ALLE ZEIT WACH
1842

H.-D. Belitz · W. Grosch

Food Chemistry

Translation from the second German Edition
by D. Hadziyev

With 345 Figures and 458 Tables

Springer Verlag Berlin Heidelberg New York
London Paris Tokyo

Professor Dr.-Ing. H.-D. Belitz

Professor Dr.-Ing. W. Grosch
Institut für Lebensmittelchemie der Technischen Universität München
and Deutsche Forschungsanstalt für Lebensmittelchemie, München
Lichtenbergstraße 4
D-8046 Garching, FRG

Translator:

Professor Dr. D. Hadziyev
Department of Food Science
University of Alberta
Edmonton, Alberta, Canada

ISBN 3-540-15043-9 Springer-Verlag Berlin Heidelberg New York
ISBN 0-387-15043-9 Springer-Verlag New York Heidelberg Berlin

Library of Congress Cataloging in Publication Data.
Belitz, H.-D. (Hans-Dieter)
Food chemistry.
Translation of: Lehrbuch der Lebensmittelchemie.
Includes bibliographies.
1. Food-Analysis. I. Grosch, W. (Werner) II. Title.
TX545.B3513 1986 664′.07 85-27642
ISBN 0-387-15043-9 (U.S.)

Typesetting: Passavia Druckerei GmbH, Passau
Offsetprinting: Saladruck, Berlin
Bookbinding: Lüderitz & Bauer, Berlin
2152/3020-543210

In memoriam
Joseph Schormüller (1903–1974)
Professor of Food Chemistry at the
Technische Universität Berlin

Preface to the First English Edition

The two German editions of the "Lehrbuch für Lebensmittelchemie" were so well accepted not only as an university textbook but also as a first comprehensive source of information for people in science, industry, official food control and administration, that the publishing house, Springer Verlag (Heidelberg), decided to edit an English version.
The first English edition is actually the second German edition which was revised for this purpose.
We are specially thankful to our colleague Prof. Dr. D. Hadziyev for the translation of the book.

Garching, December 1986 H.-D. Belitz, W. Grosch

Preface to the Second German Edition

Appreciative critiques and rapid sales have indicated that the first edition of our books has been well received by many readers.
A second print run was dictated therefore, earlier than expected, so that in this second edition we have concentrated on correcting errors and updating statistical data. Nevertheless, we have done our best to adapt the text to include the most important, recent developments and, where appropriate, the references have also been brought up to date. With respect to the latter it is important to emphasize that a textbook should, apart from discussing the state of the art, provide an incentive for more intensive study.
Our thanks are due to all readers who have assisted us in preparing this second edition by pointing out errors and proofreading by their constructive criticism. For the preparation of the manuscript and we are indebted to Mrs. A. Mödel (food chemist), Mrs. R. Berger, Mrs. J. Hahn, Mrs. I. Hofmeier, Mrs. H. Troesch and Mrs. K. Wuest. Once again, we would like to acknowledge the pleasant cooperation of our publishers.

Garching, September 1984 H.-D. Belitz, W. Grosch

Preface to the First German Edition

The very rapid development of food chemistry and technology over the last two decades, due to a remarkable augmentation to the analytical and manufacturing possibilities, make the complete lack of a comprehensive, teaching or reference text particularly noticeable. It is hoped that this textbook of food chemistry will help to fill this gap. In writing this volume we were able to draw on our experience from the lectures which we have given, covering various scientific subjects, over the last fifteen years at the Technical University of Munich.

Since a separate treatment of the important food constituents (proteins, lipids, carbohydrates, flavor compounds etc.,) and of the important food groups (milk, meat, eggs, cereals, fruits, vegetables, etc.,) has proved successful in our lectures, the subject matter is also organized in the same way in this book.

Compounds which are found only in particular foods are discussed where they play a distinctive role while food additives and contaminants are treated in their own chapters. The physical and chemical properties of the important constituents of foods are discussed in detail where these form the basis for understanding either the reactions which occur, or can be expected to occur, during the production, processing, storage and handling of foods or the methods used in analyzing foods. An attempt has also been made to clarify the relationship between the structure and properties at the level of individual food constituents and at the level of the whole food system.

The book focuses on the chemistry of foodstuffs and does not consider national or international food regulations. We have also omitted a broader discussion of aspects related to the nutritional value, the processing and the toxicology of foods. All of these are an essential part of the training of a food chemist but, because of the extent of the subject matter and the consequent specialization, must, today, be the subject of separate books. Nevertheless, for all important foods we have included brief discussions of manufacturing processes and their parameters since these are closely related to the chemical reactions occurring in foods.

Commodity and production data, of importance to food chemists, are mainly given in tabular form. Each chapter includes some references which are not intended to form an exhaustive list and no preference or judgement should be inferred from the choice of references; they are given simply to encourage further reading. Additional literature of more general nature is given at the end of the book.

This book is primarily aimed both at students of food and general chemistry but also at those students of other disciplines who are required, or choose to study food chemistry as a subsidiary subject. We also hope that this comprehensive text will prove useful to both food chemists and chemists generally who have completed their formal education.

We thank sincerely Mrs. A. Mödl (food chemist), Mrs. R. Berger, Mrs. I. Hofmeier, Mrs. E. Hortig, Mrs. F. Lynen and Mrs. K. Wüst for their help during the preparation of the manuscript and its proofreading. We are very grateful to Springer Verlag for their consideration of our wishes and for the agreeable cooperation.

Garching, July 1982 H.-D. Belitz, W. Grosch

Foreword of the Translator

Providing basic and applied information on the state of knowledge within food science and technology is a constant challenge. This translation presents an intellectually digested overview of the ever provoking field of food chemistry.

The translation is a textbook in the German sense which means it is both a textbook and a reference. It is a handbook in the North American sense, which means a comprehensive, one volume reference. Also, it exemplifies how to write a chemically oriented text without excessive repetition of basic disciplines and food processing, and showing that many food science disciplines are mature enough to be dealt with separately. On the basis of the revised second German edition, the book was additionally revised, upgraded and supplemented with English references all in a joint pursuit to develop an English edition.

Often being asked why I embarked on a "linguistic exercise" rather than writing my own book, the answer is intimately connected with the way humane deal with knowledge. If it is good, our duty is to remove its language barrier and let it be disseminated. Furthermore there were other attributes.

The authors are my personal friends. One is administrator of the Institute of Food Chemistry, Technical University Munich, where both are wonderful teachers and scientists. They are highly regarded by the scientific community for their work on the extent and mechanism of lipid oxidation, carotenoid cooxidation, and for the field of proteinase enzymes and their natural inhibitors in cereals and vegetables. Also, the correlation of sensory properties of peptides and other compounds with their structures and/or extent of protein hydrolysis is a highlight of their research. The authors reliably recount the vast heritage and progress in food chemistry in post-war Germany. Admittedly, the latter developed separately rather than closely interrelated with other countries. Hence, the translation reflects this process and sheds light on the present status of their teaching within their University degree programs.

The translation aims to serve the needs of senior undergraduate and graduate students and to serve as a handbook for teaching staff and graduates employed by baking, brewing, dairy, meat and other food industries, Agriculture and Government agencies. This is especially so since it meets the revised minimum curriculum standards set by the Institute of Food Technologists.

The translation generally follows the American style and spelling (odor or flavor rather than odour or flavour) and for compound nomenclature follows the recommendations of the Merck Index, tenth edition, with some exceptions (e.g. an "e" in flavine, but not in riboflavin, or often prolamines rather than exclusively prolamin).

It is a pleasure to acknowledge the commendable cooperation of the authors, and the skill and assistance of Leonard Steele and Judy Nuss. Also valuable was the release from active teaching, administrative and extension work by being granted a study leave – special thanks are due to my Department of Food Science and the Faculty of Agriculture and Forestry.

Edmonton, December 1986 D. Hadziyev

Table of Contents

Introduction

Foods are materials which, in their naturally occurring, processed or cooked forms, are consumed by humans as nourishment and for enjoyment.

The terms "nourishment" and "enjoyment" introduce two important properties of foods: the nutritional value and the enjoyment value. The former is relatively easy to quantify since all the important nutrients are known and their effects are defined. Furthermore, there are only a limited number of nutrients. Defining the enjoyment value of a food is more difficult because such a definition must take account of all those properties of a food, such as visual appeal, smell, taste and texture, which interact with the senses. These properties can be influenced by a large number of compounds which in part have not even been identified.

As well as for its nutritional and enjoyment values food is increasingly being judged according to properties which determine its value in use. Thus, the term "convenience foods". An obvious additional requirement of a food is that it be free from toxic materials.

Food chemistry is involved, not only in elucidating the composition of the raw materials and end-products, but also with the changes which occur in food during its production, processing, storage and cooking. The highly complex nature of food results in a multitude of desired and undesired reactions which are controlled by a variety of parameters. To gain a meaningful insight into these reactions, it is necessary to dissolve the food into model systems. Thus, starting from compositional analyses (detection, isolation and structural characterisation of food constituents), the reactions of a single constituent or of a simple mixture are followed. Subsequently, an investigation of a food in which an individual reaction dominates can be made. Inherently, such a study starts with a given compound and is thus not restricted to any one food or group of foods. Such general studies of reactions involving food constituents are supplemented by special investigations which focus on chemical processes in individual foods. Research of this kind is, from the very begining, closely associated with economic and technological aspects and contributes, by understanding the basics of the chemical processes occurring in foods, both to resolving specific technical problems and to process optimization.

A comprehensive evaluation of foods requires that analytical techniques keep pace with the available technology so that a major part of food chemistry is concerned with the application and continual development of analytical methods. This aspect is particularly important with respect to possible contamination of foods with substances which may involve a health risk. Thus, there are close links with environmental problems.

Food chemistry research is aimed at establishing objective scales by which the criteria mentioned above – nutritional value, enjoyment value, absence of toxic compounds and convenience – can be evaluated. These are a prerequisite for the production of high quality food in sufficient quantity.

The brief outline given here makes clear that food chemistry, unlike other branches of chemistry which are concerned either with particular classes of compounds or with particular methods, is a subject which, both in terms of the actual chemistry and the methods involved, has a very broad field to cover.

0 Water

0.1 Foreword

Water (moisture) is the predominant constituent in many foods (Table 0.1). As a medium water supports chemical reactions, and it is a direct reactant in hydrolytic processes. Therefore, removal of water from food or binding it by increasing the concentration of common salt or sugar retards many reactions and inhibits the growth of microorganisms, thus improving the shelf lives of a number of foods. Through physical interaction with proteins, polysaccharides, lipids and salts, water contributes significantly to the texture of food.

Table 0.1. Moisture contents of some food

Food	Moisture content (weight-%)	Food	Moisture content (weight-%)
Meat	65–75	Cereal flour	12–14
Milk	87	Coffee beans, roasted	5
Fruits, vegetables	70–90	Milk powder	4
Bread	35	Edible oil	0
Honey	20		
Butter, margarine	16–18		

The function of water is better understood when its structure and its state in a food system are clarified. Special aspects of binding of water by individual food constituents (cf. 1.4.3.3, 3.5.2 and 4.4.3) and meat (cf. 12.5) are discussed in the indicated sections.

0.2 Structure

0.2.1 Water Molecule

The six valence electrons of oxygen in a water molecule are hybridized to four sp^3 orbitals that are elongated to the corners of a somewhat deformed, imaginary tetrahedron (Figure 0.1). Two hybrid orbitals form O–H covalent bonds with a bond angle of 105° for H–O–H, whereas the other 2 orbitals hold the nonbonding electron pairs (n-electrons). The O–H covalent bonds, due to the highly electronegative oxygen, have a partial (40%) ionic character.

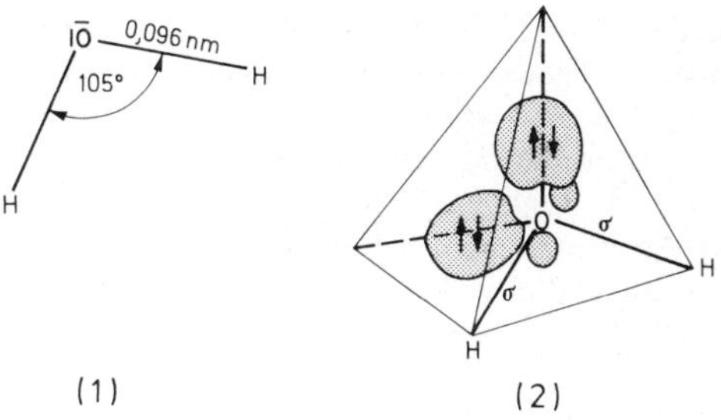

Fig. 0.1. Water. **1** Molecular geometry, **2** orbital model

Each water molecule is tetrahedrally coordinated with four other water molecules through hydrogen bonds. The two unshared electron pairs (n-electrons or sp^3 orbitals) of oxygen act as H-bond acceptor sites and the H–O bonding orbitals act as hydrogen bond donor sites (Figure 0.2). The dissociation energy of this hydrogen bond is about 25 kJ $mole^{-1}$.

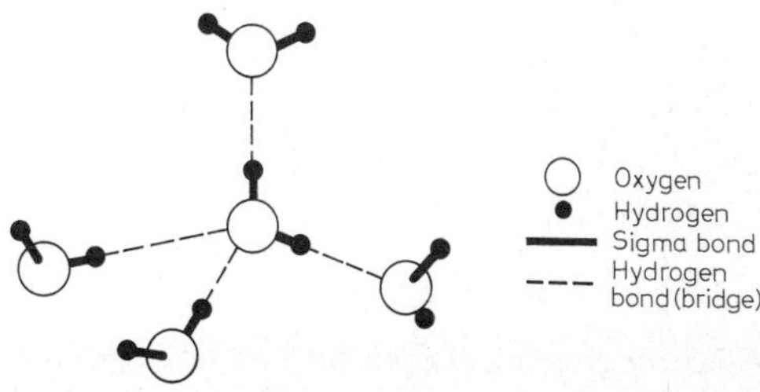

Fig. 0.2. Tetrahedral coordination of water molecules

The simultaneous presence of two acceptor sites and two donor sites in water permits association in a three-dimensional network stabilized by H-bridges. This structure which explains the special physical properties of water is unusual for other small molecules. For example, alcohols and compounds with isoelectric dipoles similar to those of water, such as HF or NH_3, form only linear or two-dimensional associations.

The partial polarization of H–O bonds is further enhanced through H-bridge formation. Therefore, the dipole moment of a complex consisting of increasing numbers of water molecules (multimolecular dipole) is higher as more molecules become associated and is certainly much higher than the dipole moment of a single molecule. Thus, the dielectric constant of water is high and surpasses the value, which can be calculated on the basis of the dipole moment of a single molecule. Proton transport takes place along the H-bridges. It is actually the jump of a proton from one water molecule to a neighboring water molecule. Regardless of whether the proton is derived from dissociation of water or originates from an acid, it will sink into the unshared electron pair orbitals of water:

$$H_2\overset{\oplus}{O}{-}H\cdots OH_2 \longrightarrow H_2O\cdots H{-}\overset{\oplus}{O}H_2 \qquad (0.1)$$

In this way a hydrated $H_3O^{\oplus}$ ion is formed with an exceptionally strong hydrogen bond (dissociation energy about 100 kJ $mole^{-1}$). A similar mechanism is valid in transport of $OH^{\ominus}$ ions, which also occurs along the hydrogen bridges:

$$H\overset{\ominus}{O}\cdots H{-}OH \longrightarrow HO{-}H\cdots \overset{\ominus}{O}H \qquad (0.2)$$

Since the crossing of a proton from one oxygen to the next occurs extremely rapidly ($\nu > 10^{12}\ s^{-1}$), proton mobility surpasses the mobilities of all other ions by a factor of 4–5, except for the stepwise movement of $OH^{\ominus}$ within the structure; its rate of exchange is only 40% less than that of a proton.

H-bridges in ice extend to a larger sphere than in water (see the following section). The mobility of protons in ice is higher than in water by a factor of 100.

0.2.2 Liquid Water and Ice

The arrangements of water molecules in "liquid water" and in ice are still under intensive investigation. The outlined hypotheses agree with existing data and are generally accepted.

Due to the pronounced tendency of water molecules to associate through H-bridges, liquid water and ice are highly structured. They differ in the distance between molecules, coordination number and time-range order (duration of stability). Stable ice-I is formed at 0 °C and 1 atm pressure. It is one of nine known crystalline polymorphic structures, each of which is stable in a certain temperature and pressure range. The coordination number in ice-I is four, the O–H···O (nearest neighbor) distance is 0.276 nm (0 °C) and the H-atom between neighboring oxygens is 0.101 nm from the oxygen to which it is bound covalently and 0.175 nm from the oxygen to which it is bound by a hydrogen bridge. Five water molecules, forming a tetrahedron, are loosely packed and kept together mostly through H-bridges.

Table 0.2. Coordination number and distance between two water molecules

	Coordination number	O–H···O Distance
Ice (0 °C)	4	2.76 Å
Water (1,5 °C)	4.4	2.9 Å
Water (83 °C)	4.9	3.05 Å

When ice melts and the resultant water is heated (Table 0.2), both the coordination number and the distance between the nearest neighbors increase. These changes have opposite influences on the density. An increase in coordination number (i.e. the number of water molecules arranged in an orderly fashion around each water molecule) increases the density, whereas an increase in distance between nearest neighbors decreases the density. The effect of increasing coordination number is predominant during a temperature increase from 0 to 4 °C. As a consequence, water has an unusual property: its density in the liquid state at 0 °C (0.9998 g cm^{-3}) is higher than in the solid state (ice-I, $\varrho = 0.9168$ g cm^{-3}). Water is a structured liquid with a short time-range order. The water molecules, through H-bridges, form short-lived polygonal structures which are rapidly cleaved and then reestablished giving a dynamic equilibrium. Such fluctuations explain the lower viscosity of water, which otherwise could not be explained if H-bridges were rigid.

The hydrogen-bound water structure is changed by solubilization of salts or molecules with polar and/or hydrophobic groups. In salt solutions the n-electrons occupy the free orbitals of the cations, forming "aqua complexes". Other water molecules then coordinate through H-bridges, forming a hydration shell around the cation and disrupting the natural structure of water.

Hydration shells are formed by anions through ion-dipole interaction and by polar groups through dipole-dipole interaction or H-bridges, again contributing to the disruption on the structured state of water.

Aliphatic groups which can fix the water molecules by dispersion forces are no less disruptive. A minimum of free enthalpy will be attained when an ice-like water structure is arranged around a hydrophobic group (tetrahedral-four-coordination) Such ice-like hydration shells around aliphatic groups contribute, for example, to stabilization of a protein, helping the protein to acquire its most thermodynamically-favorable conformation in water.

The highly structured, three-dimensional hydrogen bonding state of ice and water is reflected in many of their unusual properties. Additional energy is required to break the structured state. This accounts for water having substantially higher melting and boiling points and heats of fusion and vaporization than methanol or dimethyl ether (cf. Table 0.3). Methanol has only one hydrogen donor site, while dimethyl ether has none but does have a hydrogen bond acceptor site; neither is sufficient to form a structured network as found in water.

Table 0.3. Some physical constants of water, methanol and dimethyl ether

	F_p (°C)	K_p (°C)
H_2O	0.0	100.0
CH_3OH	−98	64.7
CH_3OCH_3	−138	−23

0.3 Water Binding

0.3.1 Sorption Isotherms

Sorption isotherms for food are presented in Figure 0.3. The data provide a relationship between the moisture content and water activity (a_w) of food. The definition of a_w is:

$$a_w = P/P_0 = ERH/100 \tag{0.3}$$

P = partial vapor pressure of food moisture at temperature T

P_0 = saturation vapor pressure of pure water at T

ERH = equilibrium relative humidity at T.

At low water content (< 50%) small changes in these parameters cause a large change in water activity. The sorption isotherm of a food with a lower water content is therefore shown with an expanded ordinate (compare Figure 0.3, a and b). Figure 0.3,b shows that the desorption isotherm curve (representing the drying process), is somewhat above the adsorption isotherm (water adsorption) curve of a food which is sensitive to moisture. This phenomenon, known as hysteresis, is explained as follows:

Capillaries close during drying of food, thus decreasing the free inner surface. To fill the remaining capillaries (i.e. rehydrate), of which the geometry differs strongly in various foods, a higher partial pressure of water vapor is required than in the case of moisture removal. Three regions of water binding are distinguished in sorption isotherm curves. In adsorption of moisture the inner dry surface is first covered by a monomolecular layer of water (segment A in Figure 0.3,b). This water interacts with food constituents as outlined under 0.2.2. In the next segment of the curve (B, Figure 0.3,b), the monomo-

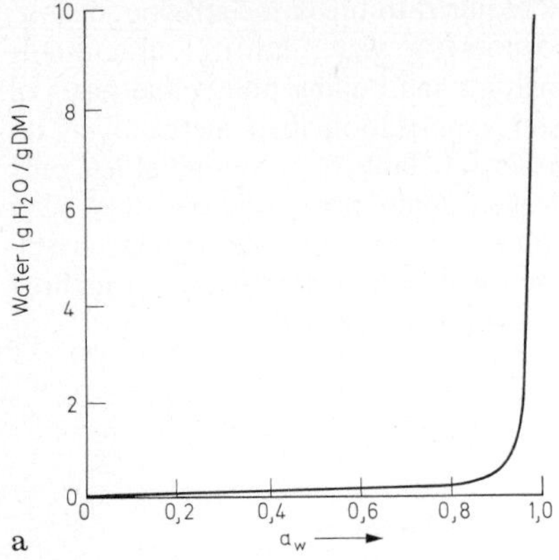

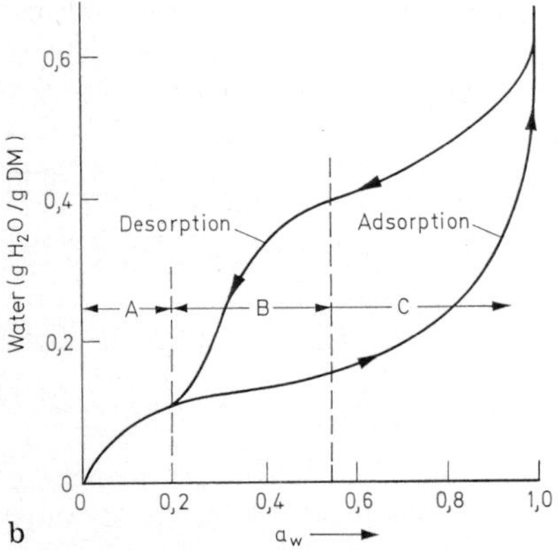

Fig. 0.3. Moisture sorption isotherm (After *T. P. Labuza* et al., 1970). **a** Food with high moisture content; **b** Food with low moisture content (DM: Dry matter)

Table 0.4. Water binding in food

Binding type	a_w Range	State of water
I	0.8–0.99	Mobile water molecules within the tissue surrounded by membranes, or in macrocapillaries (∅ > 1 µm); Freezing point somewhat decreased
II	0.25–0.8	Water molecules with restricted mobility e.g. in microcapillaries (∅ < 1 µm); Freezing point significantly decreased.
III	< 0.25	"Non freezeable" water

lecular layer coordinates additional water molecules until a hydration shell is formed. These shell molecules are fixed, i.e. have no mobility and are not freezable at low temperatures (type III binding in Table 0.4). At higher water activity (segment C, Figure 0.3,b) water condenses and increasingly fills the capillaries; this water (type I), unlike types II or III, is mobile (Table 0.4).
The attainment of a complete monomolecular coating of water on the inner surface of a food is marked by the BET-point (according to *Brunauer, Emmett* and *Teller,* who developed a method for measurement of the so-called "monolayer value" which can be considered as equivalent of the amount of water held adsorbed on specific sites of a food). The values (examples listed in Table 0.5) correspond to the amount of water required to form a monolayer coating.

Table 0.5. Moisture content of some food at BET-point

Food	Moisture content (g/100 g Dry matter)	Food	Moisture content (g/100 g Dry matter)
Starch	11	Potato	
Gelatin	11	sliced	6
Lactose,		Whey	
amorphous	6	powder	3
Dextran	9	Beef,	
Saccharose,		freeze-dried	4
granulated	0.4		

However, the values listed are approximate since the analytical procedure applied included a number of simplifications.
The "nonfreezable" type III water, which is 2- to 4-times the amount of water at the BET-point, can be readily determined by differential scanning calorimetry. Type II water can be assayed by ^{1}H-NMR spectroscopy.

0.3.2 Influence on Reaction Rate

The rates of many reactions are influenced by the extent of water binding in foods in which the water content is lower than the content of total solids.

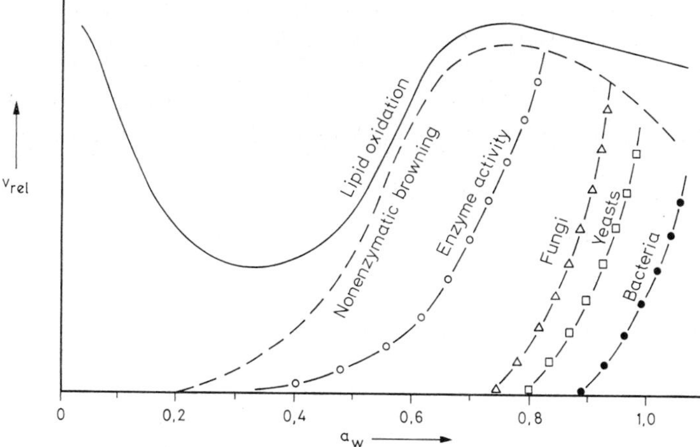

Fig. 0.4. Food shelf life (storage stability) as a function of water activity (After *T. P. Labuza*, 1971)

The effect of water activity on processes that can influence food quality is presented in Figure 0.4. Decreased water activity retards the growth of microorganisms, slows enzyme catalyzed reactions (particularly involving hydrolases; cf. 2.7) and, lastly, retards nonenzymatic browning. In contrast, the rate of lipid autoxidation increases in dried food systems (cf. 3.7.2.1.4).

The storage stability of food with an a_w between 0.2 and 0.4 is the highest. This a_w range obviates the need for preservatives against microbial spoilage and food quality is unaffected by nonenzymatic browning and lipid autoxidation since these reactions are essentially prevented.

Table 0.6. Water activity of some food

Food		Food	
Liver sausage	0.96	Marmalades	0.82–0.94
Salami	0.82–0.85	Honey	0.75
Dried fruits	0.72–0.80		

Foods with a_w values between 0.6 and 0.9 (examples in Table 0.6) are known as "intermediate moisture foods" (IMF). These foods must be protected extensively against microbial spoilage.

Table 0.7. Moisture content of some food or food ingredients at a water activity of 0.8

	Moisture content (%)		Moisture content (%)
Peas	16	Glycerol	108
Casein	19	Sorbitol	67
Starch (potato)	20	Saccharose	56
		Common salt	332

One of the options to decreasing water activity, and thus improve the shelf life of food, is to use additives with high water binding capacities (humectants). Table 0.7 shows that, in addition to common salt, glycerol, sorbitol and sucrose have potential as humectants. However, they are also sweeteners and would be objectionable from a consumer standpoint in many foods in the concentrations required to regulate water activity.

0.4 Literature

Fennema, O. R.: Water and ice. In: Principles of food science, part I (Ed.: Fennema, O. R.), p. 13, Marcel Dekker, Inc.: New York–Basel. 1976

Franks, F.: Water, ice and solutions of simple molecules. In: Water relations of foods (Ed.: Duckworth, R.B.), p. 3, Academic Press: London–New York–San Francisco. 1975

Heiss, R.: Haltbarkeit und Sorptionsverhalten wasserarmer Lebensmittel. Springer-Verlag: Berlin–Heidelberg–New York. 1968

Karel, M.: Water activity and food preservation. In: Principles of food science, part II (Eds.: Karel, M., Fennema, O.R., Lund, D.B.), p. 237, Marcel Dekker, Inc.: New York–Basel, 1975

Labuza, T.P.: Sorption phenomena in foods, Food Technol. *22,* 263 (1968)

Labuza, T.P.: Kinetics of lipid oxidation in foods. Crit. Rev. Food Technol. *2,* 355 (1971)

Labuza, T.P., Tannenbaum, S.R., Karel, M.: Water content and stability of low-moisture and intermediate-moisture foods. Food Technol. *24,* 543 (1970)

Rockland, L.B.: Water activity and storage stability. Food Technol. *23,* 1241 (1969)

1 Amino Acids, Peptides, Proteins

1.1 Foreword

Amino acids, peptides and proteins are important constituents of food. They supply the required building blocks for protein biosynthesis. In addition, they directly contribute to the flavor of food and are precursors for aroma compounds and colors formed during thermal or enzymatic reactions in production, processing and storage of food. Other food constituents, e.g. carbohydrates, take part in such reactions. Proteins also contribute significantly to the physical properties of food through their ability to build or stabilize gels, foams, doughs, emulsions and fibrillar structures.

1.2 Amino Acids

1.2.1 General Remarks

There are about 20 amino acids in a protein hydrolysate. With a few exceptions, their general structure is:

$$R{-}CH(NH_2){-}COOH \quad (1.0)$$

In the simplest case, R = H (aminoacetic acid or glycine). In other amino acids R is an aliphatic, aromatic or heterocyclic residue and may incorporate other functional groups. Table 1.1 shows the most important "building blocks" of proteins. There are about 200 amino acids found in nature (Figure 1.1). Some of the more uncommon ones, which occur mostly in plants in free form, are covered in Chapter 17 on vegetables.

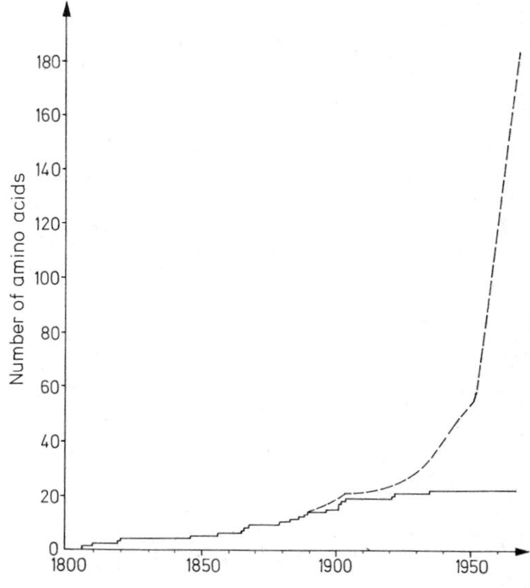

Fig. 1.1. Discovery of naturally occurring amino acids (After *Meister*, 1965). --- Amino acids, total; — protein constituents

1.2.2 Classification, Discovery and Occurrence

1.2.2.1 Classification

There are a number of ways of classifying amino acids. Since their side chains are the deciding factors for intra- and intermolecular interactions in proteins and, hence, for protein properties, amino acids can be classified as:

- Amino acids with nonpolar, uncharged side chains: e.g. glycine, alanine, valine, leucine, isoleucine, proline, phenylalanine, tryptophan and methionine.
- Amino acids with uncharged, polar side chains: e.g. serine, threonine, cysteine, tyrosine, asparagine and glutamine.
- Amino acids with charged side chains: e.g. aspartic acid, glutamic acid, histidine, lysine and arginine.

Table 1.1. Amino acids (protein building blocks) with their corresponding three ond one letter symbols

Structure	Name	Structure	Name	Structure	Name
COOH H_2N—CH_2	Glycine (Gly. G)	COOH H_2N—CH CH_2 CH_2 S CH_3	L-Methionine (Met. M)	COOH H_2N—CH CH_2 COOH	L-Aspartic acid (Asp. D)
COOH H_2N—CH CH_3	L-Alanine (Ala. A)	COOH H_2N—CH CH_2OH	L-Serine (Ser. S)	COOH H_2N—CH CH_2 CH_2 COOH	L-Glutamic acid (Glu. E)
COOH H_2N—CH CH H_3C CH_3	L-Valine (Val. V)	COOH H_2N—CH HC—OH CH_3	L-Threonine (Thr. T)	COOH H_2N—CH CH_2 CH_2 CH_2 CH_2NH_2	L-Lysine (Lys. K)
COOH H_2N—CH CH_2 CH H_3C CH_3	L-Leucine (Leu. L)	COOH H_2N—CH CH_2SH	L-Cysteine (Cys. C)	COOH H_2N—CH CH_2 CH_2 HO—CH CH_2NH_2	L-5-Hydroxy-lysine
COOH H_2N—CH H_3C—CH CH_2 CH_3	L-Isoleucine (Ile. I)	COOH HN OH	L-4-Hydroxy-proline	COOH H_2N—CH CH_2 N HN	L-Histidine (His. H)
COOH HN	L-Proline (Pro. P)	COOH H_2N—CH CH_2 OH	L-Tyrosine (Tyr. Y)	COOH H_2N—CH CH_2 CH_2 CH_2 NH C HN NH_2	L-Arginine (Arg. R)
COOH H_2N—CH CH_2	L-Phenylalanine (Phe. F)	COOH H_2N—CH CH_2 $CONH_2$	L-Asparagine[a] (Asn. N)		
COOH H_2N—CH CH_2 NH	L-Tryptophan (Trp. W)	COOH H_2N—CH CH_2 CH_2 $CONH_2$	L-Glutamine[a] (Gln. Q)		

[a] When no distinction exists between the acid and its amide then the symbols (Asx, B) and (Glx, Z) are valid.

Based on their nutritional/physiological roles, amino acids can be differentiated as:

- Essential amino acids:
  Valine, leucine, isoleucine, phenylalanine, tryptophan, methionine, threonine, histidine (essential for infants), and lysine and arginine ("semi-essential").
- Nonessential amino acids:
  Glycine, alanine, proline, serine, cysteine, tyrosine, asparagine, glutamine, aspartic acid and glutamic acid.

1.2.2.2 Discovery and Occurrence

Alanine was isolated from silk fibroin by *Th. Weyl* in 1888. It is present in most proteins and is particularly enriched in silk fibroin (35%). Gelatin and zein contain about 9% alanine, while its content in other proteins is 2–7%. Alanine is considered nonessential for humans.

Arginine was first isolated from lupin seedlings by *E. Schulze* and *E. Steiger* in 1886. It is present in all proteins at an average level of 3–6%, but is particularly enriched in protamines. The arginine content of peanut protein is relatively high (11%). Biochemically, arginine is of great importance as an intermediary product in urea synthesis. Arginine is a semiessential amino acid for humans. It appears to be required under certain metabolic conditions.

Asparagine from asparagus was the first amino acid isolated by *Vauguelin* and *Robiquet* in 1806. Its occurrence in proteins (edestin) was confirmed by *Damodaran* in 1932. In glycoproteins the carbohydrate component may be bound N-glycosidically to the protein moiety through the amide group of asparagine (cf. 11.2.3.1.1 and 11.2.3.1.3).

Aspartic acid was isolated from legumes by *H. Ritthausen* in 1868. It occurs in all animal proteins, primarily in albumins at a concentration of 6–10%. Alfalfa and corn proteins are rich in aspartic acid (14.9% and 12.3%, respectively) while its content in wheat is low (3.8%). Aspartic acid is nonessential.

Cystine was isolated from bladder calculi by *W.H. Wolaston* in 1810 and from horns by *L. Moerner* in 1899. Its content is high in keratins (9%). Cystine is very important since the peptide chains of many proteins are connected by two cysteine residues, i.e. by disulfide bonds. A certain conformation may be fixed within a single peptide chain by disulfide bonds. Most proteins contain 1–2% cystine. Although it is itself nonessential, cystine can partly replace methionine which is an essential amino acid.

Glutamine was first isolated from sugar beet juice by *Schulze* and *Bosshard* in 1883. Its occurrence in protein (edestin) was confirmed by *Damodaran* in 1932. Glutamine is readily converted into pyrrolidone carboxylic acid, which is stable between pH 2.2 and 4.0, but is readily cleaved to glutamic acid at other pH's:

$$\begin{array}{l} CH_2-CH_2-CONH_2 \\ | \\ CH-NH_2 \\ | \\ COOH \end{array} \longrightarrow \text{HOOC-(pyrrolidone ring, N-H, =O)} + NH_3 \quad (1.1)$$

Glutamic acid was first isolated from wheat gluten by *H. Ritthausen* in 1866. It is abundant in most proteins, but is particularly high in milk proteins (21.7%), wheat (31.4%), corn (18.4%) and soya (18.5%). Molasses also contains relatively high amounts of glutamic acid. Monosodium glutamate is used in numerous food products as a flavor enhancer.

Glycine is found in high amounts in structural protein. Collagen contains 25–30% glycine. It was first isolated from gelatin by *H. Braconnot* in 1820. Glycine is a nonessential amino acid although it does act as a precursor of many compounds formed by various biosynthetic mechanisms.

Histidine was first isolated in 1896 independently by *A. Kossel* and by *S. G. Hedin* from protamines. Most proteins contain 2–3% histidine. Blood proteins contain about 6%. Histidine is essential in infant nutrition.

5-Hydroxylysine was isolated by *van Slyke et al.* (1921) and *Schryver et al.* (1925). It occurs in collagen. The carbohydrate component of glycoproteins may be bound O-glycosidically to the hydroxyl group of the amino acid (cf. 12.3.2.3.1).

4-Hydroxyproline was first obtained from gelatin by *E. Fischer* in 1902. Since it is abundant in collagen (12.4%), the determination of hydroxyproline is used to detect the presence of connective tissue in comminuted meat products. Hydroxyproline is a nonessential amino acid.

Isoleucine was first isolated from fibrin by *P. Ehrlich* in 1904. It is an essential amino acid. Meat and cereal proteins contain 4–5% isoleucine; egg and milk proteins, 6–7%.

Leucine was isolated from wool and from muscle tissue by *H. Braconnot* in 1820. It is an essential amino acid and its content in most proteins is 7–10%. Cereal proteins contain variable amounts (corn 12.7%, wheat 6.9%). During alcoholic fermentation, fusel oil is formed from leucine and isoleucine.

Lysine was isolated from casein by *E. Drechsel* in 1889. It makes up 7–9% of meat, egg and milk proteins. The content of this essential amino acid is 2–4% lower in cereal proteins in which prolamin is predominant. Crab and fish proteins are the richest sources (10–11%). Along with threonine and methionine, lysine is a limiting factor in the biological value of many proteins, mostly those of plant origin. The processing of foods results in losses of lysine since its ε-amino group is very reactive (cf. *Maillard* reaction).

Methionine was first isolated from casein by *J. H. Mueller* in 1922. Animal proteins contain 2–4% and plant proteins contain 1–2% methionine. Methionine is an essential amino acid and in many biochemical processes its main role is as a methyl-donor. It is very sensitive to oxygen and heat treatment. Thus, losses occur in many food processing operations such as drying, kiln-drying, puffing, roasting or treatment with oxidizing agents. In the bleaching of flour with NCl_3 (nitrogen trichloride), methionine is converted to the toxic methionine sulfoximide:

$$H_3C-\overset{\overset{O}{\|}}{\underset{\underset{NH}{\|}}{S}}-CH_2-CH_2-\underset{\underset{NH_2}{|}}{CH}-COOH \qquad (1.1a)$$

Phenylalanine was isolated from lupins by *E. Schulze* in 1881. It occurs in almost all proteins (averaging 4–5%) and is essential for humans. It is converted *in vivo* into tyrosine, so phenylalanine can replace tyrosine nutritionally.

Proline was discovered in casein and egg albumen by *E. Fischer* in 1901. It is present in numerous proteins at 4–7% and is abundant in wheat proteins (10.3%), gelatin (12.8%) and casein (12.3%). Proline is nonessential.

Serine was first isolated from sericin by *E. Cramer* in 1865. Most proteins contain about 4–8% serine. In phosphoproteins (casein, phosvitin) serine, like threonine, is a carrier of phosphoric acid in the form of O-phosphoserine. The carbohydrate component of glycoproteins may be bound O-glycosidically through the hydroxyl group of serine and/or threonine [cf. 10.1.2.1.1 (κ-casein) and 13.1.4.2.4].

Threonine was discovered by *W. C. Rose* in 1935. It is an essential amino acid, present at 4.5–5% in meat, milk and eggs and 2.7–4.7% in cereals. Threonine is often the limiting amino acid in proteins of lower biological quality. The "bouillon" flavor of protein hydrolysates originates partly from a lactone derived from threonine (cf. 12.8, Reaction 12.18).

Tryptophan was first isolated from casein hydrolysates, prepared by hydrolysis using pancreatic enzymes, by *F. G. Hopkins* in 1902. It occurs in animal proteins in relatively low amounts (1–2%) and in even lower amounts in cereal proteins (about 1%). Tryptophan is exceptionally abundant in lysozyme (7.8%). It is completely destroyed during acidic hydrolysis of protein. Biologically, tryptophan is an important essential amino acid, primarily as a precursor in the biosynthesis of nicotinic acid.

Tyrosine was first obtained from casein by *J. Liebig* in 1846. Like phenylalanine, it is found in almost all proteins at levels of 2–6%. Silk fibroin can have as much as 10% tyrosine. It is converted through dihydroxyphenylalanine by enzymatic oxidation into brown-black colored melanins.

Valine was first isolated by *P. Schutzenberger* in 1879. It is an essential amino acid and is present in meat and cereal proteins (5–7%) and in egg and milk proteins (7–8%). Elastin contains notably high concentrations of valine (15.6%).

1.2.3 Physical Properties

1.2.3.1 Dissociation

In water amino acids are present, depending on pH, as cations, zwitter ions or anions:

$$R{-}CH(NH_3^{\oplus}){-}COOH \underset{+H^{\oplus}}{\overset{-H^{\oplus}}{\rightleftharpoons}} R{-}CH(NH_3^{\oplus}){-}COO^{\ominus} \underset{+H^{\oplus}}{\overset{-H^{\oplus}}{\rightleftharpoons}} R{-}CH(NH_2){-}COO^{\ominus} \quad (1.2)$$

With the cation denoted as ^+A, the dipolar zwitter ion as $^+A^-$ and the anion as A^-, the dissociation constant can be expressed as:

$$\frac{[^{\oplus}A^{\ominus}][H^{\oplus}]}{[^{\oplus}A]} = K_1 \qquad \frac{[A^{\ominus}][H^{\oplus}]}{[^{\oplus}A^{\ominus}]} = K_2 \quad (1.3)$$

At a pH where only dipolar ions exist, i.e. the isoelectric point, pI, $[^+A]=[A^-]$:

$$[^{\oplus}A] = \frac{[^{\oplus}A^{\ominus}][H^{\oplus}]}{K_1} = [A^{\ominus}] = \frac{[^{\oplus}A^{\ominus}]K_2}{[H^{\oplus}]}$$

$$[H^{\oplus}] = (K_1 \cdot K_2)^{0,5}$$
$$pI = 0{,}5\,(pK_1 + pK_2) \quad (1.4)$$

The dissociation constants of amino acids can be determined, for example, by titration of the acid. Figure 1.2 shows titration curves for glycine, histidine and aspartic acid. Table 1.2 lists the dissociation constants for some amino acids. In amino acids the acidity of the carboxyl group is higher and the basicity of the amino group lower than in the corresponding carboxylic acids and amines (cf. pK values for propionic acid, 2-propylamine and alanine). As illustrated by the comparison of pK values of 2-aminopropionic acid (alanine) and 3-aminopropionic acid (β-alanine), the pK is influenced by the distance between the two functional groups.

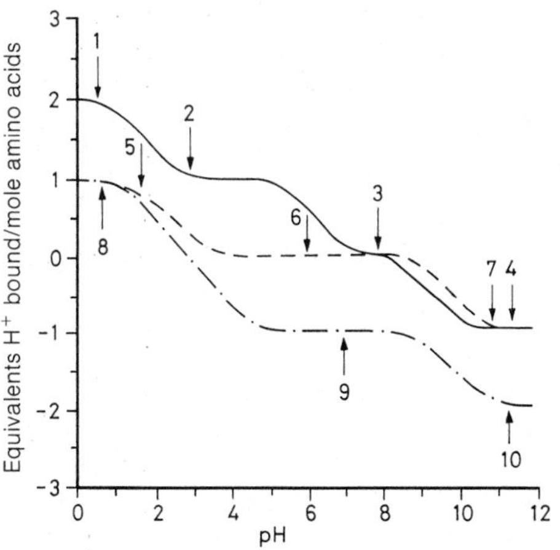

Fig. 1.2. Calculated titration curves for glycine (---), histidine (—) and aspartic acid (-·-·-). Numerals on curves are related to charge of amino acids in respective pH range: 1 $^{++}$His, 2 $^{++}$His$^-$, 3 $^+$His$^-$, His$^-$, 5 $^+$Gly, 6 $^+$Gly$^-$, 7 Gly$^-$, 8 $^+$Asp. 9 $^+$Asp^{--}, 10 Asp^{--}

Table 1.2. Amino acids: dissociation constants and isoelectric points at 25 °C

Amino acid	pK_1	pK_2	pK_3	pK_4	p^I
Alanine	2.34	9.69			6.0
Arginine	2.18	9.09	12.60		10.8
Asparagine	2.02	8.80			5.4
Aspartic acid	1.88	3.65	9,60		2.8
Cysteine	1.71	8.35	10.66		5.0
Cystine	1.04	2.10	8.02	8.71	5.1
Glutamine	2.17	9.13			5.7
Glutamic acid	2.19	4.25	9.67		3.2
Glycine	2.34	9.60			6.0
Histidine	1.80	5.99	9.07		7.5
4-Hydroxyproline	1.82	9.65			5.7
Isoleucine	2.26	9.62			5.9
Leucine	2.36	9.60			6.0
Lysine	2.20	8.90	10.28		9.6
Methionine	2.28	9.21			5.7
Phenylalanine	1.83	9.13			5.5
Proline	1.99	10.60			6.3
Serine	2.21	9.15			5.7
Threonine	2.15	9.12			5.6
Tryptophan	2.38	9.39			5.9
Tyrosine	2.20	9.11	10.07		5.7
Valine	2.32	9.62			6.0
Propionic acid	4.87				
2-Propylamine	10.63				
β-Alanine	3.55	10.24			6.9
γ-Aminobutyric acid	4.03	10.56			7.3

1.2.3.2 Configuration and Optical Activity

Amino acids, except for glycine, have at least one chiral center and, hence, are optically active. All amino acids found in proteins have the same configuration on the α-C-atom: they are considered L-amino acids or (S)-amino acids* in the *Cahn-Ingold-Prelog* system (with L-cysteine an exception; it is in the (R)-series). D-amino acids (or (R)-amino acids) also occur in nature, for example in a number of peptides of microbial origin:

COOH | H_2N—C—H | R — L-Amino acid (S)-Amino acid

COOH | H—C—NH_2 | R — D-Amino acid (R)-Amino acid (1.5)

Isoleucine, threonine and 4-hydroxyproline have two asymmetric C-atoms, thus each has four isomers:

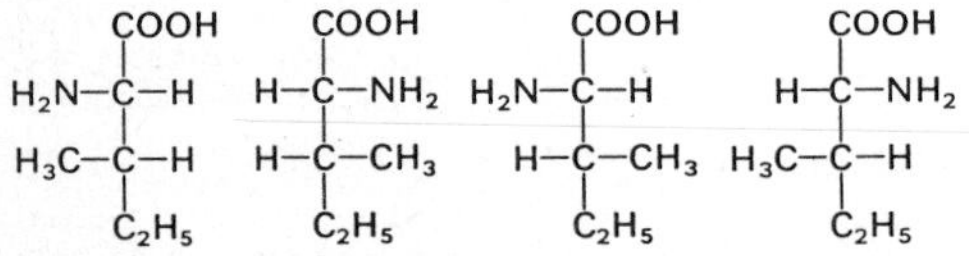

COOH	COOH	COOH	COOH
H_2N—C—H	H—C—NH_2	H_2N—C—H	H—C—NH_2
H_3C—C—H	H—C—CH_3	H—C—CH_3	H_3C—C—H
C_2H_5	C_2H_5	C_2H_5	C_2H_5
L-Isoleucine (2S:3S)-Isoleucine (Common in proteins)	D-Isoleucine (2R:3R)-Isoleucine	L-allo-Isoleucine (2S:3R)-Isoleucine	D-allo-Isoleucine (2R:3S)-Isoleucine

(1.6)

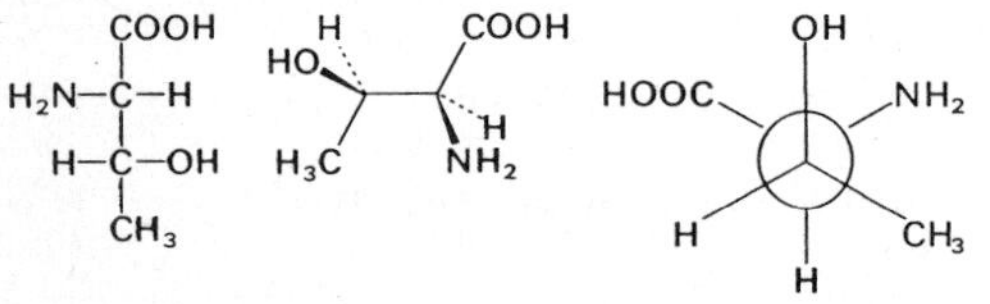

COOH | H_2N—C—H | H—C—OH | CH_3 (Fischer-projection)

H, HO, COOH, H, H_3C, NH_2 (dotted line-wedge)

OH, HOOC, NH_2, H, CH_3, H (Newman-projection)

L-Threonine, (2S:3R)-Threonine (Common in proteins) (1.7)

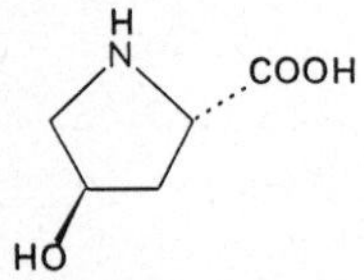

H N, COOH, HO (1.8)

L-4-Hydroxyproline, (2S:4R)-Hydroxyproline (Common in proteins)

* As with carbohydrates, D,L-nomenclature is preferred with amino acids.

The specific rotation of amino acids in aqueous solution is strongly influenced by pH. It passes through a minimum in the neutral pH range and rises after addition of acids or bases (Table 1.3).

Table 1.3. Amino acids: specific rotation ($[\alpha]^t_D$)

Amino acid	Solvent system	Temperature (°C)	$[\alpha]_D$
L-Alanine	0.97 M HCl	15	+14.7°
	water	22	+2.7°
	3 M NaOH	20	+3.0°
L-Cystine	1.02 M HCl	24	−214.4°
L-Glutamic acid	6.0 M HCl	22.4	+31.2°
	water	18	+11.5°
	1 M NaOH	18	+10.96°
L-Histidine	6.0 M HCl	22.7	+13.0°
	water	25.0	−39.01°
	0.5 M NaOH	20	−10.9°
L-Leucine	6.0 M HCl	25.9	+15.1°
	water	24.7	−10.8°
	3.0 M NaOH	20	+7.6°

1.2.3.3 Solubility

The solubilities of amino acids in water are highly variable. Besides the extremely soluble proline, hydroxyproline, glycine and alanine are also quite soluble. Other amino acids (cf. Table 1.4) are significantly less soluble, with cystine and tyrosine having particularly low solubilities. Addition of acids or bases improves the solubility through salt formation. The presence of other amino acids, in general, also brings about an increase in solubility. Thus, the extent of solubility of amino acids in a protein hydrolysate is different than that observed for the individual components.

The solubility in organic solvents (given as g amino acid/100 g solvent) is not very good because of the polar characteristics of the amino acids. All amino acids are insoluble in ether. Only cysteine and proline are relatively soluble in ethanol (1 and 5 g/100 g, respectively, at 19 °C). Methionine, arginine, leucine (0.0217 g/100 g; 25 °C), glutamic acid (0.00035 g/100 g; 25 °C), phenylalanine, hydroxy-proline, histidine and tryptophan are sparingly soluble in ethanol. The

Table 1.4. Solubility of amino acids in water (g/100 g H_2O)

Amino acid	Temperature (°C) 0	25	50	75	100
L-Alanine	12.73	16.51	21.79	28.51	37.30
L-Asparatic acid	0.209	0.500	1.199	2.875	6.893
L-Cystine	0.005	0.011	0.024	0.052	0.114
L-Glutamic acid	0.341	0.843	2.186	5.532	14.00
Glycine	14.18	24.99	39.10	54.39	67.17
L-Histidine	–	4.29	–	–	–
L-Hydroxy-proline	28.86	36.11	45.18	51.67	–
L-Isoleucine	3.791	4.117	4.818	6.076	8.255
L-Leucine	2.270	2.19	2.66	3.823	5.638
D,L-Methionine	1.818	3.381	6.070	10.52	17.60
L-Phenylalanine	1.983	2.965	4.431	6.624	9.900
L-Proline	127.4	162.3	206.7	239.0	–
D,L-Serine	2.204	5.023	10.34	19.21	32.24
L-Tryptophan	0.823	1.136	1.706	2.795	4.987
L-Tyrosine	0.020	0.045	0.105	0.244	0.565
L-Valine	8.34	8.85	9.62	10:24	–

solubility of isoleucine is relatively high in hot ethanol (0.09 g/100 g at 20 °C; 0.13 g/100 g at 78–80 °C).

1.2.3.4 UV-Absorption

Aromatic amino acids such as phenylalanine, tyrosine and tryptophan absorb in the UV-range of the spectrum with absorption maxima at 200–230 nm and 250–290 nm (Figure 1.3). Dissociation of the phenolic HO-group of tyrosine shifts the absorption curve by about 20 nm towards longer wavelengths (Figure 1.4). Absorption readings at 280 nm are used for the determination of proteins and peptides. Histidine, cysteine and methionine absorb between 200 and 210 nm.

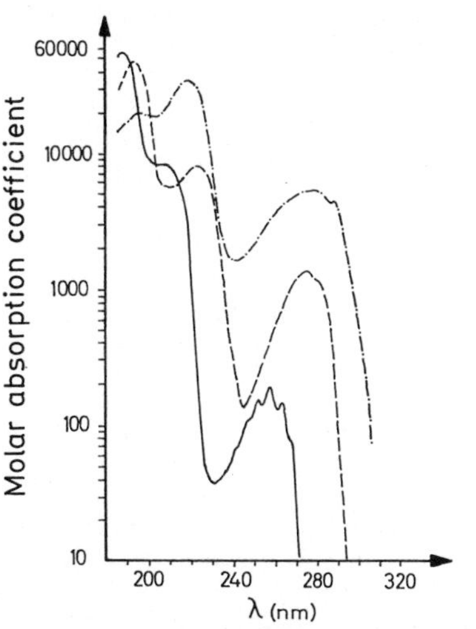

Fig. 1.3. Ultraviolet absorption spectra of some amino acids. (After *Luebke, Schroeder,* and *Kloss,* 1975). –·–·– Trp. --- Tyr. —— Phe

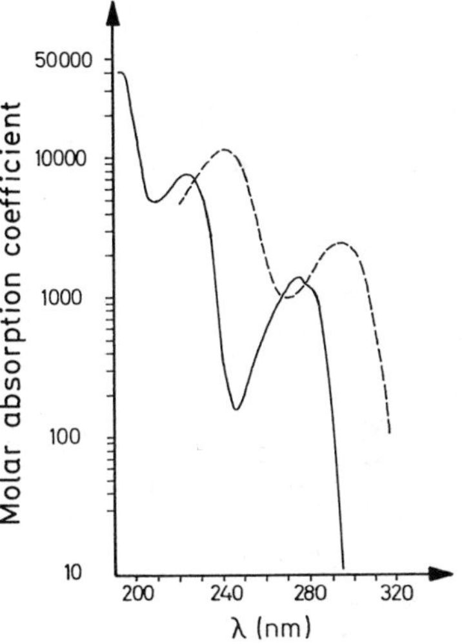

Fig. 1.4. Ultraviolet absorption spectra of tyrosine as affected by pH. (After *Luebke, Schroeder* and *Kloss,* 1975) —— 0.1 N HCl, --- 0.1 N NaOH

1.2.4 Chemical Reactions

Amino acids show the usual reactions of both carboxylic acids and amines. Reaction specificity is due to the presence of both carboxyl and amino groups and, occasionally, of other functional groups. Reactions occurring at 100–220 °C, such as in cooking, frying and baking, are particularly relevant to food chemistry.

1.2.4.1 Esterification of Carboxyl Groups

Amino acids are readily esterified by acid-catalyzed reactions. An ethyl ester hydrochloride is obtained in ethanol in the presence of HCl:

$$\underset{NH_3^{\oplus}Cl^{\ominus}}{R{-}CH{-}COOH} + R'{-}OH \xrightarrow{H^{\oplus}} \underset{NH_3^{\oplus}Cl^{\ominus}}{R{-}CH{-}COOR'} + H_2O \quad (1.9)$$

The free ester is released from its salt by the action of alkali. A mixture of free esters can then be separated by distillation without decomposition. Fractional distillation of esters is the basis of a method introduced by *Emil Fischer* for the separation of amino acids:

$$R{-}CH(NH_3^{\oplus}X^{\ominus}){-}COOR' \xrightarrow{B} R{-}CH(NH_2){-}COOR' + BH^{\oplus}X^{\ominus} \quad (1.10)$$

Free amino acid esters have a tendency to form cyclic dipeptides or open-chain polypeptides:

(1.11a)

$$-NH{-}CH(R){-}C(=O){-}OR' + H_2N{-}CH(R){-}C(=O){-} \longrightarrow -NH{-}CH(R){-}CO{-}NH{-}CH(R){-}CO{-} + R'OH \quad (1.11b)$$

Tert-butyl esters, which are readily split by acids, or benzyl esters, which are readily cleaved by HBr/glacial acetic acid or catalytic hydrogenation, are used as protective groups in peptide synthesis.

1.2.4.2 Reactions of Amino Groups

1.2.4.2.1 Acylation

Activated acid derivatives, e.g. acid halogenides or anhydrides, are used as acylating agents:

$$R'{-}COX + H_2N{-}CH(R){-}COO^{\ominus} + OH^{\ominus} \longrightarrow R'{-}CO{-}NH{-}CH(R){-}COO^{\ominus} + X^{\ominus} + H_2O \quad (1.12)$$

N-acetyl amino acids are being considered as ingredients in chemically-restricted diets and for fortifying plant proteins to increase their biological value. Addition of free amino acids to food which must be heat treated is not problem free. For example, methionine in the presence of a reducing sugar can form methional by a *Strecker* degradation mechanism, imparting an off-flavor to food. Other essential amino acids, e.g. lysine or threonine, can lose their biological value through similar reactions. Feeding tests with rats have shown that N-acetyl-L-methionine and N-acetyl-L-threonine have nutritional values equal to those of the free amino acids (this is true also for humans with acetylated methionine). The growth rate of rats is also increased significantly by the α- or ε-acetyl or α,ε-diacetyl derivatives of lysine.

Some readily-cleavable acyl residues are of importance as temporary protective groups in peptide synthesis.

The trifluoroacetyl residue is readily removed by mild base-catalyzed hydrolysis:

$$F_3C{-}C(=O){-}NH{-}R \xrightarrow{OH^{\ominus}} F_3C{-}C(O^{\ominus})(OH){-}NH{-}R \longrightarrow F_3C{-}COO^{\ominus} + H_2N{-}R \quad (1.13)$$

The phthalyl residue can be readily cleaved by hydrazinolysis:

$$\xrightarrow{H_2N{-}NH_2} + H_2N{-}R \quad (1.14)$$

The benzyloxycarbonyl group can be readily removed by catalytic hydrogenation or by hydrolysis with HBr/glacial acetic acid:

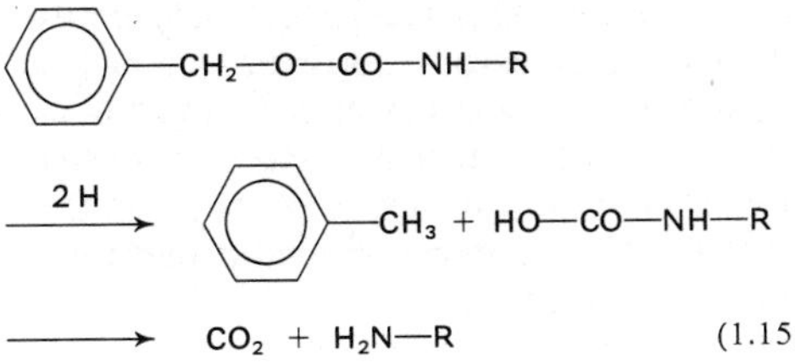

$$C_6H_5{-}CH_2{-}O{-}CO{-}NH{-}R \xrightarrow{2\,H} C_6H_5{-}CH_3 + HO{-}CO{-}NH{-}R \longrightarrow CO_2 + H_2N{-}R \quad (1.15a)$$

$$C_6H_5{-}CH_2{-}O{-}CO{-}NH{-}R \xrightarrow[\text{Glac. acetic acid}]{HBr} C_6H_5{-}CH_2{-}O{-}C^{\oplus}(OH){-}NH{-}R \;(Br^{\ominus}) \longrightarrow C_6H_5{-}CH_2Br + CO_2 + H_2N{-}R \quad (1.15b)$$

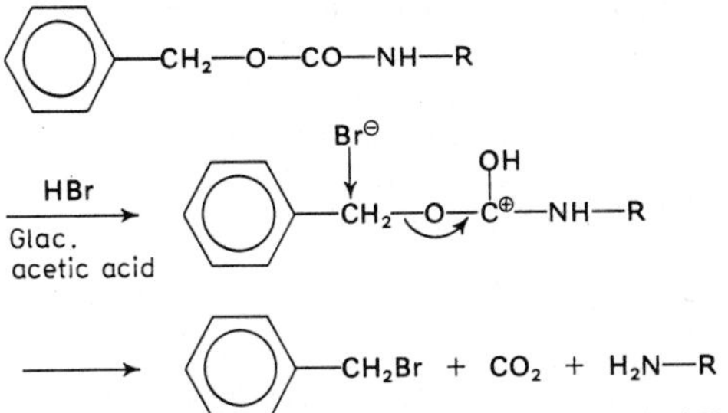

The tert-alkoxycarbonyl residues, e.g. the tert-butyloxycarbonyl groups, are cleaved under acid-catalyzed conditions:

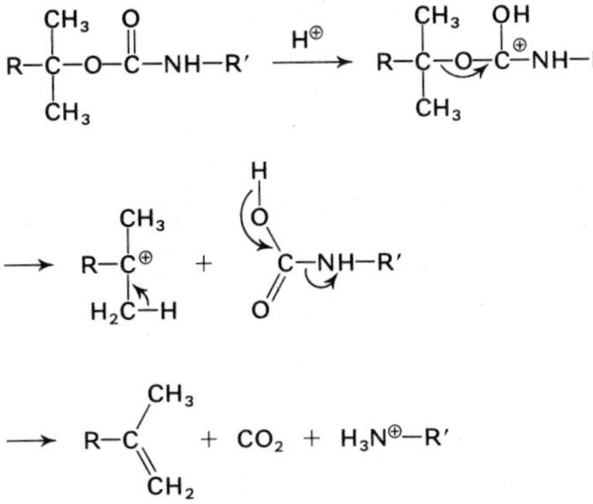

$$R{-}C(CH_3)_2{-}O{-}CO{-}NH{-}R' \xrightarrow{H^{\oplus}} R{-}C(CH_3)_2{-}O{-}C^{\oplus}(OH){-}NH{-}R' \longrightarrow R{-}C^{\oplus}(CH_3)(CH_2{-}H) + HO{-}CO{-}NH{-}R' \longrightarrow R{-}C(CH_3){=}CH_2 + CO_2 + H_3N^{\oplus}{-}R' \quad (1.16)$$

N-acyl derivatives of amino acids are transformed into oxazolinones (azlactones) by elimination of water:

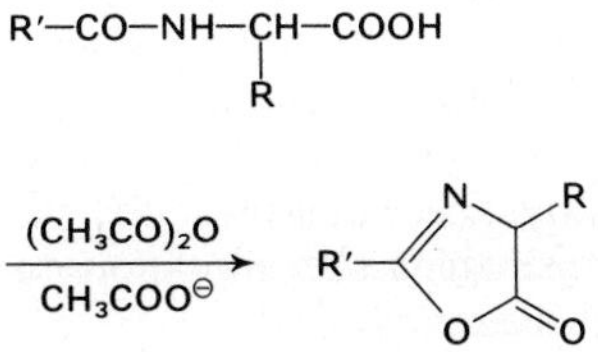

$$R'{-}CO{-}NH{-}CH(R){-}COOH \xrightarrow[CH_3COO^{\ominus}]{(CH_3CO)_2O} \text{azlactone (R', R)} \quad (1.17)$$

These are highly reactive intermediary products which form a mesomerically-stabilized anion. The anion can then react, for example, with aldehydes. This reaction is utilized in amino acid synthesis with glycine azlactone as a starting compound:

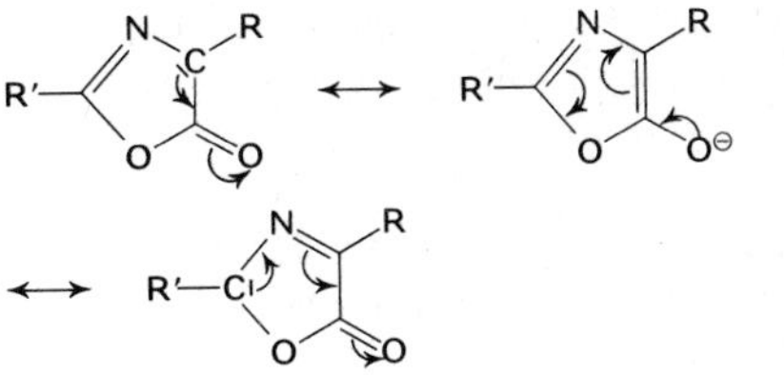

(1.18a)

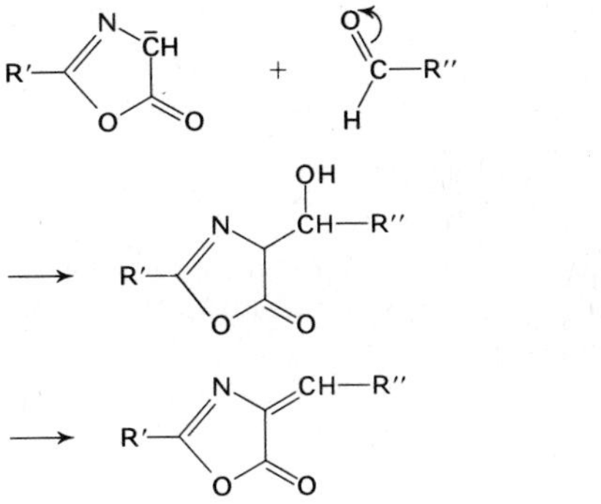

$$\text{glycine azlactone (R')} + O{=}C(H){-}R'' \longrightarrow \text{4-[CH(OH)}{-}R''\text{]-azlactone} \longrightarrow \text{4-(}CH{-}R''\text{)-azlactone} \quad (1.18b)$$

Acylation of amino acids with 5-dimethylaminonaphthalene-1-sulfonyl chloride (dansyl chloride) is of great analytical importance:

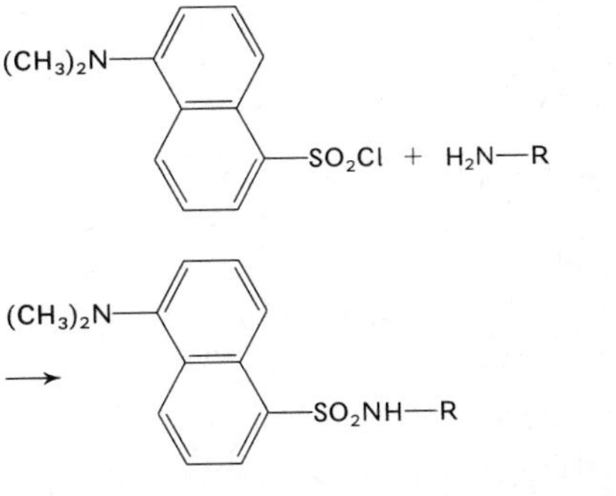

$$(CH_3)_2N{-}C_{10}H_6{-}SO_2Cl + H_2N{-}R \longrightarrow (CH_3)_2N{-}C_{10}H_6{-}SO_2NH{-}R \quad (1.19)$$

The aryl sulfonyl derivatives are very stable against acidic hydrolysis. Therefore, they are suitable for the determination of free N-terminal amino groups or free ε-amino groups of peptides or proteins. Dansyl derivatives which fluoresce in UV-light have a detection limit in the nanomole range, which is lower than that of 2,4-dinitrophenyl derivatives by a factor of 100.

1.2.4.2.2 Alkylation and arylation

N-methyl amino acids are obtained by reaction of the N-tosyl derivative of the amino acid, with methyl iodide followed by removal of the tosyl substituent with HBr:

$$H_3C{-}C_6H_4{-}SO_2{-}NH{-}CH(R){-}COOH \xrightarrow{CH_3J} H_3C{-}C_6H_4{-}SO_2{-}N(CH_3){-}CH(R){-}COOH \xrightarrow{HBr} H_3C{-}NH{-}CH(R){-}COOH \quad (1.20)$$

The N-methyl compound can also be formed by methylating the benzyl derivative of the amino acid, formed initially by reaction of the amino acid with benzaldehyde, with HCHO/HCOOH. The benzyl group is then eliminated by hydrogenolysis:

$$C_6H_5{-}CHO + H_2N{-}R \rightarrow C_6H_5{-}CH{=}N{-}R \xrightarrow{H_2\ catalyst} C_6H_5{-}CH_2{-}NH{-}R \xrightarrow{HCHO/HCOOH} C_6H_5{-}CH_2{-}N(CH_3){-}R \xrightarrow{H_2\ catalyst} HN(CH_3){-}R \quad (1.21)$$

Dimethyl amino acids are obtained by reaction with formaldehyde followed by reduction with sodium borohydride:

$$2\,HCHO + H_2N{-}R \xrightarrow[pH\ 9,\ 0\,^{\circ}C]{NaBH_4} (CH_3)_2N{-}R \quad (1.22)$$

The corresponding reactions with proteins are considered to be used as a means of protecting the ε-amino groups and, thus, of avoiding their destruction in food through the *Maillard* reaction (cf. 1.4.6.2.2).

Direct reaction of amino acids with methylating agents, e.g. methyl iodide or dimethyl sulfate, proceeds through monomethyl and dimethyl compounds to trimethyl derivatives (or generally to N-trialkyl derivatives) denoted as betaines:

$$H_2N{-}CH(R){-}COOH \xrightarrow{CH_3J} (CH_3)_3N^{\oplus}{-}CH(R){-}COO^{\ominus} \quad (1.23)$$

As shown in Table 1.5, betaines are widespread in both the animal and plant kingdoms.

Table 1.5. Occurrence of trimethyl amino acids $(CH_3)_3N^+-CHR-COO^-$ (Betaines)

Amino acid	Betaine	Occurence
β-Alanine	Homobetaine	Meat extract
γ-Amino-butyric acid	Actinine	Mollusc (shell-fish)
Glycine	Betaine	Sugar beet, other samples of animal and plant origin
Histidine	Hercynine	Mushrooms
β-Hydroxy-γ-amino-butyric acid	Carnitine	Mammals muscle tissue, yeast, wheat germ, fish, liver, whey, mollusc (shell-fish)
4-Hydroxy-proline	Betonicine	Jack beans
Proline	Stachydrine	Stachys, orange leaves, lemon peel, alfalfa, Aspergillus oryzae

Derivatization of amino acids by reaction with 1-fluoro-2,4-dinitrobenzene (FDNB) yields N-2,4-dinitrophenyl amino acids (DNP-amino acids), which are yellow compounds and crystallize readily. The reaction is important for labeling N-terminal amino acid residues and free ε-amino groups present in peptides and proteins; the DNP-amino acids are stable under conditions of acidic hydrolysis (cf. Reaction 1.24).

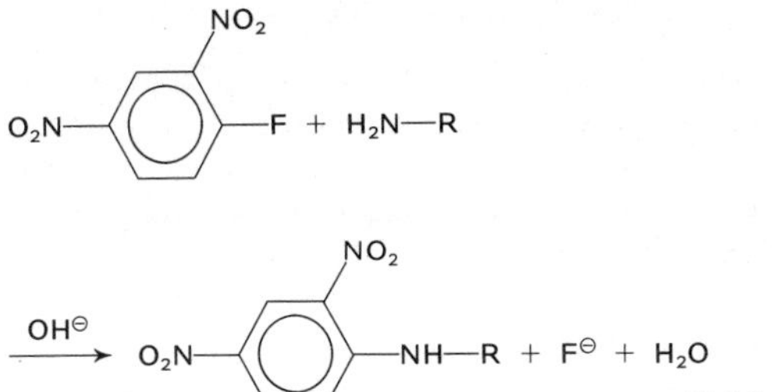

(1.24)

Reaction of amino acids with triphenylmethyl chloride (tritylchloride) yields N-trityl derivatives, which are alkali stable. However, the derivative is cleaved in the presence of acid, giving a stable triphenylmethyl cation and free amino acid:

$$(C_6H_5)_3C{-}Cl + H_2N{-}R \xrightarrow{OH^{\ominus}} (C_6H_5)_3C{-}NH{-}R \xrightarrow{H^{\oplus}} (C_6H_5)_3C^{\oplus} + H_2N{-}R \quad (1.25)$$

The reaction with trinitrobenzene sulfonic acid is also of analytical importance. It yields a yellow-colored derivative that can be used for the spectrophotometric determination of protein:

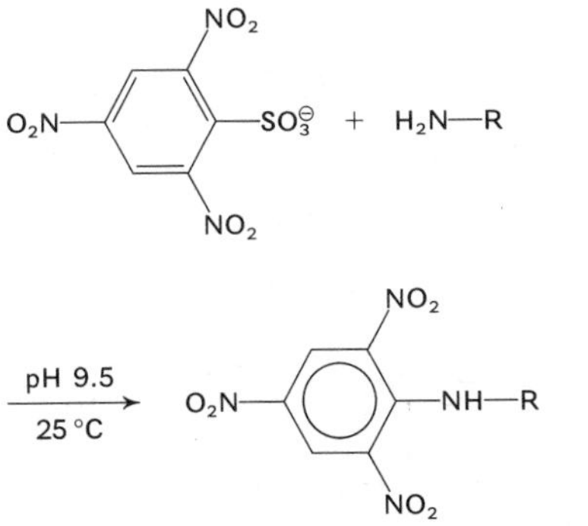

(1.26)

The reaction is a nucleophilic aromatic substitution proceeding through an intermediary addition-product (*Meisenheimer*-complex). It occurs under mild conditions only when the benzene ring structure is stabilized by electron-withdrawing substituents on the ring (cf. Reaction 1.27).

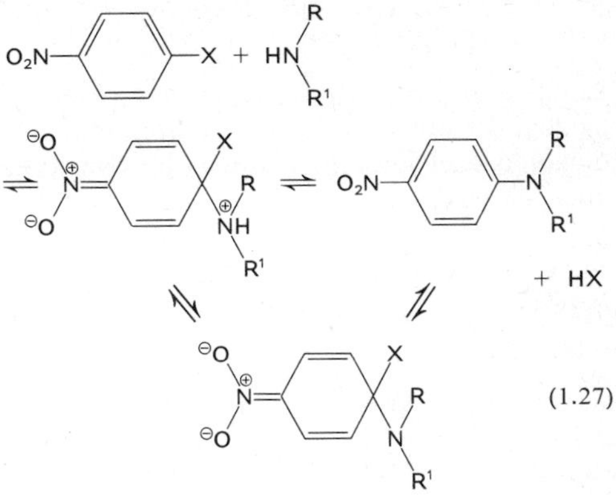

(1.27)

The formation of the *Meisenheimer*-complex has been verified by isolating the addition product from the reaction of 2,4,6-trinitroanisole with potassium ethoxide (cf. Reaction 1.28).

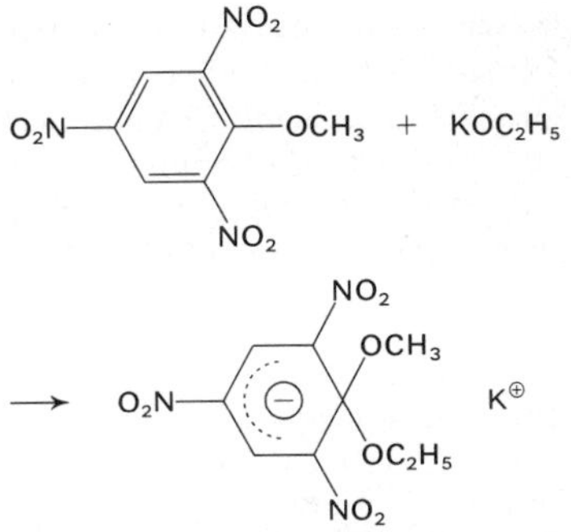

(1.28)

An analogous reaction occurs with 1,2-naphthoquinone-4-sulfonic acid (*Folin* reagent) but, instead of a yellow color (cf. Formula 1.26), a red color develops:

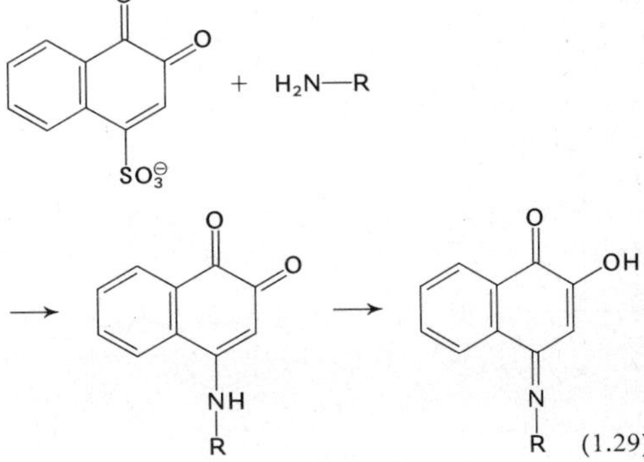

(1.29)

1.2.4.2.3 Carbamoyl and thiocarbamoyl derivatives

Amino acids react with isocyanates to yield carbamoyl derivatives which are cyclized into 2,4-dioxoimidazolidines (hydantoins) by boiling in an acidic medium:

$$R'{-}N{=}C{=}O + H_2N{-}CH(R){-}COOH \rightarrow \ldots \xrightarrow{H^{\oplus}} \ldots \qquad (1.30)$$

A corresponding reaction with phenylisothiocyanate can degrade a peptide in a stepwise fashion (*Edman* degradation). The reaction is of great importance for revealing the amino acid sequence in a peptide chain. The initial reaction involves the formation of an anilino-thiazolinone. This is unstable in an acidic medium and hydrolyzes to the phenylthiocarbamoyl amino acid, which then cyclizes into the corresponding 3-phenyl-2-thiohydantoin under suitable conditions:

$$C_6H_5{-}N{=}C{=}S + H_2N{-}CH(R){-}CO{-}NH{-}R' \rightarrow \ldots \xrightarrow{H^{\oplus}} \ldots + H_3N^{\oplus}{-}R' \xrightarrow{H_2O} \ldots \rightarrow \ldots \qquad (1.31)$$

1.2.4.2.4 Reactions with carbonyl compounds

Amino acids react with carbonyl compounds, forming azomethines. If the carbonyl compound has an electron-withdrawing group, e.g. a second carbonyl group, transamination and decarboxylation occur:

$$\begin{matrix}-C{=}O\\-C{=}O\end{matrix} + H_2N{-}CHR{-}COOH \rightarrow \begin{matrix}-C{=}N{-}CHR{-}COOH\\-C{=}O\end{matrix} \rightarrow \begin{matrix}-C{-}N{=}CHR\\-C{-}OH\end{matrix} + CO_2 \xrightarrow{H_2O} \begin{matrix}-CH{-}NH_2\\-C{=}O\end{matrix} + O{=}CHR \qquad (1.32)$$

$$\text{(I)} + H_2N{-}CHR{-}COOH \rightarrow \ldots{-}N{=}CHR \rightleftharpoons \ldots{-}N{=}CHR \xrightarrow{H_2O} \ldots{-}NH_2 + O{=}CHR \xrightarrow{I} \ldots \qquad (1.33)$$

The reaction is known as the *Strecker* degradation and plays a role in food since food can be an abundant source of dicarbonyl compounds generated by the *Maillard* reaction (cf. 4.2.4.4). The aldehydes which are formed from amino acids (*Strecker* aldehydes) are aroma compounds.

The ninhydrin reaction is a special case of the *Strecker* degradation. It is an important reaction for the quantitative determination of amino acids using spectrophotometry (cf. Reaction 1.33).

The resultant blue-violet color has an absorption maximum at 570 nm, except for reaction with proline, which yields a yellow-colored compound with $\lambda_{max} = 440$ nm:

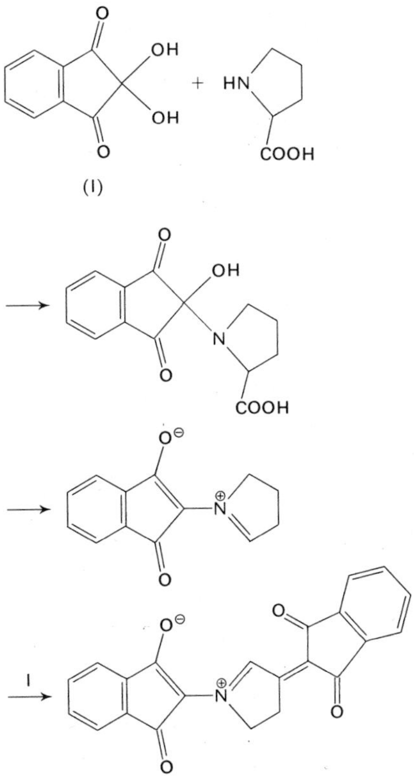

(I)

(1.34)

1.2.4.3 Reactions Involving Other Functional Groups

The most interesting of these reactions are those in which α-amino and α-carboxyl groups are blocked, that is, reactions occurring with peptides and proteins. These reactions will be covered in detail in sections dealing with modification of proteins (cf. 1.4.4 and 1.4.6.2). A number of reactions of importance to free amino acids will be covered in the following sections.

1.2.4.3.1 Lysine

A selective reaction may be performed with either of the amino groups in lysine. Selective acylation of the ε-amino group is possible using the lysine-Cu^{2+} complex as a reactant:

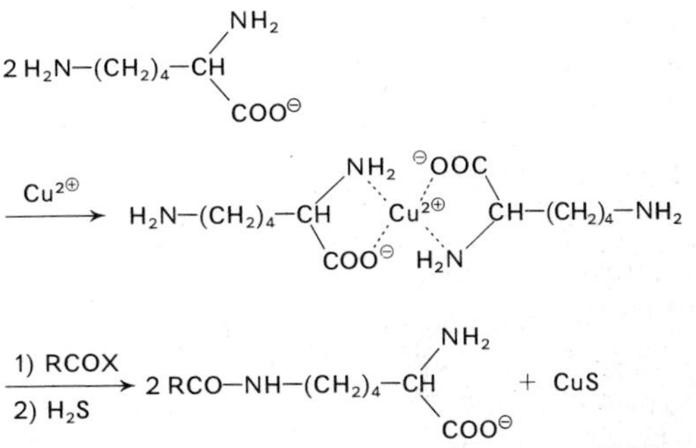

$$2\,H_2N{-}(CH_2)_4{-}CH(NH_2){-}COO^{\ominus} \xrightarrow{Cu^{2\oplus}} [H_2N{-}(CH_2)_4{-}CH(NH_2)COO^{\ominus}]_2Cu^{2\oplus} \xrightarrow[2)\ H_2S]{1)\ RCOX} 2\,RCO{-}NH{-}(CH_2)_4{-}CH(NH_2){-}COO^{\ominus} + CuS \quad (1.35)$$

Selective reaction with the α-amino group is possible using a benzylidene derivative:

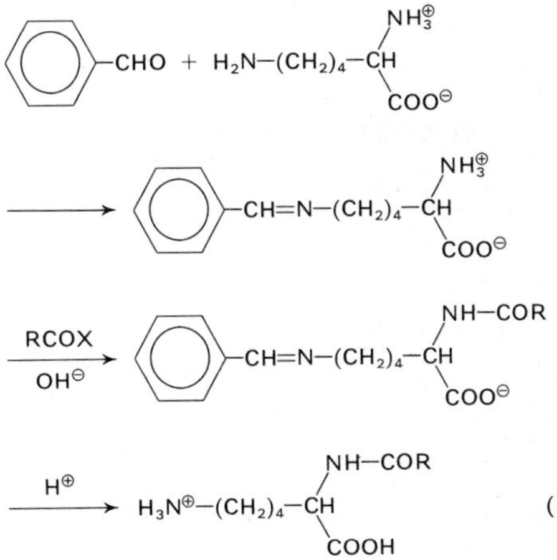

$$C_6H_5{-}CHO + H_2N{-}(CH_2)_4{-}CH(NH_3^{\oplus}){-}COO^{\ominus} \longrightarrow C_6H_5{-}CH{=}N{-}(CH_2)_4{-}CH(NH_3^{\oplus}){-}COO^{\ominus} \xrightarrow[OH^{\ominus}]{RCOX} C_6H_5{-}CH{=}N{-}(CH_2)_4{-}CH(NH{-}COR){-}COO^{\ominus} \xrightarrow{H^{\oplus}} H_3N^{\oplus}{-}(CH_2)_4{-}CH(NH{-}COR){-}COOH \quad (1.36)$$

ε-N-benzylidene-L-lysine and ε-N-salicylidene-L-lysine are as effective as free lysine in growth feeding tests with rats. Browning reactions of these derivatives are strongly retarded, hence they are of interest for lysine fortification of food.

1.2.4.3.2 Arginine

In the presence of α-naphthol and hypobromite, the guanidyl group of arginine gives a red compound with the following structure:

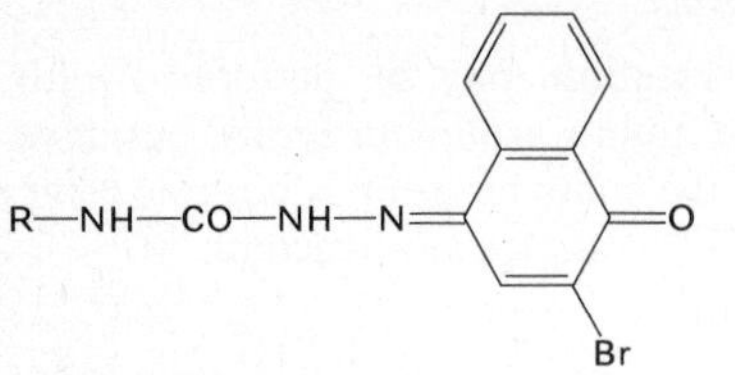

R—NH—CO—NH—N= [naphthoquinone, 3-Br] =O (1.37)

1.2.4.3.3 Aspartic and glutamic acids

The higher esterification rate of β- and γ-carboxyl groups can be used to advantage for selective reactions. On the other hand the β- and γ-carboxyl groups are more rapidly hydrolyzed in acid-catalyzed hydrolysis since protonation is facilitated by having the ammonium group further away from the carboxyl group. Alkali-catalyzed hydrolysis of methyl or ethyl esters of aspartic or glutamic acids bound to peptides can result in the formation of isopeptides:

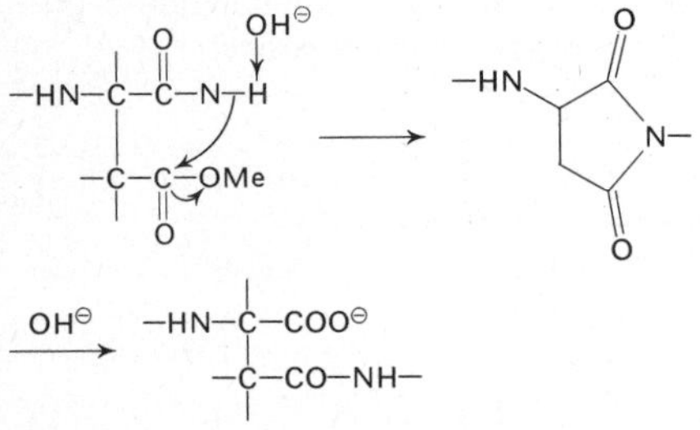

$$-HN-C(-C(=O)-NH-) \, ; \, -C-C(=O)-OMe \xrightarrow{OH^{\ominus}} \text{(succinimide)} \xrightarrow{OH^{\ominus}} -HN-C-COO^{\ominus} \, ; \, -C-CO-NH- \quad (1.38)$$

1.2.4.3.4 Serine and threonine

Acidic or alkaline hydrolysis of protein can yield α-keto acids through β-elimination of a water molecule:

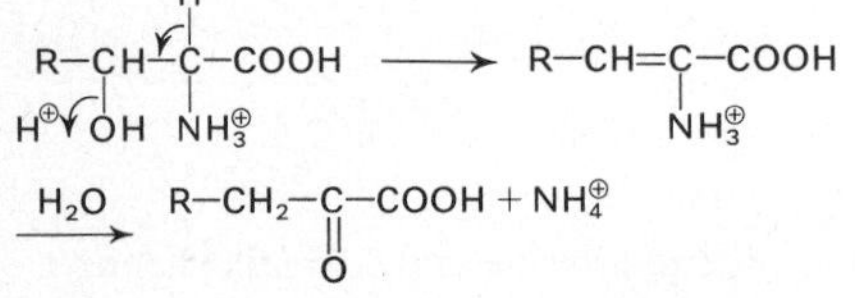

$$R-CH(OH)-CH(NH_3^{\oplus})-COOH \longrightarrow R-CH=C(NH_3^{\oplus})-COOH \xrightarrow{H_2O} R-CH_2-C(=O)-COOH + NH_4^{\oplus} \quad (1.39)$$

In this way, α-ketobutyric acid formed from threonine can yield another amino acid, α-aminobutyric acid, via a transamination reaction. Reaction 1.39 is responsible for losses of hydroxy amino acids during protein hydrolysis.

Reliable estimates of the occurrence of these amino acids are obtained by hydrolyzing protein for varying lengths of time and extrapolating the results to zero time.

1.2.4.3.5 Cysteine and cystine

Cysteine is readily converted to the corresponding disulfide, cystine, even under mild oxidative conditions, such as treatment with I_2 or potassium hexacyanoferrate(III). Reduction of cystine to cysteine is possible using sodium borohydride or thiol reagents (mercaptoethanol, dithiothreitol):

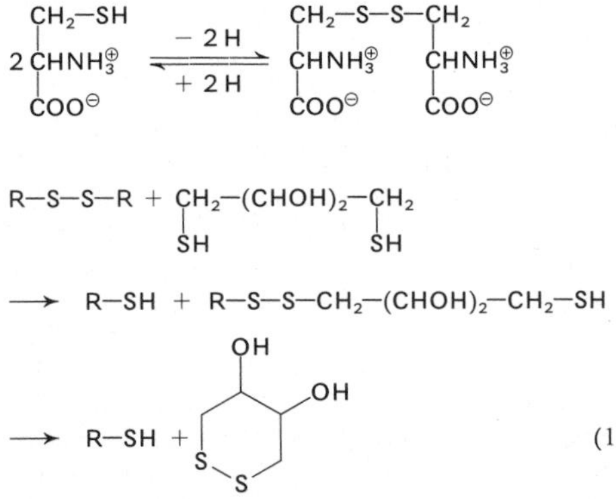

$$2\,HS-CH_2-CHNH_3^{\oplus}-COO^{\ominus} \underset{+2H}{\overset{-2H}{\rightleftharpoons}} COO^{\ominus}-CHNH_3^{\oplus}-CH_2-S-S-CH_2-CHNH_3^{\oplus}-COO^{\ominus}$$

$$R-S-S-R + CH_2(SH)-(CHOH)_2-CH_2(SH)$$

$$\longrightarrow R-SH + R-S-S-CH_2-(CHOH)_2-CH_2-SH$$

$$\longrightarrow R-SH + \text{(4,5-dihydroxy-1,2-dithiane)} \quad (1.40)$$

The equilibrium constants for the reduction of cystine at pH 7 and 25 °C with mercaptoethanol or dithiothreitol are 1 and 10^4, respectively.

Stronger oxidation of cysteine, e.g. with performic acid, yields the corresponding sulfonic acid, cysteic acid:

$$R-SH,\; R-S-S-R \xrightarrow{HCOOOH} R-SO_3H \quad (1.41)$$

Reaction of cysteine with alkylating agents yields thioethers. Iodoacetic acid, iodoacetamide, ethylenimine and vinylpyridine are the most commonly used alkylating agents:

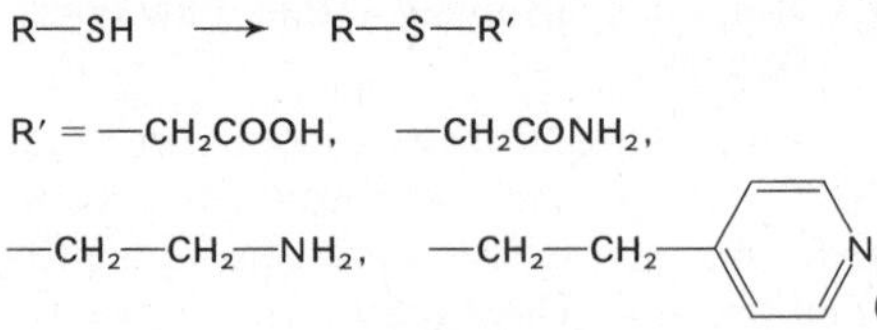

$$R-SH \longrightarrow R-S-R'$$

$$R' = -CH_2COOH, \quad -CH_2CONH_2, \quad -CH_2-CH_2-NH_2, \quad -CH_2-CH_2-\text{(4-pyridyl)} \quad (1.42)$$

1.2.4.3.6 Methionine

Methionine is readily oxidized to the sulfoxide and then to the sulfone. This reaction can result in losses of this essential amino acid during food processing:

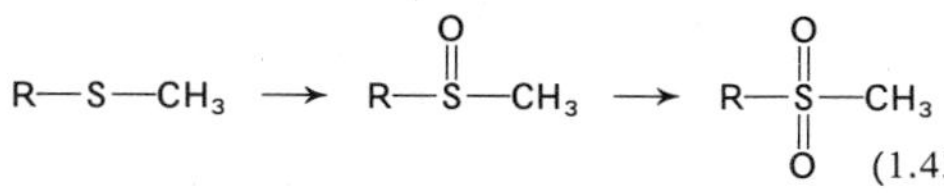

$$R{-}S{-}CH_3 \longrightarrow R{-}\overset{O}{\overset{\|}{S}}{-}CH_3 \longrightarrow R{-}\underset{O}{\underset{\|}{\overset{O}{\overset{\|}{S}}}}{-}CH_3 \quad (1.43)$$

1.2.4.3.7 Tyrosine

Tyrosine reacts, like histidine, with diazotized sulfanilic acid (*Pauly* reagent). The coupled-reaction product is a red azo compound:

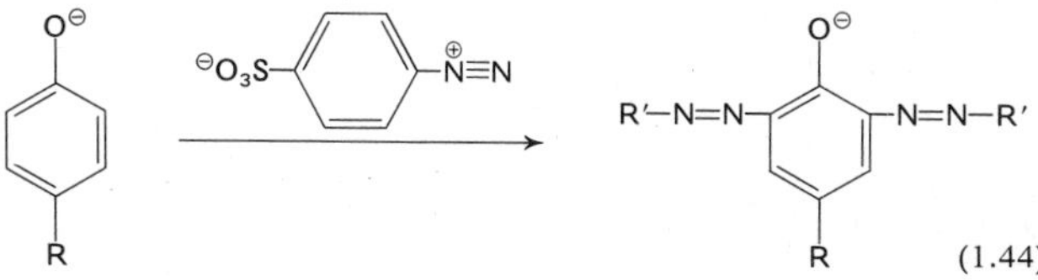

$$\text{(1.44)}$$

1.2.4.4 Thermal Degradation of Amino Acids

Reactions at elevated temperatures are important since they can occur during the preparation of food. Frying, roasting, boiling and baking develop the typical aromas of many foods which reflect the degree of heat treatment. Sulfur-containing compounds are often of importance. The presence of more than 100 volatile sulfur compounds (thiols, sulfides, disulfides, thiophenes, thiazoles and other S-heterocyclics) has been confirmed in heated meat, and the patent literature shows that heating sulfur-containing amino acids with a number of other components generates a meat-like aroma (Table 1.6).

Table 1.6. Examples of patents covering the aroma of processed meat

Inventor	Company	Patent number	Year	Reactions involved
May and Morton	Unilever	Brit. 858.660	1961	Amino acid source (must contain cysteine) + aldehyde + pentoses in water and refluxed.
Hack and Königsdorf	Corn Products Co.	US 3.480.447	1969	Amino acids (devoid of cysteine) + hydrolyzed plant proteins + reducing sugar + taurine: heated at 110 °C for 15–20 h.
Giacino	IFF	US 3.519.437	1970	Taurine + thiamine + hydrolyzed plant proteins + water or fat + heating or "flash heating".
O'Hara, Ota, Enei, Eguchi, Okamura	Ajinomoto	US 3.524.747	1970	Amino acids (devoid of cysteine) + lactic acid + phosphate + nucleoside-5′-phosphate
Kitada et al.	Ajinomoto	US 3.620.772	1971	Cysteine + reducing sugar heated at 50–120 °C and 70–200 kg/cm^2 for 1–10 h.
Tonsbeek	Unilever	Can. 862.685	1971	Amino acids + lactic acid + 5′-nucleotide, heated at 110–150 °C.
van Pottelsberghe, de la Potterie	Nestlé	US 3.716.380	1973	Protein hydrolyzate + water + methionine + sugar + by choice a carboxylic acid + heat.
van Pottelsberghe, de la Potterie	Nestlé	US 3.716.379	1973	Hydrolyzed plant protein (devoid of cysteine) + thiamine + a mono- or polysaccharide + water + heat.

Thermal decomposition of amino acids has been studied in detail in model systems. Table 1.7 is a compilation of the compounds obtained by pyrolysis of cystine and methionine under nitrogen at 270–300 °C. The following reaction pathway has been postulated:

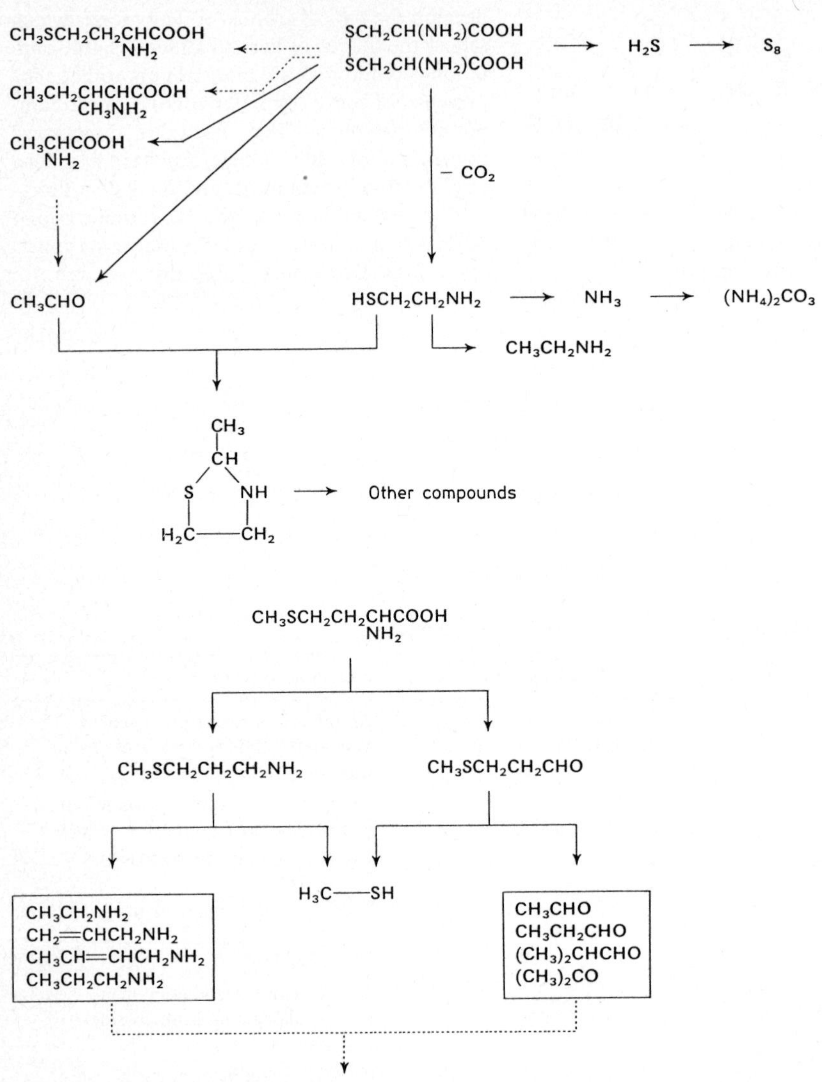

(1.45)

As a model reaction for the frying of food in fat, cysteine was heated in tributyrin at 200–220 °C for 2 h. Figure 1.5 presents the separation of the reaction products by gas chromatography. The following reaction pathways are assumed to occur (the numerals in Reaction 1.46 refer to Figure 1.5):

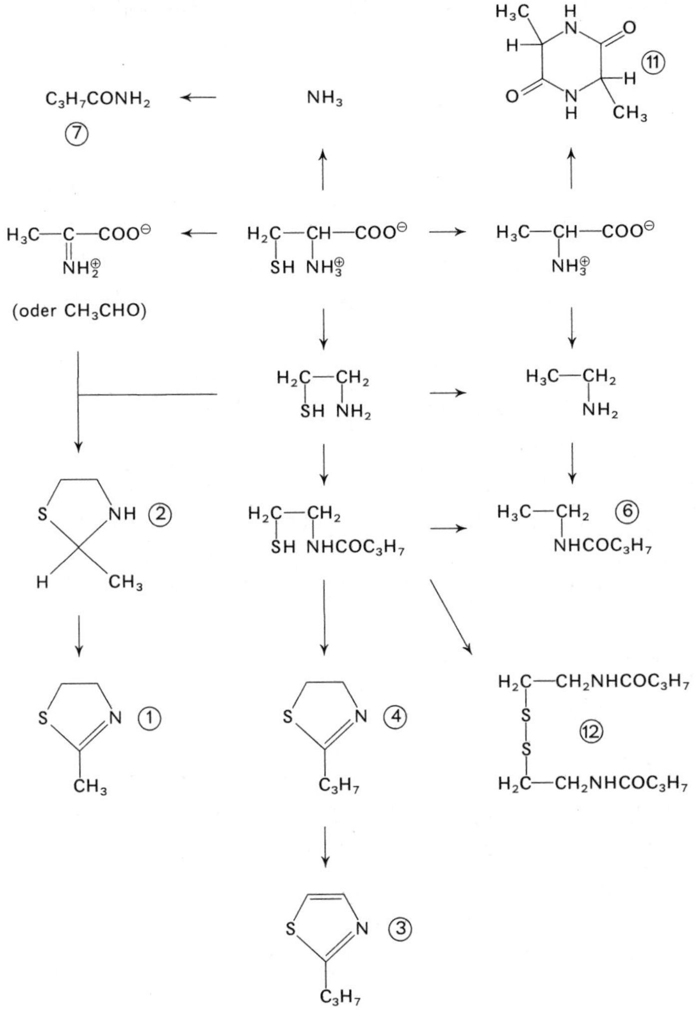

$$H_2C(SH)-CH(NH_3^{\oplus})-COO^{\ominus} \rightarrow NH_3 \rightarrow C_3H_7CONH_2\ (7)$$

$$H_2C(SH)-CH(NH_3^{\oplus})-COO^{\ominus} \rightarrow H_3C-C(=NH_2^{\oplus})-COO^{\ominus}\ (\text{oder } CH_3CHO)$$

$$H_2C(SH)-CH(NH_3^{\oplus})-COO^{\ominus} \rightarrow H_3C-CH(NH_3^{\oplus})-COO^{\ominus} \rightarrow (11)\ \text{3,6-dimethyl-2,5-dioxopiperazine}$$

$$H_2C(SH)-CH(NH_3^{\oplus})-COO^{\ominus} \rightarrow H_2C(SH)-CH_2(NH_2) \rightarrow H_3C-CH_2(NH_2) \rightarrow H_3C-CH_2(NHCOC_3H_7)\ (6)$$

$$H_2C(SH)-CH_2(NH_2) + CH_3CHO \rightarrow (2)\ \text{2-methylthiazolidine} \rightarrow (1)\ \text{2-methyl-2-thiazoline}$$

$$H_2C(SH)-CH_2(NH_2) \rightarrow H_2C(SH)-CH_2(NHCOC_3H_7) \rightarrow (4)\ \text{2-propyl-2-thiazoline} \rightarrow (3)\ \text{2-propylthiazole}$$

$$H_2C(SH)-CH_2(NHCOC_3H_7) \rightarrow (12)\ H_2C(-CH_2NHCOC_3H_7)-S-S-H_2C(-CH_2NHCOC_3H_7) \tag{1.46}$$

Table 1.7. Pyrolysis products of cystine[a] and methionine (at 270–300 °C under nitrogen at 30 mm Hg for 1 h)

Compound	Formed[b] from	
	Cystine	Methionine
1) Hydrogen sulfide	+++	
2) Methanethiol		+++
3) 2-Methylthiazolidine	+++	
4) Thiopropylamine	++	
5) 3-Methylthiopropylamine		++
6) Ethylamine	++	+
7) 2-Propenylamine		++
8) 2-Butenylamine		+
9) Ethanal	++	++
10) Propanal		++
11) Methylthiopropanal		++
12) Acetone		+
13) Isobutanal		+
14) Ammonium carbonate	+++	
15) Ammonia	+	
16) Sulfur	+	
17) Alanine	+	
18) Isoleucine	(+)	
19) Methionine	(+)	

[a] With the exceptions of sulfur and methionine all the listed compounds are formed also from cysteine
[b] Amounts +++: high; ++: medium; +: low; and (+) negligible

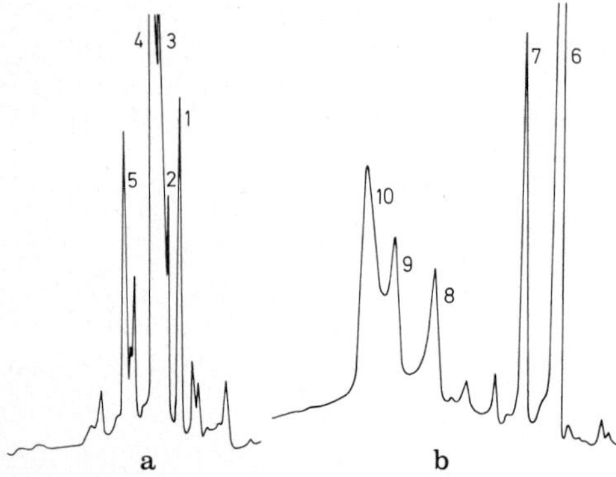

Fig. 1.5. Reaction products for cysteine heated in tributyrin (reflux at 200–220 °C for 2 h) (After *Severin* and *Ledl,* 1975). Gas chromatographic separation of a light (a) and a less volatile (b) fraction. Separation conditions: 20% carbowax on kieselgel, column lenght 5 m, diameter 1 cm; column temperature 70° (a), 80° (b), and after 15 min programming with 5 °C/min till 220 °C. 1 2-Methylthiazoline, 2 2-methylthiazolidine, 3 2-propylthiazole, 4 2-propylthiazoline, 5 butyric acid, 6 N-ethylbutyric acid amide, 7 butyric acid amide, 8, 9, 10 butyric acid mono- and diglycerides, 11 2,5-dimethyldioxopiperazine, 12 N,N′-dibutyrylcystamine.

Fat acts as an acylating agent at high temperatures. For example, 2-propylthiazoline is obtained as the main reaction product from the acylation of cysteamine (generated by the decarboxylation of cysteine), followed by cyclization. Dehydrogenation to the corresponding thiazole can occur in the presence of oxygen or free radicals. The formation of N-ethylbutyramide in relatively high yield is explained by a radical-type elimination of sulfur during pyrolysis of cystine. The occurrence of alanine and ethylamine in the pyrolysis reaction products is explained in a similar fashion.

1.2.5 Synthetic Amino Acids Utilized for Increasing the Biological Value of Food (Food Fortification)

The daily requirements of humans for essential amino acids and their occurrence in some important food proteins are presented in Table 1.8. The biological value of a protein (g protein formed in the body/100 g food protein) is determined by the absolute content of essential amino acids, by the relative proportions of essential amino acids, and by their ratios to nonessential amino acids. The highest biological value observed is for a blend of 35% egg and 65% potato proteins.

Table 1.8. Adult requirement for essential amino acids and their occurence in various food

Amino acids	1	2	3	4	5	6	7	8	9
Isoleucine	10–11	3.5	4.0	4.6	3.9	3.6	3.4	5.0	3.5
Leucine	11–14	4.2	5.3	7.1	4.3	5.1	6.5	8.2	5.4
Lysine	9–12	3.5	3.7	4.9	3.6	4.4	2.0	3.6	5.4
Methionine + Cystine	11–14	4.2	3.2	2.6	1.9	2.1	3.8	3.4	1.9
Methionine		2.0	1.9	1.9	1.2	0.9	1.4	2.2	0.8
Phenylalanine + Tyrosine	13–14	4.5	6.1	7.2	5.8	5.5	6.7	8.9	6.0
Phenylalanine		2.4	3.5	3.5	3.1	3.3	4.6	4.7	2.5
Threonine	6– 7	2.2	2.9	3.3	2.9	2.7	2.5	3.7	3.8
Tryptophan	3	1.0	1.0	1.0	1.0	1.0	1.0	1.0	1.0
Valine	11–14	4.2	4.3	5.6	3.6	3.3	3.8	6.4	4.1
Tryptophan[a]			1.7	1.4	1.4	1.5	1.1	1.0	1.3

1: Daily requirement in mg/kg body weight
2–8: Relative value related to Trp = 1 (pattern)
2: Daily requirements, 3: eggs, 4: bovine milk, 5: potato, 6: soya, 7: wheat flour, 8: rice, and 9: *Torula*-yeast
[a] Tryptophan (%) in raw protein

The biological value of a protein is generally limited by:

- Lysine: deficient in proteins of cereals and other plants
- Methionine: deficient in proteins of bovine milk and meat
- Threonine: deficient in wheat and rye
- Tryptophan: deficient in casein, corn and rice.

Since food is not available in sufficient quantity or quality in many parts of the world, increasing its biological value by addition of essential amino acids is gaining in importance. Illuminating examples are rice fortification with L-lysine and L-threonine, supplementation of bread with L-lysine and fortification of soya and peanut protein with methionine. Synthetic amino acids are used also for chemically defined diets.
The fortification of animal feed with amino acids (0.05–0.2%) is of great significance.
These demands have resulted in increased production of amino acids. Table 1.9 gives data for world production in 1978. The production of L-glutamic acid, used to a great extent as a flavor enhancer, is exceptional. Production of methionine and lysine is also significant.

Three main processes are distinguished in the production of amino acids: chemical synthesis, isolation from protein hydrolysates and microbiological methods of production, which is currently the most important. The following sections will further elucidate the important industrial processes for a number of amino acids.

Table 1.9. World production of amino acids, 1979

Amino acid	t/year	Process[a]					Mostly utilized as
		1	2	3			
				a	b	c	
L-Ala	130					+	Flavoring compound
D,L-Ala	700	+		(+)			Flavoring compound
L-Arg	500		(+)	+			Infusion Therapeutics
L-Asp	250					+	Therapeutics Flavoring compound
L-Asn	50	+					Therapeutics
L-CySH	700		+				Baking additive Antioxidant
L-Glu	270.000			+			Flavoring compound, flavor enhancer
L-Gln	500			+			Therapeutics
Gly	6.000	+					Sweetener
L-His	200			+			Therapeutics
L-Ile	150			+			Infusion
L-Leu	150		+				Infusion
L-Lys	32.000			+			Feed ingredient
D,L-Met	180.000	+					Feed ingredient
L-Phe	150	+					Infusion
L-Pro	100			+			Infusion
L-Ser	50				+		Cosmetics
L-Thr	160	(+)		+			Food additive
L-Trp	200	+					Infusion
L-Tyr	100		+				Infusion
L-Val	150			+			Infusion

[a] 1: Chemical synthesis, 2: protein hydrolysis, 3: microbiological procedure (a: direct synthesis from glucose or other simple C-source, b: synthesis from precursors, and c: enzymatic process).

1.2.5.1 Glutamic Acid

Acrylnitrile is catalytically formylated with CO/H_2 and the resultant aldehyde is transformed through a *Strecker* reaction into glutamic acid dinitrile which yields D,L-glutamic acid after alkaline hydrolysis. Separation of the racemate is achieved by preferential crystallization of the L-form from an oversaturated solution after seeding with L-glutamic acid:

$$H_2C{=}CH{-}CN \xrightarrow[{[Co(CO)_4]_2}]{CO/H_2} OHC{-}CH_2{-}CH_2{-}CN \xrightarrow{HCN/NH_3} NC{-}CH(NH_2){-}CH_2{-}CH_2{-}CN \xrightarrow{OH^{\ominus}} \text{D,L-Glu} \quad (1.47)$$

A fermentation procedure with various selected strains of microorganisms *(Brevibacterium flavum, Brev. roseum, Brev. saccharolyticum)* provides L-glutamic acid in yields of 50 g/l of fermentation liquid:

$$CH_3COONH_4\ (20\ g/l) \xrightarrow[pH\ 7.5]{MO} \text{L-Glu}\ (50\ g/l) \quad (1.48)$$

1.2.5.2 Aspartic Acid

Aspartic acid is obtained in 90% yield from fumaric acid using the aspartase enzyme:

$$\text{Fumaric acid} \xrightarrow[NH_3]{\text{Aspartase}} \text{L-Asp} \quad (1.49)$$

1.2.5.3 Lysine

A synthetic procedure starts with caprolactam, which possesses all the required structural features, except for the α-amino group, which is introduced in several steps:

$$\text{Caprolactam} \xrightarrow{2\,COCl_2} \text{N-COCl (Cl) derivative} \xrightarrow[\text{fl. } SO_2]{HNO_3/SO_3} \text{NO}_2\text{-N-COCl derivative} \rightarrow \text{NO}_2\text{-caprolactam} \xrightarrow[\text{Raney-Ni}]{H_2} \text{NH}_2\text{-caprolactam} \quad (1.50)$$

Separation of isomers is done at the α-amino caprolactam (Acl) step through the sparingly-soluble salt of the L-component with L-pyrrolidone carboxylic acid (Pyg):

D,L-Acl + L-Pyg → D-Acl + L,L-salt (D-Acl recycled: Base)

$$\rightarrow \text{L-Acl} \xrightarrow{HCl} \text{L-Lys} \cdot HCl \quad (1.51)$$

$$NC{-}CH{=}CH_2 + H_3C{-}CHO \xrightarrow{\text{Cyclohexylamine}} NC{-}CH_2{-}CH_2{-}CH_2{-}CHO \xrightarrow[NH_3]{HCN,\ CO_2} \text{NC-(CH}_2)_3\text{-hydantoin} \xrightarrow{H_2 \text{ cat.}} H_2N{-}CH_2{-}(CH_2)_3\text{-hydantoin}$$

$$\xrightarrow{OH^\ominus} \text{D,L-Lysine} \xrightarrow{\text{Sulfanilic acid}} \text{D-salt} + \text{L-salt} \quad \text{(D-salt recycled: Heating)} \quad (1.52)$$

In another procedure, acrylnitrile and ethanal react to yield cyanobutyraldehyde which is then transformed by a *Bucherer* reaction into cyanopropylhydantoin. Catalytic hydrogenation of the nitrile group followed by alkaline hydrolysis yields D,L-lysine. The isomers can be separated through the sparingly-soluble L-lysine sulfanilic acid salt (cf. Reaction 1.52).

Fermentation with a pure culture of *Brevibacterium lactofermentum or Micrococcus glutamicus* produces L-lysine directly:

$$\underset{(<15\ g/l)}{CH_3COONH_4} \xrightarrow[\text{pH } 7-8.5]{MO} \text{L-Lys } (40-90\ g/l) \quad (1.53)$$

1.2.5.4 Methionine

Interaction of methanethiol with acrolein produces an aldehyde which is then converted to the corresponding hydantoin through a *Bucherer* reaction. The product is hydrolyzed by alkaline catalysis. Separation of the resultant racemate is usually not carried out since the D-form of methionine is utilized by humans via transamination:

$$CH_3SH + H_2C{=}CH{-}CHO \rightarrow H_3C{-}S{-}CH_2CH_2{-}CHO \xrightarrow[2)\ OH^\ominus]{1)\ HCN,\ CO_2,\ NH_3} \text{D,L-Met} \quad (1.54)$$

1.2.5.5 Phenylalanine

Benzaldehyde is condensed with hydantoin, then hydrogenation using a chiral catalyst gives a product which is about 90% L-phenylalanine:

$$C_6H_5{-}CHO + \text{hydantoin} \rightarrow C_6H_5{-}CH{=}\text{hydantoin} \xrightarrow[2)\ \text{Hydrolysis}]{1)\ H_2/\text{chiral catalyst}} \text{L-Phe } (90\%) \quad (1.55)$$

1.2.5.6 Threonine

Interaction of a copper complex of glycine with ethanal yields the threo and erythro isomers in the ratio of 2:1. They are separated based on their differences in solubility:

$$\text{Cu}^{2\oplus}\text{-glycine complex} + 2\,CH_3CHO \xrightarrow{OH^{\ominus}} \text{Cu}^{2\oplus}\text{-complex of } H_3C{-}CH(OH){-}CH(NH_2){-}COO^{\ominus} \quad (1.56)$$

D,L-threonine is separated into its isomers through its N-acetylated form with the help of an acylase enzyme.

1.2.5.7 Tryptophan

Tryptophan is obtained industrially by a variation of the *Fischer* indole synthesis. Addition of hydrogen cyanide to acrolein gives 3-cyanopropanal which is converted to hydantoin through a *Bucherer* reaction. The nitrile group is then reduced to an aldehyde group. Reaction with phenylhydrazine produces an indole derivative. Lastly, hydantoin is saponified with alkali (cf. Reaction 1.58).

$$HCN + H_2C{=}CH{-}CHO \longrightarrow NC{-}CH_2{-}CH_2{-}CHO$$

$$\xrightarrow[NH_3]{HCN,\ CO_2} \text{5-(NC–CH}_2\text{–CH}_2\text{)-hydantoin} \xrightarrow{H_2\ cat.} OHC{-}(CH_2)_2{-}R$$

$$\xrightarrow{C_6H_5{-}NH{-}NH_2} \text{phenylhydrazone} \longrightarrow \text{ene-hydrazine} \longrightarrow \text{ortho-aminophenyl enamine} \longrightarrow \text{3-(R–CH}_2\text{)-indole} \xrightarrow{OH^{\ominus}} \text{D,L-Trp} \quad (1.57)$$

L-Tryptophan is also produced through enzymatic synthesis from indole and serine with the help of tryptophan synthase:

$$\text{Indole} + HOH_2C{-}CH(NH_2){-}COOH \longrightarrow \text{L-Trp} \quad (1.58)$$

1.2.6 Sensory Properties

Free amino acids can contribute to the flavor of protein-rich foods in which hydrolytic processes occur (e.g. meat, fish or cheese).

Table 1.10 provides data for taste quality and taste intensity of amino acids. Taste quality is influenced by the molecular configuration: sweet amino acids are primarily found among members of the D-series, whereas bitter amino acids are generally within the L-series. In consequence amino acids with a cyclic side chain (1-aminocycloalkane-1-carboxylic acids) are sweet and bitter.

The taste intensity of a compound is reflected by its recognition threshold value. The recognition threshold value is the lowest concentration needed to recognize the compound reliably, as assessed by a taste panel. Table 1.10 shows that the taste intensity of amino accids is dependent on the hydrophobicity of the side chains.

L-Tryptophan has a threshold value of $c_{tbitter}$ = 4–6 mmole/l and, is second only to tyrosine in its bitterness. D-tryptophan, with c_{tsweet} = 0.2–0.4 mmole/l, is the sweetest amino acid. A comparison of these threshold values with those of caffeine (c_{tbi} = 1–1.2 mmole/l) and sucrose (c_{tsw} = 10–12 mmole/l) shows that caffeine is about 5 times as bitter as L-tryptophan and that D-tryptophan is about 37 times as sweet as sucrose.

L-Glutamic acid has an exceptional value. In higher concentrations it has an imitation meat broth flavor, while in lower concentrations, it enhances the characteristic flavor of a given food (flavor enhancer, cf. 8.6.1). L-Methionine has a sulfur-like flavor.

The bitter taste of the L-amino acids can interfere with the utilization of these acids, e.g. in chemically defined diets.

Table 1.10. Taste of amino acids in aqueous solution (at pH 6–7)
sw – sweet, bi – bitter, neu – neutral

Amino acid	Taste			
	L-Compound		D-Compound	
	Quality	Intensity[a]	Quality	Intensity[a]
Alanine	sw	12–18	sw	12–18
Arginine	bi		neu	
Asparagine	neu		sw	3– 6
Aspartic acid	neu		neu	
Cystine	neu		neu	
Glutamine	neu		neu	8–12
Glutamic acid	meat broth like		neu	
Glycine[b]	sw	25–35		
Histidine	bi	45–50	sw	2– 4
Isoleucine	bi	10–12	sw	8–12
Leucine	bi	11–13	sw	2– 5
Lysine	sw		sw	
	bi	80–90		
Methionine	sulphurous		sulphurous	
Phenylalanine			sw	4– 7
	bi	5–7	sw	1– 3
Proline	sw	25–40	neu	
	bi	25–27		
Serine	sw	25–35	sw	30–40
Threonine	sw	35–45	sw	40–50
Tryptophan	bi	4– 6	sw	0,2– 0,4
Tyrosine	bi	4– 6	sw	1– 3
1-Aminocycloalkane-1-carboxylic acid[b]				
Cyclobutane derivative	sw	20–30		
Cyclopentane	sw	3-6		
derivative	bi	95–100		
Cyclohexane	sw	1-3		
derivative	bi	45–50		
Cyclooctane	sw	2-4		
derivative	bi	2-5		
Caffeine	bi	1–1,2		
Saccharose	sw	10–12		

[a] Recognition threshold value (mmole/l).
[b] Compounds not optically active

1.3 Peptides

1.3.1 General Remarks, Nomenclature

Peptides are formed by binding amino acids together through an amide linkage.
On the other hand peptide hydrolysis results in free amino acids:

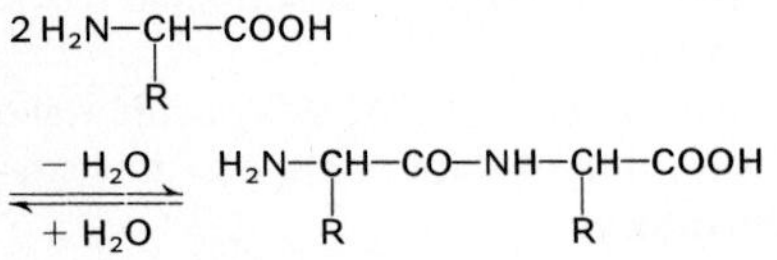

$$2\,H_2N{-}\underset{\displaystyle R}{CH}{-}COOH \underset{+H_2O}{\overset{-H_2O}{\rightleftharpoons}} H_2N{-}\underset{\displaystyle R}{CH}{-}CO{-}NH{-}\underset{\displaystyle R}{CH}{-}COOH \qquad (1.58a)$$

Functional groups not involved in the peptide synthesis reaction should be blocked. The protecting or blocking groups must be removed after synthesis under conditions which retain the stability of the newly-formed peptide bonds:

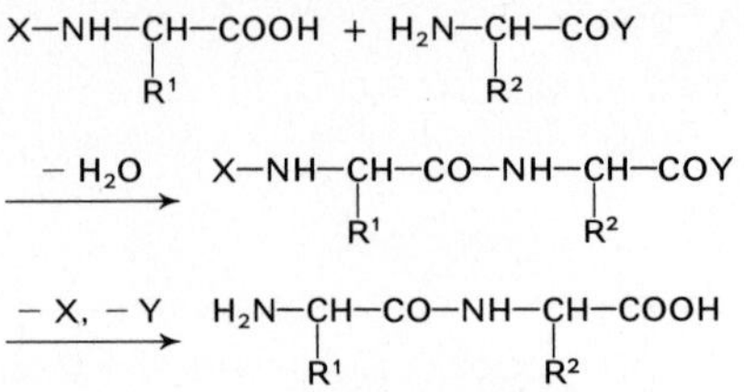

$$X{-}NH{-}\underset{\displaystyle R^1}{CH}{-}COOH + H_2N{-}\underset{\displaystyle R^2}{CH}{-}COY$$

$$\xrightarrow{-H_2O} X{-}NH{-}\underset{\displaystyle R^1}{CH}{-}CO{-}NH{-}\underset{\displaystyle R^2}{CH}{-}COY$$

$$\xrightarrow{-X,\,-Y} H_2N{-}\underset{\displaystyle R^1}{CH}{-}CO{-}NH{-}\underset{\displaystyle R^2}{CH}{-}COOH \qquad (1.59)$$

Peptides are denoted by the number of amino acid residues as di-, tri-, tetrapeptides, etc., and the term "oligopeptides" is used for those with 10 or less amino acid residues. Higher molecular weight peptides are called polypeptides. The transition of "polypeptide" to "protein" is rather undefined, but the limit is commonly assumed to be at a molecular weight of about 10 kdal, i.e. about 100 amino acid residues are needed in the chain for it to be called a protein.
Peptides are interpreted as acylated amino acids:

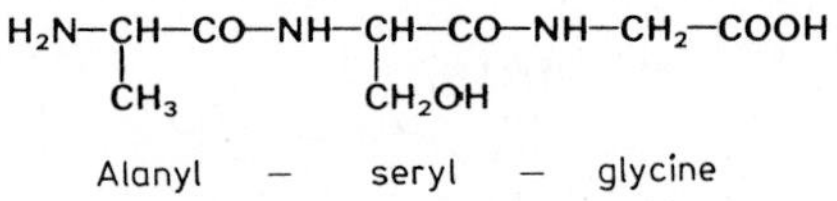

$$H_2N{-}\underset{\displaystyle CH_3}{CH}{-}CO{-}NH{-}\underset{\displaystyle CH_2OH}{CH}{-}CO{-}NH{-}CH_2{-}COOH \qquad (1.60)$$

Alanyl – seryl – glycine

The first three letters of the amino acids are used as symbols to simplify designation of peptides (cf. Table 1.1)*. Thus, the peptide shown above can also be given as:

Ala—Ser—Gly (1.61)

D-Amino acids are denoted by the prefix D-. In compounds in which a functional group of the side chain is involved, the bond is indicated by a perpendicular line. The tripeptide glutathione (γ-glutamyl-cysteinyl-glycine) is given as an illustration along with its corresponding disulfide, oxidized glutathione:

* One-letter symbols have also been established for an even more abbreviated presentation.

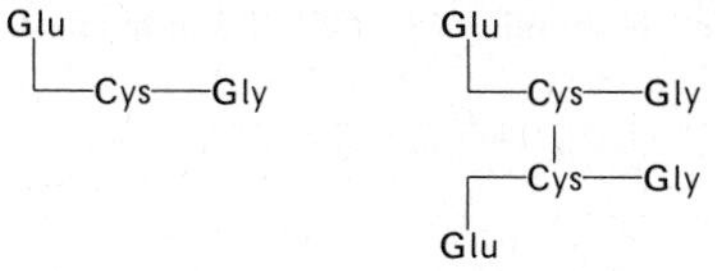

Glu–Cys–Gly (branched: Glu linked to Cys)

Glu–Cys–Gly | Cys–Gly–Glu (disulfide-linked: Cys–Cys) (1.62)

By convention, the amino acid residue with the free amino group is always placed on the left. The amino acids of the chain ends are denoted as N-terminal and C-terminal amino acid residues. The peptide linkage direction in cyclic peptides is indicated by an arrow, i.e. $-CO \rightarrow NH-$.

1.3.2 Physical Properties

1.3.2.1 Dissociation

The pK values and isoelectric points for some peptides are listed in Table 1.11. The acidity of the free carboxyl groups and the basicity of the free amino groups are lower in peptides than in the corresponding free amino acids. The amino acid sequence also has an influence (e.g. Gly-Asp/Asp-Gly).

Table 1.11. Dissociation constants and isoelectric points of various peptides (25 °C)

Peptide	pK_1	pK_2	pK_3	pK_4	pK_5	pI
Gly-Gly	3.12	8.17				5.65
Gly-Gly-Gly	3.26	7.91				5.59
Ala-Ala	3.30	8.14				5.72
Gly-Asp	2.81	4.45	8.60			3.63
Asp-Gly	2.10	4.53	9.07			3.31
Asp-Asp	2.70	3.40	4.70	8.26		3.04
Lys-Ala	3.22	7.62	10.70			9.16
Ala-Lys-Ala	3.15	7.65	10.30			8.98
Lys-Lys	3.01	7.53	10.05	11.01		10.53
Lys-Lys-Lys	3.08	7.34	9.80	10.54	11.32	10.93
Lys-Glu	2.93	4.47	7.75	10.50		6.10
His-His	2.25	5.60	6.80	7.80		7.30

1.3.3 Sensory Properties

While the taste quality of amino acids does depend on configuration, peptides, except for the sweet dipeptide esters of aspartic acid (see below), are neutral or bitter in taste, with no relationship to configuration (Table 1.12). As with amino acids, the taste intensity is influenced by the hydrophobicity of the side chains (Table 1.13). The taste intensity does not appear to be dependent on amino acid sequence (Table 1.12).

Table 1.12. Taste-threshold values of various peptides: effect of configuration and amino acid sequence (tested in aqueous solution at pH 6–7); bi – bitter

Peptide[a]	Taste	
	Quality	Intensity[b]
Gly-Leu	bi	19–23
Gly-D-Leu	bi	20–23
Gly-Phe	bi	15–17
Gly-D-Phe	bi	15–17
Leu-Leu	bi	4– 5
Leu-D-Leu	bi	5– 6
D-Leu-D-Leu	bi	5– 6
Ala-Leu	bi	18–22
Leu-Ala	bi	18–21
Gly-Leu	bi	19–23
Leu-Gly	bi	18–21
Ala-Val	bi	60–80
Val-Ala	bi	65–75
Phe-Gly	bi	16–18
Gly-Phe	bi	15–17
Phe-Gly-Phe-Gly	bi	1.0–1,5
Phe-Gly-Gly-Phe	bi	1,0–1,5

[a] L-Configuration if not otherwise designated.
[b] Recognition threshold value in mmole/l.

Bitter tasting peptides can occur in food after proteolytic reactions. For example, the bitter taste of cheese is a consequence of faulty ripening. The wide use of proteolytic enzymes to achieve well-defined modifications of food proteins without producing a bitter taste therefore causes some problems. Removal of the bitter taste of a partially-hydrolyzed protein is outlined in the section dealing with proteins modified with enzymes (cf. 1.4.6.3.2).

The sweet taste of aspartic acid dipeptide esters (I) was discovered by chance in 1969 for α-L-aspartyl-L-phenylalanine methyl ester ("Aspartame", "NutraSweet"). The corresponding peptide ester of L-aminomalonic acid (II) is also sweet.

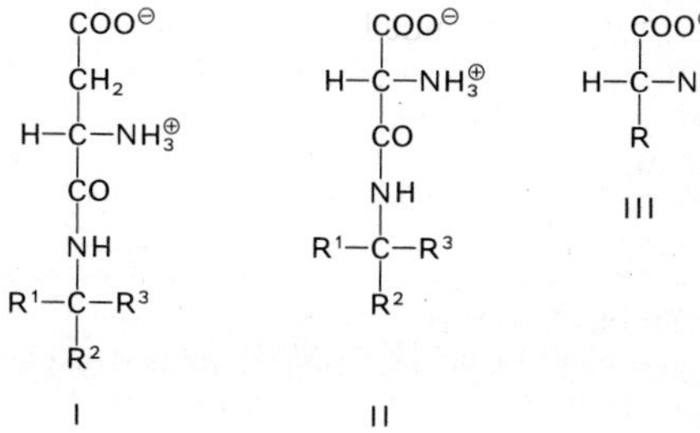

I: $COO^{\ominus}$–CH_2–$H{-}C{-}NH_3^{\oplus}$–CO–NH–$R^1{-}C{-}R^3$ (R^2)

II: $COO^{\ominus}$–$H{-}C{-}NH_3^{\oplus}$–CO–NH–$R^1{-}C{-}R^3$ (R^2)

III: $COO^{\ominus}$–$H{-}C{-}NH_3^{\oplus}$–R

(1.63)

Table 1.13. Bitter taste of dipeptide A–B: dependence of recognition threshold value (mmole/l) on side chain hydrophobicity (0: sweet or neutral taste)

A / B		Asp	Glu	Asn	Gln	Ser	Thr	Gly	Ala	Lys	Pro	Val	Leu	Ile	Phe	Tyr	Trp
		0	0	0	0	0	0	0	0	85	26	21	12	11	6	5	5
Gly	0	–	–	–	–	–	–	0	0	–	45	75	21	20	16	17	13
Ala	0	–	–	–	–	–	–	0	0	–	–	70	20	–	–	–	–
Pro	26	–	–	–	–	–	–	–	–	–	–	–	6	–	–	–	–
Val	21	–	–	–	–	–	–	65	70	–	–	20	10	–	–	–	–
Leu	12	–	–	–	–	–	–	20	20	–	–	–	4.5	–	–	3.5	0.4
Ile	11	43	43	33	33	33	33	21	21	23	4	9	5.5	5.5			0.9
Phe	6	–	–	–	–	–	–	17	–	–	2	–	1.4	–	0.8	0.8	–
Tyr	5	–	–	–	–	–	–	–	–	–	–	–	4	–	–	–	–
Trp	5	–	28	–	–	–	–	–	–	–	–	–	–	–	–	–	–

A comparison of structures I, II and III reveals a relationship between sweet dipeptides and sweet D-amino acids. The required configuration of the carboxyl and amino groups and the side chain substituent, R, is found only in peptide types I and II.

Since the discovery of the sweetness of compounds of type I there has been a systematic study of the structural prerequisites for a sweet taste.

The presence of L-aspartic acid was shown to be essential, as was the peptide linkage through the α-carboxyl group.

Table 1.14. Taste of aspartic acid dipeptide esters[a]

R^2	R^3	Taste[b]
$COOCH_3$	H	8
n-C_3H_6	$COOCH_3$	4
n-C_4H_7	$COOCH_3$	45
n-C_4H_9	$COOC_2H_5$	5
n-C_6H_{13}	CH_3	10
n-C_7H_{15}	CH_3	neutral
$COOCH(CH_3)_2$	n-C_3H_7	17
$COOCH(CH_3)_2$	n-C_4H_9	neutral
$COOCH_3$	$CH_2C_6H_5$	bitter
$CH(CH_3)C_2H_5$	$COOCH_3$	bitter
$CH_2CH(CH_3)_2$	$COOCH_3$	bitter
$CH_2C_6H_5$	$COOCH_3$	140
COO-2-methyl-cyclohexyl	$COOCH_3$	5–7,000
COO-fenchyl	$COOCH_3$	22–33,000

[a] Formula 1.63 I, $R^1 = H$.
[b] For sweet compounds the threshold value is given, related to the threshold value of saccharose ≙ 1.

R^1 may be an H or CH_3 group*, while the R^2 and R^3 groups are variable within a certain range. Several examples are presented in Table 1.14. The sweet taste intensity passes through a maximum with increasing length and volume of the R^2 residue (e.g. COO-fenchyl ester is 22–23 × 10^3 times sweeter than sucrose). The size of the R^3 substituent is limited to a narrow range. Obviously, the R^2 substituent has the greatest influence on taste intensity.

Some peptides exhibit a salty taste, e.g. ornithyl-β-alanine hydrochloride, and may be used as substitutes for sodium chloride.

1.3.4 Individual Peptides

Peptides are widespread in nature. They are often involved in specific biological activities (peptide hormones, peptide toxins, peptide antibiotics). A number of peptides of interest to food chemists are outlined in the following sections.

1.3.4.1 Glutathione

Glutathione (γ-L-glutamyl-L-cysteinyl-glycine) is widespread in animals, plants and microorganisms. A noteworthy feature is the binding of glutamic acid through its γ-carboxyl group. The peptide is the coenzyme of glyoxalase.

(1.64)

* Data are not yet available for compounds with $R^1 > CH_3$.

It is involved in active transport of amino acids and, due to its ready oxidation, is also involved in many redox-type reactions. It influences the rheological properties of wheat flour dough through thiol-disulfide interchange with wheat gluten. High concentrations of reduced glutathione in flour bring about reduction of protein disulfide bonds and a corresponding decrease in molecular weight of some of the protein constituents of dough gluten (cf. 15.4.1.4.1).

1.3.4.2 Carnosine, Anserine and Balenine

These peptides are noteworthy since they contain a β-amino acid, β-alanine, bound to L-histidine or 1-methyl- or 3-methyl-L-histidine, and are present in meat extract and in muscle of vertebrates (cf. Formula 1.65).

$H_3N^{\oplus}$–CH_2–CH_2–CO–NH–CH(R)–$COO^{\ominus}$

R: Carnosine, Anserine, Balenine (imidazole side chains: Carnosine – CH_2-imidazole with HN; Anserine – CH_2-imidazole with N–H_3C; Balenine – CH_2-imidazole with N–CH_3)

Carnosine Anserine Balenine

(1.65)

Data on the amounts of these peptides present in meat are given in Table 1.15. Carnosine is predominant in beef muscle tissue, while anserine is predominant in chicken meat. Balenine is a characteristic constituent of whale muscle, although it appears that sperm whales do not have this dipeptide. The amounts found in commercial sperm whale meat extract are probably due to the presence of meat from other whale species. These peptides are used analytically to identify the meat extract. Their physiological roles are not clear. Their buffering capacity in the pH range of 6–8 may be of some importance. They may also be involved in revitalizing exhausted muscle, i.e. in the muscle regaining its excitability and ability to contract. Carnosine may act as a neurotransmitter for nerves involved in odor perception.

Table 1.15. Occurence of carnosine, anserine and balenine (%) in meat[a]

Meat	Carnosine	Anserine	Balenine	Σ[b]
Beef muscle tissue	0.15–0.35	0.01–0.05		0.2–0.4
Beef meat extract	3.1 –5.7	0.4 –1.0		4.4–6.2
Chicken meat[c]	0.01–0.1	0.05–0.25		
Chicken meat extract	0.7 –1.2	2.5 –3.5		
Whale meat				ca. 0.3
Whale meat extract a[d]	3.1 –5.9	0.2 –0.6	13.5–23.0	16–30
Whale meat extract b[e]	2.5–4.5	1.2 –3.0	0 – 5.2	3.5–12

[a] The results are expressed as % of the moist tissue weight, or for commercially available extracts containing 20% moisture.
[b] β-Alanine peptide sum.
[c] Lean and deboned chicken meat.
[d] Commercial extract mixture of various whales.
[e] Commercial extract mixture, with sperm whale prevailing.

1.3.4.3 Nisin

This peptide is formed by several strains of *Streptococcus lactis* (Langfield-N-group). It contains

Ile–Dhb–D-Ala–Ile–Dha–Leu–Ala–D-Abu–Pro–Gly–Ala–Lys–D-Abu–Gly–Ala–Leu–Met–Gly–Ala–Asn–
Met–Lys–D-Abu–D-Ala–Abu–Ala–His–Ala–Ser–Ile–His–Val–Dha–Lys

(S bridges as shown; residue positions 10, 20, 30 marked)

H_2N–C(=CHR)–COOH

R = H: 2-Amino acrylic acid (Dehydroalanine, Dha)
R = CH_3: 2-Amino crotonic acid (Dehydro-α-amino butyric acid, Dhb)

H_2N–CH(COOH)–CHR–S–CH_2–CH(NH_2)–COOH

R = H: Lanthionine (Ala–S–Ala)
R = CH_3: β-Methyllanthionine (Abu–S–Ala)

(1.66)

a number of unusual amino acids, namely dehydroalanine, dehydro-β-methylalanine, lanthionine, β-methyl-lanthionine, and therefore also five thioether bridges (cf. Formula 1.66).
The peptide subtilin is related to nisin. Nisin is active against Gram-positive microorganisms (lactic acid bacteria, *Streptococci, Bacilli, Clostridia* and other anaerobic spore-forming microorganisms). Nisin begins to act against the cytoplasmic membrane as soon as the spore has germinated. Hence, its action is more pronounced against spores than against vegetative cells. Nisin is permitted as a preservative in several countries. It is used to suppress anaerobes in cheese and cheese products, especially in hard cheese and processed cheese to inhibit butyric acid fermentation. The use of nisin in the canning of vegetables allows mild sterilization conditions.

1.3.4.4 Lysine Peptides

A number of peptides, such as:

Gly–Lys, Ala–Lys, Glu–Lys, Gly–(ε)Lys, Glu–(ε)Lys, Lys(ε)–Glu, Gly–Lys(ε)–Gly (1.67)

have been shown to be as good as lysine in rat growth feeding tests.These peptides substantially retard the browning reaction with glucose (Figure 1.6), hence they are suitable for lysine fortification of foods in which sugars are present during heat treatment.

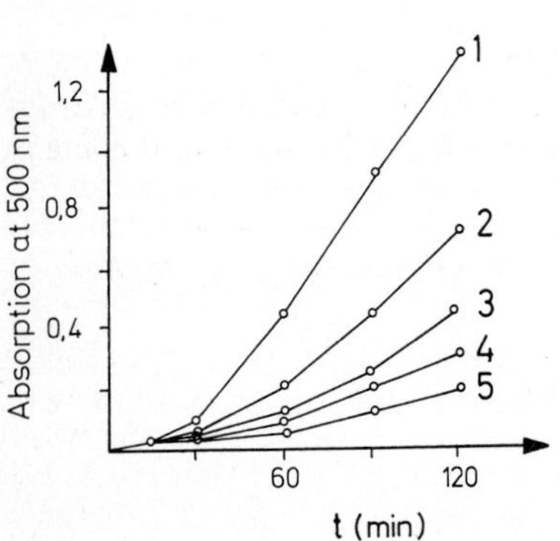

Fig. 1.6. Browing of some lysine derivatives (0.1 M lysine or lysine derivative, 0.1 M glucose in 0.1 M phosphate buffer pH 6.5 at 100 °C in sealed tubes. (After *Finot* et al., 1978.) 1 Lys, 2 Ala-Lys, 3 Gly-Lys, 4 Glu – Lys(ε)–Gly, 5 Lys(ε)–Glu

1.3.4.5 Other Peptides

Other peptides occur commonly and in variable levels in protein rich food as degradation products of proteolytic processes.

1.4 Proteins

Like peptides, proteins are formed from amino acids through amide linkages. Covalently-bound hetero constituents can also be incorporated into proteins. For example, phosphoproteins such as milk casein (cf. 10.1.2.1.1) or phosvitin of egg yolk (cf. 11.2.4.1.2) contain phosphoric acid esters of serine and threonine residues.
Glycoproteins, such as κ-casein (cf. 10.1.2.1.1), various components of egg white (cf. 11.2.3.1) and egg yolk (cf. 11.2.4.1.2), collagen from connective tissue (cf. 12.3.2.3.1) and serum proteins of some species of fish (cf. 13.1.4.2.4), contain one or more monosaccharide or oligosaccharide units bound O-glycosidically to serine, threonine or δ-hydroxylysine or N-glycosidically to asparagine:

Sugar(NHAc)–NH–CO–CH_2–CH(NH–)(CO–)

Sugar–O–CH(CH_2NH_2)–$(CH_2)_2$–CH(NH–)(CO–)

Sugar(NHAc)–O–CHR–CH(NH–)(CO–)

H_2PO_3O–RHC–CH(NH–)(CO–) (1.67a)

R : H, CH_3 ; R[1] : H, Sugar residue; Ac : Acetyl

The structure of a protein is dependent on the amino acid sequence (the primary structure) which determines the molecular conformation (secondary and tertiary structures). Proteins sometimes occur as molecular aggregates which are arranged in an orderly geometric fashion (quaternary structure). The histograms in Figure 1.7 provide data on the proteins with known amino acid sequences (1953–1976) and conformations (1960–1974).

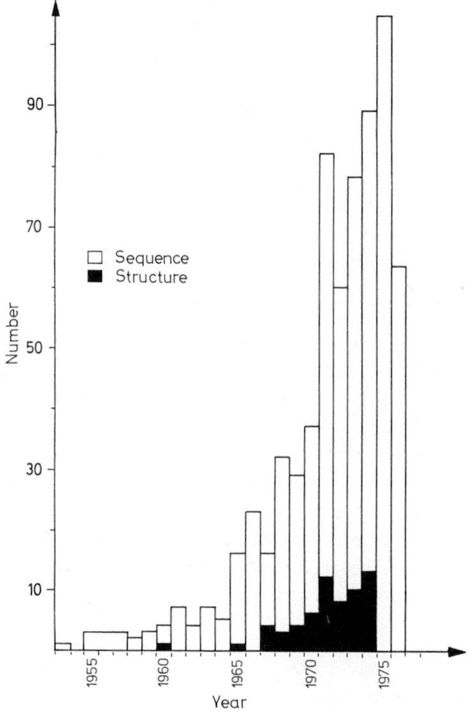

Fig. 1.7. Number of elucidate amino acid sequences (1953–1976) and protein structures (years 1960–1974)

1.4.1 Amino Acid Sequence

1.4.1.1 Amino Acid Composition, Subunits

Sequence analysis can only be conducted on a pure protein. First, the amino acid composition is determined after acidic hydrolysis. The procedure (separation on a single cation-exchange resin column and color development with ninhydrin reagent) has been standardized and automated (amino acid analyzers). Figure 1.7a shows a typical amino acid chromatogram.

It is also necessary to know the molecular weight of the protein. This is determined by gel column chromatography, ultracentrifugation or SDS-PAG electrophoresis. Furthermore, it is necessary to determine whether the protein is a single molecule or consists of a number of identical or different polypeptide chains (subunits) associated through disulfide bonds or noncovalent forces. Dissociation into subunits can be accomplished by a change in pH, by chemical modification of the protein, such as by succinylation, or with denaturing agents (urea, guanidine hydrochloride, sodium dodecyl sulfate). Disulfide bonds, which are also found in proteins which consist of only one peptide chain, can be cleaved by oxidation of cystine to cysteic acid or by reduction to cysteine with subsequent alkylation of the thiol group (cf. 1.2.4.3.5) to prevent reoxidation. Separation of subunits is achieved by chromatographic or electrophoretic methods.

1.4.1.2 Terminal Groups

N-terminal amino acids can be determined by treating a protein with 1-fluoro-2,4-dinitrobenzene (*Sanger's* reagent; cf. 1.2.4.2.2) or 5-dimethylaminonaphthalene-1-sulfonyl chloride (dansyl chloride; cf. 1.2.4.2.1). Another possibility is the reaction with cyanate, followed by elimination of the N-terminal amino acid in the form of hydantoin, and separation and recovery of the amino acid by cleavage of hydantoin (cf. 1.2.4.2.3). The N-terminal amino acid (and the amino acid sequence close to the N-terminal) is accessible by hydrolysis with aminopeptidase, in which case it should be remembered that the hydrolysis rate is dependent on amino acid side chains and that proline residues are not cleaved. A special procedure is required when the N-terminal residue is acylated (N-formyl- or N-acetyl amino acids, or pyroglutamic acid).

Determination of C-terminal amino acids is possible via the hydrazinolysis procedure recommended by *Akabori:*

$$H_2N{-}\underset{R_1}{CH}{-}CO{-}(HN{-}\underset{R_{2-n}}{CH}{-}CO{-})HN{-}\underset{R_m}{CH}{-}COOH \xrightarrow[100\,^{\circ}C]{H_2N{-}NH_2} H_2N{-}\underset{R_{1-n}}{CH}{-}CO{-}NH{-}NH_2 + H_2N{-}\underset{R_m}{CH}{-}COOH \quad (1.68)$$

The C-terminal amino acid is then separated from the amino acid hydrazides, e.g. by a cation-exchange resin, and identified. The C-terminal amino acids can be removed enzymatically by carboxypeptidase A which preferentially cleaves amino acids with aromatic and large aliphatic side chains; carboxypeptidase B which preferentially cleaves lysine, arginine and amino acids with neutral side chains or carboxypeptidase C which cleaves with less specificity and cleaves proline.

1.4.1.3 Partial Hydrolysis

Longer peptide chains are usually fragmented. The fragments are then separated and analyzed individually for amino acid sequences. Selective enzymatic cleavage of peptide bonds is accomplished primarily with trypsin, which cleaves exclusively Lys-X- and Arg-X-bonds, and chymotrypsin, which cleaves peptide bonds with less specificity (Tyr-X, Phe-X, Trp-X and Leu-X). The enzymatic attack can be influenced by modification of the protein. For example, acylation of the ε-amino group of lysine limits tryptic hydrolysis to Arg-X (cf. 1.4.4.1.3 and 1.4.4.1.4), whereas substitution of the SH-group of a cysteine residue with an aminoethyl group introduces a new cleavage position for trypsin into the molecule ("pseudolysine residue"):

$$-NH-CH(CH_2-SH)-CO-NH-CH(R)-CO-$$

$$\xrightarrow{H_2C-CH_2\ (NH)} -NH-CH(CH_2-S-CH_2-CH_2-NH_2)-CO-NH-CH(R)-CO-$$

$$\xrightarrow{\text{Trypsin}} -NH-CH(CH_2-S-CH_2-CH_2-NH_2)-COOH + H_2N-CH(R)-CO- \quad (1.69)$$

The most important chemical method for selective cleavage uses cyanogen bromide (BrCN) to attack Met-X-linkages (Reaction 1.70).

Hydrolysis of proteins with strong acids reveals a difference in the rates of hydrolysis of peptide bonds depending on the adjacent amino acid side chain. Bonds involving amino groups of serine and threonine are particularly susceptible to hydrolysis. This effect is due to N → O-acyl migration via the oxazoline and subsequent hydrolysis of the ester bond (Reaction 1.71).

$$-NH-CH(CH_2-CH_2-S-CH_3)-C(=O)-HN-CH(R)-CO-$$

$$\xrightarrow{BrCN} -NH-CH(CH_2-CH_2-S^{\oplus}(CN)-CH_3)-C(=O)-NH-CH(R)-CO- \quad Br^{\ominus}$$

$$\longrightarrow -NH-CH-C(=N-CH(R)-CO-)-O-CH_2-CH_2 \text{ (ring)}$$

$$\xrightarrow{H_2O} -NH-CH-C(=O)-O-CH_2-CH_2 \text{ (ring)} + H_2N-CH(R)-CO- \quad (1.70)$$

Hydrolysis of proteins with dilute acids preferentially cleaves aspartyl-X-bonds.

Separation of peptide fragments is achieved by gel and ion-exchange column chromatography using a volatile buffer as eluent (pyridine, morpholine acetate) which can be removed by freeze-drying of the fractions collected. Recently the

$$-NH-CH(R)-C(=O)-NH-CH(CO-)-CH_2-OH \quad (H^{\oplus})$$

$$\xrightarrow{-H_2O} -NH-CH(R)-C{=}N-CH(CO-)-CH_2-O \text{ (oxazoline ring)}$$

$$\xrightarrow{+H_2O} -CH(R)-CO-O-CH_2-CH(NH_2)-CO-$$

$$\xrightarrow{+H_2O} -NH-CH(R)-COOH + H_2N-CH(CH_2OH)-CO- \quad (1.71)$$

separation of peptides and proteins by reversed-phase HPLC has gained great importance, using volatile buffers mixed with organic, water-soluble solvents as the mobile phase.
The fragmentation of the protein is performed by different enzymic and/or chemical techniques, at least by two enzymes of different specifity. The arrangement of the obtained peptides in the same order as they occur in the intact protein is accomplished with the aid of overlapping sequences. The principle of this method is illustrated for subtilisin BPN′ as an example in Figure 1.7b.

1.4.1.4 Sequence Analysis

The *Edman* degradation is by far the most important method in sequence analysis. It involves stepwise degradation of peptides with phenylisothiocyanate (cf. 1.2.4.2.3). The resultant phenylthiohydantoin is either identified directly or the amino acid is recovered. The stepwise reactions are performed in solution or on peptide bound to a carrier, i.e. to a solid phase. Both approaches have been automated ("sequencer"). Carriers used include resins containing amino groups (e.g. amino polystyrene) or glass beads treated with amino alkylsiloxane:

$$\text{Glass}(\text{OH})_3 + (CH_3O)_3Si(CH_2)_n\text{—}NH_2$$

$$\longrightarrow \text{Glass}(\text{—O—})_3Si\text{—}(CH_2)_n\text{—}NH_2 \quad (1.72)$$

The peptides are then attached to the carrier by carboxyl groups (activation with carbodiimide or carbonyl diimidazole, as in peptide synthesis) or by amino groups. For example, a peptide segment from the hydrolysis of protein by trypsin has lysine as its C-terminal amino acid. It is attached to the carrier with p-phenylene-diisothiocyanate through the α- and ε-amino groups. Mild acidic treatment of the carrier under conditions of the *Edman* degradation splits the first peptide bond. The *Edman* procedure is then performed on the shortened peptide through second, third and subsequent repetitive reactions (Reaction 1.73).
Methods other than the *Edman* degradation can provide additional information. These include determination of terminal residues with amino- and carboxypeptidases, as already discussed, or mass spectrometric analysis of suitable volatile peptide derivatives.

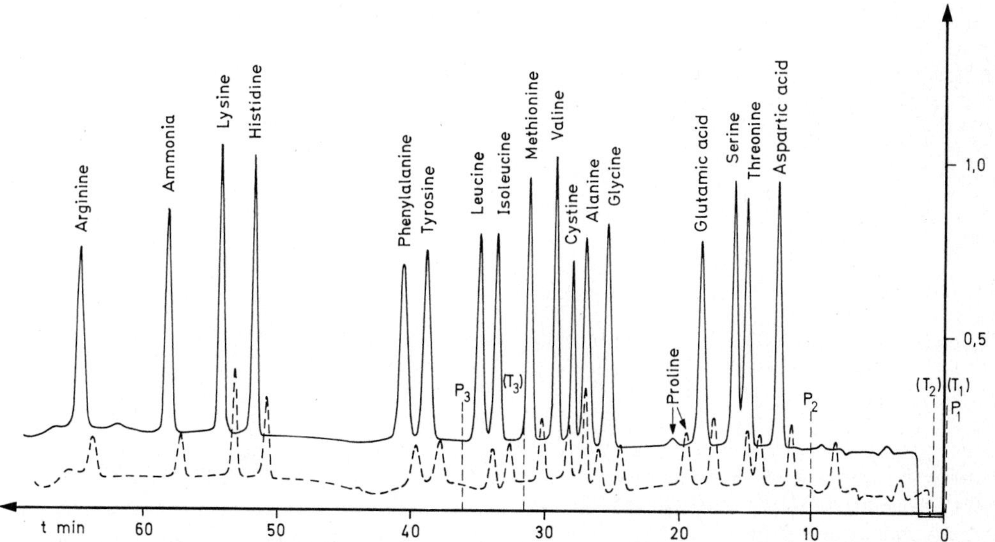

Fig. 1.7a. Amino acid chromatogram. Separation of a mixture of amino acids (10 nmole/amino acid) by an amino acid analyzer. Applied is a single ion exchange column: Durum DC-4A, 295 × 4 mm, buffers $P_1/P_2/P_3$: 0.2 N Na-citrate pH 3.20/0.2 N Na-citrate pH 4.25/1.2 N Na-citrate and NaCl of pH 6.45. Temperatures $T_1/T_2/T_3$: 48/56/80 °C. Flow rate: 25 ml/h; Absorbance reading after color development with ninhydrin at 570/440 nm: ——/----

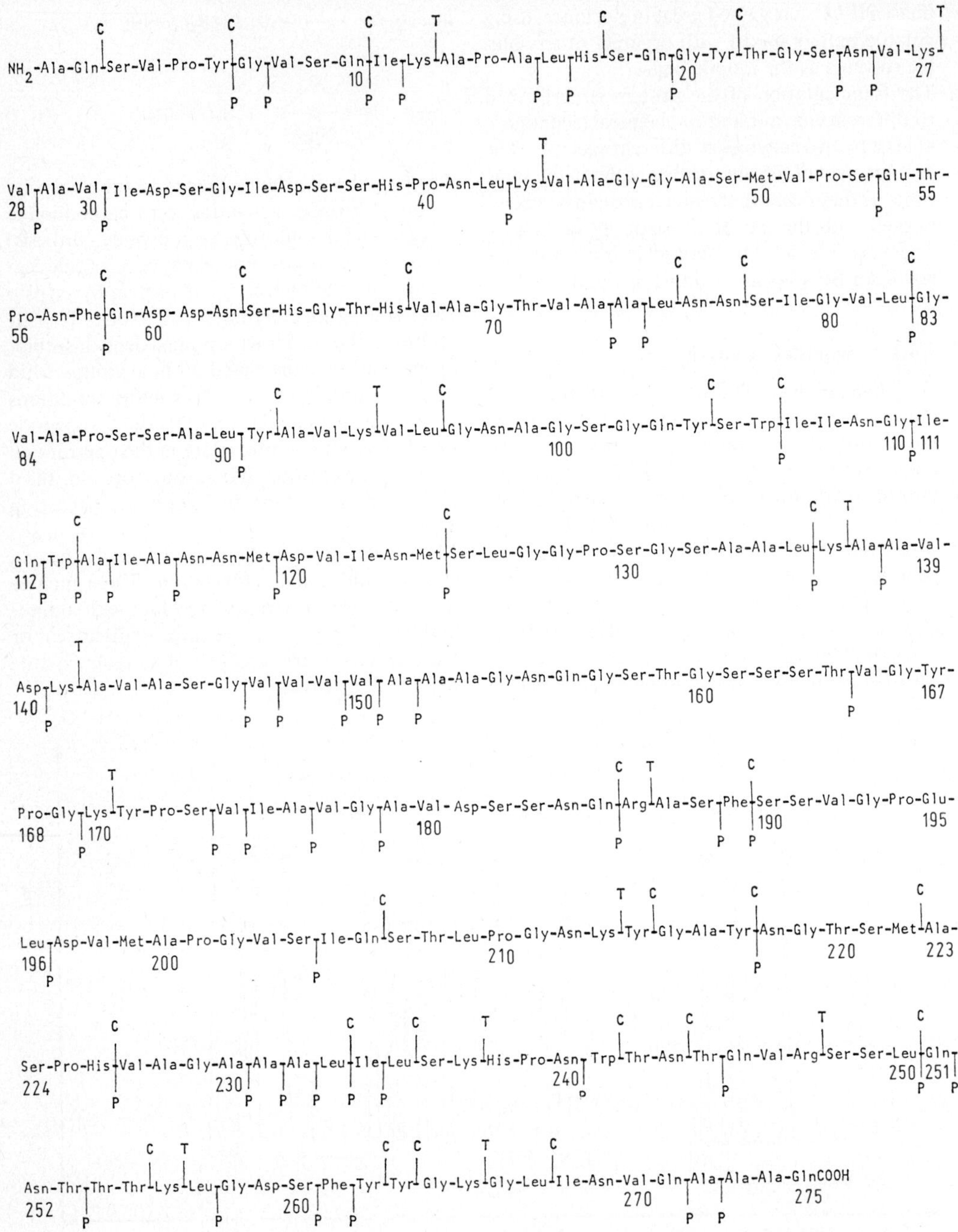

Fig. 1.7b. Subtilisin BPN′; peptide bonds hydrolyzed by trypsin (T), chymotrypsin (C), and pepsin (P).

H₂N—CH(R)—CO—B—C—D—Lys(NH₂)—COOH

↓ H₂N—Carrier—NH₂ | SCN—C₆H₄—NCS

HN—Carrier—NH (each —CS—NH—C₆H₄—NH—CS—NH—)
CHR—CO—B—C—D—Lys—COOH

↓ H⊕

HN—Carrier—NH
(thiazolinone with S, N, O, R) + H₂N—B—C—D—Lys—COOH (1.73)

1.4.2 Conformation

Information about conformation is available through X-ray crystallographic analysis of protein crystals. This assumes that, in many cases, the conformation of the protein in crystalline form is similar to that of the protein in solution. In 1960 *Kendrew et al.* succeeded in elucidating the structure of myoglobin (17.8 kdal) with a resolution of 0.2 nm. As of 1975 about 70 proteins have been analyzed with resolutions of 0.2–0.3 nm. An example of the calculated electron density distributions of 2,5-dioxopiperazine based on various degrees of resolution is presented in Figure 1.8. Individual atoms are well revealed at 0.11 nm. Such a resolution has not been achieved with proteins. Reliable localization of the C_α-atom of the peptide chain requires a resolution of less than 0.3 nm.

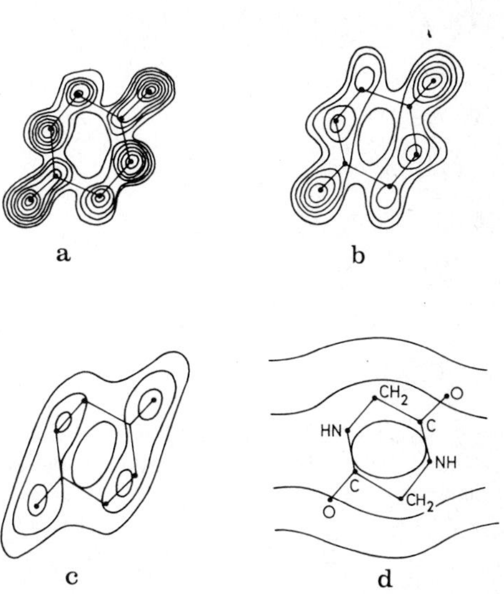

Fig. 1.8. Electron density distribution patterns for 2,5-dioxopiperazine with varying resolution extent. **a** 0.11 nm, **b** 0.15 nm **c** 0.20 nm, **d** 0.60 nm (After *Perutz,* 1962)

1.4.2.1 Extended Peptide Chains

X-ray structural analysis and other physical measurements of a fully-extended peptide chain reveal the lengths and angles of bonds (see the "ball and stick" representation in Figure 1.9). The peptide bond has partial (40%) double bond character with π electrons shared between the $C'-O$ and $C'-N$ bonds. The resonance energy is about 83.6 kJ/mole:

$$-C_\alpha-C'(=O)-N(H)-C_\alpha- \longleftrightarrow -C_\alpha-C'(-O^{\ominus})=N^{\oplus}(H)-C_\alpha- \qquad (1.74)$$

Normally the bond has a trans-configuration, i.e. the oxygen of the carbonyl group and the hydrogen of the NH group are in the trans-position; a cis-configuration occurs only in exceptional cases (e.g. in small cyclic peptides or in proteins before proline residues).

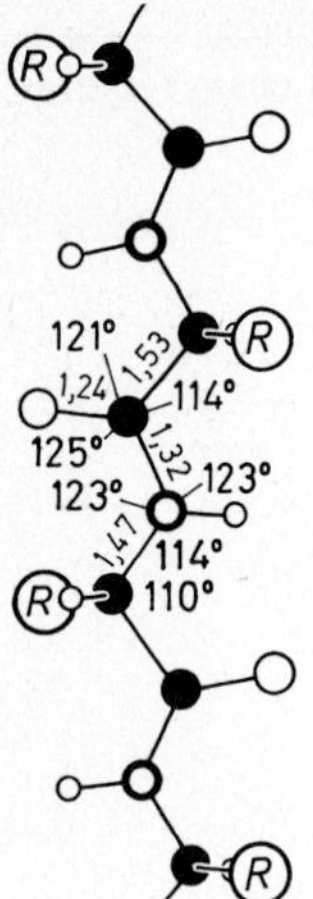

Fig. 1.9. Structure of an elongated peptide chain. • Carbon, ○ oxygen, **O** nitrogen, ∘ hydrogen and Ⓡ side chain

Six atoms of the peptide bonds, C_i^α, C_i', O_i, N_{i+1}, C_{i+1}^α and H_{i+1}, lie in one plane (cf. Figure 1.10). For a trans-peptide bond, ω_i is 180°. The position of two neighboring planes is determined by the numerical value of the angles ψ_i (rotational bond between a carbonyl carbon and an α-carbon) and ϕ_i (rotational bond between an amide-N and an α-carbon). For an extended peptide chain, $\psi_i = 180°$ and $\phi_i = 180°$. The position of side chains can also be described by a series of angles χ_i^{1-n}.

1.4.2.2 Secondary Structure (Regular Structural Elements)

The primary structure gives the sequence of amino acids in a protein chain while the secondary structure reveals the arrangement of the chain in space. The peptide chains are not in an extended or unfolded form (ψ_i, $\phi_i \neq 180°$). It can be shown with models that ψ_i and ϕ_i, at a permissible minimum distance between nonbonding atoms (Table 1.16), can assume only particular angles. Figure 1.11 presents the permissible degree ranges for amino acids other than glycine ($R \neq H$). The range is broader for glycine ($R = H$). Figure 1.12 demonstrates that most of 13 different proteins with a total of about 2,500 amino acid residues have been shown empirically to have values of ψ,ϕ-pairs within the permissible range. When a multitude of equal ψ,ϕ-pairs occurs consecutively in a peptide chain, the chain acquires regular repeating structural elements. The types of structural elements are compiled in Table 1.17.

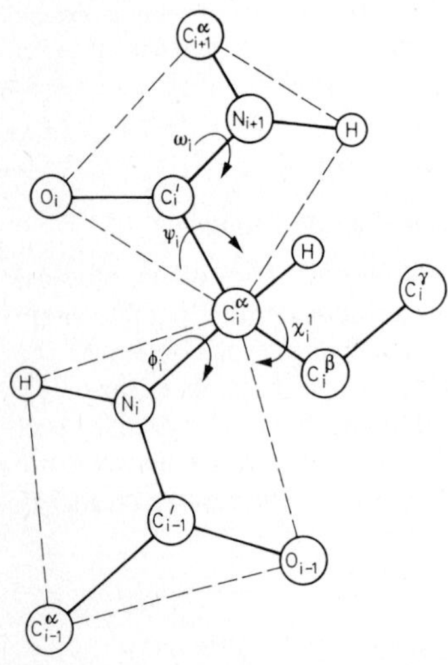

Fig. 1.10. Definitions for dihedral angles in a peptide chain

$\omega_i = 0°$ for $C_i^\alpha - C_i'/N_{i+1} - C_{i+1}^\alpha \rightarrow$ cis,
$\psi_i = 0°$ for $C_i^\alpha - N_i/C_i' - O_i \rightarrow$ trans,
$\varphi_i = 0°$ for $C_i^\alpha - C_i'/N_i - H_i \rightarrow$ trans,
$\chi_i = 0°$ for $C_i^\alpha - N_iC_i^\beta - C_i \rightarrow$ cis;

The angles are positive when the rotation is clockwise and viewed from the N-terminal side of a bond or (for X) from the atom closer to the main chain respectively. (After *Schulz* and *Schirmer*, 1979)

Table 1.16. Minimal distances for nonbonded atoms (Å)

	C	N	O	H
C	3.20[a] (3.00)[b]	2.90 (2.80)	2.80 (2.70)	2.40 (2.20)
N		2.70 (2.60)	2.70 (2.60)	2.40 (2.20)
O			2.70 (2.60)	2.40 (2.20)
H				2.00 (1.90)

[a] Normal values
[b] Extreme values

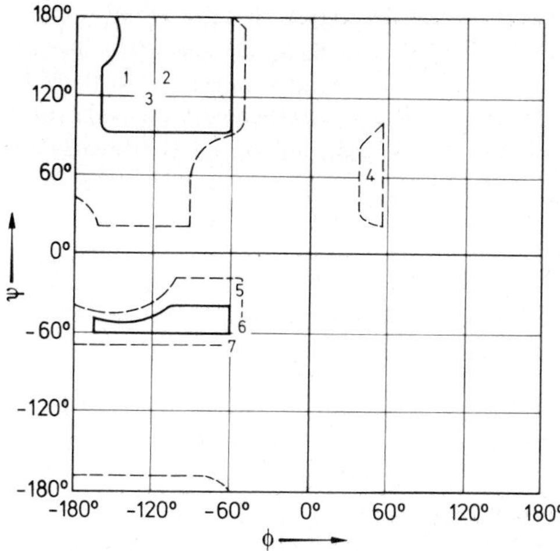

Fig. 1.11. φ,ψ-Diagram (*Ramachandran* plot). Allowed conformations for amino acids with a C^{β}-atom obtained by using normal (—) and lower limit (---) contact distances for non-bonded atoms, from Table 1.16. β-Sheet structures: antiparallel (1); parallel (2), twisted (3). Helices: α-, left-handed (4), 3_{10} (5), α, right-handed (6), π (7).

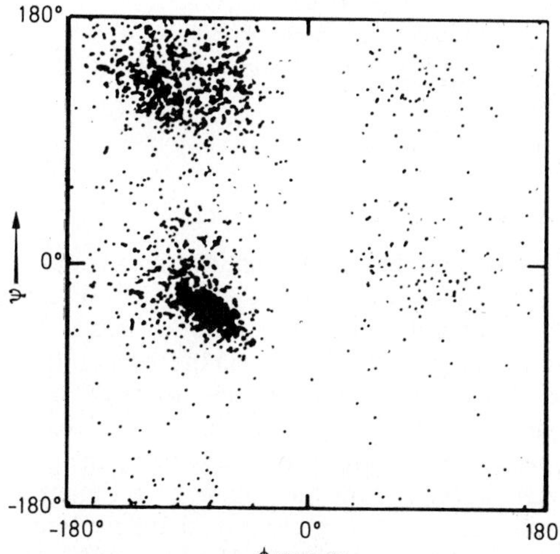

Fig. 1.12. φ,ψ-Diagram for observed values of 13 various proteins containing a total of 2,500 amino acids. (After *Schulz* and *Schirmer,* 1979)

Table 1.17. Regular structure elements (secondary structure) present in protein

	K	ψ	n[a]	d[b]	r[c]
	(°)	(°)		(Å)	(Å)
β-sheet, parallel	−119	+113	2.0	3.2	1.1
β-sheet, antiparallel	−139	+135	2.0	3.4	0.9
Twisted sheet			2.3	3.3	1.0
3_{10}-Helix	−49	−26	3.0	2.0	1.9
α-Helix (right-handed)	−57	−47	3.6	1.5	2.3
α-Helix (left-handed)	+57	+47	3.6	1.5	2.3
π-Helix	−57	−70	4.3	1.1	2.8
Poly-L-proline helix I	−83	+158	3.3	1.9	
Poly-L-proline helix II	−78	+149	3.0	3.1	
Polyglycine helix II (left-handed)	−80	+150	3,0	3,1	

[a] Amino acid residues per turn.
[b] The rise along the axis direction, per residue.
[c] The radius of the helix.

1.4.2.2.1 *β-sheet*

Three regular structural elements (pleated-sheet structures) have values in the range of $\phi = -120°$ and $\psi = +120°$. The peptide chain is always lightly folded on the C_{α} atom (cf. Figure 1.13), thus the R side chains extend perpendicularly to the extension axis of the chain, i.e. the side chains change their projections alternately from +z to −z. Such a pleated structure is stabilized when more chains are present. Subsequently, adjacent chains interact along the x-axis by hydrogen bonding, thus providing the cross-linking required for stability.

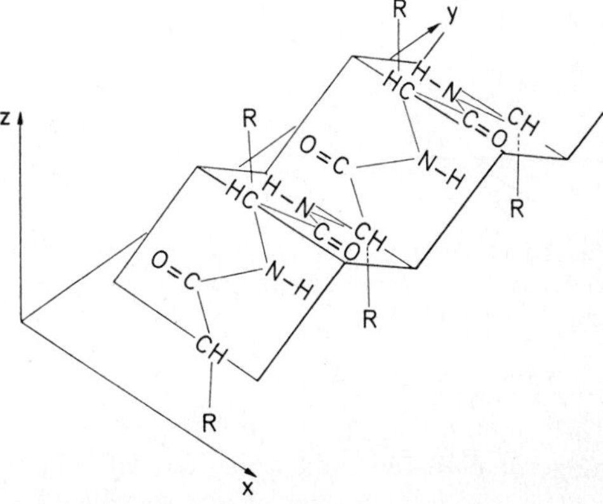

Fig. 1.13. A pleated sheet structure of a peptide chain

When adjacent chains run in the same direction, the peptide chains are parallel. This provides a stabilized, planar, parallel sheet structure. When the chains run in opposite directions, a planar, antiparallel sheet structure is stabilized (Figure 1.14). The lower free energy, twisted sheet structures, in which the main axes of the neighboring chains are arranged at an angle of 25° (Figure 1.14a), are more common than planar sheet structures.

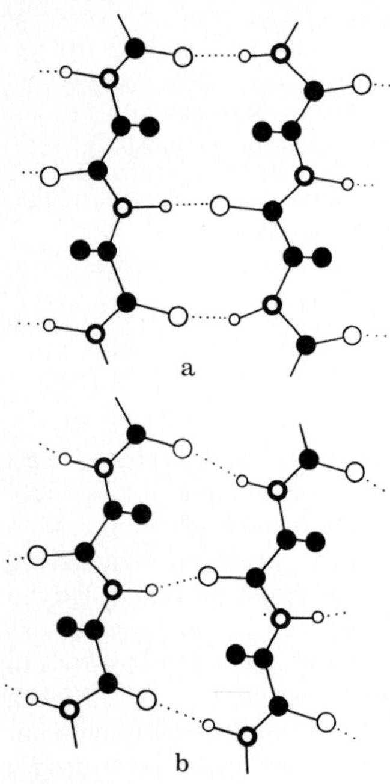

Fig. 1.14. Diagrammatic presentation of antiparallel (**a**) and parallel (**b**) peptide chain arrangements

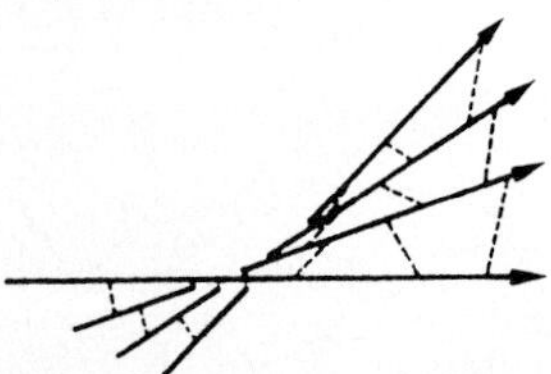

Fig. 1.14a. Diagrammatic presentation of a twisted sheet structure of parallel peptide chains (After *Schulz* and *Schirmer*, 1979)

1.4.2.2.2 Helical structures

There are three regular structural elements in the range of $\phi = -60°$ and $\psi = -60°$ (cf. Figure 1.11) in which the peptide chains are coiled like a threaded screw. These structures are stabilized by intrachain hydrogen bridges which extend almost parallel to the chain axis, cross-linking the CO and NH groups, specifically, the CO group of amino acid residue i with the NH group of residue i + 3 (3_{10}-helix), i + 4 (α-helix) or i + 5 (π-helix).

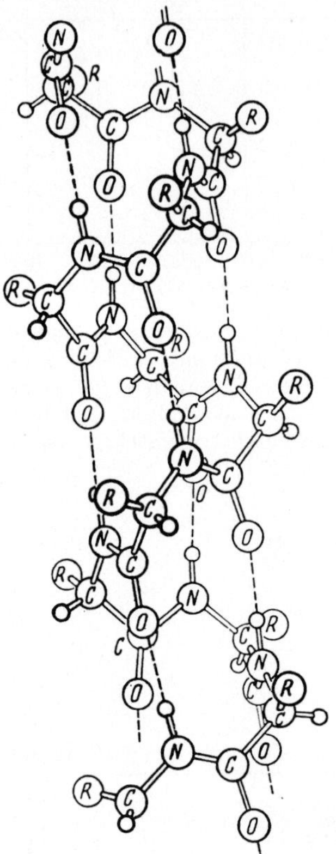

Fig. 1.15. Right-handed α-helix

The α-helix, and for polypeptides from L-amino acids exclusively the right-handed α-helix (Figure 1.15), is the most common. The 3_{10}-helix is observed only in a few proteins in the form of short segments of the peptide chains, while the existence of a π-helix is hypothetical. A helix is characterized by the angles ϕ and ψ, or by the parameters derived from these angles: n, the number of amino acid residues per turn; d, the rise along the main axis per amino acid residue; and r, the radius of the helix. Thus, the equation for the pitch, p, is $p = n \cdot d$. The parameters n and d are presented within a ϕ, ψ plot in Figure 1.16.

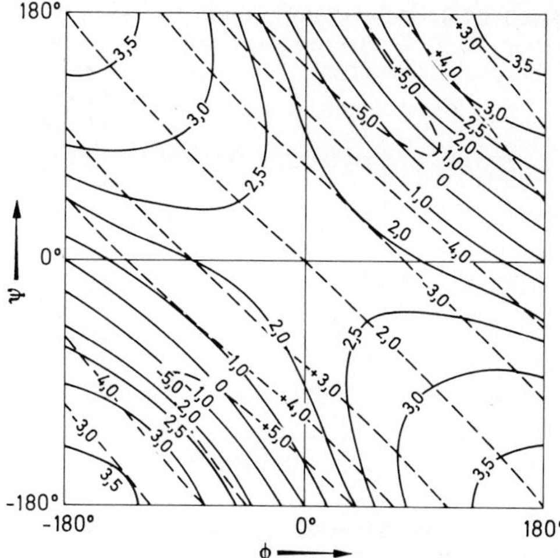

Fig. 1.16. ϕ,ψ-Diagram with marked helix parameters n (---) and d (—). (After *Schulz* and *Schirmer,* 1979)

Table 1.16a. β-Bends of egg white lysozyme

Residue Number	Sequence
20–23	Tyr Arg *Gly* Tyr
36–39	Ser Asn Phe Asn
39–42	Asn Thr Gln Ala
47–50	Thr Asp *Gly* Ser
54–57	Gly Ile Leu Glu
60–63	Ser Arg Trp Trp
66–69	Asp Gly Arg Thr
69–72	Thr Pro *Gly* Ser
74–77	Asn Leu Cys Asn
85–88	Ser Ser Asp Ile
100–103	Ser Asp *Gly* Asp
103–106	Asp Gly Met Asn

1.4.2.2.3 Reverse turns

An important conformational feature of globular proteins are the reverse turns or β-bends. They occur at "hairpin" corners, where the peptide chain changes direction abruptly. Such corners involve four amino acid residues. Glycine is favoured in the third position of β-bends for energetical reasons. The sequences of the β-bends of lysozyme are listed in Table 1.16a as an example.

1.4.2.2.4 Super-secondary structures

Analysis of known protein structures has demonstrated that regular elements can exist in combined forms. Examples are the coiled-coil α-helix (Figure 1.17,a), chain segments with antiparallel β-structures (β-meander structure; Figure 1.17,b) and combinations of α-helix and β-structure (e.g. βαβαβ; Figure 1.17,c).

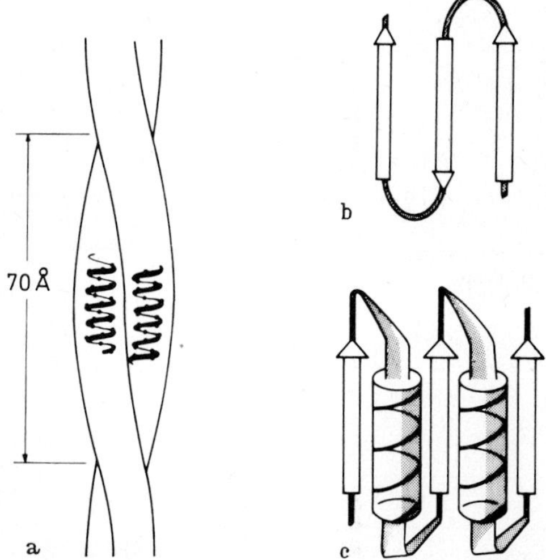

Fig. 1.17. Superhelix secondary structure (after *Schulz* and *Schirmer,* 1979). **a** coiled-coil α-helix, **b** β-meander, **c** βαβαβ-structure

1.4.2.3 Tertiary and Quaternary Structures

Proteins can be divided into two large groups on the basis of conformation: (a) fibrillar (fibrous) or scleroproteins, and (b) folded or globular proteins.

1.4.2.3.1 Fibrous proteins

The entire peptide chain is packed or arranged within a single regular structure for a variety of fibrous proteins. Examples are wool keratin (α-helix), silk fibroin (β-sheet structure) and collagen (a triple helix). Stabilization of these structures is achieved by intermolecular bonding (electrostatic interaction and disulfide linkages, but primarily hydrogen bonds and hydrophobic interactions).

1.4.2.3.2 Globular proteins

Regular structural elements are mixed with randomly-extended chain segments (randomly-coiled structures) in globular proteins. The proportion of regular structural elements is highly variable: 20–30% in casein, 45% in lysozyme and 75% in myoglobin (Table 1.18). Five structural subgroups are known in this group of proteins: (1) α-helices occur only; (2) β-structures occur only; (3) α-helical and β-structural portions occur in separate segments on the peptide chain; (4) α-helix and β-structures alternate along the peptide chain; and (5) α-helix and β-structures do not exist.

Table 1.18. Proportion of "regular structure elements" present in various globular proteins

Protein	a-Helix	β-Structure	n_G	n	%
Myoglobin	3– 16[a]			14	
	20– 34			15	
	35– 41			7	
	50– 56			7	
	58– 77			20	
	85– 93			9	
	99–116			18	
	123–145			23	
			151	113	75
Lysozyme	5– 15			11	
	24– 34			11	
		41–54		14	
	80– 85			6	
	88– 96			9	
	97–101			5	
	109–125			7	
			129	63	49
α_{S1}-Casein			199	ca	30
β-Casein			209	ca	20

[a] Position number of the amino acid residue in the sequence.

n_G: Total number of amino acid residues.

n: Amino acid residues within the regular structure.

%: Percentage of the amino acid residues present in regular structure.

The process of peptide chain folding is not yet fully understood. It occurs spontaneously, probably arising from one center or from several centers of high stability in larger proteins.

Folding of the peptide chain packs it densely, by formation of a large number of intermolecular noncovalent bonds. Data about the nature of the bonds involved are provided in Table 1.19.

Table 1.19. Bond-types in proteins

Type	Examples	Bond strength (kJ/mole)
Covalent bonds	$-S-S-$	ca. −230
Electrostatic bonds	$-COO^- H_3N^+-$ $\rangle C=O\ O=C\langle$	−21 +1.3
Hydrogen bonds	$-O-H\cdots O\langle$ $\rangle N-H\cdots O=C\langle$	−4 −3
Hydrophobic bonds	$-CH\langle^{CH_3}_{CH_3}\ {}^{H_3C}_{H_3C}\rangle CH-$	0.025[b]
	$-Ala\cdots Ala-$	−3
	$-Val\cdots Val-$	−8
	$-Leu\cdots Leu-$	−9
	$-Phe\cdots Phe-$	−13
	$-Trp\cdots Trp-$	−19

[a] For $\varepsilon = 4$.

[b] Per Å^2-surface area.

The H-bonds formed between main chains, main and side chains and side-side chains are of particular importance for folding. The portion of polar groups involved in H-bond build-up in proteins of >8.9 kdal appears to be fairly constant at about 50%.

The hydrophobic interaction of the nonpolar regions of the peptide chains also plays an important role in protein folding. These interactions are responsible for the fact that nonpolar groups are folded to a great extent towards the interior of the protein globule. The surface areas accessible to water molecules have been calculated for both unfolded and native folded forms for a number of monomeric proteins with known conformations. The proportion of the accessible surface of a stretched state which tends to be fixed in the interior of the globule as a result of folding is a simple linear function of the molecular weight (M). The gain in free energy for the folded surface is 10 kJ nm^{-2}. Therefore, the total hydrophobic contribution to free energy due to folding is:

$$\Delta G_{HP} = 88\ M + 79 \cdot 10^{-5}\ M^2\ [J \cdot mole^{-1}] \qquad (1.75)$$

This relation is valid for a range of $6{,}108 \leq M \leq 34{,}409$, but appears to be also valid for larger molecules since they often consist of several loose associations of independent globular portions, denoted as structural domains (Figure 1.18).

Abb. 1.18. Globular protein with two-domain structure (After *Schulz* and *Schirmer*,1979)

Proteins with disulfide bonds fold at a significantly slower rate than those without disulfide bonds. Folding is not limited by the reaction rate of disulfide formation. Therefore the folding process of disulfide-containing proteins seems to proceed in a different way. The reverse process, protein unfolding, is very much slowed down by the presence of disulfide bridges which generally impart great stability to globular proteins. This stability is particularly effective against denaturation. An example is the *Bowman–Birk* inhibitor from soybean (Figure 1.19) which inhibits the activity of trypsin and chymotrypsin. Its tertiary structure is stabilized by seven disulfide bridges. The reactive sites of inhibition are Lys^{16}-Ser^{17} and Leu^{43}-Ser^{44}, i.e. both sites are located in relatively small rings, each of which consists of nine amino acid residues held in ring form by a disulfide bridge. The thermal stability of this inhibitor is high.

1.4.2.3.3 Quaternary structures

In addition to the free energy gain by folding of a single peptide chain, association of more than one peptide chain (subunit) can provide further gains in free energy. For example, hemoglobin (4 associated peptide chains) $\Delta G^0 = -46$ kJ $mole^{-1}$ and the trypsin-trypsin inhibitor complex (association of 2 peptide chains) $\Delta G^0 = -75.2$ kJ $mole^{-1}$. In principle such associations correspond to the folding of a larger peptide chain around several structural domains without covalently binding the subunits. Table 1.19 a lists some proteins which partially exhibit quaternary structures.

Table 1.19 a. Examples of globular proteins

Name	Origin	Molecular weight (Kdal)	Number of subunits
Lysozyme	Chicken egg	14.6	1
Papain	*Papaya latex*	20,7	1
a-Chymotrypsin	Pancreas (beef)	23	1
Trypsin	Pancreas (beef)	23,8	1
Pectinesterase	Tomato	27.5	
Chymosin	Stomach (calves)	31	
b-Lactoglobulin	Milk	35	2
Pepsin A	Stomach (swine)	35	1
Peroxidase	Horseradish	40	1
Hemoglobin	Blood	64.5	4
Avidin	Chicken egg	68.3	4
Alcohol-dehydrogenase	Liver (horse)	80	2
	Yeast	150	4
Hexokinase	Yeast	104	2
Lactate dehydrogenase	Heart (swine)	135	4
Glucose oxidase	*P. notatum*	152	
Pyruvate kinase	Yeast	161	8
	A. niger	186	
β-Amylase	Sweet potato	215	4
Catalase	Liver (beef)	232	4
	M. lysodeikticus	232	
Adenosine triphosphatase	Heart (beef)	284	6
Urease	Jack-beans	483	6
Glutamine synthetase	*E. coli*	592	12
Arginine decarboxylase	*E. coli*	820	10

1.4.2.4 Denaturation

The term denaturation denotes a reversible or irreversible change of native conformation (tertiary structure) without cleavage of covalent bonds (except for disulfide bridges). Denaturation is possible with any treatment that cleaves hydrogen bridges, ionic or hydrophobic bonds. This can be accomplished by: changing the temperature, adjusting the pH, increasing the interface area, or adding organic solvents, salts, urea, guanidine hydrochloride or detergents such as sodium dodecyl sulfate. Denaturation is gen-

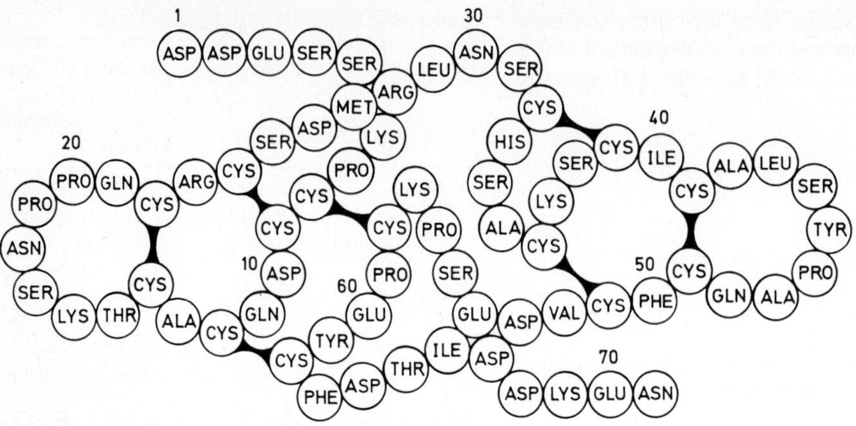

Fig. 1.19. *Bowman-Birk* inhibitor from soybean (After *Ikenaka* et al., 1974)

erally reversible when the peptide chain is stabilized in its unfolded state by the denaturing agent and the native conformation can be reestablished after removal of the agent. Irreversible denaturation occurs when the unfolded peptide chain is stabilized by interaction with other chains (as occurs for instance with egg proteins during boiling). During unfolding reactive groups, such as thiol groups, that were covered or blocked, may be exposed. Their participation in the formation of disulfide bonds may also cause an irreversible denaturation.

Denaturation of biologically-active proteins is usually associated with loss of activity. The fact that denatured proteins are more readily digested by proteolytic enzymes is also of interest.

1.4.3 Physical Properties

1.4.3.1 Dissociation

Proteins, like amino acids, are amphoteric. Depending on pH, they can exist as polyvalent cations, anions or zwitter ions. Proteins differ in their α-carboxyl and α-amino groups – since these groups are linked together by peptide bonds, the uptake or release of protons is limited to free terminal groups. Therefore, most of the dissociable functional groups are derived from side chains. Table 1.20 lists pK values of some protein groups. In contrast to free amino acids, these values fluctuate greatly for proteins since the dissociation is influenced by neighboring groups in the macromolecule. For example, in lysozyme the γ-carboxyl group of Glu^{35} has a pK of 6–6.5, while the pK of the β-carboxyl group of Asp^{66} is 1.5–2, of Asp^{52} is 3–4.6 and of Asp^{101} is 4.2–4.7.

Table 1.20. pK values of protein side chains

Group	pK (25 °C)	Group	pK (25 °C)
α-Carboxyl-	3– 4	Imidazolium-	4– 8
β,γ-Carboxyl-	3– 5	Hydroxy-	
α-Ammonium-	7– 8	(aromatic)	9–12
ε-Ammonium-	9–11	Thiol-	8–11
Guanidinium-	12–13		

The total charge of a protein, which is the absolute sum of all positive and negative charges, is differentiated from the so-called net charge which, depending on the pH, may be positive, zero or negative. By definition the net charge is zero and the total charge is maximal at the isoelectric point. Lowering or raising the pH tends to increase the net charge towards its maximum, while the total charge always becomes less than at the isoelectric point.

Since proteins interact not only with protons but also with other ions, there is a further differentiation between an isoionic and an isoelectric point. The isoionic point is defined as the pH of a protein solution at infinite dilution, with no other ions present except for H^+ and HO^-. Such a protein solution can be acquired by extensive

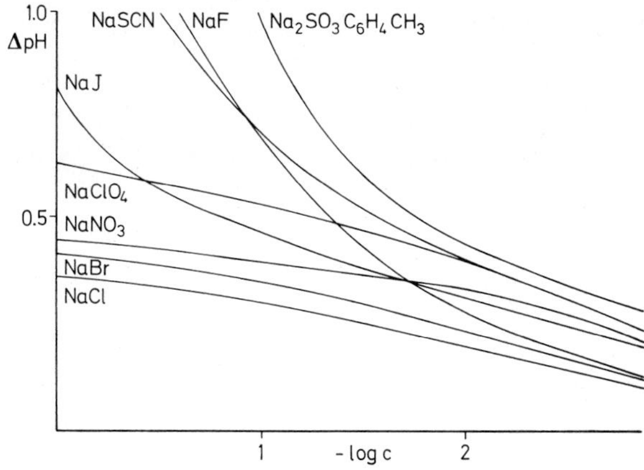

Fig. 1.20. pH-shift of isoionic serum albumin solutions by added salts. (After *Edsall* and *Wyman*, 1958)

dialysis (or, better, electrodialysis) against water. The isoionic point is constant for a given substance while the isoelectric point is variable depending on the ions present and their concentration. In the presence of salts, i.e. when binding of anions is stronger than that of cations, the isoelectric point is lower than the isoionic point. The reverse is true when cationic binding is dominant. Figure 1.20 shows the shift in pH of an isoionic serum albumin solution after addition of various salts. The shift in pH is consistently positive, i.e. the protein binds more anions than cations.

The titration curve of β-lactoglobulin at various ionic strengths (Figure 1.21) shows that the isoelectric point of this protein, at pH 5.18, is independent of the salts present. The titration curves are, however, steeper with increased ionic strength, which indicates greater suppression of the electrostatic interaction between protein molecules.

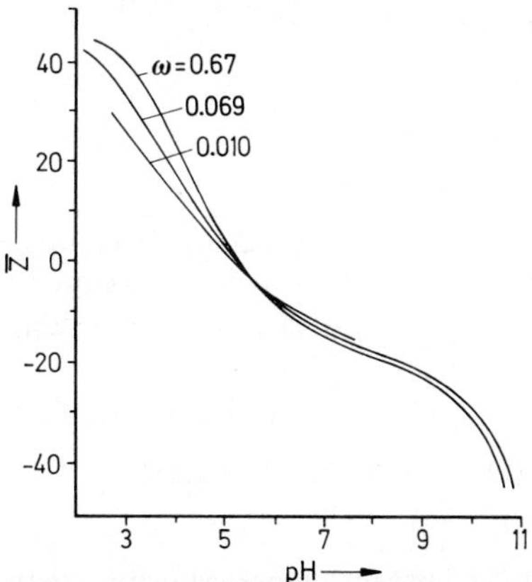

Fig. 1.21. Titration curves for β-lactoglobulin at various ionic strengths ω. (After *Edsall* and *Wyman*, 1958)

At its isoelectric point a protein is the least soluble and the most likely to precipitate ("isoelectric precipitation") and is at its maximal crystallization capacity. The viscosity of solubilized proteins and the swelling power of insoluble proteins are at a minimum at the isoelectric point.

1.4.3.2 Optical Activity

The optical activity of proteins is due not only to asymmetry of amino acids but also to the chirality resulting from the arrangement of the peptide chain. Information on the conformation of protein can be obtained from recording of the optical rotatory dispersion (ORD) or the circular dichroism (CD), especially in the range of peptide bond absorption wavelengths (190–200 nm). The *Cotton* effect occurs in this range and reveals quantitative information on secondary structure. An α-helix or a β-structure gives a negative *Cotton* effect, with absorption maxima at 199 and 205 nm, while a randomly-coiled conformation shifts the maximum to shorter wavelengths, i.e. results in a positive *Cotton* effect (Figure 1.22).

1.4.3.3 Solubility, Hydration and Swelling Power

Protein solubility is variable and is influenced by the number of polar and apolar groups and their arrangement along the molecule. Generally, proteins are soluble only in strongly polar solvents such as water, glycerol, formamide, dimethylfor-

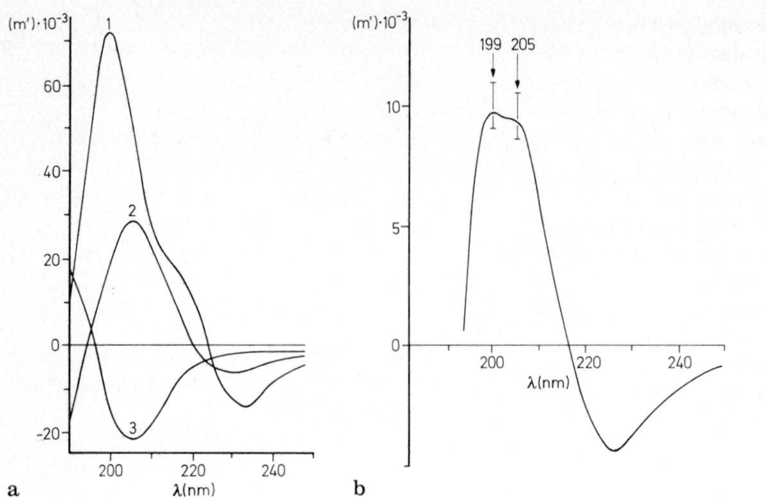

Fig. 1.22. *Cotton* effect. **a** Polylysine α-helix (1, pH 11–11.5) β-sheet structure (2, pH 11–11.3 and heated above 50 °C) and random coiled (3, pH 5–7). **b** Ribonuclease with 20% α-helix, 40% β-sheet structure and 40% random coiled region. (After *Luebke, Schroeder,* and *Kloss,* 1975)

mamide or formic acid. In a less polar solvent such as ethanol, proteins are rarely noticeably soluble (e.g. prolamins).

The solubility in water is dependent on pH and on salt concentration. Figure 1.23 shows these relationship for β-lactoglobulin.

At low ionic strength, the solubility rises with increase in ionic strength and the solubility minimum (isoelectric point) is shifted from pH 5.4 to pH 5.2. This shift is due to preferential binding of anions to the protein.

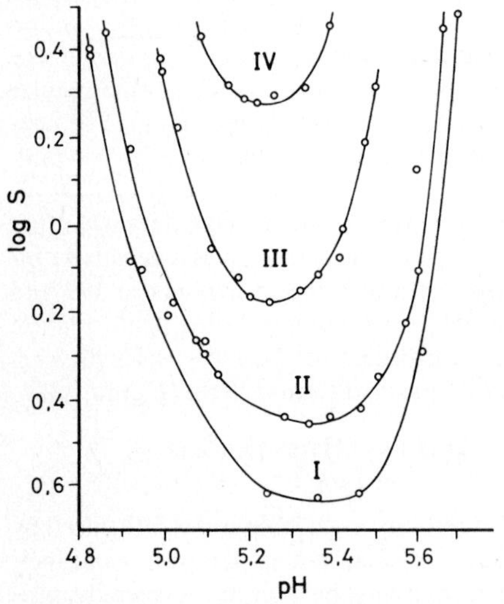

Fig. 1.23. β-Lactoglobulin solubility as affected by pH and ionic strength I. 0.001, II. 0.005, III. 0.01, IV. 0.02

As a rule, neutral salts have a two-fold effect on protein solubility. At low concentrations they increase the solubility ("salting in" effect) by suppressing the electrostatic protein-protein interaction (binding forces).

The log of the solubility (S) is proportional to the ionic strength (μ) at low concentrations (cf. Figure 1.23):

$$\log S = K \cdot \mu$$

Protein solubility is decreased ("salting out" effect) at higher salt concentrations due to the ion hydration tendency of the salts. The following relationship applies (S_0: solubility at $\mu = 0$; K: salting out constant):

$$\log S = \log S_0 - K \cdot \mu$$

Cations and anions in the presence of the same counter ion can be arranged in the following orders (*Hofmeister* series) based on their salting out effects:

$K^+ > Rb^+ > Na^+ > Cs^+ > Li^+ > NH_4^+$;
$SO_4^{2-} >$ citrate$^{2-} >$ tartrate$^{2-} >$ acetate$^-$
$> Cl^- > NO_3^- > Br^- > J^- > CNS^-$

Multivalent anions are more effective than monovalent anions, while divalent cations are less effective than monovalent cations.

Since proteins are polar substances, they are hydrated in water. The degree of hydration (g water of hydration/g protein) is variable. It is 0.22 for ovalbumin (in ammonium sulfate), 0.06 for edestin (in ammonium sulfate), 0.8 for β-lactoglobulin and 0.3 for hemoglobin.
The swelling of insoluble proteins corresponds to the hydration of soluble proteins in that insertion of water between the peptide chains results in an increase in volume and other changes in the physical properties of the protein. The amount of water taken up during swelling can exceed the dry weight of the protein by several times. For example, muscle tissue contains 3.5–3.6 g water per g protein dry matter.

1.4.4 Chemical Reactions

The chemical modification of protein is of importance for a number of reasons. It provides derivatives suitable for sequence analysis, identifies the reactive groups in catalytically-active sites of an enzyme, enables the binding of protein to a carrier (protein immobilization) and provides changes in protein properties which are important in food processing. In contrast to free amino acids and, except for the relatively small number of functional groups on the terminal amino acids, only the functional groups on protein side chains are available for chemical reactions.

1.4.4.1 Lysine Residue

Reactions involving the lysine residue can be divided into several groups: (a) reactions leading to a positively-charged derivative; (b) reactions eliminating the positive charge; (c) derivatizations introducing a negative charge; and (d) reversible reactions. The latter are of particular importance.

1.4.4.1.1 Reactions which retain the positive charge

Alkylation of the free amino group of lysine with aldehydes and ketones is possible, with a simultaneous reduction step:

$$\text{Prot–NH}_2 + \text{R–CO–R}^1 \xrightarrow{\text{NaBH}_4,\ \text{pH 9, 0 °C, 30 min}} \text{Prot–NH–CH(R)(R}^1) \qquad (1.76)$$

$(R = R^1 = CH_3;\ R = H,\ CH_3,\ R^1 = H)$

A dimethyl derivative [Prot $-N(CH_3)_2$] can be obtained with formaldehyde ($R = R_1 = H$) (cf. 1.2.4.2.2).
Guanidination can be accomplished by using O-methylisourea as a reactant. α-Amino groups react at a much slower rate than ε-amino groups:

$$\text{Prot–NH}_2 + \text{H}_3\text{C–O–C(=NH)NH}_2 \xrightarrow{\text{pH 10.6, 4 °C, 4 days}} \text{Prot–NH–C(=NH}_2^{\oplus})\text{–NH}_2 \qquad (1.77)$$

This reaction is used analytically to assess the amount of biologically-available ε-amino groups.
Derivatization with imido esters is also possible. The reactant is readily accessible from the corresponding nitriles:

$$\text{Prot–NH}_2 \xrightarrow[\text{pH 9.2, 0 °C, 20 h}]{\text{R–C(=NH}_2^{\oplus})\text{OR}'} \text{Prot–NH–C(=NH}_2^{\oplus})\text{–R}$$

$$\text{R–CN} \xrightarrow[\text{H}^{\oplus}]{\text{R'OH}} \text{R–C(=NH}_2^{\oplus})\text{–OR}' \qquad (1.78)$$

Proteins can be cross-bonded with the use of a bifunctional imido ester (cf. 1.4.4.10).
Treatment of the amino acid residue with amino acid carboxyanhydrides yields a polycondensation reaction product:

$$\text{Protein–NH}_2 + \text{(amino acid carboxyanhydride, R)} \longrightarrow$$

$$\text{Protein–NH–[CO–CHR–NH]}_n\text{–CO–CHR–NH}_2 \qquad (1.79)$$

The value n depends on reaction conditions. The carboxyanhydrides are readily accessible through interaction of the amino acid with phosgene:

$$\text{R–CH(NH}_2)\text{–COOH} \xrightarrow{\text{COCl}_2} \text{R–CH(NH–CO–Cl)–COOH} \xrightarrow{-\text{HCl}} \text{(carboxyanhydride, R)} \qquad (1.80)$$

1.4.4.1.2 Reactions resulting in a loss of positive charge

Acetic anhydride reacts with lysine, cysteine, histidine, serine, threonine and tyrosine residues. Subsequent treatment of the protein with hydroxylamine (1 M, 2 h, pH 9, 0 °C) leaves only the acetylated amino groups intact:

$$\text{Prot—NH}_2 \xrightarrow[\text{pH 7–9.5, 0 °C}]{(\text{CH}_3\text{CO})_2\text{O}} \text{Prot—NH—CO—CH}_3 \quad (1.81)$$

Carbamylation with cyanate attacks α- and ε-amino groups as well as cysteine and tyrosine residues. However, their derivatization is reversible under alkaline conditions:

$$\text{Prot—NH}_2 \xrightarrow[\text{pH 8, 37 °C, 12–24 h}]{\text{1 M KOCN}} \text{Prot—NH—C(=O)—NH}_2 \quad (1.82)$$

Arylation with 1-fluoro-2,4-dinitrobenzene (*Sanger's* reagent; FDNB) and trinitrobenzene sulfonic acid was outlined in Section 1.2.4.2.2. FDNB also reacts with cysteine, histidine and tyrosine.

4-Fluoro-3-nitrobenzene sulfonic acid, a reactant which has good solubility in water, is also of interest for derivatization of protein:

$$\text{Prot—NH}_2 + \text{F—C}_6\text{H}_3(\text{O}_2\text{N})\text{—SO}_3\text{H} \xrightarrow{\text{pH 6–9, 25 °C}} \text{Prot—NH—C}_6\text{H}_3(\text{O}_2\text{N})\text{—SO}_3^{\ominus} \quad (1.83)$$

Deamination can be accomplished with nitrous acid:

$$\text{Prot—NH}_2 \xrightarrow[\text{pH 4.35, 0 °C}]{\text{HNO}_2} \text{Prot—OH} + \text{N}_2 \quad (1.84)$$

This reaction involves α- and ε-amino groups as well as tryptophan, tyrosine, cysteine and methionine residues.

1.4.4.1.3 Reactions resulting in a negative charge

Acylation with dicarboxylic acid anhydrides, e.g. succinic acid anhydride, introduces a carboxyl group into the protein:

$$\text{Prot}\begin{cases}\text{—C}_6\text{H}_4\text{—OH}\\ \text{—NH}_2\\ \text{—SH}\end{cases} \xrightarrow[\text{pH 8–9}]{\text{succinic anhydride}} \text{Prot}\begin{cases}\text{—C}_6\text{H}_4\text{—O—CO—(CH}_2)_2\text{—COOH}\\ \text{—NH—CO—(CH}_2)_2\text{—COOH}\\ \text{—S—CO—(CH}_2)_2\text{—COOH}\end{cases} \xrightarrow{\text{H}_2\text{O}} \text{Prot}\begin{cases}\text{—C}_6\text{H}_4\text{—OH}\\ \text{—NH—CO—(CH}_2)_2\text{—COOH}\\ \text{—SH}\end{cases} \quad (1.85)$$

Introduction of a fluorescent acid group is possible by interaction of the protein with pyridoxal phosphate followed by reduction of the intermediary *Schiff* base:

$$\text{Prot—NH}_2 + \text{OHC—(pyridinium: }{}^{\ominus}\text{O, CH}_3\text{, NH}^{\oplus}\text{, CH}_2\text{OPO}_3\text{H}^{\ominus}) \xrightarrow[\text{pH 6, 25 °C}]{\text{NaBH}_4} \text{Prot—NH—CH}_2\text{—(pyridinium: }{}^{\ominus}\text{O, CH}_3\text{, NH}^{\oplus}\text{, CH}_2\text{OPO}_3\text{H}^{\ominus}) \quad (1.86)$$

1.4.4.1.4 Reversible reactions

N-Maleyl derivatives of proteins are obtained at alkaline pH by reaction with maleic anhydride. The acylated product is cleaved at pH < 5, regenerating the protein:

$$\text{Prot—NH}_2 + \text{maleic anhydride} \xrightarrow{\text{pH>7}} \text{Prot—NH—CO—CH=CH—COO}^{\ominus} \xrightarrow{\text{pH<5}} \text{Prot—NH}_2 \quad (1.87)$$

The half-life (τ) of ε-N-maleyl lysine is 11 h at pH 3.5 and 37 °C. More rapid cleavage is observed with the 2-methyl-maleyl derivative (τ < 3 min at pH 3.5 and 20 °C) and the 2,2,3,3-tetrafluoro-succinyl derivative (τ very low at pH 9.5 and 0 °C). Cysteine binds maleic anhydride through an addition reaction. The S-succinyl derivative is quite stable. This side reaction is, however, avoided when protein derivatization is done with exo-cis-3,6-endoxohexahydrophthalic acid anhydride:

(1.88)

For ε-N-acylated lysine, τ = 4–5 h at pH 3 and 25 °C.

Acetoacetyl derivatives are obtained with diketene:

$$\text{Prot–NH}_2 + \text{(diketene)} \longrightarrow \text{Prot–NH–CO–CH}_2\text{–CO–CH}_3 \quad (1.89)$$

This type of reaction also occurs with cysteine and tyrosine residues. The acyl group is readily split from tyrosine at pH 9.5. Complete release of protein from its derivatized form is possible by treatment with phenylhydrazine or hydroxylamine at pH 7.

1.4.4.2 Arginine Residue

The arginine residue of protein reacts with α- or β-dicarbonyl compounds to form cyclic derivatives:

$$\text{Protein–NH–C(=NH)NH}_2 + (\text{OHC})_2\overset{\ominus}{\text{C}}\text{–NO}_2\ \text{Na}^{\oplus} \xrightarrow[0-5\,°\text{C},\ 15\ \text{min}]{\text{pH } 12-14} \text{Protein–NH–(5-nitropyrimidin-2-yl)} \xrightarrow{\text{NaBH}_4} \text{Protein–NH–(tetrahydro-5-nitropyrimidin-2-yl)} \quad (1.90)$$

$$\text{R–NH–C(=NH}_2^{\oplus})\text{NH}_2 + \text{R'–CO–CO–R'} \rightarrow \text{R–N=C(HN–C(R')(OH))(HN–C(R')(OH))} \rightarrow \text{R–N=(imidazolidone, R', R', =O)} \quad (1.91)$$

$$\text{R–NH–C(=NH}_2^{\oplus})\text{NH}_2 + \text{1,2-cyclohexanedione} \xrightarrow[25-40\,°\text{C}]{\text{pH } 8-9} \text{R–N=(bicyclic diol adduct)} \xrightarrow[\text{pH } 7]{\text{H}_2\text{N–OH}} \text{R–NH–C(=NH}_2^{\oplus})\text{NH}_2 \quad (1.92)$$

The nitropyrimidine derivative absorbs at 335 nm. The arginyl bond of this derivative is not cleaved by trypsin but it is cleaved in its tetrahydro form, obtained by reduction with $NaBH_4$ (cf. Reaction 1.90). In the reaction with benzil, an iminoimidazolidone derivative is obtained after a benzilic acid rearrangement (cf. Reaction 1.91).

Reaction of the arginine residue with 1,2-cyclohexanedione is highly selective and proceeds under mild conditions. Regeneration of the arginine residue is again possible with hydroxylamine (cf. Reaction 1.92).

1.4.4.3 Glutamic and Aspartic Acid Residues

These amino acid residues are usually esterified with methanolic HCl. There can be side reactions, such as methanolysis of amide derivatives or N,O-acyl migration in serine or threonine residues:

$$\text{Protein–COOH} \xrightarrow[0\,°\text{C}]{\text{CH}_3\text{OH/HCl}} \text{Protein–COOCH}_3 \quad (1.93)$$

Diazoacetamide reacts with a carboxyl group and also with the cysteine residue:

$$\text{Protein—COOH} + N_2CH\text{—}CONH_2 \longrightarrow R\text{—}COOCH_2CONH_2 \quad (1.94)$$

Amino acid esters or other similar nucleophilic compounds can be attached to a carboxyl group of a protein with the help of a carbodiimide:

$$\text{Protein—COOH} + H_2N\text{—}CH_2\text{—}COOCH_3 \xrightarrow{R-N=C=N-R} \text{Protein—CO—NH—}CH_2\text{—}COOCH_3 \quad (1.95)$$

1.4.4.4 Cystine Residue (cf. also Section 1.2.4.3.5)

Cleavage of cystine is possible by a nucleophilic attack:

$$\text{Protein—S—S—Protein} + Y^{\ominus} \longrightarrow \text{Protein—S—Y} + \text{Protein—}S^{\ominus} \quad (1.96)$$

The nucleophilic reactivity of the reagents decreases in the series: hydride > arsenite and phosphite > alkanethiol > aminoalkanethiol > thiophenol and cyanide > sulfite > OH^- > p-nitrophenol > thiosulfate > thiocyanate. Cleavage with sodium borohydride and with thiols was covered in Section 1.2.4.3.5. Complete cleavage with sulfite requires that oxidative agents (e.g. Cu^{2+}) be present and that the pH be higher than 7:

$$\begin{aligned} RSSR + SO_3^{2\ominus} &\longrightarrow RSSO_3^{\ominus} + RS^{\ominus} \\ 2\,RS^{\ominus} &\xrightarrow{Cu^{2\oplus}} RSSR \end{aligned} \quad (1.97)$$

The resultant S-sulfo derivative is quite stable in neutral and acidic media and is fairly soluble in water. The S-sulfo group can be eliminated with an excess of thiol reagent.
Cleavage of cystine residues with cyanides (nitriles) is of interest since the thiocyanate formed in the reaction is cyclized into a 3-acyl-2-iminothiazolidine under cleavage of the N-acyl bond:

$$\text{R—CO—NH—CH(CH}_2\text{—S—S—R'')—CO—NH—R'} \xrightarrow{CN^{\ominus}} \text{R—CO—NH—CH(CH}_2\text{—S—CN)—CO—NH—R'} + R''\text{—}S^{\ominus}$$

$$\longrightarrow \text{R—CO—N} \overset{\text{(HN=C—S—CH}_2\text{ ring)}}{—} \text{CH—CO—NH—R'} \longrightarrow \text{R—COOH} + \text{HN—CH(—CO—NH—R')—CH}_2\text{—S—C(=NH)— (ring)} \quad (1.98)$$

This reaction can be utilized for the selective cleavage of peptide chains. Initially, all the disulfide bridges are reduced with dithiothreitol, and then are converted to mixed disulfides through reaction with 5,5′-dithio-bis-(2-nitrobenzoic acid). These mixed disulfides are then cleaved by cyanide at pH 7.
Electrophilic cleavage occurs with Ag^+ and Hg^+ or Hg^{2+} as follows:

$$\begin{aligned} 2\,Ag^{\oplus} + 2\,RSSR &\longrightarrow 2\,RSAg + 2\,RS^{\oplus} \\ 2\,RS^{\oplus} + 2\,OH^{\ominus} &\longrightarrow 2\,RSOH \longrightarrow RSO_2H + RSH \\ RSH + Ag^{\oplus} &\longrightarrow RSAg + H^{\oplus} \\ \hline 3\,Ag^{\oplus} + 2\,RSSR + 2\,OH^{\ominus} &\longrightarrow 3\,RSAg + RSO_2H + H^{\oplus} \end{aligned} \quad (1.99)$$

Electrophilic cleavage with H^+ is possible only in strong acids (e.g. 10 N HCl). The sulfenium cation which is formed can catalyze a disulfide exchange reaction:

$$\begin{aligned} RSSR + H^{\oplus} &\longrightarrow RSH + RS^{\oplus} \\ RS^{\oplus} + R^1SSR^1 &\longrightarrow RSSR^1 + R^1S^{\oplus} \end{aligned} \quad (1.100)$$

In neutral and alkaline solutions a disulfide exchange reaction is catalyzed by the thiolate anion:

$$\begin{aligned} RSSR + OH^{\ominus} &\rightleftharpoons R\text{—}SOH + RS^{\ominus} \\ R^1SSR^1 + RS^{\ominus} &\rightleftharpoons R^1SSR + R^1S^{\ominus} \end{aligned} \quad (1.101)$$

1.4.4.5 Cysteine Residue (cf. also Section 1.2.4.3.5)

A number of alkylating agents yield derivatives which are stable under the conditions for acidic hydrolysis of protein. The reaction with ethylene imine giving an S-aminoethyl derivative and, hence, an additional linkage position in the protein for hydrolysis by trypsin, was mentioned in Section 1.4.1.3. Iodoacetic acid, depending on the pH, can react with cysteine, methionine, lysine and histidine residues:

$$\text{Protein—SH} \xrightarrow{JCH_2COOH} \text{Protein—S—}CH_2\text{—COOH} \quad (1.102)$$

Maleic acid anhydride and methyl-p-nitrobenzene sulfonate are also alkylating agents:

$$\text{Protein—SH} + \text{(maleic anhydride)} \longrightarrow \text{Protein—S—CH(COOH)—}CH_2\text{—COOH} \quad (1.103a)$$

$$\text{Protein—SH} + CH_3OSO_2\text{—}C_6H_4\text{—}NO_2 \longrightarrow \text{Protein—S—}CH_3 \quad (1.103b)$$

A number of reagents make it possible to measure thiol group content spectrophotometrically. The molar absorption coefficient, ε, for the derivative of azobenzene-2-sulfenylbromide, ε_{353}, is 16,700 m^{-1} cm^{-1} at pH 1. 5,5′-dithiobis-(2-nitrobenzoic acid) has a somewhat lower ε_{412} of 13,600 at pH 8 for its product, a thionitrobenzoate anion. The derivative of p-mercuribenzoate has an ε_{250} of 7,500 at pH 7, while the derivative of N-ethylmaleic imide has an ε_{300} of 620 at pH 7:

$$\text{Protein—SH} + \text{BrS—}C_6H_4\text{—N=N—}C_6H_5 \longrightarrow \text{Protein—S—S—}C_6H_4\text{—N=N—}C_6H_5 \quad (1.104a)$$

$$\text{Protein—SH} + O_2N\text{—}C_6H_3(COOH)\text{—S—S—}C_6H_3(COOH)\text{—}NO_2 \longrightarrow \text{Protein—S—S—}C_6H_3(COOH)\text{—}NO_2 + HS\text{—}C_6H_3(COOH)\text{—}NO_2 \quad (1.104b)$$

$$\text{Protein—SH} + X\text{—Hg—}C_6H_4\text{—COOH} \longrightarrow \text{Protein—S—Hg—}C_6H_4\text{—COOH} \quad (1.104c)$$

$$\text{Protein—SH} + \text{(N-ethylmaleimide, N—}C_2H_5) \longrightarrow \text{Protein—S—(N-ethylsuccinimide, N—}C_2H_5) \quad (1.104d)$$

1.4.4.6 Methionine Residue

Methionine residues are oxidized to sulfoxides with hydrogen peroxide. The sulfoxide can be reduced, regenerating methionine, using an excess of thiol reagent (cf. 1.2.4.3.6). α-Halogen carboxylic acids, β-propiolactone and alkyl halogenides convert methionine into sulfonium derivatives, from which methionine can be regenerated in an alkaline medium with an excess of thiol reagent:

$$\text{Protein—S—}CH_3 \xrightarrow{JCH_2COOH} \text{Protein—}S^{\oplus}\text{(—}COCH_2J\text{)—}CH_3$$

$$\text{Protein—S—}CH_3 \xrightarrow{CH_3J} \text{Protein—}S^{\oplus}\text{(—}CH_3\text{)—}CH_3$$

$$\text{Protein—S—}CH_3 + \text{(β-propiolactone)} \longrightarrow \text{Protein—}S^{\oplus}\text{(—}CH_2\text{—}CH_2\text{—COOH)—}CH_3 \quad (1.105)$$

Reaction with cyanogen bromide (BrCN), which splits the peptide bond on the carboxyl side of the methionine molecule, was outlined in Section 1.4.1.3.

1.4.4.7 Histidine Residue

Selective modification of histidine residues present on active sites of serine proteinases is possible. Substrate analogues such as halogenated methyl ketones inactivate such enzymes (for example, 1-chloro-3-tosylamido-7-aminoheptan-2-one inactivates trypsin and 1-chloro-3-tosylamido-4-phenylbutan-2-one inactivates chymotrypsin) by N-alkylation of the histidine residue:

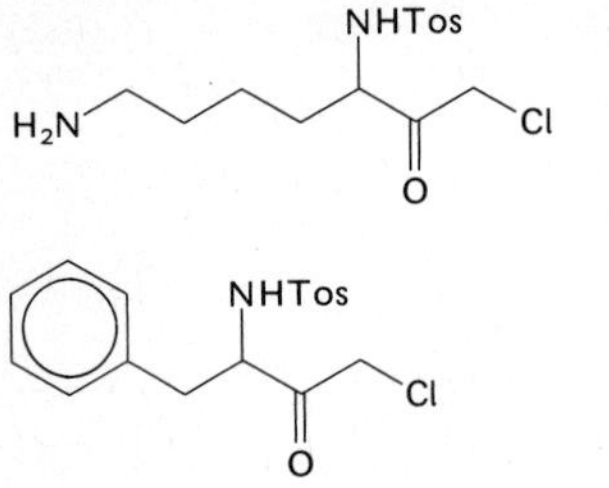

(1.106)

1.4.4.8 Tryptophan Residue

N-bromosuccinimide oxidizes the tryptophan side chain and also tyrosine, histidine and cysteine:

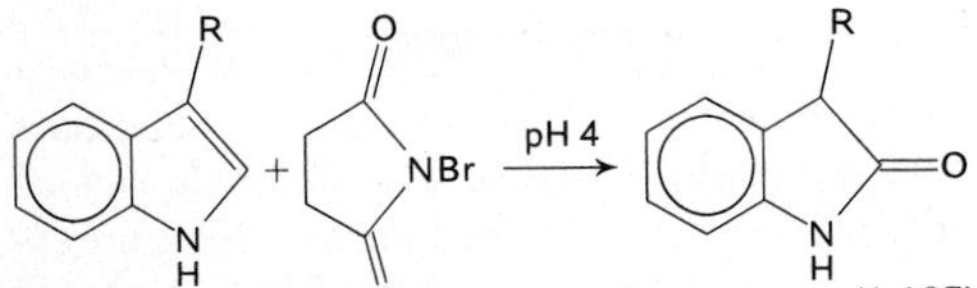

(1.107)

The reaction is used for the selective cleavage of peptide chains and the spectrophotometric determination of tryptophan.

1.4.4.9 Tyrosine Residue

Selective acylation of tyrosine can occur with acetylimidazole as a reagent:

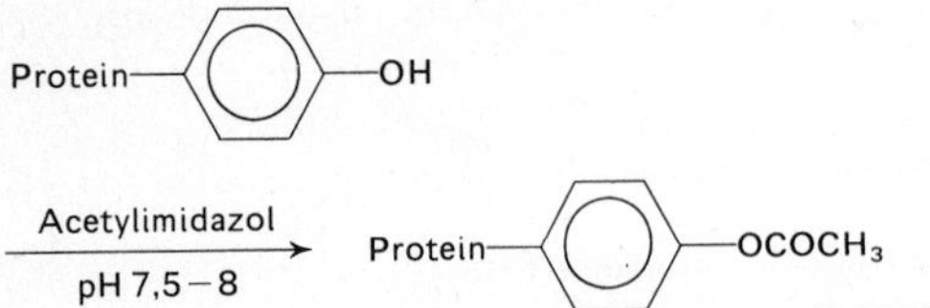

(1.108)

Diazotized arsanilic acid reacts with tyrosine and with histidine, lysine, tryptophan and arginine:

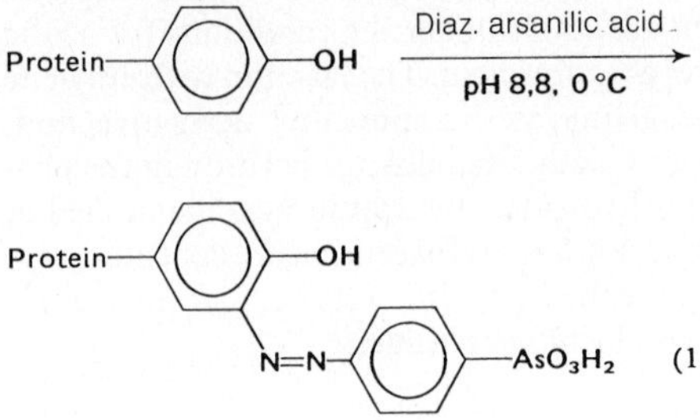

(1.109)

Tetranitromethane introduces a nitro group into the ortho position:

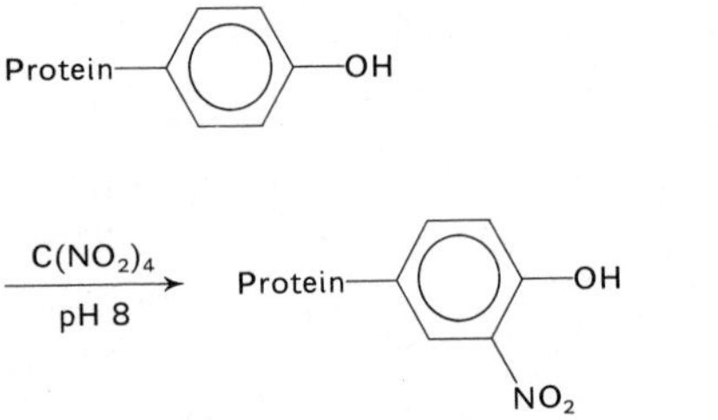

(1.110)

1.4.4.10 Bifunctional Reagents

Bifunctional reagents enable intra- and intermolecular cross-linking of proteins. Examples are bifunctional imidoester, fluoronitrobenzene and isocyanate derivatives:

$$H_3C-O-\underset{\underset{NH_2^{\oplus}}{\|}}{C}-(CH_2)_n-\underset{\underset{NH_2^{\oplus}}{\|}}{C}-O-CH_3 \quad (1.111a)$$

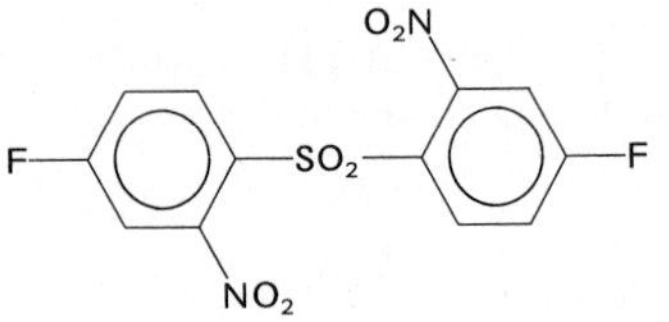

(1.111b)

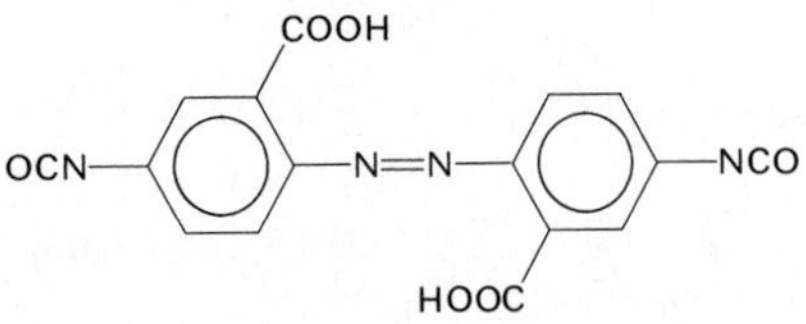

(1.111c)

1.4.4.11 Reactions Involved in Food Processing

The nature and extent of the chemical changes induced in proteins by food processing depends on a number of parameters, for example, composition of the food and processing conditions, such as temperature, pH or the presence of oxygen. As a consequence of these reactions, the biological value of proteins may be decreased by:

- Destruction of essential amino acids
- Conversion of essential amino acids into derivatives which are not metabolizable
- Decrease in the digestibility of protein as a result of intra- or interchain cross-linking.

Formation of toxic degradation products is also possible. The nutritional/physiological and toxicological assessment of changes induced by processing of food is a subject of some controversy and opposing opinions.

The *Maillard* reaction of the ε-amino group of lysine prevails in the presence of reducing sugars, for example, lactose or glucose, which yield protein-bound ε-N-desoxylactulosyl-1-lysine or ε-N-desoxyfructosyl-1-lysine, respectively. Lysine is not biologically available in these forms. Acidic hydrolysis of such primary reaction products yields lysine as well as the degradation products furosine and pyridosine in a constant ratio (cf. 4.2.4.4):

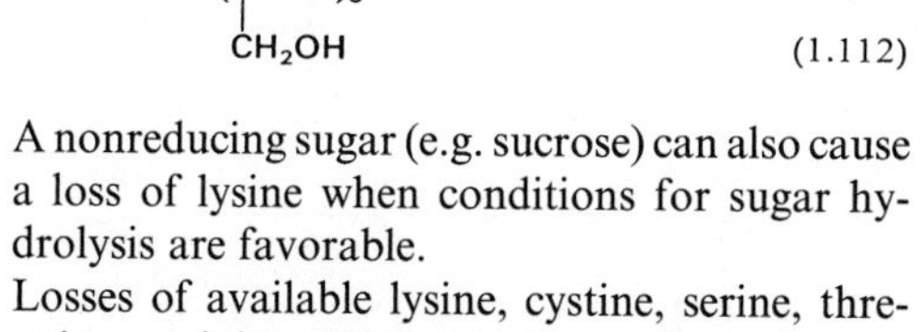

$$R^1{-}CO{-}HN{-}CH(CO{-}HNR){-}(CH_2)_4{-}NH{-}CH_2{-}CO{-}(CHOH)_3{-}CH_2OH \xrightarrow[100\,°C,\ 24\,h]{6\,N\,HCl} \text{L-Lysine } 50\% + \text{Furosine } 20\% + \text{Pyridosine } 10\% \quad (1.112)$$

A nonreducing sugar (e.g. sucrose) can also cause a loss of lysine when conditions for sugar hydrolysis are favorable.

Losses of available lysine, cystine, serine, threonine, arginine and some other amino acids occur at higher pH values. Hydrolysates of alkali-treated proteins often contain some unusual compounds, such as ornithine, β-aminoalanine, lysinoalanine, ornithinoalanine, lanthionine, methyl lanthionine and D-alloisoleucine, as well as other D-amino acids.

The formation of these compounds is based on the following reactions: 1,2-elimination in the case of hydroxy amino acids and thio amino acids results in 2-amino-acrylic acid (dehydroalanine) and 2-aminocrotonic acid (dehydroaminobutyric acid), respectively:

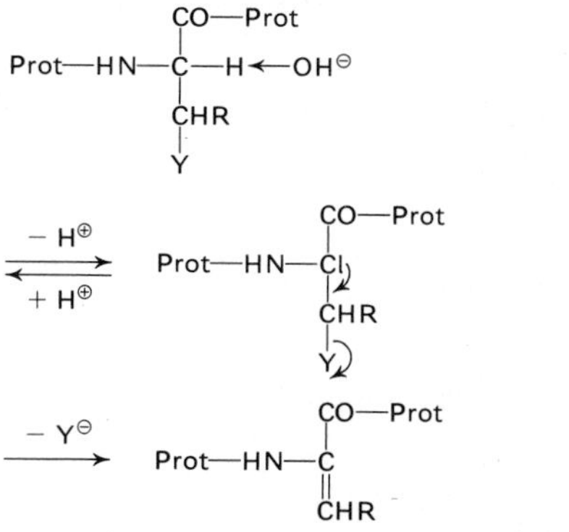

$$\text{Prot{-}HN{-}C(CO{-}Prot)(H \leftarrow OH^{\ominus}){-}CHR{-}Y} \underset{+H^{\oplus}}{\overset{-H^{\oplus}}{\rightleftharpoons}} \text{Prot{-}HN{-}C^{\ominus}(CO{-}Prot){-}CHR{-}Y} \xrightarrow{-Y^{\ominus}} \text{Prot{-}HN{-}C(CO{-}Prot){=}CHR} \quad (1.113)$$

$R = H, CH_3$; $Y = OH, OPO_3H_2, SH, SR^1, SSR^1$

In the case of cystine, the eliminated thiolcysteine can form a second dehydroalanine residue:

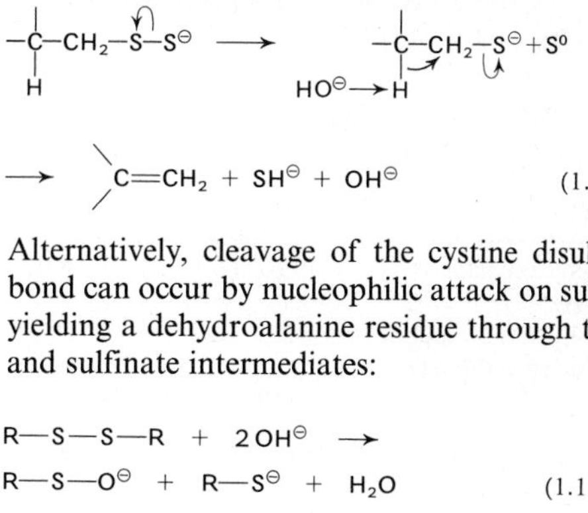

$$-\overset{|}{\underset{H}{C}}{-}CH_2{-}S{-}S^{\ominus} \longrightarrow -\overset{|}{\underset{HO^{\ominus}\rightarrow H}{C}}{-}CH_2{-}S^{\ominus} + S^0$$

$$\longrightarrow \;\rangle C{=}CH_2 + SH^{\ominus} + OH^{\ominus} \quad (1.114)$$

Alternatively, cleavage of the cystine disulfide bond can occur by nucleophilic attack on sulfur, yielding a dehydroalanine residue through thiol and sulfinate intermediates:

$$R{-}S{-}S{-}R + 2\,OH^{\ominus} \longrightarrow R{-}S{-}O^{\ominus} + R{-}S^{\ominus} + H_2O \quad (1.115a)$$

$$R{-}S{-}O^{\ominus} + R{-}S{-}S{-}R \longrightarrow R{-}\underset{\underset{O}{\|}}{S}{-}S{-}R + R{-}S^{\ominus} \quad (1.115b)$$

$$R{-}\underset{\underset{O}{\|}}{S}{-}S{-}R + 2\,OH^{\ominus} \longrightarrow R{-}SO_2^{\ominus} + R{-}S^{\ominus} + H_2O \quad (1.115c)$$

Intra- and interchain cross-linking of proteins can occur in dehydroalanine reactions involving additions of amines and thiols. Ammonia may also react via an addition reaction:

$$\text{PROTEIN}(-CH{=}CHR)_3 + H_2N{-}\text{PROTEIN} + HS{-}\text{PROTEIN} + NH_3 \longrightarrow \text{PROTEIN}\begin{cases} -CH{-}CH_2{-}HN{-}\text{PROTEIN} \\ -CH{-}CH_2{-}S{-}\text{PROTEIN} \\ -CH{-}CH_2{-}NH_2 \end{cases} \quad (1.116)$$

Acidic hydrolysis of such a cross-linked protein yields the unusual amino acids listed in Table 1.21. Ornithine is formed during cleavage of arginine (Reaction 1.17).

Table 1.21. Formation of unusual amino acids by alkali treatment of protein

Name	Formula
3-N^6-Lysinoalanine ($R{=}H$) 3-N^6-Lysino-3-methyl-alanine ($R{=}CH_3$)	$HOOC{-}CH(NH_2){-}CHR{-}NH{-}(CH_2)_4{-}CH(NH_2){-}COOH$
3-N^5-Ornithinoalanine ($R{=}H$) 3-N^5-Ornithino-3-methylalanine ($R{=}CH_3$)	$HOOC{-}CH(NH_2){-}CHR{-}NH{-}(CH_2)_3{-}CH(NH_2){-}COOH$
Lanthionine ($R{=}H$) 3-Methyllanthionine ($R{=}CH_3$)	$HOOC{-}CH(NH_2){-}CHR{-}S{-}CH_2{-}CH(NH_2){-}COOH$
3-Aminoalanine ($R{=}H$) 2,3-Diamino butyric acid ($R{=}CH_3$)	$HOOC{-}CH(NH_2){-}CHRNH_2$

$$HOOC{-}CHNH_2{-}(CH_2)_3{-}NH{-}C({=}NH){-}NH_2 \xrightarrow{OH^{\ominus}} HOOC{-}CHNH_2{-}(CH_2)_3{-}NH_2 + O{=}C(NH_2)_2 \quad (1.117)$$

Formation of D-amino acids occurs through abstraction of a proton via a C2-carbanion. The reaction with L-isoleucine is particularly interesting. L-Isoleucine is isomerized to D-alloisoleucine which, unlike other D-amino acids, is a diastereoisomer and so has a retention time different from L-isoleucine, making its determination possible directly from an amino acid chromatogram:

$$\underset{\text{L-Isoleucine}}{-HN{-}CH(CO{-}){-}CH(CH_3){-}C_2H_5} \underset{+H^{\oplus}}{\overset{-H^{\oplus}}{\rightleftharpoons}} -HN{-}C^{\ominus}(CO{-}){-}CH(CH_3){-}C_2H_5 \underset{-H^{\oplus}}{\overset{+H^{\oplus}}{\rightleftharpoons}} \underset{\text{D-allo-Isoleucine}}{H{-}C(CO{-})(NH{-}){-}CH(CH_3){-}C_2H_5} \quad (1.118)$$

Heating proteins in a dry state at neutral pH results in the formation of isopeptide bonds between lysine residues and the β- or γ-carboxamide groups of asparagine and glutamine residues:

$$\text{PROTEIN}\begin{cases} -NH_2 \quad H_2NOC{-}CH_2{-} \\ -NH_2 \quad H_2NOC{-}(CH_2)_2{-} \end{cases}\text{PROTEIN} \xrightarrow{-NH_3} \text{PROTEIN}\begin{cases} -NH{-}OC{-}H_2C{-} \\ -NH{-}OC{-}(H_2C)_2{-} \end{cases}\text{PROTEIN} \quad (1.119)$$

These isopeptide bonds are cleaved during acidic hydrolysis of protein and, therefore, do not contribute to the occurrence of unusual amino acids. A more intensive heat treatment of proteins in the presence of water leads to a more extensive degradation.

Oxidative changes in proteins primarily involve methionine, which relatively readily forms methionine sulfoxide:

$$\begin{array}{c} CO{-} \\ | \\ {-}HN{-}C{-}H \\ | \\ (CH_2)_2 \\ | \\ S \\ | \\ CH_3 \end{array} \longrightarrow \begin{array}{c} CO{-} \\ | \\ {-}HN{-}C{-}H \\ | \\ (CH_2)_2 \\ | \\ S{=}O \\ | \\ CH_3 \end{array} \qquad (1.120)$$

After *in vivo* reduction to methionine, protein-bound methionine sulfoxide is apparently biologically available.

Figure 1.24 shows the effect of alkaline treatment of a protein isolate of sunflower seeds. Serine, threonine, arginine and isoleucine concentrations are markedly decreased with increasing concentrations of NaOH. New amino acids (ornithine and alloisoleucine) are formed. Initially, lysine concentration decreases, but increases at higher concentrations of alkali. Lysinoalanine behaves in the opposite manner.

The extent of formation of D-amino acids as a result of alkaline treatment of proteins is shown in Table 1.21 a.

Table 1.21 a. Formation of D-amino acids by alkali treatment of proteins[a] (1% solution in 0.1 N NaOH, pH ~ 12.5, temperature 65 °C)

Protein	Heating time (h)	D-Asp (%)	D-Ala	D-Val	D-Leu	D-Pro	D-Glu	D-Phe
Casein	0	2.2	2.3	2.1	2.3	3.2	1.8	2.8
	1	21.8	4.2	2.7	5.0	3.0	10.0	16.0
	3	30.2	13.3	6.1	7.0	5.3	17.4	22.2
	8	32.8	19.4	7.3	13.6	3.9	25.9	30.5
Wheat gluten	0	3.3	2.0	2.1	1.8	3.2	2.1	2.3
	3	29.0	13.5	3.9	5.6	3.2	25.9	23.3
Promine D (soya protein)	0	2.3	2.3	2.6	3.3	3.2	1.8	2.3
	3	30.1	15.8	6.6	8.0	5.8	18.8	24.9
Lactalbumin	0	3.1	2.2	2.9	2.7	3.1	2.9	2.3
	3	22.7	9.2	4.8	5.8	3.6	12.2	16.5

[a] Results in % correspond to D- + L-amino acids = 100%.

Data presented in Figures 1.25 and 1.26 clearly show that the formation of lysinoalanine is influenced not only by pH but also by the protein source. An extensive reaction occurs in casein even at pH 5.0 due to the presence of phosphorylated serine residues, while noticeable reactions occur in gluten from wheat or in zein from corn only in the pH range of 8–11. Figure 1.27 illustrates the dependence of the reaction on protein concentration.

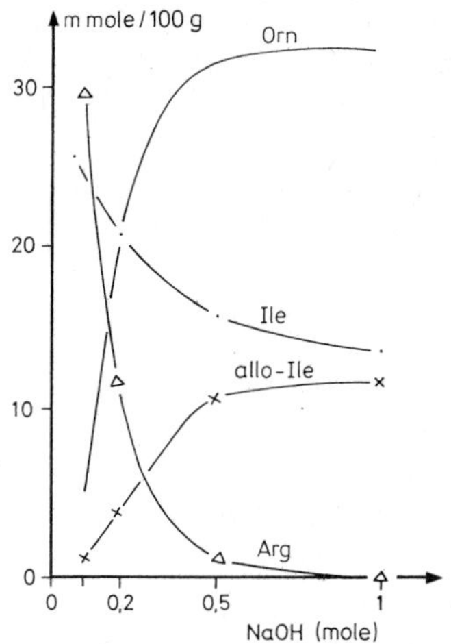

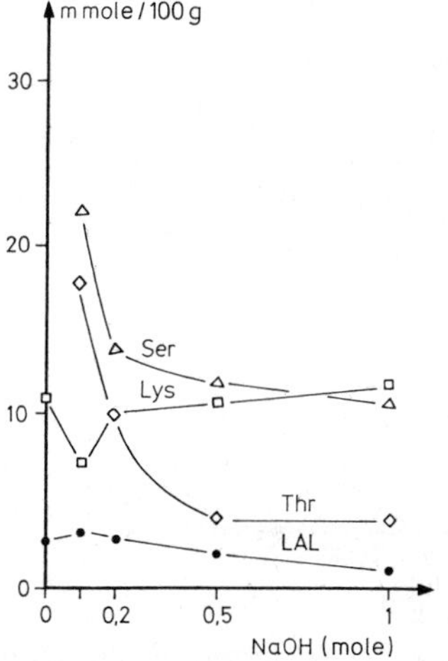

Fig. 1.24. Amino acid contents for a sunflower seed protein isolate heated in sodium hydroxide solutions at 80 °C for 16 h. (After *Mauron*, 1975)

Table 1.22 lists the contents of lysinoalanine in food products processed industrially or prepared under the "usual household conditions". The contents are obviously affected by the food type and by the processing conditions.

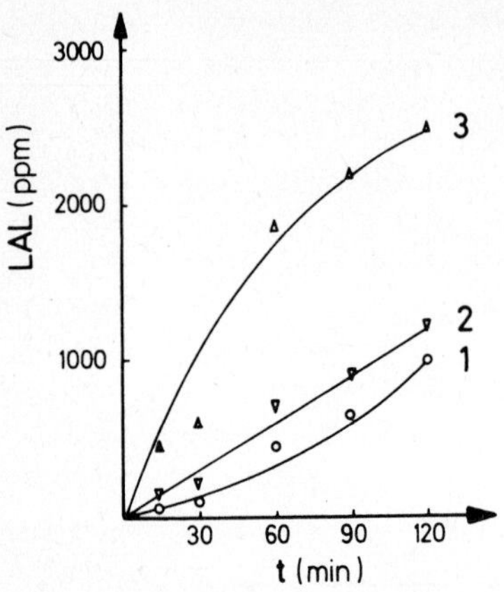

Abb. 1.25. Formation of lysinoalanine (LAL) by heating casein (5% solution at 100 °C) (After *Sternberg* and *Kim*, 1977) 1 pH 5.0, 2 pH 7.0, 3 pH 8.0

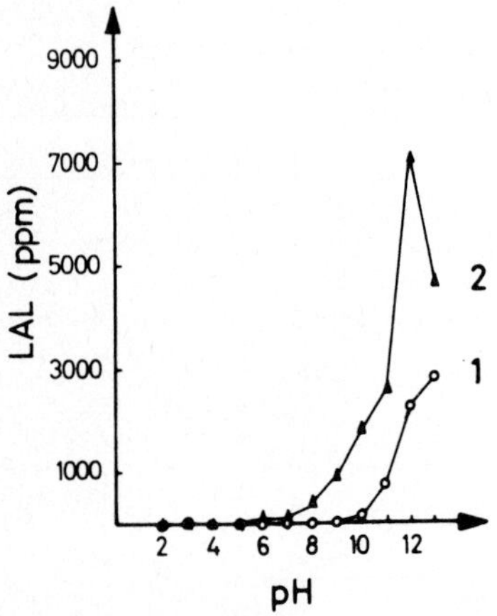

Fig. 1.26. Lysinoalanine (LAL) formation from wheat gluten (2) and corn gluten (1). Protein contents of the glutens: 70%; heated as 6.6% suspensions at 100 °C for 4 h. (After *Sternberg* and *Kim*, 1977)

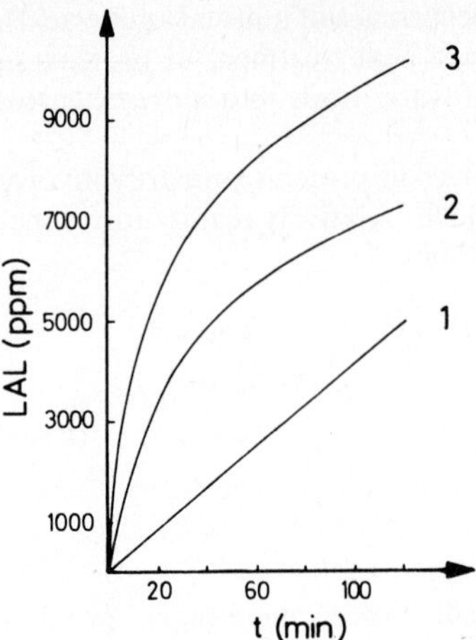

Fig. 1.27. Lysinoalanine (LAL) formation as influenced by casein concentration. (1): 5%, (2): 15%, and (3) 20% all at pH 12.8. (After *Sternberg* and *Kim*, 1977)

Table 1.22. Lysinoalanine content of various foods

Food	Origin/Treatment		Lysinoalanine (mg/kg protein)
Frankfurter	CP[a]	Raw	0
		Cooked	50
		Roasted in oven	170
Chicken drums	CP	Raw	0
		Roasted in oven	110
		Roasted in micro wave oven	200
Egg white, fluid	CP		15
Egg white		Boiled	
		(3 min)	140
		(10 min)	270
		(30 min)	370
		Baked	
		(10 min/150 °C)	350
		(30 min/150 °C)	1,100
Dried egg white	CP		160–1,820[b]
Condensed milk, sweetened	CP		360– 540
Condensed milk, unsweetened	CP		590– 860
Milk product for infants	CP		150– 640
Infant food	CP		<55– 150
Soya protein isolate	CP		0– 370
Hydrolyzed vegetable protein	CP		40– 500
Cocoa powder	CP		130– 190
Na-caseinate	CP		45– 560
Na-caseinate	CP		430–6,900
Ca-caseinate	CP		250–4,320

[a] Commercial product.
[b] Variation range for different brand name products.

1.4.5 Enzyme-Catalyzed Reactions

1.4.5.1 Foreword

A great number and variety of enzyme-catalyzed reactions are known with protein as a substrate. These include hydrolytic reactions (cleavage of peptide bonds or other linkages, e.g. the ester linkage in a phosphoprotein), transfer reactions (phosphorylation, transfer of sugar residues and methyl groups) and redox reactions (thiol oxidation, disulfide reduction, amino group oxidation or incorporation of hydroxyl groups). Table 1.23 is a compilation of some examples.

Some of these reactions are covered in Section 1.4.6.3 or in the sections related to individual

Table 1.23. Enzymatic reactions affecting proteins

Hydrolysis
– Endopeptidases
– Exopeptidases
Proteolytic induced aggregation
– Collagen biosynthesis
– Blood coagulation
– Plastein reaction
Cross-linking
– Disulfide bonds
Protein disulfide isomerase
Protein disulfide reductase (NAD(P)H)
Protein disulfide reductase (Glutathione)
Sulfhydryloxidase
Lipoxygenase
Peroxidase
– Aldol-, Aldimine condensation and subsequent reactions (connective tissue)
Lysyloxidase
Phosphorylation, dephosphorylation
– Protein kinase
– Phosphoprotein phosphatase
Hydroxylation
– Proline hydroxylase
– Lysine hydroxylase
Glycosylation
– Glycoprotein-β-galactosyltransferase
Methylation and demethylation
– Protein(arginine)-methyl-transferase
– Protein(lysine)-methyl-transferase
– Protein-O-methyl-transferase
Acetylation, deacetylation
– ε-N-Acetyl-lysine

foodstuffs. Only enzymes that are involved in hydrolysis of peptide bonds (proteolytic enzymes, peptide hydrolases) will be covered in the following sections.

1.4.5.2 Proteolytic Enzymes

Processes involving proteolysis play a role in the production of many foods. Proteolysis can occur as a result of proteinases in the food itself, e.g. autolytic reactions in meat, or with microbial proteinases, such as the addition of pure cultures of selected microorganisms during the production of cheese.
This large group of enzymes is divided up as shown in Table 1.23 a. The two subgroups formed are: peptidases (exopeptidases) that cleave amino acids or dipeptides stepwise from the terminal ends of proteins, and proteinases (endopeptidases) that hydrolyze the linkages within the peptide chain, not attacking the terminal peptide bonds. Further division is possible, for example, by taking into account the presence of a given amino acid residue in the active site of the enzyme. The most important types of proteolytic enzymes are presented in the following sections.

1.4.5.2.1 Serine proteinases

Enzymes of this group, in which activity is confined to the pH range of 7–11, are denoted as alkaline proteinases. Typical representatives from animal sources are trypsin, chymotrypsin, elastase, plasmin and thrombin. Serine proteinases are produced by a great number of bacteria and fungi, e.g. *Bacillus cereus, B. firmus, B. licheniformis, B. megaterium, B. subtilis, Serratia marcescens, Streptomyces fradiae, S. griseus, Tritirachium album, Aspergillus flavus, A. oryzae* and *A. sojae*.
These enzymes have in common the presence of a serine and a histidine residue in their active sites (cf. 2.4.2.2). Cleavage of protein occurs through an acyl-enzyme (cf. Formula 1.121 and Figure 2.15) formed as an intermediate.

R^2—NH_2 R^1—COOH

(1.121)

Inactivation of these enzymes is possible with reagents such as diisopropylfluorophosphate

Table 1.23 a. Classification of proteolytic enzymes (peptide hydrolases)

EC-No.[a]	Enzyme group	Comments	Examples
	Peptidases (Exopeptidases)	Cleave proteins/peptides stepwise from N- or C-terminals	
3.4.11.	α-Aminoacylpeptide hydrolases	Cleave amino acids from N-terminal	Various aminopeptidases
3.4.13.	Dipeptide hydrolases	Cleave dipeptides	Various dipeptidases (carnosinase, anserinase)
3.4.14.	Dipeptidylpeptide hydrolases	Cleave dipeptides from N-terminal	Cathepsin C
3.4.15.	Peptidyldipeptide hydrolases	Cleave dipeptides from C-terminal	Carboxycathepsin,
3.4.16.	Serine carboxypeptidases	Cleave amino acids from C-terminal, serine in the active site	Carboxypeptidase C Cathepsin A
3.4.17.	Metalocarboxypeptidases	Cleave amino acids from C-terminal, Zn^{2+} or Co^{2+} in the active site	Carboxypeptidases A and B
3.4.18.	Cysteine carboxypeptidases	Cleave amino acids from C-terminal, cysteine in the active site	Lysosomal carboxypeptidase B
	Proteinases	Cleave proteins/peptides bonds other than terminal ones	
3.4.21.	Serine proteinases	Serine in the active site	Chymotrypsins A, B and C, α- and β-trypsin, microbial alkaline proteinases
3.4.22.	Cysteine proteinases	Cysteine in the active site	Papain, ficin, bromelain, cathepsin B
3.4.23.	Aspartic proteinases	Aspartic acid (2 residues) in the active site	Pepsin, cathepsin D, rennin (chymosin)
3.4.24.	Metaloproteinases	Metal ions in the active site	Collagenase, microbial neutral proteinases

[a] cf. 2.2.6

(DIFP) or phenylmethanesulfonylfluoride (DMSF). These reagents irreversibly acylate the serine residue in the active site of the enzymes (cf. Reaction 1.122).

$$E{-}CH_2OH + FY \longrightarrow E{-}CH_2OY + HF \quad (1.122)$$
$$(Y: {-}PO(i\,C_3H_7O)_2,\ {-}SO_2{-}CH_2C_6H_5)$$

Irreversible inhibition can also occur in the presence of halogenated methyl ketones which alkylate the active histidine residue (cf. 2.4.1.1), or as a result of the action of proteinase inhibitors, which are also proteins, by interaction with the enzymes to form inactive complexes. These natural inhibitors are found in the organs of animals and plants (pancreas, colostrum, egg white, potato tuber and seeds of many legumes; cf. 16.2.3). The specificity of serine proteinases varies greatly (cf. Table 1.24). Trypsin exclusively cleaves linkages of amino acid residues with a basic side chain (lysyl or arginyl bonds) and chymotrypsin preferentially cleaves bonds of amino acid residues which have aromatic side chains (phenylalanyl, tyrosyl or tryptophanyl bonds). Enzymes of microbial origin often are less specific.

1.4.5.2.2 Cysteine proteinases

Typical representatives of this group of enzymes are: papain (from the sap of a tropical, melon-like fruit tree, *Carica papaya*), bromelain (from sap and stem of pineapples, *Ananas comosus*), ficin (from *Ficus latex* and other *Ficus spp.*) and a Streptococcus proteinase. The range of activities of these enzymes is very wide and, depending on the substrate, is pH 4.5–10, with a maximum at pH 6–7.5.

The mechanism of enzyme activity appears to be similar to that of serine proteinases. A cysteine residue is present in the active site. A thioester is formed as a covalent intermediary product. The enzymes are highly sensitive to oxidizing agents. Therefore, as a rule they are used in the presence of a reducing agent (e.g. cysteine) and a chelating agent (e.g. EDTA).

Table 1.24. Specificity of proteolytic enzymes [based on cleavage of oxidized B chain of bovine insulin; strong cleavage: ↓, weak cleavage (↓)]

No.[a]	Phe	Val	Asn	Gln	His	Leu	Cys	Gly	Ser	His	Leu	Val	Glu	Ala	Leu	Tyr	Leu	Val	Cys	Gly	Glu	Arg	Gly	Phe	Phe	Tyr	Thr	Pro	Lys	Ala
Serine proteinases																														
1																						↓							↓	
2															(↓)	↓									↓	↓				
3				↓		↓					↓				↓	(↓)	↓							↓	↓	(↓)				
4		↓	↓					↓				↓								↓										
5																						↓							↓	
6			↓						↓						↓	↓									↓					
7				(↓)								(↓)			↓	↓			↓						↓	↓			↓	
8			↓	↓			↓					(↓)	↓		↓	↓			↓					(↓)	↓				↓	
Cysteine proteinases																														
9			↓			(↓)	(↓)	(↓)					↓	(↓)			↓		↓						↓	↓				
10								↓	↓				↓			↓								↓	↓					
11			↓				(↓)						↓	↓		↓			↓		↓	↓		↓	(↓)	(↓)				
Metaloproteinases																														
12	↓				↓					↓				↓	↓	↓	↓						↓	↓	↓				↓	
13					↓					↓				↓		↓	↓						↓	↓						
14					↓					↓				↓		↓	↓						↓	↓	↓				↓	
Aspartic proteinases																														
15	(↓)		(↓)	(↓)					(↓)		↓		(↓)	(↓)	(↓)	↓							(↓)	↓	↓					
16	(↓)										(↓)		↓		↓	(↓)	↓							↓	↓					
17		↓									↓	↓		↓				↓							↓					
18	↓														↓	↓								↓	↓	(↓)				
19	(↓)				(↓)					↓	↓	↓		(↓)	↓	↓				(↓)			↓	↓	↓					

[a] Enzymes.

1) Trypsin (bovine)
2) Chymotrypsin A (bovine)
3) Chymotrypsin C (swine)
4) Aspergillopeptidase C (*Aspergillus oryzae*)
5) Proteinase from *Streptomyces griseus* (Trypsin-like)
6) Subtilisin BPN′
7) Proteinase from *Aspergillus oryzae*
8) Proteinase from *Aspergillus flavus*
9) Papain (*Papaya carica*)
10) Ficin III (*Ficus glabrata*)
11) Chymopapain (*Charica papaya*)
12) Proteinase II from *Aspergillus oryzae*
13) Proteinase from *Bacillus subtilis*
14) Thermolysin (*Bacillus thermoproteolyticus Rokko*)
15) Pepsin A (swine)
16) Rennin (calf)
17) Proteinase from *Candida albicans*
18) Proteinase from *Mucor miehei*
19) Proteinase from *Rhizopus chinensis*

Inactivation of the enzymes is possible with oxidative agents, metal ions or alkylating reagents (cf. 1.2.4.3.5 and 1.4.4.5). In general these enzymes are not very specific (cf. Table 1.24).

1.4.5.2.3 Metalo proteinases

This group includes exopeptidases, carboxypeptidases A and B, aminopeptidases, dipeptidases, prolidase and prolinase, and proteinases from bacteria and fungi, such as *Bacillus cereus, B. megaterium, B. subtilis, B. thermoproteolyticus* (thermolysin), *Streptomyces griseus* (pronase; it also contains carboxy- and aminopeptidases) and *Aspergillus oryzae.*

Most of these enzymes contain one mole of Zn^{2+} per mole of protein, but prolidase and prolinase contain one mole of Mn^{2+}. The metal ion acts as a *Lewis* acid in carboxypeptidase A, establishing contact with the carbonyl group of the peptide bond which is to be cleaved. Figure 1.28 shows the arrangement of other participating residues in the active site, as revealed by X-ray structural analysis of the enzyme-substrate complex.

Inhibition of these enzymes is achieved with chelating agents (e.g. EDTA) or sodium dodecyl sulfate.

1.4.5.2.4 Aspartic proteinases

Typical representatives of this group are enzymes of animal origin, such as pepsin and rennin (called Lab-enzyme in Europe), active in the pH range of 2–4, and cathepsin D, which has a pH optimum between 3 and 5, depending on the substrate and on the source of the enzyme. At pH 6–7 rennin cleaves a bond of ϰ-casein with great specificity, thus causing curdling of milk (cf. 10.1.2.1.1).

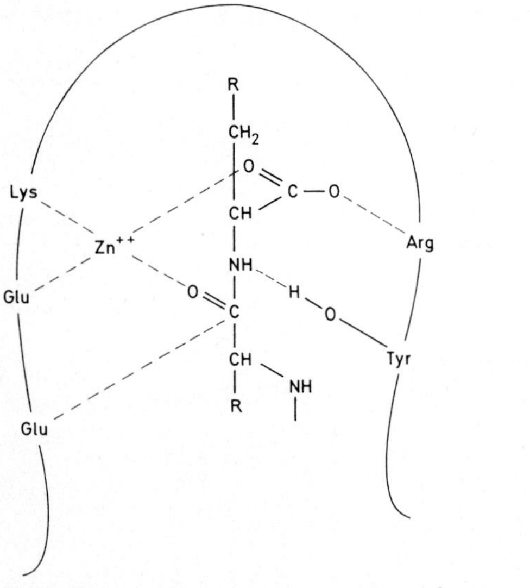

Aspartic proteinases of microbial origin can be classified as pepsin-like or rennin-like enzymes. The latter are able to coagulate milk. The pepsin-like enzymes are produced by, for example *Aspergillus awamori, A. niger, A. oryzae, Penicillium spp.* and *Trametes sanguinea.* The rennin-like enzymes are produced, for example, by *Aspergillus usamii* and *Mucor spp.,* such as *M. pusillus.*

There are two carboxyl groups, one in undissociated form, in the active site of aspartic proteinases. The mechanism postulated for cleavage of peptide bonds is illustrated in Reaction 1.123. The nucleophilic attack of a water molecule on the carbonyl carbon atom of the peptide bond is catalyzed by the side chains of Asp-32 (basic

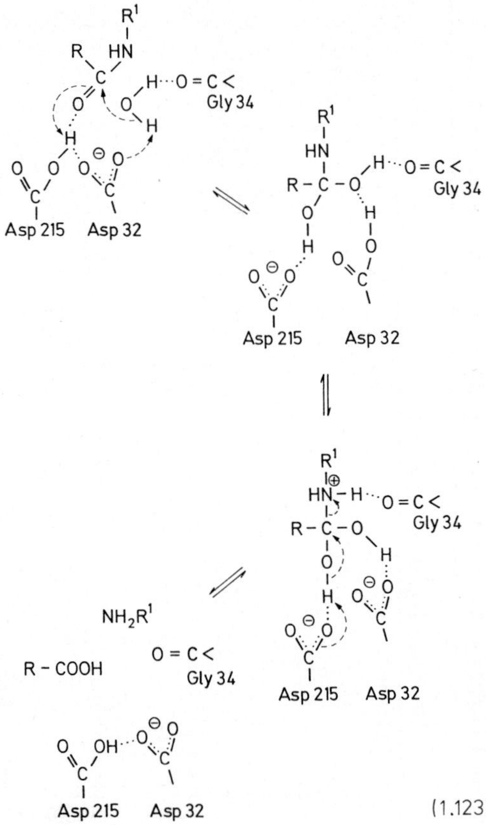

Fig. 1.28. Carboxypeptidase A active site. (After *Lowe* and *Ingraham,* 1974)

(1.123)

catalyst) and Asp-215 (acid catalyst). The numbering of the amino acid residues in the active site applies to the aspartic proteinase from *Rhizopus chinensis*.
Inhibition of these enzymes is achieved with various diazoacetylamino esters, which apparently react with carboxyl groups on the active site, and with pepstatin. The latter is isolated from various *Streptomycetes* as a peptide mixture with the general formula (R: isovaleric or n-caproic acid; AHMHA: 4-amino-3-hydroxy-6-methyl heptanoic acid):

$$\text{R—Val—Val—AHMHA—Ala—AHMAH} \quad (1.124)$$

The specifity of aspartic proteinases is given in Table 1.24.

1.4.6 Chemical and Enzymatic Reactions of Interest to Food Processing

1.4.6.1 Foreword

Standardization of food properties to meet nutritional/physiological and toxicological demands and requirements of food processing operations is a perennial endeavor. Food production is similar to a standard industrial fabrication process: on the one hand is the food commodity with all its required properties, on the other hand are the components of the product, each of which supplies a distinct part of the required properties. Such considerations have prompted investigations into the relationships in food between macroscopic physical and chemical properties and the structure and reactions at the molecular level. Reliable understanding of such relationships is a fundamental prerequisite for the design and operation of a process, either to optimize the process or to modify the food components to meet the desired properties of the product.
Modification of proteins is still a long way from being a common method in food processing, but it is increasingly being recognized as essential, for two main reasons:
Firstly, proteins fulfill multipurpose functions in food. Some of these functions can be better served by modified than by native proteins.
Secondly, persistent nutritional problems the world over necessitate the utilization of new raw materials.
Modifying reactions can ensure that such new raw materials (e.g. proteins of plant or microbial origin) meet stringent standards of food safety, palatability and acceptable biological value. A review will be given here of several protein modifications that are being used or are being considered for use. They involve chemical or enzymatic methods or a combination of both. Examples have been selected to emphasize existing trends. Table 1.25 presents some protein properties which are of interest to food processing. These properties are related to the amino acid composition and sequence and the conformation of proteins. Modification of the properties of proteins is possible by changing the amino acid composition or the size of the molecule, or by removing or inserting hetero constituents. Such changes can be accomplished by chemical and/or enzymatic reactions.
From a food processing point of view, the aims of modification of proteins are:

- Blocking the reactions involved in deterioration of food (e.g. the *Maillard* reaction)
- Improving some physical properties of proteins (e.g. texture, foam stability, whippability, solubility)
- Improving the nutritional value (increasing the extent of digestibility, inactivation of toxic or other undesirable constituents, introducing essential ingredients such as some amino acids).

Table 1.25. Properties of protein in food

Properties with	
nutritive/physiological relevance	processing relevance
Amino acid composition Availability of amino acids	Solubility, dispersibility Ability to coagulate Water binding/holding capacity Gel formation Dough formation, extensibility, elasticity Viscosity, adhesion, cohesion Whippability Foam stabilization Emulsifying ability Emulsion stabilization

1.4.6.2 Chemical Modification

Table 1.26 presents a selection of chemical reactions of proteins that are pertinent to and of current importance in food processing.

Table 1.26. Chemical reactions of proteins significant in food

Reactive group	Reaction	Product
$-NH_2$	Acylation	$-NH-CO-R$
$-NH_2$	Reductive alkylation with HCHO	$-N(CH_3)_2$
$-CONH_2$	Hydrolysis	$-COOH$
$-COOH$	Esterification	$-COOR$
$-OH$	Esterification	$-O-CO-R$
$-SH$	Oxidation	$-S-S-$
$-S-S-$	Reduction	$-SH$
$-CO-NH-$	Hydrolysis	$-COOH + H_2N-$

1.4.6.2.1 Acylation

Treatment with succinic anhydride (cf. 1.4.4.1.3) generally improves the solubility of protein. For example, succinylated wheat gluten is quite soluble at pH 5 (cf. Figure 1.29). This effect is related to disaggregation of high molecular weight gluten fractions (cf. Figure 1.30). In the case of succinylated casein it is obvious that the modification shifts the isoelectric point of the protein (and thereby the solubility minimum) to a lower pH (cf. Figure 1.31). Succinylation of leaf proteins improves the solubility as well as the flavor and emulsifying properties.

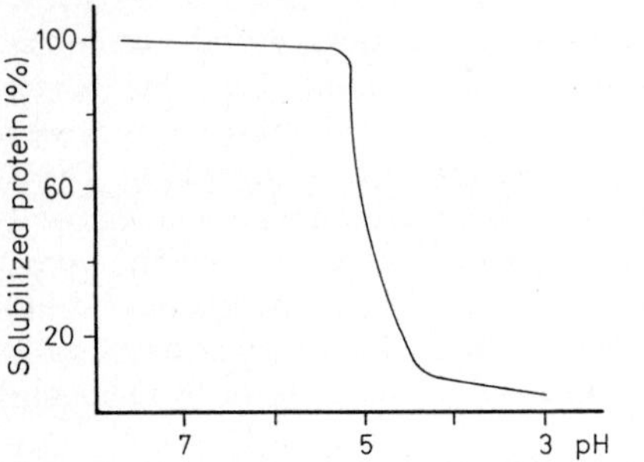

Fig. 1.29. Solubility of succinylated wheat protein in dependence on pH (0.5% solution in water). (After *Grant*, 1973)

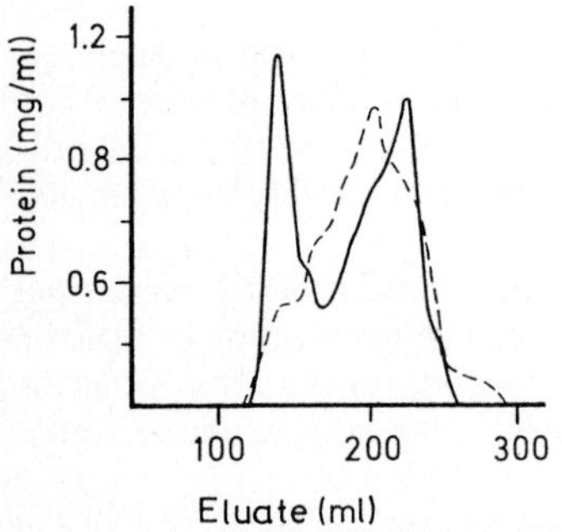

Fig. 1.30. Gel column chromatography of an acetic acid (0.2 M) wheat protein extract. Column: Sephadex G-100 (— before and --- after succinylation). (After *Grant*, 1973)

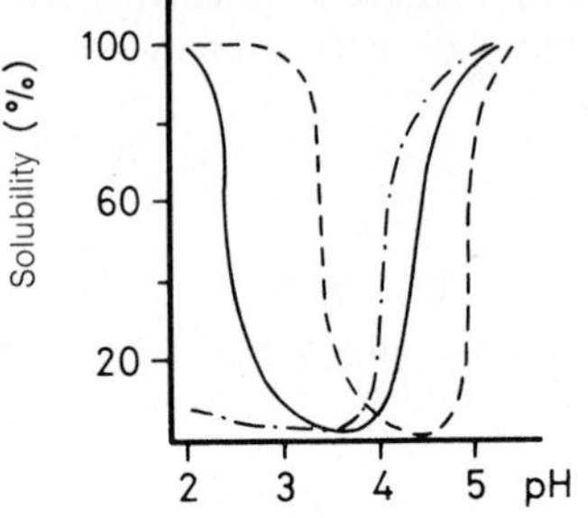

Fig. 1.31. Solubilities for native (---) and succinylated casein (— 50% and -·-·- 76%) in dependence on pH. (After *Schwenke* et al., 1977)

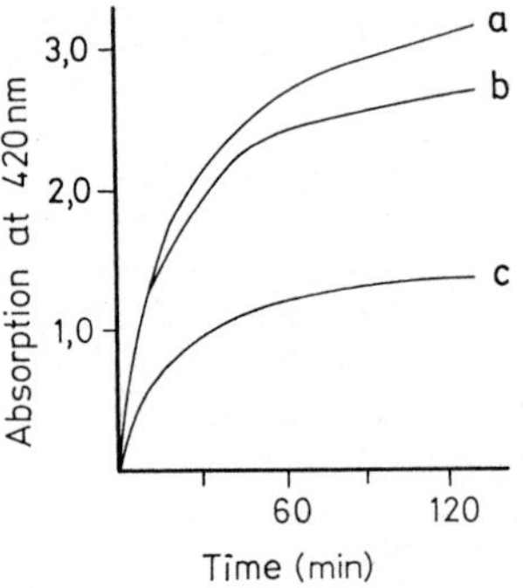

Fig. 1.32. Hydrolysis of a reductive methylated casein by bovine α-chymotrypsin. Modification extents: a 0%, b 33%, and c 52%. (After *Galembeck* et al., 1977)

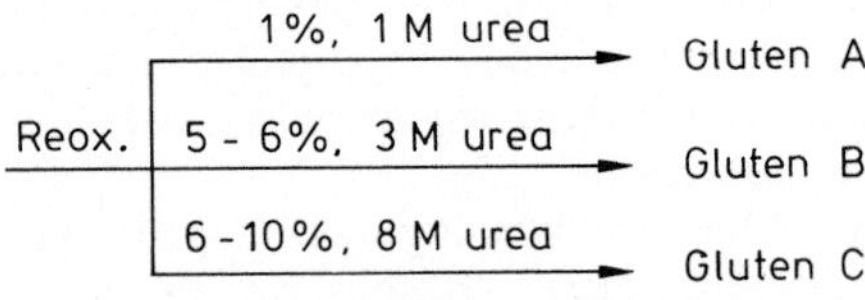

A: readily soluble, soft, adhesive, non-elastic
B: cohesive, elastic
C: sparingly soluble, strong, cohesive and non-elastic

Fig. 1.33. Properties of modified wheat gluten. (After *Lasztity*, 1975)

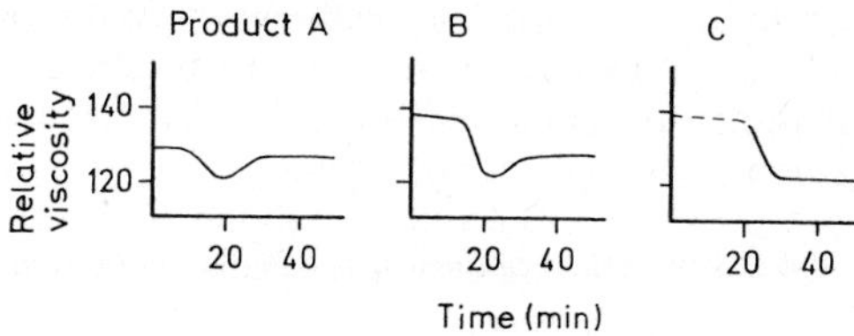

Fig. 1.34. Viscosity curves during reduction of different wheat glutens. For samples designation see Fig. 1.33. (After *Lasztity*, 1975)

Succinylated yeast protein has not only an increased solubility in the pH range of 4–6, but is more heat stable above pH 5. It has better emulsifying properties, surpassing many other proteins (Table 1.27), and has increased whippability.

Gliadin (1.4% Ala)

Edestin (3.8% Ala)

+ H_3C—CH (D,L-alanine carboxy anhydride: ring with N–H, C=O, O, C=O)

pH 6.8, 48 h → Polyalanyl-gliadin (23.6% Ala)

pH 6.8, 0.005 M SDS 72 h → Polyalanyl-edestin (6.2% Ala)

Fig. 1.35. Reaction of proteins with D,L-alanine carboxy anhydride. (After *Sela* et al., 1962 and *St. Angelo* et al., 1966)

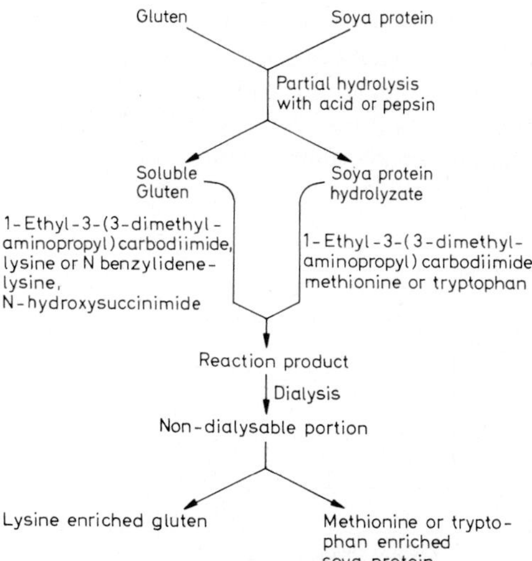

Fig. 1.36. Covalent binding of lysine to gluten (After *Li-Chan* et al., 1979) and of methionine or tryptophan to soya protein (After *Voutsinas* and *Nakai*, 1979), by applying a carbodiimide procedure

Table 1.27. Emulsifying property of various proteins[a]

Protein	Emulsifying Activity Index ($m^2 \times g^{-1}$) pH 6.5	pH 8.0
Yeast protein (88%) succinylated	322	341
Yeast protein (62%) succinylated	262	332
Sodium dodecyl sulfate (0.1%)	251	212
Bovine serum albumin	–	197
Sodium caseinate	149	166
β-Lactoglobulin	–	153
Whey protein powder A	119	142
Yeast protein (24%) succinylated	110	204
Whey protein powder B	102	101
Soya protein isolate A	41	92
Hemoglobin	–	75
Soya protein isolate B	26	66
Yeast protein (unmodified)	8	59
Lysozyme	–	50
Egg albumen	–	49

[a] Protein concentration: 0.5% in phosphate buffer of pH 6.5.

Introduction of aminoacyl groups into protein can be achieved by reactions involving amino acid carboxy anhydrides (Figure 1.35), amino acids and carbodiimides (Figure 1.36) or by BOC-amino acid hydroxysuccinimides with subsequent removal of the aminoprotecting group (BOC):

BOC – NH – CH(R) – CO – O – N(succinimide) + H_2N–[protein]–(HO–C_6H_4)

$\xrightarrow{\text{pH 9}}$ BOC – NH – CH(R) – CO – HN–[protein]; BOC – NH – CH(R) – CO – O–C_6H_4–[protein]

$\xrightarrow[\text{NH}_2\text{OH}]{\text{pH 8}}$ BOC – NH – CH(R) – CO – HN–[protein]–(HO–C_6H_4) (1.125)

$\xrightarrow[\text{TFA}]{\text{Anhydrous}}$ H_2N – CH(R) – CO – HN–[protein]–(HO–C_6H_4)

H_2N – CH(R) – COOH : Ala, Trp, Gly, Met

Feeding tests with casein with attached methionine, as produced by the above method, have demonstrated a satisfactory availability of methionine (Table 1.28). Such covalent attachment of essential amino acids to a protein may avoid the problems associated with food supplementation with free amino acids: losses in processing, development of undesired aroma due to methional, etc.

Table 1.29, using β-casein as an example, shows to what extent the association of a protein is affected by its acylation with fatty acids of various chain lengths.

1.4.6.2.2 Alkylation

Modification of protein by reductive methylation of amino groups with formaldehyde/$NaBH_4$ retards *Maillard* reactions. The resultant methyl derivative, depending on the degree of substitution, is less accessible to proteolysis (Figure 1.32). Hence, its value from a nutritional/physiological point of view is under investigation.

Table 1.28. Feeding trial with modified casein (rats): free amino acid concentration in plasma and PER value

Diet	µmole/100 ml plasma				
	Lys	Thr	Ser	Gly	Met
Casein	101	19	34	32	5
Met-Casein[a]	96	17	33	27	39
					PER[b]
Casein (10%)					2.46
Casein (10%) + Met (0.2%)					3.15
Casein (5%) + Met-Casein[a] (5%)					2.92

[a] Covalent binding of methionine to ε-NH_2-groups of casein.

[b] Protein Efficiency Ratio.

Table 1.29. Association of acylated β-casein A

Protein	SG[a] (%)	Mono-mer (%)	Poly-mer (%)	$S^0_{20,w}$ ($S \cdot 10^{-13}$)	$S^{1\%}_{20,w}$ ($S \cdot 10^{-13}$)
β-Casein A (I)	–	11	89	12.6	6.3
Acetyl-I	96	41	59	4.8	4.7
Propionyl-I	97	24	76	10.5	5.4
n-Butyryl-I	80	8	92	8.9	8.3
n-Hexanoyl-I	85	0	100	7.6	11.6
n-Octanoyl-I	89	0	100	6.6	7.0
n-Decanoyl-I	83	0	100	5.0	6.5

[a] Substitution degree.

1.4.6.2.3 Redox reactions involving cysteine and cystine

Disulfide bonds have a strong influence on the properties of proteins. Wheat gluten can be modified by reduction of its disulfide bonds to sulfhydryl groups and subsequent reoxidation of these groups under various conditions (Figure 1.33). Reoxidation of a diluted suspension in the presence of urea results in a weak, soluble, adhesive product (gluten A), whereas reoxidation of a concentrated suspension in the presence of a higher concentration of urea yields an insoluble, stiff, cohesive product (gluten C). Additional viscosity data have shown that the disulfide bridges in gluten A are mostly intramolecular while those in gluten C are preferentially intermolecular (Figure 1.34).

1.4.6.3 Enzymatic Modification

Of the great number of enzymatic reactions with protein as a substrate (cf. 1.4.5), only a small number have so far been found to be suitable for use in food processing.

1.4.6.3.1 Dephosphorylation

Figure 1.37 uses β-casein as an example to show that the solubility of a phosphoprotein in the presence of calcium ions is greatly improved by enzymatic dephosphorylation.

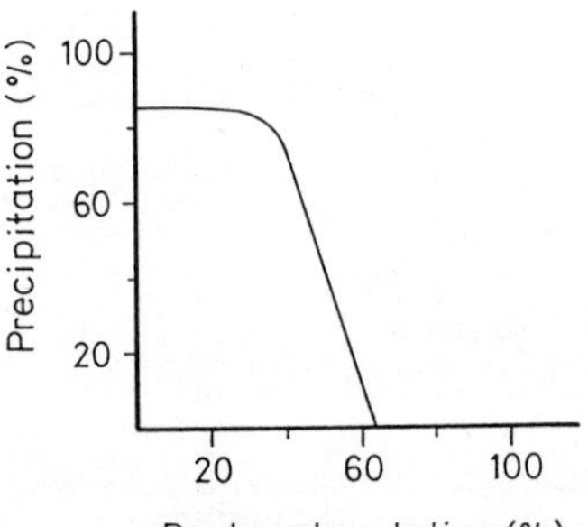

Fig. 1.37. Solubility of β-casein, partially dephosphorylated by phosphoprotein phosphatase: Precipitation: pH 7.1: 2.5 mg/ml protein: 10 mM $CaCl_2$: 35 °C; 1 h. (After *Yoshikawa* et al., 1974)

1.4.6.3.2 Plastein reaction

The plastein reaction enables peptide fragments of a hydrolysate to join enzymatically through peptide bonds, forming a larger polypeptide of about 3 kdal:

$$\text{R—CO—NH—R}^1 + \text{E—OH}$$
$$\rightleftharpoons \text{R—CO—O—E} + \text{H}_2\text{N—R}^1$$
$$\xrightarrow{\text{H}_2\text{O}} \text{R—COOH} + \text{E—OH}$$
$$\xrightarrow{\text{H}_2\text{N—R}^2} \text{R—CO—NH—R}^2 + \text{E—OH} \quad (1.126)$$

The reaction rate is affected by, among other things, the nature of the amino acid residues. Hydrophobic amino acid residues are preferably linked together (Figure 1.38). Incorporation of amino acid esters into protein is affected by the alkyl chain length of the ester. Short-chain alkyl esters have a low rate of incorporation, while the long-chain alkyl esters have a higher rate of incorporation. This is especially important for the incorporation of amino acids with a short side chain, such as alanine (cf. Table 1.30).
The plastein reaction can help to improve the biological value of a protein. Figure 1.39 shows the plastein enrichment of zein with tryptophan, threonine and lysine. The amino acid composi-

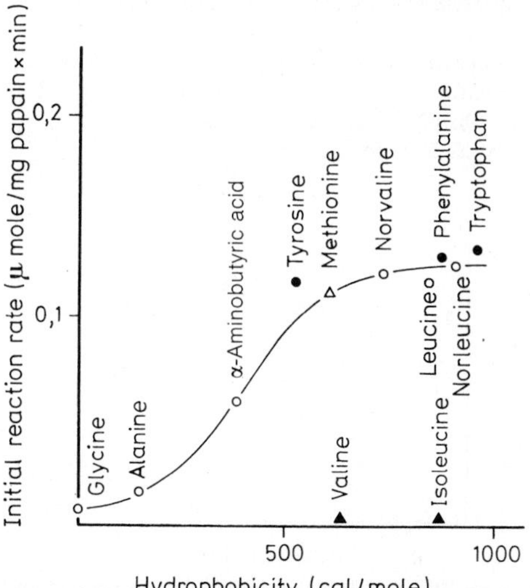

Fig. 1.38. Plastein reaction with papain: incorporation rates of amino acid esters in dependence on side chain hydrophobicity. (After *Arai* et al., 1978)

Table 1.30. Plastein reaction catalyzed by papain: rate of incorporation of amino acid esters[a]

Aminoacyl residue	OEt	OnBu	OnHex	OnOct
L-Ala	0.016	0.054	0.133	0.135
D-Ala	0.0	–	0.0	–
α-Methylala	0.0	–	0.0	–
L-Val	0.005	–	0.077	–
L-Norval	0.122	–	0.155	–
L-Leu	0.119	–	0.140	–
L-Norleu	0.125	–	0.149	–
L-Ile	0.005	–	0.048	–

[a] μmole × mg Papain^{-1} × min^{-1}.

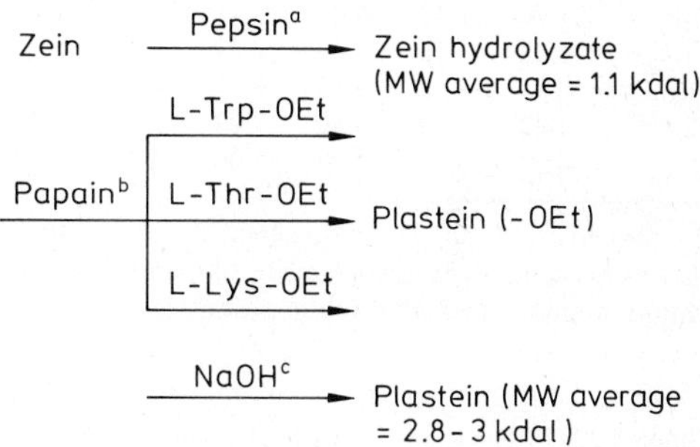

Fig. 1.39. Zein enrichment with Trp, Thr, and Lys by a plastein reaction (After *Aso* et al., 1974)
a 1% Substrate, E/S = 1/50, pH 1.6 at 37 °C for 72 h
b 50% substrate, hydrolyzate/AS-OEt = 10/1, E/S = 3/100 at 37 °C for 48 h
c 0.1 N in 50% ethanol at 25 °C for 5 h.

tion of such a zein-plastein product is given in Table 1.31.
Enrichment of a protein with selected amino acids can be achieved with the corresponding amino acid esters or, equally well, by using suitable partial hydrolysates of another protein.
Figure 1.40 presents the example of soya protein enrichment with sulfur-containing amino acids through "adulteration" with the partial hydrolysate of wool keratin. The PER (protein efficiency ratio) values of such plastein products are significantly improved, as is seen in Table 1.32.
Figure 1.41 shows that the production of plastein with an amino acid profile very close to that recommended by FAO/WHO can be achieved from very diverse proteins.
The plastein reaction also makes it possible to improve the solubility of a protein, as for exam-

Table 1.31. Amino acid composition of various plasteins (weight-%)

	1	2	3	4	5	6
Arg	1.56	1.33	1.07	1.06	1.35	1.74
His	1.07	0.95	0.81	0.75	0.81	1.06
Ile	4.39	6.39	6.58	5.49	6.23	5.67
Leu	20.18	23.70	23.05	23.75	25.28	23.49
Lys	0.20	0.20	0.24	2.14	3.24	0.19
Phe	6.63	7.26	6.82	7.34	7.22	6.98
Thr	2.40	2.18	9.23	2.36	2.46	2.13
Trp	0.38	9.71	0.25	0.40	0.42	0.33
Val	3.62	5.23	5.77	5.53	6.18	6.20
Met	1.58	1.87	1.67	1,89	2.06	2.04
Cys	1.00	0.58	0.88	0.81	0.78	0.92
Ala	7.56	7.51	8.05	7.97	7.93	8.77
Asp	4.61	3.38	3.42	3.71	3.60	3.91
Glu	21.70	12.48	14.03	14.77	12.95	13.02
Gly	1.48	1.15	1.23	1.29	1.27	1.52
Pro	10.93	8.42	9.10	9.73	9.14	9.37
Ser	4.42	3.40	3.89	3.93	3.74	4.28
Tyr	4.73	5.35	4.97	5.00	6.08	5.54

1) Zein hydrolyzate; 2) Trp-plastein; 3) Thr-plastein; 4) Lys-plastein; 5) Ac-Lys-plastein; 6) Control, an addition of amino acid ethyl esters is omitted.

Table 1.32. PER-values for various proteins and plasteins

Protein	PER value (rats)
Casein	2.40
Soya protein (I)	1.20
Plastein SW[a] + I (1:2)	2.86
Plastein-Met[b] + I (1:3)	3.38

[a] From hydrolyzate I and wool keratin hydrolyzate.
[b] From hydrolyzate I and Met-OEt.

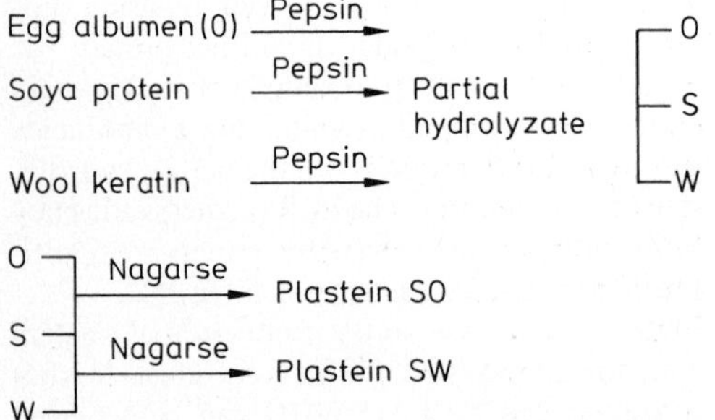

Fig. 1.40. Protein enrichment with sulfur amino acids applying plastein reaction. (After *Yamashita* et al., 1971)

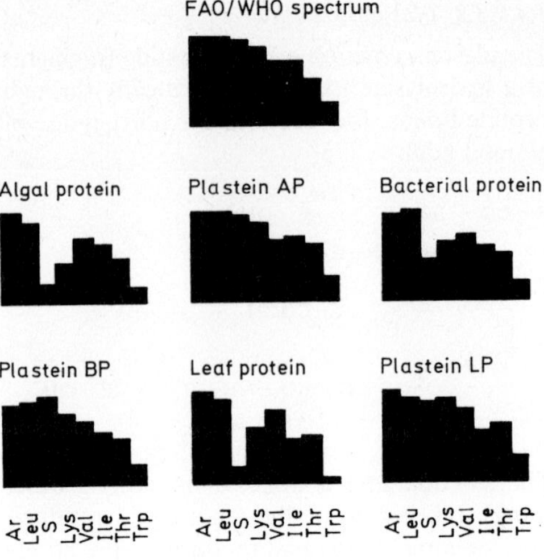

Fig. 1.41. Amino acid patterns of some proteins and their corresponding plasteins. (After *Arai* et al., 1978)

Soya globulin $\xrightarrow{\text{Pepsin}^a}$ Partial hydrolyzate $\xrightarrow[\text{Papain}^b]{\text{Glu-}\alpha,\gamma\text{-(OEt)}_2}$ Plastein (-OEt) $\xrightarrow{\text{NaOH}^c}$ Plastein

Fig. 1.42. Soy globulin enrichment with glutamic acid by a plastein reaction. (After *Yamashita* et al., 1975)
a pH 1.6
b partial hydrolyzate/Glu-α-γ-$(OEt)_2$ = 2:1, substrate concentration: 52.5%, E/S = 1/50, pH 5.5 at 37 °C for 24 h; sample contains 20% acetone
c 0.2 N, at 25 °C for 2 h.

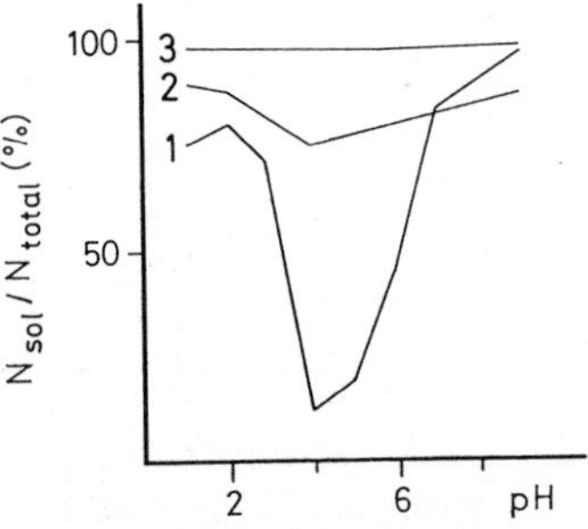

Fig. 1.43. Effect of pH on solubility of soy protein and modified products (1 g/100 ml water). 1 Soy protein, 24.1% Glu; 2 Plastein 24.8% Glu; 3 Glu-plastein with 41.9% Glu. (After *Yamashita* et al., 1975)

ple, by increasing the content of glutamic acid (Figure 1.42). A soya protein with 25% glutamic acid yields a plastein with 42% glutamic acid. Soya protein has a pronounced solubility minimum in the pH range of 3–6. The minimum is much less pronounced in the case of the unmodified plastein, whereas the glutamic acid-enriched soya plastein has a satisfactory solubility over the whole pH range (Figure 1.43) and is also resistant to thermal coagulation (Figure 1.44). Proteins with an increased content of glutamic acid show an interesting sensory effect: partial hydrolysis of modified plastein does not result in a bitter taste, rather it generates a pronounced "meat broth" flavor (Table 1.33).

Elimination of the bitter taste from a protein hydrolysate is also possible without incorporation of hydrophilic amino acids: bitter-tasting peptides, such as Leu-Phe, which are released by partial hydrolysis of protein, react preferentially in the subsequent plastein reaction and are incorporated into higher molecular weight peptides with a neutral taste.

The versatility of the plastein reaction is also demonstrated by examples wherein undesired amino acids are removed from a protein. A phenylalanine-free diet, which can be satisfied by mixing amino acids, is recommended for certain metabolic defects. However, the use of a phenylalanine-free higher molecular weight peptide is more advantageous from sensory and osmotic aspects. Such peptides can be prepared from protein by the plastein reaction. First, the protein is partially hydrolyzed with pepsin. Treatment with pronase under suitable conditions then preferentially releases amino acids with long hydrophobic side chains. The remaining peptides are separated by gel chromatography and are then subjected to the plastein reaction in the presence of added tyrosine and tryptophan (Figure 1.45). This yields a plastein that is practically phenylalanine-free and has a predetermined ratio of other amino acids, including tyrosine (Table 1.34).

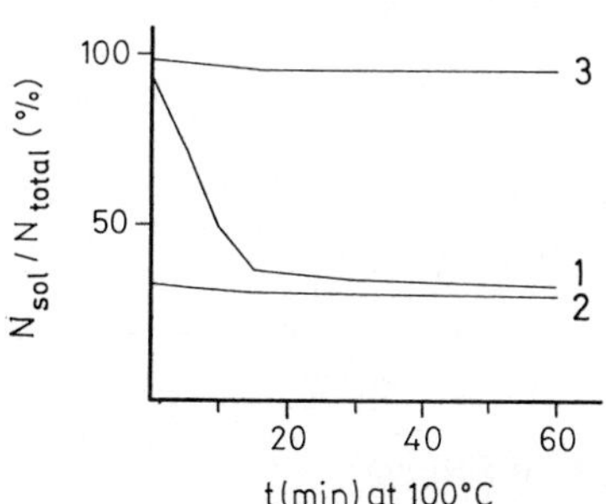

Fig. 1.44. Solubility of soy protein and modified products (800 mg/10 ml water) in dependence on heating time at 100 °C. 1 Soy protein 24.1% Glu; 2 Plastein 24.8% Glu; 3 Glu-plastein, 41.9% Glu. (After *Yamashita* et al., 1975)

Table 1.33. Taste of glutamic acid enriched plasteins

Enzyme	pH	Substrate[a]	Hydrolysis[b]	Taste[c]	
				bitter	meat broth type
Pepsin	1.5	G	67	1	1.3
		P	73	4.5	1.0
α-Chymotrypsin	8.0	G	48	1	1.0
		P	72	4.5	1.0
Molsin	3.0	G	66	1.0	5.0
		P	74	1.3	1.3
Pronase	8.0	G	66	1.0	4.3
		P	82	1.3	1.2

[a] G: Glu-plastein, P: plastein; 1 g/100 ml
[b] N_{sol} (10% TCA)/N_{total} (%)
[c] 1: no taste, 5: very strong taste.

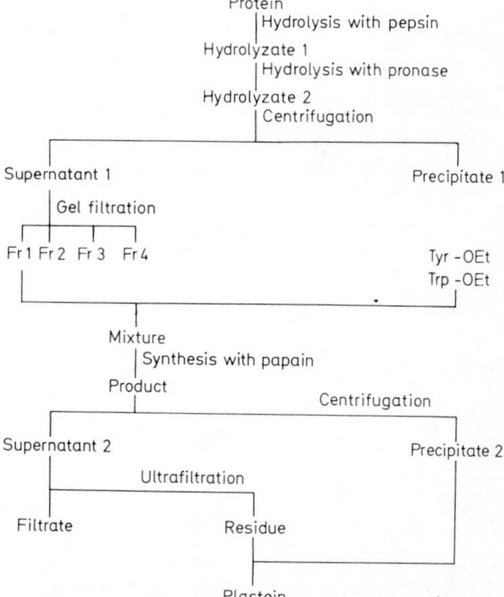

Table 1.34. Amino acid composition (weight-%) of plasteins with high tyrosine and low phenylalanine contents from fish protein concentrate (FPC) and soya protein isolate (SPI)

Amino acid	FPC	FPC-Plastein	SPI	SPI-Plastein
Arg	7.05	4.22	7.45	4.21
His	2.31	1.76	2.66	1.41
Ile	5.44	2.81	5.20	3.83
Leu	8.79	3.69	6.73	2.43
Lys	10.68	10.11	5.81	3.83
Thr	4.94	4.20	3.58	4.39
Trp	1.01	2.98	1.34	2.80
Val	5.88	3.81	4.97	3.24
Met	2.80	1.90	1.25	0.94
Cys	0.91	1.41	1.78	1.82
Phe	4.30	0.05	4.29	0.23
Tyr	3.94	7.82	3.34	7.96
Ala	6.27	4.82	4.08	2.56
Asp	11.13	13.67	11.51	18.00
Glu	17.14	27.17	16.94	33.56
Gly	4.42	3.94	4.88	3.89
Pro	3.80	4.25	6.27	2.11
Ser	4.59	3.58	5.45	4.67

The plastein reaction can also be carried out as a one-step process (Figure 1.46), thus putting these reactions to economic, industrial-scale use.

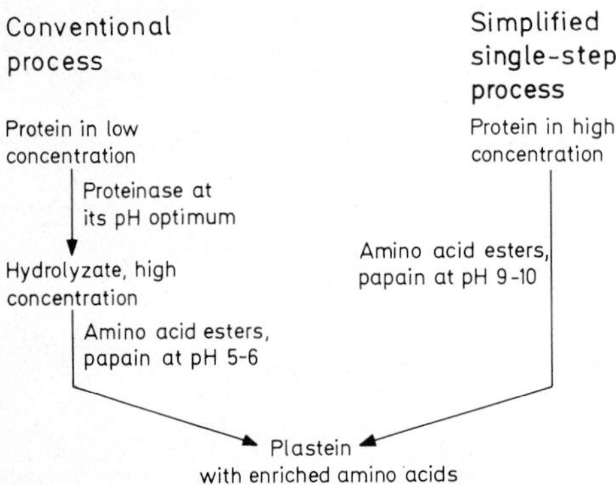

Fig. 1.46. An outline for two-and single-step plastein reactions. (After *Yamashita* et al., 1979)

1.4.6.3.3 Associations involving cross-linking

Cross-linking between protein molecules is achieved with peroxidase. The cross-linking occurs between tyrosine residues when a protein is incubated with peroxidase/H_2O_2 (cf. Reaction 1.127).

$$\text{OC}\backslash\text{CH-CH}_2\text{-C}_6\text{H}_4\text{-OH (HN)} + \text{HO-C}_6\text{H}_4\text{-CH}_2\text{-CH (CO, NH)} \xrightarrow[\text{pH 9,5}]{H_2O_2 / \text{Peroxidase}} \text{HO-C}_6\text{H}_4(\text{-CH}_2\text{-CH(CO)(NH)})\text{-C}_6\text{H}_3(\text{-CH}_2\text{-HC(OC)(HN)})\text{-OH} \quad (1.127)$$

Incubation of protein with peroxidase/H_2O_2/catechol also results in cross-linking. The reactions in this case are the oxidative deamination of lysine residues, followed by aldol and aldimine condensations, i.e. reactions analogous to those catalyzed by lysyl oxidase in connective tissue.

$$\text{-C(=O)-NH-CH(-CH}_2\text{-CH}_2\text{-CH}_2\text{-CH}_2\text{-NH}_2\text{)-C(=O)-NH-} \longrightarrow \text{-C(=O)-NH-CH(-CH}_2\text{-CH}_2\text{-CH}_2\text{-CH=O)-C(=O)-NH-} + NH_3 \quad (1.128)$$

Table 1.35 presents some of the proteins modified by peroxidase/H_2O_2 treatment and includes their dityrosine contents.

Table 1.35. Content of dityrosine in some proteins after their oxidation with horseradish peroxidase/H_2O_2 (pH 9.5, 37°C, 24 h. Substrate/enzyme = 20 : 1)

Protein	Tyrosine content prior to oxidation (g/100 g protein)	Tyrosine decrease (%)	Dityrosine content (g/100 g protein)
Casein	6.3	21.8	1.37
Soyamine[a]	3.8	11.5	0.44
Bovine serum albumin	4.56	30.7	1.40
Gliadin	3.2	5.4	0.17

[a] Protein preparation from soybean.

1.4.7 Texturized Proteins

1.4.7.1 Foreword

The protein produced for nutrition in the world is currently about 30% from animal sources and 70% from plant sources. The plant proteins are

primarily from cereals (50%) and oilseed meal (20%). Some nonconventional sources of protein (single cell proteins, leaves) have also acquired some importance.
Proteins are responsible for the distinct physical structure of a number of foods, e.g. the fibrous structure of muscle tissue (meat, fish), the porous structure of bread and the gel structure of some dairy and soya products.
Many plant proteins have a globular structure and, although available in large amounts, are used to only a limited extent in food processing. In an attempt to broaden the use of such proteins, a number of processes were developed in the mid-1950's which confer a fiber-like structure to globular proteins. Suitable processes give products with cooking strength and a meat-like structure. They are marketed as meat extenders and meat analogues and can be used whenever a lumpy structure is desired.

1.4.7.2 Starting Material

The following protein sources are suitable for the production of texturized products: soya; casein; wheat gluten; oilseed meals such as from cottonseed, groundnut, sesame, sunflower, safflower or rapeseed; zein (corn protein); yeast; whey; blood plasma; or packing plant offal such as lungs or stomach tissue.
The required protein content of the starting material varies and depends on the process used for texturization. The starting material is often a mixture such as soya with lactalbumin, or protein plus acidic polysaccharide (alginate, carrageenan or pectin).
The suitability of proteins for texturization is variable, but the molecular weight should be in the range of 10–50 kdal. Proteins of less than 10 kdal are weak fiber builders, while those higher than 50 kdal are disadvantageous due to their high viscosity and tendency to gel in the alkaline pH range. The proportion of amino acid residues with polar side chains should be high in order to enhance intermolecular binding of chains. Bulky side chains obstruct such interactions, so that the amounts of amino acids with these structures should be low.

1.4.7.3 Texturization

The globular protein is unfolded during texturization by breaking the intramolecular binding forces. The resultant extended protein chains are stabilized through interaction with neighboring chains. In practice, texturization is achieved in one of two ways:

- The starting protein is solubilized and the resultant viscous solution is extruded through a spinning nozzle into a coagulating bath (spin process).
- The starting protein is moistened slightly and then, at high temperature and pressure, is extruded with shear force through the orifices of a die (extrusion process).

1.4.7.3.1 Spin process

The starting material (protein content >90%, e.g. a soya protein isolate) is suspended in water and solubilized by the addition of alkali. The 20% solution is then aged at pH 11 with constant stirring. The viscosity rises during this time as the protein unfolds. The solution is then pressed through the orifices of a die (5,000–15,000 orifices, each with a diameter of 0.01–0.08 mm) into a coagulating bath at pH 2–3. This bath contains an acid (citric, acetic, phosphoric, lactic or hydrochloric) and, usually, 10% NaCl. Spinning solutions of protein and acidic polysaccharide mixtures also contain earth alkali salts. The protein fibers are extended further (to about 2- to 4-times the original length) in a "winding up" step and are bundled into thicker fibers with diameters of 10–20 nm. The molecular interactions are enhanced during stretching of the fiber, thus increasing the mechanical strength of the fiber bundles.
The adherent coagulating solvent is then removed by pressing the fibers between rollers, then placing them in a neutralizing bath ($NaHCO_3 + NaCl$) of pH 5.5–6 and, occasionally, also in a hardening bath (conc. NaCl).
The fiber bundles may be combined into larger aggregates with diameters of 7–10 cm.
Additional treatment involves passage of the bundles through a bath containing a binder and other additives (a protein which coagulates when heated, such as egg protein; modified starch or other polysaccharides; aroma compounds; lipids). This treatment produces bundles with improved thermal stability and aroma. A typical bath for fibers which are to be processed into a meat analogue might consist of 51% water, 15% ovalbumin, 10% wheat gluten, 8% soya flour,

7% onion powder, 2% protein hydrolysate, 1% NaCl, 0.15% monosodium glutamate and 0,5% pigments.

Finally, the soaked fiber bundles are heated and sliced.

1.4.7.3.2 Extrusion process

The moisture content of the starting material (protein content about 50%, e.g. soya flour) is adjusted to 30–40% and additives (NaCl, buffers, aroma compounds, pigments) are incorporated. Aroma compounds are added in fat as a carrier when necessary after the extrusion step to compensate for aroma losses. The protein mixture is fed into the extruder (a thermostatically-controlled cylinder or conical body which contains a polished, rotating screw with a gradually decreasing pitch) which is heated to 120–180 °C and develops a pressure of 30–40 bar. Under these conditions the mixture is transformed into a plastic, viscous state in which solids are dispersed in the molten protein. Hydration of the protein takes place after partial unfolding of the globular molecules and stretching and rearrangement of the protein strands along the direction of mass transfer.

The process is affected by the rotation rate and shape of the screw and by the heat transfer and viscosity of the extruded material and its residence time in the extruder.

As the molten material exits from the extruder, the water vaporizes, leaving behind vacuoles in the ramified protein strands.

The extrusion process is more economical than the spin process. However, it yields fiber-like particles rather than well-defined fibers. A great number and variety of extruders are now in operation. As with other food processes, there is a trend toward developing and utilizing high-temperature/short-time extrusion cooking.

1.5 Literature

Aeschbach, R., Amado, R., Neukom, H.: Formation of dityrosine cross-links in proteins by oxidation of tyrosine residues. Biochim. Biophys. Acta *439*, 292 (1976)

Arai, S., Yamashita, M., Fujimaki, M.: Nutritional improvement of food proteins by means of the plastein reaction and its novel modification. Adv. Exp. Med. Biol. *105*, 663 (1978)

Ariyoshi, Y.: The structure-taste relationships of aspartyl dipeptide esters. Agric. Biol. Chem. *40*, 983 (1976)

Aso, K., Yamashita, M., Arai, S., Fujimaki, M.: Tryptophan-, threonine-, and lysine-enriched plasteins from zein. Agric. Biol. Chem. *38*, 679 (1974)

Aso, K., Yamashita, M., Arai, S., Suzuki, J., Fujimaki, M.: Specificity for incorporation of α-amino acid esters during the plastein reaction by papain. J. Agric. Food Chem. *25*, 1138 (1977)

Belitz, H.-D., Wieser, H.: Zur Konfigurationsabhängigkeit des süßen oder bitteren Geschmacks von Aminosäuren und Peptiden. Z. Lebensm. Unters. Forsch. *160*, 251 (1976)

Boggs, R.W.: Bioavailability of acetylated derivatives of methionine, threonine and lysine. Adv. Exp. Med. Biol. *105*, 571 (1978)

Brussel, L.B.P., Peer, H.G., van der Heijden, A.: Structure-taste relationship of some sweet-tasting dipeptide esters. Z. Lebensm. Unters. Forsch. *159*, 337 (1975)

Cherry, J.P. (Ed.): Protein functionality in foods. ACS Symposium Series 147, American Chemical Society: Washington, D.C. 1981

Croft, L.R.: Introduction to protein sequence analysis, 2nd edn., John Wiley and Sons, Inc.: Chichester. 1980

Edsall, J.T., Wyman, J.: Biophysical chemistry. Vol. I, Academic Press: New York. 1958

Evans, M.T.A., Irons, L., Petty, J.H.P.: Physicochemical properties of some acyl derivatives of β-Casein. Biochim. Biophys. Acta *243*, 259 (1971)

Finot, P.-A., Mottu, F., Bujard, E., Mauron, J.: N-substituted lysines as sources of lysine in nutrition. Adv. Exp. Med. Biol. *105*, 549 (1978)

Fujimaki, M., Kato, S., Kurata, T.: Pyrolysis of sulfur containing amino acids. Agric. Biol. Chem. *33*, 1144 (1969)

Fujino, M., Wakimasu, M., Tanaka, K., Aoki, H., Nakajima, N.: L-aspartyl-aminomalonic acid diesters. Naturwissenschaften *60*, 351 (1973)

Galembeck, F., Ryan, D.S., Whitaker, J.R., Feeney, R.E.: Reactions of proteins with formaldehyde in the presence and absence of sodium borohydride. J. Agric. Food Chem. *25*, 238 (1977)

Glazer, A.N.: The chemical modification of proteins by group-specific and site-specific reagents. In: The proteins (Eds.: Neurath, H., Hill, R.L., Boeder, C.-L.), 3rd edn., Vol. II, p. 1, Academic Press: New York–London. 1976

Grant, D.R.: The modification of wheat flour proteins with succinic anhydride. Cereal Chem. *50*, 417 (1973)

Gross, E., Morell, J.L.: Structure of nisin. J. Am. Chem. Soc. *93*, 4634 (1971)

Haagsma, N., Slump, P.: Evaluation of lysinoalanine determinations in food proteins. Z. Lebensm. Unters. Forsch. *167*, 238 (1978)

Hudson, B.J.F. (Ed.): Developments in food proteins-1 ff, Applied Science Publ.: London. 1982

Ikenaka, T., Odani, S., Koide, T.: Chemical structure and inhibitory activities of soybean proteinase inhibitors. Bayer-Symposium V "Proteinase inhibitors", p. 325, Springer-Verlag: Berlin–Heidelberg. 1974

Kinsella, J. E., Shetty, K. J.: Yeast proteins: Recovery and functional properties. Adv. Exp. Med. Biol. *105,* 797 (1978)

Kinsella, J. E.: Functional properties of proteins in foods: A survey. Crit. Rev. Food Sci. Nutr. *7,* 219 (1976)

Kinsella, J. E.: Texturized proteins: Fabrication, flavoring, and nutrition. Crit. Rev. Food Sci. Nutr. *10,* 147 (1978)

Lasztity, R.: Rheologische Eigenschaften von Weizenkleber und ihre Beziehungen zu molekularen Parametern. Nahrung *19,* 749 (1975)

Li-Chan, E., Helbig, N., Holbek, E., Chau, S., Nakai, S.: Covalent attachment of lysine to wheat gluten for nutritional improvement. J. Agric. Food Chem. *27,* 877 (1979)

Lottspeich, F., Henschen, A., Hupe, K.-P. (Eds.): High performance liquid chromatography in protein and peptide chemistry. Walter de Gruyter: Berlin. 1981

Lowe, J. N., Ingraham, L. L.: An introduction to biochemical reaction mechanisms. Prentice Hall, Inc.: Englewood Cliffs, N. J. 1974

Lübke, K., Schröder, E., Kloss, G.: Chemie und Biochemie der Aminosäuren, Peptide und Proteine. Georg Thieme Verlag: Stuttgart. 1975

Masters, P. M., Friedman, M.: Racemization of amino acids in alkali-treated food proteins. J. Agric. Food Chem. *27,* 507 (1979)

Mauron, J.: Ernährungsphysiologische Beurteilung bearbeiteter Eiweißstoffe. Dtsch. Lebensm. Rundsch. *71,* 27 (1975)

Mazur, R. H., Goldkamp, A. H., James, P. A., Schlatter, J. M.: Structure-taste relationships of aspartic acid amides. J. Med. Chem. *13,* 1217 (1970)

Mazur, R. H., Reuter, J. A., Swiatek, K. A., Schlatter, J. M.: Synthetic sweeteners 3. Aspartyl dipeptide esters from L- and D-alkylglycines. J. Med. Chem. *16,* 1284 (1973)

Meister, A.: Biochemistry of the amino acid. 2nd edn., Vol. I, Academic Press: New York–London. 1965

Needleman, S. B. (Ed.): Protein sequence determination. Springer-Verlag: Berlin–Heidelberg. 1970

Needleman, S. B. (Ed.): Advanced methods in protein sequence determination. Springer-Verlag: Berlin–Heidelberg. 1977

Perutz, M. F.: Proteins and nucleic acids. Elsevier Publ. Co.: Amsterdam. 1962

Puigserver, A. J., Sen, L. C., Clifford, A. J., Feeney, R. E., Whitaker, J. R.: A method for improving the nutritional value of food proteins: Covalent attachment of amino acids. Adv. Exp. Med. Biol. *105,* 587 (1978)

Schulz, G. E., Schirmer, R. H.: Principle of protein structure. Springer-Verlag: Berlin–Heidelberg. 1979

Schwenke, K. D.: Beeinflussung funktioneller Eigenschaften von Proteinen durch chemische Modifizierung. Nahrung *22,* 101 (1978)

Sela, M., Lupu, N., Yaron, A., Berger, A.: Water-soluble polypeptidyl gliadins. Biochim. Biophys. Acta *62,* 594 (1962)

Severin, Th., Ledl, F.: Thermische Zersetzung von Cystein in Tributyrin. Chem. Mikrobiol. Technol. Lebensm. *1,* 135 (1972)

Sheppard, R. C. (Ed.): Amino-acids, peptides, and proteins, Vol. 1 ff, The Chemical Society, Berlington House, London. 1970

St. Angelo, A. J., Conkerton, E. J., Dechary, J. M., Altschul, A. M.: Modification of edestin with N-carboxyl-D,L-alanine anhydride. Biochim. Biophys. Acta *121,* 181 (1966)

Sternberg, M., Kim, C. Y.: Lysinoalanine formation in protein food ingredients. Adv. Exp. Med. Biol. *86 B,* 73 (1977)

Sulser, H.: Die Extraktstoffe des Fleisches. Wissenschaftliche Verlagsgesellschaft mbH: Stuttgart. 1978

Traub, W., Piez, K. A.: The chemistry and structure of collagen. Adv. Protein Chem. *25,* 267 (1971)

Treleano, R., Belitz, H.-D., Jugel, H., Wieser, H.: Beziehungen zwischen Struktur und Geschmack bei Aminosäuren mit cyclischen Seitenketten. Z. Lebensm. Unters. Forsch. *167,* 320 (1978)

Voutsinas, L. P., Nakai, S.: Covalent binding of methionine and tryptophan to soy protein. J. Food Sci. *44,* 1205 (1979)

Walton, A. G.: Polypeptides and protein structure, Elsevier North Holland, Inc., New York–Oxford. 1981

Whitaker, J. R., Fujimaki, M. (Eds.): Chemical deterioration of proteins, ACS Symposium Series 123, American Chemical Society: Washington D.C. 1980

Wieser, H., Belitz, H.-D.: Zusammenhänge zwischen Struktur und Bittergeschmack bei Aminosäuren und Peptiden. I. Aminosäuren und verwandte Verbindungen. Z. Lebensm. Unters. Forsch. *159,* 65 (1975)

Wieser, H., Belitz, H.-D.: Zusammenhänge zwischen Struktur und Bittergeschmack bei Aminosäuren und Peptiden. II. Peptide und Peptidderivate. Z. Lebensm. Unters. Forsch. *160,* 383 (1976)

Wieser, H., Jugel, H., Belitz, H.-D.: Zusammenhänge zwischen Struktur und Süßgeschmack bei Aminosäuren. Z. Lebensm. Unters. Forsch. *164,* 277 (1977)

Yamashita, M., Arai, S., Fujimaki, M.: A low-phenylalanine, high-tyrosine plastein as an acceptable dietetic food. J. Food Sci. *41,* 1029 (1976)

Yamashita, M., Arai, S., Amano, Y., Fujimaki, M.: A novel one-step process for enzymatic incorporation of amino acids into proteins: Application to soy protein and flour for enhancing their methionine levels. Agric. Biol. Chem. *43,* 1065 (1979)

Yamashita, M., Arai, S., Tsai, S.-J., Fujimaki, M.: Plastein reaction as a method for enhancing the sulfur-containing amino acid level of soybean protein. J. Agric. Food Chem. *19,* 1151 (1971)

Yamashita, M., Arai, S., Kokubo, S., Aso, K., Fujimaki, M.: Synthesis and characterization of a glutamic acid enriched plastein with greater solubility. J. Agric. Food Chem. *23*, 27 (1975)

Yoshikawa, M., Tamaki, M., Sugimoto, E., Chiba, H.: Effect of dephosphorylation on the self-association and the precipitation of b-Casein. Agric. Biol. Chem. *38*, 2051 (1974)

2 Enzymes

2.1 Foreword

Enzymes are proteins with enormous catalytic activity. They are synthesized by biological cells and, in all organisms, they are involved in chemical reactions related to metabolism. Therefore, enzyme-catalyzed reactions also proceed in many foods and thus enhance or deteriorate food quality. Relevant to this phenomenon are the ripening of fruits, vegetables, meat and dairy products, and the processing steps involved in the making of dough from wheat or rye flours and the production of alcoholic beverages by fermentation technology.

Enzyme inactivation or changes in the distribution patterns of enzymes in subcellular particles of a tissue can occur during storage or thermal treatment of food. Since such changes are readily detected by analytical means, enzymes often serve as suitable indicators for revealing such treatment of food. Examples are the detection of pasteurization of milk, beer or honey, and differentiation between fresh and deep frozen meat or fish.

Enzyme properties are of interest to the food chemist since enzymes are available in increasing numbers for enzymatic food analysis or for utilization in industrial food processing. Examples of both aspects of their use are provided in this chapter in section 2.5.4 on food analysis and in section 2.8, which covers food processing.

Details of enzymes which play a role in food science are restricted in this chapter to only those enzyme properties which are able to provide an insight into the build-up or functionality of enzymes or can contribute to the understanding of enzyme utilization in food analysis or food processing and storage.

2.2 General Remarks, Isolation and Nomenclature

2.2.1 Catalysis

Let us consider the catalysis of an energy-consuming reaction:

$$A \underset{k_{-1}}{\overset{k_1}{\rightleftharpoons}} P \quad (2.1)$$

with a most frequently occurring case – the reaction does not proceed spontaneously. Reactant A is metastable, since the activation energy, E_A, required to reach the activated transition state in which the chemical bonds are formed or cleaved in order to yield product P, is exceptionally high (Figure 2.1).

The reaction is accelerated by the addition of a suitable catalyst. It transforms reactant A into intermediary products (EA and EP in Figure 2.1), the transition states of which are at a lower energy level than the transition state of a non-catalyzed reaction ($A^{\ddagger}$ in Figure 2.1). The molecules of the species A contain enough energy to combine with the catalyst and, thus, to attain the "activated state" and to form or break the

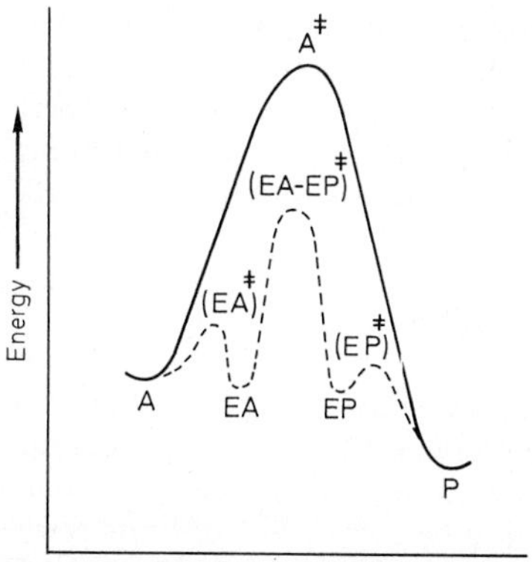

Fig. 2.1. Energy profile of an exergonic reaction A → P
— Without and --- with catalyst E

Table 2.1. Examples of catalyst activity

No. Reaction	Catalyst	Activation energy (kJ · mole^{-1})	k_{rel} (25°)
1. $H_2O_2 \rightarrow H_2O + 1/2\,O_2$	Absent	75	1.0
	$J^{\ominus}$	56.5	$\sim 2.1 \cdot 10^3$
	Catalase	26.8	$\sim 3.5 \cdot 10^8$
2. Casein + n H_2O → (n + 1) Peptide	$H^{\oplus}$	86	1.0
	Trypsin	50	$\sim 2.1 \cdot 10^6$
3. Ethylbutyrate + H_2O → butyric acid + ethanol	$H^{\oplus}$	55	1.0
	Lipase	17.6	$\sim 4.2 \cdot 10^6$
4. Saccharose + H_2O → Glucose + Fructose	$H^{\oplus}$	107	1.0
	Invertase	46	$\sim 5.6 \cdot 10^{10}$
5. Linoleic acid + O_2 → Linoleic acid hydroperoxide	Absent	150–270	1.0
	Cu^{2+}	30–50	$\sim 10^2$
	Lipoxygenase	16.7	$\sim 10^7$

covalent bond that is necessary to form the intermediary product which is then released as product P along with free, unchanged catalyst. The reaction rate constants, k_{+1} and k_{-1}, are therefore increased in the presence of a catalyst. However, the equilibrium constant of the reaction, i.e. the ratio $k_{+1}/k_{-1} = K$, is not altered.
Activation energy levels for several reactions and the corresponding decreases of these energy levels in the presence of chemical or enzymatic catalysts are provided in Table 2.1. Changes in their reaction rates are also given.
In contrast to reactions 1 and 5 (Table 2.1) which proceed at measurable rates even in the absence of catalysts, hydrolysis reactions 2, 3 and 4 occur only in the presence of protons as catalysts. However, all reaction rates observed in the case of inorganic catalysts are increased by a factor of at least several orders of magnitude in the presence of suitable enzymes. Because of the powerful activity of enzymes, their presence at levels of 10^{-8} to 10^{-6} M is sufficient for *in vitro* experiments. However, the enzyme concentrations found in living cells are often substantially higher.

2.2.2 Specificity

In addition to an enzyme's ability to substantially increase reaction rates, there is a unique enzyme property related to its high specificity for both the compound to be converted (substrate specificity) and for the type of reaction to be catalysed (reaction specificity).
The activities of allosteric enzymes (cf. 2.5.1.3) are affected by specific regulators or effectors. Thus, the activities of such enzymes show an additional regulatory specificity.

2.2.2.1 Substrate Specificity

The substrate specificity of the enzymes shows the following differences. The occurrence of a distinct functional group in the substrate is the only prerequisite for a few enzymes, such as some hydrolases. This is exemplified by nonspecific lipases (cf. Table 3.16) or proteinases (cf. 1.4.5.2.1) which act generally on an ester or peptide covalent bond.

Table 2.2. Substrate specificity of a legume α-glucosidase

Substrate	Relative activity (%)	Substrate	Relative activity (%)
Maltose	100	Cellobiose	0
Isomaltose	4.0	Saccharose	0
Maltotriose	41.5	Phenyl-α-glucoside	3.1
Panose	3.5	Phenyl-α-maltoside	29.7
Amylose	30.9		
Amylopectin	4.4		

More restricted specificity is found in other enzymes, the activities of which require that the substrate molecule contains a distinct structural feature in addition to the reactive functional group. Examples are the proteinases trypsin and chymotrypsin which cleave only ester or peptide bonds with the carbonyl derived from lysyl or arginyl (trypsin) or tyrosyl, phenylalanyl or tryptophanyl residues (chymotrypsin). Many enzymes activate only one single substrate or preferentially catalyze the conversion of one substrate while other substrates are converted into products with a lower reaction rate (cf. examples in Tables 2.2 and 3.17). In the latter cases a reliable assessment of specificity is possible only when the enzyme is available in purified form, i.e. all other accompanying enzymes, as impurities, are completely removed.
An enzyme's substrate specificity for stereoisomers is remarkable. When a chiral center is present in the substrate in addition to the group to be activated, only one enantiomer will be converted to the product. Another example is the specificity for diastereoisomers, e.g. for cis-trans geometric isomers.

Enzymes with high substrate specificity are of special interest for enzymatic food analysis. They can be used for the selective analysis of individual food constituents, thus avoiding the time consuming separation techniques required by chemical analyses, which can result in losses.

2.2.2.2 Reaction Specificity

The substrate is specifically activated by the enzyme so that, among the several thermodynamically permissible reactions, only one occurs. This is illustrated by the following example: L(+)-lactic acid is recognized as a substrate by four enzymes, as shown in Figure 2.2, although only lactate-2-monooxygenase decarboxylates the acid oxidatively to acetic acid. Lactate dehydrogenase and lactate-malate transhydrogenase form a common reaction product, pyruvate, but by different reaction pathways (Figure 2.2). This may suggest that reaction specificity should be ascribed to the different cosubstrates, such as NAD^+ or oxalacetate. But this is not the case since a change in cosubstrates stops the reaction. Obviously, the enzyme's reaction specificity as well as the substrate specificity are predetermined by the structure and chemical properties of the protein moiety of the enzyme.

Of the four enzymes considered, only the lactate racemase reacts with either of the enantiomers of lactic acid, yielding a racemic mixture.

Therefore, enzyme reaction specificity rather than substrate specificity is considered as a basis for enzyme classification and nomenclature (cf. 2.2.6).

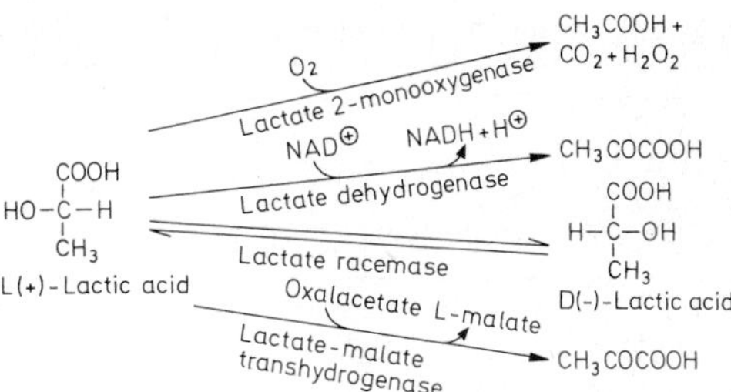

Fig. 2.2. Examples of reaction specificity of some enzymes

2.2.3 Structure

Enzymes are globular proteins with greatly differing particle sizes (cf. Table 1.19 a). As outlined in section 1.4.2, the protein structure is determined by its amino acid sequences and from its conformation, both secondary and tertiary, derived from this sequence. Larger enzyme molecules often consist of two or more peptide chains (subunits or protomers, cf. Table 1.19 a) arranged into a specified quaternary structure (cf. 1.4.2.3.3). Section 2.4.1 will show that the three dimensional shape of the enzyme molecule is actually responsible for its specificity and its effective role as a catalyst. On the other hand, the protein nature of the enzyme restricts its activity to a relatively narrow pH range (for pH optima, cf. 2.5.3) and heat treatment leads readily to loss of activity by denaturation (cf. 2.6.4).

Some enzymes are complexes consisting of a protein moiety bound firmly to a nonprotein component which is involved in catalysis, e.g. a "prosthetic" group (cf. 2.3.2). The activities of other enzymes require the presence of a cosubstrate (older designation: coenzyme) which is reversibly bound to the protein moiety (cf. 2.3.1).

2.2.4 Isolation and Purification

Most of the enzyme properties are clearly and reliably revealed only with purified enzymes. As noted under enzyme isolation, prerequisites for the isolation of a pure enzyme are selected protein chemical separation methods carried out at 0–4 °C since enzymes are often not stable at higher temperatures.

Tissue disintegration and extraction. Disintegration and homogenization of biological tissue requires special precautions: procedures should be designed to rupture the majority of the cells in order to release their contents so that they become accessible for extraction. The tissue is usually homogenized in the presence of an extraction buffer which often contains an ingredient to protect the enzymes from oxidation and traces of heavy metal ions. Particular difficulty is encountered during the isolation of enzymes which are bound tenaciously to membranes which are not readily solubilized. Extraction in the presence of tensides may help to isolate such enzymes. As a rule, large amounts of tissue have to be homogenized because the enzyme content in proportion to the total protein isolated is low and is usually further diminished by the additional purification of the crude enzyme isolate (cf. example in Table 2.3).

Table 2.3. Isolation of a glucosidase from beans *(Phaseolus vidissimus)*

No. / Isolation step	Protein (mg)	α-Glucosidase			
		Activity (μcat)	Specific activity (μcat/mg)	Enrichment (-fold)	Yield (%)
1. Extraction with 0.01 M acetate buffer of pH 5.3					
2. Saturation to 90% with ammonium sulfate followed by solubilization in buffer under step 1	44,200	3,840	0.087	1	100
3. Precipitation with polyethylene glycol (20%). Precipitate is then solubilized in 0.025 M Tris-HCl buffer of pH 7.4	7,610	3,590	0.47	5.4	93
4. Chromatography on DEAE-cellulose column, a cation exchanger	1,980	1,650	0.83	9.5	43
5. Chromatography on SP-Sephadex C-50, an anion exchanger	130	845	6.5	75	22
6. Preparative isoelectric focusing	30	565	18.8	216	15

Enzyme purification. Removal of protein impurities, usually by a stepwise process, is essentially the main approach in enzyme purification. Often, as a first step, fractional precipitation, e.g. by ammonium sulfate saturation, is used or the extracted proteins are fractionated by molecular weight e.g. column gel chromatography. The fractions containing the desired enzyme activity are collected and, for example, are purified further by ion-exchange chromatography. Other supplemental options are also available, such as various forms of preparative electrophoresis, e.g. disc gel electrophoresis or isoelectric focusing. The purification procedure can be substantially shortened by using affinity column chromatography. In this case, the column is packed with a stationary phase to which is attached the substrate or a specific inhibitor of the enzyme. The enzyme is then selectively and reversibly bound and, thus, in contrast to the other inert proteins, its elution is delayed.

Control of purity. Previously, the complete removal of protein impurities was confirmed by crystallization of the enzyme. This "proof" of purity can be circumstantial and is open to criticism. Today, electrophoretic methods of high separation efficiency are primarily used.

The behavior of the enzyme during chromatographic separation is an additional proof of purity. A purified enzyme is characterized by a symmetrical elution peak in which the positions of the protein absorbance and enzyme activity coincide and the specific activity (expressed as units per amount of protein) remains unchanged during repeated elutions.

During a purification procedure, the enzyme activities are recorded as shown in Table 2.3. They provide data which show the extent of purification achieved after each separation step and show the enzyme yield. Such a compilation of data readily reveals the undesired separation steps associated whith loss of activity and suggests modifications or adoption of other steps.

2.2.5 Multiple Forms of Enzymes

Chromatographic or electrophoretic separations of an enzyme can occasionally result in separation of the enzyme into "isoenzymes", i.e. forms of the enzyme which catalyze the same reaction although they differ in their protein structure. The occurrence of multiple enzyme forms can be the result of the following:

a) Different compartments of the cell produce genetically independent enzymes with the same substrate activity specificity, but which differ in their primary structure. An example is glutamate-oxalacetate transaminase occurring in mitochondria and also in muscle tissue sarcoplasm. This is the indicator enzyme used to differentiate fresh from frozen meat (cf. 12.9.1.2).
b) Protomers associate to form polymers of differing size. An example is the glutamate dehy-

drogenase occurring in tissue as an equilibrium mixture of molecular weights ranging from 250 to 1,000 kdal.

c) Two or more protomers combine in various amounts to form the enzyme. For example, lactate dehydrogenase is structured from a larger number of subunits with the reaction specificity given in Figure 2.2. It consists of five forms (A_4, A_3B, A_2B_2, AB_3 and B_4), all derived from two protomers, A and B.

2.2.6 Nomenclature

The members of the "International Union of Pure and Applied Chemistry" (IUPAC) and of the "International Union of Biochemistry" (IUB) adopted rules in 1978 for the systematic classification and designation of enzymes based on activity specificity. All enzymes are classified into six major classes according to the nature of the chemical reaction catalyzed:

1. Oxidoreductases.
2. Transferases.
3. Hydrolases.
4. Lyases (cleave C–C, C–O, C–N, and other groups by elimination, leaving double bonds, or conversely adding groups to double bonds).
5. Isomerases (involved in the catalysis of isomerizations within one molecule).
6. Ligases (involved in the biosynthesis of a compound with the simultaneous hydrolysis of a pyrophosphate bond in ATP or a similar triphosphate).

Each class is then subdivided into subclasses which more specifically denote the type of reaction, e.g. naming the electron donor of an oxidation-reduction reaction or naming the functional group carried over by a transferase or cleaved by a hydrolase enzyme.

Each subclass is further divided into sub-subclasses. For example, sub-subclasses of oxidoreductases are denoted by naming the acceptor which accepts the electron from its respective donor.

Each enzyme is classified by adopting this system. An example will be analyzed. The enzyme ascorbic acid oxidase catalyzes the following reaction:

$$\text{L-Ascorbic acid} + \tfrac{1}{2}O_2 \rightleftharpoons \text{L-Dehydroascorbic acid} + H_2O \quad (2.2)$$

Hence, its systematic name is L-ascorbate: oxygen oxidoreductase, and its systematic number is E.C. 1.10.3.3 (cf. Formula 2.3). The systematic names are often quite long. Therefore, the short, trivial names along with the systematic numbers are often convenient for enzyme designation. Since enzymes of different biological origin often differ in their properties, the source and, when known, the subcellular fraction used for isolation are specified in addition to the name of the enzyme preparation; for example, "ascorbate oxidase" (E.C. 1.10.3.3) from cucumber. When known, the subcellular fraction of origin (cytoplasmic, mitochondrial or peroxisomal) is also specified.

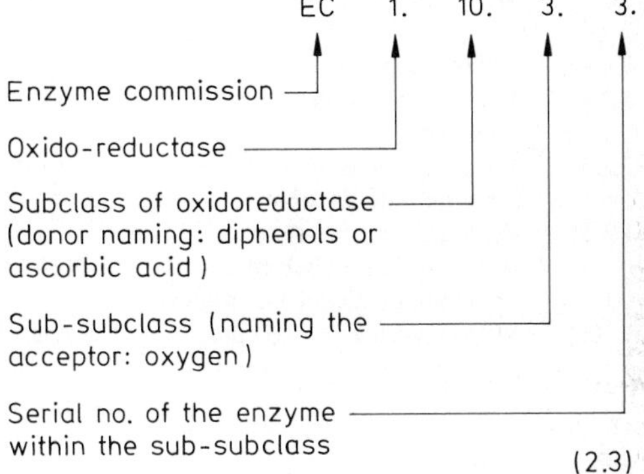

(2.3)

A number of enzymes of interest to food chemistry are described in Table 2.4. The number of the section in which an enzyme is dealt with is given in the last column.

2.2.7 Activity Units

The catalytic activity of enzymes is exhibited only under specific conditions, such as pH, ionic strength, buffer type, presence of cofactors and suitable temperature. Therefore, the rate of substrate conversion or product formation can be measured in a test system designed to follow the enzyme activity. The International System of Units (SI) designation is mole s^{-1} and its recommended designation is the "katal" (kat*). Decimal units are formed in the usual way, e.g.:

$$\mu kat = 10^{-6}\ kat = \mu mole\ s^{-1} \quad (2.4)$$

* The old definition in the literature may also be used: 1 enzyme unit (U) $\triangleq$ 1 μmole min^{-1} of substrate converted.

Table 2.4. Systematic classification of some enzymes of importance to food chemistry

Class/subclass		Enzyme	EC-Number	In text found under
1	*Oxidoreductases*			
1.1	CH–OH as a donor			
1.1.1	With $NAD^{\oplus}$ or $NADP^{\oplus}$ as an acceptor	Alcohol dehydrogenase	1.1.1.1	2.5.4
		Butanediol dehydrogenase	1.1.1.4	2.8.2.1.5
		L-Iditol dehydrogenase	1.1.1.14	2.5.4
		Lactate dehydrogenase	1.1.1.27	2.5.4
		Malate dehydrogenase	1.1.1.37	2.5.4
		Galactose dehydrogenase	1.1.1.48	2.5.4
		Glucose-6-phosphate dehydrogenase	1.1.1.49	2.5.4
1.1.3	With oxygen as an acceptor	Glucose oxidase	1.1.3.4	2.5.4 and 2.8.2.1.1
1.2	Aldehyde group as a donor			
1.2.1	With $NAD^{\oplus}$ or $NADP^{\oplus}$ as an acceptor	Aldehyde dehydrogenase	1.2.1.3	2.8.2.1.4
1.2.3	With oxygen as an acceptor	Xanthine oxidase	1.2.3.2	3.7.2.1.4
1.8	S-Compound as a donor			
1.8.5	With quinone or related compound as an acceptor	Glutathione dehydrogenase (ascorbate)	1.8.5.1	15.2.2.7
1.10	Diphenol or dienol as a donor			
1.10.3	With oxygen as an acceptor	Ascorbate oxidase	1.10.3.3	2.6.4
1.11	Hydroperoxide as a donor	Catalase	1.11.1.6	2.8.2.1.2
		Peroxidase	1.11.1.7	2.5.4. and 2.6.4
1.13	Acting on single donors			
1.13.11	Incorporation of molecular oxygen	Lipoxygenase	1.13.11.12	2.6.4 and 2.8.2.13
1.14	Acting on paired donors			
1.14.18	Incorporation of one oxygen atom	Monophenol monooxygenase (Phenol oxidase)	1.14.18.1	2.6.4
2.	*Transferases*			
2.7	Transfer of a phosphate group			
2.7.1	HO-group as an acceptor	Hexokinase	2.7.1.1	2.5.4
		Glycerol kinase	2.7.1.30	2.5.4
		Pyruvate kinase	2.7.1.40	2.5.4
2.7.3	N-group as an acceptor	Creatine kinase	2.7.3.2	2.5.4
3.	*Hydrolases*			
3.1	Cleavage of ester bonds			
3.1.1	Carboxylester hydrolases	Carboxylesterase	3.1.1.1	
		Triacylglycerol lipase	3.1.1.3	2.6.4 and 3.7.1.1
		Phospholipase A_2	3.1.1.4	3.7.1.2
		Acetylcholinesterase	3.1.1.7	2.5.4
		Pectinesterase	3.1.1.11	2.8.2.2.12
		Phospholipase A_1	3.1.1.32	3.7.1.2

Table 2.4. Continued

Class/subclass		Enzyme	EC-Number	In text found under
3.1.3	Phosphoric acid monoester hydrolases	Alkaline phosphatase	3.1.3.1	2.6.4
3.1.4	Phosphoric acid diester hydrolases	Phospholipase C	3.1.4.3	3.7.1.2
		Phospholipase D	3.1.4.4	3.7.1.2
3.2	Hydrolysing O-glycosyl compounds			
3.2.1	Glycosidases	α-Amylase	3.2.1.1	2.8.2.2.2
		β-Amylase	3.2.1.2	2.8.2.2.3
		Exo-1,4-α-D-glucosidase (glucoamylase)	3.2.1.3	2.8.2.2.4
		Cellulase	3.2.1.4	2.8.2.2.10
		Polygalacturonase	3.2.1.15	2.6.4 and 2.8.2.2.12
		Lysozyme	3.2.1.17	2.8.2.2.11
		α-D-Glucosidase (maltase)	3.2.1.20	2.5.4
		β-D-Glucosidase	3.2.1.21	2.6.4
		α-D-Galactosidase	3.2.1.22	
		β-D-Galactosidase (Lactase)	3.2.1.23	2.8.2.2.7
		β-D-Fructofuranosidase (Invertase, Saccharase)	3.2.1.26	2.8.2.2.8
		Endo-1,3-β-D-Xylanase	3.2.1.32	2.8.2.2.10
		α-L-Rhamnosidase	3.2.1.40	2.8.2.2.9
		Pullulanase	3.2.1.41	2.8.2.2.5
		Exopolygalacturonase	3.2.1.67	
3.2.3	Hydrolysing S-glycosyl compounds	Thioglucosidase (Myrosinase)	3.2.3.1	2.8.2.2.12
3.4	Peptide hydrolases[a]			
3.4.21	Serine proteinases[a]	Microbial serine proteinases e.g. subtilisin	3.4.21.14	2.8.2.2.1
3.4.22	Thiol proteinases[a]	Papain	3.4.22.1	2.8.2.2.1
		Ficin	3.4.22.3	2.8.2.2.1
		Bromelain	3.4.22.5	2.8.2.2.1
3.4.23	Carboxyl proteinases[a]	Chymosin (Rennin)	3.4.23.4	2.8.2.2.1
3.4.24	Metalloproteinases[a]	Microbial metalloproteinases	3.4.24.4	2.8.2.2.1
3.5	C–N bond cleaving (peptide hydrolases are excepted)			
3.5.2	In cyclic amides	Creatininase	3.5.2.10	2.5.4
4.	*Lyases*			
4.2	C–O-Lyases			
4.2.2	Acting on polysaccharides	Pectate lyase	4.2.2.2	2.8.2.2.12
		Exopolygalacturonate lyase	4.2.2.9	2.8.2.2.12
		Pectin lyase	4.2.2.10	2.8.2.2.12
4.4	C–S-Lyases	Alliin lyase	4.4.1.4	
5.	*Isomerases*			
5.3	Intramolecular oxidoreductases			
5.3.1	Interconverting aldoses and ketoses	Xylose isomerase	5.3.1.5	2.8.2.3
		Glucosephosphate isomerase	5.3.1.9	2.5.4
5.3.99	Other intramolecular oxidoreductases	Hydroperoxide isomerase	5.3.99.1	3.7.2.3

[a] cf. Table 1.23a.

Concentration of enzymatic activity is given as µkat l^{-1}. The following activity units are derived from this:

a) The *specific catalytic activity,* i.e. the activity of the enzyme preparation in relation to the protein concentration.
b) The *molar catalytic activity.* This can be determined when the pure enzyme with a known molecular weight is available. It is expressed as "katal per mole of enzyme" (kat $mole^{-1}$). When the enzyme has only one active site or center per molecule, the molar catalytic activity equals the "exchange number", which is defined as the number of substrate molecules converted per unit time by each active site of the enzyme molecule.

2.3 Enzyme Cofactors

Rigorous analysis has demonstrated that numerous enzymes are not pure proteins. In addition to protein, they contain metal ions and/or low molecular weight nonprotein organic molecules. These nonprotein hetero constituents are denoted as cofactors. In order to exhibit activity, many enzymes require the presence of a cofactor. Cofactors are divided into:

- prosthetic groups
- cosubstrates (older designation: coenzyme).

The difference is based on the observation that the prosthetic group is bound firmly to the enzyme. It can not be removed by, e.g. dialysis, and during enzyme catalysis it remains attached to the enzyme molecule. Often, two substrates are converted by such enzymes, one substrate followed by the other, returning the prosthetic group to its original state. On the other hand, during metabolism, the cosubstrate reacts with at least two enzymes. It transfers the hydrogen or the functional group to another enzyme and, hence, is denoted as a "transport metabolite" or as an "intermediary substrate". It is distinguished from a true substrate by being regenerated in a subsequent reaction. Therefore the concentration of the intermediary substrates can be very low.

Only those cofactors with enzymatic activities of importance in enzymatic analysis of food and/or in food processing will be presented. Some cofactors are related to the water-soluble vitamins (cf. 6.3). The metal ions are dealt with separately in section 2.3.3.

2.3.1 Cosubstrates

2.3.1.1 Nicotinamide Adenine Dinucleotide

Transhydrogenases (e.g. lactate dehydrogenase, alcohol dehydrogenase) dehydrogenate or hydrogenate their substrates with the help of a pyridine cosubstrate (Figure 2.3); its nicotinamide residue accepts or donates a hydride ion (H^-) at position 4:

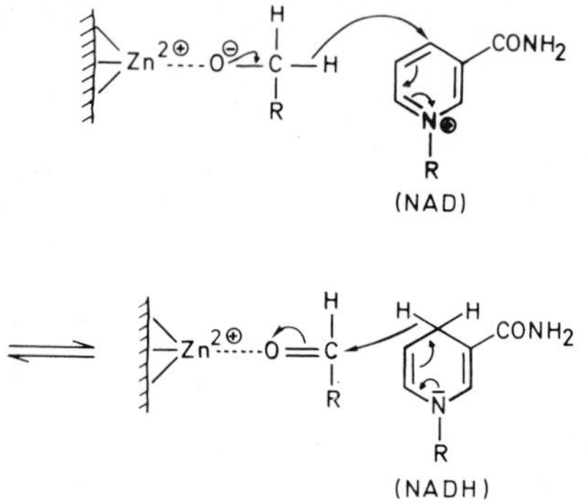

Zn$^{2\oplus}$ ··· O$^{\ominus}$–C(H)(R)–H + NAD (pyridinium, $CONH_2$, N$^{\oplus}$–R) ⇌ Zn$^{2\oplus}$ ··· O=C(H)(R) + NADH (H, H, $CONH_2$, N–R) (2.5)

The reaction proceeds stereospecifically (cf. 2.4.1.2.3); ribose phosphate and the $-CONH_2$ group force the pyridine ring of the cosubstrate to become planar on the enzyme surface. The role of Zn^{2+} ions in this catalysis is outlined in section 2.3.3.1. The transhydrogenases differ according to the site on the pyridine ring involved in or accessible to H-transfer. For example, alcohol and lactate dehydrogenases transfer the pro-R-hydrogen from the A* side, whereas glutamate or glucose dehydrogenases transfer the pro-S-hydrogen from the B* side.

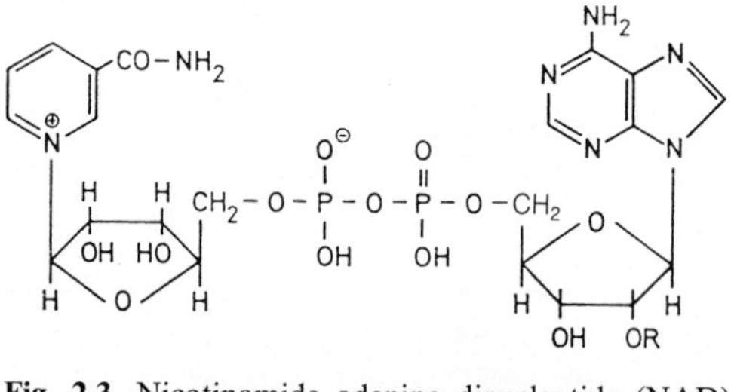

Fig. 2.3. Nicotinamide adenine dinucleotide (NAD) and nicotinamide adenine dinucleotide phosphate (NADP); R = H:NAD; R = PO_3H_2:NADP

* Until the absolute configuration of the chiral center is determined, the two sides of the pyridine ring are denoted as A and B.

The oxidized and reduced forms of the pyridine cosubstrate are readily distinguished by absorbance readings at 340 nm (Figure 2.4). Therefore, whenever possible, enzymatic reactions which are difficult to measure directly are coupled with an NAD(P)-dependent indicator reaction (cf. 2.5.4.1.1) for food analysis.

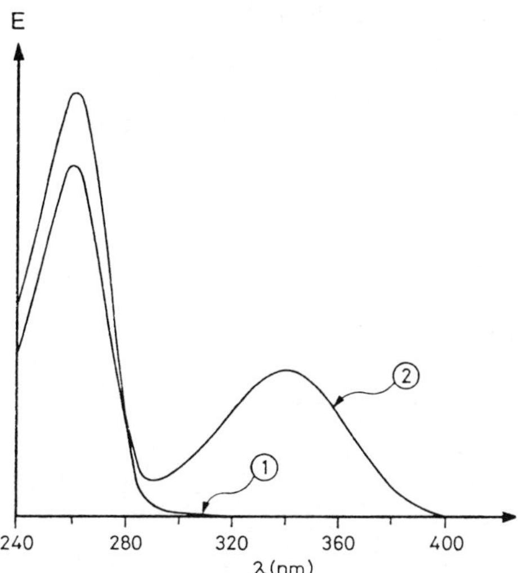

Fig. 2.4. Electron excitation spectra of NAD (1) and NADH (2)

2.3.1.2 Adenosine Triphosphate

The nucleotide adenosine triphosphate (ATP; cf. Formula 2.6) is an energy-rich compound (cf. Reaction 2.7). Various groups are cleaved and transferred to defined substrates during metabolism in the presence of ATP. One possibility, the transfer of orthophosphates by kinases, is utilized in the enzymatic analysis of food (cf. Table 2.13).

(2.6)

$$ATP + H_2O \longrightarrow ADP + H_3PO_4 \quad (2.7)$$

(ΔG^0 at pH 7 = -50 kJ mole^{-1})

2.3.2 Prosthetic Groups

2.3.2.1 Flavins

Riboflavin (7,8-dimethyl-10-ribityl-isoalloxazine), known as vitamin B_2 (cf. 6.3.2), is the building block of flavin mononucleotide (FMN) and flavin adenine dinucleotide (FAD). Both act as prosthetic groups for electron transfer reactions in a number of enzymes.

Riboflavine
FMN

Flavine adenine dinucleotide (FAD) (2.8)

An example is glucose oxidase, an enzyme often used in food processing to trap residual oxygen (cf. 2.8.2.1.1). The enzyme isolated and purified from *Aspergillus niger* is a dimer (168 kdal) with two noncovalently bound FAD molecules. In contrast to xanthine oxidase (cf. 2.3.2.2), for example, this enzyme has no heavy metal ion. During oxidation of a substrate, such as the oxidation of β-D-glucose to δ-D-gluconolactone, the flavoquinone is probably reduced by two single electron transfers:

(FAD) (2.9)

$H^{\oplus}; e^{\ominus}$ $H^{\oplus}; e^{\ominus}$

(Flavo semiquinone) ($FADH_2$)

Like glucose oxidase, many flavo-enzymes transfer the electrons to molecular oxygen, forming H_2O_2. The following intermediary products appear in this reaction:

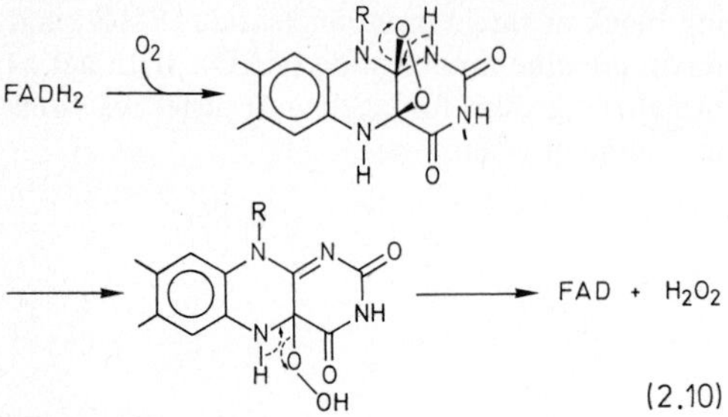

FADH2 → (O2) → … → FAD + H2O2 (2.10)

2.3.2.2 Hemin

Peroxidases from food of plant origin and several catalases (cf. 2.8.2.1.2) contain ferri-protoporphyrin III (hemin) as their prosthetic group and as the chromophore responsible for the brown color of the enzymes:

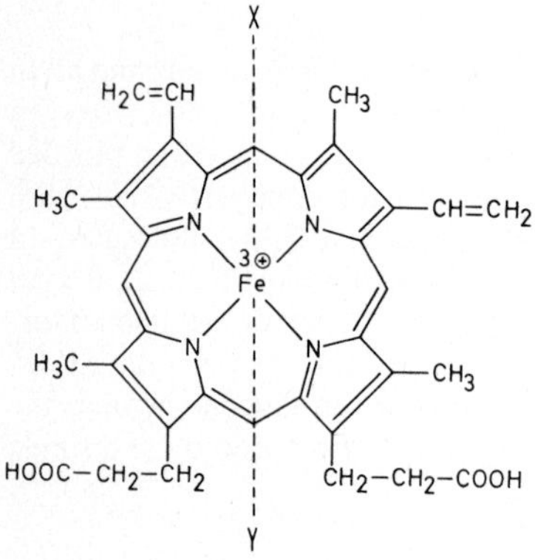

(2.11)

In catalytic reactions there is a change in the electron excitation spectra of the peroxidases (Figure 2.5a) which is caused by a valence change of the iron ion (Figure 2.6b). Intermediary compounds I (green) and II (pale red) are formed during this change.

Some verdoperoxidases, which are green in color (as suggested by their name) and found in various foods of animal origin, e.g. milk, contain an unidentified Fe-protoporphyrin as their prosthetic group.

2.3.2.3 Pyridoxal Phosphate

Pyridoxal phosphate (I) and pyridoxamine (II), derived from it, are designated as vitamin B_6 (cf. 6.3.3) and are essential ingredients of food:

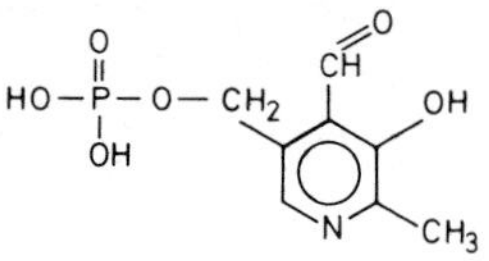

(2.12a)

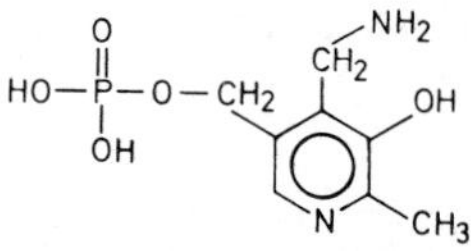

(2.12b)

Coupled to the enzyme as a prosthetic group through a lysyl residue, pyridoxal phosphate is involved in conversion reactions of amino acids. In the first catalyzed reaction step, the amino group of the amino acid substrate displaces the 6-amino group of lysine from the aldimine Schiff base-like linkage (cf. Reaction 2.13).

The positively-charged pyridine ring then exerts an electron shift towards the α-C-atom of the amino acid substrate; the shift being supported by the release of one substituent of the α-C-atom. In Figure 2.6 is shown how the ionization of the proton attached to the α-C-atom leads to *transamination* of the amino acid with formation of an α-keto acid. The reaction may also proceed through a *decarboxylation* (Figure 2.6) and yield

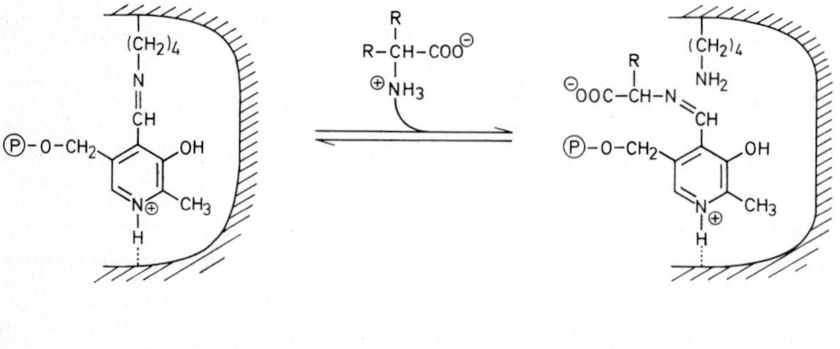

(2.13)

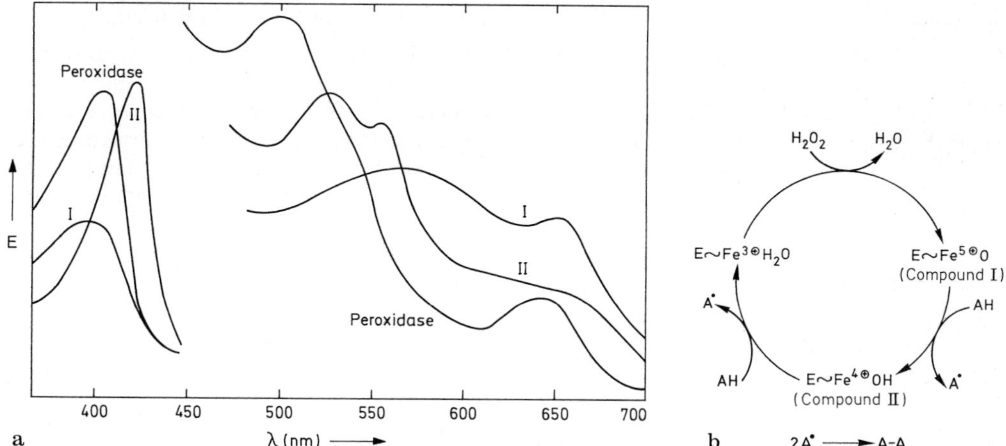

Fig. 2.5. Peroxidase reactions with H_2O_2 and a hydrogen donor (AH). **a** Electron excitation spectra of peroxidase and intermediates I and II; **b** Change in valence of the central Fe-atom of the enzyme during catalysis

an amine. Which of these two pathway options will prevail is decided by the structure of the protein moiety of the enzyme.

2.3.3 Metal Ions

Metal ions are indispensible cofactors for many enzymes. They can contribute to the stabilization of the enzyme conformation or can be involved in substrate binding to participate in catalytic reactions in the form of a *Lewis* acid or play the role of an electron carrier. Only the most important ions will be discussed.

2.3.3.1 Magnesium, Calcium and Zinc

Mg^{2+} ions activates some enzymes involved in the hydrolysis of phosphoric acid ester bonds (e.g. phosphatases; cf. Table 2.4) or is involved in the transfer of phosphate residues from ATP to a suitable acceptor (e.g. kinases; cf. Table 2.4). In both cases Mg^{2+} ions acting as an electrophilic *Lewis* acid, polarize the P−O-linkage of the phosphate residue of the substrate or cosubstrate and, thus, facilitate a nucleophilic attack (water with hydrolases; ROH in the case of kinases). An example is the hexokinase enzyme (cf. Table 2.13) which, in glycolysis, is involved in catalyzing the phosphorylation of glucose to glucose-6-phosphate with ATP as cosubstrate. The effect of a Mg^{2+} ion within the enzyme-substrate complex is obvious from the following formulation:

Adenosin−P−O−P−O−P ··· $Mg^{2\oplus}$ ··· glucose (H←B enzyme) (2.14)

Ca^{2+} ions are weaker *Lewis* acids than Mg^{2+} ions. Therefore, the replacement of Mg^{2+} by Ca^{2+} may result in an inhibition of the kinase enzymes. Enhancement of the activity of other enzymes by Ca^{2+} is based on the ability of the ion to interact with the negatively-charged site of the amino acid residues and, thus, to bring about stabilization of the enzyme conformation (e.g. α-amylase; cf. 2.8.2.2.2). The activation of the enzyme may be also caused by the involvement of the Ca^{2+} ion in substrate binding (e.g. lipase; cf. 3.7.1.1).

The Zn^{2+} ion, from the series of transition metals, is a cofactor which is not involved in redox reactions under physiological conditions. As a *Lewis* acid similar in strength to Mg^{2+}, Zn^{2+} participates in similar reactions. Hence, substituting the Zn^{2+} ion for the Mg^{2+} ion in some enzymes is possible without loss of enzyme activity.

Both metal ions can function as stabilizers of enzyme conformation and their direct participa-

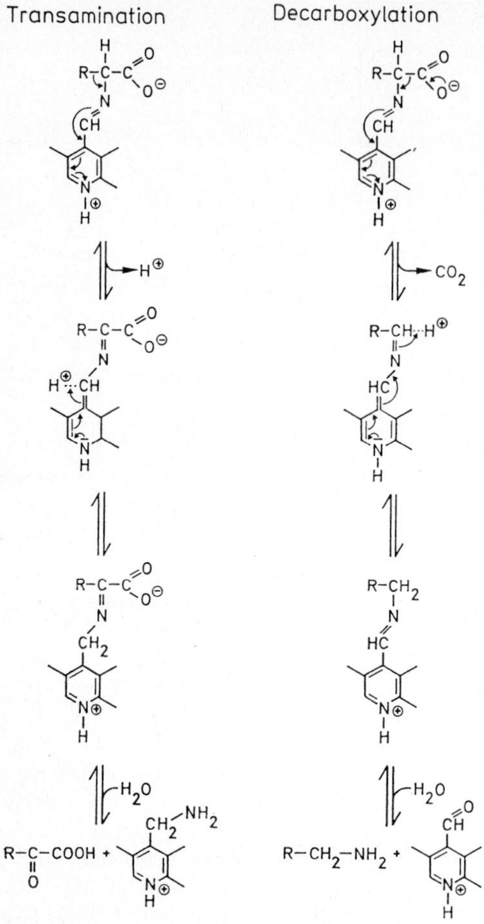

Transamination Decarboxylation

Fig. 2.6. The role of pyridoxal phosphate in transamination and decarboxylation of amino acids

tion in catalysis is readily revealed in the case of alcohol dehydrogenase. This enzyme isolated from horse liver consists of two identical polypeptide chains, each with one active site (cf. Table 1.19a). Two of the four Zn^{2+} ions in the enzyme readily dissociate. Although this dissociation has no effect on the quaternary structure, the enzyme activity is lost. As described under section 2.3.1.1, both of these Zn^{2+} ions are involved in the formation of the active site. In catalysis they polarize the substrate's C–O linkage and, thus, facilitate the transfer of hydride ions from or to the cosubstrate. Unlike the dissociable ions, removal of the two residual Zn^{2+} ions is possible only under drastic conditions, namely disruption of the enzyme's quaternary structure which is maintained by these two ions.

2.3.3.2 Iron, Copper and Molybdenum

The redox system of Fe^{3+}/Fe^{2+} covers a wide range of potentials (Table 2.5) depending on the attached ligands. Therefore, the system is exceptionally suitable for bridging the large potential differences in a stepwise electron transport system. Such an example is encountered in the transfer of electrons by the cytochromes as members of the respiratory chain (cf. textbook of biochemistry) or in the biosynthesis of unsaturated fatty acids (cf. 3.2.4), and by some individual enzymes.

The Fe-containing enzymes are attributed either to the heme (examples in 2.3.2.2) or to the nonheme Fe-containing proteins. The latter case is exemplified by lipoxygenase, for which the mechanism of activity is illustrated in section 3.7.2.2, or by xanthine oxidase.

Xanthine oxidase from milk (MW = 275 kdal) reacts with many electron donors and acceptors. However, this enzyme is most active with substrates such as xanthine or hypoxanthine as electron donors and molecular oxygen as the electron acceptor. The enzyme is assumed to have two active sites per molecule, with each having 1 FAD moiety, 4 Fe-atoms and 1 Mo-atom. During the oxidation of xanthine to uric acid:

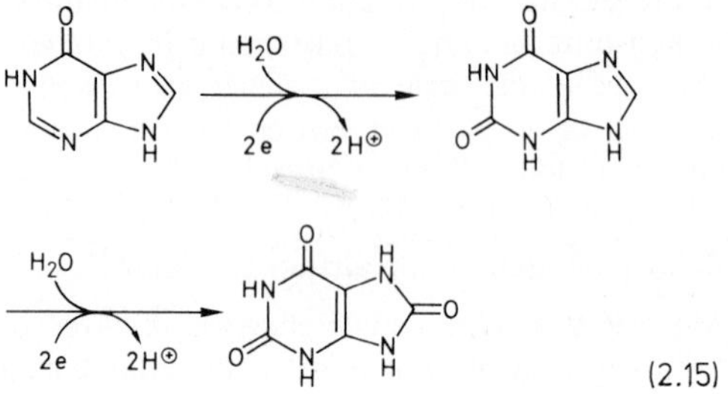

(2.15)

oxygen is reduced by two one-electron steps to H_2O_2 by an electron transfer system in which the following valence changes occur:

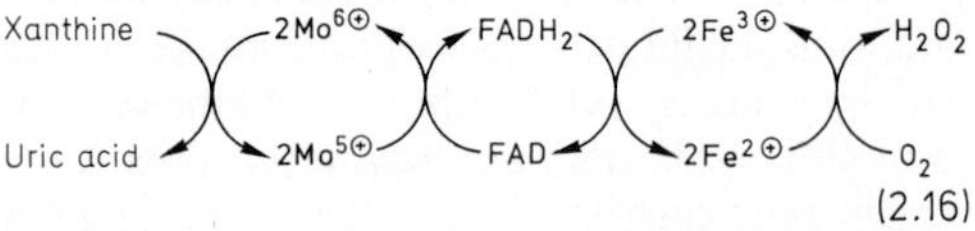

Xanthine / Uric acid ⇄ $2Mo^{6\oplus}$ / $2Mo^{5\oplus}$ ⇄ $FADH_2$ / FAD ⇄ $2Fe^{3\oplus}$ / $2Fe^{2\oplus}$ ⇄ H_2O_2 / O_2 (2.16)

Under certain conditions the enzyme releases a portion of the oxygen when only one electron transfer has been completed. This yields $O_2^{\ominus}$, the

Table 2.5. Redox potentials of Fe^{3+}/Fe^{2+} complex compounds at pH 7 (25 °C) as affected by the ligand

Redox-System	E_0' (Volt)
$[Fe^{III}(o\text{-}phen^{a})_3]^{3+}/[Fe^{II}(o\text{-}phen)_3]^{2+}$	+1.10
$[Fe^{III}(OH_2)_6]^{3+}/[Fe^{II}(OH_2)_6]^{2+}$	+0.77
$[Fe^{III}(CN)_6]^{3-}/[Fe^{II}(CN)_6]^{4-}$	+0.36
Cytochrome a (Fe^{3+})/Cytochrome a (Fe^{2+})	+0.29
Cytochrome c (Fe^{3+})/Cytochrome c (Fe^{2+})	+0.26
Hemoglobin (Fe^{3+})/Hemoglobin (Fe^{2+})	+0.17
Cytochrome b (Fe^{3+})/Cytochrome b (Fe^{2+})	+0.04
Myoglobin (Fe^{3+})/Myoglobin (Fe^{2+})	0.00
$(Fe^{III}EDTA)^{1-}/(Fe^{II}EDTA)^{2-}$	−0.12
$(Fe^{III}(oxin^{b})_3)/(Fe^{II}(oxin)_3)^{1-}$	−0.20
Ferredoxin (Fe^{3+})/Ferredoxin (Fe^{2+})	−0.40

[a] o-phen: o-Phenanthroline.
[b] oxin: 8-Hydroxyquinoline.

superoxide ion, with one unpaired electron. This ion can initiate lipid peroxidation by a chain reaction (cf. 3.7.3.1.4); hence, participation of xanthine oxidase in the generation of an "oxidation" flavor in milk is under investigation.
Phenol oxidases (cf. 18.1.2.5.7) and ascorbic acid oxidase, which occur in food, are known to have a Cu^{2+}/Cu^{1+} redox system as a prosthetic group.

2.4 Theory of Enzyme Catalysis

It has been illustrated with several examples (Table 2.1) that enzymes are substantially better catalysts than are protons or other ionic species used in nonenzymatic reactions. Enzymes invariably surpass all chemical catalysts in relation to substrate and reaction specificities.
Theories have been developed to explain the exceptional efficiency of enzyme activity. They are based on findings which provide only indirect insight into enzyme catalysis. Examples are the identification of an enzyme's functional groups involved in catalysis, elucidation of their arrangement within the tertiary structure of the enzyme, and the detection of conformational changes induced by substrate binding. Complementary studies involve low molecular weight model substrates, the reactions of which shed light on the active sites or groups of the enzyme and their coordinated interaction with other factors affecting enzymatic catalysis.

2.4.1 Active Site

An enzyme molecule is, when compared to its substrate, often larger in size by a factor of several orders of magnitude. For example, glucose oxidase is 150 kdal, while glucose is only 180 dal. This strongly suggests that in catalysis only a small locus of an active site has direct contact with the substrate. The active site is formed from amino acid residues which bind the substrate and, if required, the cofactors which assist in conversion of substrate to product. Correspondingly, the active site is composed of a binding locus and a transforming locus.

2.4.1.1 Active Site Localization

Several methods are generally used for the identification of amino acid residues present at the active site since data are often equivocal. Once obtained, the data must still be interpreted with a great deal of caution and insight.
The influence of pH on the activity assay (cf. 2.5.3) provides the first direct answer as to whether dissociable amino acid side chains, in charged or uncharged form, assist the catalysis. The data readily obtained from this assay must again be interpreted cautiously since neighboring charged groups, hydrogen bonds or the hydrophobic environment on the active site can affect the extent of dissociation of the amino acid residues and, thus, can shift their pK values (cf. 1.4.3.1).
Selective labeling of side chains which form the active site is also possible by chemical modification. When an enzyme is incubated with reagents such as iodoacetic acid (cf. 1.2.4.3.5) or dinitrofluorobenzene (cf. 1.2.4.2.2), resulting in a decrease of activity, and subsequent analysis of the modified enzyme shows that only one of the several available functional groups is bound to reagent (e.g. one of several −SH groups), then this group is most probably part of the active site.
Selective labeling data when an inhibiting substrate analogue is used are more convincing. Because of its similarity to the chemical structure of the substrate, the analogue will be bound covalently to the enzyme but not converted into product. We will consider the following examples:
N-tosyl-L-phenylalanine ethyl ester (Formula 2.17) is a suitable substrate for the proteinase chymotrypsin which hydrolyzes ester bonds.

When the ethoxy group is replaced by a chloromethyl group, an inhibitor whose structure is similar to the substrate is formed (N-tosyl-L-phenylalanine chloromethylketone, TPCK; cf. Formula 2.18).

$$C_6H_5-CH_2-CH(NH-SO_2-C_6H_4-CH_3)-CO-O-CH_2-CH_3 \qquad (2.17)$$

$$C_6H_5-CH_2-CH(NH-SO_2-C_6H_4-CH_3)-CO-CH_2-Cl \qquad (2.18)$$

Thus, the substrate analogue binds specifically and irreversibly to the active site of chymotrypsin. Analysis of the enzyme inhibitor complex reveals that, of the two histidine residues present in chymotrypsin, only His^{57} is alkylated at one of its ring nitrogens. Hence, the modified His residue is part of the active site (cf. mechanism of chymotrypsin catalysis, Figure 2.15). TPCK binds highly specifically, thus the proteinase trypsin is not inhibited. The corresponding inhibiting substrate analogue, which binds exclusively to trypsin, is N-tosyl-L-lysine chloromethylketone (TLCK):

$$H_2N-(CH_2)_4-CH(NH-SO_2-C_6H_4-CH_3)-CO-CH_2-Cl \qquad (2.19)$$

Reaction of diisopropylfluorophosphate (DIFP)

$$[(CH_3)_2CH-O]_2P(=O)F \qquad (2.20)$$

with a number of proteinases and esterases alkylates the unusually reactive serine residue at the active site. Thus, of the 28 serine residues present in chymotrypsin, only Ser^{195} is alkylated, while the other 27 residues are unaffected by the reagent. It appears that the reactivity of Ser^{195} is enhanced by its interaction with the neighboring His^{57} (cf. mechanism of catalysis in Figure 2.15). The participation of a carboxyl group at the active site in β-glucosidase catalysis has been confirmed with the help of conduritol B-epoxide, an inhibiting substrate analogue:

(2.21) [Conduritol B-epoxide (HO, OH substituted cyclohexane epoxide) attacked by the carboxylate ($-O^{\ominus}$, C=O) of the enzyme (Enzym) → ring-opened product esterified to the enzyme carboxyl group: cyclitol–O–C(=O)–Enzym]

A lysine residue is involved in enzyme catalysis in a number of lyase enzymes and in enzymes in which pyridoxal phosphate is the cosubstrate. An intermediary *Schiff* base product is formed between an ε-amino group of the enzyme and the substrate or pyridoxal phosphate (cf. 2.3.2.3). The reaction site is then identified by reduction of the *Schiff* base with $NaBH_4$.

An example of a "lysine" lyase is the aldolase enzyme isolated from rabbit muscle, the intermediary product of which is detected with dihydroxyacetone phosphate (cf. mechanism in Figure 2.17) as follows:

$$\text{ENZYME}-NH_2 + O=C(CH_2-O-PO_3H_2)(CH_2-OH) \rightleftharpoons \text{ENZYME}-N=C(CH_2-O-PO_3H_2)(CH_2OH)$$

$$\xrightarrow{NaBH_4} \text{ENZYME}-NH-CH(CH_2-O-PO_3H_2)(CH_2OH) \qquad (2.22)$$

$$\xrightarrow{\text{Hydrolysis}} HOOC(H_2N)CH-(CH_2)_4-NH-CH(CH_2-OH)(CH_2-OH)$$

2.4.1.2 Substrate Binding

2.4.1.2.1 "Lock and key" hypothesis

To explain substrate specificity, *E. Fischer* proposed a hypothesis a century ago in which he depicted the substrate as being analogous to a key and the enzyme as its lock. According to this model, the active site has a geometry which is complementary only to its substrate (Figure 2.7). In contrast, there are many possibilities for a "bad" substrate to be bound to the enzyme, but only one provides the properly positioned enzyme-substrate complex as illustrated in Figure 2.7, which is converted to the product. The rate of enzyme catalysis decreases.

The proteinases chymotrypsin and trypsin are two of the few enzymes for which secondary and tertiary structures have been elucidated by x-ray analysis and which have structures supporting the lock and key hypothesis to a certain extent. The binding site in chymotrypsin and trypsin is a three-dimensional hydrophobic pocket (Figure 2.8). Bulky amino acid residues such as aromatic amino acids fit neatly into the pocket (chymotrypsin, Figure 2.8,a), as do substrates with lysyl or arginyl residues (trypsin, Figure 2.8,b). Instead of Ser[189], the trypsin peptide chain has Asp[189] which is present in the deep cleft in the form of a carboxylate anion and which attracts the positively-charged lysyl or arginyl residues of the substrate. Thus, the substrate is stabilized and realigned by its peptide bond to face the enzyme's Ser[195] which participates in hydrolysis (transforming locus).

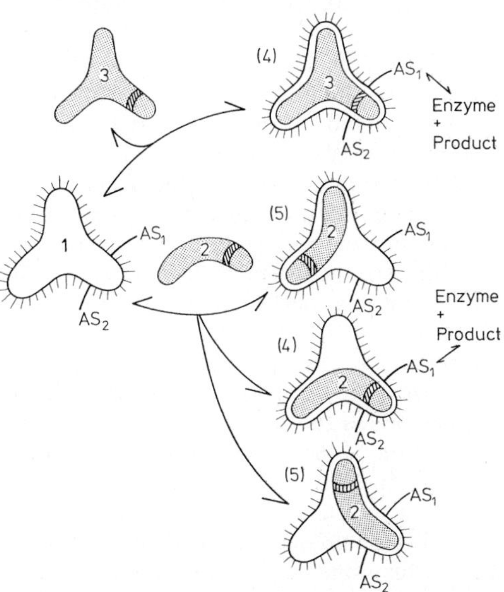

Fig. 2.7. Binding of a good (3) and of bad substrate (2) by the active site (1) of the enzyme. (After *W. P. Jencks*, 1969) (4) A productive enzyme-substrate complex; (5) a nonproductive enzyme substrate complex. As_1 and As_2: reactive amino acid residues of the enzyme involved in conversion of substrate to product

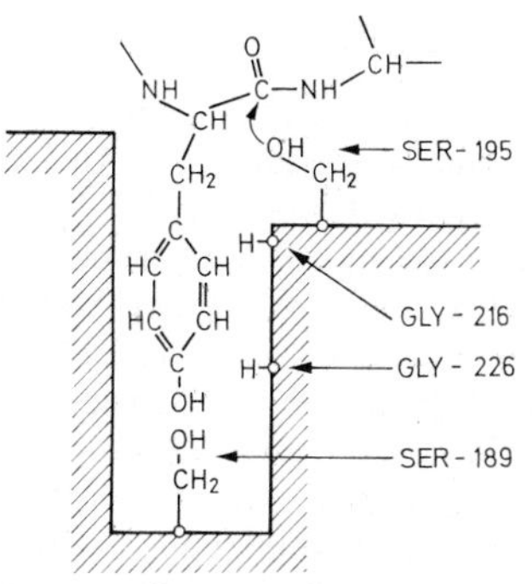

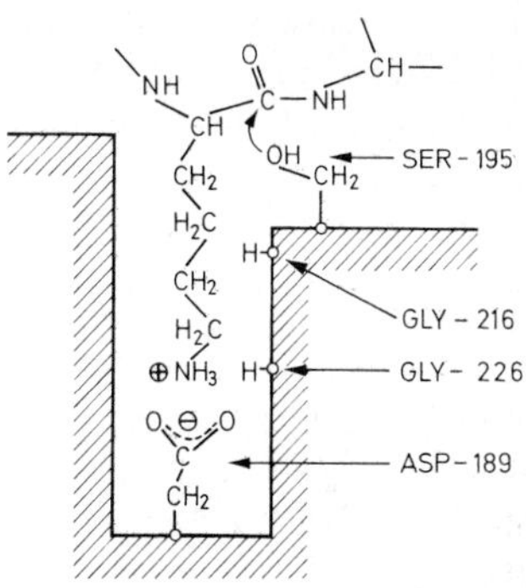

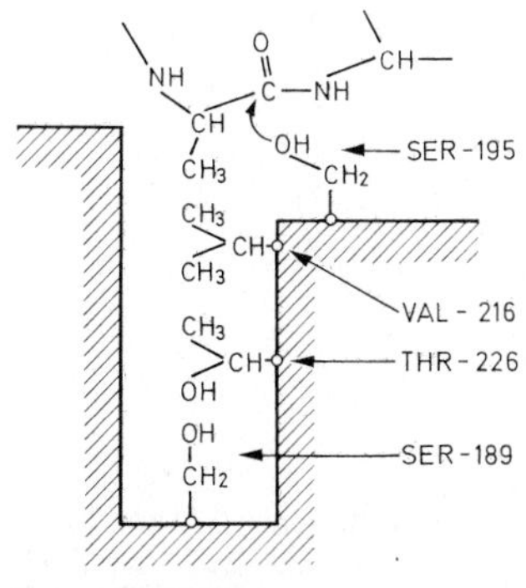

Fig. 2.8. A hypothesis about substrate binding by α-chymotrypsin, trypsin and elastase enzymes. (After *D. Shotton*, 1971)

The peptide substrate is hydrolyzed by the elastase enzyme by the same mechanism as for chymotrypsin. However, here the pocket is closed to such an extent by the side chains of Val^{216} and Thr^{226} that only the methyl group of alanine can enter the cleft (Figure 2.8,c). Therefore, elastase has specificity for alanyl peptide bonds or alanyl ester bonds.

2.4.1.2.2 Induced-fit model

The conformation of a number of enzymes is changed by the binding of the substrate. An example is carboxypeptidase A, in which the Try^{248} located in the active site moves approximately 12 Å towards the substrate, glycyl-L-phenylalanine, to establish contact. This and other observations support the dynamic induced-fit model proposed by *Koshland* (1964). Here, only the substrate has the power to induce a change in the tertiary structure of the enzyme. Thus, as the substrate molecule approaches the enzyme surface, the amino acid residues A and B change their positions to conform closely to the shape of the substrate (I, in Figure 2.9). Groups A and B are then in the necessary position for reaction with the substrate.

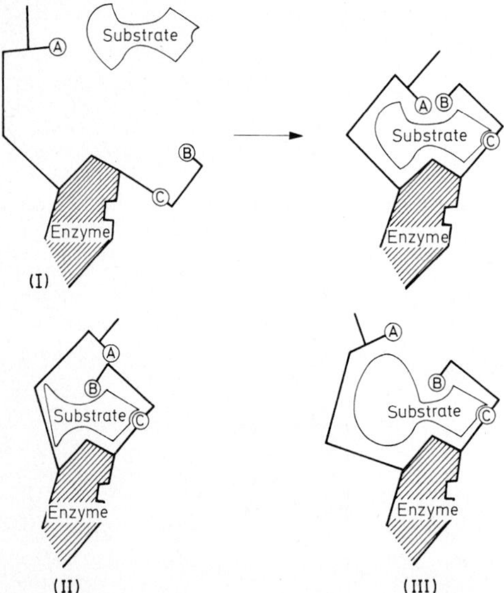

Fig. 2.9. A schematic presentation of "induced-fit model" for an active site of an enzyme. (After *D. E. Koshland,* 1964)
— polypeptide chain of the enzyme with catalytically active residues of amino acids, A and B; the residue C binds the substrate

Diagrams II and III (Figure 2.9) illustrate the case when the added compound is not suitable as substrate. Although group C positioned the substrate correctly at its binding site, the shape of the compound prevents groups A and B from being aligned properly in their active positions and, thus, from generating the product.

By accepting the working theories outlined above, one theory suitable for enzymes following the lock and key mechanism and the other theory for enzymes operating with the dynamic induced-fit model, the substrate specificity of any enzyme-catalyzed reaction can be explained satisfactorily.

In addition, the relationship between enzyme conformation and its catalytic activity thus outlined also accounts for the extreme sensitivity of the enzyme as catalyst. Even slight interferences imposed on their tertiary structure which affect the positioning of the functional groups result in loss of catalytic activity.

2.4.1.2.3 Stereospecificity

Enzymes react stereospecifically. Before being bound to the binding locus, the substrates are distinguished by their cis, trans-isomerism and also by their optical antipodes. The latter property was illustrated by the reaction of L(+)-lactic acid (Figure 2.2). There are distinct recognition areas on the binding locus. Alcohol dehydrogenase will be used to demonstrate this. This enzyme removes two hydrogen atoms, one from the methylene group and the other from the hydroxyl group, to produce acetaldehyde. However, the enzyme recognizes the difference between the two methylene hydrogens since it always stereospecifically removes the same hydrogen atom. For example, yeast alcohol dehydrogenase always removes the pro-R-hydrogen from the C-1 position of a stereospecifically deuterated substrate and transfers it to the C-4 position of the nicotinamide ring of NAD (cf. Reaction 2.23).

To explain the stereospecificity, it has been assumed that the enzyme must bind simultaneously to more than one point of the molecule. Thus, when two substituents (e.g. the methyl and hydroxyl groups of ethanol; Figure 2.10) of the prochiral site are attached to the enzyme surface at positions A and B, the position of the third substituent is fixed. Therefore, the same substituent will always be bound to reactive position

$$\mathrm{CH_3{-}C(D_R)(H_S){-}OH} + \text{(NAD)} \rightleftharpoons \mathrm{CH_3{-}CHO} + \text{(NADH)} + \mathrm{H^{\oplus}} \quad (2.23)$$

C, e.g. one of the two methylene hydrogens in ethanol. In other words, the two equal substituents in a symmetrical molecule are differentiated by asymmetric binding to the enzyme.

2.4.1.3 Effect of Substrate Binding on Reaction Rate

2.4.1.3.1 Steric effect – orientation effect

The specificity of substrate binding contributes substantially to the rate of an enzyme-catalyzed reaction. The following hypotheses, supported by several model assays, will be discussed.

Binding to the active site of the enzyme concentrates the reactant in comparison to its dilute solution. In addition, the reaction is now the favored one since binding places the substrate's susceptible reactive group in the proximity of the catalytically-active group of the enzyme.

Therefore the contribution of substrate binding to the reaction rate is partially due to a change in the molecularity of the reaction. The intermolecular reaction of the two substrates is replaced by an intramolecular reaction of an enzyme-substrate complex. The consequences can be clarified by using model compounds which have all the reactive groups within their molecules and, thus, are subjected to an intramolecular reaction. Their reactivity can then be compared with the corresponding bimolecular systems and the results expressed as a ratio of the reaction rates of the intramolecular (k_1) to the intermolecular (k_2) reactions and, based on their dimensions, are denoted as "effective molarity". As an example, let us consider the cleavage of p-bromophenylacetate in the presence of acetate ions, yielding acetic acid anhydride:

$$\mathrm{Br{-}C_6H_4{-}O{-}CO{-}CH_3 + CH_3COO^{\ominus} + Na^{\oplus} \longrightarrow Br{-}C_6H_4{-}O^{\ominus}Na^{\oplus} + (CH_3CO)_2O} \quad (2.24)$$

Intramolecular hydrolysis is substantially faster than the intermolecular reaction (Table 2.6). The effective molarity sharply increases when the reactive carboxylate anion is in close proximity to the ester carbonyl group and, by its presence, retards the mobility of the carbonyl group. Thus, the effective molarity increases (Table 2.6) as the C–C bond mobility decreases. Two bonds can rotate in a glutaric acid ester, whereas only one can rotate in a succinic acid ester. The free rotation is effectively blocked in a bicyclic system. Hence, the reaction rate is sharply increased. Here, the rigid steric arrangement of the acetate ion and of the ester group provides a configuration that imitates that of a transitional state.

In contrast to the examples given in Table 2.6, examples should be mentioned in which substrates are not bound covalently by their enzymes. The following model will demonstrate that other interactions can also promote close positioning of the two reactants.

Hydrolysis of p-nitrophenyldecanoic acid ester is catalyzed by an alkylamine:

$$\mathrm{C_9H_{19}{-}CO{-}O{-}C_6H_4{-}NO_2 + H_2O \xrightarrow{(R-NH_2)} C_9H_{19}{-}COOH + HO{-}C_6H_4{-}NO_2}$$

$$\mathrm{R{-}NH_2:\ C_{10}H_{21}{-}NH_2;\ C_2H_5{-}NH_2} \quad (2.25)$$

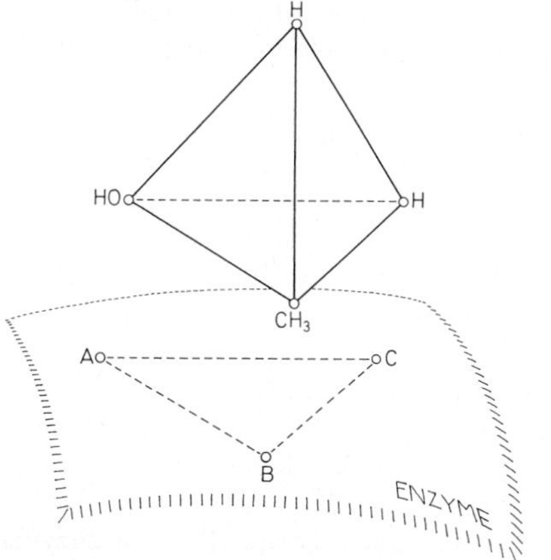

Fig. 2.10. A model for binding of a prochiral substrate (ethanol) by an enzyme

Table 2.6. Relative reaction rate for the formation of acid anhydrides

Model		k_1/k_2
I.	CH_3—COOR + CH_3—COO⊖	$9.5 \cdot 10^2$ M
II.	H_2C (CH_2—COOR) (CH_2—COO⊖)	$2.2 \cdot 10^5$ M
III.	H_2C—COOR, H_2C—COO⊖	$5 \cdot 10^7$ M
IV.	O, COOR, COO⊖	
R: Br—C_6H_4—		

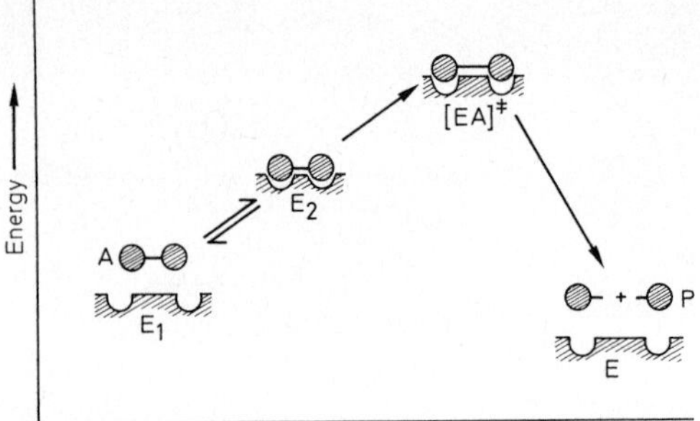

Fig. 2.11. Covalent bond deformation in an enzyme-substrate complex. (After *W. P. Jencks*, 1969)
E Enzyme in states 1 and 2; [EA]‡ enzyme-substrate-complex in transition state, A substrate, P product

The reaction rate in the presence of decylamine is faster by a factor of 700 than that in the presence of ethylamine. This implies that the reactive amino group has been oriented close to the vicinity of the susceptible carbonyl group of the ester by the establishment of a maximal number of hydrophobic contacts. Correspondingly, there is a decline in the reaction rate as the alkyl amine group is lengthened further.

2.4.1.3.2 Bond distortion

Distortion of the substrate's bonds once the substrate is bound to the active site of the enzyme enhances the reaction rate. As shown in Figure 2.11, and as outlined by one of the hypotheses, the binding site geometry of the enzyme (E_1) in its most stable conformation does not match the form of the substrate molecule. When the substrate binds, the enzyme assumes the induced-fit conformation (E_2) which is energy enriched and, thus, thermodynamically less favored, but is more suitable to bind the substrate. The binding forces between substrate and enzyme supply the energy required to reach the transition state. In this state the tendency of the enzyme is to return to its more stable, lower energy ground state. Therefore, it distorts the substrate's bonds and forms the intermediate transition state of the enzyme-substrate complex.

A chemical model which amply demonstrates the effect of strain in enhancing reaction rates is the hydrolysis of the following phosphoric acid diesters:

O=P(OH)(O—CH_2—CH_2—O) (I)   O=P(OH)(O—CH_3)(O—CH_3) (II)   (2.26)

Compound I hydrolyzes approximately 10^8 times faster than Compound II.

2.4.1.3.3 Entropy effect

An interpretation in thermodynamic terms takes into account that a loss of entropy due to the loss of freedom of rotation and translation of the reactants occurs during catalysis. This entropy effect is probably quite large in the case of the formation of an enzyme-substrate complex since the reactants are fairly rigidly positioned before the transition state is reached. In consequence the conversion of the enzyme-substrate complex to the transition state is accompanied by little or no change of the entropy. As an example a reaction running at 27 °C with a decrease in entropy of 140 J K^{-1} $mole^{-1}$ is considered. Calculations indicate that the difference in free energy between the ground and transition states is reduced by about 43 kJ. This value falls in the range of the amount by which the activation energy of a reaction is lowered by an enzyme (cf. Table 2.1) and which can have the effect of increasing the reaction rate by a factor of 10^8.

The catalysis by chymotrypsin, for example, shows how powerful the entropy effects can be. In section 2.4.2.2 we will see that this catalysis is a two-step event proceeding through an acylated-enzyme intermediate. Here we will consider only the second step, deacylation, thereby distinguishing the following intermediates:

a) N-acetyl-L-tyrosyl-chymotrypsin
b) Acetyl-chymotrypsin

In case a) deacylation is faster by a factor of 3540 since the carbonyl group is immobilized by insertion of the bulky N-acetyl-L-tyrosyl group into a hydrophobic pocket on the enzyme (Figure 2.12,a) at the correct distance from the attacking nucleophilic OH^- ion derived from water (cf. 2.4.2.2). In case b) the immobilization of the small acetyl group is not possible (Figure 2.12,b) so that the difference between the ground and transition states is very large. The closer the ground state is to the transition state, the more positive will be the entropy of the transition state, $\Delta S^{\#}$; a fact that as mentioned before lead to a considerable increase in reaction rate. The thermodynamic data in Table 2.7 show that the difference in reaction rates depends, above all, on an entropy effect; the enthalpies of the transition states scarcely differ.

Table 2.7. Thermodynamic data for transition states of two acyl-chymotrypsins

Acyl-enzyme	$\Delta G^{\#}$ (kJ·mole^{-1})	$\Delta H^{\#}$ (kJ·mole^{-1})	$\Delta S^{\#}$ (J·K^{-1}·mole^{-1})
N-Acetyl-L-tyrosyl	59.6	43.0	−55.9
Acetyl	85.1	40.5	−149.7

2.4.2 Reaction Mechanisms

Reaction steps which lead to the conversion of an enzyme-substrate complex to an enzyme-product complex are generally classified as acid-base catalysis, covalent-type catalysis or redox catalysis.

2.4.2.1 General Acid-Base Catalysis

When the reaction rate is affected by the concentration of hydronium ($H_3O^{\oplus}$) or $OH^{\ominus}$ ions from water, the reaction is considered to be specifically

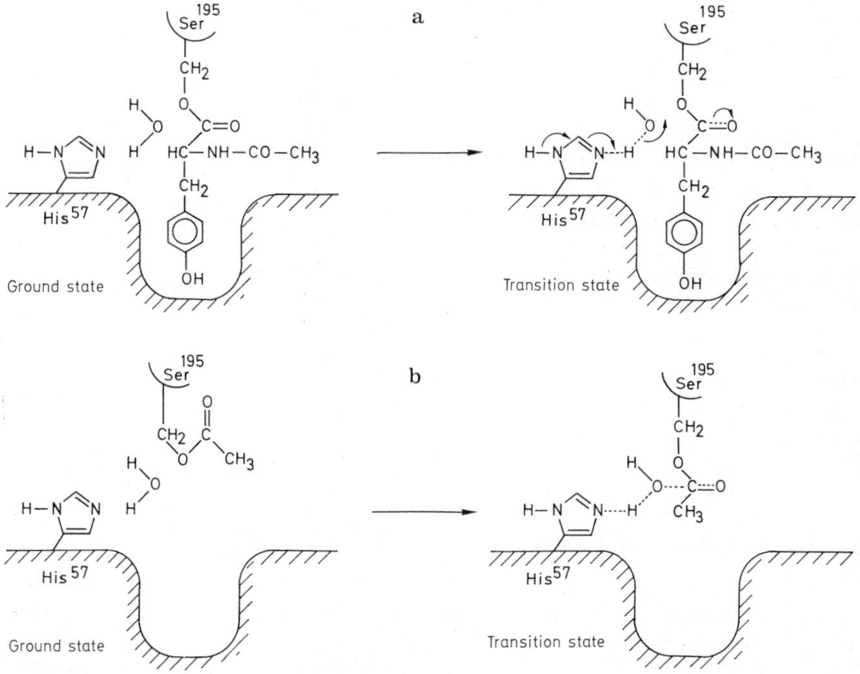

a

Ground state

Transition state

b

Ground state

Transition state

Fig. 2.12. Influence of the steric effect on deacylation of two acyl-chymotrypsins. (After *M. L. Bender* et al., 1964). **a** N-acetyl-L-tyrosyl-chymotrypsin, **b** acetyl-chymotrypsin

acid or base catalyzed. In the so-called general acid or base catalysis the reaction rate is affected by prototropic groups located on the side chains of the amino acid residues. These groups involve proton donors (denoted as general acids) and proton acceptors (general bases). Most of the amino acids located on the active site of the enzyme influence the reaction rate by general acid-base catalysis.

The effect of such general acid-base catalysis is elegantly illustrated by the rate of mutarotation, a non-enzyme-catalyzed reaction. The mutarotation of a glucose derivative is examined in benzene instead of in water in order to exclude the influence of specific acid-base catalysis:

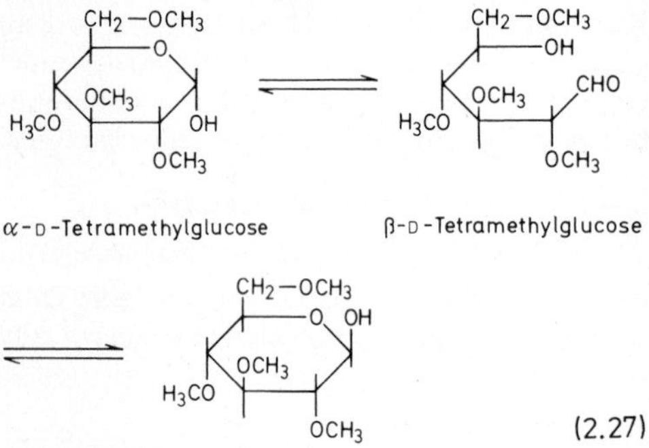

α-D-Tetramethylglucose ⇌ β-D-Tetramethylglucose ⇌ (2.27)

As seen from the data given in Table 2.8, the rate of mutarotation in benzene is slow. When a general base (pyridine) or a general acid (p-cresol) is added, the rate is increased. However, a substantial increase in mutarotation rate is obtained when both are added. The increase is larger than the sum of the individual effects. It is a concerted general acid and general base catalysis that is involved in opening the pyranose ring:

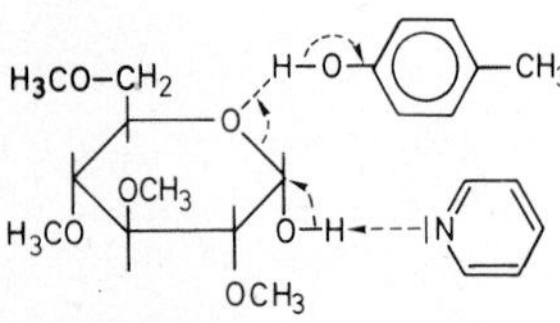

(2.28)

Table 2.8. Relative rate of mutarotation of tetramethylglucose (0.09 M) in benzene

Catalyst	k_{rel}	Catalyst	k_{rel}
Absent	1.0	0.1 M Pyridine + 0.1 M p-cresol	101
0.1 M Pyridine	5.0		
0.1 M p-Cresol	5.4		

As already mentioned, the amino acid residues in enzymes have prototropic groups which have the potential to act as a general acid or as a general base. Of these, the imidazole ring of histidine is of special interest since it can perform both functions simultaneously:

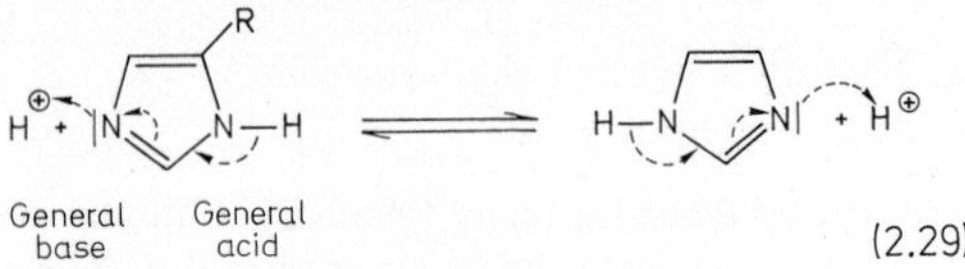

General base — General acid (2.29)

The imidazole ring ($pK_2 = 6.1$) can cover the pH range optima of many enzymes.

Thus, two histidine residues are involved in the catalytic activity of ribonuclease, a phosphodiesterase. The enzyme hydrolyzes pyrimidine-2′,3′-cyclic phosphoric acids. As shown in Figure 2.13, cytidine-2′,3′-cyclic phosphoric acid is positioned between two imidazole groups at the binding locus of the active site. His[12] serves as a general base, removing the proton from a water molecule. This is followed by nucleophilic attack of the intermediary $OH^{\ominus}$ ions on the electrophilic phosphate group. This attack is supported by the concerted action of the general acid His[119].

Another concerted general acid-base catalysis is illustrated by triose phosphate isomerase, an enzyme involved in glycolysis. Here, the concerted action involves the carboxylate anion of a glutamic acid residue as a general base with a general acid which has not yet been identified:

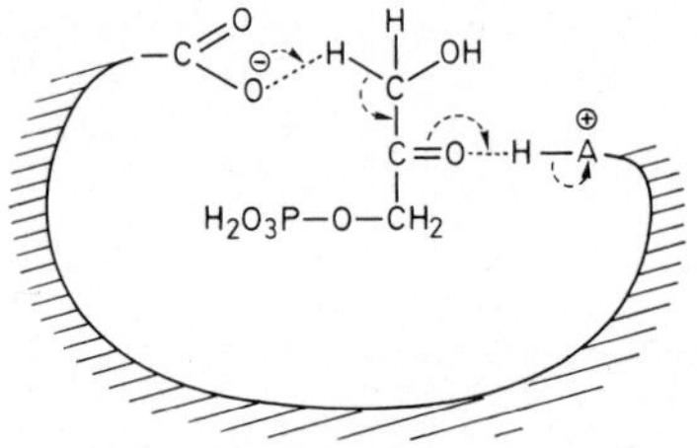

(2.30)

The endiol formed from dihydroxyacetone-3-phosphate in the presence of enzyme isomerizes into glyceraldehyde-3-phosphate. A characteristic feature for general acid-base catalysis is obvious from these two examples. The rate limiting step of the reaction is the rate of proton transfer between the prototropic groups involved in the reaction. Moreover, covalent bond formation is not involved between the enzyme and its substrate.

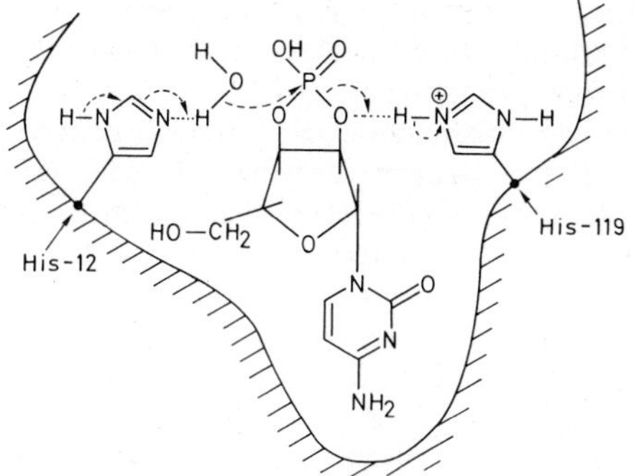

Fig. 2.13. Hydrolysis of cytidine-2′,3′-phosphate by ribonuclease. (Reaction mechanism after *D. Findlay*, 1962)

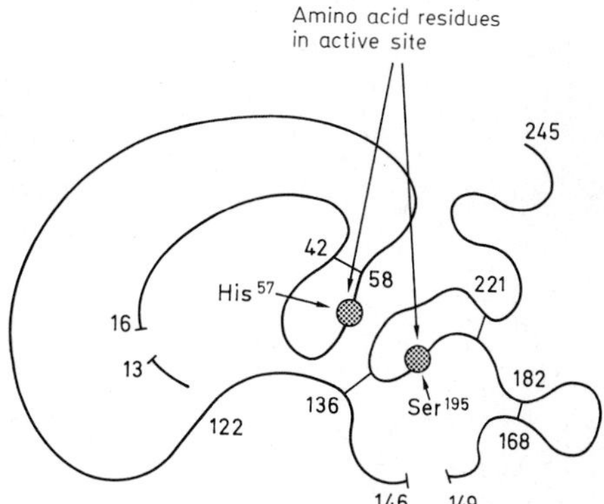

Fig. 2.14. Polypeptide chain conformation in the chymotrypsin molecule. (After *A. L. Lehninger*, 1977)

2.4.2.2 Covalent Catalysis

Studies aimed at identifying the active site of an enzyme (cf. 2.4.1.1) have shown that, during catalysis, a number of enzymes bind the substrate by covalent linkages.

Examples of enzyme functional groups which are involved in covalent bonding and are responsible for the transient intermediates of an enzyme-substrate complex are compiled in Table 2.9. Nucleophilic catalysis is dominant (examples 1–6. Table 2.9), since amino acid residues are present in the active site of these enzymes, which react only with substrate by donating an electron pair (nucleophilic catalysis). Electrophilic reactions occur mostly by involvement of carbonyl groups (example 7, Table 2.9) or with the help of metal ions.

Table 2.9. Examples of covalent enzyme substrate linked intermediates

Enzyme	Reactive functional group	Intermediate
1. Chymotrypsin	HO-(Serine)	Acylated enzyme
2. Papain	HS-(Cysteine)	Acylated enzyme
3. β-Amylase	HS-(Cysteine)	Maltosyl-enzyme
4. Aldolase	ε-H_2N-(Lysine)	*Schiff* base
5. Alkaline phosphatase	HO-(Serine)	Phosphoenzyme
6. Glucose-6-phosphatase	Imidazole-(Histidine)	Phosphoenzyme
7. Histidine decarboxylase	O=C (Pyruvate)	*Schiff* base

A number of proteinase and esterase enzymes react covalently in substitution reactions by a two-step nucleophilic mechanism. In the first step, the enzyme is acylated; in the second step, it is deacylated. Chymotrypsin will be discussed as an example of this reaction mechanism. Its activity is dependent on His^{57} and Ser^{195}, which are positioned in close proximity within the active site of the enzyme because of folding of the peptide chain (Figure 2.14).

Because Asp^{102} is located in hydrophobic surroundings, it can polarize the functional groups in close proximity to it. Thus, His^{57} acts as a strong general base and abstracts a proton from the OH-group of the neighboring Ser^{195} residue (step ‘a’, Figure 2.15). The oxygen remaining on Ser^{195} thus becomes a strong nucleophile and attacks the carbon of the carbonyl group of the peptide bond of the substrate. At this stage an amine (the first product) is released (step ‘b’, Figure 2.15) and the transient covalently-bound acyl enzyme is formed. A deacylation step follows. The previous position of the amine is occupied by a water molecule. Again, His^{57}, through support from Asp^{102}, serves as a general base, abstracting the proton from water (step ‘c’, Figure 2.15). This is followed by nucleophilic attack of the resultant $OH^{\ominus}$ ion on the carbon of the carbonyl group of the acyl enzyme (step ‘d’, Figure 2.15), resulting in free enzyme and the second product of the enzymic conversion.

An exceptionally reactive serine residue has been identified in a great number of hydrolase enzymes, e.g. trypsin, subtilisin, elastase, acetylcholine esterase and some lipases. These enzymes

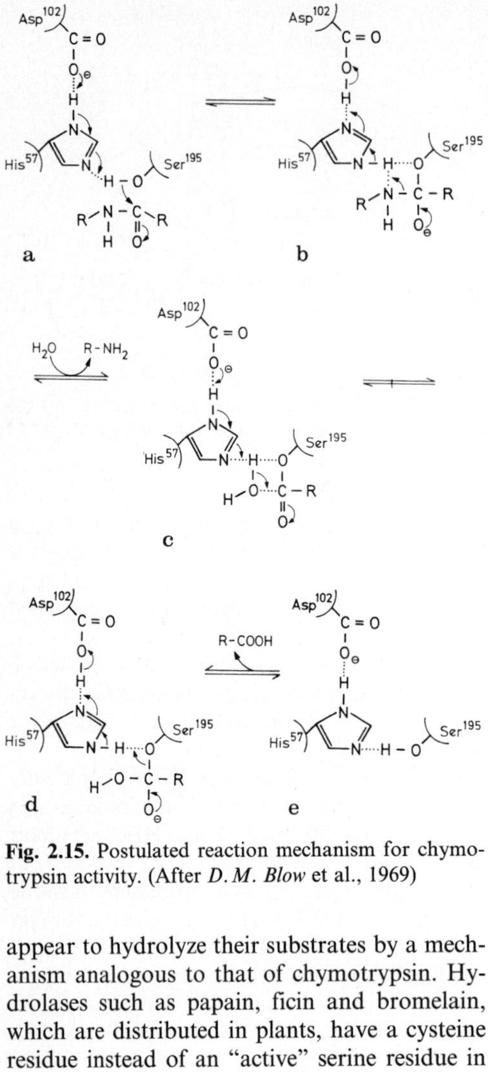

Fig. 2.15. Postulated reaction mechanism for chymotrypsin activity. (After *D. M. Blow* et al., 1969)

appear to hydrolyze their substrates by a mechanism analogous to that of chymotrypsin. Hydrolases such as papain, ficin and bromelain, which are distributed in plants, have a cysteine residue instead of an "active" serine residue in their active sites, thus, the transient intermediates are thioesters.

Enzymes involved in the cleavage of carbohydrates can also function by the above mechanism. Figure 2.16 shows that amylose hydrolysis by β-amylase occurs with the help of four functional groups in the active site. The enzyme-substrate complex is subjected to a nucleophilic attack by an SH-group on the carbon involved in the α-glycosidic bond. This transition step is facilitated by the carboxylate anion in the role of a general base and by the imidazole ring as an acid which donates a proton to glycosidic oxygen. In the second transition state the imidazole ring, as a general base in the presence of a water molecule, helps to release maltose from the maltosylenzyme intermediate.

Lysine is another amino acid residue actively involved in covalent enzyme catalysis (cf. 2.4.1.1). Many lyases react covalently with a substrate containing a carbonyl group. They catalyze, for example, aldol or retroaldol condensations important for the conversion and cleavage of monosaccharides or for decarboxylation reactions of β-keto acids. An example of the details of the reaction involved will be considered for aldolase (Figure 2.17). The enzyme-substrate complex is first stabilized by electrostatic interaction between the phosphate residues of the substrate and the charged groups present on the enzyme. A covalent intermediate, a *Schiff* base, is then formed by nucleophilic attack of the ε-amino group of the "active" lysine on a carbonyl group of the substrate. The *Schiff* base cation facilitates the retroaldol cleavage of the substrate, whereas a negatively-charged group on the enzyme (e.g. a thiolate or carboxylate anion) acts as a general base, i.e. binds the free proton (step 'b' in Figure 2.15). Thus, the first product, glyceraldehyde-3-phosphate, is released. An enamine rearrangement into a ketimine structure is followed by release of dihydroxyacetone phosphate.

This is the mechanism of catalysis by aldolases which occur in plant and animal tissues (lysine aldolases or class I aldolases). A second group of these enzymes often produced by microorganisms contains a metal ion (metallo-aldolases). This group is involved in accelerating retroaldol condensations through electrophilic reactions with carbonyl groups.

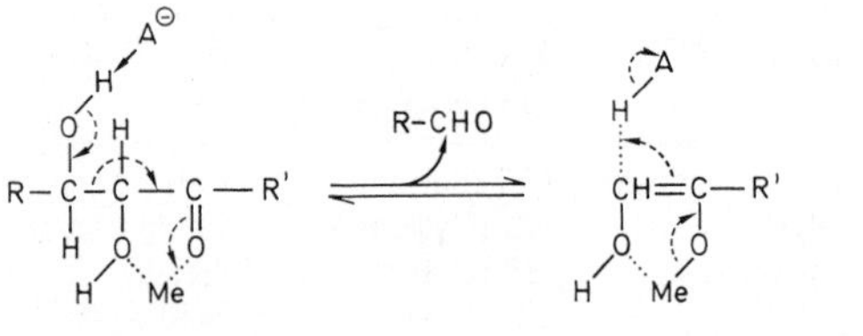

Me: Metal ion, probably $Zn^{2\oplus}$ (2.31)

Other examples of electrophilic metal catalysis are given under section 2.3.3.1. Electrophilic reactions are also carried out by enzymes which have an α-keto acid (pyruvic acid or α-keto butyric acid) at the transforming locus of their ac-

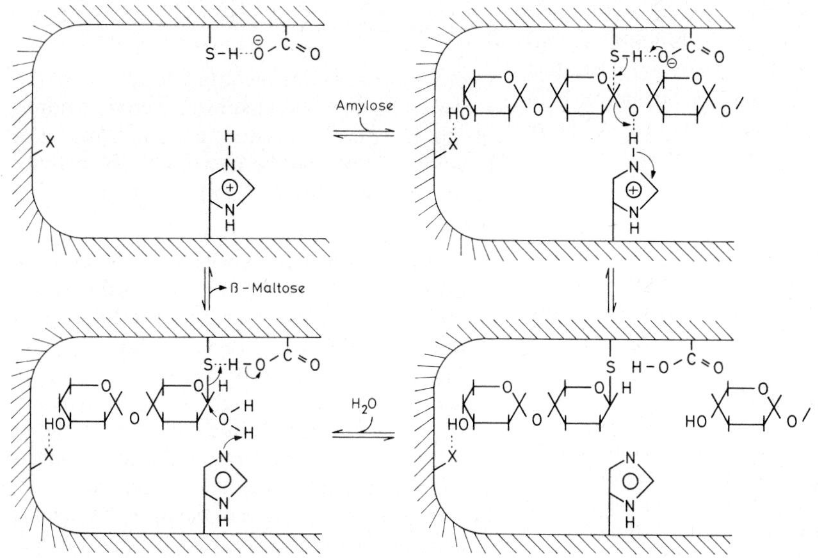

Fig. 2.16. Postulated mechanism for hydrolysis of amylose by β-amylase

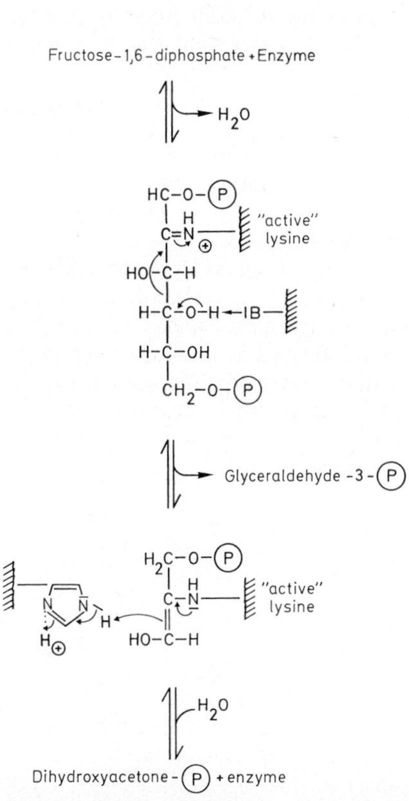

Fructose-1,6-diphosphate + Enzyme

H_2O

"active" lysine

Glyceraldehyde-3-(P)

"active" lysine

H_2O

Dihydroxyacetone-(P) + enzyme

Fig. 2.17. Aldolase of rabbit muscle tissue. A model for its activity; P: PO_3H_2

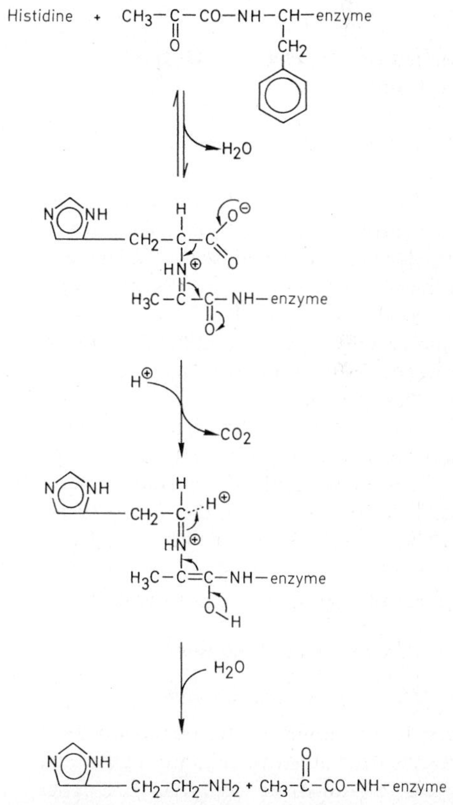

Histidine + $CH_3-C(=O)-CO-NH-CH(CH_2C_6H_5)-$enzyme

H_2O

$H^{\oplus}$

CO_2

H_2O

$CH_2-CH_2-NH_2$ + $CH_3-C(=O)-CO-NH-$enzyme

Fig. 2.18. A proposed mechanism for reaction of histidine decarboxylase

tive site. One example of such an enzyme is histidine decarboxylase in which the N-terminal amino acid residue is bound to pyruvate. Histidine decarboxylation is initiated by formation of a *Schiff* base by the reaction mechanism in Figure 2.18.

2.4.2.3 Redox Catalysis

The most important mechanisms involved in a redox-type enzyme catalysis have been dealt with under sections 2.3.1.1, 2.3.2.1, 2.3.2.2 and 2.3.3.2.

2.4.3 Closing Remarks

The hypotheses discussed here allow some understanding of the fundamentals involved in the action of enzymes. However, the knowledge is far away from the point where the individual or combined effects which regulate the rates of enzyme-catalyzed reactions can be calculated.

2.5 Kinetics of Enzyme-Catalyzed Reactions

Enzymes in food are detected only indirectly by measuring their catalytic activity and, in this way, their activities are differentiated from those of other enzymes. This is the rationale for acquiring knowledge needed to analyze the parameters which influence or determine the rate of an enzyme-catalyzed reaction.
The reaction rate is dependent on the concentrations of the components involved in the reaction. Here we mean primarily the substrate and the enzyme. Also, the reaction can be influenced by the presence of activators and inhibitors. Finally, the pH, the ionic strength of the reaction medium, the dielectric constant of the solvent (usually water) and the temperature exert an affect.

2.5.1 Effect of Substrate Concentration

2.5.1.1 Single-Substrate Reactions

2.5.1.1.1 Michaelis-Menten equation

Let us consider a single-substrate reaction. Enzyme E reacts with substrate A to form an intermediary enzyme-substrate complex, EA. The complex then forms the product P and releases the free enzyme:

$$E + A \underset{k_{-1}}{\overset{k_1}{\rightleftharpoons}} EA \xrightarrow{k_2} E + P \qquad (2.37)$$

In order to determine the catalytic activity of the enzyme, the decrease in substrate concentration or the increase in product concentration as a function of time can be measured. The activity curve obtained (Figure 2.19) has the following regions:

a) The maximum activity which occurs for a few msec until an equilibrium is reached between the rate of enzyme-substrate formation and rate of breakdown of this complex.
   Measurements in this pre-steady state region which provide an insight into the reaction steps and mechanism of catalysis are difficult and time consuming. Hence, further analysis of the pre-steady state will be ignored.
b) The usual procedure is to measure the enzyme activity when a steady state has been reached. In a steady state the intermediary complex concentration remains constant while the concentration of the substrate and end product are changing. For this state, the following is valid:

$$\frac{dEA}{dt} = -\frac{dEA}{dt} \qquad (2.38)$$

c) The reaction rate continuously decreases in this region in spite of an excess of substrate. The decrease in the reaction rate can be considered to be a result of:
   Enzyme denaturation which can readily occur; hence the enzyme concentration in the reaction system is decreasing continuously; or the product formed increasingly inhibits enzyme activity or, after the concentration of the product increases, the reverse reaction takes place, converting the product back into the initial reactant.

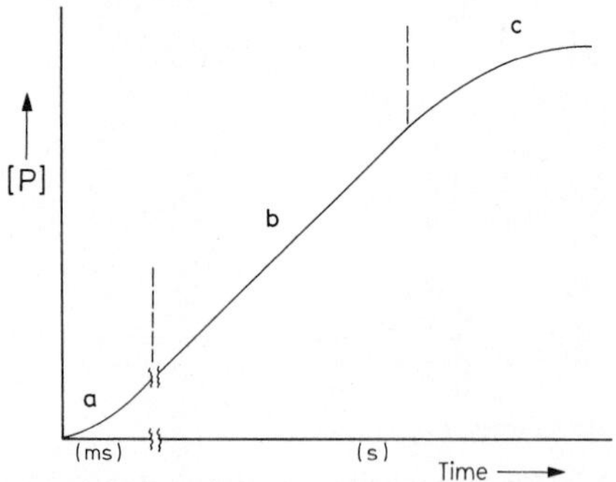

Fig. 2.19. Progress of an enzyme-catalyzed reaction

Since such unpredictable effects should be avoided during analysis of enzyme activities, as a rule the initial reaction rate, v_0, is measured as soon as possible after the start of the reaction.
The basics of the kinetic properties of enzymes in the steady state were given by *G. E. Briggs* and *J. B. S. Haldane* (1925) and are supported by earlier mathematical models proposed by *L. Michaelis* and *M. L. Menten* (1913).
The following definitions and assumptions should be introduced in relation to the reaction in Equation 2.37:

$[E_0]$ = total enzyme concentration available at the start of the catalysis

$[E]$ = concentration of free enzyme not bound to the enzyme-substrate complex, EA, i.e. $[E] = [E_0] - [EA]$.

$[A_0]$ = total substrate concentration available at the start of the reaction. Under these conditions, $[A_0] \gg [E_0]$. Since in catalysis only a small portion of A_0 reacts, the substrate concentration at any time, [A], is approximately equal to $[A_0]$.

When the initial reaction rate, v_0, is considered, the concentration of the product, [P], is 0. Thus, the reaction in Equation 2.37 takes the form:

$$\frac{dP}{dt} = v_0 = k_2(EA) \qquad (2.39)$$

The concentration of enzyme-substrate complex, [EA], is unknown and can not be determined experimentally for Equation 2.39. Hence, it is calculated as follows:
The rate of formation of EA, according to Equation 2.37, is:

$$\frac{dEA}{dt} = k_1(E)(A_0) \qquad (2.40)$$

and the rate of EA breakdown is:

$$-\frac{dEA}{dt} = k_{-1}(EA) + k_2(EA) \qquad (2.41)$$

Under steady-state conditions the rates of breakdown and formation of EA are equal (cf. Equation 2.38):

$$k_1(E)(A_0) = (k_{-1} + k_2)(EA) \qquad (2.42)$$

Also, the concentration of free enzyme, [E], can not be readily determined experimentally. Hence, free enzyme concentration from the above relationship ($[E] = [E_o] - [EA]$) is substituted in Equation 2.42:

$$k_1[(E_0) - (EA)](A_0) = (k_{-1} + k_2)(EA) \qquad (2.43)$$

Solving Equation 2.43 for the concentration of the enzyme-substrate complex, [EA], yields:

$$(EA) = \frac{(E_0)(A_0)}{\frac{k_{-1} + k_2}{k_1} + (A_0)} \qquad (2.44)$$

The quotient of the rate constants in Equation 2.44 can be simplified by defining a new constant, K_m, called the *Michaelis* constant:

$$(EA) = \frac{(E_0)(A_0)}{K_m + (A_0)} \qquad (2.45)$$

Substituting the value of [EA] from Equation 2.45 in Equation 2.39 gives the *Michaelis–Menten* equation for v_0 (initial reaction rate):

$$v_0 = \frac{k_2(E_0)(A_0)}{K_m + (A_0)} \qquad (2.46)$$

Equation 2.46 contains a quantity, $[E_0]$, which can be determined only when the enzyme is present in purified form. In order to be able to make kinetic measurements using impure enzymes, *Michaelis* and *Menten* introduced an approximation for Equation 2.46 as follows.
In the presence of a large excess of substrate, $[A_0] \gg K_m$ in the denominator of Equation 2.46. Therefore, K_m can be neglected compared to $[A_0]$:

$$v_0 = \frac{k_2(E_0)\cancel{(A_0)}}{\cancel{(A_0)}} = V \qquad (2.47)$$

Thus, a zero order reaction rate is obtained. It is characterized by a rate of substrate breakdown or product formation which is independent of substrate concentration, i.e. the reaction rate, V, is dependent only on enzyme concentration. This rate, V, is denoted as the maximum velocity.
From Equation 2.47 it is obvious that the catalytic activity of the enzyme must be measured in the presence of a large excess of substrate.
To eliminate the $[E_0]$ term, V is introduced into Equation 2.46 to yield:

$$v_0 = \frac{V(A_0)}{K_m + (A_0)} \qquad (2.48)$$

If $[A_0] = K_m$, the following is derived from Equation 2.48:

$$v_0 = \frac{V}{2} \quad (2.49)$$

Thus, the *Michaelis* constant, K_m, is equal to the substrate concentration at which the reaction rate is half of its maximal value. K_m is independent of enzyme concentration. The lower the value of K_m, the higher the affinity of the enzyme for the substrate, i.e. the substrate will be bound more tightly by the enzyme and most probably will be more efficiently converted to product. Usually, the values of K_m are within the range of 10^{-2} to 10^{-5} M.
From the definition of K_m:

$$K_m = \frac{k_{-1} + k_2}{k_1} \quad (2.50)$$

it follows that K_m is identical to the enzyme-substrate dissociation constant, K_s, only if $k_{+2} \ll k_{-1}$.

$$k_2 << k_{-1} \frown K_m \approx \frac{k_{-1}}{k_1} = K_s \quad (2.51)$$

Some values for the constants k_{+1}, k_{-1}, and k_0 are compiled in Table 2.10. In cases in which the catalysis proceeds over more steps as shown in Equation 2.37 the constant k_{+2} is replaced by k_0. The rate constant, k_{+1}, for the formation of the enzyme-substrate complex has values in the order of 10^6 to 10^8: in a few cases it approaches the maximum velocity ($\sim 10^9$ M^{-1} s^{-1}), especially when small molecules such as substrate can readily diffuse through the solution to the active site of the enzyme. The values for k_{-1} are substantially lower in most cases, whereas k_0 values are in the range of 10^1 to 10^6 s^{-1}.

Table 2.10. Rate constants for some enzyme catalyzed reactions

Enzyme	Substrate	k_1 ($M^{-1}s^{-1}$)	k_{-1} (s^{-1})	k_0 (s^{-1})
Fumarase	Fumarate	$>10^9$	$4.5 \cdot 10^4$	10^3
Acetylcholine esterase	Acetylcholine	10^9		10^3
Alcohol dehydrogenase (liver)	NAD	$5.3 \cdot 10^5$	74	
	NADH	$1.1 \cdot 10^7$	3.1	
	Ethanol	$>1.2 \cdot 10^4$	>74	10^3
Catalase	H_2O_2	$5 \cdot 10^6$		10^7
Peroxidase	H_2O_2	$9 \cdot 10^6$	<1.4	10^6
Hexokinase	Glucose	$3.7 \cdot 10^6$	$1.5 \cdot 10^3$	10^3
Urease	Urea	$>5 \cdot 10^6$		10^4

A third extreme case to be considered is if $[A_0] \ll K_m$, which occurs at about $[A_0] < 0.05$ K_m. In that case, $[A_0]$ in the denominator of Equation 2.46 can be neglected:

$$v_0 = \frac{k_2 (E_0)(A_0)}{K_m} \quad (2.52)$$

and, considering that $k_2[E_0] = V$, it follows that:

$$v_0 = \frac{V}{K_m} (A_0) \quad (2.53)$$

In this case the *Michaelis–Menten* equation reflects a first-order reaction in which the rate of substrate breakdown depends on substrate concentration. In using a kinetic method for the determination of substrate concentration (cf. 2.5.4.1.3), the experimental conditions must be selected such that Equation 2.53 is valid.

2.5.1.1.2 Determination of K_m *and* V

In order to determine values of K_m und V, the catalytic activity of the enzyme preparation is measured as a function of substrate concentration. Very good results are obtained when $[A_0]$ is in the range of 0.1 K_m to 10 K_m.
A graphical evaluation of the result is obtained by inserting the data into Equation 2.48. As can be seen from a plot of the data in Figure 2.20, the equation corresponds to a rectangular hyperbola. This graphical approach yields correct values for K_m only when the maximum velocity, V, can be accurately determined.

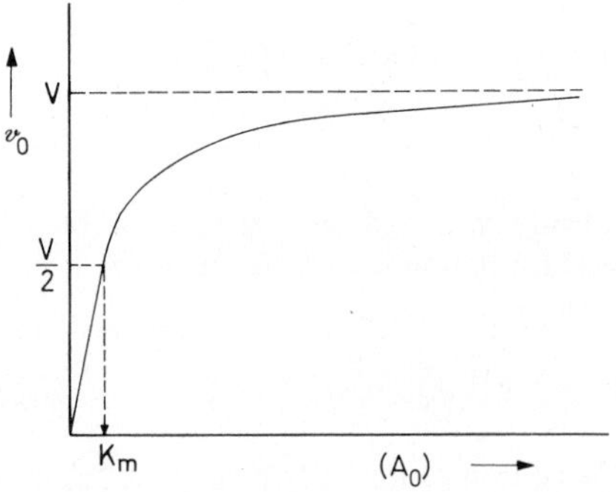

Fig. 2.20. Determination of *Michaelis* constant, K_m, after equation (2.48)

For a more reliable extrapolation of V, Equation 2.48 is transformed into a straight-line equation. Most frequently, the *Lineweaver–Burk* plot is used which is the reciprocal form of Equation 2.48:

$$\frac{1}{v_0} = \frac{K_m}{V} \cdot \frac{1}{(A_0)} + \frac{1}{V} \tag{2.54}$$

Figure 2.21 graphically depicts a plot in this case of $1/v_0$ versus $1/[A_0]$. The values V and K_m are obtained from the intercepts of the ordinate (1/V) and of the abscissa ($-1/K_m$), respectively. If the data do not fit a straight line, then the system deviates from the required steady-state kinetics; e.g. there is inhibition by excess substrate or the system is influenced by allosteric effects (cf. 2.5.1.3; allosteric enzymes do not obey *Michaelis–Menten* kinetics).

A disadvantage of the *Lineweaver–Burk* plot is the possibility of departure from a straight line since data taken in the region of saturating substrate concentrations or at low substrate concentrations can be slightly inflated and, thus, values taken from the straight line may be somewhat overestimated.

A procedure which yields a more uniform *distribution* of the data on the straight line is that proposed by *Hoftstee* (the *Eadie–Hofstee* plot). In this procedure the *Michaelis–Menten* equation, 2.48, is algebraically rearranged into:

$$\begin{aligned} &v_0 (A_0) + v_0 K_m = V \cdot (A_0) && \text{(a)} \\ &v_0 + \frac{v_0}{(A_0)} \cdot K_m = V && \text{(b)} \\ &v = -K_m \frac{v}{(A_0)} + V && \text{(c)} \end{aligned} \tag{2.55}$$

When Equation 2.55c is plotted using the substrate-reaction velocity data, a straight line with

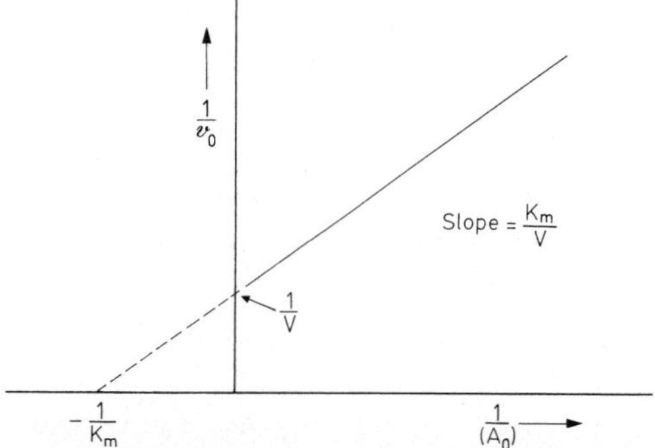

Fig. 2.21. Determination of K_m and V. (After *Lineweaver* and *Burk*)

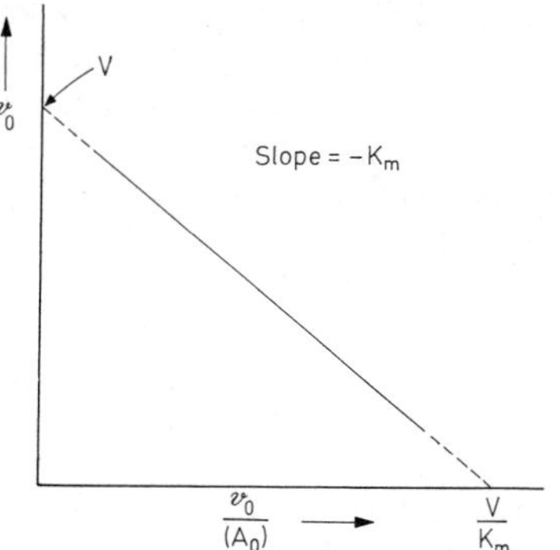

Fig. 2.22. Determination of K_m and V. (After *Hofstee*)

a negative slope is obtained (Figure 2.22) where y is v_0 and x is $v_0/[A_0]$. The y and x intercepts correspond to V and V/K_m, respectively.

Single-substrate reactions, for which the kinetics outlined above (with some exceptions, cf. 2.5.1.3) are particularly pertinent, are those catalyzed by lyase enzymes and certain isomerases. Hydrolysis by hydrolase enzymes can also be considered a single-substrate reaction when the water content remains unchanged, i.e., when it is present in high concentration (55.6 M). Thus, water, as a reactant, can be disregarded.

Characterization of an enzyme-substrate system by determining values for K_m and V is important in enzymatic food analysis (cf. 2.5.4.1) and for assessment of enzymatic reactions occurring in food (e.g. enzymatic browing of sliced potatoes, cf. 2.5.1.2.1) and for utilization of enzymes in food processing, e.g. aldehyde dehydrogenase (cf. 2.8.2.1.4).

2.5.1.2 Two-Substrate Reactions

For many enzymes, for examples oxidoreductase and ligase-catalyzed reactions, two or more substrates or cosubstrates are involved.

2.5.1.2.1 Order of substrate binding

In the reaction of an enzyme with two substrates, the binding of the substrates can occur sequentially in a specific order. Thus, the binding mechanism can be divided into catalysis which proceeds through a ternary adsorption complex (enzyme + two substrates) or through a binary complex (enzyme + one substrate), i.e. when the enzyme binds only one of the two available substrates at a time.

A ternary enzyme-substrate complex can be formed in two ways. The substrates are bound to the enzyme in a random fashion ("random mechanism") or they are bound in a well-defined order ("ordered mechanism").
Let us consider the reaction

$$A + B \underset{}{\overset{E}{\rightleftharpoons}} P + Q \qquad (2.56)$$

If the enzyme reacts by a "random mechanism", substrates A and B form the ternary enzyme-substrate complex, EAB, in a random fashion and the P and Q products dissociate randomly from the ternary enzyme-product complex, EPQ:

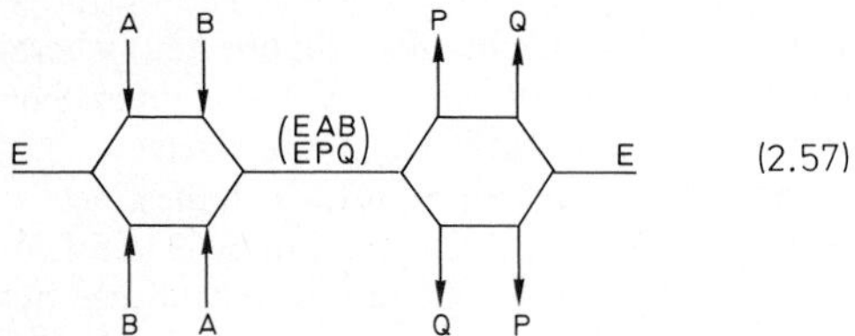

Creatine kinase from muscle (cf. 12.3.6) is an example of an enzyme which reacts by a random mechanism:

$$\text{Creatine} + \text{ATP} \rightleftharpoons \text{Creatine-phosphate} + \text{ADP} \qquad (2.58)$$

In an "ordered mechanism" the binding during the catalyzed reaction is as follows:

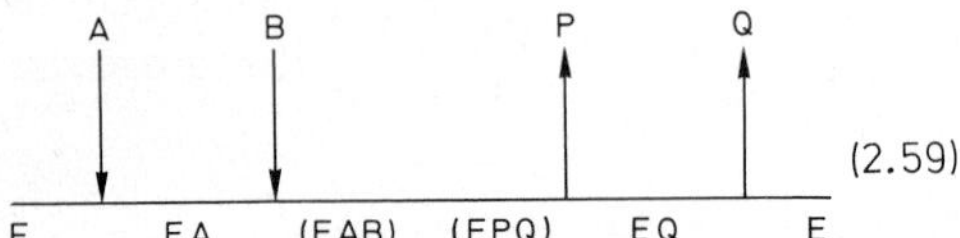

Alcohol dehydrogenase reacts by an "ordered mechanism", although the order of the binding of substrates NAD^+ and ethanol is decided by the ethanol concentration. NAD^+ is absorbed first at low concentrations (<4 mM):

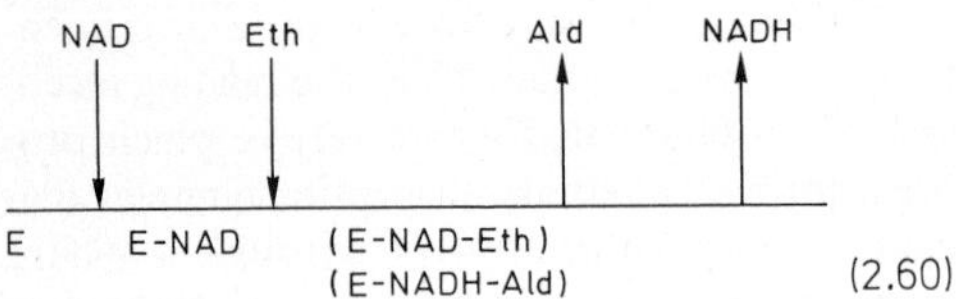

When the ethanol concentration is increased to 7–8 mM, ethanol is absorbed first, followed by the cosubstrate. The order of removal of products (acetaldehyde and NADH) is, however, not altered.
Phenol oxidase from potato tuber also reacts by an "ordered mechanism". Oxygen is adsorbed first, followed by phenolic substrates. The main substrates are chlorogenic acid and tyrosine. Enzyme affinity for tyrosine is greater and the reaction velocity is higher than for chlorogenic acid. The ratio of chlorogenic acid to tyrosine affects enzymatic browning to such an extent that it is considered to be the major problem in potato processing. The deep-brown colored melanoidins are formed quickly from tyrosine but not from chlorogenic acid. In assessing the processing quality of potato cultivars, the differences in phenol oxidase activity and the content of ascorbic acid in the tubers should also be considered in relation to "enzymatic browning". Ascorbic acid retards formation of melanoidins by its ability to reduce o-quinone, the initial product of enzymatic oxidation (cf. 18.1.2.5.7).
In enzymatic reactions where functional group transfers are involved, as a rule only binary enzyme-substrate complexes are formed by the so-called "ping pong mechanism".
A substrate is adsorbed by enzyme, E, and reacts during alteration of the enzyme (a change in oxidation state of the prosthetic group, a conformational change, or only a change in covalent binding of a functional group). The modified enzyme which, is denoted F, binds the second substrate and the second reaction occurs, which regenerates the initial enzyme, E, and releases the second product:

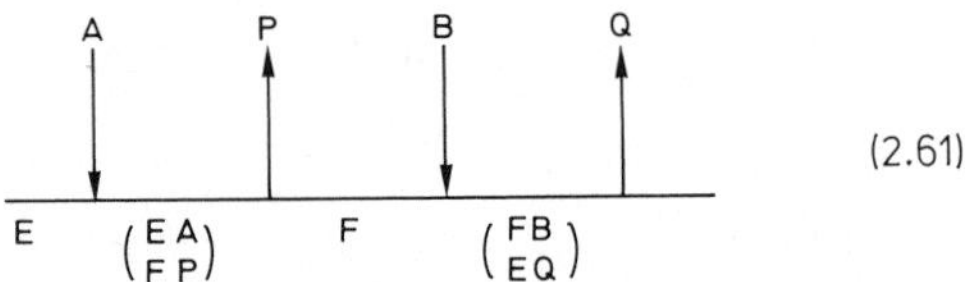

The glycolytic enzyme hexokinase reacts by a "ping pong mechanism" (cf. Formula 2.62).

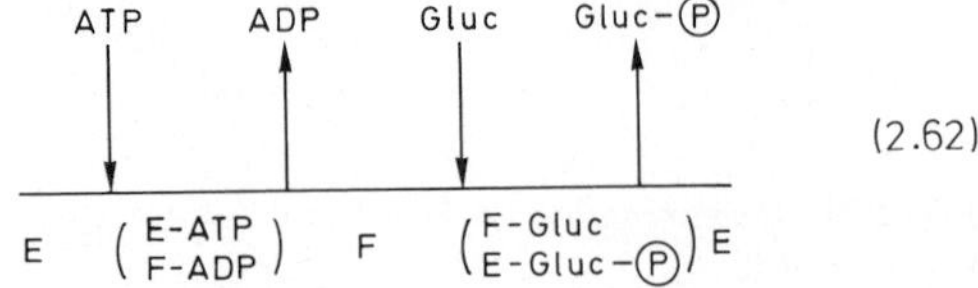

2.5.1.2.2 Rate equations for a two-substrate reaction

Here the reaction rate is distinguished by dependence of the rate on two reactants, either two molecules of the same compound or two different compounds. The rate equations can be derived by the same procedures as used for single-substrate catalysis. Only the final forms of the equations will be considered.
When the catalysis proceeds through a ternary enzyme-substrate complex, EAB, the general equation is:

$$v_0 = \frac{V}{1 + \frac{K_a}{(A_0)} + \frac{K_b}{(B_0)} + \frac{K_{ia} \cdot K_b}{(A_0)(B_0)}} \quad (2.63)$$

When compared to the rate equation for a single-substrate reaction (Equation 2.48), the difference becomes obvious when the equation for a single-substrate reaction is expressed in the following form:

$$v_0 = \frac{V}{1 + \frac{K_a}{(A_0)}} \quad (2.64)$$

The constants K_a and K_b in Equation 2.63 are defined analogously to K_m, i.e. they yield the concentrations of A or B for $v_0 = V/2$ assuming that, at any given moment, the enzyme is saturated by the other substrate (B or A). Each of the constants, like K_m (cf. Equation 2.50), is composed of several rate constants. K_{ia} is the inhibitor constant for A.
When the binding of one substrate is not influenced by the other, each substrate occupies its own binding locus on the enzyme and the substrates form a ternary enzyme-substrate complex in a defined order ("ordered mechanism"), the following is valid:

$$K_{ia} \cdot K_b = K_a \cdot K_b \quad (2.65)$$

or from Equation 2.63:

$$v_0 = \frac{V}{1 + \frac{K_a}{(A_0)} + \frac{K_b}{(B_0)} + \frac{K_a \cdot K_b}{(A_0)(B_0)}} \quad (2.66)$$

However, when only a binary enzyme-substrate complex is formed, i.e. one substrate or one product is bound to the enzyme at a time by a "ping pong mechanism", the denominator term $K_{ia} \cdot K_b$ must be omitted since no ternary complex exists. Thus, Equation 2.63 is simplified to:

$$v_0 = \frac{V}{1 + \frac{K_a}{(A_0)} + \frac{K_b}{(B_0)}} \quad (2.67)$$

For the determination of rate constants, the initial rate of catalysis is measured as a function of the concentration of substrate B (or A) for several concentrations of A (or B). Evaluation can be done using the *Lineweaver–Burk* plot. Reshaping Equation 2.63 for a "random mechanism" leads to:

$$\frac{1}{v_0} = \left[\frac{K_b}{V} + \frac{K_{ia} \cdot K_b}{(A_0)V}\right]\frac{1}{(B_0)} + \left[1 + \frac{K_a}{(A_0)}\right]\frac{1}{V} \quad (2.68)$$

A plot is first made of $1/v_0$ versus $1/[B_0]$. The corresponding slopes and ordinate intercepts are taken from the straight lines obtained at various values for $[A_0]$ (Figure 2.23):

$$\text{Slope} = \frac{K_b}{V} + \frac{K_{ia}\,K_b}{V} \cdot \frac{1}{(A_0)} \quad (2.69)$$

$$\text{Ordinate intercept} = \frac{1}{V} + \frac{K_a}{V} \cdot \frac{1}{(A_0)}$$

and are then plotted against $1/[A_0]$. In this way two straight lines are obtained (Figure 2.24,a and b), with slopes and ordinate intercepts which provide data for calculating constants K_a, K_b and the maximum velocity, V. If the catalysis proceeds through a "ping pong mechanism", then plotting $1/v_0$ versus $1/[B_0]$ yields a family of parallel lines (Figure 2.25) which are then subjected to the calculations described above.
A comparison of Figures 2.23 and 2.25 leads to the conclusion that the dependence of the initial catalysis rate on substrate concentration allows the differentiation between a ternary and a binary enzyme-substrate complex. However, it is not possible to differentiate an "ordered" from a "random" reaction mechanism by this means.

2.5.1.3 Allosteric Enzymes

We are already acquainted with some enzymes consisting of several protomers (cf. Table 1.19 a). When the protomer activities are independent of each other in catalysis, the *Michaelis–Menten* kinetics, as outlined under sections 2.5.1.1 and

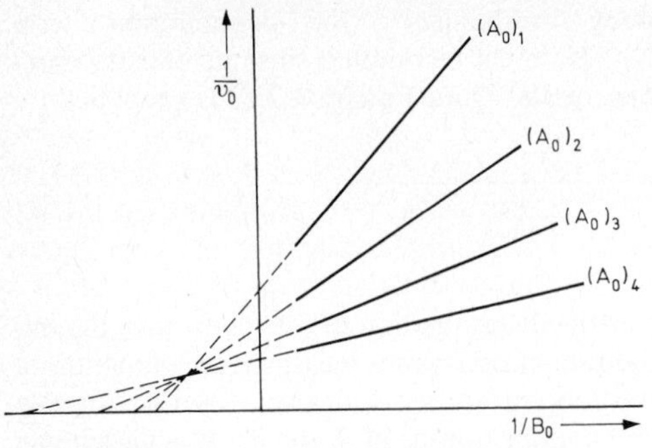

Fig. 2.23. Evaluation of a two-substrate reaction, proceeding through a ternary enzyme-substrate-complex. (After *Lineweaver* and *Burk*) $[A_0]_4 > [A_0]_3 > [A_0]_2 > [A_0]_1$

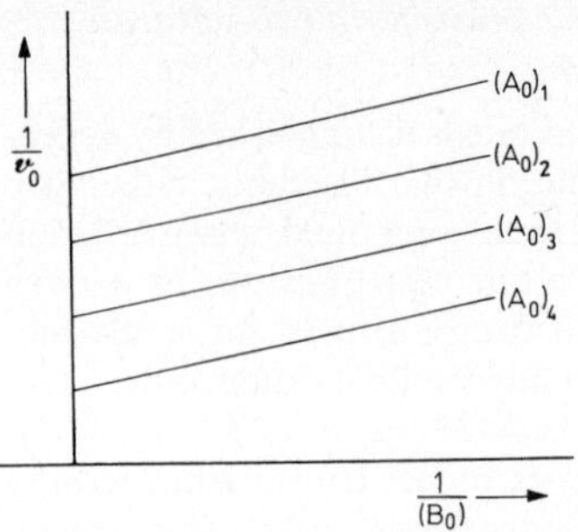

Fig. 2.25. Evaluation of a two-substrate reaction, proceeding through a binary enzyme-substrate-complex. (After *Lineweaver* and *Burk*) $[A_0]_4 > [A_0]_3 > [A_0]_2 > [A_0]_1$

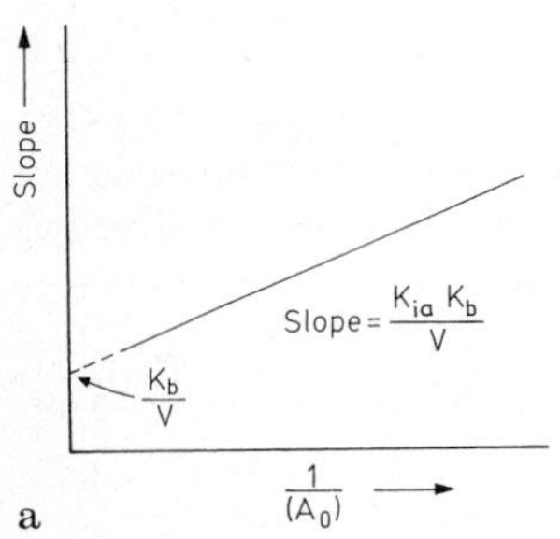

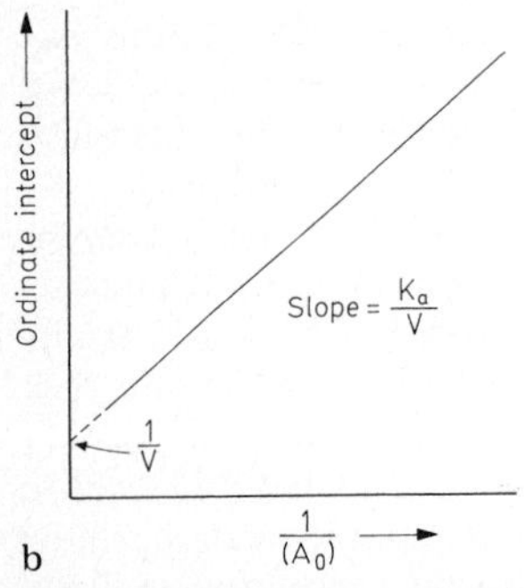

Fig. 2.24. Plotting slopes (**a**) and ordinate intercepts (**b**) from Fig. 2.23 versus $1/[A_0]$

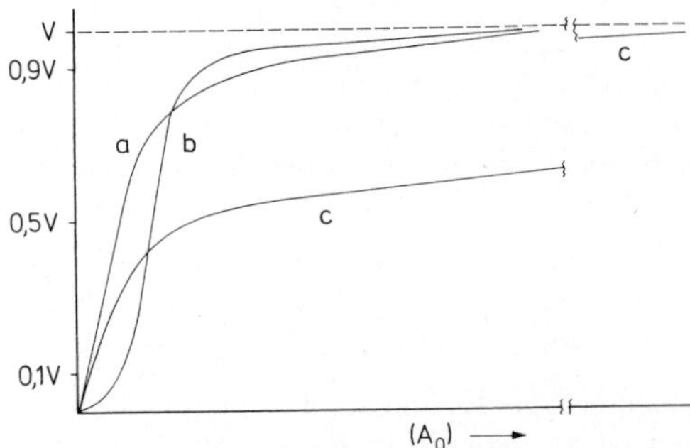

Fig. 2.26. The effect of substrate concentration on the catalysis reaction rate. **a** Enzyme obeying *Michaelis-Menten* kinetics; **b** allosterically regulated enzyme with positive cooperativity; **c** allosterically regulated enzyme with negative cooperativity

2.5.1.2, are valid. However, when the subunits cooperate, the enzymes deviate from these kinetics. This is particularly true in the case of positive cooperation when the enzyme is activated by substrate. For this kind of plot, v_0 versus $[A_0]$ yields not a hyperbolic curve but a saturation curve with a sigmoidal shape (Figure 2.26).

Thus, enzymes which do not obey the *Michaelis–Menten* model of kinetics are allosterically regulated. These enzymes have a site which reversibly binds the allosteric regulator (substrate, cosubstrate or low molecular weight compound) in addition to an active site with a binding and transforming locus. Allosteric enzymes are, as a rule, engaged at control sites of metabolism. An example is tetrameric phosphofructokinase, the key enzyme in glycolysis, an enzyme which, in glycolysis and alcoholic fermentation, catalyzes the phosphorylation of fructose-6-phosphate to fructose-1,6-diphosphate. The enzyme is activated by its substrate in the presence of ATP. The prior binding of a substrate molecule which enhances the binding of each succeeding substrate molecule is called positive cooperation.

The two enzyme-catalyzed reactions, one which obeys *Michaelis–Menten* kinetics and the other which is regulated by allosteric effects, can be reliably distinguished experimentally by comparing the ratio of the substrate concentration needed to obtain the observed value of 0.9 V to that needed to obtain 0.1 V. This ratio, denoted as R_s, is a measure of the cooperativity of the interaction.

$$R_S = \frac{(A_0)_{0,9V}}{(A_0)_{0,1V}} \tag{2.70}$$

For all enzymes which obey *Michaelis–Menten* kinetics, $R_s = 81$ regardless of the value of K_m or V. The value of R_s is either lower or higher than 81 for allosteric enzymes. $R_s < 81$ is indicative of positive cooperation. Each substrate molecule, often called an effector, accelerates the binding of succeeding substrate molecules, thereby increasing the catalytic activity of the enzyme (case b in Figure 2.26). When $R_s > 81$, the system shows negative cooperation. The effector (or allosteric inhibitor) decreases the binding of the next substrate molecule (case c in Figure 2.26).

Various models have been developed in order to explain the allosteric effect. Only the symmetry model proposed by *J. Monod, J. Wyman* and *J.P. Changeux* (1965) will be described in its simplified form; specifically, when the substrate acts as a positive allosteric regulator or effector. Based on this model, the protomers of an allosteric enzyme exist in two conformations, one with a high affinity (R-form) and the other with a low affinity (T-form) for the substrate. These two forms are interconvertible. There is an interaction between protomers. Thus, binding of the allosteric regulator by one protomer induces a conformational change of all the subunits and greatly increases the activity of the enzyme.

Let us assume that the R- and T-forms of an enzyme consisting of four protomers are in equilibrium which lies completely on the side of the T-form:

$$\text{(T-Form)} \overset{A}{\rightleftharpoons} \text{(R-Form)} \tag{2.71}$$

Addition of substrate, which here is synonymous to the allosteric effector, shifts the equilibrium from the low affinity T-form to the substantially more catalytically active R-form. Since one substrate molecule activates four catalytically active sites, the steep rise in enzyme activity after only a slight increase in substrate concentration is not unexpected. In this model it is important that the RT conformation is not permitted. All subunits must be in the same conformational state at one time to conserve the symmetry of the protomers.

The equation given by *A. V. Hill* in 1913, derived from the sigmoidal absorption of oxygen by hemoglobin, is also suitable for a quantitative description of allosteric enzymes with sigmoidal behavior:

$$v_0 = \frac{V\,(A_0)^n}{K' + (A_0)^n} \tag{2.72}$$

The equation says that the catalysis rate increases by the nth power of the substrate concentration when $[A_0]$ is small in comparison to K. The *Hill* coefficient, n, is a measure of the sigmoidal character of the curve and, therefore, of the extent of the enzyme's cooperativity. For n = 1 (Equation 2.72) the reaction rate is transformed into the *Michaelis–Menten* equation, i.e. in which no cooperativity factor exists.

In order to assess the experimental data, Equation 2.72 is rearranged into an equation of a straight line:

$$\log \frac{v_0}{V - v_0} = n \log (A_0) - \log K' \tag{2.73}$$

The slope of the straight line obtained by plotting, the substrate concentration as $\log[A_0]$ versus $\log [v_0/(V - v_0)]$ is the *Hill* coefficient, n (Figure 2.27). The constant K incorporates all the individual K_m values involved in all the steps

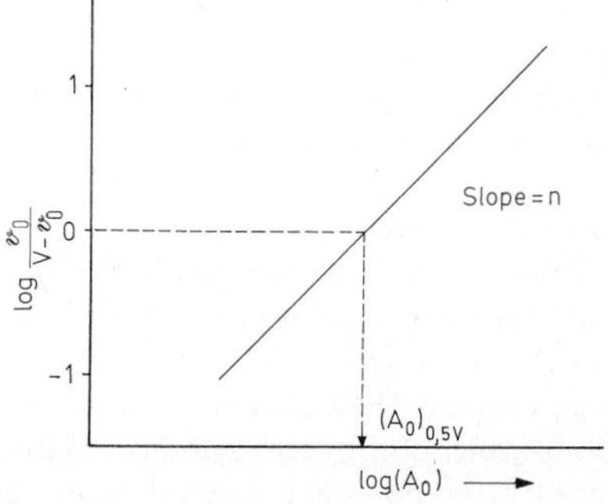

Fig. 2.27. Linear presentation of *Hill's* equation

of substrate binding and transformation. The value of K_m is obtained by using the substrate concentration, denoted as $[A_0]_{0.5V}$, at which $v_0 = 0.5$ V. Under these conditions, the following is derived from Equation 2.73:

$$\log \frac{0.5\,V}{0.5\,V} = 0 = n \cdot \log (A_0)_{0.5\,V} - \log K' \quad (a)$$
$$K' = (A_0)^n_{0.5\,V} \quad (b) \qquad (2.74)$$

2.5.2 Effect of Inhibitors

The catalytic activity of an enzyme, in addition to substrate concentration, is affected by the type and concentration of inhibitors, i.e. compounds which decrease the rate of catalysis, and activators, which have the opposite effect. Metal ions and compounds which are active as prosthetic groups or which provide stabilization of the enzyme's conformation or of the enzyme-substrate complex (cf. 2.3.2 and 2.3.3) are activators. The effect of inhibitors will be discussed in more detail in this section.

Inhibitors are found among food constituents. Proteins which specifically inhibit the activity of certain proteinases (cf. 16.2.3), amylases or β-fructofuranosidase are examples. Furthermore, food contains substances which nonselectively inhibit a wide spectrum of enzymes. Phenolic constituents of food (cf. 18.1.2.5) and mustard oil (cf. 14.3.2.2.5) belong to this group. In addition, a food might be contaminated with pesticides, heavy metal ions and other chemicals from a polluted environment (cf. Chapter 9) which can become inhibitors under some circumstances. These possibilities should be taken into account when an enzymatic food analysis is performed.

A food is usually heat treated (cf. 2.6) to suppress the undesired enzymatic reactions. As a rule, no inhibitors are used in food processing. An exception is the addition of, for example, SO_2 to inhibit the activity of phenolase (cf. 8.12.6).

Much data concerning the mechanism of action of enzyme inhibitors has been compiled by recent biochemical research. These data cover elucidation of the effect of inhibitors on functional groups of an enzyme, their effect on the active site and the clarification of the general mechanism involved in an enzyme-catalyzed reaction (cf. 2.4.1.1).

Based on kinetic considerations, inhibitors are divided into two groups: inhibitors bound *irreversibly* to enzyme and those bound *reversibly*.

2.5.2.1 Irreversible Inhibition

In an irreversible inhibition the inhibitor binds mostly covalently to the enzyme; the EI complex formed does not dissociate:

$$E + I \xrightarrow{k_1} EI \qquad (2.75)$$

The rate of inhibition depends on the reaction rate constant k_{+1} in Equation 2.75, the enzyme concentration, [E], and the inhibitor concentration, [I]. Thus, irreversible inhibition is a function of reaction time. The reaction can not be reversed by diluting the reaction medium. These criteria serve to distinguish irreversible from reversible inhibition.

Examples of irreversible inhibition are the reactions of SH-groups of an enzyme with iodoacetic acid:

$$\text{Enz-SH} + \text{JCH}_2\text{COOH} \longrightarrow \text{Enz-S-CH}_2\text{-COOH} + \text{HJ} \qquad (2.76)$$

and other reactions with inhibitors (cf. section 2.4.1.1).

2.5.2.2 Reversible Inhibition

Reversible inhibition is characterized by an equilibrium between enzyme and inhibitor:

$$E + I \rightleftharpoons EI \quad (a)$$
$$\frac{[E] \cdot [I]}{[EI]} = K_i \quad (b) \qquad (2.77)$$

The equilibrium constant or dissociation constant of the enzyme-inhibitor complex, K_i, also known as the inhibitor constant, is a measure of the extent of inhibition. The lower the value of K_i, the higher the affinity of the inhibitor for the enzyme.

Kinetically, three kinds of reversible inhibition can be distinguished: competitive, non-competitive and uncompetitive inhibition (examples in Table 2.11). Other possible cases, such as allosteric inhibition and partial competitive or partial non-competitive inhibition, are omitted in this treatise.

Table 2.11. Examples of reversible enzyme inhibition

Enzyme	EC-Number	Substrate	Inhibitor	Inhibition type[a]	K_i(M)
Glucose dehydrogenase	1.1.1.47	Glucose/NAD	Glucose-6-phosphate	C	$4.4 \cdot 10^{-5}$
Glucose-6-phosphate dehydrogenase	1.1.1.49	Glucose-6-phosphate/NADP	Phosphate	C	$1 \cdot 10^{-1}$
Succinate dehydrogenase	1.3.99.1	Succinate	Fumarate	C	$1.9 \cdot 10^{-3}$
Creatine kinase	2.7.3.2	Creatine/ATP	ADP	NC	$2 \cdot 10^{-3}$
Glucokinase	2.7.1.2	Glucose/ATP	D-Mannose	C	$1.4 \cdot 10^{-2}$
			2-Deoxyglucose	C	$1.6 \cdot 10^{-2}$
			D-Galactose	C	$6.7 \cdot 10^{-1}$
Fructose-biphosphatase	3.1.3.11	D-Fructose 1,6-biphosphate	AMP	NC	$1.1 \cdot 10^{-4}$
α-Glucosidase	3.2.1.20	p-Nitrophenyl-α-D-glucopyranoside	Saccharose	C	$3.7 \cdot 10^{-2}$
			Turanose	C	$1.1 \cdot 10^{-2}$
Cytochrome c oxidase	1.9.3.1	Ferrocytochrome c	Azide		

[a] C: competitive, NC: noncompetitive, and UC: uncompetitive.

2.5.2.2.1 Competitive inhibition

Here the inhibitor binds to the active site of the free enzyme, thus preventing the substrate from binding. Hence, there is competition between substrate and inhibitor:

$$E + I \rightleftharpoons EI \ (a) \quad E + A \rightleftharpoons EA \ (b) \quad (2.78)$$

According to the steady-state theory for a single-substrate reaction, we have:

$$v_0 = \frac{V(A_0)}{K_m\left(1 + \frac{[I]}{K_i}\right) + (A_0)} \quad (2.79)$$

In the presence of inhibitors, the *Michaelis* constant is apparently increased by the factor:

$$1 + \frac{[I]}{K_i} \quad (2.80)$$

Such an effect can be useful in the case of enzymatic substrate determinations (cf. 2.5.4.1.3). When inhibitor activity is absent, i.e. $[I] = 0$, Equation 2,79 is transformed into the *Michaelis–Menten* equation (Equation 2.48).
The *Lineweaver–Burk* plot (Figure 2.28,a) shows that the intercept of $1/V$ versus $1/[A_0]$ with the ordinate is the same in the presence and in the absence of the inhibitor, i.e. the value of V is not affected although the slopes of the lines differ. This shows that the inhibitor can be fully dislodged by the substrate from the active site of the enzyme when the substrate is present in high concentration. In other words, inhibition can be overcome at high substrate concentrations (see application in Figure 2.46). The inhibitor constant, K_i, can be calculated from the corresponding intercepts with the abscissa in Figure 2.28,a by calculating the value of K_m from the abscissa intercept when $[I] = 0$.

2.5.2.2.2 Non-competitive inhibition

The non-competitive inhibitor is not bound to the active site of the enzyme but to some other site. Therefore, the inhibitor can react equally with free enzyme or with enzyme-substrate complex. Thus, three processes occur in parallel:

$$E + A \rightleftharpoons EA \ (a) \quad E + I \rightleftharpoons EI \ (b) \quad EA + I \rightleftharpoons EAI \ (c) \quad (2.81)$$

Postulating that EAI and EI are catalytically inactive and the dissociation constants K_i and K_{EAi} are numerically equal, the following equation is obtained by rearrangement of the equation for a single-substrate reaction into its reciprocal form:

$$\frac{1}{v_0} = \frac{K_m}{V}\left(1 + \frac{[I]}{K_i}\right)\frac{1}{(A_0)} + \frac{1}{V}\left(1 + \frac{[I]}{K_i}\right) \quad (2.82)$$

The double-reciprocal plot (Figure 2.28,b) shows that, in the presence of a noncompetitive inhibitor; K_m is unchanged whereas, the values of V are decreased such that V becomes $V/(1 + [I]/K_i)$, i.e. non-competitive inhibition can not be overcome by high concentrations of substrate.

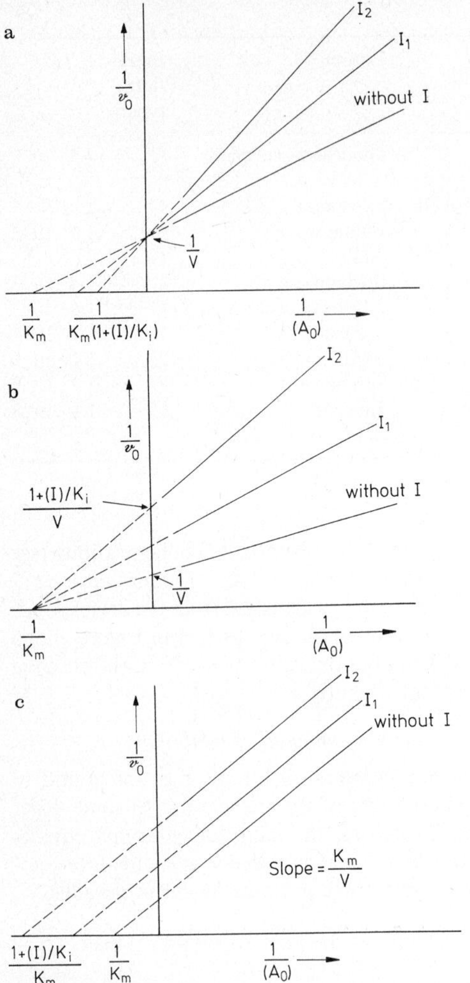

Fig. 2.28. Evaluation of inhibited enzyme-catalyzed reaction after *Lineweaver* and *Burk*, $[I_1] < (I_2)$. **a** Competitive inhibition, **b** noncompetitive inhibition, **c** uncompetitive inhibition

This also indicates that, in the presence of inhibitor, the amount of enzyme available for catalysis is decreased.

2.5.2.2.3 Uncompetitive inhibition

In this case the inhibitor reacts only with enzyme-substrate complex:

$$E + A \rightleftharpoons EA \begin{array}{l} \nearrow E + P \\ \underset{I}{\overset{I}{\rightleftharpoons}} EAI \end{array} \qquad (2.83)$$

Rearranging Equation 2.83 into an equation for a straight line, the reaction rate becomes:

$$\frac{1}{v_0} = \frac{K_m}{V} \frac{1}{(A_0)} + \frac{1}{V}\left(1 + \frac{[I]}{K_i}\right) \qquad (2.84)$$

The double reciprocal plot (Figure 2.28,c) shows that, in the presence of an uncompetitive inhibitor, both the maximum velocity, V, and K_m are changed but not the ratio of K_m/V, hence the slopes of the lines are equal and, in the presence of increasing amounts of inhibitor, the lines plotted are parallel. Uncompetitive inhibition is rarely found in single-substrate reactions. It occurs more often in two-substrate reactions.

In conclusion, it can be stated that the three types of reversible inhibitions are kinetically distinguishable by plots of reaction rate versus substrate concentration using the procedure developed by *Lineweaver* and *Burk* (Figure 2.28).

2.5.3 Effect of pH on Enzyme Activity

Each enzyme is catalytically active only in a narrow pH range and, as a rule, each has a sharp pH optimum which is often between pH 5.5 and 7.5 (Table 2.12).

The optimum pH is affected by the type and ionic strength of the buffer used in the assay. The reasons for the sensitivity of the enzyme to changes in pH are two-fold:

a) sensitivity is associated with a change in protein structure leading to irreversible denaturation,

Table 2.12. pH-optima of various enzymes

Enzyme	Source	Substrate	pH-Optimum
Pepsin	Stomach	Protein	2
Chymotrypsin	Pancreas	Protein	7.8
Papain	Tropical plants	Protein	7–8
Lipase	Microorganisms	Olive oil	5–8
α-Glucosidase (maltase)	Microorganisms	Maltose	6.6
β-Amylase	Malt	Starch	5.2
β-Fructofuranosidase (invertase)	Tomato	Saccharose	4.5
Pectin lyase	Microorganisms	Pectic acid	9.0–9.2
Xanthine oxidase	Milk	Xanthine	8.3
Lipoxygenase, type I[a]	Soybean	Linoleic acid	9.0
Lipoxygenase, type II[a]	Soybean	Linoleic acid	6.5

[a] See 3.7.2.2

b) the catalytic activity depends on the quantity of electrostatic charges on the enzyme's active site contributed by prototropic groups of the enzyme (cf. 2.4.2.1).

In addition the ionization of dissociable substrates as affected by pH can be of importance to the reaction rate. However, such effects should be determined separately. Here, only the influences under b) will be considered, with some simplifications.
An enzyme, E, its substrate, A, and the enzyme-substrate complex formed, EA, depending on pH, form the following equilibria:

$$\begin{array}{c} E^{n+1} \overset{-H^{\oplus}}{\rightleftharpoons} E^{n} \overset{-H^{\oplus}}{\rightleftharpoons} E^{n-1} \\ E^{n+1}A \overset{-H^{\oplus}}{\rightleftharpoons} E^{n}A \overset{-H^{\oplus}}{\rightleftharpoons} E^{n-1}A \end{array} \quad (2.85)$$

Which of the charged states of E and EA are involved in catalysis can be determined by following the effect of pH on V and K_m.

a) Plotting K_m versus pH reveals the type of prototropic groups involved in substrate binding and which support or maintain the conformation of the enzyme. The results of such a plot, as a rule, resemble one of the four diagrams shown in Figure 2.29.
Figure 2.29,a: K_m is independent of pH in the range of 4–9. This means that the forms

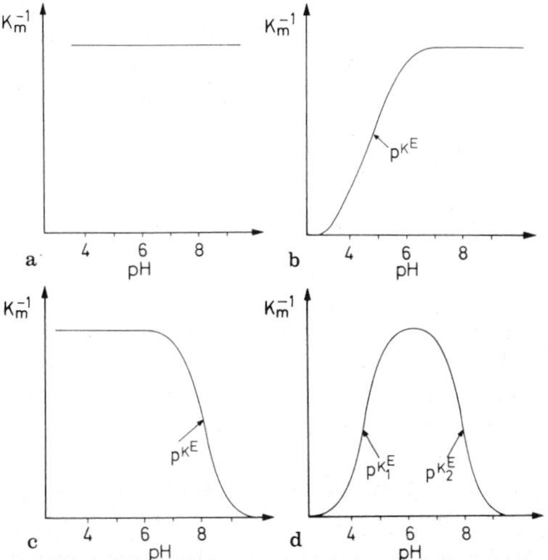

Fig. 2.29. The possible effects of pH on the *Michaelis* constant, K_m

E^{n+1}, E^n and E^{n-1}, i.e. enzyme forms which are neutral or positively or negatively charged on the active site, can bind substrate.
Figures 2.29,b and c: here, K_m is dependent on one prototropic group, the pK value of which is below (Figure 2.29,b) or above (Figure 2.29,c) neutrality. In the former case, E^n and E^{n-1} are the active forms, while in the latter, E^{n+1} and E^n are the active enzyme forms in substrate binding.
Figure 2.29,d: here K_m is dependent on two prototropic groups; the active form in substrate binding is E^n.

b) The involvement of prototropic groups in the conversion of an enzyme-substrate complex into product occurs when the enzyme is saturated with substrate, i.e. when Equation 2.47 is valid ($[A_0] \gg K_m$), by which V is defined. Thus, a plot of V versus pH provides essentially the same four possibilities presented in Figure 2.29, the difference being that, here, the prototropic groups of EA, which are involved in the conversion to product, are revealed.
In order to better understand the form of the enzyme involved in catalysis, a hypothetical enzyme-substrate system will be assayed and interpreted. We will start from the assumption that data are available for v_0 (initial velocity) as a function of substrate concentration at several pH's, e.g. plotted by the *Lineweaver* and *Burk* method. The values for K_m and V are obtained from the family of straight lines (Figure 2.30) and plotted against pH. The diagram of $K_m^{-1} = f(\text{pH})$ depicted in Figure 2.31,a corresponds to Figure 2.29,c which

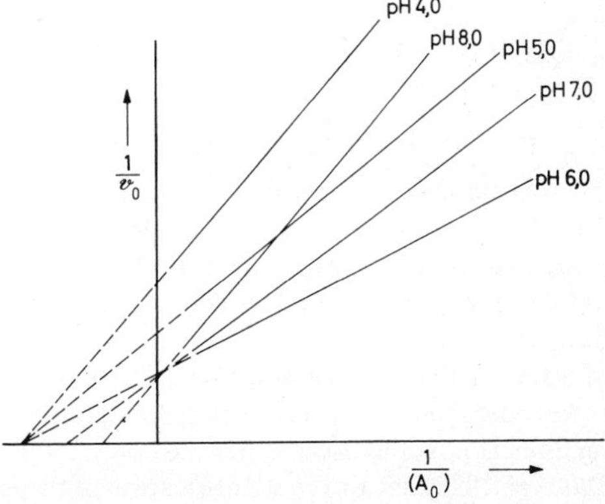

Fig. 2.30. Determination of V and K_m at different pH's

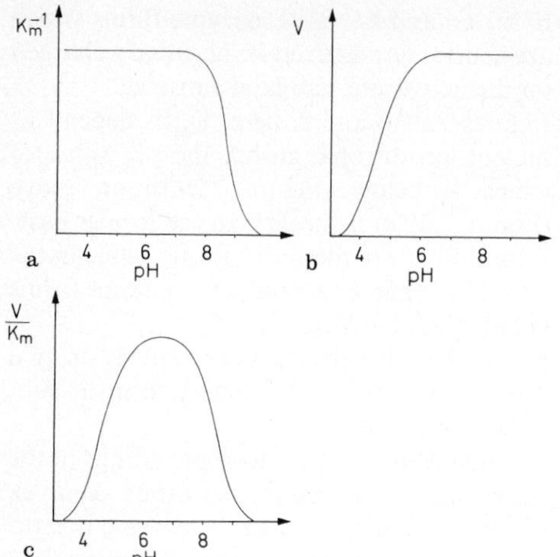

Fig. 2.31. Evaluation of K_m and V versus pH for a hypothetical case

implies that neutral (E^n) and positively-charged (E^{n+1}) enzyme forms are active in binding the substrate.

Figure 2.31 b: V is dependent on one prototropic group, the pK value of which is below neutrality. Therefore, of the two enzyme-substrate complexes, $E^{n+1}A$ and E^nA, present in the equilibrium state, only the latter complex is involved in the conversion of A to the product.

In the example given above, the overall effect of pH on enzyme catalysis can be illustrated as follows:

$$\begin{array}{ccccc} E^{n+1} & \overset{A}{\rightleftharpoons} & E^{n+1}A & & E^{n+1} \\ \updownarrow & & \updownarrow & & \updownarrow \\ E^{n} & \overset{A}{\rightleftharpoons} & E^{n}A & \xrightarrow{\quad} P & E^{n} \\ \updownarrow & & & & \updownarrow \\ E^{n-1} & & & & E^{n-1} \end{array} \tag{2.86}$$

This schematic presentation is also in agreement with the diagram of $V/K_m = f(\text{pH})$ (Figure 2.31,c) which reveals that, overall, two prototropic groups are involved in the enzyme-catalyzed reaction.

An accurate determination of the pK values of prototropic groups involved in enzyme-catalyzed reactions is possible using other assays (cf. *J. R. Whitaker,* 1972). However, identification of these groups solely on the basis of pK values is not possible since the pK value is often strongly influenced by surrounding groups. Pertinent to this claim is our recollection that the pH of acetic acid in water is 4.75, whereas in 80% acetone it is about 7. Therefore, the enzyme activity data as related to pH have to be considered only as preliminary data which must be supported and verified by supplementary investigations.

2.5.4 Enzymatic Analysis

Enzymatic food analysis involves the determination of food constituents which can be both substrates or inhibitors of enzymes, and determination of enzyme activity in food.

2.5.4.1 Substrate Determination

2.5.4.1.1 Principles

Qualitative and quantitative analysis of food constituents using enzymes can be rapid, highly sensitive, selective and accurate (examples in Table 2.13). Prior purification and separation steps, as a rule, are not necessary in the enzymatic analysis of food.

In an enzymatic assay, spectrophotometric or electrochemical determination of the reactant or the product is the preferred approach. When this is not applicable, the determination is performed by a coupled enzyme assay. The coupled reaction includes an auxiliary reaction* in which the food constituent is the reactant to be converted to product, and an indicator reaction which involves an indicator enzyme and its reactant or product, the formation or breakdown of which can be readily followed analytically. In most cases, the indicator reaction follows the auxiliary reaction:

$$\begin{aligned} A + B &\xrightleftharpoons{\text{Auxiliary enzyme}} P + Q \\ P + C &\xrightleftharpoons{\text{Indicator enzyme}} R + S \end{aligned} \tag{2.87}$$

Reactant A is the food constituent which is being analyzed. C or R or S is measured. The equlibrium state of the coupled indicator reaction is concentration dependent. The reaction has to be adjusted in some way in order to remove, for example, P from the auxiliary reaction before an

* The term "auxiliary reaction" for the main reaction is an unfortunate choice. Nevertheless, it has been generally accepted.

Table 2.13. Examples of enzymatic analysis of food constituents[a]

Constituent	Auxiliary reaction	Indicator reaction
Glucose	β-D-Glucose[b] + O_2 $\xrightleftharpoons[\text{oxidase}]{\text{Glucose}}$ ∂-D-Gluconolactone + H_2O_2 (a_H)	o-Dianisidine + H_2O_2 $\xrightleftharpoons{\text{Peroxidase}}$ Oxid. o-Dianisidine (a_I)
	Glucose + ATP $\xrightleftharpoons[\text{kinase}]{\text{Hexo-}}$ Glucose-6P (b_H)	Glucose-6P + $NADP^{\oplus}$ $\xrightleftharpoons[\text{dehydrogenase}]{\text{Glucose-6P}}$ Gluconate-6P + NADH + $H^{\oplus}$ (b_I)
Fructose	Fructose + ATP $\xrightleftharpoons[\text{kinase}]{\text{Hexo-}}$ Fructose-6P	Fructose-6-P $\xrightleftharpoons[\text{isomerase}]{\text{Glucosephosphate}}$ Glucose-6P
Sorbitol	D-Sorbitol + NAD $\xrightleftharpoons[\text{drogenase}]{\text{Sorbitol dehy-}}$ Fructose + NADH + $H^{\oplus}$	
Maltose	Maltose + H_2O $\xrightleftharpoons{\alpha\text{-Glucosidase}}$ 2 Glucose	As Glucose ($b_H + b_I$)
Starch	Starch + (n − 1) H_2O $\xrightleftharpoons[\text{glucosidase}]{\text{Amylo-}}$ n-Glucose	As Glucose ($b_H + b_I$)
Galactose	β-D-Galactose + $NAD^{\oplus}$ $\xrightleftharpoons[\text{dehydrogenase}]{\text{Galactose}}$ D-Galactono-γ-lactone + NADH + $H^{\oplus}$	
Ethanol	Ethanol + NAD $\xrightleftharpoons[\text{dehydrogenase}]{\text{Alcohol}}$ Acetaldehyde + NADH + $H^{\oplus}$	
Glycerol	Glycerol + ATP $\xrightleftharpoons[\text{kinase}]{\text{Glycerol}}$ sn-Glycerol-3P + ADP	ADP + Phosphoenolpyruvate $\xrightleftharpoons[\text{kinase}]{\text{Pyruvate}}$ ATP + Pyruvate (c)
		Pyruvate + NADH + $H^{\oplus}$ $\xrightleftharpoons[\text{dehydrogenase}]{\text{Lactate}}$ Lactate + $NAD^{\oplus}$ (d)
Lactate	L-Lactate assay is achieved by a reversed reaction under d), and D-lactate assay with a dehydrogenase specific for D-isomer.	
Creatinine and Creatine	Creatinine + H_2O $\xrightleftharpoons{\text{Creatininase}}$ Creatine	Creatine + ATP $\xrightleftharpoons[\text{kinase}]{\text{Creatine}}$ Creatine-P + ADP;
Individual amino acids	R − CH(NH_2)COOH $\xrightleftharpoons[\text{decarboxylase}^{d}]{\text{Amino acid}}$ R − CH_2 − NH_2 + CO_2	ADP is determined through c) and d)
L-Malate	L-Malate + $NAD^{\oplus}$ $\xrightleftharpoons[\text{dehydrogenase}]{\text{Malate}}$ Oxalacetate + NADH + $H^{\oplus}$	

[a] For saccharose and lactose see Fig. 2.32.
[b] The content of α-anomeric form is accessible through mutarotation.
[c] After hydrolysis this method is suitable for the assay of acylglycerols.
[d] Specific decarboxylases are available as exemplified by those for L-tyrosine, L-lysine, L-glutamic acid, L-aspartic acid, or L-arginine

$$\begin{array}{lcll} \text{Glucose} + \text{ATP} & \xrightarrow{\text{HK}} & \text{Glucose-6P} & (a) \\ \text{Glucose-6P} + \text{NADP}^{\oplus} & \xrightarrow{\text{G6P-DH}} & \text{6-Phosphogluconate} + \text{NADPH} + \text{H}^{\oplus} & (b) \\ \text{Lactose} & \xrightarrow{\beta\text{-Ga}} & \text{Glucose} + \text{Galactose} & (c) \\ \text{Saccharose} & \xrightarrow{\beta\text{-F}} & \text{Glucose} + \text{Fructose} & (d) \end{array} \quad (2.88)$$

equilibrium is achieved. By using several sequential auxiliary reactions with one indicator reaction, it is possible to simultaneously determine several constituents in one assay. An example is the analysis of glucose, lactose and saccharose (cf. Reaction 2.88).

First, glucose is phosphorylated with ATP in an auxiliary reaction (a). The product, glucose-6-phosphate, is the substrate of the NADP-dependent indicator reaction (b). Addition of β-galactosidase starts the lactose analysis (c) in which the released glucose, after phosphorylation, is again measured through the indicator reaction [(b) of Reaction 2.88 and also Figure 2.32]. Finally, after addition of β-fructosidase, saccharose is cleaved (d) and the released glucose is again measured through reactions (a) and (b) as illustrated in Figure 2.32.

2.5.4.1.2 End-point method

This procedure is reliable when the reaction proceeds virtually to completion. If the substrate is only partly consumed, the equilibrium is displaced in favor of the products by increasing the concentration of reactant or by removing one of the products of the reaction. If it is not possible to achieve this, a standard curve must be prepared. In contrast to kinetic methods (see below), the concentrations of substrates which are to be analyzed in food must not be lower than the *Michaelis* constant of the catalyzing enzyme for the auxiliary reaction. The reaction time is readily calculated: it occurs when the reaction rate follows first-order kinetics for the greater part of the enzymatic reaction.

In a two-substrate reaction the enzyme is saturated with the second substrate. Since Equation 2.48 is valid under these conditions, the catalytic activity of the enzyme needed for the assay can be determined for both one- and two-substrate reactions. The examples shown in Table 2.14 suggest that enzymes with low K_m values are desirable in order to handle the substrate concentrations for the end-point method with greater flexibility.

Data for K_m and V are needed in order to calculate the reaction time required (cf. *H. U. Bergmeyer* and *K. Gawehn,* 1977). A prerequisite is a reaction in which the equilibrium state is displaced toward formation of product with a conversion efficiency of 99%.

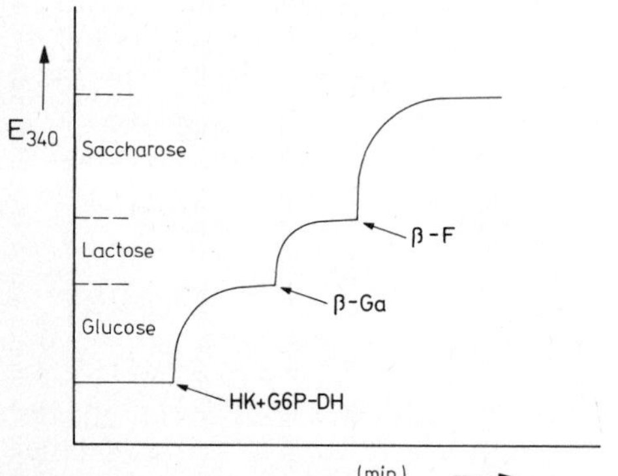

Fig. 2.32. Enzymatic determination of glucose, saccharose and lactose in one run. After adding cosubstrates, ATP and NADP, the enzymes are added in the order: hexokinase (HK), glucose-6-phosphate dehydrogenase (G6P-DH), β-galactosidase (β-Ga) and β-fructosidase (β-F)

Table 2.14. Enzyme concentrations used in the end-point method of enzymatic food analysis

Substrate	Enzyme	K_m (M)	Enzyme concentration (μcat/l)
Glucose	Hexokinase	$1.0 \cdot 10^{-4}$ (30 °C)	1.67
Glycerol	Glycerol kinase	$5.0 \cdot 10^{-5}$ (25 °C)	0.83
Uric acid	Urate oxidase	$1.7 \cdot 10^{-5}$ (20 °C)	0.28
Fumaric acid	Fumarase	$1.7 \cdot 10^{-6}$ (21 °C)	0.03

2.5.4.1.3 Kinetic method

Substrate concentration is obtained using a method based on kinetics by measuring the reaction rate. To reduce the time required per assay, the requirement for the quantitative conversion of substrate is abandoned. Since kinetic methods are less susceptible to interference than the endpoint method, they are advantageous for automated methods of enzymatic analysis.

The determination of substrate using kinetic methods is possible only as long as Equation 2.53 is valid. Hence, the following is required to perform the assay:

a) For a two-substrate reaction, the concentration of the second reactant must be high so that the rate of reaction depends only on the concentration of the substrate which is being analyzed.
b) Enzymes with high *Michaelis* constants are required; this enables relatively high substrate concentrations to be determined.
c) If enzymes with high *Michaelis* constants are not available, the apparent K_m is increased by using competitive inhibitors.

In order to explain requirement c), let us consider the example of the determination of glycerol as given in Table 2.13. This reaction allows the determination of only low concentrations of glycerol since the K_m values for participating enzymes are low: 6×10^{-5} M to 3×10^{-4} M.

In the reaction sequence the enzyme pyruvate kinase is competitively inhibited by ATP with respect to ADP. The expression $K_m(1 + (I)/K_I)$ (cf. 2.5.2.2.1) may in these circumstances assume a value of 6×10^{-3} M, for example. This corresponds to an apparent increase by a factor of 20 for the K_m of ADP (3×10^{-4} M). The ratio (S)/ $K_m(1 + [I]/K_I)$ therefore becomes 1×10^{-3} to 3×10^{-2}. Under these conditions, the auxilary reaction with pyruvate kinase follows pseudo-first-order kinetics with respect to ADP over a wide range of concentration and, as a result of the inhibition by ATP, it is also the rate-determining step of the overall reaction. It is then possible to kinetically determine higher concentrations of glycerol.

2.5.4.2 Determination of Enzyme Activity

In the foreword of this chapter it was emphasized that enzymes are suitable indicators for identifying heat-treated food. However, the determination of enzyme activity reaches far beyond this possibility: it is being used to an increasing extent for the evaluation of the quality of raw food and for optimizing the parameters of particular food processes. In addition, the activities of enzyme preparations have to be controlled prior to use in processing or in enzymatic food analysis.

The measure of the catalytic activity of an enzyme is the rate of the reaction catalyzed by the enzyme. The conditions of an enzyme activity assay are optimized with relation to: type and ionic strength of the buffer, pH, and concentrations of substrate, cosubstrate and activators used. The closely-controlled assay conditions, including the temperature, are critical because, in contrast to substrate analysis, the reliability of the results in this case often can not be verified by using a weighed standard sample.

Temperature is a particularly important parameter which strongly influences the enzyme assay. Temperature fluctuations significantly affect the reaction rate (cf. 2.6); e.g. a 1 °C increase in temperature results in about a 10% increase in activity. Whenever possible, the incubation temperature should be maintained at 25 °C.

The substrate concentration in the assay is adjusted ideally so that Equation 2.47 is valid, i.e. $[A_0] \gg K_m$. Difficulties often arise while trying to achieve this condition: the substrate's solubility is limited; spectrophotometric readings become unreliable because of high light absorbance by the substrate; or the high concentration of the substrate inhibits enzyme activity. For such cases procedures exist to assess the optimum substrate concentration which will support a reliable activity assay (cf. *H. U. Bergmeyer* and *K. Gawehn*, 1977).

2.6 Temperature Effects

The enzymatic process is influenced the most by the temperature to which food is exposed during processing and storage. The desired reactions such as ripening or other processes in which enzymes are added as outlined under section 2.8.2 are promoted by adjusting and maintaining the optimal incubation temperature. Reactions which negatively affect the aroma, nutritional value or texture of food are slowed down by low

temperature storage, or are stopped by enzyme inactivation during heat treatment.
In some foods the enzymes which must be inactivated in order to preserve food quality have been identified. Examples are listed in Table 2.15. For many foods such knowledge is still lacking; what is known is that food must be sterilized, pasteurized or blanched in order to ensure shelf life stability by thermal inactivation of enzymes.
The following section describes parameters which characterize the effect of temperature on an enzyme. This is followed by a discussion of the phenomena involved in the thermal treatment of enzymes or of enzyme-active food.

2.6.1 Q_{10} Value

In biology and food chemistry the effect of temperature on the rate of a complex process is often described by the Q_{10} value which is defined as the increase in reaction rate for a 10 °C increase in temperature:

$$Q_{10} = \frac{v_{T+10°}}{v_T} \qquad (2.89)$$

v_T = Reaction rate at temperature T

The Q_{10} value for most chemical and enzymatic reactions is 2–3. It is related to the reaction rate v through an exponential function:

$$\log (v_{T_1} - v_{T_2}) = \frac{\Delta T}{10°} \log Q_{10} \qquad (2.90)$$

Temperature difference $\Delta T = T_1 - T_2$

Consequently, a small change in Q_{10} is reflected by a large difference in reaction rate.

2.6.2 Activation Energy

An empirical relation exists for the dependence of reaction rate on temperature. It is expressed by a general equation of *S. Arrhenius:*

$$\log k = \log A - \frac{E_a}{2{,}3RT} \qquad (2.91)$$

k: Rate constant for the reaction at temperature T (°K)
E_a: Activation energy
T: Temperature (K)
R: Gas constant (8.3143 J K^{-1} $mole^{-1}$)
A: Arrhenius factor (the frequency factor)

Table 2.15. Thermal inactivation of enzymes to prevent deterioration of food quality

Food product	Enzyme	Quality loss
Potato products, apple	Monophenol oxidase	Enzymatic browning
Semi-ripe peas	Lipoxygenase, peroxidase	Flavor defects; bleaching
Fish products	Proteinase, thiaminase	Texture (liquefaction), loss of vitamine B_1
Tomato purée	Polygalacturonase	Texture (liquefaction)
Apricot products	β-Glucosidase	Color defects
Oat flakes	Lipase, lipoxygenase	Flavor defects (bitter taste)

In the case of enzyme reactions, the rate constant from Equation 2.47 can be replaced by V when the temperature dependence is measured at correspondingly high substrate concentrations ($[A_0] \gg K_m$).
The activation energy, i.e. the minimum amount of energy which a molecule must possess in order to be converted to product, can be best determined graphically by plotting the experimental data in the form of log V or log k versus 1/T (cf. example in Figure 2.34).
Which step of catalysis requires the ascertained activation energy and, thus, is rate determining, can be clarified only after the reaction mechanism is known. However, such investigations are difficult and time consuming. Detailed energy profiles have been determined for only a few enzymes (e.g. fumarase, α-chymotrypsin; cf. *J. R. Whitaker,* 1972).
It can be concluded from the data in Table 2.1 that the activation energies (E_a) of enzyme-catalyzed reactions fall in the range of 10–60 kJ $mole^{-1}$.
For enzymes which are able to convert more than one substrate or compound into product, the activation energy many be dependent on the substrate. For example, alcohol dehydrogenase, an important enzyme for aroma formation in peas, dehydrogenates alcohols with different activation energies (Table 2.16). On the other hand, the activation energies for the reverse reaction of this enzyme (aldehydes to alcohols) are only slightly influenced by substrate (Table 2.16)

The Q_{10} value and the activation energy (E_a) are related by the equation:

$$Q_{10} = e^{10 E_a / RT^2} \qquad (2.92)$$

Table 2.16. Alcohol dehydrogenase from pea seeds: activation energy for alcohol dehydrogenation and aldehyde reduction

Alcohol	E_a ($kJ \cdot mole^{-1}$)	Aldehyde	E_a ($kJ \cdot mole^{-1}$)
Ethanol	20		
n-Propanol	37	n-Propanal	20
2-Propenol	18		
n-Butanol	40	n-Butanal	21
n-Hexanol	35	n-Hexanal	18
2-trans-hexenol	15	2-trans-Hexenal	19
		2-trans-Heptenal	18

Hence, Q_{10} values are dependent on temperature, whereas E_a is independent of temperature. The *Arrhenius* equation (Equation 2.91) is explained with the help of the collision theory of chemical reactions and the theory of the transition state (cf. textbook of physical chemistry).

2.6.3 Temperature Optimum

An enzyme-catalyzed reaction behaves initially as any other chemical reaction when the temperature is increased: the reaction rate increases (Figure 2.33). However, the increase in reaction rate departs from this behavior after reaching a "temperature optimum". Beyond this point, the rate of catalysis declines rapidly (Figure 2.33). The "temperature optimum" (example in Figure 2.44) results from the overlapping of two counter effects:

- Enzyme catalysis
- Enzyme inactivation

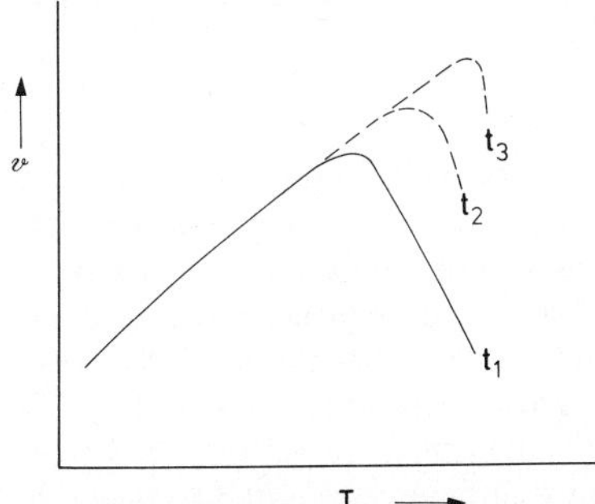

Fig. 2.33. Catalytic activity of an enzyme as influenced by temperature. Times of thermal treatment during incubation differ $t_1 > t_2 > t_3$

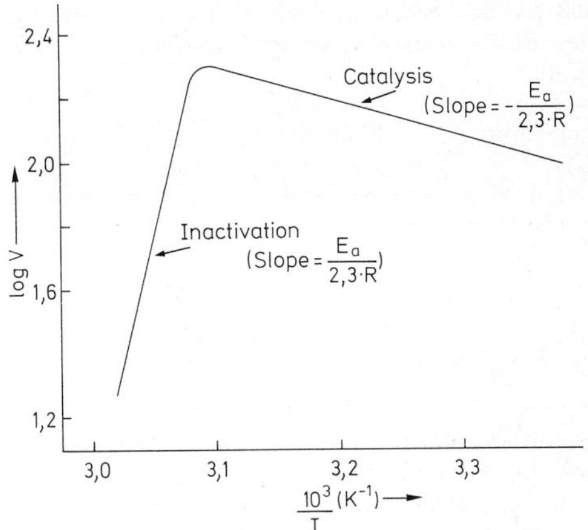

Fig. 2.34. Fungal α-amylase. Amylose hydrolysis versus temperature. *Arrhenius* diagram for assessing the activation energy of enzyme catalysis and enzyme inactivation; V = total reaction rate

and both reaction rates increase with increasing temperature.

In the following example, the two effects are far apart with respect to their activation energies. This is the case for the hydrolysis of amylose catalyzed by a preparation of microbial α-amylase as affected by temperature. By making an *Arrhenius* plot of log V or log k versus 1/T (Figure 2.34), the following values of E_a are calculated from the both slopes of the curve:

E_a (hydrolysis): 20 kJ $mole^{-1}$
E_a (inactivation): 295 kJ $mole^{-1}$

These values are within the limits for general enzyme catalysis (E_a: 10–60 kJ $mole^{-1}$; cf. Table 2.1) and enzyme inactivation (E_a: 200–750 kJ $mole^{-1}$).

As a consequence of the difference in activation energies, the rate of enzyme inactivation is substantially faster with increasing temperature than the rate of enzyme catalysis. Based on activation energies for our example (Table 2.17), the following relationship can be derived: increasing the temperature from 0 to 60 °C increases the hydrolysis rate by a factor of 5, while the rate of enzyme inactivation is accelerated by more than a power of 10.

The "temperature optimum" can be used only conditionally to characterize the enzyme since its position depends on the assay conditions (Figure 2.33). The longer the enzyme is exposed to higher

Table 2.17. α-Amylase activity as affected by temperature: relative rates of hydrolysis and enzyme inactivation

Temperature (°C)	Relative rates[a]	
	hydrolysis	inactivation
0	1.0	1.0
10	1.35	$1.0 \cdot 10^2$
20	1.8	$0.7 \cdot 10^4$
40	3.0	$1.7 \cdot 10^7$
60	4.8	$1.5 \cdot 10^{10}$

[a] Activation energies of 20 kJ · mole^{-1} for hydrolysis and 295 kJ · mole^{-1} for enzyme inactivation were used for calculation after *J. R. Whitaker* (1972).

temperatures, the greater is the amount of enzyme being inactivated or denatured; hence, the lower is the observed "temperature optimum". Therefore, the term "temperature optimum" is more an operational parameter than a reliable characteristic of the enzyme.

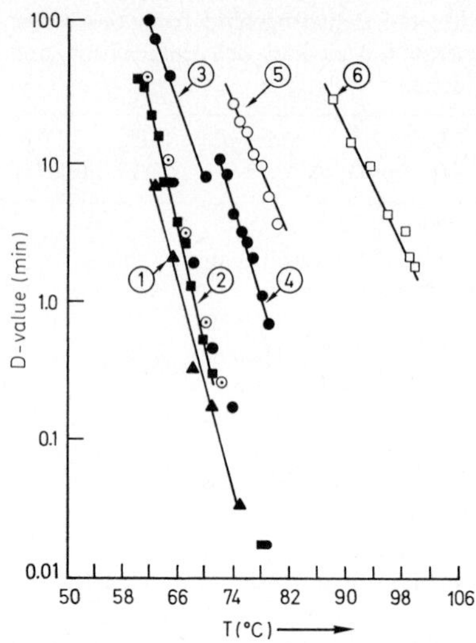

Fig. 2.35. Thermal inactivation of enzymes of milk. 1 Lipase (inactivation extent, 90%), 2 alkaline phosphatase (90%), 3 catalase (80%), 4 xanthine oxidase (90%), 5 peroxidase (90%), and 6 acid phosphatase (99%)

2.6.4 Thermal Stability

The thermal stability of enzymes is quite variable. Only a few enzymes lose their catalytic activity at 0 °C, while some are even resistant to high temperatures, at least for a short time. In a few cases enzyme stability is lower at 0–10 °C than in a medium temperature range (20–40 °C). Lipase and alkaline phosphatase in milk are thermolabile (Figure 2.35), whereas acid phosphatase is relatively stable. Therefore, alkaline phosphatase (its activity is easier to determine than that of lipase) is used to distinguish raw from pasteurized milk. Peroxidase is the last enzyme in the potato tuber to be thermally inactivated (Figure 2.36).

Such inactivation patterns are often found among enzymes in vegetables. In such cases, peroxidase is a suitable indicator for controlling the total inactivation of all the enzymes as is required in assessing the adequacy of a blanching process.

All the changes which occur in proteins outlined in section 1.4.2.4 also occur during the heating of enzymes. In the case of enzymes the consequences are even more readily observed since a slight conformational change at the active site can result in total loss of activity.

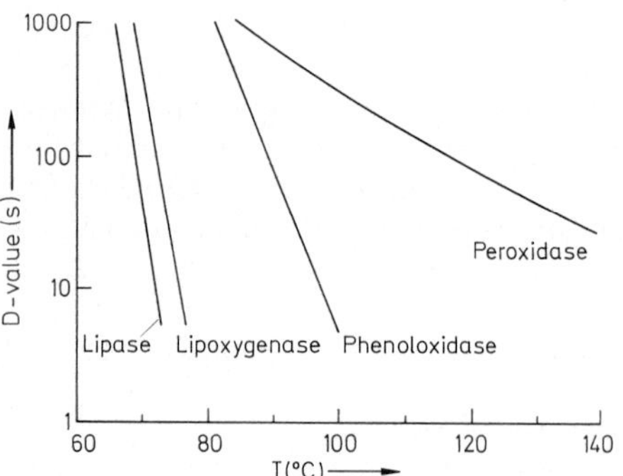

Fig. 2.36. Thermal inactivation (90%) of enzymes present in potato tuber

In the graphs depicted in Figures 2.35 and 2.36 the so-called "D-value" is plotted against temperature. This value represents the time (min, s) needed to reduce the catalytic activity to a preselected residual activity (usually 10% but also 50% which corresponds to D = 1/2 t). The "Z-value" is derived from these values. This gives the temperature difference required to decrease the D-value from 100% to 10%. The higher the Z-value, the higher the thermal stability of the enzyme (cf. Figure 2.36 and Table 2.18).

Table 2.18. Thermal inactivation of various enzymes from potato tuber

Enzyme	Z-Value (°C)
Lipase	3.1
Lipoxygenase	3.6
Monophenol oxidase	7.8
Peroxidase	35.0

The inactivation rate depends on several factors. The pH effect is significant. Lipoxygenase isolated from pea seeds (Figure 2.37) denatures most slowly at its isoelectric point. This property of lipoxygenase coincides with that of many other enzymes.

Some thermal stability data for proteinases of intereast in food processing are compiled in Table

Table 2.18 a. Examples for the use of microbial enzymes in food.

EC-Number	Enzyme[a]	Biological Origin	Application[b]
Oxidoreductases			
1.1.1.39	Malate dehydrogenase (decarboxylating)	*Leuconostoc oenos*	10
1.1.3.4	Glucose oxidase	*Aspergillus niger*	7, 10, 16
1.11.1.6	Catalase	*Micrococcus lysodeicticus*	
		Aspergillus niger	1, 2, 7, 10, 16
Hydrolases			
3.1.1.1	Carboxylesterase	*Mucor miehei*	2, 3
3.1.1.3	Triacylglycerol lipase	*Aspergillus niger, A. oryzae, Candida lipolytica, Mucor javanicus, M. miehei, Rhizopus arrhizus, R. niveus*	2, 3
3.1.1.11	Pectinesterase	*Aspergillus niger*	9, 10, 17
3.1.1.20	Tannase	*Aspergillus niger, A. oryzae*	10
3.2.1.1	α-Amylase	*Bacillus licheniformis, B. subtilis, Aspergillus oryzae*	3, 8, 9, 10, 12, 14, 15
		Aspergillus niger, Rhizopus delemar, R. oryzae	8, 9, 10, 12, 14, 15
3.2.1.2	β-Amylase	*Bacillus cereus, B. magatherium, B. subtilis*	8, 10
3.2.1.3	Exo-1,4-α-D-glucosidase	*Aspergillus oryzae*	3, 9, 10, 12, 14, 15, 18
		Aspergillus niger, Rhizopus arrhizus, R. delemar, R. niveus, R. oryzae, Trichoderma reesei	9, 10, 12, 14, 15, 18
3.2.1.4	Cellulase	*Aspergillus niger, A. oryzae, Rhizopus delemar, R. oryzae, Sporotrichum dimorphosporum, Thielavia terrestris, Trichoderma reesei*	9, 10, 18
3.2.1.6	Endo-1,3(4)-β-D-glucanase	*Bacillus circulans, B. subtilis, Aspergillus niger, A. oryzae, Penicillium emersonii, Rhizopus delemar, R. oryzae*	10
3.2.1.7	Inulinase	*Kluyveromyces fragilis*	12
3.2.1.11	Dextranase	*Klebsiella aerogenes, Penicillium funicolosum, P. lilacinum*	12
3.2.1.15	Polygalacturonase	*Aspergillus niger, Penicillium simplicissimum, Trichoderma reesei*	3, 9, 10, 17
		A. oryzae, Rhizopus oryzae	3, 9, 10
		Aspergillus niger	9, 10, 17
3.2.1.20	α-D-Glucosidase	*Aspergillus niger, A. oryzae, Rhizopus oryzae*	8
3.2.1.21	β-D-Glucosidase	*Aspergillus niger, Trichoderma reesei*	9
3.2.1.22	α-D-Galactosidase	*Aspergillus niger, Mortierella vinacea sp., Saccharomyces carlsbergensis*	12

2.20. However, data for isolated and additionally purified enzymes may not be transferrable or pertinent to the same enzymes in intact tissue in which they are protected by other constituents in the cell. Additional studies mostly related to heat transfer in food, have developed, in a few cases, successful procedures to calculate the degree of enzyme inactivation based on thermal stability data of isolated enzymes. The extent of agreement between the calculated and the experimental results for pea lipoxygenase is presented in Figure 2.38.

Peroxidase activity can partially reappear during storage of vegetables in which the enzymes have been inactivated by blanching. The reason for this recurrence, which is also observed for alkaline phosphatase of milk, is not yet clear.

Enzymes behave differently below the freezing point. The changes observed depend on the type of enzyme and on a number of other factors. The activity usually parallels the increase in enzyme or substrate concentration due to ice crystallization, it either increases or decreases according to shifts in pH and it decreases with a rise in viscosity of the medium. The latter phenomenon is due to the restricted diffusion of substrate. In completely frozen food, a state reached only during deep-freezing, the catalytic activity slows down or stops temporarily. Relatively few enzymes are irreversibly destroyed by freezing.

Table 2.18 a: Continued

EC-Number	Enzyme[a]	Biological Origin	Application[b]
3.2.1.23	β-D-Galactosidase	*Aspergillus niger, A. oryzae, Kluyveromyces fragilis, K. lactis*	1, 2, 4, 18
3.2.1.26	β-D-Fructofuranosidase	*Aspergillus niger, Saccharomyces carlsbergensis, S. cerevisiae*	14
3.2.1.32	Endo-1,3-β-D-xylanase	*Streptomyces sp., Aspergillus niger, Sporotrichum dimorphosporum*	8, 10, 13
3.2.1.41	Pullulanase	*Bacillus acidopullulyticus*	8, 10, 12, 14, 15
		Klebsiella aerogenes	8, 10, 12
3.2.1.55	α-L-Arabinofuranosidase	*Aspergillus niger*	9, 10, 17
3.2.1.58	Exo-1,3-β-D-glucosidase	*Trichoderma harzianum*	10
3.2.1.68	Isoamylase	*Bacillus cereus*	8, 10
3.2.1.78	Endo-1,4-β-D-mannanase	*Bacillus subtilis, Aspergillus oryzae, Rhizopus delemar, R. oryzae, Sporotrichum dimorphosporum, Trichoderma reesei*	13
		Aspergillus niger	13, 17
3.4.21.14	Microbial serine proteinase[c]	*Bacillus licheniformis*	5, 6, 10, 11
3.4.23.6	Microbial carboxyl proteinases	*Aspergillus melleus, Endothia parasitica, Mucor miehei, M. pusillus*	2
		Aspergillus oryzae	2, 5, 6, 8, 9, 10, 11, 15, 18
3.4.24.4	Microbial metalloproteinases	*Bacillus cereus, B. subtilis*	10, 15
Lyases			
4.2.2.10	Pectin lyase	*Aspergillus niger*	9, 10, 17
Isomerases			
5.3.1.5	Xylose isomerase[d]	*Actinoplanes missouriensis, Arthrobacter sp., Bacillus coagulans, Streptomyces albus, S. olivaceus, S. olivochromogenes, S. rubiginosus*	8, 9, 10, 12

[a] Principal activity

[b] 1) Milk, 2) Cheese, 3) Fats and oils, 4) Edible ice, 5) Meat, 6) Fish, 7) Egg, 8) Cereal and starch, 9) Fruit and vegetables, 10) Beverages (soft drinks, beer, wine), 11) Soups and broths, 12) Sugar and honey, 13) Cacao, chocolate, coffee, tea, 14) Confectionery, 15) Bakery, 16) Salads, 17) Spices and flavors, 18) Dietary food.

[c] Similar to Subtilisin

[d] Same enzymes also convert D-glucose to D-fructose

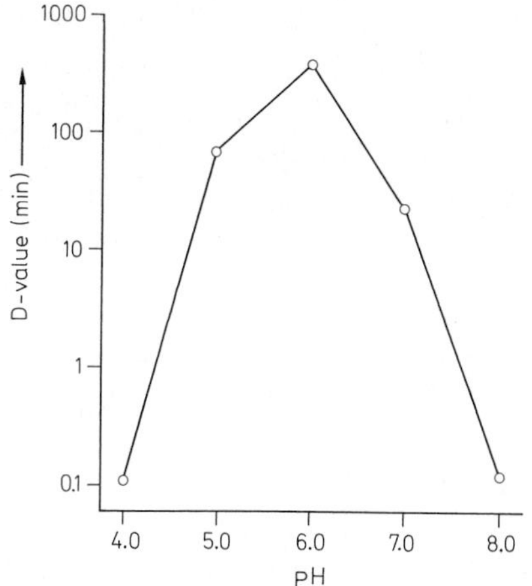

Fig. 2.37. Pea seed lipoxygenase. Inactivation extent at 65 °C as affected by pH

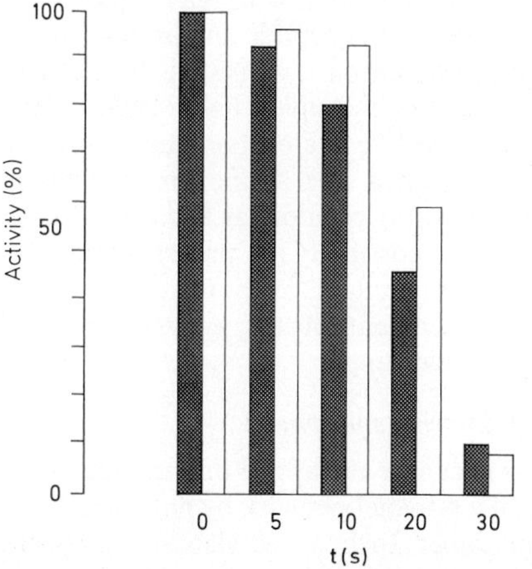

Fig. 2.38. Blanching of semiripened peas at 95 °C; lipoxygenase inactivation. (After *S. Svensson*, 1977). ■ Experimentally found, □ calculated

2.7 Effect of Water Activity

The rates of enzymatic reactions decrease with lowering of the water activity (cf. Figure 0.4). Hydrolases require water since it acts as a substrate. Furthermore water may enhance substrate diffusion to the enzyme. Therefore, the rate of enzyme catalysis is decreased when food is dried (example in Figure 2.39).

Some enzymatic reactions are possible even below the BET-point (definition in 0.3.1). In this case the substrate can reach the enzyme without

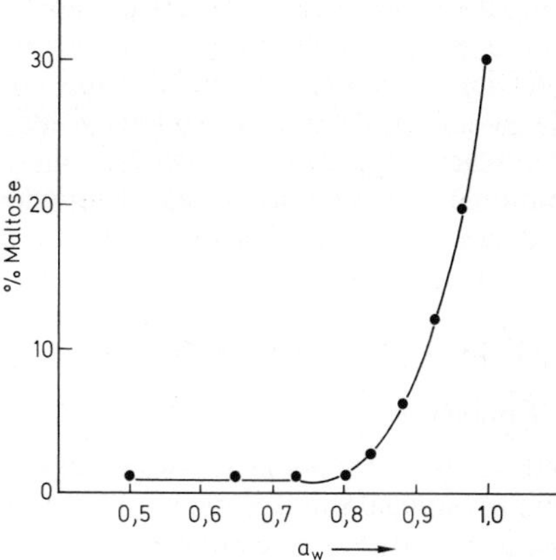

Fig. 2.39. Starch hydrolysis by β-amylase as affected by water activity. (After *J. A. Troller* and *J. B. Christian*, 1978)

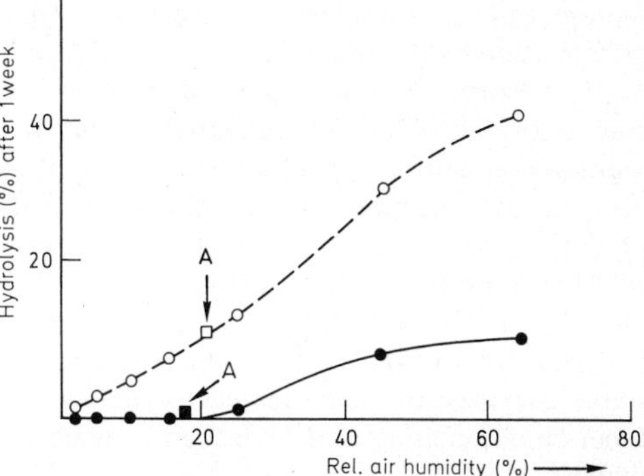

Fig. 2.40. Storage stability of mixtures of cellulose, lipase and trilaurine (●–●) and cellulose, lipase and triolein (–○–○–). (After *L. Acker* and *R. Wiese*, 1972) A: BET-point

the help of water. So, it is possible to hydrolyze liquid triolein but not solid trilaurin below the BET – point with a lipase enzyme (Figure 2.40).

2.8 Enzyme Utilization in the Food Industry

Enzyme-catalyzed reactions in food processing have been used unintentionally from ancient times. The enzymes are either an integral part of the food or are obtained from microorganisms. Addition of enriched or purified enzyme preparations of animal, plant or, especially, microbial origin is a recent practice. Such intentionally-used additives provide a number of advantages in food processing: exceptionally pronounced substrate specificity (cf. 2.2.2); high reaction rate under mild reaction conditions (temperature, pH); permits a fast and continuous, readily controlled reaction process with generally modest operational costs and investment. Examples for the application of microbial enzymes in food processing are given in Table 2.18 a.

2.8.1 Technical Enzyme Preparations

2.8.1.1 Production

The methods used for industrial-scale enzyme isolation are outlined in principle under section 2.2.4. In contrast to production of highly purified enzymes for analytical use, the production of enzymes for technical purposes is directed to removing the interfering activities which would be detrimental to processing and to staying within economically acceptable costs. Selective enzyme precipitation by changing the ionic strength and/or pH, adsorption on inorganic gels such as calcium phosphate gel or hydroxyapatite, chromatography on porous gel columns and ultrafiltration through membranes are among the fractionation methods commonly used. Ion-exchange affinity chromatography (cf. 2.2.4) and preparative electrophoresis are relatively expensive and are seldom used. A few temperature-stable enzymes are heat treated to remove the other contaminating and undesired enzyme activities.

Commercial enzyme preparations are available with defined catalytic activity. The activity is usually adjusted by the addition of suitable inert fillers such as salts or carbohydrates. The amount of active enzyme is relatively low, e.g. proteinase preparations contain 5–10% proteinase, whereas amylase preparations used for treatment of flour contain only 0,1% pure fungal α-amylase.

2.8.1.2 Immobilized Enzymes

Enzymes in solution are usually used only once. The repeated use of enzymes fixed to a carrier is more economical. The use of enzymes in a continuous process, for example, immobilized enzymes used in the form of a stationary phase which fills a reaction column where the reaction can be controlled simply by adjustment of the flow rate, is the most advanced technique. Immobilized enzymes are produced by various methods (Figure 2.41).

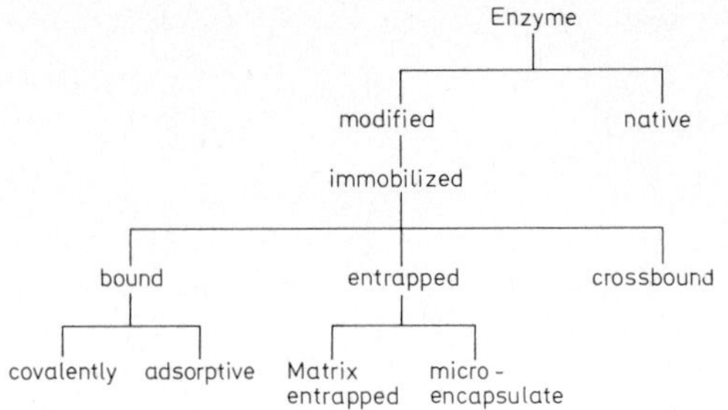

Fig. 2.41. Forms of immobilized enzyme

2.8.1.2.1 Bound enzymes

An enzyme can be bound to a carrier by covalent chemical linkages or, in many cases, by physical forces such as adsorption, by charge attraction, H-bridge formation and/or hydrophobic interactions. The covalent attachment to carrier, in this case an activated matrix, is usually achieved by methods employed in peptide and protein chemistry. First, the matrix is activated. In the next step, the enzyme is coupled under mild conditions to the reactive site on the matrix, usually by reaction with a free amino group. This is illustrated by using cellulose as a matrix (Figure 2.42). Another possibility is a process of copolymerization with suitable monomers. Generally, covalent attachment of the enzyme prevents leaching or "bleeding".

2.8.1.2.2 Enzyme entrapment

An enzyme can be entrapped or enclosed in the cavities of a polymer network by polymerization of a monomer such as acrylamide or N,N′-methylene-bis-acrylamide in the presence of enzyme, and remain accessible to substrate through the network of pores. Furthermore, suitable processes can bring about enzyme encapsulation in a semipermeable membrane (microencapsulation) or confinement in hollow fiber bundles.

2.8.1.2.3 Cross-linked enzymes

Derivatization of enzymes using a bifunctional reagent, e.g. glutaraldehyde, can result in cross-linking of the enzyme and, thus, formation of large, still catalytically-active insoluble complexes. Such enzyme preparations are relatively unstable for handling and, therefore, are used mostly for analytical work.

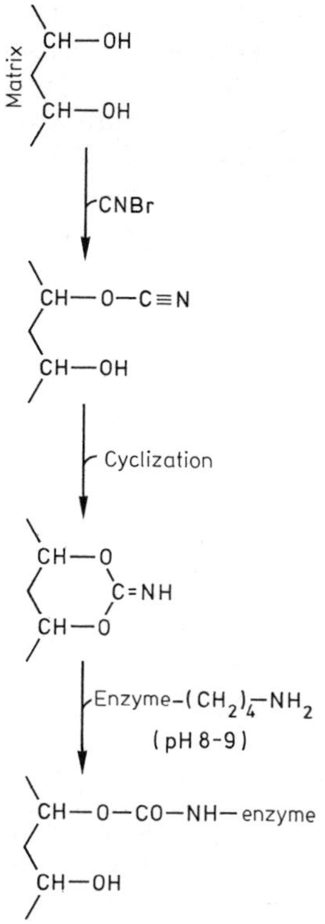

Matrix
CH—OH
CH—OH
↓ CNBr
CH—O—C≡N
CH—OH
↓ Cyclization
CH—O
C=NH
CH—O
↓ Enzyme–$(CH_2)_4$–NH_2 (pH 8–9)
CH—O—CO—NH—enzyme
CH—OH

Fig. 2.42. Enzyme immobilization by covalent binding to a cellulose matrix

2.8.1.2.4 Properties

The properties of an immobilized enzyme are often affected by the matrix and the methods used for immobilization.

Kinetics. As a rule, higher substrate concentrations are required for saturation of an entrapped enzyme than for a free, native enzyme. This is due to a decrease in the concentration gradient which takes place in the pores of the polymer network. Also, there is an increase in the "apparent" *Michaelis* constant for an enzyme bound covalently to a matrix carrying an electrostatic charge. This is also true when substrate and the functional groups of the matrix carry the same charge. On the other hand, opposite charges bring about an increase of substrate affinity towards the matrix. Consequently, this decreases the "apparent" K_m.

pH Optimum. Negatively-charged groups on a carrier matrix shift the pH optimum of the covalently-bound enzyme to the alkaline region, wherease positive charges shift the pH optimum towards lower pH values. The change in pH optimum of an immobilized enzyme can amount to one to two pH units in comparison to that of a free, native enzyme.

Thermal inactivation. Unlike native enzymes, the immobilized forms are often more heat stable (cf. example for β-D-glucosidase, Figure 2.43). Heat stability and pH optima changes induced by immobilization are of great interest in the industrial utilization of enzymes.

2.8.2 Individual Enzymes

2.8.2.1 Oxidoreductases

Broader applications for the processing industry, besides the familiar use of glucose oxidase, are found primarily for catalase and lipoxygenase among the many enzymes of this group. A number of oxidoreductases have been suggested or are in the experimental stage of utilization, particularly for aroma improvement (examples under 2.8.1.4 and 2.8.2.1.5).

2.8.2.1.1 Glucose oxidase

The enzyme produced by fungi such as *Apergillus niger* (cf. 2.3.2.1) and *Penicillium notatum* catalyzes glucose oxidation by consuming oxygen from the air (Table 2.13); hence, it is used for the

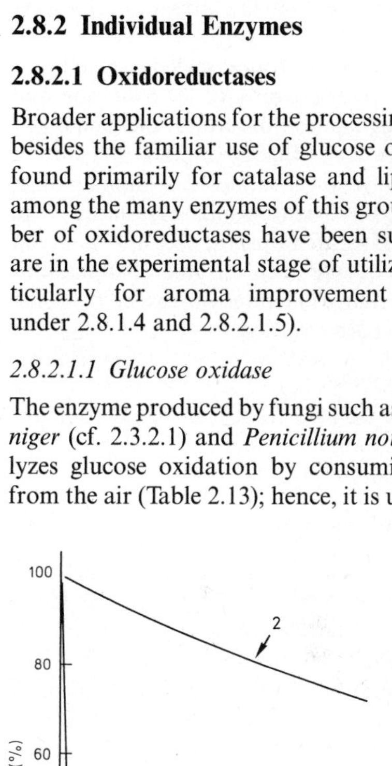

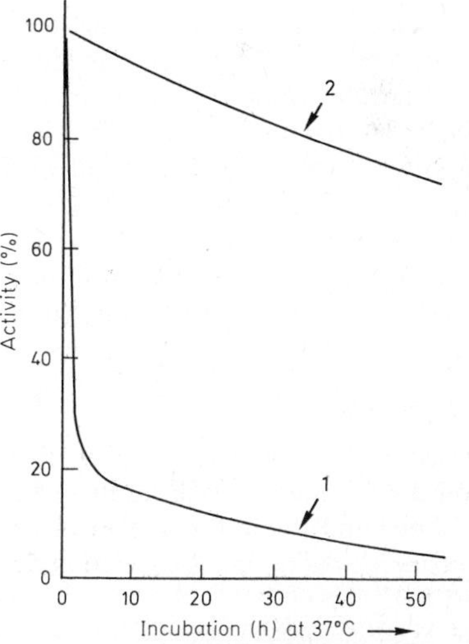

Fig. 2.43. Thermal stabilities of free and immobilized enzymes. (After *O. R. Zaborsky*, 1973). 1 β-D-glucosidase, free, 2 β-D-glucosidase, immobilized

removal of either glucose or oxygen. The H_2O_2 formed in the reaction is occasionally used as an oxidizing agent, but it is usually degraded by catalase.

Removal of glucose during the production of egg powder using glucose oxidase (cf. 11.4.3) prevents the *Maillard* reaction responsible for discoloration of the product and deterioration of its whippability. Similar uses of glucose oxidase for some meat and protein products would be to enhance the golden-yellow color rather than the brown color of potato chips or French fries which is obtained in the presence of excess glucose.

Removal of oxygen from a sealed package system results in suppression of fat oxidation and oxidative degradation of natural pigments. For example, the color change of crabs and shrimp from pink to yellow is hindered by dipping them into a glucose oxidase/catalase solution. The shelf life of citrus fruit juices, beer and wine can be prolonged with such enzyme combinations since the oxidative reactions which lead to aroma deterioration are retarded.

2.8.2.1.2 Catalase

The enzyme isolated from liver or microorganisms is important as an auxiliary enzyme for the decomposition of H_2O_2:

$$2\,H_2O_2 \rightleftharpoons 2\,H_2O + O_2 \qquad (2.93)$$

Hydrogen peroxide is a by-product in the treatment of food with glucose oxidase or is added to food in some specific canning procedures. An example is the pasteurization of milk with H_2O_2, which is important when the thermal process is shut down by technical problems. Milk thus stabilized is also suitable for cheesemaking since the sensitive casein system is spared from heat damage. The excess H_2O_2 is then eliminated by catalase.

2.8.2.1.3 Lipoxygenase

The properties of this enzyme are described under section 3.7.2.2 and its utilization in the bleaching of flour and the improvement of the rheological properties of dough is covered under section 15.4.1.4.3.

2.8.2.1.4 Aldehyde dehydrogenase

During soya processing, volatile degradation compounds (hexanal, etc.) with a "bean-like" aroma defect are formed because of the enzymatic oxidation of unsaturated fatty acids. These defects can be eliminated by the enzymatic oxidation of the resultant aldehydes to carboxylic acids. Since the flavor threshold values of these acids are high, the acids generated do not interfere with the aroma improvement process.

$$\text{n-Hexanal} + NAD^{\oplus} \rightleftharpoons \text{Caproic acid} + NADH + H^{\oplus} \qquad (2.94)$$

Of the various aldehyde dehydrogenases, the enzyme from beef liver mitochondria has a particularly high affinity for n-hexanal (Table 2.19); hence its utilization in the production of soya milk is recommended.

Table 2.19. *Michaelis* constants for aldehyde dehydrogenase (ALD) from various sources

Substrate	K_m (μM)			
	ALD (bovine liver)			ALD
	Mitochondria	Cytosol	Microsomes	Yeast
Ethanal	0.05	440	1,500	30
n-Propanal	–	110	1,400	–
n-Butanal	0.1	<1	–	–
n-Hexanal	0.075	<1	<1	6
n-Octanal	0.06	<1	<1	–
n-Decanal	0.05	–	–	–

2.8.2.1.5 Butanediol dehydrogenase

Diacetyl formed during the fermentation of beer can be a cause of a flavor defect. The enzyme from *Aerobacter aerogenes*, for example, is able to correct this defect by reducing the diketone to the flavorless 2,3-butanediol:

$$CH_3{-}CO{-}CO{-}CH_3 + NADH + H^{\oplus} \rightleftharpoons CH_3{-}CH(OH){-}CH(OH){-}CH_3 + NAD^{\oplus} \qquad (2.95)$$

Such a process is improved by the utilization of yeast cells which, in addition to the enzyme and NADH, contain a system able to regenerate the cosubstrate. In order to prevent contamination of beer with undesirable cell constituents, the yeast cells are encapsulated with gelatin.

2.8.2.2 Hydrolases

Most of the enzymes used in the food industry are derived from the class of hydrolase enzyme (cf. Table 2.18 a).

2.8.2.2.1 Proteinases

The mixture of proteolytic enzymes used in the food industry contains primarily endopeptidases (specificity and classification under section 1.4.5.2). These enzymes are isolated from animal organs, higher plants or microorganisms, i.e. from their fermentation media (Table 2.20). Examples of their utilization are:

Proteinases are added to wheat flour in the production of some bakery products to modify rheological properties of dough and, thus, the firmness of the end-product. During such dough treatment, the firm or hard wheat gluten is partially hydrolyzed to a soft-type gluten (cf. 15.4.1.4.5).

In the dairy industry the formation of casein curd is achieved with chymosin or rennin (cf. Table 2.20) by a reaction mechanism described under section 10.1.2.1.1. Casein is also precipitated through the action of other proteinases by a mechanism which involves secondary proteolytic activity resulting in diminished curd yields and lower curd strength. Rennin is essentially free of other undesirable proteinases and is, therefore, especially suitable for cheesemaking. However, there is a shortage of rennin since it is isolated from the stomach of a suckling calf. Hence, substitutes have been sought in recent decades. Microbial proteinases, especially those of *Mucor* spp., e.g. *Mucor pusillus,* appear to meet the demand.

Plant proteinases (cf. Table 2.20) and also those of microorganisms are utilized for ripening and tenderizing meat. The practical problem to be solved is how to achieve uniform distribution of the enzymes in muscle tissue. An optional method appears to be injection of the proteinase into the blood stream immediately before slaughter, or rehydration of the freeze-dried meat in enzyme solutions.

Cold turbidity in beer is associated with protein sedimentation. This can be eliminated by hydrolysis of protein using plant proteinases (cf. Table 2.20). Utilization of papain was suggested by *Wallerstein* in 1911. Production of complete or partial protein hydrolysates by enzymatic methods is another example of an industrial use of proteinases. This is the process used in the liquefaction of fish proteins to make products with good flavors.

One of the concerns in the enzymatic hydrolysis of proteins is to avoid the release of bitter-tasting peptides and/or amino acids (cf. 1.3.3). Their occurrence in the majority of proteins treated (an exception is collagen) can not be ignored, especially when the extent of hydrolysis yields peptide fragments of less than 6 kdal.

Table 2.20. Proteinases utilized in food processing

Name	Source	pH-optimum	Optimal stability pH-range
	A. Proteinases of animal origin		
Pancreatic proteinase[a]	Pancreas	9.0[b]	3–5
Pepsin	Gastric lining of swine or bovine	2	
Chymosin	Stomach lining of calves	6–7	5.5–6.0
	B. Proteinases of plant origin		
Papain	Tropical melon tree (*Carica papaya*)	7–8	4.5–6.5
Bromelain	Pineapple (fruit and stalk)	7–8	
Ficin	Figs (*Ficus carica*)	7–8	
	C. Bacterial proteinases		
Alkaline proteinase e.g. subtilisin	*Bacillus subtilis*	7–11	7.5–9.5
Neutral proteinase e.g. thermolysin	*Bacillus thermoproteolyticus*	6–9	6–8
Pronase	*Streptomyces griseus*		
	D. Fungal proteinases		
Acid proteinase	*Aspergillus oryzae*	3.0–4.0[d]	5
Neutral proteinase	*Aspergillus oryzae*	5.5–7.5[d]	7.0
Alkaline proteinase	*Aspergillus oryzae*	6.0–9.5[d]	7–8
Proteinase	*Mucor pusillus*	3.5–4.5[d]	3–6
Proteinase	*Rhizopus chinensis*	5.0	3.8–6.5

[a] A mixture of trypsin, chymotrypsin, and various peptidases with amylase and lipase as accompanying enzymes.
[b] With casein as a substrate.
[c] A mixture of various proteinases including amino- and carboxypeptidases.
[d] With hemoglobin as a substrate.

2.8.2.2.2 α-Amylase

This enzyme, as do the following enzymes (up to and including section 2.8.2.2.13), belongs to the class of hydrolases and the subclass glycosidases (Table 2.4). α-Amylase hydrolyzes starch, glycogen and other 1,4-α-glucans. The enzymatic attack occurs inside the molecule, not at the terminal ends of the molecule. Hence, the enzyme is comparable to an endopeptidase. The enzyme releases oligosaccharide fragments with 6–7 glucose units from amylose. Obviously, the enzyme attacks the amylose helix (cf. 4.4.3.12.1) and hydrolyzes the aligned neighboring glycoside linkages of each helical turn. Unlike amylose, amylopectin is cleaved randomly; the branching points (cf. 4.4.3.12.3) being skipped over. Ca^{2+} ions activate α-amylase (cf. 2.3.3.1).

The viscosity of a starch solution drops rapidly ("starch liquefaction") during hydrolysis with α-amylase and the typical starch iodine-blue color disappears. After prolonged hydrolysis, the initially-formed dextrins degrade further, yielding reducing sugars and, finally, the disaccharide α-maltose. The enzyme activity drops rapidly as the degree of polymerization of the substrate decreases.

Gelatinization of starch (cf. 4.4.3.12) enhances enzyme catalysis. Thus, the swollen starch substrate is degraded 300 times faster by bacterial amylase and 10^5 times faster by fungal amylase than the native, unswollen and ungelatinized starch.

During industrial-scale production, α-amylase is isolated from the pancreatic gland of swine and cattle and from microbial cultures e.g. the bacterium *Bacillus subtilis* or the fungi: *Aspergillus oryzae*. The high temperature-resistant bacterial amylases, particularly that of *Bacillus licheniformis* (Figure 2.44) are of interest for the hydrolysis of corn starch (gelatinization at 105–110 °C). The hydrolysis rate of this enzyme can be enhanced further by the addition of Ca^{2+} ions.

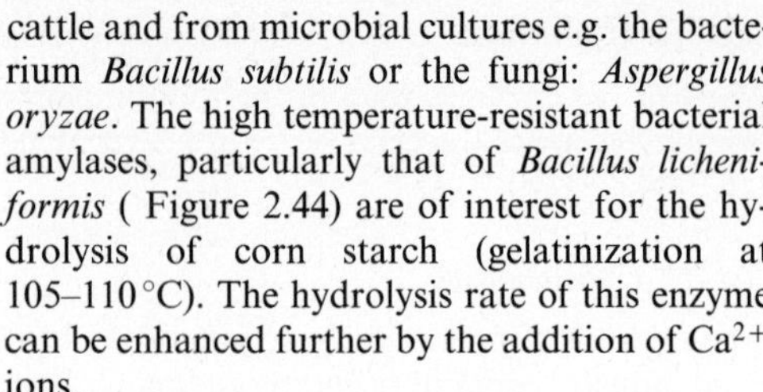

α-Amylase is utilized in the baking industry (cf. 15.4.1.4.8) and in the production of starch derivatives (cf. 2.8.2.2.4). In addition, studies are being conducted in several countries in which malt will be partially replaced by α-amylase and other enzymes of microbial origin in the production of beer and spirits.

2.8.2.2.3 β-Amylase

The enzyme hydrolyzes the 1,4-α-D-glycosidic linkage in polysaccharides (reaction mechanism in Figure 2.16). The reaction proceeds from the non-reducing end of the polysaccharide chain by the sequential release of maltose units. During hydrolysis a *Walden* inversion occurs on C-1 and so, instead of the α-form, β-maltose is obtained. In contrast to amylose, amylopectin is not completely hydrolyzed. When the enzyme reaches a branching point of the molecule, its activity ceases. During the malting of cereals, such as barley, β-amylase is biosynthesized to a greater extent than its counterpart, α-amylase. Because of the presence of these enzymes, malt is used extensively in processes where saccharification of starch is required (cf. Figure 2.45).

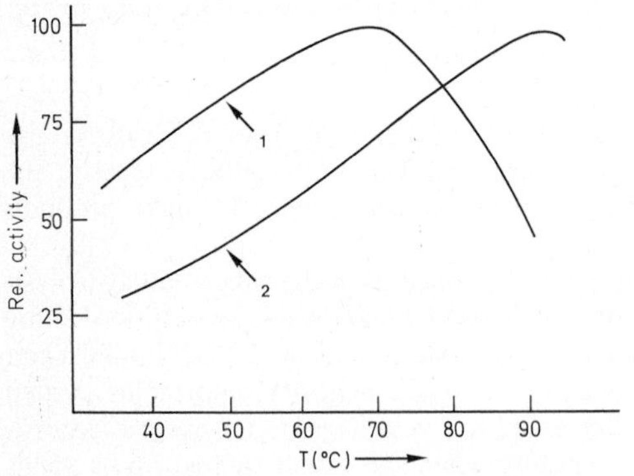

Fig. 2.44. The activity of α-amylase as influenced by temperature. 1 α-amylase from *Bacillus subtilis*, 2 from *Bacillus licheniformis*

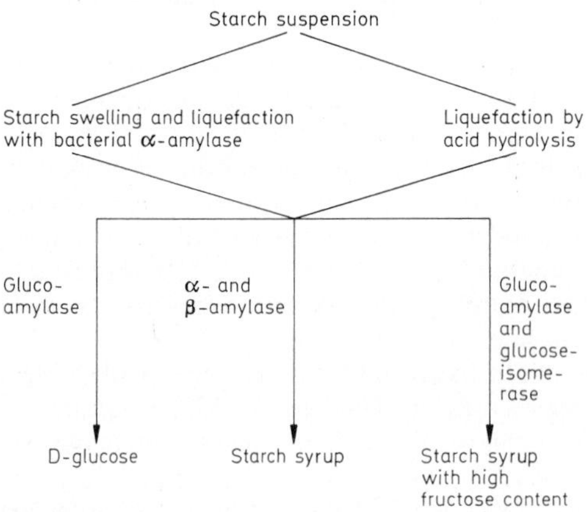

2.45. Enzymatic starch degradation

2.8.2.2.4 *Exo-1,4-α-D-glucosidase (glucoamylase)*

Glucoamylase cleaves β-D-glucose units from the non-reducing end of an 1,4-α-D-glucan. The α-1,6-branching bond present in amylopectin is cleaved at a rate about 30 times slower than the α-1,4-linkages occurring in straight chains. The enzyme preparation is produced from bacterial and fungal cultures. The removal of transglucosidase enzymes which catalyze, for example, the transfer of glucose to maltose, thus lowering the yield of glucose in the starch saccharification process, is important in the production of glucoamylase.

The starch saccharification process is illustrated in Figure 2.45. In a purely enzymatic process (left side of the Figure), the swelling and gelatinization and liquefaction of starch can occur in a single step using heat-stable bacterial α-amylase (cf.2.8.2.2.2). The action of amylases yields starch syrup which is a mixture of glucose, maltose and dextrins (cf. 19.1.4.3.2).

2.8.2.2.5 *Pullulanase and isoamylase*

These enzymes hydrolyze the 1,6-α-D-glucosidic linkage present in polysaccharides such as amylopectin and glycogen. Pullulan is only hydrolyzed by pullulanase. Both enzymes are used in the food industry for the hydrolysis of amylopectin, which yields linear fragments, i.e. amylose chains. These enzymes find application in the brewing industry and in starch hydrolysis. In combination with β-amylase, these enzymes provide a starch syrup enriched with maltose.

2.8.2.2.6 *α-D-Galactosidase*

This and the following enzymes (up to and including section 2.8.2.2.9) attack the non-reducing ends of di-, oligo- and polysaccharides and cleave the terminal monosaccharide. The substrate specificity is revealed by the name of the enzyme, e.g. α-D-galactosidase (cf. Reaction 2.96).

In the production of sucrose from sugar beets (cf. 19.1.4.1.2), the enzymatic preparation from *Mortiella vinacea* hydrolyzes raffinose and, thus, improves the yield of granular sugar in the crystallization step. Raffinose in amounts >8% effectively prevents crystallization of sucrose.

Gas production (flatulence) in the stomach or intestines produced by legumes originates from

CH$_2$–OH, HO, O, OH, O–R, OH
$\xrightarrow{H_2O}$
CH$_2$OH, HO, O, OH, ~OH + HO–R, OH
(2.96)

the sugar stachyose (cf. 16.2.5). When this tetrasaccharide is cleaved by α-D-galactosidase, flatulency from this source is eliminated.

2.8.2.2.7 *β-D-Galactosidase (lactase)*

This enzyme preparation from fungi *(Aspergillus niger)* or from yeast is of importance in the dairy industry. The preparation is used to hydrolyze lactose, the solubility of which is low, and, hence, interferes with the production of skim milk concentrate or of ice cream. The immobilized form of the enzyme is also used successfully.

2.8.2.2.8 *β-D-Fructofuranase (invertase)*

Enzyme preparations isolated from special yeast strains are used for saccharose (sucrose) inversion in the confectionery or candy industry. Invert sugar is more soluble and, because of the presence of free fructose, is sweeter than saccharose.

2.8.2.2.9 *α-L-Rhamnosidase*

Some citrus fruit juices and purees (especially those of grapes) contain naringin (a dihydrochalcone), with a very bitter taste. Treatment of naringin with combined preparations of α-L-rhamnosidase and β-D-glucosidase yields the nonbitter aglycone compound naringenin (cf. 18.1.2.5.4).

2.8.2.2.10 *Glycosidase mixtures*

The baking quality of rye flour and the shelf life of rye bread can be improved by partial hydrolysis of the rye pentosans. Technical pentosanase preparations are mixtures of β-glycosidases (1,3- and 1,4-β-D-xylanases, etc.).

Solubilization of plant constituents by soaking in an enzyme preparation (maceration) is a mild and sparing process. Such preparations usually contain exo- and endo- cellulases, α- and β-mannosidases and pectolytic enzymes (cf. 2.8.2.2.13). Examples of the utilization of glycosidase mixtures are: production of fruit and vegetable purées (mashed products), disintegration of tea leaves, or production of dehydrated mashed potatoes. Some of these enzymes are used to prevent mechanical damage to cell walls during mashing and, thus, to prevent excessive leaching of gelatinized starch from the cells, which would make the purée too sticky.

Glycosidases (cellulases and amylases from *Aspergillus niger*) in combination with proteinases are recommended for removal of shells from shrimp. The shells are loosened and then washed off in a stream of water.

2.8.2.2.11 Lysozyme

The cell walls of bacteria are formed from peptidoglycan (synonymous with murein). Peptidoglycan consists of repeating units of the disaccharide N-acetylglucosamine (NAG) and N-acetylmuramic acid (NAM) connected by β-1,4-glycosidic linkages, a tetrapeptide and a pentaglycine peptide bridge. The NAG and NAM residues in peptidoglycan alternate and form the linear polysaccharide chain.

Lysozyme (cf. 11.2.3.1.4) solubilizes peptidoglycan by cleaving the 1,4-β-linkage between NAG and NAM. In order to decrease the spore count in cheeses and, thus, to prevent later undesired puffing by *Clostridia,* addition of lysozyme to cheeses has been suggested.

2.8.2.2.12 Thioglucosidase

Proteins from seeds of the mustard family *(Brassicaceae),* such as turnip, rapeseed or brown or black mustard, contain glucosinolates which can be enzymatically decomposed into pungent mustard oils (esters of isothiocyanic acid, R−N=C=S). The oils are usually isolated by steam distillation. The reactions of thioglycosidase and a few glucosinolates occurring in *Brassicaceae* are covered in section 14.3.2.2.5.

2.8.2.2.13 Pectolytic enzymes

Enzyme preparations which break down pectins (cf. 4.4.3.11) in plant foods contain enzymes of the following groups:

a) Pectin esterase (EC 3.1.1.11) is present in plants and microorganisms. It cleaves the methoxyl group from methylated pectic substances (pectin), generating free acids (pectic acid; cf. Reaction 2.99).

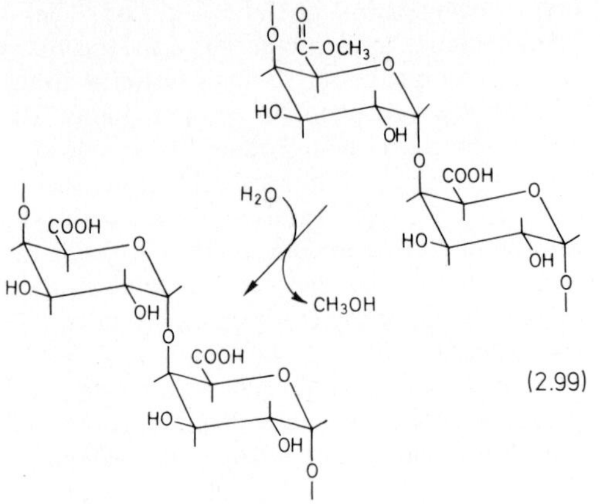

(2.99)

Pectic acid flocculates in the presence of Ca^{2+} ions. This reaction is responsible for the undesired "cloud" flocculation in citrus juices. After thermal inactivation of the enzyme at about 90 °C, this reaction is not observable. However, such treatment brings about deterioration of the aroma of the juice. Investigations of the pectin esterase of orange peel have shown that the enzyme activity is affected by competitive inhibitors: oligogalacturonic acid and pectic acid (cf. Figure 2.46). Thus, the increase in turbidity of citrus juice can be prevented by the addition of such compounds.

b) Enzymes which hydrolyze the glycosidic bonds in polygalacturonides (Table 2.21). Hy-

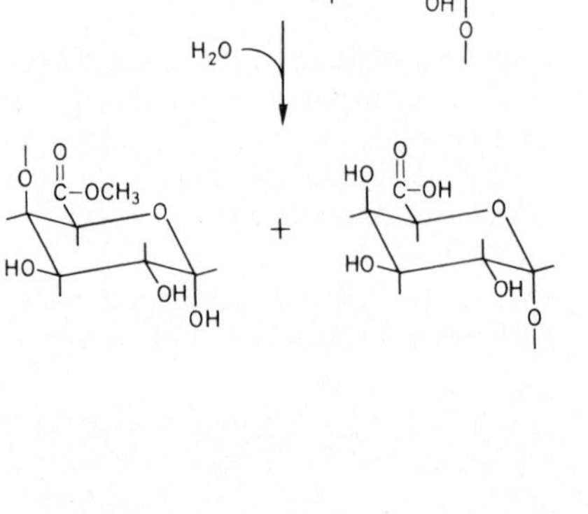

(2.97)

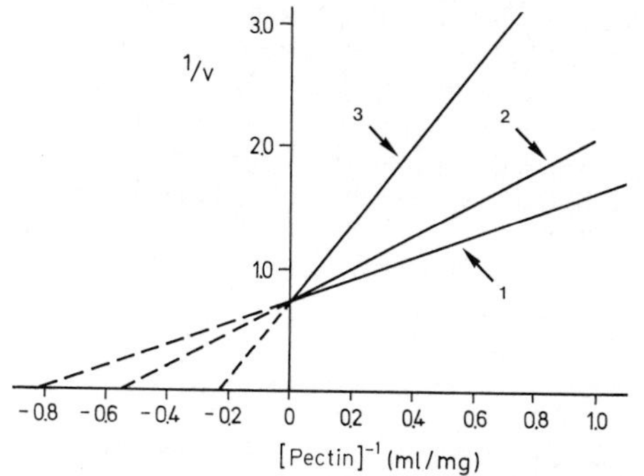

Fig. 2.46. Pectin esterase (orange) activity as affected by inhibitors (After *F. Termote*, 1977). 1 Without inhibitor, 2 hepta- and octagalacturonic acids, 3 pectic acid

drolases and lyases, with the latter catalyzing a transelimination reaction (cf. Reaction 2.98), belong to this group.

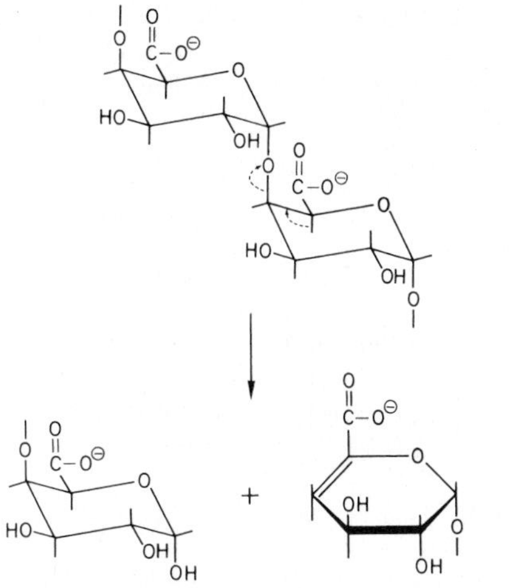

(2.98)

The double bond present in the product of trans-elimination catalysis causes an increase in absorbance at 235 nm.

According to the data in Table 2.21, the enzymes which hydrolyze polygalacturonides ("polygalacturonases") differ in specificity for hydrolyzing pectic acid or pectin; or in the position of cleavage. These enzymes can be subdivided into endo-enzymes, which hydrolyze the inner portion of the polysaccharide, or exo-enzymes, which cleave the terminal units of the long-chain molecule. The two enzymes can be readily distinguished. The endo-enzyme brings about a rapid decrease in the viscosity of a pectin solution.

Table 2.21. Enzymes involved cleaving pectin or pectic acids

Enzyme	EC-Number	Substrate
Polygalacturonase	3.2.1.15	
Endo-polymethylgalact-uronase		Pectin
Endo-polygalacturonase		Pectic acid
Exo-polygalacturonase	3.2.1.67	
Exo-polymethylgalact-uronase		Pectin
Exo-polygalacturonase		Pectic acid
Pectin lyase	4.2.2.10	Pectin
Pectate lyase	4.2.2.2.	
Endo-polygalacturonate lyase		Pectic acid
Exo-polygalacturonate lyase	4.2.2.9	Pectic acid

Polygalacturonases occur in plants and microorganisms. They are activated by the presence of NaCl and some by Ca^{2+} ions.

Pectin and pectate lyases are produced only by microorganisms. In all cases, these enzymes are activated by Ca^{2+} ions and are clearly differentiated from polygalacturonases by their pH optima (pH 8.5–9.5 versus pH 5.0–6.5, respectively).

Pectinolytic enzymes are used for the clarification of fruit and vegetable juices. The mechanism of clarification is as follows: the core of the turbidity-causing particles consists of carbohydrates and proteins (35%). The prototropic groups of these proteins acquire a positive charge at the pH of fruit juice (3.5). Negatively-charged pectin molecules form the outer shell of the particle. Partial pectinolysis exposes the positive core. Aggregation between the polycations and the polyanions then follows, resulting in flocculation. Clarification of juice by gelatin (at pH 3.5 gelatin is positively charged) and the inhibition of clarification by alginates which are polyanions at pH 3.5 support this suggested model.

In addition, pectinolytic enzymes play an important role in food processing, increasing the yield of fruit and vegetable juices and the yield of oil from olive fruits.

2.8.2.2.14 Lipases

The mechanism of lipase activity is described under section 3.7.1.1. Lipase from microbial sour-

ces (e.g. *Candida lipolytica*) is utilized for enhancement of aromas in cheesemaking.
Limited hydrolysis of milk fat is also of interest in the production of chocolate milk. It enhances the "milk character" flavor. The utilization of lipase for this commodity is also possible.
Staling of bakery products is retarded by lipase, presumably through the release of mono- and diacylglycerols (cf. 15.4.4). The defatting of bones, which has to be carried out under mild conditions in the production of gelatin, is facilitated by using lipase-catalyzed hydrolysis.

2.8.2.2.15 Tannases

Tannases hydrolyze polyphenolic compounds (tannins):

$$\text{Digallate} \xrightarrow{H_2O} 2\ \text{Gallate} \qquad (2.100)$$

For example, preparations from *Aspergillus niger* prevent development of turbidity in cold tea extracts.

2.8.2.3 Isomerases

Of this group of enzymes, glucose isomerase, which is used in the production of starch syrup with a high content of fructose (cf. 2.8.2.2.4 and 19.1.4.3.5), is very important. The enzyme used industrially is of microbial origin. Since its activity for xylose isomerization is higher than for glucose, the enzyme is classified under the name "xylose isomerase" (cf. Table 2.18a).

2.9 Literature

Acker, L., Wiese, R.: Über das Verhalten der Lipase in wasserarmen Systemen. Z. Lebensm. Unters. Forsch. *150,* 205 (1972)

Bender, M. L., Kezdy, F. J., Gunter, C. R.: The anatomy of an enzymatic catalysis. α-Chymotrypsin. J. Am. Chem. Soc. *86,* 3714 (1964)

Bergmeyer, H. U., Bergmeyer, J., Graßl, M.: Methods of enzymatic analysis. 3rd edn., Vol. 1 ff, Verlag Chemie: Weinheim–Deerfield Beach, Florida–Basel. 1983 ff

Bergmeyer, H. U., Gawehn, K.: Grundlagen der enzymatischen Analyse. Verlag Chemie: Weinheim. 1977

Betz, A.: Enzyme. Verlag Chemie: Weinheim. 1974

Birch, G. G., Blakebrough, N., Parker, K. J.: Enzymes and food processing. Applied Science Publ.: London. 1981

Blow, D. M., Birktoft, J. J., Harley, B. S.: Role of a buried acid group in the mechanism of action of chymotrypsin. Nature *221,* 337 (1969)

Eriksson, C. E.: Enzymic and non-enzymic lipid degradation in foods. In: Industrial aspects of biochemistry (Ed.: Spencer, B.), Federation of European Biochemical Societies, Vol. 30, p. 865, North Holland/American Elsevier: Amsterdam. 1974

Fennema, O.: Activity of enzymes in partially frozen aqueous systems. In: Water relations of foods (Ed.: Duckworth, R. B.), p. 397, Academic Press: London–New York–San Francisco. 1975

Gray, C. J.: Enzyme-catalysed reactions. Van Nostrand Reinhold Co: London. 1971

International Union of Biochemistry: Enzyme nomenclature 1978. Academic Press: New York–San Francisco–London. 1979

Jencks, W. P.: Catalysis in chemistry and enzymology. McGraw-Hill: New York. 1969

Karel, M.: Physico-chemical modification of the state of water in foods – a speculative survey. In: Water relations of foods (Ed.: Duckworth, R. B.), p. 639, Academic Press: London–New York–San Francisco. 1975

Kilara, A., Shahani, K. A.: The use of immobilized enzymes in the food industry: a review. Crit. Rev. Food Sci. Nutr. *12,* 161 (1979)

Koshland, D. E.: Conformation changes at the active site during enzyme action. Fed. Proc. *23,* 719 (1964)

Koshland, D. E., Neet, K. E.: The catalytic and regulatory properties of proteins. Ann. Rev. Biochem. *37,* 359 (1968)

Lehninger, A. L.: Biochemie. Verlag Chemie: Weinheim. 1977

Matheis, G., Belitz, H.-D.: Untersuchungen zur enzymatischen Bräunung bei Kartoffeln. Z. Lebensm. Unters. Forsch. *163,* 186 und 191 (1977)

Phipps, D. A.: Metals and metabolism. Clarendon Press: Oxford. 1978

Potthast, K., Hamm, R., Acker, L.: Enzymic reactions in low moisture foods. In: Water relations of foods (Ed.: Duckworth, R. B.), p. 365, Academic Press: London–New York–San Francisco. 1975

Reed, G.: Enzymes in food processing. 2nd edn., Academic Press: New York–London. 1975

Richardson, T.: Enzymes. In: Principles of food science, Part I (Ed.: Fennema, O. R.), Marcel Dekker, Inc.: New York–Basel. 1977

Schwimmer, S.: Source book of food enzymology. AVI Publ. Co.: Westport, Conn. 1981

Scrimgeour, K. G.: Chemistry and control of enzyme reactions. Academic Press: London–New York. 1977

Segel, I. H.: Biochemical calculations. John Wiley and Sons, Inc.: New York. 1968

Shotton, D.: The molecular architecture of the serine

proteinases. In: Proceedings of the international research conference on proteinase inhibitors (Eds.: Fritz, H., Tschesche, H.), p. 47, Walter de Gruyter: Berlin–New York. 1971

Svensson, S.: Inactivation of enzymes during thermal processing. In: Physical, chemical and biological changes in food caused by thermal processing (Eds.: Hoyem, T., Kvalle, O.), p. 202, Applied Science Publ.: London. 1977

Termote, F., Rombouts, F. M., Pilnik, W.: Stabilization of cloud in pectinesterase active orange juice by pectic acid hydrolysates. J. Food Biochem. *1*, 15 (1977)

Troller, J. A., Christian, J. H. B.: Water activity and food. Academic Press: New York–San Francisco–London. 1978

Whitaker, J. R.: Principles of enzymology for the food sciences. Marcel Dekker, Inc.: New York. 1972

Zaborsky, O. R.: Immobilized enzymes. CRC-Press: Cleveland, Ohio. 1973

3 Lipids

3.1 Foreword

Lipids are formed from structural units with a pronounced hydrophobicity. This solubility character, rather than a common structural feature, is unique for this class of compounds. Lipids are soluble in organic solvents but not in water. Water insolubility is the analytical property used as the basis for their facile separation from proteins and carbohydrates. Some lipids are surface-active since they are amphipathic molecules (contain both hydrophilic and hydrophobic moieties). Hence, they are polar and thus distinctly different from neutral lipids. The two approaches generally accepted for lipid classification are presented in Table 3.1

The majority of lipids are derivatives of fatty acids. In these so-called acyl lipids the fatty acids are present as esters and in some minor lipid groups in amide form (Table 3.1). The acyl residue influences strongly the hydrophobicity and the reactivity of the acyl lipids

Some lipids act as building blocks in the formation of biological membranes which surround cells and subcellular particles. Such lipids occur in food, but usually at less than 2% (cf. Table 3.11). Nevertheless, even as minor food constituents they deserve particular attention, since their high reactivity may strongly influence the organoleptic quality of the food.

Primarily triacylglycerols (also called triglycerides) are deposited in some animal tissues and organs of some plants. Lipid content in such storage tissues can rise to 15–20% or higher and so serve as a commercial source for isolation of triacylglycerols. When this lipid is refined, it is available to the consumer as an edible oil or fat. The nutritive/physiological importance of lipids is based on their role as fuel molecules (39 kJ/g triacylglycerols) and as a source of essential fatty acids and vitamins. Apart from these roles, some other lipid properties are indispensable in food handling or processing. These include the pleasant creamy or oily mouthfeel, and the ability to solubilize many taste and aroma constituents of food. These properties are of importance for food to achieve the desired texture, specific mouthfeel and aroma, and a satisfactory aroma retention. In addition, some foods are prepared by deep frying, i.e. by dipping the food into fat or oil heated to a relatively high temperature.

The lipid class of compounds also includes some important food aroma substances or precursors

Table 3.1. Lipid classification

A. Classification after "acyl residue" characteristics	
I. Simple lipids (not saponifiable)	
Free fatty acids, isoprenoid lipids (sterols, carotenoids, monoterpenes), tocopherols	
II. Acyl lipids (saponifiable)	Constituents
Mono-, di-, triacylglycerols	Fatty acid, glycerol
Phospholipids (phosphatides)	Fatty acid, glycerol or sphingosine, phosphoric acid, organic base
Glycolipids	Fatty acid, glycerol or sphingosine, mono-, di- or oligosaccharide
Diol lipids	Fatty acid, ethane-, propane-, or butane diol
Waxes	Fatty acid, fatty alcohol
Sterol esters	Fatty acid, sterol
B. Classification after the characteristics "neutral-polar"	
Neutral lipids	Polar (amphiphilic) lipids
Fatty acids ($>C_{12}$)	Glycerophospholipid
Mono-, di-, triacylglycerols	Glyceroglycolipid
Sterols, sterol esters	Sphingophospholipid
Carotenoids	Sphingoglycolipid
Waxes	
Tocopherols[a]	

[a] Tocopherols and quinone lipids are often considered as "redox lipids".

which degrade, providing the aroma compounds. Some lipid compounds are indispensable as food emulsifiers, while others are important as fat- or oil-soluble pigments or food colorants.

3.2 Fatty Acids

3.2.1 Nomenclature and Classification

Acyl lipid hydrolysis releases aliphatic carboxylic acids which differ in chemical structure. They can be divided into groups by chain length, by number, position and configuration of their double bonds, and by the occurrence of additional functional groups along the chains. The fatty acid distribution pattern in food is another criterion for differentiation.

Table 3.2 compiles the major fatty acids which occur in food. Palmitic, oleic and linoleic acids frequently occur in higher amounts, while the other acids listed, though widely distributed, as a rule occur only in small amounts (major vs minor fatty acids). Percentage data of acid distribution make it obvious that unsaturated fatty acids are the predominant form in nature.

Fatty acids are usually denoted in the literature by a "shorthand description", e.g. 18:2 (9, 12) for linoleic acid. Such an abbreviation shows the number of carbon atoms in the acid chain and the number, positions and configurations of the double bonds. By default, the bonds are considered to be cis; when trans-bonds are present, an additional "tr" is shown. As will be outlined later in a detailed survey of lipid structure, the carbon skeleton of lipids should be shown as a *zigzag* line (Table 3.2).

3.2.1.1 Saturated Fatty Acids

Unbranched, straight-chain molecules with an even number of carbon atoms are dominant among the saturated fatty acids (Table 3.3). The short-chain, low molecular weight fatty acids (< 14:0) are triglyceride constituents only in fat and oil of milk, coconut and palmseed. In free form or esterified with low molecular weight alcohols, they occur in nature only in small amounts, particularly in food processed with the aid of microorganisms, in which they are aroma substances.

The odor threshold values for fatty acids in water vary from 1–10 ppm. When water is the solvent, these values are affected by pH since only the undissociated molecules are odor active. Fatty acids with > 14 carbon atoms are considered to be flavorless and odorless.

Some high molecular weight fatty acids (> 18:0) are found in legumes (peanut butter). They can be used, like lower molecular weight homologues, for identification of sources of triglycerides (cf. 14.5.2.3). Fatty acids with odd numbers of carbon atoms, such as valeric (5:0) or enanthic (7:0) acids (Table 3.3) are present in food only in traces. Some of these short-chain homologues are important as food aroma constituents. Pentadecanoic and heptadecanoic acids are odd-numbered fatty acids present in milk and a number of plant oils. The common name "margaric acid" for 17:0 is an erroneous designation. *M. E. Chevreul* (1786–1889), who first discovered that fats are glycerol esters of fatty acids, coined the word "margarine" to denote a product from oleomargarine (a fraction of edible beef tallow), believing

Table 3.2. Structures of the major fatty acids

Abbreviated designation	Structure[a]	Common name	Proportion (%)[b]
14:0	 COOH	Myristic acid	2
16:0	 COOH	Palmitic acid	11
18:0	 COOH	Stearic acid	4
18:1 (9)	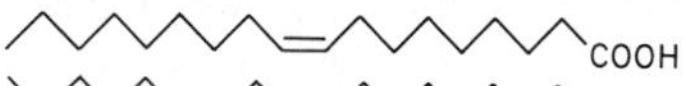 COOH	Oleic acid	34
18:2 (9, 12)	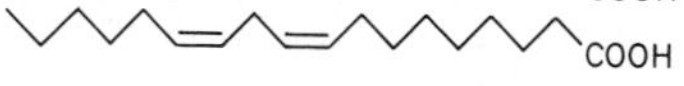 COOH	Linoleic acid	34
18:3 (9, 12, 15)	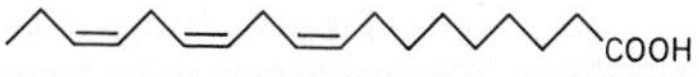 COOH	Linolenic acid	5

[a] Numbering of carbon atoms starts with carboxyl group-C as number 1.
[b] A percentage estimate based on world production of edible oils.

Table 3.3. Saturated fatty acids

Abbreviated designation	Structure	Systematic name	Common name	Melting point (°C)	Odor threshold value (ppm)[a]
A. Even numbered straight chain fatty acids					
4:0	$CH_3(CH_2)_2COOH$	Butanoic acid	Butyric acid	−7.9	0.5–10
6:0	$CH_3(CH_2)_4COOH$	Hexanoic acid	Caproic acid	−3.9	3
8:0	$CH_3(CH_2)_6COOH$	Octanoic acid	Caprylic acid	16.3	3
10:0	$CH_3(CH_2)_8COOH$	Decanoic acid	Capric acid	31.3	10
12:0	$CH_3(CH_2)_{10}COOH$	Dodecanoic acid	Lauric acid	44.0	10
14:0	$CH_3(CH_2)_{12}COOH$	Tetradecanoic acid	Myristic acid	54.4	
16:0	$CH_3(CH_2)_{14}COOH$	Hexadecanoic acid	Palmitic acid	62.9	
18:0	$CH_3(CH_2)_{16}COOH$	Octadecanoic acid	Stearic acid	69.6	
20:0	$CH_3(CH_2)_{18}COOH$	Eicosanoic acid	Arachidic acid	75.4	
22:0	$CH_3(CH_2)_{20}COOH$	Docosanoic acid	Behenic acid	80.0	
24:0	$CH_3(CH_2)_{22}COOH$	Tetracosanoic acid	Lignoceric acid	84.2	
26:0	$CH_3(CH_2)_{24}COOH$	Hexacosanoic acid	Cerotic acid	87.7	
B. Odd numbered straight chain fatty acids					
5:0	$CH_3(CH_2)_3COOH$	Pentanoic acid	Valeric acid	−34.5	3–10
7:0	$CH_3(CH_2)_5COOH$	Heptanoic acid	Enanthoic acid	−7.5	3
9:0	$CH_3(CH_2)_7COOH$	Nonanoic acid	Pelargonic acid	12.4	3
15:0	$CH_3(CH_2)_{13}COOH$	Pentadecanoic acid		52.1	
17:0	$CH_3(CH_2)_{15}COOH$	Heptadecanoic acid	Margaric acid	61.3	
C. Branched chain fatty acids					
	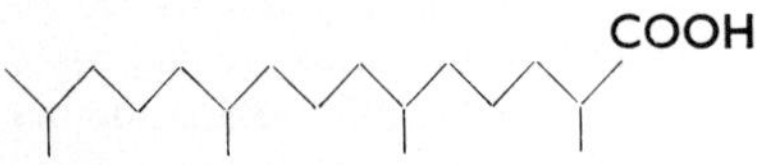 COOH	2,6,10,14-Tetra-methyl-penta-decanoic acid	Pristanic acid		
	 COOH	3,7,11,15-Tetra-methyl-hexa-decanoic acid	Phytanic acid		

[a] In water.

that the product contained a new fatty acid, 17:0. Only later was it clarified that such margarine or "17:0 acid" was a mixture of palmitic and stearic acids.

Branched-chain acids, such as iso- (with an isopropyl terminal group) or anteiso- (a secondary butyl terminal group) are rarely found in food. Pristanic and phytanic acids have been detected in milk fat (Table 3.3) They are isoprenoid acids obtained from the degradation of the phytol side chain of chlorophyll.

3.2.1.2 Unsaturated Fatty Acids

The unsaturated fatty acids, which dominate lipids, contain one, two or three allyl groups in their acyl residues (Table 3.4). Acids with isolated double bonds (a methylene group inserted between the two cis-double bonds) are usually denoted as isolenic or nonconjugated fatty acids. The structural relationship that exists among the unsaturated, nonconjugated fatty acids derived from a common biosynthetic pathway is distinctly revealed when the double bond position is determined by counting from the methyl end of the chain (it should be emphasized that position designation using this method of counting requires the suffix "ω"). Acids with the same methyl ends are then combined into groups. Thus, three family groups exist: ω3 (linolenic type), ω6 (linoleic type) and ω9 (oleic acid type; Table 3.4). Using this classification, the common structural features abundantly found in C_{18} fatty acids

Table 3.4. Unsaturated fatty acids

Abbreviated designation	Structure	Common name	Melting point (°C)
A. Fatty acids with nonconjugated cis couble bonds			
	ω9-Family		
18:1 (9)	$CH_3-(CH_2)_7-CH=CH-CH_2-(CH_2)_6-COOH$	Oleic acid	13.4
22:1 (13)	$-(CH_2)_{10}-COOH$	Erucic acid	34.7
24:1 (15)	$-(CH_2)_{12}-COOH$	Nervonic acid	42.5
	ω6-Family		
18:2 (9, 12)	$CH_3-(CH_2)_4-(CH=CH-CH_2)_2-(CH_2)_6-COOH$	Linoleic acid	−5.0
18:3 (6, 9, 12)	$-(CH=CH-CH_2)_3-(CH_2)_3-COOH$	γ-Linolenic acid	
20:4 (5, 8, 11, 14)	$-(CH=CH-CH_2)_4-(CH_2)_2-COOH$	Arachidonic acid	−49.5
	ω3-Family		
18:3 (9, 12, 15)	$CH_3-CH_2-(CH=CH-CH_2)_3-(CH_2)_6-COOH$	α-Linolenic acid	−11.0
16:3 (7, 10, 13)	$-(CH_2)_4-COOH$		
	Δ9-Family		
18:1 (9)	$CH_3-(CH_2)_7-CH=CH-CH_2-(CH_2)_6-COOH$	Oleic acid	13.4
16:1 (9)	$CH_3-(CH_2)_5-$	Palmitoleic acid	0.5
14:1 (9)	$CH_3-(CH_2)_3-$	Myristoleic acid	
B. Fatty acids with nonconjugated trans-double bonds			
18:1 (tr9)	$CH_3-(CH_2)_7-CH\overset{tr}{=}CH-(CH_2)_7-COOH$	Elaidic acid	46
18:2 (tr9, tr12)	$CH_3-(CH_2)_4-CH\overset{tr}{=}CH-CH_2-CH\overset{tr}{=}CH-(CH_2)_7-COOH$	Linolelaidic acid	28
C. Fatty acids with conjugated double bonds			
18:3 (9, tr11, tr13)	$CH_3-(CH_2)_3-CH\overset{tr}{=}CH-CH\overset{tr}{=}CH-CH\overset{c}{=}CH-(CH_2)_7-COOH$	α-Eleostearic acid	48
18:3 (tr9, tr11, tr13)	$CH_3-(CH_2)_3-CH\overset{tr}{=}CH-CH\overset{tr}{=}CH-CH\overset{tr}{=}CH-(CH_2)_7-COOH$	β-Eleostearic acid	71.5
18:4 (9, 11, 13, 15)[a]	$CH_3-CH_2-(CH=CH)_4-(CH_2)_7-COOH$	Parinaric acid	85

[a] Geometry of the double bond was not determined.

(Table 3.2) are readily revealed with less frequently occurring fatty acids. Thus, erucic acid (20 : 1), occurring only in the mustard family of seeds (*Brassicaceae*, cf. 14.3.2.2.5), belongs to the ω9 group, arachidonic acid (20 : 4), occurring in meat, liver, lard and lipids of chicken eggs, belongs to the ω6 group, while hexadecatrienoic acid, found in rape leaves and some lower plants, belongs to the ω3 group.

Nervonic acid (24 : 1, ω9) is a constituent of sphingolipids. Fatty acids C_{20}–C_{22} with five and six double bonds are characteristic features of fish oils (cf. 14.3.2).

Linoleic acid can not be synthesized by the human body. This acid and other members of the ω6 family are considered as essential fatty acids required as building blocks for biologically-active membranes. In contrast to humans, the ω6 acids can be synthesized in mammals from linoleic acid. Whether α-linolenic acid, which belongs to the ω3 family and which is synthesized only by plants, plays a nutritional role as an essential fatty acid is disputed.

A formal relationship exists in some olefinic unsaturated fatty acids with regard to the position of the double bond when counted from the carboxyl end of the chain. Oleic, palmitoleic and myristoleic acids belong to such a Δ9 family (cf. Table 3.4); the latter two fatty acids are minor in food of animal or plant origin.

Unsaturated fatty acids with an unusual structure are those with one trans-double bond and/or conjugated double bonds (Table 3.4). Such trans-unsaturated acids are formed as artifacts in the industrial processing of oil or fat (heat treatment, oil hardening). However, several occur in nature. The trans-analogue of oleic acid is found in mutton tallow, while that of linoleic acid is found in *Chilopsis linearis* seeds. Conjugated fatty acids with diene, triene or tetraene systems occur frequently in several seed oils, but do not play a role in human nutrition. Table 3.4 presents, as an example, two naturally occurring acids with conjugated triene systems which differ in the configuration of one double bond in position 9 (cis, trans).

3.2.1.3 Substituted Fatty Acids

Hydroxy fatty acids: Ricinoleic acid is the best known of the straight-chain hydroxy fatty acids. Its structure is 12-OH, 18 : 1 (9). It is an optically active acid with a D(+)-configuration:

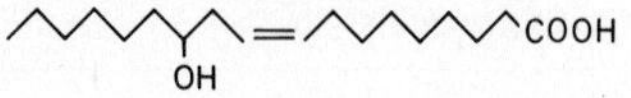

[skeletal formula: OH on C-12, double bond, COOH] (3.1)

Ricinoleic acid is the main acid of castor bean oil, comprising up to 90% of the total acids. Hence, it can serve as an indicator for the presence of this oil in edible oil blends.

D-2-Hydroxy saturated 16 : 0 to 25 : 0 fatty acids with both even and odd numbers of carbons in their chains occur in lipids in green leaves of a great number of vegetables.

γ- or δ-Lactones are obtained from 4- and 5-hydroxycarboxylic acids (C_8 to C_{16}) by the elimination of water. δ-lactones have been found in milk fat and in apricots and peaches.

Oxo fatty acids: Natural oxo (or keto) acids are less common than hydroxy acids. About 1% of milk fat consists of saturated (C_{10}–C_{24}) oxo fatty acids with an even number of carbon atoms in which the carbonyl groups is located on C-5 to C-13. One of 47 identified compounds of this group has the following structure (Formula 3.2):

$$CH_3-(CH_2)_4-CH=CH-CH_2-CH_2-\overset{\displaystyle O}{\overset{\|}{C}}-(CH_2)_7-COOH \quad (3.2)$$

Furanoid fatty acids: These occur in fish liver oil in a range of 1–6% and up to 25% in some freshwater fish. These acids contain a furan ring. The eight furanoid acids identified so far differ in position and nature of substituents on the furan ring. One of these acids has the formula:

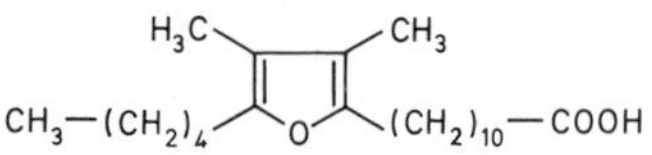

[furan ring with H_3C and CH_3 substituents; $CH_3-(CH_2)_4$ and $(CH_2)_{10}-COOH$ side chains] (3.3)

Substituted fatty acids are also derived by autoxidation or enzymatic peroxidation of unsaturated fatty acids, which will be dealt with in more detail in 3.7.2.3 and 3.7.2.4.1.

3.2.2 Physical Properties

3.2.2.1 Carboxyl Group

Carboxylic acids have a great tendency to form dimers which are stabilized by hydrogen bonds:

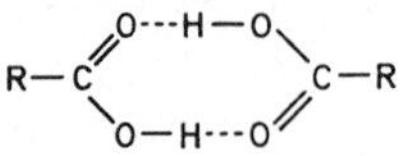

[R–C(=O···H–O)(–O–H···O=)C–R dimer] (3.4)

The binding energy of the acid dimer dissolved in hexane is 38 kJ/mole. Also, the fatty acid

molecules are arranged as dimers in the crystalline lattice (cf. Figure 3.2).
The acidic character of the carboxyl group is based on proton dissociation and on the formation of the resonance-stabilized carboxylate anion:

$$R-C{\overset{O}{\underset{OH}{}}} \rightleftharpoons H^{\oplus} + \left[R-C{\overset{O}{\underset{O^{\ominus}}{}}} \longleftrightarrow R-C{\overset{O^{\ominus}}{\underset{O}{}}} \right] \quad (3.5)$$

The pK_s values for the C_2–C_9 short-chain acid homologues range from 4.75–4.95. The pK_s of 7.9 for linoleic acid deviates considerably from this range. This unexpected and anomalous behavior, which has not yet been clarified, is clearly illustrated in the titration curves for propionic, caproic and linoleic acids recorded under identical conditions (Figure 3.1).

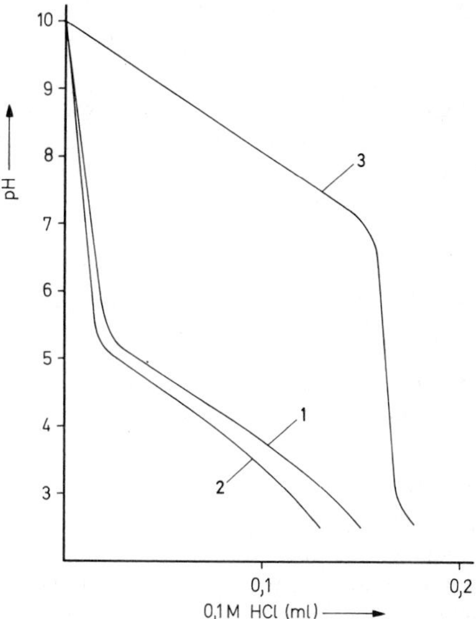

Fig. 3.1. Fatty acid titration curves (After *G. S. Bild* et al., 1977). Aqueous solutions (0.1 M) of Na-salts of propionic (1), caprylic (2) and linoleic acids (3) were titrated with 0.1 M HCl

3.2.2.2 Crystalline Structure, Melting Points

Melting properties of fats are determined by the arrangement of the acyl residues in the crystal lattice in addition to other factors attributed solely to the structure of triglycerides.
Calculations of the energy content of the carbon chain conformation have revealed that at room temperature 75% of the C–C bonds with a saturated fatty acid chain are present in a fully staggered *zigzag* or "trans" conformation and only 25% in the energetically a little less favorable skew conformation.
The unsaturated fatty acids, because of their double bonds, are not free to rotate and hence have rigid kinks along the carbon chain skeleton. However, a molecule is less bent by a trans than by a cis double bond. Thus, the cis-configuration in oleic acid causes a bending of about 40°:

COOH (3.6)

The corresponding elaidic acid, with a trans-configuration, has a slightly shortened C-chain, but is still similar to the linear form of stearic acid:

COOH (3.7)

The extent of molecular crumpling is also increased by an increase in the number of cis double bonds. Thus, the four cis double bonds in arachidonic acid increase the deviation from a straight line by 165°:

COOH (3.8)

When fatty acids crytallize, the saturated acids are oriented as depicted by the simplified pattern in Figure 3.2. The dimer molecular arrangement is thereby retained. The principal reflections of the X-ray beam are from the planes (c) of high electron density in which the carboxyl groups are situated. The length of the fatty acid molecule can be determined from the "main reflection" site intervals (distance d in Figure 3.2). For stearic acid (18:0), this distance is 2.45 nm.
The crystalline lattice is stabilized by hydrophobic interactions along the acyl residues. Correspondingly, the energy and therefore the temperature required to melt the crystal increase with an increased number of carbons in the chain.
Odd-numbered as well as unsaturated fatty acids can not be uniformly packed into a crystalline lattice as can the saturated and even-numbered acids. The odd-numbered acids are interfered with to a certain extent by their terminal methyl groups.

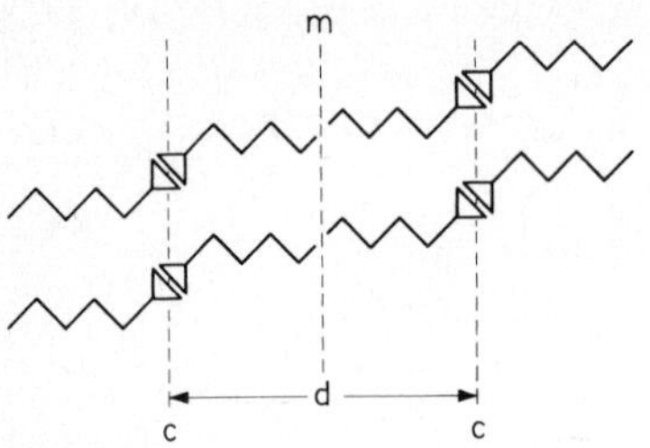

Fig. 3.2. Arrangement of caproic acid molecules in crystal. (After *J. M Mead* et al., 1965). Results of a X-ray diffraction analysis reveal a strong diffraction in the plane of carboxyl groups (c) and a weak diffraction at the methyl terminals: d (m); d: identity period

The consequence of less symmetry within the crystal is that the melting points of even-numbered acids (C_n) exceed the melting points of the next higher odd-numbered (C_{n+1}) fatty acids (cf. Table 3.3).

The molecular arrangement in the crystalline lattice of unsaturated fatty acids is not strongly influenced by trans double bonds, but is strongly influenced by cis double bonds. This difference, due to steric interference as mentioned above, is reflected in a decrease in melting points in the fatty acid series 18:0, 18:1 (tr9) and 18:1 (9). However, this ranking should be considered as reliable only when the double bond positions within the molecules are fairly comparable. Thus, when a cis double bond is at the end of the carbon chain, the deviation from the form of a straight extended acid is not as large as in oleic acid. Hence, the melting point of such an acid is higher. The melting point of cis-2-octadecenoic acid is in agreement with this rule; it even surpasses the 9-trans isomer of the same acid (Table 3.5).

Table 3.5. The effect of number, configuration and double bond position on melting points of fatty acids

Fatty acid		Melting point (°C)
18:0	Stearic acid	69
18:1 (tr9)	Elaidic acid	46
18:1 (2)	cis-2-Octadecenoic acid	51
18:1 (9)	Oleic acid	13.4
18:2 (9, 12)	Linoleic acid	−5
18:2 (tr9, tr12)	Linolelaidic acid	28
18:3 (9, 12, 15)	α-Linolenic acid	−11
20:0	Arachidic acid	75.4
20:4 (5, 8, 11, 14)	Arachidonic acid	−49.5

3.2.2.3 Urea Adducts

When urea crystallizes, channels with a diameter of 0.8–1.2 nm are formed within its crystals and can accomodate long-chain hydrocarbons. The stability of such inclusion compounds with fatty acids parallels the geometry of the acid molecule. Any deviation from a straight-chain arrangement brings about weakening of the adduct. A tendency to form inclusion compounds decreases in the series 18:0 > 18:1 (9) > 18:2 (9, 12).

A substitution on the acyl chain prevents adduct formation. Thus, it is possible to separate branched or oxidized fatty acids or their methyl esters from the corresponding straight-chain compounds on the basis of formation of urea adducts. This principle is used as a method for preparative-scale enrichment and separation of branched or oxidized acids from a mixture of fatty acids.

3.2.2.4 Solubility

Long-chain fatty acids are practically insoluble in water; instead, they form a floating film on the water surface. The polar carboxyl groups in this film are oriented toward the water, while the hydrophobic tails protrude into the gaseous phase. The solubility of the acids increases with decreasing carbon number; butyric acid is completely soluble in water.

Ethyl ether is the best solvent for stearic acid and other saturated long-chain fatty acids since it is sufficiently polar to attract the carboxyl groups. A truly nonpolar solvent, such as petroleum ether, is not suitable for free fatty acids.

The solubility of fatty acids increases with an increase in the number of cis double bonds. This is illustrated in Figure 3.3 with acetone as a solvent. The observed differences in solubility can be utilized for separation of saturated from unsaturated fatty acids. The mixture of acids is dissolved at room temperature and cooled stepwise to −80 °C. However, the separation efficiency of such a fractional crystallization is limited since, for example, stearic acid is substantially more soluble in acetone containing oleic acid than in pure acetone. This mutual effect on solubility has not been considered in Figure 3.3

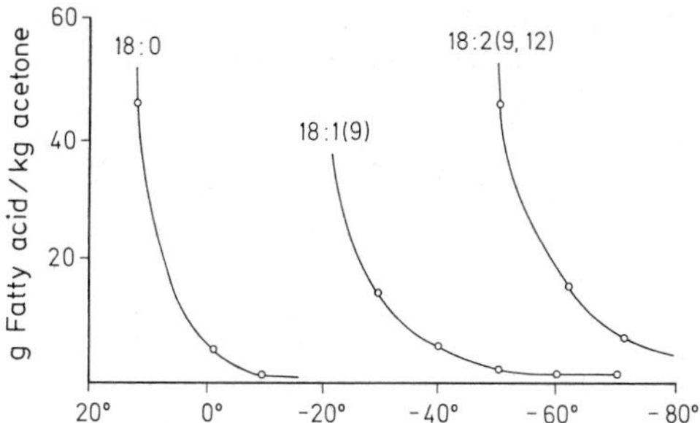

Fig. 3.3 Fatty acid solubility in acetone (After *J. M. Mead* et al., 1965)

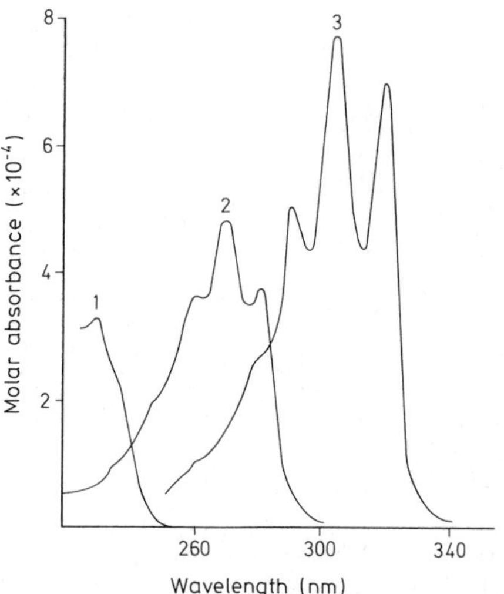

Fig. 3.4. Electron excitation spectra of conjugated fatty acids. (After *H. Pardun*, 1976). 1 9,11-isolinoleic acid, 2 α-elaeostearic acid, 3 parinaric acid

3.2.2.5 UV-Absorption

All unsaturated fatty acids which contain an isolated cis double bond absorb UV light at a wavelength close to 190 nm. Thus, the acids can not be distinguished spectrophotometrically. Conjugated fatty acids absorb light at various wavelengths depending on the length of conjugation and configuration of the double bond system. Figure 3.4 illustrates such behavior for several fatty acids. See 3.2.3.2.2 for the conversion of an isolene-type fatty acid into a conjugated fatty acid.

3.2.3 Chemical Properties

3.2.3.1 Methylation of Carboxyl Groups

The carboxyl group of a fatty acid must be depolarized by methylation in order to facilitate gas chromatographic separation or separation by fractional distillation. Reaction with diazomethane is preferred for analytical purposes. Diazomethane is formed by alkaline hydrolysis of N-nitroso-N-methyl-p-toluene sulfonamide. The gaseous CH_2N_2 released by hydrolysis is swept by a stream of nitrogen into a receiver containing the fatty acid solution in ether-methanol (9:1 v/v). The reaction:

$$R\text{-}COOH + CH_2N_2 \longrightarrow R\text{-}COOCH_3 + N_2 \qquad (3.9)$$

proceeds under mild conditions without formation of by-products. Further possibilities for methylation include: esterification in the presence of excess methanol and a *Lewis* acid (BF_3) as a catalyst; or the reaction of a fatty acid silver salt with methyl iodide:

$$R\text{-}COOAg + CH_3J \longrightarrow R\text{-}COOCH_3 + AgJ \qquad (3.10)$$

3.2.3.2 Reactions of Unsaturated Fatty Acids

A number of reactions which are known for olefinic hydrocarbons play an important role in the analysis and processing of lipids containing unsaturated fatty acids.

3.2.3.2.1 Halogen addition reactions

The number of double bonds present in an oil or fat can be determined through their iodine number (cf. 14.5.2.1). The fat or oil is treated with a halogen reagent which reacts only with existing double bonds. Substitution reactions generating hydrogen halides must be avoided. IBr in an inert solvent, such as glacial acetic acid, is a suitable reagent:

$$>C=C< + JBr \longrightarrow \underset{\overset{\oplus}{J}}{>C-C<}\ Br^{\ominus} \longrightarrow -\underset{J}{\underset{|}{\overset{|}{C}}}-\overset{Br}{\overset{|}{\underset{|}{C}}}- \qquad (3.11)$$

The number of double bonds is calculated by titrating the reagent with thiosulfate and after the fat (oil) is added to the reaction mixture.

3.2.3.2.2 Transformation of isolene-type fatty acids to conjugated fatty acids

Allyl systems are labile and are readily converted to a conjugated double bond system in the presence of base (KOH or K-tert-butylate):

$$-CH=CH-\overset{H\ \ H}{C}-CH=CH- \xrightarrow{B^{\ominus},\ H^{\oplus}} -CH=CH-CH=CH-CH_2- \qquad (3.12)$$

During this reaction, an equilibrium is established between the isolene and the conjugated forms of the fatty acid, the equilibrium state being dependent on the reaction conditions. This isomerization is used analytically since it provides a way to simultaneously determine linoleic, linolenic and arachidonic acids in a fatty acid mixture. The corresponding conjugated diene, triene and tetraene systems of these fatty acids have a maximum absorbance at distinct wavelengths (cf. Figure 3.4). The assay conditions can be selected to isomerize only the naturally-occurring cis double bonds and to ignore the trans fatty acids formed, for instance, during oil hardening (cf. 14.4.2).

3.2.3.2.3 Formation of a π-complex with Ag^+ ions

Unsaturated fatty acids or their triacylglycerols, as well as unsaturated aldehydes obtained through autoxidation of lipids (cf. 3.7.2.1.5), can be separated by "argentation chromatography". The separation is based on the number, position and configuration of the double bonds present. The separation mechanism involves interaction of the π-electrons of the double bond with Ag^+ions, forming a reversible π-complex of variable stability:

$$>C=C< + Ag^{\oplus} \rightleftharpoons >C \underset{Ag^{\oplus}}{-} C< \qquad (3.13)$$

The complex stability increases with increasing number of double bonds. This means a fatty acid with two cis double bonds will not migrate as far as a fatty acid with one double bond on a thin-layer plate impregnated with a silver salt. The R_f values increase for the series 18:2 (9, 12) < 18:1 (9) < 18:0. Furthermore, fatty acids with isolated double bonds form a stronger Ag^+ complex than those with conjugated bonds. Also, the complex is stronger with a cis- than with a trans-configuration. The complex is also more stable the further the double bond is from the end of the chain. Finally, a separation of nonconjugated from conjugated fatty acids and of isomers that differ only in their double bond configuration is possible by argentation chromatography.

3.2.3.2.4 Hydrogenation

In the presence of a suitable catalyst, e.g. Ni, hydrogen can be added to the double bond of an acyl-lipid. This heterogenous catalytic hydrogenation occurs stereo-selectively as a cis-addition. Catalyst-induced isomerization from an isolene-type fatty acid to a conjugated fatty acid occurs with fatty acids with more double bonds:

(3.14)

Since diene fatty acids form a more stable complex with a catalyst than do monoene fatty acids, the former are preferentially hydrogenated. Since Nature is not an abundant source of the solid fats which are required in food processing, the partial and selective hydrogenation, just referred to, plays an important role in the industrial processing of fats and oils (cf. 14.4.2).

3.2.4 Biosynthesis of Unsaturated Fatty Acids

The biosynthetic precursors of unsaturated fatty acids are saturated fatty acids in an activated form (cf. a biochemistry textbook). These are aerobically and stereospecifically dehydrogenated by dehydrogenase action in plant as well as animal tissues. A flavoprotein and ferredoxin are involved in plants in the electron transport system which uses oxygen as a terminal electron acceptor (cf. Reaction 3.15).

To obtain polyunsaturated fatty acids, the double bonds are introduced by a stepwise process. A fundamental difference exists between mammals and plants: in the former, oleic acid synthesis is possible and, also, additional double bonds can be inserted towards the carboxyl end of the fatty acid molecule. For example, γ-linolenic acid can be formed from the essential acid, linoleic acid and, also, arachidonic acid (Figure 3.5) can be formed by chain elongation of γ-linolenic acid. In a diet deficient in linoleic acid, oleic acid is dehydrogenated to isolinoleic acid and its deriva-

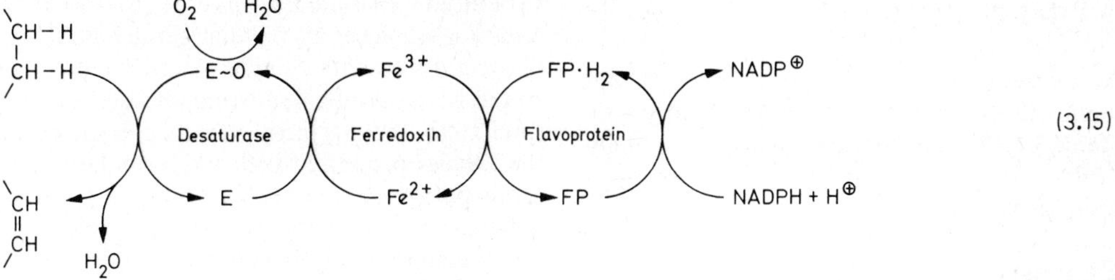

(3.15)

tives (Figure 3.5), but none of these acquire the physiological function of an essential acid such as linoleic acid.

Plants can introduce double bonds into fatty acids in both directions: towards the terminal CH_3-group or towards the carboxyl end. Oleic acid (oleoyl-CoA ester or β-oleoyl-phosphatidylcholine) is thus dehydrogenated to linoleic and then to linolenic acid. However, synthesis of the latter can be achieved by another pathway involving stepwise dehydrogenation of lauric acid with chain elongation reactions involving C_2 units (Figure 3.5).

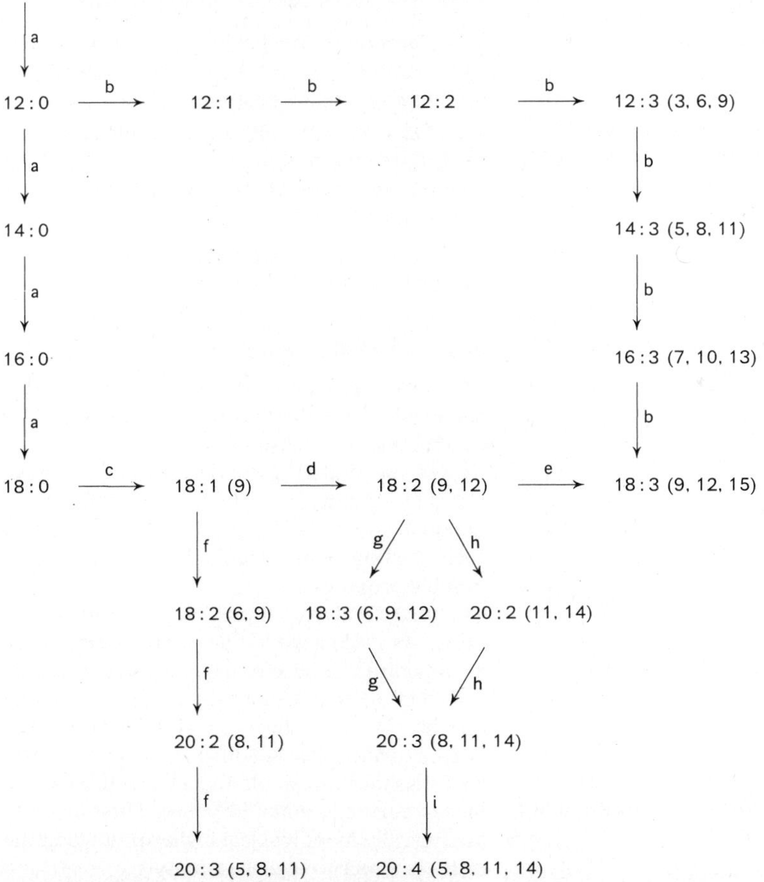

Fig. 3.5. Biosynthesis of unsaturated fatty acids
Synthesis pathways: a, b, and a, c, d in higher plants; a, c, d, e and a, c, d, g, i in algae; a, c, f and g, i (main pathway for arachidonic acid) or h, i in mammals

3.3 Acylglycerols

Acylglycerols (or acylglycerides) comprise the mono-, di- or triesters of glycerol with fatty acids (Table 3.1). They are designated as neutral lipids. Edible oils or fats consist nearly completely of triacylglycerols.

3.3.1 Triacylglycerols (TG)

3.3.1.1 Nomenclature, Classification

Glycerol, as a trihydroxylic alcohol, can form triesters with one, two or three different fatty acids. In the first case a triester is formed with three of the same acyl residues (e.g. tripalmitin; P_3), while the mixed esters involve two or three different acyl residues, e.g. dipalmito-olein (P_2O) and palmito-oleo-linolein (POL). The rule of this shorthand designation is that the acid with the shorter chain or, in the case of an equal number of carbons in the chain, the chain with fewer double bonds, is mentioned first. The Z number gives the possible different triacylglycerols which can occur in a fat (oil), where n is the number of kinds of fatty acids identified in that fat (oil):

$$Z = \frac{n^3 + n^2}{2} \tag{3.16}$$

For n = 3, the possible number of triglycerols (Z) is 18. However, such a case when a fat (oil) contains only three fatty acids is rarely found in nature. One exception is Borneo tallow (cf. 14.3.2.2.4), which contains essentially only 16:0, 18:0 and 18:1 (9) fatty acids.

Naturally, the Z value also takes into account the number of possible positional isomers within a molecule, for example, by the combination of POS, PSO and SOP. When only positional isomers are considered and the rest disregarded, Z is reduced to Z′:

$$Z' = \frac{n^3 + 3n^2 + 2n}{6} \tag{3.17}$$

Thus, when n = 3, Z′ = 10.

A chiral center exists in a triacylglycerol when the acyl residues in positions 1 and 3 are different:

```
          CH2-O-CO-R1
           |*
R1-CO-O-CH                *Chirality
           |                center
          CH2-O-CO-R2
```
(3.18)

In addition enantiomers may be produced by 1-monoglycerides, all 1,2-diglycerides and 1,3-diglycerides containing unlike substituents.

In the stereospecific numbering of acyl residues (prefix sn), the L-glycerol molecule is shown in the *Fischer* projection with the secondary HO-group pointing to the left. The top carbon is then denoted C-1. Actually, in a *Fischer* projection, the horizontal bonds denote bonds in front and the vertical bonds those behind the plane of the page:

```
      CH2-OH      sn-1
       |
HO—C—H          sn-2
       |
      CH2-OH      sn-3
```
(3.19)

For example, the nomenclature for a triacylglycerol which contains P, S and O:

sn-POS = sn-1-Palmito-2-oleo-3-stearin.

This assertion is only possible when a stereospecific analysis (cf. 3.3.1.4) provides information on the fatty acids in positions 1, 2 and 3.

rac-POS = sn-POS and sn-SOP in the molar ratio 1 : 1, i.e. the fatty acid in position 2 is fixed while the other two acids are equally distributed in positions 1 and 3.

POS = mixture of sn-POS, sn-OPS, sn-SOP, sn-PSO, sn-OSP and sn-SPO

3.3.1.2 Melting Properties

TG melting properties are affected by fatty acid composition and their distibution within the glyceride molecule (Table 3.6).

Mono-, di- and triglycerides are polymorphic, i.e. they crystallize in different modifications, denoted as γ, α, β' and β. These forms differ in their melting points (Table 3.6) and crystallographic properties.

The glassy and short-lived γ-form is usually obtained by chilling the molten triglycerides but, at a temperature close to its melting point, is rapidly forced into a transition to the α-form. Upon further heating, the α-form is transformed into the β'-form, which then solidifies in the β-form. The β-form is the most stable and, hence, also has the highest melting point (Table 3.6). These changes are typically monotropic, i.e. they proceed in the order of lower to higher-stability.

Crystallization of triglycerides from a solvent system generally yields β-form crystals. Tristea-

Table 3.6. Triacylglycerols and their polymorphic forms

Compound	Melting point (°C) of polymorphic form			
	γ	α	β′	β
Tristearin		55	65	72.5
Triolein		−32	−12	5.5
Tripalmitin		45	56.5	65.5
1,2-Dipalmitoolein	18.5	28.9		34.8
1,3-Dipalmitoolein	18	26.5		38.5
1-Palmito-3-stearo-2-olein		18.2	33	39
1-Palmito-2-stearo-3-olein		26.3	40.2	
2-Palmito-1-stearo-3-olein		25.3	40.2	
Trilinolein				−10
1,2-Dipalmitolinolein				26 to 27
1-Palmito-dilinolein				−4 to −3
1-Stearo-dilinolein				5 to 6
1,2-Diacetopalmitin		22.4		42.3
1,2-Distearo-olein	30.4	43.5		
1,3-Distearo-olein	22.4	37	41.5	44.3
Trimyristin		33	46.5	57.0
Trilaurin		15	35	46.5

rin is known to form hexagonal crystals in its α-form, orthorhombic crystals in the β′-form and triclinic crystals in the β-form. Thus, polymorphism is a reflection of different patterns of molecular packing in fat crystals.

Electron density measurements have revealed that saturated triacylglycerols in the β-form exist in the "chair" (B) configuration, while the β′-form exists in a "tuning fork" configuration (A; cf. Formula 3.20).

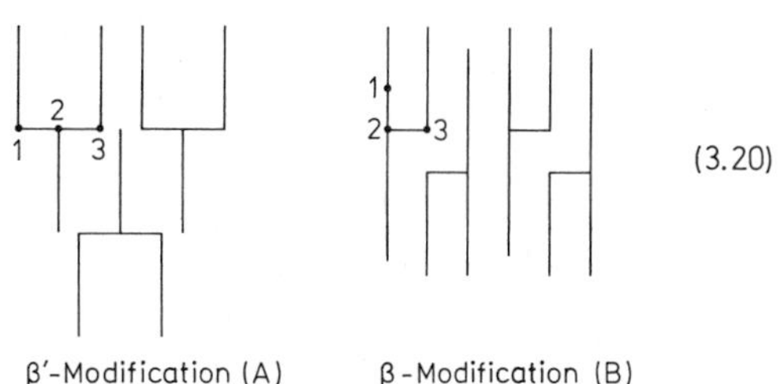

(3.20)

Unsaturated fatty acids interfere with the orderly packing of molecules in the crystalline lattice, thereby causing a decrease in the melting point of the crystals.

TG such as 1,3-diaceto-palmitin, i.e. a triglyceride with one long and two short-chain fatty acids, exists in the exceptionally stable α-form. Since films of such TG's can expand by 200 to 300 times their normal length, they are of interest for application as protective coatings for fat-containing foods. The edible oils and fats can also be classified by the dominant crystalline forms they acquire after solidification (Table 3.7).

Table 3.7. Crystallization patterns of edible fats or oils

β-Type	β′-Type	β-Type	β′-Type
Coconut oil	Cottonseed oil	Peanut butter	Tallow
Corn germ oil	Butter	Sunflower oil	Whale oil
Olive oil	Palm oil	Lard	
Palm seed oil	Rapeseed oil		

3.3.1.3 Chemical Properties

Hydrolysis, methanolysis and interesterification are the most important chemical reactions for TG's.

Hydrolysis. The fat or oil is cleaved or saponified by treatment with alkali (e.g. alcoholic KOH):

$$\text{R-CO-O}\left[\begin{array}{l}\text{O-CO-R}\\ \\ \text{O-CO-R}\end{array}\right. + 3\,\text{KOH} \longrightarrow \text{HO}\left[\begin{array}{l}\text{OH}\\ \\ \text{OH}\end{array}\right. + 3\,\text{RCOOK} \quad (3.21)$$

After acidification and extraction, the free fatty acids are recovered as alkali salts (commonly called soaps). This procedure is of interest for analysis of fat or oil samples. Commercially, the free fatty acids are produced by cleaving triglycerides with steam under elevated pressure and temperature and by increasing the reaction rate in the presence of an alkaline catalyst (ZnO, MgO or CaO) or an acidic catalyst (aromatic sulfonic acid).

Methanolysis. The fatty acids in TG are usually analyzed by gas liquid chromatography, not as free acids, but as methyl esters. The required transesterification is most often achieved by Na-methylate (methoxide) in methanol and in the presence of 2,2-dimethoxypropane to bind the released glycerol. Thus, the reaction proceeds rapidly and quantitatively even at room temperature.

Interesterification. This reaction is of industrial importance (cf. 14.4.3) since it can change the

$$R{-}CO{-}O{-}CH(CH_2{-}O{-}CO{-}R)_2 \xrightarrow{NaOCH_3/CH_3OH} 3\,RCOOCH_3 + \text{glycerol (HO–, –OH, –OH)}$$

$$\text{glycerol} + (CH_3)_2C(OCH_3)_2 \rightarrow 2\,CH_3OH + \text{isopropylidene glycerol (HO–CH}_2\text{–dioxolane with } C(CH_3)_2) \quad (3.22)$$

physical properties of fats or oils or their mixtures without the necessity of altering the chemical structure of the fatty acids. Both intra- and intermolecular acyl residue exchanges occur in the reaction until an equilibrium is reached which depends on the structure and composition of the TG molecules. The usual catalyst for interesterification is Na-methylate.

The principle of the reaction will be elucidated by using a mixture of tristearin (SSS) and triolein (OOO) or stearo-olein (OSO). Two types of interesterification are recognized:

a) A single-phase interesterification where the acyl residues are randomly distributed (cf. Formula 3.23).

$$\underset{(50\%)}{\text{St St St}} + \underset{(50\%)}{\text{O O O}} \xrightarrow{(NaOCH_3)} \underset{(12{,}5\%)}{\text{St St St}}\; \underset{(12{,}5\%)}{\text{St O St}}\; \underset{(25\%)}{\text{O St St}}\; \underset{(25\%)}{\text{St O O}}\; \underset{(12{,}5\%)}{\text{O St O}}\; \underset{(12{,}5\%)}{\text{O O O}} \quad (3.23)$$

b) A directed interesterification in which the reaction temperature is lowered so that the higher melting and least soluble TG molecules in the mixture crystallize. These molecules cease to participate in further reactions, thus the equilibrium is continuously changed. Hence, a fat (oil) can be divided into high and low melting point fractions, e.g.:

$$\text{O St O} \xrightarrow{(NaOCH_3)} \underset{(33{,}3\%)}{\text{St St St}}\; \underset{(66{,}7\%)}{\text{O O O}} \quad (3.24)$$

3.3.1.4 Structural Determination

Apart from identifying a fat or oil from an unknown source (cf. 14.5.2), TG structural analysis is important for the clarification of the relationship existing between the chemical structure and the melting or crystallization properties, i.e. the consistency.

An introductory example: cocoa butter and beef tallow, the latter used during the past century for adulteration of cocoa butter, have very similar fatty acid compositions, especially when the two main saturated fatty acids, 16:0 and 18:0, are considered together (Table 3.8). In spite of their compositions, the two fats differ significantly in their melting properties. Cocoa butter is hard and brittle and melts in a narrow temperature range (28–36 °C). Edible beef tallow, on the other hand, melts at a higher temperature (approx. 45 °C) and over a wider range and has a substantially better plasticity. The melting property of cocoa butter is controlled by the presence of the patterns of triglycerols: SSS, SUS and SSU (cf. Table 3.8). The chemical composition of Borneo tallow (Tenkawang fat) is so close to that of cocoa butter that the TG distribution patterns shown in Table 3.8 are practically indistinguishable. Also, the melting properties of the two fats are similar, consequently, Borneo tallow is currently used as an important substitute for cocoa butter. Analysis of the TG's present in fat (oil) could be a tedious task, when numerous TG compounds

Table 3.8. Average fatty acid and triacylglycerol composition (weight-%) of cocoa butter, tallow and borneo tallow (a cocoa butter substitute)

	Cocoa butter	Edible beef tallow	Borneo tallow[a]
16:0	25	36	20
18:0	37	25	42
20:0	1		1
18:1 (9)	34	37	36
18:2 (9, 12)	3	2	1
SSS[b]	2	29	4
SUS	81	33	80
SSU	1	16	1
SUU	15	18	14
USU		2	
UUU	1	2	1

[a] cf. 14.3.2.2.3

[b] S: Saturated, and U: unsaturated fatty acids.

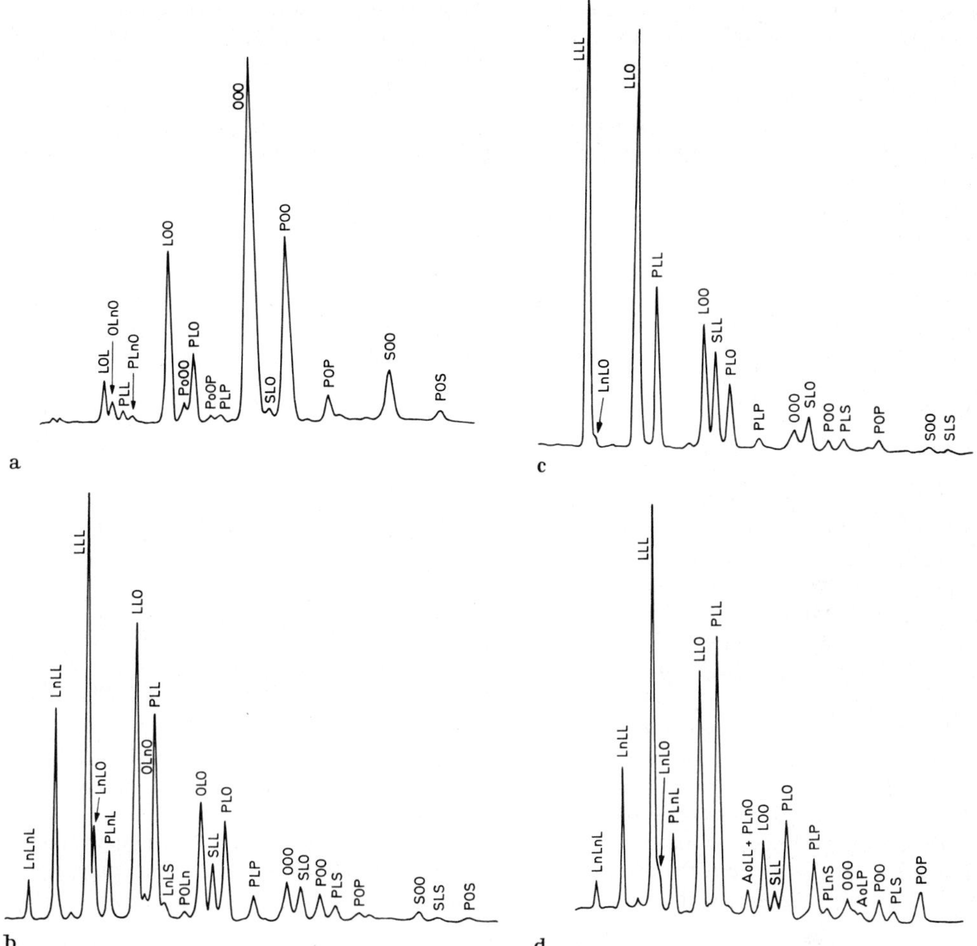

Fig. 3.6. Composition of triacylglycerols present in edible fats or oils as determined by HPLC
a olive oil, **b** soybean oil, **c** sunflower oil, **d** wheat germ oil
Fatty acids: P palmitic, S stearic, O oleic, L linoleic., Ln linolenic, Ao eicosenoic

have to be separated. The composition of milk fat is particularly complex. It contains more than 150 types of TG molecules.

The separation by HPLC using reverse phases is the first step in TG analysis. It is afforded by the chain length and the degree of unsaturation of the TG's. As shown in Fig. 3.6 the oils from different plant sources yield characteristic pattern in which distinct TG's predominate.

Various hypotheses have been advanced, supported by results of TG biosynthesis to predict the TG composition of a fat or oil when all the fatty acids occurring in the sample are known. The values calculated with the aid of the *1,3-random-2-random* hypothesis agree well with values found experimentally for plant oil or fat. The hypothesis starts with two separated fatty acid pools. The acids in both pools are randomly distributed and used as such for TG biosynthesis. The primary HO-groups (positions 1 and 3 of glycerol) from the first pool are esterified, while the secondary HO-group is esterified in the second pool. The proportion of each TG is then determined (as mole %):

$$\beta\text{-XYZ (mol-\%)} = 2 \cdot \begin{bmatrix} \text{mol-\% X in} \\ \text{1,3-Position} \end{bmatrix} \cdot \begin{bmatrix} \text{mol-\% Y in} \\ \text{2-Position} \end{bmatrix} \cdot \begin{bmatrix} \text{mol-\% Z in} \\ \text{1,3-Position} \end{bmatrix} \cdot 10^{-4} \qquad (3.25)$$

The data required in order to apply the formula are obtained as follows: after partial hydrolysis of fat (oil) with pancreatic lipase (cf. 3.7.1.1), the fatty acids bonded at positions 1 and 3 are determined. The fatty acids in position 2 are calculated from the difference between the total acids and those acids in positions 1 and 3.
Table 3.9 illustrates the extent of agreement for TG composition of sunflower oil obtained experimentally and by calculation using the 1,3-random-2-random hypothesis. However, both approaches disregarded the differences between positions 1 and 3. In addition, the hypothesis is directed to plants, of which the fats and oils consist of only major fatty acids.

Stereospecific analysis. Biochemically, the esterified primary HO-groups of glycerol are differentiated from each other; thus, the determination of fatty acids in positions 1, 2 and 3 is possible. The reaction sequence of one of the many procedures designed to carry out a stereospecific analysis is presented in Figure 3.7. First, the TG (I) is hydrolyzed under controlled conditions to a diacylglycerol using pancreatic lipase (cf. 3.7.1.1). Phosphorylation with a diacylglycerol kinase follows. The enzyme reacts stereospecifically since it phosphorylates only the 1,2- but not the 2,3-diglyceride. Additionally, compound I will be hydrolyzed to a monoacylglycerol (III). The distribution of the acyl residues in positions 1, 2 and 3 is calculated from the results of the fatty acid analysis of compounds I, II and III.

Table 3.9. Triacylglycerols composition (mole-%) of a sunflower oil. A comparison of experimental values with calculated values based on a 1,3-random-2-random hypothesis

Triacylglycerol[a]	Found	Calculated	Triacylglycerol[a]	Found	Calculated
β-StOSt	0.3	0.5	β-OStL	0.5	0.2
β-StStO	0.2	trace	β-OOL	8.1	6.5
β-StOO	2.3	1.6	β-OLO	3.1	4.2
β-OStO	0.1	trace	β-StLL	13.2	14.0
β-StStL	0.3	0.2	β-LStL	1.3	0.3
β-StLSt	2.2	1.7	β-OLL	20.4	21.9
OOO	1.3	1.2	β-LOL	8.4	8.7
β-StOL	4.4	4.2	LLL	28.1	28.9
β-StLO	4.0	5.3	Others	0.9	0.9

St: Stearic, O: oleic, and L: linoleic acid

[a] Prefix β: The middle fatty acid is esterified at the β- or sn-2-position, the other two acids are at the sn-1 or sn-3 positions.

Individual TG's or their mixtures can be analyzed with this procedure. Based on these results (some are presented in Table 3.10), general rules for fatty acid distribution in plant oils or fats can be deduced:

- The primary HO-groups in positions 1 and 3 of glycerol are preferentially esterified with saturated acids.
- Oleic and linolenic acids are equally distributed in all positions, with some exceptions, such as cocoa butter (cf. Table 3.10).

Fig. 3.7. Stereospecific analysis of triacylglycerols

Table 3.10. Results of stereospecific analysis of some fats and oils[a]

Fat/Oil	Posi-tion	16:0	18:0	18:1 (9)	18:2 (9, 12)	18:3 (9, 12, 15)
Peanut	1	13.6	4.6	59.2	18.5	–
	2	1.6	0.3	58.5	38.6	–
	3	11.0	5.1	57.3	18.0	–
Soya	1	13.8	5.9	22.9	48.4	9.1
	2	0.9	0.3	21.5	69.7	7.1
	3	13.1	5.6	28.0	45.2	8.4
Cocoa	1	34.0	50.4	12.3	1.3	–
	2	1.7	2.1	87.4	8.6	–
	3	36.5	52.8	8.6	0.4	–
Liver (rat)	1	19.0	2.53	6.2	5.3	–
	2	1.2	0.2	13.0	18.0	–
	3	6.0	0.8	14.0	10.0	–

[a] In order to simplify the Table other fatty acids present in fat/oil are not listed.

- The remaining free position, 2, is then filled with linoleic acid.

Results compiled in Table 3.10 show that, for oil or fat of plant origin, there is little difference in acyl residues between positions 1 and 3. Therefore, the 1,3-random-2-random hypothesis provides results that agree well with experimental findings.

The fatty acid pattern in animal fats is strongly influenced by the fatty acid composition of animal feed. A steady state is established only after 4–6 months of feeding with the same feed composition. The examples given in Table 3.10 show that positions 1 and 3 in triglycerides of animal origin show much greater variability than in fats or oils of plant origin. Therefore, any prediction of TG types in animal fat should be calculated from three separate fatty acid pools (*1-random-2-random-3-random* hypothesis).

3.3.1.5 Biosynthesis

A TG molecule is synthesized in the fatty cells of mammals and plants from L-glycerol-3-phosphate and fatty acid-CoA esters (Figure 3.8). The L-glycerol-3-phosphate supply is provided by the reduction of dihydroxy acetone phosphate by NAD$^+$-dependent glycerol phosphate dehydrogenase. The dihydroxy acetone phosphate originates from glycolysis.

The lipid bodies (oleosomes, spherosomes) synthesized are surrounded by a membrane and are deposited in storage tissues.

The TG fatty acid composition within a plant species depends on environment; especially temperature. A general rule is that plants in cold climates produce a higher proportion of unsaturated fatty acids. Obviously, the mobility of TG's is thus retained. In the sunflower (cf. Figure 3.9), this rule is highly pronounced; whereas, in safflower, only a weak response to temperature variations is observed (Figure 3.9).

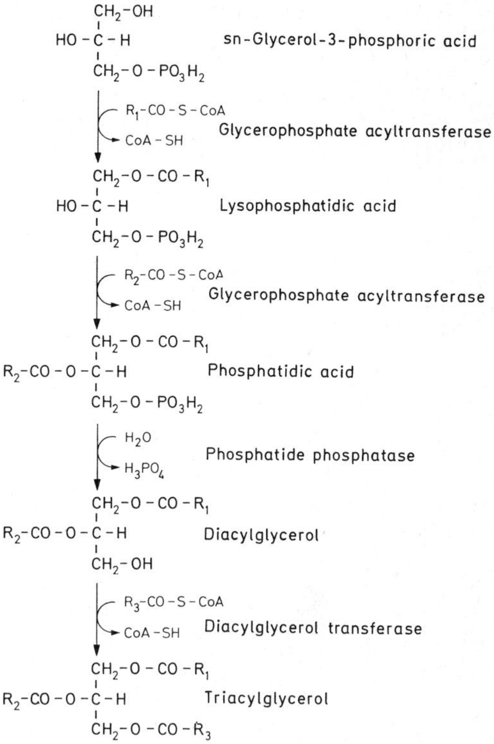

CH_2-OH / $HO-C-H$ / $CH_2-O-PO_3H_2$ — sn-Glycerol-3-phosphoric acid

↓ $R_1-CO-S-CoA$ → $CoA-SH$ — Glycerophosphate acyltransferase

$CH_2-O-CO-R_1$ / $HO-C-H$ / $CH_2-O-PO_3H_2$ — Lysophosphatidic acid

↓ $R_2-CO-S-CoA$ → $CoA-SH$ — Glycerophosphate acyltransferase

$CH_2-O-CO-R_1$ / $R_2-CO-O-C-H$ / $CH_2-O-PO_3H_2$ — Phosphatidic acid

↓ H_2O → H_3PO_4 — Phosphatide phosphatase

$CH_2-O-CO-R_1$ / $R_2-CO-O-C-H$ / CH_2-OH — Diacylglycerol

↓ $R_3-CO-S-CoA$ → $CoA-SH$ — Diacylglycerol transferase

$CH_2-O-CO-R_1$ / $R_2-CO-O-C-H$ / $CH_2-O-CO-R_3$ — Triacylglycerol

Fig. 3.8. Biosynthesis of triacylglycerols

3.3.2 Mono- and Diacylglycerols (MG, DG)

The occurrence of MG and DG in edible oils or fats or in raw food is very low. However, their levels may be increased by the action of hydrolases during food storage or processing.

MG and DG are produced commercially by fat glycerolysis:

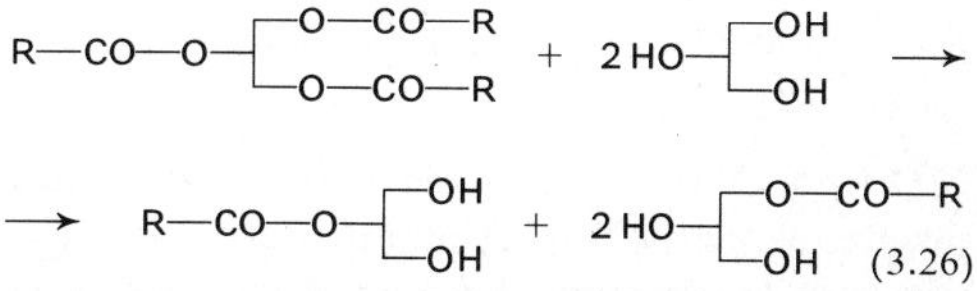

$$R{-}CO{-}O{-}C(CH_2{-}O{-}CO{-}R)_2 + 2\,HO{-}C(CH_2{-}OH)_2 \longrightarrow R{-}CO{-}O{-}C(CH_2{-}OH)_2 + 2\,HO{-}C(CH_2{-}O{-}CO{-}R)(CH_2{-}OH) \quad (3.26)$$

Physical properties: MG and DG crystallize in different forms (polymorphism; cf. 3.3.1.2). The

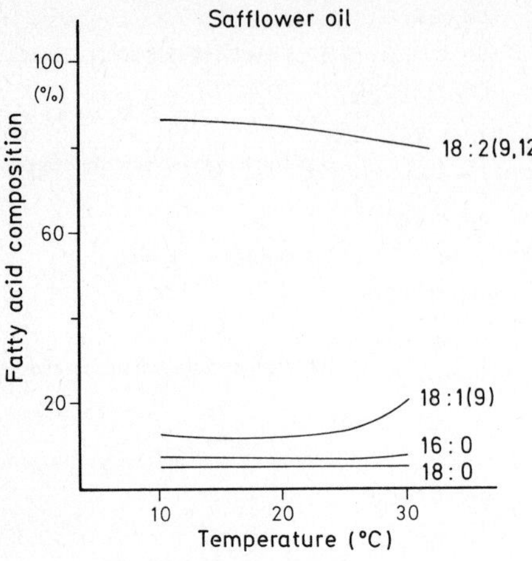

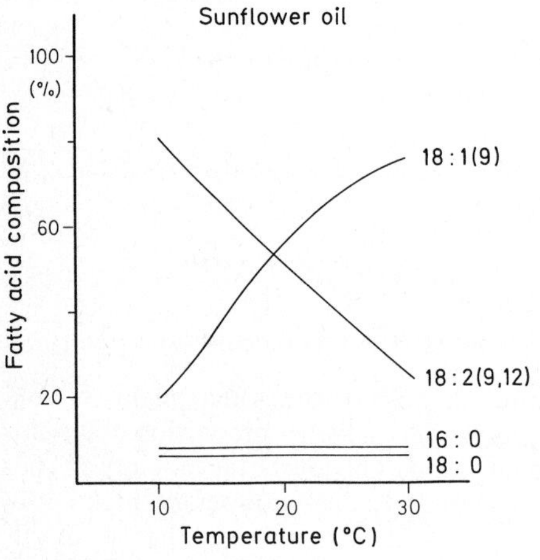

Fig. 3.9. The effect of climate (temperature) on fatty acid composition of triacylglycerols

melting point of an ester of a given acid increases for the series 1,2-DG < TG < 2-MG < 1,3-DG < 1-MG:

	Melting Point (°C) β-form
Tripalmitin	65.5
1,3-Dipalmitin	72.5
1,2-Dipalmitin	64.0
1-Palmitin	77.0
2-Palmitin	68.5

MG and DG are surface-active agents. Their properties can be further modified by esterification with acetic, lactic, fumaric, tartaric or citric acids. These esters play a singificant role as emulsifiers in food processing (cf. 8.15).

3.4 Phospho- and Glycolipids

3.4.1 Classes

Phospho- and glycolipids, together with proteins, are the building blocks of biological membranes. Hence, they invariably occur in all foods of animal and plant origin. Examples are compiled in Table 3.11. As surface-active compounds, phospho- and glycolipids contain hydrophobic moieties (acyl residue, N-acyl sphingosine) and hydrophilic portions (phosphoric acid, carbohydrate). Therefore, they are capable of forming orderly structures (micelles or planar layers) in aqueous media; the bilayer structures are found in all biological membranes. Examples for the composition of membrane lipids are listed in Table 3.12.

3.4.1.1 Phosphatidyl Derivatives

The following phosphoglycerides are derived from phosphatidic acid:

$$\begin{array}{l} (1)\ CH_2-O-CO-R_1 \\ R_2-CO-O-CH\ (2) \\ (3)\ CH_2-O-P(=O)(O^{\ominus})-O-CH_2-CH_2-N^{\oplus}(CH_3)_3 \end{array} \quad (3.27)$$

Phosphatidyl choline or lecithin (phosphate group esterified with the HO-group of choline).

$$\begin{array}{l} CH_2-O-CO-R_1 \\ R_2-CO-O-CH \\ CH_2-O-P(=O)(O^{\ominus}\ X^{\oplus})-O-CH_2-CH(COO^{\ominus})-\overset{\oplus}{N}H_3 \end{array} \quad (3.28a)$$

Table 3.11. Composition of lipids of various foods[a]

	Milk	Soya	Wheat	Apple
Total lipids	3.6	23.0	1.5	0.088
Triacylglycerols	94	88	41	5
Mono-, and diacylglycerols	1.5		1	
Sterols	<1		1	15
Sterol esters			1	2
Phospholipids	1.5	10	20	47
Glycolipids		1.5	29	17
Sulfolipids				1
Others		0.54	7	15

[a] Total lipids as %, while lipid fractions are expressed as percent of the total lipids.

Table 3.12. Lipid composition of biomembranes

	Lipid content of membrane (%)	Lipid composition (%)			
		Neutral lipids	Glycero-glycolipids	Glycero-phospholipids	Sphingo-lipids
Chloroplast (spinach)	52	29	45	9	no data
Mitochondria (bovine heart)	26	8	no data	92	0
Endoplasmic reticulum (heart, swine)	25	32	no data	55	11
Myelin (brain of mammal animals)	78	25	no data	32	31

Phosphatidyl serine (phosphate group esterified with the HO-group of the amino acid serine).

```
              CH2-O-CO-R1
              |
R2-CO-O-CH         O
              |          ||                        ⊕
              CH2-O-P-O-CH2-CH2-NH3
                         |
                         O⊖
```
(3.28b)

Phosphatidyl ethanolamine (phosphate group esterified with ethanolamine).

```
              CH2-O-CO-R1
              |
R2-CO-O-CH         O
              |          ||
              CH2-O-P-O-[inositol ring: OH, OH, OH, HO, OH]
                         |
                 ⊕X  ⊖O
```
(3.28c)

Phosphatidyl inositol (phosphate group esterified with inositol).

A mixture of phosphatidyl serine and phosphatidyl ethanolamine was once referred to as cephalin.

Only one acyl residue is cleaved by hydrolysis (cf. 3.7.1.2.1) with phospholipase A. This yields the corresponding lyso-compounds from lecithin or phosphatidyl ethanolamine. Some of these lyso-derivatives occur in nature, e.g. in cereals.

Phosphatidyl glycerol is invariably found in green plants, particularly in chloroplasts:

```
              CH2-O-CO-R1       CH2-OH
              |                 |
R2-CO-O-CH         O            CH-OH
              |          ||           |
              CH2-O-P-O-CH2
                         |
                         O⊖ X⊕
```
(3.29a)

Cardiolipin, first identified in beef heart, is also a minor constituent of green plant lipids. Its chemical structure is diphosphatidyl glycerol:

```
                                                O
                                                ||
              CH2-O-CO-R1       CH2-O-P-O-CH2
              |                 |         |        |
R2-CO-O-CH         O            CH-OH     O⊖ X⊕   CH-O-CO-R3
              |          ||          |                 |
              CH2-O-P-O-CH2              R4-CO-O-CH2
                         |
                         O⊖ X⊕
```
(3.29b)

The plasmalogens occupy a special place in the class of phospho-glycerides. They are phosphatides in which position 1 of glycerol is linked to a straight-chain aldehyde with 16 or 18 carbons. The linkage is an enol-ether type with a double bond in the cis-configuration. Plasmalogens of 1-0-(1-alkenyl)-2-0-acylglycerophospholipid type occur in small amounts in animal muscle tissue and also in milk fat. The enol-ether linkage, unlike the ether bonds of a 1-0-alkylglycerol (cf. 3.6.2), is readily hydrolyzed even by weak acids.

```
              CH2-O-CH=CH-R1
              |
R2-CO-O-CH         O
              |          ||                        ⊕
              CH2-O-P-O-CH2-CH2-NH3
                         |
                         O⊖
```
(3.30)

Plasmalogen

Phospholipids are sensitive to autoxidation since they contain an abundance of linoleic acid. The other acid commonly present is palmitic acid.

Phospholipids are soluble in chloroform-methanol and poorly soluble in water-free acetone. The pK_s value of the phosphate group is between 1 and 2. Phosphatidyl choline and phosphatidyl ethanolamine are zwitter-ions at pH 7.

Phospholipids can be hydrolyzed stepwise by alcoholic KOH. Under mild conditions, only the fatty acids are cleaved, whereas, with strong alkalies, the base moiety is released. The bonds between phosphoric acid and glycerol or phosphoric acid and inositol are stable to alkalies, but are readily hydrolyzed by acids. Phosphatidyl derivatives, together with triacyglycerols and sterols, constitute the lipid fraction of the lipoprotein complex (cf. 3.5.1).

Lecithin: Lecithin plays a significant role as a surface-active substance in the production of emulsions. "Raw lecithin", especially that of soya and that isolated from egg yolk, is available for use on a commercial scale.
The major phospholipids of raw soya lecithin are given in Table 3.13. The manufacturer often separates lecithin into ethanol-soluble and ethanol-insoluble fractions.
Pure lecithin is a W/O emulsifier with a HLB-value (cf. 8.15.3) of about 3. Since the commercial used lecithins are complex mixtures of lipids their HBL-values lie within a wide range. The ethanol-insoluble fraction (Table 3.13) is suitable for stabilization of W/O emulsions and the ethanol-soluble fraction for O/W emulsions. To increase the HLB-value "hydroxylated lecithins" are produced by a partial oxidation of the unsaturated acyl residues with hydrogen peroxide or benzoyl peroxide.

3.4.1.2 Glyceroglycolipids

These lipids consist of 1,2-diacylglycerols and a mono-, di- or less frequently, tri- or tetrasaccharide bound in position 3 of glycerol. Galactose is predominant as the sugar component among plant glycerolipids. The chloroplasts are particularly abundant in these lipids (Table 3.12).

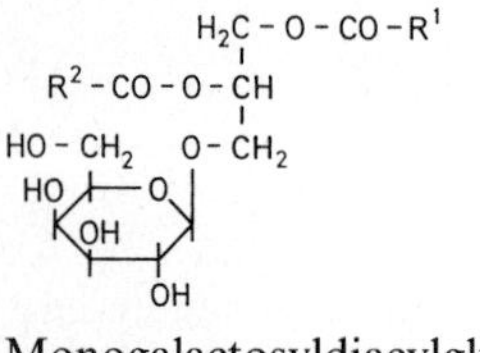

$H_2C-O-CO-R^1$, $R^2-CO-O-CH$, $HO-CH_2$ $O-CH_2$ (3.31a)

Monogalactosyldiacylglycerol (MGDG)
(1,2-diacyl-3-β-D-galactopyranosyl-L-glycerol)

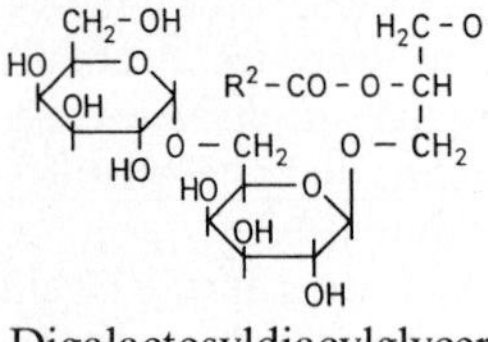

CH_2-OH, $H_2C-O-CO-R^1$, $R^2-CO-O-CH$, $O-CH_2$ $O-CH_2$ (3.31b)

Digalactosyldiacylglycerol (DGDG)
[1,2-diacyl-3-(α-D-galactopyranosyl-1,6-β-D-galactopyranosyl)-L-glycerol]

Table 3.13. Composition of soya "raw lecithin"

	Weight-% of raw lecithin		
	Unfractionated	Ethanol soluble fraction	Ethanol insoluble fraction
Phosphatidyl ethanolamine	32.6	32.5	32.6
Phosphatidyl choline	32.6	65.1	4.6
Phosphatidyl inositol	34.8	2.4	62.8

6-O-acyl-MGDG and 6-O-acyl-DGDG are minor components of plant lipids.
Sulfolipids are glyceroglycolipids which are highly soluble in water since they contain a sugar moiety esterified with sulfuric acid. The sugar moiety is 6-sulfochinovose. Sulfolipids occur in chloroplasts but are also detected in potato tubers:

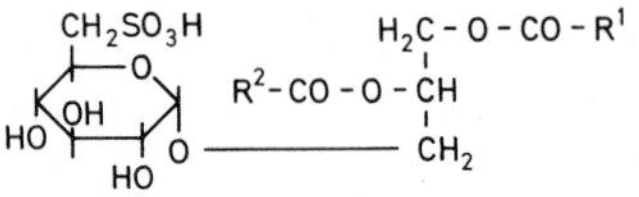

CH_2SO_3H, $H_2C-O-CO-R^1$, $R^2-CO-O-CH$, CH_2 (3.32)

Sulfolipid
[1,2-diacyl-(6-sulfo-α-D-chinovosyl-1,3)-L-glycerol]

3.4.1.3 Sphingolipids

Sphingolipids contain sphingosine, an amino alcohol with a long unsaturated hydrocarbon chain (D-*erythro*-1,3-dihydroxy-2-amino-trans-4-octadecene) instead of glycerol:

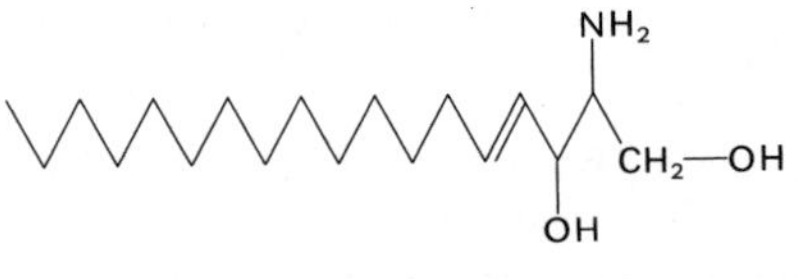

NH_2, CH_2-OH, OH (3.33)

Sphingolipids, which occur in plants, e.g. wheat contain phytosphingosines:

$$CH_3(CH_2)_n-\underset{OH}{CH}-\underset{OH}{CH}-\underset{NH_2}{CH}-\underset{OH}{CH_2} \quad (3.34)$$

n: 13, 14, 15, 16

In the sphingolipids, the amino group is linked to a fatty acid by an amide bond. The primary HO-group (on C-1) is either esterified with phosphoric acid (sphingophospholipid) or is bound glycosidically to a mono-, di- or oligosaccharide (sphingoglycolipid).

Sphingophospholipids: Sphingomyelin is one example of a sphingophospholipid. It is the most abundant sphingolipid and is found in myelin, the fatty substance of the sheath around nerve fibers. The structure of sphingomyelin is:

$$H_3C-\overset{\oplus}{N}(CH_3)_2-CH_2-CH_2-O-\overset{O}{\overset{\|}{P}}(O^{\ominus})-O-CH_2-CH(NH-CO-R)-CH(OH)-CH=CH-\ldots \quad (3.35)$$

Sphingoglycolipids (Cerebrosides) are found in abundance in membranes of nerve cells, brain cells and the myelin sheath. They also occur in plants, particularly cereals.

$$\text{Glc}-O-CH_2-CH(NH-CO-R^1)-CH(OH)-CH(OH)-(CH_2)_{13}-CH_3 \quad (3.36)$$

Cerebroside from wheat flour

Cerebrosides from plants and also from liver and spleen contain glucose, whereas those from brain contain galactose.

Sphingosine is attached to a large oligosaccharide in gangliosides that are present in various organs, such as brain, spleen or red blood cells. This oligosaccharide contains sialic acids (the trivial name for the N-acetyl- and N-glucosyl neuramic acids; cf. Formula 3.37) in addition to glucose and galactose.

Phytosphingolipids: These lipids have an even more complex structure than cerebrosides. Total hydrolysis yields phytosphingosine, inositol, phosphoric acid and various monosaccharides (galactose, arabinose, mannose, glucosamine, glucuronic acid).

Phytosphingolipids are isolated from soya and peanuts (cf. Formula 3.38).

3.4.2 Analysis

Extraction and removal of nonlipids. A solvent mixture of chloroform/methanol (2 + 1 v/v) is suitable for a quantitative extraction of lipids. Addition of a small amount of BHA (cf. 3.7.3.2.2) is recommended for the stabilization of lipids against autoxidation.

Nonlipid impurities were earlier removed by shaking the extracts with a special salt solution under rather demanding conditions. An improved procedure, by which emulsion formation is avoided, is based on column chromatography with dextran gels.

Separation and identification of classes of lipids. Isolated and purified lipids can be separated into classes by thin layer chromatography using developing solvents of different polarity. Figure 3.10 shows an example of the separation of neutral and polar lipids. Identification of polar lipids is based on using spraying reagents which react on the plates with the polar moiety of the lipid molecules. For example, phosphoric acid is identified by the molybdenum-blue reaction, monosaccharides with orcinol-$FeCl_3$, choline by

$$CH_3-(CH_2)_{12}-CH=CH-CH(OH)-CH(NH-CO-R)-CH_2-O-\text{Glucose-Galactose-N-Acetylgalactosamine-Galactose} \quad (3.37)$$

N-Acetylneuraminic acid

$$HO-CH_2-CH(OH)-CH(OH)-\text{[pyranose ring with } H_2N, OH, COOH, OH]$$

Neuraminic acid

$$CH_3-(CH_2)_{13}-CH(OH)-CH(OH)-CH(NH-CO-R_1)-CH_2-O-\overset{O}{\overset{\|}{P}}(O^{\ominus}\,X^{\oplus})-O-\text{Inositol-Glucuronic acid-Glucosamine}\left\{\begin{array}{l}\text{Galactose}\\\text{Arabinose}\\\text{Mannose}\end{array}\right. \quad (3.38)$$

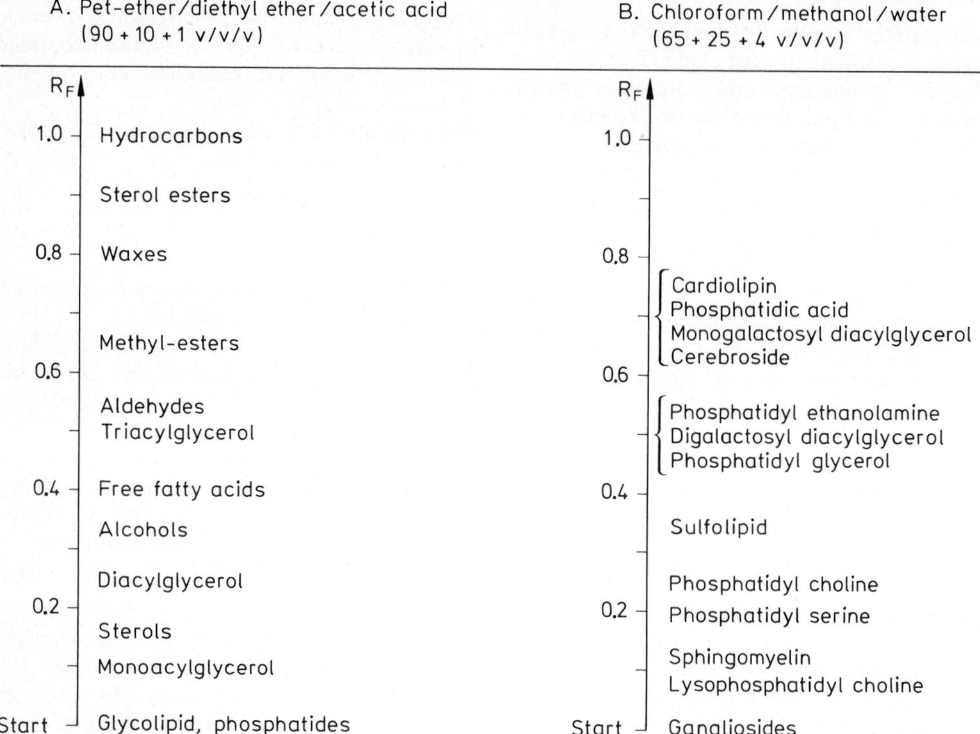

Fig. 3.10. Separation of lipid classes by thin layer chromatography using Kieselgel as an adsorbent. R_f values in two solvent systems

bismuth iodide (*Dragendorff* reagent), ethanolamine and serine by the ninhydrin reaction, and sphingosine by a chlorine-benzidine reagent.

When sufficient material is available, it is advisable to perform a preliminary separation of lipids by column chromatography on magnesium silicate (florisil), silicic acid, hydrophobic dextran gel or a cellulose-based ion-exchanger, such as DEAE-cellulose.

Analysis of lipid components. Fatty acid composition is determined after methanolysis of the lipid. For positional analysis of acyl residues (positions 1 or 2 in glycerol), phosphatidyl derivatives are selectively hydrolyzed with phospholipases (cf. 3.7.1.2.1) and the fatty acids liberated are analyzed by gas chromatography.

The sphingosine base can also be determined by gas chromatography after trimethylsilyl derivatization. The length of the carbon skeleton, of interest for phytosphingosine, can be determined by analyzing the aldehydes released after the chain has been cleaved by periodate:

$$\begin{array}{c} CH_3-(CH_2)_n-\underset{OH}{\underset{|}{CH}}-\underset{OH}{\underset{|}{CH}}-\underset{NH_2}{\underset{|}{CH}}-CH_2-OH \\ \downarrow JO_4^{\ominus} \\ CH_3-(CH_2)_n-CHO + NH_3 + 3\,CH_2O \end{array} \quad (3.39)$$

The monosaccharides in glycolipids can also be determined by gas chromatography. The lipids are hydrolyzed with 2 N trifluoroacetic acid and then derivatized to an acetylated glyconic acid nitrile. By using this sugar derivative, the chromatogram is simplified because of the absence of sugar anomers (cf. 4.2.4.6).

3.5 Lipoproteins, Membranes

3.5.1 Lipoproteins

3.5.1.1 Definition

Lipoproteins are aggregates, consisting of proteins, polar lipids and triacylglycerols, which are water soluble and can be separated into protein and lipid moieties by an extraction procedure using suitable solvents. This indicates that only noncovalent bonds are involved in the formation of lipoproteins. The aggregates are primarily stabilized by hydrophobic interactions between the apolar side chains of hydrophobic regions of the protein and the acyl residues of the lipid. In addition, there is a contribution to stability by ionic forces between charged amino acid residues and charges carried by the phosphatides. Hydrogen bonds, important for stabilization of the secondary structure of protein, play a small role in binding lipids since phosphatidyl derivatives have only a few sites available for such linkages. Hydrogen bonds can exist to a greater extent between proteins and glycolipids; however, such lipids have not yet been found as lipoprotein components, but rather as building blocks of biological membranes. An exception may be their occurrence in wheat flour, where they are responsible for gluten stability which develops in dough. Here, the lipoprotein complex consists of prolamine and glutelin attached to glycolipids by hydrogen bonds and hydrophobic forces. Although the presence of covalent bonds between lipids and proteins can not be completely excluded, experimental results do not support such an assumption.

3.5.1.2 Classification

Lipoproteins exist as globular particles in an aqueous medium. They are solubilized from biological sources by buffers with high ionic strength, by a change of pH or by detergents in the isolating medium. The latter, a more drastic approach, is usually used in the recovery of lipoproteins from membranes.

Lipoproteins are characterized by ultracentrifugation. Since lipids have a lower density (0.88–0.9 g/ml) than proteins (1.3–1.35 g/ml), the separation is possible because of differences in the ratios of lipid to protein within a lipoprotein complex. The lipoproteins of blood plasma have been thoroughly studied. They are separated by a stepwise centrifugation in solutions of NaCl into three fractions with different densities (Figure 3.11). The "very low density lipoproteins" (VLDL; density < 1.006 g/ml), the "low density lipoproteins" (LDL; 1.063 g/ml) and the "high density lipoproteins" (HDL; 1.21 g/ml) float, and the sediment contains the plasma proteins. The VLDL fraction can be separated further by electrophoresis into chylomicrons (the lightest lipoprotein, density < 1.000 g/ml) and pre-β-lipoprotein.

Lipoproteins in the LDL fraction from an electrophoretic run have a mobility close to that of blood plasma β-globulin, therefore, the LDL fraction is denoted as β-lipoprotein. An analogous designation of α-lipoprotein is assigned to the HDL fraction.

Chylomicrons, the diameters of which range from 1,000–10,000 Å, are small droplets of triacylglycerol stabilized in the aqueous medium by a membrane-like structure composed of protein, phosphatides and cholesterol. The role of chylomicrons in blood is to transport triacylglycerols

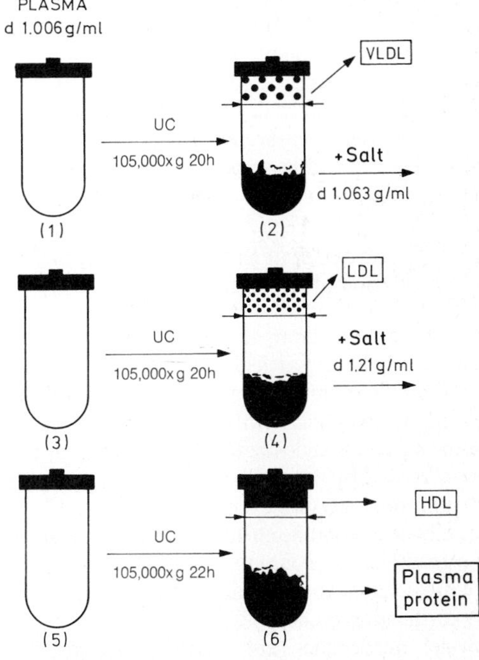

Fig. 3.11. Plasma protein fractionation by a preparative ultracentrifugation method. (After *D. Seidel,* 1971)

Table 3.14. Composition of typical lipoproteins

Source	Lipoprotein	Particle weight (kdal)	Protein (%)	Glycero-phospholipids (%)	Cholesterol free (%)	Cholesterol esterified (%)	Triacyl-glycerols (%)
Human blood serum	Chylomicron	10^6–10^7	1–2	4	2.5–3	3–4	85–90
	Pre-β-lipoprotein	$5–100 \cdot 10^3$	8.3	19.2	7.4	11.1	54.2
	LDL (β-lipoprotein)	$2.3 \cdot 10^3$	22.7	27.9	8.5	28.8	10.5
	HDL (α-lipoprotein)	$1–4 \cdot 10^2$	58.1	24.7	2.9	9.2	5.9
Egg yolk (chicken)	β-Lipovitellin	$4 \cdot 10^2$	78	12	0.9	0.1	9
	LDL	$2–10 \cdot 10^3$	18	22	3.5	0.2	58
Bovine milk	LDL	$3.9 \cdot 10^3$	12.9	52	0	0	35.1

to various organs, but preferentially from the intestines to adipose tissue and the liver. The milk fat globules (cf. 10.1.2.3) have a structure similar to that of chylomicrons. The composition of plasma lipoprotein is presented in Table 3.14.
Some diseases related to fat metabolism can be clinically diagnosed by the content and composition of the plasma lipoprotein fractions.
Two lipoproteins have been isolated from egg yolk (Table 3.14). An LDL similar to plasma lipoprotein occurs in a soluble form. β-lipovitellin, the composition of which corresponds to an HDL (Table 3.14), has the properties of a membrane constituent.
Electron microscopy studies have revealed that the fat globules in milk have small particles attached to their membranes; these are detached by detergents and have been identified as LDL (cf. Table 3.14).

3.5.2 Involvement of Lipids in the Formation of Biological Membranes

Membranes that compartmentalize the cells and many subcellular particles are formed from two main building blocks: proteins and lipids (phospholipids and cholesterol). Differences in membrane structure and function are reflected by the proteins and by the compositional differences of membrane lipids (see examples in Table 3.12).
Studies of membrane structure are difficult since the methods for isolation and purification profoundly change the organization and functionality of the membrane.
Model membranes are readily formed. The major forces in such events are the hydrophobic interactions between the acyl tails of phospholipids, providing a bilayer arrangement. In addition, the amphipathic character of the lipid molecules makes membrane formation a spontaneous process. The acyl residues are sequestered and oriented in the nonpolar interior of the bilayer, whereas the polar hydrophilic head groups are oriented toward the outer aqueous phase.
Another arrangement in water that satisfies both the hydrophobic acyl tails and the hydrophilic polar groups is a globular micelle. Here, the hydrocarbon tails are sequestered inside, while the polar groups are on the surface of the sphere. There is no bilayer in this arrangement.
The favored structure for most phospho- and glycolipids in water is a bimolecular arrangement, rather than a micelle. Two model systems can exist for such bimolecular arrangements. The first is a lipid vesicle, known as a liposome, the core of which is an aqueous compartment surrounded by a lipid bilayer, and the second is a planar, bilayer membrane. The latter, together with the micellular model, is presented in Figure 3.12. Globular proteins, often including enzymes, are found in animal cell membranes and are well embedded or inserted into the bimolecular layer. Some of these so-called integral membrane proteins protrude through both sides of

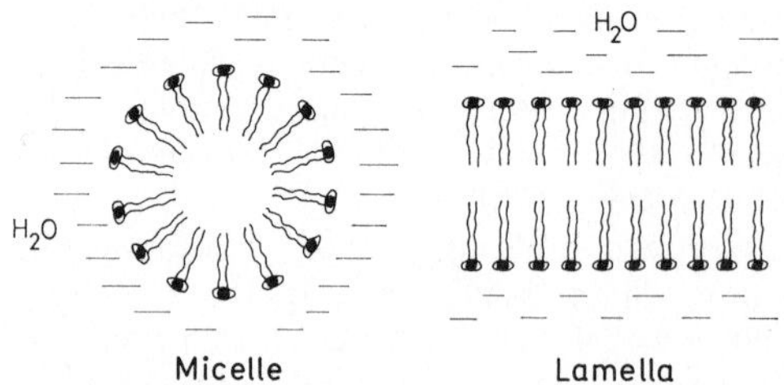

Fig. 3.12. Arrangement of polar acyl lipids in aqueous medium. ⫰ Polar lipid tails; ≈ hydrophobic lipid tails

the membrane (fluid mosaic model, Figure 3.13). Although integral proteins interact extensively with the hydrophobic acyl tails of membrane lipids, they are mobile within the lipid membrane. Older membrane models which have been proposed but, so far, not proven, considered the membrane as an inner lipid bilayer coated on both sides with membrane proteins. It was assumed that these proteins were attached to the surface of the polar lipids by ionic and/or hydrogen bonds.

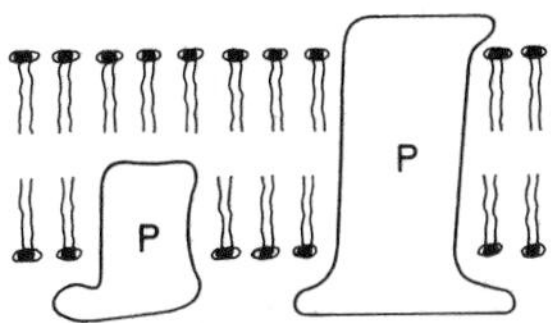

Fig. 3.13. Fluid mosaic model of a biological membrane. The protein (P) is not fixed but is mobile in the phospholipid phase

3.6 Diol Lipids, Higher Alcohols, Waxes and Cutin

3.6.1 Diol Lipids

The diol lipids which occur in both plant and animal tissues are minor lipid constituents. The diol content is about 1% of the content of glycerol. Exceptions are sea stars, sea urchins and mollusks, the lipids of which in summer contain 25–40% diol lipids. This proportion decreases sharply in winter and spring. Neutral and polar lipids derived from ethylene glycol, propane-(1,2 and 1,3)-diol and butane-(1,3; 1,4 and 2,3)-diol have been identified in the diol lipid fraction. Several of those isolated from corn oil have the following structures:

$$\begin{array}{l} H_2C-O-CO-R \\ | \\ H_2C-O-CO-R \end{array} \qquad \begin{array}{l} H_3C \\ | \\ HC-O-CO-R \\ | \\ HC-O-CO-R \\ | \\ H_3C \end{array} \tag{3.40}$$

$$\text{(sugar ring with } CH_2\text{-OH, HO, OH, OH)}-O-CH_2-CH_2-O-CO-R \tag{3.41}$$

Diol lipids with structures analogous to phosphatidyl choline or plasmalogen have also been identified.

3.6.2 Higher Alcohols and Derivatives

Higher alcohols occur either free or bound in plant and animal tissues. Free higher alcohols are abundant in fish oil and include:

Cetyl alcohol $C_{16}H_{33}OH$
Stearyl alcohol $C_{18}H_{37}OH$
Oleyl alcohol $C_{18}H_{35}OH$

Waxes are important derivatives of higher alcohols. They are higher alcohols esterified with long-chain fatty acids. Plant waxes are usually found on leaves or seeds. Thus, cabbage leaf wax consists of the primary alcohols C_{12} and C_{18}–C_{28} esterified with palmitic acid and other acids. The dominant components are stearyl and ceryl alcohol ($C_{26}H_{53}OH$). In addition to primary alcohols, esters of secondary alcohols, e.g. esters of nonacosane-15-ol, are present:

$$H_3C-(CH_2)_{13}-\underset{\displaystyle OH}{\underset{|}{CH}}-(CH_2)_{13}-CH_3 \tag{3.42}$$

The role of waxes is to protect the surface of plant leaves, stems and seeds from dehydration and infections by microorganisms. Waxes are removed together with oils by solvent extraction of nondehulled seeds. Waxes are oil-soluble at elevated temperatures but crystallize at room temperature, causing undesired oil turbidity. Cerotyl cerotate (ceryl alcohol esterified with cerotic acid, $C_{25}H_{51}COOH$)

$$H_3C-(CH_2)_{24}-CO-O-CH_2-(CH_2)_{24}-CH_3 \tag{3.43}$$

is removed from seed hulls during extraction of sunflower oil. Waxes are removed by an oil refining winterization step during the production of clear edible oil.

Waxes are present in fish oils, especially in sperm whale blubber and the whale's large head, which contain a "reservoir" of spermaceti wax. The main constituent of this wax is cetyl alcohol esterified with palmitic acid.

The higher alcohols, 16:0, 18:0 and 18:1 (9), form mono- and diethers with glycerol. Such alkoxy-lipids are widely distributed in small amounts in mammals and sea animals. Examples of confirmed structures are shown in Formula 3.44.

The elucidation of ether lipid structure is usually accomplished by cleavage by concentrated HI at elevated temperatures.

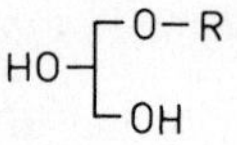

Alkyl glycerol ether (HO–, –O–R, –OH) 1,2-Dialkyl-glycerol ether (R–O–, –O–R, –OH)

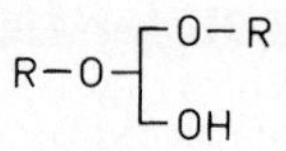

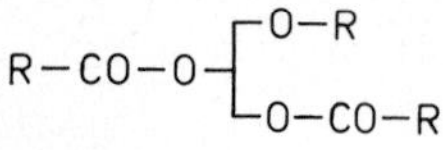

O-Alkyl-diacylglycerol (R–CO–O–, –O–R, –O–CO–R) O,O-Dialkyl-monoacylglycerol (R–O–, –O–R, –O–CO–R) (3.44)

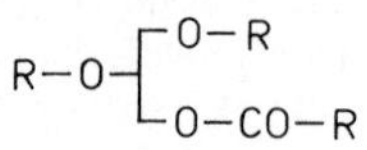

It is possible to differentiate edible beef tallow from lard by alkoxy lipid analysis. The alkoxy lipids are in a three-fold higher concentration (approx. 370 μg/g) in tallow than in lard. This ratio is independent of the composition of animal feed. For qualitative and quantitative analysis, the deacylated alkoxy lipids are recovered from the "unsaponifiables" and are then enriched by extraction and column chromatography procedures. The isolates are collected and, after silylation of one or two of the free HO-groups of glycerol, the mono- and diether concentrations are determined.

Common names of some deacylated alkoxy lipids (1-0-alkylglycerol) are the following:

$$CH_3-(CH_2)_{14}-CH_2-O-CH_2-\underset{OH}{CH}-\underset{OH}{CH_2} \quad (3.45a)$$

Chimyl alcohol

$$CH_3-(CH_2)_{16}-CH_2-O-CH_2-\underset{OH}{CH}-\underset{OH}{CH_2} \quad (3.45b)$$

Batyl alcohol

$$CH_3-(CH_2)_7-CH=CH-(CH_2)_7-CH_2-O-CH_2-\underset{OH}{CH}-\underset{OH}{CH_2}$$

Selachyl alcohol (3.45c)

3.6.3 Cutin

Plant epidermal cells are protected by a suberized or waxy cuticle. An additional layer of epicuticular waxes is deposited above the cuticle in many plants. The waxy cuticle consists of cutin. This is a complex, high molecular weight polyester which is readily solubilized by alkali. The structural units of the polymer are hydroxy fatty acids. The latter are similar in structure to the compounds given in 3.7.2.4.1. A segment of the postulated structure of cutin is presented in Figure 3.14.

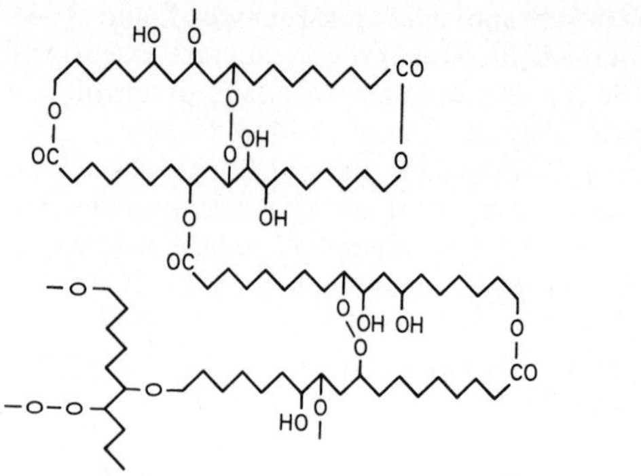

Fig. 3.14. A structural segment of cutin. (After *C. Hitchcock* and *B. W. Nichols,* 1971)

3.7 Changes in Acyl Lipids of Food

3.7.1 Enzymatic Hydrolysis

Hydrolases, which cleave acyl lipids, are present in food and microorganisms. The release of short-chain fatty acids ($< C_{14}$), e.g. the hydrolysis of milk fat, has a direct effect on food aroma. Lypolysis is undesirable in fresh milk since the free C_4–C_{12} fatty acids (cf. Table 3.3 for odor threshold values) are responsible for the rancid aroma defect. On the other hand, lypolysis occurring during the ripening of cheese is a desired and favorable process since, here, the short-chain fatty acids are involved in the build-up of specific cheese aromas. Likewise, slight hydrolysis of milk fat is advantageous in the production of chocolate.

Linoleic and linolenic acid freed by hydrolysis and present in emulsified form affect the flavor of food even at low concentrations. They cause a bitter-burning sensation. In addition, they decompose by autoxidation (cf. 3.7.2.1) or enzymatic oxidation (cf. 3.7.2.2) into compounds with an intensive odor. In fruits and vegetables enzymatic oxidation in conjunction with lipolysis occur, as a rule, at a high reaction rate, especially when tissue is sliced or homogenized (an example for rapid lypolysis is shown in Table 3.15). Also, enzymatic hydrolysis of a small amount of the acyl lipids present can not be avoided during disintegration of oil seeds. Release of higher fatty acids promotes foaming and so they are removed during oil refining (cf. 14.4.1).

Table 3.15. Lipid hydrolysis occurring during potato tuber homogenization

	µmoles/g[a]	
	Acyl lipids	Free fatty acids
Potato	2.34	0.70
Homogenate[b]	2.04	1.40
Homogenate[b] kept for 10 min at 0 °C	1.72	1.75
Homogenate kept for 10 min at 25 °C	0.54	2.90

[a] Potato tissue fresh weight.
[b] Sliced potatoes were homogenized for 30 sec at 0 °C.

Enzymes with lypolytic activity belong to the carboxylic-ester hydrolase group of enzymes (cf. Table 2.4).

3.7.1.1 Triacylglycerol Hydrolases (Lipases)

Lipases (cf. Table 2.4) hydrolyze only emulsified acyl lipids; they are active on a water/lipid interface. Lipases differ from esterase enzymes since the latter cleave only water-soluble esters, such as triacetylglycerol.
Lipase activity is detected in, for example, milk, oilseeds (soybean, peanut), cereals (oats, wheat), in fruits and vegetables and in the digestive tract of mammals. Many microorganisms release lipase-type enzymes into their culture media.
As to specificity, fat-splitting enzymes, which preferentially cleave primary HO-group esters are distinguished from those which indiscriminantly hydrolyze all three ester bonds of acyl glycerols (Table 3.16).
The lipase secreted by the swine pancreas is the most studied. Its molecular weight is 48 kdal. The enzyme cleaves the following types of acyl glycerols, with a decreasing rate of hydrolysis: triacyl- > diacyl- ≫ monoacylglycerols. Table 3.16 shows that pancreatic lipase reacts with acyl residues at positions 1 and 3. The third acyl residue of a triacylglycerol is cleaved (cf. Reaction 3.46) only after acyl migration, which requires a longer incubation time.

$$\text{(O–H, O–C(=O)R, O–H glycerol)} \xrightarrow[\ H_2O]{\ominus OH,\ H^{\oplus}} \text{(O–CO–R, OH, OH)} \xrightarrow{\text{Lipase (Pancreas)}} \text{(OH, OH, OH)} + RCOOH \qquad (3.46)$$

The smaller the size of the oil droplet, the larger the oil/water interface and, therefore, the higher the lipase activity. This relationship should not be ignored when substrate emulsions are prepared for the assay of enzyme activities.
A model for pancreatic lipase has been suggested to account for the enzyme's property to be active on the oil/water interface (Figure 3.15). The lipase's "hydrophobic head" is bound to the oil droplet by hydrophobic interactions, while the enzyme's active site aligns with and binds to the substrate molecule. The active site resembles that of serine proteinase. The splitting of the ester bond occurs with the involvement of Ser, His and Asp residues on the enzyme by a mechanism analogous to that of chymotrypsin (cf. 2.4.2.2). The dissimilarity of pancreatic lipase to serine proteinase is in the active site: lipase has a leucine residue within this site in order to establish hydrophobic contact with the lipid substrate and to align it with the activity center.

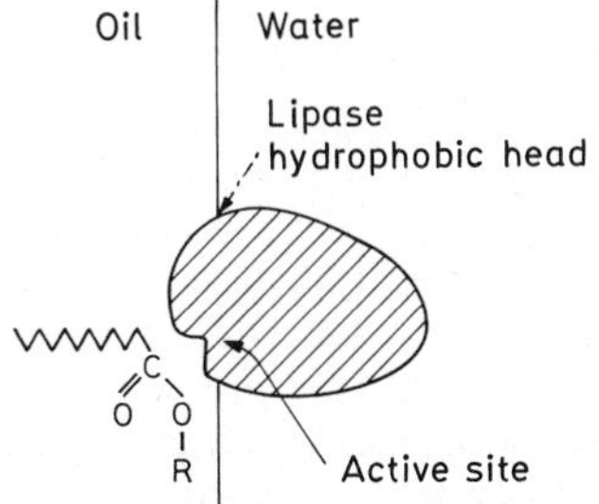

Fig. 3.15. A hypothetical model of pancreatic lipase fixation on an oil/water interphase. (After *H. Brockerhoff*, 1974)

Table 3.16. Examples for lipase specificity

From a triacylglycerol hydrolyzed are	Lipase source
Acyl residues in positions 1 and 3	Pancreas, milk, *Pseudomonas fragi*, *Penicillium roqueforti*
Acyl residues in positions 1, 2 and 3	Oats, Castor bean, *Aspergillus flavus*
Oleic and linoleic acids in position 1, 2 and 3	*Geotrichum candidum*

Lipase-catalyzed reactions are accelerated by Ca^{2+} ions since the freed fatty acids are precipitated as insoluble Ca-salts.
The properties of milk lipase closely resemble those of pancreatic lipase.
A lipase of microbial origin with exceptional specificity (Table 3.16) has been detected. It hydrolyzes fatty acids only when they have a double bond in position 9. It is used to elucidate triacylglyceride structure. The use of lipases in food processing was outlined in section 2.8.

3.7.1.2 Polar-Lipid Hydrolases

These enzymes are denoted as phospholipases or glycolipid hydrolases, depending on the substrate.

3.7.1.2.1 Phospholipases

Some of the enzymes that cleave phosphatides react with specificity. The action sites of various phospholipases will be clarified by the illustration of lecithin degradation. In general, these enzymes have been accepted as invaluable tools for the elucidation of the structure of the phosphatides.

Phospholipases A_1 and A_2: Both enzymes occur in many mammals. They are located in the lysosomes. Phospholipase A_1 cleaves the acyl residue in position 1 of lecithin and phospholipase A_2 hydrolyzes the acyl residue in position 2. In both cases lysolecithins are formed.

```
                  P-Lip. B
             ┌───────┴───────┐
             ↓               ↓
            H2C—O—┊—CO—R1
                  ←───────────── P-Lip. A1
R2—CO—┊—O—CH        O                   CH3
                    ‖                 ⊕/
            H2C—O—┊—P—┊—O—CH2—CH2—N—CH3
                    |                  \
P-Lip. A2           O⊖                  CH3

          P-Lip. C        P-Lip. D
```
(3.47)

Phospholipase B: This enzyme is specific for removing the remaining acyl residue in position 1 or 2 of a lysophosphatide. It has been isolated from rat brain and purified from contaminating phospholipase A activity. Occurrence of the enzyme in plants has not been confirmed. In the older literature an enzyme is described that cleaves both acyl residues of a phosphatide and it is denoted there as "phospholipase B". This hydrolase enzyme occurs in barley.

Phospholipase C: It hydrolyzes lecithin to a 1,2-diacylglyceride and phosphoryl choline. The enzyme is found in snake venom and in bacteria.

Phospholipase D: This enzyme cleaves the choline group in the presence of water or an alcohol, such as methanol, ethanol or glycerol, yielding free or esterified phosphatidic acid. For example:

$$\text{Phosphatidylcholine} + \text{ROH} \longrightarrow \text{Phosphatidyl-OR} + \text{Choline}$$

$$R: H, CH_3, CH_3CH_2, CH_2(OH)—CH(OH)—CH_2 \quad (3.48)$$

Phospholipase D can not cleave phosphatidyl inositol. The enzyme is present in cereals, such as rye and wheat, and in legumes. It was isolated and purified from peanuts.

3.7.1.2.2 Glycolipid hydrolases

Enzymes that cleave the acyl residues from mono- and digalactosyl-diglycerides are localized in green plants. A substrate specificity study for such a hydrolase from potato (Table 3.17) shows that plants also contain enzymes able to hydrolyze polar lipids in general. The potato enzyme preferentially cleaves the acyl residue from monoacylglycerols and lysolecithin, whereas triacylglycerols, such as triolein, are not affected.

Table 3.17. Purified potato acyl hydrolase: substrate specificity

Substrate	Relative activity (%)	Substrate	Relative activity (%)
Monolein	100	Lecithin	13
Diolein	21	Monogalactosyl-diacylglycerol	31
Triolein	0.2		
Methyloleate	28	Digalactosyl-diacylglycerol	17
Lysolecithin	72		

3.7.2 Peroxidation of Unsaturated Acyl Lipids

Acyl lipid constituents, such as oleic, linoleic and linolenic acids, have one or more allyl groups within the fatty acid molecule (cf. Table 3.4) and thus are readily oxidized to hydroperoxides. The latter, after subsequent degradation reaction, yield a great number of other compounds. Therefore, under the usual conditions of food storage, unsaturated acyl lipids can not be considered as stable food constituents.

Autoxidation should be distinguished from *lipoxygenase catalysis* in the process denoted as *lipid peroxidation*. Both oxidations provide hydroperoxides, but the latter occurs only in the presence of the enzyme.

Lipid peroxidation provides numerous volatile and nonvolatile compounds. Since the volatiles are exceptionally odorous compounds, lipid peroxidation is detected even in food with unsaturated acyl lipids present as minor constituents, or in food in which only a small portion of lipid was subjected to oxidation.

Induced changes in food aroma are continually assessed by consumers as objectionable, for example, as rancid, fishy, metallic or cardboard-like, or as an undefined aged or stale flavor. On the other hand, the fact that some volatile compounds, at a level below their off-flavor threshold values, contribute to the pleasant aroma of many fruits and vegetables and to rounding-off the aroma of many fat- or oil-containing foods should not be neglected.

3.7.2.1 Autoxidation

Autoxidation is quite complex and involves a great number of interrelated reactions of intermediates. Hence, autoxidation of food is usually imitated by the study of a model system in which, for example, changes of one unsaturated fatty acid or one of its intermediary oxidation products are recorded in the presence of oxygen under controlled experimental conditions.

Model system studies have revealed that the rate of autoxidation is affected by fatty acid composition, degree of unsaturation, the presence and activity of pro- and antioxidants, partial pressure of oxygen, the nature of the surface being exposed to oxygen and the storage conditions (temperature, light exposure, moisture content, etc.) of fat/oil-containing food. Thus, the autoxidation rate can vary considerably.

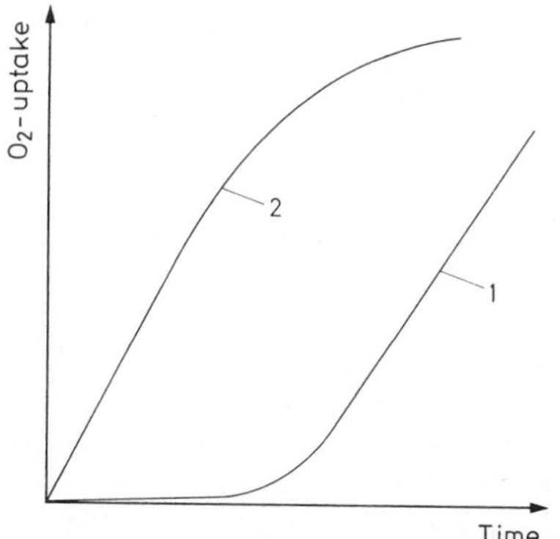

Fig. 3.16. Autoxidation rate of some unsaturated acyl lipids. Prooxidant concentration: 1 low, 2 high

An extreme case (cf. Figure 3.16-1) demonstrates what has invariably been found in food: the initial oxidation products are detectable only after a certain elapsed storage time. When this *induction period,* which is typical for a given autoxidation process, has expired, a steep rise occurs in the reaction rate. The prooxidant concentration is high in some foods. In these cases, illustrated in Figure 3.16-2, the induction period may be nonexistent.

3.7.2.1.1 Fundamental steps of autoxidation

The length of the induction period and the rate of oxidation depend, among other things, on the fatty acid composition of the lipid (Table 3.18); the more allyl groups present, the shorter the induction period and the higher the oxidation rate.

Table 3.18. Induction period and relative rate of oxidation for fatty acids at 25 °C

Fatty acid	Number of allyl groups	Induction period (h)	Oxidation rate (relative)
18:0	0		1
18:1 (9)	1	82	100
18:2 (9, 12)	2	19	1,200
18:3 (9, 12, 15)	3	1.34	2,500

Both phenomena, the induction period and the rise in reaction rate in the series, oleic, linoleic and linolenic acid can be explained by the following assumption. Oxidation proceeds by a sequential free radical chain-reaction mechanism. Relatively stable radicals that can abstract H-atoms from the activated methylene groups in an olefinic compound are formed. *Farmer and coworkers* (1942) and *Bolland* (1949) proposed an autoxidation mechanism for olefinic compounds and thus, also for unsaturated fatty acids, based on this assumption and, in addition, on the fact the oxidation rate is exponential. This mechanism has several fundamental steps. As shown in Figure 3.17, the oxidation process is essentially a radical-induced chain reaction divided into initiation (start), propagation, branching and ter-

Start: Formation of peroxy ($RO_2\cdot$), alkoxy ($RO\cdot$) or alkyl ($R\cdot$) radicals

Chain propagation:

(1) $R^\cdot + O_2 \longrightarrow RO_2^\cdot$ k_1: $10^9\ M^{-1}s^{-1}$

(2) $RO_2^\cdot + RH \longrightarrow ROOH + R^\cdot$ k_2: $10-60\ M^{-1}s^{-1}$

(3) $RO^\cdot + RH \longrightarrow ROH + R^\cdot$

Chain branching:

(4) $ROOH \longrightarrow RO^\cdot + {}^\cdot OH$

(5) $2\,ROOH \longrightarrow RO_2^\cdot + RO^\cdot + H_2O$

Chain termination:

(6) $2\,R^\cdot \longrightarrow$
(7) $R^\cdot + RO_2^\cdot \longrightarrow$ } Stable products
(8) $2\,RO_2^\cdot \longrightarrow$

Fig. 3.17. Basic steps in autoxidation of olefins

mination steps. Autoxidation is initiated by free radicals frequently of unknown origin.

Measuring and calculating the reaction rate constants for the different steps of radical chain reaction shows that due to the stability of the peroxy free radicals ($ROO^\cdot$) the whole process is limited by the conversion of these free radicals into monohydroperoxide molecules (ROOH). This reaction is achieved by abstraction of an H-atom from a fatty acid molecule [reaction step 2 (RS-2 in Figure 3.17)]. The H-abstraction is the slowest and, hence, the rate limiting step in radical ($R^\cdot$) formation.

Peroxidation of unsaturated fatty acids is accelerated autocatalytically by radicals generated from the degradation of hydroperoxides by a monomolecular reaction mechanism (RS-4 in Figure 3.17). This reaction is promoted by heavy metal ions or heme(in)-containing molecules (cf. 3.7.2.1.4). Also, degradation of hydroperoxides is considered as a starting point in discussions pertinent to volatile reaction products (cf. 3.7.2.1.5).

After a while the hydroperoxide concentration reaches a level at which it begins to generate free radicals by a bimolecular degradation mechanism (RS-5 in Figure 3.17). Reaction RS-5 is exothermic, unlike the endothermic monomolecular decomposition of hydroperoxides (RS-4 in Figure 3.17) which needs approx. 150 kJ/mol. However, in most foods, RS-5 is of no relevance since fat (oil) oxidation makes a food unpalatable well before reaching the necessary hydroperoxide level for the RS-5 reaction step to occur. RS-4 and RS-5 (Figure 3.17) are the branching reactions of the free radical chain.

At room temperature a radical may intitiate the formation of 100 hydroperoxide molecules before chain termination occurs. In the presence of air (oxygen partial pressure > 130 mbar) all acyl radicals are transformed into peroxy radicals through the rapid radical chain reaction 1 (RS-1, Figure 3.17). Therefore, chain termination occurs through collision of two peroxy radicals (RS-8, Figure 3.17).

Termination reactions RS-6 and RS-7 in Figure 3.17 play a role when, for example, the oxygen level is low, e.g. in the inner portion of a fatty food.

The hypothesis presented in Figure 3.17 is valid only for the initiation phase of autoxidation. The process becomes less and less clear with increasing reaction time since, in addition to hydroperoxides, secondary products appear that partially autoxidize into tertiary products. The stage from which the process starts to become difficult to survey depends on the stability of the primary products. It is instructive here to compare the difference in the structures of monohydroperoxides derived from linoleic and linolenic acids.

3.7.2.1.2 Monohydroperoxides

The peroxy radical formed in RS-1 (Figure 3.17) is slow reacting and therefore it selectively abstracts the most weakly bound H-atom from a fat molecule. It differs in this property from, for example, the substantially more reactive hydroxy ($HO^\cdot$) and alkoxy ($RO^\cdot$) radicals (cf. 3.7.2.1.4). RS-2 in Figure 3.17 has a high reaction rate only when the energy for H-abstraction is clearly lower than the energy released in binding H to O during formation of hydroperoxide groups (about 376 kJ $mole^{-1}$).

Table 3.19 lists the energy inputs needed for H-abstraction from the carbon chain segments or groups occurring in fatty acids. The peroxy radical abstracts hydrogen more readily from a methylene group of a 1,4-pentadiene system than from a single allyl group. In the former case the 1,4-diene radical that is generated is more effectively stabilized by resonance, i.e. electron delocaliza-

Table 3.19. Energy requirement for a an H-atom abstraction

	D_{R-H} (kJ/mole)
$-CH_2-$ (H)	422
CH_3-CH- (H)	410
$-CH-CH=CH-$ (H)	322
$-CH=CH-CH-CH=CH-$ (H)	272

tion over 5 C-atoms. Such considerations explain the difference in rates of autoxidation for unsaturated fatty acids and show why, at room temperature, the unsaturated fatty acids are attacked very selectively by peroxy radicals while the saturated acids are stable.

The general reaction steps shown in Figure 3.17 are valid for all unsaturated fatty acids. In the case of oleic acid, H-atom abstraction occurs on the methylene group adjacent to the double bond, i.e. positions 8 and 11 (Figure 3.18). This would give rise to four hydroperoxides. In reality, they have all been isolated and identified as autoxidation products of oleic acid. The newly-formed double bond configuration of the hydroperoxides is affected by temperature. This configuration is 33% cis and 67% of the more stable trans-configuration at room temperature.

Oxidation of the methylene group in position 11 of linoleic acid is activated especially by the two neighboring double bonds. Hence, this is the initial site for abstraction of an H-atom (Figure 3.19). The pentadienyl radical generated is stabilized by formation of two hydroperoxides, each retaining a conjugated diene system. These hydroperoxides have a UV maximum absorption at 235 nm and can be separated by high performance liquid chromatography as methyl esters, either directly or after reduction to hydroxydienes (Figure 3.20).

The monoallylic groups in linoleic acid (positions 8 and 14 in the molecule), in addition to the bisallylic group (position 11), also react to a small extent, giving rise to four hydroperoxides (8-, 10-, 12- and 14-OOH), each isomer having two

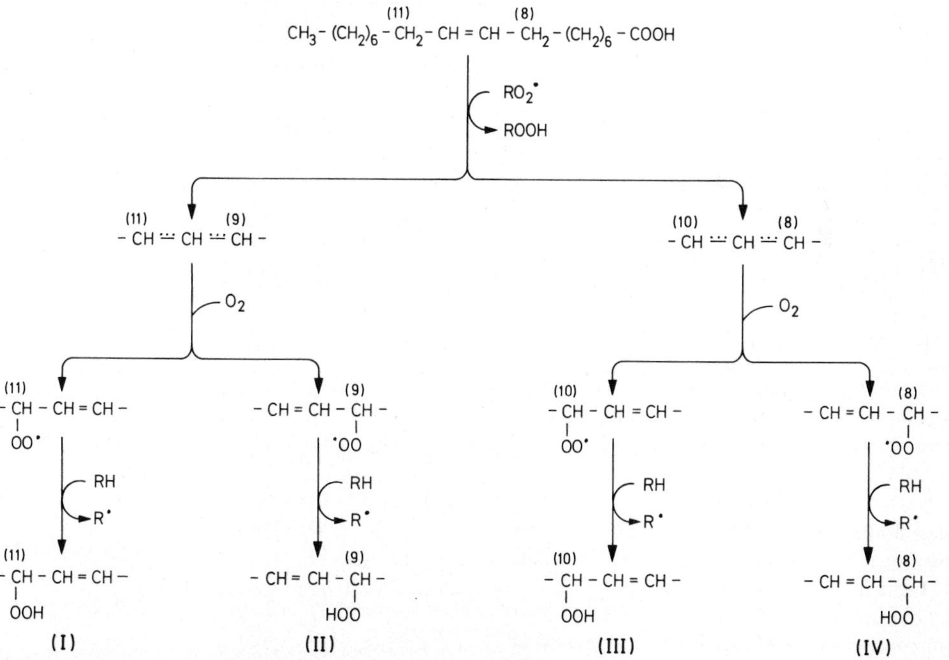

Fig. 3.18. Autoxidation of oleic acid. Primary reaction products:
I 11-Hydroperoxyoctadec-9-enoic acid; II 9-hydroperoxyoctadec-10-enoic acid, III 10-hydroperoxyoctadec-8-enoic acid, IV 8-hydroperoxyoctadec-9-enoic acid

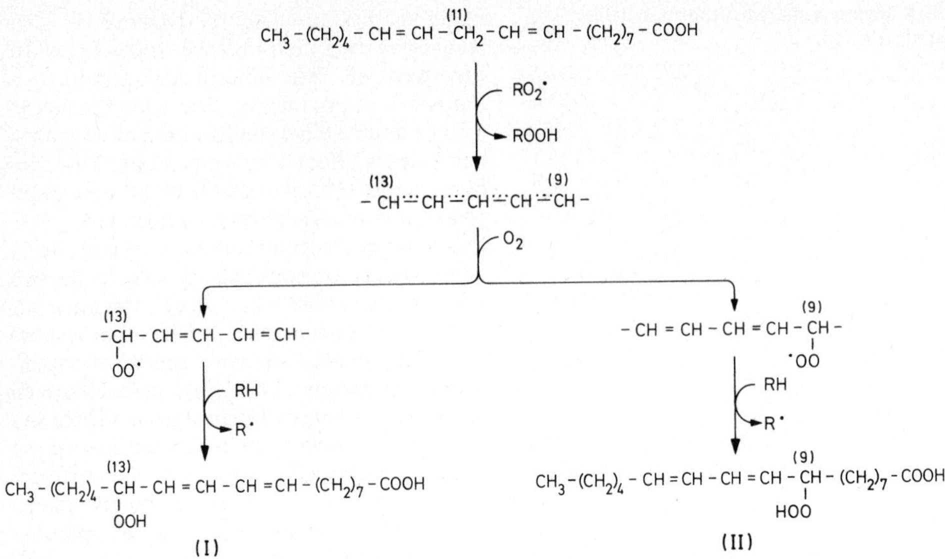

Fig. 3.19. Autoxidation of linoleic acid. Primary reaction products. I 13-Hydroperoxyoctadeca-9,11-dienoic acid, II 9-hydroperoxyoctadeca-10,12-dienoic acid

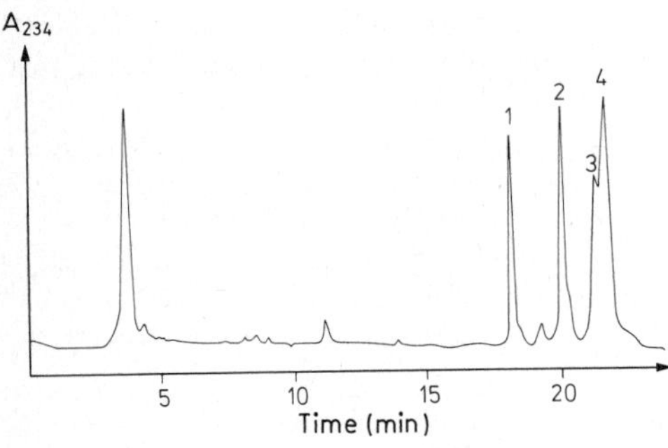

Fig. 3.20. Autoxidation of linoleic acid methyl ester. Analysis of primary products by HPLC (After *H. W. S. Chan* and *G. Levett,* 1977). **1** 13-Hydroperoxy-cis-9, trans-11-octadecadienoic acid methyl ester, **2** 13-hydroperoxy-trans-9, trans-11-octadecadienoic acid methyl ester, **3** 9-hydroperoxy-trans-10, cis-12-octadecadienoic acid methyl ester, **4** 9-hydroperoxy-trans-10, trans-12-octadecadienoic acid methyl ester

isolated double bonds. The proportion of these minor monohydroperoxides is about 4% of the total (Table 3.20).

Autoxidation of linolenic acid yields four monohydroperoxides (Table 3.20). Formation of the monohydroperoxides is easily achieved by H-abstraction from the bis-allylic groups in positions 11 and 14. The resultant two pentadiene radicals then stabilize analogously to linoleic acid oxidation (Figure 3.19); each radical corresponds to two monohydroperoxides. However,

Table 3.20. Monohydroperoxides formed by autoxidation ($^3O_2$) and photoxidation ($^1O_2$) of unsaturated fatty acids

Fatty acid	Monohydroperoxide			
	Position of		Proportion (%)	
	HOO-group	double bond	$^3O_2$	$^1O_2$
Oleic acid	8	9	27	
	9	10	23	48
	10	8	23	52
	11	9	27	
Linoleic acid	8	9, 12	1.5	
	9	10, 12	46.5	32
	10	8, 12	0.5	17
	12	9, 13	0.5	17
	13	9, 11	49.5	34
	14	9, 12	1.5	
Linolenic acid	9	10, 12, 15	31	23
	10	8, 12, 15		13
	12	9, 13, 15	11	12
	13	9, 11, 15	12	14
	15	9, 12, 16		13
	16	9, 12, 14	46	25

the four isomers are not formed in equimolar amounts; the 9- and 16-isomers predominate (Table 3.20). The configuration of the conjugated double bonds again depends on reaction conditions. Cis-hydroperoxides are the main products at temperatures $<40\,°C$.

Competition exists between conversion of the peroxy radical to monohydroperoxide and reactions involving β-fragmentation and cyclization. Allyl peroxy radicals can undergo β-fragmentation which results, after a new oxygen molecule is attached, in a peroxy radical positional isomer, e.g. rearrangement of an oleic acid peroxy radical:

(11) (9) (11) (9)
-CH-CH=CH- ⟶ -CH=CH-CH-
O-O· ·O-O

(10) (8) (10) (8)
-CH-CH=CH- ⟶ -CH=CH-CH-
O-O· ·O-O (3.49)

However, hydroperoxides can also be isomerized by such a reaction pathway. When they interact with free radicals (H-abstraction from -OOH group) or with heavy metal ions (cf. Reaction 3.58 b), they are again transformed into peroxy radicals. Thus, the 13-hydroperoxide of linoleic acid isomerizes into the 9-isomer and vice versa:

(13) (9)
—CH—CH=CH—CH=CH—
OOH

RO· → ROH

—CH—CH=CH—CH=CH— O·
OO· ·O

RH → R·

(13) (9)
—CH=CH—CH=CH—CH—
HOO (3.50)

3.7.2.1.3 *Hydroperoxide-epidioxides*

Peroxy radicals which contain isolated β, γ double bonds are prone to cyclization reactions in competition with reactions leading to monohydroperoxides. A hydroperoxide-epidioxide results through attachment of a second oxygen molecule and abstraction of a hydrogen atom:

·O-O ⟶ O-O

O_2; RH, R· ⟶ HOO O-O (3.51)

Peroxy radicals with isolated β, γ double bonds are formed as intermediary products after autoxidation and photooxidation (reaction with singlet O_2) of unsaturated fatty acids having two or more double bonds.

For this reason the 10- and 12-peroxy radicals obtained from linoleic acid readily form hydroperoxy-epidioxides. While such radicals are only minor products during autoxidation, in photooxidation they are generated as intermediary products yields similar to the 9- and 13-peroxy radicals, which do not cyclize. Ring formation by 10- and 12-peroxy radicals decreases accumulation of the corresponding monohydroperoxides (Table 3.20; reaction with $^1O_2$).

Among the peroxy radicals of linolenic acid which are formed by autoxidation, the isolated β, γ double bond system exists only for the 12- and 13-isomers, and not for the 9- and 16-isomers. Also, the inclination of the 12- and 13-peroxy radicals of linolenic acid to form hydroperoxy-epidioxides has, as a consequence, the accumulation of less monohydroperoxide of the corresponding isomers, as opposed to the 9- and 16-isomers (Table 3.20).

Peroxy radicals interact rapidly with antioxidants which may be present to give monohydroperoxides (cf. 3.7.3.1). Thus, it is not only the chain reaction which is inhibited by antioxidants, but also β-fragmentation and peroxy radical cyclization.

R^1 ... R^2
O-O·

⟶ R^1 ... R^2
O-O O-OH

⟶ R^1 ... CHO + OCH-R^2 + ··· (3.52)

R^1: CH_3-$(CH_2)_3$; R^2: $(CH_2)_7$-COOH

Fragmentation occurs when a hydroperoxide-epidioxide is heated, resulting in formation of aldehydes and aldehydo-acids. For example, hydroperoxide-epidioxide fragments derived from the 12-peroxy radical of linoleic acid are formed as shown in Reaction 3.52.

Peroxy radicals formed from fatty acids with three or more double bonds can form bicycloendoperoxides with an epidioxide radical as intermediate. This is illustrated by Reaction 3.68.

3.7.2.1.4 Initiation of a radical chain reaction

Since autoxidation of unsaturated acyl lipids frequently results in deterioration of food quality, an effort is made to at least decrease the rate of this deterioration process. However, pertinent measures are only possible when better knowledge is acquired about the reactions involved during the induction period of autoxidation and how they trigger the start of autoxidation.

In recent decades model system studies have revealed that two fundamentally different groups of reactions are involved in initiating autoxidation.

The first group is confined to the initiating reactions which overcome the energy barrier required for the reaction of molecular oxygen with an unsaturated fatty acid. This group of reactions includes photosensitized oxidation (photooxidation) and lipid oxidation by lipoxygenase catalysis (cf. 3.7.2.2). Both reactions provide the "first" primary hydroperoxides. These are then converted further into radicals by the second group of reactions. Heavy metal ions and heme(in) proteins are involved in this second reaction group. Some enzymes which generate the superoxide radical anion can be placed in between these two delineated reaction groups since at least H_2O_2 is in addition necessary as reactant for the formation of radicals.

Photooxidation: In order to understand photooxidation and to differentiate it from autoxidation, the electronic configuration of the molecular orbital energy levels for oxygen should be known. As presented in Figure 3.21, the allowed energy levels correspond to $^3\Sigma^-g$, $^1\Delta^+g$ and $^1\Sigma^+g$.

The notation for the molecular orbital of O_2 is $(\sigma\, 2s)^2\, (\sigma^*\, 2s)^2\, (\sigma\, 2p)^2\, (\pi\, 2p)^4\, (\pi^*\, 2p)^2$

In the ground state oxygen is a triplet ($^3O_2$). As seen from the above notation, the term $(\pi^*\, 2p)^2$ accounts for two unpaired electrons in the oxygen molecule. These are the two antibonding π orbitals available: π^*2p_y and π^*2p_z. The two electrons occupy these orbitals alone. The net

Electrons in molecular orbitals: $(\sigma_1)^2\, (\sigma_1^*)^2\, (\sigma_2)^2\, (\pi)^4\, (\pi^*)^2$

		π^*-molecular orbital[a] $2p_y$	$2p_x$	Lifetime (s) Gas phase	Liquid phase
	2. Singlet state ($^1\Sigma_g^+$)	↑	↓	7–12	10^{-9}
155 kJ/mole					
	1. Singlet state ($^1\Delta_g$)	↑↓		$3 \cdot 10^3$	10^{-1}–10^{-3} b)
92 kJ/mole					
	Ground state ($^3\Sigma_g^-$)	↑	↑	∞	∞

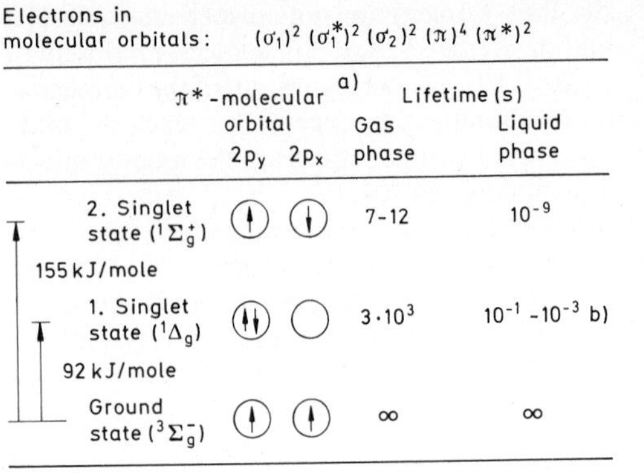

Fig. 3.21. Configuration of electrons in an oxygen molecule

[a] Electrons in $2p_x$ and $2p_y$ orbitals

[b] Dependent on solvent, e. g. 2 µs in water, 20 µs in D_20 and 200 µs in methanol

angular momentum of the unpaired electrons has three components, hence the term "triplet". When the electrons are paired, the angular momentum can not be split into components and this represents a singlet state. In the triplet state oxygen reacts preferentially with radicals, i.e. molecules having one unpaired electron. In contrast direct reactions of triplet-state oxygen with molecules which have all electrons paired, as in the case with fatty acids, are prevented by spin barriers. For this reason the activation energy of the reaction

$$RH + {}^3O_2 \longrightarrow ROOH \qquad (3.53)$$

is so high (146–273 kJ/mole) that it does not occur without some assistance.

Oxygen goes from the ground state to the short-lived 1-singlet-state ($^1O_2$) by the uptake of 92 kJ/mole of energy (Figure 3.21). The previously unpaired single electrons are now paired on the π^*2p_y antibonding orbital. The reactivity of this molecule resembles ethylenic or general olefinic π electron pair reactions, but it is more electrophilic. Hence, in the reaction with oleic acid, the 1-singlet-state oxygen attacks the 9–10 double bond, generating two monohydroperoxides, the 9- and 10-isomers (cf. Table 3.20). The second singlet-state of oxygen ($^1\Sigma^+g$) has a much shorter life than the 1-singlet-state and plays no role in the oxidation of fats or oils.

For a long time it has been recognized that the stability of stored fat (oil) drops in the presence of light. Light triggers lipid autoxidation. Low amounts of some compounds participate as sensitizers.

According to *Schenk* and *Koch* (1960), there are two types of sensitizers. Type I sensitizers are those which, once activated by light (sen*), react directly with substrate, generating substrate radicals. These then trigger the autoxidation process. Type II sensitizers are those which activate the ground state of oxygen to the $^1O_2$ singlet state.

Type I and II photooxidation compete with each other. Which reaction will prevail essentially depends on the structure of the sensitizer but also on the concentration and the structure of the substrate available for oxidation.

Table 3.20 shows that the composition of hydroperoxide isomers derived from an unsaturated acid by autoxidation ($^3O_2$) differs from that obtained in the reaction with $^1O_2$. The isomers can be separated by analysis of hydroperoxides using high performance liquid chromatography and, thus, one can distinguish Type I from Type II photooxidation. Such studies have revealed that sensitizers, such as chlorophylls a and b, pheophytins a and b and riboflavin, present in food, promote the Type II oxidation of oleic and linoleic acids.

As already stated, the Type II sensitizer, once activated, does not react with substrate but with ground state triplet oxygen, transforming it with an input of energy into 1-singlet-state oxygen:

$$\mathrm{Sen} \xrightarrow{h\nu} \mathrm{Sen}^*$$
$$\mathrm{Sen}^* + O_2 \longrightarrow \mathrm{Sen} + {}^1O_2 \quad (3.55)$$

The singlet $^1O_2$ formed now reacts directly with the unsaturated fatty acid by a mechanism of "cyclo-addition":

$$\mathrm{O{=}O} + \mathrm{RHC{=}CH{-}CH(H){-}R^1} \longrightarrow \mathrm{RHC(O{-}O{-}H){-}CH{=}CH{-}R^1} \quad (3.56)$$

The formation of hydroperoxides in numbers which double the number of isolated double bonds present in the fatty acid molecule is in agreement with the above reaction mechanism. The reaction is illustrated in Figure 3.22 for the oxidation of linoleic acid. In addition to the two hydroperoxides with a conjugated diene system already mentioned (Figure 3.19), two hydroperoxides are obtained with isolated double bonds.

Formation of 1-singlet oxygen ($^1O_2$) is inhibited by carotenoids (car), which deplete the $^1O_2$ of its excess energy, dissipating it as radiation and forming the ground state triplet ($^3O_2$):

$$^1\mathrm{Car} \xrightarrow{^1O_2 \;\to\; ^3O_2} {}^3\mathrm{Car} \xrightarrow{\text{(Energy dissipation by radiation)}} {}^1\mathrm{Car} \quad (3.57)$$

The quenching effect of carotenoids (transition of $^1O_2$ to $^3O_2$) is very fast ($k = 3 \times 10^{10}$ mole^{-1} s^{-1}). They also prevent energy transfer from excited-state chlorophyll to $^3O_2$. Therefore, carotenoids are particularly suitable for protecting fat (oil)-containing food from Type II photooxidation.

Heavy metal ions: These ions are involved in the second group of initiation reactions, namely, in the decomposition of initially-formed hydroperoxides into radicals which then propel the radical chain reaction of the autooxidation process. Fats, oils and foods always contain traces of heavy metals, the complete removal of which in a refining step would be uneconomic. The metal ions, primarily Fe, Cu and Co, may originate from:

- Raw food. Traces of heavy metal ions are present in many enzymes and other metal-bound proteins. For example, during the crushing and solvent extraction of oilseeds metal bonds dissociate and the freed ions bind to fatty acids.
- From processing and handling equipment. Traces of heavy metals are solubilized during the processing of fat (oil). Such traces are inactive physiologically but active as prooxidants.
- From packaging material. Traces of heavy metals from metal foils or cans or from wrapping paper can contaminate food and diffuse into the fat or oil phase.

The concentration of heavy metal ions that results in fat (oil) shelf life instability is dependent on the nature of the metal ion and the fatty acid composition of the fat (oil). Edible oils of the

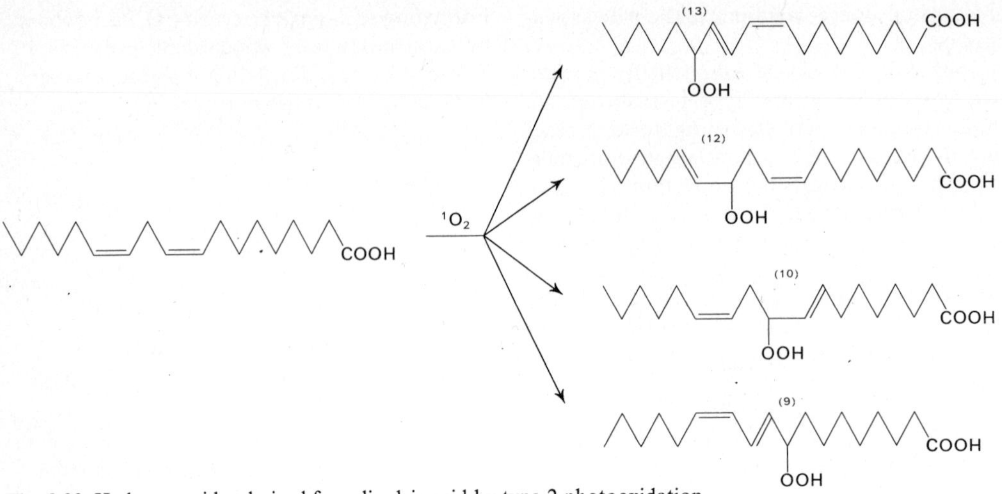

Fig. 3.22. Hydroperoxides derived from linoleic acid by type-2 photooxidation

linoleic acid type, such as sunflower and corn germ oil, should contain less than 0.03 ppm Fe and 0.01 ppm Cu to maintain their stability. The concentration limit is 5 ppm for both Cu and Fe in fat with a high content of oleic and/or stearic acids, e.g. butter.

Heavy metal ions trigger the autoxidation of unsaturated acyl lipids only when they contain hydroperoxides. That is, the presence of a hydroperoxide group is a prerequisite for metal ion activity, which leads to decomposition of the hydroperoxide group into a free radical:

$$Me^{n\oplus} + ROOH \longrightarrow Me^{(n+1)\oplus} + RO^{\bullet} + OH^{\ominus} \qquad (3.58\,a)$$

$$Me^{(n+1)\oplus} + ROOH \longrightarrow RO_2^{\bullet} + H^{\oplus} + Me^{n\oplus} \qquad (3.58\,b)$$

Me: Heavy metal ion

Reaction rate constants for the decomposition of linoleic acid hydroperoxide are given in Table 3.21. As seen with iron, the lower oxidation state (Fe^{2+}) provides a ten-fold faster decomposition rate than the higher state (Fe^{3+}). Correspondingly, Reaction 3.58 a proceeds must faster than Reaction 3.58 b in which the reduced state of the metal ion is regenerated. The start of autoxidation then is triggered from hydroperoxide generated radicals.

The decomposition rates for hydroperoxides emulsified in water depend on pH (Table 3.21). The optimal activity for Fe and Cu ions is in the pH range of 5.5–6.0. The presence of ascorbic acid, even in traces, accelerates the decomposition. Apparently, the acid sustains the reduced state of the metal ions.

The direct oxidation of fatty acids to an acyl radical by a heavy metal ion

$$RH + Me^{(n-1)\oplus} \longrightarrow R^{\bullet} + H^{\oplus} + Me^{n\oplus} \qquad (3.59)$$

proceeds similarly, but with an exceptionally slow rate. It seems to be without significance for the initiation of autoxidation.

The autoxidation of acyl lipids is also influenced by the moisture content of food. The reaction rate is high for both dehydrated and water-containing food, but is minimal at a water activity (a_w) of 0.3 (Figure 0.4). The following hypotheses are discussed to explain these differences: The high reaction rate in dehydrated food is due to

Table 3.21. Linoleic acid hydroperoxides: decomposition by heavy metal or heme compounds at 23 °C. Relative reaction rates k_{rel} are given at two pH's[a]

Heavy metal ion[b]	k_{rel} pH 7	k_{rel} pH 5.5	Heme compound[b]	k_{rel} pH 7	k_{rel} pH 5.5
Fe^{3+}	1	10^2	Hematin	$4 \cdot 10^3$	$4 \cdot 10^4$
Fe^{2+}	14	10^3	Methemoglobin	$5 \cdot 10^3$	$7.6 \cdot 10^3$
Cu^{2+}	0.2	1,5	Cytochrome C	$2.6 \cdot 10^3$	$3.9 \cdot 10^3$
Co^{3+}	$6 \cdot 10^2$	1	Oxyhemoglobin	$1.2 \cdot 10^3$	
Mn^{2+}	0	0	Myoglobin	$1.1 \cdot 10^3$	
			Catalase	1	
			Peroxidase	1	

[a] Linoleic acid hydroperoxide is emulsified in respective buffers.
[b] Reaction rate constant is related to reaction rate in presence of Fe^{3+} at pH 7 ($k_{rel} = 1$).

metal ions with depleted hydration shells. In addition, ESR spectroscopic studies show that food drying promotes the formation of free radicals which might initiate lipid peroxidation. As the water content starts to increase, the rate of autoxidation decreases. It is assumed that this decrease in rate is due to hydration of ions and also of radicals. Above an a_w of 0.3, free water is present in food in addition to bound water. Free water appears to enhance the mobility of prooxidants, thus accounting for the renewed increase in autoxidation rate that is invariably observed at high moisture levels in food.

Heme(in) compounds. Heme (Fe^{2+}) and hemin (Fe^{3+}) proteins are widely distributed in food. Lipid peroxidation in animal tissue is accelarated by hemoglobin, myoglobin and cytochrome C. These reactions are often responsible for rancidity or aroma defects occurring during storage of fish, poultry and cooked meat. The most important heme(in) proteins are in plant food peroxidase and catalase. Cytochrome P_{450} is a particularly powerful catalyst for lipid peroxidation, although it is not yet clear to what extent the compound affects food shelf life "in situ".
Unlike heavy metal ion catalysis, decomposition of hydroperoxides in the presence of Fe-porphyrin protein does not involve a change in the valence of Fe ions. This is shown by, for example, the fact that ascorbic acid does not promote the reaction.
The role of myoglobin will be examined in more detail to clarify the mechanism of hydroperoxide decomposition. The central Fe^{2+}-atom in myoglobin binds the ligands, adopting the complex geometry of a regular octahedron. Four ligands are derived from the protoporphyrin ring system and the other two are a histidine residue of the protein and a water molecule. The outer coordination sphere of the complex can be presented schematically as:

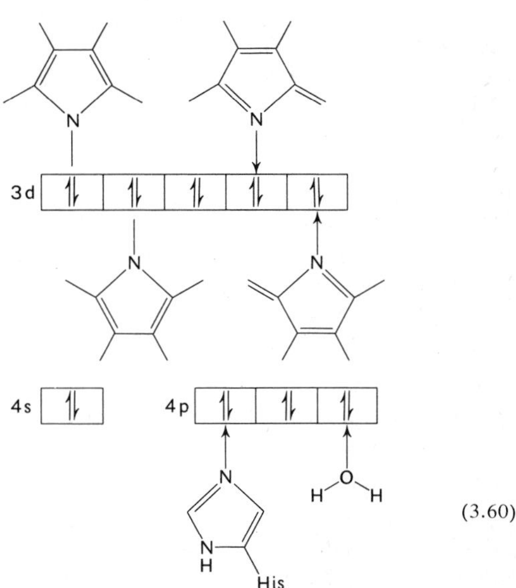

(3.60)

The water (bound to the complex with two electrons paired in the 4p orbitals) is readily displaced by a hydroperoxide molecule, thus augmenting the stability of the ligand field. Interaction in the next stage with the porphyrin ring system, details of which are still unknown, brings about hydroperoxide cleavage into an alkoxy and a hydroxy radical. This suggest that the porphyrin ring system performs a reaction which is identical with one of the branching reactions (cf. RS-4 in Figure 3.17) of the free radical chain of autoxidation. In comparison to iron ions some heme(in) compounds degrade the hydroperoxides more rapidly by orders of magnitude (cf. Table 3.21). Therefore they are more effective as initiators of lipid peroxidation. Their activity is also negligibly influenced by a decrease in the pH-value.
However, the activity of a heme(in) protein for hydroperoxides is influenced by its steric accessibility to fatty acid hydroperoxides. Hydroperoxide binding to the Fe-porphyrin moiety of native catalase and peroxidase molecules is obviously not without interferences. The prosthetic group is free to promote hydroperoxide decomposition only after heat denaturation of the enzymes. Results of a model study involving peroxidase are given in Table 3.22. As the data show, heating the enzyme results in an increase of prooxidative

Table 3.22. Rates of linoleic acid peroxidation in the presence of heat-treated horseradish peroxidase

Heat treatment 2 min at °C	Linoleic acid peroxidation (μmole O_2/min)	Enzyme activity (%)
25	8.7	100
53	10.5	100
90	47.5	80
120	79.5	50
140	96.0	14

activity by a factor of 10 and, as expected, in a concomitant drop of enzyme activity. Similar results are obtained in reaction systems containing catalase.

Suppression of peroxidase and catalase activity is of importance for the shelf life of heat-processed food. As long as the protein moiety has not been denatured, it is the lipoxygenase enzyme which is the most active for lipid peroxidation (cf. 3.7.2.2). After lipoxygenase activity is destroyed by heat denaturation, its role is replaced by the heme(in) proteins. As already suggested, an assay of heme(in) protein enzyme activity does not necessarily reflect its prooxidant activity.

Activated oxygen from enzymatic reactions.

In enzymatic reactions oxygen can form three intermediates, which differ greatly in their activities and which are all ultimately reduced to water:

$$O_2 \xrightarrow{e} O_2^{\ominus} \xrightarrow[2H^{\oplus}]{e} H_2O_2 \xrightarrow[H^{\oplus}\ H_2O]{e} OH^{\bullet} \xrightarrow[H^{\oplus}]{e} H_2O \quad (3.61)$$

e: Electron

Oxygen takes up one electron to form a superoxide radical anion [O_2^-, $(\sigma 2s)^2$ $(\sigma^* 2s)^2$ $(\sigma 2p)^2$ $(\pi 2p)^4$ $(\pi^* 2p)^3$; i.e. one electron-pair exists in the antibonding π molecular orbital]. This anion radical is a reducing agent with chemical properties dependent on pH, according to the equilibrium:

$$O_2^{\ominus} + H^{\oplus} \rightleftharpoons HO_2^{\bullet} \ (pK_s: 4,8) \quad (3.62)$$

Based on its pK_s value under physiological conditions, this activated oxygen species occurs as an anion with its radical character suppressed. It acts as a nucleophilic reagent (e.g. it promotes phospholipid hydrolysis within the membranes) under such conditions, but is not directly able to abstract an H-atom and to initiate lipid peroxidation. The free radical activity of the superoxide anion appears only in acidic media, wherein the perhydroxy radical form ($HO_2^{\bullet}$) prevails. The O_2^- has a finite stability since it slowly dismutates ($k = 0.35\ m^{-1}\ s^{-1}$):

$$2O_2^{\ominus} + 2H^{\oplus} \longrightarrow H_2O_2 + O_2 \quad (3.63)$$

An enzyme with superoxide dismutase activity which significantly accelerates ($k = 2 \times 10^9\ m^{-1}\ s^{-1}$) Reaction 3.63 occurs in numerous animal and plant tissues.

The superoxide radical anion, O_2^-, is generated by flavin enzymes, such as xanthine oxidase (cf. 2.3.3.2). The involvement of this enzyme in the development of milk oxidation flavor has been questioned for a long time.

Hydrogen peroxide, H_2O_2, is the second intermediate of oxygen reduction. In the absence of heavy metal ions, energy-rich radiation including UV light, or of elevated temperatures, H_2O_2 is a rather indolent and sluggish reaction agent. On the other hand, the hydroxy radical ($HO^{\bullet}$) derived from it is exceptionally active, since during the abstraction of an H-atom,

$$R{-}H + HO^{\bullet} \longrightarrow R^{\bullet} + H_2O \quad (3.64)$$

the energy input in the HO-bond formed is 497 kJ/mole, thus exceeding the dissociation energy for abstraction of hydrogen from each C−H bond by at least 75 kJ/mole (cf. Table 3.19). Therefore, the $HO^{\bullet}$ radical reacts nonselectively with all organic constituents of food. Consequently, it can directly initiate lipid peroxidation. However, in a complex system such as food, the following question is always pertinent: "Has the $HO^{\bullet}$ radical actually reached the unsaturated acyl lipid, or was it trapped prior to lipid oxidation by some other food ingredient?".

The reaction of the superoxide radical anion with hydrogen peroxide should be emphasized in relation to initiation of autoxidation. This is the so-called *Fenton* reaction, which occurs especially in the presence of an Fe-complex:

$$O_2^{\ominus} \rightarrow O_2;\quad ADP\text{-}Fe^{3\oplus} \rightleftharpoons ADP\text{-}Fe^{2\oplus};\quad H_2O_2 \rightarrow HO^{\bullet} + OH^{\ominus} \quad (3.65)$$

The Fe-complex (e.g. with ADP) occurs in food of plant and animal origin. The Fe^{2+} obtained by reduction with O_2^- can then reduce the H_2O_2 present and generate free $HO^{\bullet}$ radicals.

3.7.2.1.5 Secondary products

The primary products of autoxidation, the monohydroperoxides, are odorless and tasteless (such as linoleic acid hydroperoxides; cf. Table 3.28). Food quality is not affected until volatile

Table 3.23. Volatile carbonyl compounds formed by autoxidation of unsaturated fatty acids (μg/g)[a]

Oleic acid		Linoleic acid		Linolenic acid	
Heptanal	50	Pentanal	55	Propanal[b]	
Octanal	320	Hexanal	5,100	1-Penten-3-one	30
Nonanal	370	Heptanal	50	2tr-Butenal	10
Decanal	80	2tr-Heptenal	450	2tr-Pentenal	35
2tr-Decenal	70	Octanal	45	2c-Pentenal	45
2tr-Undecenal	85	1-Octen-3-one	2	2tr-Hexenal	10
		2c-Octenal	990	3tr-Hexenal	15
		2tr-Octenal	420	3c-Hexenal	90
		3c-Nonenal	30	2tr-Heptenal	5
		3tr-Nonenal	30	2tr,4c-Heptadienal	320
		2tr-Nonenal	30	2tr,4tr-Heptadienal	70
		2c-Decenal	20	2c,5c-Octadienal	20
		2tr,4tr-Nonadienal	30	3,5-Octadien-2-one	30
		2tr,4c-Decadienal	250	2tr,6c-Nonadienal	10
		2tr,4tr-Decadienal	150	2,4,7-Decatrienal	85

[a] Each fatty acid in amount of 1 g was autoxidized at 20 °C by an uptake of 0.5 mole oxygen/mole fatty acid.
[b] Major compound of autoxidation.

compounds are formed. The latter are usually powerfully odorous compounds and, even in the very small amounts in which they occur, affect the odor and flavor of food.
Carbonyl compounds are the main aroma carriers in lipid peroxidation. A lesser role is ascribed to hydrocarbons, such as alkanes, alkenes and alkylfurans, and to alcohols.

Volatile monocarbonyl compounds. Examples of the great number of carbonyl compounds identified are presented in Table 3.23. The corresponding odor threshold values, because of the poor solubility of these compounds, are lower in water than in a hydrophobic medium (Table 3.24). As seen from the threshold values, several of these compounds, at a level of a few mg/tonne of food, can stimulate odor perception, with an aroma note as shown in Table 3.24.
Particularly powerful aroma compounds are formed by the autoxidation of α-linolenic acid (cf. Tables 3.23 and 3.24). Therefore, rapid deterioration of food containing linolenic acid should not be ascribed solely to the preferential reactivity of this acid, but also to the low odor threshold values of its aldehydes, such as 3-cis-hexenal and 2-trans,6-cis-nonadienal. Carbonyl compounds with exceptionally strong aromas can be released in food by the autoxidation of some fatty acids, even if they are present in low amounts. An example is octadeca-cis-11,cis-15-dienoic acid (the precursor for 4-cis-heptenal), which occurs in beef and mutton and often in butter. The odor threshold of 4-cis-heptenal is 10 μg/kg.
Also, the processing of oil and fat can provide an altered fatty acid profile. These can then provide new precursors for a new set of carbonyls. For example, 6-trans-nonenal, the precursor of which is octadeca-cis-9-trans-15-dienoic acid, is a product of the partial hydrogenation of linolenic acid. This aldehyde can be formed during storage of partially-hardened soya and linseed oils. The aldehyde, together with other compounds, is responsible for an off-flavor denoted as "hardened flavor".
Several reaction mechanisms have been suggested to explain the formation of volatile carbonyl compounds. The most probable mechanism is the β-scission of monohydroperoxides with formation of an intermediary short-lived alkoxy radical (Figure 3.23). Such β-scission is catalyzed by heavy metal ions or heme(in) compounds (cf. 3.7.2.1.4).
There are two possibilities for β-scission of each hydroperoxide fatty acid (Figure 3.23). Option "B", i.e. the cleavage of the C–C bond located further away from the double bond position, is the energetically preferred one since it leads to resonance-stabilized "oxo-ene" or "oxo-diene" compounds. Applying this β-scission mechanism ("B") to both major monohydroperoxide iso-

Table 3.24. Sensory properties of aliphatic aldehydes and vinyl ketones

Compound	Odor description									Odor threshold value (ppb) in	
	sharp	oily fatty	tallowy	frying odor	green leafy	cucumber-like	fishy	orange peel-like	metallic	paraffin oil	water
Aldehydes											
5:0	+									100	10
6:0			+		+					150	4.5
7:0		+								45	30
8:0		+								50	40
9:0			+							250	40
10:0								+		900	5
5:1 (2tr)		+			+					700	–
6:1 (2tr)		+			+					1,500	17
6:1 (3c)					+					100	0.3
7:1 (2tr)		+								14,000	50
8:1 (2tr)		+								7,000	4
9:1 (2tr)			+			+				3,500	0.08
7:2 (2tr, 4c)		+		+						50	–
7:2 (2tr, 4tr)		+								10,000	–
9:2 (2tr, 4tr)		+								460	90
9:2 (2tr, 6c)						+				2	0.05
9:2 (2tr, 6tr)			+		+					240	1
10:2 (2tr, 4c)				+						20	–
10:2 (2tr, 4tr)				+						200	0.07
10:3 (2tr, 4c, 7c)							+			–	–
Vinyl ketones											
1-Penten-3-one							+			3	1
1-Octen-3-one									+	0.1	0.1
1,cis-5-Octadien-3-one		+							+	0.1	0.001

mers of linoleic acid gives the following fragmentation products:

$H_3C-(CH_2)_4$ … OOH … $(CH_2)_7-COOH$

(13-LOOH)

$$\rightarrow H_3C-(CH_2)_3-\dot{C}H_2 \xrightarrow[R^\bullet]{RH} CH_3-(CH_2)_3-CH_3$$

$$\rightarrow OHC \ldots (CH_2)_7-COOH \qquad (3.67a)$$

From the volatile autoxidation products which contain the methyl end of the linoleic acid molecule, the formation of 2,4-decadienal and pentane can be explained in this way.

$H_3C-(CH_2)_4$ … OOH … $(CH_2)_7-COOH$

(9-LOOH)

$$\rightarrow H_3C-(CH_2)_4-CH{=}CH-CH{=}CH-CHO$$

$$\rightarrow H_2C^\bullet-(CH_2)_6-COOH \xrightarrow[R^\bullet]{RH} CH_3-(CH_2)_6-COOH \qquad (3.67b)$$

The formation of hexanal among the main volatile compounds derived from linoleic acid (cf. Table 3.23) is still an open question. The preferential formation of hexanal in aqueous systems can be explained with an ionic mechanism. As shown in Figure 3.24, the heterolytic cleavage is initiated by the protonation of the hydroperoxide

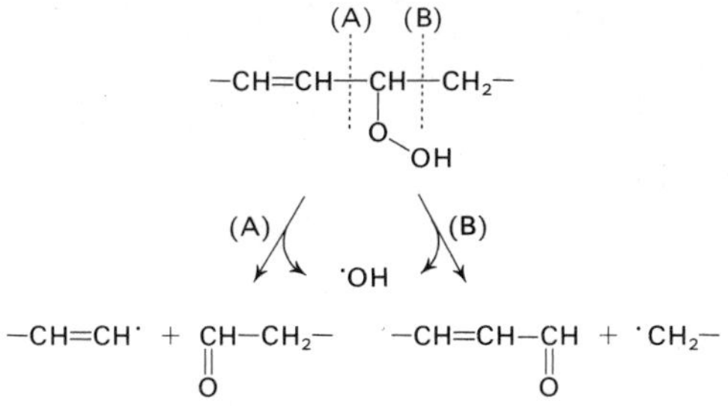

Fig. 3.23. β-Scission of monohydroperoxides. (After *H. T. Badings*, 1970)

group. After elimination of a water molecule, the oxo-cation formed is subjected to an insertion reaction exclusively on the C–C linkage adjacent to the double bond. The carbonium ion then splits into an oxo-acid and hexanal. The fact that linoleic acid 9-hydroperoxide gives rise to 2-nonanal is in agreement with this outline.

However, in the water-free fat or oil phase of food the homolytic cleavage of hydroperoxides, presented above, is the predominant reaction mechanism. Since option "A" of the cleavage reaction is excluded (Figure 3.23), some other reactions should be assumed to occur to account for formation of hexanal and other aldehydes from linoleic acid. The further oxidation reactions of monohydroperoxides and carbonyl compounds are among the possibilities.

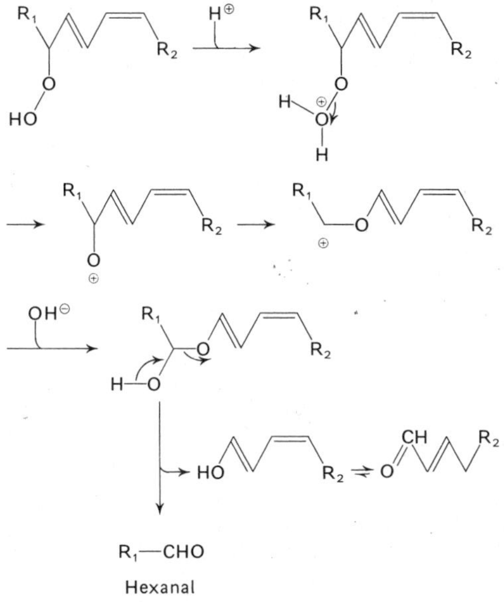

The above assumption is supported by the finding that 2-alkenals and 2,4-alkadienals are oxidized substantially faster than the unsaturated fatty acids (Figure 3.25). In addition, the autoxidation of 2,4-decadienal yields hexanal and other volatiles which coincide with those obtained from linoleic acid. Since saturated aldehydes oxidize slowly, as demonstrated by nonanal (Figure 3.25), they will enrich the oxidation products and become predominant.

Other studies elucidating the multitude of aldehydes which arise suggest that decomposition of minor hydroperoxides formed by autoxidation of linoleic acid (cf. Table 3.20) is a contributing source to the profile of aldehydes. This suggestion is supported by pentanal, which originates from the 14-hydroperoxide.

The occurrence of 2,4-heptadienal (from the 12-hydroperoxide isomer) and of 2,4,7-decatrienal (from the 9-hydroperoxide isomer) as oxidation products is, thereby, readily explained by accepting the fragmentation mechanism outlined above (option "B" in Figure 3.23) for the autoxidation of α-linolenic acid. The formation of other volatile carbonyls can then follow by autoxidation of these two aldehydes or from the further oxidation of labile monohydroperoxides.

R_1: $CH_3(CH_2)_4$ R_2: $(CH_2)_7$—COOH

Fig. 3.24. Proton-catalyzed cleavage of linoleic acid 13-hydroperoxide. (After *G. Ohloff*, 1973)

Malonaldehyde. This dialdehyde is preferentially formed by autoxidation of fatty acids with three or more double bonds. The compound is odor-

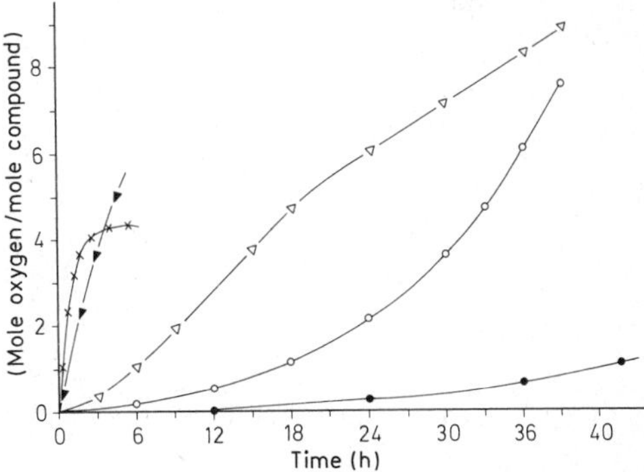

Fig. 3.25. Reaction rate of an autoxidation process. (After *D. A. Lillard* and *E. A. Day*, 1964). –▽–▽– Linolenic acid methyl ester, –○–○– linoleic acid methyl ester, × – × – 2-nonenal, ▼–▼–▼ 2,4-heptadienal, –●–●– nonanal

less. In food it may be bound to proteins by a double condensation, crosslinking the proteins (cf. 3.7.2.4.3). Malonaldehyde is formed from α-linolenic acid by a modified reaction pathway, as outlined under the formation of hydroperoxide-epidioxide (cf. 3.7.2.1.3). However, a bicyclic compound is formed here as an intermediary product that readily fragments to malonaldehyde (cf. Reaction 3.68).

Malonaldehyde, even at trace levels, is often used as an indicator for fat or oil oxidation (cf. thiobarbituric acid test, 14.5.3.2.1).

Furan derivatives. Several furan derivatives occur among the autoxidation products of linoleic and linolenic acids (Table 3.25). These derivatives should be responsible for an aroma defect in soybean oil and green beans. The odor is removed by refining of the oil (cf. 14.4.1), but it returns if the oil is stored under unfavorable conditions ("reversion flavor").

Alkanes, Alkenes. The main constituents of the volatile hydrocarbon fraction are ethane and pentane. Since these hydrocarbons are readily quantitated by gas chromatography using head space analysis, they can serve as suitable indicators for *in situ* detection of lipid peroxidation. Pentane is probably formed from the 13-hydroperoxide of linoleic acid by the β-scission mechanism. The corresponding pathway for 16-hydroperoxide of linolenic acid should then yield ethane.

(11) $COOCH_3$ → ($RO_2^{\cdot}$, O_2; ROOH) → $COOCH_3$ → $COOCH_3$ → $COOCH_3$ → (O_2, RH; $R^{\cdot}$) → $COOCH_3$, OOH → (Heat or $H^{\oplus}$) → + $COOCH_3$, OOH (3.68)

3.7.2.2 Lipoxygenase: Occurrence and Properties

A lipoxygenase (linoleic acid oxygen oxidoreductase, EC 1.13.11.12) enzyme occurs in many plants and also in erythrocytes and leucocytes. It catalyzes the oxidation of some unsaturated fatty

Table 3.25. Volatile furan derivatives

Compound	Precursor	Aroma description
$CH_3-(CH_2)_4-$(furan) 2-n-Pentylfuran	18:2 (9, 12)	Liquorice
2-(cis-2′-Pentenyl)furan	18:3 (9, 12, 15)	Grass, butter note
2-(trans-2′-Pentenyl)furan	18:3 (9, 12, 15)	Fatty-oily, metallic
$CH_3-CH_2-CH{=}CH-CH_2-$(furan)$-CHO$ 5-(Pentenyl)-2-furaldehyde	18:3 (9, 12, 15)	Liquorice

acids to their corresponding monohydroperoxides. These hydroperoxides have the same structure as those obtained by autoxidation. Unlike autoxidation, reactions catalyzed by lipoxygenase are characterized by all the features of enzyme catalysis: substrate specificity, peroxidation selectivity, occurrence of a pH optimum, susceptibility to heat treatment and a high reaction rate in the range of 0–20 °C. Also, the activation energy for linoleic acid peroxidation is rather low; 17 kJ/mole (for comparison with the activation energy of a noncatalyzed reaction, see 3.7.2.1.4).

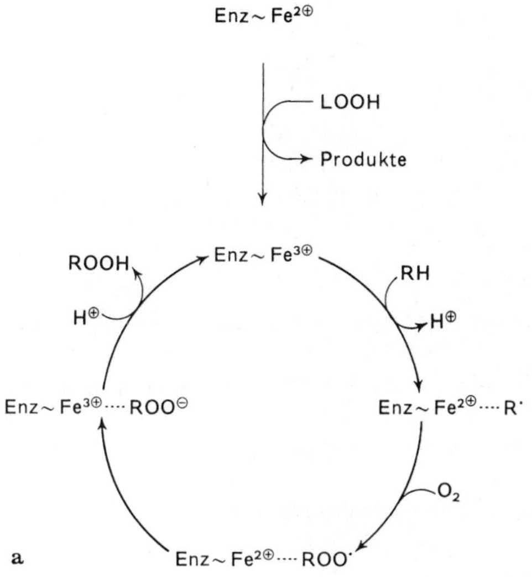

Lipoxygenase oxidizes only fatty acids which contain a 1-cis,4-cis-pentadiene system. Therefore, the preferred substrates are linoleic and linolenic acids for the plant enzyme, and arachidonic acid for the animal enzyme; oleic acid is not oxidized.

Lipoxygenase is a metal-bound protein with an Fe-atom in its active center. The enzyme is activated by hydroperoxide and, during activation, Fe^{2+} is oxidized to Fe^{3+}. The catalyzed oxidation pathway is assumed to have the following reaction steps (cf. Figure 3.26a): abstraction of a methylene H-atom from the substrate's 1,4-pentadiene system and oxidation of the H-atom to a proton. The pentadienyl radical bound to enzyme is then rearranged into a conjugated diene system, followed by uptake of oxygen. The peroxy radical formed is then reduced by the enzyme and, after attachment of a proton, the hydroperoxide formed is released.

It should be emphasized that enzyme-catalyzed oxidation is initiated even in the absence of monohydroperoxides. This means the enzyme alone is able to overcome the energy barrier of this reaction (cf. 3.7.2.1.4).

With respect to the plant enzyme properties, two types of enzymes exist (although a transitional form may also exist). The enzyme that peroxidizes only free fatty acids with a high stereo- and regioselectivity is a type I lipoxygenase. It gives rise to an optically-active hydroperoxide with a cis,trans-diene system (Figure 3.26b). This type

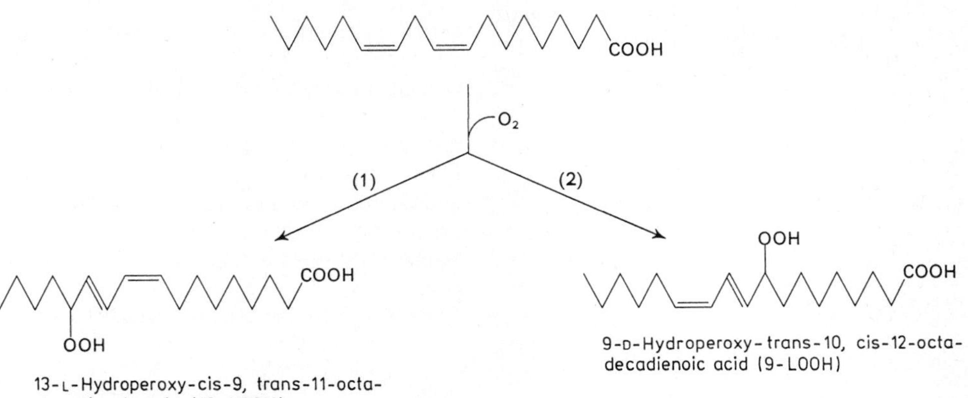

Fig. 3.26. Reactions of lipoxygenase – type I

a Proposed mechanism of reaction (After *G.A. Veldink*, 1977); RH: Linoleic acid; LOOH: linoleic acid hydroperoxide

b Specificity for linoleic acid oxidation. 1 Lypoxygenase from soybean (L-1; cf. Table 3.26); 2 lipoxygenase from tomato (cf. Table 3.26)

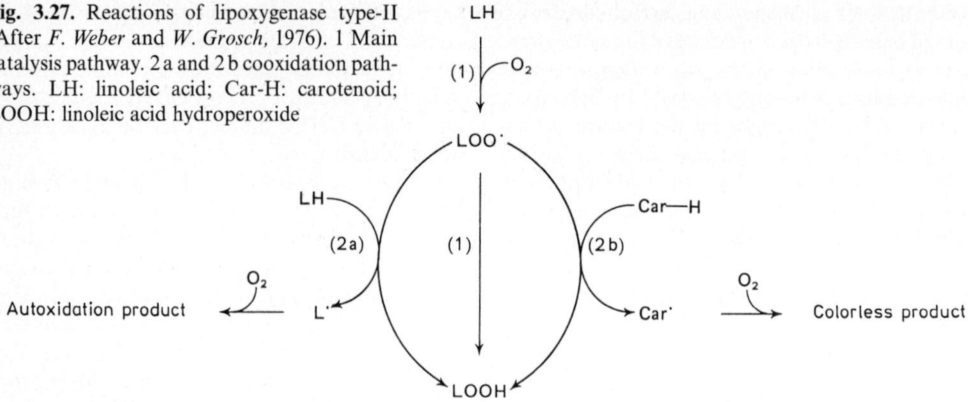

Fig. 3.27. Reactions of lipoxygenase type-II (After *F. Weber* and *W. Grosch,* 1976). 1 Main catalysis pathway. 2a and 2b cooxidation pathways. LH: linoleic acid; Car-H: carotenoid; LOOH: linoleic acid hydroperoxide

I enzyme forms preferentially either 9- or 13-hydroperoxides from free linoleic acid as the substrate (Table 3.26 and Figure 3.26b).

Type II lipoxygenase, on the other hand, acts more like a catalyst of autoxidation with much less reaction specificity for linoleic acid, from which it produces both 9- and 13-hydroperoxides in equal amounts (as in the noncatalyzed autoxidation) and some other products, such as ketodiene fatty acids. Moreover, lipoxygenase type II also reacts with an esterified substrate, thus not requiring prior release of fatty acids by a lipase enzyme for activity in food.

The type II lipoxygenase can cooxidize carotenoids and chlorophyll and thus can degrade these pigments to colorless products. The property of the enzyme is utilized in flour "bleaching" (cf. 15.4.1.4.3). The involvement of the enzyme in cooxidation reactions can be explained by the possibility that the peroxy radicals are not fully converted to their hydroperoxides as in the reaction of the type I enzyme. Thus, a fraction of the free peroxy radicals are released by the enzyme. It can abstract an H-atom either from the unsaturated fatty acid present (pathway 2a in Figure 3.27) or from a polyene (pathway 2b in Figure 3.27).

The type II enzyme present in legumes brings about a wide spectrum of volatile aldehydes from lipid substrates. These aldehydes, indentical to those of a noncatalyzed autoxidation, can be further reduced into their alcohols, depending on the state of $NADH\text{-}NAD^+$ (cf. 5.3.2.1).

Table 3.26. Occurrence and properties of various lipoxygenases

Food	pH optimum	Peroxidation specificity[a]		
		9-LOOH (%)	13-LOOH (%)	Type
Soybean, L-1	9.0	5	95	I
Soybean, L-2	6.5	50	50	II
Peas, L-2	6.5	50	50	II
Peanut	6.0	0	100	I
Potato	5.5	95	5	I
Tomato	5.5	95	5	I
Wheat	6.0	90	10	I
Cucumber	5.5	75	25	
Apple	6.0	10	90	
Strawberry	6.5	23	77	
Gooseberry	6.5	45	55	II

[a] Against linoleic acid; 9- or 13-LOOH, cf. Fig. 3.26b.

3.7.2.3 Enzymatic Degradation of Hydroperoxides

In a number of fruits and vegetables a hydroperoxide lyase occurs which degrade the hydroperoxides to volatile C_6 and C_9 aldehydes. The cucumber has been well studied (Figure 3.28).

The main cucumber aroma compound, 2-trans,6-cis-nonadienal, is formed from the combined activity of a hydroperoxide lyase and of an isomerase. In mushrooms linoleic acid is enzymatically oxidized to 10-(S)-hydroperoxy-trans-8,cis-12-octadecadienoic acid which in turn is cleaved by a hydroperoxide lyase into 1-octen-3-(R)-ol and 10-oxo-trans-8-decenoic acid:

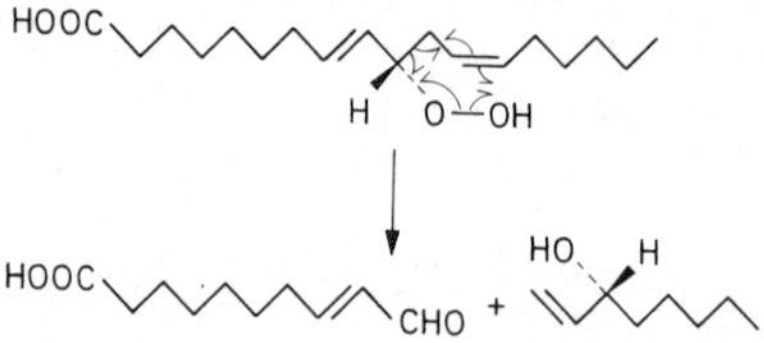

(3.68a)

1-octen-3-(R)-ol is the character impact flavor compound of fresh mushrooms.

The presence of a hydroperoxide-isomerase has been confirmed in barley, corn and linseed. It catalyzes the formation of α- or γ-ketol fatty acids, for example:

R: CH_3—$(CH_2)_4$

R′: $(CH_2)_7$—COOH

(3.69)

Linoleic acid 9-hydroperoxide formed by lipoxygenase in potato is changed enzymatically by elimination of water into a fatty acid with a dienyl-ether structure:

H_2O

(3.70)

(1′-trans, 3′-cis-Nonadienyloxa)-trans-8-nonenoic acid

In addition to lipoxygenase, lipoperoxidase activity has been observed in oats. The 9-hydroperoxide formed initially is reduced to 9-hydroxy-trans-10,cis-12-octadecadienoic acid.

Since hydroxy but not hydroperoxy acids taste bitter, this reaction should contribute to the bitter taste generated during the storage of oats (cf. 15.2.2.3).

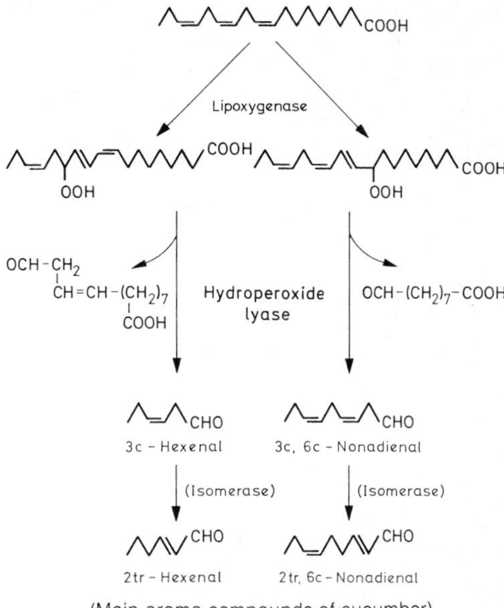

Fig. 3.28. Formation of aroma compounds in cucumber. (After *T. Galliard* et al., 1976)

3.7.2.4 Hydroperoxide-Protein Interactions

3.7.2.4.1 Products formed from hydroperoxides

Hydroperoxides formed enzymatically in food are usually degraded further. This degradation can also be of a nonenzymatic nature. In nonspecific reactions involving heavy metal ions, heme(in) compounds or proteins, hydroperoxides are transformed into oxo, epoxy, mono-, di- and trihydroxy acids (Table 3.27). Unlike hydroperoxides, i.e. the primary products of autoxidation, some of these derivatives are characterized as having a bitter taste (Table 3.28). Such compounds are detected in legumes and cereals. They may play a role in other foods rich in unsaturated fatty acids and proteins, such as fish and fish products.

In order to clarify the formation of the compounds presented in Table 3.27, the reaction sequences given in Figure 3.29 have been assumed to occur. The start of the reaction is from the alkoxydiene radical generated from the 9- or 13-hydroperoxide by the catalytic action of heavy metal ions or heme(in) compounds (cf. 3.7.2.1.4). The alkoxydiene radical may disproportionate into a hydroxydiene and an oxodiene fatty acid. Frequently this reaction is only of secondary importance since the alkoxydiene radical rearranges immediately to an epoxyallylic radical which is susceptible to a variety of radical combination reactions. Under aerobic conditions the epoxyallylic radical combines preferentially with molecular oxygen. The epoxyhydroperoxides formed are, in turn, subject to homolysis via an

Table 3.27. Products obtained by non-enzymic degradation of linoleic acid hydroperoxides

Product[a]	Hydroperoxide interaction with				
	Fe^{3+} cysteine	Hemo-globin	Soya homogenate	Pea homogenate	Wheat flour
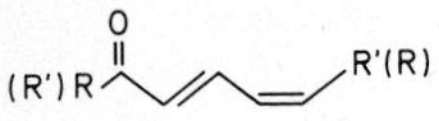 O ‖ (R')R R'(R)	+	+	+	+	
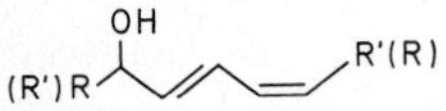 OH (R')R R'(R)	+	+	+	+	+
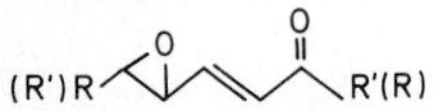 O O ‖ (R')R R'(R)	+		+		
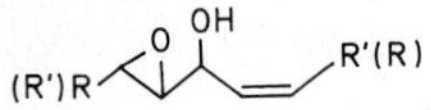 O OH (R')R R'(R)	+	+	+		
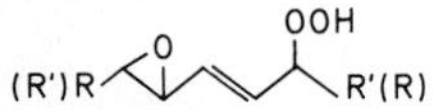 OOH O (R')R R'(R)	+		+		
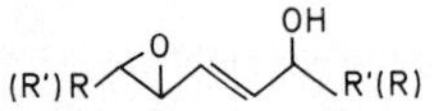 OH O (R')R R'(R)		+		+	+
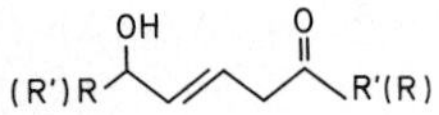 OH O ‖ (R')R R'(R)	+				
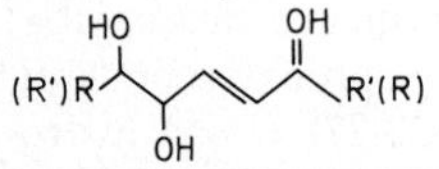 HO OH ‖ (R')R R'(R) OH	+		+	+	+

[a] As a rule a mixture of two isomers are formed with a R: $CH_3(CH_2)_4$ and R': $(CH_2)_7COOH$.

Table 3.28. Taste of oxidized fatty acids

Compound	Threshold value for bitter taste (mmole/l)
13-Hydroperoxy-cis-9,trans-11-octa-decadienoic acid	not bitter[a]
9-Hydroperoxy-trans-10,cis-12-octa-decadienoic acid	not bitter[a]
13-Hydroxy-cis-9,trans-11-octa-decadienoic acid	7.6–8.5[a]
9-Hydroxy-trans-10,cis-12-octa-decadienoic acid	6.5–8.0[a]
9,12,13-Trihydroxy-trans-10-octa-decadienoic acid }	0.6–0.9[b]
9,10,13-Trihydroxy-trans-11-octa-decadienoic acid }	

[a] A burning taste sensation.
[b] A blend of the two trihydroxy fatty acids was assessed.

oxyradical. A disproportionation reaction leads to epoxyoxo and epoxyhydroxy compounds. Under anaerobic conditions the epoxyallylic radical combines with other radicals, e.g. hydroxy radicals (Fig. 3.29) or thiyl radicals (Fig. 3.30).

Of the epoxides produced the allylic epoxides are known to be particularly susceptible to hydrolysis in the presence of protons. As shown in Fig. 3.29 trihydroxy fatty acids may result from the hydrolysis of an allylic epoxyhydroxy compound.

3.7.2.4.2 *Lipid-protein complexes*

Studies related to the interaction of hydroperoxides with proteins have shown that, in the absence of oxygen, linoleic acid 13-hydroperoxide reacts with N-acetylcysteine, yielding an adduct of which one isomer is shown:

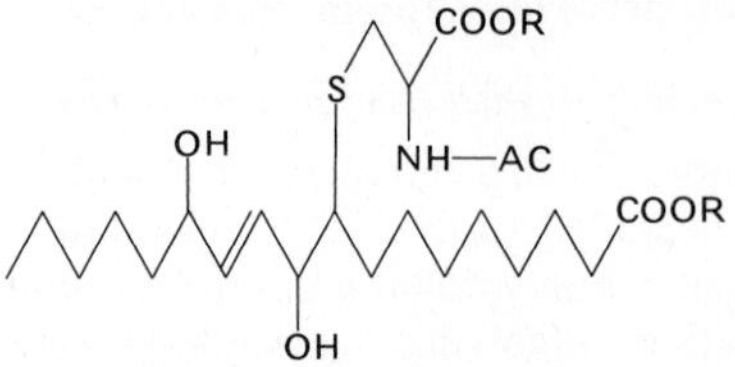

COOR
S
OH NH—AC
COOR (3.71)
OH

However, in the presence of oxygen, covalently-bound amino acid-fatty acid adduct formation is significantly suppressed; instead, oxidized fatty acids are formed as listed in Table 3.27.
The difference in reaction products is explained in the reaction scheme shown in Figure 3.30 which gives an insight into the different reaction pathways. The thiyl radical, derived from cysteine by abstraction of an H-atom, is added to the epoxyallyllic radical only in the absence of oxygen (pathway 2 in Figure 3.30). Otherwise, in the presence of oxygen, oxidation of cysteine to cysteine oxide and of fatty acids to their more oxidized forms (Figure 3.29) both occur with a higher reaction rate than the previous reaction. As a consequence, a large portion of the oxidized lipid from a protein-containing food stored in air does not have lipid-protein covalent bonds and,

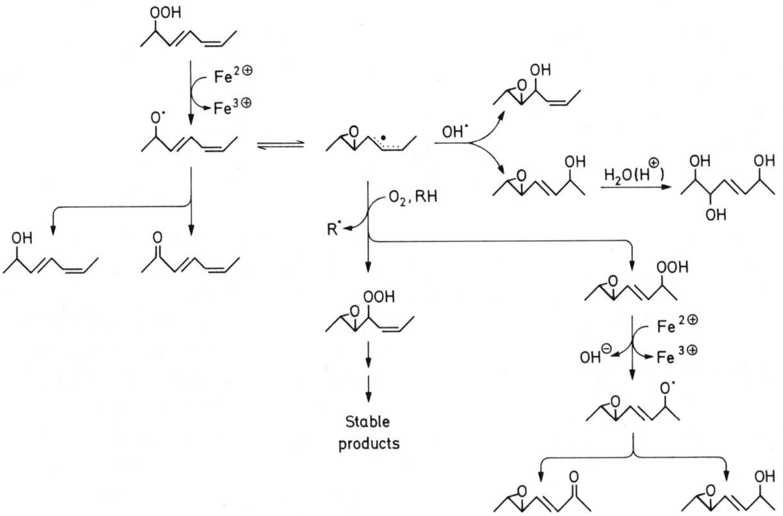

Fig. 3.29. Degradation of linoleic acid hydroperoxides to hydroxy-, epoxy- and oxo-fatty acids. The postulated reaction sequence explains the formation of identified products. Only segments of the structures are presented. (After *H. W. Gardner*, 1985)

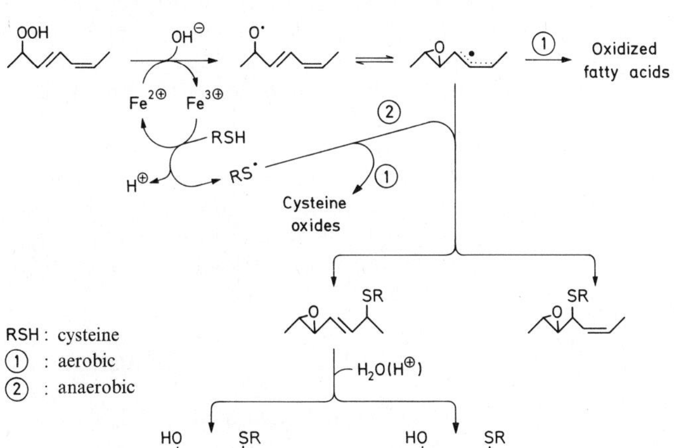

Fig. 3.30. Interaction of linoleic acid hydroperoxides with cysteine. A hypothesis to explain the reaction products obtained. Only segments of the structures are presented. (After *H. W. Gardner*, 1985)

hence, is readily extracted with a lipid solvent such as chloroform/methanol (2:1).

3.7.2.4.3 Protein changes

Some properties of proteins are changed when they react with hydroperoxides and their degradation products. This is reflected by changes in food texture, decreases in protein solubility (formation of cross-linked proteins), color (browning) and changes in nutritive value (loss of essential amino acids).

As shown in Figure 3.30, the radicals generated from hydroperoxides can abstract H-atoms from protein (PH), preferentially from the amino acids Trp, Lys, Tyr, Arg, His, cysteine and cystine, wherein the phenolic HO-, S- or N-containing group reacts:

$$RO^{\bullet} + PH \longrightarrow P^{\bullet} + ROH \qquad (3.72a)$$

$$2\,P^{\bullet} \longrightarrow P{-}P \qquad (3.72b)$$

In Reaction 3.27 b, protein radicals combine with each other, resulting in the formation of a protein network. Malonaldehyde is generated (cf. 3.7.2.1.5) under certain conditions during lipid peroxidation. As a bifunctional reagent, malonaldehyde can crosslink proteins through a *Schiff* base reaction with the ε-NH_2 group of lysine:

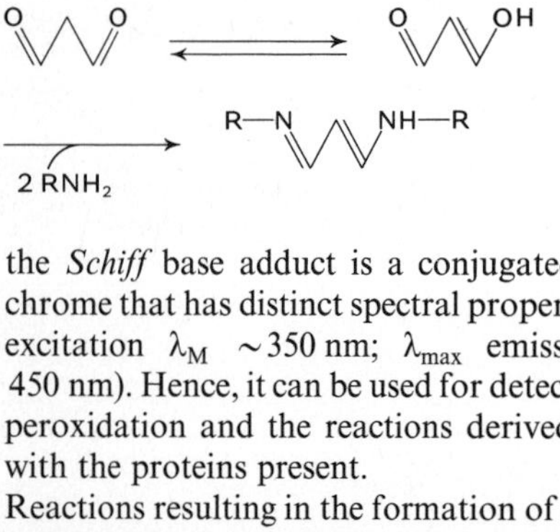

$$OHC{-}CH_2{-}CHO \rightleftharpoons OHC{-}CH{=}CH{-}OH \xrightarrow{2\,RNH_2} R{-}N{=}CH{-}CH{=}CH{-}NH{-}R \qquad (3.73)$$

the *Schiff* base adduct is a conjugated fluorochrome that has distinct spectral properties (λ_{max} excitation λ_M ~350 nm; λ_{max} emission λ = 450 nm). Hence, it can be used for detecting lipid peroxidation and the reactions derived from it with the proteins present.

Reactions resulting in the formation of a protein network like that outlined above also have practical implications; e.g. they are responsible for the decrease in solubility of fish protein during frozen storage.

Also, the monocarbonyl compounds derived from autoxidation of unsaturated fatty acids readily condense with protein-free NH_2 groups, forming *Schiff* bases that can provide brown

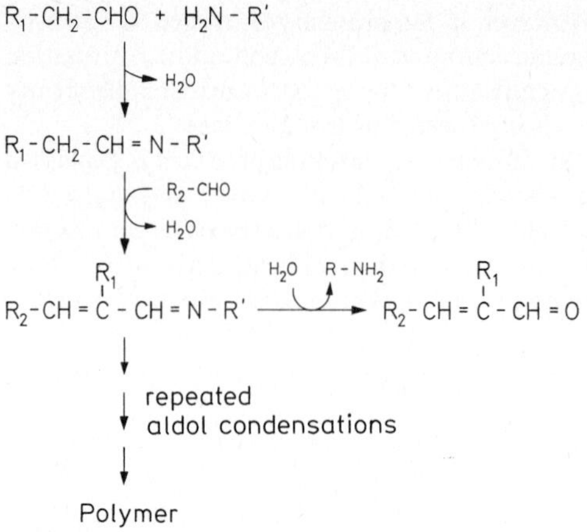

$$R_1{-}CH_2{-}CHO + H_2N{-}R'$$
$$\downarrow -H_2O$$
$$R_1{-}CH_2{-}CH{=}N{-}R'$$
$$\downarrow +R_2{-}CHO,\ -H_2O$$
$$R_2{-}CH{=}C(R_1){-}CH{=}N{-}R' \xrightarrow[H_2O]{-R{-}NH_2} R_2{-}CH{=}C(R_1){-}CH{=}O$$
$$\downarrow \text{repeated aldol condensations}$$
$$\text{Polymer}$$

Fig. 3.31. Reaction of volatile aldehydes with protein amino groups

polymers by repeated aldol condensations (Figure 3.31). The brown polymers are often N-free since the amino compound can be readily eliminated by hydrolysis. When hydrolysis occurs in the early stages of aldol condensations (after the first or second condensation; cf. Figure 3.31) and the released aldehyde, which has a powerful odor, does not reenter the reaction, the condensation process results not only in discoloration (browning) but also in a change in aroma.

3.7.2.4.4 Decomposition of amino acids

Studies of model systems have revealed that protein cleavage and degradation of side chains, rather than formation of protein networks, are the preferred reactions when the water content of protein/lipid mixtures decreases. Several examples of the extent of losses of amino acids in a protein in the presence of an oxidized lipid are presented in Table 3.29. The strong dependence of this loss on the nature of the protein and reaction conditions is obvious. Degradation products obtained in model systems of pure amino acids and oxidized lipids are described in Table 3.30.

3.7.3 Inhibition of Lipid Peroxidation

Autoxidation of unsaturated acyl-lipids can be retarded by:

- Exclusion of oxygen. Possibilities are packaging under a vacuum or addition of glucose oxidase (cf. 2.8.2.1.1).

Table 3.29. Amino acid losses occurring in protein reaction with peroxidized lipids

Reaction system		Reaction conditions		Amino acids loss
protein	lipid	time	T(°C)	(% loss)
Cytochrome C	Linolenic acid	5 h	37	His (59), Ser (55), Pro (53), Val (49), Arg (42), Met (38), Cys (35)[a]
Trypsin	Linoleic acid	40 min	37	Met (83), His (12)[a]
Lysozyme	Linoleic acid	8 days	37	Trp (56), His (42), Lys (17), Met (14), Arg (9)
Casein	Linoleic acid ethyl ester	4 days	60	Lys (50), Met (47), Ile (30), Phe (30), Arg (29), Asp (29), Gly (29), His (28), Thr (27), Ala (27), Tyr (27)[a,b]
Ovalbumin	Linoleic acid ethyl ester	24 h	55	Met (17), Ser (10), Lys (9), Ala (8), Leu (8)[a,b]

[a] Trp analysis was not performed.
[b] Cystine analysis was omitted.

Table 3.30. Amino acid products formed in reaction with peroxidized lipid

Reaction system		Compounds formed from amino acids
amino acid	lipid	
His	Methyl linoleate	Imidazolelactic acid, Imidazoleacetic acid
Cys	Ethyl arachidonate	Cystine, H_2S, cysteic acid, alanine, cystine-disulfoxide
Met	Methyl linoleate	Methionine-sulfoxide
Lys	Methyl linoleate	Diaminopentane, aspartic acid, glycine, alanine, α-aminoadipic acid, pipecolinic acid, 1,10-diamino-1,10-dicarboxydecane

- Storage at low temperature in the dark. The autoxidation rate is thereby decreased substantially. However, in fruits and vegetables with the lipoxygenase enzyme, these precautions are not applicable. Food deterioration is prevented only after inactivation of the enzyme by a blanching process (cf. 2.6.4).
- Addition of antioxidants to food.

3.7.3.1 Antioxidant Activity

The peroxy and oxy free radicals formed during the propagation and branching steps of the autoxidation radical chain (cf. Figure 3.17) are scavenged by antioxidants (AH; cf. Figure 3.32). Antioxidants containing a phenolic group play the major role in food. In reactions 1 and 2 in Figure 3.32 they form radicals which are stabilized by an aromatic resonance system. In contrast to the acyl peroxy and oxy free radicals they are not able to abstract a H-atom from an unsaturated fatty acid and therefore cannot initiate lipid peroxidation. The end-products formed in reactions 3 and 4 in Figure 3.32 are relatively stable and in consequence the autoxidation radical chains are shortened.
The reaction scheme (Fig. 3.32) shows that one antioxidant molecule combines with two radicals. Therefore, the maximum achievable stoichiometric factor is n = 2. In practice, the value of n is between 1 and 2 for the antioxidants used. Antioxidants, in addition to their main role as radical scavengers, are able to in part reduce hydroperoxides to hydroxy compounds.

3.7.3.2 Antioxidants in Food

3.7.3.2.1 Natural antioxidants

The unsaturated lipids in living tissue are relatively stable. Plants and animals have the necessary complement of antioxidants and of enzymes, for instance, glutathione peroxidase and superoxide dismutase, to effectively prevent lipid oxidation.
During isolation of oil from plants (cf. 3.8.3), tocopherols are also isolated. A sufficient level is retained in oil even after refining, thus, tocopherols secure the stability of the oil end-product. Soya oil, due to its relatively high level of linolenic acid (cf. Table 14.9), is an exception. The

$$RO_2^{\cdot} + AH \longrightarrow ROOH + A^{\cdot} \quad (1)$$

$$RO^{\cdot} + AH \longrightarrow ROH + A^{\cdot} \quad (2)$$

$$RO_2^{\cdot} + A^{\cdot} \longrightarrow ROOA \quad (3)$$

$$RO^{\cdot} + A^{\cdot} \longrightarrow ROA \quad (4)$$

Fig. 3.32. Activity of an antioxidant as a radical scavenger AH: Antioxidant

tocopherol content of animal fat is influenced by animal feed.

In vivo the relative antioxidant activity of the tocopherols yields α-T > β T > γ T > δ T. This corresponds to both the vitamin E activity of the tocopherols (cf. 6.2.3) and to the rate constants found at 30 °C for reaction 1 in Figure 3.32: 23.5, 16.6, 15.9 and $6.5 \times 10^5\ M^{-1} \times s^{-1}$, respectively. α-T has a larger rate constant than any synthetic antioxidant. However, in food the antioxidant activity of the tocopherols is reverse: δ-T > γ-T > β-T > α-T. The following effects are discussed to explain the differences of the tocopherols in vivo and in food.

- α-T reacts with radicals as shown in formula 3.74. Product I is stable whereas product II is a radical which can initiate an autoxidation reaction. Possibly the formation of product II is surpressed in vivo.
- In aqueous media δ-T is the most effective in protecting hydroperoxides of linoleic acid from decomposition and α-T is the least. A higher stability of the traces of hydroperoxides formed in reaction 1 of Figure 3.32 would prolong the induction period of lipid autoxidation. In vivo the hydroperoxides formed in reaction 1 are rapidly converted into other compounds, e.g. by enzymatic reduction to hydroxy fatty acids which cannot degrade into radicals.
- α-T is easily oxidized by gaseous oxygen while γ-T and especially δ-T are stable. It is assumed that the oxidation of α-T leads to radicals which initiate lipid peroxidation. In vivo α-T is protected against gaseous oxygen.

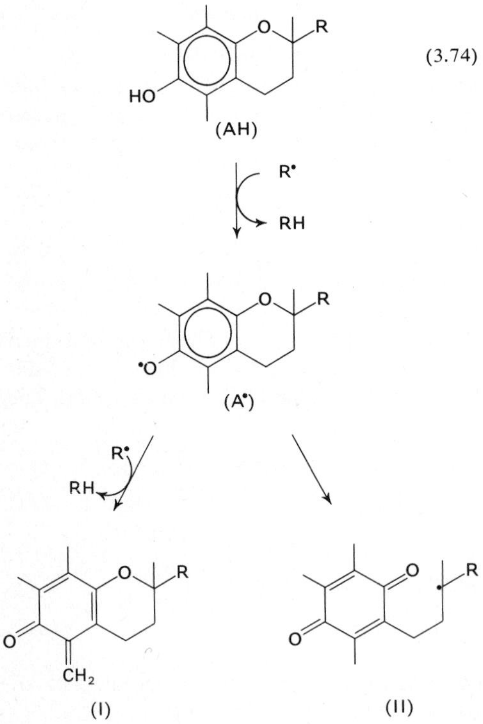

(3.74)

An increase in the concentration of α-T from 0.04% to 4% (by weight to linoleic acid) caused a conversion of its antioxidant activity into a prooxidant activity. γ- and δ-T do not show this conversion; they exhibit antioxidant activity also at the higher concentration. The prooxidant activity of α-T is explained with the formation of radicals during oxidation by gaseous oxygen.

Ascorbic acid (cf. 6.3.9) is active as an antioxidant in aqueous media, but only at higher concentrations ($\sim 10^{-3}$ M). A prooxidant activity is also observed at lower levels (10^{-5} M), especially in the presence of heavy metal ions.

Tocopherols and vitamin C are used as additives to increase food stability. In order to improve the solubility of vitamin C in the fat or oil phase, it is applied in the form of ascorbyl palmitate.

Flavanones (cf. 18.1.2.5.4) and flavonols (cf. 18.1.2.5.5) are phenolic compounds which are widely distributed in plant tissues, where they act as natural antioxidants. The protective effect of several herbs and spices (e.g. sage or rosemary) against fat (oil) oxidation is based on the presence of such natural antioxidants (cf. 22.1.1.5).

Polyphenols, such as lignin, undergo thermal cracking, resulting in volatile phenols, during the generation of smoke by burning wood or, even more so, sawdust. These phenols deposit on the food surface during smoking and then penetrate into the food, thus acting as antioxidants.

A very powerful antioxidant has been isolated from the desert plant commonly called the creosote bush. It is an alkyl-substituted catechol derivative, nordihydroguaiaretic acid (NDGA):

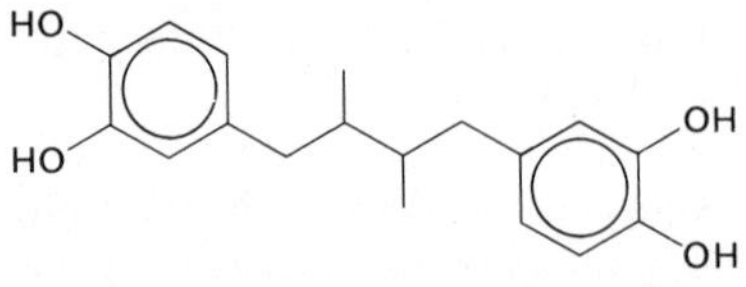

(3.75)

Utilization of this compound in food processing is still open to question. Finally, some of the *Maillard* reaction products, such as reductones (cf. 4.2.4.4), should be considered as naturally active antioxidants.

3.7.3.2.2 Synthetic antioxidants

In order to be used as an antioxidant, a synthetic compound has to meet the following requirements: it should not be toxic; it has to be highly active at low concentrations (0.01–0,02%); it has to concentrate on the surface of the fat or oil phase. Therefore, strongly lipophilic antioxidants are particularly suitable (with low HLB values, e.g. BHA, BHT or α-tocopherol) for o/w emulsions. In addition, antioxidants should be stable under the usual food processing conditions. This stability is denoted as the "carry through" effect.
Some of the synthetic antioxidants used worldwide are:

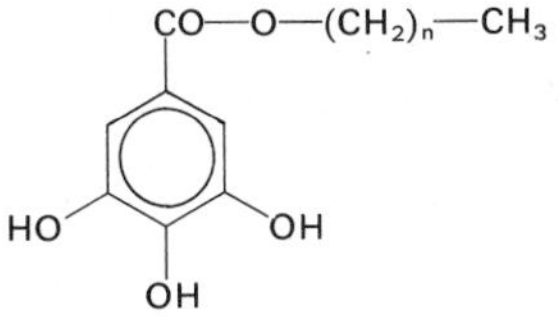

(3.76)

Propyl (n = 2), octyl (n = 7) and dodecyl (n = 11) gallate

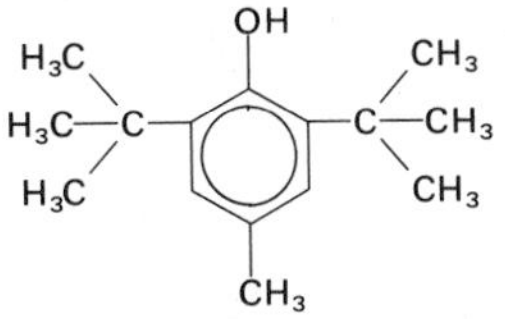

(3.77)

2,6-di-tert-butyl-p-hydroxytoluene (BHT)

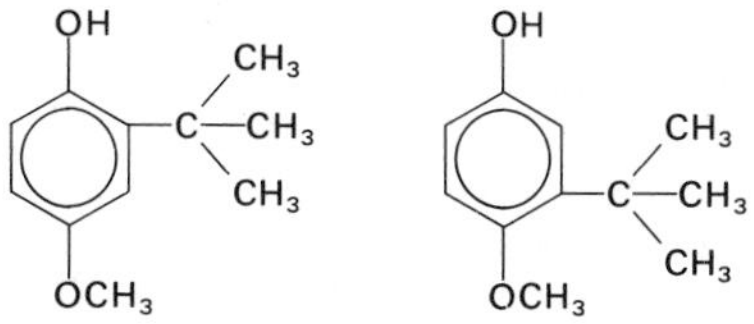

(3.78)

Tert-butyl-4-hydroxyanisole (BHA)

Commercial BHA is a mixture of two isomers, 2- and 3-tert-butyl-4-hydroxyanisole

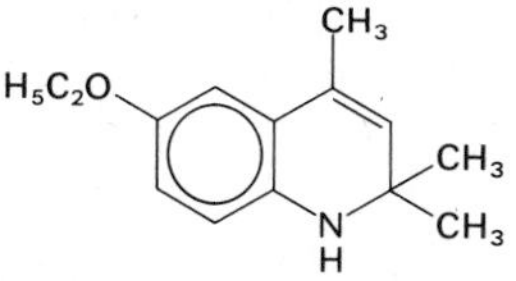

(3.79)

6-Ethoxy-1,2-dihydro-2,2,4-trimethylquinoline (ethoxyquin)

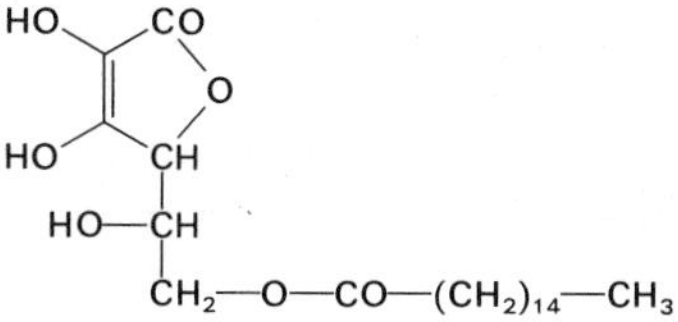

(3.80)

Ascorbyl palmitate

ESR spectroscopy has demonstrated that a large portion of ethoxyquin is present in oil as a free radical:

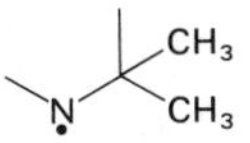

(3.81)

and stabilization by dimerization of the radical occurs. The radical, and not the dimer, is the active antioxidant.
tert-Butylhydroquinone (TBHQ) is a particularly powerful antioxidant used, for example, for stabilization of soya oil. The "carry through" properties are of importance for using BHA, TBHQ and BHT in food processing. All three antioxidants are water steam distillable at higher temperatures. Utilization of antioxidants is often regulated by governments through controls on the use of food additives. In North America incorporation of antioxidants is permitted at a maximum level of 0.01% of any one antioxidant, and a maximum of 0.02% for any combination. The regulations related to permitted levels often vary from country to country.
The efficiency of an antioxidant can be evaluated by a comparative assay, making use of an "antioxidative factor" (AF):

$$AF = I_A/I_0 \tag{3.82}$$

where I_A = oxidation induction period for a fat or oil (cf. 3.7.2.1.1) in the presence of an antioxidant and I_0 = oxidation induction period of a fat or oil without an antioxidant.
Hence, the efficiency of an antioxidant increases with an increase in the AF value. As illustrated

Table 3.31. Antioxidative factor (AF) values of some antioxidants (0.02%) in refined lard

Antioxidant	AF	Antioxidant	AF
d-α-Tocopherol	5	Octyl gallate	6
dl-γ-Tocopherol	12	Ascorbyl palmitate	4
BHA	9.5	BHA and	
BHT	6	BHT[a]	12

[a] Each compound is added in amount of 0.01%.

by the data in Table 3.31, BHA in comparison to BHT shows a higher efficiency in a lard sample. This result is understandable since in BHT both tertiary butyl substituents sterically hinder the reaction with radicals to a certain extent (reaction 1 in Figure 3.32). The effect of antioxidants depends not only on the origin of fat or oil but, also, on the processing steps used in the isolation and refining procedures. Hence, data in Table 3.31 serve only as an illustration.

BHA and BHT together at a given total concentration are more effective in extending shelf-life of a fat or oil than either antioxidant alone at the same level of use (Table 3.31).

To explain this, it is suggested that BHA, by participating in reaction 1 (Figure 3.32), provides a phenoxy radical (I):

$$RO_2^{\bullet} + \text{(BHA)} \longrightarrow ROOH + \text{(I)} \quad (3.83a)$$

which is then regenerated into the original molecule by rapid interaction with BHT:

$$\text{(I)} + \text{(BHT)} \longrightarrow \text{BHA} + \text{(II)} \quad (3.83b)$$

On the other hand, the phenoxy radical (II) derived from BHT can react further with an additional peroxy radical:

$$\text{(II)} + RO_2^{\bullet} \longrightarrow \text{(4-methyl-4-OOR-2,6-di-tert-butylcyclohexadienone)} \quad (3.84)$$

Propyl gallate (PG) increases BHA efficiency, but not that of BHT. Ascorbyl palmitate, which is by itself a rather weak antioxidant, substantially sustains the antioxidative activity of γ-D,L-tocopherol.

3.7.3.2.3 Synergists

Substances which enhance the activity of antioxidants are called synergists, and the main examples are citric, phosphoric, citraconic and fumaric acids, i.e. compounds which complex heavy metal ions (chelating agents, sequesterants or scavengers of trace metals). Thus, initiation of heavy metal-catalyzed lipid autoxidation can be prevented (cf. 3.7.2.1.4). Results compiled in Table 3.32 demonstrate the synergistic activities of citric and phosphoric acids in combination with lauryl gallate. Whereas citric acid enhances the antioxidant effectiveness in the presence of all three metal ions, phosphoric acid is able to do so with copper and nickel, but not with iron. Also, use of citric acid is more advantageous since phosphoric acid promotes polymerization of fat or oil during deep frying.

3.7.4 Fat or Oil Heating (Deep Frying)

Deep frying is one of the methods of food preparation used both in the home and in industry.

Table 3.32. Synergistic action of citric (C) and phosphoric acids (P) in combination with lauryl gallate (LG) on oxidation of fats and oils

To fat/oil added	AF value after addition of				
	0.01% C	0.01% P	0.01% LG	0.01% LG + 0.01% C	0.01% LG + 0.01% P
0.2 ppm Cu	0.3	0.2	0.9	4.7	4.1
2 ppm Fe	0.6	0.5	0.1	5.7	0.2
2 ppm Ni	0.5	0.6	3.0	7.0	4.4

Table 3.33. Chemical changes in cottonseed oil heated at 225 °C

	Time (h)		
	0	72	194
Average molecular weight (dalton)	850	1,080	1,510
Viscosity (Poise)	0.6	2,1	18.1
Iodine number	110	91	73
Peroxide value	2.5	1.5	0.0
Urea adduct filtrate (%)[a]	2.3	17.5	31.0

[a] Reaction products (polymeric and oxidized fatty acids) not able to form urea adducts.

Meat, fish, doughnuts, potato chips or french fries are dipped into fat (oil) heated to about 180 °C. After several minutes of frying, the food is sufficiently tender to be served.

The frying fat or oil changes substantially in its chemical and physical properties after prolonged use. Data for cottonseed oil compiled in Table 3.33 indicate that heating of oil causes reactions involving double bonds (a decrease in iodine number, an increase in urea adduct filtrate) which bring about the formation of higher molecular weight compounds and, consequently, compounds with higher intrinsic viscosity. Peroxides formed at elevated temperatures fragment immediately. Therefore, determination of peroxide values to evaluate the quality of fat or oil in deep frying is not appropriate.

A great number of volatile and nonvolatile products are obtained during deep frying of oil or fat. The types of reactions involved in and responsible for changes in oil and fat structures are compiled as a review in Table 3.34. Several types of reactions presented will be outlined in more detail.

3.7.4.1 Autoxidation of Saturated Acyl-Lipids

Autoxidation selectivity decreases above 60 °C since the hydroperoxides formed are subjected to homolysis into hydroxy and alkoxy radicals (Reaction RS-4 in Figure 3.17) which, due to their high reactivity, can abstract H-atoms even from saturated fatty acids.

A multitude of compounds result from these reactions. For example, Table 3.35 lists a series of aldehydes and methyl ketones derived preferentially from tristearin. Both classes of compounds are also formed by thermal degradation of free fatty acids. The presence of these acids is the result of triglyceride hydrolysis or the acids are formed by the oxidation of aldehydes.

Methyl ketones are obtained by thermally-induced β-oxidation followed by a decarboxylation reaction (Figure 3.33), whereas aldehydes

Table 3.34. A review of reactions occurring in heat treated fats and oils

Fat/oil heating	Reaction	Products
1. Deep frying without food	Autoxidation Isomerization Polymerization	Volatile acids aldehydes esters alcohols Epoxides Branched chain fatty acids Dimeric fatty acids Mono- and bicyclic compounds Aromatic compounds Compounds with trans double bonds Hydrogen, CO_2.
2. Deep frying with food added	As under 1. and in addition hydrolysis	As under 1. and in addition free fatty acids, mono- and diacylglycerols and glycerol

Table 3.35. Volatile compounds formed from heat-treated tristearin[a]

Class of compound	Portion	C-number	Major compounds
Alcohols	2.7	4–14	n-Octanol n-Nonanol n-Decanol
γ-Lactones	4.1	4–14	γ-Butyrolactone γ-Pentalactone γ-Heptalactone
Alkanes	8.8	4–17	n-Heptadecane n-Nonane n-Decane
Acids	9.7	2–12	Caproic acid Valeric acid Butyric acid
Aldehydes	36.1	3–17	n-Hexanal n-Heptanal n-Octanal
Methyl ketones	38.4	3–17	2-Nonanone 2-Heptanone 2-Decanone

[a] Tristearin is heated in air at 192 °C.

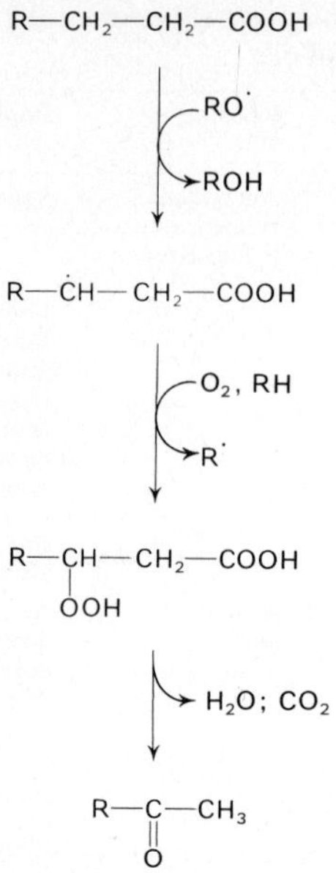

$$R{-}CH_2{-}CH_2{-}COOH \xrightarrow{RO^{\cdot} \to ROH} R{-}\dot{C}H{-}CH_2{-}COOH \xrightarrow{O_2,\ RH \to R^{\cdot}} R{-}CH(OOH){-}CH_2{-}COOH \xrightarrow{\to H_2O;\ CO_2} R{-}C(=O){-}CH_3$$

Fig. 3.33. Autoxidation of saturated fatty acids. Postulated reaction steps involved in formation of methyl ketones

are obtained from the fragmentation of hydroperoxides by a β-scission mechanism (Figure 3.34) occurring nonselectively at elevated temperatures (compare the difference with 3.7.2.1.5). Some volatiles are important odorous compounds. Some unsaturated lactones, such as the γ-lactone of 4-hydroxy-2-nonenoic acid

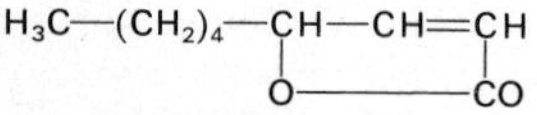

$$H_3C{-}(CH_2)_4{-}\underset{|}{C}H{-}CH{=}\underset{|}{C}H \quad O{-}CO \tag{3.85}$$

should contribute to the pleasant deep-fried flavor. Since most of such compounds are formed by thermal degradation of linoleic acid, fats or oils containing this acid provide a better aroma during deep frying. However, when a fat is used for a prolonged period of time, the unpleasant aroma of volatile compounds becomes noticeable. This is a sign of the advanced stage of fat or oil deterioration.

3.7.4.2 Polymerization

Under deep frying conditions the isolenic fatty acids are isomerized into conjugated fatty acids which in turn interact by a 1,4-cycloaddition, yielding so-called *Diels-Alder* adducts (cf. Reaction 3.86).

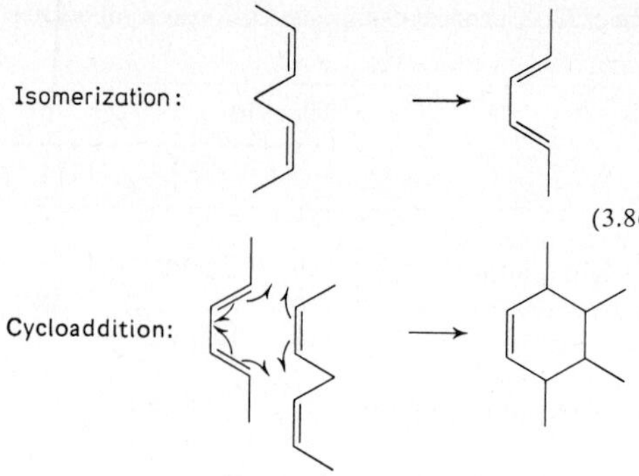

Isomerization:

Cycloaddition:

(3.86)

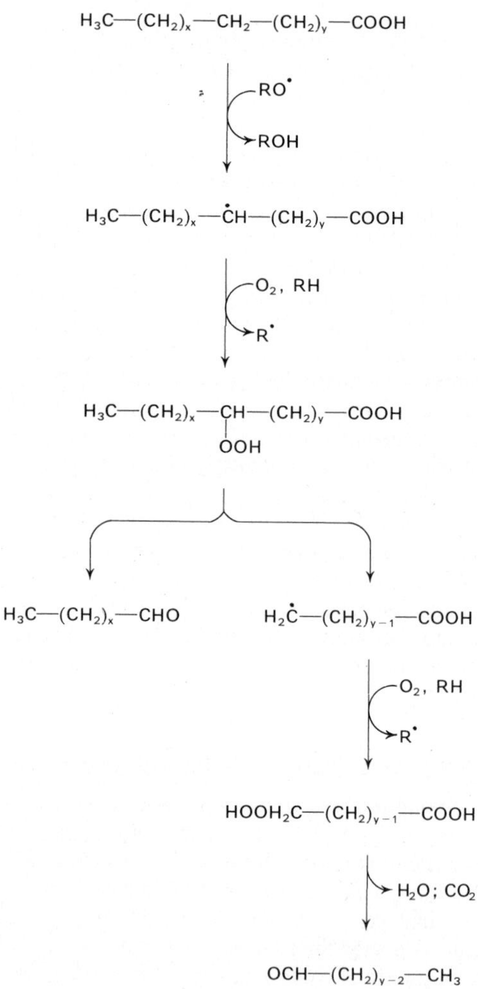

$$H_3C{-}(CH_2)_x{-}CH_2{-}(CH_2)_y{-}COOH \xrightarrow{RO^{\cdot} \to ROH} H_3C{-}(CH_2)_x{-}\dot{C}H{-}(CH_2)_y{-}COOH \xrightarrow{O_2,\ RH \to R^{\cdot}} H_3C{-}(CH_2)_x{-}CH(OOH){-}(CH_2)_y{-}COOH$$

$$\to H_3C{-}(CH_2)_x{-}CHO \quad + \quad H_2\dot{C}{-}(CH_2)_{y-1}{-}COOH$$

$$H_2\dot{C}{-}(CH_2)_{y-1}{-}COOH \xrightarrow{O_2,\ RH \to R^{\cdot}} HOOH_2C{-}(CH_2)_{y-1}{-}COOH \xrightarrow{\to H_2O;\ CO_2} OCH{-}(CH_2)_{y-2}{-}CH_3$$

Fig. 3.34. Autoxidation of saturated fatty acids. Hypothetical reactions involved in formation of volatile aldehydes

The side chains of the resultant tetra-substituted cyclohexene derivatives are shortened by oxidation to oxo, hydroxy or carboxyl groups. In addition, the cyclohexene ring is readily dehydrogenated to an aromatic ring, hence compounds related to benzoic acid can be formed.

The fatty acids or triacylglycerol radicals formed by H-abstraction in the absence of oxygen can dimerize and then form a ring structure:

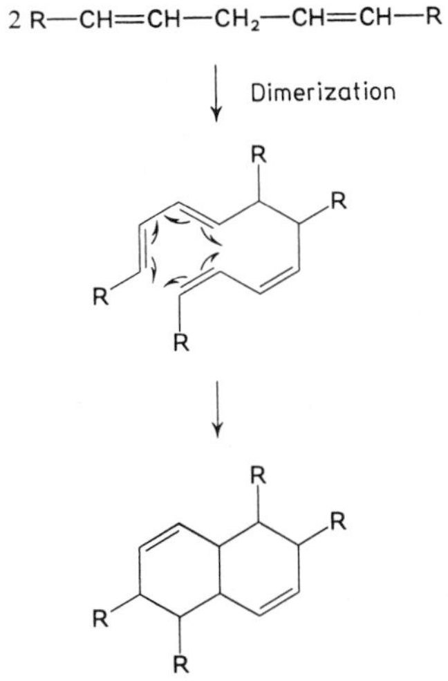

$$2\ R{-}CH{=}CH{-}CH_2{-}CH{=}CH{-}R \xrightarrow{\text{Dimerization}} \ldots \quad (3.87)$$

On the other hand, polymers with ether and peroxide linkages are formed in the presence of oxygen. They also may contain hydroxy, oxo or epoxy groups. The following structures, among others, have been identified:

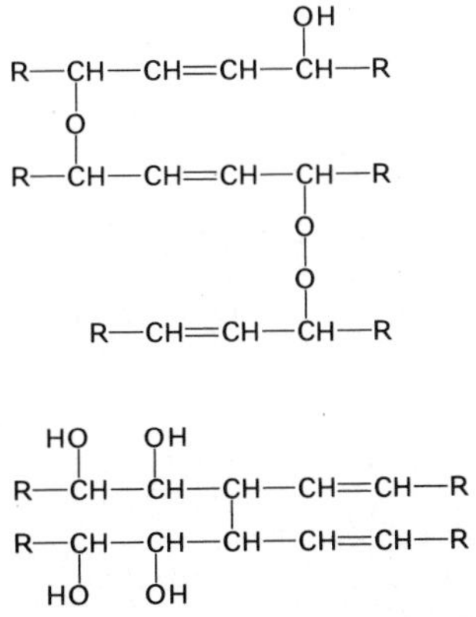

$$R{-}CH(O{-})\,{-}CH{=}CH{-}CH(OH){-}R$$
$$R{-}CH(O{-})\,{-}CH{=}CH{-}CH(O{-}O{-}){-}R$$
$$R{-}CH{=}CH{-}CH(O{-}O{-}){-}R \quad (3.88)$$
$$R{-}CH(OH){-}CH(OH){-}CH{-}CH{=}CH{-}R$$
$$R{-}CH(OH){-}CH(OH){-}CH{-}CH{=}CH{-}R$$

Such compounds are undesirable in deep-fried oil or fat since they persistently diminish the flavoring characteristics of the oil or fat and, because of their HO-groups, behave like surface-active agents, i.e. they foam.

Table 3.36. Relative stability of various fats and oils in deep-frying (RSDF)

Oil/fat	RSDF	Oil/fat	RSDF
Sunflower	1.0	Coconut	2.4
Rapeseed	1.0	Edible beef tallow	2.4
Soya	1.0	Soya oil, hydrogenated	2.3
Peanut	1.2	Peanut oil hydrogenated	4.4
Palm	1.5		
Lard	2.0		
Butter fat	2.3		

Disregarding the odor or taste deficiencies developed in a fat or oil heated for a prolonged period of time, the oil is considered spoiled when its petroleum ether-insoluble oxidized fatty acids reach a level $\geq 1\%$ (or $\geq 0{,}72\%$ at the decreased smoke-point temperature of $\leq 170\,°C$). The fats or oils differ in their heat stability (Table 3.36). The stability is increased by hydrogenation of the double bonds.

3.7.5 Microbial Degradation of Acyl Lipids to Methyl Ketones

The fatty acids with short and medium chain lengths present in milk fat, coconut and palm oils are degraded to methyl ketones by some fungi. A number of *Penicillium* and *Aspergillus* species, as well as several *Ascomycetes, Phycomycetes* and *Fungi imperfecti* are able to do this. The microorganisms first hydrolyze the triglycerides enzymatically (cf. 3.7.1) and then they degrade the free acids by a β-oxidation pathway mechanism (Figure 3.35). The fatty acids $< C_{14}$ are transformed to methyl ketones, the C-skeletons of which are those of the fatty acids shortened by one C-atom. Apparently, the thiohydrolase activity of these fungi is higher than the β-ketothiolase activity. Hence, ester hydrolysis occurs instead of thioclastic cleavage of the thioester of a β-keto acid (see a textbook of biochemistry). The β-keto acid released is rapidly decarboxylated enzymatically; a portion of the methyl ketones is reduced to the corresponding secondary alcohols.

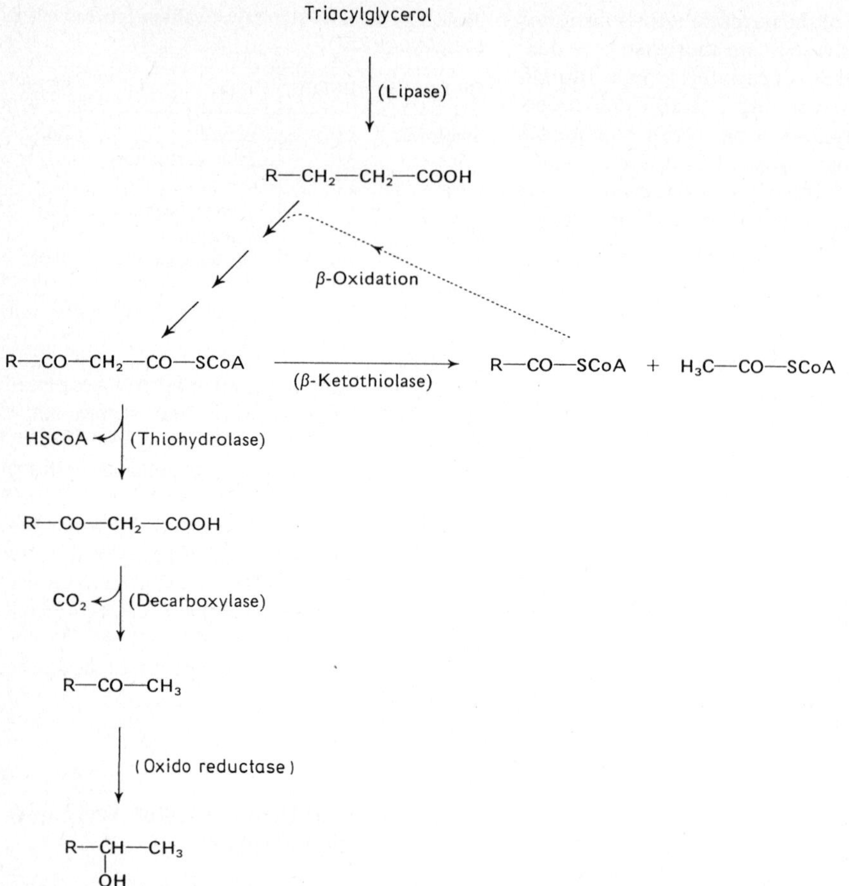

Fig. 3.35. Fungal degradation of triacylglycerols to methyl ketones. (After *J. E. Kinsella* and *D. H. Hwang*, 1976)

The odor threshold values for methyl ketones are substantially higher than those for aldehydes (cf. Tables 3.24 and 3.37). Nevertheless, they act as aroma constituents, particularly in flavors of mold-ripened cheese (cf. 10.2.7.2). However, methyl ketones in coconut or palm oil or in milk fat provide an undesirable, unpleasant odor denoted as "perfume rancidity".

3.8 Unsaponifiable Constituents*

Disregarding a few exceptions, fats and oils contain an average of 0.2–1.5% unsaponifiable compounds (Table 3.38). They are isolated from a soap solution (alkali salts of fatty acids) by extraction using an organic solvent.

The unsaponifiable matter contains hydrocarbons, steroids, tocopherols and carotenoids. In

* The free higher alcohols described under 3.62 and the deacylated alkoxy-lipids belong to this class.

Table 3.37. Sensory properties of methyl ketones

Compound	Odor description	Odor threshold (ppb; in water)
2-Pentanone	Fruity, after bananas	2,300
2-Hexanone		930
2-Heptanone	Fragrant, herbaceous	650
2-Octanone	Flowery, refreshing	190
2-Nonanone	Flowery, fatty	190

Table 3.38. Content of unsaponifiables in various fats and oils

Fat/Oil	Unsaponifiable (weight-%)	Fat/Oil	Unsaponifiable (weight-%)
Soya	0.6–1.2	Rapeseed	0.7–1.1
Sunflower	0.3–1.2	Shea	3.6–10.0
Cocoa	0.2–0.3	Lard	0.1–0.2
Peanut	0.2–4.4	Fish liver	0.7–9.0
Olive	0.4–1.1	Herring	1.2–2.3
Palm	0.3–0.9		

addition, contaminants or fat or oil additives, such as mineral oil, plasticizers or pesticide residues, can be found.

Each class of compounds in the unsaponifiable matter is represented by a number of components, the structures and properties of which have been thoroughly elucidated in the past decade or two, thus reflecting the advance in the analytical chemistry of fats and oils.

Studies aimed at elucidating the constituents and their structures of unsaponifiable matter are motivated by a desire to find compounds which can serve as a reliable indicator for the identity of a fat or an oil.

3.8.1 Hydrocarbons

All edible oils contain hydrocarbons with an even or an odd C-number (C_{11} to C_{35}). Olive, rice and fish oils are particularly rich in this class of compounds. The main hydrocarbon constituent of olive oil (1–7 g/kg) and rice oil (3.3 g/kg) is a linear triterpene (C_{30}), squalene:

(3.89)

This compound is used as an analytical indicator for olive oil (cf. Table 14.20).

Squalene is present in a substantially higher concentration in fish liver oil. For example, shark liver oil has up to 30% squalene, and 7% pristane (2,6,10,14-tetramethylpentadecane) and some phytane (3,7,11,15-tetramethylhexadecane).

Apart from traces of polycyclic hydrocarbons (cf. 9.7), low levels of alkylbenzenes are detected. Such compounds occurring in olive oil are shown in Formula 3.90.

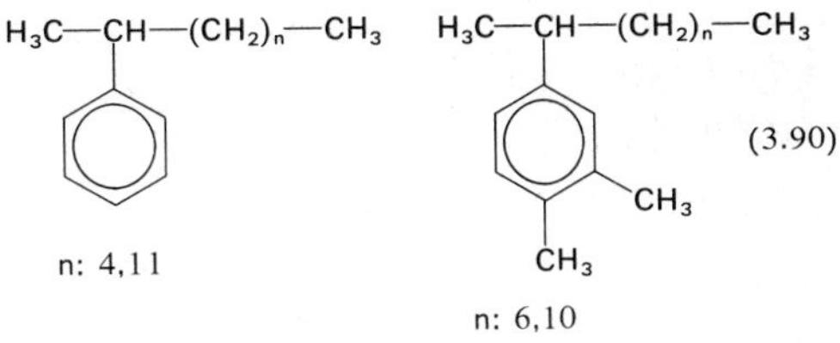

$H_3C—CH—(CH_2)_n—CH_3$ (phenyl) n: 4,11 $H_3C—CH—(CH_2)_n—CH_3$ (dimethylphenyl, CH_3, CH_3) n: 6,10

(3.90)

3.8.2 Steroids and Steroid Derivatives

3.8.2.1 Structure, Nomenclature

The steroid skeleton is made up of four condensed rings; A, B, C and D. The first three are in the chair conformation, whereas ring D is usually planar. While rings B and C, and C and D are fused in a trans-conformation, rings A and B can be fused in a trans- or in a cis-conformation. A characteristic of steroids is the presence of an alcoholic HO-group in position C-3.

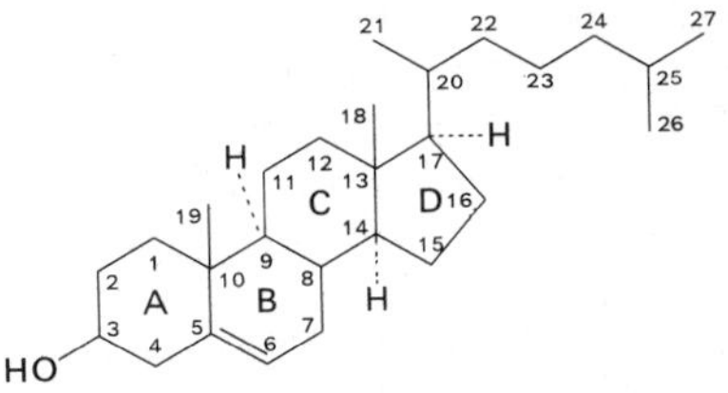

(3.91)

Conformational isomers introduced by fusing rings A and B in cholest-5-ene-3-β-ol (cholesterol; cf. Formula 3.91) ar not possible since the C-5 position has a double bond.

By convention, the steric arrangement of substituents and H-atoms is related to the angular methyl group attached at C-10. When the plane containing the four rings is assumed to be the plane of this page, the substituent at C-10, by definition, is above the plane; all substituents below the plane are denoted by dashed or dotted lines. They are said to be α-oriented and have a trans-conformation. Substituents above the plane are termed β-oriented and are shown by solid line bonds and, in relation to the angular C-10 methyl group, are of cis-conformation.

In cholesterol (cf. Formula 3.91) the HO-group, the angular methyl group at C-13, the side chain on C-17 and the H-atom on C-8 are β-oriented (cis), whereas the H-atoms at C-9, C-14 and

C-17 are α-oriented (trans). Steroids that are not methylated at position C-4 are denoted as desmethyl steroids.

3.8.2.2 Steroids of Animal Food

3.8.2.2.1 Cholesterol

Cholesterol (cf. Formula 3.91) is obtained biosynthetically from squalene (see a textbook of biochemistry). It is the main steroid of mammals and occurs in lipids in free form or esterified with saturated and unsaturated fatty acids. The content of cholesterol in some foods is illustrated by the data in Table 3.38 a.

Autoxidation of cholesterol, which is accelerated manyfold by 18:2 and 18:3 fatty acid peroxy radicals, proceeds through the intermediary 3β-hydroxycholest-5-en-7α- and 7β-hydroperoxides, of which the 7β-epimer is more stable because of its quasi-equatorial conformation and, hence, is formed predominantly. Unlike autoxidation, the photosensitized oxidation (reaction with a singlet oxygen) of cholesterol yields 3β-hydroxycholest-6-en-5α-hydroperoxide. Among the many derivatives obtained by the further degradation of the hydroperoxides, cholest-5-en-3β,7α-diol; cholest-5-en-3β,7β-diol; 3β-hydroxycholest-5-en-7-one; 5,6α-epoxy-5α-cholestan-3β-ol; 5,6β-epoxy-5β-cholestan-3β-ol and cholestan-3β,5α,6β-triol have been identified as major products. These so-called "oxycholesterols" have been detected especially in spray dried whole eggs and spray dried egg yolks. The are also formed during deep frying of butter oil and tallow.

Cholesterol is the precursor in animal organisms for the biosynthesis of other steroids, such as sex hormones and bile acids. The latter, due to their interface activity, play a role in the emulsification and absorption of triacyl lipids in the intestine. Cholic acid is a bile acid with three α-oriented HO-groups:

COOH
OH
HO OH

(3.92)

Bile acids present in bile or in intestines are almost always bound to glycine e.g. cholic acid as glycocholic acid, and to taurine e.g. taurocholic acid.

Table 3.38 a. Cholesterol content of some food

Food	Amount (mg/100 g)
Calf brain	2,000
Egg yolk[a]	1,010
Pork kidney	410
Pork liver	340
Butter	240
Pork meat, lean	70
Beef, lean	60
Fish (Halibut; *Hypoglossus vulgaris*)	50

[a] Egg white is devoid of cholesterol

3.8.2.2.2 Vitamin D

Cholecalciferol (vitamin D_3) is formed by photolysis of 7-dehydrocholesterol, a precursor in cholesterol biosynthesis. As shown in Figure 3.36, UV radiation opens the B-ring. The precalciferol formed is then isomerized to vitamin D_3 by a rearrangement of the double bond which is influenced by temperature. Side-products, such as lumi- and tachisterol, have no vitamin D activity. Cholecalciferol is converted into the active hormone, 1,25-dihydroxy-cholecalciferol, by hydroxylation reactions in liver and kidney.

7-Dehydrocholesterol, the largest part of which is supplied by food intake and which accumulates in human skin, is transformed by UV light into vitamin D_3. The occurrence and the physiological significance of the D vitamins are covered in Chapter 6.22.

The main steroid of yeasts, ergosterol (ergosta-5,7,22-trien-3β-ol, i.e. two conjugated double bonds in ring B and a third in the side chain) is known as provitamin D_2.

3.8.2.3 Plant Steroids (Phytosterols)

3.8.2.3.1 Desmethylsterols

Cholesterol, long considered to be an indicator of the presence of animal fat, also occurs in small amounts in plants (Table 3.39). Campe-, stigma- and sitosterol, which are predominant in the sterol fraction of some plant oils are structurally related to cholesterol; only the side chain on C-17 is changed. The following structural segments (only ring D and the side chain) show these differences:

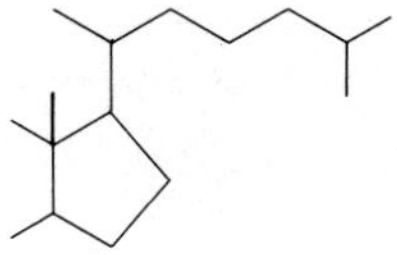

(3.93 a)

Cholesterol

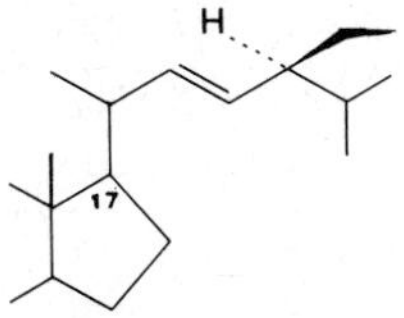

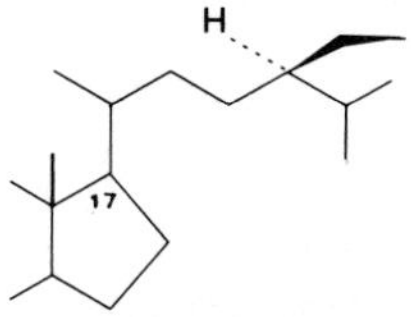

(3.93 b)

Sitosterol (24-α-ethyl-cholesterol)

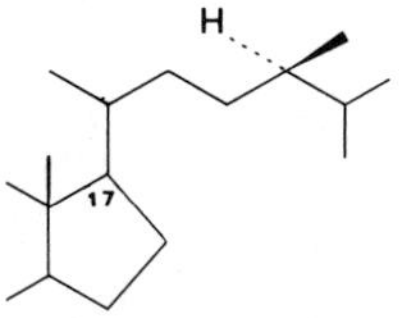

(3.93 c)

Stigmasterol (double bond at C-22)

(3.93 d)

Campesterol (24-α-methyl-cholesterol)

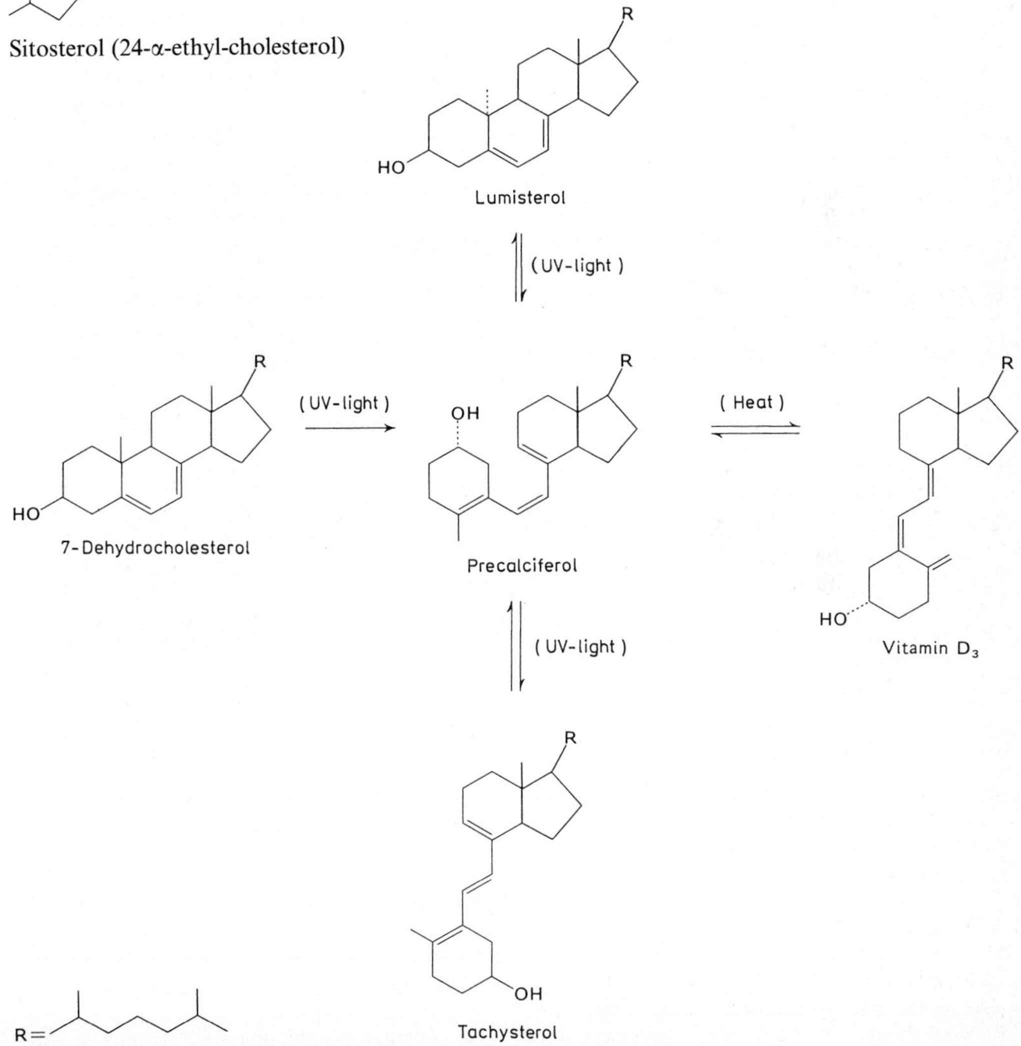

Lumisterol

(UV-light)

7-Dehydrocholesterol

(UV-light)

Precalciferol

(Heat)

Vitamin D_3

(UV-light)

Tachysterol

Fig. 3.36. Photochemical conversions of provitamin D_3

Table 3.39. Composition of various plant oils' sterol fraction (%)[a]

Oil	Sterol fraction[b]	Chole-sterol	Campe-sterol	Stigma-sterol	Sito-sterol	Δ^5-Avena-sterol	Δ^7-Stigma-sterol	Δ^7-Avena-sterol	Other sterols
Cottonseed	0.35	trace	9.2	2.5	88.2	trace	–	–	–
Peanut	0.25	–	12.8	10.8	74.6	–	1.8	–	–
Corn	0.90	–	22.1	7.2	62.2	3.5	trace	–	–
Rapeseed	0.84	0.8	29.4	–	54.9	5.7	trace	–	9.9
Olives	0.16	trace	3.9	2.1	85.4	8.6	trace	–	–
Soya	0.37	trace	21.3	19.1	53.5	2.4	2.0	1.4	–
Cocoa butter	0.25	0.7	8.7	27.9	62.6	trace	–	–	–

[a] Relative to sum of sterol fractions taken as ≙ 100%.
[b] As percent of oil weight.

Δ^5-Avenasterol is a sitosterol derivative:

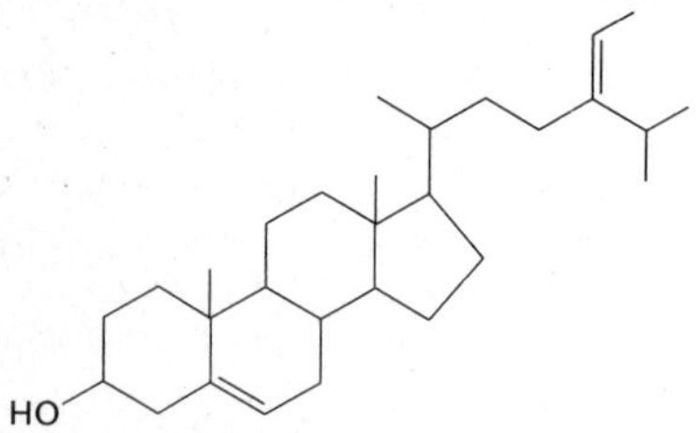

(3.94)

Δ^5-Avenasterol

Δ^7- and Δ^5-sterols occur in plant lipids; for example:

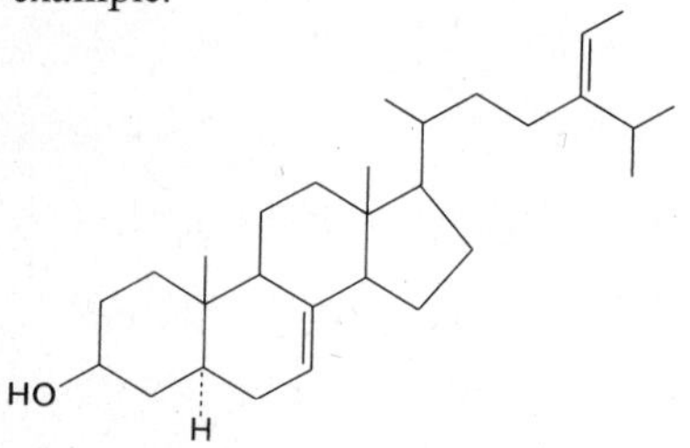

(3.95 a)

Δ^7-Avenasterol

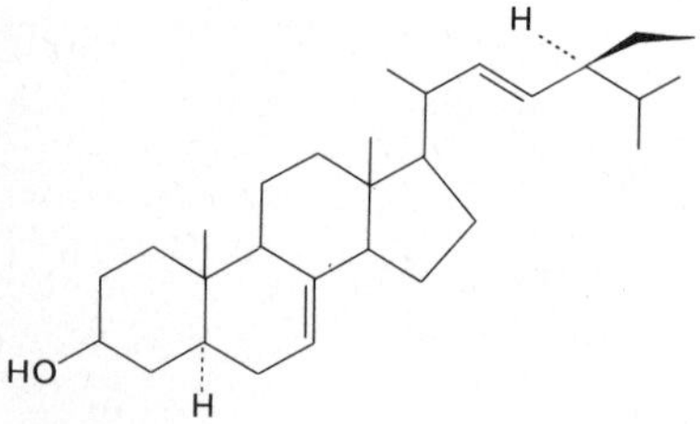

(3.95 b)

Δ^7-Stigmasterol

Plant lipids contain 0.15–0.9% sterols, with sitosterol as the main component (Table 3.39).
In order to identify blends of fats (oils), the data on the predominant steroids are usually expressed as a quotient. For example, the ratio of stigmasterol/campesterol is determined in order to detect adulteration of cocoa butter. As seen from Table 3.40, this ratio is significantly lower in a number of cocoa butter substitutes than in pure cocoa butter. The phytosterol fraction (e.g. sito- and campesterol) has to be determined in order to detect the presence of plant fats in animal fats.

3.8.2.3.2 Methyl- and dimethylsterols

Steroids with α-oriented C-4 methyl groups occur in oils of plant origin. The main compounds are:

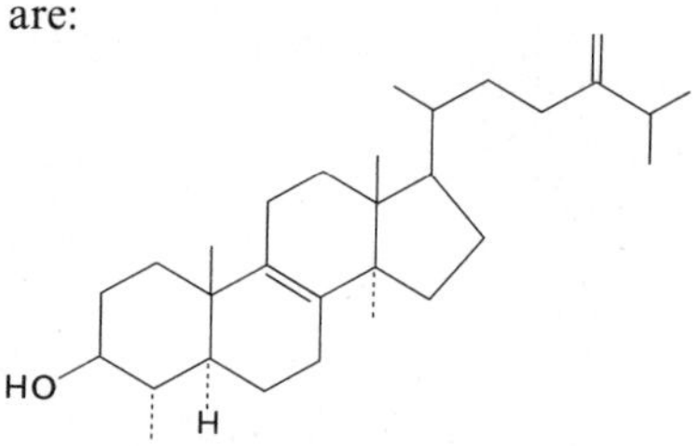

(3.96 a)

4α,14α-dimethyl-24-methylene-5α-cholest-8-en-3β-ol (Obtusifoliol)

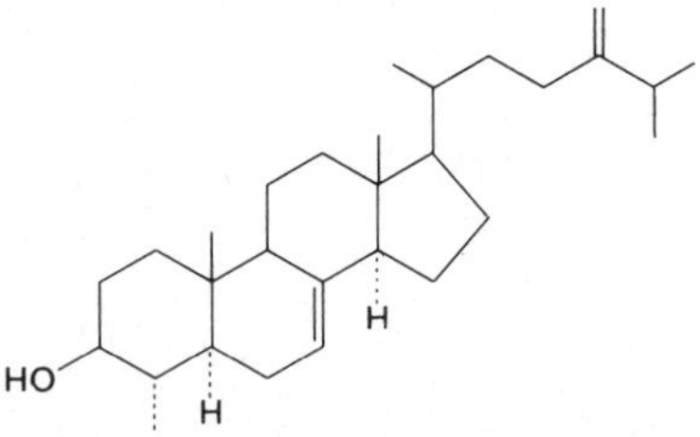

(3.96 b)

4α-methyl-24-methylene-5α-cholest-7-en-3β-ol (Gramisterol)

Gas chromatographic-mass spectrometric studies have also revealed the presence of 4,4-dimeth-

Table 3.40. Ratio of stigmasterol to campesterol in various fats

Fat	% Stigmasterol[a] / % Campesterol
Cocoa	2.8–3.5
Tenkawang	0.42–0.55
Coconut oil	1.5
Peanut oil, hydrogenated	0.72
Coberine[b]	0.31–0.60
Calvetta[b]	0.58–0.61

[a] The ratio is calculated from peak area after the unsaponifiables (sterol fraction) were separated by gas chromatography.
[b] Cocoa butter substitutes.

ylsterols in the steroid fraction of many plant oils:

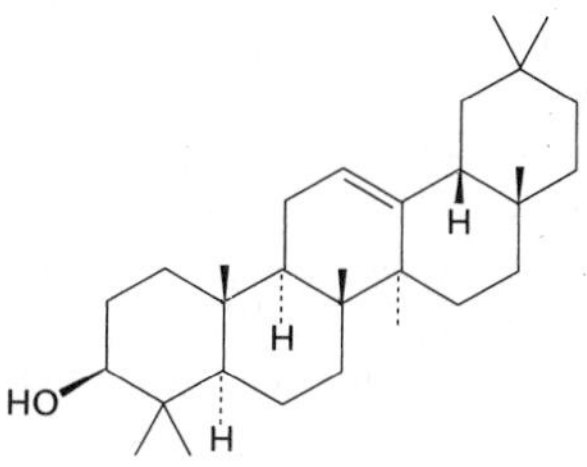

(3.97a)

β-Amyrine

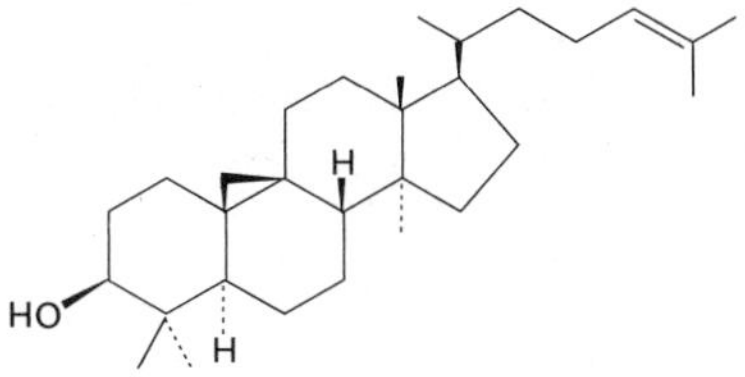

(3.97b)

Cycloartenol

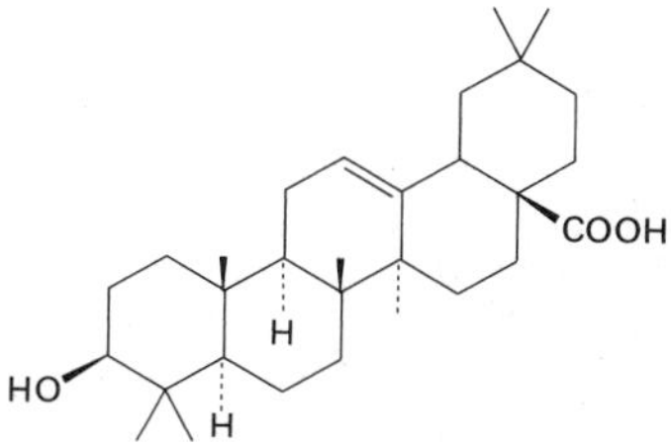

(3.97c)

Oleanolic acid

Oleanolic acid has long been known as a constituent of olive oil. Methyl- and dimethylsterols are important in identifying fats and oils (cf. Figure 3.38).

3.8.2.4 Analysis

Qualitative determination of sterols is done using the *Liebermann-Burchard* reaction, in which a mixture of glacial acetic and concentrated sulfuric acids reacts directly with the fat or oil or the unsaponifiable fraction. Several modifications of this basic assay have been developed which, depending on the steroid and the oxidizing agent used, result in the production of a green or red color. The reaction is more sensitive when the SO_3 oxidizing agent is replaced by the Fe^{3+} ion. The conversion of sterols into a chromophore is based on the reaction sequence given in Figure 3.37. As seen there, the assay is applicable only to sterols containing a double bond, such as in the B ring of cholesterol.

Sterols are separated as 3,5-dinitrobenzoic acid derivatives by thin layer chromatography and, after reaction with 1,3-diaminopropane, are determined quantitatively with high sensitivity in the form of a *Meisenheimer* adduct. Today, gas

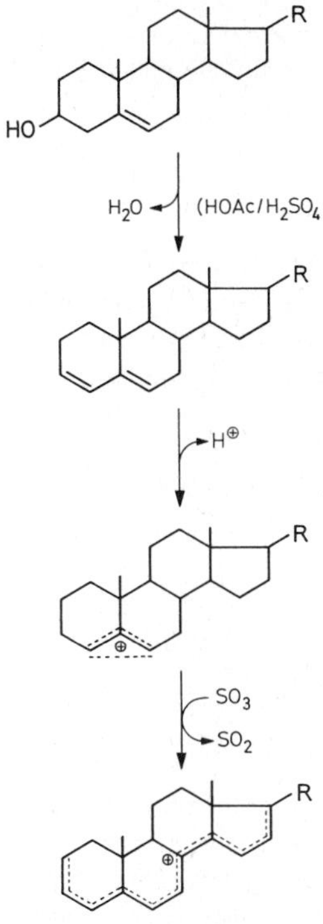

Pentaenyl cation (λ_M: 620 nm)

Fig. 3.37. Sterol detection after *Lieberman-Burchard.* Reactions involved in color development

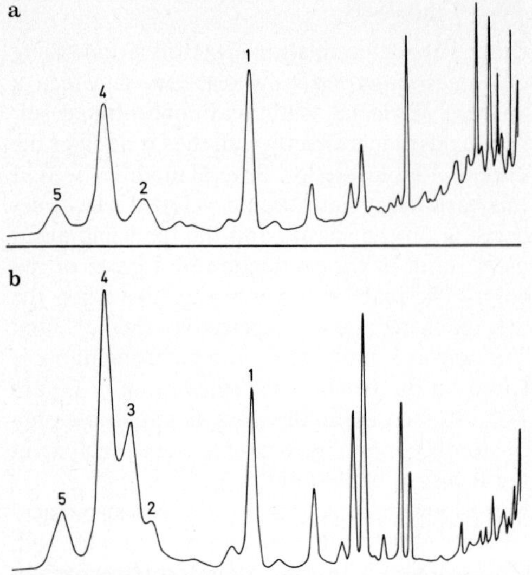

3.38. Gas chromatographic separation of the triterpene alcohol fraction after silylation: cocoa butter **a** and cocoa butter +2% coberine **b**. (After *A. Fincke*, 1976). Peak 1: cholesterol added as a marker compound; peak 2: β-amyrin; peak 3: butyrospermol and α-amyrin; peak 4: cycloartenol; peak 5: 24-methylenecycloartenol

chromatographic analysis of silylated sterols is used more often. One application of this method is illustrated by the detection of 2% coberine in cocoa butter (Figure 3.38). Coberine is a cocoa butter substitute made by blending palm oil and shea butter (the shea is an African tree with seeds that yield a thick white fat, shea butter).

The content of egg (more accurately, the yolk) in pasta products or cookies can be calculated after the cholesterol content has been determined, usually by gas chromatography or HPLC.

Vitamin D determination requires specific procedures in which precautions are taken with regard to the compound's sensitivity to light. A chemical method uses thin layer chromatographic separation of unsaponifiables, elution of vitamin D from the plate and photometric reading of the color developed by antimony (III) chloride. An alternative method recommends the use of HPLC.

3.8.3 Tocopherols and Tocotrienols

3.8.3.1 Structure, Importance

All four tocopherols and tocotrienols, with the chemical structures given in Figure 3.39, are found in nature, primarily in cereals (especially wheat germ oil), nuts and rapeseed oils. These redox-type lipids are of nutritional/physiological and analytical interest. As antioxidants (cf. 3.7.3.2.1), they prolong the shelf lives of many foods containing fat or oil. The significance of tocopherols such as vitamin E was outlined in 6.2.3.

About 60–70% of the tocopherols in oilseeds are retained during the oil extraction and refining process (cf. 14.4.1). Some oils with very similar fatty acid compositions can be distinguished by their distinct tocopherol spectrum. To illustrate this, two examples are provided. The amount of

HO (5) (7) (8) O (1) CH_3 R

Tocol $R = H_2C{-}CH_2{-}CH_2{-}CH(CH_3){-}CH_2{-}CH_2{-}CH_2{-}CH(CH_3){-}CH_2{-}CH_2{-}CH_2{-}CH(CH_3){-}CH_3$

Tocotrienol $R = H_2C{-}CH_2{-}CH{=}C(CH_3){-}CH_2{-}CH_2{-}CH{=}C(CH_3){-}CH_2{-}CH_2{-}CH{=}C(CH_3){-}CH_3$

Substitution	*Tocopherols* (T)	*Tocotrienols* (T-3)
5,7,8-Trimethyl	α-T	α-T-3
5,8-Dimethyl	β-T	β-T-3
7,8-Dimethyl	γ-T	γ-T-3
8-Methyl	δ-T	δ-T-3

Fig. 3.39. Tocopherols and tocotrienols present in food

Table 3.41. Tocopherol and tocotrienol contents in edible oils (mg/100 g)

Oil	α-T	α-T-3	β-T	β-T-3	γ-T	γ-T-3	δ-T	δ-T-3
Corn	11.2	tr	5.0		60.2	tr	1.8	
Cottonseed	38.9				38.7			
Palm	25.6	14.3		3.2	31.6	28.6	7.0	6.9
Rapeseed	18.4				38.0		1.2	
Sunflower	48.7	5			5.1		0.8	
Soya	10.1				59.3		26.4	
Wheat germ	133.0	2.6	71.0	18.1	26.0		27.1	

tr: Traces; T: tocopherol; T-3: tocotrienol.

β-tocopherol in wheat germ oil is quite high (Table 3.41), hence it serves as an indicator of that oil. The blending of soya oil with sunflower oil is detectable by an increase in the content of linolenic acid (cf. Table 14.19). However, it is possible to make a final conclusive decision about the presence and quantity of soya oil in sunflower oil only after an analysis of the composition of the tocopherols.

3.8.3.2 Analysis

Isolation of tocopherols is accompanied by losses due to oxidation. Therefore, the edible oil is dissolved in acetone at 20–25 °C in the presence of ascorbyl palmitate as an antioxidant. The major portion of triacylglycerols is separated by crystallization at −80 °C. Tocopherols remaining in solution are then analyzed by thin layer or gas chromatography (after silylation of phenolic HO-groups) or by HPLC (cf. Figure 3.40). UV spectrophotometry is also possible. However, the fluorometric method based on an older colorimetric procedure developed by *Emmerie* and *Engel* is even more sensitive. It involves reduction of the Fe (III) ion to Fe (II) by tocopherols and the reaction of the reduced iron with 2,2′-bipyridyl to form an intensive red colored complex.

3.8.4 Carotenoids

Carotenoids are polyene hydrocarbons biosynthesized from eight isoprene units (tetraterpenes) and, correspondingly, have a 40-C skeleton.

Table 3.42. Carotenoids in various food

Food	Concentration (ppm)[a]	Food	Concentration (ppm)[a]
Carrots	54	Peaches	27
Spinach	26–76	Apples	0.9–5.4
Tomatoes	51	Peas	3–7
Apricots	35	Lemons	2–3

[a] On dry weight basis.

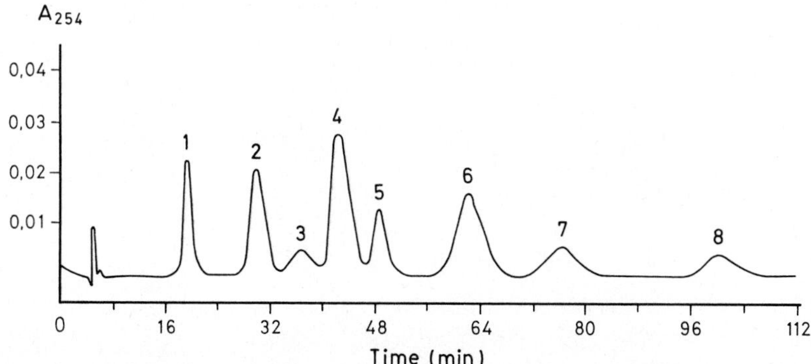

Fig. 3.40. Tocopherol and tocotrienol analysis by HPLC (After *J. F. Cavins* and *G. E. Inglett*, 1974). 1 α-Tocopherol, 2 α-tocotrienol, 3 β-tocopherol, 4 γ-tocopherol, 5 β-tocotrienol, 6 γ-tocotrienol, 7 δ-tocopherol, and 8 δ-tocotrienol

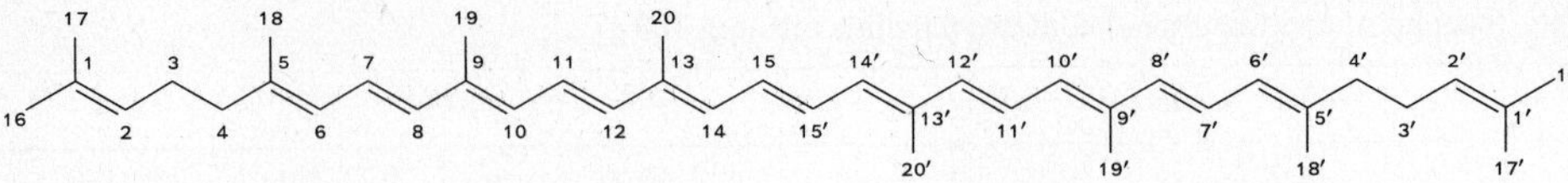

(3.97 d)

They provide the intensive yellow, orange or red color to a great number of foods of plant origin (Table 3.42). They are synthesized only by plants (see a textbook of biochemistry). However, they reach by feed (pasture, fodder) animal tissues, which has the ability to absorb, modify and deposit these compounds.

A well known example is the chicken egg yolk, which is colored by carotenoids. The carotenoids in green plants are masked by chlorophyll. When the latter is degraded, the presence of carotenoids is readily revealed (e.g. the green pepper becomes red after ripening).

3.8.4.1 Chemical Structure, Occurrence

Other carotenoids are derived by hydrogenation, dehydrogenation and/or cyclization of the basic structure of the 40-C carotenoids (cf. Formula 3.97 d). The latter reaction can occur at one or both end groups. The differences in 9-C end groups are denoted by Greek letters (cf. Formula 3.97 e).

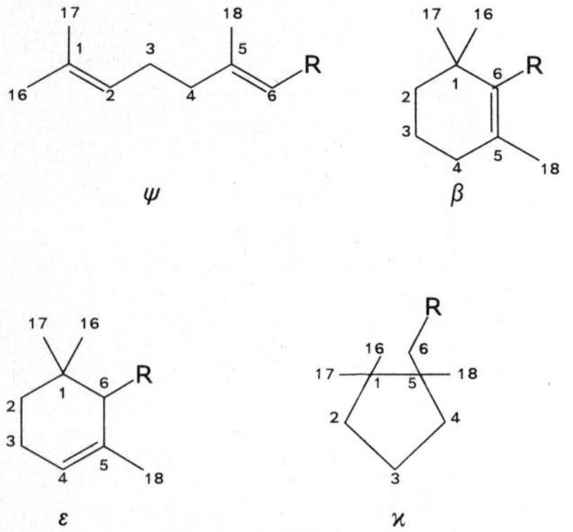

(3.97 e)

A semisystematic nomenclature used at times has two Greek letters as a prefix for the generic name "carotene", denoting the structure of both C_9-end groups (e.g. formulas III, IV, VI or X; cf. Formulas 3.98 c, 3.98 d, 3.99 a and 3.103, respectively). Designations such as α-, β- or γ-carotene are common names.

Carotenoids are divided into two main classes: carotenes and xanthophylls. In contrast to carotenes, which are pure polyene hydrocarbons, xanthophylls contain oxygen in the form of hydroxy, epoxy or oxo groups. Some carotenoids of importance to food are presented in the following sections.

3.8.4.1.1 Carotenes

Acyclic or aliphatic carotenes

Carotenes I, II and III (cf. Formulas 3.98 a–c) are intermediary or precursor compounds which, in biosynthesis after repeated dehydrogenizations, provide lycopene (IV; see a textbook of biochemistry). Lycopene is the red color of the tomato (and also of wild rose hips). In yellow tomato cultivars, lycopene precursors are present, probably together with β-carotene (Table 3.43).

Table 3.43. Carotenes (ppm) in some tomato cultivars

Cultivar	Phytoene (I)	Phytofluene (II)	β-Carotene (VII)	ξ-Carotene (III)	γ-Carotene (V)	Lycopene (IV)
Campbell	24.4	2.1	1.4	0	1.1	43.8
Ace Yellow	10.0	0.2	trace	0	0	0
High Beta	32.5	1.7	35.6	0	0	0
Jubilee	68.6	9.1	0	12.1	4.3	5.1

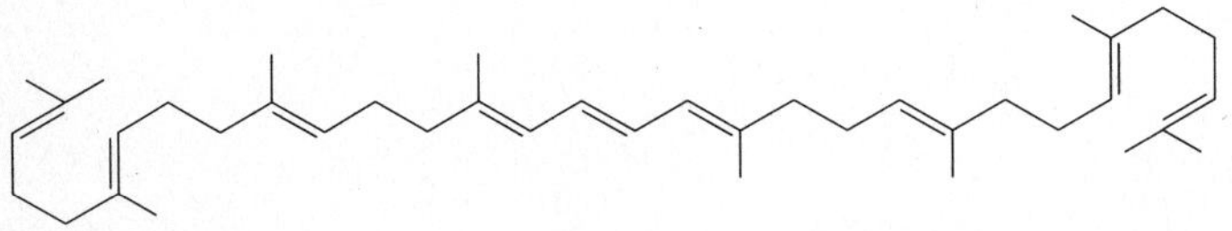

(3.98 a)

Phytoene (I)

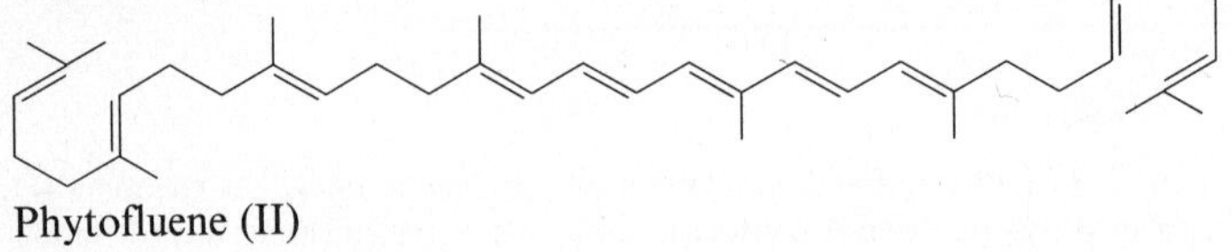

(3.98 b)

Phytofluene (II)

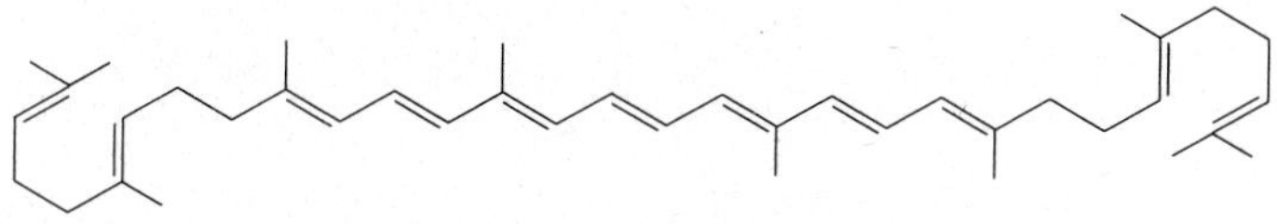

(3.98 c)

ξ-carotene (7,8,7′,8′-tetrahydro-ψ,ψ-carotene) (III)

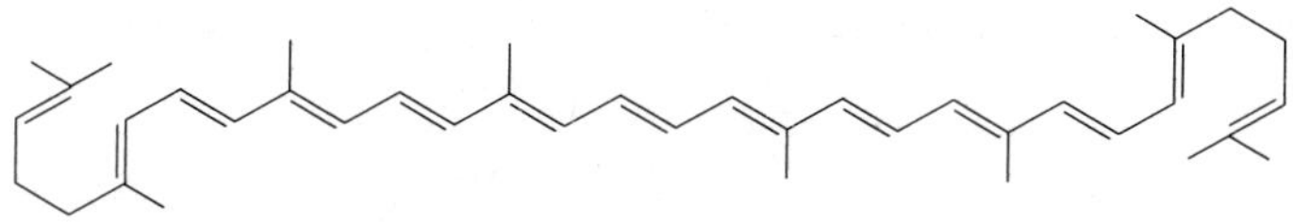

(3.98 d)

Lycopene (ψ,ψ-carotene) (IV)

Monocyclic carotenes

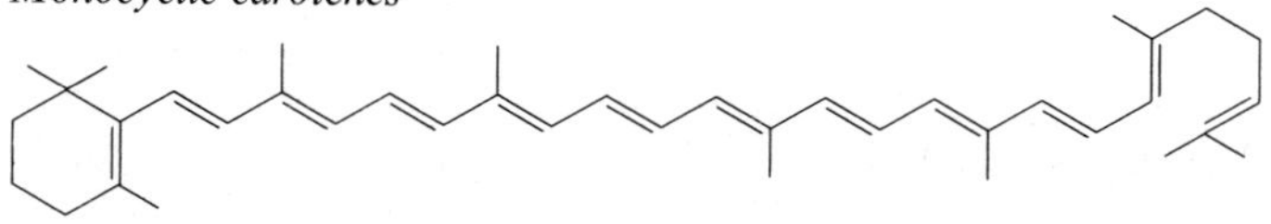

(3.99 a)

γ-carotene (V)

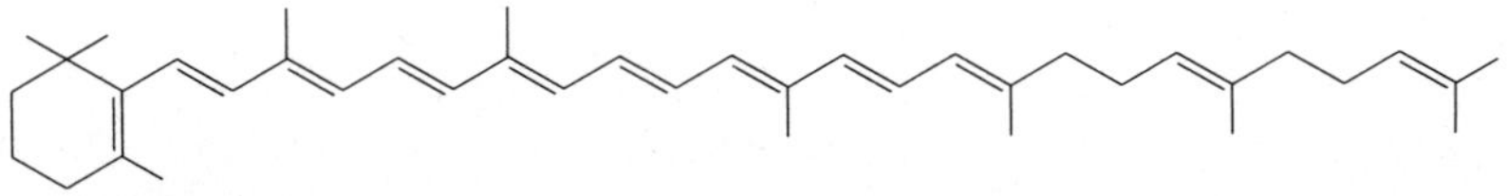

(3.99 b)

β-carotene (VI)

Bicyclic carotenes

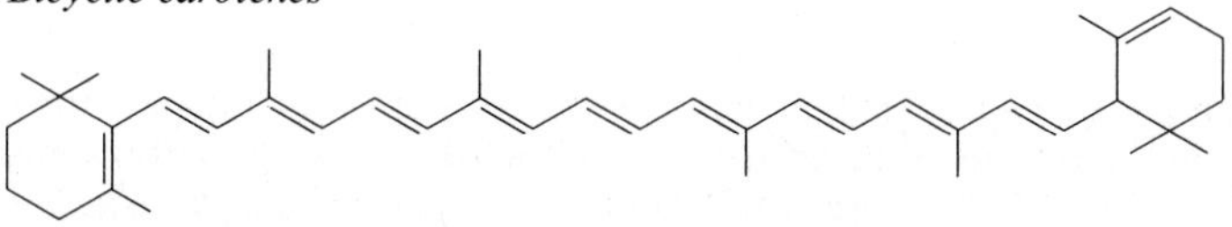

(3.100 a)

α-carotene (β,ε-carotene) (VI)

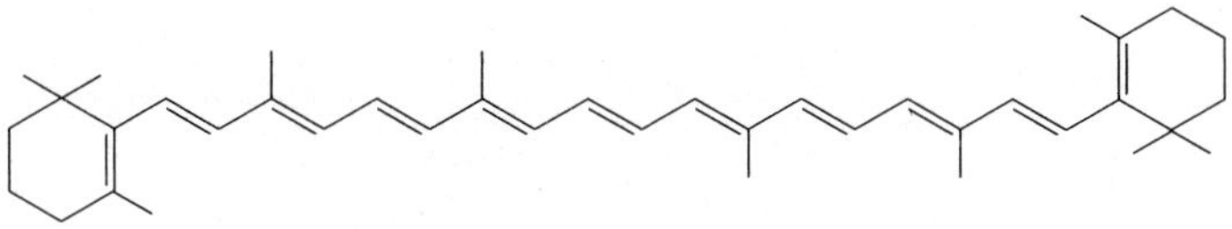

(3.100 b)

β-carotene (β,β-carotene) (VII)

The importance of β-carotene as provitamin A is covered under 6.2.1.

3.8.4.1.2 Xanthophylls
Hydroxy compounds

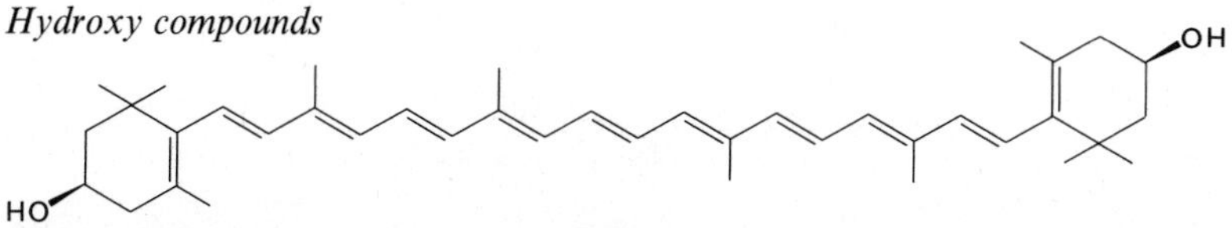

OH

HO

(3.101)

Zeaxanthin (β,β-carotene-3,3′-diol) (VIII)

This xanthophyll is present in corn *(Zea mays)*.

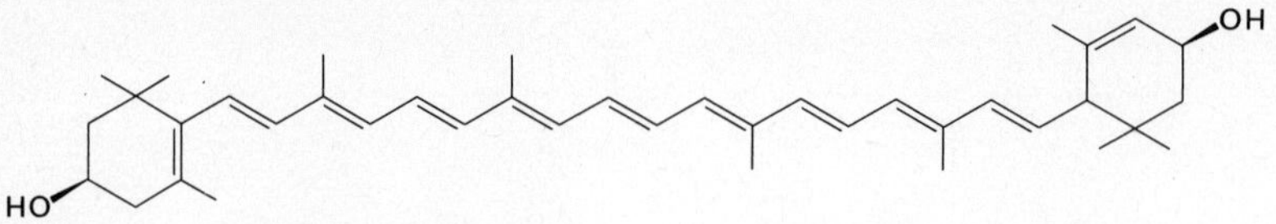

(3.102)

Lutein (β,ε-carotene-3,3′-diol (IX)

This xanthophyll occurs in green leaves and in egg yolk.

Keto compounds

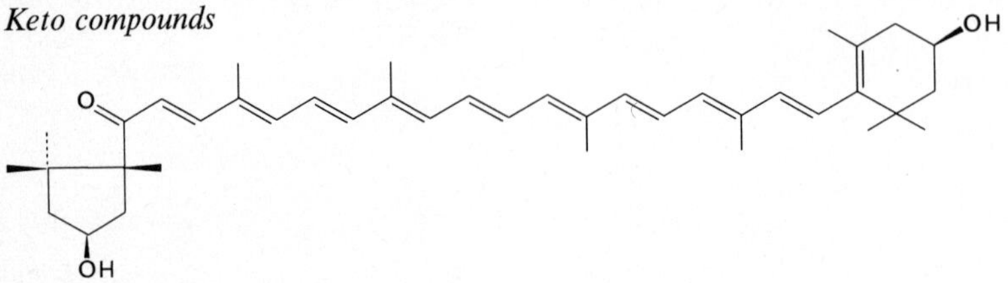

(3.103)

Capsanthin (3,3′-dihydroxy-β,ϰ-carotene-6′-one) (X)

This xanthophyll is the major carotene of paprika peppers.

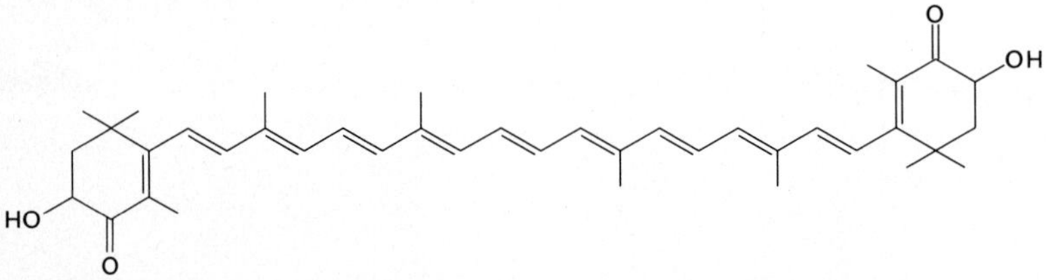

(3.104)

Astaxanthin (XI)

Astaxanthin is present in crab and lobster shells and, in combination with proteins, provides three blue hues (α-, β- and γ-crustacyanin) and one yellow pigment. During the cooking of crabs and lobsters, the red astaxanthin is released from a green carotenoid-protein complex. Astaxanthin usually occurs in lobster shell as an ester, e.g. dipalmitic ester.

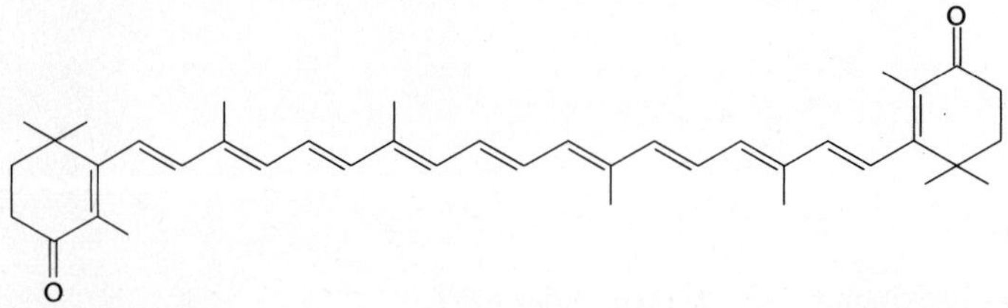

(3.105)

Canthaxanthin (XII)

This xanthophyll is used as a food colorant (cf. 3.8.4.5).

Epoxy compounds

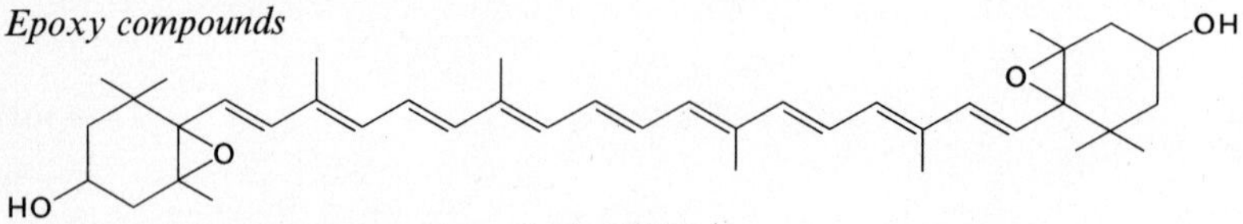

(3.106)

Violaxanthin (zeaxanthin-diepoxide) (XIII)

Violaxanthin is present in orange juice (cf. Table 3.44) and it also occurs in green leaves.

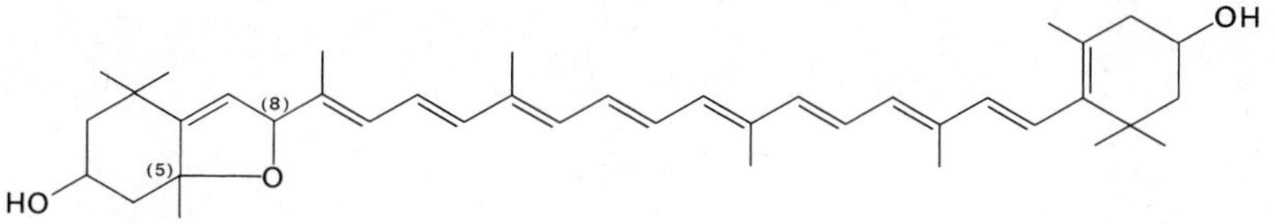

(3.107)

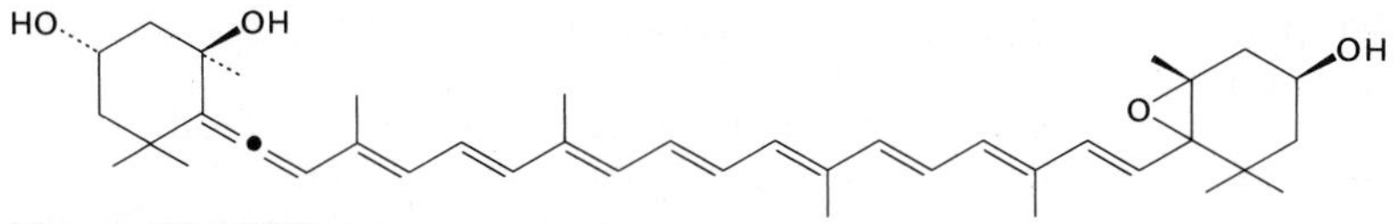

(3.107a)

Neoxanthin (XX)

Mutatoxanthin (5,8-epoxy-5,8-dihydro-β,β-carotene-3,3′-diol) (XVI)

This epoxy carotenoid is present in oranges (cf. Table 3.44).

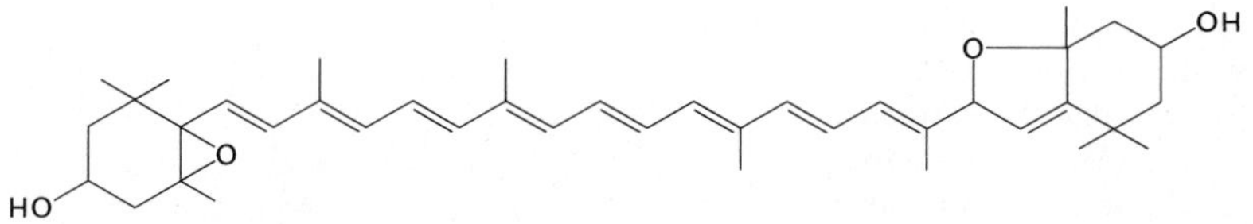

(3.108)

Luteoxanthin (XIV)

Luteoxanthin is the major carotenoid of oranges (cf. Table 3.44).

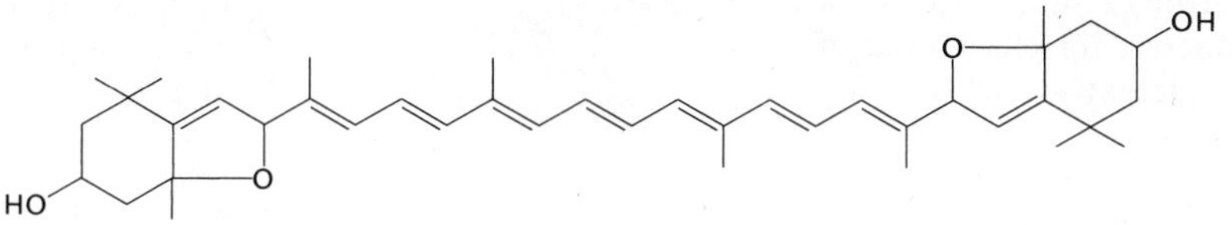

(3.109)

Auroxanthin (XV)

This carotenoid is a furanoid rearrangement product of violaxanthin. It is also a constituent of oranges (cf. Table 3.44) and is the main color of the yellow flowers of *Viola tricolor.*

Dicarboxylic acids and esters

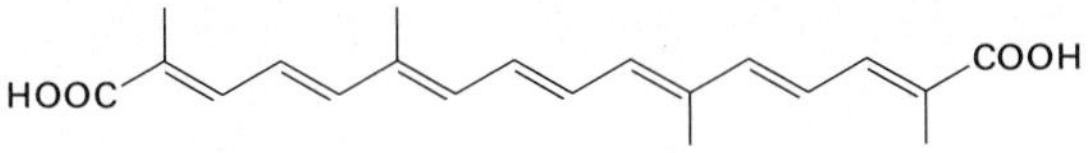

(3.110)

Crocetin (XVII)

This carboxylic acid carotenoid is the yellow pigment of saffron. It occurs in plants as a diester, i.e. glycoside with the disaccharide gentiobiose. The diester, called crocin, is therefore water-soluble.

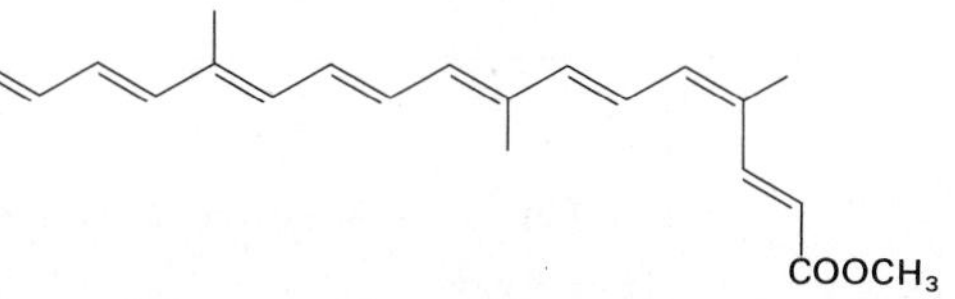

(3.111)

Bixin (XVIII)

Bixin is the main pigment of annato extract. Annato originates from the West Indies and the pigment is isolated from the seed pulp of the tropical bush *Bixa orellana.* Bixin is the monomethyl ester of norbixin, a dicarboxylic acid homologous to crocetin.

(3.112)

β-apo-8′-carotenal*

Carotenoids are, as a rule, present in plants as a complex mixture. For example, the orange has more than 50 well characterized compounds, of which only those that exceed 5% of the total carotenoids are presented in Table 3.44.

3.8.4.2 Physical Properties

Carotenoids are very soluble in apolar solvents, including edible fats and oils, but they are not soluble in water. Hence, they are denoted "lipochromes". Carotenoids are readily extracted from plant sources with petroleum ether, ether or benzene. Ethanol and acetone are also suitable solvents.

The color of carotenoids is the result of the presence of a conjugated double bond system in the molecules. The electron excitation spectra of such systems are of interest for elucidation of their structure and for qualitative and quantitative analyses.

Carotenoids show three distinct maxima in the visible spectrum, with wavelength position dependent on the number of conjugated double bonds (Table 3.45). The fine structure of the spectrum is better distinguished with acyclic lycopene (IV) than bicyclic β-carotene, since the latter is no longer a fully planar molecule due to the interferences of methyl groups positioned on the rings and also to the polyenic chain. Such steric effects prevent the total overlapping of π orbitals; consequently, a hypsochromic shift (a shift to a longer wavelength) is observed for the major absorption bands (Figure 3.41,a).

Oxo groups in conjugation with the polyene system shift the major absorption bands to shorter wavelengths (a bathochromic effect) with a si-

Table 3.44. Major carotenoid components in orange juice

Carotenoid	As percent of total carotenoids
Phytoene (I)	13
ξ-Carotene (III)	5.4
Cryptoxanthin (3-Hydroxy-β-carotene)	5.3
Antheraxanthin (5,6-Epoxyzeaxanthin)	5.8
Mutatoxanthin (XVI)	6.2
Violaxanthin (XIII)	7.4
Luteoxanthin	17.0
Auroxanthin	12.0

* The prefix "apo" indicates a compound derived from a carotenoid by removing part of its structure.

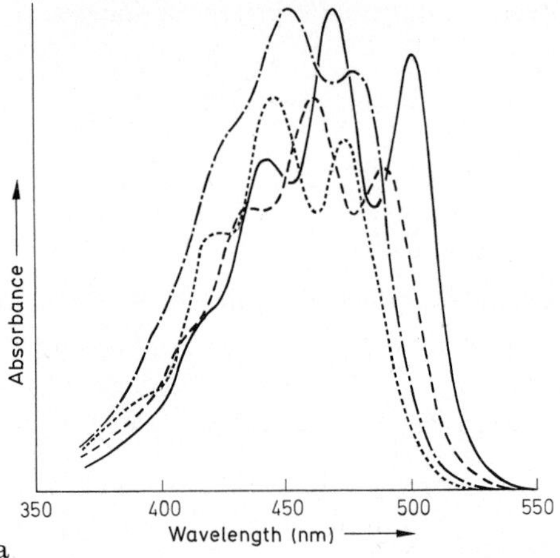

a

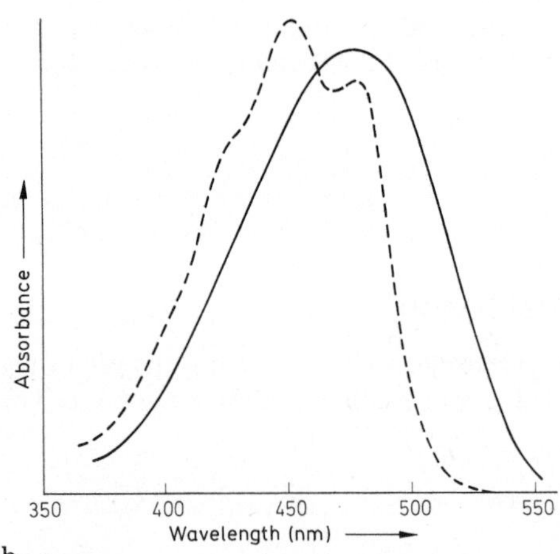

b

Fig. 3.41. Electron excitation spectra of carotenoids. (After *O. Isler*, 1971)
a —— Lycopene (IV), ---- γ-carotene (V), ······ α-carotene (VI), — · — · — · β-carotene (VII); **b** Canthaxanthin (XII) before —— and after ---- oxo groups reduction with $NaBH_4$

Table 3.45. Absorption wavelength maxima for some carotenoids

Compound	Conjugated double bonds	Wavelength, nm (petroleum ether)		
A. Effect of the number of conjugated double bonds				
Phytoene (I)	3	275	285	296
Phytofluene (II)	5	331	348	367
ξ-Carotene (III)	7	378	400	425
Neurosporene	9	416	440	470
Lycopene (IV)	11	446	472	505
B. Effect of the ring structure				
γ-Carotene (V)	11	431	462	495
β-Carotene (VII)	11	425[a]	451	483

[a] Maximum absorption wavelength is not unequivocal (cf. Fig. 3.41).

multaneous enhancement of the fine structure of the spectrum (Figure 3.41,b). The hydroxyl groups in the molecule have no influence on the spectra.

A change of solvent system alters the position of absorption maxima. For example, replacing hexane with ethanol leads to a bathochromic shift.

The carotenoids in nature and, thus, in food are all of the trans-double bond configuration. When a mono-cis- or di-cis-compound occurs, the prefix "neo" is used. When one bond of all trans-double bonds is rearranged into the cis-configuration, there is a small shift in absorption maxima with a new minor "cis band"shoulder on the side of the shorter wavelength.

3.8.4.3 Chemical Properties

Carotenoids are highly sensitive to oxygen and light. When these factors are excluded, carotenoids in food are stable even at high temperatures. Their degradation is, however, accelerated by intermediary radicals occurring in food due to lipid peroxidation (cf. 3.7.2). The cooxidation phenomena in the presence of lipoxygenase (cf. 3.7.2.2) are particularly visible.

Changes in extent of coloration often observed with dehydrated paprika and tomato products are related to oxidative degradation of carotenoids. Such discoloration is desirable in flours (flour bleaching; cf. 15.4.1.4.3).

The color change in paprika from red to brown, as an example, is due partly to a slow *Maillard* reaction, but primarily to oxidation of capsanthin (Figure 3.42) and to some as yet unclear polymerization reactions.

3.8.4.4 Precursors of Aroma Compounds

Aroma compounds are formed during the oxidative degradation of carotenoids. Such compounds, their precursors and the foods in which they occur are listed in Table 3.46. Among them are β-ionone, significant for raspberry aroma and with a low threshold value (14 ppb; water). The formation of β-ionone in dehydrated carrots cause the undesired off-flavor "odor of violets". Unsaturated ketones derived from degradation of carotenoids are readily further oxidized. For example, dihydroactinidiolide (Table 3.47) is a decomposition product of β-ionone, whereas 1,2-dihydro-1,1,6-trimethyl-naphthalene (Table 3.47) is derived from β-ionone epoxide or from 3-oxo-β-ionone.

Other volatile compounds presented in Table 3.47 are derived from carotenoid metabolism by

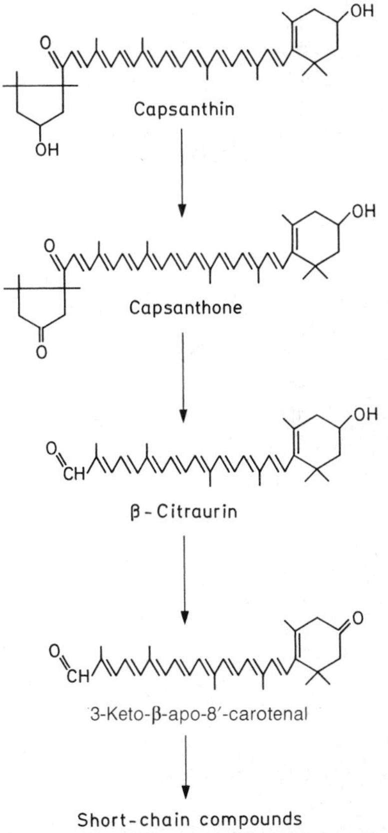

Fig. 3.42. Oxidative degradation of capsanthin during storage of paprika. (After *T. Philip* and *F.J. Francis*, 1971)

Table 3.46. Major aroma compounds formed in oxidative degradation of carotenoids

Precursor	Aroma compound	Occurrence
Lycopene (I)	6-Methyl-5-hepten-2-one	Tomato
	6,10-Dimethyl-5,9-undecadien-2-one	Tomato
	6,10,14-Trimethyl-5,9,13-pentadecadien-2-one	Tomato
Dehydrolycopene	6-Methyl-3,5-heptadien-2-one	Tomato
β-Carotene (VII)	β-Ionone[a]	
β-Carotene-5,6-epoxide	β-Ionone-5,6-epoxide	Tomato, raspberry, blackberry, passion fruit, black tea
α-Carotene (VI)	α-Ionone[a]	

[a] Odor threshold values are given in Table 5.1.

pathways which are not yet clarified. They are also related structurally to ionones.

Damascone and damascenone contribute significantly to the tobacco aroma. Theaspirone (15 μg/kg green tea leaves) is recommended as an aroma enhancer for black fermented tea.

3.8.4.5 Use of Carotenoids in Food Processing

Carotenoids are utilized as food pigments to color margarine, ice creams, various cheese products, beverages, sauces, meat, and confectionery and bakery products. Plant extracts and/or individual compounds are used.

3.8.4.5.1 Plant extracts

Annato is a yellow oil or aqueous alkaline extract of fruit pulp of *Raku* or *Orleans* shrubs or brushwood *(Bixa orellana)*. The major pigments of annato are bixin (XVIII) and norbixin, which both give dicarboxylic acids upon hydrolysis.

Table 3.47. Volatile compounds with a structure related to α- or β-ionone

Compound	Occurrence
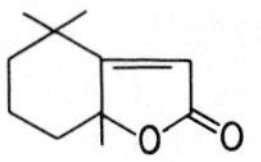 Dihydroactinidiolide	Black (fermented) tea, tomato
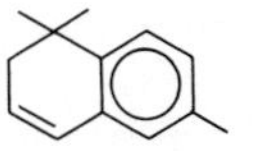 1,2-Dihydro-1,1,6-trimethylnaphthalene	Peaches, strawberries, red wine
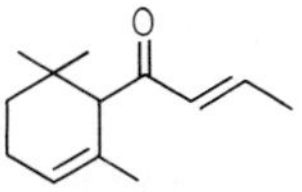 α-Damascone	Tea leaves
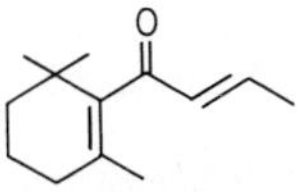 β-Damascone	Tea leaves
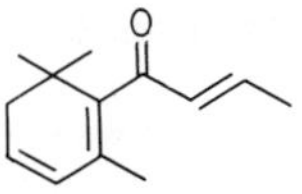 β-Damascenone[a]	Tea leaves, raspberry, cooked apple, grapes, beer, coffee
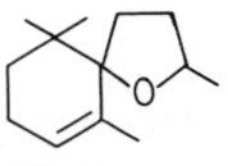 Theaspiran	Raspberry, passion fruit, tea leaves
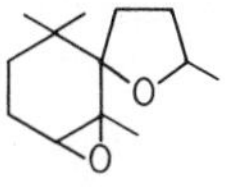 Theaspiran-4,5-epoxide	Black (fermented) tea
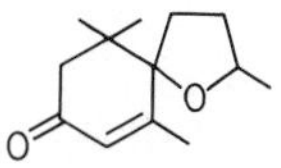 Theaspirone	Tea leaves, passion fruit

[a] Odor threshold value is given in Table 5.1.

Oleoresin from paprika is a red, oil extract containing about 50 different pigments. The aqueous extract of saffron (more accurately, from the pistils of the flower *Crocus sativus*) contains crocin (XVII) as its main constituent. It is used for coloring beverages and bakery products.

Raw, unrefined palm oil contains 0.05–0.2% carotenoids, with α- and β-carotenes, in a ratio of 2:3, as the main constituents. It is of particular use as a colorant for margarine.

3.8.4.5.2 Individual compounds

β-Carotene (VII), canthaxanthin (XII), β-apo-8′-carotenal (XIX) and the carboxylic acid ethyl ester derived from the latter are synthesized for use as colorants for edible fats and oils. These carotenoids, in combination with surface-active agents, are available as microemulsions (cf. 8.15.1) for coloring foods with a high moisture content.

3.8.4.6 Analysis

The total lipids are first extracted from food with isopropanol/petroleum ether (3:1 v/v) or by acetone. Alkaline hydrolysis follows, removing the extracted acyl-lipids and the carotenoids from the unsaponifiable fraction. This is the usual procedure when alkali-stable carotenoids are analyzed. Although carotenoids are generally alkali-stable, there are exceptions. When alkali-labile carotenoids are present, the acyl lipids are removed instead by a saponification method using column chromatography as the separation technique.

A preliminary separation of the lipids into classes of carotenoids is carried out when a complex mixture of carotenoids is present. For example, column chromatography is used with Al_2O_3 as an adsorbent (Table 3.48). Additional separation into classes or individual compounds is achieved by thin layer chromatography. Thin layers made of MgO or $ZnCO_3$ are suitable. These adsorbent layers permit separation of carotenoids into

Table 3.48. Separation of carotenoids into classes by column chromatography using neutral aluminum oxide (6% moisture) as an adsorbent

P: Petroleum ether, D: diethyl ether

Elution with	Carotenoids in effluent
100% P	Carotenes
5% D in P	Carotene-epoxides
20–59% D in P	Monohydroxy-carotenoids
100% D	Dihydroxy-carotenoids
5% Ethanol in D	Dihydroxy-epoxy-carotenoids

classes by the number, position and configuration of double bonds.

Identification of carotenoids is based on chromatographic data and on electron excitation spectra (cf. 3.8.4.2), supplemented when necessary with tests specific to each group. For example, a hypsochromic effect after addition of $NaBH_4$ suggests the presence of oxo or aldehyde groups, whereas the same effect after addition of HCl suggests the presence of a 5,6-epoxy group. The latter "blue hue shift" is based on a rearrangement reaction:

(3.112a)

(5,6-Epoxide) (5,8-Epoxide)

Such rearrangements can also occur during chromatographic separations of carotenoids on silicic acid. Hence, this adsorbent is a potential source of artifacts.

Epoxy group rearrangement in the carotenoid molecule can also occur during storage of food with a low pH, such as orange juice.

Elucidation of the structure of carotenoids requires, in addition to VIS/UV spectrophotometry, supplemental data from mass spectrometry and IR spectroscopy. Carotenoids are determined photometrically with high sensitivity based on their high molar absorbancy coefficients. This is often used for simultaneous qualitative and quantitative analysis. New separation methods based on high performance liquid chromatography have also been proven advantageous for the qualitative and quantitative analysis of carotenoids present as a highly complex mixture in food.

3.9 Literature

Axelrod, B.: Lipoxygenases. In: Food related enzymes (Ed.: Whitaker, J. R.), p. 324, Advances in Chemistry Series 136, American Chemical Society: Washington, D.C. 1974

Badings, H. T.: Cold storage defects in butter and their relation to the autoxidation of unsaturated fatty acids. Ned. Melk Zuiveltijdschr. *24,* 147 (1970)

Bergelson, L. D.: Diol lipids. New types of naturally occuring lipid substances. Fette Seifen Anstrichm. *75,* 89 (1973)

Biermann, U., Wittmann, A., Grosch, W.: Vorkommen bitterer Hydroxyfettsäuren in Hafer und Weizen. Fette Seifen Anstrichm. *82,* 236 (1980)

Bild, G.S., Ramadoss, C.S., Axelrod, B.: Effect of substrate polarity on the acitvity of soybean lipoxygenase isoenzymes. Lipids *12,* 732 (1977)

Britton, G., Goodwin, T.W.: Biosynthesis of carotenoids. In: Methods in enzymology (Eds.: Colowick, S.P., Kaplan, N.O.), Vol. XVIII, part C, p. 654, Academic Press: New York–London. 1971

Brockerhoff, H.: Lipolytic enzymes. In: Food related enzymes (Ed.: Whitaker, J.R.), p. 131, Advances in Chemistry Series 136, American Chemical Society: Washington, D.C. 1974

Burton, G.W., Ingold, K.U.: Autoxidation of biological molecules. 1. The antioxidant activity of vitamin E and related chain-breaking phenolic antioxidants in vitro. J. Am. Chem. Soc. *103,* 6472 (1981)

Cavins, J.F., Inglett, G.E.: High-resolution liquid chromatography of vitamin E isomers. Cereal Chem. *51,* 605 (1974)

Chan, H.W.-S., Levett, G.: Autoxidation of methyl linoleate. Separation and analysis of isomeric mixtures of methyl linoleate hydroperoxides and methyl hydroxylinoleates. Lipids *12,* 99 (1977)

Chang, S.S., Peterson, R.J., Ho, C.-T.: Chemistry of deep fat fried flavor. In: Lipids as a source of flavor (Ed.: Supran, M.K.), p. 18, ACS Symposium Series 75, American Chemical Society: Washington, D.C. 1978

Christie, W.W.: Lipid analysis. 2nd Ed. Pergamon Press: Oxford. 1982

Dugan, L.R., jr.: Lipids. In: Principles of food science (Ed.: Fennema, O.R.), part I, p. 139, Marcel Dekker, Inc.: New York–Basel. 1976

Emken, E.A.: Commercial and potential utilization of lipoxygenase. J. Am. Oil Chem. Soc. *55,* 416 (1978)

Eriksson, C.E.: Enzymic and non-enzymic lipid degradation in foods. In: Industrial aspects of biochemistry (Ed.: Spencer, B.), Federation of European Biochemical Societies, Vol. 30, p. 865, North Holland/American Elsevier: Amsterdam. 1974

Fideli, E.: Lipids of olives. Prog. Chem. Fats Other Lipids *15,* 57 (1977)

Fiebig, H.-J.: HPLC-Trennung von Triglyceriden. Fette Seifen Anstrichm. *87,* 53 (1985)

Fincke, A.: Kakaobutter und Ersatzfette-Chemie und Analytik. Dtsch. Lebensm.Rundsch. *72,* 6 (1976)

Foote, C.S.: Photosensitized oxidation and singlet oxygen: Consequences in biological systems. In: Free radicals in biology (Ed.: Pryor, W.A.), Vol. II, p. 85, Academic Press: New York–San Francisco–London. 1976

Forss, D.A.: Odour and flavour compounds from lipids. Prog. Chem. Fats Other Lipids *13,* 177 (1973)

Frankel, E.N.: Lipid oxidation. Prog. Chem. Fats Other Lipids *19,* 1 (1980)

Frankel, E.N., Neff, W.E.: Formation of malonaldehyde from lipid oxidation products. Biochim. Biophys. Acta *754*, 264 (1983)

Galliard, T., Mercer, E.I. (Eds.): Recent advances in the chemistry and biochemistry of plant lipids. Academic Press: London–New York–San Francisco. 1975

Galliard, T., Phillips, D.R., Reynolds, J.: The formation of cis-3-nonenal, trans-2-nonenal and hexanal from linoleic acid hydroperoxide isomers by a hydroperoxide cleavage enzyme system in cucumber (Cucumis sativus) fruits. Biochim. Biophys. Acta *441*, 181 (1976)

Gardner, H.W.: Decomposition of linoleic acid hydroperoxides. Enzymic reactions compared with nonenzymic. J. Agric. Food Chem. *23*, 129 (1975)

Gardner, H.W.: Lipid hydroperoxide reactivity with proteins and amino acids: A review. J. Agric. Food Chem. *27*, 220 (1979)

Gardner, H.W., Plattner, R.D., Weisleder, D.: The epoxyallylic radical from homolysis and rearrangement of methyl linoleate hydroperoxide combines with the thiyl radical of N-acetylcysteine. Biochem. Biophys. Acta *834*, 65 (1985)

Grosch, W.: Ablauf und Analytik des oxydativen Fettverderbs. Z. Lebensm. Unters. Forsch. *157*, 70 (1975)

Grosch, W.: Reaktionen in Lebensmitteln pflanzlicher Herkunft, die durch das Enzym Lipoxygenase (EC 1.13.11.12) beschleunigt werden. Mitteilungsbl. GDCh Fachgruppe Lebensmittelchem. Gerichtl. Chem. *30*, 1 (1976)

Grosch, W.: Lipid degradation products and flavours. In: Food flavours (Eds.: Morton, I.D., MacLeod, A.J.), Part A, p. 325, Elsevier Publ. Co.: Amsterdam–Oxford–New York. 1982

Gunstone, F.D., Norris, F.A.: Lipids in foods. Pergamon Press: Oxford. 1983

Hitchcock, C., Nichols, B.W.: Plant lipid biochemistry. Academic Press: London–New York. 1971

Homberg, E.: Gaschromatographische Trennung pflanzlicher Sterine. Fette Seifen Anstrichm. *79*, 234 (1977)

Isler, O. (Ed.): Carotenoids. Birkhäuser Verlag: Basel–Stuttgart. 1971

Jeong, T.M., Itoh, T., Tamura, T., Matsumoto, T.: Analysis of methylsterol fractions from twenty vegetable oils. Lipids *10*, 634 (1975)

Johnson, A.R., Davenport, J.B.: Biochemistry and methodology of lipids. John Wiley and Sons: New York. 1971

Keppler, J.G.: Twenty-five years of flavor research in a food industry. J. Am. Oil Chem. Soc. *54*, 474 (1977)

Kindl, H., Wöber, G.: Biochemie der Pflanzen. Springer-Verlag: Berlin–Heidelberg–New York. 1975

Kinsella, J.E., Hwang, D.H.: Enzymes of penicillium roqueforti involved in the biosynthesis of cheese flavour. Crit. Rev. Food Sci. Nutr. *8*, 191 (1976)

Korycka-Dahl, M.B., Richardson, T.: Activated oxygen species and oxidation of food constituents. Crit. Rev. Food Sci. Nutr. *10*, 209 (1978)

Koskas, J.P., Cillard, J., Cillard, P.: Autoxidation of linoleic acid and behavior of its hydroperoxides with and without tocopherols. J. Am. Oil Chem. Soc. *61*, 1466 (1984)

Labuza, T.P.: Kinetics of lipid oxidation in foods. Crit. Rev. Food Technol. *2*, 355 (1971)

Lillard, D.A., Day, E.A.: Degradation of monocarbonyls from autoxidizing lipids. J. Am. Oil Chem. Soc. *41*, 549 (1964)

Litchfield, C.: Analysis of tryglycerides. Academic Press: New York–London. 1972

Mahadevan, V.: Fatty alcohols: Chemistry and metabolism. Prog. Chem. Fats Other Lipids *15*, 255 (1977)

Mead, J.F., Howton, D.R., Nevenzel, J.C.: Fatty acids, long-chain alcohols and waxes. In: Comprehensive biochemistry (Eds.: Florkin, M., Stotz, E.H.), Vol. 6, p. 1, Elsevier Publ. Co.: Amsterdam–London–New York. 1965

Meijboom, P.W., Jongenotter, G.A.: Flavor perceptibility of straight chain, unsaturated aldehydes as a function of couble-bond position and geometry. J. Am. Oil Chem. Soc. *58*, 680 (1981)

Nawar, W.W., Bradley, S.J., Lamanno, S.S., Richardson, G.G., Whiteman, R.C.: Volatiles from frying fats: A comparative study. In: Lipids as a source of flavor (Ed.: Supran, M.K.), p. 42, ACS Symposium Series 75, American Chemical Society: Washington, D.C. 1978

Ohloff, G.: Fette als Aromavorstufen. In: Fette als funktionelle Bestandteile von Lebensmitteln (Hrsg.: Solms, J.), S. 119, Forster Verlag AG: Zürich. 1973

Ohloff, G.: Recent developments in the field of naturally-occuring aroma components. Fortschr. Chem. Org. Naturst. *35*, 431 (1978)

Perkins, E.G. (Ed.): Analysis of lipids and lipoproteins. American Oil Chemists' Society: Champaign, Ill. 1975

Philip, T., Francis, F.J.: Oxidation of capsanthin. J. Food Sci. *36*, 96 (1971)

Plattner, R.D.: High performance liquid chromatography of triglycerides: Controlling selectivity with reverse phase columns. J. Am. Oil Chem. Soc. *58*, 638 (1981)

Porter, N.A., Lehman, L.S., Weber, B.A., Smith, K.J.: Unified mechanism for polyunsaturated fatty acid autoxidation. Competition of peroxy radical hydrogen atom abstraction, b-scission, and cyclization. J. Am. Chem. Soc. *103*, 6447 (1981)

Pryde, E.H. (Ed.): Fatty acids. American Oil Chemists' Society: Champaign, Ill. 1979

Pryor, W.A., Stanley, J.P., Blair, E.: Autoxidation of polyunsaturated fatty acids. II. A suggested mechanism for the formation of TBA-reactive materials from prostaglandin – like endoperoxides. Lipids *11*, 370 (1976)

Schieberle, P., Haslbeck, F.: Laskawy, G., Grosch, W.: Comparison of sensitizers in the photooxidation of unsaturated fatty acids and their methyl esters. Z. Lebensm. Unters. Forsch. *179,* 93 (1984)

Seher, H., Vogel, H.: Untersuchung von Steringemischen. V. Die Sterine in Sonnenblumen und anderen Pflanzenölen. Fette Seifen Anstrichm. *78,* 301 (1976)

Seidel, D.: Plasmalipoproteine – Funktion und Charakterisierung. In: Fettstoffwechselstörungen (Hrsg.: Schettler, G.), S, 24, Georg Thieme Verlag: Stuttgart. 1971

Sessa, D.J., Rackis, J.J.: Lipid – derived flavors of legume protein products. J. Am. Oil Chem. Soc. *54,* 468 (1977)

Sherwin, E.R.: Oxidation and antioxidants in fat and oil processing. J. Am. Oil Chem. Soc. *55,* 809 (1978)

Simic, M.G., Karel, M. (Eds.): Autoxidation in food and biological systems. Plenum Press: New York–London. 1980

Slower, H.L.: Tocopherols in foods and fats. Lipids *6,* 291 (1971)

Smith, L.L.: Cholesterol autoxidation. Plenum Press: New York. 1981

Smouse, T.H.: A review of soy bean oil reversion flavor. J. Am. Oil Chem. Soc. *56,* 747 A (1979)

Thiele, O.W.: Lipide, Isoprenoide mit Steroiden. Georg Thieme Verlag: Stuttgart. 1979

Veldink, G.A., Vliegenthart, J.F.G., Boldingh, J.: Plant lipoxygenases. Prog. Chem. Fats Other Lipids *15,* 131 (1977)

Wachs, W.: Fette und Lipoide (Lipide), Wachse, Harze. In: Handbuch der Lebensmittelchemie, Bd. I (Hrsg.: Schormüller, J.), S. 308, Springer-Verlag: Berlin–Heidelberg–New York. 1965

Warner, K., Evans, C.D., List, G.R., Dupuy, H.P., Wadsworth, J.I., Goheen, G.E.: Flavor score correlation with pentanal and hexanal contents of vegetable oil. J. Am. Oil Chem. Soc. *55,* 252 (1978)

Wurzenberger, M., Grosch, W.: Stereochemistry of the cleavage of the 10-hydroperoxide isomer of linoleic acid to 1-octen-3-ol by a hydroperoxide lyase from mushrooms (Psalliota bispora). Biochim. Biophys. Acta *795,* 163 (1984)

4 Carbohydrates

4.1 Foreword

Carbohydrates are the most widely distributed and abundant organic compounds on earth. They have a central role in the metabolisms of animals and plants. Carbohydrate biosynthesis by green plants from carbon dioxide and water in the presence of light energy, i. e. photosynthesis, supports the existence of all other organisms.
Carbohydrates are a basic food, accounting for a large portion of total nutrient intake. Even the nondigestible carbohydrates, acting as ballast material, are of importance in a balanced daily nutrition. Other important functions in food are fulfilled by carbohydrates. They act for instance as sweetening, gel- or paste-forming and thickening agents, stabilizers and are also precursors for aroma and coloring substances generated within the food by a series of reactions during handling and processing.
The term carbohydrates goes back to times when it was thought that all compounds of this class were hydrates of carbon, on the basis of their empirical formula, e. g. glucose, $C_6H_{12}O_6$ ($6C + 6H_2O$). Meanwhile, many compounds were identified which deviated from this general formula, but retained common reactions and, hence, were also classed as carbohydrates. These are exemplified by deoxysugars, amino sugars and sugar carboxylic acids. Carbohydrates are commonly divided into monosaccharides, oligosaccharides and polysaccharides. Monosaccharides are polyhydroxy-aldehydes or -ketones, generally with an unbranched C-chain. Well known representatives are glucose, fructose and galactose. Oligosaccharides are carbohydrates which are obtained from monosaccharides by elimination of water, e. g. by the reaction:

$$n\ C_6H_{12}O_6 \xrightarrow{-(n-1)\ H_2O} C_{6n}H_{10n+2}O_{5n+1} \qquad (4.1)$$

Well known representatives of disaccharides are saccharose (sucrose), maltose and lactose, and of trisaccharides, raffinose, and tetrasaccharides, stachyose.

In polysaccharides, consisting of n monosaccharides, the number n is rather large. Hence, the properties of these high molecular weight polymers differ greatly from other carbohydrates. Unlike mono- or oligosaccharides, polysaccharides are in many cases insoluble or at best not readily soluble in water. They do not have a sweet taste and are essentially inert. Well known representatives are starch, cellulose and pectin.

4.2 Monosaccharides

4.2.1 Structure and Nomenclature

4.2.1.1 Nomenclature

Monosaccharides are polyhydroxy-aldehydes (aldoses), formally considered to be derived from glyceraldehyde, or polyhydroxyketones (ketoses), derived from dihydroxyacetone by inserting CHOH units into the carbon chains. The resultant compounds in the series of aldoses are denoted by the total number of carbons as trioses, for the starting glyceraldehyde, and tetroses, pentoses, hexoses, etc. The ketose series begins with the simplest ketose, dihydroxyacetone, a triulose, followed by tetruloses, pentuloses, hexuloses, etc. The position of the keto group is designated by a numerical prefix, e. g. 2-pentulose, 3-hexulose.
When a monosaccharide carries a second carbonyl group, it is denoted as a –dialdose (2 aldehyde groups), –osulose (aldehyde and keto groups) or –diulose (2 keto groups). Substitution of an HO-group by an H-atom gives rise to a deoxy sugar, and by an H_2N-group to amino-deoxy compounds (cf. Formula 4.2).
Monosaccharides can form cyclic five- and six-membered rings starting with tetroses and 2-pentuloses. These are intramolecular hemiacetals or lactols. With the exception of erythrose, monosaccharides are crystallized in these cyclic forms and, even in solution, there is a equilibrium between the open chain carbonyl form and cyclic

CHO | CH$_2$OH | CHO | CH$_2$OH | CHO
CHOH | CO | CHOH | CHOH | CHOH
CHOH | CHOH | CHOH | CO | CHOH
CH$_2$OH | CH$_2$OH | CHOH | CHOH | CHOH
 | | CHOH | CHOH | CHOH
 | | CH$_2$OH | CH$_2$OH | CHO

Tetrose, 2-Tetrulose, Hexose, 3-Hexulose, Hexo-dialdose

2-Hexos-ulose, 2,4-Hexo-diulose, 2-Desoxy-pentose, 2-Amino-2-desoxy-hexose (4.2)

hemiacetals, with the latter predominating. The tendency to cyclize is pronounced in simple ω-hydroxyaldehydes, and is even more pronounced in monosaccharides, as shown by ΔG°-values and equilibrium concentrations in 75% aqueous ethanol (cf. Formula 4.3).

4-Hydroxybutanal (11.4%) ⇌ ($\Delta G^0 = -1.2$ kcal/Mol) 2-Hydroxytetrahydrofuran (88.6%)

5-Hydroxypentanal (6.1%) ⇌ ($\Delta G^0 = -1.6$ kcal/Mol) 2-Hydroxytetrahydropyran (93.9%) (4.3)

Glucose (open-chain structure) (0.0026%) ⇌ ($\Delta G^0 = -6.3$ kcal/Mol) Glucose (pyranose form) (99.9974%; sum of all cyclic structures)

Lactols can be considered as tetrahydrofuran or tetrahydropyran derivatives, hence, they are also denoted as furanoses or pyranoses.

4.2.1.2 Configuration

Glyceraldehyde, a triose, has one chiral center, so it exists as an enantiomer pair, i.e. in D- and L-forms. It is possible by cyanhydrin synthesis to obtain from each enantiomer a pair of diastereomeric tetroses (with a new chiral center at carbon position 2):

Glyceraldehyde —HCN→ D-Erythronic acid nitrile + D-Threonic acid nitrile

—H⁺→ D-Erythronic acid lactone + D-Threonic acid lactone (4.4)

—NaHg/NaBH$_4$, pH 4–8→ D-Erythrose + D-Threose

Correspondingly, L-erythrose and L-threose are obtained from L-glyceraldehyde:

L-Erythrose L-Threose (4.5)

After repeated cyanhydrin reactions, four tetroses will provide a total of eight pentoses (each tetrose provides a pair of new diastereomers with one more chiral center), which can then yield sixteen stereoisomeric hexoses. The compounds derived from D-glyceraldehyde are designated as D-aldoses and those from L-glyceraldehyde as L-aldoses.

Figure 4.1 shows the formulas and names for D-aldoses using simplified *Fischer* projections. The occurrence of aldoses of importance in food is compiled in Table 4.1. Epimers are monosaccharides which differ in configuration at only one chiral C-atom. D-glucose and D-mannose are 2-epimers. D-glucose and D-galactose are 4-epimers.

The enantiomers D- and L-tetrulose, by formally inserting additional CHOH-groups between the

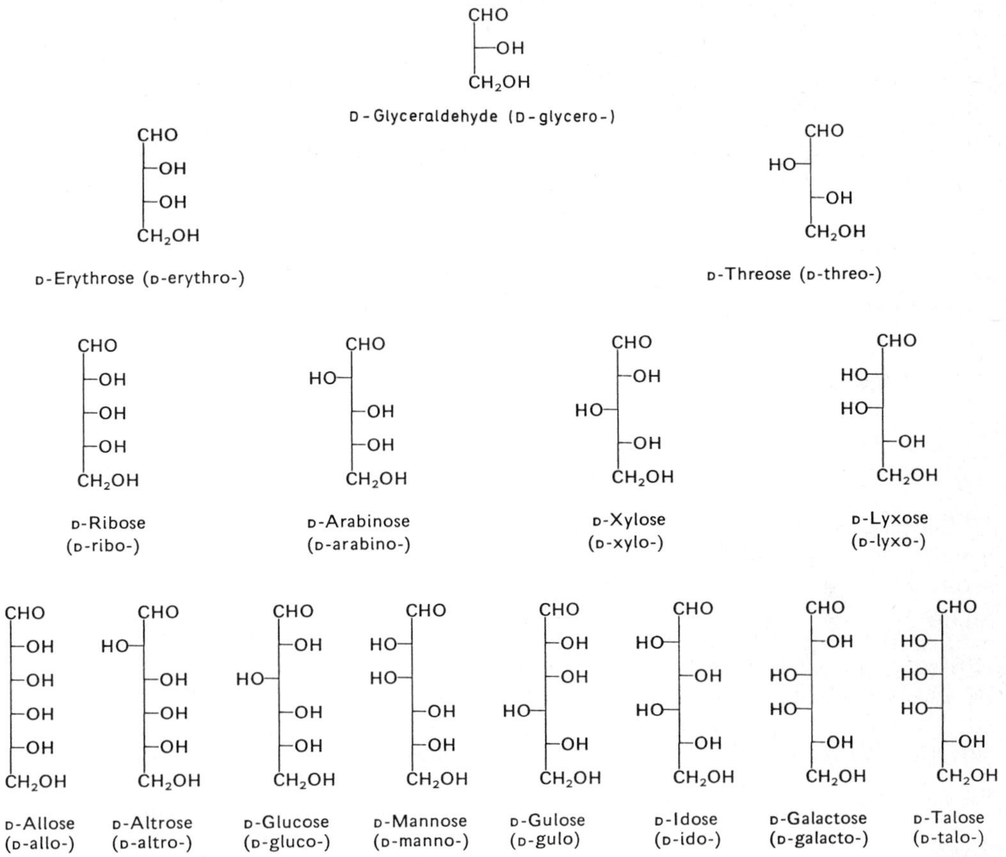

D-Glyceraldehyde (D-glycero-)

D-Erythrose (D-erythro-)   D-Threose (D-threo-)

D-Ribose (D-ribo-)   D-Arabinose (D-arabino-)   D-Xylose (D-xylo-)   D-Lyxose (D-lyxo-)

D-Allose (D-allo-)   D-Altrose (D-altro-)   D-Glucose (D-gluco-)   D-Mannose (D-manno-)   D-Gulose (D-gulo)   D-Idose (D-ido-)   D-Galactose (D-galacto-)   D-Talose (D-talo-)

Fig. 4.1. D-Aldoses in *Fischer* projection

Table 4.1. Occurrence of aldoses

Name, structure	Occurrence
Pentoses	
D-Apiose (3-C-Hydroxy-methyl-D-glycero-tetrose)	Parsley, celery seed
L-Arabinose	Plant gums, hemicelluloses pectins, glycosides
2-Deoxy-D-ribose	Deoxyribonucleic acid
D-Lyxose	Yeast-nucleic acid
2-O-Methyl-D-xylose	Hemicelluloses
D-Ribose	Ribonucleic acid
D-Xylose	Xylanes, hemicelluloses plant gums, glycosides
Hexoses	
L-Fucose (6-Deoxy-L-galactose)	Human milk, seaweed (algae) plant gums and mucilage
D-Galactose	Widespread in oligo- and polysaccharides
D-Glucose	Widespread in plants and animals
D-Mannose	Widespread as polysaccharide building blocks
L-Rhamnose (6-Deoxy-L-mannose)	Plant gums and mucilage, glycosides

keto and existing CHOH-groups, form a series of D- and L-2-ketoses. Figure 4.2 gives D-2-ketoses in their simplified *Fischer* projections.

Data are provided in Table 4.2 on the occurrence of ketoses of interest in food.

For simplified presentation of oligo- and polysaccharide structures, abbreviations are used which, as with amino acids, are of three characters. Usually, the first three letters are from the name of the monosaccharide. Figure 4.1 gives the configuration prefix derived from the trivial names, representing a specified configuration applied in monosaccharide classification. Thus, systematic names for D-glucose and D-fructose are D-gluco-hexose and D-arabino-2-hexulose. Such nomenclature makes it possible to systematically denote all monosaccharides that contain more than four chiral centers. In this procedure the portion of the molecule adjacent to the carbonyl group is given as the maximal possible

CH2OH
=O
-OH
CH2OH

D-Tetrulose

CH2OH
=O
-OH
-OH
CH2OH

D-Ribulose

CH2OH
=O
HO-
-OH
CH2OH

D-Xylulose

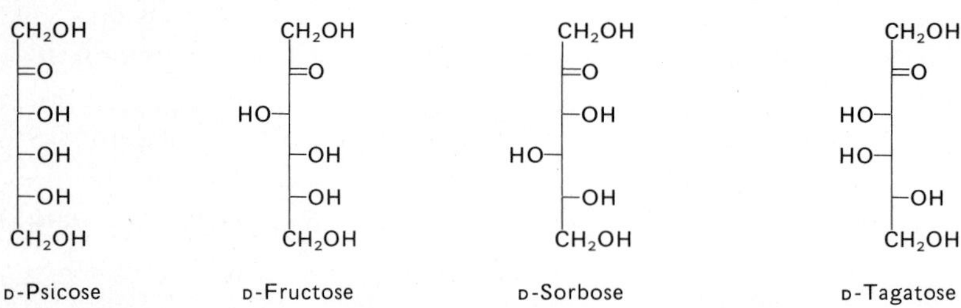

Fig. 4.2. D-Ketoses in *Fischer* projection

Table 4.2. Occurrence of ketoses

Name, structure	Occurrence
Hexulose	
D-Fructose	Present in plants and honey
D-Psicose	Found in residue of fermented mollases
Heptulose	
D-manno-2-Heptulose	Avocado fruit
Octulose	
D-glycero-D-manno-2-Octulose	Avocado fruit
Nonulose	
D-erythro-L-gluco-2-Nonulose	Avocado fruit

prefix, while the portion furthest from the carbonyl group is denoted first. In the case of ketoses, the two portions of the molecule separated by the keto group are given. In a combined prefix designation, as with aldoses, the portion which has the C-atom furthest from the keto group is mentioned first. However, when a monosaccharide does not have more than four chiral centers, a designation in the ketose series may omit the two units separated by the keto group. The examples of formula 4.6 illustrate the rule.

Hemiacetal or lactol formation provides a new chiral center on C-1. Thus, there are two additional diastereomers for each pyranose and furanose. These isomers are called anomers and are denoted as α and β-forms. Formula 4.7 illustrates the two anomeric D-glucose molecules in both *Fischer* and *Haworth* projections.

Based on the cis-arrangement of the two adjacent HO-groups in positions C-1 and C-2, α-D-glucopyranose, unlike its β-anomer, increases the conductivity of boric acid. A borate complex is formed which is a stronger acid than boric acid itself (cf. Formula 4.8).

In *Haworth* projections the HO-group of α/β-anomers of the D-series usually occurs below/above the pyranose or furanose ring planes, while in the L-series the reverse is true (cf. Formula 4.9).

Each monosaccharide can exist in solution together with its open chain molecule in a total of five forms. Due to the strong tendency for an open chain molecule to be in a cyclic form, the

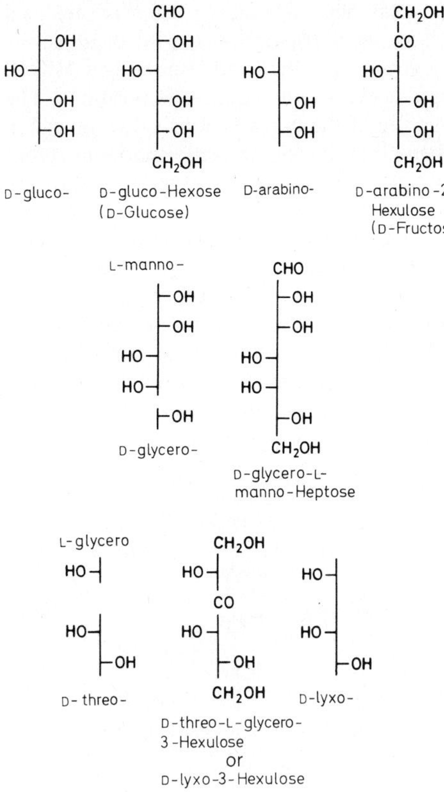

D-gluco- D-gluco-Hexose (D-Glucose) D-arabino- D-arabino-2-Hexulose (D-Fructose)

L-manno- D-glycero- D-glycero-L-manno-Heptose (4.6)

L-glycero D-threo- D-lyxo- D-threo-L-glycero-3-Hexulose or D-lyxo-3-Hexulose

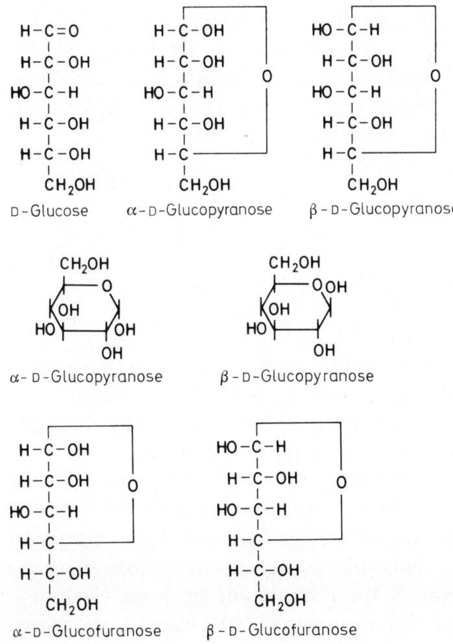

D-Glucose α-D-Glucopyranose β-D-Glucopyranose

α-D-Glucopyranose β-D-Glucopyranose (4.7)

α-D-Glucofuranose β-D-Glucofuranose

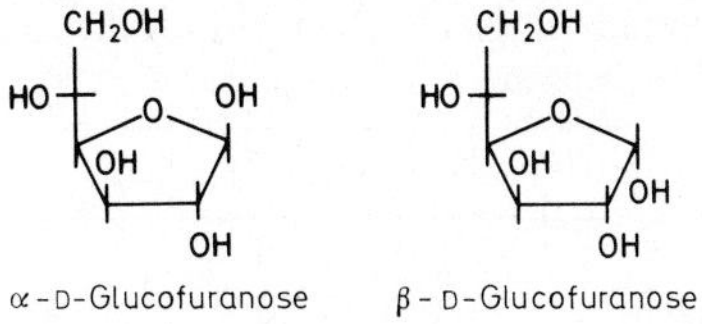

α-D-Glucofuranose β-D-Glucofuranose (4.7 cont.)

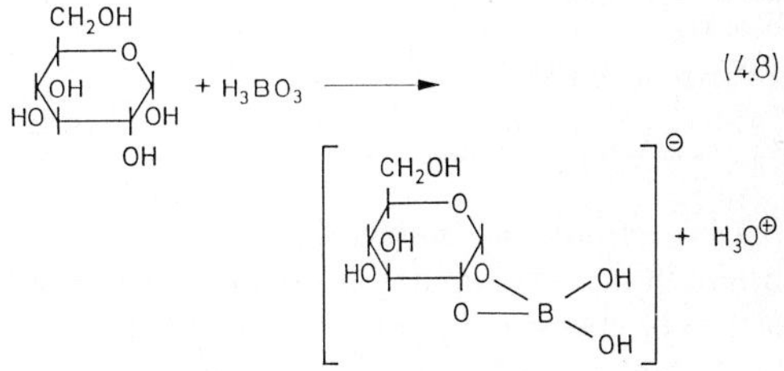

$+ H_3BO_3 \longrightarrow$ $+ H_3O^{\oplus}$ (4.8)

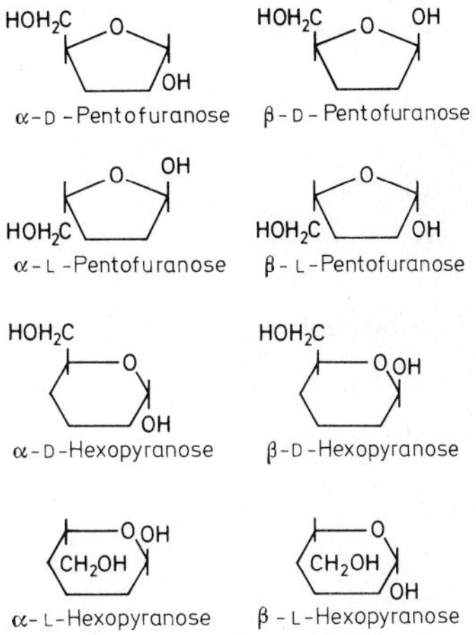

α-D-Pentofuranose β-D-Pentofuranose

α-L-Pentofuranose β-L-Pentofuranose (4.9)

α-D-Hexopyranose β-D-Hexopyranose

α-L-Hexopyranose β-L-Hexopyranose

amount of the open chain form is negligible. The contribution of the different cyclic forms to the equilibrium state in a solution depends on the configuration pattern of the sugar. An aqueous D-glucose solution is nearly exclusively the two pyranoses, with 36% α- and 64% β-anomer, while the furanose ring form is less than 1%. The equilibrium state varies greatly among sugars (cf. Table 4.6).

The transition into various forms occurs through the open-chain carbonyl compound. The acid- and base-catalyzed ring opening is the rate limiting step of the reaction:

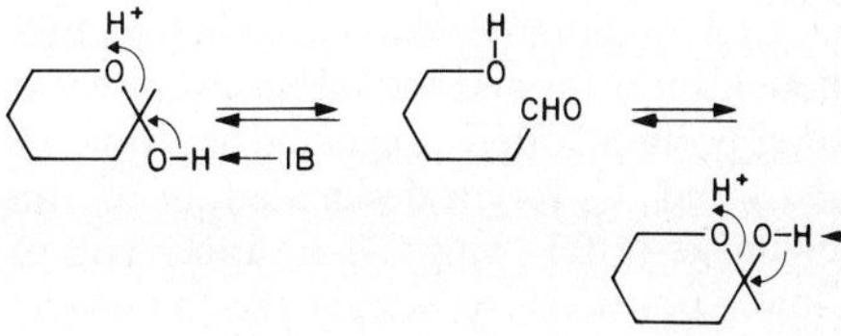

(4.10)

Table 4.3. Mutarotation rate of 2,3,4,6-tetramethyl-D-glucose (0.09 M) in benzene

Catalyst	k (min^{-1})	k_{rel}
–	7.8×10^{-5}	1.0
Pyridine (0.1 M)	3.7×10^{-4}	4.7
p-Cresol (0.1 M)	4.2×10^{-4}	5.4
Pyridine + p-cresol (0.1 M)	7.9×10^{-3}	101
2-Pyridone (0.1 M)	1.8×10^{-1}	2,307
Benzoic acid (0.1 M)	2.2	28,205

2,3,4,6-Tetramethyl-D-glucose reaches equilibrium in benzene rapidly through the concerted action of cresol and pyridine as acid-base catalysts (Table 4.3). A bifunctional reagent, 2-pyridone, can also be efficient as an acid-base catalyst in both polar and nonpolar solvents:

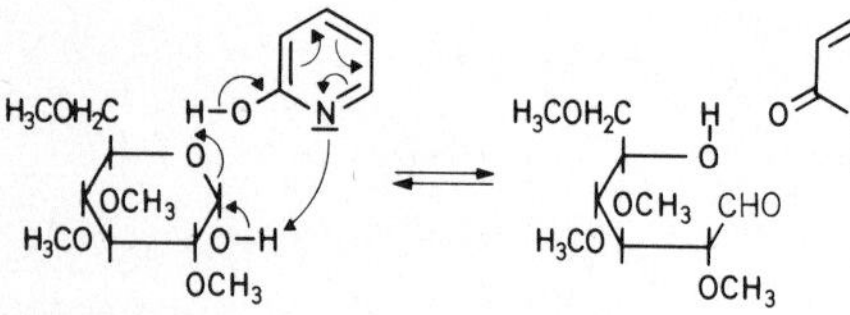

(4.11)

Water can also be a bifunctional catalyst:

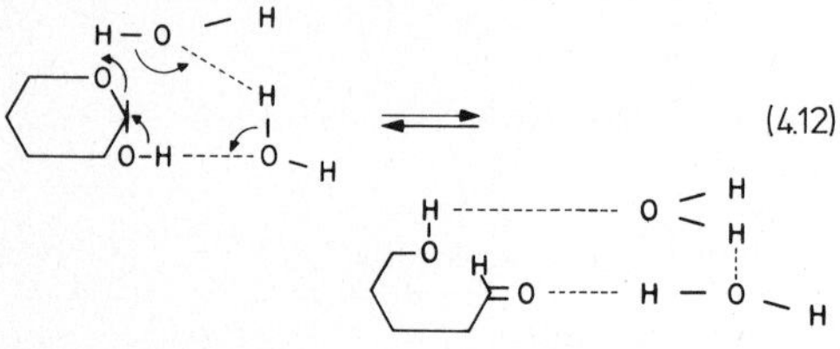

(4.12)

The reaction rate for the conversion of the α- and β-forms (so-called mutarotation) has a wide minimum in an aqueous medium and in a pH range of 2–7, as illustrated in Figure 10.6 with lactose, and the rate increases beyond this pH range.

4.2.1.3 Conformation

The preferred conformation for a pyranose is the so-called chair conformation and not the twisted-boat conformation, since the former has the highest thermodynamic stability. The two chair C-conformations are $^4C_1$ (the superscript corresponds to the number of the C-atom in the upper position of the chair and the subscript to that in the lower position; often designated as C1 or "O-outside") and $^1C_4$ (often designated as 1C, the mirror image of C1, with C-1 in upper and C-4 in lower positions, or simply the "O-inside" conformer). The $^4C_1$-conformation is preferred in the series of D-pyranoses, with most of the bulky groups, e.g. HO and, especially, CH_2OH, occupying the roomy equatorial positions. The interaction of the bulky groups is low in such a conformation, hence the conformational stability is high. This differs from the $^1C_4$-conformation, in which most of the bulky groups are crowded into axial positions, thus imparting a thermodynamic instability to the molecule.

β-D-glucopyranose in $^4C_1$-conformation is an exception. All substituents are arranged equatorially, while in $^1C_4$ all are axial. α-D-Glucopyranose in $^4C_1$-conformation has one axial group and is also thermodynamically more stable than $^1C_4$ (cf. Table 4.5):

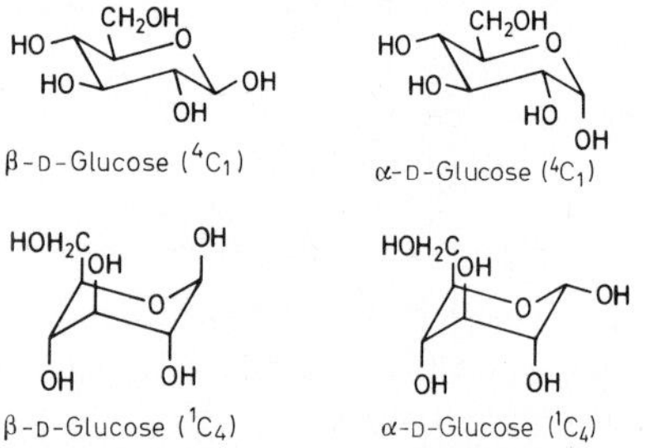

β-D-Glucose ($^4C_1$) α-D-Glucose ($^4C_1$)

β-D-Glucose ($^1C_4$) α-D-Glucose ($^1C_4$) (4.13)

The arrangement of substituents differs in α-D-idopyranose. Here, all the substituents are in axial positions in the $^4C_1$-conformation (axial HO-groups at 1, 2, 3, 4), except for the CH_2OH-group, which is equatorial. However, the $^1C_4$-conformation is thermodynamically more favorable despite the fact that the CH_2OH-group is axial (cf. Table 4.5):

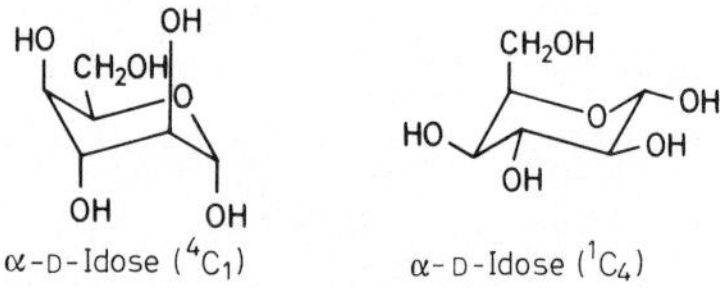

α-D-Idose ($^4C_1$) α-D-Idose ($^1C_4$) (4.14)

A second exception (or rather an extreme case) is α-D-altropyranose. Both conformations (O-outside and O-inside) have practically the same stability in this sugar (cf. Table 4.5).

The free energy of the conformers in the pyranose series can be calculated from partial interaction energies (derived from empirical data). Only the 1,3-diaxial interactions (with exception of the interactions between H-atoms), 1,2-gauche or staggered (60°) interactions of two

HO-groups and that between HO-groups and the CH_2OH-group will be considered. The partial interaction energies are compiled in Table 4.4 and conformational free energies, calculated from these data for various conformers, are presented in Table 4.5. In addition to the interaction energies an effect is considered which destabilizes the anomeric HO-group in equatorial position, while it stabilizes this group in axial position. This is called the *anomeric effect*. The effect is attributed to repulsion forces between the parallel dipoles resulting from the polarized bonds C5–O5 and C1–O1 (equatorial). The repulsion forces the anomeric HO-group to take up the more stable axial or α-position:

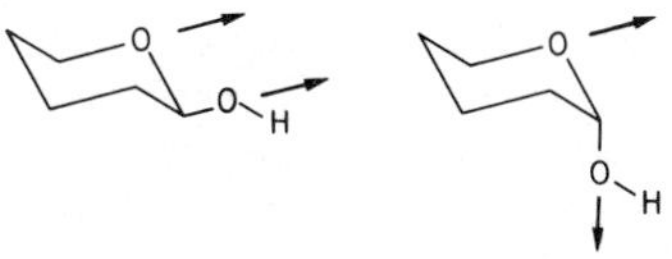

(4.15)

Table 4.4. Free energies of unfavorable interactions between substituents on tetrahydropyrane rings

Interaction	Energy kJ/mole[a]
$H_{ax} - O_{ax}$	1.88
$H_{ax} - C_{ax}$	3.76
$O_{ax} - O_{ax}$	6.27
$O_{ax} - C_{ax}$	10.45
$O_{eq} - O_{eq}/O_{ax} - O_{eq}$	1.46
$O_{eq} - C_{eq}/O_{ax} - C_{eq}$	1.88
Anomeric effect[b]	
for O^{c2}_{eq}	2.30
for O^{c2}_{ax}	4.18

[a] Aqueous solution, room temperature.
[b] To be considered only for an equatorial position of the anomeric HO-group.

The other substituents influence the anomeric effect, particularly the HO-group in C-2 position. Here, due to an antiparallel dipole formation, the axial position enhances stabilization better than does the equatorial position. Correspondingly, in an equilibrium state in water, D-mannose is 67% in its α-form, while α-D-glucose or α-D-galactose are only 36% and 27%, respectively (Table 4.6). The anomeric effect (dipole-dipole interaction) increases as the dielectric constant of the solvent system decreases.

Table 4.5. Calculated conformational free energies ($E_{conf.}$) for hexopyranoses

Hexopyranose	Conformation	$E_{conf.}$ (kJ/mole)
α-D-Glucose	$^4C_1$	10.03
	$^1C_4$	27.38
β-D-Glucose	$^4C_1$	8.57
	$^1C_4$	33.44
α-D-Mannose	$^4C_1$	10.45
β-D-Mannose	$^4C_1$	12.33
α-D-Galactose	$^4C_1$	11.91
β-D-Galactose	$^4C_1$	10.45
α-D-Idose	$^4C_1$	18.18
	$^1C_4$	16.09
β-D-Idose	$^4C_1$	16.93
	$^1C_4$	22.36
α-D-Altrose	$^4C_1$	15.26
	$^1C_4$	16.09

Table 4.6. Aldoses and ketoses: isomers in water at equilibrium state[a]

Sugar	α-Pyranose	β-Pyranose	α-Furanose	β-Furanose
D-Glucose	36	64		< 1
D-Mannose	67	33		< 1
D-Galactose	27	73		< 1
D-Idose	31	37	16	16
D-Ribose	21.5	58.5	6.5	13.5
D-Xylose	35	65		< 1
D-Fructose	3	57	9	31

[a] Results are expressed as % of all sugar forms present in solution.

Alkylation of the anomeric HO-group also enhances the anomeric effect (Table 4.7).

Table 4.7. Methylglucoside isomers in methanol (1% HCl) at equilibrium state[a]

Compound	α-Pyranoside	β-Pyranoside	α-Furanoside	β-Furanoside
Methyl-D-glucoside	66	32.5	0.6	0.9
Methyl-D-mannoside	94	5.3	0.7	0
Methyl-D-galactoside	58	20	6	16
Methyl-D-xyloside	65	30	2	3

[a] Values in %.

Conformational isomers of furanose also occur since its ring is not planar. There are two basic conformations, the envelope (E) and the twist (T), which are the most stable; in solution a mixture exists of conformers similar in energy (cf. Formula 4.16).

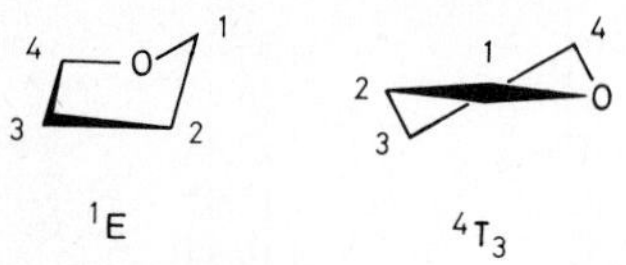

^1E $\quad$ $^4T_3$ (4.16)

An anomeric effect forces the anomeric HO-group preferentially into the axial position. This is especially the case when the HO-group attached to C-2 is also axial. When pyranose ring formation is prevented or blocked, as in 5-O-methyl-D-glucose, the twisted $^3T_2$-conformer becomes the dominant form:

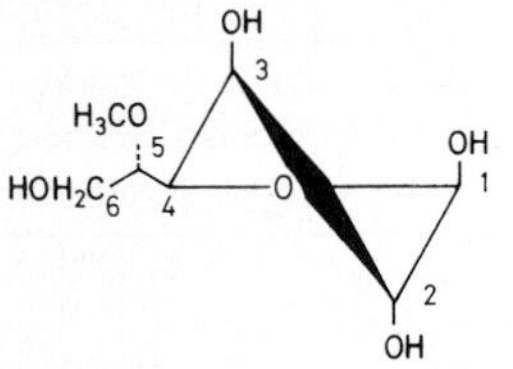

(4.17)

5-O-Methyl-D-glucose ($^3T_2$)

A pyranose is generally more stable than a furanose, hence, the former and not the latter conformation is predominant in most monosaccharides (Table 4.6).
The composition of isomers in aqueous solution, after equilibrium is reached, is compiled for a number of monosaccharides in Table 4.6. Evidence for such compositions is obtained by polarimetry, by oxidation with bromine, which occurs with a higher reaction rate with β- than α-pyranose and, above all, by nuclear magnetic resonance spectroscopy.
In proton magnetic resonance (PMR) spectroscopy of sugars, the hydrogen atoms bound to oxygen, which complicate the spectrum, are replaced by derivatization (O-acetyl derivatives) or are exchanged by deuterium when the sugar is in D_2O solution. The chemical shift of the retained H-atoms bound covalently to carbon is distinct. Protons on an anomeric carbon, due to less shielding than other protons, appear at a lower magnetic field in the order pyranose > furanose and in the range of $\tau = 5.5 - 4.5$ (free monosaccharides). Furthermore, an axial proton (β-form of D-series) appears at higher field than an equatorial proton (α-form of D-series). The sugar conformation is elucidated by coupling effects of neighboring protons: equatorial-equatorial, equatorial-axial (small coupling constants) or axial-axial (larger coupling constants).

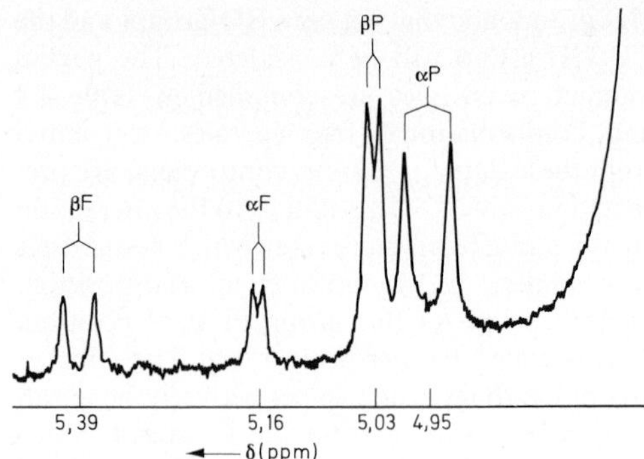

Fig. 4.3. Proton resonance spectrum of D-idose. (After *Angyal*, 1969)

Figure 4.3 shows the PMR spectrum for D-idose, beginning with the pyranose (P) and followed by the furanose (F) ring conformations, in decreasing τ-value order. The high coupling constant recorded at highest τ-value indicates the presence of a diaxial H-atom arrangement on C-1 and C-2, which is incompatible with either β-pyranose conformer but is compatible with the α-pyranose $^1C_4$-conformation. Hence, the second signal belongs to β-pyranose which, in both $^1C_4$- and $^4C_1$-conformations, has a small coupling constant. In furanose the signal at lower field with higher coupling constant is due to the cis-arrangement of protons at C-1 and C-2 (β-form), while at higher field and with lower coupling constant the signal is due to trans-arrangement (α-form).
Elucidation of sugar conformation can also be achieved by ^{13}C-nuclear magnetic resonance spectroscopy. Like PMR, the chemical shifts differ for different C-atoms and are affected by the spatial arrangement of ring substituents.

4.2.2 Physical Properties

4.2.2.1 Hygroscopicity and Solubility

The moisture uptake of sugars is variable and depends, for example, on the sugar structure, isomers present and sugar purity. Solubility decreases as the sugars cake together, as often hap-

pens in sugar powders or granulates. On the other hand, the retention of food moisture by concentrated sugar solutions, e.g. glucose syrup, is utilized in the baking industry.
The solubility of mono- and oligosaccharides in water is good. Anomers may differ substantially in their solubility, as exemplified by α- and β-lactose (cf. 10.1.2.2 and Table 10.14). Monosaccharides are soluble to a small extent in ethanol and are insoluble in organic solvents such as ether, chloroform or benzene.

4.2.2.2 Optical Rotation, Mutarotation

Specific rotation constants, designated as [α] for sodium D-line light at 20–25 °C, are listed in Table 4.8 for some important mono- and oligosaccharides. The specific rotation constant $[\alpha]^t_\lambda$ at a selected wavelength and temperature is calculated from the angle of rotation, α, by the equation:

$$[\alpha]^t_\lambda = \frac{100 \cdot \alpha}{l \cdot c} \quad (4.18)$$

where l is the polarimeter tube length in decimeters and c the number of grams of the optically-active sugar in 100 ml of solution. The molecular rotation, [M], is suitable for comparison of the rotational values of compounds with differing molecular weights:

$$[M]^t_\lambda = \frac{M[\alpha]^t_\lambda}{100} \quad (4.19)$$

where M represents the compound's molecular weight. Since the rotational value differs for anomers and also for pyranose and furanose conformations, the angle of rotation for a freshly-prepared solution of an isomer changes until an equilibrium is established. This phenomenon is known as mutarotation. When an equilibrium exists only between two isomers, as with glucose (α- and β-pyranose forms), the reaction rate follows first order kinetics:

$$-\frac{dc_\alpha}{dt} = k \cdot c_\alpha - k' \cdot c_\beta \quad (4.20)$$

A simple mutarotation exists in this example, unlike complex mutarotations of other sugars, e.g. idose which, in addition to pyranose, is also to a substantial extent in the furanose form. Hence, the order of its mutarotation kinetics is more complex.

Table 4.8. Specific rotation of various mono- and oligosaccharides

Compound	$[\alpha]_D$[a]	Compound	$[\alpha]_D$[a]
Monosaccharides		*Oligosaccharides* (continued)	
L-Arabinose	+105	Kestose	+28
α-	+55.4	Lactose	+53.6
β-	+190.6	β-	+34.2
D-Fructose	−92	Maltose	+130
β-	−133.5	α-	+173
D-Galactose	+80.2	β-	+112
α-	+150.7	Maltotriose	+160
β-	+52.8	Maltotetraose	+166
D-Glucose	+52.7	Maltopentaose	+178
α-	+112	Maltulose	+64
β-	+18.7	Manninotriose	+167
D-manno-2-Heptulose	+29.4	Melezitose	+88.2
D-Mannose	+14.5	Melibiose	+143
α-	+29.3	β-	+123
β-	−17	Palatinose	+97.2
D-Rhamnose	−7.0	Panose	+154
D-Ribose	−23.7	Raffinose	+101
D-Xylose	+18.8	Saccharose	+66.5
α-	+93.6	*α-Schardinger-*Dextrin	+151
Oligosaccharides (including disaccharides)		*β-Schardinger-*Dextrin	+162
Cellobiose	+34.6	*γ-Schardinger-*Dextrin	+180
β-	+14.2	Stachyose	+146
Gentianose	+33.4		
Gentiobiose	+10		
α-	+31		
β-	−3		

[a] Temperature: 20–25 °C.

4.2.3 Sensory Properties

Mono- and oligosaccharides and their corresponding sugar alcohols, with a few exceptions, are sweet. β-D-Mannose has a sweet-bitter taste, and some oligosaccharides are bitter, e.g. gentiobiose.
The most important sweeteners are saccharose (sucrose), starch syrup (a mixture of glucose, maltose and malto-oligosaccharides) and glucose. Invert sugar, fructose-containing glucose syrups, fructose, lactose and sugar alcohols, such as sorbitol, mannitol and xylitol, are also of importance.

The sugars differ in quality of sweetness and taste intensity. Saccharose is distinguished from other sugars by its pleasant taste even at high concentrations. The taste intensity of oligosaccharides drops regularly as the chain length increases.

The taste intensity can be measured by determining the recognition threshold of the sugar (the lowest concentration at which sweetness is still perceived) or by comparison with a reference substance (isosweet concentrations). The threshold value is related to the affinity of sweet-taste chemoreceptors for the sweet substance and is of importance in elucidation of relationships between the chemical structure of a compound and its taste. For practical purpose, the use of a reference substance is of greater importance; taste intensity is dependent on concentration and varies greatly among sweet compounds.

Saccharose is the reference substance usually chosen. Tables 4.9, 4.10a and 4.10b list some sugar sweetness threshold values and relative sweetness values. Only mean values are given, with deviations omitted. The recognition threshold values for saccharose cited in the literature vary from 0.01 to 0.037 mole/l.

Taste quality and intensity are dependent not only on a compound's structure but on other taste reception parameters: temperature, pH and the presence of additional sweet or nonsweet compounds.

Furthermore, an interrelationship exists between the sugar content of a solution and the sensory assessment of the volatile aroma compounds present. Even the color of the solution might affect taste evaluation. Figures 4.4–4.7 clarify these phenomena, with fruit juice and canned fruits as selected food samples.

The overall conclusion is that the composition and concentration of a sweetening agent has to be adjusted for each food formulation to provide optimum sensory perception.

A prerequisite for a compound to be sweet is the presence in its structure of a proton donor/

Table 4.9. Taste threshold value of sugars in water

Sugar	Recognition threshold		Detection threshold	
	mole/l	%	mole/l	%
Fructose	0.052	0.94	0.02	0.24
Glucose	0.090	1.63	0.065	1.17
Lactose	0.116	4.19	0.072	2.60
Maltose	0.080	2.89	0.038	1.36
Saccharose	0.024	0.81	0.011	0.36

Table 4.10a. Relative sweetness of sugars and sugar alcohols to sucrose[a]

Sugar/ sugar alcohol	Relative sweetness	Sugar/ sugar alcohol	Relative sweetness
Saccharose	100	D-Mannitol	69
Galactitol	41	D-Mannose	59
D-Fructose	114	Raffinose	22
D-Galactose	63	D-Rhamnose	33
D-Glucose	69	D-Sorbitol	51
Invert sugar	95	Xylitol	102
Lactose	39	D-Xylose	67
Maltose	46		

[a] 10% aqueous solution.

Table 4.10b. Concentration (%) of iso-sweet aqueous solutions of sugars and sugar alcohols

D-Fructose	D-Glucose	Lactose	Saccharose	D-Sorbitol	Xylitol
0.8	1.8	3.5	1.0		
1.7	3.6	6.5	2.0		
4.2	8.3	15.7	5.0	9.3	8.5
8.6	13.9	25.9	10.0	17.2	9.8
13.0	20.0	34.6	15.0		
16.1	27.8		20.0	28.2	18.5
	39.0		30.0		25.4
	48.2		40.0	48.8	

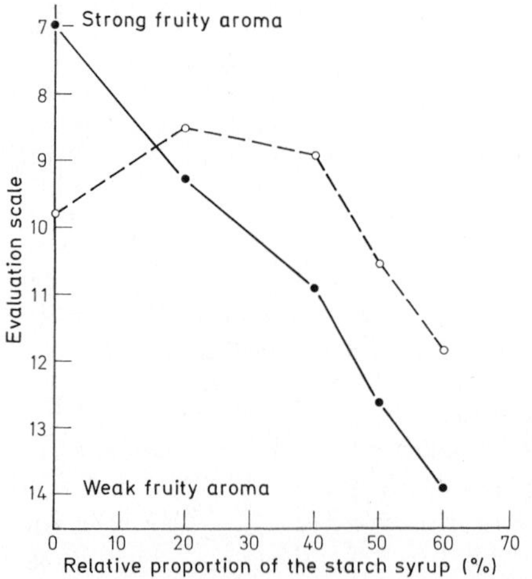

Fig. 4.4. Sensory evaluation of the "fruity flavor" of canned peaches at different ratios of saccharose/starch syrup (●—● 60° Brix, ○--○ 50° Brix). (After *Pangborn*, 1959)

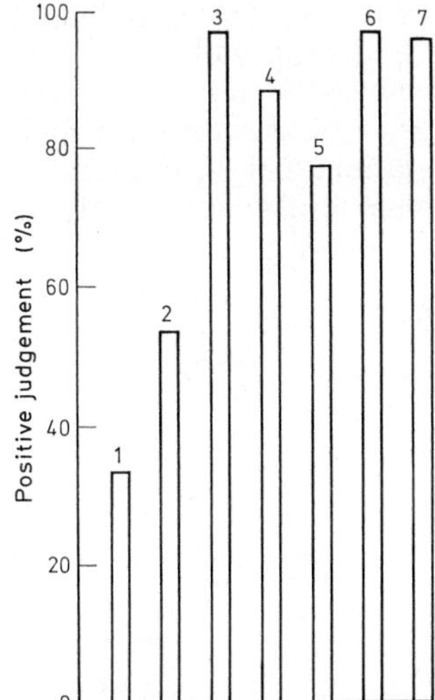

Fig. 4.5. Sensory evaluation of canned cherries prepared with different sweeteners 1, 2, 3: 60, 50, 40% saccharose, 4: 0.15% cyclamate, 5: 0.05% saccharin, 6: 10% saccharose + 0.10% cyclamate, 7: 10% saccharose + 0.02% saccharin. (After *Salunkhe,* 1963)

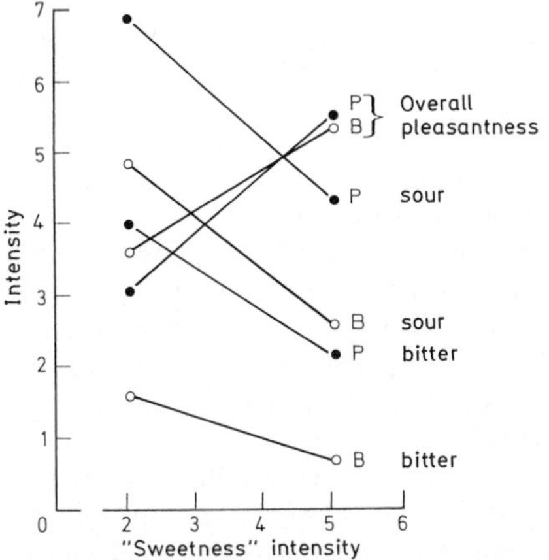

Fig. 4.6. Sensory evaluation of the categories "overall pleasantness", "sour" and "bitter" versus sweetness intensity. B bilberry (○—○) and P (•—•) cranberry juice. (After *Sydow,* 1974)

acceptor system (AH/B-system), which may be supplemented by a hydrophobic site X (cf. 8.8.1 and Figure 8.3). This AH/B/X-system interacts with a complementary system of the taste receptor site located on the taste buds. Based on studies of the taste quality of sugar derivatives and deoxy sugars, the following AH/B/X-systems

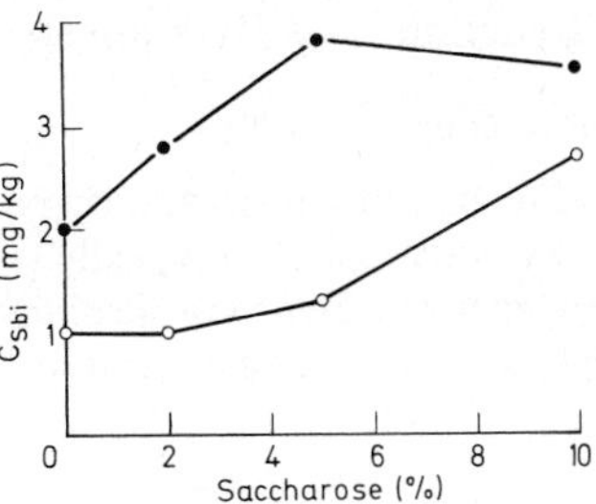

Fig. 4.7. Bitter taste threshold values of limonin (○—○) and naringin ($\times\ 10^{-1}$ •—•) in aqueous saccharose solution. (After *Guadagni,* 1973)

have been proposed for β-D-glucopyranose and β-D-fructopyranose respectively:

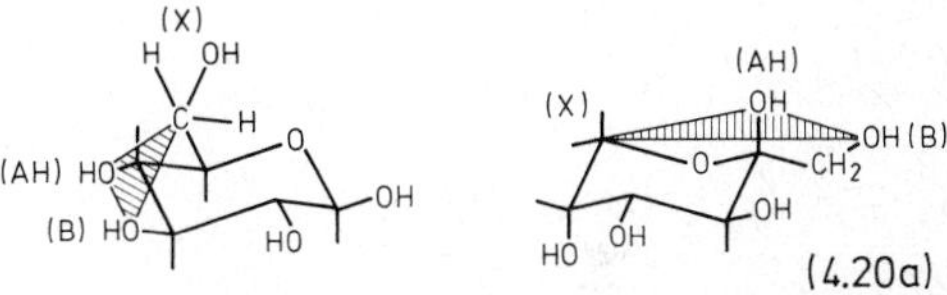

(4.20a)

β-D-glucopyranose and β-L-glucopyranose are sweet. Molecular models show that the AH/B/X-system of both sugar components fits equally well with the complementary receptor system $AH_R/B_R/X_R$:

β-D-Glucose β-L-Glucose

(4.20b)

With asparagine enantiomers, the D-form is sweet, while the L-form is tasteless. Here, unlike D- and L-glucose, only the D-form interacts with the complementary $AH_R/B_R/X_R$-system:

D-Asparagine L-Asparagine

(4.20c)

These examples demonstrate that the AH/B/X-model, as a prerequisite for eliciting sweet taste responses, is correct in principle.

4.2.4 Chemical Reactions and Derivatives

4.2.4.1 Reduction to Sugar Alcohols

Monosaccharides can be reduced to the corresponding alcohols by $NaBH_4$, electrolytically or via catalytic hydrogenation. Two new alcohols are obtained from ketoses since a new chiral center is formed:

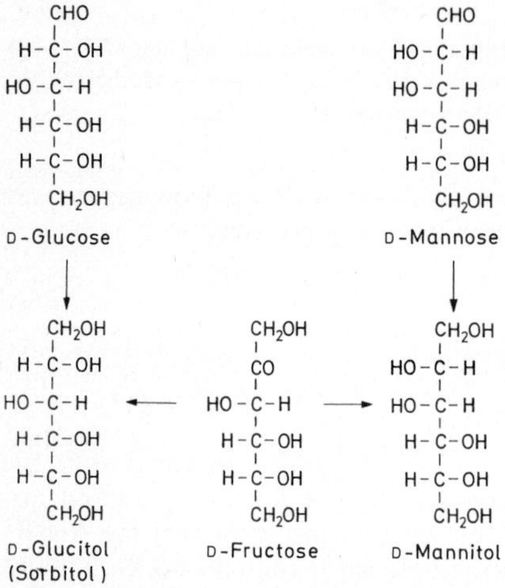

D-Glucose → D-Glucitol (Sorbitol); D-Mannose → D-Mannitol; D-Glucitol ← D-Fructose → D-Mannitol (4.21)

The alcohol name is derived from the sugar name by replacing the suffix -ose or -ulose with the suffix -itol. The sugar alcohols of importance in food processing are xylitol, the only one of the four pentitols (meso-ribitol, D,L-arabitol, meso-xylitol) used, and only D-glucitol (sorbitol) and D-mannitol of the ten stereoisomeric forms of hexitols [meso-allitol, meso-galactitol (dulcitol), D,L-glucitol (sorbitol), D,L-iditol, D,L-mannitol and D,L-altritol]. They are used as sugar substitutes in dietetic food formulations, to decrease water activity in "intermediate moisture foods", as softeners, as crystallization inhibitors and for improving the rehydration characteristics of dehydrated food. Sorbitol is found in nature in many fruits, e. g. pears, apples and plums. Maltitol, the reduction product of the disaccharide maltose, is being considered for wider use in food formulations.

4.2.4.2 Oxidation to Aldonic, Dicarboxylic, and Uronic Acids

Under mild conditions, e. g. with bromine water in buffered neutral or alkaline media, aldoses are oxidized to aldonic acids. Oxidation involves the lactol group exclusively. β-Pyranose is oxidized more rapidly than the α-form. Since the β-form is more acidic (cf. 4.2.1.3), it can be considered that the pyranose anion is the reactive form. The oxidation product is a δ-lactone which is in equilibrium with the γ-lactone and the free form of aldonic acid. The latter form prevails at pH > 3.

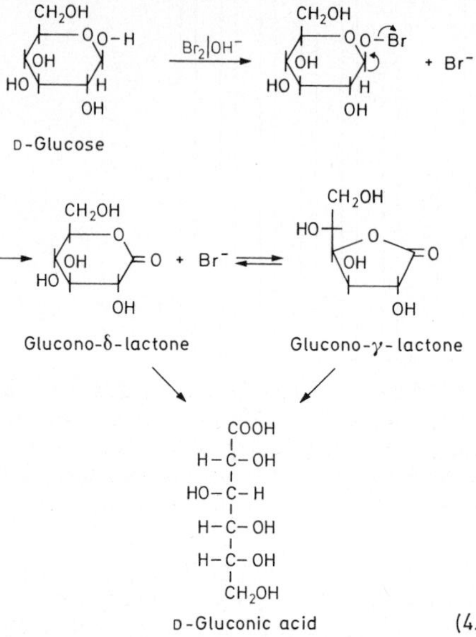

D-Glucose → (Br_2/OH^-) → Glucono-δ-lactone + Br^- ⇌ Glucono-γ-lactone → D-Gluconic acid (4.22)

The transition of lactones from δ- to γ-form and vice versa probably proceeds through an intermediary bicyclic form.

The acid name is obtained by adding the suffix -onic acid (e. g. aldose → aldonic acid).

Glucono-δ-lactone is utilized in food when a slow acid release is required, as in baking powders, raw fermented sausages or dairy products.

Treatment of aldose with more vigorous oxidizing agents, such as nitric acid, brings about oxidation of the C-1 aldehyde group and the CH_2OH-group, resulting in formation of a dicarboxylic acid (nomenclature: stem name of the parent sugar + the suffix -aric acid, e. g. aldose → aldaric acid). Thus, galactaric acid (common or trivial name: mucic acid) is obtained from galactose:

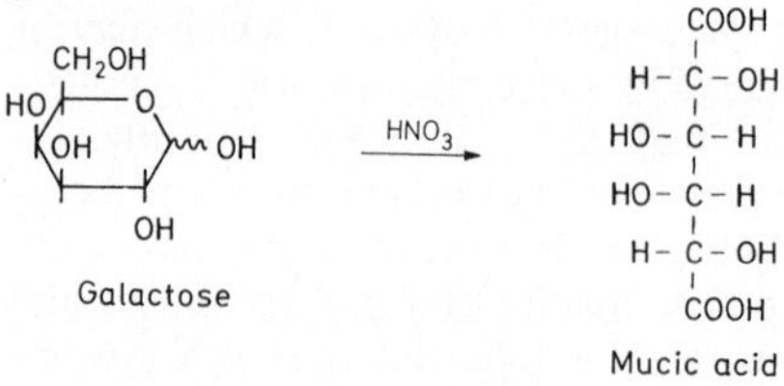

Galactose → (HNO_3) → Mucic acid (4.23)

The dicarboxylic acid can, depending on its configuration, form mono- or dilactones.

Oxidation of the CH_2OH-group by retaining the carbonyl function at C-1, with the aim of obtaining uronic acids (aldehydocarboxylic acids), is possible only by protecting the carbonyl group during oxidation. A suitable way is to temporarily block the vicinal HO-groups by ketal formation which, after the oxidation at C-6 is completed, are deblocked under mild acidic conditions:

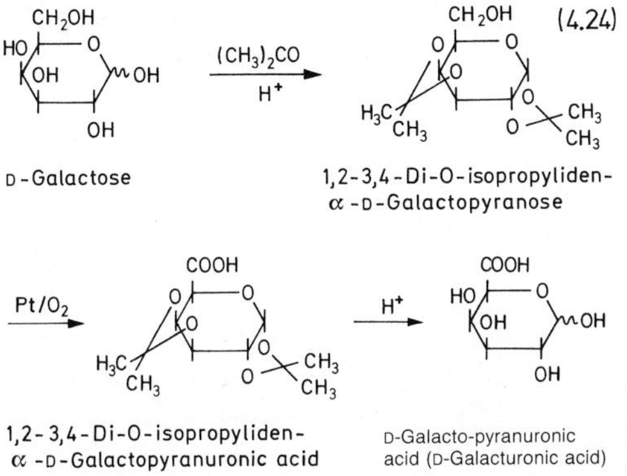

(4.24)

$(CH_3)_2CO$, H^+

D-Galactose → 1,2-3,4-Di-O-isopropyliden-α-D-Galactopyranose

Pt/O_2 → 1,2-3,4-Di-O-isopropyliden-α-D-Galactopyranuronic acid

H^+ → D-Galacto-pyranuronic acid (D-Galacturonic acid)

An additional possibility for uronic acid synthesis is the reduction of monolactones of the corresponding aldaric acids:

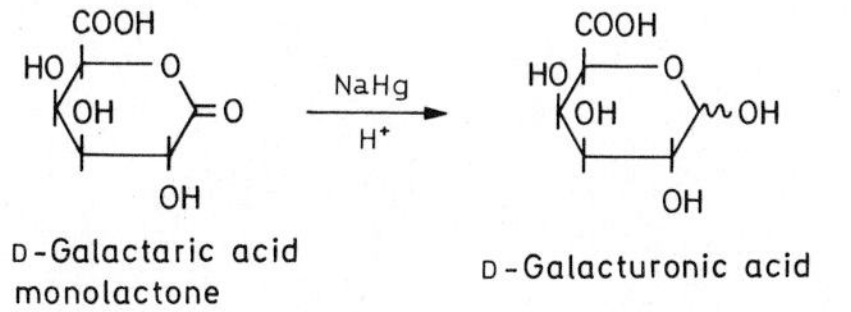

D-Galactaric acid monolactone —NaHg, H^+→ D-Galacturonic acid (4.25)

Another industrially-applied glucuronic acid synthesis involves first oxidation then hydrolysis of starch:

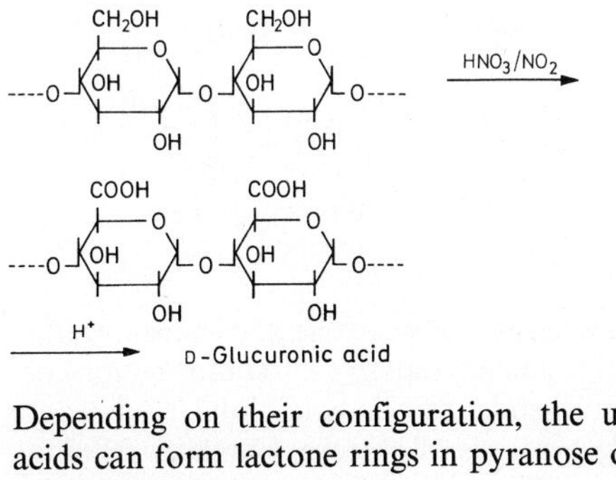

HNO_3/NO_2 → H^+ → D-Glucuronic acid (4.26)

Depending on their configuration, the uronic acids can form lactone rings in pyranose or furanose forms.

Uronic acid biosynthesis also occurs from the monosaccharide blocked at C-1 (cf. Reaction Sequence 4.27).

A number of uronic acids occur fairly abundantly in nature. Some are constituents of polysaccharides of importance in food processing, such as gel-forming and thickening agents, e.g. pectin (D-galacturonic acid) and sea weed-derived alginic acid (D-mannuronic acid, L-guluronic acid).

4.2.4.3 Reactions in the Presence of Acids and Alkalis

Monosaccharides are fairly stable in a pH range of 3–7. However, at both ends of the pH limits, depending on conditions, extensive conversion can occur. Water elimination is the predominant reaction, retaining the C-chain length in an acidic medium, but treatment with alkali causes enolization and even fragmentation of the sugar molecule, followed by additional secondary reactions.

4.2.4.3.1 Reactions in strongly acidic media

The reverse of glycoside hydrolysis (cf. 4.2.4.5), i.e. formation of glycosides, occurs in dilute mineral acids. All the possible disaccharides and higher oligosaccharides, but preferentially isomaltose and gentiobiose, are obtained from glucose:

(4.28)

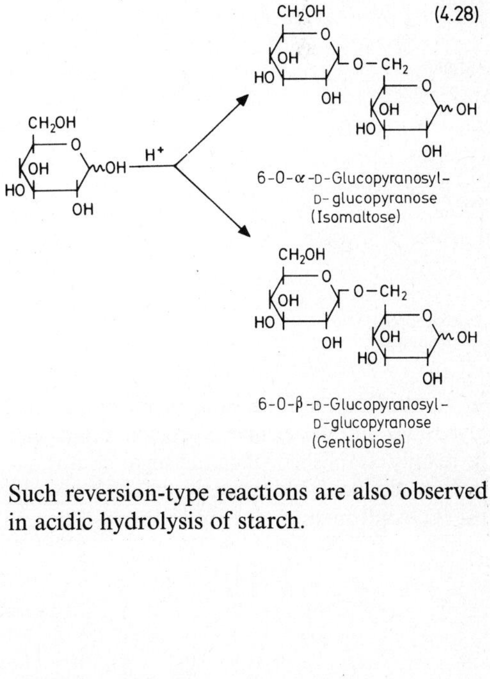

H^+ → 6-O-α-D-Glucopyranosyl-D-glucopyranose (Isomaltose)

→ 6-O-β-D-Glucopyranosyl-D-glucopyranose (Gentiobiose)

Such reversion-type reactions are also observed in acidic hydrolysis of starch.

D-Galactose
D-Glucose
D-Glucose-6-phosphate
D-Galactose-1-phosphate
D-Glucose-1-phosphate
UTP
Uridindiphosphate-D-galactose
Uridindiphosphate-D-glucose (UDPG)
Uridindiphosphate-D-galacturonic acid
Uridindiphosphate-D-glucuronic acid
CO_2
Uridindiphosphate-L-arabinose
Uridindiphosphate-D-xylose

(4.27)

In addition to the formation of intermolecular glycosides, intramolecular glycosidic bonds can be readily established when the sugar conformation is suitable. β-Idopyranose, which occurs in the $^1C_4$-conformation, is readily changed to 1,6-anhydroidopyranose, while the same reaction with β-D-glucopyranose ($^4C_1$-conformation) occurs only under drastic conditions, for instance during pyrolysis of glucose, starch or cellulose. Heating glucose syrup above 100 °C can also form 1,6-anhydroglucopyranose, but only in traces:

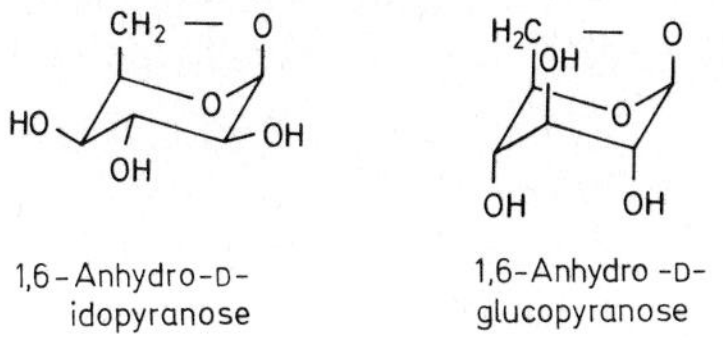

(4.29)

1,6-Anhydro-D-idopyranose

1,6-Anhydro-D-glucopyranose

Tricyclic compounds can be formed through intermolecular glycosidic bonds. For example, small amounts of di-D-fructopyranose-1,2′:2,1′-dianhydride are detected in molten saccharose (cf. Formula 4.30).

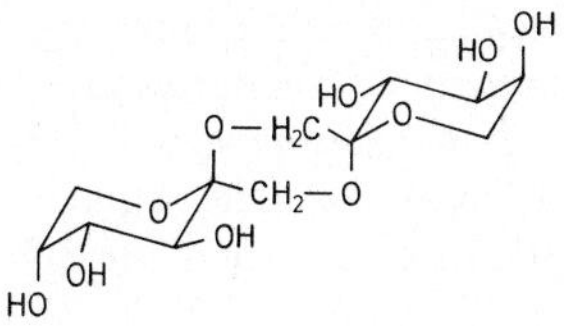

(4.30)

Heating monosaccharides in weakly acidic media, and more intensively at higher acid concentrations, leads, after a slow enolization step, to rapid proton-catalyzed β-eliminations of water which then, through several reaction steps, lead to furan derivatives. Dehydration of pentoses forms furfural and of hexoses, 5-hydroxymethyl

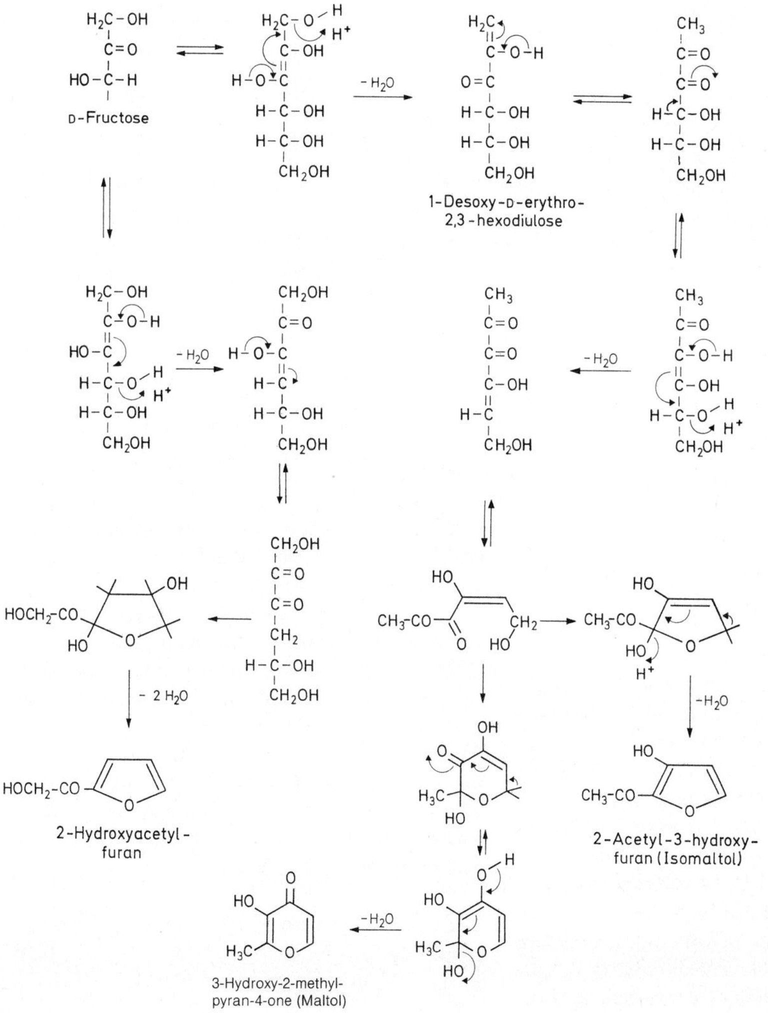

D-Fructose

1-Desoxy-D-erythro-2,3-hexodiulose

2-Hydroxyacetyl-furan

2-Acetyl-3-hydroxy-furan (Isomaltol)

3-Hydroxy-2-methyl-pyran-4-one (Maltol)

(4.32)

furfural, with 3 molecules of water eliminated in each case. The reactivity of 2-ketohexoses is higher because the 1,2-enolization step occurs more easily than with an aldose. The sequence of reactions is shown with glucose and fructose taken as examples:

D-Glucose

D-Fructose

1,2-Endiol

3-Deoxy-D-gluco-sulose

trans- cis-

3,4-Dideoxy-D-glycero-3-hexenulose

5-Hydroxymethyl furfural

(4.31)

Fructose, in addition to the predominant 1,2-endiol, can form the 2,3-endiol, hence the spectrum of its degradation products is wider than with glucose. Elimination of an HO-group on C-1, and corresponding intermediary steps, form 2-acetyl-3-hydroxyfuran (isomaltol) or 3-hydroxy-2-methyl-pyran-4-one (maltol), whereas HO-group elimination from C-4 forms hydroxyacetylfuran (cf. Reaction 4.32).

Formation of a number of other products can also be explained. Thus, 2,4-dihydroxy-2,5-dimethyl-3-furanone (diacetylformosine), a compound with a strong caramel-like odor, can be formed from 1-deoxy-2,3-hexodiulose through 1,6-di-deoxy-2,4,5-hexotriulose as an intermediate (the latter, in cyclic form, is actually the end-product):

2,4-Dihydroxy-2,5-dimethyl 3-furanone (Diacetylformosine)

(4.33)

Compounds, such as diacetylformosine and several intermediary products of the reaction steps described above, which retain a carbonyl group in the vicinity of an endiol group, are called reductones. Ascorbic acid belongs to this class of compounds, a class characterized by strong reducing power in acidic media, even at low temperatures.

Thus, Ag^{+}, Au^{3+} and Pt^{4+} are reduced to the metal state, Cu^{2+} to Cu^{+}, Fe^{3+} to Fe^{2+} and Br_2 and I_z to their respective anions, Br^{-} and I^{-}. The reductone is oxidized into its dehydro compound in these reactions:

(4.34)

Reductones are stable at pH < 6, as resonance-stabilized monoanions, while at higher pH the dianion is unstable in the presence of oxygen:

C=O / C−OH / C−O$^{\ominus}$ ⟷ C−O$^{\ominus}$ / C−OH / C=O (4.35)

The reactions outlined above take place even under mild conditions (cf. 4.2.4.4) in the presence of amino compounds. Reductones formed in this way in food act as natural antioxidants.

4.2.4.3.2 Reactions in strongly alkaline solution

Aldoses and ketoses are readily enolized in the presence of alkali. Thus, in the presence of alkali, glucose, mannose and fructose are in equilibrium through the common 1,2-endiol, along with a small amount of psicose, which is derived from fructose by 2,3-enolization:

D-Glucose ⇌ 1,2-Endiol ⇌ D-Mannose

D-Fructose ⇌ 2,3-Endiol ⇌ D-Psicose (4.36)

In this isomerization reaction, known as the *Lobry de Bruyn-van Ekenstein* rearrangement, one type of sugar can be transformed to another sugar in widely differing yields. The reaction is also applicable to disaccharides, transforming an aldose to a ketose. For example, in the presence of sodium aluminate as a catalyst, lactose is rearranged into lactulose:

Lactose (4-O-β-D-Galacto pyranosido-D-glucopyranose) → Lactulose (4-O-β-D-Galacto pyranosido-d-fructofuranose and -fructopyranose) (4.37)

Lactulose utilization in infant nutrition is under consideration, since it acts as a bifidus factor and prevents obstipation.

Cleavage of the double bond of an endiol occurs in the presence of oxygen or other oxidizing agents, e. g. Cu^{2+}, resulting in two corresponding carboxylic acids. In such a reaction with glucose, the main products are D-arabinonic and formic acids. Depending on reaction conditions, particularly the type of alkali present, other hydroxyacids are also formed, due to enolization occurring along the molecule:

H−C=O / H−C−OH / HO−C−H / H−C−OH / H−C−OH / CH_2OH ⇌ H−C−OH / C−OH / HO−C−H / H−C−OH / H−C−OH / CH_2OH —Ox.→ H−COOH + COOH / HO−C−H / H−C−OH / H−C−OH / CH_2OH (4.38)

The nonstoichiometric sugar oxidation process in the presence of alkali is used for both qualitative and quantitative determination of reducing sugars (*Fehling's* reaction with alkaline cupric tartrate; *Nylander's* reaction with alkaline trivalent bismuth tartrate; or using *Benedict's* solution, in which cupric ion complexes with citrate ion).

Likewise, hydroxyaldehydes and hydroxyketones arise from cleavage of the carbon chain of a sugar at the endiol double bond under nonoxidative conditions using dilute alkalies at elevated temperatures or concentrated alkalies even at room temperature. Since enolization is not restricted to any part of the molecule and, since water elimination is not restricted in amount, even the spectrum of primary cleavage products is great. These primary products are highly reactive and provide a great number of secondary products by aldol condensations and intramolecular *Cannizzaro* reactions (cf. Reaction 4.39).

The compounds formed in fructose syrup of pH 8–10 heated for 3 h are listed in Table 4.11. Some, e. g. the cyclopentenolones, are typical caramel-like aroma substances. Their formation is assumed to follow the reaction sequence depicted under Reaction 4.39.

Saccharic acids are specific reaction products of monosaccharides in strong alkalies, particularly of alkaline-earth metals. They are obtained as diastereomeric pairs by benzilic acid rearrangement of 1,2- and 2,3-dicarbonyl compounds, mentioned in 4.2.4.3.1 as derivatives obtained from sugar by enolization and water elimination (cf. Reaction 4.40).

(4.39)

Metasaccharic acids

D-Glucose

Saccharic acids

Isosaccharic acids (4.40)

Table 4.11. Volatile reaction products obtained from fructose by an alkali degradation (pH 8–10)

Acetic acid
Hydroxyacetone
1-Hydroxy-2-butanone
3-Hydroxy-2-butanone
4-Hydroxy-2-butanone
Furfuryl alcohol
5-Methyl-2-furfuryl alcohol
2,5-Dimethyl-4-hydroxy-3-(2H)-furanone
2-Hydroxy-3-methyl-2-cyclopenten-1-one
3,4-Dimethyl-2-hydroxy-2-cyclopenten-1-one
3,5-Dimethyl-2-hydroxy-2-cyclopenten-1-one
3-Ethyl-2-hydroxy-2-cyclopenten-1-one
γ-Butyrolactone

Ammonia catalyzes the same reactions. However, N-containing compounds are also formed, e.g. 1-amino-1-deoxyglycoses (glycosylamines), diglycosylamines and 1-amino-1-deoxyglyculoses. The mechanisms involved in their formation are covered in the section on N-glycosides (4.2.4.4). Reactive intermediary products can polymerize further into high molecular weight, brown pigments, or form a number of imidazole, pyrazine and pyridine derivatives.

4.2.4.3.3 Caramelization

Brown-colored products with a typical caramel aroma are obtained by melting sugar or by heating sugar syrup in the presence of acidic and/or alkaline catalysts. The reactions involved were covered in the previous two sections. The process can be directed more towards aroma formation or more towards brown pigment accumulation. Heating of saccharose syrup in a buffered solution enhances molecular fragmentation and, thereby, formation of aroma substances. Primarily dihydrofuranones, cyclopentenolones, cyclohexenolones and pyrones are formed (cf. 4.2.4.3.2). On the other hand, heating glucose syrup with sulfuric acid in the presence of ammonia provides intensively-colored polymers ("sucre couleur"). The stability and solubility of these polymers are enhanced by bisulfite anion addition to double bonds:

$$>C=C< \xrightarrow{SO_3H^{\ominus}} >CH-C(-SO_3^{\ominus})< \quad (4.41)$$

4.2.4.4 Reactions with Amino Compounds (N-Glycosides, Maillard Reaction)

The interaction of amino compounds with monosaccharides initially probably involves addition of a carbonyl group to a primary amino group of an amino acid, peptide or protein, followed by water elimination, leading to an intermediary imine, which cyclizes to a glycosylamine (N-glycoside):

$$H-C(R)=O + H_2N-R_1 \longrightarrow H-C(OH)(R)-NH-R_1 \xrightleftharpoons[+H_2O]{-H_2O} H-C(R)=N-R_1 \rightleftharpoons \text{D-Glucosylamine} \quad (4.42)$$

D-Glucose

D-Glucosylamine

Unlike O-glycosides, N-glycosides exhibit mutarotation. This results from acid-catalyzed isomerization of the compound through an intermediary open-chain immonium ion:

(4.43)

N-glycosides are widely distributed in nature (nucleic acids, NAD^+, Coenzyme A). They are generated in food whenever a reducing sugar occurs together with a compound containing an amino group, e.g. free amino acids, peptides, proteins or amines. They are obtained more readily at higher temperature or in food with lower water activity.

Aldosylamine, which is the initial product from aldose, will rearrange to a 1-amino-1-deoxyketose (an *Amadori* rearrangement). Ketosylamine can yield 2-amino-2-deoxyaldose via a *Heyns* rearrangement. Both reactions correspond to alkali-catalyzed isomerization of aldoses and ketoses, and both start with an immonium ion and proceed through an endiol-like enaminol stage:

$$H-C(=\overset{\oplus}{N}H-R)-C(H)(OH)- \rightleftharpoons H-C(-NH-R)=C(-OH)- \rightleftharpoons H-C(H)(-NH-R)-C(=O)-$$

1-Amino-1-desoxyketose

(4.44 a)

$$H-C(H)(OH)-C(=\overset{\oplus}{N}H-R)- \rightleftharpoons H-C(-OH)=C(-NH-R)- \rightleftharpoons H-C(=O)-C(H)(-NH-R)-$$

2-Amino-2-desoxy aldose

(4.44 b)

1-amino-1-deoxyketoses are found in various foods, such as dried fruits (peaches, apricots), dehydrated vegetables, milk powder or liver extracts.

In milk powder, the reaction occurs between the ε-amino group of protein and the disaccharide lactose (cf. Reaction 4.45). The initially-formed lactosylamine is changed by an *Amadori* rearrangement into a protein-bound N-alkyl-1-amino-1-deoxylactulose which, after acidic hydrolysis, yields fructoselysine. This is then transformed into the unusual amino acids furosine and pyridosine, which occur in milk powder hydrolysate and other food hydrolysates. Both acids serve as reliable indicator compounds for revealing sugar-amine interaction.

Glycosylamines and *Amadori* rearrangement products are, however, only intermediary compounds in a sequence of reactions denoted as "*Maillard* reaction" or as "nonenzymatic browning". These reactions actually correspond to those already outlined for acid/base-catalyzed conversions of monosaccharides. Starting with N-intermediary products and therefore containing a basic nitrogen atom, which act as catalysts within the molecule, these reactions proceed at

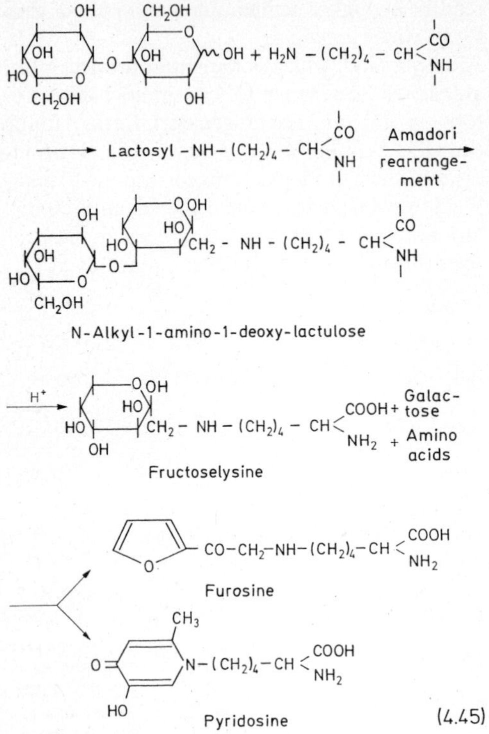

Lactosyl – NH – $(CH_2)_4$ – CH... Amadori rearrangement

N-Alkyl-1-amino-1-deoxy-lactulose

Fructoselysine

Furosine

Pyridosine

(4.45)

substantially higher rates and under substantially milder conditions present in many foods. These reactions provide brown pigments, known as melanoidins, which are nitrogenous polymers and copolymers with variable contents of nitrogen and differing molecular weights and solubilities in water.

An insight into their structures is provided by studies involving model systems. Heating of pentoses with amino compounds (amino acids, primary and secondary amines) in nearly neutral aqueous solutions yields a significant amount of 4-hydroxy-5-methyl-2,3-dihydrofuran-3-one (I). The CH-acid compound condenses with aldehydes and ketones (which also originate from sugar rearrangement reactions) to form colored compounds (II, III; cf. Reaction 4.45a).

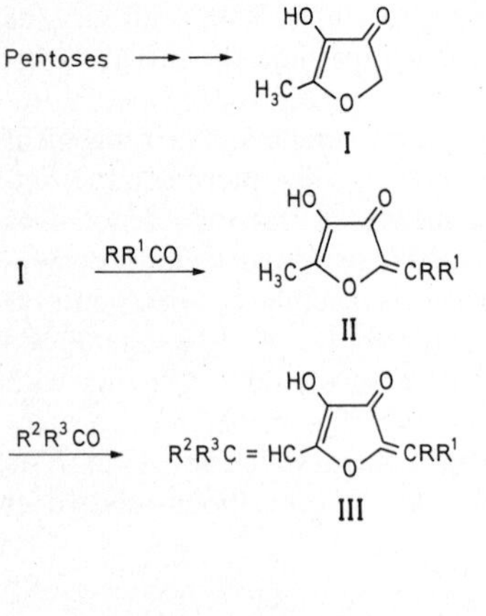

Pentoses → I; I + RR'CO → II; + R^2R^3CO → III (4.45a)

The orange-colored compound, IV, is obtained from xylose and glycine in aqueous solution and compound V from xylose and diethylamine in ethanol:

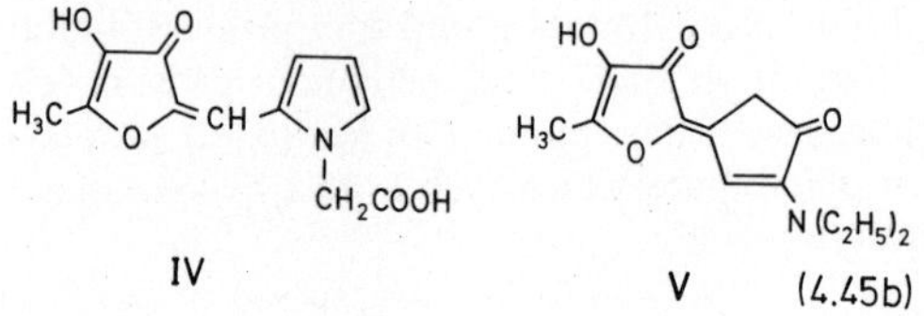

IV V (4.45b)

Maillard reaction of hexoses yields, among others, methylene reductonic acid (VI) and the dihydro-γ-pyrone VIII, which also can condense to give colored products (VII, IX):

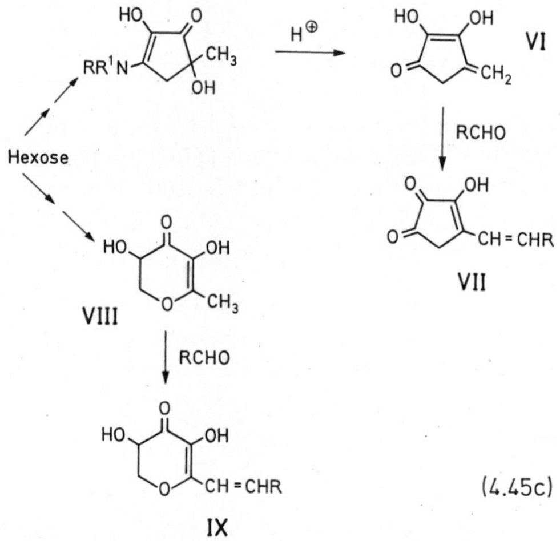

Hexose → VI (RCHO) → VII; Hexose → VIII (RCHO) → IX (4.45c)

While the pathway to the lactam X (λ_{max} = 406 nm, in methanol) is yet unknown, the acylpyrrole XI, may be formed via sugar retroaldol cleavage and recombination of the fragments of the primary amine. The orange-colored compound, XII (λ_{max} = 478 nm, in methanolic HCl) is obtained by heating glucose/piperidinium acetate in ethanol. It contains the furanone ring as a CH-acid component which is characteristic for pentoses. Formaldehyde elimination is probably followed by a retroaldol reaction.

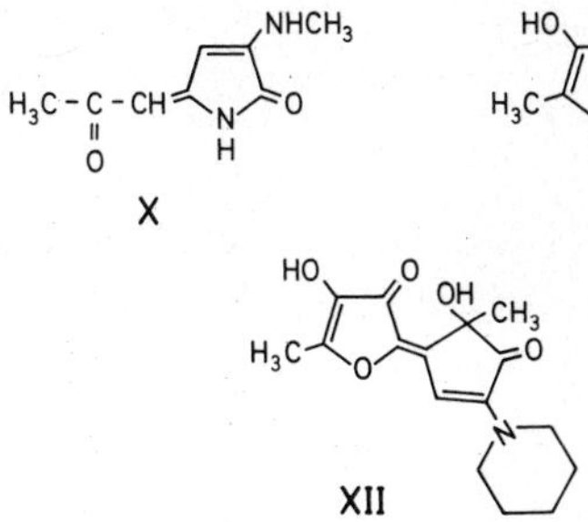

X XI XII (4.45d)

It is evident that color formation by the *Maillard* reaction involves both C–C covalent bond cleavages and subsequent recombinations.

Furthermore, volatile compounds with strong odor intensity are formed. Browning and aroma formation accompanying cooking, frying, baking or roasting are essentially caused by the *Maillard* reaction; as are the losses in essential amino acids (lysine and methionine) and undesired discoloration and off-flavor development in food, mostly in its dehydrated form during storage or heat treatment, such as in pasteurization, sterilization or roasting.

A pathway, which is outlined in Reaction 4.46, leads from glycosylamine through 3-deoxy-glycosulose to hydroxymethyl furfural.

Parallel to the above reaction, but to a lesser extent, 1-amino-1-deoxyketose, through a 2,3-enolization and subsequent amine elimination, provides 1-deoxy-2,3-glycodiulose. This compound is transformed into maltol and isomaltol and, in addition, provides degradation fragments which might enter into a number of secondary reactions. However maltol is formed under normal conditions of the *Maillard* reaction only from disaccharides and not from monosaccharides (cf. Reaction 4.46 a).

In both pathways the amino compound is eliminated in the initial stages of the reaction, but it can later interact with intermediary products.

Reaction 4.46 a summarizes important products and probable pathways of the *Maillard* reaction. The compounds mentioned were isolated from model systems or from food:

Sugars (mono- and disaccharides) react with amino compounds via glycosyl amines and *Amadori* compounds to 1-, 3- and 4-deoxy compounds (I–III).

Dehydration of the *1-deoxy compounds* of mono- and disaccharides yields 1,6-di-deoxy compounds (IV a, IV b), which can react with secon-

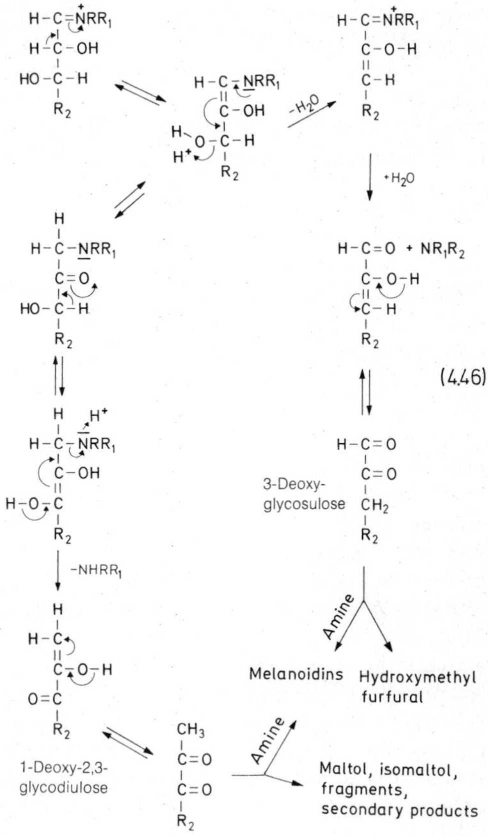

3-Deoxy-glycosulose

Melanoidins Hydroxymethyl furfural

1-Deoxy-2,3-glycodiulose

Maltol, isomaltol, fragments, secondary products

(4.46)

dary amines to aminohexose reductones (V) and with primary amines to pyrrolinone reductones (VI), respectively. Until now a formation of maltol and of isomaltol derivatives via 1,6-dideoxy compounds from 1-deoxy compounds was observed only for disaccharides and not for monosaccharides. Lactose, for instance, yields mainly β-galactosyl isomaltol, while from maltose mainly maltol and not α-glucosylisomaltol is obtained. Pyridones (X) are formed in the presence of primary amines. The corresponding lysine derivative (X: $R^3 = (CH_2)_4\text{–}CHNH_2\text{–}COOH$) occurs, bound to protein, in heated milk products and may be determined after acid hydrolysis during amino acid analysis, together with pyridosin (cf. formula 4.45), which has the same retention time.

A typical *Maillard* compound from monosaccharides is the dihydropyranone XI, which can be formed only if $R^1 = H$ and which has been isolated from food. Under normal *Maillard* reac-

tion conditions no dehydration of XI to yield maltol (VIII) was observed.
Condensation of 1-deoxy-compounds (in their open and cyclic forms, respectively) with aldehydes yields colored compounds (XII, XIII). Further colored compounds, for instance VII in Reaction 4.45c, can be formed in an acidic medium via methylene reductinic acid (VI in Reaction 4.45c). The colored compounds XIV and XV were obtained from pentoses.
3-Deoxy compounds (II) yield furfural (XVIII, R=H) or hydroxymethyl furfural (XVIII, R=CH_2OH) and pyrrol (XIX) or pyridine derivatives (XX) in the presence of primary amines. The 3,4-dideoxy compound (XVII) can condense with aldehydes to give compound XXI, which may be stabilized by formation of the acetal XXII. Further possible reactions of XXI are rearrangement (XVIIb) and additional condensation with aldehydes to yield XXIII or (in the presence of primary amines) XXIV.
The *4-deoxy compounds* (III) yield hydroxyacetyl furan (XXV), and (in the presence of primary amines) the pyrrole and pyridine derivatives XXVI and XXVII, respectively.
A further reaction sequence, the *Strecker* degradation, starts from α-dicarbonyl compounds, which occur as intermediary- or end-products of other decomposition reactions, and amino acids. Their interaction, involving transamination, provides an aminoketone, an aldehyde and CO_2 (cf. 1.2.4.2.4):

$$R-CH(NH_2)-COOH \xrightarrow[-H_2O]{>C(=O)-C(=O)<} \;>C=N-C(R)(H)-COOH,\ >C=O \longrightarrow\; >C(OH)=C<-N=C(R)-COOH \xrightarrow{H_2O} H_2N-C=C-OH + R-C(=O)-COOH \xrightarrow{-CO_2} R-CHO \qquad (4.47)$$

This reaction pathway occurs in food at higher concentrations of free amino acids and under more drastic reaction conditions, e.g. higher temperatures or pressures. The aldehydes formed, often called *Strecker* aldehydes, can act

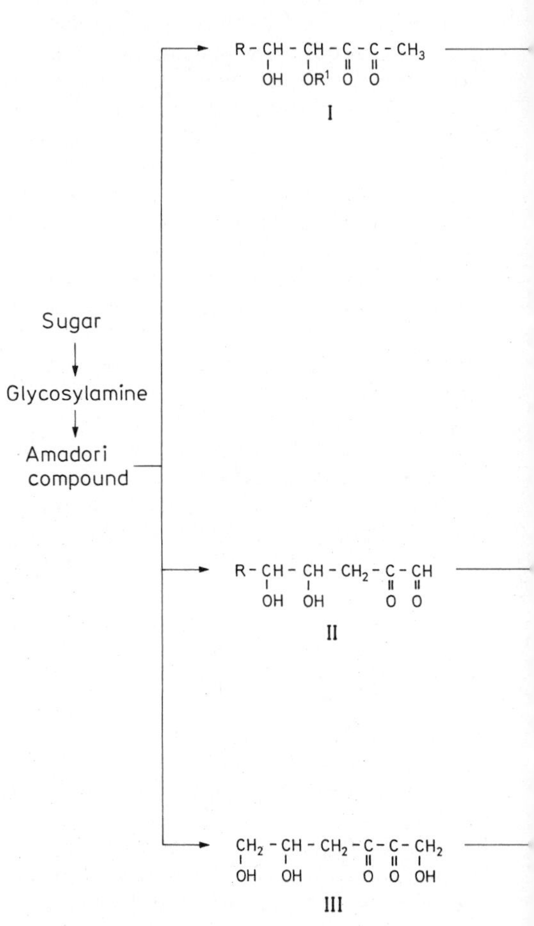

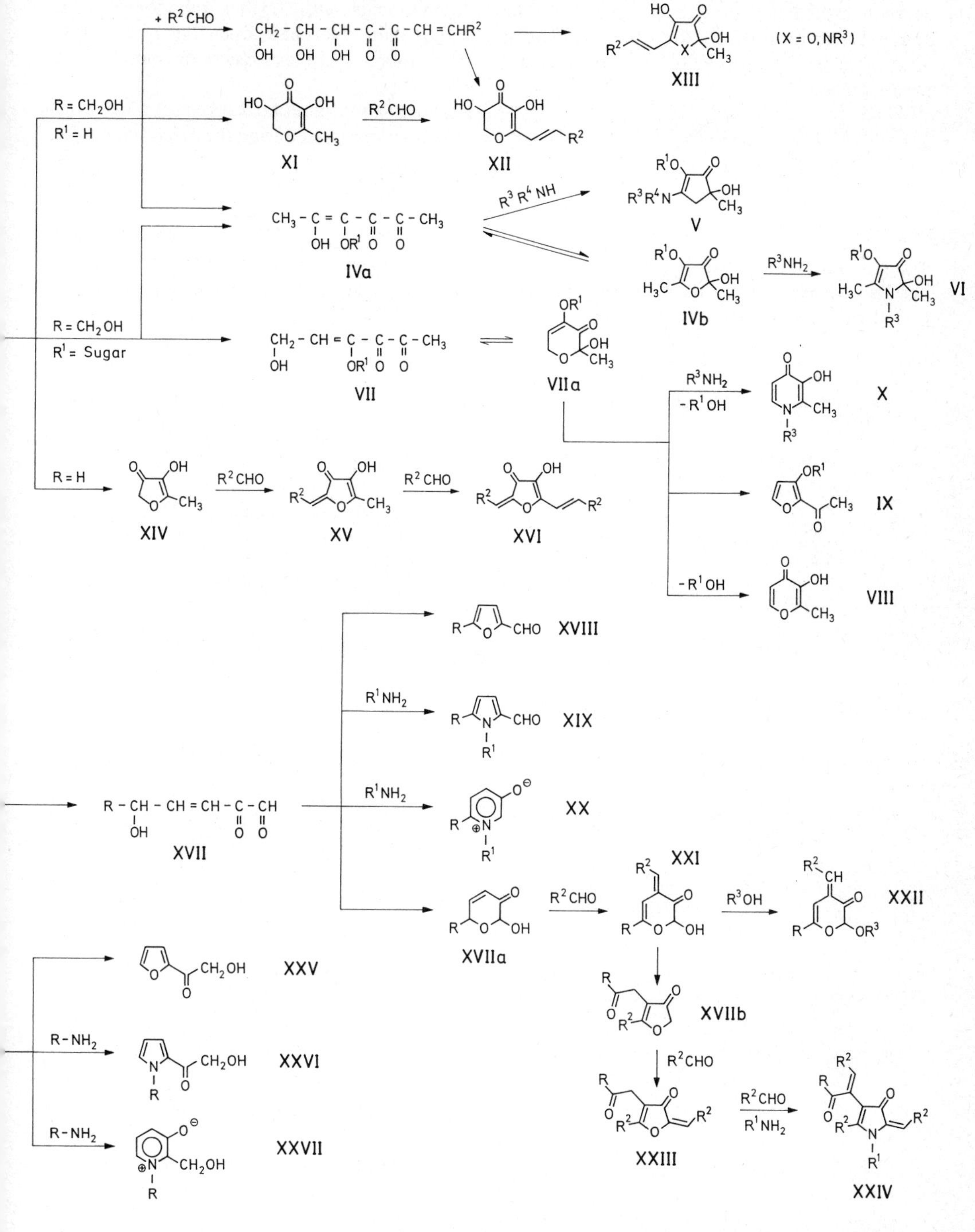
+ R2 CHO
(X = O, NR3)
XIII
R = CH2OH
R1 = H
R2 CHO
XI
XII
R3 R4 NH
V
IVa
R3NH2
IVb
VI
R = CH2OH
R1 = Sugar
VII
VIIa
R3NH2
- R1 OH
X
IX
VIII
R = H
R2 CHO
XIV
XV
XVI
XVIII
R1NH2
XIX
XX
XVII
XXI
R3OH
XXII
XVIIa
XVIIb
XXIII
R1NH2
XXIV
XXV
R-NH2
XXVI
XXVII

(4.46 a)

as food aroma constituents (cf. 5.3.1.1). The aminoketone formed, on the other hand, can yield pyrazine derivatives (also powerful aroma constituents):

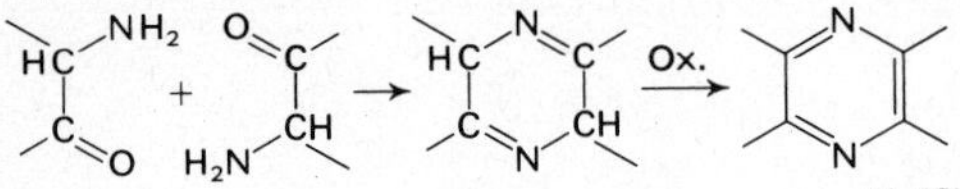

(4.48)

Transformation into pyrrole compounds is also possible by reaction of an α-dicarbonyl compound with a 1-amino-1-deoxyketose:

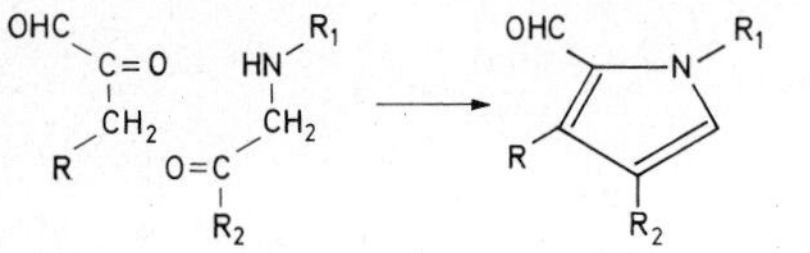

(4.49)

Amino acids with further functional groups within their side chain open the possibility of even more complex reactions (cf. 5.3.1.6–5.3.1.10).

Reaction 4.49 a summarizes the pathways, which have been observed during the *Strecker* reaction: The *3-deoxy compound* (I) is formally reduced to the endiol II and yields in the reaction with primary amino acids (pathway a) the products III to V. Maltoxazin (VI) ist formed via the reaction with proline (pathway b), while the reaction with hydroxyproline yields the furan derivative VII (pathway c).

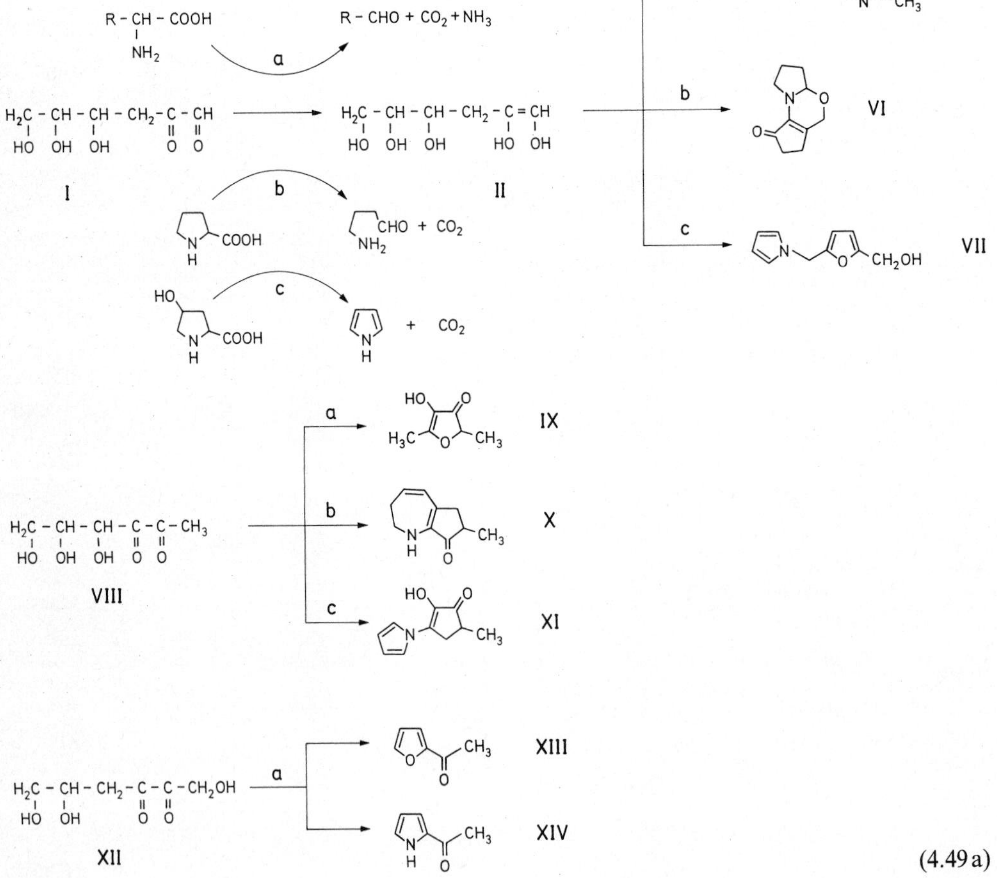

(4.49 a)

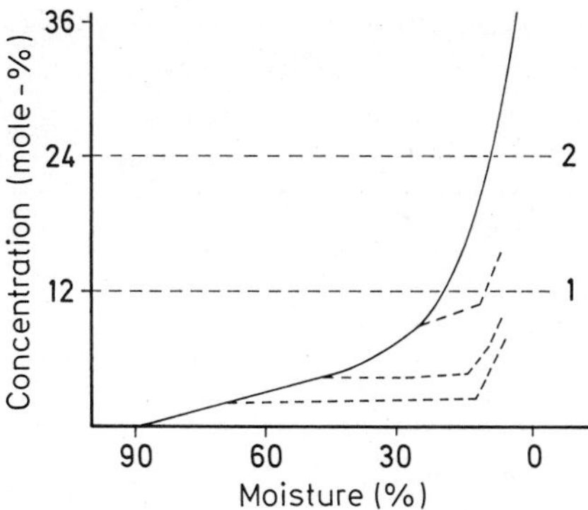

Fig. 4.7a. Concentration increase of Amadori compounds in two stage air drying of carrots as influenced by carrot moisture content. (— 10, 20, 30 min at 110° C; --- 60 °C; sensory assessment: 1) detection threshold 2) quality limit. (After *Eichner* and *Wolf,* in *Waller* and *Feather,* 1983)

The products IX to XI are related to the *1-deoxy compound* (VIII) and the products XIII and XIV to the *4-deoxy compound,* respectively.

Measures to curtail or inhibit the *Maillard* reaction, when it is undesirable in food, involve lowering of pH, handling of food at low temperatures, avoiding the critical range of water activity (cf. Figure 0.4) during food processing and storage, use of nonreducing sugars, and use of sulfite as an additive. Figure 4.7a demonstrates, by the example of carrot dehydration the advantages of running a two-stage process to curtail the *Maillard* reaction.

4.2.4.5 Reactions with Hydroxy Compounds (O-Glycosides)

The lactol group of monosaccharides heated in alcohol in the presence of an acid catalyst is substituted by an alkoxy or aryloxy group, denoted as an aglycone (*Fischer* synthesis), to produce alkyl- and arylglycosides. It is assumed that the initial reaction involves the open form. With the majority of sugars, the furanosides are formed in the first stage of reaction. They then equilibrate with the pyranosides. The transition from furanoside to pyranoside occurs most probably through an open carboxonium-ion, whereas pyranoside isomerization is through a cyclic one (cf. Reaction 4.50).

Furanosides are obtainable by stopping the reaction at a suitable time. The equilibrium state in

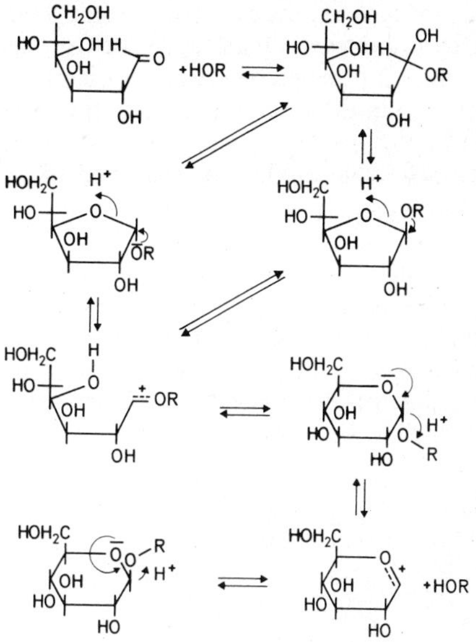

(4.50)

alcohol is, as in water, dependent on conformational factors. The alcohol as solvent and its R-moiety both increase the anomeric effect and thus α-pyranoside becomes a more favorable form than was α-pyranose in aqueous free sugar solutions (Table 4.7). In the system D-glucose/methanol in the presence of 1% HCl, 66% of the methylglucoside is present as α-pyranoside, 32.5% as β-pyranoside, and only 0.6% and 0.9% are in α- and β-furanoside forms. Under the same conditions, D-mannose and D-galactose are 94% and 58% respectively, in α-pyranoside forms.

A highly stereospecific access to glycosides is possible by C-1 bromination of acetylated sugars.

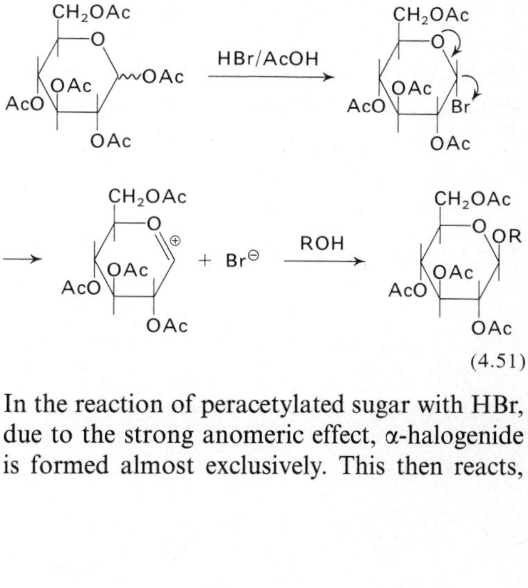

(4.51)

In the reaction of peracetylated sugar with HBr, due to the strong anomeric effect, α-halogenide is formed almost exclusively. This then reacts, probably through its glycosyl cation form. Due to the steric influence of the acetylated group on C-2, the 1,2-trans-glycoside is preferentially obtained, e.g. in the case of D-glucose, β-glucoside results.

Acetylglycosyl halogenides are also used for a highly stereoselective synthesis of α-glycosides. The compound is first dehalogenated into a glycal. Then, addition of nitrosylchloride follows, giving rise to 2-deoxy-2-nitrosoglycosylchloride. The latter, in the presence of alcohol, eliminates HCl and provides a 2-deoxy-2-oximino-α-glycoside. Reaction with ethanal yields the 2-oxo compound, which is then reduced to α-glycoside:

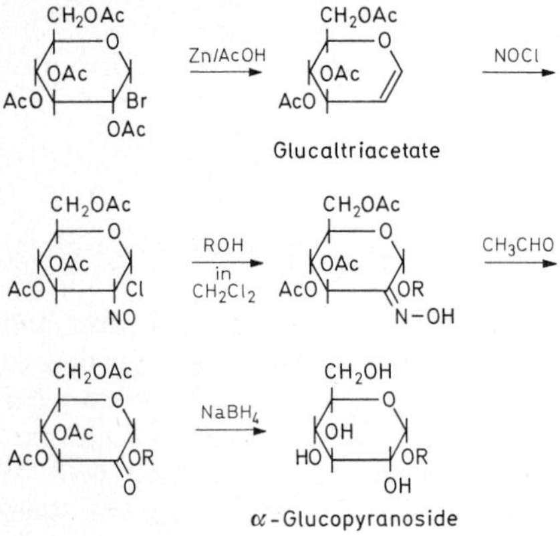

Zn/AcOH; Glucaltriacetate; NOCl; ROH in CH_2Cl_2; CH_3CHO; $NaBH_4$

α-Glucopyranoside (4.52)

O-Glycosides are widely distributed in nature and are the constituents, such as glycolipids, glycoproteins, flavanoid glycosides or saponins, of many foods.

O-Glycosides are readily hydrolyzed by acids. Hydrolysis by alkalies is achieved only under drastic conditions which simultaneously decompose monosaccharides.

The acid hydrolysis is initiated by glycoside protonation. Alcohol elimination is followed by addition of water:

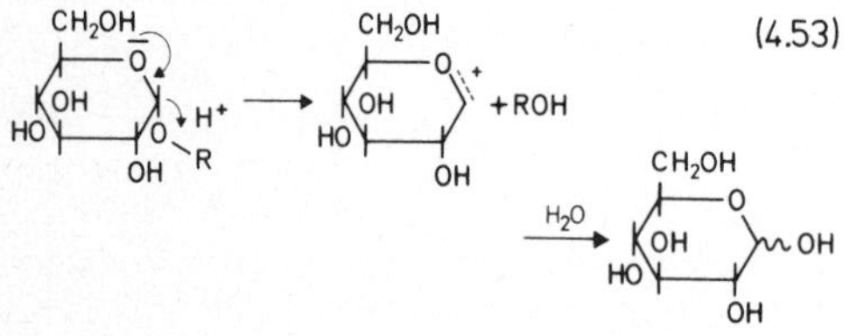

(4.53)

The hydrolysis rate is dependent on aglycone and the monosaccharide itself. The most favored form of alkylglycoside, α-pyranoside, usually is

Table 4.12. Relative rate of hydrolysis of glycosides (a: 2 M HCl, 60 °C; b: 0.5 M HCl, 75 °C)

Compound	Hydrolysis condition	k_{rel}
Methyl-α-D-glucopyranoside	a	1.0
Methyl-β-D-glucopyranoside	a	1.8
Phenyl-α-D-glucopyranoside	a	53.7
Phenyl-β-D-glucopyranoside	a	13.2
Methyl-α-D-glucopyranoside	b	1.0
Methyl-β-D-glucopyranoside	b	1.9
Methyl-α-D-mannopyranoside	b	2.4
Methyl-β-D-mannopyranoside	b	5.7
Methyl-α-D-galactopyranoside	b	5.2
Methyl-β-D-galactopyranoside	b	9.2

the isomer most resistant to hydrolysis. This is also true for arylglycosides, however, due to steric effects, the β-pyranoside isomer is synthesized preferentially and so the β-isomer better resists hydrolysis.

The influence of the sugar moiety on the rate of hydrolysis is related to the conformational stability. Glucosides with high conformational stability are hydrolyzed more slowly (cf. data compiled in Table 4.12).

4.2.4.6 Esters

Esterification of monosaccharides is achieved by reaction of the sugar with an acyl halide or an acid anhydride. Acetylation, for instance with acetic anhydride, is carried out in pyridine solution:

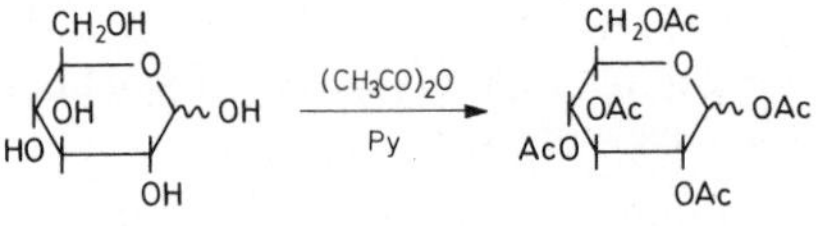

$(CH_3CO)_2O$, Py (4.54)

Acyl groups have a protective role in some synthetic reactions. Aldonitrile acetates are analytically suitable sugar derivatives for gas chromatographic separation and identification of sugars. An advantage of these compounds is that they simplify a chromatogram since there are no anomeric peaks:

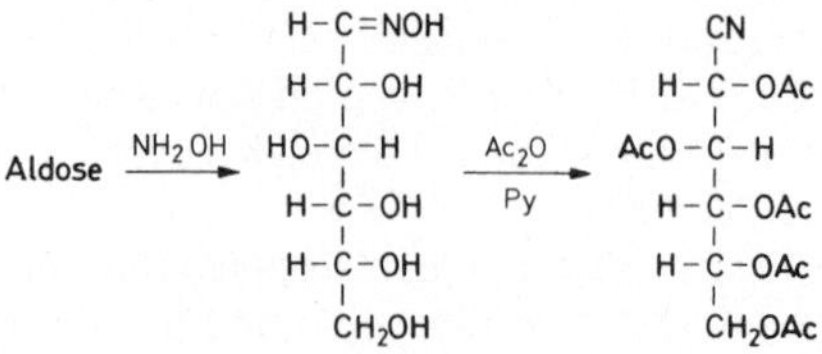

Aldose $\xrightarrow{NH_2OH}$ H−C=NOH / H−C−OH / HO−C−H / H−C−OH / H−C−OH / CH_2OH $\xrightarrow{Ac_2O,\ Py}$ CN / H−C−OAc / AcO−C−H / H−C−OAc / H−C−OAc / CH_2OAc (4.54a)

Selective esterification of a given HO-group is also possible. For example, glucose can be selectively acetylated in position 3 by reacting 1,2,5,6-di-O-isopropylidene-α-D-glucofuranose with acetic acid anhydride, followed by hydrolysis of the diketal:

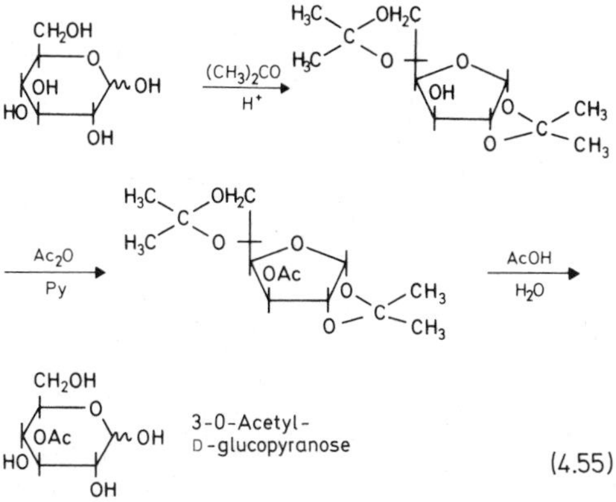

3-O-Acetyl-D-glucopyranose (4.55)

Hydrolysis of acyl groups can be achieved by interesterification or by an ammonolysis reaction:

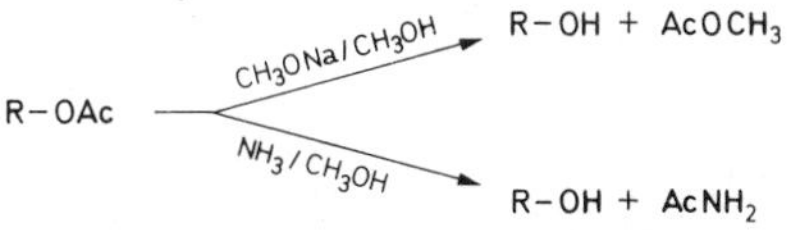

$$R{-}OAc \xrightarrow{CH_3ONa/CH_3OH} R{-}OH + AcOCH_3 \qquad R{-}OAc \xrightarrow{NH_3/CH_3OH} R{-}OH + AcNH_2 \quad (4.56)$$

Sugar esters are also found widely in nature. Phosphoric acid esters are important intermediary products of metabolism, while sulfuric acid esters are constituents of some polysaccharides. Examples of organic acid esters are vacciniin in blueberry (6-benzoyl-D-glucose) and the tannin-type compound, corilagin (1,3,6-trigalloyl-D-glucose):

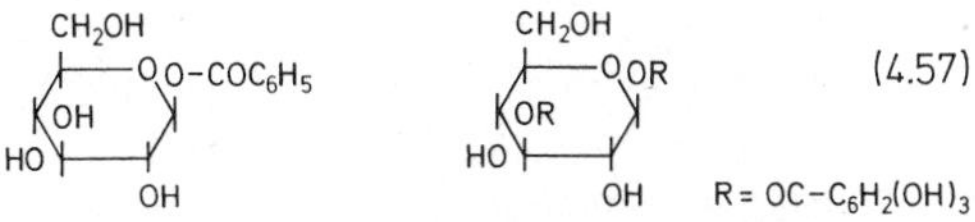

(4.57) $R = OC{-}C_6H_2(OH)_3$

Sugar esters or sugar alcohol esters with long chain fatty acids (lauric, palmitic, stearic and oleic) are produced industrially and are very important as surface-active agents. These include sorbitan fatty acid esters (cf. 8.15.4.3) and those of saccharose (cf. 8.15.4.2), which have diversified uses in food processing (cf. Table 8.15).

4.2.4.7 Ethers

Methylation of sugar HO-groups is possible using dimethylsulfate or methyliodide as the methylating agent. Methyl esters are of importance in analysis of sugar structure since they provide data about ring size and linkage positions. Permethylated saccharose, for example, after acid hydrolysis provides 2,3,4,6-tetra-O-methyl-D-glucose and 1,3,4,6-tetra-O-methyl-D-fructose. This suggests the presence of a 1,2-linkage between the two sugars and the pyranose and furanose structures for glucose and fructose, respectively:

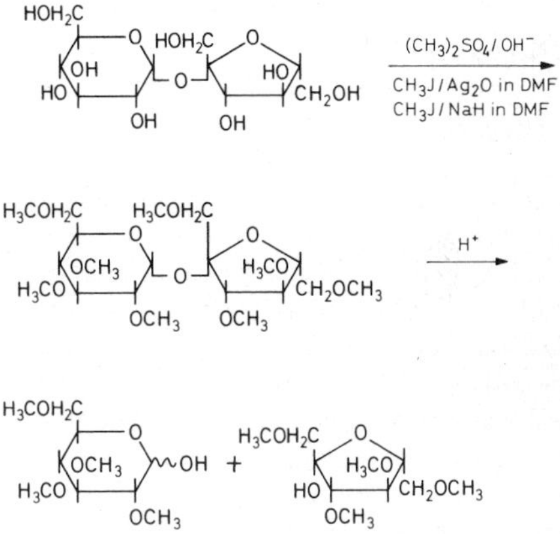

$(CH_3)_2SO_4/OH^-$; CH_3J/Ag_2O in DMF; CH_3J/NaH in DMF; H^+ (4.58)

Trimethylsilyl ethers (TMS-ethers) are unstable against hydrolysis and alcoholysis, but have remarkable thermal stability and so are suitable for gas chromatographic sugar analysis. Treatment of a sugar with hexamethyldisilazane and trimethylchlorosilane, in pyridine as solvent, provides a sugar derivative with all HO-groups silylated:

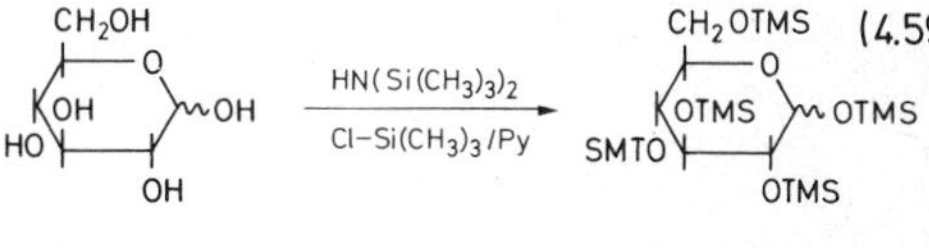

$HN(Si(CH_3)_3)_2$; $Cl{-}Si(CH_3)_3/Py$ (4.59)

4.2.4.8 Oxidative Cleavage of 1,2-Diols

Oxidative cleavage of vicinal dihydroxy groups or hydroxy-amino groups of a sugar with lead tetraacetate or periodate is of importance for structural elucidation. Fructose, in a 5-membered furanose form, consumes 3 moles of periodate (splitting of each α-glycol group requires 1 mole of oxidant) while, in a pyranose ring form, it consumes 4 moles of periodate.

Saccharose consumes 3 moles and maltose 4 moles of periodate (cf. Reaction 4.60).
The final conclusion as to sugar linkage positions and ring structure is drawn from the periodate consumption, the amount of formic acid produced (in the case of saccharose, 1 mole; maltose, 2 moles) and the other carbonyl fragments which are oxidized additionally by bromine to stable carboxylic acids and then released by hydrolysis. The glycol splitting reaction should be considered an optional or complementary method to the permethylation reaction applied in structural elucidation of carbohydrates.

4.3 Oligosaccharides

4.3.1 Structure and Nomenclature

Monosaccharides are able to form glycosides (cf. 4.2.5.4). When this occurs between the lactol group of one monosaccharide and any HO-group of a second monosaccharide, a disaccharide results.
Compounds with up to about 10 monosaccharide residues are designated as oligosaccharides. When a glycosidic linkage is established only between the lactol groups of two monosaccharides, then a *nonreducing disaccharide* is formed, and when one lactol group and one alcoholic HO-group are involved, a *reducing disaccharide* results. The former is denoted as a glycosylglycoside, the latter as a glycosylglycose, with additional data for linkage direction and positions. Examples are saccharose and maltose:

β-D-Fructofuranosyl-α-D-glucopyranoside (Saccharose)

O-α-D-Glucopyranosyl-(1 → 4)-D-glucopyranose (Maltose) (4.61)

$3\,JO_4^-$; $+HCOOH$; Br_2, H_2O ; H^+

D-Glyceric acid + Hydroxypyruvic acid + Glycoxalic acid + D-Glyceric acid

(4.60)

$4\,JO_4^-$; $+HCOOH$; $+HCOOH$; Br_2, H_2O ; H^+

CO_2 + Erythronic acid + Glycoxalic acid + D-Glyceric acid

An abbreviated method of nomenclature is to use a three letter designation or symbol for a monosaccharide and a suffix *f* or *p* for furanose or pyranose. For example, saccharose and maltose can be written as O-β-D-Fru*f*(2 → 1)α-D-Glc*p* and O-α-D-Glc*p*(1 → 4)D-Glcp, respectively.
Branching also occurs in oligosaccharides. It results when one monosaccharide is bound to two

glycosyl residues. The name of the second glycosyl residue is inserted into square brackets. A trisaccharide which represents a building block of the branched chain polysaccharides amylopectin and glycogen is given as an example:

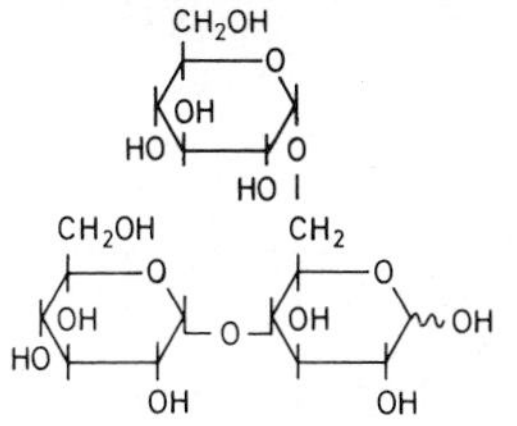

(4.62)

O-α-D-Glucopyranosyl-(1→4)-O-[α-D-glucopyranosyl-(1→6)]-D-glucopyranose

The abbreviated formula for this trisaccharide is as follows:

αD Glc*p* (1→4) Glc*p*
(6
↑
1) αGlc*p* (4.63)

The conformations of oligo- and polysaccharides, like peptides, can be described by providing the angles Φ and ψ:

(4.64)

A calculation of conformational energy for all conformers with allowed Φ, ψ pairs provides a Φ, ψ diagram with lines corresponding to isoconformational energies. The low-energy conformations calculated in this way agree with data obtained experimentally (X-ray diffraction, NMR, ORD) for oligo- and polysaccharides.
H-bonds fulfill a significant role in conformer stabilization. Cellobiose and lactose conformations are well stabilized by an H-bond formed between the HO-group of C-3 in the glycose residue and the ring oxygen of the glycosyl residue. Conformations in aqueous solutions appear to be similar to those in the crystalline state:

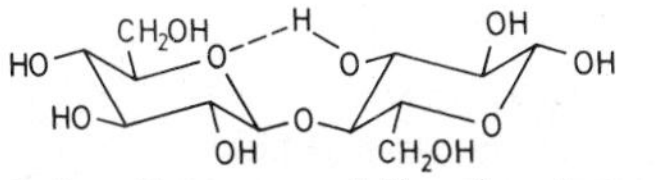

(4.65)

O-β-D-Glucopyranosyl-(1→4)-D-glucopyranose (Cellobiose)

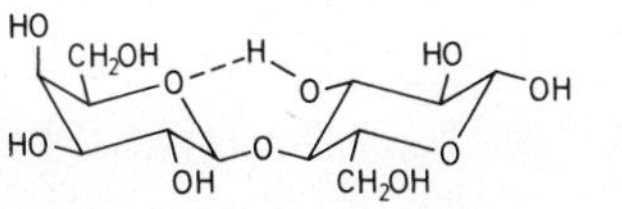

O-β-D-Galactopyranosyl-(1→4)-D-glucopyranose (Lactose)

In crystalline maltose and in nonaqueous solutions of this sugar, a hydrogen bond is established between the HO-groups on C-2 of the glucosyl and on C-3 of the glucose residues (4.66a), while in aqueous solution a conformer partially present is stabilized by H-bonds established between the CH_2OH-group of the glucosyl residue and the HO-group of C-3 on the glucose residue (4.66b). Both conformers (a, b in Formula 4.66) correspond to the two energy minima in the Φ, ψ diagram.

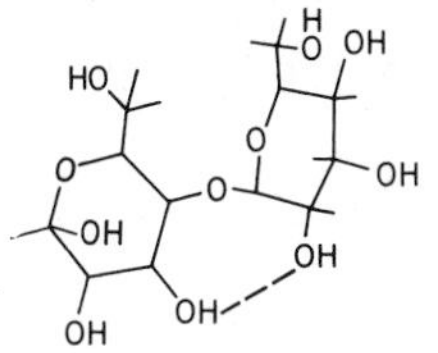

(4.66a)

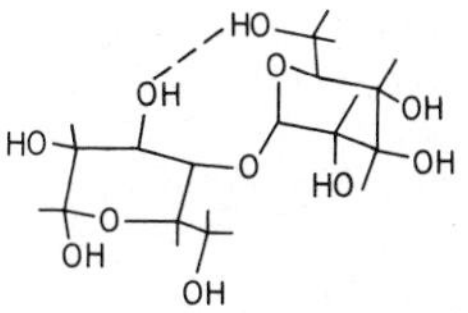

(4.66b)

Two H-bonds are possible in saccharose, the first between the HO-groups on the C-1 of the fructose and the C-2 of the glucose residues, and the second between the HO-group on the C-6 of the fructose residue and the ring oxygen of the glucose residue:

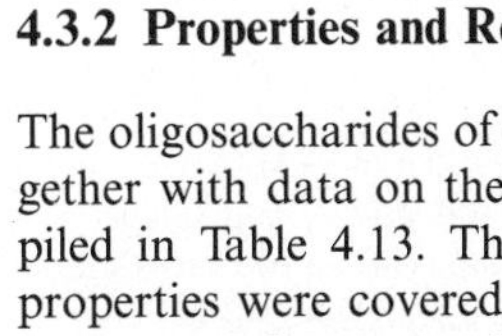

(4.67)

β-D-Fructofuranosyl-α-D-glucopyranoside (Saccharose)

4.3.2 Properties and Reactions

The oligosaccharides of importance to food, together with data on their occurrence, are compiled in Table 4.13. The physical and sensory properties were covered with monosaccharides, as were reaction properties, though the difference between reducing and nonreducing oligosaccharides should be mentioned. The latter do not have a free lactol group and so lack reducing properties, mutarotation and the ability to react with alcohols and amines.

Table 4.13. Structure and occurrence of oligosaccharides

Name	Structure	Occurrence
Disaccharides		
Cellobiose	O-β-D-Glc*p*-(1 → 4)-D-Glc*p*	Building block of cellulose
Gentiobiose	O-β-D-Glc*p*-(1 → 6)-D-Glc*p*	Glycosides (amygdalin)
Isomaltose	O-α-D-Glc*p*-(1 → 6)-D-Glc*p*	Found in mother liquor during glucose production from starch
Lactose	O-β-D-Gal*p*-(1 → 4)-D-Glc*p*	Milk
Lactulose	O-β-D-Gal*p*-(1 → 4)-D-Fru*f*	Conversion product of lactose
Maltose	O-α-D-Glc*p*-(1 → 4)-D-Glc*p*	Building block of starch, sugar beet, honey
Maltulose	O-α-D-Glc*p*-(1 → 4)-D-Fru*f*	Conversion product of maltose, honey, beer
Melibiose	O-α-D-Gal*p*-(1 → 6)-D-Glc*p*	Cacao beans
Neohesperidose	O-α-L-Rha*p*-(1 → 2)-D-Glc*p*	Glycosides (naringin, neohesperidin)
Neotrehalose	O-α-D-Glc*p*-(1 → 1)-β-D-Glc*p*	Koji extract
Nigerose	O-α-D-Glc*p*-(1 → 3)-D-Glc*p*	Honey, beer
Palatinose	O-α-D-Glc*p*-(1 → 6)-D-Fru*f*	Microbial product of saccharose
Rutinose	O-α-L-Rha*p*-(1 → 6)-D-Glc*p*	Glycosides (hesperidin)
Saccharose	O-β-D-Fru*f*-(2 → 1)-α-D-Glc*p*	Sugar beet, sugar cane, spread widely in plants
Trehalose	O-α-D-Glc*p*-(1 → 1)-αD-Glc*p*	Ergot *(Claviceps purpurea)*, young mushrooms
Trisaccharides		
Fucosidolactose	O-α-D-Fuc*p*-(1 → 2)-O-β-α-Gal*p*-(1 → 4)-D-Gal*p*	Human milk
Gentianose	O-β-D-Glc*p*-(1 → 6)-O-α-D-Glc*p*-(1 → 2)-β-D-Fru*f*	Gentian rhizome
Isokestose (1-Kestose)	O-α-D-Glc*p*-(1 → 2)-O-β-D-Fru*f*-(1 → 2)-β-D-Fru*f*	Product of saccharase action on saccharose as a substrate
Kestose (6-Kestose)	O-α-D-Glc*p*-(1 → 2)-O-β-D-Fru*f*-(6 → 2)-β-D-Fru*f*	Saccharose subjected to yeast saccharase activity, honey
Maltotriose	O-α-D-Glc*p*-(1 → 4)-O-α-D-Glc*p*-(1 → 4)-D-Glc*p*	Degradation product of starch, starch syrup
Manninotriose	O-α-D-Gal*p*-(1 → 6)-O-α-D-Gal*p*-(1 → 6)-D-Glc*p*	Manna
Melezitose	O-α-D-Glc*p*-(1 → 3)-O-β-D-Fru*f*-(2 → 1)-α-D-Glc*p*	Manna, nectar
Neokestose	O-β-D-Fru*f*-(2 → 6)-O-α-D-Glc*p*-(1 → 2)-β-D-Fru*f*	Product of saccharase action on saccharose as a substrate
Panose	O-α-D-Glc*p*-(1 → 6)-O-α-D-Glc*p*-(1 → 4)-D-Glc*p*	Degradation product of amylopectin, honey
Raffinose	O-α-D-Gal*p*-(1 → 6)-O-α-D-Glc*p*-(1 → 2)-β-D-Fru*f*	Sugar beet, sugar cane, widely distributed in plants
Umbelliferose	O-α-D-Gal*p*-(1 → 2)-O-α-D-Glc*p*-(1 → 2)-β-D-Fru*f*	Umbelliferae roots
Tetrasaccharides		
Maltotetraose	O-α-D-Glc*p*-(1 → 4)-O-α-D-Glc*p*-(1 → 4)-O-α-D-Glc*p*-(1 → 4)-D-Glc*p*	Starch syrup
Stachyose	O-α-D-Gal*p*-(1 → 6)-O-α-D-Gal*p*-(1 → 6)-O-α-D-Glc*p*-(1 → 2)-β-D-Fru*f*	Widespread in plants (artichoke, soybean)
Higher oligosaccharides		
Maltopentaose	[O-α-D-Glc*p*-(1 → 4)]$_4$-D-Glc*p*	Starch syrup
α-*Schardinger*-Dextrin, Cyclohexaglucan (α,1 → 4)		Growth of
β-*Schardinger*-Dextrin, Cycloheptaglucan (α,1 → 4)		*Bacillus macerans*
γ-*Schardinger*-Dextrin, Cyclooctaglucan (α,1 → 4)		on starch syrup

As other glycosides, oligosaccharides are readily hydrolyzed by acids, while they are relatively stable against alkalies. Saccharose hydrolysis is denoted as an inversion and the resultant equimolar mixture of glucose and fructose is called invert sugar. The term is based on a change of specific rotation during hydrolysis. In saccharose the rotation is positive, while it is negative in the hydrolysate, since D-glucose rotation to the right (hence its name dextrose) is surpassed by the value of the left-rotating fructose (levulose):

$$\underset{[\alpha]_D = +66.5°}{\text{Saccharose}} \xrightarrow{H^{\oplus}} \underset{[\alpha]_D = +52.7°}{\text{D-Glucose}} + \underset{[\alpha]_D = -92.4°}{\text{D-Fructose}} \quad [\alpha]_D = -19.8° \tag{4.68}$$

Conclusions can be drawn from mutarotation, which follows hydrolysis of reducing disaccharides, for the configuration on the anomeric C-atom. Since the α-anomer has a higher specific rotation in the D-series than the β-anomer, cleavage of β-glycosides increases the specific rotation while cleavage of α-glycosides decreases it:

$$\underset{[\alpha]_D = +34.6°}{\text{Cellobiose } (\beta)} \longrightarrow \underset{[\alpha]_D = +52.7°}{2\ \text{D-Glucose}}$$

$$\underset{[\alpha]_D = +130°}{\text{Maltose } (\alpha)} \longrightarrow \underset{[\alpha]_D = +52.7°}{2\ \text{D-Glucose}} \tag{4.69}$$

$$\underset{[\alpha]_D = +52.3°}{\text{Lactose } (\beta)} \longrightarrow \underset{[\alpha]_D = +80.2°}{\text{D-Galactose}} + \underset{[\alpha]_D = +52.7°}{\text{D-Glucose}} \quad [\alpha]_D = +66.3°$$

Enzymatic cleavage of the glycosidic linkage is specified by the configuration on anomeric C-1 and also by the whole glycosyl moiety, while the aglycone residue may vary within limits.

The methods used to elucidate the linkage positions in an oligosaccharide (methylation, oxidative cleavage of glycols) were outlined under monosaccharides.

4.4 Polysaccharides

4.4.1 Classification, Structure

Polysaccharides, like oligosaccharides, consist of monosaccharides bound to each other by glycosidic linkages. Their acidic hydrolysis yields monosaccharides. Partial chemical and enzymatic hydrolysis, in addition to total hydrolysis, are of importance for structural elucidation. Enzymatic hydrolysis provides oligosaccharides, the analysis of which elucidates monosaccharide sequences and the positions and types of linkages.

Polysaccharides (glycans) can consist of one type of sugar structural unit (homoglycans) or of several types of sugar units (heteroglycans). The monosaccharides may be joined in a linear pattern (as in cellulose and amylose) or in a branched fashion (amylopectin, glycogen, guaran). The frequency of branching sites and the length of side chains can vary greatly (glycogen, guaran). The monosaccharide residue sequence may be periodic, one period containing one or several alternating structural units (cellulose, amylose or hyaluronic acid), the sequence may contain shorter or longer segments with periodically-arranged residues separated by nonperiodic segments (alginate, carrageenans, pectin), or the sequence may be nonperiodic all along the chain (as in the case of carbohydrate components in glycoproteins).

4.4.2 Conformation

The monosaccharide structural unit conformation and the positions and types of linkages in the chain determine the chain conformation of a polysaccharide. In addition to irregular conformations, regular conformations are known which reflect the presence of at least a partial periodic sequence in the chain. Some typical conformations will be explained in the following discussion, with examples of glucans and some other polysaccharides.

An *extended or stretched ribbon-type conformation* is typical for 1,4-linked β-D-glucopyranosyl residues (Figure 4.8, a), as occur, for instance, in cellulose fibers:

CH_2OH O HO HO O O (4.70)

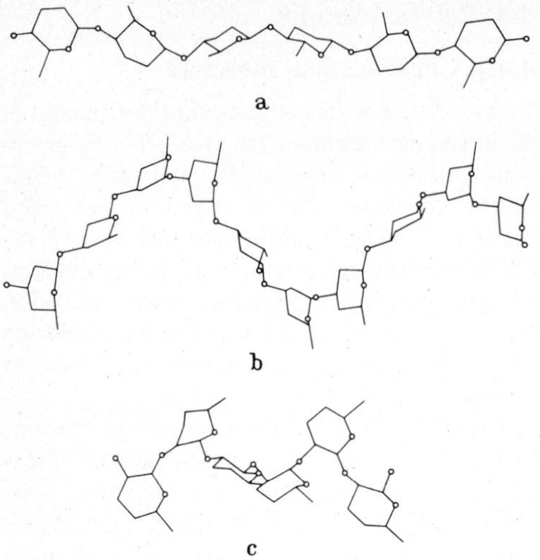

a

b

c

Fig. 4.8. Conformations of some β-D-glucans. Linkages: **a** 1 → 4, **b** 1 → 3, **c** 1 → 2. (After *Rees*, 1977)

This formula shows that the stretched chain conformation is due to a *zigzag* geometry of monomer linkages involving oxygen bridging. The chain may be somewhat shortened or compressed to enable formation of H-bonds between neighboring residues and thus contribute to conformational stabilization. In the ribbon-type, stretched conformation, with the number of monomers in a turn denoted as n and the pitch (advancement) in the axial direction per monomer unit as h, the range of n is from 2 to ± 4, while h is the length of a monomer unit. Thus, the chain given in Figure 4.8 a has $n = -2.55$ and $h = 5.13$ Å.

A strongly pleated, ribbon-type conformation might also occur, as shown by a segment of a pectin chain (1,4-linked α-D-galactopyranosyluronate units):

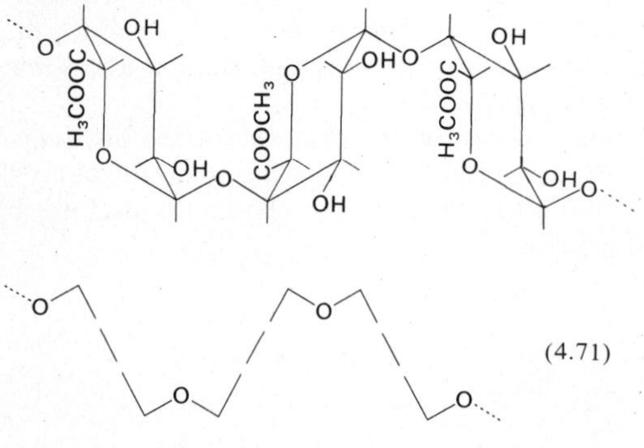

(4.71)

and the same pleated conformation is shown by an alginate chain (1,4-linked α-L-gulopyranosyluronate units):

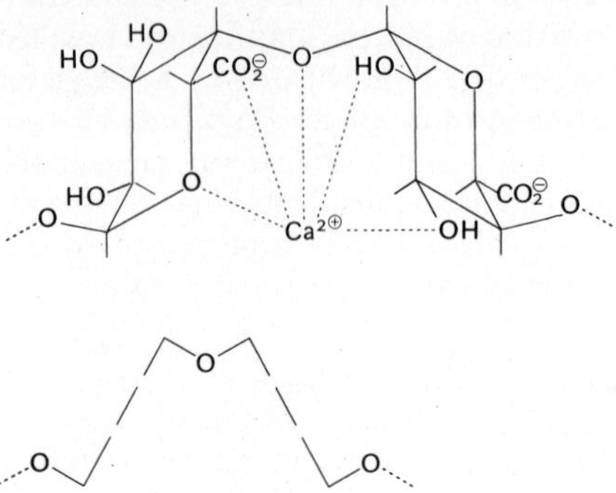

(4.72)

Ca^{2+} ions can be involved to stabilize the conformation. In this case, two alginate chains are assembled in a conformation which resembles an egg box *(egg box type of conformation)*:

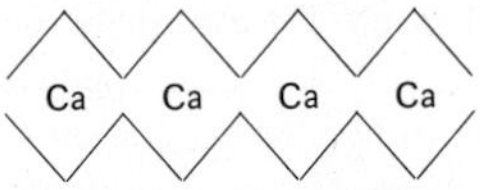

Ca Ca Ca Ca (4.73)

It should be emphasized that in all examples the linear, ribbon-type conformation has a zigzag geometry as a common feature.

A *hollow helix-type conformation* is typical for 1,3-linked β-D-glucopyranose units (Figure 4.8, b), as occur in the polysaccharide lichenin, found in moss-like plants (lichens):

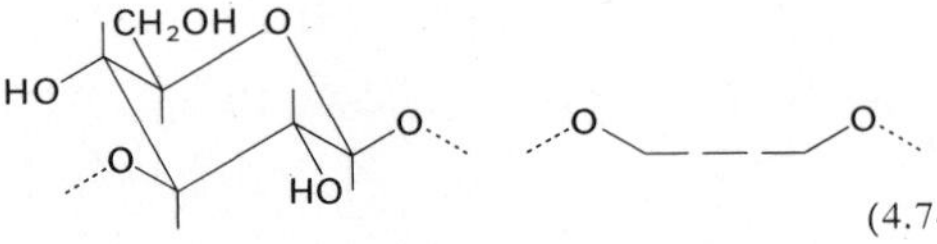

(4.74)

The formula shows that the helical conformation of the chain is imposed by a U-form geometry of the monomer linkages. Amylose (1,4-linked α-D-glucopyranosyl residues) also has such a geometry, and hence a helical conformation:

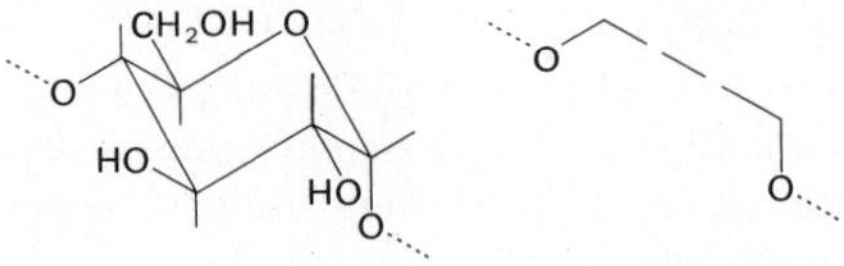

(4.75)

The number of monomers per turn (n) and the pitch in the axial direction per residue (h) is highly variable in a hollow helical conformation.

The value of n is between 2 and ± 10, whereas h can be near its limit value of 0. The conformation of an $\beta(1 \rightarrow 3)$-glucan, with $n = 5.64$ and $h = 3.16$ Å, is shown in Figure 4.8, b.

The helical conformation can be stabilized in various ways. When the helix diameter is large, inclusion (clathrate) compounds can be formed (Figure 4.9, a; cf. 4.4.4.12.2). More extended or stretched chains, with smaller helix diameter, can form double or triple stranded helices (Figure 4.9, b; cf. 4.4.4.3.2 and 4.4.4.12.2), while strongly-stretched chains, in order to stabilize the conformation, have a zigzag, pleated association and are not stranded (Figure 4.9, c).

A *crumpled type of conformation* occurs with, for example, 1,2-linked β-D-glucopyranosyl residues (Figure 4.8, c). This is due to the wrinkled geometry of the monomer O-bridge linkages:

(4.76)

Here, the n value varies from 4 up to -2 and h is 2–3 Å. The conformation reproduced in Figure 4.8, c has $n = 2.62$ and $h = 2.79$ Å. The likelihood of such a disorderly form associating into more orderly conformations is low. Polysaccharides of this conformational type play only a negligible role in nature.

Glycans with 1,6-linked β-D-glucopyranosyl units exhibit a particularly great variability in conformation. These are *loosely-jointed conformations*:

(4.77)

The great flexibility of this glycan-type conformation is based on the nature of the connecting bridge between the monomers. The bridge has three free rotational bonds and, furthermore, the sugar residues are further apart.

The examples considered so far have demonstrated that a prediction is possible for a homoglycan conformation based on the geometry of the bonds of the monomer units which maintain the oxygen bridges. It is more difficult to predict the conformation of a heteroglycan with a periodic sequence of several monomers, which implies different types of conformations. Such a case is shown by ι-carrageenan, in which the β-D-galactopyranosyl-4-sulfate units have a U-form geometry, while the 3,6-anhydro-α-D-galactopyranosyl-2-sulfate residues have a zigzag geometry:

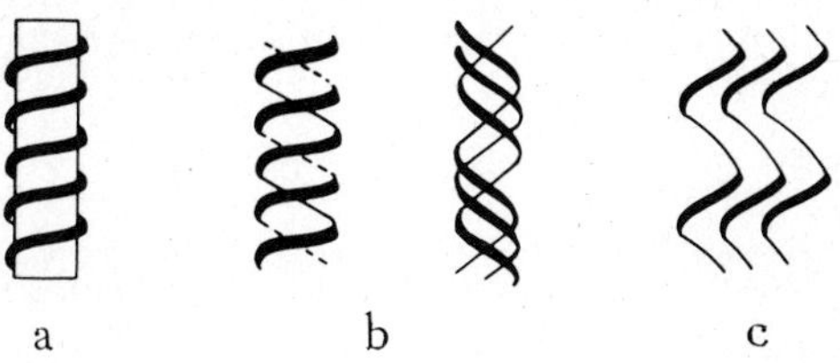

Fig. 4.9. Stabilization of helical conformations. **a** Clathrate compounds, **b** coiled double or triple helices, **c** "nesting". (After *Rees*, 1977)

(4.78)

Calculations have shown that conformational possibilities vary from a shortened, compressed ribbon band type to a stretched helix type. X-ray diffraction analyses have proved that a stretched helix exists, but as a double stranded helix in order to stabilize the conformation (cf. 4.4.4.3.2 and Figure 4.14).

It was outlined in the introductory section (cf. 4.4.1) that the periodically-arranged monosaccharide sequence in a polysaccharide can be interrupted by nonperiodic segments. Such sequence interferences result in conformational disorders. This will be explained in more detail on ι-carrageenan, mentioned above, since it will shed light on the gel-setting mechanism of macromolecules in general.

Initially, a periodic sequence of altering units of β-D-galactopyranose-4-sulfate (I, conformation $^4C_1$) and α-D-galactopyranose-2,6-disulfate (II, conformation $^4C_1$) is built up in carrageenan biosynthesis:

$^{\ominus}O_3SO$ CH$_2$OH O HO I

O CH$_2$OSO$_3^{\ominus}$ HO $^{\ominus}O_3SO$ II

OSO$_3^{\ominus}$ III

(4.79)

When the biosynthesis of the chain is complete, an enzyme-catalyzed reaction eliminates sulfate from most of α-D-galactopyranose-2,6-disulfate (II), transforming the unit to 3,6-anhydro-α-D-galactopyranose-2-sulfate (III, conformation $^1C_4$). This transformation is associated with a change in linkage geometry. Some II-residues remain in the sequence, acting as interference sites. While the undisturbed, ordered segment of one chain can associate with the same segment of another chain, forming a double helix, the nonperiodic or disordered segments can not participate in such associations (Figure 4.10).

In this way, a gel is formed with a three-dimensional network in which the solvent is immobilized. The gel properties, e.g. its strength, are influenced by the number and distribution of α-D-galactopyranosyl-2,6-disulfate residues, i.e. by a structural property regulated during polysaccharide biosynthesis.

The example of the ι-carrageenan gel-building mechanism, involving a chain-chain interaction of sequence segments of orderly conformation, interrupted by randomly-coiled segments corresponding to a disorderly chain sequence, can be applied generally to gels of other macromolecules. Besides a sufficient chain length, the structural prerequisite for gel-setting ability is interruption of a periodic sequence and its orderly conformation. The interruption is achieved by insertion into the chain of a sugar residue of a different linkage geometry (carrageenans, alginates, pectin), by a suitable distribution of free and esterified carboxyl groups (glycuronans) or by insertion of side chains. The interchain associations during gelling (network formation), which involve segments of orderly conformation, can then occur in the form of a double helix (Figure 4.11, a); a multiple bundle of double helices (Figure 4.11, b); an association between stretched ribbon-type conformations, such as an egg box model (Figure 4.11, c); some other similar associations (Figure 4.11, d); or, lastly, forms consisting of double helix and ribbon-type combinations (Figure 4.11, e).

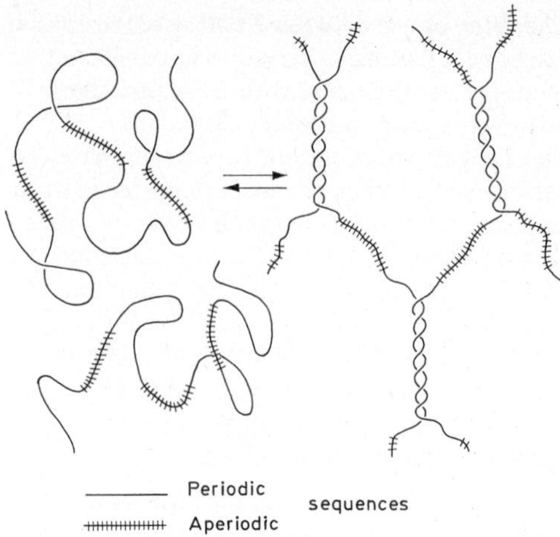

Fig. 4.10. Schematic representation of a gel setting process. (After *Rees*, 1977)

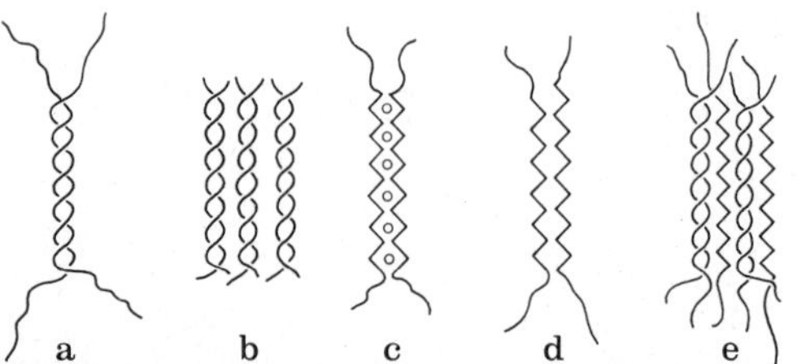

Fig. 4.11. Interchain aggregation between regular conformations. **a** Double helix, **b** double helix bundle, **c** egg-box, **d** ribbon-ribbon, and **e** double helix, ribbon interaction

4.4.3 Properties

4.4.3.1 General Remarks

Polysaccharides are widely and abundantly distributed in nature, fulfilling roles as:

- Structure-forming skeletal substances (cellulose, hemicellulose and pectin in plants; chitin, mucopolysaccharides in animals)
- Assimilative reserve substances (starch, dextrins, inulin in plants; glycogen in animals)
- Water-binding substances (agar, pectin and alginate in plants; mucopolysaccharides in animals).

As a consequence, polysaccharides occur in many food products and even then they often retain their natural role as skeletal substances (fruits and vegetables) or assimilative nutritive substances (cereals, potatoes, legumes). Isolated polysaccharides are utilized to a great extent in food processing, either in native or modified form, as: thickening or gel-setting agents (starch, alginate, pectin, guaran gum); stabilizers for emulsions and dispersions; film-forming, coating substances to protect sensitive food from undesired change; and inert fillers to increase the proportion of indigestible ballast substances in a diet (cellulose).

The outlined functions of polysaccharides are based on their highly variable properties. They vary from insoluble forms (cellulose) to those with good swelling power and solubility in hot and cold water (starch, guaran gum). The solutions may exhibit low viscosities even at very high concentrations (gum arabic), or may have exceptionally high viscosities even at low concentrations (guaran gum). Some polysaccharides, even at a low concentration, set into a thermoreversible gel (alginates, pectin). While most of the gels melt at elevated temperatures, some cellulose derivatives set into a gel.

These properties and their utilization in food products are described in more detail in section 4.4.4, where individual polysaccharides are covered. Here, only a brief account will be given to relate their properties to their structures in a general way.

4.4.3.2 Perfectly-Linear Polysaccharides

Compounds with a *single* neutral monosaccharide structural unit and with *one* type of linkage (as occurs in cellulose or amylose) are denoted as perfectly-linear polysaccharides. They are usually insoluble in water and can be solubilized only under drastic conditions, e. g. at high temperature, or by cleaving H-bonds with alkalies or other suitable reagents. They readily precipitate from solution (example: starch retrogradation). The reason for these properties is the existence of an optimum structural prerequisite for setting an orderly conformation within the chain and also for chain-chain interaction. Often, the conformation is so orderly that a partial crystallinity state develops. Large differences in properties are found within these group of polysaccharides when there is a change in structural unit, linkage type or molecular weight. This is shown by properties of cellulose, amylose or β-1,3-glucan macromolecules.

4.4.3.3 Branched Polysaccharides

Branched polysaccharides (amylopectin, glycogen) are more soluble in water than their perfectly-linear counterparts since the chain-chain interaction is less pronounced and there is a greater extent of solvation of the molecules. Solutions of branched polysaccharides, once dried, are readily rehydrated. Compared to their linear counterparts of equal molecular weights and equal concentrations, solutions of branched polysaccharides have a lower viscosity. It is assumed that the viscosity reflects the "effective volume" of the macromolecule. The "effective volume" is the volume of a sphere with diameter determined by the longest linear extension of the molecule. These volumes are generally larger for linear than for branched molecules (Figure 4.12). Exceptions are found with highly pleated linear chains. The tendency of branched polysaccharides to precipitate is low. They form a sticky paste at higher concentrations, probably due to side chain-side chain interactions (interpenetration, entanglement). Thus, branched polysaccharides are suitable as binders or adhesives.

4.4.3.4 Linearly-Branched Polysaccharides

Linearly-branched polysaccharides, i.e. polymers with a long "backbone" chain and with many short side chains, such as guaran or alkyl cellulose, have properties which are a combination of those of perfectly-linear and of branched molecules. The long "backbone" chain is responsible for high solution viscosity. The presence of numerous short side chains greatly weakens interactions between the molecules, as shown by the good solubility and rehydration rates of the molecules and by the stability even of highly concentrated solutions.

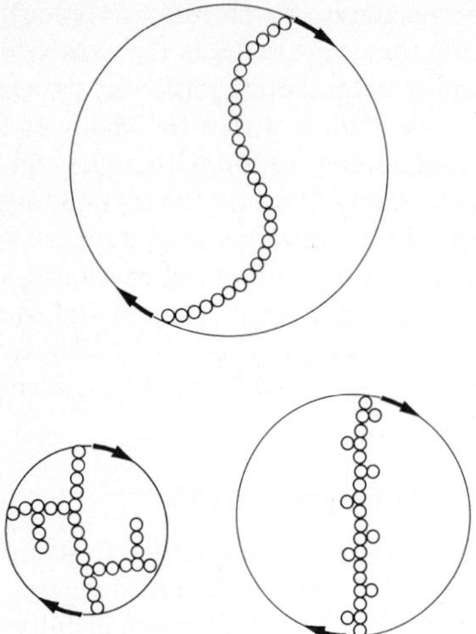

Fig. 4.12. Schematic representation of the "effective volumes" of linear, branched and linear-side chain types of polysaccharides

4.4.3.5 Polysaccharides with Carboxyl Groups

Polysaccharides with carboxyl groups (pectin, alginate, carboxymethyl cellulose) are very soluble as alkali salts in the neutral or alkaline pH range. The molecules are negatively charged due to carboxylate anions and, due to the repulsive charge forces, the molecules are relatively stretched and resist intermolecular associations. The solution viscosity is high and is pH-dependent. Gel setting or precipitation occur at pH ≤ 3 since electrostatic repulsion ceases to exist. In addition, undissociated carboxyl groups dimerize through H-bridges. However, a divalent cation is needed to achieve gel setting in a neutral solution.

4.4.3.6 Polysaccharides with Strongly Acidic Groups

Polysaccharides with strongly acidic residues, present as esters along the polymer chains (sulfuric, phosphoric acids, as in furcellaran, carrageenan or modified starch), are also very soluble in water and form highly viscous solutions. Unlike polysaccharides with carboxyl groups, in strongly acidic media these solutions are distinctly stable.

4.4.3.7 Modified Polysaccharides

Modification of polysaccharides, even to a low substitution degree, brings about substantial changes in their properties.

4.4.3.7.1 Derivatization with neutral substituents

The solubility in water, viscosity and stability of solutions are all increased by binding neutral substituents to linear polysaccharide chains. Thus the properties shown by methyl, ethyl and hydroxypropyl celluloses correspond to those of guaran and locust bean gum. The effect is explained by interference of the alkyl substituents in chain interactions, which then facilitates hydration of the molecule. An increased degree of substitution increases the hydrophobicity of the molecules and, thereby, increases their solubility in organic solvents.

4.4.3.7.2 Derivatization with acidic substituents

Binding acid groups to a polysaccharide (carboxymethyl, sulfate or phosphate groups) also results in increased solubility and viscosity for reasons already outlined. Some derivatized polysaccharides, when moistened, have a pasty consistence.

4.4.4 Individual Polysaccharides

4.4.4.1 Agar

Agar is a gelatinous product isolated from seaweed (red algae class, *Rhodophyceae*), e. g. *Gelidium spp., Pterocladia spp.* and *Gracilaria spp.*, by a hot water extraction process. Purification is possible by congealing the gel.

Agar is a heterogenous complex mixture of related polysaccharides having the same backbone chain structure. The main components of the chain are β-D-galactopyranose and 3,6-anhydro-α-L-galactopyranose, which alternate through $1 \rightarrow 4$ and $1 \rightarrow 3$ linkages:

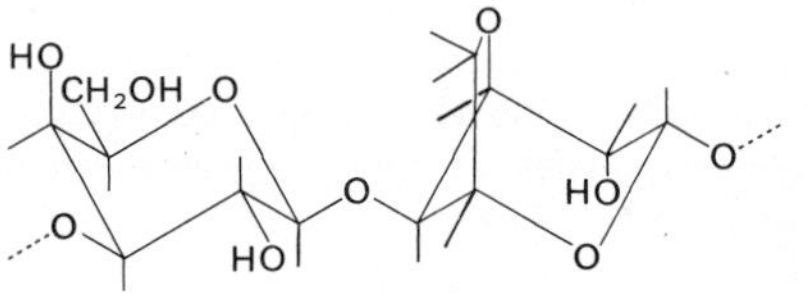

(4.80)

The chains are esterified to a low extent with sulfuric acid. The sulfate content differentiates the agarose fraction (the main gelling component of agar), in which close to every tenth galactose unit of the chain is esterified, and the agaropectin fraction, which has a higher sulfate esterification degree and, in addition, has pyruvic acid bound in ketal form [4,6-(1-carboxyethylidene)-D-galactose]. The ratio of the two polymers can vary greatly. Uronic acid, when present, does not exceed 1%.

Agar is insoluble in cold water, slightly soluble in ethanolamine and soluble in formamide. Agar precipitated by ethanol from a warm aqueous dispersion is, in its moist state, soluble in water at 25 °C, while in the dried state it is soluble only in hot water. Gel setting occurs upon cooling. Agar is a most potent gelling agent as gelation is perceptible even at 0.04%. Gel setting and stability are affected by agar concentration and its average molecular weight. A 1.5% solution sets to a gel at 32–39 °C, but does not melt below 60–97 °C. The great difference between gelling and melting temperatures, due to hysteresis, is a distinct and unique feature of agar.

Utilization. Agar is widely used, for instance in preparing nutritive media in microbiology. Its application in the food industry is based on its main properties: it is essentially indigestible, forms heat resistant gels, and has emulsifying and stabilizing activity. Agar is added to sherbets (frozen desserts of fruit juice, sugar, water or milk) and ice creams (at about 0.1%), often in combination with gum tragacanth or locust (carob) bean gum or gelatin. An amount of 0.1–1% stabilizes yoghurt, some cheeses and candy and bakery products (pastry fillings). Furthermore, agar retards bread staling and provides the desired gel texture in poultry and meat canning. Lastly, agar has a role in vegetarian diets (meat substitute products) and in desserts and pretreated instant cereal products.

4.4.4.2 Alginates

Alginates occur in all brown algae *(Phaeophyceae)* as a skeletal component of their cell walls. The major source of industrial production is the giant kelp, *Macrocystis pyrifera.* Some species of *Laminaria, Ascophyllum* and *Sargassum* are also used. Algae are extracted with alkalies. The polysaccharide is usually precipitated from the extract by acids or calcium salts.

Alginate building blocks are β-D-mannuronic and α-L-guluronic acids, joined by $1 \rightarrow 4$ linkages:

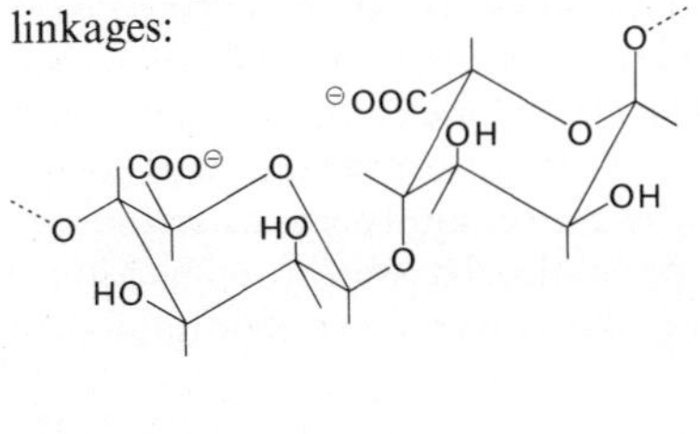

(4.81)

The ratio of the two sugars (mannuronic/guluronic acids) is generally 1.5, with some deviation with source. Alginates extracted from *Laminaria hyperborea* have ratios of 0.4–1.0. Partial hydrolysis of alginate yields chain fragments which consist predominantly of mannuronic or guluronic acid, and also fragments where the two uronic acid residues alternate in a 1 : 1 ratio. Alginates are linear copolymers consisting of the following structural units:

$$[\rightarrow 4)\text{-}\beta\text{-D-ManpA}(1 \rightarrow 4)\text{-}\beta\text{-D-ManpA}(1 \rightarrow]_n$$
$$[\rightarrow 4)\text{-}\alpha\text{-L-GulpA}(1 \rightarrow 4)\text{-}\alpha\text{-L-GulpA}(1 \rightarrow]_m$$
$$[\rightarrow 4)\text{-}\beta\text{-D-ManpA}(1 \rightarrow 4)\text{-}\alpha\text{-L-GulpA}(1 \rightarrow]_p \quad (4.82)$$

The molecular weights of alginates are 32–200 kdal. This corresponds to a degree of polymerization of 180–930. The carboxyl group pK-values are 3.4–4.4. Alginates are water soluble in the form of alkali, magnesium, ammonia or amine salts. The viscosity of alginate solutions is influenced by molecular weight and the counter ion of the salt. In the absence of di- and trivalent cations or in the presence of a chelating agent, the viscosity is low ("long flow" property). However, with a rise in multivalent cation levels (e. g. calcium) there is a parallel rise in viscosity ("short flow"). Thus, the viscosity can be adjusted as desired. Freezing and thawing of an Na-alginate solution containing Ca^{2+} ions can result in a further rise in viscosity. The curves in Figure 4.13 show the effect on viscosity of the concentrations of three alginate preparations: low, moderate and high viscosity types. These data reveal that a 1% solution, depending on the type of alginate, can have a viscosity range of 20–2,000 cps. The viscosity is unaffected in a pH range of 4.5–10. It rises at a pH below 4.5, reaching a maximum at pH 3–3.5.

Gels, fibers or films are formed by adding Ca^{2+} or acids to Na-alginate solutions. A slow reaction is needed for uniform gel formation. It is achieved by a mixture of Na-alginate, calcium phosphate and glucono-δ-lactone, or by a mixture of Na-alginate and calcium sulfate.

Propylene glycol alginate is a derivative of economic importance. This ester is obtained by the reaction of propylene oxide with partially-neutralized alginic acid. It is soluble down to pH 2 and, in the presence of Ca^{2+} ions, forms soft, elastic, less brittle and syneresis-free gels.

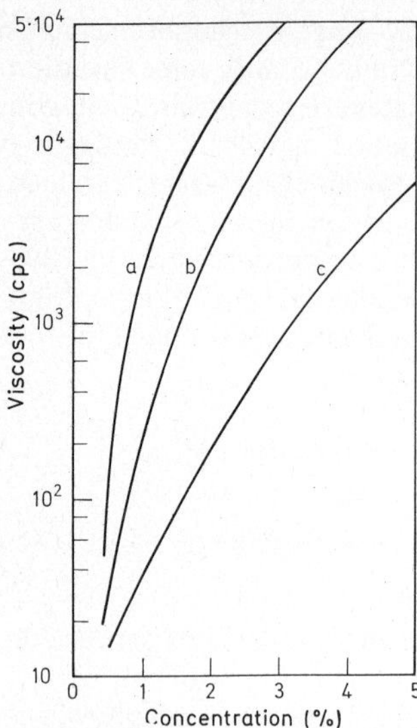

Fig. 4.13. Viscosity of aqueous alginate solutions. Alginate with (a) high, (b) medium, and (c) low viscosity

Utilization. Alginate is a powerful thickening, stabilizing and gel-forming agent. At a level of 0.25–0.5% it improves and stabilizes the consistency of fillings for baked products (cakes, pies), salad dressings and milk chocolates, and prevents formation of larger ice crystals in ice creams during storage. Furthermore, alginates are used in a variety of gel products (cold instant puddings, fruit gels, dessert gels, onion rings, imitation caviar) and are applied to stabilize fresh fruit juice and beer foam.

4.4.4.3 Carrageenans

Red sea weeds *(Rhodophyceae)* produce two types of galactans: agar and agar-like polysaccharides composed of D-galactose and 3,6-anhydro-L-galactose residues; and carrageenan and related polysaccharides, composed of D-galactose and 3,6-anhydro-D-galactose, which are partially sulfated as 2-, 4- and 6-sulfates and 2,6-disulfates. Galactose residues are alternatively linked by 1 → 3 and 1 → 4 linkages. Carrageenans are isolated from *Chondrus* (*Chondrus crispus,* the Irish moss), *Eucheuma, Gigartina, Gloiopeltis* and *Iridaea* species by hot water extraction under mild alkaline conditions, followed by drying or isolate precipitation.

Carrageenans are a complex mixture of various polysaccharides. They can be separated by fractional precipitation with potassium ions. Table 4.14 compiles data on these fractions and their monosaccharide constituents. Two major fractions are ϰ (gelling and K^+-insoluble fraction) and λ (nongelling, K^+-soluble).

Table 4.14. Building blocks of carrageenans

Carrageenan	Monosaccharide-building block
ι-Carrageenan	D-Galactose-4-sulfate, 3,6-Anhydro-D-galactose-2-sulfate
ϰ-Carrageenan	D-Galactose-4-sulfate, 3,6-Anhydro-D-galactose
λ-Carrageenan	D-Galactose-2-sulfate, D-Galactose-2,6-disulfate
μ-Carrageenan	D-Galactose-4-sulfate, D-Galactose-6-sulfate, 3,6-Anhydro-D-galactose
ν-Carrageenan	D-Galactose-4-sulfate, D-Galactose-2,6-disulfate, 3,6-Anhydro-D-galactose
Furcellaran	D-Galactose-D-Galactose-2-sulfate, D-Galactose-4-sulfate, D-Galactose-6-sulfate, 3,6-Anhydro-D-galactose

ϰ-Carrageenan is composed of D-galactose, 3,6-anhydro-D-galactose and ester-bound sulfate in a molar ratio of 6:5:7. The galactose residue are essentially fully sulfated in position 4, whereas the anhydrogalactose residues can be sulfated in position 2 or substituted by α-D-galactose-6-sulfate or 2,6-disulfate. A typical sequence of ϰ-(or ι-)carrageenan is:

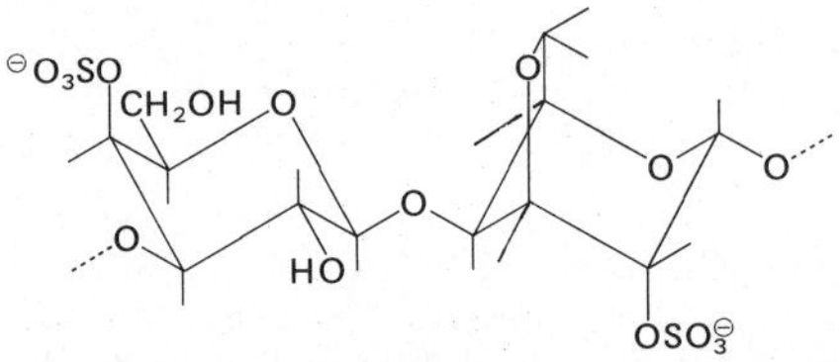

(4.83)

The sequence favors the formation of a double-stranded helix (Figure 4.14). λ-Carrageenan is characterized by a higher sulfate content, which favors a zigzag, ribbon-shaped conformation. The molecular weights of ϰ- and λ-carrageenans are 200–800 kdal. The water solubility is higher as the carrageenan sulfate content is higher and as the content of anhydrosugar residue is lower. The viscosity of the solution depends on the carrageenan type, molecular weight, temperature, ions present and carrageenan concentration.

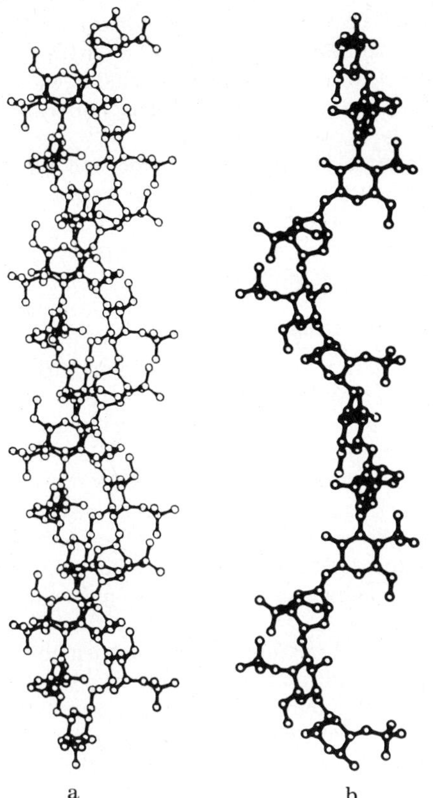

Fig. 4.14. ι-Carrageenan conformation. **a** Double helix, **b** single coil is presented to clarify the conformation. (After *Rees,* 1977)

As observed in all linear macromolecules with charges along the chain, the viscosity increases exponentially with the concentration (Figure 4.15). Aqueous ϰ-carrageenan solutions, in the presence of ammonium, potassium, rubidium or caesium ions, form thermally-reversibly gels, which is not so with lithium and sodium ions.

This strongly suggests that gel-setting ability is highly dependent on the radius of the hydrated counter ion. The latter is about 0.23 nm for the former group of cations, while hydrated lithium (0.34 nm) and sodium ions (0.28 nm) exceed the limit. The action of cations is visualized as a zipper arrangement between aligned segments of linear polymer sulfates, with low ionic radius cations locked between alternating sulfate residues. Gel-setting ability is probably also due to a mechanism based on formation of partial double helix structures between various chains. The ex-

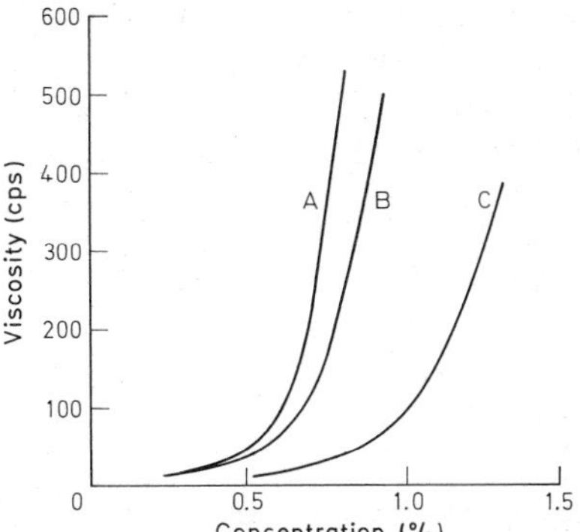

Fig. 4.15. Viscosity curves of carrageenan aqueous solutions. A: *Eucheuma spinosum,* C: *Chondrus crispus,* B: A and C in a ratio of 2:1, 40 °C, 20 rpm. (After *Whistler,* 1973)

tent of intermolecular double helix formation, and thus the gel strength, is greater as the chain sequence is more uniform. Each substitution of a 3,6-anhydrogalactose residue by another residue, e.g. galactose-6-sulfate, results in a kink within the helix and, thereby, a decrease in gelling strength. The helical conformation is also affected by the position of sulfate groups. The effect is more pronounced with sulfate in the 6-position, than in 2- or 4-positions. Hence, the gel strength of ϰ-carrageenan is dependent primarily on the content of esterified sulfate groups in the 6-position.

The 6-sulfate group can be removed by heating carrageènans with alkali to yield 3,6-anhydrogalactose residues. This elimination results in a significantly increased gel strength.

Carrageenans and other acidic polysaccharides coagulate proteins when the pH of the solution is lower than the proteins' isoelectric points. This can be utilized for separating protein mixtures.

Utilization. Carrageenan utilization in food processing is based on the ability of the polymer to gel, to increase solution viscosity and to stabilize emulsions and various dispersions. A level as low as 0.03% in chocolate milk prevents fat droplet separation and stabilizes the suspension of cocoa particles. Carrageenans prevent syneresis in fresh cheese and improve dough properties and enable a higher amount of milk powder incorporation in baking. The gelling property in the presence of K^+ salt is utilized in desserts and canned meat. Protein fiber texture is also improved. Protein sedimentation in condensed milk is prevented by carrageenans which, like κ-casein, prevent milk protein coagulation by calcium ions. Carrageenans are also used to stabilize ice cream and clarify beverages.

4.4.4.4 Furcellaran

Furcellaran (Danish agar) is produced from red sea weed (algae *Furcellaria fastigiata*). Production began in 1943 when Europe was cut off from its agar suppliers. After alkali pretreatment of algae, the polysaccharide is isolated using hot water. The extract is then concentrated under vacuum and seeded with 1–1.5% KCl solution. The separated gel threads are concentrated further by freezing, the excess water is removed by centrifugation or pressing and, lastly, the polysaccharide is dried. The product is a K-salt and contains, in addition, 8–15% occluded KCl.

Furcellaran is composed of D-galactose (46–53%), 3,6-anhydro-D-galactose (30–33%) and sulfated portions of both sugars (16–20%).

The structure of furcellaran is similar to κ-carrageenan. The essential difference is that κ-carrageenan has one sulfate ester per two sugar residues, while furcellaran has one sulfate ester residue per three to four sugar residues. Sugar sulfates identified are: D-galactose-2-sulfate, -4-sulfate and -6-sulfate, and 3,6-anhydro-D-galactose-2-sulfate. Branching of the polysaccharide chain can not be excluded. Furcellaran forms thermally-reversible aqueous gels by a mechanism involving double helix formation, similar to κ-carrageenan (cf. 4.4.4.3).

The gelling ability is affected by polysaccharide polymerization degree, amount of 3,6-anhydro-D-galactose, and by the radius of the cations present. K^+, NH_4^+, Rb^+ and Cs^+ form very stable, strong gels. Ca^{2+} has a lesser effect, while Na^+ prevents gel setting. Addition of sugar affects the gel texture, which goes from a brittle to a more elastic texture.

Utilization. Furcelleran, with milk, provides good gels and therefore it is used as an additive in puddings. It is also suitable for cake fillings and icings. In the presence of sucrose, it gels rapidly and retains good stability, even against food grade acids. Furcelleran has the advantage over pectin in marmalades since it allows stable gel setting at a sugar concentration even below 50–60%. The required amount of polysaccharide is 0.2–0.5%, depending on the marmalade's sugar content and the desired gel strength. To keep the hydrolysis extent low, a cold aqueous 2–3% solution of furcellaran is mixed into a hot, cooked slurry of fruits and sugar.

Furcellaran is also utilized in processed meat products, such as spreadable meat pastes and pastry fillings. It facilitates protein precipitation during brewing of beer and thus improves the final clarification of the beer.

4.4.4.5 Gum Arabic

Gum arabic is a tree exudate of various *Acacia* species, primarily *Acacia senegal,* and is obtained as a result of tree bark injury. It is collected as air-dried droplets with diameters from 2–7 cm. The annual yield per tree averages 0.9–2.0 kg. The major producer is Sudan, with 50–60,000 t/annum, followed by several other African countries. Gum arabic has been known since ancient Egypt as "kami", an adhesive for pigmented paints.

Gum arabic is a mixture of closely-related polysaccharides, with an average molecular weight range of 260–1,160 kdal. The main structural units, with molar proportions for the gum exudate *A. senegal* given in brackets, are L-arabinose (3.5), L-rhamnose (1.1), D-galactose (2.9) and D-glucuronic acid (1.6). The proportion varies sig-

```
                      D-GlcpA
                         1
                         ↓
                         6
                  R→3)-D-Galp
                         1
                         ↓
                         6
  →3)-D-Galp(1→3)-D-Galp(1→3)-D-Galp(1→
         6                          6
         ↑                          ↑
         1                          1
  R→3)-D-Galp               R→3)-D-Galp
         6                          6
         ↑                          ↑
         1                          1
  R→3)-D-Galp               R→3)-D-Galp
         6                          6
         ↑                          ↑
         1                          1
  R→4)-D-GlcpA              R→4)-D-GlcpA

R: L-Rhap(1→, L-Araf(1→,
   D-Galp(1→3)-L-Araf(1→,
   L-Arap(1→3)-L-Araf(1→.                    (4.84)
```

Table 4.15. Viscosity (cps) of polysaccharides in aqueous solution as affected by concentration (25 °C)

Concentration (%)	Gum arabic	Tragacanth	Carrageenan	Sodium alginate	Methyl cellulose 1,500 cps	Locust bean gum	Guaran gum
1		54	57	214	38.9	58.5	3,025
2		906	397	3,760	512	1,114.3	25,060
3		10,605	4,411	29,400	3,850	8,260	111,150
4		44,275	25,356		12,750	39,660	302,500
5	7.3	111,000	51,425		67,575	121,000	510,000
6		183,500					
10	16.5						
20	40.5						
30	200.0						
40	936.3						
50	4,162.5						

nificantly with Acacia species. Gum arabic has a major core chain built of β-D-galactopyranosyl residues linked by 1 → 3 bonds, in part carrying side chains attached at position 6 (cf. Formula 4.84).

Gum arabic occurs neutral or as a weakly acidic salt. Counter ions are Ca^{2+}, Mg^{2+} and K^+. Solubilization in 0.1 N HCl and subsequent precipitation with ethanol yields free acid.

Gum arabic is very soluble in water and solutions up to 50% gum can be prepared. The solution viscosity starts to rise steeply only at high concentrations (Figure 4.16). This property is unlike that of many other polysaccharides, which provide highly viscous solutions even at low concentrations (Table 4.15).

Utilization. Gum arabic is used as a emulsifier and stabilizer, e. g. in baked products. It retards sugar crystallization and fat separation in confectionery products and large ice crystal formation in ice creams, and can be used as a foam stabilizer in beverages. Gum arabic is also applied as a flavor fixative in the production of encapsulated, powdered aroma concentrates. For example, essential oils are emulsified with gum arabic solution and then spray-dried. In this process, the polysaccharide forms a film surrounding the oil droplet, which then protects the oil against oxidation and other changes.

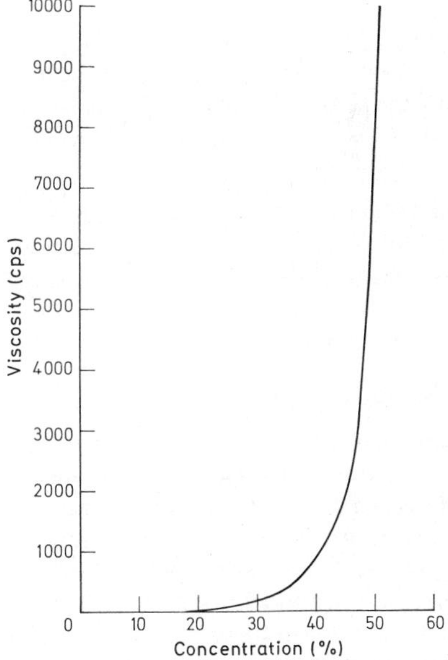

Fig. 4.16. Viscosity curve of an aqueous gum arabic solution. (After *Whistler,* 1973) (25.5 °C, Brookfield viscometer)

4.4.4.6 Gum Tragacanth

Gum tragacanth is a tasteless and odorless plant exudate collected from *Astragalus* species shrubs grown in the Middle East (Iran, Syria, Turkey).

Gum tragacanth consists of a water-soluble fraction, the so-called tragacanthic acid, and the insoluble swelling component, bassorin. The soluble fraction is a complex mixture of various acidic polysaccharides. Their basic units are D-galacturonic acid, D-galactose, L-fucose, D-xylose and L-arabinose. Most carboxyl groups in the insoluble fraction are esterified with methanol; their hydrolysis yields tragacanthic acid. Its molecular weight is about 840 kdal. The mole-

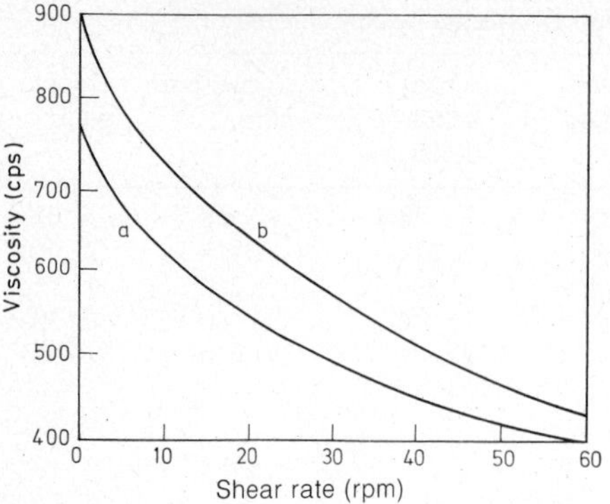

Fig. 4.17. The effect of shear rate on viscosity of aqueous tragacanth solutions. **a** Flake form tragacanth, 1%; **b** tragacanth, ribbon form, 0.5%. (After *Whistler,* 1973)

cules are highly elongated (450 × 1.9 nm) in aqueous solution and are responsible for the high viscosity of the solution (Table 4.15). As shown in Figure 4.17, the viscosity is highly dependent on shear rate.

Utilization. Gum tragacanth is used as a thickening agent and a stabilizer in salad dressings (0.4–1.2%) and in fillings and icings in baked goods. As an additive in ice creams (0.5%), it provides a soft texture.

4.4.4.7 Guaran Gum

Guar flour is obtained from the seed endosperm of the leguminous plant *Cyamopsis tetragonoloba.* The seed is decoated and the germ removed. In addition to the polysaccharide guaran, guar flour contains 10–15% moisture, 5–6% protein, 2.5% crude fiber and 0.5–0.8 ash. The plant is cultivated for forage in India, Pakistan and the United States (Texas).
Guaran gum consists of a chain of β-D-mannopyranosyl units joined with 1 → 4 linkages. Every

(4.85)

second residue has a side chain, a D-galactopyranosyl residue that is bound to the main chain by an α (1 → 6) linkage (cf. Formula 4.85). Guaran gum forms highly viscous solutions (Table 4.15), the viscosity of which is shear rate dependent (Figure 4.18).

Utilization. Guaran gum is used as a thickening agent and a stabilizer in salad dressings and ice creams (application level 0.3%). In addition to the food industry, it is widely used in paper, cosmetic and pharmaceutical industries.

4.4.4.8 Locust Bean Gum

The locust bean (carob bean; St. John's bread) is from an evergreen cultivated in the Mediterranean area since ancient times. Its long, edible, fleshy seed pod is also used as fodder. The dried seeds were called "carat" by Arabs and served as a unit of weight (approx. 200 mg). Even today, the carat is used as a unit of weight for precious stones, diamonds and pearls, and as a measure of gold purity (1 carat = 1/24 part of pure gold). The locust bean seeds consist of 30–33% hull material, 23–25% germ and 42–46% endosperm. The seeds are milled and the endosperm is separated and utilized like the guar flour described above. The commercial flour contains 88% galactomannoglucan, 5% other polysaccharides, 6% protein and 1% ash.

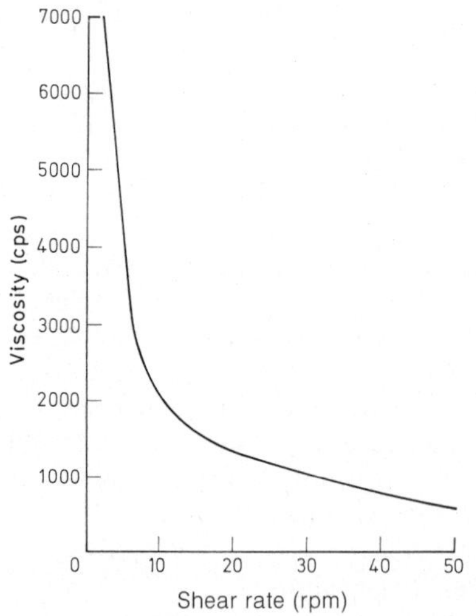

Fig. 4.18. Viscosity of 1% aqueous guar solution at 25 °C versus shear rate (rpm.). Viscometer: Haake rotovisco. (After *Whistler,* 1973)

The main locust bean polysaccharide is similar to that of guaran gum: a linear chain of 1 → 4 linked β-D-mannopyranosyl units, with α-D-galactopyranosyl residues 1 → 6 joined as side chains. The ratio mannose/galactose is 3 to 6; this indicates that, instead of every second mannose residue, as in guaran gum, only every 4th to 5th is substituted at the C-6 position with a galactose molecule.

The molecular weight of the galactomannan is close to 310 kdal. Physical properties correspond to those of guar gum, except the viscosity of the solution is not as high (cf. Table 4.15).

Utilization. Locust bean flour is used as a thickener, binder and stabilizer in meat canning, salad dressings, sausages, soft cheeses and ice creams. It also improves the water binding capacity of dough, especially when flour of low gluten content is used.

4.4.4.9 Tamarind Flour

Tamarind is one of the most important and widely grown trees of India (*Tamarindus indica;* date of India). Its brown pods contain seeds which are rich in a polysaccharide that is readily extracted with hot water and, after drying, recovered in a powdered form.

The polysaccharide consists of D-galactose (1), D-xylose (2) and D-glucose (3), with respective molar ratios given in brackets. L-Arabinose is also present. The suggested structure is presented in Formula 4.86.

The polysaccharide forms a stable gel over a wide pH range. Less sugar is needed to achieve a desired gel strength than in corresponding pectin gels (Figure 4.19). The gels exhibit only a low syneresis phenomenon.

Utilization. The tamarind seed polysaccharide is a suitable substitute for pectin in the production of marmalades and jellies. It can be used as a thickening agent and stabilizer in ice cream and mayonnaise production.

4.4.4.10 Arabinogalactan from Larch

Coniferous larch-related woods (*Larix* species; similar to pine, but shed their needle-like leaves) contain 5–35% of the dry weight of the wood of a water-soluble arabinogalactan. It can be isolated from chipped wood by a counter-current extraction process, using water or dilute acids, and the extract is then usually drum dried.

The polysaccharide consists of straight chain β-D-galactopyranosyl units joined by 1 → 3 linkages and, in part, has side chains of galactose and arabinose residues bound to positions 4 and 6. The suggested structure is given in Formula 4.87.

In general, the polysaccharide is highly branched. The molecular weight is 50–70 kdal. The molecule is nearly spherical in shape, so its aqueous solution behaves like a *Newton*ian fluid. The viscosity is exceptionally low. At a temperature of 20 °C, the viscosity of a 10% solution is 1.74 cps, a 30% solution 7.8 cps at pH 4 or 8.15

[→4)-β-D-Glcp(1→4)-β-D-Glcp(1→4)-β-D-Glcp(1→4)-β-D-Glcp(1→]ₙ
with side chains (1→6): α-D-Xylp (on 1st Glcp); α-D-Xylp (on 2nd Glcp), itself carrying β-D-Galp (1→2); L-Araf (on 3rd Glcp)

(4.86)

β-L-Arap(1→3)L-Araf (1→6); L-Araf(1→3)D-Galp (1→6); D-Galp(1→6)D-Galp (1→6)

→3)-β-D-Galp(1→3)-β-D-Galp(1→3)-β-D-Galp(1→3)-β-D-Galp(1→3)-β-D-Galp(1→3)-β-D-Galp(1→

β-D-Galp(1→6)D-Galp (1→4)

(4.87)

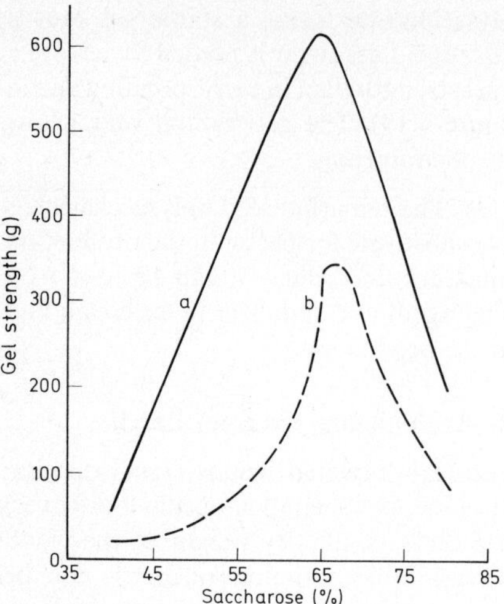

Fig. 4.19. Gel strength of (a) tamarind flour and (b) pectin from lemons versus saccharose concentration. (After *Whistler,* 1973)

cps at pH 11, and a 40% solution 23.5 cps. These data show that the viscosity is practically unaffected by pH. The solution acquires a thick paste consistency only at concentrations exceeding 60%.

Utilization. Arabinogalactan, due to its good solubility and low viscosity, is used as an emulsifier and stabilizer, and as a carrier substance in essential oils, aroma formulations, and sweeteners.

4.4.4.11 Pectin

Pectin is widely distributed in plants. It is produced commercially from peels of citrus fruits and from apple pomace (crushed and pressed residue). It is 20–40% of the dry matter content in citrus fruit peel and 10–20% in apple pomace. Extraction is achieved at pH 1.5–3 at 60–100 °C. The process is carefully controlled to avoid hydrolysis of glycosidic and ester linkages. The extract is concentrated to a liquid pectin product or is dried by spray- or drum-drying into a powdered product. Purified preparations are obtained by precipitation of pectin with ions which form insoluble pectin salts (e. g. Al^{3+}), followed by washing with acidified alcohol to remove the added ions, or by alcoholic precipitation using isopropranol and ethanol.

Pectin is a chain-like polymer consisting of α-D-galacturonic structural units joined by 1 → 4 linkages:

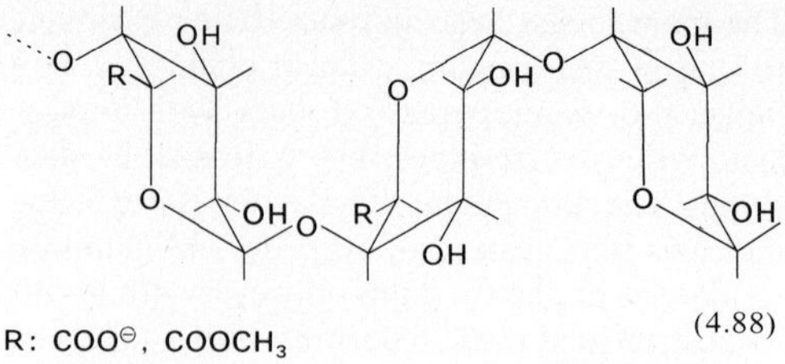

R: $COO^{\ominus}$, $COOCH_3$ (4.88)

The main chain, however, is one-tenth rhamnose residues. In segments in which rhamnose is enriched, rhamnose units may be in adjacent or alternate positions. Pectin also contains small amounts of D-galactan and arabinan in its extended side chains and, to a lesser extent, fucose and xylose sugars in its short side chains (1 to 3 unit chains). These short side chains are not regarded as typical pectin constituents. The galacturonic acid carboxyl groups along the main chain are esterified to a variable extent with methanol, while the HO-groups in 2- and 3-positions may be acetylated to a small extent. Pectin stability is highest at pH 3–4. The glycosidic linkage hydrolyzes in a stronger acidic medium. In an alkaline medium, both linkages, ester and glycosidic, are split to the same extent, the latter by an elimination reaction:

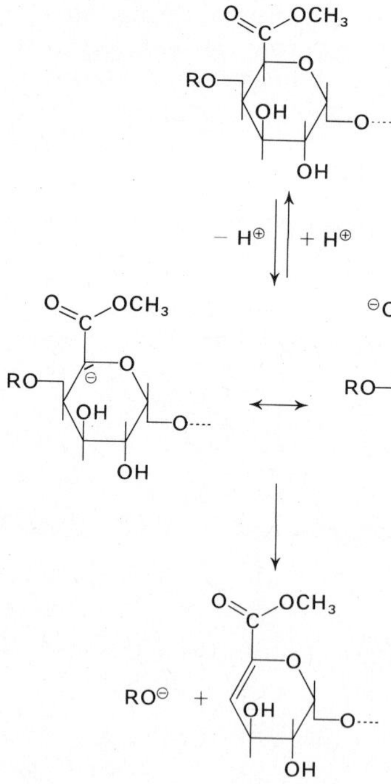

(4.89)

The elimination reaction occurs more readily with galacturonic acid units having an esterified carboxyl group, since the H-atom on C-5 is more acidic than with residues having free carboxyl groups.
At a pH of about 3, and in the presence of Ca^{2+} ions also at higher pH's, pectin forms a thermally-reversible gel. The gel-forming ability, under comparable conditions, is directly proportional to the molecular weight and inversely proportional to the esterification degree. For gel formation, low ester pectins require very low pH values and/or calcium ions, but they set into a gel in the presence of a relatively low sugar content. High ester pectins require an increasing amount of sugar with rising esterification degree. The gel-setting time for high ester pectins is longer than that for pectin products of low esterification degree (Table 4.16).

Utilization. Since pectin can set into a gel, it is widely used in marmalade and jelly production. Standard conditions to form a stable gel are, for instance: pectin content < 1%, sucrose 58–75% and pH 2.8–3.5. In low sugar products, low ester pectin is used in the presence of Ca^{2+} ions. Pectin use as a stabilizer for beverages (to emulsify the essential oil components) or ice creams is also of importance.

Table 4.16. Gelling time of pectins with differing degrees of esterification

Pectin type	Esterification degree (%)	Gelling time[a] (s)
Fast gelling	72–75	20–70
Normal	68–71	100–135
Slow gelling	62–66	180–250

[a] Difference between the time when all the prerequisites for gelling are fulfilled and the time of actual gel setting.

4.4.4.12 Starch

4.4.4.12.1 General remarks

Starch is widely distributed in various plant organs as a storage carbohydrate. As an ingredient of many foods, it is also the most important carbohydrate source in human nutrition. In addition, starch and its derivatives are important industrially, for example, in the paper and textile industries.

Starch production is based on its isolation from corn, millet, potatoes, wheat, cassava (also called manioc) and sweet potatoes and, often, from rice, arrowroot and sago palm.
In some cases, e. g. potato tubers, starch granules occur free, deposited in cell vacuoles; hence, their isolation is relatively simple. The plant material is disintegrated, the starch granules are washed out with water, and then sedimented from the "starch milk" suspension and dried. In other cases, such as cereals, the starch is embedded in the endosperm protein matrix, hence granule isolation is a more demanding process. Thus, a counter-current process with water at 50 °C for 36–48 h is required to soften corn (steeping step of processing). The steeping water contains 0.2% SO_2 in order to loosen the protein matrix and, thereby, to accelerate the granule release and increase the starch yield. The corn grain is then disintegrated. The germ, due to its high oil content, has a low density and is readily separated by flotation. It is the source for corn oil isolation (cf. 14.3.2.2.4). The protein and starch are then separated in hydrocyclones. The separation is based on density difference (protein < starch).

Table 4.17. Amylose content, gelatinization temperature and swelling power of various starches

Starch from	Amylose (%)	Gelatinization temperature (°C)	Swelling power[a] at 95 °C
Wheat	26	53–65	21
Rye		57–70	
Barley	22	56–62	
Corn	28	62–70	24
Amylomaize	51–65	67–87	
Waxy corn (maize)	1	63–72	64
Oats	27	56–62	
Rice	18	61–78	19
Waxy rice	1	55–65	56
Sorghum	25	69–75	22
Waxy sorghum	1	68–74	49
Potatoes	23	58–66	
Beans	24	64–67	
Peas	35	57–70	
Cassava (trade name tapioca)		52–64	

[a] Swollen starch weight increase per starch dry matter basis; values were corrected for leached out polysaccharides.

The protein by-product is marketed as animal feed or used for production of a protein hydrolysate (seasoning). The recovered starch is washed and dried.

Starch granules are formed from two glucans, amylose and amylopectin. Most starches contain 20–39% amylose (Table 4.17). New corn cultivars (amylomaize) have been developed which contain 50–80% amylose. The amylose can be isolated from starch, e.g. by crystallization of a starch dispersion, usually in the presence of salts ($MgSO_4$) or by precipitation with a polar organic compound (alcohols, such as n-butanol, or lower fatty acids, such as caprylic or capric), which form a complex with amylose and thus enhance its precipitation.

Normal starch granules contain 70–80% amylopectin, while some corn cultivars and millet, denoted as waxy maize or waxy millet, contain almost 100% amylopectin.

4.4.4.12.2 Amylose structure and properties

Amylose is a chain polymer of α-D-glucopyranosyl residues linked 1 → 4:

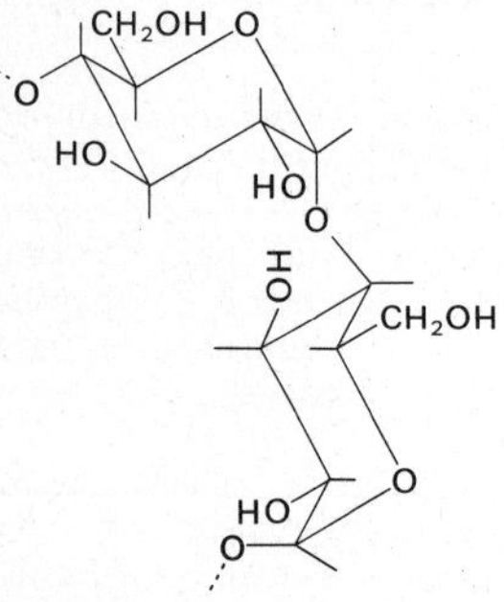

(4.90)

Enzymatic hydrolysis of the chain is achieved by α-amylase, β-amylase and glucoamylase. Often, β-amylase does not degrade the molecule completely into maltose, since a very low branching is found along the chain with α (1 → 6) linkages. The molecular size of amylose is variable. The polymerization degree in wheat starch is 1,000–2,000, while in potatoes it can rise up to 4,500. This corresponds to a molecular weight of 150–750 kdal. Amylose is not readily soluble or dispersible in cold water. When intact starch granules are heated in water, they swell and provide a pasty and sticky starch suspension, which consists of highly-swollen intact granules dispersed in a solution of free starch molecules that leached from the granules during swelling. The temperature range at which granule swelling occurs, with a simultaneous destruction of its internal structure, is called the gelatinization temperature. It is specific for a starch type and, hence, varies for starches of different botanical origin (Table 4.17).

A gelatinization curve for potato starch is reproduced in Figure 4.20 a. The number of fully-gelatinized starch granules in an aqueous granule suspension is determined by polarized light microscopy.

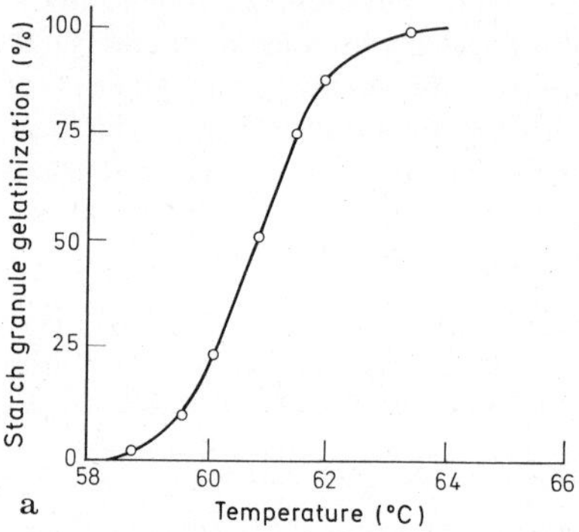

Fig. 4.20 a. Potato starch gelatinization curve. (After *Banks* and *Muir,* 1980)

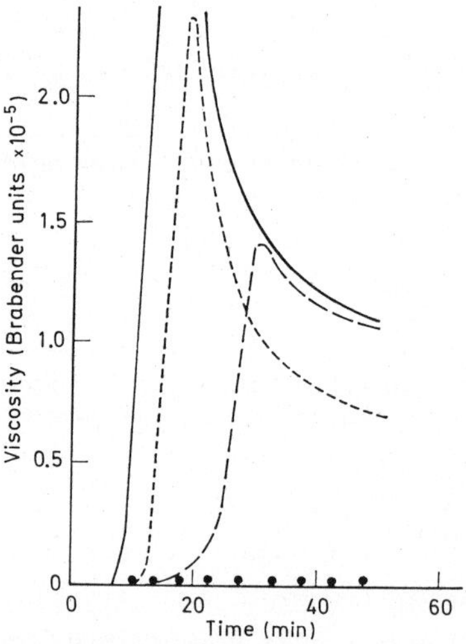

Fig. 4.20 b. Gelatinization properties of various starches. (After *Banks* and *Muir,* 1980). Brabender visco-amylograph. 40 g starch/460 ml water, temperature programming: start at 50 °C, heated to 95 °C at a rate of 1.5 °C/min. Held at 95 °C for 30 min —— potato, - - - waxy corn, — — — normal corn, and ●●● amylomaize starch

Another way to monitor gelatinization versus temperature is to use a rotoviscometer to continuously measure the viscosity of a starch suspension. The viscosity curves (Figure 4.20b) show that, initially, a rise in temperature leads to a steep rise in viscosity due to starch granule swelling. Then the intact swollen granules burst and disintegrate, causing a drop in viscosity. The shape of the curve varies significantly for different starches. Potato starch shows a very high viscosity maximum (~4,000 Brabender units, BU), which is followed by a steep viscosity drop. Waxy corn shows a similar behavior, except that its maximum is not as high. In normal corn starch the maximum is lower, but the viscosity drop is still lower, which indicates a greater stability of its starch granules to disintegration. Under these conditions, amylomaize starch does not swell, instead about 35% of the material solubilizes.

When a continuously-mixed (Brabender rotovisco-amylograph) gelatinized starch paste is cooled rapidly, the viscosity generally rises; however, without mixing, the starch paste sets to a gel. Amylose gels tend to retrograde. This term usually denotes an irreversible transition from the solubilized or dispersed state to an insoluble microcrystalline state (mostly by realignment of amylose molecules, cf. Figure 4.20c). A similar retrogradation results when starch paste is cooled slowly. The tendency for retrogradation is enhanced at lower temperatures, especially near 0 °C, at a neutral pH and by high starch concentration and the absence of surface-active agents. The extent of retrogradation is also dependent on the molecular weight of the starch molecules (chain length or polymerization degree) and the origin of the starch. It increases in the series: potato – corn – wheat.

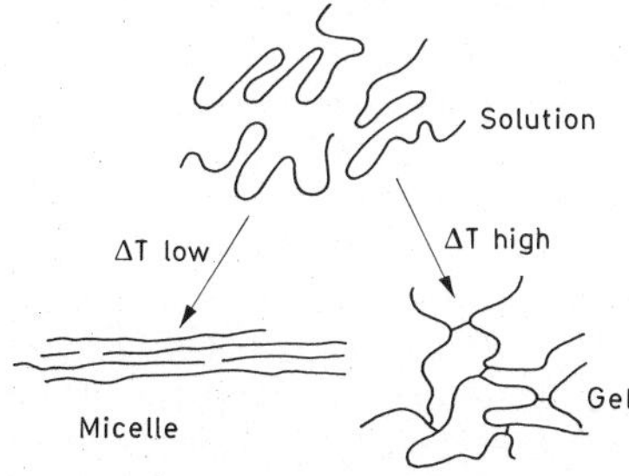

Fig. 4.20c. Behavior of amylose molecules during cooling of their concentrated aqueous solutions

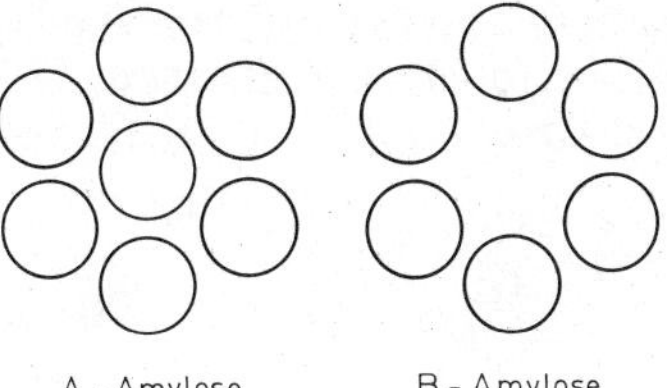

Fig. 4.20d. Arrangement of double helices (o) in A-type and B-type amylose (cross-sections)

The transitions of a water-deficient starch granule, a highly swollen water-enriched granule, molecules in solution, and of a crystallized shrunken state are related to conformational changes of amylose chains which have not yet been fully elucidated. The conformation is affected by many factors, even by the presence of low molecular weight compounds.

With the aid of X-ray diffraction analysis, native starch granules reveal three crystallographic patterns, denoted as A, B and C patterns. An additional pattern occurs in swollen granules and it is designated as the V form.

While A and B patterns are genuine crystalline modifications, the C pattern appears to be a mixed A and B form. The A pattern is usually present in cereal starches, while the B pattern is in potatoes, amylomaize and in retrograded starch. The C pattern is observed in mixtures of corn and potato starches and is also found in legume starches.

The structural element of the B pattern is a double helix (cf. Figure 4.22) packed in an antiparallel, hexagonal mode. The central channel is filled with water. The A pattern is very similar, except that the central channel is occupied by another double helix. The water molecules are distributed between double helices (Figure 4.20d). A double helix structure is shown in Figure 4.21, a.

The double helix, depending on conditions, can change into other helical conformations. In the presence of KOH, for instance, a more extended helix results, with 6 glucose residues per helical turn (Figure 4.21, b) while, in the presence of KBr, the helix is even more stretched, to 4 residues per turn (Figure 4.21, c). Inclusion (clathrate) compounds are formed in the presence of small molecules and stabilize the V starch conformation (Figure 4.21, d); it also has 6 glucose residues per helical turn. Stabilization may be achieved by

H-bridges between O-2 and O-3 of neighboring residues within the same chain and between O-2 and O-6 of the residues i and i + 6 neighbored

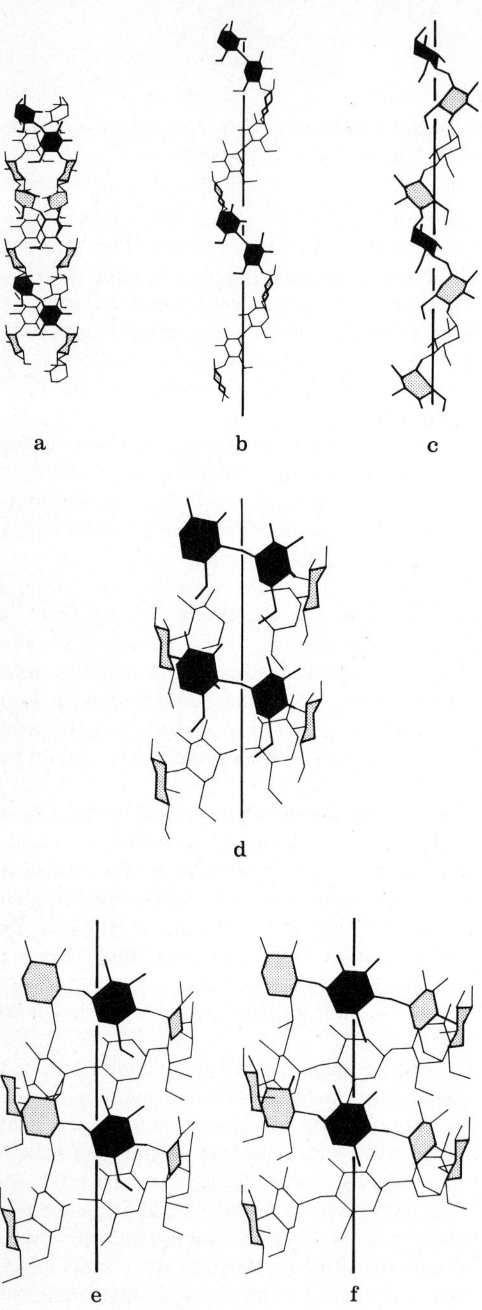

Fig. 4.21. Amylose conformation (for explanation see text). (After *Rees*, 1977)

on the helix surface. Many molecules, such as iodine, monoacylglycerides, phenols, arylhalogenides, n-butanol, t-butanol or cyclohexane, form clathrate compounds with amylose molecules. The helix diameter, to a certain extent, conforms to the size of the enclosed guest molecule; it varies from 13.7 Å to 16.2 Å. While the iodine complex and that of n-butanol have 6 glucose residues per turn in a V conformation, in a complex with t-butanol the helix turn is enlarged to 7 glucose residues/turn (Figure 4.21, e). It is shown by an α-naphthol clathrate that up to 8 residues are allowed (Figure 4.21, f). Since the helix is internally hydrophobic, the enclosed "guest" has also to be lipophilic in nature. The enclosed molecule contributes significantly to the stability of a given conformation. For example, it is observed that the V conformation, after "guest" compound removal, slowly changes in a humid atmosphere to a more extended B conformation. Such a conformational transition also occurs during staling of bread or other bakery products. Freshly-baked bread shows a V spectrum of gelatinized starch, but aged bread typically has the retrograded starch B spectrum. Figure 4.22 illustrates both conformations in the form of cylinder projections. While in V amylose, as already outlined, O-2 of residues i and O-6 of residues i + 6 come into close contact through H-bridges, in the B pattern the inserted water molecules increase the double-strand distance along the axis of progression (h) from 0.8 nm for the V helix to 1 nm for the B helix.

Cereal starches are stabilized by the enclosed lipid molecules, so their swelling power is low. The swelling is improved in the presence of alcohols (ethanol, amyl alcohol, tert-amyl alcohol). Obviously, these alcohols are dislodging and removing the "guest" lipids from the helices.

As already suggested, the three main starch conformations (A, B and V) can interchange under suitable conditions. Moistened potato starch, when heated, changes from a B into a A pattern conformation. Amylose triacetate fibers saponified with methanolic KOH, followed by alkaline amylose extraction with absolute methanol and steeping in 75% ethanol, provide V pattern starch. When alkaline amylose is kept for several days in a moist atmosphere (80% relative humidity), it gives an A starch; and, when treated with boiling water, a B starch. Only the V conformation is suitable for clathrate formation.

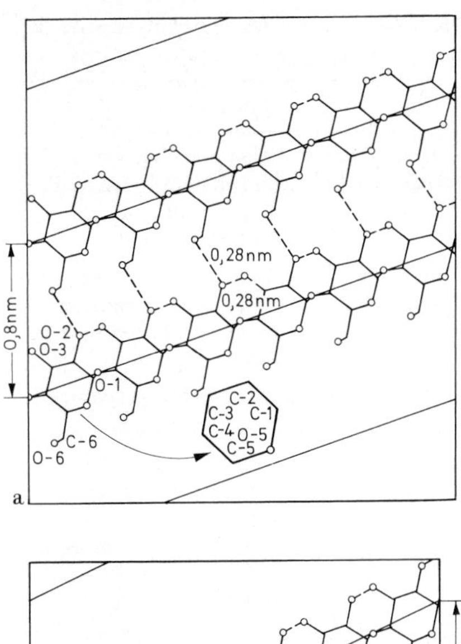

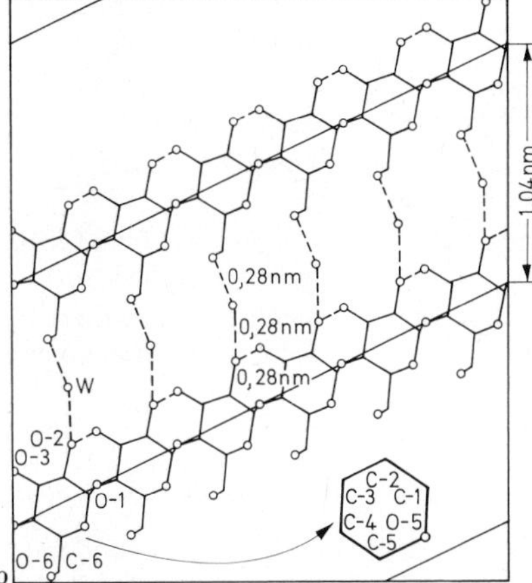

Fig. 4.22. Amylose: V-conformation (**a**) and B-conformation (**b**) in a cylinder projection. (After *Ebert*, 1980)

4.4.4.12.3 Amylopectin structure and properties

Amylopectin is a branched glucan with side chains attached in the 6-position of the glucose residues of the principal chain:

(4.91)

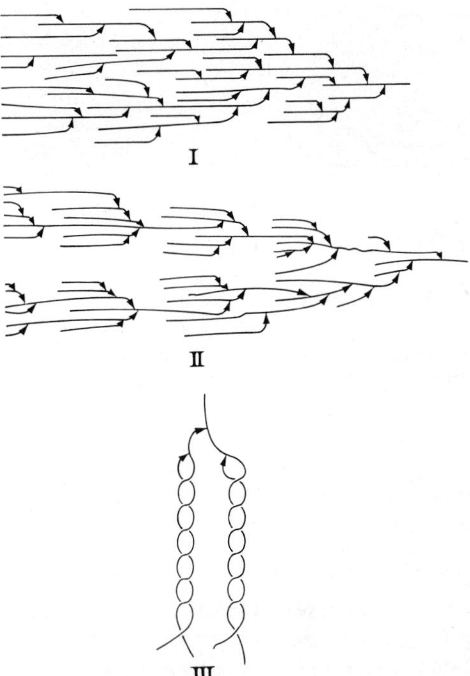

Fig. 4.23. Structural models (I, II) for amylopectin with parallel double helices. III is an enlarged segment of I or II. (After *Banks* and *Muir*, 1980)

An average of 15–30 glucose residues are present in short chain branches and each of these branch chains is joined by linkage of C-1 to C-6 of the next chain. The proposed structural models (Figure 4.23) suggest that amylopectin also has double helices organized in parallel. Probably, the main portion of a starch granule's crystallinity is derived from amylopectin. This is supported by findings that a waxy maize starch shows crystallinity coinciding with that of a normal maize starch and that gelatinized starch granules retain amylopectin, but lose amylose by leaching or diffusion.

The molecular weight of amylopectin is very high, 100–200 Mdal. One phosphoric acid residue is found for an average of 400 glucose residues.

The organization of amylopectin molecules in starch granules is shown in Figure 4.24.

Enzymatic degradation of amylopectin is similar to that of amylose. The enzyme β-amylase degrades the molecule up to the branching points. The remaining resistant core is designated as "limit-dextrin".

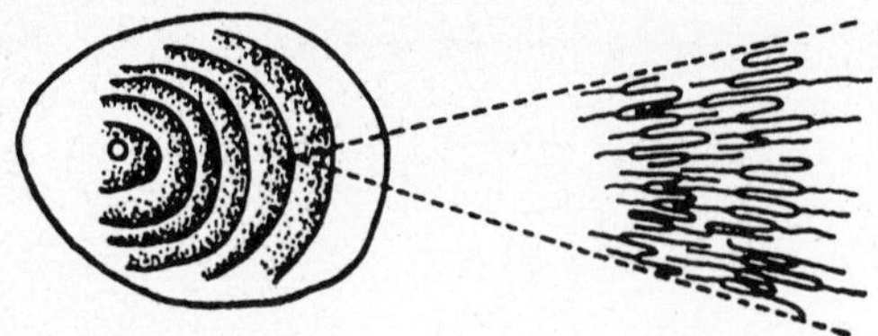

Fig. 4.24. Arrangement of amylopectin molecules in a starch granule

Amylopectin, when heated in water, forms a transparent, highly viscous solution, which is ropy, sticky and coherent. Unlike with amylose, there is no tendency for retrogradation. There are no staling or aging phenomena and no gelling, except at very high concentrations. However, there is a rapid viscosity drop in acidic media and by autoclaving or applying stronger mechanical shear force.

4.4.4.12.4 Utilization

Starch is an important thickening and binding agent and is used extensively in the production of puddings, soups, sauces, salad dressings, diet food preparations for infants, pastry filling, mayonnaise, etc.

A layer of amylose can be used as a protecting cover for fruits (dates or figs) and dehydrated and candied fruits, preventing their sticking together. Amylose treatment of French fries decreases their susceptibility to oxidation. The good gelling property of a dispersable amylose makes it a suitable ingredient in instant puddings or sauces. Amylose films can be used for food packaging, as edible wrapping or tubing, as exemplified by a variety of instant coffee or tea products. Amylopectin utilization is also diversified. It is used to a large extent as a thickener or stabilizer and as an adhesive or binding agent. Table 4.18 lists the range of its applications.

Table 4.18. Utilization of amylopectin and its derivatives

Starch	Utilization
Unmodified waxy starch (also in blend with normal starch and flours)	Salad dressing, sterilized canned and frozen food, soups, broth, puffed cereals, and snack food
Pregelatinized waxy starch or isolated amylopectin	Baked products, paste (pâté) fillings, sterilized bread, salad dressing, pudding mixtures
Thin boiling waxy starch	Protective food coatings
Cross-linked waxy starch	Paste fillings, white and brown sauces, broth, sterilized or frozen canned fruit, puddings, salad dressing, soups, spreadable cream products for sandwiches, infant food
Waxy starch hydroxypropyl ether	Sterilized and frozen canned food
Waxy starch carboxymethyl ether	Emulsion stabilizer
Waxy starch acetic acid ester	Sterilized and frozen canned food, infant food
Waxy starch succinic- and adipic acid esters	Sterilized and frozen canned food, aroma encapsulation
Waxy starch sulfuric acid ester	Thickenig agent, emulsion stabilizer, ulcer treatment (pepsin inhibitor)

4.4.4.13 Modified Starches

Starch properties and those of amylose and amylopectin can be improved or "tailored" by physical and chemical methods to fit or adjust the properties to a particular application or food product.

Pregelatinized starch. Heating of a starch suspension above its gelatinization temperature, followed by suspension drying, provides a starch product that is soluble in cold water and that gels. These products are used in instant foods, e.g. pudding, and as baking aids (cf. Table 4.18).

Thin-boiling starch. Partial acidic hydrolysis yields a starch product which is not very soluble in cold water but is readily soluble in boiling water. The solution has a lower viscosity than the untreated starch, and remains fluid after cooling. Retrogradation is low. These starches are utilized as thickeners and as protective films (cf. Table 4.18).

Starch ethers. When a 30–40% starch suspension is reacted with ethylene oxide or propylene oxide in the presence of hydroxides of alkali and/or alkali earth metals (pH 11–13), hydroxyethyl- or hydroxypropyl-derivates are obtained ($R' = H, CH_3$):

$$R{-}OH + \underset{O}{\triangle}{-}R^1 \xrightarrow{OH^{\ominus}} R{-}O{-}CH_2{-}\underset{\underset{OH}{|}}{C}HR^1 \quad (4.92)$$

These derivatives are also obtained in reaction with the corresponding epichlorohydrins. The substitution degree can be controlled over a wide range by adjusting process parameters. Low substitution products contain up to 0.1 mole alkyl group/mole glucose, while those with high substitution degree have 0.8–1 mole/mole glucose. Introduction of hydroxylalkyl groups, often in combination with a small extent of cross-linking (see below) greatly improves starch swelling power and solubility, lowers the gelatinization temperature and substantially increases the freeze-thaw stability and the paste clarity of highly-viscous solutions. Therefore, these products are utilized as thickeners for refrigerated foods (apple and cherry pie fillings, etc.), and heat-sterilized canned food (cf. Table 4.18).
Reaction of starch with monochloroacetic acid in an alkaline solution yields carboxymethyl starch:

$$R{-}OH + ClCH_2COO^{\ominus} \xrightarrow{OH^{\ominus}} R{-}O{-}CH_2{-}COO^{\ominus} \quad (4.93)$$

These products swell instantly, even in cold water and in ethanol. Dispersions of 1–3% carboxymethyl starch have an ointment-like (pomade) consistency, whereas a 3–4% dispersions provide a gel-like consistency. These products are of interest as thickeners and gel-forming agents.

Starch esters. Starch monophosphate ester is produced by dry heating of starch with alkaline orthophosphate or alkaline tripolyphosphate at 120–175 °C:

$$R{-}OH \xrightarrow[POCl_3\,/\,Alkali\ phosphates]{OH^{\ominus}} R{-}OPO_3H^{\ominus} \quad (4.94)$$

Starch organic acid esters, such as those of acetic acid, longer chain fatty acids (C_6–C_{26}), succinic, adipic or citric acids, are obtained in reactions with the reactive derivatives (e. g. vinyl acetate) or by heating the starch with free acids or their salts. The thickening and paste clarity properties of the esterified starch are better than in the corresponding native starch. In addition, esterified starch has an improved freeze-thaw stability. These starches are utilized as thickeners and stabilizers in bakery products, soup powders, sauces, puddings, refrigerated food, heat-sterilized canned food and in margarines. The starch esters are also suitable as protective coatings, e. g. for dehydrated fruits or for aroma trapping or encapsulation (cf. Table 4.18).

Cross-linked starches. Cross-linked starches are obtained by the reaction of starch (R–OH) with bi- or polyfunctional reagents, such as sodium trimetaphosphate, phosphorus oxychloride, epichlorohydrin or mixed anhydrides of acetic and dicarboxylic acids (e. g. adipic acid):

$$2\,R{-}OH + R^1CO{-}O{-}CO{-}(CH_2)_n{-}CO{-}O{-}COR^1 \longrightarrow R{-}O{-}CO{-}(CH_2)_n{-}CO{-}O{-}R \quad (4.95a)$$

$$R{-}OH + \text{(epichlorohydrin, } CH_2Cl\text{)} \xrightarrow{OH^{\ominus}} R{-}O{-}CH_2{-}CH(OH){-}CH_2Cl \xrightarrow{OH^{\ominus}} R{-}O{-}CH_2{-}\text{(epoxide)} \xrightarrow[OH^{\ominus}]{ROH} R{-}O{-}CH_2{-}CH(OH){-}CH_2{-}O{-}R \quad (4.95b)$$

$$2\,R{-}OH \xrightarrow{(NaPO_3)_3} R{-}O{-}P(=O)(O^{\ominus}){-}O{-}R \quad (4.95c)$$

The starch granule gelatinization temperature increases in proportion to the extent of cross-linking, while the swelling power decreases (Figure 4.25). Starch stability remains high at extreme pH values (as in the presence of food acids) and under conditions of shear force. Cross-linked

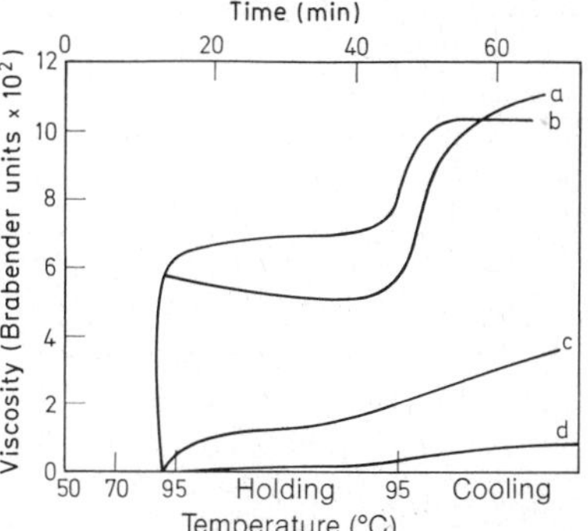

Fig. 4.25. Corn starch viscosity curves as affected by crosslinking degree. Instruments: Brabender amylograph. a Control, b crosslinked 0.05%, c 0.10%, d 0.15% epichlorohydrin. (After *Pigman,* 1970)

starch derivatives are generally used when high starch stability is demanded.

Oxidized starches. Starch hydrolysis and oxidation occur when aqueous starch suspensions are treated with sodium hypochlorite at a temperature below the starch gelatinization temperature range. The products obtained have an average of one carboxyl group per 25–50 glucose residues:

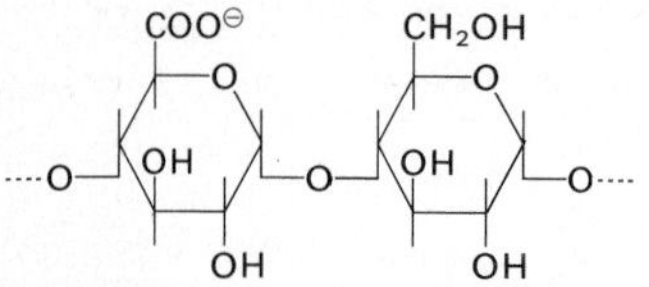

(4.96)

Oxidized starch is used as a lower-viscosity filler for salad dressings and mayonnaise. Unlike thin-boiling starch, oxidized starch does not retrograde nor does it set to an opaque gel.

4.4.4.14 Cellulose

Cellulose is the main constituent of plant cell walls, where it usually occurs together with hemicelluloses, pectin and lignin. Since cellulase enzymes are absent in the human digestive tract, cellulose, together with some other inert polysaccharides, constitute the indigestible carbohydrate of plant food (vegetables, fruits or cereals), referred to as dietary fiber. Cellulases are also absent in the digestive tract of animals, but herbivorous animals can utilize cellulose because of the rumen microflora (which hydrolyze the cellulose). The importance of dietary fiber in human nutrition appears mostly to be the maintenance of intestinal motility (peristalsis).

Cellulose consists of β-glucopyranosyl residues joined by 1 → 4 linkages (cf. Formula 4.97).

Cellulose crystallizes as monoclinic, rod-like crystals. The chains are oriented parallel to the fiber direction and form the long b-axis of the unit cell (Figure 4.26). The chains are probably somewhat pleated to allow intrachain hydrogen bridge formation between O-4 and O-6, and between O-3 and O-5 (cf. Formula 4.98).

Intermolecular hydrogen bridges (stabilizing the parallel chains) are present in the direction of the a-axis, while hydrophobic interactions exist in the c-axis direction. The crystalline sections comprise an average of 60% of native cellulose. These sections are interrupted by amorphous gel regions, which can become crystalline when moisture is removed. The acid- or alkali-labile bonds also apparently occur in these regions. Microcrystalline cellulose is formed when these bonds are hydrolyzed. This partially-depolymerized cellulose product with a molecular weight of 30–50 kdal, is still water insoluble, but does not have a fibrous structure.

Cellulose has a variable degree of polymerization (denoted as DP; number of glucose residues per chain) depending on its origin. The DP can range from 1,000 to 14,000 (with corresponding molecular weights of 162 to 2,268 kdal). Because of its high molecular weight and crystalline structure, cellulose is insoluble in water. Also, its swelling power or ability to absorb water, which depends partly on cellulose source, is poor or negligible.

Utilization. Microcrystalline cellulose is used in low-calorie food products and in salad dressings, desserts and ice creams. Its hydration capacity and dispersibility are substantially enhanced by adding it in combination with small amounts of carboxymethyl cellulose.

4.4.4.15 Cellulose Derivatives

Cellulose can be alkylated into a number of derivatives with good swelling properties and im-

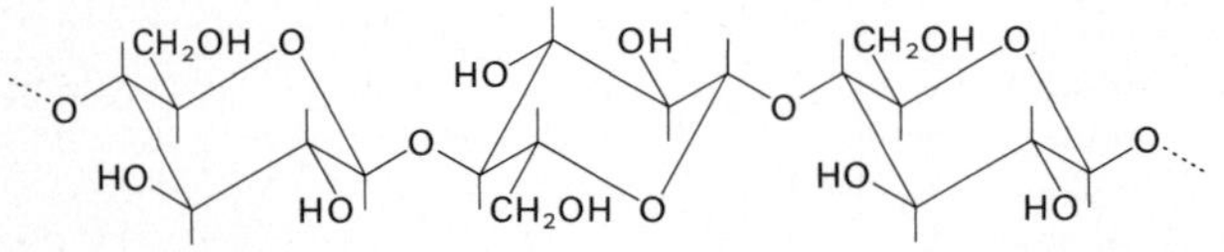

(4.97)

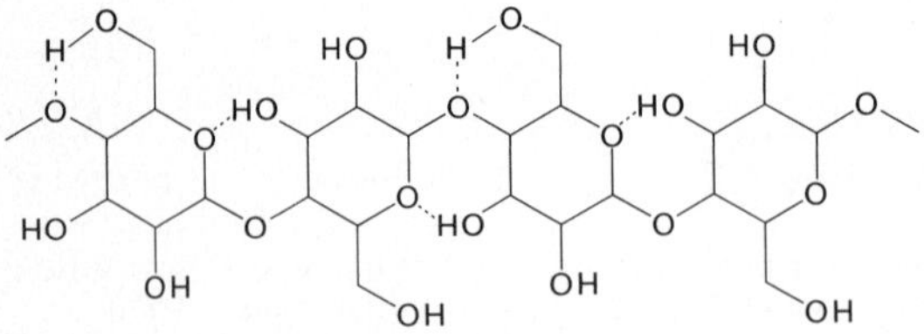

(4.98)

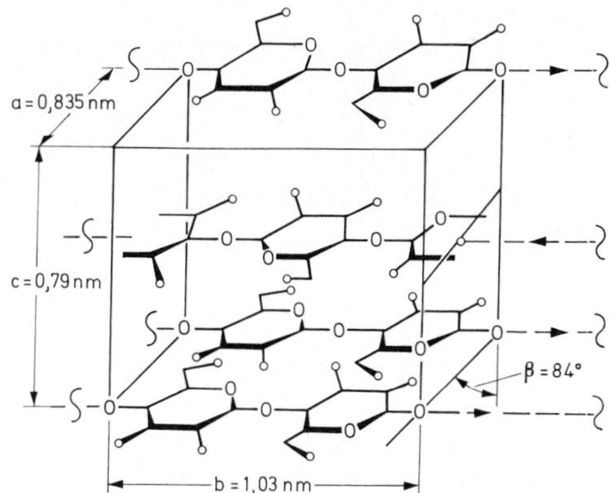

Fig. 4.26. Unit cell of cellulose. (After *Meyer* and *Misch*)

proved solubility. Such derivatives have a wide field of application.

Alkyl cellulose, hydroxyalkyl cellulose. The reaction of cellulose with methylchloride or propylene oxide in the presence of a strong alkali introduces methyl or hydroxypropyl groups into cellulose. The degree of substitution (DS) is dependent on reaction conditions (cf. Reaction 4.99).

Mixed substituted products are also produced, e.g. methylhydroxypropyl cellulose or methylethyl cellulose. The substituents interfere in the normal crystalline packing of the cellulose chains, thus facilitating chain solvation. Depending on the nature of the substituent (methyl, ethyl, hydroxymethyl, hydroxyethyl or hydroxypropyl) and the substitution degree, products are obtained with variable swelling powers and water solubilities. A characteristic property for methyl cellulose and double-derivatized methylhydroxypropyl cellulose is their initial viscosity drop with rising temperature, setting to a gel at a specific temperature. Gel setting is reversible. Gelling temperature is dependent on substitution type and degree. Figure 4.27 shows the dependence of gelling temperature on the substitution type and the concentration of the derivatives in water. Hydroxyalkyl substituents stabilize the hydration layer around the macromelecule and, thereby, increase the gelling temperature. Changing the proportion of methyl to hydroxypropyl substituents can vary gel setting within a wide temperature range.

The above properties of cellulose derivatives permit their diversified application (Table 4.19). In baked products obtained from gluten-poor or gluten-free flours, such as those of rice, corn or rye, the presence of methyl and methylhydroxypropyl celluloses decreases the crumbliness and friability of the product, enables a larger volume of water to be worked into the dough and, thus, improves the extent of starch swelling during oven baking. Since differently-substituted celluloses offer a large choice of gelling temperatures, each application can be met by using a most suitable derivative. Their addition to batter or a coating mix for meats (panure) decreases oil uptake in frying. Their addition to dehydrated fruits and vegetables improves rehydration char-

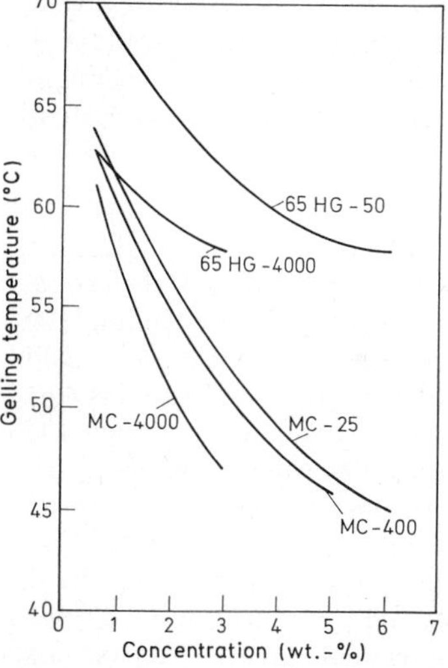

Fig. 4.27. Gelling behavior of alkyl celluloses. (After *Balser,* 1975). MC: methyl cellulose, HG: hydroxypropylmethyl cellulose with a hydroxypropyl content of about 6.5%. The numerical suffix is the viscosity (cps) of a 2% solution

$$\text{Cellulose}{-}OH + CH_3Cl \xrightarrow{OH^\ominus} \text{Cellulose}{-}O{-}CH_3$$

$$\text{Cellulose}{-}OH + CH_3{-}\underset{O}{CH{-}CH_2} \xrightarrow{OH^\ominus} \text{Cellulose}{-}O{-}CH_2{-}\underset{OH}{\underset{|}{CH}}{-}CH_3 \quad (4.99)$$

Table 4.19. Utilization of cellulose derivatives (in amounts of 0.01 to 0.8%)

Food product	Cellulose derivative[a]			Effect								
	1	2	3	A[b]	B	C	D	E	F	G	H	I
Baked products	+		+		+		+		+			
Potato products	+	+			+		+					+
Meat and fish	+		+	+		+						+
Mayonnaise, dressings	+		+	+	+			+				
Fruit jellies	+			+	+	+						
Fruit juices	+			+								
Brewery	+	+								+	+	
Wine	+	+								+	+	
Ice cream, cookies	+			+	+			+				
Diet food	+	+	+		+							

[a] 1: Carboxymethyl cellulose, Na-salt; 2: methyl cellulose; 3: hydroxypropyl methyl cellulose.
[b] A: Thickening effect; B: water binding/holding; C: gel setting at cold; D: gel setting at higher temperatures; E: emulsifier; F: suspending effect; G: surface activity; H: adsorption; and I: film-forming property.

acteristics and texture upon reconstitution. Sensitive foods can be preserved by applying alkyl cellulose as a protective coating or film. Cellulose derivatives can also be used as thickening agents in low calorie diet foods. Hydroxypropyl cellulose is a powerful emulsion stabilizer, while methylethyl cellulose has the property of a whipping cream: it can be whipped into a stable foam consistency.

Carboxymethyl cellulose. Carboxymethyl cellulose is obtained by treating alkaline cellulose with chloroacetic acid. The properties of the product depend on the degree of substitution (DS; 0.3–0.9) and of polymerization (DP; 500–2,000). Low substitution types (DS $\leq$ 0.3) are insoluble in water but soluble in alkali, whereas higher DS types (>0.4) are water soluble. Solubility and viscosity are dependent on pH.
Carboxymethyl cellulose is an inert binding and thickening agent used to adjust or improve the texture of many food products, such as jellies, paste fillings, spreadable process cheeses, salad dressings and cake fillings and icings (Table 4.19). It retards ice crystal formation in ice cream, stabilizing the smooth and soft texture. It retards the undesired saccharose crystallization in candy manufacturing and inhibits starch retrogradation or the undesired staling in baked goods. Lastly, carboxymethyl cellulose improves the stability and rehydration characteristics of many dehydrated food products.

4.4.4.16 Xanthan Gum

Xanthan gum, the extracellular polysaccharide from *Xanthomonas campestris* and some related microorganisms, is produced on a nutritive medium containing glucose, NH_4Cl, a mixture of amino acids, and minerals. The polysaccharide is recovered from the medium by isopropanol precipitation in the presence of KCl.
Xanthan gum is a heteroglycan consisting of D-glucose, D-mannose and D-glucuronic acid in a molar ratio of 2.8:2:2. Some sugar residues are acetylated, while some are present as ketals, formed by pyruvate [4,6-O-(1-carboxyethylidine)-D-glucopyranose]. The molecule consists of a "backbone" made of 1,4-linked β-glucopyranosyl residues. In the average every other clucose residue bears a trisaccharide side chain:

→4)-β-D-Glc*p*-(1→4)-β-D-Glc*p*-(1→4)-β-D-Glc*p*-(1→
| 3 ↑ 1 |
β-D-Man*p*-(1→4)-β-D-GlcA*p*-(1→2)-α-D-Man*p*-6-OAc
4 6 C
H_3C COOH (4.100)

The molecular weight of xanthan gum is $>10^6$ Mdal. In spite of this weight, it is quite soluble in water. The highly viscous solution exhibits a pseudoplastic behavior (Figure 4.28). The viscosity is, to a great extent, independent of tempera-

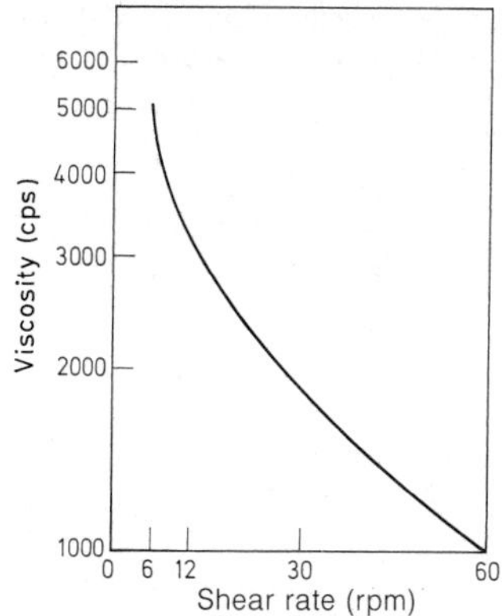

Fig. 4.28. Viscosity of aqueous xanthan gum solution as affected by shear rate. (After *Whistler*, 1973). Viscometer: Brookfield Model LVF

ture. Solutions, emulsions and gels, in the presence of xanthan gums, acquire a high freeze-thaw stability.

Utilization. The practical importance of xanthan gum is based on its emulsion-stabilizing and particle-suspending abilities (turbidity problems, essential oil emulsions in beverages). Due to its high thermal stability, it is useful as a thickening agent in food canning. Xanthan gum addition to starch gels substantially improves their freeze-thaw stability.

Xanthan gum properties might also be utilized in instant puddings: a mixture of locust bean flour, Na-pyrophosphate and milk powder with xanthan gum as an additive provides instant jelly after reconstitution in water. The pseudoplastic thixotropic properties, due to intermolecular association of single-stranded xanthan gum molecules, are of interest in the production of salad dressings: that is, a high viscosity in the absence of a shear force and a drop in viscosity to a fluid state under a shear force.

4.4.4.17 Scleroglucan

Sclerotium species, e.g. *S. glucanicum,* produce scleroglucan on a nutritive medium of glucose, nitrate as N-source and minerals. The polysaccharide is recovered from the nutritive medium by alcoholic precipitation.

The "backbone" of scleroglucan is a β-1,3-glucan chain that, on the average, has an attached glucose as a side chain on every third sugar residue (cf. Formula 4.101).

The polysaccharide has a molecular weight of about 130 kdal and is very soluble in water. Solutions have high viscosities and exhibit pseudoplastic thixotropic properties.

Utilization. Scleroglucan is used as a food thickener and, on the basis of its good film-forming property, is applied as a protective coating to dried foods.

4.4.4.18 Dextran

Leuconostoc mesenteroides, Streptobacterium dextranicum and some other bacteria produce dextran from a nutrient medium containing saccharose. Dextran is a β-1,6-glucan with several glucose side chains, bound primarily to the main chain of the macromolecule through 1,3-linkages but, in part, also by 1,4- and 1,2-linkages:

(4.102)

On the average, 95% of the glucose residues are present in the main chain. Dextran is very soluble in water.

Utilization. Dextran is used mostly in medicine as a blood substitute. In the food industry it is

(4.101)

used as a thickening and stabilizing agent, as exemplified by its use in baking products, confections, beverages and in the production of ice creams.

4.4.4.19 Polyvinyl Pyrrolidone (PVP)

This compound is used as if it were a polysaccharide-type additive. Therefore, it is described here. The molecular weight of PVP can range from 10–360 kdal.

$$\left[-CH(N\text{-pyrrolidone})-CH_2- \right]_n \quad (4.103)$$

It is quite soluble in water and organic solvents. The viscosity of the solution is related to the molecular weight. PVP forms insoluble complexes with phenolic compounds and, therefore, is applied as a clarifying agent in the beverage industry (beer, wine, fruit juice). Furthermore, it serves as a binding and thickening agent, and as a stabilizer, e. g. of vitamin preparations. Its tendency to form films is utilized in protective food films (particle solubility enhancement and aroma fixation in instant tea and coffee production).

4.5 Literature[a]

Angyal, S. J.: Zusammensetzung und Konformation von Zuckern in Lösung. Angew. Chem. *81,* 172 (1969)

Balser, K.: Celluloseäther. In: Ullmanns Encyklopädie der technischen Chemie, 4. Aufl., Bd. 9, S. 192. Verlag Chemie: Weinheim. 1975

Banks, W., Muir, D. D.: Structure and chemistry of the starch granule. In: The biochemistry of plants (Eds.: Stumpf, P. K., Conn, E. E.), Vol. 3, p. 321, Academic Press: New York–London. 1980

Birch, G. G.: Structural relationships of sugars to taste. Crit. Rev. Food Sci. Nutr. *8,* 57 (1967)

Birch, G. G. (Ed.): Developments in food carbohydrate-1, Applied Science Publ.: London. 1977 ff.

Birch, G. G., Parker, K. J. (Eds.): Nutritive sweeteners. Applied Science Publ.: London. 1982

Brimacombe, J. C. (Ed.): Carbohydrate chemistry, Vol. 1, The Chemical Society, Berlington House: London. 1969.

Davidson, R. L. (Ed.): Handbook of water-soluble gums and resins. McGraw-Hill Book Co.: New York. 1980

Ebert, G.: Biopolymere. Dr. Dietrich Steinkopff Verlag: Darmstadt. 1980

Eriksson, C. (Ed.): Maillard reactions in food. Chemical, physiological and technological aspects. Pergamon Press: Oxford. 1981

Guadagni, D. G., Maier, V. P., Turnbaugh, J. G.: Effect of some citrus juice constituents on taste thresholds for limonin and naringin bitterness. J. Sci. Food Agric. *24,* 1277 (1973)

Ledl, F., Severin, T.: Untersuchungen zur Maillard-Reaktion. XIII. Bräunungsreaktion von Pentosen mit Aminen. Z. Lebensm. Unters. Forsch. *167,* 410 (1978)

Ledl, F., Severin, T.: Untersuchungen zur Maillard-Reaktion. XVI. Bildung farbiger Produkte aus Hexosen. Z. Lebensm. Unters. Forsch. *175,* 262 (1982)

Ledl, F., Krönig, U., Severin, T., Lotter, H.: Untersuchungen zur Maillard-Reaktion. XVIII. Isolierung N-haltiger farbiger Verbindungen. Z. Lebensm. Unters. Forsch. *177,* 267 (1983)

Ledl, F., Fritsch, G., Hiebl, J., Pachmayr, O., Severin, T.: Degradation of Maillard Products. In: Amino-Carbonyl Reactions in Food and Biological Systems (Eds.: Fujimaki, M., Namiki, M., Kato, H.). Developments in Food Science 13, Elsevier: Amsterdam. 1986

Lehmann, J.: Chemie der Kohlenhydrate. Georg Thieme Verlag: Stuttgart. 1976

Loewus, F. A., Tanner, W. (Eds.): Plant carbohydrates I, II. Springer Verlag: Berlin–Heidelberg. 1981/82

Pangborn, R. M., Leonard, S., Simone, M., Luh, B. S.: Freestone peaches. I. Effect of sucrose, citric acid and corn syrup on consumer acceptance. Food Technol. *13,* 444 (1959)

Pigman, W., Horton, D. (Eds.): The Carbohydrates. 2nd edn., Academic Press: New York–London. 1970–1980

Radley, J. A. (Ed.): Starch production technology. Applied Science Publ.: London. 1976

Rees, D. A.: Polysaccharide shapes. Chapman and Hall: London. 1977

Salunkhe, D. K., McLaughlin, R. L., Day, S. L., Merkley, M. B.: Preparation and quality evaluation of processed fruits and fruit products with sucrose and synthetic sweeteners. Food Technol. *17,* 203 (1963)

Shallenberger, R. S.: Advanced sugar chemistry, principles of sugar stereochemistry. Ellis Horwood Publ.: Chichester. 1982

Von Sydow, E., Moskowitz, H., Jacobs, H., Meiselman, H.: Odor – Taste interaction in fruit juices. Lebensm. Wiss. Technol. *7,* 18 (1974)

Waller, G. R., Feather, M. S. (Eds.): The Maillard reaction in foods and nutrition. ACS Symposium Series 215, American Chemical Society: Washington, D.C. 1983

Whistler, R. L. (Ed.): Industrial gums. 2nd edn., Academic Press: New York–San Francisco–London. 1973

[a] Cf. 19.3.

5 Aroma Substances

5.1 Foreword

5.1.1 Concept Delineation

When food is consumed, the interaction of taste, odor and textural feeling provides an overall sensation which is best defined by the English word "flavor". German and some other languages do not have an adequate expression for such a broad and comprehensive term. Flavor results from compounds that are divided into two broad classes: Those *responsible for taste* and those *responsible for odors,* the latter often designated as aroma substances. However, there are compounds which provide both sensations.
Compounds *responsible for taste* are generally nonvolatile at room temperature. Therefore, they interact only with taste receptors located in taste buds of the tongue. The four important basic taste perceptions are provided by: sour, sweet, bitter and salty compounds. They are covered in separate sections (cf., for example, 8.10, 22.3, 1.2.6, 1.3.3, 4.2.3, 8.8 and 19.22.2).

Aroma substances are volatile compounds which are perceived by the odor receptor sites of the smell organ, i.e. the olfactory tissue of the nasal cavity. The concept of aroma substances, like the concept of taste substances, should be used loosely, since a compound might contribute to the typical odor or taste of one food, while in another food it might cause a faulty odor or taste, or both, resulting in an off-flavor.

5.1.2 Threshold Value

The lowest concentration of a compound that can still be directly recognized by its odor is designated as an odor threshold. Threshold concentration data allow comparison of the intensity or potency of odorous substances. The examples in Table 5.1 illustrate that great differences exist between individual aroma compounds, with an odor potency range of several orders of magnitude.

Table 5.1. Odor threshold values in water of some aroma compounds (20 °C)

Compound	Threshold value (mg/l)
Pyrazine	300
Ethanol	100
Maltol	35
Hexanol	0.7
Butyric acid	0.2
Vanillin	0.02
Limonene	0.01
Linalool	0.006
Hexanal	0.0045
2-Phenylethanal	0.004
α-Ionone	0.004
2-Methylpropanal	0.001
Ethylbutyrate	0.001
(+)-Nootkatone	0.001
(−)-Nootkatone	1.0
2-Methylbutyric acid ethyl ester	0.0001
4-Hydroxy-2,5-dimethyl-3(2H)-furanone	0.00004
4-Methoxy-2,5-dimethyl-3(2H)-furanone	0.00003
Methylmercaptan	0.00002
β-Damascone	0.000009
β-Ionone	0.000007
2-Isobutyl-3-methylpyrazine	0.000002
1-p-Menthen-8-thiol	0.00000002

In an example provided by nootkatone, an essential aroma compound of grapefruit oil (cf. 18.1.2.6.3), it is obvious that the two enantiomers (optical isomers) differ significantly in their aroma intensity (cf. 5.3.2.3) and, occasionally, in aroma quality or character.
The threshold concentrations (values) for aroma compounds are dependent on their vapor pressure, which is affected by both temperature and medium (cf. 5.4). The values are also influenced by the assay procedure and/or performance of the sensory panel. The frequent discrepancies in threshold values in the literature are basically due to such differences.

5.1.3 Impact Compounds of Natural Aroma

The amount of volatile substances in food is exceptionally low (Table 5.2). They include numerous components (Table 5.2), not all of which are important to food aroma. To be considered an aroma compound, a component of the volatile fraction must be present in food in higher concentration than its threshold value (cf. 5.2.4). The discrimination of the aroma constituents from the other volatile compounds is often very difficult, and usually provides only approximate values at best (cf. 5.2.4).

Table 5.2. Occurrence of volatile compounds in various foods

Food product	Content (ppm)	Number of compounds	
		identified	unidentified
Meat (beef)	50	270	250
Coffee	50	468	> 500
Onion	900	96	?
Raspberry	1.7	95	120
Strawberry	10	324	?
Pineapple	11.3	59	?
Passion fruit	12	194	50

Particularly important aroma constituents are those compounds which bear the characteristic aroma of the food, the so-called "character impact compounds". With regard to the occurrence of such key compounds, food can be divided into four groups:

- Group 1: The aroma is decisively carried by one compound. The presence of other aroma compounds serves only to round-off the characteristic aroma of the food.
- Group 2: Several compounds, one of which may play a major role, create or determine the typical aroma of the food.
- Group 3: The aroma may be closely simulated or reproduced only with a large number of compounds. Usually, a character impact compound is not present.
- Group 4: The food aroma can not be satisfactorily reproduced even with a large number of volatile compounds.

Examples of all four groups outlined above exist among fruit and vegetable aromas (cf. 17.9 and Table 18.24). Butter and blue cheese aroma are decisively formed by 2,3-butanedione and supported by acetaldehyde and dimethylsulfide or by 2-heptanone and 2-nonanone, respectively, and are suitable examples of Group 2.
Food aroma obtained by a thermal process, alone or in combination with a fermentation process, is highly complex in its composition. This aroma should belong to Group 3 (processed meat, roasted coffee, tea or bread) or to Group 4 (cocoa or beer).

5.1.4 Off-Flavors

A strange, extraneous type of aroma normally not present in a food may arise through loss of "impact compounds" or a shift in aroma concentration or change in composition of the individual components of the aroma. This is designated as an aroma defect, or simply as an off-flavor. Some common off-flavors which can arise during food processing and/or storage are listed in Table 5.3.
Examples of microbial metabolites which may be involved in pigsty-like and earthy-muddy off-flavors are skatole (I; faecal-like, 10 μg/kg*), 2-methylisoborneol (II; earthy-muddy, 0.03 μg/kg*) and geosmin (III; earthy, 0.01 μg/kg*):

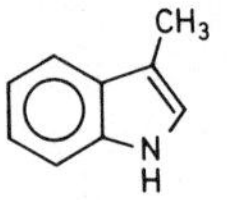

CH_3 N H OH

I 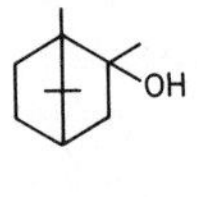 II

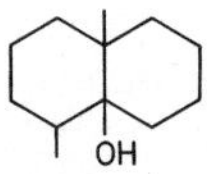

(5.1)

OCH_3 Cl Cl OH Cl

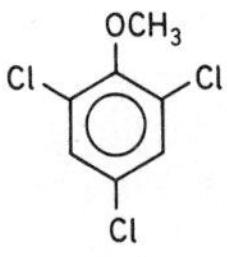

III IV

2,4,6-trichloroanisole with an extreme low odor threshold (mouldy-like; 3.10^{-5} μg/kg, water) is an example of an off-flavor substance (cf. 20.2.7) which is produced by fungal degradation and methylation of pentachlorophenol fungicides.

---

* Odor threshold in water.

Table 5.3. "Off-flavors" in food products

Food product	Off-flavor	Cause
Milk	Sunlight flavor	Photooxidation of methionine to methional (with riboflavin as a sensitizer)
Milk powder	Bean-like	The level of O_3 in air too high: ozonolysis of 8,15- and 9,15-isolinoleic acid to 6-trans-nonenal
Milk fat	Metallic	Autoxidation of pentaene- and hexaene fatty acids to octa-1,cis-5-dien-3-one
Milk products	Malty	Faulty fermentation by *Streptococcus lactis, var. maltigenes;* formation of phenylacetaldehyde and 2-phenylethanol from phenylalanine
Milk powder	Gluey, glutinous	Degradation of tryptophan to o-amino-acetophenone
Mutton-meat	Sweet, acidic	4-Methyloctanoic acid, 4-methylnonanoic acid
Peas, deep frozen	Hay-like	Saturated and unsaturated aldehydes, octa-3,5-dien-2-one, 3-alkyl-2-methoxy-pyrazines, hexanol
Orange juice	Grapefruit note	Metal-catalyzed oxidation or photooxidation of valencene to nootkatone 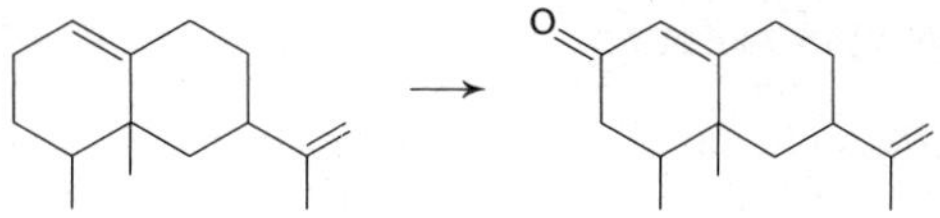
Orange juice	Terpene note	d-Limonene oxidation to carvone and carveol 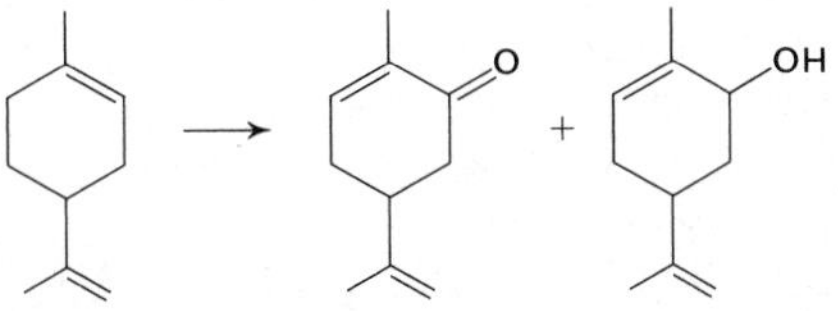
Passion fruit juice	Aroma flattening during pasteurization	Oxidation of (6-trans-2′-trans)-6-(but-2′-enyliden)-1,5,5-trimethylcyclohex-1-ene to 1,1,6-trimethyl-1,2-dihydronaphthalene: 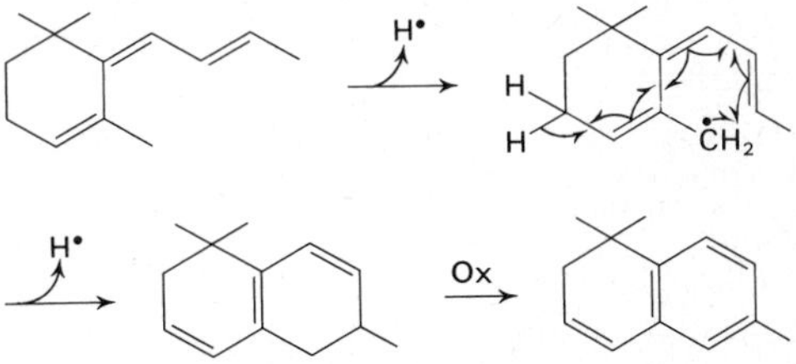
Beer	Sunlight flavor	Photolysis of humulon: reaction of one degradation product with hydrogen sulfide yielding 3-methyl-2-buten-1-thiol
Beer	Phenolic note	Faulty fermentation: hydrocinnamic acid decarboxylation by *Hafnia protea*

5.2 Aroma Analysis

The aroma substances consist of highly diversified classes of compounds, some of them being highly reactive, and almost all are present in food in extremely low concentrations. The difficulties usually encountered in qualitative and quantitative analysis of aroma compounds are based on these features. Other difficulties are associated with identification of aroma compounds, elucidation of their chemical structure and characterization of sensory properties. The results of an aroma analysis can serve as an objective guide in food processing for assessing the suitability of individual processing steps, and to assess the quality of raw material, intermediate- and end-products. In addition, investigation of food aroma broadens the possibility of food flavoring with substances that are prepared synthetically, but that are chemically identical to those found in nature, i.e. the so-called "nature identical flavors" (cf. 5.5).

5.2.1 Aroma Isolation

Before the isolation of volatile compounds, the question is raised as to whether all the food aroma constituents present are to be analyzed or only that part in the vapor phase above the food, which reaches the olfactory tissue of the nasal cavity as a vapor. The latter compounds are determined by *head space analysis,* while the former, the total aroma compounds of the food, are analyzed by separation of an aroma concentrate. Such concentrates should be prepared by several methods, since each method used for the isolation of volatiles has its own drawbacks.

Additional difficulties are encountered in foods which retain fully-active enzymes, which can alter the aroma. For example, during the homogenization of fruits and vegetables, hydrolases split the aroma ester constituents, while lipoxygenase, together with hydroperoxide lyase, enrich the aroma with newly-formed volatile compounds (cf. 3.7.2.2 and 3.7.2.3). To avoid such interferences, tissue disintegration is done in the presence of enzyme inhibitors or, when possible, by rapid sample preparation.

An additional aspect of aroma isolation not to be neglected is the ability of the aroma substances to bind to the solid food matrix. Such binding ability differs for many aroma constituents (cf. 5.4).

5.2.1.1 Distillation, Extraction

The volatile aroma compounds, together with some water, are removed by vacuum distillation from an aqueous food suspension. The highly volatile compounds are condensed in an efficiently-cooled trap. The aqueous distillate is then extracted with an organic solvent to separate the organic compounds present.

An extraction combined with distillation can be achieved using an apparatus designed by *Likens-*

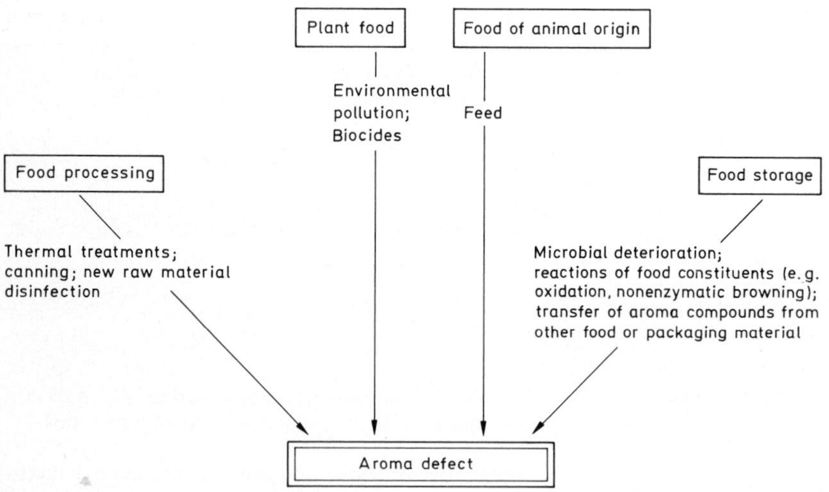

Fig. 5.1. The causes for aroma defects in food

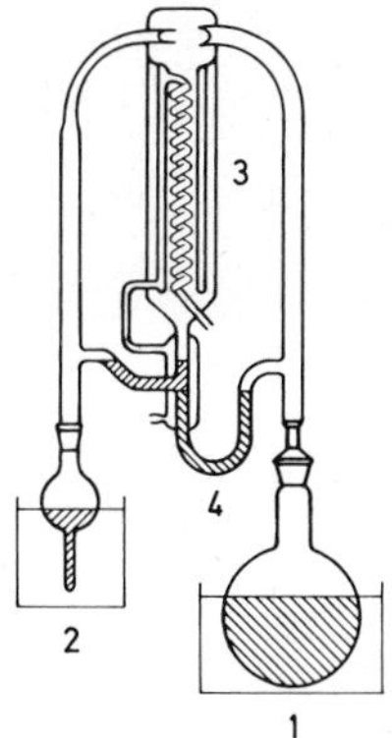

Fig. 5.2. Apparatus after *Likens* and *Nickerson* used for simultaneous extraction and distillation of volatile compounds.
1 Flask with heating bath containing the aqueous sample, 2 flask with heating bath containing the solvent (e. g. pentane), 3 cooler, 4 condensate separator: extract is the upper and water the lower phase

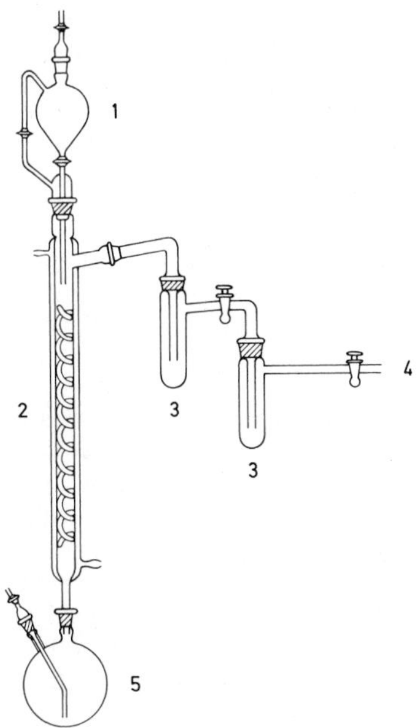

Fig. 5.3. Apparatus for separation of volatile compounds from fats and oils or other high boiling point solvents. (After *C. Weurman*, 1969). 1 Sample, 2 glass column with heating jacket (40–60 °C) and a rotating spiral to disperse the sample over a large surface, 3 condensing traps cooled in liquid nitrogen or acetone/dry ice, 4 connection to vacuum pump, 5 receiving flask for sample depleted of volatile compounds

Nickerson (Figure 5.2). Experiments with the classes of compounds listed in Table 5.4 provide high recoveries for the C_5–C_{11} homologues. However, recovery is incomplete for polar compounds with pronounced water solubility, i. e. the low molecular weight homologues (Table 5.4). On the other hand, the volatility of the compounds is reduced for molecular weights above 150 dal and, thus, their recovery is greatly decreased.
Recovery of the volatiles from fats and oils in a cold trap (Figure 5.3) provides concentrates free of water.

5.2.1.2 Gas Extraction

Volatile compounds can be isolated from a solid or liquid food sample by purging the sample with an inert gas (e. g. N_2, CO_2, He) and absorbing the volatiles on a porous, granulated polymer (Tenax GC, Porapak Q, Chromosorb 105), followed by recovery of the compounds. Water is retarded to only a negligible extent by these polymers (Table 5.5). The desorption of volatiles is usually achieved stepwise in a temperature gradient. At low temperatures, the traces of water are removed by elution, while at elevated temperatures, the volatiles are released and flushed out by a carrier gas into a cold trap, usually connected to a gas chromatograph.

5.2.1.3 Headspace Analysis

The headspace analysis procedure is simple: the food is sealed in a container, then brought to the desired temperature and left for a while to establish an equilibrium between volatiles bound

Table 5.4. Yield of volatile compounds (%) obtained by their isolation from highly diluted aqueous solutions (0.6 ppm) using the distillation and extraction (pentane) apparatus after *Likens* and *Nickerson*

C-Number	1-Alkanol	2-Alkanone	Alkanal	Alkane
3	Trace			
4	Trace	Trace	Trace	
5	93	79	101	
6	97	104	91	64
7	101	101	101	94
8	102	94	94	103
9	99	97	83	94
10		102		90
11		101		94
12				104

Table 5.5. Relative retention time (t_{rel}) of some compounds separated by gas chromatography using Porapak Q as stationary phase (Porapak: styrene divinylbenzene polymer; T: 55 °C)

Compound	t_{rel}	Compound	t_{rel}
Water	1.0	Ethyl-mercaptan	20.2
Methanol	2.3	Dimethylsulfide	19.8
Ethanol	8.1	Formic acid ethyl ester	6.0
Acetaldehyde	2.5		
Propanal	15.8		
Methyl-mercaptan	2.6		

to the food matrix and those present in the vapor phase. A given volume of the headspace is withdrawn with a gas syringe and then injected into a gas chromatograph equipped with a suitable separation column. Since the presence of water and a large volume of headspace provides inferior separations, the sample volume has to be moderate and therefore only the major volatiles are found.

An increase in analysis sensitivity is possible if the headspace volatiles are flushed out, absorbed and thus concentrated in a polymer, as outlined in the previous section. However, it is difficult to obtain a representative sample by this flushing procedure, a sample that would match the original headspace composition. A model system assay (Figure 5.4) might clarify the problems. Samples (e) and (f) were obtained by absorption on different polymers. They are different from each other and differ from sample (b), which was obtained directly for headspace analysis. The results might agree to a greater extent by varying the gas flushing parameters (gas flow, time), but substantial differences would still remain.

A comparison of samples (a) and (g) in Figure 5.4 shows that the results obtained by the distillation-extraction procedure, as applied to a model mixture, are fairly reproducible, with ethanol the only exception.

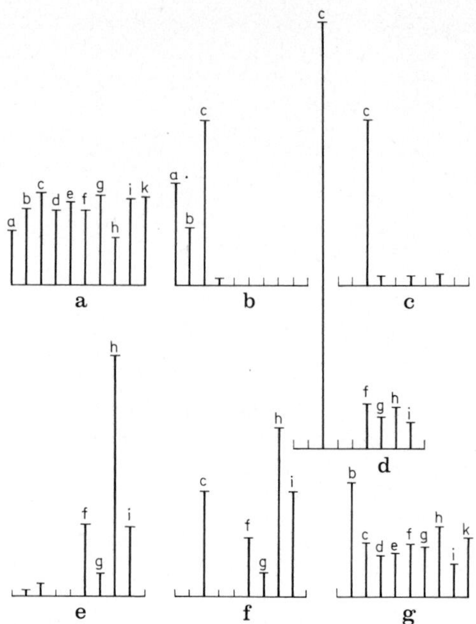

Fig. 5.4. A comparison of some methods used for aroma compound isolation. (After *W. G. Jennings* and *M. Filsoof,* 1977). **a** a Ethanol, b 2-pentanone, c heptanone, d pentanol, e hexanol, f hexyl formate, g 2-octanone, h d-limonene, i heptyl acetate and γ-heptalactone. **b** Headspace analysis of aroma mixture **a**. **c** From aroma mixture 10 μl is dissolved in water and the headspace is analyzed. **d** As in **c** but the water is saturated with 80% NaCl. **e** As in **c** but purged with nitrogen and trapped by Porapak Q. **f** As in **c** but purged with nitrogen and trapped by Tenax GC. **g** As in **e** but distilled and extracted according *Nickerson* and *Likens* (cf. Fig. 5.2)

5.2.2 Separation

When an aroma concentrate contains phenols, organic acids and or bases, preliminary separation of these compounds from neutral volatiles by extraction with alkali or acids is advantageous.

The acidic, basic and neutral fractions are individually analyzed. The neutral fraction by itself consists of so many compounds that in most cases not even a gas chromatographic column with the highest resolving power is able to separate them into individual peaks. Thus, separation of the neutral fraction is advisable and is usually achieved by fractional distillation, or preparative gas or high pressure liquid chromatography. An example given in Figure 5.5, which deals with an analysis of cognac aroma, may meet all the requirements for such an analysis.

5.2.3 Chemical Structure

Mass spectrometry has become an indispensable tool in the structural elucidation of volatile compounds. The sample separation requirement is adequately met by the efficiency of a capillary gas chromatographic column or other microscale column chromatography. Usually, a versatile mass spectrometer inlet system is connected directly to the effluent exit tube of a gas chromatograph, through an enrichment device or sample splitter. In such a combination, the gas chromatography column as a rule is a long capillary, the walls of which are coated with the stationary phase (wall coated open tubular column).

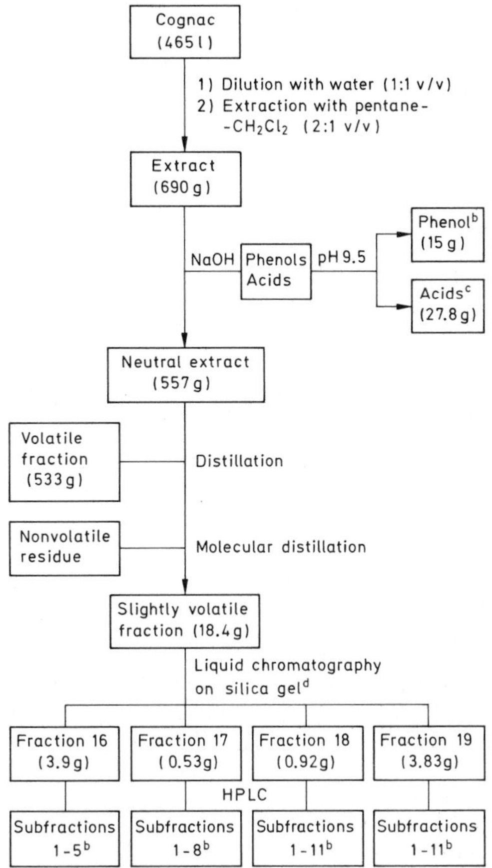

Fig. 5.5. Volatile compounds of cognac[a]. (Analysis scheme after *Ter Heide et al.*, 1978)

[a] The analysis is limited to fractions that significantly contribute to cognac aroma.

[b] GC/MS analysis identified the presence of 18 acetals, 59 alcohols, 28 aldehydes, 70 esters, 35 ketones, 3 lactones, 8 phenols, and 44 other compounds.

[c] GC/MS analysis of acids as methyl esters provided 27 compounds

[d] Of 22 fractions collected, four fractions were further separated by high pressure liquid chromatography

Mass spectral data, together with gas chromatography retention indices, are sufficient for the elucidation of the structure of compounds having only one functional group and low molecular weight. Usually, the conclusions are supported by IR spectroscopy data and, if necessary, by derivatization analysis.

With more complicated compounds, e.g. heterocyclics, the mass spectral data are often ambiguous. As an example, to be considered are the two compounds shown in Figure 5.6. Their mass spectra are very similar. Differentiation is possible only from the ^{1}H-NMR spectra (Figure 5.6). Wider use of ^{1}H-NMR in the elucidation of aroma compound structure is now possible since new recording techniques have been developed which are suitable for compounds also present in relatively small amounts.

"Reaction gas chromatography" is another aid for qualitative aroma analysis. A sample of only several µg is fixed to a porous carrier in a capillary column, and the carrier, if necessary, is impregnated with a catalyst. The sample is then hydrogenated or is split by ozonization. Mass spectral analysis is used to determine the structure based on the reaction products separated by gas chromatography.

It is obvious that identification of a compound is considered final only when the proposed structure coincides in all aspects with a synthetic reference compound.

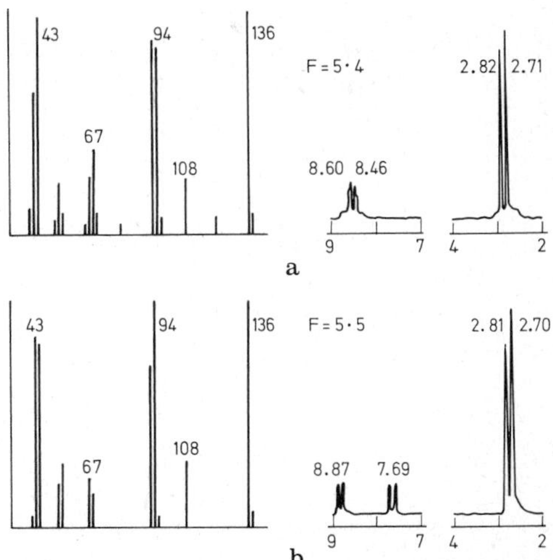

Fig. 5.6. Mass spectra and ^{1}H-NMR spectra. Excerpts from 2-acetyl-3-methylpyrazine (**a**) and 4-acetyl-2-methylpyrimidine (**b**) recordings (After *R. Tressl*, 1980)

5.2.4 Sensory Properties

5.2.4.1 Aroma Value

The significance of the volatile compound identified as an aroma substance of the food being studied is given by the aroma value of the compound, calculated from the ratio:

Aroma value, $A_x = c_x/a_x$

where c_x is the concentration of compound "x" in the food and a_x is the odor threshold concentration (cf. 5.1.2) of compound "x" in the food. Thus, if the actual concentration of compound "x" in food is higher than its threshold concentration, then it is expected to contribute significantly to the food aroma. Obtaining the experimental data needed to calculate the aroma value is not so simple. Quantitative determination of volatiles present in trace amounts in food is associated with difficulties which are not readily overcome, hence the experimental data are mostly only approximations.

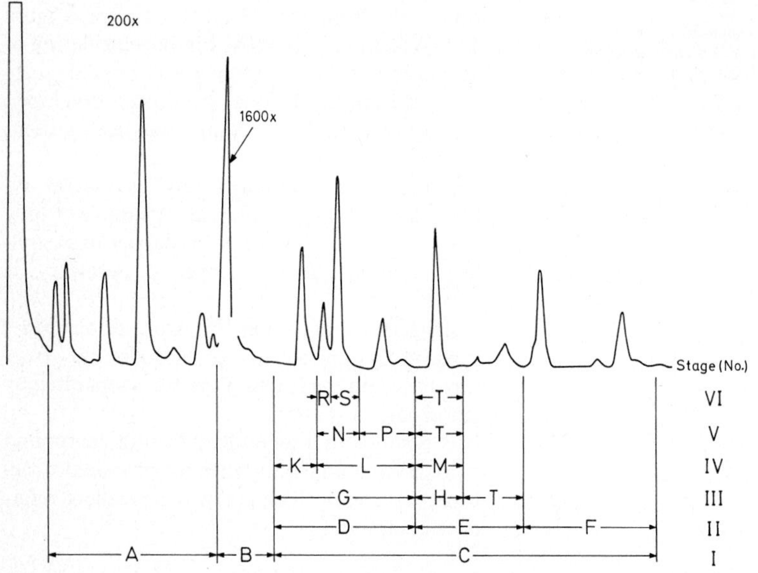

Fig. 5.6a. Blueberry aroma analysis. (After *Parliment* and *Scarpellino,* 1977)

Fractions collected after gas chromatographic separation were tested individually and in combination

Step	GC-fraction	"Blueberry" Aroma Note	
		yes	no
I	A		+
	B		+
	C	+	
II	D		+
	E		+
	F		+
	D + E	+	
	D + F		+
	E + F		+
⋮	⋮		⋮
VI	R		+
	S		+
	T		+
	R + T	+	

An even greater problem is to obtain experimentally the threshold value based on not a simple model system, but on a complex medium which has to be the same as the composition and polarity of the solid matrix in the food. The threshold value is significantly affected by the medium, as illustrated in Table 5.6.

The volatile compounds in potato chips are listed in Table 5.7 according to their aroma values. As can be seen, great differences exist in the contributions of the individual compounds to the aroma. Additional sensory panel analysis has confirmed that methional, with aroma values higher than all others, is also one of the aroma constituents, and its odor quality is the closest to the aroma of potato chips.

However, assessment of a volatile compound solely on the basis of its aroma value may give an overestimate since the above considerations are based on many simplifications. Thus, the fact that the dependence of odor intensity on compound concentration does not follow the same linear quantitative relationship for various aroma compounds is disregarded. Additionally, it is not a single compound but all of the compounds in food which contribute to aroma in an additive way, i.e. the desirable aroma is exhibited only by a mixture of these compounds. Also disregarded are the interactions which might result in intensification or weakening of an aroma. There are indications that synergistic and antagonistic effects exist in many food systems.

Table 5.6. Ratios of odor threshold values of aroma compounds in water and white bread

Compound	Factor [a]	Compound	Factor [a]
Diacetyl	2,500	Vanillin	375
Maltol	1,330	3-Methyl-butanol	26
Ethanol	1,230		
Ethylmaltol	750		

[a] A factor by which the threshold value is higher in white bread than water.

Table 5.7. Aroma compounds of potato chips

Compound	Presence in concentrate (%)	Odor threshold value (μg/kg oil)	Aroma value
Methional	2.0	0.2	100,000
2-Phenylethanal	18.0	22	8,180
3-Methylbutanal	5.0	13	3,850
2-Ethyl-3,6-dimethyl-pyrazine	7.4	24	2,720
2-trans,4-trans-Decadienal	7.5	135	560
2-Methylbutanal	7.4	140	530
2-Ethyl-5-methylpyrazine	6.0	320	190
1-Penten-3-one	0.1	5.5	180
Hexanal	2.1	120	175
2-Methylpropanal	0.5	43	120
2-trans-Nonenal	1.5	150	100

5.2.4.2 Sensory Analysis of GC-Peaks

The relative importance of the various fractions of aroma substances eluted from a gas chromatographic column can be determined by a trained sensory panel by sniffing the aroma-active regions of the effluent ("aromagram"; cf. Figure 21.2) or by sniffing the aroma of fractions collected from the effluent. Additional separation of these fractions can result in loss of typical food aroma. However, when aroma can be restored by a combination of certain peaks, then the typical aroma constituents are within these peaks. A subsequent chemical identification would then require recovery of peaks in larger amounts. This procedure of analysis is shown in Figure 5.6a by using a simplified version of blueberry aroma analysis to determine its "character impact compounds".

In the first stage of analysis, the gas chromatographic peaks are collected as three fractions, A, B and C. Then each is assessed by a sensory panel. The aroma in fraction C resembles that of blueberries. When fraction C is additionally separated into subfractions D, E and F, the aroma disappears. When subfraction D (with a green, harsh aroma note) and E (with a note of perfume or lemon) are mixed together, the blueberry aroma note is reconstituted. Further localization reveals that chromatographic peaks R and T are "character impact compounds", identified chemically as 3-cis-hexenol (R) and linalool (T).

5.2.4.3 Statistical Analysis

With a complex aroma, such as that formed during food processing, a relationship between the sensory data for odor intensity and quality, on the one hand, and the chemically-identified volatile compounds, on the other, can be formulated and analyzed statistically using a data processing system.

Attention will be given to the following example: analysis of an off-flavor which might develop during canning of meat. Headspace analysis has revealed 95 compounds. In order to correlate this analysis with the results of sensory evaluation, 23 compounds were selected with an aroma value (A_x) much higher than 1. The concentrations of a maximum of four of these compounds were algebraically combined by linear or vectorial summation into a so-called S-value. The sensory assessment was based on 15 odor quality perceptions (e.g. "sulfur-like", "cooked", "vegetable-like" "or sickly") by using an intensity scale of 1 to 10. Thus for the odor note "burnt, smoky", a linear relationship between the sensory R data and the concentrations of acetaldehyde and 2-methylthiophene was found: S = [(ethanal) (2-methylthiophene]$^{1/2}$ (cf. Figure 5.7).

In addition, it can be shown that hydrogen sulfide, methyl mercaptan, dimethyl sulfide, 2-methyl propanal and 2- and 3-methyl butanals are responsible for the canned beef off-flavor. The observations that the addition of lysine, which binds to aldehydes, and sodium fumarate, which binds hydrogen sulfide and methyl mercaptan, reduces the perception of the off-flavor of canned beef are in agreement with this finding.

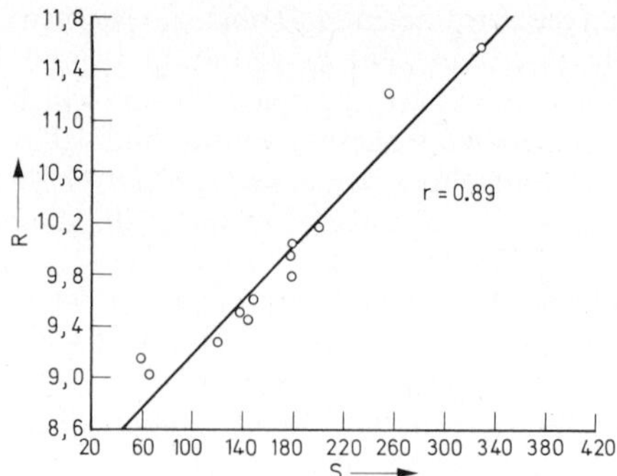

Fig. 5.7. Aroma note "burnt, smoky" in canned meats: sensory data (R + 10) in dependence on stimulant concentration. S = [(ethanal) (2-methylthophene)]$^{1/2}$ (After *E. von Sydow*, 1975)

5.3 Individual Aroma Compounds

A comprehensive survey of information related to the occurrence of aroma compounds in food recognizes, on the one hand, the vast diversity of the classes of compounds present and, on the other hand, reveals the frequent similarity of various food aromas. Thus, some simple volatiles, such as alcohols, aldehydes, amines and monocarboxylic acids, are identified in almost all kinds of food. Other aroma compounds also occur in a number of different foods, e.g. of the 400 compounds identified in cocoa aroma, 260 also occur in other foods.

The diversity of chemical structures suggests the involvement of numerous reactions in aroma formation. These reactions occur in many foods and are responsible for overlapping aroma spectra. Fruits and vegetables, herbs and spices contain aroma substances derived primarily as secondary products of plant metabolism. Common qualitative features of aroma are revealed in many unrelated plant families, suggesting that, in spite of their diversity, a common pattern of metabolic pathways operates in these plant cells. Thus, the wide spectrum of terpenes is found not only in citrus fruits but also in herbs and spices, red currants, cranberries, carrots and celery. The differences in aroma between these products are due to deviations which occur in terpene biosynthesis or to a dissimilarity in the secondary metabolism of these compounds.

Meat, fish, milk and cereals are, by their nature, a poor source of aroma. The specific aroma of the product is generated only through procesing, such as heating and/or fermentation. When common types of carbohydrates, amino acids, peptides and/or lipids are present in food, similar or equivalent aromas are formed during food processing if the processing parameters are comparable (temperature, duration, presence of water, oxygen). Aromas obtained by fermentation are similar when the same aroma precursors in the raw material are utilized by microorganisms which correspond in their secondary metabolism operational for the biosynthesis of volatiles. Examples are the synthesis of higher alcohols and esters by yeast in alcoholic beverages or breadmaking, or the microbial degradation of milk fat and coconut oil into 2-alkanones and 2-alkanols.

A selection of aroma compounds, grouped according to their origin into those formed by nonenzymatic or enzymatic reactions and listed by classes of compounds, is presented in the following sections. Some aroma compounds formed by both enzymatic and nonenzymatic reactions are covered in sections 5.3.1 and 5.3.2. It should be noted that the reaction pathways for each aroma compound are differentially established. In many cases they are dealt with by using hypothetical reaction pathways which lead from the precursor to the flavor compound. The reaction steps and the intermediates on the pathway are postulated by using the general knowledge of organic chemistry or biochemistry. So far, only in a few cases have the intermediary products been isolated and identified, the enzymes involved proven, and the proposed reaction pathway verified in a model system. Obviously, these subjects of aroma research are especially difficult since they involve, in most cases, elucidation of the side pathways occurring in chemical or biochemical reactions, which quantitatively are often not much more than negligible.

5.3.1 Nonenzymatic Reactions

Aroma changes at room temperature caused by nonenzymatic reactions are observed only after prolonged storage of food. Lipid peroxidation (cf. 3.7.2.1.5), *Strecker* degradation (cf. 4.2.4.4) of amino acids, carbohydrate heterolysis and the further interactions between their intermediary products, involving mostly aldehydes (cf. Figure 3.31), all play a part. These processes are greatly accelerated during heat treatment of food.
The diversity of aroma is enriched at the higher temperatures used during roasting or frying. The food surface, once dehydrated, is subjected to pyrolysis of its main constituents: carbohydrates, proteins, lipids and other constituents and, as a consequence, a diversified aroma spectrum results.
The large number of volatile compounds formed by the degradation of only one or two constituents is characteristic of nonenzymatic reactions. For example, 41 sulfur-containing compounds, including 20 thiazoles, 11 thiophenes, 2 dithiolanes and 1 dimethyltrithiolane, are obtained by heating cysteine and xylose in tributyrin at 200 °C.

5.3.1.1 Carbonyl Compounds

The most important reactions which provide volatile carbonyl compounds were presented in sections 3.7.2.1.5 (lipid peroxidation), 4.2.4.3.2 (caramelization) and 4.2.4.4 (amino acid decomposition by the *Strecker* degradation mechanism). Some *Strecker* aldehydes found in many foods are listed in Table 5.8 together with the corresponding aroma quality data. Data for carbonyls derived from fatty acid degradation are found in Table 3.24. Carbonyls are also obtained by degradation of carotenoids (cf. 3.8.4.4).

Table 5.8. Some *Strecker* degradation aldehydes

Amino acid precursor	*Strecker*-aldehyde			Odor threshold value
	name	structure	aroma description	(μg/l; water)
Gly	Formaldehyde	CH_2O	Mouse-urine, ester-like	$50 \cdot 10^3$
Ala	Ethanal	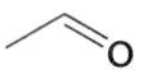	Sharp, penetrating, fruity	25
Val	2-Methylpropanal	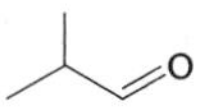	Penetrating, green	2
Leu	3-Methylbutanal	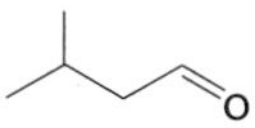	Malty, green	3
Ile	2-Methylbutanal	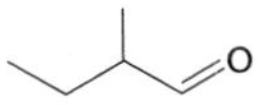	Etheral, bitter almond, green	4
Phe	2-Phenylethanal	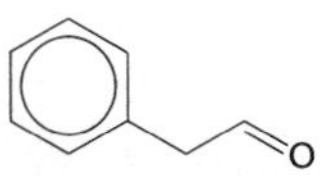	Flowery, honey-like	4

5.3.1.2 Pyrones

Maltol (3-hydroxy-2-methyl-4H-pyran-4-one) is obtained from carbohydrates as outlined in 4.2.4.3.1 and has a caramel-like odor (threshold value in Table 5.1). Its occurrence in food is given in Table 5.9.

Table 5.9. Occurrence of maltol in food

Food product	mg/kg	Food product	mg/kg
Malt coffee	292	Chocolate	3.3
Biscuit	19.7	Beer	0–3.4

Maltol enhances the sweet taste of food, especially sweetness produced by sugars (cf. 8.6.3), and is able to mask the bitter flavor of hops and cola.

Ethyl maltol [3-hydroxy-2-ethyl-4H-pyran-4-one] enhances the aroma but is 4- to 6-times more powerful than maltol. It has not been detected as a natural constituent in food. Nevertheless, it is used for food aromatization.

5.3.1.3 Furans

Among the great number of products obtained from carbohydrate degradation, 3(2H)-fura-

Table 5.10. 3(2H)-Furanones in food

Structure	Substituent	Aroma description	Occurrence
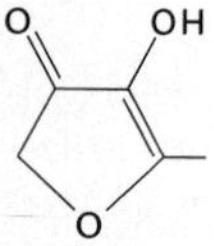 (I)	4-Hydroxy-5-methyl	Roasted chicory-like caramel	Meat broth
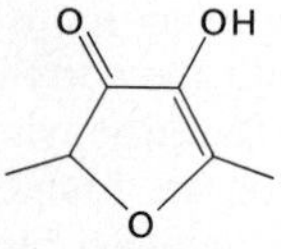 (II)	4-Hydroxy-2,5-dimethyl[a]	Heat-treated pineapple-like caramel	Pineapple, strawberry, burnt almond, meat broth, popcorn
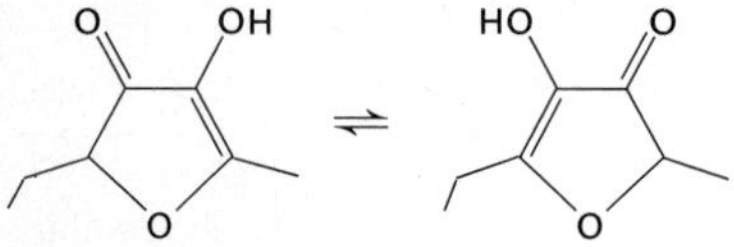 (III)	4-Hydroxy-2-ethyl-5-methyl	Sweet, pastry, caramel	Soya souce[b]
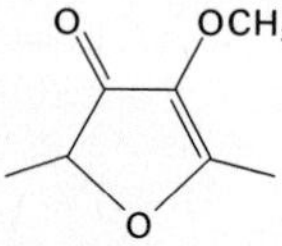 (IV)	4-Methoxy-2,5-dimethyl[a]	Sherry-like	Strawberry, raspberry[c]
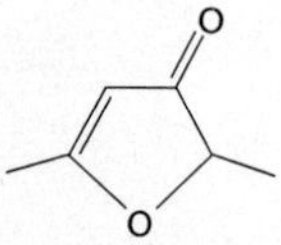 (V)	2,5-Dimethyl	Bread-like	White bread, coffee

[a] Odor threshold values are given in Table 5.1.
[b] Both tautomeric forms presented are identified.
[c] Arctic bramble *(Rubus arcticus)*.

nones belong to the most striking aroma compounds (Table 5.10).

Compounds I–III in Table 5.10, as well as maltol and cyclopentenolone (cf. 4.2.4.3.2), have a planar enol-oxo-configuration

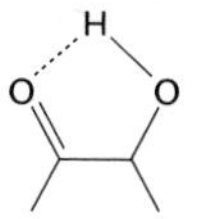

(5.2)

and a caramel-like odor. When the hydroxy group is absent (V) or is methylated (IV), this aroma note disappears.

The conversion of fructose into a furanone was presented in section 4.2.4.3.1. An alternative reaction is the involvement of a retroaldol condensation (Figure 5.8), which explains the formation and occurrence of compound I (Table 5.10). L-Rhamnose has been identified as the precursor for compound II (Table 5.10). Figure 5.9 explains the formation of compound II by the *Maillard* reaction mechanism. Whether the furanones detected in fruit are the favored reaction products of low pH of fruit and are obtained exclusively by a nonenzymatic reaction is not yet clear.

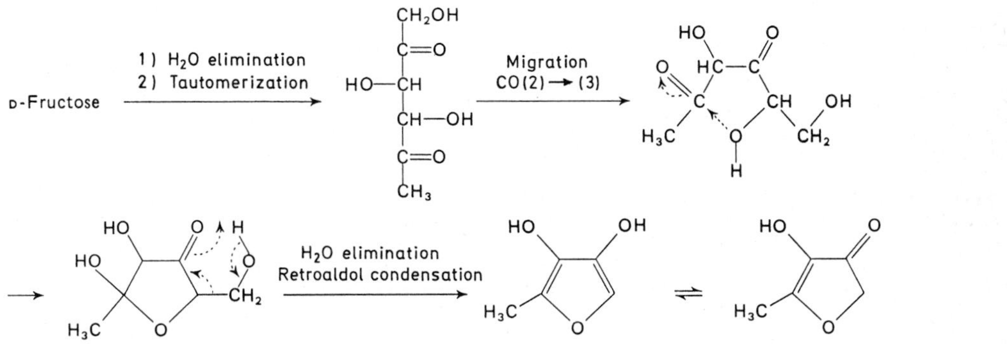

D-Fructose → 1) H_2O elimination 2) Tautomerization → Migration CO(2) → (3) → H_2O elimination Retroaldol condensation

Fig. 5.8. Formation of 4-hydroxy-5-methyl-3-(2H)-furanone

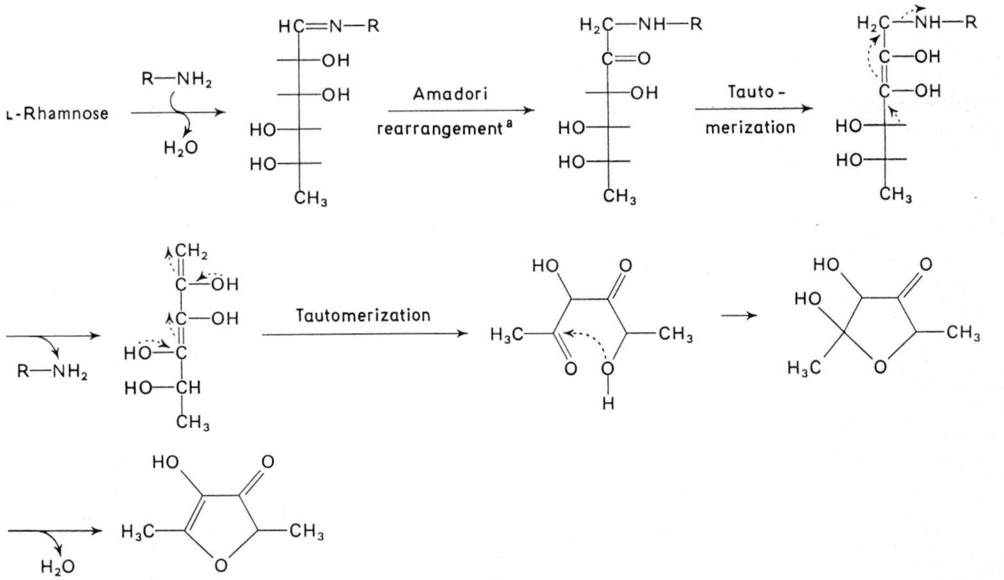

L-Rhamnose → R—NH_2 / H_2O → Amadori rearrangement[a] → Tautomerization → R—NH_2 → Tautomerization → H_2O

Fig. 5.9. Formation of 4-hydroxy-2.5-dimethyl-3(2H)-furanone by *Maillard* reaction

[a] cf. 4.2.4.4

Table 5.11. Lactones in food

Name	Structure	Aroma description	Occurrence
4-Butanolide (γ-Butyrolactone)	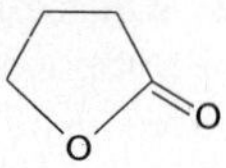 (I)	Sweet, aromatic, slightly after butter	Dried mushrooms, popcorn, roasted nuts, crispbread, pineapple
3-Penten-4-olide (α-Angelica Lactone)	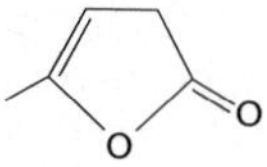 (II)	Sweet, herbaceous	White bread, soybeans, raisins
4-Nonanolide[a]	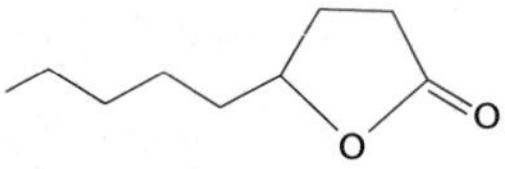 (III)	Reminiscent of coconut oil	Fat-containing food, crispbread, peaches
4-Decanolide[b] (γ-Decalactone)	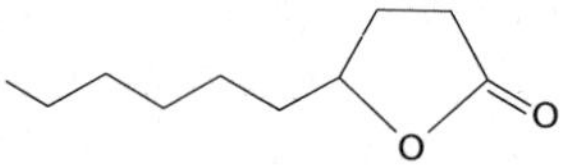 (IV)	Fruity, after peaches	Fat-containing food, strawberries
5-Decanolide[c] (δ-Decalactone)	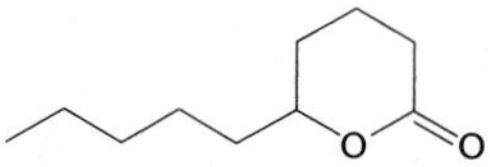 (V)	Oily, after peaches	Fat-containing food
5-Oxo-4-hexanolide (Solerone)	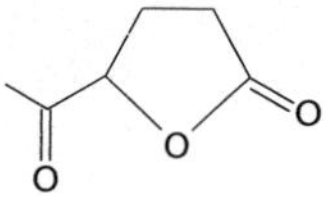 (VI)	Wine-like	Wine
5-Hydroxy-4-hexanolide (Sherry-Lactone)	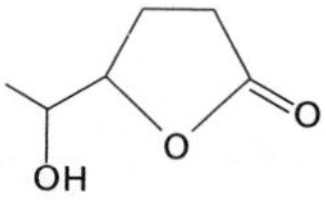 (VII)		Sherry
3-Methyl-4-octanolide (Whiskey-Lactone)	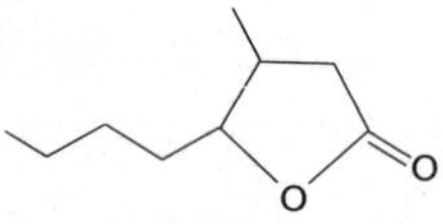 (VIII)		Whiskey

Odor threshold value (ppm): [a] 0.065 (H_2O), 2.4 (Oil);
[b] 0.088 (H_2O);
[c] 0.16 (H_2O).

6-Methyl-2,3-dihydro-thieno(2,3-c)-furan (I) and 2-mercaptomethylfuran (II) are significant contributors to coffee aroma:

(I)

(5.3a and 5.3b)

(II)

Compound I, also known as kahweofuran, mediates a "smoky burnt" note when greatly diluted, while the aroma of compound II resembles that of coffee.

5.3.1.4 Lactones

Many lactones with differing structures and odor qualities occur in food. Table 5.11 presents a short list. Since some of the aroma of the lactones is very pleasant, they are also of interest for commercial aromatization of food. Long-chain lactones (e.g. III–V in Table 5.11) found in fatty food (e.g. milk fat, meat, beef bouillon and some fruit, such as peach and coconut; cf. 18.1.2.6.7) are obtained from the corresponding hydroxy fatty acids (cf. 3.2.1.3).
Lactones VI–VII, formed by routes not yet clarified, are of importance to the aroma of some alcoholic beverages (Table 5.11). Two diastereoisomers exist for the whiskey lactone (VII). Thus cis-lactone, with an odor threshold value about ten-fold higher than the trans-lactone (0.79 ppm and 0.067 ppm, respectively) also has a more pleasant aroma.

5.3.1.5 Thiols, Thioethers, Di- and Trisulfides

An abundance of sulfurous compounds is obtained from cysteine, cystine and methionine by heating food. Some are very powerful aroma compounds (Table 5.12) and are involved in the generation of some delightful but also some irritating, unpleasant odor notes. A number of sulfur compounds obtained by nonenzymatic browning are recognized by their aroma note, which imitates that of processed meat.

Table 5.12. Odor threshold values (water) of some volatile sulfur compounds

Compound	Threshold value (µg/l)
Hydrogen sulfide	5
Methylmercaptan	0.02
Ethylmercaptan	0.008
Dimethylsulfide	1.0
Dimethyldisulfide	7.6
Dimethyltrisulfide	0.01
Methional	0.2
2-Acetylthiazole	10
2-Isobutylthiazole	2
3,4-Dimethylthiophene	1.3
5-Methyl-2-formylthiophene	1.0
3,5-Dimethyl-1,2,4-trithiolane	10

Numerous reaction systems which provide such an aroma are described in the patent literature (Table 1.6).
Thiols are important constituents of food aroma because of their intensive odor and their occurrence as intermediary products which can react with other volatiles by addition to carbonyl groups or to double bonds.
Hydrogen sulfide and 2-mercaptoacetaldehyde are obtained during the course of the *Strecker* degradation of cysteine (Figure 5.10). In a similar way, methionine gives rise to methional, which releases methylmercaptan by β-elimination (Figure 5.11). Dimethylsulfide is obtained by methylation during heating in the presence of pectin, or directly from methionine (cf. Reaction 5.4).

Met → $(CH_3)_2S^{\oplus}$–CH₂CH₂CH(NH₂)COOH $X^{\ominus}$ →(H_2O, −HX) CH_3SCH_3 + HO–CH₂CH₂CH(NH₂)COOH (5.4)

The sulfur compounds mentioned above occur in practically all protein-containing foods when they are heated or stored for a prolonged period of time.
The sensory properties of dimethylsulfide are of interest. Dimethylsulfide, even at very low levels close to the odor threshold value, is an important constituent of coffee and tea aroma. However, in other food, it is responsible for off-flavors

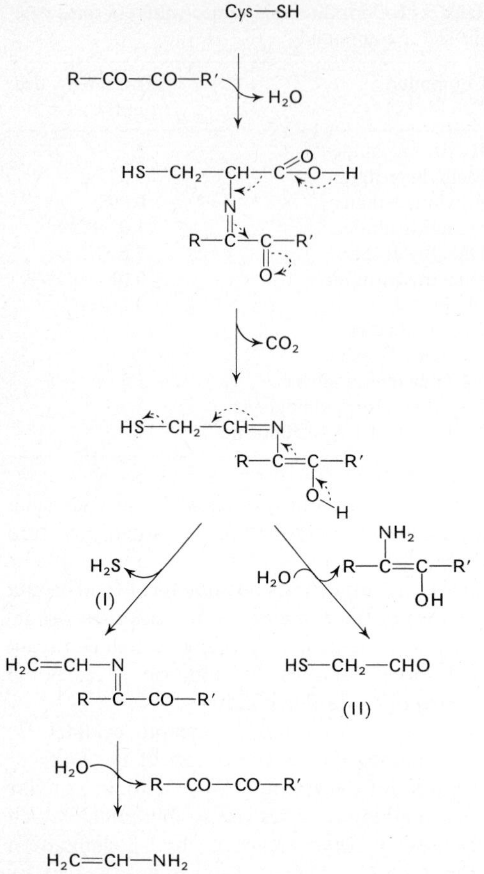

Fig. 5.10. Cysteine decomposition by a *Strecker* degradation mechanism: formation of H_2S (I) or 2-mercaptoethanal (II)

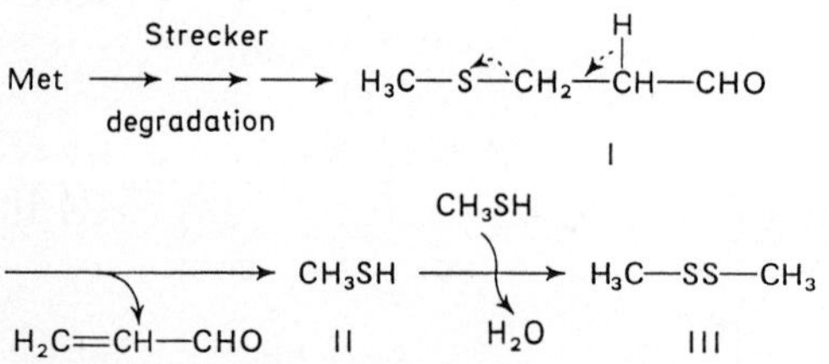

Fig. 5.11. Methionine degradation to methional, methylmercaptan and dimethylsulfide

designated as "crude oil" (frozen crustaceans), "onion" flavor (beer) or "feed smell" (such as occurs in milk). Bacteria are involved in the formation of dimethylsulfide in beer.

Methional is responsible for the "sunlight" flavor of milk (cf. Table 5.3) and for the typical flavor of processed potatoes (Table 5.7).

The dithiosemiacetal, 1-methylthioethanethiol, is obtained by reaction 5.5. It has a meat-like aroma and is found in meat broth.

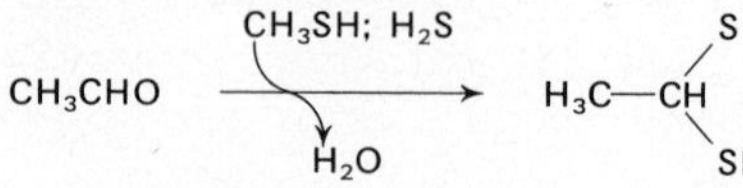

$$CH_3CHO \xrightarrow[H_2O]{CH_3SH;\ H_2S} H_3C{-}CH(S{-}CH_3)(SH) \quad (5.5)$$

Thiols are responsible for marked aroma defects. Besides the "sunlight" flavor defect of beer (Table 5.3), the "cat urine" taint of canned beef should be mentioned. Here, the reaction components are mesityl oxide, probably derived from solvent contamination, and hydrogen sulfide:

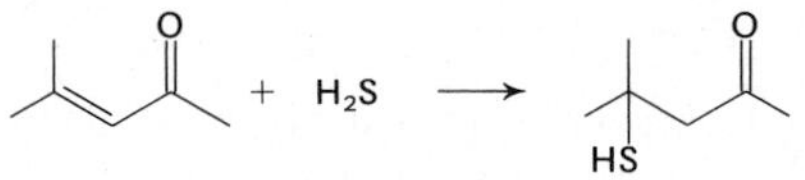

(5.6)

Thiols are readily oxidized to disulfides (Figure 5.11), which can disproportionate to trisulfides:

(5.7)

The exceptionally aroma-active, dimethyltrisulfide (Table 5.12) contributes to the flavor of poultry meat. In addition, it influences cooked white cabbage and cauliflower aroma.

Interaction of acetaldehyde and hydrogen sulfide provides heterocyclic compounds (Figure 5.12).

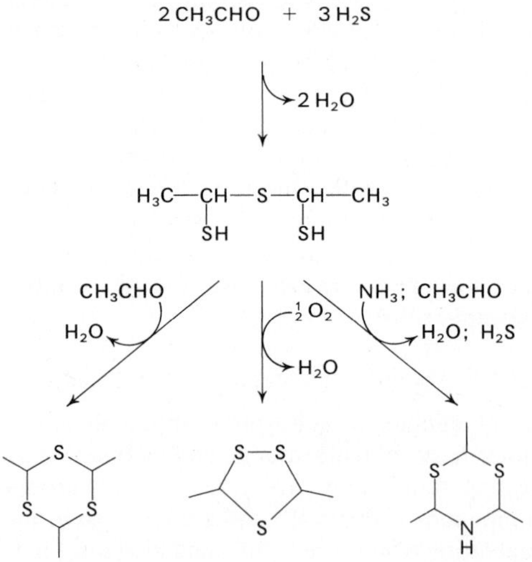

Fig. 5.12. Formation of 2,4,6-trimethyl-s-trithiane (I), 3,5-dimethyl-1,2,4-trithiolane (II) and 2,4,6-trimethyl-5,6-dihydro-1,3,5-dithiazine (III)

If ammonia is present, the profile of the heterocyclic compound is broadened (Figure 5.12).
Sulfides I and II in Figure 5.12 and trithioacetone, analogous to trithioacetaldehyde (I), are formed by cooking meat. Their significance to aroma is being discussed.

5.3.1.6 Thiophenes

2-Mercaptoethanal (II, Figure 5.10), obtained by the *Strecker* degradation of cysteine, can react with unsaturated aldehydes, such as acrolein or 2-butenal (obtained from acetaldehyde by aldol condensation), to form thiophene derivatives (Figure 5.13). These are usually found in cooked and roasted meat, roasted coffee and nuts. The contribution of 2-methylthiophene to the flavor of canned meat is covered in section 5.2.4.3.
Glucose syrup acquires a honey-like aroma due to the presence of 2.5 ppm of 2-acetyl-3-methylthiophene. In coffee, this compound at 1.1 ppm creates a nut-like note somewhat reminiscent of starch aroma.
Cysteine, when heated with 3(2H)-furanone (e.g. compound I, Table 5.10), develops a meat-like aroma. The following reactions are assumed to occur:

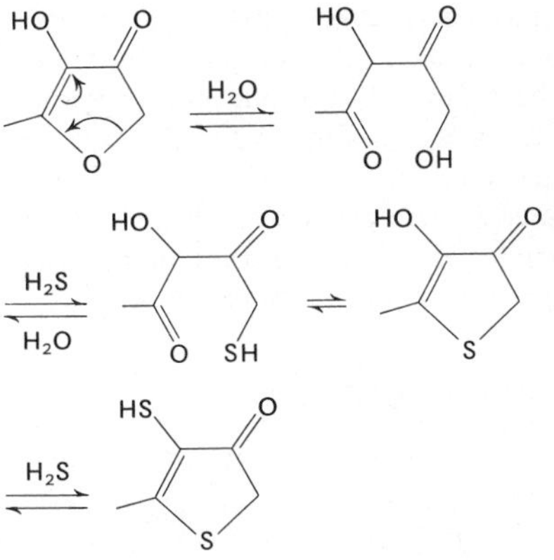

(5.8)

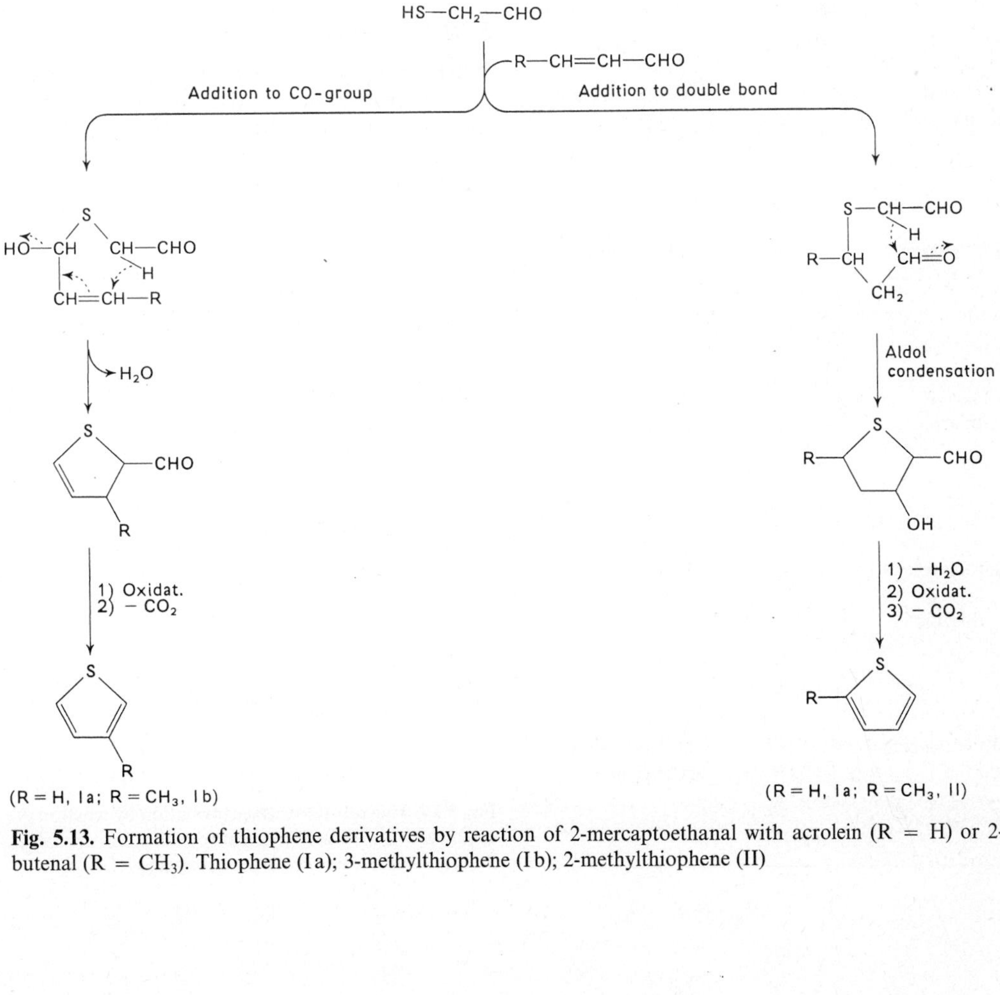

($R = H$, Ia; $R = CH_3$, Ib) ($R = H$, Ia; $R = CH_3$, II)

Fig. 5.13. Formation of thiophene derivatives by reaction of 2-mercaptoethanal with acrolein ($R = H$) or 2-butenal ($R = CH_3$). Thiophene (Ia); 3-methylthiophene (Ib); 2-methylthiophene (II)

Whether 4-mercapto-5-methyl-3(2H)-thiophenone, which has a sweet, meat-like odor, plays a significant role in the aroma of cooked meat has yet to be clarified.

5.3.1.7 Thiazoles

Thiazole and its derivatives are detected in foods such as coffee, fried meat, fried potatoes, heated milk and beer. Several of about 30 compounds identified are listed in Table 5.13. The nut-like aroma typical of many thiazoles is strikingly demonstrated by compound V, which is found in fried meat (Table 5.13). The compound is used in the aromatization of processed food.

Table 5.13. Thiazoles in food

Name	Structure	Aroma description
Thiazole	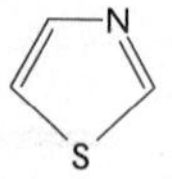 (I)	Pyridine-like
2-Methyl-thiazole	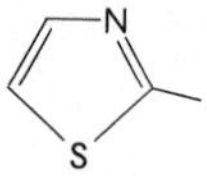 (II)	Green vegetable
2-Isobutyl-thiazole	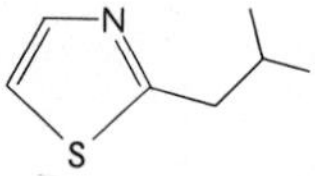 (III)	Green tomato leaves, wine (2 ppb)
2-Acetyl-thiazole	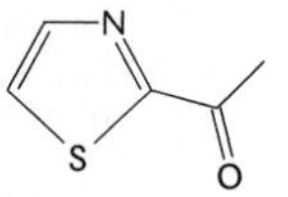 (IV)	Nutty, cereal, popcorn
2,4-Dimethyl-5-vinyl-thiazole	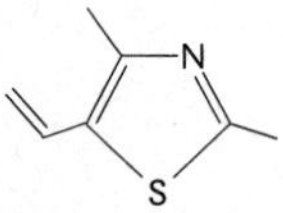 (V)	Nutty
Benzothiazole	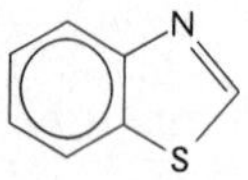 (VI)	Quinoline-like, rubber-like

Thiazole IV (Table 5.13) occurs widely in heat-treated food. However, its presence in beer is considered as an off-flavor. The formation of this thiazole by heating food is presented in the reaction sequence outlined in Figure 5.14. Its precursor (I in Figure 5.14) has an intensive, fresh, bread crust-like odor. In spite of this fact, it is not an important aroma constituent of bread. The compound is readily synthesized and is utilized for commercial aromatization. Thiazole VI (Table 5.13) can occur in milk when it is heated, and is responsible for a "stale" off-flavor. Thiazole III (Table 5.13) is a significant constituent of tomato aroma. The aroma of tomato products is usually enhanced by the addition of 20–50 ppb of thiazole III (for the biosynthesis of the compound, see section 5.3.2.4).

Thiazoles are also obtained from the thermal degradation of thiamine (Figure 5.15). Thiamine can yield furan and thiophene derivatives by an alternative pathway.

5.3.1.8 Oxazoles

When protein-containing food is heated, oxazoles or oxazolines can arise from serine or threonine by an alternative *Strecker* degradation mechanism (Figure 5.16). Several oxazolines, e. g. 2,4,5-trimethyl-3-oxazoline, the aroma of

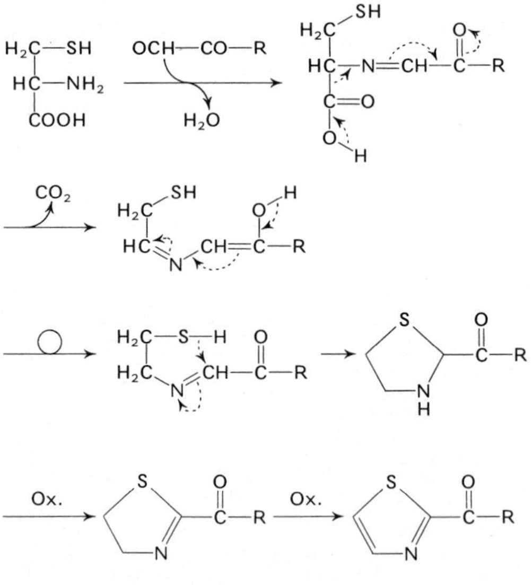

Fig. 5.14. Thiazole derivatives formation by reaction of cysteine with dicarbonyl compounds. R = CH_3: 2-acetyl-2-thiazoline (I), 2-acetylthiazole (II)

which is described as "woody" (odor threshold value in water is 1 ppm), are present in meat aroma:

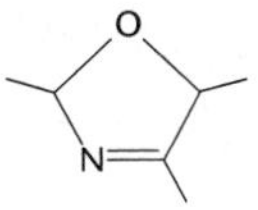

(5.9)

5.3.1.9 Pyrroles

So far, 40 pyrrole derivatives have been identified in the aroma of cooked or roasted food. Some are shown in Table 5.14. They are obtained primarily by the *Maillard* reaction, as outlined under section 4.2.4.4, or from the reaction of an amino acid with 2-acyl furan:

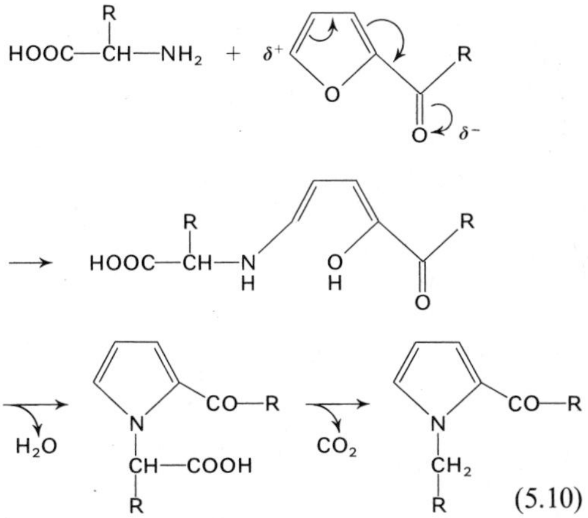

(5.10)

5.3.1.10 Pyrazines

Pyrazines are powerful aroma compounds. More than 50 pyrazines have been found in food. Some are presented in Table 5.15. Pyrazines are formed by the *Maillard* reaction and by pyrolysis of some amino compounds. Accordingly, they are widely distributed in heat-treated food, for example, bread, meat, roasted coffee, cocoa and roasted nuts. The examples in Table 5.15 show the aroma notes of several pyrazines (a wide aroma spectrum indeed: paprika, chocolate, coffee, potato, etc.). The comprehensive patent literature (several examples are provided in Table 5.16) serves as a convincing illustration of the efforts put into the production of pyrazines and their utilization as flavoring compounds. The odor potency can vary within this single class of compounds in a range of eight orders of magnitude. The potency is greatly affected by the nature of the molecule's side chains and positions in the ring.

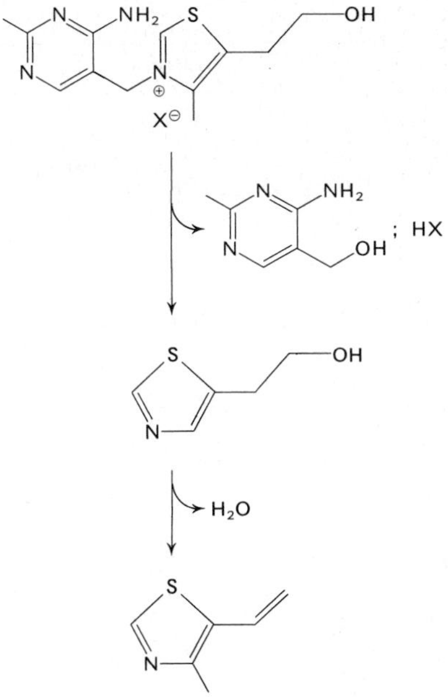

Different pyrazines are formed by the *Maillard* reaction and by pyrolysis. The diversity of com-

Fig. 5.15. Thermal degradation of thiamine at pH 6; I: 4-methyl-5-vinylthiazole

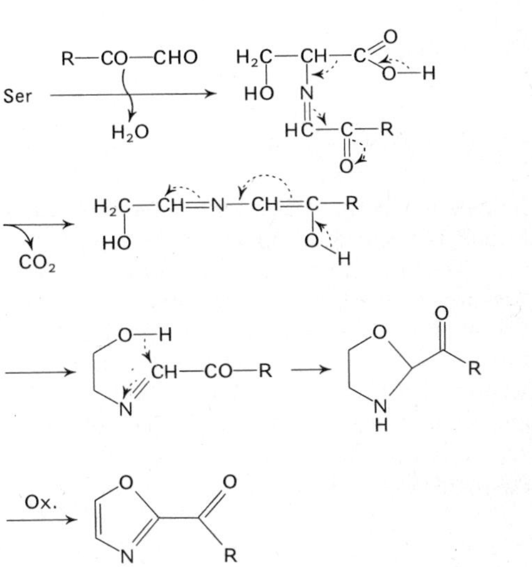

Fig. 5.16. *Strecker* degradation of serine, providing 2-acyloxazoles

Table 5.14. Pyrroles in food

Name	Structure	Aroma description	Occurrence
2-Formylpyrrole	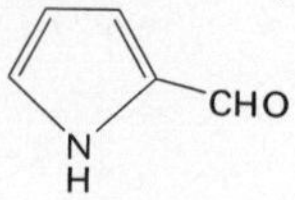	}	Coffee, roasted peanut, cocoa
2-Acetylpyrrole	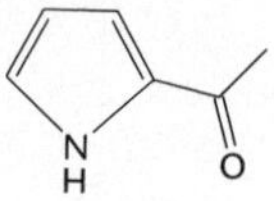	}	Coffee, roasted peanut, cocoa
2-Formyl-1-methylpyrrole	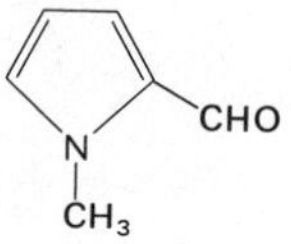	}	Coffee, roasted peanut, cocoa
Pyrrolidine	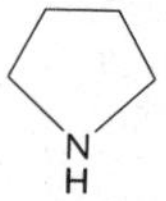	Amine-like	Caviar, cheese, beer
2-Acetyl-1-methyl-pyrrolidine	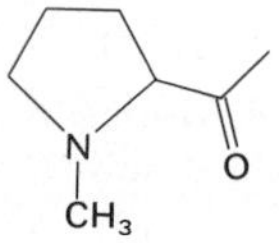	Bread-like	Bread
1-Furfurylpyrrole	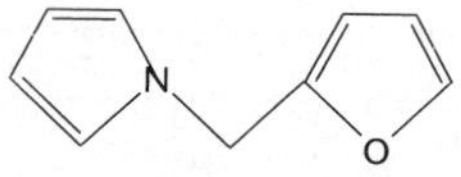	Hay-like	Beer

pounds has been studied with reference to the available N-source. Products of heating glucose with N-supplying reaction components, such as various amino acids and ammonium chloride, are shown in Figure 5.17. The pyrazines were formed by the reaction mechanism outlined in section 4.2.4.4. Hydroxy-amino compounds like the amino acid threonine are the preferred precursors in pyrazine synthesis (cf. Table 5.17) during pyrolysis (cf. Reaction 5.11).

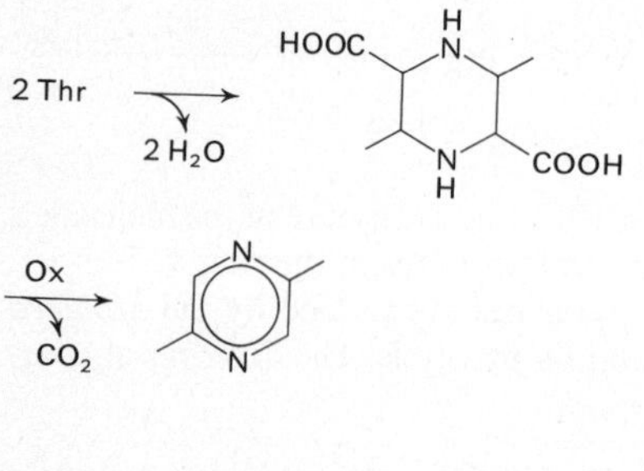

$$2\,Thr \xrightarrow{-2\,H_2O} \text{(dimethylpiperazine dicarboxylic acid)} \xrightarrow[-CO_2]{Ox} \text{(2,5-dimethylpyrazine)} \quad (5.11)$$

The reaction of cyclopentenolone (cf. 4.2.4.3.2) with an aminoketone (cf. 4.2.4.4) and ammonia in heated food can give rise, for example, to 6,7-dihydro-5H-cyclopenta(b)pyrazine (cf. Reaction 5.12), a compound with a bread crust odor note.

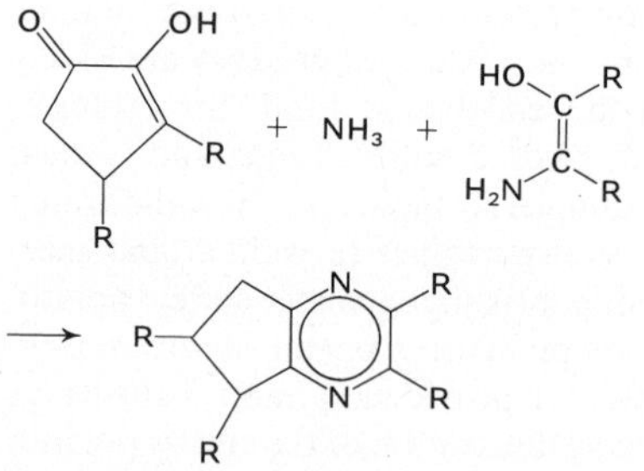

$$\text{cyclopentenolone} + NH_3 + HO{-}C(R){=}C(R){-}NH_2 \longrightarrow \text{6,7-dihydro-5H-cyclopenta(b)pyrazine (R-substituted)} \quad (5.12)$$

The powerfully odorous pyrazines VI–VII (Table 5.15) appear as metabolic by-products in some molds and microorganisms (cf. 5.3.2.5).

Table 5.15. Pyrazines in food

Structure	Substituent	Aroma description	Odor threshold value (µg/l; water)
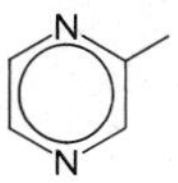 (I)	2-Methyl-	Roasted peanuts	10^5
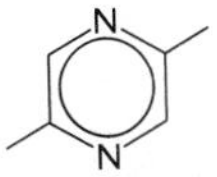 (II)	2,5-Dimethyl-	Roasted peanuts	1.8×10^3
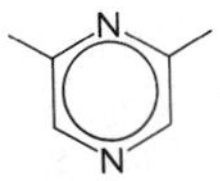 (III)	2,6-Dimethyl-	Chocolate	10^4
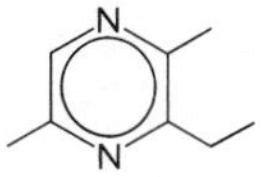 (IV)	2,5-Dimethyl-3-ethyl-	Lard	4.3×10^4
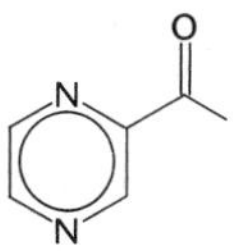 (V)	Acetyl-	Roasted corn	62
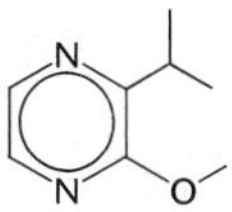 (VI)	2-Isopropyl-3-methoxy-	Potatoes	0.002
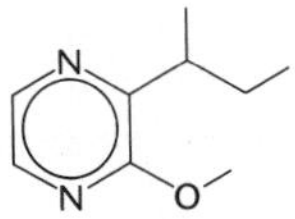 (VII)	2-sec-Butyl-3-methoxy-		0.001
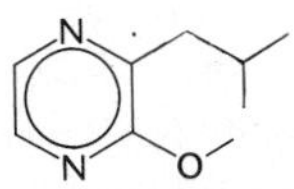 (VIII)	2-Isobutyl-3-methoxy-	Hot paprika (red pepper)	0.002

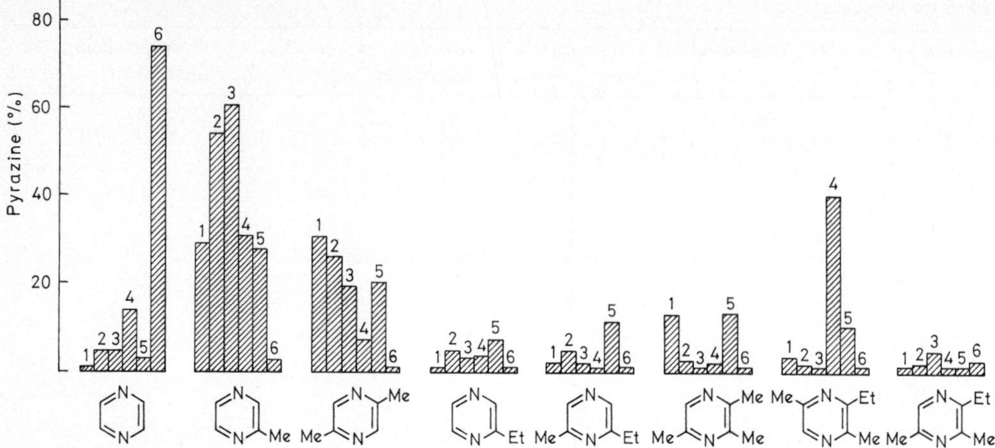

Fig. 5.17. Pyrazine formation in roasted peanuts and in *Maillard* reaction model systems (After *P. E. Koehler* et al, 1969). Roasted peanuts (I); glucose reacted at 120 °C with asparagine (2), glutamine (3), glutamic acid (4), aspartic acid (5) and ammonium chloride (6)

Table 5.16. Food flavoring with pyrazines

Compound (mg/kg)	Food product	Aroma description
2-Ethyl-3-vinyl-pyrazine (6)	Instant coffee	Earthy
2-Ethyl-3,5-dimethyl-pyrazine (50)	Glucose syrup	Burnt almond
2-Ethyl-3,6-dimethyl-pyrazine (20)	Glucose syrup	Hazelnut
Formylpyrazine (12.5)	Instant coffee	Roasted note
2-Ethoxy-3-methyl-pyrazine	Ice cream	Roasted nuts
2-Ethyl-3-methoxy-pyrazine	Potato products	Potatoes

Table 5.17. Pyrazine formation by pyrolysis. Amounts obtained are recorded as very high (4), high (3), medium (2), low (1), and not detectable (0)

Pyrazine	Precursor				
	Ser	Thr	Ethanolamine	Glucoseamine	Ala
Pyrazine	3	0	4	1	0
Methylpyrazine	2	1	3	4	0
2,3-Dimethylpyrazine	1	0	0	1	0
2,5-Dimethylpyrazine	0	4	1	3	0
Ethylpyrazine	4	0	2	0	0
2-Ethyl-5-methylpyrazine	0	0	0	1	0
2-Ethyl-6-methylpyrazine	1	0	0	0	0
2,6-Dimethylpyrazine	2	0	0	0	0
3-Ethyl-2,5-dimethylpyrazine	1	3	2	0	0
Trimethylpyrazine	0	3	0	2	0

5.3.1.11 Phenols

Phenolic acids and lignin are degraded thermally or decomposed by microorganisms into phenols, which are then detected in food. Some of these compounds are listed in Table 5.18.

Smoke generated by burning wood (lignin pyrolysis) is used for cold or hot smoking of meat and fish products. This is a phenol enrichment process since phenol vapors penetrate the meat or fish muscle tissue. Also, some alcoholic beverages, such as Scotch whiskey, and also butter (cf. 5.19) have low amounts of some phenols, the presence of which is needed to round-off their typical aromas. Model system studies involving pyrolysis of single phenolic acids (Table 5.20) have verified the formation of large numbers of phenols. To explain such a reaction which, for example, accompanies the process of roasting coffee or the kiln drying of malt, it has to be assumed that thermally-formed free radicals

Table 5.18. Phenols in food

Name	Structure	Aroma description	Occurrence
Phenol	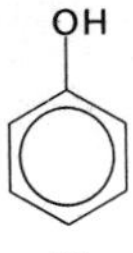 OH (I)	Smoky	Coffee, beer, sherry, milk, roasted peanuts, tomatoes
m-Cresol	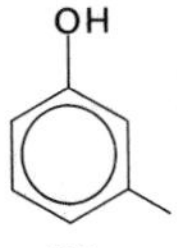 OH (II)	Smoky	Coffee, sherry, milk, roasted peanuts, asparagus
4-Ethylphenol	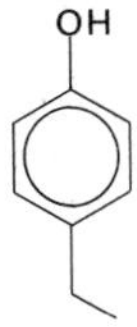 OH (III)	Woody	Milk, soya souce, roasted peanuts, tomatoes
Guaiacol	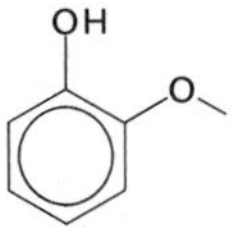 OH, O (IV)	Smoky, burnt, sweet	Coffee, milk crispbread
4-Vinylphenol	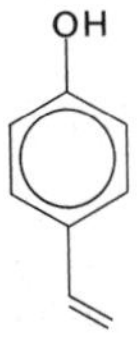 OH (V)	Harsh, smoky	Beer, milk, roasted peanuts
2-Methoxy-4-vinylphenol	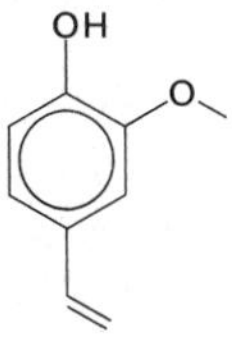 OH, O (VI)	Clove-like	Coffee, beer
Isoeugenol	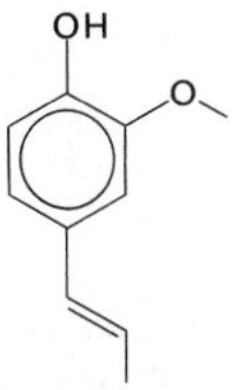 OH, O (VII)	Clove-like	Coffee, beer

regulate the decomposition pattern of phenolic acids (cf., for example, heat decomposition of ferulic acid, Figure 5.18).

Table 5.19. Phenols in butter[a]

Compound	Concentration found		Aroma threshold[b]	Optimal concentration[c]
	sample 1	sample 2		
Phenol	16	9	10	50
m-Cresol	3.3	3	10	50
p-Cresol	5	2	2	5
Guaiacol	0.8	2	2	

[a] All results are expressed as µ/kg.
[b] Medium: butter.
[c] Adjustment to this concentration provides a higher valued butter aroma.

Table 5.20. Pyrolysis products of some phenolic acids (T: 200 °C; air)

Phenolic acid	Product	Distribution (%)
Ferulic acid	4-Vinylguaiacol	79.9
	Vanillin	6.4
	4-Ethylguaiacol	5.5
	Guaiacol	3.1
	3-Methoxy-4-hydroxy-acetophenone (Acetovanillone)	2.6
	Isoeugenol	2.5
Sinapic acid	2,6-Dimethoxy-4-vinylphenol	78.5
	Syringaldehyde	13.4
	2,6-Dimethoxyphenol	4.5
	2,6-Dimethoxy-4-ethylphenol	1.8
	3,5-Dimethoxy-4-hydroxy-acetophenone (Acetosyringone)	1.1

Fig. 5.18. Thermal degradation of ferulic acid. 4-Vinylguaiacol (I), vanillin (II), and guaiacol (III)

5.3.2 Enzymatic Reactions

Aroma compounds are formed by numerous reactions which occur as part of the normal metabolism of animals, plants and microorganisms. The enzymatic reactions triggered by tissue disruption, as experienced during disintegration or slicing of fruits and vegetables, are of particular importance. Enzymes can also be involved indirectly in aroma formation by setting up the preliminary stage of the process, e. g. by releasing amino acids from available proteins, sugars from polysaccharides, or ortho-quinones from phenolic compounds, which are then converted into aroma compounds by further nonenzymatic reactions. In this way, the enzymes enhance the aroma of bread, meat, beer, tea and cacao.

5.3.2.1 Carbonyl Compounds, Alcohols

Fatty acids and amino acids are precursors of a great number of volatile aldehydes, while carbohydrate degradation is the source of ethanal only.

Long chain aldehydes (C_{13}–C_{17}) are formed from fatty acids by an α-oxidation mechanism:

$$R{-}CH_2{-}COOH \xrightarrow{\alpha\text{-Hydroxylation}} R{-}CH(OH){-}COOH \xrightarrow{O_2,\ H_2O} R{-}CO{-}COOH \xrightarrow{-CO_2} R{-}CHO \quad (5.13)$$

Such a degradation mechanism operating in plant metabolism has been confirmed in cucumbers and other vegetables.

Linoleic and linolenic acids in fruits and vegetables are subjected to oxidative degradation by lipoxygenase alone or in combination with a hydroperoxide lyase, as outlined in sections 3.7.2.2 and 3.7.2.3. The oxidative cleavage yields oxo acids, aldehydes and allyl alcohols. Among the aldehydes formed, hexanal, 2-trans-hexenal, 3-cis-hexenal and/or 2-trans-nonenal, 3-cis-nonenal, 2-trans,6-cis-nonadienal and 3-cis,6-cis-nonadienal are important for aroma.

Frequently, these aldehydes appear along with some of the alcohols derived from them soon after the disintegration of the tissue in the presence of oxygen.

Aldehydes formed by the *Strecker* degradation (cf. 5.3.1.1; Table 5.8) can also be obtained as metabolic by-products of the enzymatic transamination or oxidative deamination of amino acids. First, the amino acids are converted enzymatically to α-keto acids and then to aldehydes by decarboxylation in a side reaction:

$$\text{R—CH(NH}_2\text{)—COOH} \longrightarrow \text{R—C(=O)—COOH} \xrightarrow{-CO_2} \text{R—CHO} \qquad (5.14)$$

Unlike other amino acids, threonine can eliminate a water molecule and, by subsequent decarboxylation, can yield propanal:

$$\text{H}_3\text{C—CH(OH)—CH(NH}_2\text{)—COOH} \xrightarrow{-H_2O} \text{H}_3\text{C—CH=C(NH}_2\text{)—COOH} \rightleftharpoons \text{H}_3\text{C—CH}_2\text{—C(=NH)—COOH} \xrightarrow[-NH_3]{H_2O} \text{H}_3\text{C—CH}_2\text{—CO—COOH} \xrightarrow{-CO_2} \text{H}_3\text{C—CH}_2\text{—CHO} \qquad (5.15)$$

Many aldehydes derived from amino acids occur in plants and fermented food.

A study involving the yeast *Saccharomyces cerevisiae* clarified the origin of 2-methylpropanal and 2- and 3-methylbutanal. They are formed to a negligible extent by decomposition but mostly as by-products during the biosynthesis of valine, leucine and isoleucine.

Figure 5.19 shows that α-ketobutyric acid, derived from threonine (cf. Reaction 5.15), can be converted into isoleucine. Butanal and 2-methylbutanal are formed by side-reaction pathways.

2-Acetolactic acid, obtained from the condensation of two pyruvate molecules, is the intermediary product in the biosynthetic pathways of valine and leucine (Figure 5.20). However, 2-acetolactic acid can be decarboxylated in a side reaction into acetoin, the precursor of diacetyl. At α-keto-3-methylbutyric acid the metabolic pathway branches to form 2-methylpropanal and branches again at α-keto-4-methyl valeric acid to form 3-methylbutanal (Figure 5.20).

The enzyme that decarboxylates the α-ketocarboxylic acids to aldehydes has been detected in oranges. Substrate specificity for this decarboxylase is shown in Table 5.21.

Alcohol dehydrogenases (cf. 2.3.1.1) can reduce the aldehydes derived from fatty acid and amino acid metabolism into the corresponding alcohols:

$$\text{R—CH}_2\text{—OH} + \text{NAD}^{\oplus} \rightleftharpoons \text{R—CHO} + \text{NADH} + \text{H}^{\oplus} \qquad (5.16)$$

Alcohol formation in plants and microorganisms is strongly favoured by the reaction equilibrium and, primarily, by the predominance of the concentration of NADH over that of NAD^+. Nevertheless, the enzyme specificity is highly variable. In most cases aldehydes $>C_5$ are only slowly

Table 5.21. Substrate specificity of a 2-oxocarboxylic acid decarboxylase from orange juice

Substrate	v_{rel} (%)
Pyruvate	100
2-Oxobutyric acid	34
2-Oxovaleric acid	18
2-Oxo-3-methylbutyric acid	18
2-Oxo-3-methylvaleric acid	18
2-Oxo-4-methylvaleric acid	15

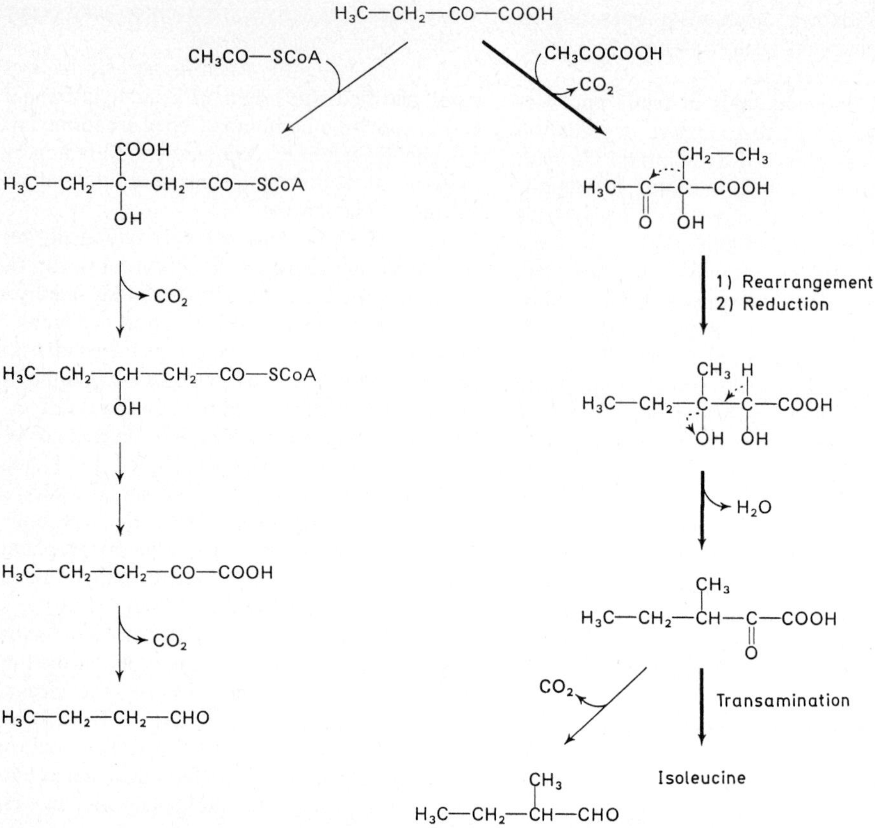

H₃C—CH₂—CO—COOH

CH₃CO—SCoA | CH₃COCOOH, CO₂

H₃C—CH₂—C(OH)(COOH)—CH₂—CO—SCoA | H₃C—C(=O)—C(OH)(CH₂—CH₃)—COOH

CO₂ | 1) Rearrangement 2) Reduction

H₃C—CH₂—CH(OH)—CH₂—CO—SCoA | H₃C—CH₂—C(OH)(CH₃)—C(OH)(H)—COOH

H₂O

H₃C—CH₂—CH₂—CO—COOH | H₃C—CH₂—CH(CH₃)—C(=O)—COOH

CO₂ | CO₂ | Transamination

H₃C—CH₂—CH₂—CHO | Isoleucine

H₃C—CH₂—CH(CH₃)—CHO

Fig. 5.19. Formation of aldehydes during isoleucine biosynthesis (After *A. Piendl,* 1969). ➞ main pathway → side pathway of the metabolism

reduced; thus, with aldehydes rapidly formed by, for example, oxidative cleavage of unsaturated fatty acids, a mixture of alcohols and aldehydes results, in which the aldehydes predominate.

5.3.2.2 Esters

Esters are significant aroma constituents of many fruits. They are synthetized only by intact cells:

$$\begin{aligned} &R{-}CO{-}SCoA + R'{-}OH \\ &\longrightarrow R{-}CO{-}O{-}R' + CoASH \end{aligned} \qquad (5.17)$$

Acyl-CoA originates from the β-oxidation of fatty acids and also occasionally from amino acid metabolism. Figure 5.21 shows an example of how ethyl 2-trans, 4-cis-decadienoate, an important aroma constituent of pears, is synthesized from linoleic acid.

When fruits are homogenized, such as in the processing of juice, the esters are rapidly hydrolyzed by the hydrolase enzymes present, and the fruit aroma flattens.

5.3.2.3 Terpenes

The mono- and sesquiterpenes in fruits (cf. 18.1.2.6) and vegetables (cf. 17.1.2.6), herbs and spices (cf. 22.1.1.1) and wine (cf. 20.2.6.9) are presented in Table 5.22. These compounds stimulate a wide spectrum of aromas, mostly perceived as very pleasant (examples in Table 5.23). Volatile terpenes in many foods play the role of "character impact compounds" and, hence, are often utilized for food aromatization.

Monoterpenes with hydroxyl groups, such as linalool, geraniol and nerol, are present in fruit juice at least in part as glycosides. Linalool-β-rutinoside (I) and linalool-6-0-α-arabinofurano-

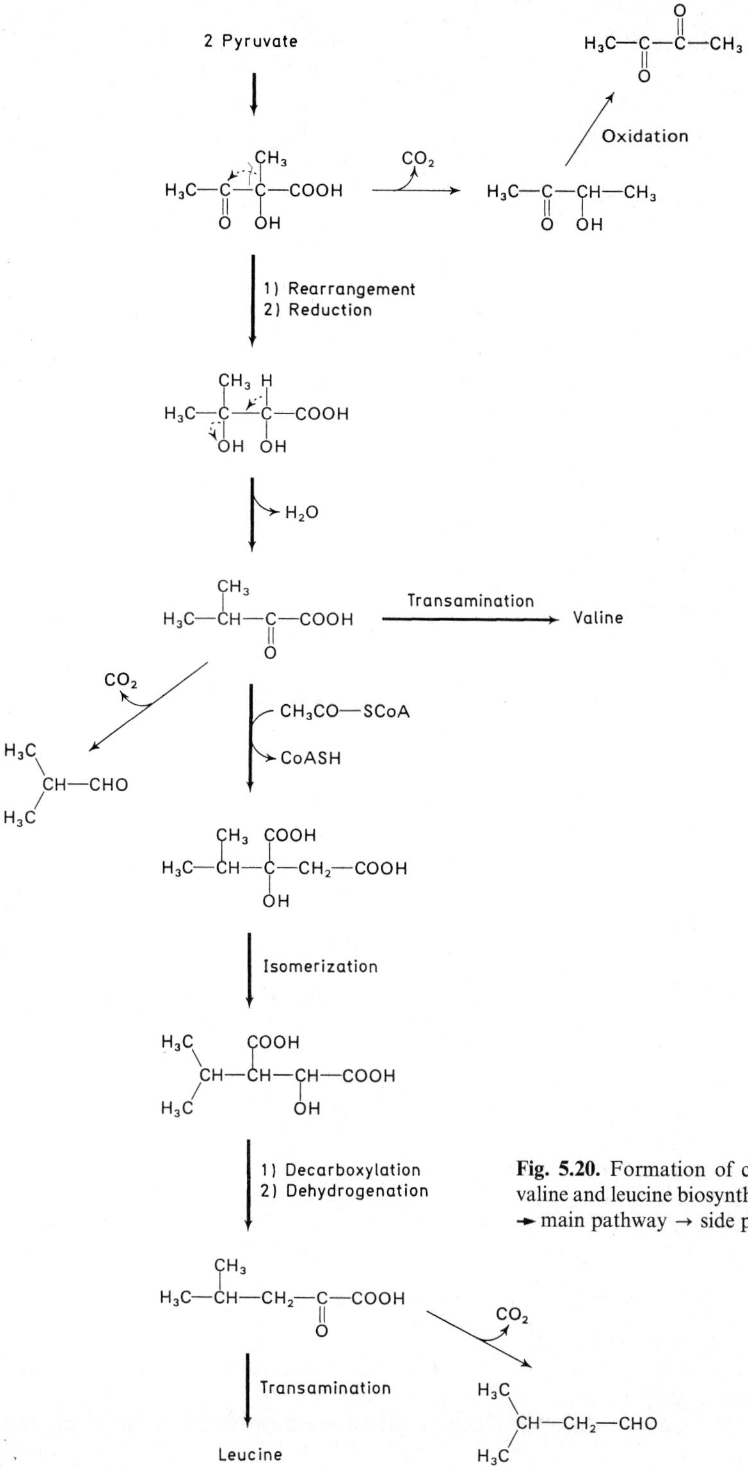

Fig. 5.20. Formation of carbonyl compounds during valine and leucine biosynthesis. (After *A. Piendl*, 1969). ➞ main pathway → side pathway of the metabolism

syl-β-D-glucopyranoside (II) have been found in wine grapes and in wine (cf. 20.2.6.9):

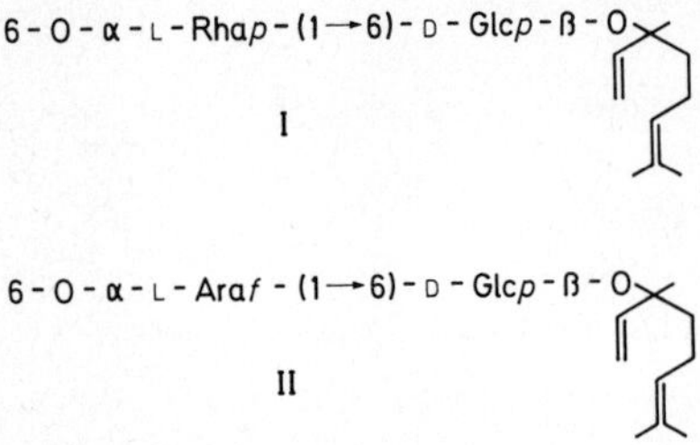

6-O-α-L-Rhap-(1→6)-D-Glcp-β-O

I

(5.18)

6-O-α-L-Araf-(1→6)-D-Glcp-β-O

II

The terpene glycosides hydrolyze even under slight warming, depending on the acidity of the fruit juice. Under these conditions, terpenes with two or three hydroxyl groups which are released undergo further reactions, forming hotrienol (IV) and neroloxide (V) from 3,7-dimethylocta-1,3-dien-3,7-diol in grape juice, or cis- and trans-linalool oxides (VIa and VIb) from 3,7-dimethylocta-1-en-3,6,7-triol in grape juice and peach sap (cf. Formulas 5.19 and 5.20).

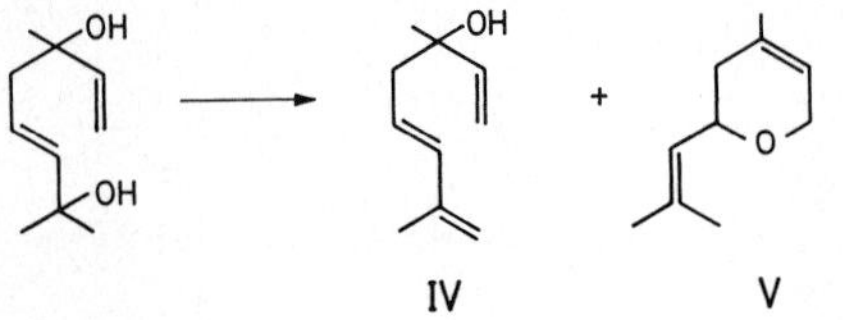

OH OH → OH IV + O V

(5.19)

OH HO OH → HO O VIa + HO O VIb

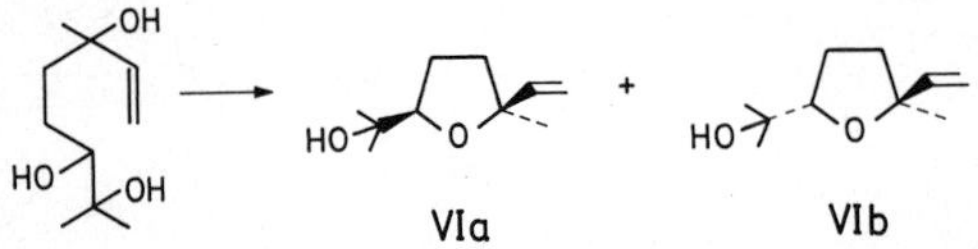

(5.20)

Most terpenes contain one or more chiral centers. Of several terpenes the optically-inactive form, and the l- or d-form occur in different plants. The enantiomers and diastereoisomers differ regularly in their odor characteristics. For example, menthol (cf. Formula 5.48) in the l-form (1R, 3R, 4S) which occurs in peppermint oil, has a clean sweet, cooling and refreshing peppermint aroma, while in the d-form (1S, 3S, 4R) it has remarkable, disagreeable notes such as phenolic, medicated, camphor and musty. Carvone (cf. Formula 5.48) in the l-form has a peppermint odor. In the d-form it has an aroma similar to caraway.

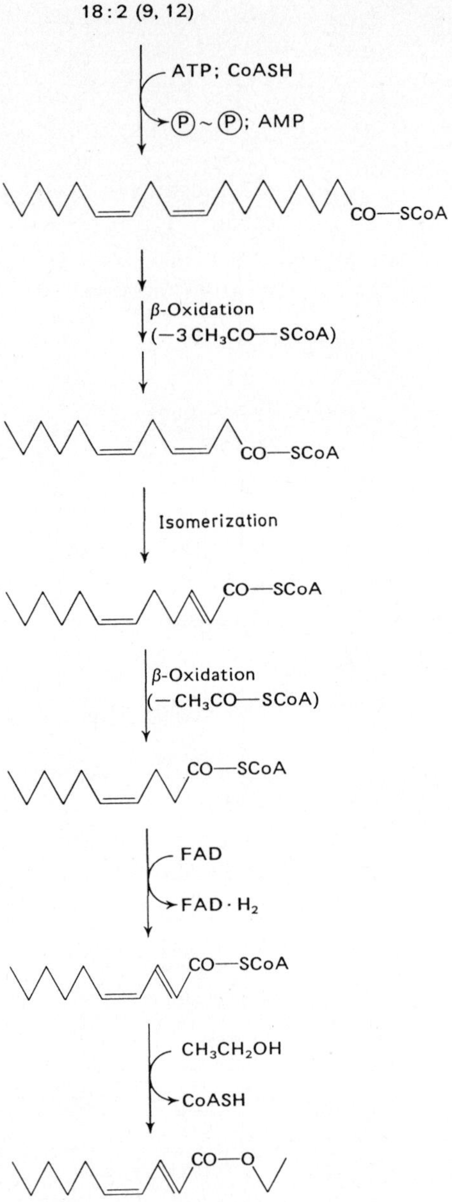

18:2 (9, 12)

ATP; CoASH

Ⓟ~Ⓟ; AMP

CO—SCoA

β-Oxidation
$(-3\,CH_3CO—SCoA)$

CO—SCoA

Isomerization

CO—SCoA

β-Oxidation
$(-CH_3CO—SCoA)$

CO—SCoA

FAD

$FAD \cdot H_2$

CO—SCoA

CH_3CH_2OH

CoASH

CO—O

Fig. 5.21. Biosynthesis of 2-trans, 4-cis-decadienoic acid ethyl ester in pears (After *W. G. Jennings* and *R. Tressl*, 1974)

Some terpenes are readily oxidized during food storage. Examples of aroma defects resulting from oxidation are provided in Table 5.3 and section 22.1.1.1.

Table 5.22. Terpenes in food

Monoterpenes

Acyclic (including cyclic derivatives)

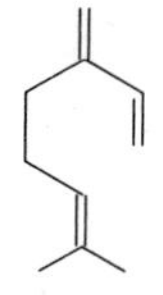

Myrcene (I)

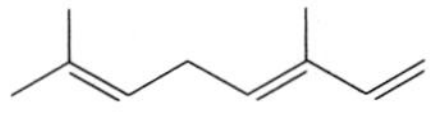

trans-Ocimene (II)

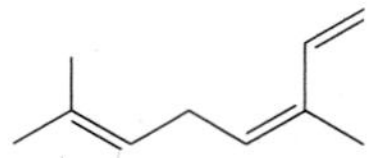

cis-Ocimene (III)

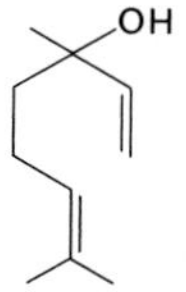

OH

Linalool (IV)

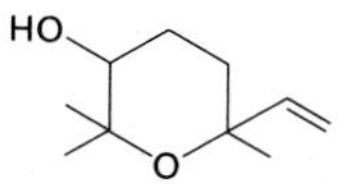

HO

O

2,6,6-Trimethyl-2-vinyl-5-hydroxytetrahydropyran[a] (IV a)

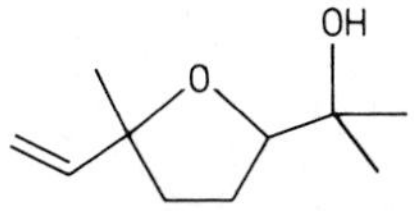

OH

O

2-Methyl-2-vinyl-5-hydroxyisopropyltetrahydrofuran[a] (IV b)

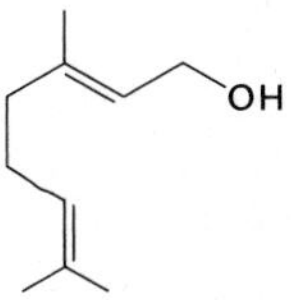

OH

Geraniol[b,c] (V)

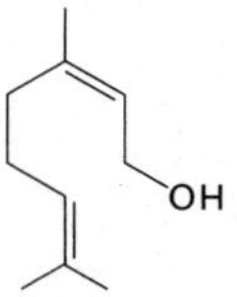

OH

Nerol[b] (VI)

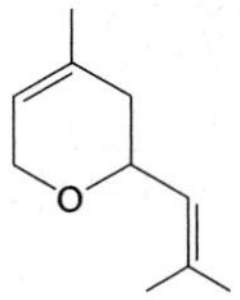

O

Neroloxide (VI a)

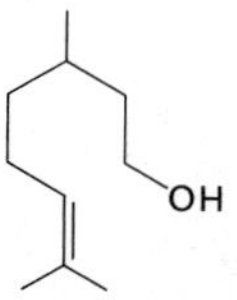

OH

Citronellol[b] (VII)

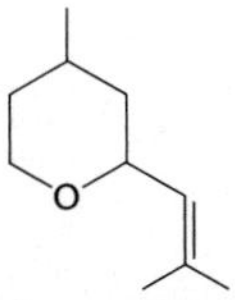

O

Rosenoxide (VII a)

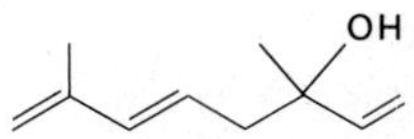

OH

Hotrienol[d] (VIII)

Monocyclic

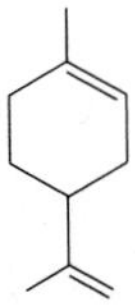

Limonene (IX)

α-Terpinene (X)

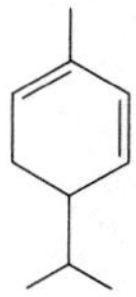

α-Phellandrene (XI)

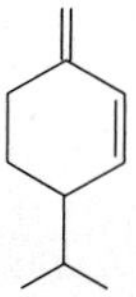

β-Phellandrene (XII)

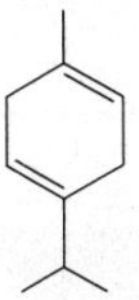

γ-Terpinene (XIII)

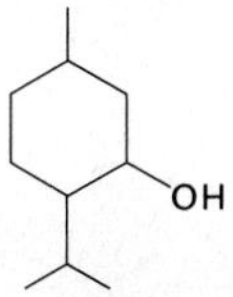

OH

Menthol (XIV)

Table 5.22. (Continued)

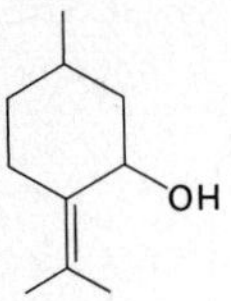

Pulegol (XV)

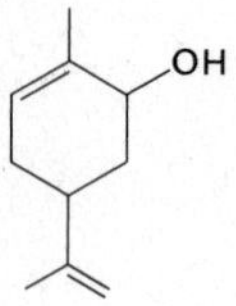

Carveol (XVI)

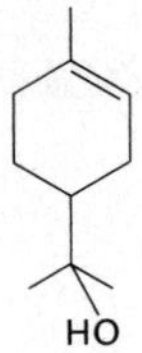

α-Terpineol (XVII)

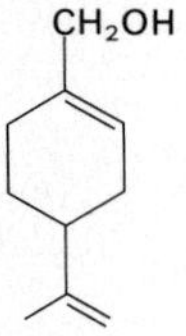

Perilla alcohol (XVIII)

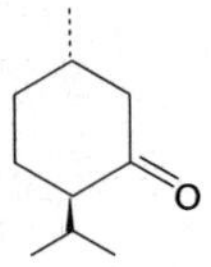

Menthone (XIX)

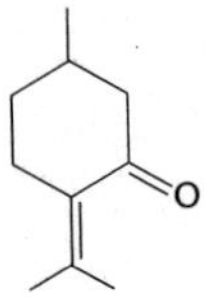

Pulegone (XX)

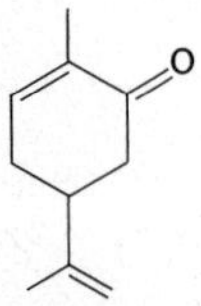

Carvone (XXI)

1,3-p-Menthadien-7-al (XXII)

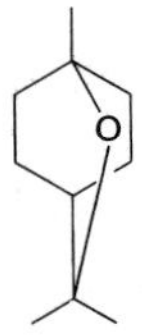

1,8-Cineole (XXIII)

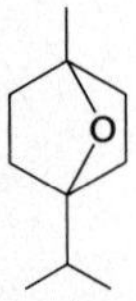

1,4-Cineole (XXIV)

Bicyclic

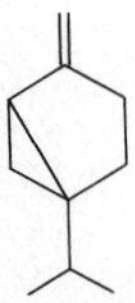

Sabinene (XXV)

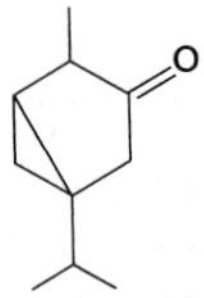

Thujone (XXVI)

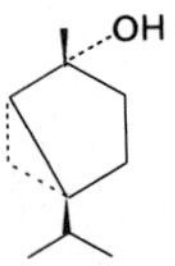

(+)-cis-Sabinene hydrate (XXVII)

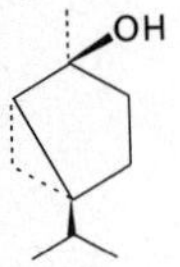

(+)-trans-Sabinene hydrate (XXVIII)

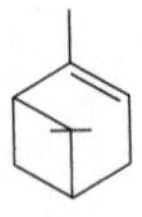

α-Pinene (XXIX)

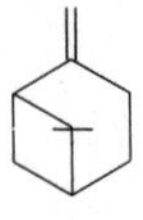

β-Pinene (XXX)

Table 5.22. (Continued)

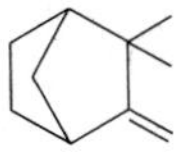

Camphene (XXXI)

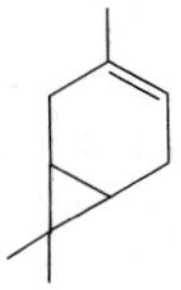

Δ^3-Carene (XXXII)

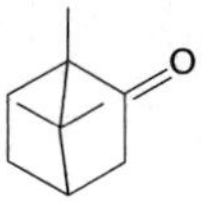

Camphor (XXXIII)

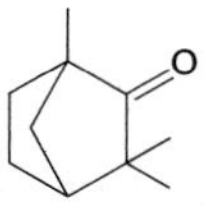

Fenchone (XXXIV)

Sesquiterpenes

Acyclic

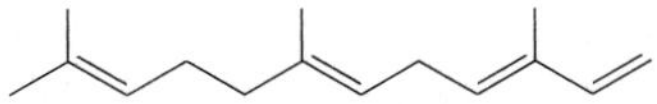

trans-α-Farnesene (XXXV)

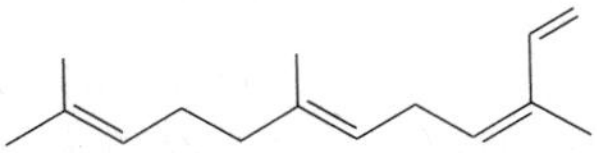

cis-α-Farnesene (XXXVI)

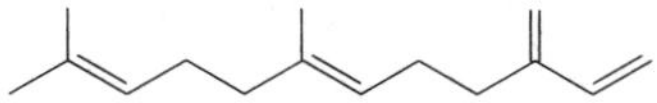

β-Farnesene (XXXVII)

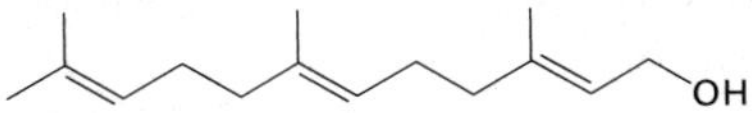

OH

Farnesol (XXXVIII)

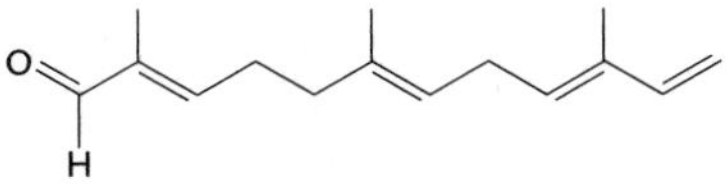

O H

(all-trans)-α-Sinensal (XXXIX)

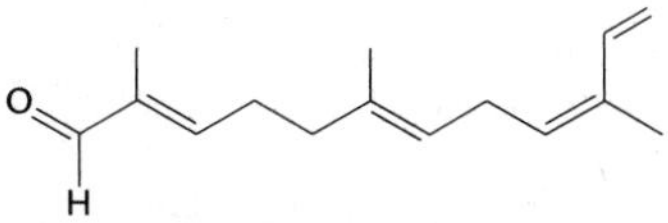

O H

(trans,trans,cis)-α-Sinensal (XL)

Monocyclic

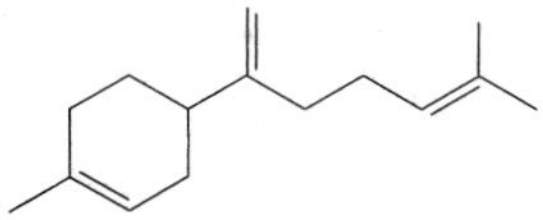

β-Bisabolene (XLI)

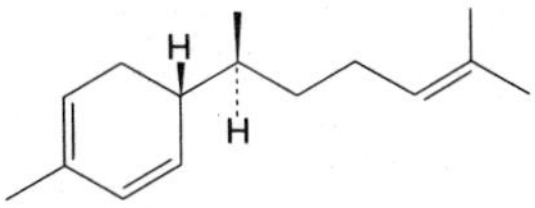

H H

(−)-Zingiberene (XLII)

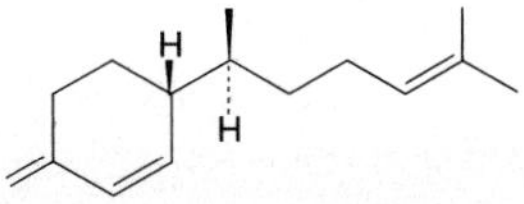

H H

(−)-Sesquiphellandrene (XLIII)

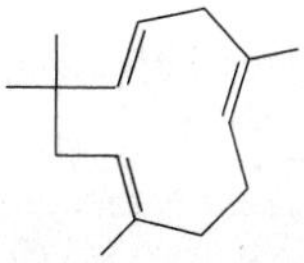

Humulene (XLIV)

Table 5.22. (Continued)

Bicyclic

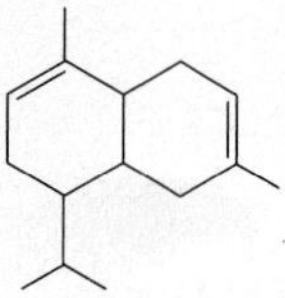

β-Cadinene (XLV) 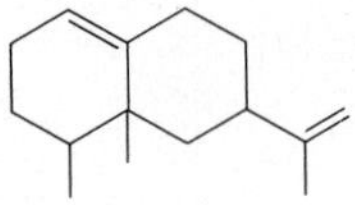 | Valencene (XLVI) 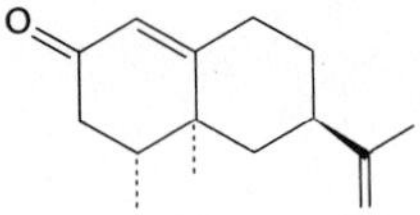 | (+)-Nootkatone (XLVII)

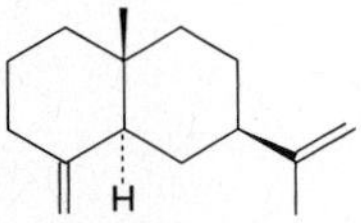

β-Selinene (XLVIII) 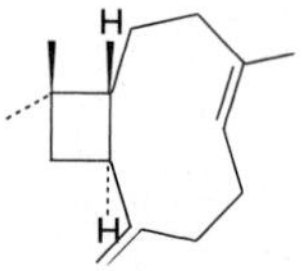 | β-Caryophyllene (XLIX)

[a] Compounds IV a and IV b, of which occur cis- and trans-isomers in wine (cf. Table 20.11), are also denoted as linalooloxides.
[b] Corresponding aldehydes geranial (V a), neral (VI b) and citronellal (VII a) also occur in food. Citral is a mixture of neral and geranial.
[c] Corresponding acid is an important aroma constituent of wine cultivars "Traminer" and "Scheurebe".
[d] (−)-3,7-Dimethyl-1,5,7-octatrien-3-ol (hotrienol) is found in grape, wine and tea aromas.

Terpene biosynthesis is carried out only by plants and some microorganisms and is initiated in both cases by acetyl-CoA.
Three molecules of acetyl-CoA condense to form 3-hydroxy-3-methylglutaryl-CoA (I) which, after hydrolysis and reduction, is converted to mevalonic acid (II). Phosphorylation, and elimination of CO_2 and water yields isopentenyl

Table 5.23. Aroma qualities of various terpenes

Compound[a]	Aroma description
Linalool[b] (IV)	Flowery
Linalooloxide (IV b)	Sweet, woody with flowery earthy undertone
Nerol (VI)	Rose-like with a green undertone
Neroloxide (VI a)	Sweet-flowery with green-herbaceous undertone
Limonene[b] (IX)[b]	Lemon-like
α-Phellandrene (XI)	Lemon-like, somewhat minty
α-Terpineol (XVII)	Lilac-like
Pulegone (XX)	Peppermint and camphor-like
α-Pinene (XXIX)	Resinous, after turpentine
Fenchone (XXXIV)	Camphor-like, sweet
Farnesol (XXXVIII)	Blossom-like

[a] Roman numerals given in brackets refer to Table 5.22.
[b] Odor threshold values are presented in Table 5.1.

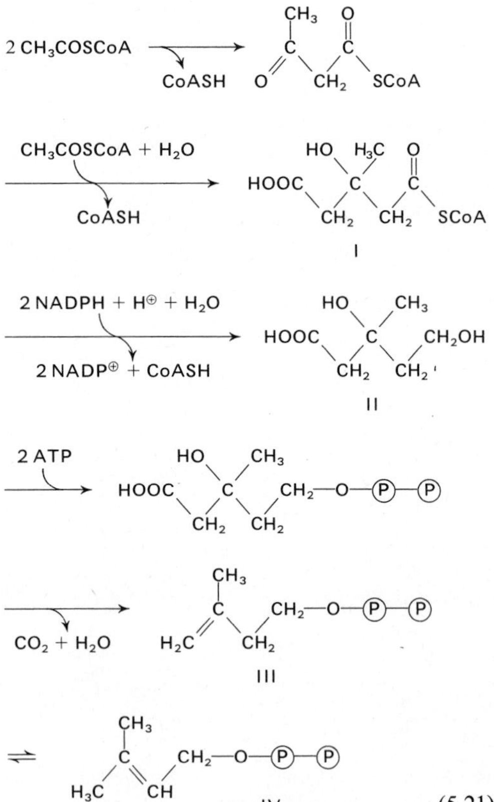

$$2\,CH_3COSCoA \xrightarrow{-\,CoASH} CH_3{-}CO{-}CH_2{-}CO{-}SCoA$$

$$\xrightarrow[CoASH]{CH_3COSCoA\,+\,H_2O} HOOC{-}CH_2{-}C(OH)(CH_3){-}CH_2{-}CO{-}SCoA \quad (I)$$

$$\xrightarrow[2\,NADP^{\oplus}\,+\,CoASH]{2\,NADPH\,+\,H^{\oplus}\,+\,H_2O} HOOC{-}CH_2{-}C(OH)(CH_3){-}CH_2{-}CH_2OH \quad (II)$$

$$\xrightarrow{2\,ATP} HOOC{-}CH_2{-}C(OH)(CH_3){-}CH_2{-}CH_2{-}O{-}(P){-}(P)$$

$$\xrightarrow{-\,CO_2\,+\,H_2O} H_2C{=}C(CH_3){-}CH_2{-}CH_2{-}O{-}(P){-}(P) \quad (III)$$

$$\rightleftharpoons H_3C{-}C(CH_3){=}CH{-}CH_2{-}O{-}(P){-}(P) \quad (IV) \qquad (5.21)$$

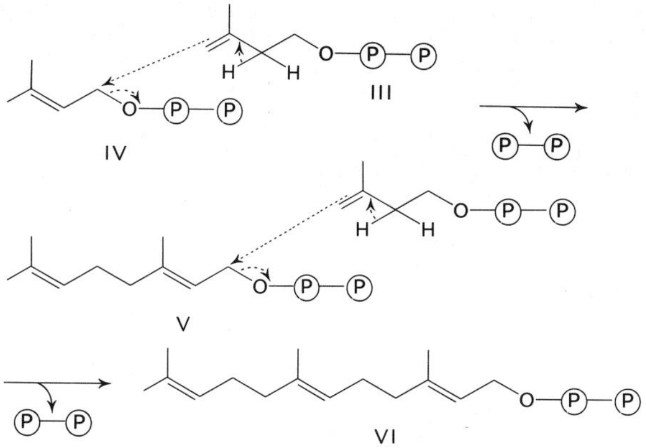

(5.22)

pyrophosphate (III), which can partly isomerize into dimethyl-allyl pyrophosphate (IV). A cation derived from IV then reacts with the double bond of compound III (cf. Reactions 5.21 and 5.22).

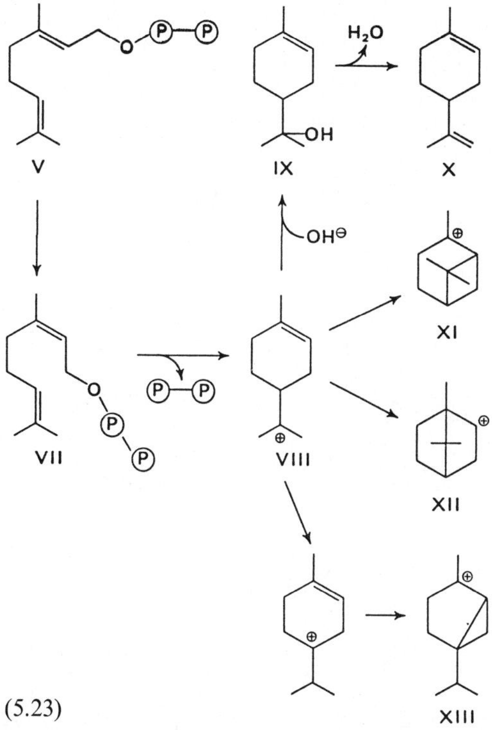

(5.23)

Geranyl diphosphate (V), the parent compound for monoterpene (C_{10}) biosynthesis, is generated by elimination of a proton. Condensation of compound V with isopentenyl diphosphate (III) leads to farnesyl diphosphate (VI), which is the parent compound for the biosynthesis of all the sesquiterpenes (C_{15}).

Although acyclic monoterpenes can be generated from V (e.g. geraniol by hydrolysis of compound V), cyclization is possible. It can only occur via isomerization of compound V from the trans- to the cis-form, i.e. formation of neryl pyrophosphate (VII, Reaction 5.23). The chain elongation reaction leading to farnesyl diphosphate is interrupted by this isomerization.

It is assumed that the cyclization mechanism of VII occurs via the cation VIII, forming terpenes with a p-menthane skeleton, such as α-terpineol (IX) and limonene (X). The cation is also involved in formation of bicyclic terpenes, such as pinane (XI), camphane (XII) and thujane (XIII). The cations present as intermediary products can attract a nucleophilic HO-group and thus provide a variety of terpene alcohols. However, the oxygen-containing moiety can also be acquired by oxidation of the complete menthane skeleton. For example, in peppermint leaves, l(−)menthol (XIV) is synthesized from cation VIII by the following pathway:

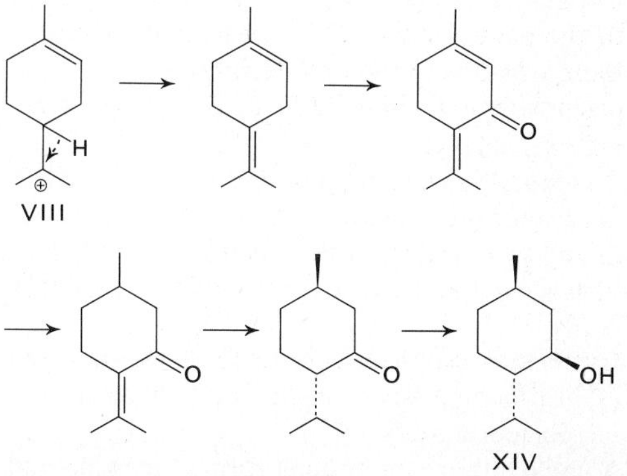

(5.24)

(5.25)

Similarly, sesquiterpene biosynthesis begins with a trans-to-cis isomerization. The intermediary cation (XVI) obtained from cis-farnesyl pyrophosphate (XV) provides a number of reaction possibilities due to the length of the carbon skeleton and the three double bonds within the skeleton. Only two metabolic pathways are presented here; one postulated for the biosynthesis of β-bisabolene (XVII) and the second for β-cadinene (XIX; cf. Reaction 5.25).

5.3.2.4 Volatile Sulfur Compounds

The aroma of many vegetables is due to volatile sulfur compounds obtained by a variety of enzymatic reactions. Examples are the vegetables of the plant families *Brassicacea* and *Liliaceae;* their aroma is formed by decomposition of glucosinolates (cf. 14.3.2.2.5) or S-alkyl-cysteine-sulfoxides (cf. 17.1.2.6.7).

2-Isobutylthiazole (compound III, Table 5.13) contributes to tomato aroma. It is probably obtained as a product of the secondary metabolism of leucine and cysteine (cf. postulated Reaction 5.26).

Cysteine is also the precursor for asparagus acid (1,2-dithiolane-4-carboxylic acid) in asparagus (cf. Reaction 5.27).

Volatile sulfur compounds formed in wine and beer production originate from methionine and

(5.26)

Cys—SH —Deamination→ CH_2—SH / C=O / COOH

—CH_3COOH, Aldol cond.→ HOOC—CH_2—C(—OH)(CH_2—SH)—COOH

—(CO_2; H_2O)→ H_2C=C(CH_2—SH)—COOH —H_2S→

HS—H_2C—CH(COOH)—CH_2—SH —Ox, (H_2O)→ (S—S ring)—COOH (5.27)

are by-products of the microorganism's metabolism. The compounds formed are methional (I), methionol (II) and acetic acid-3-(methylthio)-propyl ester (III, cf. Reaction 5.28).

Met —1) Deamination 2) Decarboxylation→ I (CHO)

—Reduction→ II (OH)

—$CH_3COSCoA$→ III (5.28)

5.3.2.5 Pyrazines

Paprika pepper *(Capsicum anuum)* and chillies *(Capsicum frutenscens)* contain high concentrations of 2-isobutyl-3-methoxypyrazine (cf. Table 5.15 for structure). Its biosynthesis from leucine is assumed to be through the following pathway:

Leu —Amidation→ (NH_2, O=C—NH_2) —(HC—CH, O O)→ (HO-pyrazine) —Methylation→ (MeO-pyrazine) (5.29)

Pyrazines are also produced by microorganisms. For example, 2-isopropyl-3-methoxypyrazine has been identified as a metabolic by-product of *Pseudomonas perolans* and *Pseudomonas taetrolens*. This pyrazine is responsible for a musty-earthy off-flavor in eggs, dairy products and fish.

5.4 Interactions with Other Food Constituents

Aroma interactions with lipids, proteins and carbohydrates affect the retention of volatiles within the food and, thereby, the levels in the gaseous phase. Consequently, the interactions affect the intensity and quality of food aroma. Since such interactions can not be clearly followed in a real food system, their study has been transferred to model systems which can, in essence, reliably imitate the real systems. A knowledge of the binding of aroma to solid food matrices, from the standpoint of food aromatization, aroma behavior and food processing and storage, is of great importance.

5.4.1 Lipids

In an o/w emulsion (cf. 8.15.1), the distribution coefficient, K, for aroma compounds is related to aroma activity:

$$K = \frac{C_o}{C_w} \quad (5.30)$$

where C_o is the concentration of the aroma compound in the oil phase, and C_w the concentration of the aroma compound in the aqueous phase.

In a homologous series, e.g. n-alkane alcohols (cf. Figure 5.22), the value of K increases with increasing chain length. The solubility in the fat or oil phase rises proportionally as the hydrophobicity imposed by chain length increases. The vapor pressure behavior is the reverse of that; it drops as the hydrophobicity of the aroma compounds increases. The vapor pressure also drops as the volume of the oil phase increases, and the odor threshold value increases at the same time. The above outlines are well clarified in Figure 5.23. The solubility of 2-heptanone is higher in whole milk than in skim milk which, in this case, behaves as an aqueous phase. When this phase is replaced by oil (Figure 5.23, experiment 5), 2-

heptanone concentration in the gas phase is the lowest.

Assays with n-alcohols demonstrate that, with increasing chain length of volatile compounds, the speed of migration of the molecules from oil to water phase increases. An increase in oil viscosity retards such migration.

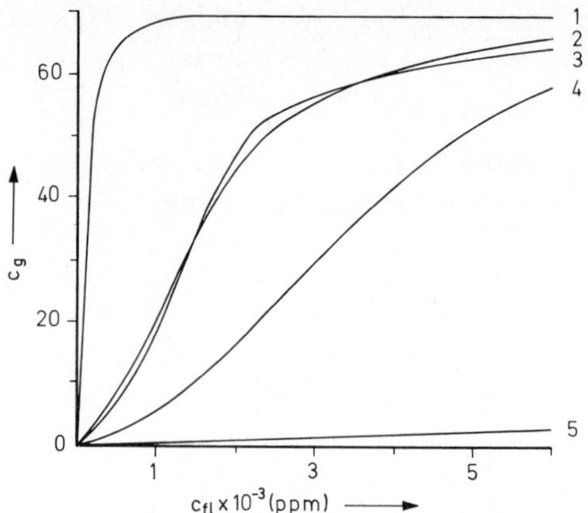

Fig. 5.23. Influence of the media on 2-heptanone concentration in the gas phase. (After *W. W. Nawar,* 1966). 2-Heptanone alone (1), in water (2), in skim milk (3), in whole milk (4), in oil (5). c_{fl}: concentration in liquid; c_g: concentration in gas phase (detection signal height from headspace analysis)

5.4.2 Proteins, Polysaccharides

The sorption characteristics of various proteins for several volatile compounds are presented in Figure 5.24. Ethanol is bound to the greatest extent, probably with the aid of hydrogen bonds. The binding of the nonpolar aroma compounds probably occurs on the hydrophobic protein surface regions. A proposal for the utilization of data on the sorption of aroma volatiles on a biopolymer (protein, polysaccharide) is based on the law of mass action. When a biopolymer, B, has a group which attracts and binds the aroma molecule, A, then the following equation is valid:

$$K = \frac{[BA]}{C_f[B]} \tag{5.31}$$

where K = a single binding constant; and C_f = concentration of free aroma compound molecules.

$$[BA] = KC_f[B] \tag{5.32}$$

To calculate the average number of aroma mole-

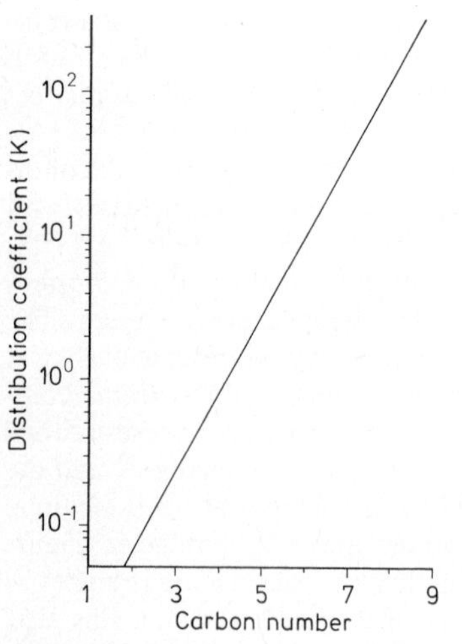

Fig. 5.22. Distribution of n-alkanols in the system oil/water (After *P. B. McNulty* and *M. Karel,* 1973)

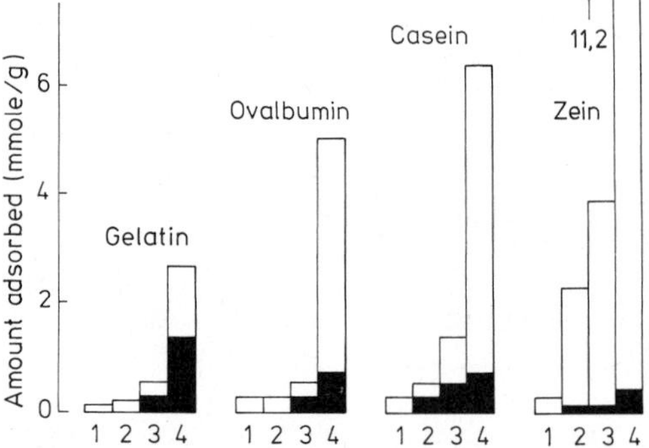

Fig. 5.24. Sorption of volatile compounds on proteins at 23 °C. (After *H. G. Maier,* 1974). Hexane (1), ethyl acetate (2), acetone (3), ethanol (4). □ plus ■: maximal sorption, ■: after desorption

cules bound to a biopolymer, the specific binding capacity, r, has to be introduced:

$$r = \frac{[BA]}{([B] + [BA])} \tag{5.33}$$

The concentration of the complex BA from Equation 5.32 is substituted in Equation 5.33:

$$r = \frac{KC_f[B]}{([B] + K[B]C_f)} = \frac{KC_f}{1 + KC_f} \tag{5.34}$$

When a biopolymer binds not only one molecular species (as A in the above case) but has a number (n) of binding groups (or sites) equal in binding ability and independent from each other, then r has to be multiplied by n, and Equation 5.34 acquires the form:

$$r = \frac{nKC_f}{1 + KC_f} \tag{5.35}$$

$$\frac{r}{C_f} = Kn - Kr = K' - Kr \tag{5.36}$$

where K′ = overall binding constant.

The utilization of data then follows Equation 5.36 presented in graphic form, i.e. a diagram of $r/C_f = f(r)$. Three extreme or limiting cases should be observed:

a) A straight line (Figure 5.25a) indicates that only one binding region on a polymer, with n binding sites (all equivalent and independent from each other) is involved. The values n and K′ are obtained from the intersection of the straight line with the abscissa and the ordinate, respectively.

b) A straight line parallel to the abscissa (Figure 5.25b) is obtained when the single binding constant, K, is low and the value of n is very high. In this special case, Equation 5.36 has the form:

$$r = K'C_f \tag{5.37}$$

c) A curve (Figure 5.25c) which in approximation is the merging of two straight lines, as shown separately (Figure 5.25d). This indicates two binding constants, K_1' and K_2', and their respective binding groups, n_1 and n_2, which are equivalent and independent from each other.

By plotting r versus C_f, values of K′ are obtained from the slope of the curve. An example for a model system with two binding regions (case c in Figure 5.25) is given by aroma binding with starch. It should be remembered that starch binds the volatiles only after gelatinization by trapping the volatiles in its helical structure, and that starch is made up of two constituents, amylose and amylopectin. The binding parameters are listed for some aroma compounds in Table 5.24. Numerous observations indicate that K_1' and binding region n_1 are related to the inner space of the helix, while K_2' and the n_2 region are related to the outer surface of the helix. K_1' is greater than K_2', which shows that, within the helix, the aroma compounds are bound more efficiently to glucose residues of the helix. The fraction 1/n is a measure of the size of the binding region. It decreases, as expected, with increasing molecular weight of alkane alcohols, but it is still larger within the helix than on the outer surface. Altogether, it should be concluded that, within a helix, the trapped compound can not fulfill an active role as an aroma constituent.

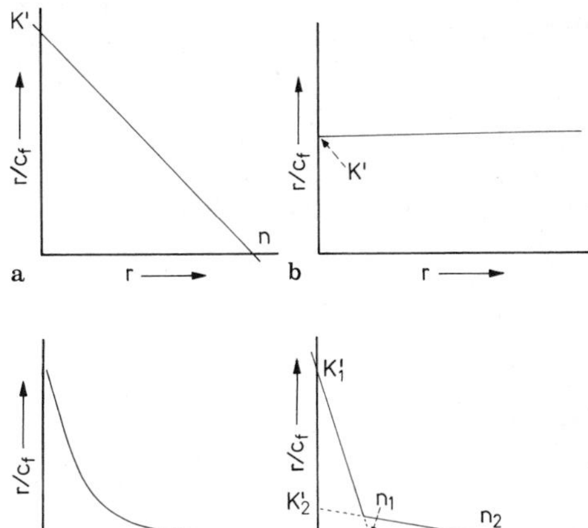

Fig. 5.25. Binding of aroma compounds by biopolymers. Graphical determination of binding parameters. (After *J. Solms,* 1975)

Table 5.24. Binding of aroma compounds by potato starch

Compounds	Binding constant K_1'	n_1	K_2'	n_2
1-Hexanol	$5.45 \cdot 10^1$	0.10	–	–
1-Octanol	$2.19 \cdot 10^2$	0.05	$2.15 \cdot 10^1$	0.11
1-Decanol	$1.25 \cdot 10^2$	0.04	$1.29 \cdot 10^1$	0.11
Capric acid	$3.30 \cdot 10^2$	0.07	$4.35 \cdot 10^1$	0.19
Menthone	$1.84 \cdot 10^2$	0.012	8.97	0.045
Menthol	$1.43 \cdot 10^2$	0.007	–	–
β-Pinene	$1.30 \cdot 10^1$	0.027	1.81	0.089

An unlimited number of binding sites exist in proteins dissolved or dispersed in water (case b, Figure 5.25). K′ values for several aroma compounds are given in Table 5.25. The value of the constant decreases in the order of aldehydes, ketones, alcohols, while compounds such as dimethylpyrazine and butyric acid are practically unable to bind. In the case of aldehydes, the assumption that they can react with free amino-

Table 5.25. Binding of aroma compounds by proteins (0.4% solutions at pH 4.5)

Aroma compound	Total binding constant $K' \cdot 10^{-3} (M^{-1})$			
	Bovine serum albumin		Soya protein	
	20 °C	60 °C	20 °C	60 °C
Butanal	9.765	11.362	10.916	9.432
Benzaldehyde	6.458	6.134	5.807	6.840
2-Butanone	4.619	5.529	4.975	5.800
1-Butanol	2.435	2.786	2.100	2.950
Phenol	3.279	3.364	3.159	3.074
Vanillin	2.070	2.490	2.040	2.335
2,5-Dimethyl pyrazine	0.494			
Butyric acid	0			

and SH-groups can not be excluded. The high values of K′ can reflect other than secondary forces.

Bovine serum albumin and soya proteins are practically the same with regard to the binding of the aroma compounds (Table 5.25). Since both proteins have a similar hydrophobicity, it is apparent that hydrophobic rather than hydrophilic interactions are responsible for aroma binding in proteins.

5.5 Food Aromatization

Aromatized food has been produced and consumed for centuries, as exemplified by confectionery and baked products, and tea or alcoholic beverages. In recent decades the number of aromatized foods has increased greatly. In FR Germany this amounts to about 15% of the total food consumed. A significant reason for such development is the rise of industrially-processed food and its share of the food market, which demands aromatization during processing and storage. In addition, introduction of new raw materials, e. g. protein isolates to diversify or expand traditional food sources, or the production of food substitutes, which are of better quality not only visually but in flavor or aroma, has contributed to food aromatization. Aroma concentrates, essences, extracts and individual compounds are used for aromatization. They are usually blended in a given proportion by a flavorist; thus, an aroma mixture is "composed". The empirically-developed "aroma formulation" is based primarily on the flavorist's experience and personal sensory assessment and is supported by the results of a physico-chemical aroma analysis. Legislative measures that regulate food aromatization differ in various countries.

5.5.1 Raw Materials for Essences

In FR Germany, up to 75% of the aromas used for food aromatization are of plant origin and, thus, designated as "natural aroma substances". The rest of the aroma compounds are synthetic, but 99% of this portion is chemically identical to their natural counterparts. Only 1% are synthetic aroma compounds not found in nature.

5.5.1.1 Essential Oils

Essential (volatile) oils are obtained preferentially by steam distillation of plants (whole or parts) such as clove buds, nutmeg (mace), lemon, caraway, fennel, and cardamon fruits (cf. 22.1). After steam distillation, the essential oil is separated from the water layer, clarified and stored. The pressure and temperature in the process are selected to incur the least possible loss of aroma substances by thermal decomposition, oxidation or hydrolysis.

Many essential oils, such as those of citrus fruits, contain terpene hydrocarbons which contribute little to aroma but are readily autooxidized and polymerized ("resin formation"). These undesirable oil constituents (for instance, limonene from orange oil) can be removed by fractional distillation. Fractional distillation is also used to enrich or isolate a single aroma compound. Usually, this compound is the dominant constituent of the essential oil. Examples of single aroma compounds isolated as the main constituent of an essential oil are: 1,8-cineole from eucalyptus, l(−)-menthol from peppermint, anethole from anise seed, eugenol from clove, or citral (mixture of geranial and neral, the pleasant odorous compounds of lemon or lime oils) from litseacuba.

5.5.1.2 Extracts

When the content of essential oil is low in the raw material, the aroma constituents are destroyed by steam distillation or the aroma is lost by their solubility in water, then the oil in the raw material is recovered by an extraction process.

Examples are certain herbs or spices (cf. 22.1.1.1) and some fruit powders. Hexane, methylene chloride, acetone, ethanol, water and/or edible oil or fat are used as solvents. Good yields are also obtained by using liquid CO_2. The volatile solvent is then fully removed by distillation. The oil extract (resin, absolue) often contains volatile aroma compounds in excess of 10% in addition to lipids, waxes, plant pigments and other substances extractable by the chosen solvent. Extraction may be followed by chromatographic or counter-current separation to isolate some desired aroma fractions. If the solvent used is not removed by distillation, the product is called an extract. The odor intensity of the extract, compared to the pure essential oil, is weaker for aromatization purposes by a factor of 10^2 to 10^3.

5.5.1.3 Distillates

The aroma constituents in fruit juice are more volatile during the distillation concentration process than is the bulk of the water. Hence, the aroma volatiles are condensed and collected (cf. 18.2.12). Such distillates yield highly concentrated aroma fractions through further purification steps.

5.5.1.4 Microbial Aromas

Cheese aroma concentrates are now offered on the market which have an aroma intensity at least 20-fold higher than that of normal cheese. They are produced by the combined action of lipases and *Penicillium roqueforti* using whey and fats/oils of plant origin as substrates. In addition to C_4–C_{10} fatty acids, the aroma is determined by the presence of 2-heptanone and 2-nonanone.

5.5.1.5 Synthetic Natural Aroma Compounds

In spite of the fact that a great number of food aroma compounds have been identified, economic factors have resulted in only a limited number of them being synthesized on a commercial scale. Synthesis starts with the natural compound available in large amounts at the right cost, or from chemical compounds from other sources. Several examples will be considered below:

A most important aroma compound worldwide, vanillin, is obtained primarily by alkaline hydrolysis of lignin (sulfite waste of the wood pulp industry), which yields coniferyl alcohol. It is converted to vanillin by oxidative cleavage:

CH₂—OH … HO … O—CH₃ —Ox.→ CHO … HO … O—CH₃ (5.38)

The most important source of citral, used in large amounts in food processing, is the steam-distilled oil of lemongrass *(Cymbopogon flexuosus)*. Citral actually consists of two geometrical isomers: geranial (I) and neral (II). They are isolated from the oil in the form of bisulfite adducts:

Lemonleaf oil —$NaHSO_3$→ CH(SO₃Na)(OH) adduct + CH(SO₃Na)(OH) adduct —$NaHSO_3$→ CHO (I) + CHO (II) (5.39)

The aroma compound menthol is primarily synthesized from petrochemically-obtained m-cresol. Thymol is obtained by alkylation and is then further hydrogenated into racemic menthol:

OH —Alkylation→ OH —H_2→ OH (5.40)

A more expensive processing step then follows, in which the racemic form is separated and l(−)-menthol is recovered. The d-optical isomer substantially decreases the quality of the aroma (cf. 5.3.2.3).

The purity requirement imposed on synthetic aroma substances is very high. The purification steps usually used are not only needed to meet the stringent legal requirements (i.e. beyond any doubt safe and harmless to health), but also to remove undesirable contaminating aroma compounds. For example, menthol has a phenolic off-flavor note even in the presence of only 0.01% thymol as an impurity. This is not surprising since the odor threshold value for thymol is lower than that of l(−)-menthol by a factor of 450.

5.5.1.6 Synthetic Aroma Compounds

Some synthetic aroma compounds used in food aromatization are compiled in Table 5.26.

Table 5.26. Synthetic aroma compounds used for flavoring (not naturally occurring in food)

Name	Structure	Aroma description
Ethyl vanillin	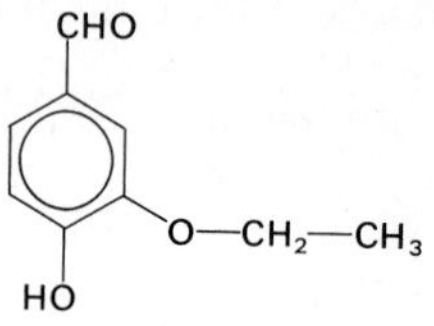 CHO; HO; O—CH$_2$—CH$_3$	Sweet after vanilla (2 to 4-times stronger than vanillin)
Ethyl maltol	cf. 5.3.1.2	Caramel-like
Musk ambrette	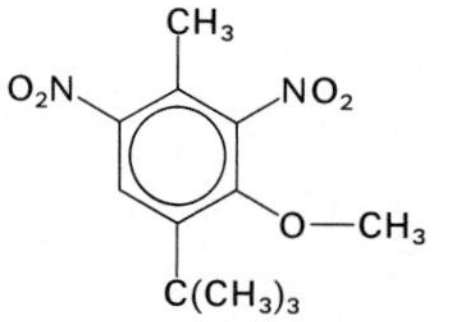 CH$_3$; O$_2$N; NO$_2$; O—CH$_3$; C(CH$_3$)$_3$	Musk-like
Allyl phenoxyacetate	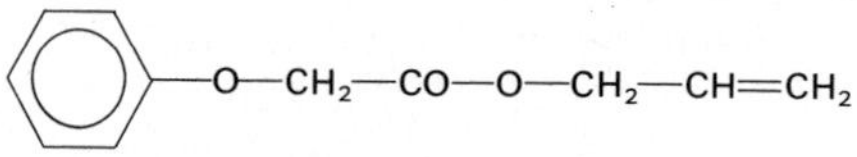 —O—CH$_2$—CO—O—CH$_2$—CH=CH$_2$	Fruity, pineapple-like
α-Amyl cinnamic-aldehyde	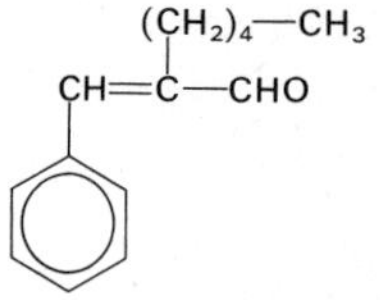 (CH$_2$)$_4$—CH$_3$; CH=C—CHO	Floral, jasmin and lilies
Hydroxycitronellal	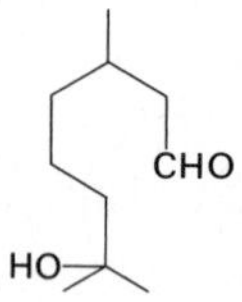 CHO; HO	Sweet, flowery, liliaceous
Resorcinol-dimethylether	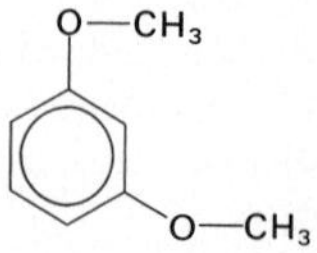 O—CH$_3$; O—CH$_3$	Sharp, penetrating, fruity

Table 5.26. (Continued)

Name	Structure	Aroma description
Anisylacetone	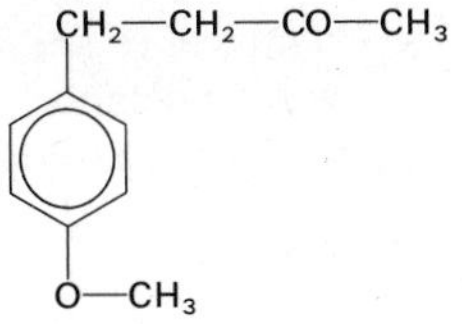 CH_2—CH_2—CO—CH_3; O—CH_3	Sweet, flowery, fruity
6-Methyl coumarin	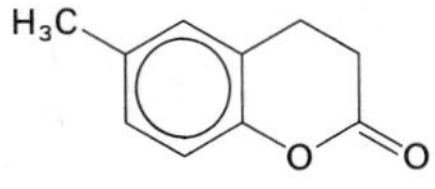 H_3C; O; O	Dried, herbaceous
β-Naphthyl methylketone	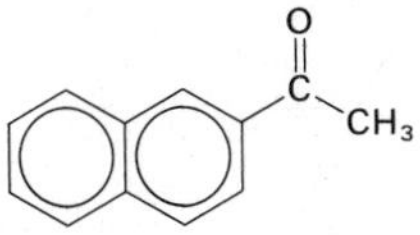 O; C; CH_3	After orange flowers, strawberry-like
Propenylguaethol (vanatrope)	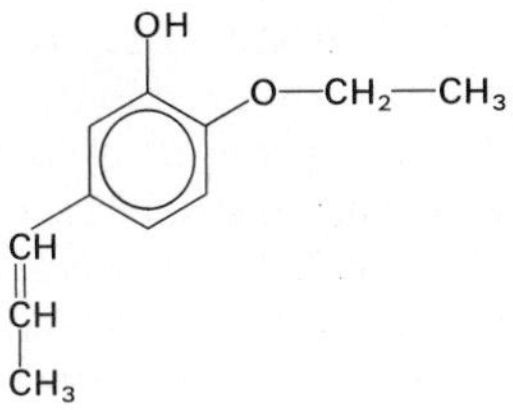 OH; O—CH_2—CH_3; CH; CH; CH_3	Phenolic, anise-like
Piperonyl isobutyrate	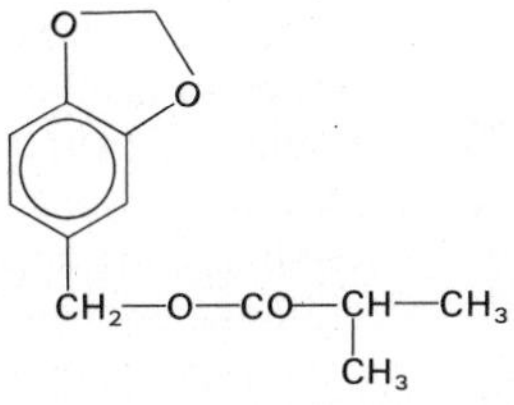 O; O; CH_2—O—CO—CH—CH_3; CH_3	Sweet, fruity, after berry fruits

5.5.2 Essences

The flavorist composes essences from raw materials. In addition to striving for an optimal aroma, the composition of the essence has to meet the food processing demands, e. g. compensate for possible losses during heating. The "aroma formulation" is an empirical one, developed as a result of long experience dealing with many problems, disappointments and failures, and is rigorously guarded after the "know-how" is acquired. Based on these facts, the example given in Table 5.27 provides a formulation for pineapple essence only in principle.

5.5.3 Aromas from Precursors

The aroma of food that has to be heated, in which the impact aroma compounds are generated by the *Maillard* reaction, can be improved by increasing the levels of precursors involved in the reaction. This is a trend in food aromatization. Some of the precursors are added directly (consider the proposal put forward in Table 1.6 for producing the meat aroma note), while some precursors are generated within the food by the preliminary release of the reaction components required for the *Maillard* reaction (cf. 4.2.4.4). This is achieved by adding protein and polysaccharide hydrolases to food.

Table 5.27. Formulations for pineapple essence

From naturally occurring raw materials:	From synthetic aroma compounds (Continued):
686 g pineapple fruit juice concentrate	20.0 g diethyl sebacate
300 g pineapple shell distillate	16.4 g allyl cyclohexyl propionate
10 g orange oil[a]	16.0 g ethyl propionate
2 g oil of wine yeast[a]	13.0 g ethyl heptanoate
2 g camomile oil (*Matricaria chamomilla*)	8.0 g butyric acid
1,000g	5.6 g vanillin
	4.0 g citronellyl butyrate
From synthetic aroma compounds:	2.5 g methyl allyl caproate
376 g ethyl acetate	2.0 g methyl-β-methyl thiopropionate
112 g amyl butyrate	2.0 g allyl phenoxy-acetate
105 g butyl acetate	1.0 g methyl caprylate
45 g ethyl butyrate	0.6 g citral
36 g ethyl isovalerate	0.3 g cinnamyl acetate
28.6 g amyl acetate	0.1 g bornyl acetate
22.5 g orange oil	162 g solvent
21.4 g allyl caproate	1,000 g

[a] Aroma compounds content approx. 1%.

5.6 Molecular Structure and Odor

When the peripheral receptors within the olfactory tissue are stimulated, the olfactory mechanism recognizes two characteristics of the odor: quality and intensity. Only the latter can be measured, e.g. by determining odor threshold values (cf. 5.1.2). Quality, on the other hand, can be described by comparison. Odor stimulants can be grouped into those of the same or similar qualities. For example, primary quality or modality is a quality which is recognized as being homogeneous, i.e. it is not reproducible by two or more nonidentical stimulants being superimposed.

The subject of many studies related to human chemoreception usually raises the same question: "Which structural elements control the receptor's perception of a compound as a specific odor or flavor?". It has to be admitted that the whole field is very perplexing. Nevertheless, a number of similarities between the structure of aroma volatiles and its relationship to odor have been observed, and this will be emphasized below in more detail.

A compound's odor properties are affected by:

- Molecular structure or geometry
- Presence of functional groups

The significance and meaning of a compound's geometry is demonstrated, for example, by several camphor-like compounds:

I II III IV (5.41)

2,2,3,3-tetramethylbutane (II) and bicyclooctane (III) have an odor very close to that of camphor (I). Also, cyclooctane (IV) is camphor-like, though not as clearly so as II and III. These examples strongly suggest that functional groups have no decisive importance for the camphor-like odor, but the molecular shape or geometry of the molecule, which is close to that of a sphere, does.

A further indication of the importance of the molecular geometry is the possible substitution of groups within the molecule by other groups having similar *van der Waals* radii (e.g. H ↔ F; CH_3 ↔ Br). This substitution has basically no effect on the odor quality (cf. Formula 5.42). The corresponding fluoro-derivatives of compound V (see compound VI) are similar in odor though slightly more herbaceous. The musky odor character is retained in compounds VII/VIII or IX/X, while it is absent in compounds XI/XII. The sandalwood odor is retained by compounds XIII/XIIIa when the cis-decalin group is replaced by a t-butyl-cyclohexyl group. Also the pairs XIV/XIVa and XV/XVI (cf. Formula 5.42) have the same odors, respectively.

The need to maintain the molecule's dipole moment in such substitutions is important since the dipole moment is responsible for the required orientation of the molecule on the odor's chemoreceptor site. cis-t-Butylcyclohexyl acetate (XVII) has an intensive specific odor, while its trans-isomer (XVIII) has only a weak and flat odor (cf. Formula 5.43).

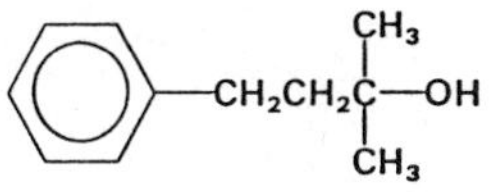

V

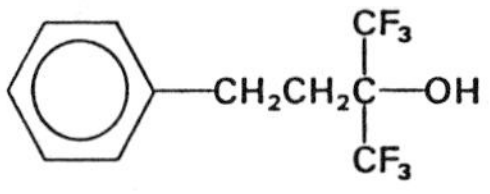

VI

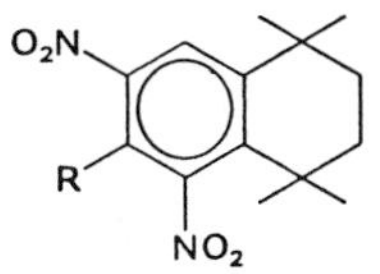

VII: R=Me
VIII: R=Br

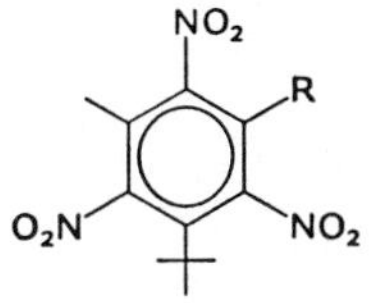

IX: R=Me
X: R=Br

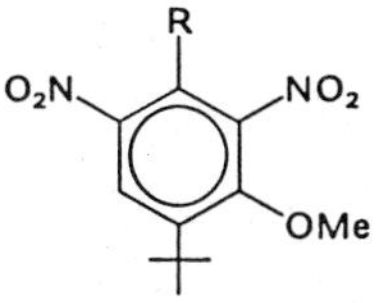

XI: R=Me
XII: R=Br

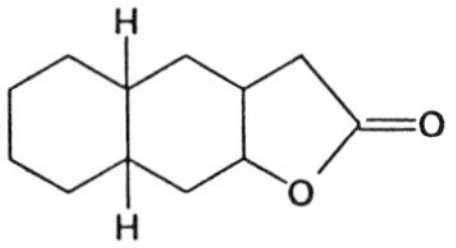

XIII

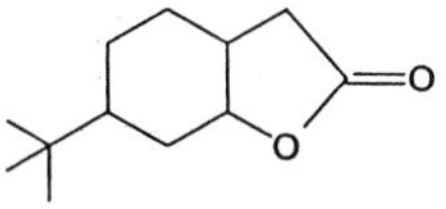

XIII a

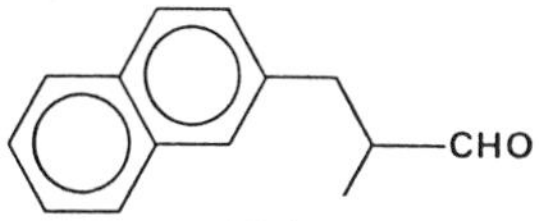

XIV

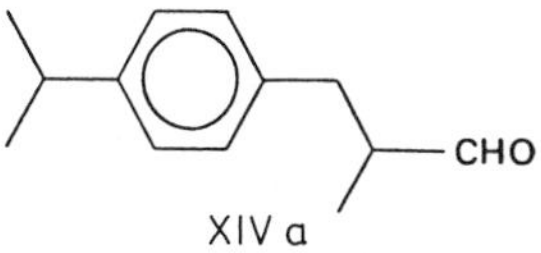

XIV a

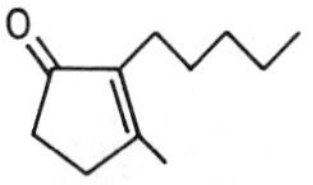

(5.42)

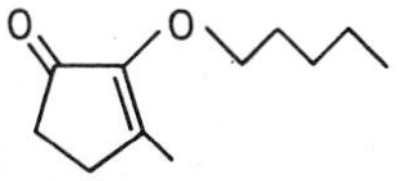

XV

XVI

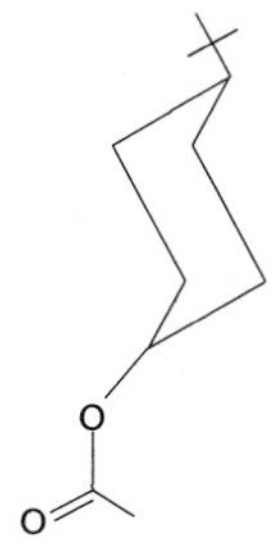

XVII 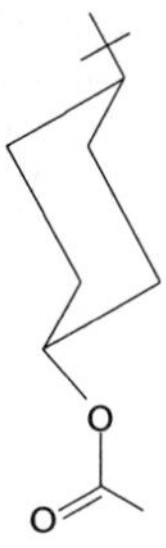 XVIII (5.43)

The fact that a similar type of molecular geometry supports a similar type of odor is demonstrated in the case of limonene (XIX). Both isomers of p-methene-8 differ clearly: XX has a characteristic orange odor (like limonene), while XXI has the flat odor of a hydrocarbon.

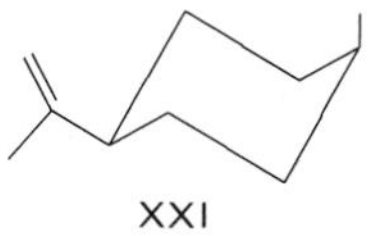

XXI 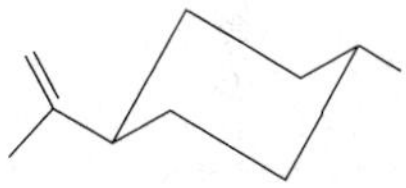 XX

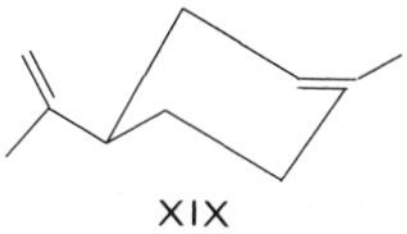

XIX (5.44)

Functional groups are not at all essential for an aroma compound, as illustrated by hydrocarbons. These compounds have a specific odor though they have no functional group, thus again illustrating the importance of molecular geometry. On the other hand, there are compounds, such as NH_3, H_2S and CH_3SH, which consist of only one functional group and their smell (unpleasant odor) is extremely intensive. In this case consideration of steric factors makes no sense; the functional group is obviously and solely the determinant.

An increase in molecular size results in a decrease of the effect of the functional group and a concomitant increase of the influence of molecular orientation, due to its dipole moment, on the perception of odor quality. Actually, both odor quality and odor intensity are influenced indirectly. This is exemplified by the following compounds: R_2S, R_2NH, R'_2O. For R = Me, the odor quality is distinctly different. For R^1 = Me, R^2 =

PhEt, a difference exists but is not distinct, as in the case of ether and thioether, which have the same odor qualities.

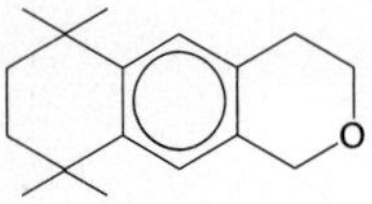

XXII 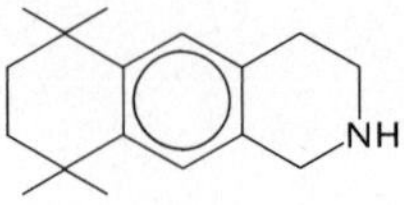 XXIII

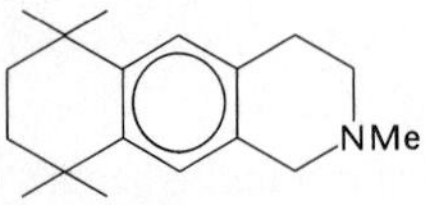

XXIV 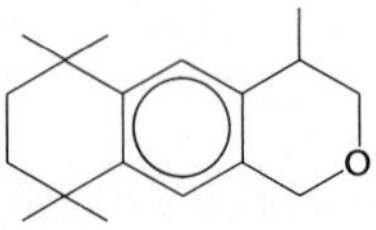 XXV

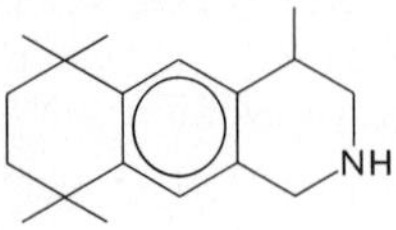

XXVI 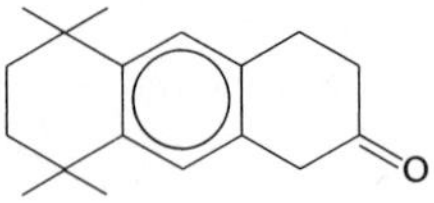 XXVII (5.45)

Further examples are found among the odorous substances of musk. It is possible to retain the original odor quality by substituting the oxygen in (XXII) with NH (XXIII) or MeN (XXIV); however, the odor intensity is decreased. The same is true for XXV and XXVI. Substitution of oxygen in XXV with CO (XXVII) results in an odor loss. The latter is the result of a negligible volatility of the new compound and change of its dipole moment.

The surrounding of functional groups within a molecule is important since it influences the accessibility of these groups. Camphor odor is quite well imitated by the compound methyl isopropyl ketone. In this case, surroundings of the functional group are very similar. Several examples of the musk-like compounds can also verify the involvement of the surroundings of the functional group as a factor in aroma perception (cf. Formula 5.46).

The CO group in XXVIII is required for a musk-like odor, since XXIX does not have such an odor. The quaternary C-atoms at positions 3 and 5 are important for odor intensity. Positions 2 and 6 interfere with accessibility to the CO group and also with maintaining the coplanarity of the molecule. Compounds XXIXa–XXIXc have a strong musk-like odor, while XXIXd and XXIXe have a very weak musk-like odor.

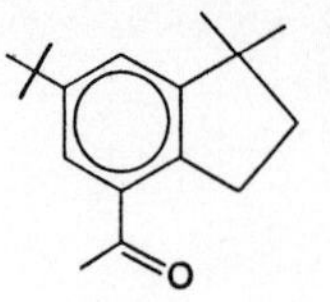

XXVIII 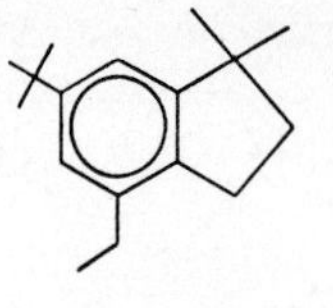 XXIX

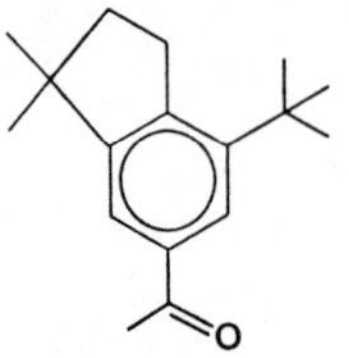

XXIX a 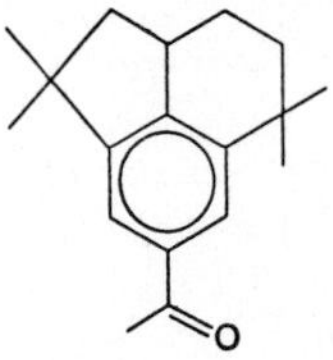 XXIX b

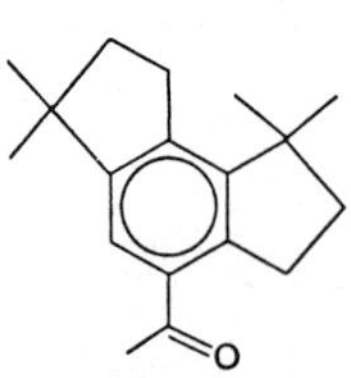

XXIX c

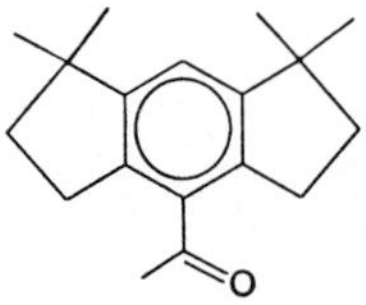

XXIX d 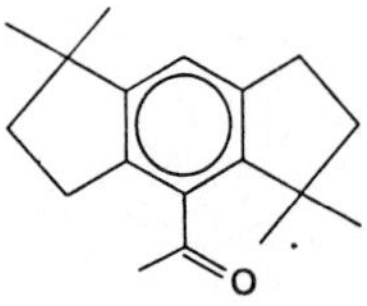 XXIX e (5.46)

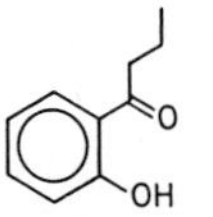

XXX 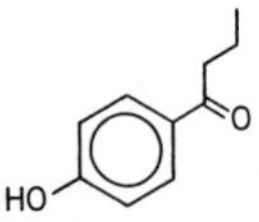 XXXI

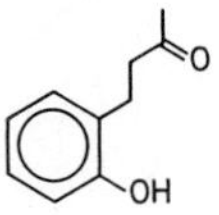

XXXII 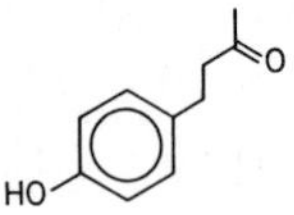 XXXIII

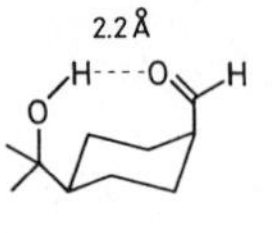

2.2 Å 6 Å

XXXIV 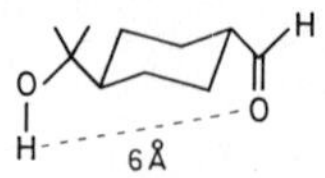 XXXV (5.47)

In compounds with several functional groups, their relative positions can also be of importance for odor quality. Compound XXX, for example, has a strong phenolic odor resembling that of salicylic acid, but this odor is weak in compound XXXIII. Compound XXXI is odorless, while XXXIII has a strong fruity odor and is known as "raspberry ketone". Of the pair of isomers, XXXIV/XXXV, the cis-compound (XXXIV) has a strong flowery-woody odor note, while the trans-isomer (XXXV) is odorless. Very probably, both functional groups interact as a bifunctional unit with the corresponding chemoreceptor site. The OH-groups, in this case, play the role of proton donors (HA-group), while the CO-group functions as a proton acceptor (B-group). In many additional compounds with differing odor qualities, it has been found that the distance between proton donor and proton acceptor must be less than 3 Å. The odor quality in these compounds appears to be determined by the hydrophobic part of the molecule which, as the third X group, establishes the hydrophobic bond with the receptor. It is of interest that this AH/B/X-system postulated for a number of aroma compounds has stood the test for more than a decade in the discussion of structure-activity relationship of sweet compounds (cf. 4.2.3 and 8.8.1).

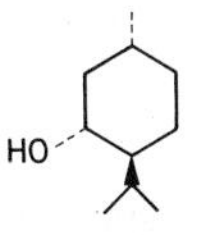

HO XXXVI OH XXXVII

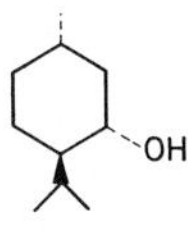

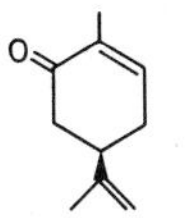

O XXXVIII O XXXIX (5.48)

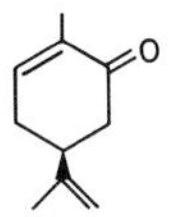

The acceptance of such a hypothesis, stimulant-receptor-interaction over at least three binding points, is also supported by compounds having chiral centers, the optical isomers of which, as a rule, have very different aroma notes. For example, l(−)-menthol (XXXVI) has a sweet peppermint odor, cooling and refreshing, while its d(+)-form (XXXVII) has in addition to the minty odor disagreeable notes such as phenolic, medicated, camphor or musty. With carvone, the l-form (XXXVIII) is also peppermint-like, while the d-form (XXXIX) has the odor of caraway.

In conclusion, the following can be said for the role of functional groups: in small molecules, they directly influence the odor quality while, in large molecules, they influence the dipole moment orientation and, thus, indirectly, the odor quality. In large molecules, the most important influence is derived from the molecular geometry. It appears that both parameters, geometry and functional groups, should not be separated in any consideration.

In some cases, the chemical *structure-odor activity relationship* can be expressed by simple rules. An example is decaline (XL), which has an amber tree-like odor.

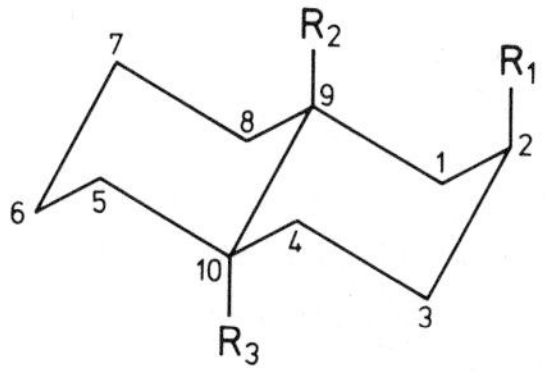

R_2 R_1 7 9 8 1 2 6 5 4 10 3 R_3 XL (5.49)

The structural requirement for the amber-tree-like odor is: trans-decaline must be axially substituted (H is counted as a substituent) in positions 2, 9 and 10 and one of the substituents must be oxygen. This is the rule established by *Ohloff* and is supported by many examples.

In conclusion, a few words will be said about individual *odor modalities*. The number of primary odor qualities is disputed. Some examples are reproduced in Table 5.28. Each primary quality or modality is illustrated by a compound(s) which well represents the specified odor quality.

Musk-like odor is a typical primary odor quality or modality. Musk is the penetrating odorous substance secreted by male deer. The main constituent of musk is a cyclic ketone with a 15-membered ring, called muscone. The chemical structure of musk-like macro rings in relation to odor activity has been thoroughly investigated. Information is also available for some other modalities. However, systematic research covering the chemical structure-odor quality relationship is almost nonexistent for other modalities.

Macro rings XLI a–XLI c, with n = 15 or 16, have good natural musk-like odor characteristics. For n > 16, the animal-like perfume note increases, while for ketones with n = 9–12 rings, the odor is like camphor and, for n = 13, like

Table 5.28. Some chemicals representing particular odour qualities

Almond-like:	Benzaldehyde; Nitrobenzene.
Animal:	Civet; Amber; Musk; Phenylacetic acid.
Aromatic:	Benzaldehyde; Cinnamyl Cinnamate.
Blood-like, raw meat:	*Absolute Musk B* (a proprietary product).
Burnt:	Methyl Benzoate; Guaiacol; Birch Tar Oil; Plexanethiol.
Camphor-like:	Camphor; Borneol.
Cool, cooling:	Camphor; (−)-Menthol; Methyl Salicylate.
Disinfectant:	o-Cresol; Phenol.
Earthy:	"Würzel Körper" (H + R) Phenyl acetaldehyde-dimethyl acetal.
Ethereal/anaesthetic:	Ethyl amyl ketone.
Ethereal:	Ethyl acetate; Di-ethyl ether.
Faecal:	Civet; Skatole; Valeric acid.
Fishy:	Trimethylamine; Hexylamine; certain Thiols (or Mercaptans).
Floral:	Hydroxycitronellal; Indole; Aurantiol; *Lilial* (Givaudan); Phenyl-ethyl alcohol.
Fragrant:	Chanel No. 5; Perfumes – various; β-Ionone.
Fruity (citrus):	Oils of orange, lemon and mandarin; Citral; Oil of Bergamot.
Fruity (other):	So-called Aldehydes C_{14}, C_{16}, and C_{18}; (+)- and (−)-Decalactone; Benzyl acetate; Hexyl butyrate; Amyl acetate; β-Ionone.
Garlic, onion:	Dimethylsulfide; Propylpropenyldisulfide; Diethylsulfide; Allylsulfide.
Green/freshly cut grass:	Phenylacetaldehyde; p-Tolylacetaldehyde; Hexenal; Methylheptine carbonate; cis-Hexyl acetate.
Heavy:	Oakmoss; Coumarin; Benzoin.
Herbal:	Carvacryl methyl ether.
Light:	Ethyl acetate; Amyl acetate; Linalyl acetate; Ylang.
Metallic:	Undecylene alcohol; n-Nonyl acetate; Allyl octoate; Amylvinylketone.
Minty, peppermint:	Menthol; Menthone; (Safrole?).
Mothball-like:	Naphthalene; Indole; *Yara-Yara*.
Musk-like:	Exaltolide; Musk tincture; Musk ambrette.
Oily fatty:	Alcohols C_9, C_{10}, C_{11}, C_{12}; 2-trans-4-trans-Decadienal.
Paint-like:	Turpentine; White Spirit.
Powdery:	Coumarin; Musk ketone; Musk ambrette.
Petrol-like:	Benzene; Ethylbenzene; Cyclohexanol.
Rancid:	Valeric acid; Butyric acid; γ-Butyrolactone.
Resinous:	Olibanum; Myrrh; Oppoponax.
Rose:	Geraniol and β-Phenylethyl alcohol 1 : 1 mixture.
Sharp, pungent:	Ammonia; Formic acid.
Soapy:	Alcohol C_{12}; Stearic acid; Octanol; Lauryl alcohol; Sodium stearate.
Sour, acid:	Citral; Citronellyl acetate; Acetic acid; Propionic acid.
Spicy:	Eugenol; Pepper; Nutmeg; Safrole; Ginger; Clove.
Sulfurous:	Dimethylsulfide; Dimethylpropanethiol; Hydrogen sulfide.
Sweaty:	iso-Amylsalicylate; iso-Valeric acid; Costus oil.
Sweet:	Vanillin; Heliotropin; β-Ionone; Glycerol.
Vanilla-like:	Vanillin; Ethyl vanillin; Balsam of Peru.
Vegetables (cooked):	Dibutylsulfide.
Warm:	Costus; Amber; Patchouli Oil.
Woody:	Cedarwood.

XLI a: ring $(CH_2)_{n-1}$ — X; X = CO, O, S, NH, NMe

XLI b: ring $(CH_2)_{n-2}$ — X—Y; X−Y = O—CO

XLI c: ring $(CH_2)_{n-3}$ — X—Y—Z; X−Y−Z = O—CO—O, CO—O—CO, $O—CH_2—O$ (5.50)

cedar. The nature of the atoms of the ring can be varied without producing any essential alteration in the odor. In the lower-membered rings, however, the odor quality depends strongly on the nature of the polar group present.

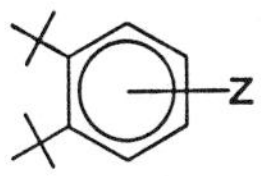

XLII a 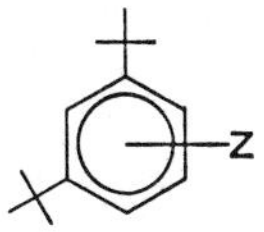 XLII b

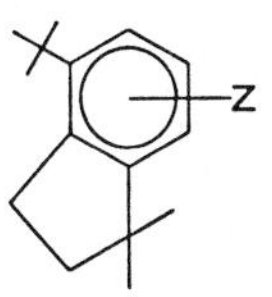

XLII c 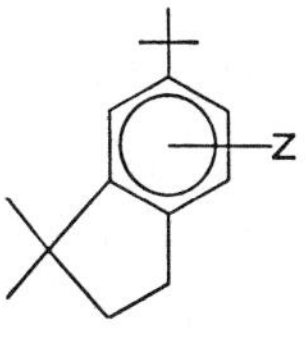 XLII d

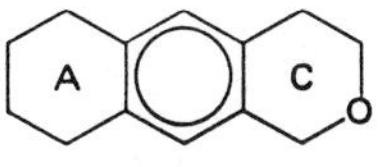

XLIII 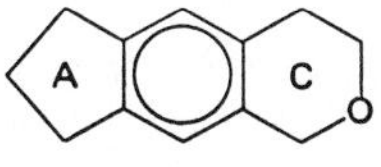 XLIV (5.51)

Aromatic ketones XLII a–XLII d (z = R – CO), as well as isochromans XLIII and XLIV (with A and C rings alkylated) and aromatic nitro compounds XLV and XLVI, likewise can have a musk-like odor.

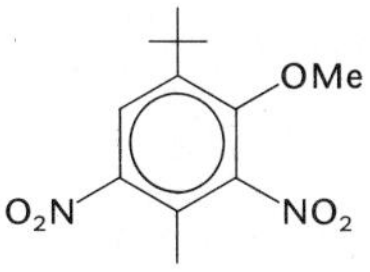

XLV 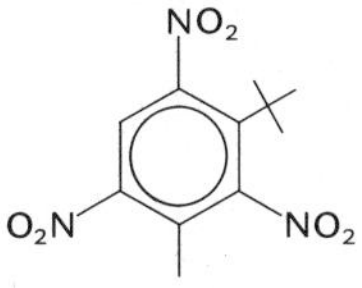 XLVI (5.52)

Camphor (I) acts as though it has a quasi-spherical molecular shape and so do hydrocarbons and their respective polar derivatives (II–IV; XLVII–L):

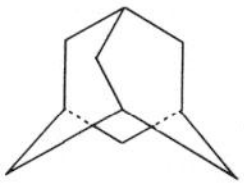

XLVII 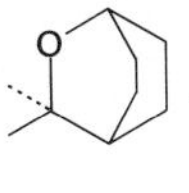 XLVIII 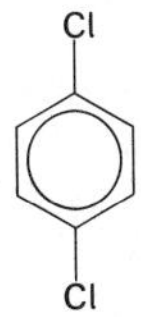 XLIX 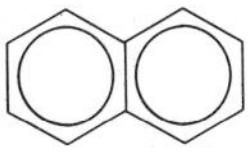 L (5.53)

The above-mentioned HA/B^--system appears to be essential for compounds with a caramel-like, sweet, nutty odor obtained by nonenzymatic browning of sugars present in food. Variations in odor quality are affected by the hydrophobic part of the molecule.

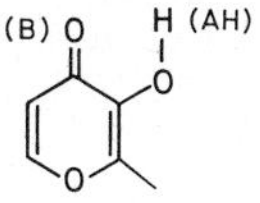

LI 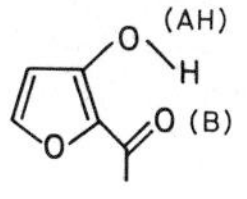 LII

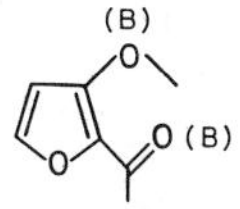

LIII (5.54)

Maltol (LI) and isomaltol (LII) are typical representatives of this class of aroma compounds. Any change of the bifunctional proton donor/proton acceptor unit results in an extensive change in odor quality; for example, O-methylisomaltol (LIII) is completely devoid of the typical maltol character, retaining only a weak, slightly fruity odor.

5.7 Literature

Angrick, M., Rewicki, D.: Die Maillard-Reaktion. Chem. Unserer Zeit *14,* 149 (1980)

Baltes, W.: Die Bedeutung der Maillard-Reaktion für die Aromabildung in Lebensmitteln. Mitteilungsbl. GDCh Fachgruppe Lebensmittelchem. Gerichtl. Chem. *34,* 39 (1980)

Beets, M.G.J.: Structure – activity relationships in human chemoreception. Applied Science Publ.: London. 1978

Beyeler, M., Solms, J.: Interaction of flavour model compounds with soy protein and bovine serum albumin. Lebensm. Wiss. Technol. *7,* 217 (1974)

Boyko, A.L., Morgan, M.E., Libbey, L.M.: Porous polymer trapping for GC/MS analysis of vegetable flavors. In: Analysis of foods and beverages – Headspace techniques (Ed.: Charalambous, G.), p. 57, Academic Press: New York–London. 1978

Charalambous, G., Inglett, G.E. (Eds.): Flavors. International symposium on aroma research held at the central institute of nutrition and food research TNO, Zeist, the Netherlands, May 26–29, 1975. Centre for Agricultural Publishing and Documentation: Wageningen. 1975

Franzen, K.L., Kinsella, J.E.: Physicochemical aspects of food flavouring. Chem. Ind. *1975,* 505

Harper, R.: Some chemicals representing particular odour qualities. Chemical Senses and Flavor *1,* 5 (1974)

Jennings, W.G., Filsoof, M.: Comparison of sample preparation techniques for gas chromatographic analyses. J. Agric. Food Chem. *25*, 440 (1977)

Maga, J.A.: The role of sulfur compounds in food flavor. I. Thiazoles. Crit. Rev. Food Sci. Nutr. *6*, 153 (1975)

Maga, J.A.: The role of sulfur compounds in food flavor. II. Thiophenes. Crit. Rev. Food Sci. Nutr. *6*, 241 (1975)

Maga, J.A.: The role of sulfur compounds in food flavor. III. Thiols. Crit. Rev. Food Sci. Nutr. *7*, 147 (1976)

Maga, J.A.: Lactones in foods. Crit. Rev. Food Sci. Nutr. *8*, 1 (1976)

Maga, J.A.: Simple phenol and phenolic compounds in food flavor. Crit. Rev. Food Sci. Nutr. *10*, 323 (1978)

Maga, J.A.: Furans in foods. Crit. Rev. Food Sci. Nutr. *11*, 355 (1979)

Maga, J.A.: Pyrazines in foods: an update. Crit. Rev. Food Sci. Nutr. *16*, 1 (1982)

Maier, H.H.: Zur Bindung flüchtiger Aromastoffe an Proteine. Dtsch. Lebensm. Rundsch. *70*, 349 (1974)

McNulty, P.B., Karel, M.: Factors affecting flavour release and uptake in O/W-emulsions. J. Food Technol. *8*, 319 (1973)

Murray, K.E., Whitfield, F.B.: The occurrence of 3-alkyl-2-methoxypyrazines in raw vegetables. J. Sci. Food Agric. *26*, 973 (1975)

Nawar, W.W.: Some considerations in interpretation of direct headspace gas chromatographic analysis of food volatiles. Food Technol. *20*, 213 (1966)

Ohloff, G.: Recent developments in the field of naturally-occuring aroma components. Progress in the Chemistry of Organic Natural Products *35*, 431 (1978)

Ohloff, G.: Bifunctional unit concept in flavour chemistry. In: Flavour '81 (Ed.: Schreier, P.), p. 757, Walter de Gruyter: Berlin–New York. 1981

Ohloff, G., Flament, I.: The role of heteroatomic substances in the aroma compounds of foodstuffs. Progress in the Chemistry of Organic Natural Products *36*, 231 (1978)

Parliment, T.H., Scarpellino, R.: Organoleptic techniques in chromatographic food flavor analysis. J. Agric. Food Chem. *25*, 97 (1977)

Piendl, A.: Brauereitechnologie und Molekularbiologie. Brauwissenschaft *22*, 175 (1969)

Reineccius, G.A.: Off-flavors in meat and fish – a review. J. Food Sci. *44*, 12 (1979)

Reymond, D., Solms, J.: Geruch und Geschmack verändernde Stoffe. In: Kosmetika, Riechstoffe und Lebensmittelzusatzstoffe (Eds.: Aebi, H., Baumgartner, E., Fiedler, H.P., Ohloff, G.), S. 193, Georg Thieme Verlag: Stuttgart. 1978

Rothe, M.: Einführung in die Aromaforschung. Akademie-Verlag: Berlin. 1978

Schlegel, H.G.: Allgemeine Mikrobiologie. 4. Aufl., Georg Thieme Verlag: Stuttgart. 1976

Schreier, P.: Aromastoff-Analytik in Lebensmitteln. Dtsch. Lebensm. Rundsch. *74*, 321 (1978)

Schreier, P. (Ed.): Flavor '81. Walter de Gruyter: Berlin–New York. 1981

Schreier, P.: Chromatographic studies of biogenesis of plant volatiles. Dr. Alfred Hüthig Verlag: Heidelberg. 1984

Schutte, L.: Precursors of sulfur-containing flavor compounds. Crit. Rev. Food Technol. *4*, 457 (1973)

Shankaranarayana, M.L., Raghavan, B., Abraham, K.O., Natarajan, C.P.: Volatile sulfur compounds in food flavors. Crit. Rev. Food Technol. *4*, 395 (1973)

Solms, J.: Aromastoffe als Liganden. In: Geruch- und Geschmackstoffe (Ed.: Drawert, F.), S. 201, Verlag Hans Carl: Nürnberg. 1975

Teranishi, R., Flath, R.A., Sugisawa, H. (Eds.): Flavor Research, Marcel Dekker: New York. 1981

Ter Heide, R., de Valois, P.J., Visser, J., Jaegers, P.P., Timmer, R.: Concentration and identification of trace constituents in alcoholic beverages. In: Analysis of foods and beverages – Headspace techniques (Ed.: Charalambous, G.), p. 249, Academic Press: New York–San Francisco–London. 1978

Tressl, R.: Probleme bei der Beurteilung von natürlichen Aromen. Mitteilungsbl. GDCh Fachgruppe Lebensmittelchem. Gerichtl. Chem. *34*, 47 (1980)

Tressl, R., Renner, R.: Bildung von Aromastoffen in Lebensmitteln. Dtsch. Lebensm. Rundsch. *72*, 37 (1976)

Van Straten, S. (Ed.): Volatile compounds in food. 4th edn., with supplements 1 and 2, Central Institute for Nutrition and Food Research TNO: Zeist, the Netherlands. 1977/1978

Vernin, G. (Ed.): Chemistry of heterocyclic compounds in flavours and aromas. Ellis Horwood Publ.: Chichester. 1982

Von Sydow, E.: Contribution for individual chemical compounds to the aroma of foods. In: Geruch- und Geschmackstoffe (Ed.: Drawert, F.), S. 187, Verlag Hans Carl: Nürnberg. 1975

Weurman, C.: Isolation and concentration of volatiles in food odor research. J. Agric. Food Chem. *17*, 370 (1969)

Whitfield, F.B., Tindale, C.R.: The role of microbial metabolites in food off-flavours. Food Technol. Australia *36*, 204 (1984)

Williams, P.J., Strauss, C.R., Wilson, B., Massy-Westropp, R.A.: Novel monoterpene disaccharide glycosides of *Vitis vinifera* grapes and wines. Phytochemistry *21*, 2013 (1982)

Williams, P.J., Strauss, C.R., Wilson, B., Massy-Westropp, R.A.: Use of C_{18}-reversed phase liquid chromatography for the isolation of monoterpene glycosides and nor-isoprenoid precursors from grape juice and wines. J. Chromatogr. *235*, 471 (1982)

Ziegler, E.: Die natürlichen und künstlichen Aromen. Dr. Alfred Hüthig Verlag: Heidelberg. 1982

6 Vitamins

6.1 Foreword

Vitamins are minor but essential constituents of food. They are required for the normal growth, maintenance and functioning of the human body. Hence, their preservation during storage and processing of food is of far-reaching importance. Data are provided in Tables 6.1 and 6.2 to illustrate vitamin losses in some preservation methods for fruits and vegetables. Vitamin losses can occur through chemical reactions which lead to inactive products, or by extraction or leaching, as in the case with water-soluble vitamins during blanching and cooking.

The vitamin requirement of the body is usually adequately supplied by a balanced diet. A deficiency can result in hypovitaminosis and, if more severe, in avitaminosis. Both can occur not only as a consequence of insufficient supply of vitamins by food intake, but can be caused by disturbances in resorption, by stress and by disease.

An assessment of the extent of vitamin supply can be made by determination of vitamin content in blood plasma, or by measuring a biological activity which is dependent on the presence of a vitamin, as are many enzyme activities.

Vitamins are usually divided into two general classes: the fat-soluble vitamins, such as A, D, E and K_1, and the water-soluble vitamins, B_1, B_2, B_6, nicotinamide, pantothenic acid, biotin, folic acid, B_{12} and C.

Data about the average human requirement of some vitamins are presented by age group in Table 6.3.

6.2 Fat-Soluble Vitamins

6.2.1 Retinol (Vitamin A)

6.2.1.1 Biological Role

Retinol (I, in Formula 6.1) is of importance in protein metabolism of cells which develop from the ectoderm (such as skin or mucous-coated linings of the respiratory or digestive systems). Lack of retinol in some way negatively affects epithelial tissue (thickening of skin, hyperkeratosis) and also causes night blindness.

Furthermore, retinol, in the form of 11-cis-retinal (II), is the chromophore component of the visual cycle chromoproteins in three types of cone cells, blue, green and red (λ_{max} 435, 540 and 565 nm, respectively) and of rods of the retina.

Table 6.1. Vitamin losses (%) through processing/canning of vegetables

Processed/ canned product	Samples of vegetable analyzed	Vitamin losses as % of freshly cooked and drained product				
		A	B_1	B_2	Niacin	C
Frozen products (cooked and drained)	10[a]	12[c] 0–50[d]	20 0–61	24 0–45	24 0–56	26 0–78
Sterilized products (drained)	7[b]	10 0–32	67 56–83	42 14–50	49 31–65	51 28–67

[a] Asparagus, lima beans, green beans, broccoli, cauliflower, green peas, potatoes, spinach, brussels sprouts, and baby corn-cobs.
[b] As under a) with the exception of broccoli and brussels sprouts; the values for potato include the cooking water.
[c] Average values.
[d] Variation range.

Table 6.2. Vitamin losses (%) through processing/canning of fruits

Processed/canned product	Fruit samples analyzed	Vitamin losses as % of fresh product				
		A	B_1	B_2	Niacin	C
Frozen products (not thawed)	8[a]	37[c]	29	17	16	18
		0–78[d]	0–66	0–67	0–33	0–50
Sterilized products (including the cooking water)	8[b]	39	47	57	42	56
		0–68	22–67	33–83	25–60	11–86

[a] Apples, apricots, bilberries, sour cherries, orange juice concentrate (calculated for diluted juice samples), peaches, raspberries and strawberries.
[b] As under a) except orange juice and not its concentrate was analyzed.
[c] Average values.
[d] Variation range.

Table 6.3. Vitamins: daily dietary requirements

Age group (years):		<1	1–4	4–10	10–18	>18
Vitamin	Unit					
A	IE[a]	1,500	2,000–2,500	2,500–3,000	4,500–5,000	5,000–6,000
D	IE[b]	400	400	400	400	400
E	IE[c]	5	10	10	15	15
C	mg	35	40	40	45	45–80
B_1	mg	0.3–0.5	0.6–0.8	0.9–1.2	1.5	1.5
B_2	mg	0.4–0.6	0.8	0.9–1.2	1.3–1.8	1.2–1.8
B_6	mg	0.3–0.5	0.7–0.9	0.9–1.2	1.6–2.0	2.0–2.5
Nicotinamide	mg	5–8	9–12	12–16	14–20	12–20
Folic acid	mg	0.05	0.1–0.2	0.2–0.3	0.4	0.4–0.8
B_{12}	µg	0.3–0.4	1–1.5	1.5–2.0	2.0–3.0	3.0–4.0

[a] Requirement covers a supply in forms of vitamin A (75%) and carotenes (25%), 1 international unit ≙ 0.3 µg vitamin A, 1.8 µg β-carotene or 3.6 µg of other carotenoid with a vitamin A activity.
[b] 1 IE ≙ 0.025 µg Vitamin D_3.
[c] 1 IE ≙ 1 mg D,L-α-Tocopherol acetate.

The chromoproteins (rhodopsins) are formed in the dark from the corresponding proteins (opsins) and 11-cis-retinal, while in the light the chromoproteins dissociate into the more stable all-trans-retinal and protein. This conformational change triggers a nerve impulse in the adjacent nerve cell. The all-trans-retinal is then converted to all-trans-retinol and through an inter-

I CH_2OH

II CHO

(6.1)

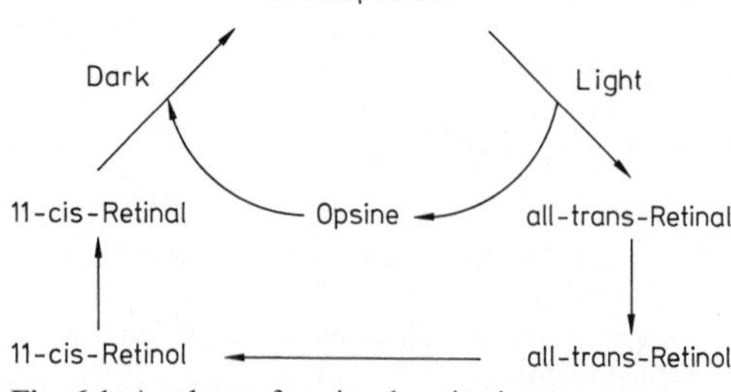

Fig. 6.1. A scheme for visual excitation process

mediary, 11-cis-retinol, is transformed back into 11-cis-retinal (see Figure 6.1 for the visual cycle reactions).

6.2.1.2 Requirement, Occurrence

The daily adult requirement of vitamin A is 1.5–1.8 mg. Approx. 75% is provided by retinol intake (as fatty acid esters, primarily retinylpalmitate), while the remaining 25% is through β-carotene and other provitamin-active carotenoids. Due to the limited extent of carotenoid cleavage, at least 6 g of β-carotene are required to yield 1 g retinol.

Vitamin A resorption and its storage in the liver occur essentially in the form of fatty acid esters. Its content in liver is 250 μg/g fresh tissue, i.e. a total of about 240–540 mg is stored. The liver supplies the blood with free retinol, which then binds to proteins in blood. Vitamin A concentration is 45–84 μg/100 ml plasma in adults; values below 15–24 μg/100 ml indicate a deficiency.

A hypervitaminosis is known, but the symptoms disappear if the intake of retinal is decreased.

Vitamin A occurs only in animal tissue; above all in fish liver oil, in livers of mammals, in milk fat and in egg yolk. Plants are devoid of vitamin A but do contain carotenoids which yield vitamin A by cleavage of the centrally-located double bond (provitamins A). Carotenoids are present in almost all vegetables but primarily in green, yellow and leafy vegetables (carrots, spinach, cress, kale, bell peppers, paprika peppers, tomatoes) and in fruit, with outstanding sources being rose hips, pumpkin, apricots, oranges and palm oil, which is often used for yellow coloring. Animal carotenoids are always of plant origin, derived from feed. Table 6.4 gives vitamin A contents of some common foods. These values can vary highly with cultivar, stage of ripeness, etc. An accurate estimate of the vitamin A content of a food must include a detailed analysis of its carotenoids.

6.2.1.3 Stability, Degradation

Food processing and storage can lead to 5–40% destruction of vitamin A and carotenoids. In the absence of oxygen and at higher temperatures, as experienced in cooking or food sterilization, the preferential reactions are isomerization and fragmentation. In the presence of oxygen, oxidative degradation leads to a series of products, some of which are volatile (cf. 3.8.4.4). This oxidation often parallels lipid oxidation (cooxidation process). The rate of oxidation is influenced by oxygen partial pressure, water activity, temperature, etc. Dehydrated foods are particularly sensitive to oxidative degradation.

6.2.2 Calciferol (Vitamin D)

6.2.2.1 Biological Role

Cholecalciferol (vitamin D_3, I) is formed from cholesterol in the skin through photolysis of 7-dehydrocholesterol (provitamin D_3) by ultraviolet light ("sunshine vitamin"; cf. 3.8.2.2.2). Similarly, vitamin D_2 (ergocalciferol, II; cf. Formula 6.2) is formed from ergosterol.

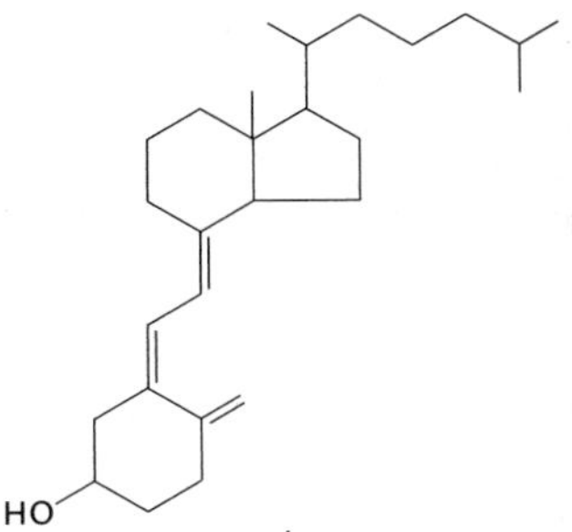

HO I

HO II (6.2)

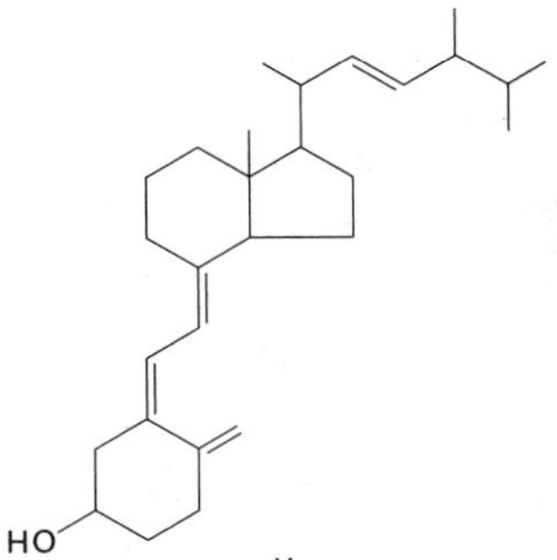

The active metabolites, 25-hydroxycholecalciferol and 1α,25-dihydroxycholecalciferol, are responsible for intestinal resorption of calcium and for calcium salt deposition into the organic matrix of bones by triggering the biosynthesis of the calcium-binding protein.

Vitamin D deficiency can result in an increased excretion of calcium and phosphate and, consequently, impairs bone formation through inadequate calcification of cartilage and bones (child-

Table 6.4. Vitamin content of some food products[a]

Food product	Carotene[b] mg	A mg	D µg	E mg	K mg	B_1 mg	B_2 mg	NAM[c] mg	PAN[d] mg	B_6 mg	BIO[e] µg	FOL[f] µg	B_{12} µg	C mg
Milk and milk products														
Bovine milk, raw	0.018	0.030	0.06	0.09		0.04	0.18	0.09	0.35	0.05	3.5	6.0	0.4	1.7
Humane milk	0.024	0.054	0.05	0.52	0.003	0.02	0.04	0.17	0.21	0.01	0.6	5.0	0.05	4.4
Butter	0.38	0.59	1.3	2.2	0.06	0.005	0.02	0.03	0.05	0.005				0.2
Cheese														
Emmental	0.14	0.32	1.1	0.35		0.05	0.34	0.18	0.40	0.07	3.0	4.0	2.2	0.5
Camembert (60% fat)		0.63				0.04	0.37	1.18	0.7	0.2	2.8			
Camembert (30% fat)	0.1	0.2	0.17	0.30		0.05	0.67	1.2	0.9	0.3	5.0	66	3.1	
Eggs														
Chicken egg yolk		1.12		3.0		0.29	0.40	0.07	3.7	0.3	50	150	2.0	0.3
Chicken egg white						0.02	0.32	0.09	0.14	0.012	7	16	0.1	
Meat and meat products														
Beef, whole carcass, lean						0.08	0.18	4.9		0.5		20	1.3	
Pork, whole carcass, lean						0.66	0.17	3.7		0.4			0.8	
Calf liver		3.92	0.33	1.2	0.15	0.28	2.61	15.0	7.9	0.9	80	240	60	35
Pork liver		5.8		0.45	0.02	0.31	3.2	15.7	6.8	0.6	30	220	40	23
Chicken liver		11.6	1.3	0.4		0.32	2.49	11.6	7.2	0.8		380	20	28
Pork kidneys						0.34	1.8	8.4	3.1	0.6			20	16
Blood sausage	0.02					0.07	0.13	1.2					50	
Fish and fish products														
Herring		0.04	30	1.5		0.04	0.22	3.8	0.9	0.5	4.5	5	8.5	0.5
Eel		0.98	13	8		0.18	0.32	2.6	0.3			13	1	1.8
Cod-liver oil		30	330	3.26										
Cereals and cereal products														
Wheat, whole kernel	0.02			3.2		0.48	0.14	5.1	1.2	0.4	6	49		
Wheat flour, type 405				2.3		0.06	0.03	0.7	0.2	0.2	1.5	10		
Wheat flour, type 550				2.0		0.11	0.08	0.5	0.4	0.1	1.1	20		
Wheat germ				27.6	0.35	2.01	0.72	4.5	1.0	3.3	17	520		
Wheat gluten				9.1	0.08	0.65	0.51	17.7	2.5	2.5	44	400		
Rye whole kernel				3.8		0.35	0.17	1.8	1.5	0.3	4.6	42		
Rye flour, type 997				3.4		0.19	0.11	0.8						
Corn whole kernel	0.37			5.8	0.04	0.36	0.20	1.5	0.7	0.4	6	26		

Corn (breakfast cereal, corn flakes)			0.43				1.4	0.2	0.07		6		
Oat flakes			3.7		0.59	0.15	1.0	1.1	0.16	20	24		
Rice, unpolished			4.5		0.41	0.09	5.2	1.7	0.68	12	16		
Rice, polished			0.4		0.06	0.03	1.3	0.6	0.15		29		
Vegetables													
Watercress	2.68				0.09	0.17	0.7						51
Mushrooms (cultivated)	0.01	1.94	0.08	0.02	0.10	0.44	5.2	2.1	0.07	20	30		4.9
Chicory	1.29				0.05	0.03	0.24		0.05		50		10.2
Endive	1.14				0.05	0.12	0.4				50		9.4
Lamb's lettuce	3.9				0.07	0.08	0.4						35
Kale	4.1				0.1	0.25	2.1		0.3	0.5	60		105
Potatoes	0.01		0.09		0.11	0.05	1.2	0.4	0.2	0.4	7		17
Kohlrabi	0.2				0.05	0.05	1.8	0.1	0.1				63.3
Head lettuce	0.8		0.4	0.2	0.06	0.08	0.3	0.1	0.06	1.9	40		13
Lentils, dried	0.1		1.3		0.43	0.26	2.2	1.4	0.6		40		
Carrots	12		0.7	0.08	0.07	0.05	0.6	0.3	0.1	5	8		7.1
Brussels sprouts	0.4		1.1	0.6	0.11	0.14	0.7		0.3	0.4	80		114
Spinach	4.2		2.5	0.4	0.11	0.23	0.6	0.3	0.22	6.9	80		52
Edible mushroom *(Boletus edulis)*		3.1	0.63		0.03	0.37	4.9	2.7					2.5
Tomatoes	0.82		0.49	0.63	0.06	0.04	0.5	0.3	0.1	4	40		24.2
White cabbage	0.04		0.02		0.05	0.04	0.3	0.3	0.1		80		45.8
Fruits													
Orange	0.09		0.24		0.08	0.04	0.3	0.2	0.05	2.3	20		50
Apricot	1.8		0.5		0.04	0.05	0.8	0.3	0.1		4		9.4
Strawberry	0.05		0.22	0.02	0.03	0.05	0.5	0.3	0.06	4	20		64
Grapefruit	0.02		0.27		0.05	0.02	0.24	0.25	0.03	0.4	10		44
Rose hips													1,250
Currants-red	0.04		0.21		0.04	0.03	0.23	0.06	0.05	2.6			36
Currants-black	0.14		1.0		0.05	0.04	0.28	0.4	0.08	2.4			177
Cherries-sour	0.3				0.05	0.06	0.4						12
Plums	0.2		0.8		0.07	0.01	0.4	0.2	0.05	0.1	2		5.4
Sea buckthorn	1.5				0.03	0.21	0.3	0.2	0.11	3.3	10		450
Yeast													
Baker's yeast, pressed					1.43	2.31	17.4	3.5	0.8	30	1,020		
Brewer's yeast, dried					12.0	3.8	44.8	7.2	4.4	20	3,200	20	

[a] Values are given in mg or µg per 100 g of edible portion. [b] Total carotenoids with vitamin A activity. [c] Nicotinamide. [d] Pantothenic acid. [e] Biotin. [f] Folic acid.

hood rickets). Vitamin D deficiency in adults leads to osteomalacia, a softening and weakening of bones. Hypercalcemia is a result of excessive intake of vitamin D, causing calcium carbonate and calcium phosphate deposition disorders involving various organs.

6.2.2.2 Requirement, Occurrence

The daily requirement is 10 µg. Indicators of deficiency are the concentration of the metabolite 25-hydroxycholecalciferol in plasma and the activity of alkaline serum phosphatase, which increases during vitamin deficiency.
Most natural foods have a low content of vitamin D_3. Fish liver oil is an exceptional source of vitamin D_2. The D-provitamins, ergosterol and 7-dehydrocholesterol, are widely distributed in the animal and plant kingdoms. Yeast, some mushrooms, cabbage, spinach and wheat germ oil are particularly abundant in provitamin D_2. Vitamin D_3 and its provitamin are present in egg yolk, butter, cow's milk, beef and pork liver, mollusks, animal fat and pork skin. However, the most important vitamin D source is fish oil, primarily liver oil. The vitamin D requirement of humans is best supplied by 7-dehydrocholesterol. Table 6.4 gives data on vitamin D occurrence in some foods. However, these values can vary widely, as shown by variations in dairy cattle milk (summer or winter), feed or frequency of pasture grazing and exposure to the ultraviolet rays of sunlight.

6.2.2.3 Stability, Degradation

Vitamin D is sensitive to oxygen and light. Its stability in food is not a problem, because adults usually obtain a sufficient supply of this vitamin.

6.2.3 α-Tocopherol (Vitamin E)

6.2.3.1 Biological Role

The various tocopherols differ in number and position of the methyl group on the ring. α-Tocopherol (Formula 6.3; Table 6.5) has the highest biological activity. Its activity is as an antioxidant, preventing or retarding lipid oxidation (cf. 3.7.3.1) and stabilizing some other biologically-active compounds (e. g. vitamin A, ubiquinone, hormones and enzymes) against oxidation. Its mechanism of action is not fully elucidated.

6.2.3.2 Requirement, Occurrence

The daily requirement is 15 mg α-tocopherol. The intake increases in diets with a higher content of unsaturated fatty acids. A normal supply results in a concentration of 0.7–1.6 mg/100 ml in blood plasma. A level less than 0.4 mg/100 ml is considered a deficiency.
Table 6.4 provides data for tocopherol content in some foods. The main sources are vegetable oils, particularly germ oils of cereals.

6.2.3.3 Stability, Degradation

Losses occur in vegetable oil processing into margarine and shortening. Losses are also encountered in intensive lipid autoxidation, particularly in dehydrated or deep fried foods (Table 6.6).

Table 6.5. Biological activity of tocopherols

Tocopherol	Activity (IE/mg)
d,l-α-Tocopherol acetate	1.0
d,l-α-Tocopherol	1.1
d-α-Tocopherol acetate	1.4
d-α-Tocopherol	1.5
l-α-Tocopherol	0.5
d,l-β-Tocopherol	0.3
d,l-γ-Tocopherol	0.15
d,l-δ-Tocopherol	0.01

Table 6.6. Tocopherol stability during deep frying

	Tocopherol total (mg/100 g)	Loss (%)
Oil before deep frying	82	
after deep frying	73	11
Oil extracted from potato chips		
immediately after frying	75	
after 2 weeks storage		
at room temperature	39	48
after 1 month storage		
at room temperature	22	71
after 2 months storage		
at room temperature	17	77
after 1 month kept at −12 °C	28	63
after 2 months kept at −12 °C	24	68
Oil extracted from French fries		
immediately after frying	78	
after 1 month kept at −12 °C	25	68
after 2 months kept at −12 °C	20	74

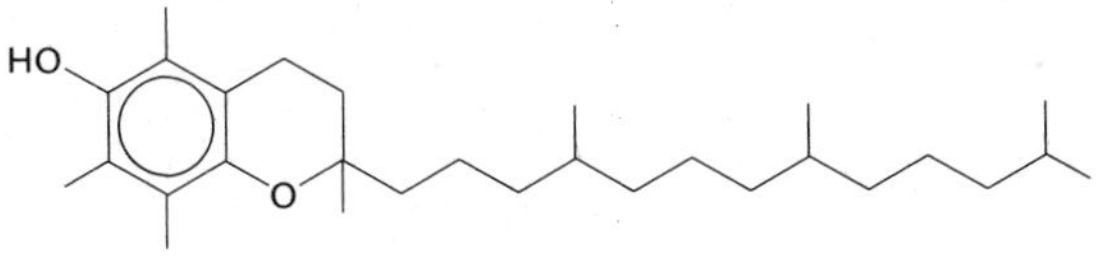

(6.3)

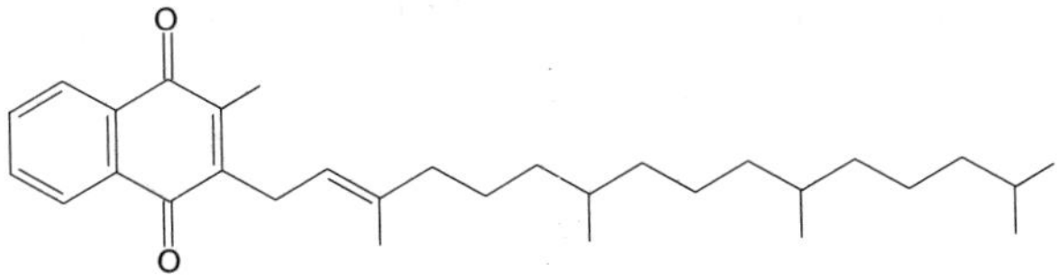

(6.4)

6.2.4 Phytomenadione (Vitamin K_1)

6.2.4.1 Biological Role

The K-group vitamins are naphthoquinone derivatives which differ in their side chains. The structure of vitamin K_1 is shown in Formula 6.4. Vitamin K_1 is involved in biosynthesis of some blood-clotting factors (prothrombin, proconvertin, Christmas or Stuart factor). Deficiency of this vitamin causes a reduced activity of prothrombin and results in hemorrhage.

6.2.4.2 Requirement, Occurrence

The daily adult requirement of vitamin K_1 is estimated to be 1–4 mg. It is supplied by food, most abundantly by green leafy vegetables, and by synthesis within the body by intestinal bacteria. Their synthesizing efficiency, 1–1.5 mg/day, is impressive as it probably can meet the daily requirement.

Vitamin K_1 occurs primarily in green leafy vegetables (spinach, cabbage, cauliflower), but liver (veal or pork) is also an excellent source (Table 6.4).

6.2.4.3 Stability, Degradation

Little is known about the metabolism of vitamin K_1 and the effects of food processing on it.

6.3 Water-Soluble Vitamins

6.3.1 Thiamine (Vitamin B_1)

6.3.1.1 Biological Role

Thiamine, in the form of its pyrophosphate,

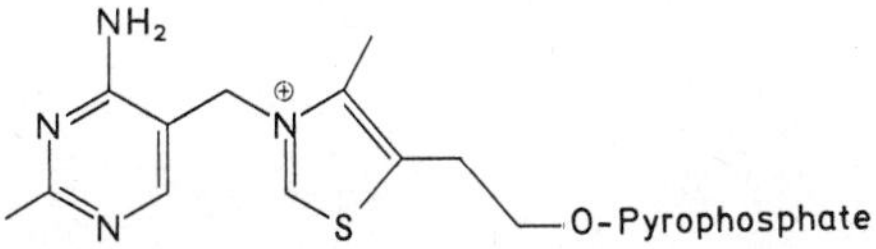

(6.5)

is the coenzyme of several important enzymes, such as pyruvate dehydrogenase, transketolase, phosphoketolase and α-ketoglutarate dehydrogenase, in reactions involving the transfer of an activated aldehyde unit (D: donor; A: acceptor):

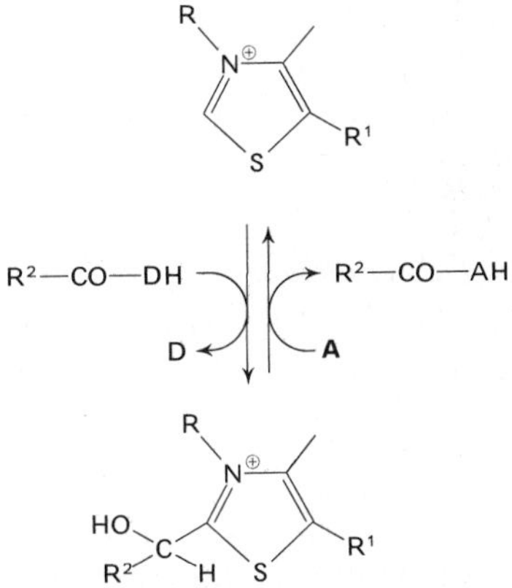

R^2—CO—DH → R^2—CO—AH

(6.6)

Vitamin B_1 deficiency is shown by a decrease in activity of the enzymes mentioned above. The disease known as beri-beri, which has neurological and cardiac symptoms, results from a severe dietary deficiency of thiamine.

6.3.1.2 Requirement, Occurrence

The daily adult requirement is 1–2 mg. Since thiamine is a key substance in carbohydrate metabolism, the requirement increases in a carbohydrate-enriched diet. The assay of transketolase activity of red blood cells or the extent of transketolase reactivation on addition of thiamine pyrophosphate can be used as indicators for sufficient vitamin intake in the diet.

Vitamin B_1 is found in many plants. It is present in the pericarp and germ of cereals, in yeast, vegetables (potatoes) and shelled fruit. It is abundant in pork, beef, fish, eggs and in animal organs such as liver, kidney, brain and heart. Human

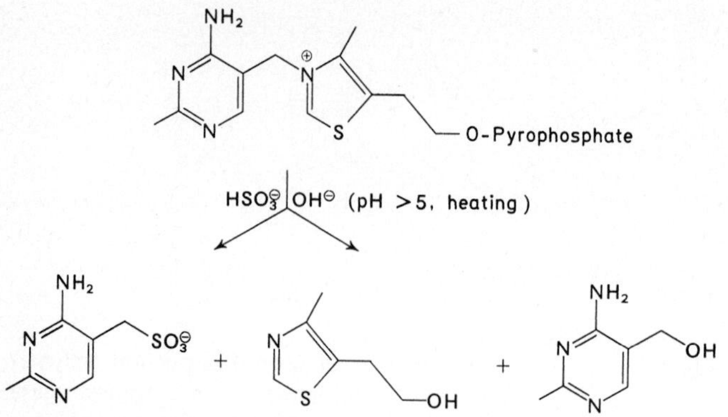

(6.7)

milk and cow's milk contain vitamin B_1. Wholegrain bread and potatoes are important dietary sources. Since vitamin B_1 is localized in the outer part of cereal grain hulls, flour milling with a low extraction grade or rice polishing remove most of the vitamin in the bran. Table 6.4 lists data on the occurrence of thiamine.

6.3.1.3 Stability, Degradation

Thiamine stability in aqueous solution is relatively low. It is influenced by pH (Figure 6.2), temperature (Table 6.7), ionic strength and metal ions. The enzyme-bound form is less stable than free thiamine (Figure 6.2). Strong nucleophilic reagents, such as HSO_3^- or OH^- cause rapid decomposition by forming 5-(2-hydroxyethyl)-4-methylthiazole and 2-methyl-4-amino-5(methylsulfonic acid)-pyrimidine, or 2-methyl-4-amino-5-hydroxymethylpyrimidine (see Reactions 6.7).

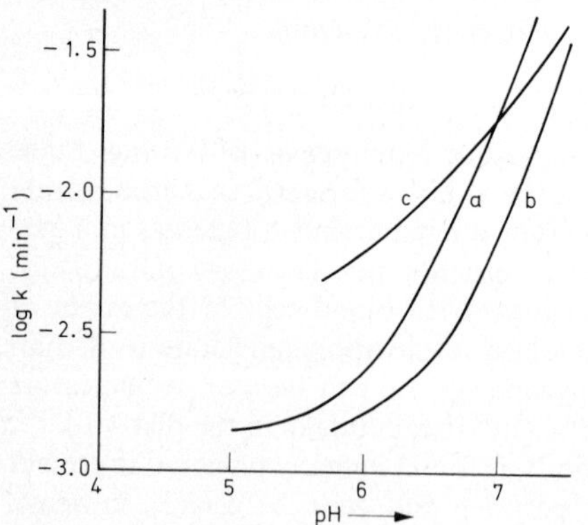

Fig. 6.2. Inactivation rate of thiamine as affected by pH **a** Thiamine in phosphate buffer, **b** thiamine in wheat or oat flours, **c** thiamine pyrophosphate in flour

Table 6.7. Thiamine losses in food during storage (12 months)

Food	Thiamine loss, %	
	1.5 °C	38 °C
Apricots	28	65
Orange juice	0	22
Peas	0	32
Green beans	24	92
Tomato juice	0	40

Thermal degradation of thiamine, which also initially yields the thiazole and pyrimidine derivatives mentioned above, is involved in the formation of meat-like aroma in cooked food. The degradation products of the thiazole, namely furan, thiophene and dihydrothiophene derivatives and free H_2S, are responsible for this aroma.

Thiamine is inactivated by nitrites, probably through reaction with the amino group attached to the pyrimidine ring.

Strong oxidants, such as H_2O_2 or potassium ferricyanide, yield the fluorescent thiochrome. This reaction is often used in chemical determination of thiamine content in food (see Reaction 6.8).

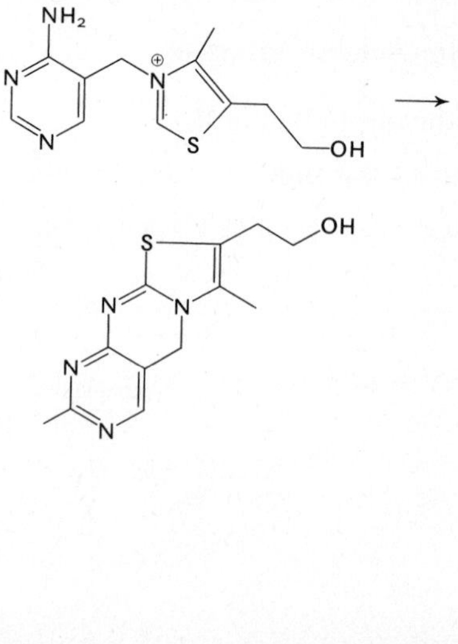

(6.8)

The following losses of thiamine can be expected: 15–25% in canned fruit or vegetables stored for more than a year; 0–60% in meat cooked under household conditions, depending on temperature and preparation method; 20% in salt brine pickling of meat and in baking of white bread; 15% in blanching of cabbage without sulfite and 40% with sulfite. Losses caused by sulfite are pH dependent. Practically no thiamine degradation occurs in a stronger acidic medium (e. g. lemon juice).

6.3.2 Riboflavin (Vitamin B_2)

6.3.2.1 Biological Role

Riboflavin (Formula 6.9) is the prosthetic group of flavine enzymes, which are of great importance in general metabolism and particularly in metabolism of protein.

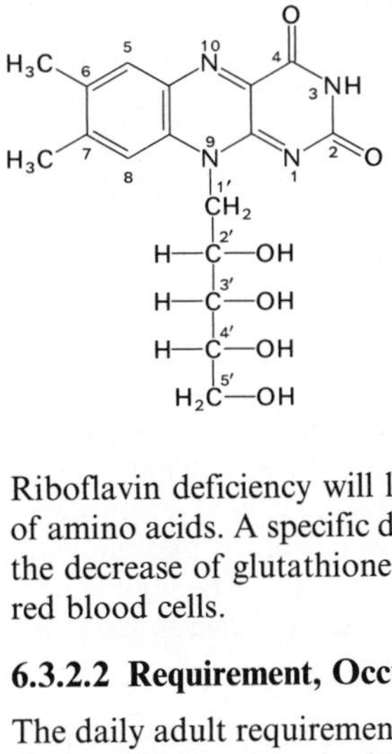

(6.9)

Riboflavin deficiency will lead to accumulation of amino acids. A specific deficiency symptom is the decrease of glutathione reductase activity in red blood cells.

6.3.2.2 Requirement, Occurrence

The daily adult requirement is 1.6–2.6 mg. Deficiency symptoms are rarely observed with a normal diet and, since the riboflavin pool in the body is very stable, even in a deficient diet it is not depleted by more than 30–50%. The riboflavin content of urine is an indicator of riboflavin supply levels. Values above 80 μg riboflavin/g creatinine are normal; 27–79 μg/g is low; and less than 27 μg/g strongly suggests a vitamin-deficient diet. Glutathione reductase activity assay can provide similar information.

The most important sources of riboflavin are milk and milk products, eggs, various vegetables, yeast, meat products, particularly variety meats such as heart, liver and kidney, and fish liver and roe. Table 6.4 provides data about the occurrence of riboflavin in some common foods.

6.3.2.3 Stability, Degradation

Riboflavin is relatively stable in normal food handling processes. Losses range from 10–15%. Exposure to light, especially in the visible spectrum from 420–560 nm, photolytically cleaves ribitol from the vitamin, converting it to lumiflavin:

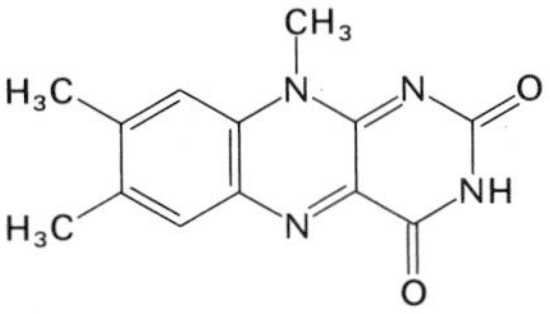

(6.10)

6.3.3 Pyridoxine (Pyridoxal, Vitamin B_6)

6.3.3.1 Biological Role

Vitamin B_6 activity involves pyridoxine or pyridoxol (R = CH_2OH), pyridoxal (R = CHO) and pyridoxamine (R = CH_2NH_2). The metabolically-active form is pyridoxalphosphate, which is a prosthetic group (cf. 2.3.2.3)

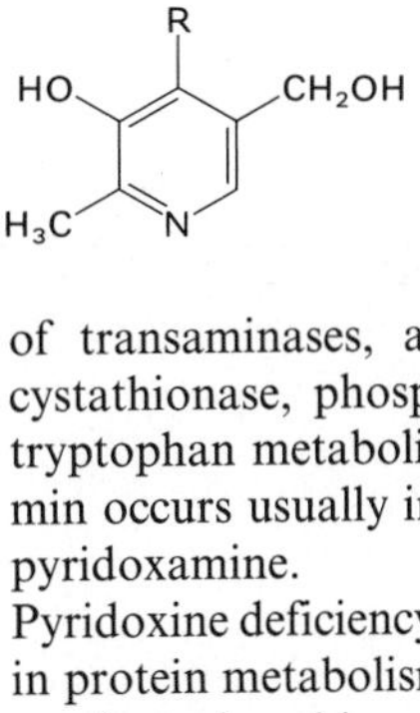

(6.11)

of transaminases, amino acid-decarboxylases, cystathionase, phosphorylases and enzymes of tryptophan metabolism. The intake of the vitamin occurs usually in the forms of pyridoxal or pyridoxamine.

Pyridoxine deficiency in the diet causes disorders in protein metabolism. Hydroxykynurenine and xanthurenic acid accumulate, since the conversion of tryptophan to nicotinic acid, a step regulated by the kynureninase enzyme, is interrupted.

6.3.3.2 Requirement, Occurrence

The daily adult requirement is 2 mg. An indicator for sufficient supply is the activity of glutamate oxalacetate transaminase, an enzyme present in red blood cells. This activity is decreased in vitamin deficiency. The occurrence of pyridoxine in food is outlined in Table 6.4.

6.3.3.3 Stability, Degradation

The most stable form of the vitamin is pyridoxal, and this form is used for vitamin fortification of food. Vitamin B_6 loss is 45% in cooking of meat and 20–30% in cooking of vegetables. During milk sterilization a reaction with cysteine transforms the vitamin into an inactive thiazolidine derivative. This reaction may account for vitamin losses in other heat-treated foods:

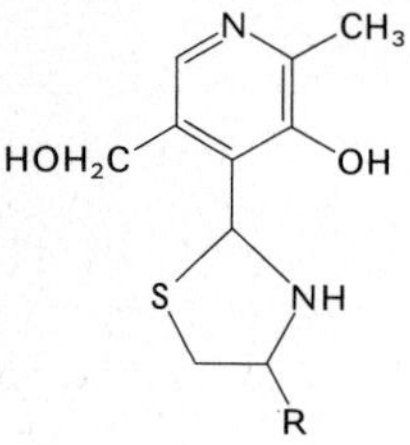

(6.12)

6.3.4 Nicotinamide (Niacin)

6.3.4.1 Biological Role

Nicotinic acid amide (I), in the form of nicotinamide adenine dinucleotide (NAD^+), or its phosphorylated form ($NADP^+$), is a coenzyme of dehydrogenases. Its excretion in urine is essentially in the form of N^1-methylnicotinamide (trigonelline amide, II), N^1-methyl-2-pyridone-5-carboxamide (III), N^1-methyl-4-pyridone-3-carboxamide (IV) and several further degradation products:

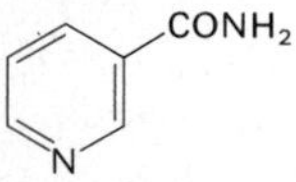

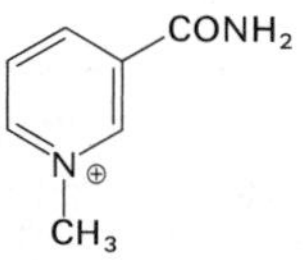

(6.13)

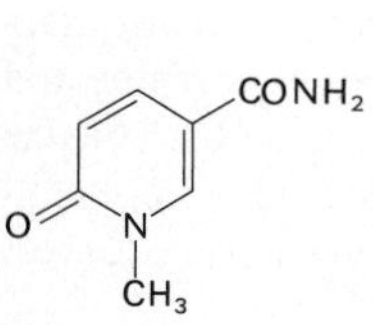

Vitamin deficiency is observed initially by a drop in concentration of NAD^+ and $NADP^+$ in liver and muscle, while levels remain normal in blood, heart and kidney. The classical deficiency disease is pellagra, which affects the skin, digestion and the nervous system (dermatitis, diarrhea and dementia). However, the initial deficiency symptoms are largely nonspecific.

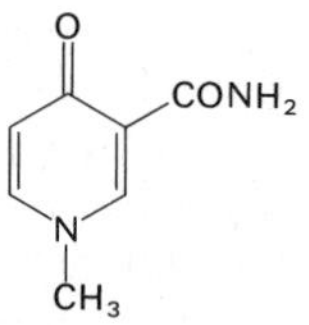

6.3.4.2 Requirement, Occurrence

The daily adult requirement is 12–20 mg, with 60–70% of that covered by tryptophan intake. Hence, milk and eggs, though they contain little niacin, are good foods for prevention of pellagra because they contain tryptophan. It substitutes for niacin in the body, with 60 mg L-tryptophan equalling 1 mg nicotinamide. Indicators for sufficient supply of niacin in the diet are the levels of metabolites, II in urine or III and IV in blood plasma.

The vitamin occurs in food as nicotinic acid, either as its amide or as a coenzyme. Animal organs, such as liver, and lean meat, cereals, yeast and mushrooms are abundant sources of niacin. Table 6.4 provides data on its occurrence in food.

6.3.4.3 Stability, Degradation

Nicotinic acid is quite stable. Moderate losses of up to 15% are observed (cf. Tables 6.1 and 6.2) in blanching of vegetables. The loss is 25–30% in the first days of ripening of meat.

6.3.5 Pantothenic Acid

6.3.5.1 Biological Role

Pantothenic acid is the building unit of coenzyme A (CoA), the main carrier of acetyl and other acyl groups in cell metabolism. Acyl groups are linked to CoA by a thioester bond. Pantothenic acid occurs in free form in blood plasma, while in organs it is present as CoA. The highest concentrations are in liver, adrenal glands, heart and kidney.

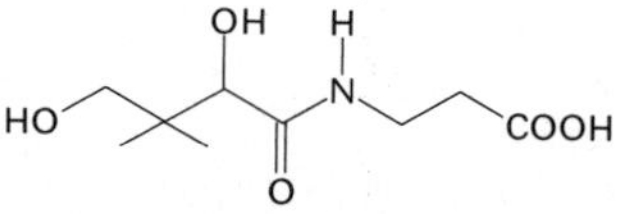

(6.14)

A normal diet provides an adequate supply of the vitamin.

6.3.5.2 Requirement, Occurrence

The daily adult requirement is 6–8 mg. The concentration in blood is 10–40 μg/100 ml and 2–7 mg/day are excreted in urine.

Table 6.4 lists data on pantothenic acid occurrence in food.

6.3.5.3 Stability, Degradation

Pantothenic acid is quite stable. Losses of 10% are experienced in processing of milk. Losses of 10–30%, mostly due to leaching, occur during cooking of vegetables.

6.3.6 Biotin

6.3.6.1 Biological Role

Biotin is the prosthetic group of carboxylating enzymes, such as acetyl-CoA-carboxylase, pyruvate carboxylase and propionyl-CoA-carboxylase, and therefore plays an important role in fatty acid biosynthesis and in gluconeogenesis. The carboxyl group of biotin forms an amide bond with the ε-amino group of a lysine residue of the enzyme protein:

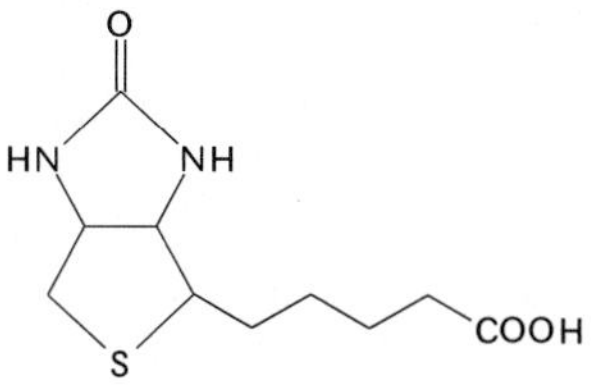

O HN NH S COOH (6.15)

Biotin deficiency rarely occurs. Consumption of large amounts of raw egg white might inactivate biotin by its specific binding to avidin (cf. 11.2.3.1.9).

6.3.6.2 Requirement, Occurrence

The daily adult requirement is 150–300 μg. An indicator for sufficient biotin supply is the excretion level in urine, which is normally 30–50 μg/day. A deficiency is indicated by a drop to 5 μg/day.

Biotin is not free in food, but is bound to proteins. Table 6.4 provides data on its occurrence in food.

6.3.6.3 Stability, Degradation

Biotin is quite stable. Losses during processing and storage of food are 10–15%.

6.3.7 Folic Acid

6.3.7.1 Biological Role

The tetrahydrofolate derivative (II) of folic acid (I) is the cofactor of enzymes which transfer single carbon units in various oxidative states, e.g. formyl or hydroxymethyl residues. In transfer reactions the single carbon unit is attached to the N^5- or N^{10}-atom of tetrahydrofolic acid (cf. Formula 6.16).

Folic acid deficiency can occur by insufficient supply in the diet, malfunction of resorptive processes, or therapy which uses a folic acid antagonist. Deficiency is detected by a decrease in folic acid concentration in red blood cells and plasma, and by a change in blood cell patterns.

6.3.7.2 Requirement, Occurrence

The daily adult requirement is estimated to be 0.4–0.8 mg. A sufficient supply can be monitored by the level of free folic acid in blood serum or red blood cells. Serum values of 5–20 ng/ml are normal, while less than 5 ng/ml is a deficiency level. When folic acid is lacking, there is increased excretion of formiminoglutamic acid, which is formed from histidine, since its conversion to glutamic acid as the final step in histidine degradation is dependent on folic acid.

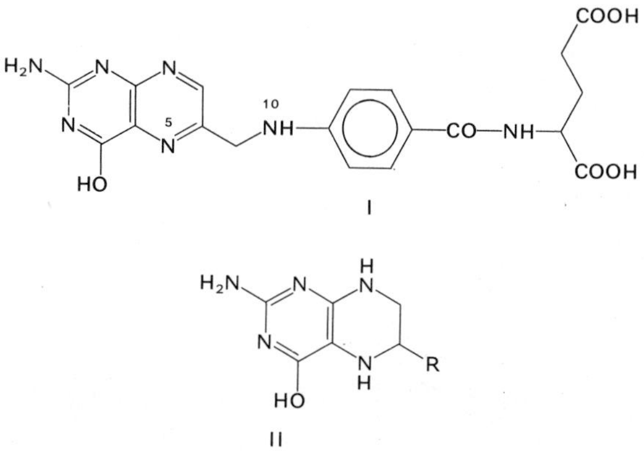

H_2N N N 5 N 10 NH CO—NH COOH COOH HO I

H_2N N H N N H N R HO II

(6.16)

In food folic acid is mainly bound to oligo-γ-L-glutamates made up of 1–8 glutamic acid residues. Unlike free folic acid, the resorption of this conjugated form is limited and occurs only after the glutamic acid moiety is cleaved by the enzyme folic acid conjugase, which occurs in intestinal mucosa. The folic acid content of foods varies. It is mainly present in conjugated form in vegetables, while it is in free form in liver. Data on folic acid occurrence in food are compiled in Table 6.4.

6.3.7.3 Stability, Degradation

Folic acid is quite stable. There is no destruction during blanching of vegetables, while cooking of meat gives only small losses. Losses in milk are apparently due to an oxidative process and parallel those of ascorbic acid. Ascorbate added to food preserves folic acid.

6.3.8 Cyanocobalamin (Vitamin B_{12})

6.3.8.1 Biological Role

The resorption of cyanocobalamin (Formula 6.17) is achieved with the aid of an "intrinsic factor" formed by the stomach mucosa. An exchange in liver of the cyano group for adenine nucleoside transforms the resorpted compound into a coenzyme. For example, 5′-desoxyadenosylcobalamin is the coenzyme for methylmalonyl-CoA-mutase, a key enzyme in the metabolism of propionic acid (cf. 10.2.7.2). Vitamin deficiency causes excretion of methylmalonic acid in urine. The B_{12}-coenzymes, together with folic acid, are involved in transfers of labile methyl groups. Furthermore, the vitamin appears to have a role in protein metabolism efficiency.

Although inadequate diets are occasionally responsible for deficiency of vitamin B_{12}, more often the deficiency is caused by impairment in resorption due to lack of glycoprotein, the "intrinsic factor", in gastric juice. This deficiency then causes pernicious anemia.

6.3.8.2 Requirement, Occurrence

The daily adult requirement of vitamin B_{12} is 3–4 μg. The plasma concentration is normally 450 pg/ml.

The ability of vitamin B_{12} to promote growth alone or together with antibiotics, for example in young chickens, suckling pigs and young hogs, is of particular importance. This effect appears to be due to the influence of the vitamin on protein metabolism, and it is practised in animal feeding. The increase in feed utilization is exceptional with underdeveloped young animals. Vitamin B_{12} is of importance also in poultry operations (enhanced egg laying and chick hatching). The use of vitamin B_{12} in animal feed vitamin fortification is obviously well justified.

Liver, kidney, spleen, thymus glands and muscle tissue are abundant sources of vitamin B_{12}. Con-

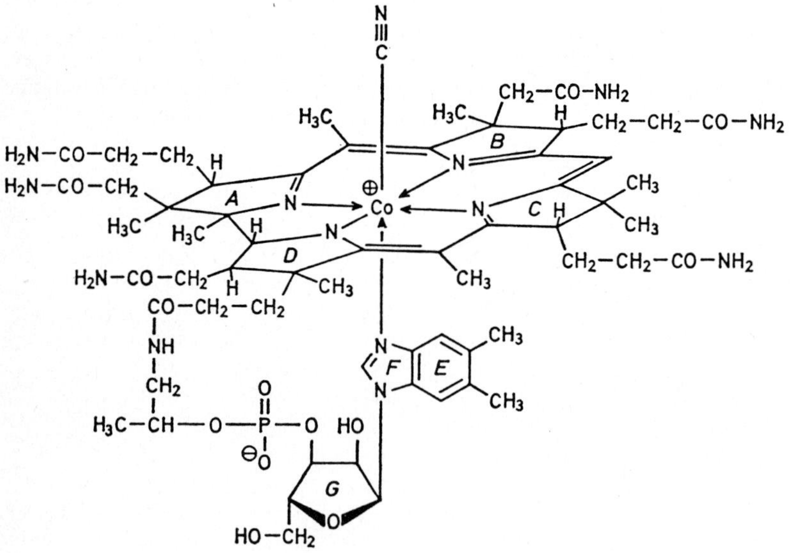

(6.17)

sumption of internal organs (variety meats) of animals is one method of alleviating vitamin B_{12} deficiency symptoms in humans.

6.3.8.3 Stability, Degradation

The stability of vitamin B_{12} is very dependent on a number of conditions. It is fairly stable at pH 4–6, even at high temperatures. In alkaline media or in the presence of reducing agents, such as ascorbic acid or SO_2, the vitamin is destroyed to a greater extent.

6.3.9 L-Ascorbic Acid (Vitamin C)

6.3.9.1 Biological Role

Ascorbic acid is involved in hydroxylation reactions, e.g. biosynthesis of catecholamines, hydroxyproline and corticosteroids (11-β-hydroxylation of deoxycorticosterone and 17-β-hydroxylation of corticosterone). Vitamin C is fully resorbed and distributed throughout the body, with the highest concentration in adrenal and pituitary glands.

About 3% of the body's vitamin C pool, which is 20–50 mg/kg body weight, is excreted in urine as ascorbic acid, dehydroascorbic acid (a combined total of 25%) and their metabolites, 2,3-diketo-L-gulonic acid (20%) and oxalic acid (55%). An increase in excreted oxalic acid occurs only with a very high intake of ascorbic acid.

Scurvy is caused by a dietary deficiency of ascorbic acid.

6.3.9.2 Requirement, Occurrence

The daily adult requirement is about 45–80 mg. An indicator for insufficient vitamin supply in the diet is a low level in blood plasma (0.4 mg/100 ml).

Vitamin C is present in all animal and plant cells, mostly in free form, and it is probably bound to protein as well. Vitamin C is particularly abundant in rose hips, black and red currants, strawberries, parsley, oranges, lemons (in peels even more so than in pulp), grapefruit, a variety of cabbages and potatoes. Vitamin C loss during storage of vegetables from winter through late spring can be as high as 70%.

Table 6.4 provides data on vitamin C occurrence in a variety of foods.

Table 6.8. Effect of pH on ultraviolet absorption maxima of ascorbic acid

pH	λ max (nm)
2	244
6 – 10	266
> 10	294

6.3.9.3 Stability, Degradation

Ascorbic acid (I) has an acidic hydroxyl group (pK_1 = 4.04, pK_2 = 11.4 at 25 °C). Its UV absorption depends on the pH value (Table 6.8). Ascorbic acid is readily and reversibly oxidized to dehydroascorbic acid (II), which is present in aqueous media as a hydrated hemiketal (IV). The biological activity is lost when the dehydroascorbic acid lactone ring is irreversibly opened, giving rise to 2,3-diketogulonic acid (III):

I

Red. Ox.

H_2O

II

III

H_2O

IV

(6.18)

The oxidation of ascorbic acid to dehydroascorbic acid and its further degradation products depends on a number of parameters. Oxygen partial pressure, pH, temperature and the presence of heavy metal ions are of great importance. Metal-catalyzed destruction proceeds at a higher rate than noncatalyzed spontaneous autoxidation. Traces of heavy metal ions, particularly Cu^{2+} and Fe^{3+}, result in high losses.

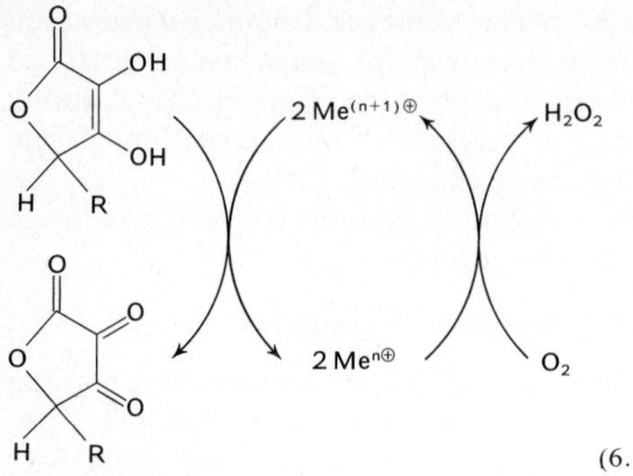

(6.19)

The principle of metal catalysis is schematically presented in Reaction 6.19.

The rate of anaerobic vitamin C degradation, which is substantially lower than that of noncatalyzed oxidation, is maximal at pH 4 and minimal at pH 2. It probably proceeds through the keto-form of ascorbate, then via a keto-anion to diketogulonic acid:

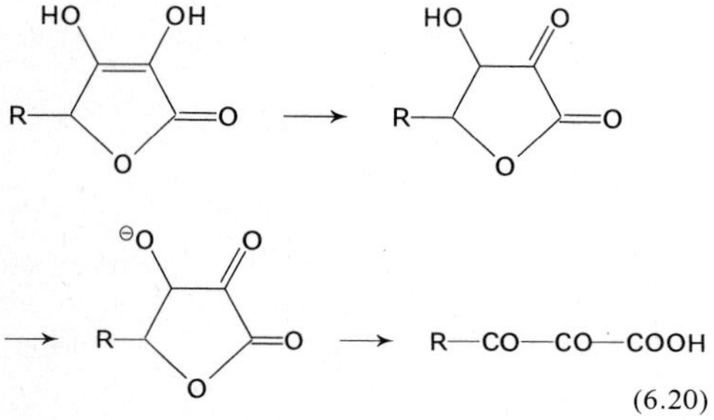

(6.20)

Diketogulonic acid degradation products, xylosone and 4-desoxypentosone, are then converted further into ethylglyoxal, various reductones (cf. 4.2.4.3.1), furfural and furancarboxylic acid:

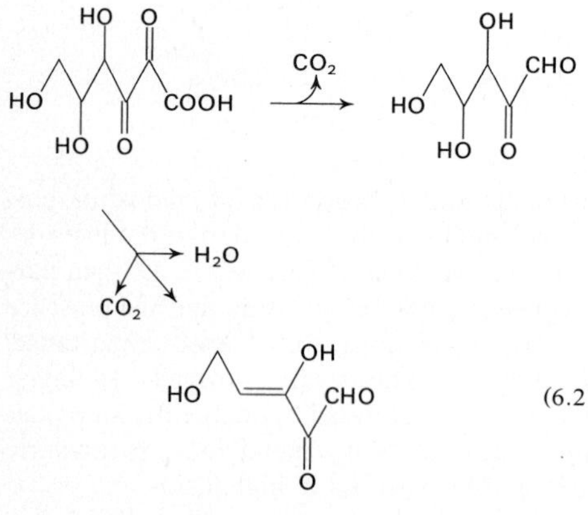

(6.21)

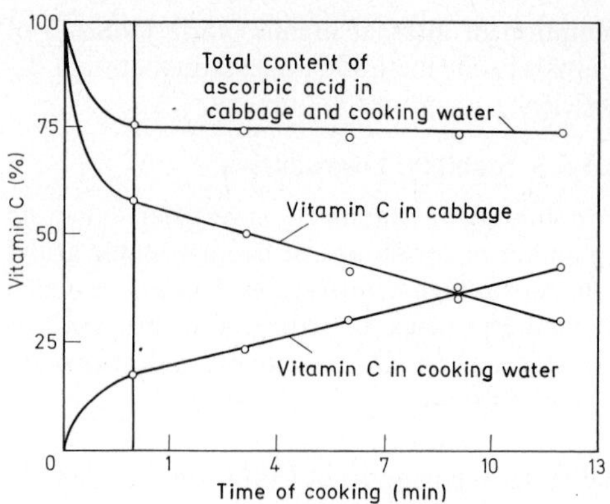

Fig. 6.3. Ascorbic acid losses as a result of cooking of cabbage. (After *Plank*, 1966)

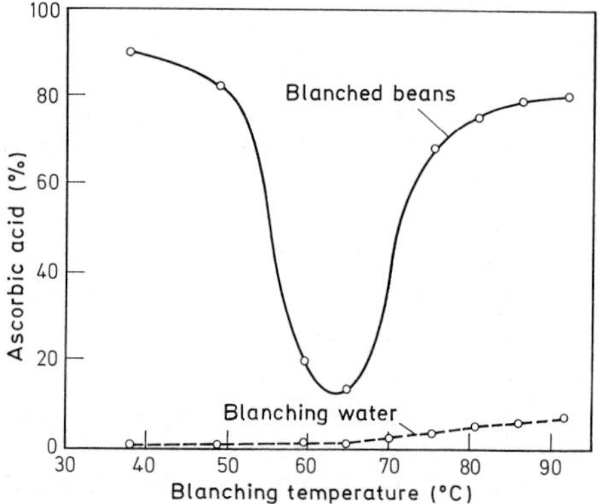

Fig. 6.4. Ascorbic acid losses in green beans versus blanching temperature. (After *Plank*, 1966)

In the presence of amino acids, ascorbic acid, dehydroascorbic acid and their degradation products might be changed further by entering into *Maillard*-type browning reactions (cf. 4.2.4.4).

A wealth of data is available about ascorbic acid losses during preservation, storage and processing of food. Tables 6.1 and 6.2 and Figures 6.3 and 6.4 present several examples. Ascorbic acid degradation is often used as a general indicator of changes occurring in food.

6.4 Literature

Bässler, K.H., Lang, K.: Vitamine. Dr. Dietrich Steinkopff Verlag: Darmstadt. 1975

Counsell, J.N., Hornig, D.H. (Eds.): Vitamin C (Ascorbic Acid). Applied Science Publ.: London 1981

Farrer, K. T. H.: The thermal destruction of vitamin B_1 in foods. Adv. Food Res. *6,* 257 (1955)

Körner, W. F., Völlm, J.: Vitamine. In: Klinische Pharmakologie und Pharmakotherapie. 3. Aufl. (Hrsg.: Kuemmerle, H. P., Garrett, E. R., Spitzy, K. H.), S. 361, Urban & Schwarzenberg: München–Berlin–Wien. 1976

Labuza, T. P., Riboh, D.: Theory and application of Arrhenius kinetics to the prediction of nutrient losses in foods. Food Technol. *36* (10), 66 (1982)

Lund, D. B.: Effect of commercial processing on nutrients. Food Technol. *33,* (2) 28 (1979)

Machlin, L. J. (Ed.): Handbook of Vitamins. Marcel Dekker: New York. 1984

Plank, R.: Handbuch der Kältetechnik. Bd. XIV, S. 475, Springer-Verlag: Berlin–Heidelberg. 1966

Seib, P. A., Tolbert, B. M. (Eds.): Ascorbic acid: Chemistry, metabolism, and uses. Advances in Chemistry Series 200, American Chemical Society: Washington, D.C. 1982

7 Minerals

7.1 Foreword

Minerals are the constituents which remain as ash after the incineration of plant and animal tissues. They may be divided into two main categories; *main elements* (Ca, P, K, Cl, Na, Mg) and *trace elements* (Fe, Zn, Cu, Mn, I, Mo, etc.). According to their biological roles, they may also be divided into *essential elements,* for which the biological roles are known; *nonessential elements,* with unknown functions, if any; and *toxic elements,* which may be ingested through food intake or from water or absorbed from the air.
Essential elements, including the main elements and a number of trace elements, fulfill various functions: as electrolytes, as enzyme constituents (cf. 2.3.3) and as building materials, e.g. in bones and teeth. Table 7.1 summarizes the mineral content of the human body.
The importance of minerals as food ingredients is not only their nutritional and physiological roles. They contribute to food flavor and activate or inhibit enzyme-catalyzed and other reactions, and they affect the texture of food.

7.2 Main Elements

7.2.1 Sodium

The sodium content of the body is 1.4 g/kg. Sodium is present mostly as an extracellular constituent and maintains the osmotic pressure of the extracellular fluid. In addition, it activates some enzymes, such as amylase. Sodium absorption is rapid; it starts 3–6 min after intake and is completed within 3 h. Daily intake of sodium ranges from 1.7–6.9 g. The intake of too little or too much sodium can result in serious disorders. From a nutritional standpoint, only excessive intake is of concern since it results in hypertension, i.e. abnormally high blood pressure. A low intake of sodium can be achieved by a nonsalty diet or by using diet salt (common salt substitutes, cf. 22.2.5). An adult's requirement for sodium averages 460 mg/day. Table 7.2 provides values for the sodium content of some foods.

Table 7.1. Mineral content of the human body

Element	Content g/kg	Element	Content mg/kg
Calcium	10–20	Iron	70–100
Phosphorus	6–12	Zinc	20–30
Potassium	2–2.5	Copper	1.5–2.5
Sodium	1–1.5	Manganese	0.15–0.3
Chlorine	1–1.2	Iodine	0.1–0.2
Magnesium	0.4–0.5	Molybdenum	0.1

7.2.2 Potassium

The concentration of potassium in the body is 2 g/kg. Potassium is localized mostly within the cells. It regulates the osmotic pressure within the cell, is involved in cell membrane transport and also in the activation of a number of glycolytic and respiratory enzymes. The potassium intake in a normal diet is 2–5.9 g/day. The minimum daily requirement is estimated to be 782 mg. Potassium deficiency is associated with a number of symptoms and may be a result of undernourishment or predominant consumption of potassium-deficient foods, e.g. white bread, fat or oil. The potassium content in food is summarized in Table 7.2. Potatoes and molasses are particularly abundant sources.

7.2.3 Magnesium

The concentration of magnesium in the body is 250 mg/kg. The daily requirement is 300–350 mg. In a normal diet, the daily intake is 300–500 mg. As a constituent and activator of many enzymes, particularly those associated with the conversion of energy-rich phosphate compounds, magnesium is a life-supporting element. Because of its indispensable role in body metabolism, magnesium deficiency causes serious disorders.

Table 7.2. Mineral content (Na, K, Ca, Fe, and P) of some foods

Food product	Na	K	Ca	Fe	P
Milk and dairy products					
Bovine milk, raw, high quality	48	157	120	0.046	92
Human milk	16	53	31	0.08	15
Butter	5	16	13	–	21
Cheese					
Emmental (45% fat)	450	107	1,020	0.31	636
Camembert (60% fat)	944	105	400	0.58	
Camembert (30% fat)	900	120	600	0.17	310
					540
Eggs					
Chicken egg yolk	51	138	140	7.2	590
Chicken egg white	170	154	11	0.2	21
Meat and meat products					
Beef, whole carcass, lean	58	342	11	2.6	170
Pork, whole carcass, lean	58	260	9	2.3	176
Calf liver	87	316	8.7	7.9	306
Pork liver	77	350	20	22.1	362
Chicken liver	68	218	18	7.4	240
Pork kidney	173	242	7	10	260
Blood sausage	680	38	6.5	6.4	22
Fish and fish products					
Herring	117	360	34	1.1	250
Eel	65	217	17	0.6	223
Cereals and cereal products					
Wheat, whole kernel	7.8	502	43.7	3.3	406
Wheat flour, type 405	2.0	108	15	1.95	–
Wheat flour, type 550	3.0	126	16	1.1	95
Wheat germ	5	837	69	8.1	1,100
Wheat gluten	2	1,390	43	3.6	1,240
Rye, whole kernel	40	530	64	–	373
Rye flour, type 997	1	240	31	2.2	–
Corn, whole kernel	6	330	15	–	256
Breakfast cereals (corn flakes)	915	139	13	2.0	59
Oat flakes	5	335	54	4.6	391
Rice, unpolished	10	150	23	2.6	325
Rice, polished	6	103	6	0.6	120
Vegetables					
Watercress	12	276	180	3.1	64
Mushrooms (cultivated)	8	422	8	1.26	123
Chicory	4.4	192	26	0.74	26
Endive	53	346	54	1.4	54
Peas, green	2	304	24	1.84	108
Lamb's lettuce	4	421	35	2.0	49
Kale	42	490	212	1.9	87
Potatoes	3.2	443	9.5	0.8	50
Kohlrabi	32	380	68	0.9	49
Head lettuce	10	224	37	1.1	.7
					33
Vegetables					
Lentils, dried	4	810	74	6.9	412
Carrots	60	290	41	0.66	35
Brussels sprout	7	411	31	1.1	84
Spinach	65	633	126	4.1	55
Edible mushroom *(Boletus edulis)*	6	486	9	1.0	115
Tomato	6	297	14	0.5	26
White cabbage	13	227	46	0.5	28
Fruits					
Apple	3	144	7	0.48	12
Orange	1.4	177	42	0.4	23
Apricots	2	278	16	0.65	21
Strawberry	2.5	147	26	0.96	29
Grapefruit	1.6	180	18	0.34	17
Rose hips	146	291	257	–	258
Currants-red	1.4	238	29	0.91	27
Currants-black	1.5	310	46	1.29	40
Cherries-sour	2	114	–	0.6	7
Plums	1.7	221	14	0.44	18
Sea buckthorn	3.5	133	42	0.44	9
Yeast					
Baker's yeast, pressed	34	649	28	4.9	605
Brewer's yeast, dried	–	1,410	–	17.6	1,900

[a] Data are in mg/100 g edible portion (average values).

7.2.4 Calcium

The total amount of calcium in the body is about 1,500 g. Because of the large amounts of calcium all over the body, it is one of the most important minerals. It is abundant in the skeleton and in some body tissues. Calcium is considered an essential nutrient because of its importance in building and maintaining bones and its role in blood clotting and muscle contraction. Therefore, calcium deficiency also causes serious disorders. The daily requirement is 0.8–1.0 g, while the average intake in a normal diet is 0.8–0.9 g. The main dietary source of calcium is milk and other dairy products. Fruits and vegetables, cereals, meat, fish and eggs have much lower amounts. Table 7.2 provides data about the occurrence of calcium in foods.

7.2.5 Chloride

The chloride content of human tissue is 1.1 g/kg body weight. Chloride intake, mostly in the form of common salt (NaCl), is 3–12 g/day. Chloride serves as a counter ion for sodium in extracellular fluid and for hydrogen ions in gastric juice. Chloride absorption is as rapid as its excretion in the urine. An accurate estimate for the daily requirement of chloride is not available.

7.2.6 Phosphate

The total phosphorus content in the body is about 700 g. The daily requirement is about 0.8–

1.2 g. The Ca/P ratio in food should be about 1. Phosphorus, in the form of phosphate, free or bound as an ester or present as an anhydride, plays an important role in metabolism and, as such, is an essential nutrient. The organic forms of phosphorus in food are cleaved by intestinal phosphatases and, thereby, absorption occurs mostly in the form of inorganic phosphate. Polyphosphates, used as food additives, are absorbed only after prior hydrolysis into orthophosphate. The extent of hydrolysis is influenced by the degree of condensation of the polyphosphates. Table 7.2 includes a compilation of the phosphorus content of some foods.

Table 7.3. Trace elements in the human body and their daily intake[a]

Element	Content (mg/kg body weight)	Intake (mg/day)
Essential		
Fe	60	15
F	37	2.5
Zn	33	6–22
Si	ca. 14	cf.[b]
Cu	1.0	3.2
V	0.3	2
Se	0.2	0.07
Mn	0.2	2–9
J	0.2	0.2
Sn	0.2	4.0
Ni	0.1	0.4
Mo	0.1	0.3
Cr	0.02	0.05–0.1
Co	0.02	0.3
Nonessential		
Rb	4.6	1–2
Br	2.9	7.5
Al	0.9	5–35
B	0.7	1.3
Ba	0.3	1.3
Ti	0.1	0.9
Au	0.1	
Sb	0.1	
Te	0.1	0.2
Li	0.03	2.0
Cs	0.02	
U	0.001	
Bi	0.0004	

[a] Average values.
[b] Silicon intake is strongly ifluenced by overall food composition (vegetarian nutrition ≫ meat based nutrition).

7.3 Trace Elements

7.3.1 General Remarks

There are 14 essential trace elements present in hormones, vitamins, enzymes and other proteins which have distinct biological roles. In addition, numerous other elements occur in the human body and their physiological roles have not yet been determined. They are usually associated with related elements, e. g. Li with Na, or Rb with K. Table 7.3 summarizes the content of essential and nonessential trace elements in the body and their average daily intake in food. Toxic elements (e. g. As, Sb, Cd, Hg, Tl and Pb) may appear in food by various routes. These toxicants are dealt with separately in Chapter 9.2. A deficiency in the essential trace elements results in metabolic disorders that are primarily associated with the absence or decreased activity of metabolic enzymes.

7.3.2 Essential Trace Elements

7.3.2.1 Iron

The iron content of the body is 4–5 g. Most is present in the hemoglobin (blood) and myoglobin (muscle tissue) pigments. The metal is also present in a number of enzymes (peroxidase, catalase, hydroxylases and flavine enzymes), hence it is an essential ingredient of the daily diet. The iron requirement depends on the age and sex of the individual,but it is about 1–2.8 mg/day. Iron supplied in the diet must be in the range of 5–28 mg/day in order to meet this daily requirement. The large variation in intake can be explained by different extents of absorption of the various forms of iron present in food (organic iron compound vs simple salts). The most utilizable source is iron in meat, for which the extent of absorption is 20–30%. The absorption is much less from liver (6.3%) and fish (5.9%), or from cereals, vegetables and milk, from which iron absorption is the lowest (1.0–1.5%). Eggs decrease and ascorbic acid increases the extent of absorption. Bran interferes with iron absorption due to the high content of phytate. Apparently, the absorption of iron present in food is, in a healthy organism, regulated by the requirement of the organism. Nevertheless, in order to provide a sufficient supply of iron to persons who require higher amounts (children, women before menopause and pregnant or nursing women), cereals (flour, bread, rice, pasta products) fortified with iron to the extent of 55–130 mg/kg are recommended. Extensive feeding tests with chickens and rats have shown that $FeSO_4$ is the most suitable form of iron, but ferrous gluconate and ferrous glycerol phosphate are also efficiently absorbed. Two food processing problems arising from mineral fortification are the increased probability that oxidation will occur and, in the case of wheat flour, decreased baking quality. Generally, iron is an undesirable element in food processing; for example, iron catalyzes the oxidation of fat or oil, increases turbidity of wine and, as a constituent of drinking water, it supports the growth of iron-requiring bacteria. The iron content of various foods is shown in Table 7.2.

7.3.2.2 Copper

The amount of copper in the body is 100–150 mg. Copper is a component of a number of oxidoreductase enzymes (cytochrome oxidase, tyrosinase, uricase, amine oxidase). In blood plasma, it is bound to ceruloplasmin, which catalyzes the oxidation of Fe^{2+} to Fe^{3+}. This reaction is of great significance since it is only the Fe^{3+} form in blood which is transported by the transferrin protein to the iron pool in the liver. The daily copper requirement is 1–2 mg and it is supplied in a normal diet. Copper is even less desirable than iron during food processing and storage since it catalyzes many unwanted reactions (e. g. oxidative destruction of ascorbic acid).

7.3.2.3 Zinc

The total zinc content in adult human tissue is 2–4 g. The daily requirement of 6–22 mg is provided by a normal diet. Zinc is a component of a number of enzymes (e. g. alcohol dehydrogenase, lactate dehydrogenase, malate dehydrogenase, glutamate dehydrogenase, carboxypeptidases A and B, and carbonic anhydrase). Other enzymes, e. g. dipeptidases, alkaline phosphatase, lecithinase and enolase, together with some other divalent metal ions, are also activated by zinc. Zinc deficiency in animals causes serious disorders, while high zinc intake by humans is toxic. Zinc poisonings have been reported as a result of consumption of soured food kept in zinc-plated metal containers (e. g. potato salad from institutional catering services).

7.3.2.4 Manganese

The body contains a total of 10–40 mg of manganese. The daily requirement, 2–48 mg/day, is met by the normal daily food intake. Manganese is the metal activator for pyruvate carboxylase and, like some other divalent metal ions, it activates various enzymes, such as arginase, amino peptidase, alkaline phosphatase, lecithinase or enolase. Manganese, even in higher amounts, is relatively nontoxic.

7.3.2.5 Cobalt

The total cobalt content of the body is 1–2 mg. Since it was discovered that vitamin B_{12} contains cobalt as its central atom, the nutritional importance of cobalt has been emphasized and it has been assigned the status of an essential element. Its requirement is met by normal nutrition.

7.3.2.6 Vanadium

The total vanadium content of the body is 17–43 mg. Feeding tests using chickens and rats have indicated a growth promoting effect. Obviously, this element has a biological role. Animal feeding tests indicate that the daily requirement for humans should be about 1–2 mg, which is provided by a normal diet.

7.3.2.7 Chromium

The chromium content of the body varies considerably depending on the region; the range is 6–12 mg. The daily intake also varies greatly from 5 to 200 μg. Chromium is important in the utilization of glucose. For instance, it activates the enzyme phosphoglucomutase and increases the activity of insulin; therefore, chromium deficiency causes a decrease in glucose tolerance. Chromium, as the chromate ion, proved to be nontoxic when used at 25 ppm in a long-term feeding experiment with rats.

7.3.2.8 Selenium

The selenium content in humans is 10–15 mg, while the daily requirement is 0.05–0.1 mg. Selenium is an antioxidant and can enhance tocopherol activity. The enzyme glutathione peroxidase contains selenium. It catalyzes the following reaction:

$$ROOH + 2GSH \rightarrow ROH + H_2O + GSSG$$

Selenium toxicity, for example, its strong carcinogenic activity, is well known from numerous animal feeding studies and from diseases of cattle grazing in pastures on selenium-enriched soil. Serious disorders are caused by as little as 2–8 ppm selenium in animal feed.

7.3.2.9 Molybdenum

The body contains 8–10 mg of molybdenum. Daily intake in food is approx. 0.3 mg. It is a component of aldehyde oxidase and xanthine oxidase. The bacterial nitrate reductase involved in meat curing and pickling processes contains molybdenum. High levels of the metal are toxic, as has been shown by cattle grazing on molybdenum-enriched soils. The grass on such soils contains 20–100 μg molybdenum/g dry matter.

7.3.2.10 Nickel

Nickel is an activator for a number of enzymes, e. g. alkaline phosphatase and oxalacetate decarboxylase, which can also be activated by some other divalent metal ions. Nickel also enhances insulin activity. The essential role of nickel has been established by inducing deficiency symptoms in feeding experiments with chickens and rats. Some of these symptoms result from respiratory changes of liver mitochondria.

7.3.2.11 Tin

Tin occurs in all human organs. It has a growth-promoting effect in rats. The natural level of tin in food is very low, but it can be increased by eating canned food when tin-plated metal sheet containers are used for canning. The contaminating tin may be substantial with canned food of low pH. Pineapple or grapefruit juice kept in faulty-plated tin cans may contain 2 g tin/l juice. The tin content in canned food is generally below 50 mg/kg and should not exceed 250 mg/kg. Tin absorption occurs only to a low extent, so the metal is only slightly toxic.

7.3.2.12 Silicon

Silicon, as soluble silicic acid, is rapidly absorbed. The silicon content of the body is approx. 1 g. The main source is cereal products. Silicon promotes growth and thus has a biological role. The toxicity of this element is apparent only at concentrations at or above 100 mg/kg.

7.3.2.13 Fluorine

The body contains 2.6 g fluorine. It plays an essential role, as indicated by feeding experiments with rats and mice – deficient diets of 2.5 ppm and 0.1–0.3 ppm, respectively, resulted in disorders in test animals in growth and reproduction. The positive effect of fluorine on teeth caries is well established. The addition to drinking water of 0.5–1.5 ppm fluorine in the form of NaF or $(NH_4)_2SiF_6$ inhibits tooth decay. Its beneficial effect appears to be in retarding solubilization of tooth enamel and inhibiting the enzymes involved in development of caries. Toxic effects of fluorine appear at a level of 2 ppm. Therefore, the beneficial effects of fluoridating drinking water are disputed by some and it is a controversial topic of mineral nutrition.

7.3.2.14 Iodine

The content of iodine in the body is about 10 mg, of which the largest portion (70–80%) is covalently bound in the thyroid gland. Iodine absorption from food occurs exclusively and rapidly as iodide and is utilized in the thyroid gland in the biosynthesis of the hormone thyroxine (tetraiodothyronine) and its less iodized form, triiodothyronine. In this process, the iodide ion is first oxidized, then iodization of the tyrosine residues of thyroglobulin occurs. Diiodotyrosine condenses with itself or with monoiodotyrosine to form thyroglobulin-bound tyroxine or triiodothyronine. Both active hormones are released from thyroglobulin by the action of a proteinase. Also released are several peptides which, however, lack activity. The iodine requirement of humans is 100–200 μg/day. Iodine deficiency results in enlargement of the thyroid gland (iodine-deficiency induced goiter). There is little iodine in most food. Good sources are milk, eggs and, above all, seafood. Drinking water contributes little to the body's iodine supply. In areas where goiter is found, the water has 0.1–2.0 μg I/l, while in goiter-free districts, 2–15 μg I/l are present in drinking water. To avoid diseases caused by low iodine supply, some countries with iodine-deficient districts employ prophylactic measures to combat the deficiency symptoms. This involves iodization of common salt with KI, with 100 μg iodine added to 1–10 g NaCl. Higher amounts of iodine are toxic and, as shown with rats, the animal's normal reproduction and lactation are affected.

7.3.3 Some Nonessential Trace Elements

7.3.3.1 Boron

Boron is present in humans and animals but, apparently, is not required. However, it is important for plants. Hollow heart and drying defects in sugar beet and the browning disease of kohlrabi are symptoms of boron deficiency. Boron is present in many foods. Expressed on a fresh weight basis, the contents in ppm are: fruits, 5–30; vegetables, 0.5–2; cereals, 0.5–3; eggs, 0.1; and milk 0.1–0.2. The amount of boron in a normal daily diet is 4–41 mg. The amount consumed is affected by drinking of wine, which contains a relatively high concentration, 10 mg/l. Boric acid accumulates at higher levels in adipose tissue and mostly in the central nervous system (brain and spinal cord). Since the implications of this storage pool are unknown, the use of boric acid in the preservation of food has been abandoned.

7.3.3.2 Aluminum

The body contains 50–150 mg of aluminum. Higher levels are found in aging organisms. The daily average intake of aluminum is 5–35 mg. It is absorbed in only negligible amounts by the gastrointestinal tract. The largest portion is eliminated in feces. Excretion of aluminum in urine is less than 0.1 mg/day. It is not secreted in milk. Animal feeding tests with high levels of aluminium in the diet through several generations showed that aluminum is nontoxic. This seems to be true also for humans. Hence, the reluctance to use aluminum cookware in food processing is unfounded. Some recent studies, however, have revealed that a pathologically-caused accumulation of aluminum in humans can cause significant damage to the cells of the central nervous system.

7.4 Minerals in Food Processing

The contribution of minerals to the nutritive/physiological value and the physical state of food has been covered in the Foreword of this Chapter and under the individual elements.

However, there are metal ions, derived from food itself or acquired during food processing and storage, which interfere with the quality and visual appearance of food. They can cause discoloration of fruit and vegetable products (cf. 18.1.2.5.7) and many metal-catalyzed reactions responsible for losses of some essential nutrients, for example, ascorbic acid oxidation (cf. 6.3.9.3). Also, they are responsible for taste defects or off-flavors, for example, as a consequence of fat oxidation (cf. 3.7.2.1.4). Therefore, the removal of many interfering metal ions by chelating agents (cf. 8.14) or by other means is of importance in food processing.

7.5 Literature

Lang, K.: Biochemie der Ernährung, 4. Aufl., Dr. Dietrich Steinkopff Verlag: Darmstadt. 1979

8 Food Additives

8.1 Foreword

A food additive is a substance (or a mixture of substances) which is added to food and is involved in its production, processing, packaging and/or storage without being a major ingredient. Additives or their degradation products generally remain in food, but in some cases they may be removed during processing. The following examples illustrate and support the use of additives to enhance the:

- *Nutritive value of food*
  Additives such as vitamins, minerals, amino acids and amino acid derivatives are utilized to increase the nutritive value of food. A particular diet may also require the use of thickening agents, emulsifiers, sweeteners, etc.
- *Sensory value of food*
  Color, odor, taste and consistency or texture, which are important for the sensory value of food, may decrease during processing and storage. Such decreases can be corrected or readjusted by additives such as pigments, aroma compounds or flavor enhancers. Development of "off flavor", for instance that derived from fat or oil oxidation, can be suppressed by antioxidants. Food texture can be stabilized by adding minerals or polysaccharides, and by many other means.
- *Shelf life of food*
  The current forms of food production and distribution, as well as the trend towards so-called convenience foods, have increased the demand for longer shelf life. Furthermore, the world food supply situation requires preservation by avoiding deterioration as much as possible. The extension of shelf life involves protection against microbial spoilage, for example by using additives that affect growth of microflora, and use of active agents which suppress and retard undesired chemical and physical changes in food. The latter is achieved by stabilization of pH using buffering additives or stabilization of texture with thickening or gelling agents, which are polysaccharides.

It is implicitly understood that food additives and their degradation products should be nontoxic at their recommended levels of use. This applies equally to acute and to chronic toxicity, particularly the potential carcinogenic, teratogenic (causing a malformed fetus) and mutagenic effects. It is generally recognized that additives are applied only when required for the nutritive or sensory value of food, or for its processing or handling. The use of additives is regulated by Food and Drug or Health and Welfare administrations in most countries. The regulations differ in part from country to country but there are endeavors under way to harmonize them on the basis of both current toxicological knowledge and the requirements of modern food technology. The most important groups of additives are outlined in the following sections. No reference is given to legislated regulations or definitions provided therein. A compilation of the relative importance of various groups of additives is presented in Table 8.1.

Table 8.1. Utilization of food additives in United States (1965 as % of total additives used)[a]

Additives, class	As % of total	Additives, class	As % of total
Aroma compounds	42.5	Chelating agents	2.6
Natural aroma substances	21	Colors	2.1
Nutritional fortifiers	6.9	Chemical preservatives	1.8
Surface active agents (tensides)	5	Stabilizers	1.8
Buffering substances, acids, bases	3.5	Antioxidants	1.7
		Maturing and bleaching agents	1.4
		Sweeteners	0.5
		Other additives	9.4

[a] In 1965 a total of 1696 substances (= 100%) were utilized.

8.2 Vitamins

Many food products are enriched or fortified with vitamins to adjust for processing losses or to increase the nutritive value. Such enrichment is important, particularly for fruit juices, canned vegetables, flour and bread, milk, margarine and infant food formulations. Table 8.2 provides an overview of vitamin enrichment of food.

Several vitamins have some desirable additional effects. Ascorbic acid is a dough improver, but can play a role similar to tocopherol as an antioxidant. Carotenoids and riboflavin are used as coloring pigments, while niacin improves the color stability of fresh and cured and pickled meat.

8.3 Amino Acids

The increase in the nutritive value of food by addition of essential amino acids and their derivatives is dealt with in sections 1.2.5 and 1.3.4.4.

8.4 Minerals

Food is usually an abundant source of minerals. Fortification is considered for iron, which is often not fully available, and for calcium, magnesium, copper and zinc. Iodization of salt is of importance in iodine deficient areas (cf. 22.2.4).

Table 8.2. Examples of vitamin fortification of food

Vitamin	Food product
B_1	Cocoa powder and its products, beverages and concentrates, confectionary and other baked products
B_2	Baked products, beverages
B_6	Baked and pasta products
B_{12}	Beverages, etc.
Pantothenic acid	Baked products
Folic acid/biotin	Not commonly used as additives
C	Fruit drinks, desserts, dairy products, flour
A	Skim milk powder, breakfast cereals (flakes), beverage concentrates, margarine, baked products, etc.
D	Milk, milk powder, etc.
E	Various food products, e.g. margarine

8.5 Aroma Substances

The use of aroma substances of natural or synthetic origin is of great importance (cf. Table 8.1). The aroma compounds are dealt with in detail in Chapter 5 and in individual sections covering some food commodities.

8.6 Flavor Enhancers

These are compounds that enhance the aroma of a food commodity, though by themselves, in the concentrations used, they do not have a distinct odor or taste. An enhancer's effect is apparent to the senses as "feeling", "volume", "body" or "freshness" (particularly in thermally-processed food) of the aroma, and also by the speed of aroma perception ("time factor potentiator").

8.6.1 Monosodium Glutamate (MSG)

Glutamic acid was isolated by *Ritthausen* in 1866. In 1908 *Ikeda* found that MSG is the beneficial active component of the algae *Laminaria japonica,* used for a long time in Japan as a flavor improver of soup and similarly prepared food. The consumption of MSG in 1978 was 200,000 tonnes worldwide.

In a pH range of 5–8 and at its usually applied level of 0.2–0.5%, MSG has a pleasant, lightly salty-sweet taste and a property often described as "mouth satisfaction". It affects and promotes sensory perception, particularly of meat-like aroma notes, and is frequently used as an additive in frozen, dehydrated or canned products of fish and meat.

The intake of larger amounts of MSG by some hypersensitive persons can trigger a "Chinese restaurant syndrome", which is characterized by temporary disorders such as drowsiness, headache, stomach ache and stiffening of joints. These disappear after a short time.

8.6.2 5′-Nucleotides

5′-Inosine monophosphate (IMP, disodium salt) and 5′-guanosine monophosphate (GMP, disodium salt) have properties similar to MSG but heightened by a factor of 10–20. Their flavor enhancing ability at 75–500 ppm is good in all food (e.g. soups, sauces, canned meat or tomato juice). However, some other specific effects, besides the "MSG effect", have been described for

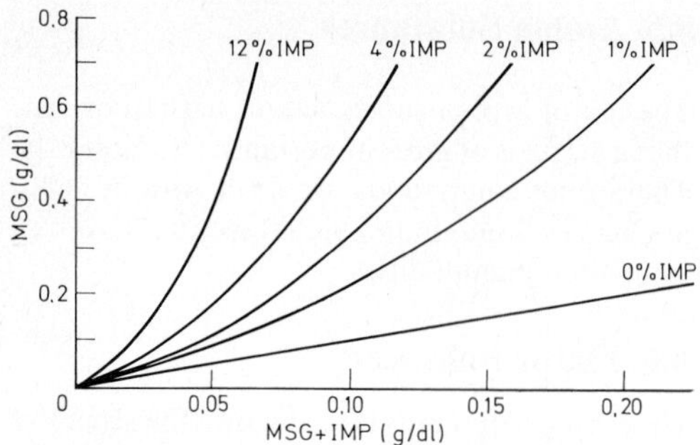

Fig. 8.1. Synergistic activities of Na-glutamate (MSG) and disodium-inosine monophosphate (IMP). The curves give the concentrations of MSG and MSG + IMP in water that are rated as being sensory equivalent by a taste panel

nucleotides. For example, they imprint a sensation of higher viscosity in liquid food. The sensation is often expressed as "freshness" or "naturalness", but the expressions "body" and "mouthfeel" are more appropriate for soups. Synergistic flavor-enhancing effects are experienced with simultaneous use of MSG and IMP or GMP (Figure 8.1).

8.6.3 Maltol

Maltol (3-hydroxy-2-methyl-4-pyrone, cf. 5.3.1.2) has a caramel-like odor (melting point 162–164 °C). It enhances the perception of sweetness in carbohydrate-rich food (e. g. fruit juices, marmalades, fruit jelly). Addition of 5–75 ppm maltol allows a decrease of sugar content by about 15%, while retaining the sweetness intensity.

8.6.4 Other Compounds

Numerous patents have been issued or are pending for other compounds that influence, improve or enhance the flavor of food or suppress or modify unpleasant flavor notes. The significance of these compounds, with some of them active at exceptionally low levels, will obviously rise. Simultaneously, their analytical determination will become a challenge.

An example is dioctyl sodium sulfosuccinate

(8.1)

which in low concentrations provides a perception of "freshness" to sterilized milk. N,N′-di-o-tolylethylenediamine, in a concentration range of 5×10^{-7} ppb to 10 ppb,

(8.2)

enhances the butter-like aroma note of margarine and the milk flavor of reconstituted milk powder.

8.7 Sugar Substitutes

Sugar substitutes are those compounds that are used like sugars (sucrose, glucose) for sweetening, but are metabolized without the influence of insulin. Important sugar substitutes are the sugar alcohols, sorbitol, xylitol and mannitol and, to a certain extent, fructose (cf. 19.1.4.5–19.1.4.9).

8.8 Sweeteners

Sweeteners are natural or synthetic compounds which provide a sweet perception, but possess no or negligible nutritional value ("nonnutritive sweeteners") in relation to the extent of sweetness.

There is considerable interest in new sweeteners. The rise in obesity in industrialized countries has established a trend for calorie-reduced nutrition. Also, there is an increased discussion about the safety of saccharin and cyclamate, the two sweeteners which were predominant for a long time. The search for new sweeteners is complicated by the fact that the relationship between chemical structure and sweetness perception is not yet satisfactorily resolved. In addition, the health safety of suitable compounds has to be certain. Some other criteria must also be met, for example, the compound must be adequately soluble and stable over a wide pH and temperature range, have a clean sweet taste without side or post-flavor effects, and provide a sweetening effect as cost-effectively as does sucrose. The use of some compounds which match most of these demands will be discussed.

The following sections describe several sweeteners, irrespective of whether or not they are approved, banned or are just being considered for future commercial use.

8.8.1 Structural Provisions for Sweetness

A sweet taste can be derived from compounds with very different chemical structures. *Shallenberger* and *Acree* consider that, for sweetness, a compound must contain a proton donor/acceptor system (AH_s/B_s-system), which has to match some steric prerequisites and which can interact with a complementary receptor system (AH_r/B_r-system) by involvement of two hydrogen bridges (Figure 8.2). The expanded model of *Kier* has an additional hydrophobic interaction with a group, X, present at a distinct position of the molecule (Figure 8.3). The examples in Figures 8.2 and 8.3 show that these models are applicable to many sweet compounds from highly different classes. An expanded model substitutes the AH_s/B_s-system with a nucleophilic/electrophilic system (n_s/e_s-system) and the localized contact with group X with an extended hydrophobic contact. Thus, a receptor for a sweetener has to be depicted schematically as a hydrophobic pocket, containing its own complementary n_r/e_r-system.

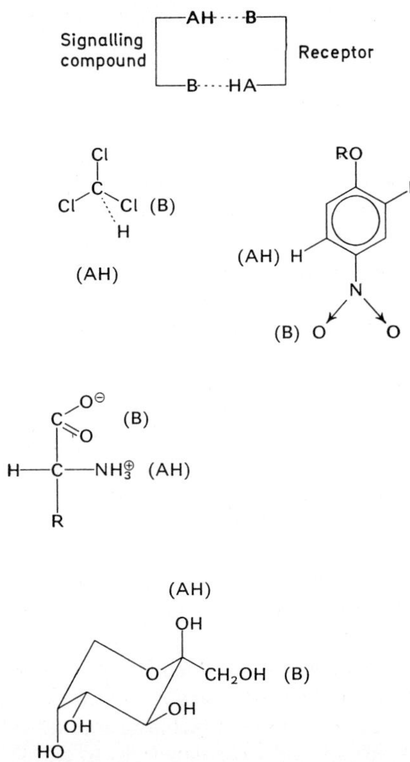

Fig. 8.2. AH/B-systems of various sweet compounds

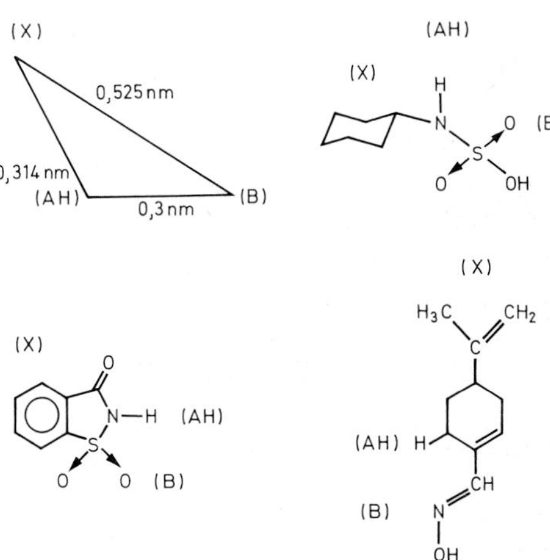

Fig. 8.3. AH/B/X-systems of various sweet compounds

While *two* polar (n_s/e_s) groups should be present in sweet compounds and, when necessary, be supplemented with a hydrophobic group, a bitter tasting compound requires the presence of only *one* polar (n_s or e_s) and one hydrophobic group. The sweetness intensity of a compound can be measured numerically and expressed as:

- Threshold detection value, c_{tsv} (e.g. the lowest concentration of an aqueous solution that can still be perceived as being sweet).
- Factor f_{sac}, which represents any aqueous concentration of a sweetener, multiplied by a factor to obtain the sucrose (saccharose) concentration which provides equal sweetness (isosweet saccharose solution). This factor is concentration dependent and can be expressed in weight ($f_{sac,\,g}$) or in mole concentration ($f_{sac,\,mole}$).

8.8.2 Saccharin

Saccharin, in addition to cyclamate, is one of the most important sweeteners and is usually sold as its Na- or Ca-salt. In higher concentrations the compound has a light, bitter after taste. For a 10% sucrose solution, $f_{sac,\,g} \sim 320$. The synthesis of saccharin starts with toluene and proceeds as shown in Reactions 8.3.

PCl_5 ; $KMnO_4$; NH_3

(separated by freezing)

(8.3)

Numerous saccharin analogues have been synthesized and tested for sweetness, but none has surpassed saccharin.

8.8.3 Cyclamate

Cyclamate is a widespread sweetener and is marketed as the Na- or Ca-salt of cyclohexane sulfamic acid (delisted in 1970 in North America as an unsafe nonnutritive sweetener). The taste intensity is substantially lower than for saccharin and is $f_{sac, g} \sim 35$ for the 10% sucrose solution. There is no bitter side taste. Overall, the sweet taste of cyclamate is not as pleasant as that of saccharin. The synthesis of the compound is based on sulfonation of cyclohexylamine:

$ClSO_3H$

(8.4)

Table 8.3 shows several homologous compounds to illustrate the dependence of sweetness intensity on cycloalkyl ring size. The larger the ring size, the higher the sweetness, i.e. the lower the sweetness threshold value.

Table 8.3. Taste threshold values of some alicyclic sulfamic acids (Na-salts)

R	c_{tsw} (mmole/l)	R	c_{tsw} (mmole/l)
Cyclobutyl	100	Cycloheptyl	0.5–0.7
Cyclopentyl	2–4	Cyclooctyl	0.5–0.8
Cyclohexyl	1–3		

8.8.4 Monellin

The pulp of *Dioscoreophyllum cumminsii* fruit contains monellin, a sweet protein with a molecular weight of 11.5 kdal. It consists of two peptide chains, A and B, which are not covalently bound. Their amino acid sequences are shown in Table 8.4. The separated individual chains are not sweet. When the chains are recombined, a sweet taste is restored slowly, but the sweetness intensity of the original native protein is not reached. This strongly suggests that peptide chain separation results in irreversible conformational changes.

The SH group of cysteine residues of the B chain reacts only slowly with SH reagents and under rather drastic conditions; hence its contribution to the protein's sweet taste is questionable. The threshold value is $f_{sac, g} = 3{,}000$. Based on its low stability, slow triggering and slow fading away of taste perception, monellin probably will not succeed as a commercial sweetener.

8.8.5 Thaumatins

The fruit of *Thaumatococcus daniellii* contains two sweet proteins: thaumatin I and II, with $f_{sac, g} \sim 2{,}000$. There are also low amounts of three other sweet proteins (thaumatin a, b and c). The complete amino acid sequence of thaumatin I, a peptide chain with 207 amino acid residues, has been established (Table 8.5). A comparison of the sequences of monellin A or B and thaumatin I shows the presence of five identical tripeptide units (Table 8.6). Successive acetylation of the ε-amino groups weakens the sweet taste and, when just four groups are acetylated, it is extinguished. However, modification of up to seven

Table 8.4. Amino acid sequences of monellin A and B chains

A-chain

5 10 15 20
Phe Arg Glu Ile Lys Gly Tyr Gly Tyr Gln Leu Tyr Val Tyr Ala Ser Asp Lys Leu Phe
25 30 35 40
Arg Ala Asp Ile Ser Glu Asp Tyr Lys Thr Arg Gly Arg Lys Leu Leu Arg Phe Asn Gly
45
Pro Val Pro Pro Pro

B-chain

5 10 15 20
Gly Glu Trp Glu Ile Ile Asp Ile Gly Pro Phe Thr Gln Asn Leu Gly Lys Phe Ala Val
25 30 35 40
Asp Glu Glu Asn Lys Ile Gly Gln Tyr Gly Arg Leu Thr Phe Asn Lys Val Ile Arg Pro
45 50
Cys Met Lys Lys Thr Ile Tyr Glu Asn Glu

Table 8.5. Amino acid sequence of thaumatin I

1 10 20
Ala Thr Phe Glu Ile Val Asn Arg Cys Ser Tyr Thr Val Trp Ala Ala Ala Ser Lys Gly
21 30 40
Asp Ala Ala Leu Asp Ala Gly Gly Arg Gln Leu Asn Ser Gly Glu Ser Trp Thr Ile Asn
41 50 60
Val Glu Pro Gly Thr Asn Gly Gly Lys Ile Trp Ala Arg Thr Asp Cys Tyr Phe Asp Asp
61 70 80
Ser Gly Ser Gly Ile Cys Lys Thr Gly Asp Cys Gly Gly Leu Leu Arg Cys Lys Arg Phe
81 90 100
Gly Arg Pro Pro Thr Thr Leu Ala Glu Phe Ser Leu Asn Gln Tyr Gly Lys Asp Tyr Ile
101 110 120
Asp Ile Ser Asn Ile Lys Gly Phe Asn Val Pro Met Asn Phe Ser Pro Thr Thr Arg Gly
121 130 140
Cys Arg Gly Val Arg Cys Ala Ala Asp Ile Val Gly Gln Cys Pro Ala Lys Leu Lys Ala
141 150 160
Pro Gly Gly Gly Cys Asn Asp Ala Cys Thr Val Phe Gln Thr Ser Glu Tyr Cys Cys Thr
161 170 180
Thr Gly Lys Cys Gly Pro Thr Glu Tyr Ser Arg Phe Phe Lys Arg Leu Cys Pro Asp Ala
181 190 200
Phe Ser Tyr Val Leu Asp Lys Pro Thr Thr Val Thr Cys Pro Gly Ser Ser Asn Tyr Arg
201 207
Val Thr Phe Cys Pro Thr Ala

Table 8.6. Identical tripeptide sequences in monellins and thaumatin

	116 120		99 100 105
Thaumatin	-Pro-Thr-*Thr-Arg-Gly*-Cys-Arg-	Thaumatin	-Tyr-Ile-*Asp-Ile-Ser*-Asn-Ile
	27 30		20 25
Monellin A	-Tyr-Lys-*Thr-Arg-Gly*-Arg-Lys-	Monellin A	-Arg-Ala-*Asp-Ile-Ser*-Glu-Asp-
	126 130		92 95
Thaumatin	-Cys-Ala-*Ala-Asp-Ile*-Val-Gly-	Thaumatin	-Leu-Asn-*Gln-Tyr-Gly*-Lys-Asp-
	19 20 25		26 30
Monellin A	-Phe-Arg-*Ala-Asp-Ile*-Ser-Glu-	Monellin B	-Ile-Gly-*Gln-Tyr-Gly*-Arg-Leu-
	98 100		
Thaumatin	-Asp-Tyr-*Ile-Asp-Ile*-Ser-Asn-		
	4 5 10		
Monellin B	-Glu-Ile-*Ile-Asp-Ile*-Gly-Pro-		

ε-amino groups by reductive methylation with $HCHO/NaBH_4$ does not decrease or modify the sweet taste intensity. Apparently, the isoelectric point of the protein plays a role in sweetness activity. Thaumatin probably will find application in several fields as a sweetener.

8.8.6 Miraculin

Miraculin is a glycoprotein present in fruits of *Synsepalum dulcificum* (a tropical fruit known as miracle berry). It has no taste but changes the taste of acidic solutions into a sweet taste note; hence it is designated as a "taste modifier". Thus, lemon juice seems sweet when the mouth is first rinsed with a solution of miraculin. The molecular weight of this taste modifier is 42–44 kdal.

8.8.7 Gymnema silvestre Extract

The extract from *Gymnema silvestre* is related to the taste modifier miraculin. It enhances sweet taste perception for several hours, without interfering in reception of other taste qualities. The active substance has not yet been identified.

8.8.8 Stevioside

Leaves of *Stevia rebaudiana* contain approx. 6% stevioside ($f_{sac,\,g}$ ~ 300). The compound could become a sweetener of commercial importance. Its structure is:

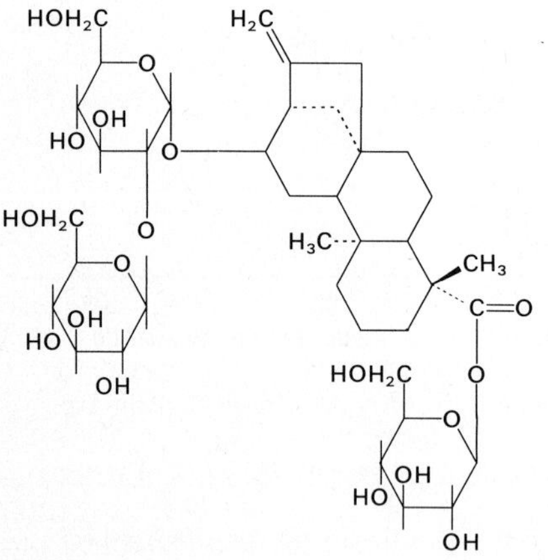

(8.5)

8.8.9 Osladin

Osladin occurs in rhizomes of *Polypodium vulgare*. It is very sweet ($f_{sac,\,g}$ ~ 3,000); however, due to its toxic effect, it is not likely to become an approved sweetener (see Formula 8.6).

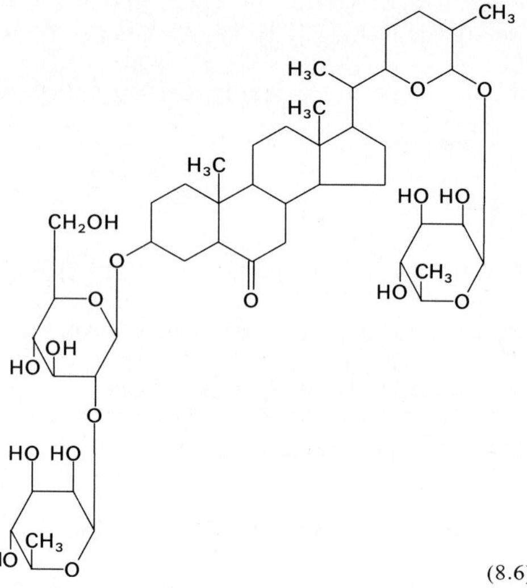

(8.6)

8.8.10 Phyllodulcin

The leaves of *Hydrangea macrophylla* contain a 3,4-dihydroisocoumarin derivate, phyllodulcin. Its sweetness matches that of dihydrochalcones and of licorice root. The taste perception builds relatively slowly and also fades away slowly. The sweetness intensity is $f_{sac,\,g}$ = 200–300. Its application is being considered in manufacturing of several brands of candy and chewing gum. A study of a number of related isocoumarin derivatives shows that taste quality and intensity are very much dependent on the substitution pattern of the molecule (cf. Table 8.7).

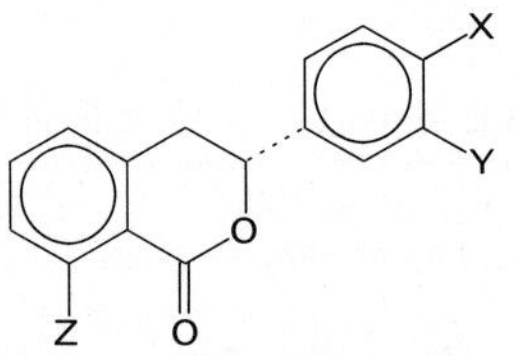

(8.7)

Z = OH, X = OMe, Y = OH

8.8.11 Glycyrrhizin

The active substance from licorice root *(Glycyrrhiza glabra)* is a β,β′-glucuronido-glucuronide of glycyrrhetic acid (cf. Formula 8.8):

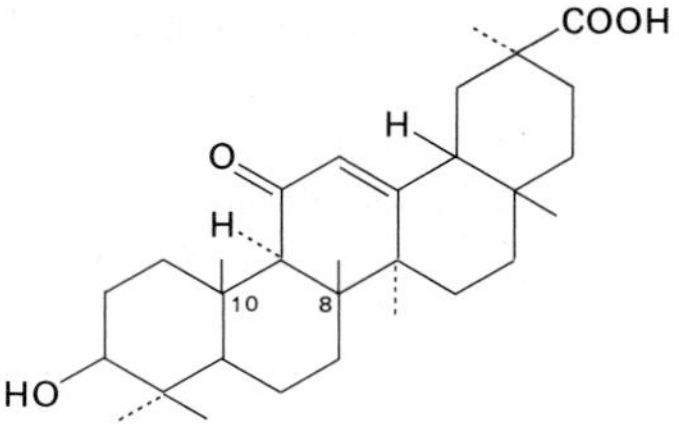

(8.8)

The sweet taste intensity is $f_{sac} = 50$. The compound is utilized for production of licorice (also spelled as liquorice). Its cortisone-like side activity limits its wide application.

8.8.12 Nitroanilines

Several m-nitroanilines are potent sweeteners:

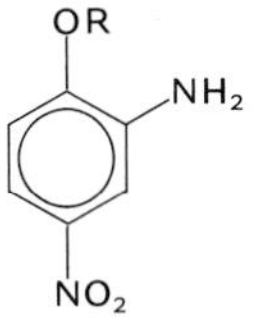

(8.9)

As shown in Table 8.8, the taste intensity is strongly dependent on the substituent R. The propoxy derivative has an $f_{sac} = 4{,}100$ and was for some time used under the brand name "Ultrasweet P-4000" but, due to its toxic side effects, is no longer important.

Table 8.7. Sensory properties of some 2,3-dihydroisocoumarins

Compound[a]			Taste
X	Y	Z	
OMe	OH	OH	very sweet
OMe	OMe	OH	bitter
OMe	OMe	OMe	no taste
OMe	OAc	OAc	light sweet
OH	OH	OH	no taste
OH	H	OH	no taste
OH	OH	H	no taste
OMe	OH	H	very sweet
OH	OMe	H	no taste

[a] Formula 8.7.

Table 8.8. Sweetness of some 1-alkoxy-2-amino-4-nitrobenzenes (Formula 8.9)

R	f_{sac}	R	f_{sac}
H	120	C_4H_9	1,000
CH_3	220	$(CH_3)_2CH$	600
C_2H_5	950	$CH_3CH{=}CH$	2,000
C_3H_7	4,100		

8.8.13 Dihydrochalcones

These sweeteners are obtained from phenolic glucosides present in citrus fruit peels. Some are derived from flavanones (cf. 18.1.2.5.4 and Table 18.22) and have a relatively clean sweet taste that is slowly perceived but persists for some time. The sweet taste intensity of β-neohesperidin-dihydrochalcone is $f_{sac,\,g} = 1{,}100$ (at the threshold level) or $f_{sac,\,g} = 600$ (10–12% saccharose solution). Its structure is (R = β-neohesperidosyl):

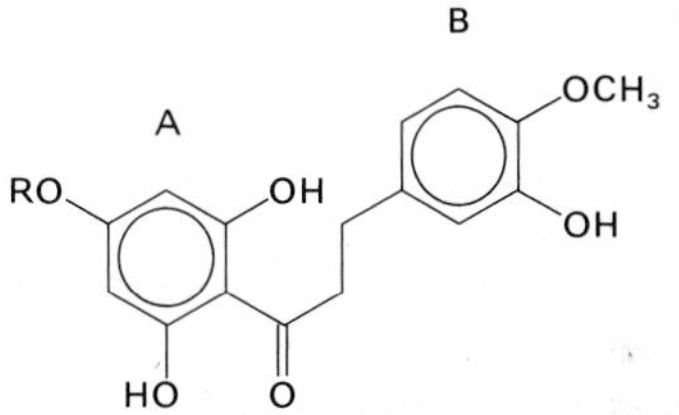

(8.10)

The compound is being considered for application in chewing gum, mouthwashes and various brands of candy. The sweet taste quality and intensity of dihydrochalcone are related particularly to the substitution pattern in ring B. The prerequsite for a sweet taste is the presence in ring B of at least one –OH group, but not three adjacent hydroxy- or alkyl-substituted phenolic groups.

8.8.14 Dulcin

Dulcin, 4-ethoxyphenylurea, is closer in taste to sucrose than saccharin. The sweetness intensity is $f_{sac,\,g} = 109$ (5% saccharose solution), i.e. lower than that of saccharin. It was used earlier as a mix with saccharin, but now has no importance due to its toxic side effects.

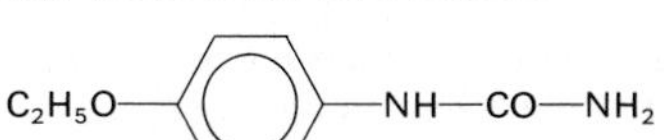

$C_2H_5O-C_6H_4-NH-CO-NH_2$ (8.11)

8.8.15 Oximes

It has long been known that the anti-aldoxime of perillaldehyde (discovered in the essential oil of *Perilla nankinensis*) has an intensive sweet taste ($f_{sac,\,g} \sim 2{,}000$). For its structure see Formula 8.12 (I).

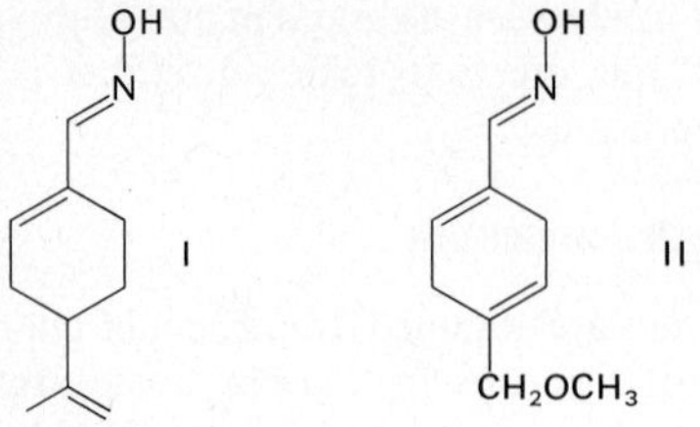

(8.12)

Its poor water solubility works against its utilization. In the meantime, a related compound, (II), with improved solubility, has been reported, but its sweetness is just moderately high ($f_{sac,g} \sim 450$).

8.8.16 Oxathiazinone Dioxides

This compound belongs to a new class of sweeteners with an AH/B-system corresponding to that of saccharin. Based on properties and toxicological data, the compound is a sweetener of potential commercial importance. The sweet taste intensity depends on substituents R^1 and R^2 (Table 8.9):

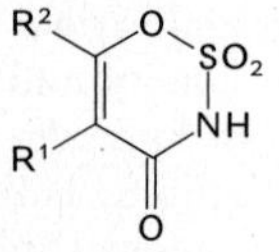

(8.13)

Table 8.9. Sweetness of some oxathiazinone dioxides (Na-salts)

R^1	R^2	$f_{sac,g}$	R^1	R^2	$f_{sac,g}$
H	H	10	Et	H	20
H	Me	130[a]	Et	Me	250
Me	H	20	Pr	Me	30
Me	Me	130	i-Pr	Me	50
H	Et	150			

[a] Acesulfame.

Table 8.10. Comparison of sweetness intensities of aspartame and saccharose (concentrations of isosweet aqueous solutions in %)

Saccharose	Aspartame	$f_{sac,g}$
0.34[a]	0.001[a]	340
4.3	0.02	215
10.0	0.075	133
15.0	0.15	100

[a] Threshold value.

Oxathiazinone dioxide is obtained from fluorosulfonyl isocyanate and alkynes, i.e. from compounds with active methylene groups, as exemplified by 1,3-diketones, 3-oxocarbonic acids and 3-oxocarbonic acid esters (see Reactions 8.14).

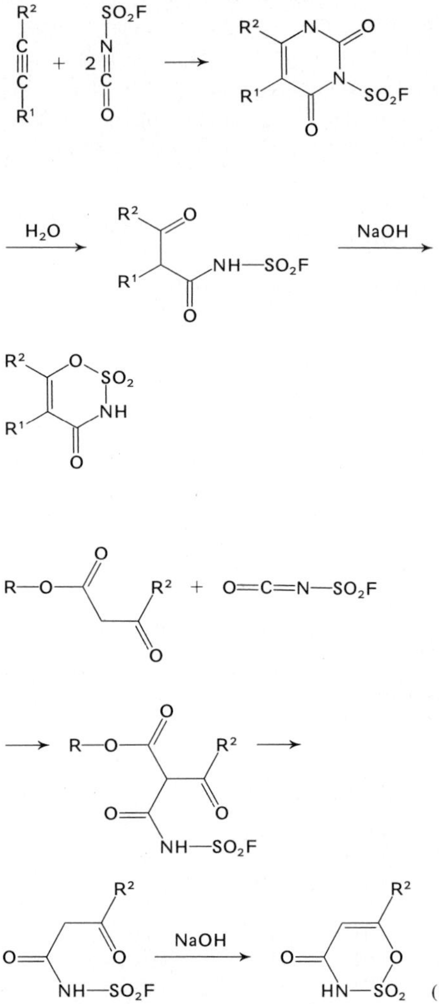

(8.14)

8.8.17 Dipeptide Esters

A dipeptide, L-aspartyl-L-phenylalanine methyl ester (L-Asp-L-Phe-OMe), has recently been approved for use as a sweetener in North America (aspartame, "NutraSweet"). It is as sweet as a number of other dipeptide esters of L-aspartic acid. The relationship between chemical structure and taste among these compounds was outlined in more detail in 1.3.3. The sweetness intensity relative to saccharose is concentration dependent (Table 8.10). Its utilization in many food products (such as diet cola drinks) is increasing, though its stability is not always satisfactory. Unlike sweetening of drinks (coffee or tea) which are drunk immediately, problems exist in the use of aspartame in food which has to be heated or in sweetened drinks which have to be stored for longer periods of time. Possible degradation reactions involve hydrolysis into amino acid constituents and cyclization into the 2,4-dioxopiperazine derivative:

L-Asp-L-Phe-OMe → L-Asp + L-Phe + MeOH

L-Asp-L-Phe-OMe → HOOC—H₂C-(2,5-dioxopiperazine, NH, HN)-CH₂C₆H₅ + MeOH (8.15)

Aspartame synthesis on a large scale is achieved by the following reactions:

L-Asp —PCl₅→ (aspartic anhydride, H_2N) —L-Phe-OMe→

L-Asp-L-Phe-OMe (I) + L-Asp(L-Phe-OMe) (II) →

diluted HCl → L-Asp-L-Phe-OMe (precipitate) (8.16)

Separation of the two dipeptide isomers (I; II) is possible since there are solubility differences between the two isomers as a consequence of their differing isoelectric points ($IP_I > IP_{II}$).

8.8.18 Hernandulcin

(+)-Hernandulcin is a sweet sesquiterpene from *Lippia dulcis Trev.* (*Verbenaceae*), with the structure 6-(1.5-dimethyl-1-hydroxy-hex-4-enyl)-3-methyl-cyclohex-2-enone:

I II

The sweet taste intensity is $f_{sac,mol} = 1{,}250$ (0,25 M sucrose). Hernandulcin is somewhat less pleasant in taste than sucrose and exhibits some bitterness.

The racemic compound was synthesized via a directed aldol-condensation reaction by adding 6-methyl-5-hepten-2-one to a mixture of 3-methyl-2-cyclohexen-1-one and lithium diisopropylamide in tetrahydrofuran, followed by chromatographic separation of (±)-hernandulcin (I, 95%) from the diastereomeric counterpart (±)-epihernandulcin (II, 5%). Whereas I is sweet, II exhibits no sweet taste.

The carbonyl and hydroxyl groups, which are located about 2.6 Å apart in the preferred conformation, are considered as an AH/B-system. The sweet taste is lost when these groups are modified (reduction of the carbonyl group to an alcohol, or acetylation of the hydroxy group).

8.9 Food Colors

A number of natural colors are available and used to adjust or correct food discoloration or color change during processing or storage. Carotenoids (cf. 3.8.4.5) are used the most, followed by red beet pigment and brown-colored caramels. The number of approved synthetic dyes is low. Table 8.11 lists the pigments of importance in food coloring. Yellow and red colors are used the most. Food products which are often colored are confections, beverages, dessert powders, cereals, ice creams and dairy products.

8.10 Acids

The acids in food fulfill a number of other functions besides flavor and antimicrobial activities. The most important acids used in food processing and storage are outlined in this section.

8.10.1 Acetic Acid and Its Higher Homologues

Acetic, propionic and sorbic acids are dealt with under antimicrobial agents (8.12). Other short

Table 8.11. Examples for food colorants (natural and synthetic)

Number	Name	FD & C (USA)	EEC No.	Color	λ max (nm) (solvent[b])	Formula[a]	Examples for utilization in food processing
1	Tartrazine	Yellow No 5	E 102,	lemon-yellow (W)	426 (W)	I	Pudding powders, confectionary and candies, ice creams, pop (effervescent) drinks
2	Riboflavin		E 101,	yellow (W)	445 (W)		Mayonnaise, soups, puddings. desserts, confectionary and candy products
3	Curcumin		E 100,	yellow-reddish (E)	426 (E)	II	Mustard
4	Zeaxanthin			yellow (oil)	455–460 (CH)		Fat, hot and cold drinks, puddings, water
5	Sunset Yellow FCF	Yellow No 6	E 110,	orange (W)	485 (W)	III	Beverages, fruit preserves, confectionary and candy products, honey-like products, sea salmon, crabs
6	β-Carotene		E 160 a,	orange (oil)	453–456 (CH)		Fat, beverages, soups, pudding, water, confectionary and candy products, yoghurt
7	Bixin		E 160 b,	orange (oil)	471/503 ($CHCl_3$)		Fat, mayonnaise
8	Lycopene		E 160 d,	orange (oil)	478 (H)		Mayonnaise, ketchup, sauces
9	Canthaxanthin	Food Orange 8	E 161 g,	orange (oil)	485 ($CHCl_3$)		Sea salmon, beverages, tomato products
10	Astaxanthin			orange (oil)	488 ($CHCl_3$)		Beverages, tomato products, confectionary and candy products
11	β-Apo-8′-carotenal		E 160 e,	orange (oil)	460–462 (CH)		Sauces, beverages, confectionary and candy products
12	Carmoisine		E 122,	red with bluish tint (W)	516 (W)	IV	Beverages, confectionary and candy products, ice cream, pudding, fruit preserves
13	Amaranth	Red No 2 (delisted)	E 123,	red with bluish tint (W)	520 (W)	V	Beverages, fruit preserves confectionary and candy products, jams
14	Ponceau 4R		E 124,	scarlet-red (W)	505 (W)	VI	Beverages, candy products (bonbons), sea salmon, cheese coatings

Table 8.11. (Continued)

Number	Name		EEC No.	Color	λ max (nm) (solvent[b])	Formula[a]	Examples for utilization in food processing
15	Carmine		E 120,	bright-red	518 (W ammonia solution)	VII	Alcoholic beverages
16	Anthocyanidin (from red grape pomace)		E 163 a–f	red-violet[c] (W)	520–546 (M + 0.01% HCl)		Jams, pop (effervescent) drinks
17	Erythrosine	Red No 3	E 127	cherry-red (W)	527 (W)	VIII	Fruits, jams, confectionary and candy products
18	Red 2 G			red with bluish tint	532 (W)	IX	Confectionary and candy products
19	Indigo Carmine	Blue No 2	E 132,	purple blue (W)	610 (W)	X	Also in combination with yellow colorant for confectionary and candy products and liqueurs
20	Patent Blue V		E 131,	blue with a greenish tint (W)	638 (W)	XI	Mostly in combination with yellow colorants for confectionary and candy products, beverages
21	Brilliant Blue FCF	Blue No 1		blue with a greenish tint (W)	630 (W)	XII	Mostly in combination with yellow colorants for confectionary and candy products, beverages
22	Chlorophyll		E 140,	green	412 ($CHCl_3$)		Edible oils
23	Chlorophyllin copper complex		E 141,	green (W)	405 (W)		Confectionary and candy products, liqueurs, jellies, cream food products
24	Green S (Brilliant Green BS)		E 142,	green (W)	632 (W)	XIII	
25	Black BN		E 151,	violet with bluish tint	570 (W)	XIV	For fish roe coloring, confectionary and candy products

[a] Formulas in Table 8.11 a; [b] solvent W: water, CH: cyclohexane, M: methanol, H: hexane, and E: ethanol; [c] color is pH dependent

Table 8.11 a. Structures of the synthetic food colorants listed in Table 8.11.

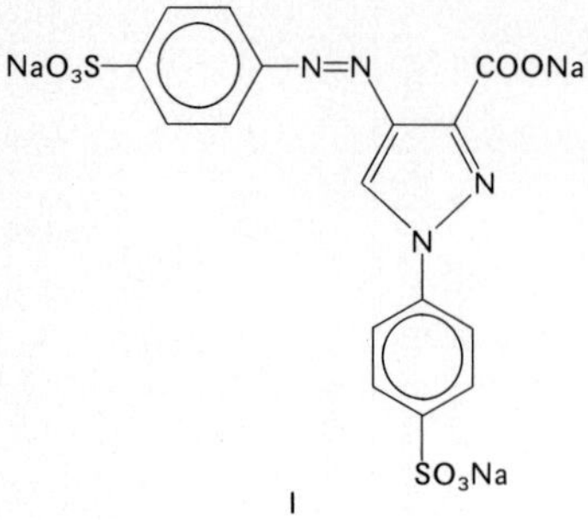

NaO3S– –N=N– COONa N N SO3Na
I

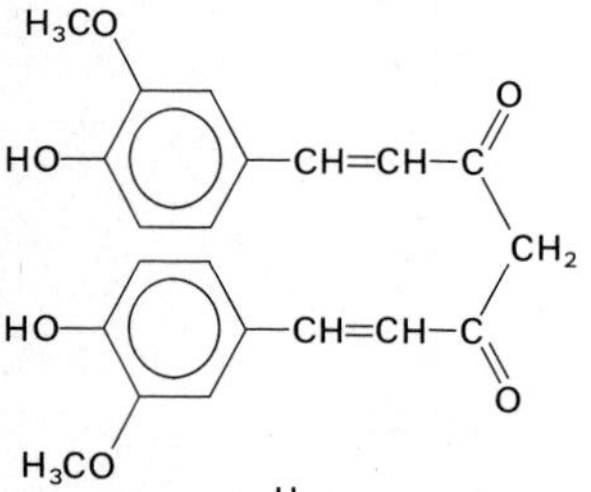

HO SO3Na NaO3S– –N=N– SO3Na
V

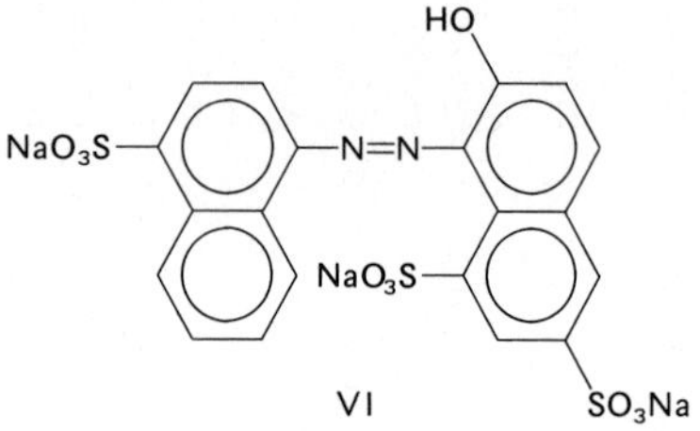

H3CO HO– –CH=CH–C O CH2 HO– –CH=CH–C O H3CO
II

HO NaO3S– –N=N– NaO3S– SO3Na
VI

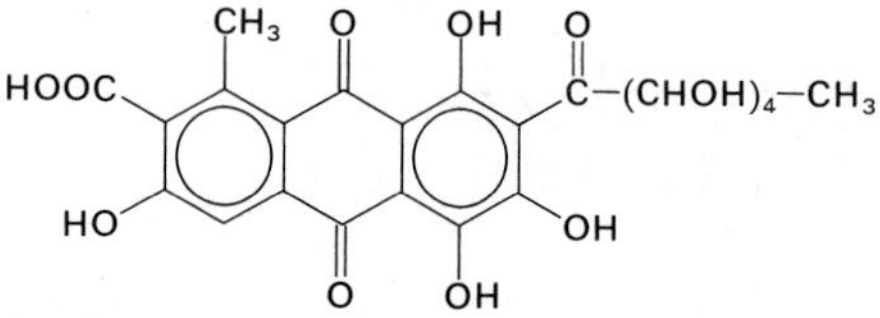

CH3 O OH O HOOC C–(CHOH)4–CH3 HO OH O OH
VII

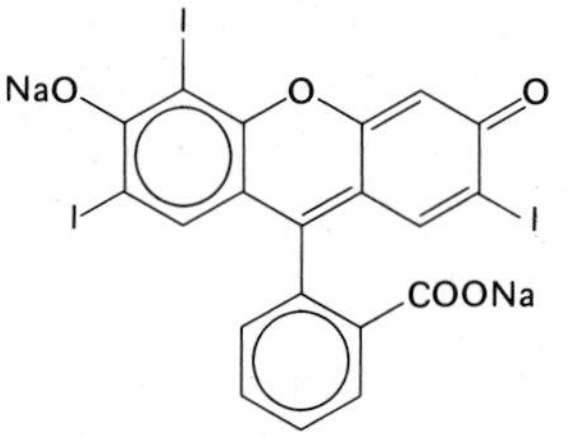

HO NaO3S– –N=N– SO3Na
III

I NaO O O I I COONa
VIII

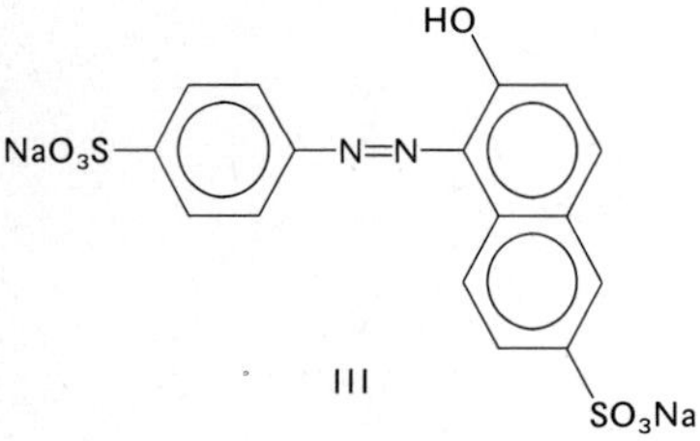

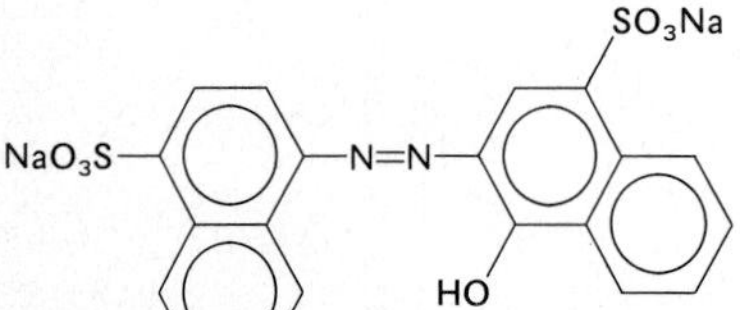

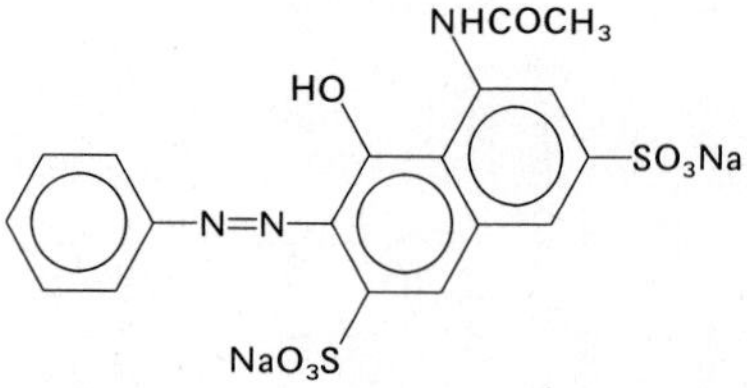

SO3Na NaO3S– –N=N– HO
IV

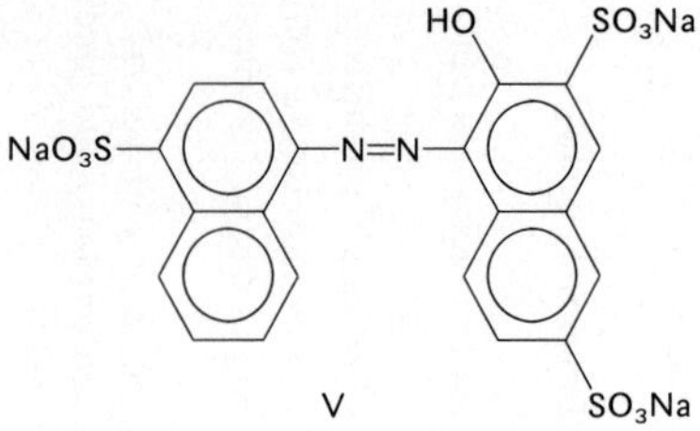

NHCOCH3 HO –SO3Na –N=N– NaO3S
IX

Table 8.11 a. (Continued)

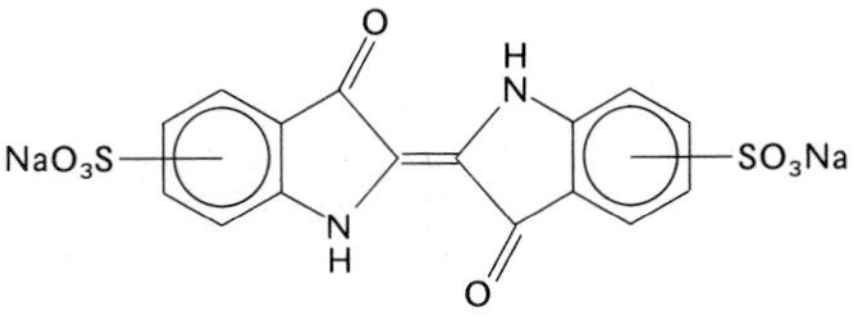

X

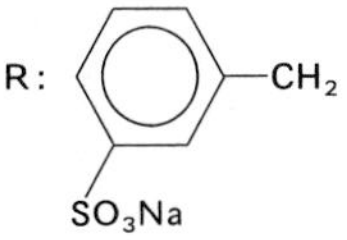

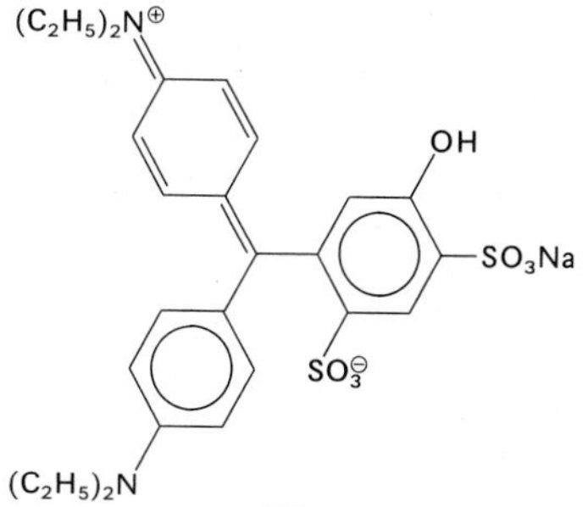

XI

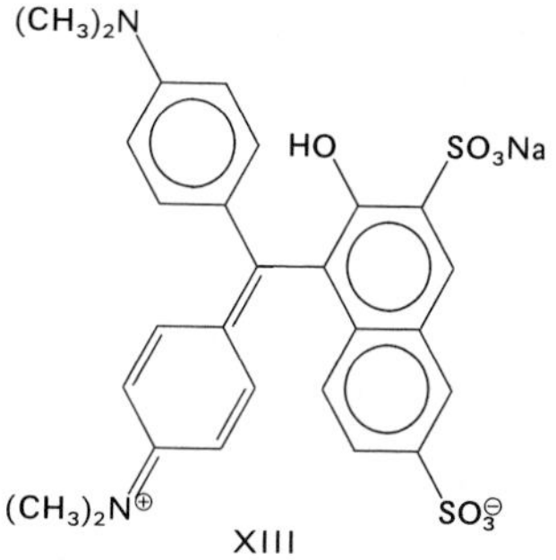

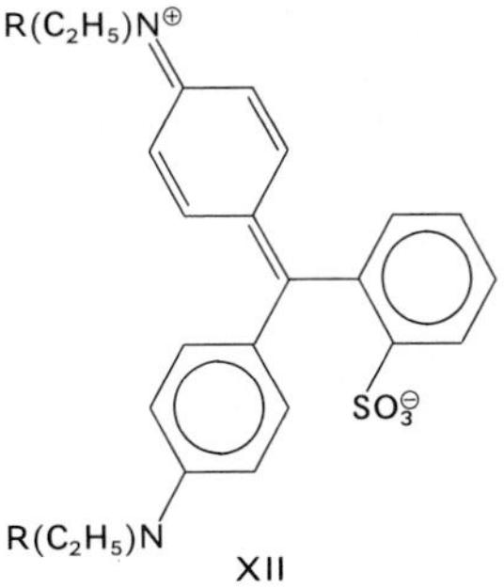

XII

XIII

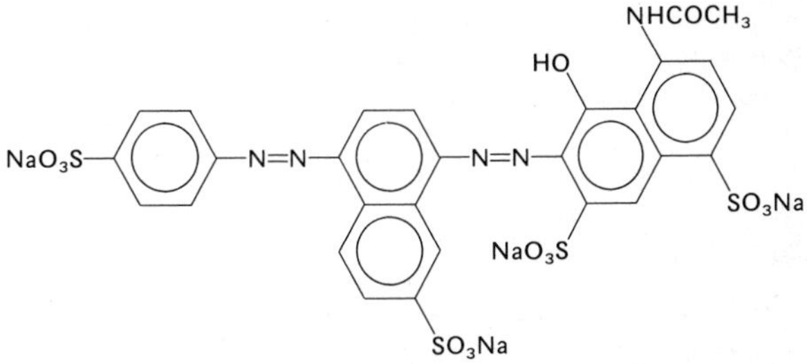

XIV

chain fatty acids, such as butyric and higher homologues, are used in aroma formulations.

8.10.2 Succinic Acid

The acid ($pK_1 = 4.19$; $pK_2 = 5.63$) is applied as a plasticizer in dough making. Succinic acid monoesters with glycerol are used as emulsifiers in the baking industry. The acid is synthesized by catalytic hydrogenation of fumaric or maleic acids.

8.10.3 Succinic Anhydride

This is the only acid anhydride used as a food additive. The hydrolysis proceeds slowly, hence the compound is suitable as an agent in baking powders and for binding of water in some dehydrated food products.

8.10.4 Adipic Acid

Adipic acid ($pK_1 = 4.43$; $pK_2 = 5.62$) is used in baking powders, in powdered fruit juice drinks, for improving the gelling properties of marmalades and fruit jellies, and for improving cheese texture. It is synthesized from phenol or cyclohexane (cf. Reactions 8.17).

8.10.5 Fumaric Acid

Fumaric acid ($pK_1 = 3.00$; $pK_2 = 4.52$) increases the shelf life of some dehydrated food products (e.g. pudding and jelly powders). It is also used

$$\text{Phenol} \xrightarrow[\text{Catalyst}]{H_2} \text{Cyclohexanol} \xrightarrow{HNO_3} HOOC{-}(CH_2)_4{-}COOH$$

$$\text{Cyclohexane} \xrightarrow[\text{Catalyst}]{O_2} \text{Cyclohexyl hydroperoxide (O–OH)} \rightarrow \text{Cyclohexanol (OH)} + \text{Cyclohexanone (=O)} \xrightarrow[O_2/\text{catalyst}]{HNO_3} HOOC{-}(CH_2)_4{-}COOH \quad (8.17)$$

to lower the pH, usually together with food preservatives (e.g. benzoic acid), and as an additive promoting gel setting. Fumaric acid is synthesized via maleic acid anhydride which is obtained by catalytic oxidation of butene or benzene (cf. Reaction 8.18) or is produced microbiologically from molasses by fermentation using *Rhizopus spp.*.

$$\text{Benzene} \xrightarrow{V_2O_5,\ 4.5\,O_2} \text{Maleic anhydride}; \quad \text{Butenes} \xrightarrow{V_2O_5,\ H_3PO_4,\ 3\,O_2} \text{Maleic anhydride} \xrightarrow[\text{Isom.}]{H_2O} HOOC{-}CH{=}CH{-}COOH \quad (8.18)$$

8.10.6 Lactic Acid

D,L- or L-lactic acid (pK = 3.86) is utilized as an 80% solution. A specific property of the acid is its formation of intermolecular esters, providing oligomers or a dimer lactide:

$$HO{-}CH(CH_3){-}COOH \rightarrow \text{Lactide (cyclic dimer)} \ /\ \text{oligomeric esters} \quad (8.19)$$

Such intermolecular esters are present in all lactic acid solutions with an acid concentration higher than 18%. More dilute solutions favor complete hydrolysis to lactic acid. The lactide can be utilized as an acid generator. Lactic acid is used for improving egg white whippability (pH adjustment to 4.8–5.1), flavor improvement of beverages and vinegar-pickled vegetables, prevention of discoloration of fruits and vegetables and, as calcium lactate, an additive in milk powders.
Lactic acid production is based on synthesis starting from ethanal, leading to racemic D,L-lactic acid, or on a homofermentation process (*Lactobacillus delbruckii, L. bulgaricus, L. leichmannii*) of carbohydrate-containing raw material, which generally provides L- but also D,L-lactic acids under the conditions of fermentation.

$$\text{Saccharose} \rightarrow \text{Lactate} \rightarrow \text{Ca-lactate} \xrightarrow{H_2SO_4} \text{Lactic acid}$$
$$CH_3CHO \xrightarrow{HCN} CH_3CHOHCN \rightarrow \text{Lactate} \quad (8.20)$$

8.10.7 Malic Acid

Malic acid ($pK_1 = 3.40$; $pK_2 = 5.05$) is widely utilized in manufacturing of marmalades, jellies and beverages and canning of fruits and vegetables (e.g. tomato). The monoesters with fatty alcohols are effective antispattering agents in cooking and frying fats and oils.
Malic acid synthesis, which provides the racemic D,L-form, is achieved by addition of water to maleic/fumaric acids.

8.10.8 Tartaric Acid

Tartaric acid ($pK_1 = 2.98$; $pK_2 = 4.34$) has a "rough", "hard" taste. It is utilized in fruit juice drinks and sour candies, and is added to baking powders as an acid component. Tartaric acid is produced from the pomace residue after must has been drained off, from maturing wine sediments, or by oxidation of maleic acid anhydride to the racemic mixture (cf. Reaction 8.21).

$$\text{Maleic anhydride} \xrightarrow{H_2O_2} \text{HOOC–CH(OH)–CH(OH)–COOH} \quad (8.21)$$

8.10.9 Citric Acid

Citric acid ($pK_1 = 3.09$; $pK_2 = 4.74$; $pK_3 = 5.41$) is utilized in candy production, fruit juice, ice cream, marmalade and jelly manufacturing, in vegetable canning and in dairy products such as processed cheese and buttermilk (aroma improver). It is also used to suppress browning in fruits and vegetables and as a synergetic compound for antioxidants. Its production is based on microbial fermentation of molasses by *Aspergillus niger*. The yield of citric acid is 50–70% of the fermentable sugar content.

8.10.10 Phosphoric Acid

Phosphoric acid ($pK_1 = 2.15$; $pK_2 = 7.1$; $pK_3 \sim 12.4$) and its salts account for 25% of all the acids used in food industries. The bulk of the acids (salts) used in the industry is citric acid (about 60%), while the use of other acids is only 15%. The main field of use of phosphoric acid is the soft drink industry (cola drinks). It is also used in fruit jellies, processed cheese and flour and as an active buffering agent or pH-adjusting ingredient in fermentation processes. Acid salts, e.g. $Ca(H_2PO_4)_2 \cdot H_2O$ (fast activity), $NaH_{14}Al_3(PO_4)_8 \cdot 4H_2O$ (slow activity) and $Na_2H_2P_2O_7$ (slow activity) are used in baking powders as components of the reaction to slowly or rapidly release the CO_2 from $NaHCO_3$.

8.10.11 Hydrochloric and Sulfuric Acids

Both acids are used in starch and sucrose hydrolyses. Hydrochloric acid is also used in protein hydrolysis in industrial production of seasonings.

8.10.12 Glucono-δ-lactone

Glucono-δ-lactone slowly hydrolyzes, releasing protons. Hence, it is applied as an additive when slow acidification is needed, as with baking powder, raw sausage ripening and several sour milk products.

8.11 Bases

NaOH and a number of alkaline salts, such as $NaHCO_3$, Na_2CO_3, $MgCO_3$, MgO, $Ca(OH)_2$, Na_2HPO_4 and Na-citrate, are used in food processing for various purposes, for example:

Ripe olive fruits are treated with 0.25–2% NaOH to eliminate the bitter flavor and to develop the desired dark fruit color.

In alkali-baked goods, molded dough pieces are dipped into 1.25% NaOH at 85–88 °C in order to form the typical flat deep brown surface during baking.

In chocolate manufacturing, $NaHCO_3$ enhances the *Maillard* reactions, providing dark bitter chocolates.

In production of molten processed cheese, the pH rise needed to improve the swelling of casein gels is achieved by addition of alkali salts.

8.12 Antimicrobial Agents

Elimination of microflora by physical methods is not always possible, therefore, antimicrobial agents are needed. The spectrum of compounds used for this purpose has hardly changed for a long time. It is not easy to find new compounds with wider biological activity, negligible toxicity for mammals and acceptable cost.

8.12.1 Benzoic Acid

Benzoic acid activity is directed both to cell walls and to inhibition of citrate cycle enzymes (α-ketoglutaric acid dehydrogenase, succinic acid dehydrogenase) and of enzymes involved in oxidative phosphorylation.

The acid is used in its alkali salt form as an additive, since solubility of the free acid is unsatisfactory low. However, the undissociated acid ($pK_a = 4.19$) is active – it is probably in this form that the acid acquires its ability to cross cell membranes.

Benzoic acid usually occurs in nature as a glycoside (in cranberry, bilberry, plum and cinnamon trees and cloves). Its activity is primarily against yeasts and molds, less so against bacteria. Figures 8.4 and 8.6 show the pH-dependent activity of the acid against *Escherichia coli, Staphylococcus aureus* and *Aspergillus niger*.

The LD_{50} (rats; orally) is 1.7–3.7 g/kg body weight; the LD_{100} (guinea pig, cat, dog, rabbit; orally) is 1.4–2 g/kg. A daily intake of <0.5 g Na-benzoate is tolerable for humans. No dangerous accumulation of the acid occurs in the body even at a dosage of as much as 4 g/day. It is eliminated by excretion in urine as hippuric acid while, at higher levels of intake, the glucuronic acid derivative is also excreted.

Benzoic acid (0.05–0.1%) is often used in combination with other preservatives and, on the basis of its higher activity at acidic pH's, it is used for preservation of sour food (pH 4–4.5 or lower), beverages with carbon dioxide, fruit salads, marmalades, jellies, fish preserves, margarine, paste (pâté) fillings and pickled sour vegetables. A change in aroma, occurring mostly in fruit products, may result as a consequence of benzoic acid esterification.

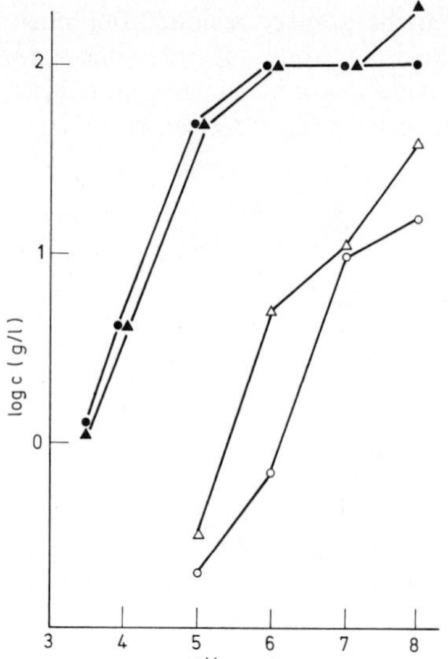

Fig. 8.4. The effect of benzoic acid on *Escherichia coli* (○ bacteriostatic. ● bactericidal activity) and *Staphylococcus aureus* (△ bacteriostatic and ▲ bactericidal activity)

8.12.2 PHB-Esters

The alkyl esters of p-hydroxybenzoic acid (PHB; parabens) are quite stable. Their solubility in water decreases with increasing alkyl chain length (methyl → butyl). The esters are mostly soluble in 5% NaOH.

The esters are primarily antifungal agents and are also active against yeasts but less so against bacteria, especially those which are gram-negative. The activity rises with increasing alkyl chain length (Figure 8.5). Nevertheless, because of better solubility, lower members of the homologous series are preferred.

The LD_{50} (mice; orally) is >8 g/kg body weight. In a feeding experiment over 96 weeks using 2% PHB-ester, no weight decrease was observed, while a slight decrease was found at the 8% level. In humans the compounds hydrolyze and are excreted in urine as p-hydroxybenzoic acid or its glycine or glucuronic acid conjugates.

Unlike benzoic acid, the esters can be used over a wide pH range since their acitivity is almost independent of pH (cf. Figure 8.6). As additives, they are applied at 0.3–0.06% as aqueous alkali solutions or as ethanol or propylene glycol solutions in fillings for baked goods, fruit juices, marmalades, syrups, preserves, olives and pickled sour vegetables.

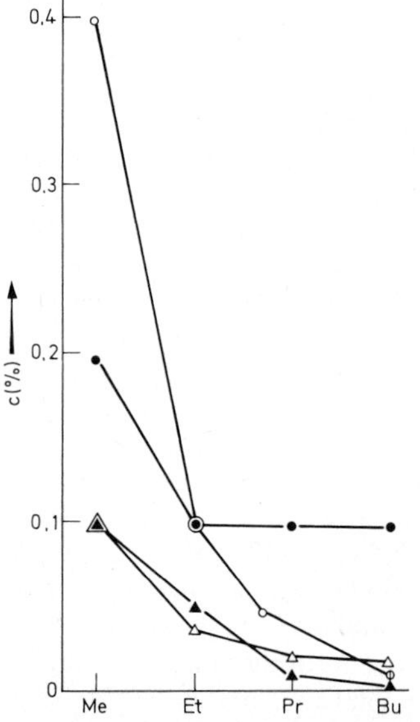

Fig. 8.5. Inhibition of *Salmonella typhosa* (●), *Aspergillus niger* (△), *Staphylococcus aureus* (○), and *Saccharomyces cerevisiae* (▲) by PHB-esters

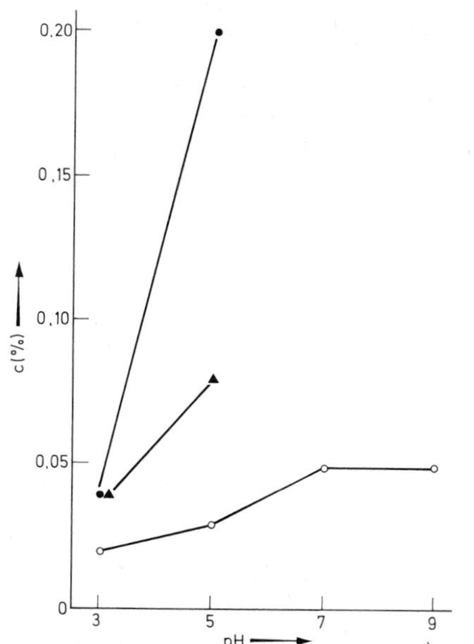

Fig. 8.6. Growth inhibition of *Aspergillus niger by benzoic acid (●), p-hydroxybenzoic acid propyl ester (○) and sorbic acid (▲).*

8.12.3 Sorbic Acid

The antimycotic effect of straight chain carboxylic acids has long been known. In particular the unsaturated acids, for example crotonic acid and its homologues, are very active. Sorbic acid (2-trans, 4-trans-hexadienoic acid; pK = 4.76) has the advantage that it is odorless and tasteless at the levels of use (0.3% or less).
The acid is obtained by several syntheses:

$$2\,CH_3CHO \longrightarrow CH_3CH(OH)CH_2CHO \xrightarrow{-H_2O} CH_3CH{=}CHCHO$$

CH_2CO → a: β-lactone (4-propenyl-2-oxetanone) → $H^{\oplus}/OH^{\ominus}$ → sorbic acid (COOH)
CH_2CO → b: polyester → 170–180° → sorbic acid (COOH)

a: BF_3, T < 0 °C b: Diethylene glycol, Zn-Isobutyrate, T > 25 °C (8.23)

- From parasorbic acid [(S)-2-hexen-5-olide); cf. Reaction 8.22]. The acid is present in berries of the mountain ash tree *(Sorbus aucuparia)*.

$$\text{parasorbic acid} \xrightarrow[\text{EtOH}]{\text{HCl in}} CH_3CH{=}CHCH{=}CHCOOEt \xrightarrow{H^{\oplus}} CH_3CH{=}CHCH{=}CHCOOH \quad (8.22)$$

- From crotonaldehyde obtained from ethanal (cf. Reaction 8.23).
- From ethanal:

$$3\,CH_3CHO \xrightarrow[\text{acetate}]{\text{Piperidine}} CH_3CH{=}CHCH{=}CHCHO + 2\,H_2O \xrightarrow[Ag/O_2/30\,°C]{pH > 12.5} CH_3CH{=}CHCH{=}CHCOOH \quad (8.24)$$

The second synthesis is the most important. The microbial activity of sorbic acid is primarily against fungi and yeasts, less so against bacteria. The activity is pH dependent (Figure 8.6). Its utilization is recommended up to pH 6.5.
The LD_{50} (rats) of sorbic acid is 4–6 g/kg body weight. Feeding experiments with rats for more than 90 days, with 1–8% sorbic acid in the diet, had no effect, while only 60% of the animals survive an 8% level of benzoic acid.
Sorbic acid is degraded biochemically like a fatty acid, i.e. by a β-oxidation mechanism. A small portion of the acid is degraded by ω-oxidation, yielding trans, trans-muconic acid (cf. Reaction 8.25).

$$\text{CH}_3\text{-CH=CH-CH=CH-COOH} \rightarrow \text{HOOC-CH=CH-CH=CH-COOH} \quad (8.25)$$

Some microorganisms, such as *Penicillium roqueforti,* have the ability to decarboxylate sorbic acid and thus convert it into 1,3-pentadiene, which has no antimicrobial activity and in addition contributes to an off-flavor in cheeses:

$$\text{CH}_3\text{-CH=CH-CH=CH-COOH} \rightarrow \text{CH}_3\text{-CH=CH-CH=CH}_2 \quad (8.26)$$

Sorbic acid or its salts are effective antifungal agents in baked products, cheeses, beverages (fruit juices, wines), marmalades, jellies, dried fruits and in margarine.

8.12.4 Propionic Acid

Propionic acid is most often found in nature where propionic acid fermentation occurs, as for example in Emmental cheese, in which it is present up to 1%.

Its antimicrobial activity is mostly against molds, less so against bacteria. Propionic acid has practically no effect against yeast. Its activity is pH dependent. It is recommended and used up to pH 5 and only occasionally up to pH 6.

Propionic acid is practically nontoxic. It is used as an additive in baked products for inhibition of molds, and to prevent ropiness caused by the action of *Bacillus mesentericus*. It is added to flour at 0.1–0.2% as its Ca-salt and is used in cheese manufacturing by dipping the cheese into an 8% solution of the acid.

8.12.5 Acetic Acid

The preserving activity of vinegar (cf. 22.3) has been known from ancient times. The acid has a two-fold importance: as a preservative and as a seasoning agent. It is more active against yeasts and bacteria than against molds. It is used as the free acid, Na- and Ca-salts, or as Na-diacetate ($CH_3COOH \cdot CH_3COONa \cdot 1/2\ H_2O$), in ketchup, mayonnaise, acid-pickled vegetables, bread and other baked products.

8.12.6 SO_2 and Sulfite

The activity of these preserving agents covers yeasts, molds and bacteria. The activity increases with decreasing pH and is mostly derived from undissociated sulfurous acid, which predominates at a pH < 3.

Toxicity is negligible at the levels usually applied. Possible mutagenic activity is under investigation. Thiamine is destroyed by SO_2. Excretion of sulfite in urine occurs after oxidation to sulfate.

SO_2 is used in the production of dehydrated fruits and vegetables, fruit juices, syrups, concentrates or purée. The form of application is SO_2, Na_2SO_3, K_2SO_3, $NaHSO_3$, $Na_2S_2O_5$ and $K_2S_2O_5$ at levels of 200 ppm or less.

SO_2 is added in the course of wine making prior to must fermentation to exclude the effect of interfering microorganisms. During wine fermentation with selected pure yeast cultures, SO_2 is used at a level of 50–100 ppm, while 50–75 ppm are used for wine storage.

SO_2 is not only antimicrobially active, but inhibits discoloration by blocking compounds with a reactive carbonyl group (*Maillard* reaction; nonenzymatic browning) or by inhibiting oxidation of phenols by phenol oxidase enzymes (enzymatic browning).

8.12.7 Diethyl (Dimethyl) Pyrocarbonate

Diethyl pyrocarbonate (DEPC or diethyl dicarbonate) is a colorless liquid of fruit-like or ester odor. Its antimicrobial acitivity covers yeasts (10–100 ppm), bacteria (*Lactobacilli:* 100–170 ppm) and molds (300–800 ppm). The levels of the compound required for a clear inhibition are given in brackets. Diethyl pyrocarbonate readily hydrolyzes to yield carbon dioxide and ethanol:

$$\text{C}_2\text{H}_5\text{O-CO-O-CO-OC}_2\text{H}_5 \rightarrow 2\,C_2H_5OH + 2\,CO_2 \quad (8.27)$$

or it reacts with food ingredients. In alcoholic beverages it yields a small amount of diethyl carbonate:

$$\text{C}_2\text{H}_5\text{O-CO-O-CO-OC}_2\text{H}_5 \xrightarrow{C_2H_5OH} \text{C}_2\text{H}_5\text{O-CO-OC}_2\text{H}_5 + CO_2 + C_2H_5OH \quad (8.28)$$

In the presence of ammonium salts, DEPC can form ethyl urethane in a pH-dependent reaction (cf. Reaction 8.29).

$$\text{EtO-CO-O-CO-OEt} + NH_3 \longrightarrow \text{EtO-CO-NH}_2 + CO_2 + C_2H_5OH \quad (8.29)$$

Since diethyl carbonate may be a teratogenic agent and ethyl urethane is a carcinogen, the use of diethyl pyrocarbonate is discussed under toxicological aspects. The compound should be replaced by dimethyl pyrocarbonate, since methyl urethane, unlike ethyl urethane, is not carcinogenic.
Both compounds are used in cold pasteurization of fruit juices, wine and beer, at a concentration of 120–300 ppm.

8.12.8 Ethylene Oxide, Propylene Oxide

These compounds are active against all microorganisms, particularly vegetative cells and spores, and also against viruses. Propylene oxide is somewhat less reactive than ethylene oxide.
The pure compounds, since they are efficient alkylating agents, are very toxic. After their application, all the residual amounts must be completely removed. The glycols resulting from their hydrolysis are not as toxic (ethylene glycol: LD_{50} for rats is 8.3 g/kg body weight). Toxic reaction products can be formed, as exemplified by chlorohydrin obtained in the presence of chloride:

$$\text{ethylene oxide} \xrightarrow{Cl^{\ominus}} \text{HO-CH}_2\text{-CH}_2\text{-Cl} \quad (8.30)$$

In addition, some essential food constituents react with formation of biologically inactive derivatives. Examples are riboflavin, pyridoxine, niacin, folic acid, histidine or methionine. However, these reactions are not of importance under the conditions of the normal application of ethylene oxide or propylene oxide.
Both compounds are used as gaseous sterilants (ethylene oxide, boiling point 10.7 °C; propylene oxide, 35 °C) against insects and for gaseous sterilization of some dehydrated foods for which other methods, e. g. heat sterilization, are not suitable. Examples are the gaseous sterilization of walnuts, starches, dehydrated foods (fruits and vegetables) and, above all, spices, in which a high spore count (and plate count in general) is often a sanitary problem. The sterilization is carried out in pressure chambers in a mixture with an inert gas (e. g. 80–90% CO_2). The need to remove the residual unreacted gas (vacuum by "gaseous rinsing") has already been emphasized. An alternative method of sterilization for the above-mentioned food products is high energy irradiation (UV-light, X-ray, or gamma irradiation).

8.12.9 Nitrite, Nitrate

These additives are used primarily to preserve the red color of meat (cf. 12.3.2.2.2). However, they also have antimicrobial activity, particularly in a mixture with common salt. Of importance is their inhibitive action, in nonsterilized meat products, against infections by *Clostridium botulinum* and, consequently, against accumulation of its toxin.
The activity is dependent on pH and is proportional to the level of free HNO_2. Acute toxicity has been found only at high levels of use (formation of methemoglobin). A problem is the possibility of the formation of nitrosamines, compounds with powerful carcinogenic activity. Numerous animal feeding tests have demonstrated tumor occurrence when the diet contained amines (sensitive to nitroso substitution) and nitrite. Consequently, the trend is to exclude or further reduce the levels of nitrate and nitrite in food. No suitable replacement has been found for nitrite in meat processing.

8.12.10 Antibiotics

The use of antibiotics in food preservation raises a problem since it might trigger development of more resistant microorganisms and thus create medical/therapeutic difficulties.
Of some importance is nisin, a polypeptide-type of antibiotic, produced by some *Streptococcus lactis* strains. It is active against Gram-positive microorganisms and all spores, but is not used in human medicine. This heat-resistant peptide is applied as an additive for sterilization of dairy products, such as cheeses or condensed or evaporated milk (cf. 1.3.4.3).
Pimaricin (Formula 8.31) is active at 5–100 ppm against yeasts and molds and is used as an additive in surface treatment of cheeses. It also finds application for suppressing the growth of molds on ripening raw sausages.

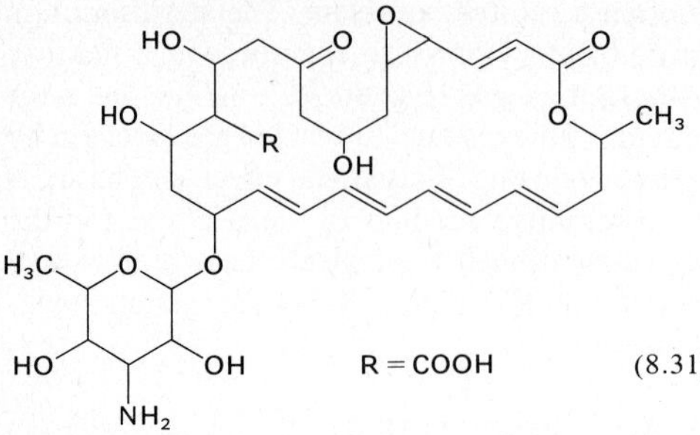

$$R = COOH \quad (8.31)$$

The possibility of incorporating the wide spectrum antibiotics chlortetracycline and oxytetracycline into fresh meat, fish and poultry, in order to delay spoilage, is still under investigation.

8.12.11 Diphenyl

Diphenyl, due to its ability to inhibit growth of molds, is used to prevent their growth on peels of citrus fruits (lemon, orange, lime, grapefruit). It is applied by impregnating the wrapping paper and/or cardboard packaging material (1–5 g diphenyl/m^2).

8.12.12 o-Phenylphenol

This compound, at a level of 10–50 ppm and a pH range of 5–8, inhibits the growth of molds. The inhibition effect, which is enhanced with a pH rise, is utilized in the preservation of citrus fruits. It is applied by dipping the fruits into a 0.5–2% solution at pH 11.7.

8.12.13 Thiabendazole, 2-(4-Thiazolyl)benzimidazole

This compound is particularly powerful against molds which cause the so-called blue mold rots, e.g. *Penicillium italicum* (blue-green-spored "contact mold") and *Penicillium digitatum* (green-spored mold). It is used for preserving the peels of citrus fruits and bananas. The application mode is by dipping or spraying the fruit with a wax emulsion containing 0.1–0.45% thiabendazole.

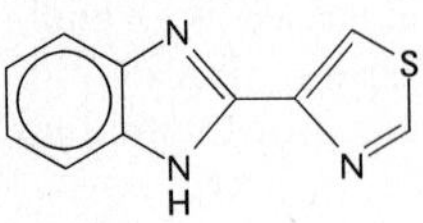

(8.32)

8.13 Antioxidants

Since lipids are widely distributed in food and since their oxidation yields degradation products of great aroma impact, their degradation is an important cause of food deterioration by generation of undesirable aroma. Lipid oxidation can be retarded by oxygen removal or by using antioxidants as additives. The latter are mostly phenolic compounds, which provide the best results often as a mixture and in combination with a chelating agent. The most important antioxidants, natural or synthetic, are tocopherols, ascorbic acid esters, gallic acid esters, tert-butylhydroxyanisole (BHA) and di-tert-butylhydroxytoluene (BHT). They are covered in 3.7.3.2.2.

8.14 Chelating Agents (Sequestrants)

Chelating agents have acquired greater importance in food processing. Their ability to bind metal ions has contributed significantly to stabilization of food color, aroma and texture. Many natural constituents of food can act as chelating agents, e.g. carboxylic acids (oxalic, succinic), hydroxy acids (lactic, malic, tartaric, citric), polyphosphoric acids (ATP, pyrophosphates), amino acids, peptides, proteins and porphyrins.
Table 8.12 lists the chelating agents utilized by the

Table 8.12. Chelating agents used as additives in food processing. (Compounds given in brackets are utilized only as salts or derivatives)

(Acetic acid)	Na-, K-, Ca-salts
Citric acid	Na-, K-, Ca-salts monoisopropyl ester monoglyceride ester triethyl ester monostearyl ester
EDTA	Na-, Ca-salts
(Gluconic acid)	Na-, Ca-salts
Oxystearin	
Orthophosphoric acid	Na-, K-, Ca-salts
(Pyrophosphoric acid)	Na-salt
(Triphosphoric acid)	Na-salt
(Hexametaphosphoric acid, 10–15 residues)	Na-, Ca-salts
(Phytic acid)	Ca-salt
Sorbitol	
Tartaric acid	Na-, K-salts
(Thiosulfuric acid)	Na-salt

food industry, while Table 8.13 gives the stability constants for some of their metal complexes.
Traces of heavy metal ions can act as catalysts for fat or oil oxidation. Their binding by chelating agents increases antioxidant efficiency and inhibits oxidation of ascorbic acid and fat-soluble vitamins. The stability of the aroma and color of canned vegetables is substantially improved.
In production of herb and spice extracts, the combination of an antioxidant and a chelating agent provides an improved extract quality. Chelating agents are also used in dairy products, wherein their deaggregating activity for the casein complexes is often utilized; in blood recovery processes, since they prevent clotting; and in the sugar industry to facilitate sucrose crystallization, a process which is otherwise retarded by sucrose-metal complexes.

8.15 Surface-Active Agents

Naturally occurring and synthetic surface-active agents (tensides), some of which are listed in Table 8.14, are used in food processing when a decrease in surface tension is required e.g. in production and stabilization of all kinds of dispersions (Table 8.15).
Dispersions include emulsions, foams, aerosols and suspensions (Table 8.15a). In all cases an *outer, continuous* phase is distinct from an *inner,* discontinuous, *dispersed* phase. Emulsions are of particular importance and they will be outlined in more detail.

8.15.1 Emulsions

Emulsions are dispersed systems, usually of two immiscible liquids. When the outer phase consists of water and the inner of oil, it is considered as an "oil in water" (o/w) type of emulsion. When this is reversed, i.e. water is dispersed in oil, a w/o emulsion exists. Examples of food emulsions are: milk (o/w), butter (w/o) and mayonnaise (o/w).
The visual appearance of an emulsion depends on droplet diameter. When 1 μm or greater, the emulsion is milky-turbid. When the diameter is close to light wavelengths (10^{-5}–10^{-6} cm), the system is optically clear and is called a microemulsion.

Table 8.13. Stability constants (pK-values) of some metal complexes

Chelating agent	Ca^{2+}	Co^{2+}	Cu^{2+}	Fe^{2+}	Fe^{3+}	Mg^{2+}	Zn^{2+}
Acetate	0.5	2.2				0.5	1.0
Glycine	1.4	5.2	8.2	4.3	10.0	3.5	5.2
Citrate	3.5	4.4	6.1	3.2	11.9	2.8	4.5
Tartrate	1.8		3.2		7.5	1.4	2.7
Gluconate	1.2		18.3			0.7	1.7
Pyrophosphate	5.0		6.7		22.2	5.7	8.7
ATP	3.6	4.6	6.1			4.0	4.3
EDTA	10.7	16.2	18.8	14.3	25.7	8.7	16.5

Table 8.14. Surfactants (surface active agents) in food

I. Naturally occurring:

A. Ionic: proteins, phospholipids (lecithin, cf. 3.4.1.1), bile acids

B. Nonionic: glycolipids (cf. 3.4.1.2), saponins

II. Synthetic:

A. Ionic: stearyl-2-lactylate

B. Nonionic: mono-, diacylglycerols and their acetic-, citric-, tartaric-, and lactic acid esters, saccharose fatty acid esters, sorbitan fatty acid esters, polyoxyethylene sorbitan fatty acid esters

Table 8.15. Examples of surfactant utilization in food industry

Utilization in production of	Effect
Margarine	Stabilization of a w/o emulsion
Mayonnaise	Stabilization of an o/w emulsion
Ice cream	Stabilization of an o/w emulsion, achievement of a "dry" consistency
Sausages	Prevention of fat separation
Bread and other baked products	Improvement of crumb structure, baked product volume, inhibition of starch retrogradation (bread staling)
Chocolate	Improvement of rheological properties, inhibition of "fat blooming"
Instant powders	Solubilization
Spice extracts	Solubilization

Table 8.15a. Dispersion systems

Type	Inner phase	Outer phase
Emulsion	liquid	liquid
Foam	gaseous	liquid
Aerosol	liquid or solid	gaseous
Suspension solid	liquid	liquid

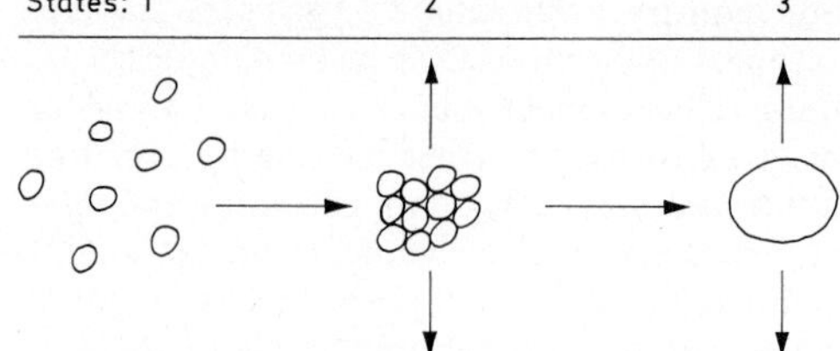

Fig. 8.7. Changes in an emulsion. 1 The droplets are dispersed in a continuous phase. 2 The droplets form aggregates. An increase in particle diameter results in acceleration of their flotation or sedimentation. 3 Coalescence: the aggregated droplets merge into large droplets. Lastly, two continuous phases are formed; the emulsion is destroyed.

Each emulsifier can disperse a limited amount of an inner phase, i. e. it has a fixed *capacity*. When the limit is reached, further addition of outer phase breaks down the emulsion. The capacity and other related parameters differ among emulsifiers and can be measured accurately under standardized conditions.

8.15.2 Emulsifier Action

Emulsions are made and stabilized with the aid of a suitable tenside, usually called an emulsifier. Its activity is based on its molecular structure. There is a lipophilic or hydrophobic part with good solubility in a nonaqueous phase, such as an oil or fat, and a polar and hydrophilic part, soluble in water. The hydrophobic part of the molecule is generally a long-chain alkyl residue, while the hydrophilic part consists of a dissociable group or of a number of hydroxyl groups.

In an immiscible system such as oil/water, the emulsifier is located on the interface, where it decreases interfacial tension. Thus, even in very low concentration, it facilitates a fine distribution of one phase within the other. The emulsifier also prevents droplets, once formed, from aggregating and coalescing, i. e. merging into a single, large drop (Figure 8.7).

Ionic tensides stabilize o/w emulsions in the following way (Figure 8.8a): on the interface, their alkyl residues are solubilized in oil droplets, while the charged end groups project into the aqueous phase. The involvement of counter ions forms an electrostatic double layer, which prevents oil droplet aggregation.

Nonionic, neutral tensides are oriented on the oil droplet's surface with the polar end of the tenside projecting into the aqueous phase. The coalescence of the droplets of an o/w emulsion is prevented by an anchored "hydrate shell" built around the polar groups.

The coalescence of water droplets in a w/o emulsion first requires that water molecules break through the double-layered hydrophobic region of emulsifier molecules (Figure 8.8b). This escape is only possible when sufficient energy is supplied to rupture the emulsifier's hydrophobic interaction.

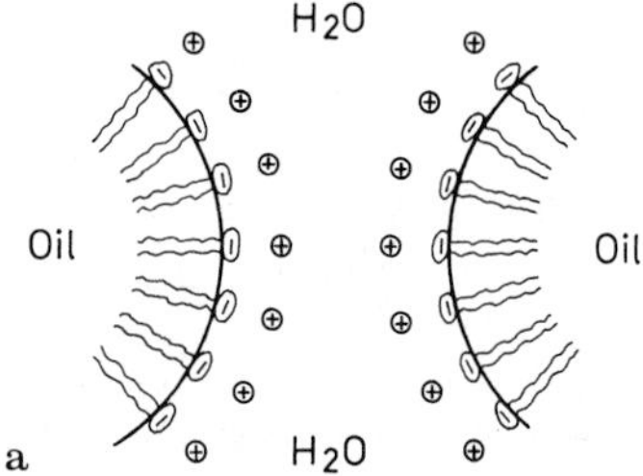

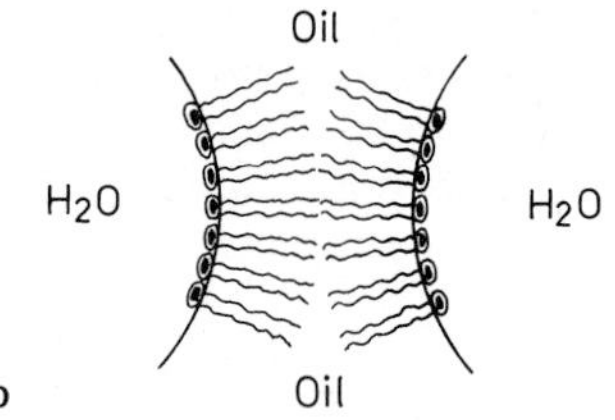

Fig. 8.8. Stabilization of an emulsion. **a** Activity of an ionic emulsifier in an o/w emulsion. **b** Activity of a nonpolar emulsifier in w/o emulsion. ○ Polar groups, ~~ apolar tails of the emulsifier

The stability of an emulsion is increased when additives are added which curtail droplet mobility. This is the basis of the stabilization effect of hydrocolloids (cf. 4.4.3) on o/w emulsions in which the stabilization is a consequence of a rise in viscosity of the outer, aqueous phase.

A rise in temperature negatively affects emulsion stability, and can be applied whenever an emulsion has to be destroyed. Elevated temperatures are used along with shaking, agitation or pressure (mechanical destruction of interfacial films as, for example, in butter manufacturing, cf. 10.2.3). Other ways of decreasing the stability of an emulsion are addition of ions which collapse the electrostatic double layer, or hydrolysis to destroy the emulsifier.

8.15.3 HLB-Value

From the models depicted above, it follows that a tenside with a relatively stronger lipophilic and a weaker hydrophilic group is predominantly soluble in the oil phase and in consequence acts preferentially as stabilizer of w/o emulsions. The same is valid for the opposite tenside and emulsion type. A numerical measure has been developed to assess the relative strength or "activity" of hydrophilic and lipophilic groups. This is the so-called *HLB-value* ("hydrophilic-lipophilic balance"). The value can be determined, for example, from dielectric constants or from chromatographic properties of the tensides. The HLB-value of a fatty acid ester of a polyhydroxy alcohol is obtained from (SV = saponification number of the tenside; AV = acid value of the separated acids):

$$\mathrm{HLB} = 20\left(1 - \frac{\mathrm{SV}}{\mathrm{AV}}\right) \quad (8.33)$$

Table 8.16 gives HLB-values for various substances. HLB-values as related to the field of application are listed in Table 8.17; low values favor w/o emulsions and higher values, o/w emulsions.

With nonionic, neutral emulsifiers it is observed that, with a rise in temperature, the extent of polar group hydration decreases and the influence of the lipophilic part increases. Consequently, at elevated temperatures a phase change from o/w to w/o might occur with some emulsifiers. The temperature at which the conversion occurs is called the "phase conversion temperature".

8.15.4 Synthetic Emulsifiers

Nonionic emulsifiers are the most important of the numerous synthetic emulsifiers used by the food industry. Unlike ionic emulsifiers, nonionic emulsifiers are not in danger of decreasing in interfacial activity by interaction with salt in food.

The utilization of emulsifiers is legislated and often differently regulated in some countries. The synthetic emulsifiers described below are used worldwide.

Table 8.16. Hydrophilic lipophilic balance (HLB) values of some surfactants

Compound	HLB-value
Oleic acid	1.0
Sorbitol tristearate	2.1
Stearyl monoglyceride	3.4
Sorbitol monostearate	4.7
Sorbitol monolaurate	8.6
Gelatin	9.8
Polyoxyethylene sorbitol tristearate	10.5
Methylcellulose	10.5
Polyoxyethylene sorbitol monostearate	14.9
Polyoxyethylene sorbitol monooleate	15.0
Sodium oleate	18.0
Potassium oleate	20.0

Table 8.17. HLB-values related to their industrial application

HLB-range	Application
3– 6	w/o-emulsifiers
7– 9	Humectants
8–18	o/w-emulsifiers
15–18	Turbidity stabilization

8.15.4.1 Mono-, Diacylglycerols

These nonionic emulsifiers were described in 3.3.2.

8.15.4.2 Sugar Esters

They are obtained, among other methods, by transesterification of fatty acid methyl esters (14:0, 16:0, 18:0 and/or 18:1, double bond position 9) with sucrose and lactose. The resultant mono – and diesters are odorless and tasteless. Depending on their structure, they cover an HLB range of 7–13, and are used in stabilization of o/w emulsions, or in stabilization of some instant dehydrated and powdered foods.

8.15.4.3 Sorbitan Fatty Acid Esters

Sorbitol is heated with fatty acids in the presence of NaOH or H_2SO_4 (180–250 °C) to produce 1,4- and 1,5-sorbitan esters:

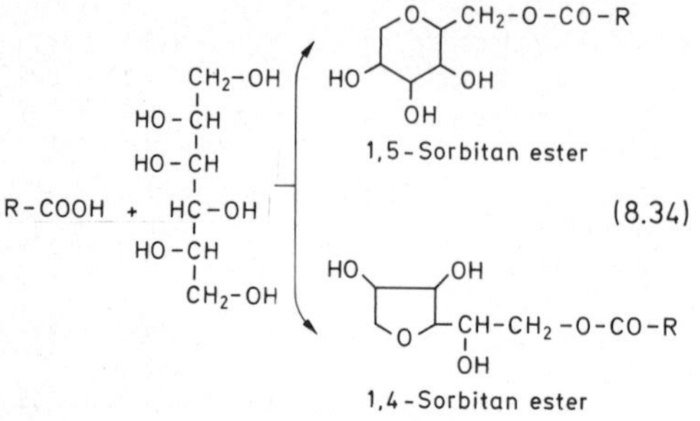

$$R\text{-}COOH + HOCH_2\text{-}(CHOH)_4\text{-}CH_2OH \rightarrow \text{1,5-Sorbitan ester} + \text{1,4-Sorbitan ester} \quad (8.34)$$

Sorbitan fatty acid esters are used, for example, to stabilize w/o emulsions (HLB-value 1.8 to about 9).

8.15.4.4 Polyoxyethylene Sorbitan Esters

Polyoxyethylene groups are introduced into the molecules to increase the hydrophilic property of sorbitan esters (cf. Formula 8.35).

Polyoxyethylene sorbitan monoesters (examples in Table 8.16) are used to stabilize o/w emulsions.

8.15.4.5 Stearyl-2-Lactylate

Stearyl-2-lactylate is the only synthetic tenside with dissociable groups which is used in food processing (cf. Formula 8.36).

The HLB-value of the sodium salt is 8–9, and that of the calcium salt, 6–7. The sodium salt is used to stabilize an o/w emulsion which is subjected to repeated cycles of freezing and thawing.

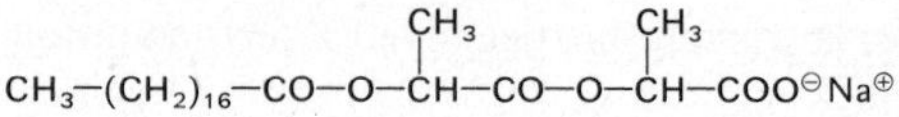

$$CH_3\text{—}(CH_2)_{16}\text{—CO—O—}\underset{}{\overset{CH_3}{CH}}\text{—CO—O—}\overset{CH_3}{CH}\text{—COO}^{\ominus}Na^{\oplus} \quad (8.36)$$

8.16 Thickening Agents, Gel Builders, Stabilizers

A number of polysaccharides and their modified forms, even at low concentrations, are able to increase a system's viscosity, to form gels and to stabilize emulsions, suspensions or foams. These compounds are also active as crystallization inhibitors (e. g. in confections, ice creams) and are suitable for aroma encapsulation, as is often needed for dehydrated food. These properties make polysaccharides important additives in food processing and storage. The compounds of importance, together with their properties and utilization, were described in detail in the chapter on carbohydrates. Among proteins, gelatin is an important gel-forming agent used widely in food products (cf. 12.3.2.3.1).

8.17 Humectants

Some polyols (1,2-propanediol, glycerol, mannitol, sorbitol) have distinct hygroscopic properties and act as humectants, i. e. additives for retaining food moisture and softness and inhibiting crystallization, as is often required in a confectionery product. When glycerol or sorbitol is added to mashed vegetables or fruits or in the production of other powdered foods before the final drying stage, the dehydrated products have improved rehydration characteristics.

8.18 Anticaking Agents

Some food products, such as common salt, seasoning salt (e. g. a mixture of onion or garlic powder with common salt), dehydrated vegetable and fruit powders, soup and sauce powders and baking powder, tend to cake or harden into a hard lump. Lumping can be avoided by using any of a number of compounds that either absorb water or provide protective hydrophobic films. Anticaking compounds include sodium-, potassium- and calcium hexacyanoferrate (II), calcium silicate, tricalcium phosphate, magnesium silicate and magnesium carbonate.

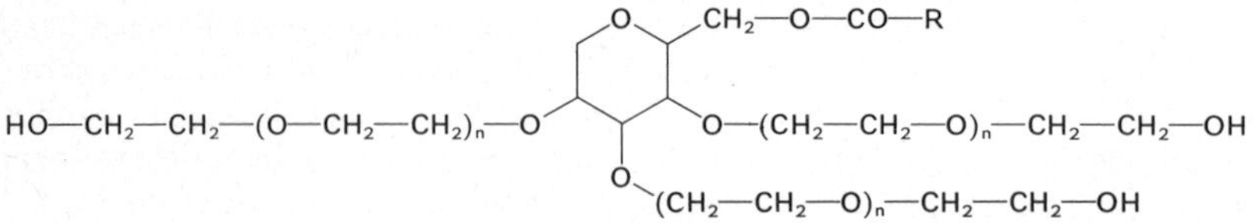

$$HO\text{—}CH_2\text{—}CH_2\text{—}(O\text{—}CH_2\text{—}CH_2)_n\text{—O—[sorbitan ring: }CH_2\text{—O—CO—R; O—}(CH_2\text{—}CH_2\text{—}O)_n\text{—}CH_2\text{—}CH_2\text{—OH; O—}(CH_2\text{—}CH_2\text{—}O)_n\text{—}CH_2\text{—}CH_2\text{—OH]} \quad (8.35)$$

8.19 Bleaching Agents

Bleaching is used primarily in flour production. The removal of yellow carotenoids by oxidation can be achieved by a number of compounds that, in addition to bleaching, improve the baking quality of flour. Examples of some approved common bleaching agents are Cl_2, ClO_2, NOCl, NO_2 and N_2O_4. Lipoxygenase enzyme also has an efficient bleaching activity.

8.20 Clarifying Agents

In some beverages, such as fruit juices, beer or wine, turbidity and sediment formation can occur with the involvement of phenolic compounds, pectins and proteins. These defects can be corrected by: (a) partial enzymatic degradation of pectins and proteins; (b) removal of phenolic compounds with the aid of gelatin, polyamide or polyvinyl pyrrolidone powders; and (c) by protein removal with bentonite (an aluminum silicate) or tannin.

8.21 Propellants, Protective Gases

Food sensitive to oxidation and/or microbial spoilage can be stored in an atmosphere of protective gas or a gas mixture (N_2, CO_2, CO, etc.; modified or controlled atmosphere storage). This is often a suitable method for lengthening the shelf life of food.

Liquid food can be filled into pressurized containers and, when needed, using a propellant, be discharged in the form of a cream or paste (e.g. cream cheese, ketchup), a foam (whipping cream) or a mist (herb or spice extracts in oil; liquid barbecue smoke). Propellants used are N_2, N_2O, CO_2, octafluorocyclobutane (Freon 318) and chloropentafluoroethane (Freon 115).

Due to its low solubility in water, fat and oil, N_2 is used preferentially as a propellant when foam formation is not desired (ketchup). On the other hand, gases such as N_2O and CO_2 are preferred at times due to their good solubility in water. The fluorinated hydrocarbons are readily soluble in fats and oils and, therefore, are used for making of instant foam (whipping cream).

8.22 Literature

Beets, M.G.J.: Structure-activity relationships in human chemoreception. Applied Science Publ.: London. 1978

Belitz, H.-D., Chen, W., Jugel, H., Treleano, R., Wieser, H., Gasteiger, J., Marsili, M.: Sweet and bitter compounds: Structure and taste relationship. In: Food taste chemistry (Ed.: Boudreau, J.C.), ACS Symposium Series 115, p. 93, American Chemical Society: Washington, D.C. 1979

Belitz, H.-D., Chen, W., Jugel, H., Stempfl, W., Treleano, R., Wieser, H.: Structural requirements for sweet and bitter taste. In: Flavour '81 (Ed.: Schreier, P.), p. 741, Walter de Gruyter: Berlin–New York. 1981

Belitz, H.-D., Chen, W., Jugel, H., Stempfl, W., Treleano, R., Wieser, H.: QSAR of bitter tasting compounds. Chem. Ind. *1983,* 23

Birch, G.G., Brennan, J.G., Parker, K.J. (Eds.): Sensory properties of foods. Applied Science Publ.: London. 1977

Compadre, C.M., Pezzuto, J.M., Kinghorn, A.D.: Hernandulcin: An intensely sweet compound discovered by review of ancient literature. Science *227,* 417 (1985)

Furia, T.E. (Ed.): Handbook of food additives. 2nd edn., CRC Press: Cleveland, Ohio. 1972

Griffin, W.C., Lynch, M.J.: Surface active agents. In: Handbook of food additives, 2nd end. (Ed.: Furia, T.E.), p. 397, CRC Press: Cleveland, Ohio. 1972

Heusch, R.: Emulsionen. In: Ullmanns Encyklopädie der technischen Chemie, 4. Aufl., Bd. 10, S. 449, Verlag Chemie: Weinheim. 1975

Iyengar, R.B., Smits, P., Van der Ouderaa, F., Van der Wel, H., Van Brouwershaven, J., Ravestein, P., Richters, G., von Wassenaar, P.D.: The complete amino-acid sequence of the sweet protein Thaumatin I. Eur. J. Biochem. *96,* 193 (1979)

Kier, L.B.: A molecular theory of sweet taste. J. Pharm. Sci. *61,* 1394 (1972)

Lange, H., Kurzendorfer, C.-P.: Zum Mechanismus der Stabilisierung von Emulsionen. Fette Seifen Anstrichm. *76,* 120 (1974)

Lück, E.: Chemische Lebensmittelkonservierung. Springer Verlag: Berlin-Heidelberg. 1977

Powrie, W.D., Tung, M.A.: Food dispersions. In: Principles of food science, part I (Ed.: Fennema, O.R.), p. 539, Marcel Dekker, Inc.: New York. 1976

Rosival, L., Engst, R., Szokolay, A.: Fremd- und Zusatzstoffe in Lebensmitteln. VEB Fachbuchverlag: Leipzig. 1978

Shallenberger, R.S., Acree, T.E.: Chemical structure of compounds and their sweet and bitter taste. In: Handbook of sensory physiology, Vol. IV, Part 2 (Ed.: Beidler, L.M.), p. 221, Springer Verlag: Berlin–Heidelberg. 1971

Souci, S.W., Mergenthaler, E.: Fremdstoffe in Lebensmitteln. J.F. Bergmann Verlag: München. 1958

Van der Wel, H., Bel, W.J.: Effect of acetylation and methylation on the sweetness intensity of Thaumatin I. Chem. Senses Flavor *2,* 211 (1976)

9 Food Contamination

9.1 General Remarks

Special attention must be paid to the possibility of contamination of food with toxic compounds. They may be present incidentally and may be derived in various ways. Examples of such contaminants are:

- Pollutants derived from burning of fossil fuels, radionuclides from fallout, or emissions from industrial processing (toxic trace elements, radionuclides, polycyclic aromatic hydrocarbons).
- Components of packaging material, and of other frequently used products (monomers, polymer stabilizers, plasticizers, polychlorinated biphenyls, cleansing/washing agents and disinfectants).
- Toxic metabolites of microorganisms (enterotoxins, mycotoxins).
- Residues of agricultural chemicals used for crop protection (pesticides, such as insecticides, fungicides and herbicides, antisprouting agents, plant hormones).
- Residues from livestock and poultry husbandry (veterinary medicinals and feed additives).

Toxic food contaminants might also be formed within the food itself or within the human digestive tract by reactions of some food ingredients and additives (e.g. nitrosamines).

Measures required to prevent contamination include:

- Extensive analytical control of food.
- Determination of the sources of contamination.
- Legislation (legal standards to permit, ban, curtail or control the use of potent food contaminants, and the processes associated with them) to establish permissible levels of contaminants.

Such efforts have paid off, as shown in a report of the German Society for Nutrition released in 1980 on chemical/toxicological aspects of food, which stated:

"When the overall situation in food processing and distribution is compared to that described in the nutrition report of 1976, we have to say that no new health hazard source has been discovered. Where contamination problems existed, appropriate measures were taken to reduce or avoid them; consequently, decreases in residues hazardous to health have already been observed. Where residue problems can not be avoided, our duty is to inform the consumer not to include the food as a regular part of the daily diet but to less frequently consume some of the more contaminated foods, such as some wild mushrooms, kidney, liver, specialty meats and freshwater fish. Where man is obviously involved in creation of residue problems, by applying pesticides improperly or which are banned, or applying illegal animal medicinals or by not observing the waiting period after use, only greater vigilance and control can be of help. This increasingly concerns the individual producers of milk and eggs, growers of fruit and vegetables and, particularly, importers of some food commodities."

Toxicological assessment of a contaminant may, for various reasons, be a difficult task. Firstly, sufficient data are not available for all compounds. Also, the possibility of synergistic effects of various substances, often including their degradation products, should not be excluded. Further, the hazard might be influenced by age, sex, state of health and by habitual consumption. Based on these considerations, any nutritional statement as to "tolerable concentration" must take sufficient safety factors into account.

Toxicity assay involves the determination of:

- Acute toxicity, designated as LD_{50} (the dose that will kill 50% of the animals in a test series).
- Subacute toxicity, determined by animal feeding tests lasting four weeks.
- Chronic toxicity, assessed by animal feeding tests lasting for 6 months to 2 years.

In chronic toxicity tests attention is especially

given to the occurrence of carcinogenic, mutagenic and teratogenic symptoms. The tests are conducted with at least two animal species, one of which is not a rodent.

The upper dosage level for a substance, fed to test animals over their life span and observed for several generations, which does not show an effect, is designated as the "No Effect Level" (NEL, mg/kg body weight of the animal tested). This level can be used as a basis for estimating the hazard for humans in all cases in which a correlation between dose and effect has been observed. The NEL is multiplied by a safety factor of 10^{-1} to 5×10^{-4}, but usually 10^{-2}, to obtain the toxicologically Acceptable Dose (AD, mg/kg human body weight). It is expressed as Acceptable Daily Intake (ADI) or as Acceptable Weekly Intake (AWI). These dose calculations also take into account the extreme deviations of data from the mean, as found with particularly sensitive organisms.

Taking into consideration the consumption habits, a toxicologically Permissible Concentration (PC) can be calculated from the toxicologically Acceptable Dose for a contaminant present in a given food.

The AD can not be exceeded, if the PC is observed.

$$PC = \frac{NEL \times FV}{SF} \times \frac{BW}{CA \times ASF}$$

In this formula, PC is the toxicological permissible concentration for a particular food (expressed in mg/kg food); NEL, no effect level (mg/kg feed); FV, daily intake of feed by test animals (kg feed/kg body weight); SF, safety factor (10–2,000, but usually 100); BW, body weight of an adult (50–80 kg); CA, amount in kg consumed per day of food for which the PC is being calculated; and ASF, additional safety factor for particularly sensitive persons, such as children or the sick (the accepted ASF is 2.5). The maximum concentrations of contaminants allowed by legislation are often still well below toxicological tolerance concentrations, because other parameters such as "good" agricultural practice", are taken into account.

Table 9.1. Mercury, lead, and cadmium levels in food (FR Germany report for 1979)[a]

Food	Mercury			Lead			Cadmium		
	1	2	3	1	2	3	1	2	3
Milk				0.001–0.0835	0.019	0.05	0.001–0.007	0.001	0.002
Chicken egg	0.0008–0.24	0.011	0.03	0.0002–0.8689	0.074	0.2	0.0005–0.0871	0.024	0.05
Beef/veal meat	0.0005–0.105	0.003	0.02	0.001–0.967	0.070	0.3	0.001–0.32	0.016	0.1
Pork meat	0.001–0.18	0.006	0.05	0.01–0.600	0.061	0.3	0.001–0.099	0.009	0.1
Beef/veal liver	0.0025–0.879	0.015	0.1	0.01–3.31	0.278	0.8	0.001–4.1	0.127	0.5
Pork liver	0.001–1.434	0.058	0.1	0.007–1.488	0.149	0.8	0.0025–1.61	0.165	0.8
Fish (fresh water)	0.0005–2.74	0.257	1.0	0.0005–1.08	0.124	0.5	0.0005–0.8035	0.020	0.05
Fish (sea)	0.0035–1.78	0.128	1.0						
Fish products	0.002–1.6	0.189	1.0						
Leafy vegetables	0.00025–0.033	0.004		0.0025–9.136	0.620	1.2	0.001–0.3875	0.044	0.1
Sprout vegetables	0.00025–0.025	0.003		0.0005–0.55	0.101	1.2	0.0005–0.09	0.019	0.1
Fruit vegetables	0.00025–0.012	0.003		0.0015–1.91	0.120	0.2	0.0005–0.166	0.020	0.1
Root vegetables				0.0015–1.24	0.205	0.5	0.001–0.104	0.023	0.05
Pomme fruits	0.00025–0.0125	0.002		0.0005–1.54	0.171	0.5	0.0005–0.116	0.010	0.05
Stone fruits	0.00025–0.0099	0.001		0.0075–1.349	0.142	0.5	0.0005–0.076	0.014	0.05
Berries	0.00025–0.00167	0.002		0.0002–2.08	0.245	0.5	0.0001–0.101	0.018	0.05
Fruit juice				0.01–0.20	0.057	0.2	0.004–0.015	0.007	0.02
Cereals	0.0005–0.0642	0.004	0.03	0.01–0.61	0.041	0.5	0.004–0.80	0.035	0.1
Potatoes	0.0005–0.0154	0.006	0.02	0.0015–0.319	0.075	0.2	0.001–0.202	0.050	0.1
Wine				0.005–3.08	0.173	0.3	0.0005–0.03	0.003	0.1
Drinking water	0.00002–0.002	0.0003	0.004	0.0021–0.0225	0.009	0.04	0.0004–0.0044	0.001	0.006

[a] All values given are in mg/kg or mg/l, column 1: variation range, column 2: an arithmetic average, column 3: a guiding value which should not be exceeded, or in some cases legally set as the highest permissible level.

9.2 Toxic Trace Elements

9.2.1 Mercury

Mercury poisoning caused by food intake is derived from organomercury compounds, e.g. dimethyl mercury ($CH_3-Hg-CH_3$), methyl mercury salts (CH_3-Hg-X; X = chloride or phosphate), and phenyl mercury salts (C_6H_5-Hg-X; X = chloride or acetate). These highly toxic compounds are lipid soluble, readily absorbed and accumulate in erythrocytes and the central nervous system. Some are used as fungicides and for treating seeds (seed dressing). Methyl mercury compounds are also synthesized by microflora from inorganic mercury salt sediments found on lake and river bottoms. Hence, the content of these compounds might rise in fish and other organisms living in water.
The natural mercury level in the environment appears to have stabilized in the last 50 years. Poisonings recorded in Japan appear to have been caused by consumption of fish caught in waters heavily contaminated by mercury-containing industrial waste water, and in Iraq by milling and consuming seed cereals dressed with mercury and which were intended for sowing. The permitted level or tolerance dose for an adult of 70 kg is 18 mg Hg per year, of which a maximum of 10 mg may be derived from the highly toxic methyl mercury. The mercury content in some foods is compiled in Table 9.1. The average mercury intake with food, most of which is by consumption of fish, is estimated at 5.7 mg/capita/year in FR Germany.

9.2.2 Lead

The contamination of the environment with lead is increased by industrialization and by emissions from cars running on leaded gasoline. Tetraethyllead [$(C_2H_5)_4Pb$], an antiknocking additive used to increase the octane value of gasoline, is converted by combustion into PbO, $PbCl_2$ and other inorganic lead compounds. The major part of these compounds is found in an approx. 30 m wide band along roads or highways; the lead level sharply decreases beyond this distance. At a distance of 100 m from a road with heavy traffic, the lead level in the atmosphere decreases by a factor of 10 and that in soil and plants by a factor of 20 from the level found at or close to the road. A decrease in the level of lead in gasoline and increased use of nonleaded gasoline has resulted in a drop in the extent of contamination. Environmental lead contamination has not, however, significantly increased the level of lead in food. The lead in soil is rather immobilized; thus the increase in the lead level of plants is not proportional to the extent of soil contamination. Vegetables with larger surface areas (spinach, cabbage) may contain higher levels of lead when cultivated near the lead emission source. When contaminated plants are fed to animals, the body does not absorb much lead since most is excreted in feces.
Further sources of contamination are lead-containing tin cookware and soldered metal cans and lead-containing enamels. This is particularly so in contact with sour food. These sources of contamination are of lesser importance.
3.5 mg of lead are considered as the temporary acceptable weekly intake for adults of 70 kg. The lead content of food is compiled in Table 9.1. The average intake of lead in FR Germany is 0.7–1.2 mg/week, of which in excess of 90% is derived from food. A decrease in lead intake is desirable.
Hair and bone analyses have revealed that lead contamination of humans in preindustrialized times was apparently higher than today. This might be due to the use in those days of lead pipes for drinking water, lead-containing tinware, and excessive use of lead salts for heavily-glazed pottery used as kitchenware.

9.2.3 Cadmium

Cadmium ions, unlike Pb^{2+} and Hg^{2+}, are readily absorbed by plants and distributed uniformly in all their tissues, thus decontamination by dehulling or by removal of outer leaves, as with lead, is not possible. Some wild mushrooms accumulate higher levels of cadmium. In food of animal origin, cadmium is found primarily in internal organs, such as liver and kidney, and in milk. The contamination sources are industrial waste water and the sludge from plant clarifiers, which is often used as fertilizer. The cadmium content of food is compiled in Table 9.1.
A prolonged intake of cadmium results in its accumulation in the human organism, primarily in liver and kidney. A level of 0.2–0.3 mg Cd/g kidney cortex causes damage of the tubuli. The

tolerance dose of cadmium is estimated at 0.5 mg/week. Its intake varies, but is estimated at 0.24 mg/week in FR Germany and some other countries. More than 95% is derived from the diet. This intake suggests that no acute health hazard exists with cadmium. Nevertheless, cadmium contamination, like lead and mercury, still presents a problem and, hence, a decrease in cadmium intake is a justified aspiration.

9.2.4 Radionuclides

It is estimated that the average radiation exposure in FR Germany in 1975 was 172 mrad, of which 21 mrad were ascribed to internal radiation by natural radionuclides incorporated in the body (about 90% from ^{40}K, the rest from ^{14}C) and less than 1 mrad by nuclides acquired as a result of atmospheric fallout from nuclear explosion tests (50% from ^{137}Cs, a radionuclide with a half life of 30 years, but quickly excreted by the body; approx. 50% from ^{90}Sr, a most dangerous radioisotope, capable of inducing leukemia and bone cancer; and traces of ^{14}C and tritium). ^{137}Cs and ^{90}Sr are the escort elements of potassium and calcium, respectively. Food contamination with radionuclides in FR Germany had its peak in 1964/65, when the intake in food per day per person was 240 pCi of ^{137}Cs and 30 pCi of ^{90}Sr. The intake is now much less (less than 10% of previous values) as a result of the moratorium on atmospheric testing of atomic weapons. Radionuclide residues in food are not now a health hazard. Only a negligible fraction of the total radioactivity in the flesh of domestic animals is derived from fallout of nuclides.

About 40% of ^{90}Sr contamination in man is obtained from milk and other dairy products. The level of tritium infiltrating the biosphere is expected to rise from 28 MCi to 250 MCi due to nuclear plant operation on a long-term basis.

Table 9.2. Food poisoning by bacterial toxins

Microorganism:	*Staphylococcus aureus*	*Clostridium botulinum*	*Bacillus cereus*	*Clostridium perfringens*
Growth conditions				
temperature range	10–45 °C	4–35 °C	10–45 °C	12–52 °C
pH range	4.5	5		5–8.5
Toxin				
type	Protein	Protein	Lipid (?)	Protein
effective amount	0.5–1 μg	0.1–1 μg	10^8 spores/g	10^6 spores/g
stability	Relatively thermostable	Thermolabile, inactivation at 80 °C/30 min or 100 °C/5 min		
Incubation time	2–6 h	1–3 days	1–12 h	8–24 h
Duration of illness	1–3 days	Death after 1–8 days, with survivors ill 6–8 months	0.5–1 day	0.5–1 day
Symptoms	Vomiting, diarrhea, abdominal pain	Paralysis of the nerve centres of the *medulla oblongata*	Abdominal pain, nausea, diarrhea, vomiting	Diarrhoea, abdominal convulsions, nausea, loss of appetite
Foods usually accounting for poisoning	Cold meat and cheese slices, mildly acidic salads (meat, poultry, sausage, cheese, potatoes), mayonnaise, cream fillings in baked products	Homemade canned meat, hind bony ham, sliced sausages, trout fillets, canned green beans	Institutional/ community catering. Heated and warmed dishes, cereal containing dishes (corn, rice)	Institutional/ communal catering. Heated and warmed meat dishes, warm desserts, puddings cream fillings in baked products

9.3 Toxic Compounds of Microbial Origin

9.3.1 Food Poisonings by Bacterial Toxins

Most (60–90%) food poisonings are bacterial in nature. They are distinguished by food intake causing:

- Intoxication (poisoning, e. g. by *Clostridium botulinum, Staphylococcus aureus*).
- Diseases by massive pollution with facultative pathogenic spores, e. g. *Clostridium perfringens, Bacillus cereus*.
- Infections by *Salmonella* spp. or *Shigella* spp.
- Diseases of unclear etiology, such as those from *Proteus* spp., *Escherichia coli, Pseudomonas* spp.

The harmful activity of these bacteria in the digestive tract is ascribed to enterotoxins, which are classified into two groups: exotoxins (toxins excreted by microorganisms into the surrounding medium) and endotoxins (retained by the microorganism cells but released when the cell disintegrates).

Exotoxins are released primarily by gram-positive bacteria during their growth. They consist mostly of proteins which are antigenic and very poisonous. They become active after a latent period. This group includes the toxins released by *Clostridium botulinum* (botulin toxin, a globular protein neurotoxin), *Cl. perfringens* and *Staphylococcus aureus*. Table 9.2 gives some important data for these microorganisms, including harmful effects. Intoxications with *St. aureus* are the most frequent food poisoning. Symptoms are vomiting, diarrhea and stomach ache and are caused primarily by food of animal origin (meat and meat products, poultry, cheese, potato salad, pastry).

Endotoxins are produced primarily by gram-negative bacteria. They act as antigens, are firmly bound to the bacterial surface and are complex in nature. They have protein, polysaccharide and lipid components. Endotoxins are relatively heat

Table 9.3. Mycotoxins

Fungus/mold	Toxin[a]	Toxicity[b]	Effect	Occurrence
Claviceps purpurea	Ergot alkaloids (I)		Ergotism (gangrenous convulsions)	Preferentially rye, to a lesser extent wheat
Aspergillus flavus *A. parasiticus*	Aflatoxins (II)	7.2 mg/kg (rat, orally)	Liver cirrhosis, liver cancer	Groundnuts and other nuts (almond, Brasil nut) corn and other cereals, animal feed, milk
Aspergillus versicolor *A. nidulans*	Sterigmatocystin (III)	120 mg/kg (rat, orally)	Liver cancer	Corn, wheat, animal feed
Penicillium expansum *P. urticae* *Byssochlamis nivea, B. fulva*	Patulin (IV)	35 mg/kg (mouse, orally)	Cellular poison	Putrifying fruits, fruit juices
Aspergillus ochraceus *A. melleus*	Ochratoxin A (V)	20 mg/kg (rat, orally)	Fatty liver and kidney damage	Barley, corn
Fusarium graminearum	Zearalenone (VI) (Fusariotoxin F_2)	0.1 mg/kg over 5 days (swine, orally[c])	Estrogen, infertility	Corn and other cereals, animal feed
Fusarium oxysporum *F. tricinctum*	Fusariotoxin T_2 (VII)	3.8 mg/kg (rat, orally)	Toxic aleukia, hemorrhagic syndrome	Cereals, animal feed

[a] Roman numerals refer to the structural formulas in Fig. 9.1.
[b] Acute toxicity (LD_{50}).
[c] Estrogenic activity.

stable and are in general active without a latent period. The toxins causing typhus and paratyphus fevers, salmonellosis and bacterial dysentery are in this group. Salmonellosis is very serious. It is an infection by toxins of about 300 different but closely-related organisms. The infection is characterized by enteric fever, gastroenteritis and salmonella septicemia. Sources of infections are egg products, frozen poultry, ground or minced beef, confectionery products and cocoa.

9.3.2 Mycotoxins

There are about 80–90 mycotoxins produced under certain conditions by about 120 fungi or molds. Table 9.3 presents data on mycotoxins of particular interest to food preservation and storage. The chemical structures of these toxins are presented in Figure 9.1.

Infections of rye and, to a lesser extent, of other cereal grains with *Claviceps purpurea* (ergot, or rooster's spur) are responsible for the disease

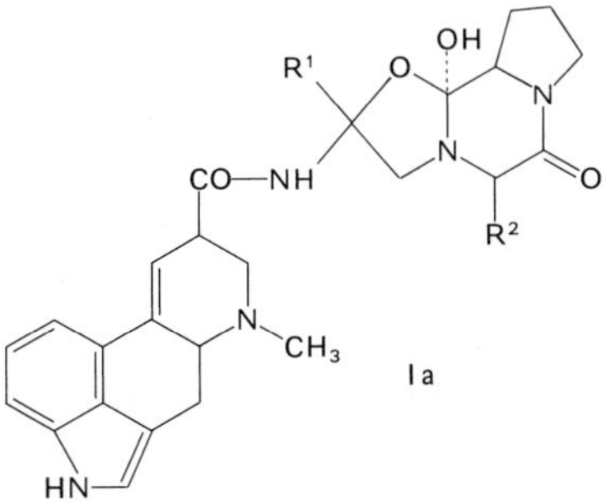

Ia

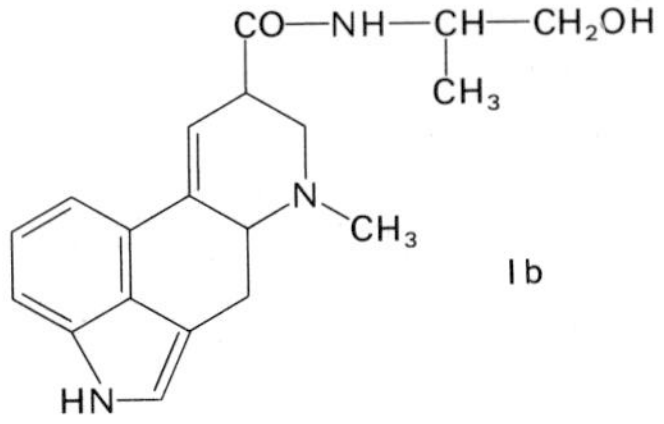

Ib

IIa

IIb

IIc

IId

I: Alkaloids of Ergot
Ia: Ergocristine ($R^1 = CH(CH_3)_2$, $R^2 = H_2C-C_6H_5$), Ergostine ($R^1 = C_2H_5$, $R^2 = H_2C-C_6H_5$), Ergotamine ($R^1 = CH_3$, $R^2 = H_2C-C_6H_5$), Ergocryptine ($R^1 = CH(CH_3)_2$, $R^2 = H_2C-CH(CH_3)_2$), Ergocornine ($R^1 = CH(CH_3)_2$, $R^2 = CH(CH_3)_2$), Ergosine ($R^1 = CH_3$, $R^2 = CH_2-CH(CH_3)_2$) Ib: Ergometrine

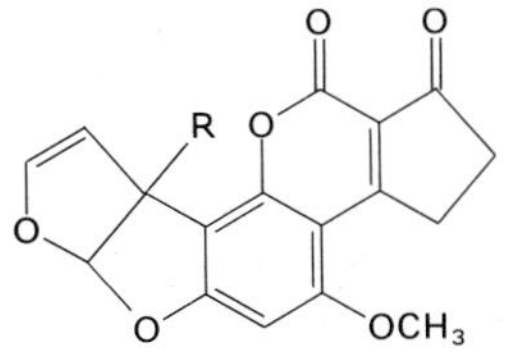

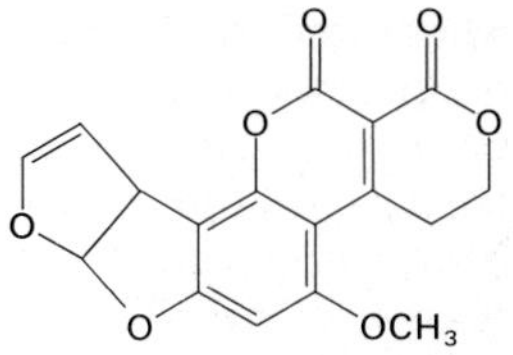

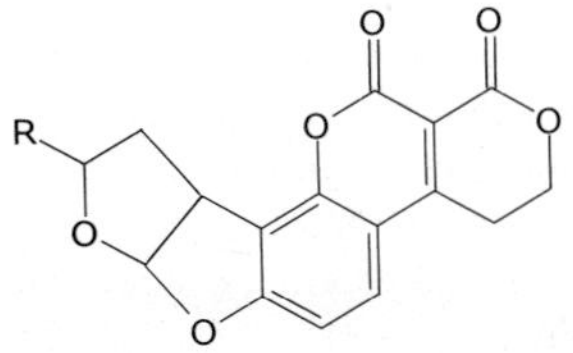

II: Aflatoxins
IIa: Aflatoxin B_1 (R=H), Aflatoxin M_1 (R=OH), IIb: Aflatoxin B_2 (R, R^1=H), Aflatoxin M_2 (R=OH, R^1=H), Aflatoxin B_{2a} (R=H, R^1=OH), IIc: Aflatoxin G_1, IId: Aflatoxin G_2 (R=H), Aflatoxin G_{2a} (R=OH)

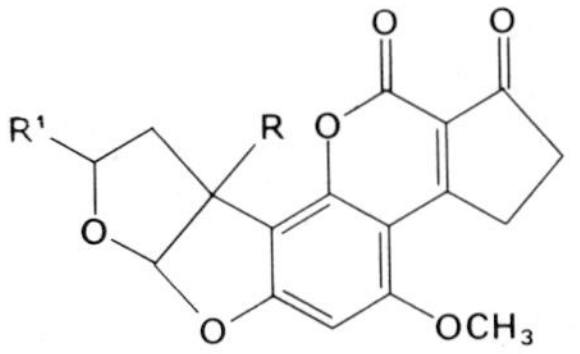

Fig. 9.1. Structures of some mycotoxins (cf. Table 9.3)

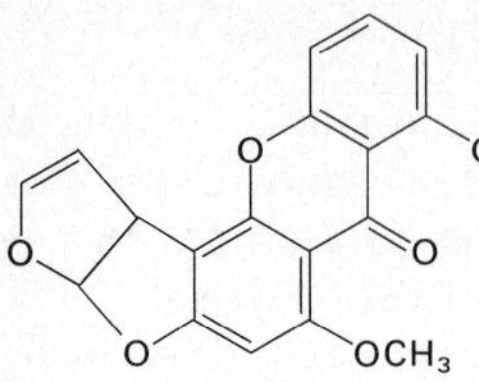

III: Sterigmatocystine
IV: Patuline

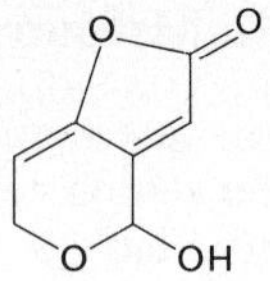

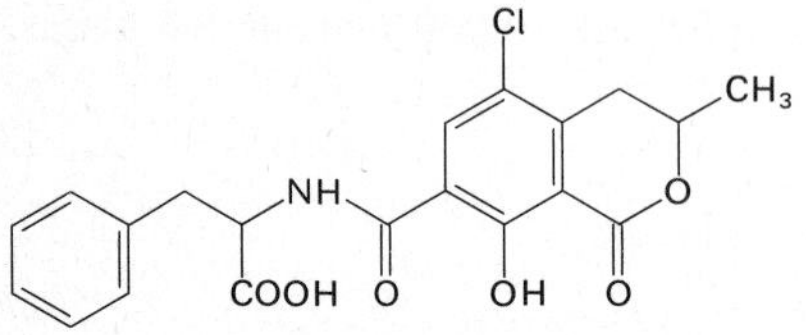

V: Ochratoxin A
VI: trans-Zearalenone
VII: Fusariotoxin T_2

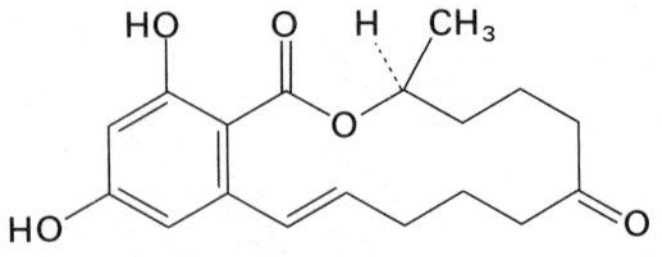

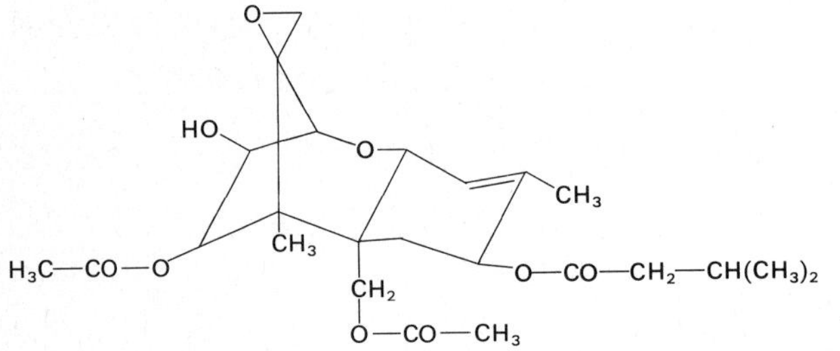

Fig. 9.1 (Continued)

called ergotism (symptoms: gangrene and convulsions). The disease was important in the past when bread from infected rye grain was eaten. It has practically ceased to exist due to seed treatment with fungistatic agents and due to grain cleaning prior to milling.

Most mycotoxin data are on the genera *Aspergillus* and the aflatoxins they produce during growth. These are the most common and highly toxic fungal toxins, e.g. Aflatoxin B_1, the most powerful carcinogen known. In animal feeding tests (rats) its carcinogenic effect is revealed at a daily dose of only 10 μg/kg body weight. In a comparative study, the carcinogenic property of the highly toxic dimethylnitrosamine was revealed only at a daily dose of 750 μg/kg body weight. It is primarily plant material (particularly peanuts, peanut butter, rice, maize) that is contaminated with aflatoxins. Aflatoxin migrates from moldy feed to animal products, primarily milk. The dairy cow's metabolism converts the B-group aflatoxins to those of the M-group ("M" stands for metabolite), which are also carcinogenic.

Thorough analytical research and strict legislative measures have significantly reduced aflatoxin contamination and kept it under control. Only *one* peanut out of 10^4 is contaminated with aflatoxin. Aflatoxins M_1 and M_2, obtained by hydroxylation of aflatoxins B_1 and B_2, respectively, average 3–8 ng/l of milk.

Substantially less analytical data are available on the occurrence of the other mycotoxins listed in Table 9.2. The most significant household source in quantity of mycotoxins is mold infections of fruit, bread and other baked products, meat and processed meat products.

Table 9.4. World utilization of pesticides

	Pesticide group[a]				Total[b]
	Insecticides	Fungicides	Herbicides	Others	
Western Europe	21	27	47	5	20.6
FR Germany	11	19	64	6	2.8
France	19	25	50	6	7.2
Italy	28	45	22	5	3.2
Eastern Europe	22	28	45	5	18.3
USA/Canada	28	6	65	2	23.1
Latin America	58	13	28	1	12.0
Africa	60	17	16	7	4.9
Asia/Middle East	49	22	26	2	19.4
Australia/ New Zealand	30	13	56		1.7

[a] Percent of the total value (wholesale price).
[b] Percent of the world market (wholesale price).

9.4 Pesticides

9.4.1 General Remarks

The term pesticides includes the compounds used in agricultural food production to protect cultivated plants from plant- and insect-caused diseases, parasites or weeds, or from detrimental microorganisms. The most important groups of pesticides are: (1) *herbicides* to protect the plant from weeds; (2) *fungicides* to suppress the growth of undesired fungi or molds; and (3) *insecticides* to protect the plants from damage caused by insects. In addition to these main groups, there are *acaricides* to control mites, *nematocides* to control worms or nematodes, *molluscicides* to protect the plant from snails and slugs, *rodenticides* to control rodents (mice or rats), and plant *growth regulators* (plant growth hormones [cf. 18.1.4] and sprout inhibitors). Table 9.4 provides data on the global share of various groups of pesticides.

Pesticide use is rewarding since it reduces losses in crop harvesting and storage. It has also contributed to the control or eradication of insect-spread diseases such as malaria. Nevertheless, crop harvest losses average 35% worldwide (Europe, 25%; Asia, 43%).

Pesticides are applied in various forms and by various means: dusting as powder, fumigation, spraying as a liquid, or pad or furrow irrigation. The strict observance of directions for use, waiting the recommended time between final application and harvest, and restricting the application to the necessary dose, is required to maintain the residual pesticide levels in food at a minimum. These requirements and recommendations, as set by regulations, are listed in an application booklet.

The contamination of food of plant origin can be direct by treating the crop for storage and distribution (fruit and vegetable treatment with fungicides, cereal treatment with insecticides). It can be indirect from uptake from the soil of residual pesticides by the subsequent crop, from the atmosphere or drifting from neighboring fields, or from a storage space pretreated with pesticide.

Contamination of food of animal origin occurs by ingestion of feed, from stall- and barn-cleansing agents (fungicides, insecticides), insecticides, wooden studs and boards preserved with fungistatic agents, veterinary medicines and, occasionally, use of disinfected corn as a fodder.

Table 9.5 and Figure 9.2 review the important pesticides and their application methods. Table 9.6 gives data on residues in FR Germany.

9.4.2 Insecticides

The most important insecticide classes are chlorinated hydrocarbons, organophosphoric acid esters (organophosphates) and carbamates (Table 9.5; Figure 9.2). The greatest attention is given to chlorinated hydrocarbons. They are very stable, persistent in the environment, and fat soluble and thus deposit in human adipose tissue and in human milk fat. Direct analysis of human fat for these residues is a good indicator of the overall contamination by this class of insecticide.

Contamination of food with chlorinated organic compounds is a declining event since their use is decreasing in favor of thiophosphoric acid esters and carbamates. Since these compounds are substantially degraded over the span of a recommended waiting period, residue problems do not arise.

Analytical data compiled in Table 9.6 show that most of the samples with identified residues contained primarily persistent organochlorine and, rarely, organophosphate insecticide, fungicide or herbicide. Foods of animal origin contained almost exclusively organochlorine residues (HCH, HCB, DDT, DDE, dieldrin and PCB); however, in comparison to previous years, as already mentioned, there is a very clear downward trend in

Table 9.5. Some selected pesticide trade names and applications

Number	Name	Application
Synthetic Insecticides		
I	Lindane (γ-HCH), BHC, Gammexane	Seed dressing (cereals, beets, legumes), vegetables and fruits
I a	β-HCH	In FR Germany not permissible (the same is the case with α-HCH)
II	p,p′-DDT	Fight against malaria, in FR Germany not permissible
III	Methoxychlor	Fruits, vegetables, potatoes, cabbage, wheat, rye
IV	Endosulfan, Thiodan	Vegetables, fruits, potatoes, rapeseed, beets, field beans, corn, animal feed
V	Aldrin, HHDN	Grape (vine) cultivation
VI	Dieldrin, HEOD	In FR Germany not permissible
VII	Heptachlor	Seed dressing (beets)
VIII	Heptachlor-epoxide (A, B)	Metabolites from VII
IX	Azinphos -ethyl Gusathion H	Potatoes, cereals, rapeseed, beets, vegetables, animal feed
X	Azinphos-methyl Gusathion	Fruits, vegetables (asparagus), vines
XI	Dimethoate, Rogor, Perfekthion	Vegetables, fruits, vines, potatoes, beets, cereals, animal feed
XII	Fenitrothion	Vegetables, fruits, potatoes, cereals, clover, beets, vines
XIII	Methidathion	Fruits, potatoes, beets, vines
XIV	Mevinphos, Phosdrin, PD 5	Fruits, vegetables, cereals, potatoes, rapeseed, beets, animal feed
XV	Malathion	Cereals, potatoes, vegetables, fruits
XVI	Parathion (-ethyl), E 605, Eftol	Vegetables, fruits cereals, beets, rapeseed, potatoes, vines, animal feed
XVII	Parathion-methyl, ME 605	As compound XVI
XVIII	Phosphamidon	Vegetables, fruits, cereals, potatoes, rapeseed, beets, clover, vines
XIX	Bromophos, Nexion	Vegetables, fruits, cereals, potatoes, beets, rapeseed, animal feed
XX	Dibrom, Naled	As compound XIX
XXI	Chlorfenvinphos, Birlane	Vegetables, potatoes, beets, corn
XXII	Dichlorvos, DDVP, Vapona	Vegetables, fruits, cereals, rapeseed, beets, potatoes, mushroom, vines, animal feed
XXIII	Carbaryl, Sevin	Fruits, vegetables, potatoes, vines
XXIV	Methomyl, Lannate	Vegetables, vines
XXV	Promecarb, Carbamult	Pomme fruits, vegetables, potatoes, rapeseed, beets
XXVI	Propoxur, Unden	Vegetables, fruits, cereals, potatoes, beets
Natural insecticides		
XXVI a	Nicotine	Vegetables, fruits
XXVI b	Pyrethrins (from *Chrysanthemum cinerariaefolium*)	Fruits, vegetables, potatoes, rapeseed, beets, mushroom, vines
XXVI c	Rotenone, Derris	Fruits, vegetables, potatoes, rapeseed, beets, vines

Number	Name	Application
Acaricides		
XXVII	Ethion	Fruits
XXVIII	Kelthane, Dicofol	Fruits, vegetables, vines
XXIX	Tetradifon, Tedion	Fruits, vines
XXX	Tetrasul, Animert	Cucumbers, tomatoes
Fungicides		
XXXI	Copperoxichloride	Fruits, vegetables, rapeseed, beets, potatoes, vines
XXXII	Sulfur	Fruits, vegetables, vines, potato seeds
XXXIII	Thiram, TMTD, Pomarsol	Pomme fruits, strawberries, chicory, head lettuce, vegetables, vines
XXXIV	Ferbam	Pomme fruits
XXXV	Ziram	Pomme fruits
XXXVI	Maneb, Dithane M-22	Potatoes, beets, vegetables, asparagus. berry fruits, vines, pomme fruits
XXXVII	Zineb, Dithane, Z-78	Pomme fruits, stone fruits, asparagus, vines
XXXVIII	Mancozeb (complex consisting of XXXVI and XXXVII), Dithane ultra	Pomme fruits, stone fruits, cucumbers, tomatoes, vegetables, vines
XXXIX	Benomyl, Benlate	Pomme fruits, cucumbers, head lettuce, strawberrie, cereals (seed dressing)
XL	Captafol, Difolatan	Pomme fruits, peaches, vegetables, legumes, potatoes, vines
XLI	Captan, Orthocide-406	Pomme fruits, cherries, peaches, vegetables, corn, legumes, head lettuce, vines
XLII	Dichlofluanid, Euparen	Head lettuce, tomatoes, berry fruits, pomme fruits, peaches, vines
XLIII	Folpet, Phaltan	Pomme fruits, cherries, strawberries, beans, peas, cucumbers, vines
XLIV	Hexachlorobenzene (HCB)	In FR-Germany not permissible
XLV	Quintozene, PCNB, Brassicol (contains up to 3% XLIV)	Seed dressing (cereals, potato seed)
XLVI	Thiabendazole, Tecto	Pomme fruits, surface treatment of fruits (bananas, citrus)
XLVII	Fentin acetate, Brestan	Carrots, potatoes
XLVIII	Methoxyethyl-mercuric chloride, Ceresan	Wheat, rye, barley oats
IL	Phenylmercuric acetate, PMA	As compound XLVIII
L	Phenylmercuric chloride PMC	As compound XLVIII
Herbicides		
LI	Alachlor	Corn, cabbages
LII	Amitrole	For all crops and pomme fruits
LIII	Atrazine	Corn, asparagus
LIV	Bromacil	Pomme fruits
LV	Buturone	Berry fruits, vines, winter wheat cultivars

Table 9.5. (Continued)

Number	Name	Application
LVI	Chlorbufam	Vegetables (carrots, red beet)
LVII	Chloropropham, CIPC	Potatoes (antisprouting agent)
LVIII	Chloroxuron, Tenoran	Vegetables, carrots
LIX	Chlortoluron	Winter wheat cultivars
LX	2,4-D	Cereals
LXI	Diquat	Potatoes (vine killing) vegetables, strawberries
LXII	Desmetryn	Cabbages
LXIII	Diallate, DATC	Peas, red beets, beets
LXIV	Diuron	Asparagus, berry fruits
LXV	Lenacil	Strawberries, spinach, beets
LXVI	Linuron	Peas, beans, carrots, celery, asparagus, potatoes, vines
LXVII	MCPA	Pomme fruits, barley, vines
LXVIII	Metobromuron	Lamb's lettuce, potatoes
LXIX	Metoxuron	Carrots, cereals
LXX	Metribuzin	Tomatoes, asparagus, potatoes
LXXI	Monalid	Carrots, parsley, cellery
LXXII	Monolinuron	Bush snap beans, asparagus, potatoes, cereals, vines
LXXIII	Paraquat, Gramoxone	Vegetables, fruits, strawberries, vines and all forage crops
LXXIV	Pentanochlor	Carrots, parsley, tomatoes, celery, peppermint
LXXV	Prometryn	Carrots, leek, celery
LXXVI	Propham, IPC	Caraway, potatoes (antisprouting agent)
LXXVII	Propyzamide	Chicory, head lettuce, pomme fruits, rapeseed
LXXVIII	Simazire, Gesatop	Pomme and stone fruits, strawberries, vino, peas, corn, field beans, sugar beet
LXXIX	2,4,5-T	Cereals, vines
LXXX	TCA	Sugar beet
LXXXI	Terbutryne, Igran	Peas, corn
LXXXII	Triallate, Avadex	Cereals, sugar beet
LXXXIII	Trifluoralin	Cauliflower, turnips, winter cereal cultivars
Nematicides		
LXXXIV	Dazomet	Fruits, vegetables, forage crops, potatoes
LXXXV	1,3-Dichloropropane, 1,2-Dichlorpropene, DD mixture	Potatoes, fruits, vegetables, vines, forage crops
LXXXVI	Methyl bromide, Terabol	Vegetables, fruits, forage crops, potatoes, beets
LX XXVII	Methyl isocyanate	As compound LXXXVI
LXXXVIII	Zinophos	Vegetables
Molluscicides		
LXXXIX	Metaldehyde	Vegetables, strawberries, cereals
XC	Mercapto-dimethur	Cereals
Rodenticides		
XCI	Endrin	Pomme- and stone fruits, currants
XCII	Toxaphene (Chlorinated camphene)	Cereals

accumulation of DDT and its metabolites, as well as of dieldrin, HCB and heptachloroepoxide. High DDT values were found in some fruits, particularly wine grapes and apples. Since the content of organochlorine compounds in human milk is higher by a factor of 10–30 than in cow's milk (Table 9.6a), the Acceptable Daily Intake (ADI) values for breast fed children are significantly exceeded for DDT, and also for dieldrin by a factor of 2–4 and HCB by a factor of 8–9. Nevertheless the expert opinion is that the advantages of infant feeding with mother's milk during a 3–6 month period outweigh the health hazard of the residue levels.

Figure 9.2a uses fresh dairy milk as an example to demonstrate the difference in results for chlorinated hydrocarbon residues in blended milk samples and in milk delivered by individual dairy farmers. The data emphasize the importance of taking random samples from individual dairy farmers.

9.4.3 Herbicides

Herbicides, used to protect the growth of cultivated plants from weeds, are divided into compounds with broad or selective activity. The former group includes chlorate, copper sulfate, calcium cyanamide and chlorinated fatty acid derivatives. The group with selective activity includes growth regulators, such as aryloxy fatty acids, carbamic acid and urea derivatives, triazines and pyridines (cf. Table 9.5 and Figure 9.2).

The main field of herbicide application is corn (maize), other cereals and beet cultivation. Residue problems are almost nonexistent. The toxicity of the compounds used is generally low in warmblooded organisms.

A possible side effect of herbicides which is not to be ignored or underestimated is their influence on the arthropods and the natural microflora of soil.

9.4.4 Fungicides

These compounds are used to protect plants against diseases caused by fungi or molds, e.g. potato and tomato rots, flour dew and fruit scabs.

Important fungicides, in addition to inorganic compounds (copper oxychloride, sulfur, sulfur-

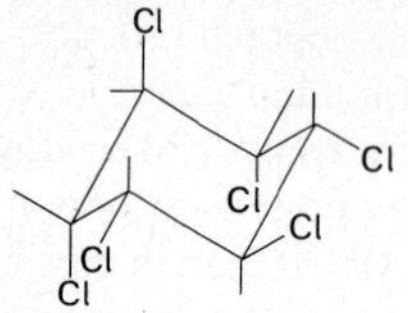

I

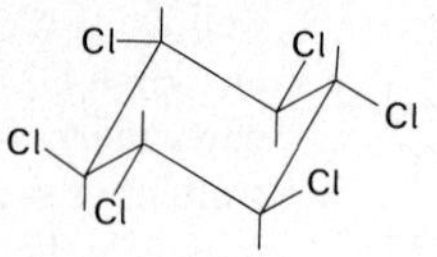

Ia

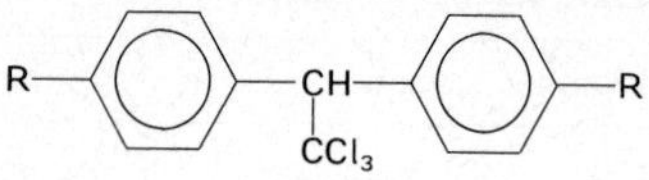

II: R = Cl, III: R = OCH_3

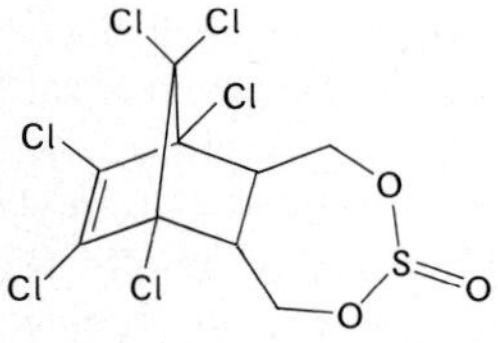

IV

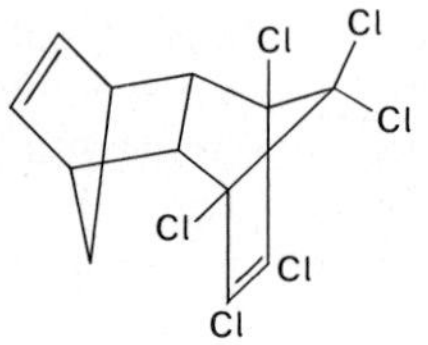

V

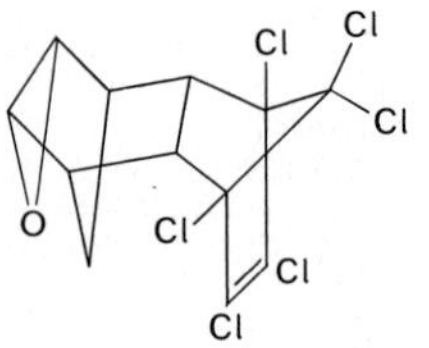

VI

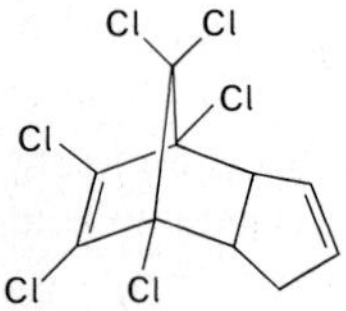

VII

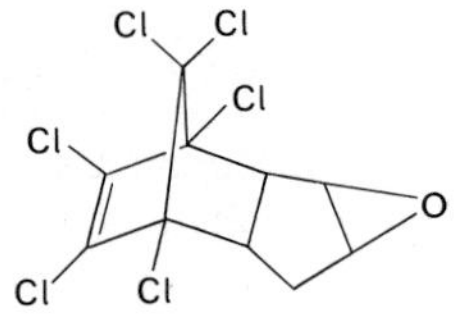

VIII

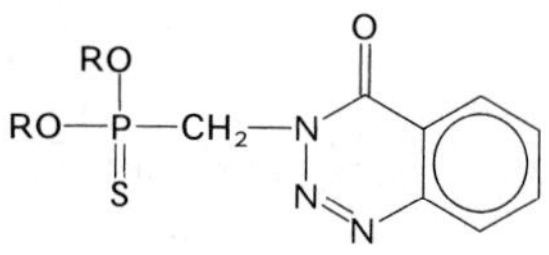

IX: R = C_2H_5, X: R = CH_3

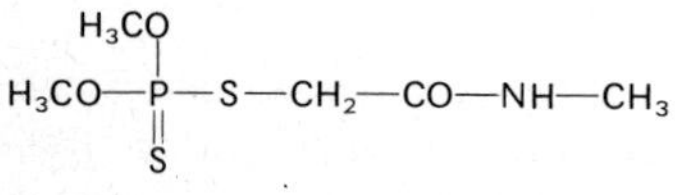

XI

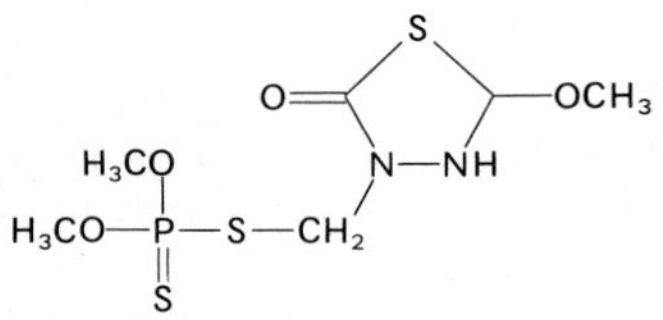

XIII

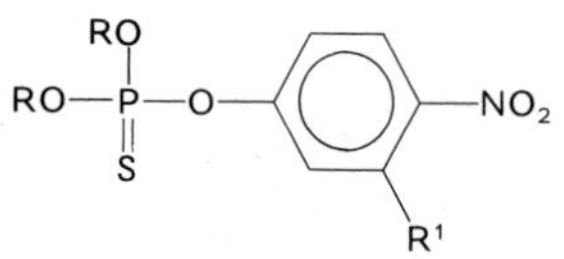

XII : R = CH_3, R^1 = CH_3
XVI : R = C_2H_5, R^1 = H
XVII: R = CH_3, R^1 = H

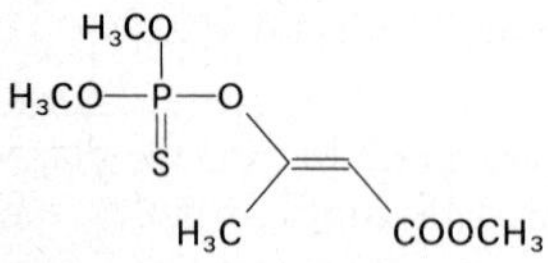

XIV

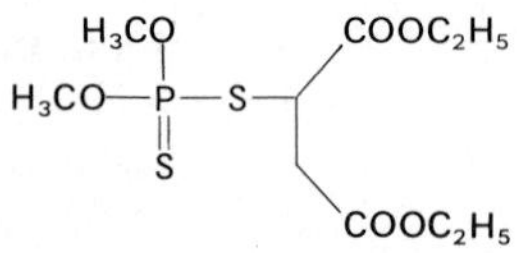

XV

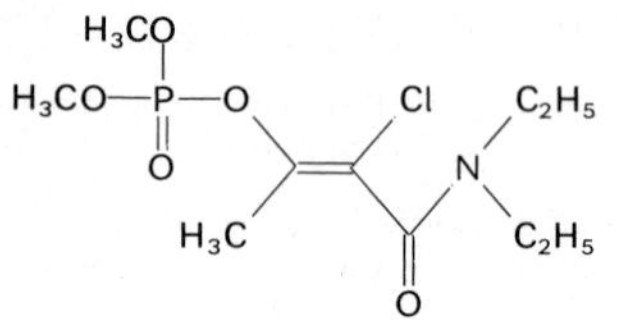

XVIII

Fig. 9.2. Structures of some selected pesticides. Part 1. The Roman numerals refer to Table 9.5

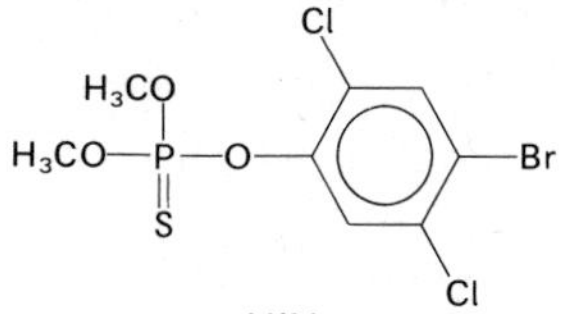

XIX

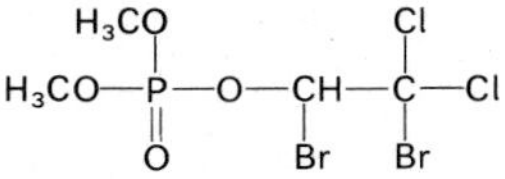

XX

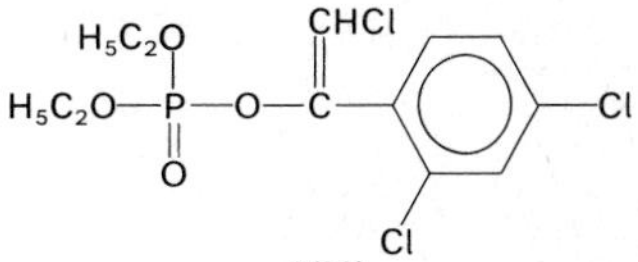

XXI

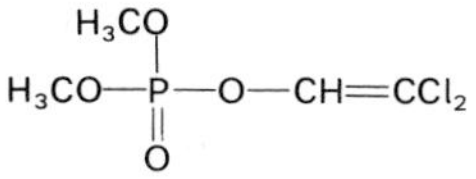

XXII

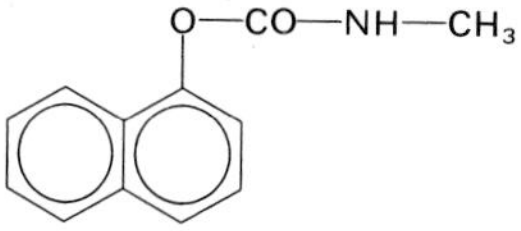

XXIII

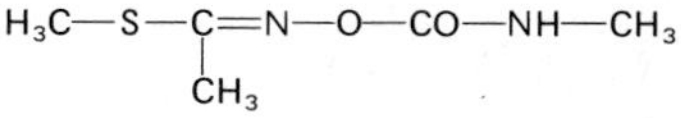

XXIV

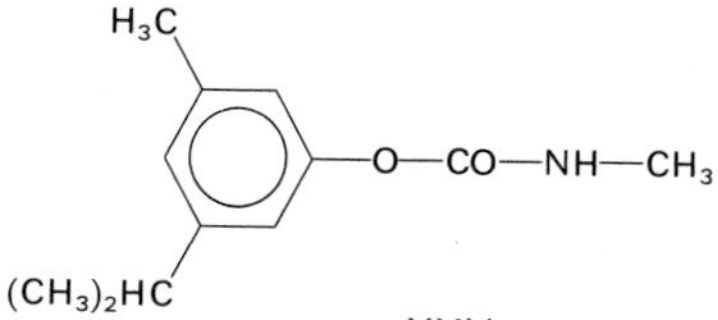

XXV

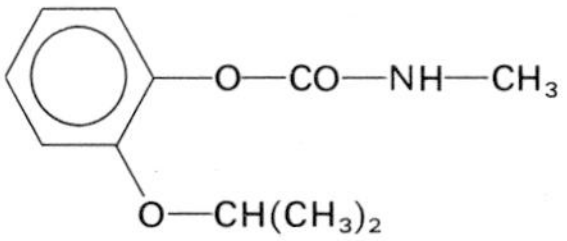

XXVI

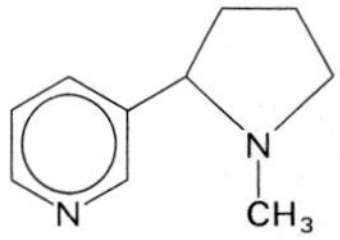

XXVIa

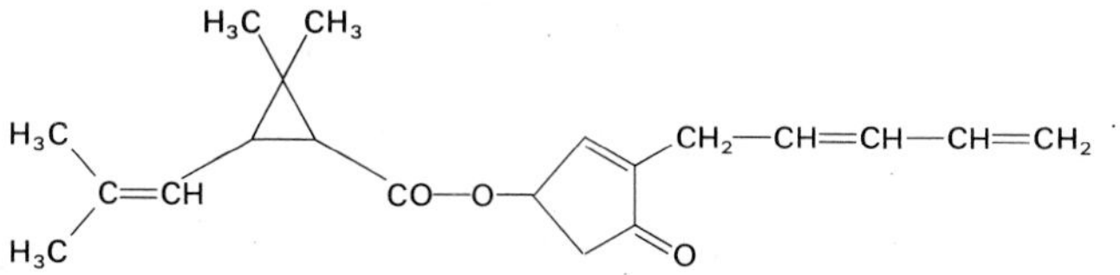

XXVIb (Pyrethrin I)

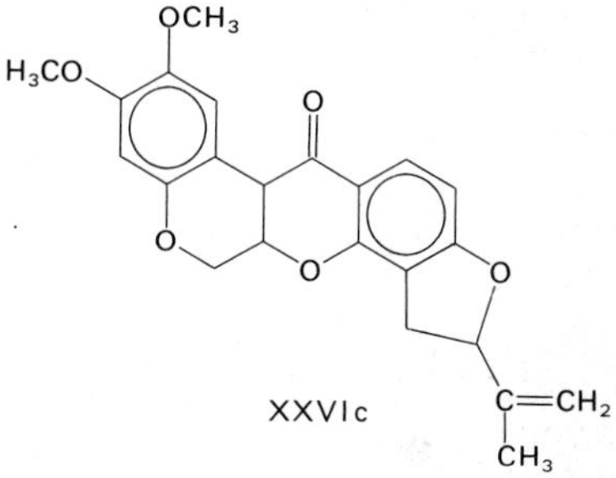

XXVIc

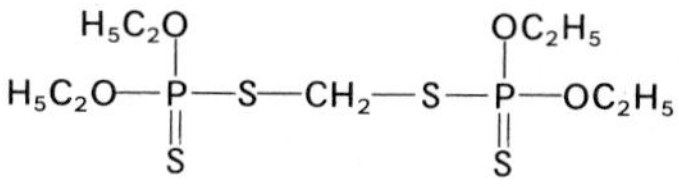

XXVII

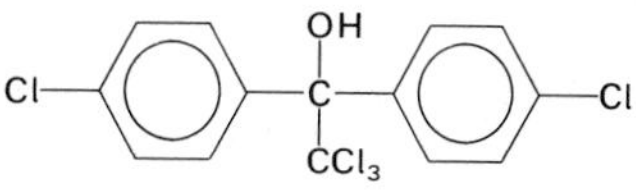

XXVIII

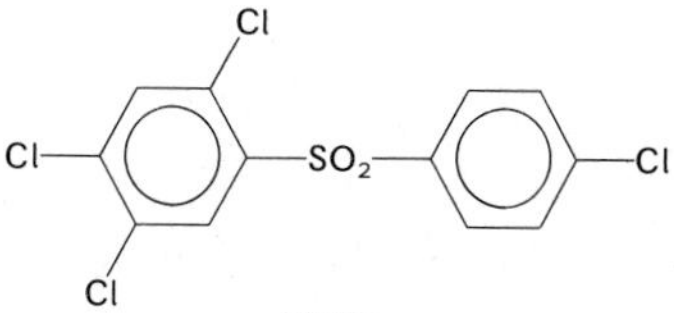

XXIX

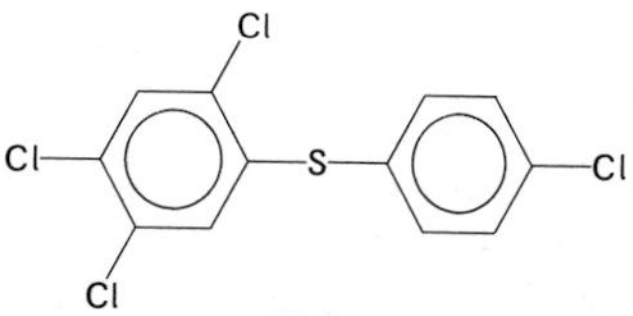

XXX

Fig. 9.2. Structures of some selected pesticides. Part 2. The Roman numerals refer to Table 9.5.

XXXIII

XXXIV: n = 3, Me = $Fe^{3\oplus}$
XXXV: n = 2, Me = $Zn^{2\oplus}$

XXXVI: Me = $Mn^{2\oplus}$
XXXVII: Me = $Zn^{2\oplus}$

XXXIX

XL: R = CCl_2—$CHCl_2$
XLI: R = CCl_3

XLII

XLIII

XLIV: R = Cl
XLV: R = NO_2

XLVI

R_nMeX

XLVII:	n = 3, Me = $Sn^{4\oplus}$, X = $OCOCH_3$, R = C_6H_5
XLVIII:	n = 1, Me = $Hg^{2\oplus}$, X = Cl, R = $CH_3O(CH_2)_2$
IL:	n = 1, Me = $Hg^{2\oplus}$, X = $OCOCH_3$, R = C_6H_5
L:	n = 1, Me = $Hg^{2\oplus}$, X = Cl, R = C_6H_5

LI

LII

LIII:	R = Cl, $R^1 = C_2H_5$, $R^2 = CH(CH_3)_2$
LXII:	R = SCH_3, $R^1 = CH_3$, $R^2 = CH(CH_3)_2$
LXXV:	R = SCH_3, R^1, $R^2 = CH(CH_3)_2$
LXXVIII:	R = Cl, R^1, $R^2 = C_2H_5$
LXXXI:	R = SCH_3, $R^1 = C_2H_5$, $R^2 = C(CH_3)_3$

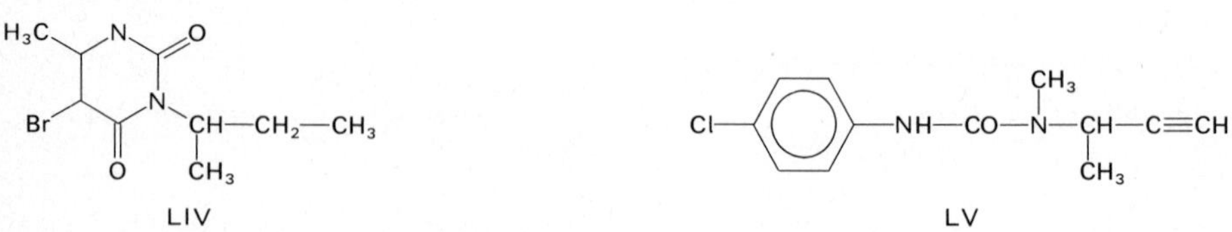

Fig. 9.3. Structures of some selected pesticides. Part 3. The Roman numerals refer to Table 9.5.

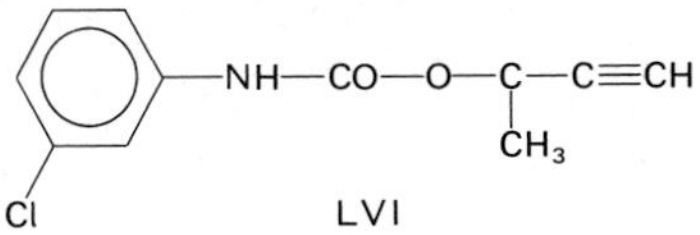

LVI

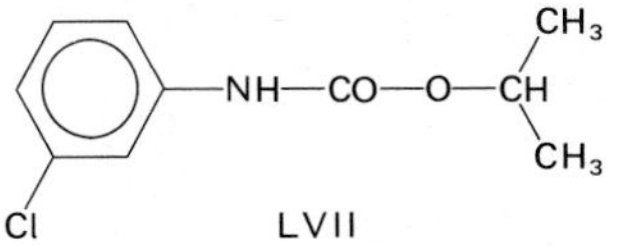

LVII

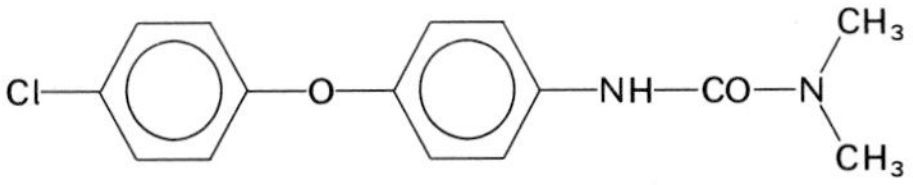

LVIII

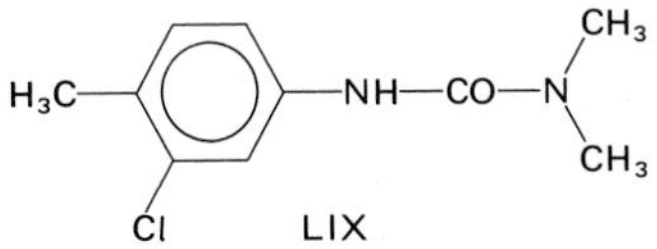

LIX

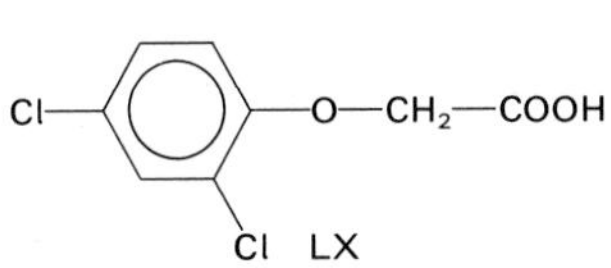

LX

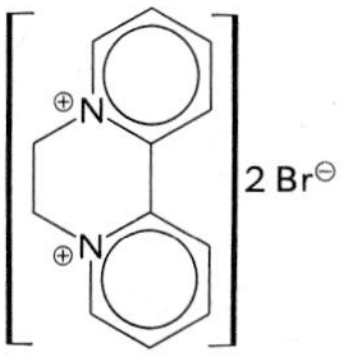

LXI

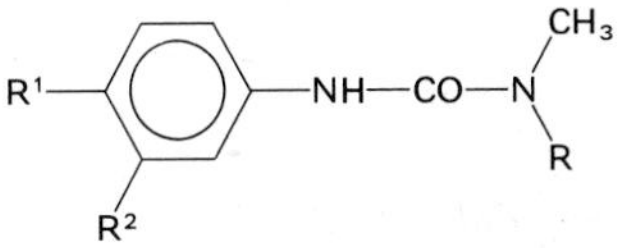

LXIV: $R^1, R^2 = Cl$, $R = CH_3$
LXVI: $R^1, R^2 = Cl$, $R = OCH_3$
LXVIII: $R^1 = Br$, $R^2 = H$, $R = OCH_3$
LXIX: $R^1 = OCH_3$, $R^2 = Cl$, $R = CH_3$
LXXII: $R^1 = Cl$, $R^2 = H$, $R = OCH_3$

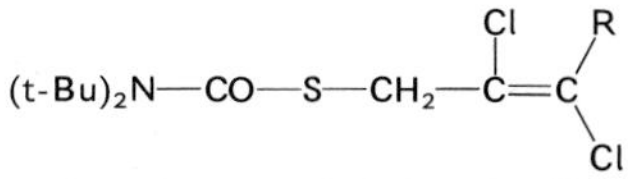

$(t\text{-}Bu)_2N{-}CO{-}S{-}CH_2{-}C(Cl){=}C(R)Cl$

LXIII: R = H, LXXXII: R = Cl

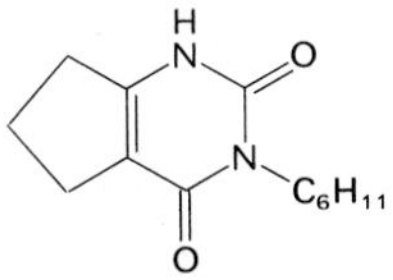

LXV

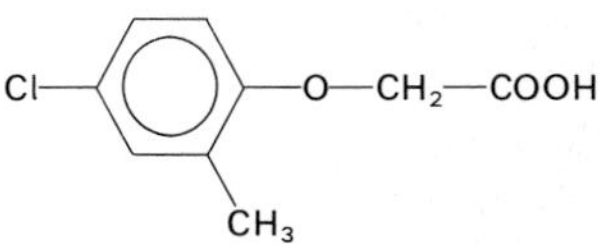

LXVII

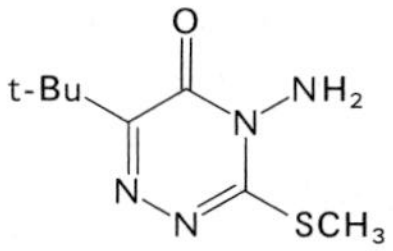

LXX

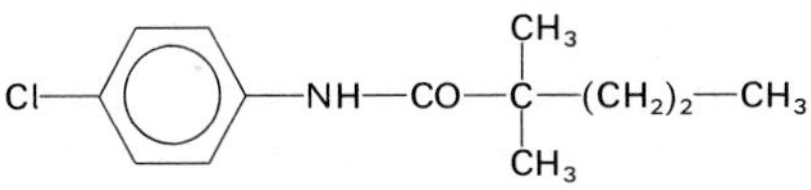

LXXI

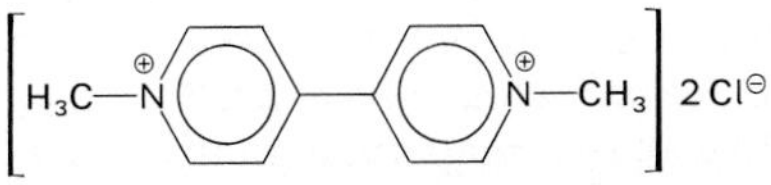

LXXIII

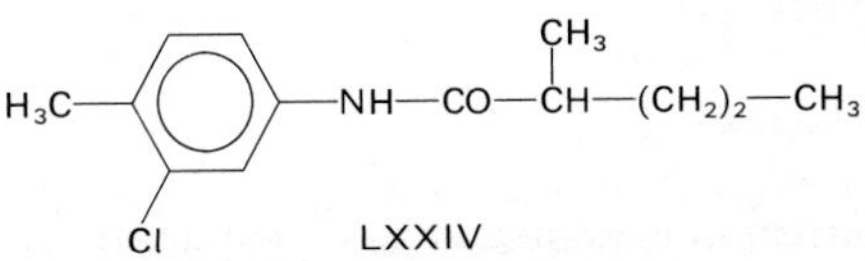

LXXIV

Fig. 9.2. Structures of some selected pesticides. Part 4. The Roman numerals refer to Table 9.5.

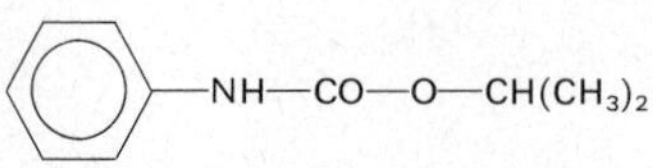

LXXVI

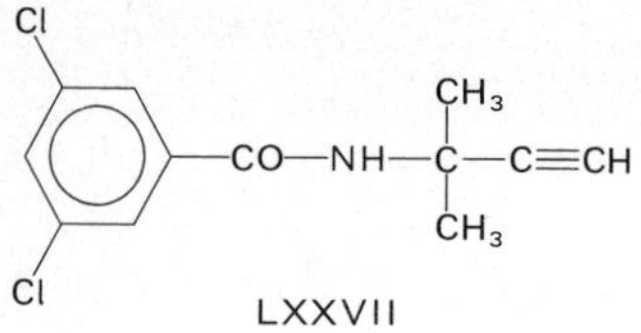

LXXVII

CCl_3COOH

LXXX

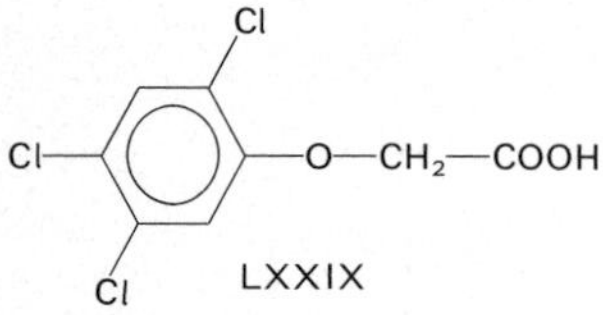

LXXIX

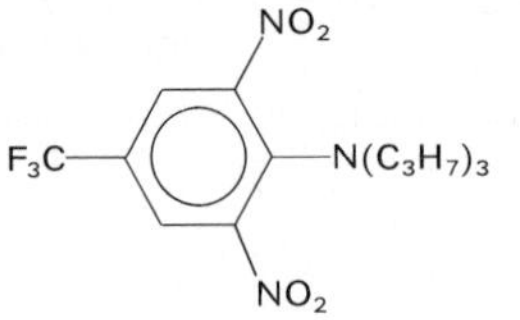

LXXXIII

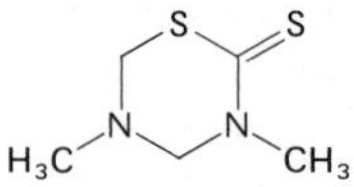

LXXXIV

$HC{=}CCl{-}CH_2Cl$

$ClCH{=}CH{-}CH_2Cl$

LXXXV

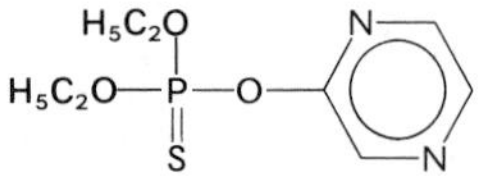

LXXXVIII

CH_3CHO

LXXXIX

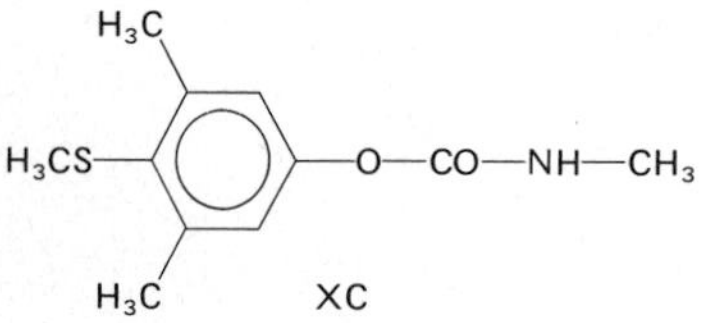

XC

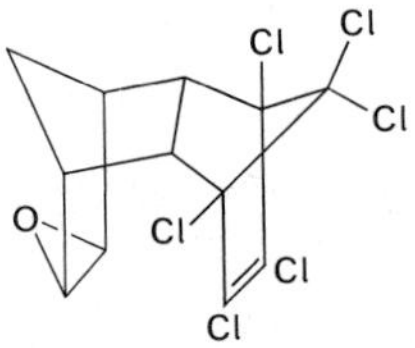

XCI

Fig. 9.2. Structures of some selected pesticides. Part 5. The Roman numerals refer to Table 9.5.

lime broth), are dithiocarbamates and organometallic compounds (cf. Table 9.5 and Figure 9.2). Residues of HCB and dithiocarbamates are observed in green vegetables, particularly in lettuce. Quintozene, which contains HCB as an impurity, also occurs as a residue.

9.5 Veterinary Medicines and Feed Additives

9.5.1 Foreword

The current practice in animal husbandry is the wide use of veterinary medicines, which serve not only for therapy, but to a large extent for prophylaxis and economic aims (e.g. to shorten animal growth or feeding time; to abate the risk of losses). Veterinary preparation residues in food are ingested by humans, though in low amounts but continuously and, hence, could be a health hazard. This possibility was, for a long time, not carefully examined. Therefore, as in the field of pesticides, supporting and maintaining appropriate measures (printing suitable directions for use, analytical control or supervision, elucidation of toxicological problems) has the ultimate aim of protecting human health.
A brief outline of some important groups of veterinary medicines follows. Table 9.7 and Figure 9.3 provide a review of their use and chemical structures.

Table 9.6. Pesticide residues in food (FR Germany, 1977 and 1978, after Nutrition Report, 1980)

Food	Year	Origin	Number of samples	Samples with residues	
				below the permissible level in %	above the permissible level in %
Foods of plant origin (fruits and vegetables)	1977	Domestic	216	19.0	1.4
		Import	594	51.5	8.2
	1978	Domestic	629	28.6	0
		Import	938	43.3	2.0
Vegetable	1977	Domestic	166	14.5	1.8
		Import	280	40.4	8.2
	1978	Domestic	223	16.6	0
		Import	312	51.9	3.5
Fruits (excluding citrus fruits)	1977	Domestic	50	28.0	0
		Import	255	41.6	8.2
	1978	Domestic	98	45.9	0
		Import	375	42.4	1.9
Citrus fruits	1977	Domestic and	59	64.4	8.5
	1978	Import	102	49.0	1.0
Foods of animal origin	1977	Domestic	649	95.7	2.5
		Import	242		
	1978	Domestic	748	92.0	1.9
		Import	189		
Milk	1977	Domestic and	115	95.7	3.5
	1978	Import	307	98.4	0.7
Dairy products	1977	Domestic and	59	100	0
	1978	Import	38	97.4	0
Butter	1977	Domestic and	80	98.6	0
	1978	Import	114	97.4	0
Cheese	1977	Domestic and	67	79.1	8.7
	1978	Import	47	87.3	10.6
Eggs	1977	Domestic and	64	84.4	9.4
	1978	Import	77	85.7	0
Meat products	1977	Domestic and	143	95.8	1.4
	1978	Import	254	83.9	3.9

Table 9.6a. Chlorinated hydrocarbons in human and bovine milk (mg/kg fat)

		α-HCH	β-HCH	γ-HCH
Bovine milk[b]	1979	0.028	0.009	0.031
Human milk[c]	1974/75	0.032	0.56	0.087
	1979/80	0.02	0.3	0.02
		HCB	DDT[a]	PCB
Bovine milk[b]	1979	0.030	0.033	0.11
Human milk[c]	1974/75	2.65	3.51	6.5
	1979/80	1.5	2.0	4.0

HCH: Hexachlorocyclohexane, HCB: hexachlorobenzene, DDT: 4,4-dichlorodiphenyltrichlorethane, DDE: 1,1-dichloro-2,2-bis(p-chlorphenyl)-ethylene, DDD: 1,1-dichloro-2,2-bis(4-chlorphenyl)-ethane, PCB: Polychlorinated biphenyl.

[a] Total DDT: DDT + degradation products (DDE, DDD).

[b] 810 samples analyzed were collected from different districts of FR Germany.

[c] Samples from Northwestern Germany.

Table 9.7. Animal medicines and feed additives (selected structural formulas are presented in Figure 9.3)

Number	Compound	Application
Antibiotics, Sulfonamides		
I	Penicillins *(Penicillium notatum)*	Therapeutics, feed additives
II	Streptomycin *(Streptomyces griseus)*	Therapeutics
III	Tetracyclines *(Streptomyces spp.)*	Therapeutics, feed additives
IV	Chloramphenicol *(Streptomyces venezuelae)*	As III
V	Oleandomycin *(Streptomyces antibioticus)*	Feed additive for poultry and swine
VI	Spiramycin *(S. ambofaciens)*	As V and calves, sheep, goats
VII	Tylosin *(S. fradiae)*	As V, for swine
VIII	Flavophospholipol	As V, for poultry, calves, swines
IX	Virginiamycin *(S. virginae)*	As VIII
X	Zinc-Bacitracin *(Bacillus subtilis)*	As VI
XI	Sulfonamide	Therapeutics
XII	Carbadox	Therapeutic for animal breeding
Steroid hormones and other compounds with estrogenic activity		
XIII	Corticosteroids	Therapeutics
XIV	17-β-Estradiol	Therapeutics, anabolic agents
XV	Oestrone	
XVI	17-α-Estradiol	
XVII	Estriol	
XVIII	Progesterone	
XIX	Testosterone	
XX	17-β-Estradiol-3-benzoate	
XXI	Estradiol-17-monopalmitate	
XXII	Testosterone-propionate	
XXIII	Trenbolone	
XXIV	Trenbolone acetate	
XXV	Diethylstilbestrol	
XXVI	Genisterin	
XXVII	Hexestrol	
XXVIII	Coumestrol	
XXIX	Dienestrol	
XXX	Zeranol	Anabolic agents
Psychopharmaceuticals		Therapeutics (Sedatives)
XXXI	Meprobamate (Aneural, Miltaun)	
XXXII	Hydroxyzine (Atarax, Marmoran)	
XXXIII	Chlordiazepoxide (Librium)	
XXXIV	Diazepam (Valium)	
XXXV	Oxazepam (Adumbran)	
XXXVI	Chlorpromazine (Megaphen)	
XXXVII	Promazine (Verophen)	

Table 9.7. (Continued)

Number	Compound	Application
XXXVIII	Azepromazine (Plegicil)	
XXXIX	Xylazine (Rompun)	
XL	Azaperone (Stresnil)	
XLI	Reserpine	
Thyreostatica		Therapeutics, anabolic agents
XLII	Methylthiouracil (Thyreostat)	
XLIII	Thiamazole (Favistan)	
XLIV	Carbimazole (Neo-Thyreostat)	
Coccidiostatica		Poultry feed additive for prophylaxis against Cocciodiose
XLV	Amprolium	
XLVI	Decoquinate	
XLVII	DOT, Dinitolmide	
Antiparasitica		
XLVIII	Niclofolan (Menichlopholan)	A remedy against liver leech
IL	Oxyclozanide	As XLVIII
L	Trichlorphon (Metrifonate)	
Other medicines		
LI	Niclosamide (Masonil)	A remedy against worms
LII	Nitarsone	Therapy (poultry, swine diarrhea), growth promoter
Antioxidants		Feed additives
LIII	Ethoxyquin	
LIV	Butylated hydroxytoluene (BHT)	

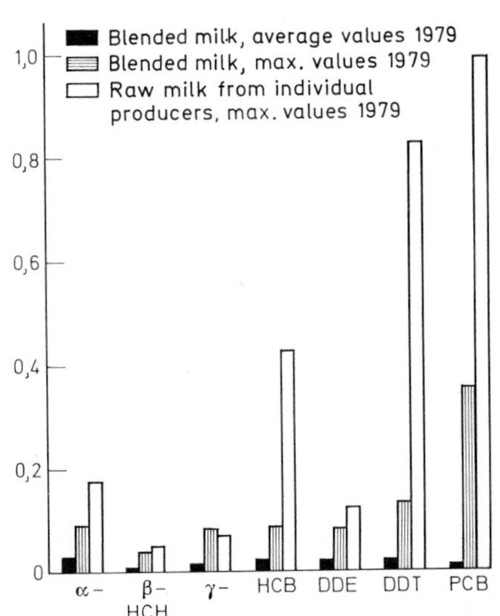

Fig. 9.2a. Chlorinated hydrocarbons in cow's milk, mg/kg fat. (After *Acker*, 1981.) For abbreviations see Table 9.6a

9.5.2 Antibiotics

Antibiotics are used for therapy and as growth-promoting agents, since they improve feed utilization and, thus, animal growth (calves, hogs and poultry). Residues may be found in eggs and milk (e.g. after treatment for mastitis). A constant intake of antibiotics, even at low doses, is a risk to human health since some microorganisms may become resistant and allergic reactions may develop. Therefore, the trend in current practice is to use only those antibiotics as feed additives, which are not used for human therapy or in treatment of animal diseases.

9.5.3 Glucocorticoides

Preparations of the hormones of the adrenal gland cortex, e.g. cortisone, are used in situations where animals are stressed. These preparations have a broad spectrum activity and are not to have uncontrolled access to the food chain.

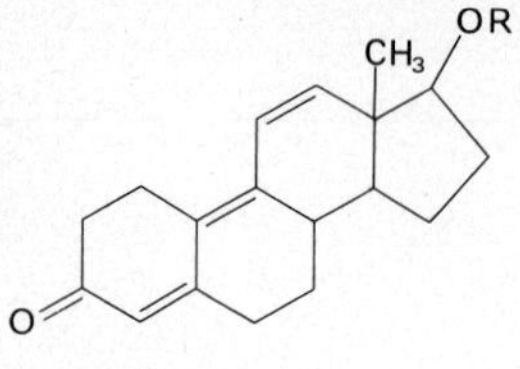

XXIII: R = H
XXIV: R = $COCH_3$

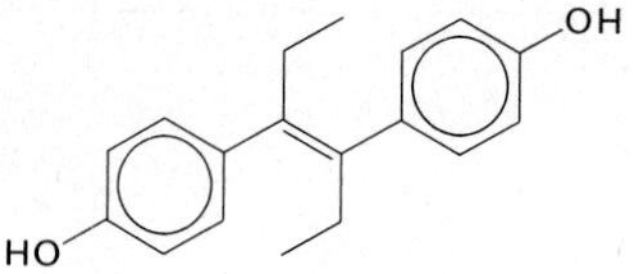

XXV

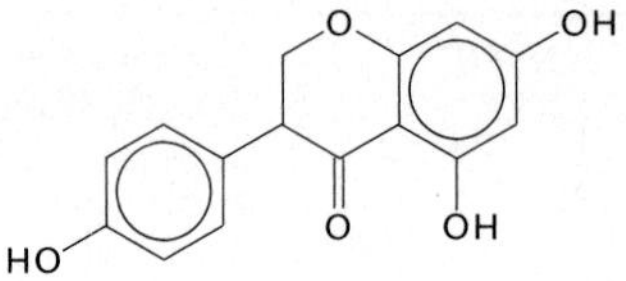

XXVI

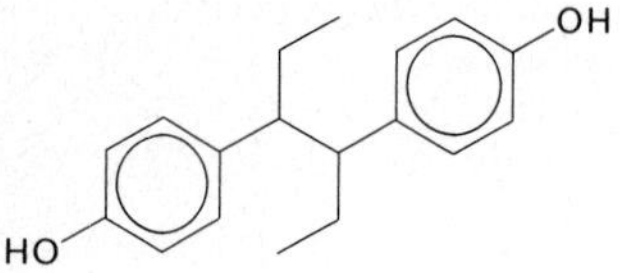

XXVII

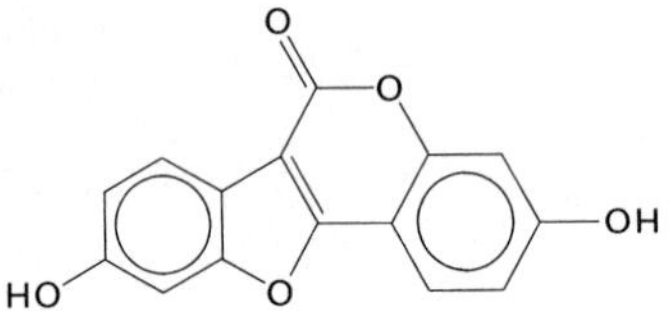

XXVIII

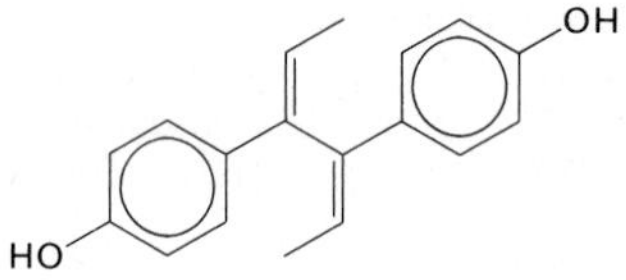

XXIX

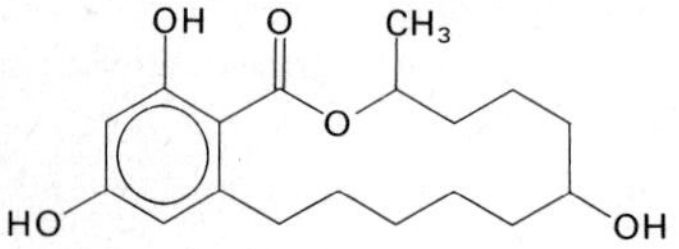

XXX

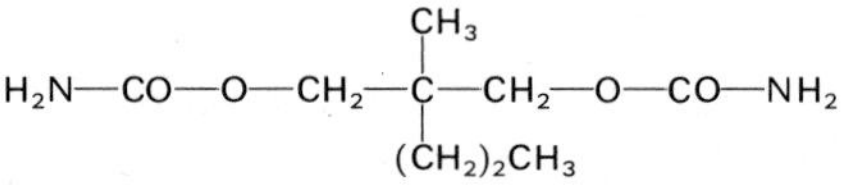

$H_2N—CO—O—CH_2—C(CH_3)((CH_2)_2CH_3)—CH_2—O—CO—NH_2$

XXXI

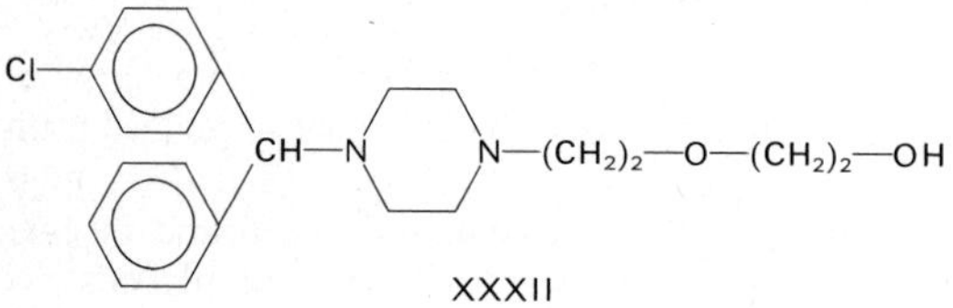

XXXII

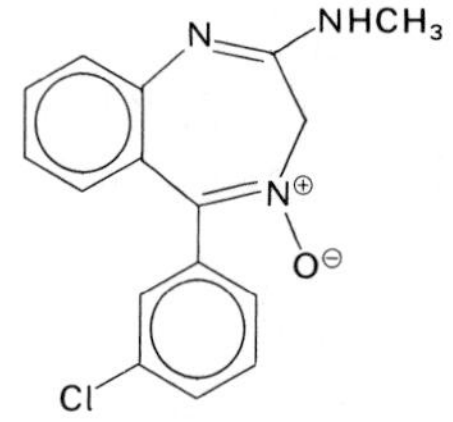

XXXIII

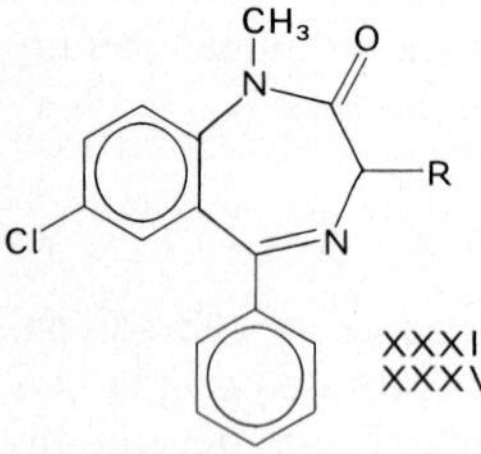

XXXIV: R = H
XXXV: R = OH

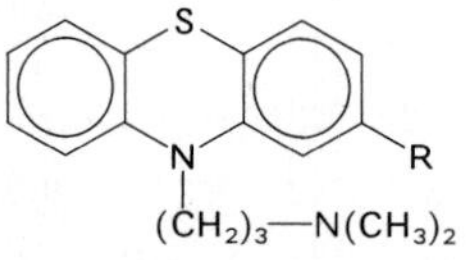

XXXVI: R = Cl
XXXVII: R = H
XXXVIII: R – $COCH_3$

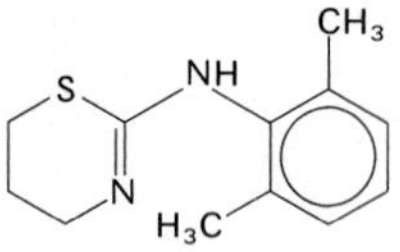

XXXIX

Fig. 9.3. Structures of some selected veterinary medicines and feed additives. The Roman numerals refer to Table 9.7

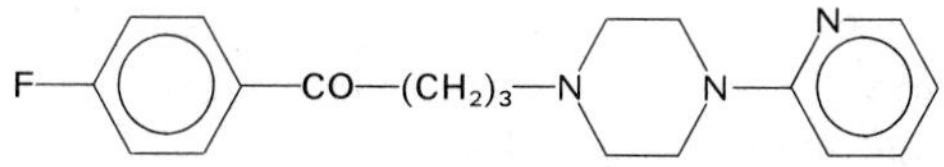

XL

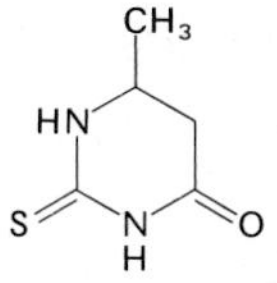

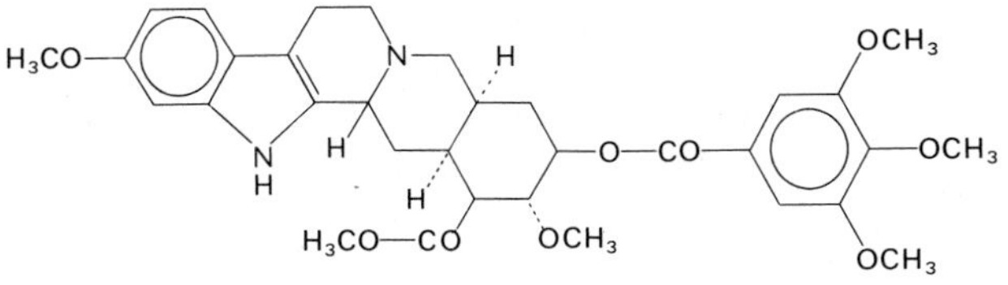

XLI

XLII

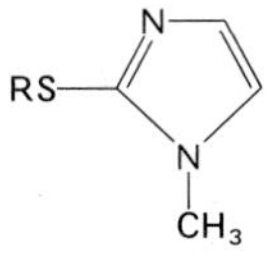

XLIII: R = H
XLIV: R = OCOCH3

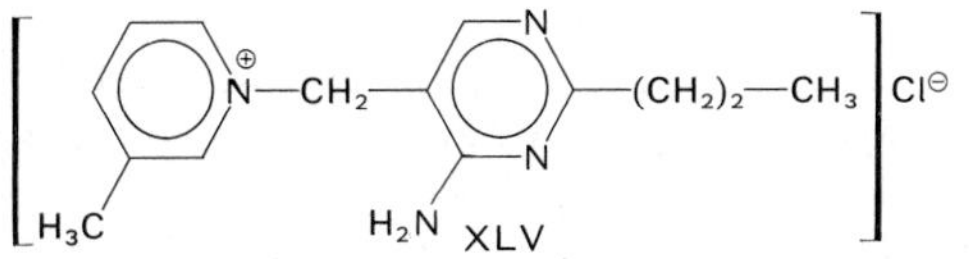

XLV

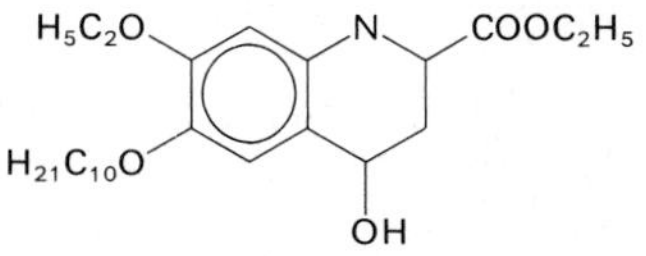

XLVI

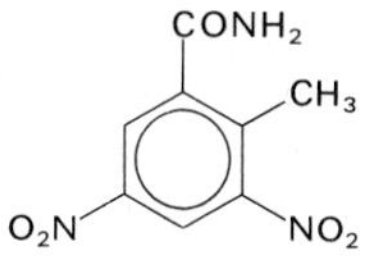

XLVII

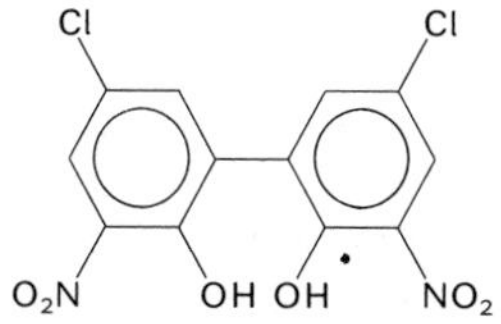

XLVIII

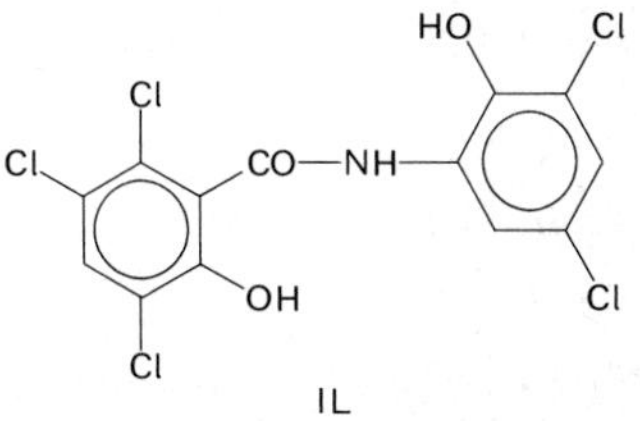

IL

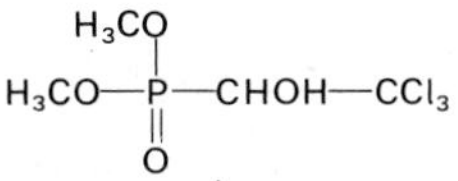

L

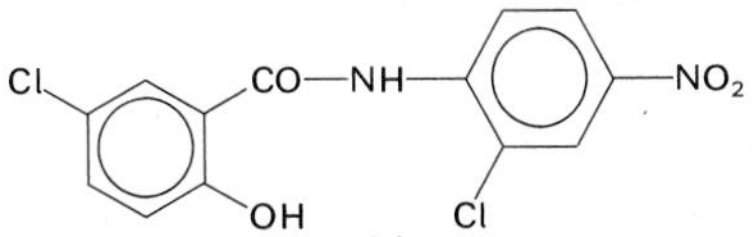

LI

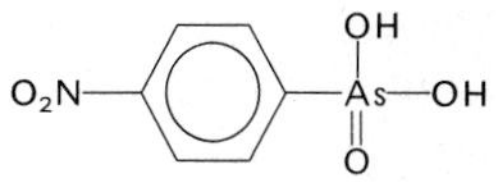

LII

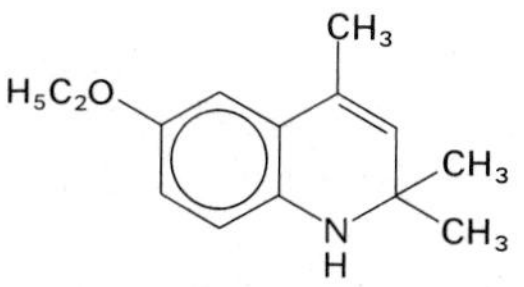

LIII

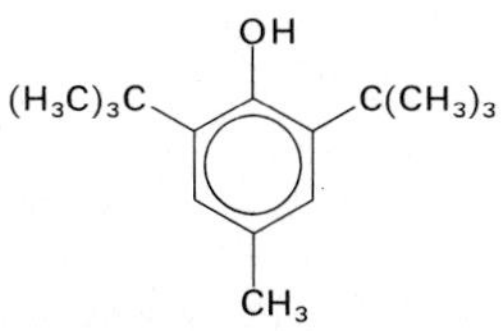

LIV

Fig. 9.3. (Continued)

9.5.4 Sex Hormones

In addition to their use in therapy, these compounds are used as growth or feed utilization promoters due to their anabolic activity. These include testicular hormones, such as testosterone and estradiol, as well as synthetic compounds with corresponding activity: esters of estradiol and testosterone, trenbolone, diethylstilbestrol [3,4- bis (p-hydroxyphenyl)-3-hexene] and zeranol.

While residues of the endogenic hormones are assumed not to present a health hazard, the other compounds mentioned are generally stable and have a long residence time in the body. Diethylstilbestrol has teratogenic and carcinogenic side effects.

9.5.5 Psychosomatic Agents

Compounds of this group (e.g. librium, valium) are used as sedatives to lessen the irritation or excitement of the animal and are given prior to vaccination or slaughter. The anabolic conversion of feed by a tranquil animal may be indirectly improved as well. However, the use of sedatives might impede the recognition of sickness in an animal.

9.5.6 Thyreostatica

These compounds lower the basal metabolic rate, thus effecting an increase in body weight (muscle and adipose tissue deposition) and improving the quality of the muscle. Thiouracile and mercaptobenzimidazole are the most commonly used. Thiourea derivatives, found among degradation products, have carcinogenic activity.

9.5.7 Coccidiostatica

The compounds of this class are added to animal feed to combat coccidiosis diseases (such as enteritis or cachexie) caused by protozoans living as parasites in intestines. Poultry and rabbits are the animals most often affected. Residues may be found in eggs.

9.5.8 Other Compounds

Antiparasitica are used, for example, against the cattle liver leech, which enters and infects the body after intake of snails or slugs during grazing. Antiparasitica residues may occur in milk.
Antioxidants, e.g. ethoxyquin (6-ethoxy-1,2-dihydro-2,2,4-trimethylquinoline), are used as an additive to protect animal feed from spontaneous combustion during transport.

9.6 Polychlorinated Biphenyls (PCB's)

Chlorination of biphenyl yields a mixture of compounds with differing chlorine contents. A widely-used commercial product, Clophen A 60, with 60% bound chlorine (corresponding in North America to Aroclor 1260, a compound with an average chlorine content of 63% and an average molecular weight of 372), consists of about 55 components of differing toxicities. The toxicity is influenced by the number of chlorine atoms and their substitution pattern in the biphenyl molecule, and by the overall composition of the product.

PCB's are widely used in industry as plasticizers in synthetic polymers, in paints, in heat exchange media, in hydraulic presses and as a dielectric fluid in transformers. Although external use has been abandoned, PCB's are still widely found as environmental contaminants (soil, atmosphere and water) and as a consequence of this as residues in food (eggs, wild game, fish, etc.). The concentrations, averaging 12 μg/kg, are indeed low but, as with all persistent lipophilic compounds, the level is constantly being increased through the nutritional chain (plant-animal-human).

9.7 Polycyclic Aromatic Hydrocarbons

Burning of organic materials, such as wood (wood smoke and its semi-dry distillation product, the wood smoke vapor phase), coal or fuel oil, results in pyrolytic reactions which yield a great number of polycyclic aromatic hydrocarbons, with more than three linearly or angularly fused benzene rings, that are carcinogenic to varying extents (1,2-benzanthracene, benzo[a]pyrene, chrysene, fluoranthene, pyrene, etc.). The quantity and diversity of compounds generated is affected by the conditions of the burning process. Benzo[a]pyrene usually serves as an indicator compound (Formula 9.1):

(9.1)

Contamination of food with polycyclic compounds can occur by deposition from the atmosphere (as often occurs with fruit and leafy vegetables in industrial districts), by direct drying of cereals with combustion gases, by smoking or roasting of food (barbecuing or charcoal broiling; smoking of sausage, ham or fish; roasting of coffee). The content in meat and processed meat products should not exceed 1 µg/kg end-product, measured as benzo[a]pyrene. Most smoked products (about 98%) are within this limit. The highest concentrations of polycyclic aromatic hydrocarbons in smoked fish have been found in eels and salmon.

9.8 Nitrosamines

Nitrosamines and nitrosamides are powerful carcinogens. They are obtained from secondary amines, N-substituted amides and nitrous acid:

$$RR^1NH + HONO \longrightarrow RR^1N{-}NO \quad (9.2a)$$

$$R{-}CO{-}NHR^1 + HONO \longrightarrow R{-}CO{-}N(NO){-}R^1 \quad (9.2b)$$

in reactions in which the nitrosonium ion, NO^+, or a nitrosyl halogenide, XNO, is the reactive component:

$$NO_2^{\ominus} \underset{}{\overset{H^{\oplus}}{\rightleftharpoons}} HONO \overset{H^{\oplus}}{\rightleftharpoons} H_2ONO^{\oplus}$$

$$H_2ONO^{\oplus} \rightleftharpoons H_2O + NO^{\oplus}$$

$$H_2ONO^{\oplus} \overset{X^{\ominus}}{\rightleftharpoons} X{-}N{=}O$$

$$H_2ONO^{\oplus} \overset{NO_2^{\ominus}}{\rightleftharpoons} H_2O + N_2O_3 \quad (9.3)$$

Nitrosamine formation is also possible from primary amines or diamines:

$$R{-}CH_2{-}NH_2 \xrightarrow[\text{1) Nitrosation, 2) Diazotization, 3) Deamination}]{HNO_2} R{-}CH_2^{\oplus}$$

$$\xrightarrow[\text{4) Dimerization}]{R-CH_2-NH_2} (R{-}CH_2)_2NH \xrightarrow[\text{5) Nitrosation}]{HNO_2} (R{-}CH_2)_2N{-}NO \quad (9.4)$$

from diamines:

$$H_2N{-}(CH_2)_n{-}CH(R){-}NH_2 \xrightarrow[-N_2]{HNO_2} H_2N{-}(CH_2)_n{-}CH^{\oplus}(R) \xrightarrow{\text{Cyclization}}$$

$$\text{cyclo}[(CH_2)_n{-}CHR{-}NH] \xrightarrow{HNO_2} \text{cyclo}[(CH_2)_n{-}CHR{-}N(NO)]$$

R = H: diamine
R = COOH: diamino carboxylic acid (9.5)

and also from tertiary amines:

$$R_2N{-}CHR^1R^2 \xrightarrow{HNO_2} R_2N^{\oplus}(NO){-}CHR^1R^2 \xrightarrow{-HNO} R_2\overset{\oplus}{N}{=}CR^1R^2 \xrightarrow{H_2O}$$

$$R_2NH_2^{\oplus} + OCR^1R^2 \xrightarrow{HNO_2} R_2N{-}NO \quad (9.6a)$$

$$2\,HNO \longrightarrow H_2N_2O_2 \longrightarrow H_2O + N_2O \quad (9.6b)$$

Nitrosamines are detected in variable amounts in many foods (Table 9.8). The most common compound is dimethylnitrosamine, which is also a most powerful carcinogen. Some activity has been ascribed to nitrosopiperidine and nitrosopyrrolidine. In meat products cured and treated with pickling salt, 30% of the samples contained nitrosodimethylamine (NDMA; 0.5–15 μg/kg) and 13% nitrosopyrrolidine (NPYR; > 0.5 μg/kg). About 25% of the cheese samples analyzed were contaminated (0.5–4.9 μg/kg).

Nitrosopyrrolidine is formed from the amino acid proline by nitrosation followed by decarboxylation at elevated temperatures, such as in roasting or frying:

$$\text{Proline} \xrightarrow{HNO_2} \text{N-Nitrosoproline} \xrightarrow{-CO_2} \text{N-Nitrosopyrrolidine} \tag{9.7}$$

The nitrosopyrrolidine (1.5 μg/kg) in meat products increases almost ten-fold (to 15.4 μg/kg) during roasting and frying. An estimate for the average daily intake of nitrosamines ranges from 0.1 μg nitrosodimethylamine and 0.1 μg nitrosopyrrolidine, to a total of 1 μg.

An endogenic dose should be included in addition to the above exogenic dose. It may result from ingestion of amines, and of nitrate ions, both of which are abundant in food. Table 9.9 presents data on the forms of amines present in food. The levels of nitrate are also given. The occurrence of nitrate is very high in some vegetables (e.g. 3,100 mg/kg spinach), and nitrate is occasionally found in drinking water.

The minimum nitrate intake is estimated at 75 mg/person/day. Nitrate is reduced to nitrite by microflora of mouth saliva. This is a prerequisite for a nitrosation reaction in the gastric acid medium. A nitrosation may also occur with a number of medicinal products. Dimethylamino or diethylamino compounds, which might be degraded under release of dimethyl or diethyl nitrosamines, are most commonly involved.

Inhibition of a nitrosation reaction is possible, e.g. with ascorbic acid, which is oxidized by nitrite to its dehydro form, while nitrite is reduced to NO. Similarly, tocopherols and some other food constituents inhibit substitution reactions. Representative suitable measures to decrease exo- and endogenic nitrosamine hazard are:

- Decreasing nitrite and nitrate incorporation into processed meat. However, to completely relinquish the use of nitrite is a great health hazard due to the danger from bacterial intoxication (especially by botulism).
- Addition of inhibitors (ascorbic acid, tocopherols).
- Decreasing the nitrate content of vegetables.

Table 9.8. Nitrosamines in food

Food product	Compound[a]	Content μg/kg	Year of analysis
Frankfurter (hot dog)	NDMA	0–84	1972
Fish (raw)	NDMA	0–4	1971, 1972
Fish, smoked and pickled with nitrites or nitrates	NDMA	4–26	1971
Fish, fried	NDMA	1–9	1972
Cheese (Danish, Blue, Gouda, Tilsiter, goatmilk cheese)	NDMA	1–4	1972
Salami	NDMA	10–80	1972
Bacon (hog's hind leg) smoked meat	NDMA	1–60	1975
Pepper-coated ham, raw and roasted	NPIP	4–67	1975
	NPYR	1–78	1975

[a] NDMA: N-Nitrosodimethylamine, NPIP: N-nitrosopiperidine, NPYR: N-nitrosopyrrolidine.

9.9 Cleansing Agents and Disinfectants

Residues introduced by large-scale animal husbandry and use of milking machines are gaining in importance. The residues in meat and processed meat products originate from the surfaces of processing equipment, while residues in milk arise from measures involved in disinfection of the udder. Iodine-containing disinfectants, including udder dipping or soaking agents, may be an additional source of contamination of milk with iodine.

Table 9.9. Amines and nitrate in food (mg/kg)

Compound	Spinach	Cabbage (kale)		Carrots	Red beet	Celery	Lettuce	Rhubarb	Herring			Cheese		
		red	green						salted	smoked	in oil	Tilsiter	Camembert	Limburger
Ammonia	18.280	11.060	15.260	3.970	8.800	19.600	10.260	6.340	2.928	270	–	164.400	–	–
Methylamine	12	22.7	16.6	3.8	30	5.1	37.5	–	3.4	–	7	–	12	3
Ethylamine	8.4	1.3	–	1	1	–	3.3	–	0.1	0.4	–	–	4	1
Dimethylamine	–	2.8	5.5	–	–	6.4	7.2	–	7.8	6.3	45	–	–	–
Methylethylamine	–	0.9	–	7	7	–	7.5	–	–	–	1	–	–	–
n-Propylamine	–	–	–	–	–	–	–	–	–	–	–	8.7	2	2
Diethylamine	15	–	–	–	–	–	–	–	1.9	5.2	–	–	–	–
n-Butylamine	–	–	7	–	–	–	–	–	–	–	–	3.7	–	–
i-Butylamine	–	–	–	–	–	–	–	–	–	0.3	–	–	0.2	0.2
Pyrroline	–	–	–	–	–	–	–	–						
n-Pentylamine	0.3	0.6	0.4	–	–	0.8	3	–	–	–	17	1.2	–	–
i-Pentylamine	3.8	–	0.5	–	–	–	–	3.9	–	–	–	–	0.2	tr
Pyrrolidine	2.5	–	–	–	–	0.4	–	–	–	–	–	19.9	1	0.1
Di-n-propylamine									–	–	–	8.4	–	–
Piperidine									0.7	0.2	–	–	–	tr
Aniline	–	1.0	0.7	30.9	–	0.7	0.6	5						
N-Methylaniline	3.4	0.3	–	0.8	–	0.5	–	–	–	–	–	37.9	–	–
N-Methylbenzylamine	–	–	–	16.5	–	–	10	–	–	–	2	–	–	–
Toluidine	–	–	1.1	7.2	–	1.1	–	–						
Benzylamine	6.1	3.3	3.8	2.8	0.1	3.4	11.5	2.9						
Phenylethylamine	1.1	8.6	3	2	0.5	–		3.2						
N-Methylphenyl-ethylamine	2.4	3.7	2	2	0.4	0.5	0.4	2.6	–	–	0.1	2.6	–	–
Nitrate	3.100	600	200	500	500	400	1.600	–	–	–	–	444	–	–

[a] Traces

9.10 Literature

Acker, L.: Die Rückstandssituation in der Bundesrepublik Deutschland – Versuch einer Bestandsaufnahme. Lebensmittelchem. Gerichtl. Chem. *35*, 1, (1981)

Benarie, M. M. (Ed.): Atmospheric pollution 1982. Proc. of the 15-th intl collog., Paris. Elsevier Science Publ., Amsterdam. 1982

Berg, H. W., Diehl, J. F., Frank, H.: Rückstände und Verunreinigungen in Lebensmitteln. Dr. Dietrich Steinkopff Verlag: Darmstadt. 1978

Bund für Lebensmittelrecht und Lebensmittelkunde e. V.: Wie sicher sind unsere Lebensmittel? Schriftenreihe Heft 102, B. Behr's Verlag: Hamburg. 1983

Clayson, D. B., Krewski, D. and Munro, I.: Toxicological risk assessment. Vol. I–II. CRC Press Inc., Boca Raton, FL: 1985

Deutsche Forschungsgemeinschaft: Rückstände in Gebäck und Getreideprodukten. Harald Boldt Verlag KG: Boppard. 1981

Deutsche Forschungsgemeinschaft: Rückstände in Lebensmitteln tierischer Herkunft. Verlag Chemie: Weinheim. 1983

Deutsche Forschungsgemeinschaft: Das Nitrosamin-Problem. Verlag Chemie: Weinheim. 1983

Deutsche Forschungsgemeinschaft: Rückstände und Verunreinigungen in alkoholfreien Getränken. Verlag Chemie: Weinheim. 1983

Dominquez, G. S. and Bartlett, K. G.: Hazardous waste management. Vol. I. CRC Press Inc., Boca Raton, FL. 1986

Eisenbrand, G.: N-Nitrosoverbindungen in Nahrung und Umwelt. Wissenschaftliche Verlagsgesellschaft mbH: Stuttgart. 1981

Ernährungsbericht 1980. Deutsche Gesellschaft für Ernährung e. V.: Frankfurt/Main. 1980

Ernährungsbericht 1984. Deutsche Gesellschaft für Ernährung e. V.: Frankfurt/Main. 1984

Expert Committee on Pesticide Use in Agriculture. Pesticide research report, Agriculture Canada, Ottawa. 1985

Follweiler, J. M. and Sherma, J.: Pesticides and related organic chemicals. CRC Press Inc., Boca Raton, FL. 1984

Hassall, K. A.: The chemistry of pesticides. Verlag Chemie: Weinheim. 1982

Hathway, D. E.: Molecular aspects of toxicology. The Royal Society of Chemistry, Burlington House, London. 1984

Heitefuß, R.: Pflanzenschutz. Georg Thieme Verlag: Stuttgart. 1975

Heyns K.: Über die endogene Nitrosamin-Entstehung beim Menschen. Landwirtschaftliche Forschung, Sonderheft 36, S. 145, Kongressband 1979 Gießen, J. D. Sauerländer's Verlag: Frankfurt/Main. 1980

Lindner, E.: Toxikologie der Nahrungsmittel. 2. Aufl., Georg Thieme Verlag: Stuttgart. 1979

Mandava, N. B.: CRC Handbook of natural pesticides Vol. I–II. CRC Press Inc., Boca Raton, FL. 1985

Neely, W. B. and Blau, G. E.: Environmental exposure from chemicals. CRC Press Inc., Boca Raton, FL. 1985

Purves, D.: Trace – element contamination of the environment. 2-nd rev. ed. Elsevier Science Publ., Amsterdam. 1985

Rechcigl, M. Jr. CRC Handbook of naturally occurring food toxicants. CRC Press Inc., Boca Raton, FL. 1983

Safety and quality in food (Proc. of DSA symposium, Brussels, 1984). Elsevier Science Publ., Amsterdam. 1984

Siewierski, M. (Ed.): Determination and assessment of pesticide exposure (Proc. of a working conf., Hershey, P. A., 1982) Elsevier Science Publ., Amsterdam. 1984

Wegler, R. (Hrsg.): Chemie der Pflanzenschutz- und Schädlingsbekämpfungsmittel. Bde. 1 bis 8, Springer-Verlag: Berlin–Heidelberg. 1970–1982

Whicker, W. and Schultz, V.: Radioecology: Nuclear energy and the environment. CRC Press Inc., Boca Raton, FL. 1982

10 Milk and Dairy Products

10.1 Milk

Milk is the secreted fluid of the mammary glands of female mammals. It contains nearly all the nutrients necessary to sustain life. Since the earliest times, mankind has used the milk of goats, sheep and cows as food. Today the term "milk" is synonymous with cow's milk. The milk of other animals is spelled out, e.g. sheep milk or goat milk, when supplied commercially.
The yield of milk per cow in Germany more than doubled within this century as a result of dairy cattle breeding and improvement in feed quality. The yield was 1,260 kg per cow in 1812, 2,163 kg in 1926, 3,800 kg in 1970 and 4,181 kg in 1977. Milk production in various countries, its processing into dairy products and its consumption are summarized in Tables 10.1–10.3.

10.1.1 Physical and Physico-Chemical Properties

Milk is a white or yellow-white, opaque liquid. The color is influenced by scattering and absorption of light by milk fat globules and protein micelles. Therefore, skim milk also retains its white color. A yellowish, i.e. yellow-green, color is derived from carotene (ingested primarily dur-

Table 10.1. World production of milk, 1981 (1,000 t)

Continent	Cow milk	Buffalo milk	Sheep milk	Goat milk
World	428,213	27,943	7,910	7,559
Africa	10,516	1,303	697	1,434
America, North-, Central-	78,593			327
America, South-	23,462		35	136
Asia	38,134	26,537	3,579	3,655
Europe, West-	134,371	72	2,568	1,470
Europe, East- + USSR	131,248	30	1,032	536
Australia + South Pacific Islands	11,889			

[a] World production ≙ 100%.

Table 10.1. (Continued)

Country	Cow milk	Country	Buffalo milk
USSR	88,000	India	17,500
USA	60,161	Pakistan	6,519
France	33,700	China	1,410
FR Germany	24,817	Egypt	1,303
UK	15,862	Nepal	489
Poland	15,259	Turkey	305
India	13,500	Italy	72
Holland	12,148	Sri Lanka	55
Brasil	10,500	Burma	54
Italy	10,490	Vietnam	50
German DR	8,500		
Canada	8,025	Σ (%)[a]	99
Mexico	6,885		
Japan	6,620		
New Zealand	6,500		
Σ (%)[a]	75		

Country	Sheep milk	Country	Goat milk
Turkey	1,200	India	948
France	1,082	Turkey	647
Iran	705	Bangladesh	495
Italy	608	France	492
Greece	580	Pakistan	441
China	495	Greece	421
Syria	476	USSR	400
Rumania	340	Sudan	392
Bulgaria	308	Spain	307
Afghanistan	225	Mexico	296
		Somalia	286
Σ (%)[a]	76	China	265
		Iran	222
		Nigeria	140
		Σ (%)[a]	76

[a] World production ≙ 100%.

Table 10.2. World production of dairy products, 1981 (1,000 t)

Continent	Cheese	Butter[a]	Condensed milk	Whole milk powder	Skim milk powder[b]	Whey powder
World	11,612	6,870	4,749	1,727	4,276	1,043
Africa	370	162	43	12	18	
America, North-, Central-	2,586	731	1,354	81	701	403
America, South-	457	168	180	364	4	
Asia	688	1,331	642	116	185	9
Europe, West-	4,481	2,396	1,632	739	2,440	589
Europe, East- + USSR	2,810	1,754	822	246	659	33
Australia + South Pacific Islands	221	328	77	169	269	8

Country	Cheese	Country	Butter[a]	Country	Condensed milk
USA	2,212	USSR	1,277	USA	825
USSR	1,511	India	670	Holland	563
France	1,185	USA	561	USSR	530
FR Germany	818	FR Germany	545	FR Germany	508
Italy	600	France	530	India	324
Holland	468	German DR	281	Canada	229
Poland	386	Poland	280	Mexico	189
Egypt	247	New Zealand	247	France	157
Denmark	247	Pakistan	221	Czechoslovakia	152
Argentina	240	Holland	183	UK	150
UK	240	UK	171	Σ (%)[c]	76
German DR	213	Czechoslovakia	127		
Canada	207	Turkey	123		
Czechoslovakia	181	Σ (%)[c]	76		
Σ (%)[c]	75				

Country	Whole milk powder	Country	Skim milk powder[b]	Country	Whey powder
USSR	205	France	807	USA	345
Holland	190	FR Germany	697	France	255
Brasil	180	USA	551	Holland	157
France	139	USSR	290	FR Germany	130
FR Germany	131	UK	252	Canada	58
New Zealand	104	New Zealand	208	UK	33
Japan	95	German DR	170	Finland	23
Denmark	87	Holland	163	Spain	13
Venezuela	73	Ireland	142	China	9
Australia	65	Canada	141	Belgium/Luxemburg	8
Argentina	62	Σ (%)[c]	80	Σ (%)[c]	99
Σ (%)[c]	77				

[a] Includes fat from buffalo milk (ghee).

[b] Includes powder from buttermilk recovered from butter churning.

[c] World production ≙ 100%.

Table 10.3. Consumption of milk and dairy products in FR Germany (in kg/capita and year)

	64/65	66/67	68/69	70/71	72/73	74/75	76/77
Fluid milk	95.7	93.4	92.0	92.5	86.5	82.9	83.8
Cream	2.5	2.8	3.2	3.5	3.7	3.9	4.1
Cheese	8.4	9.1	9.4	10.2	11.1	11.7	12.4
Butter	8.5	8.6	8.5	8.3	7.3	7.0	6.4

ing pasture grazing) present in the fat phase and from riboflavin present in the aqueous phase. Milk tastes mildly sweet, while its odor and flavor are normally quite faint.

Milk fat occurs in the form of droplets or globules, surrounded by a membrane and emulsified in milk serum (also called whey). The fat globules (called cream) separate after prolonged storage or after centrifugation. The fat globules float on the skim milk. Homogenization of milk so finely divides and emulsifies the fat globules that cream separation does not occur even after prolonged standing.

Proteins of various sizes are dispersed in milk serum. They are called micelles and consist mostly of calcium salts of casein molecules.

Various proteins, carbohydrates, minerals and other ingredients are solubilized in milk serum. The specific density of milk decreases with increasing fat content, and increases with increasing amounts of protein, milk sugar and salts. The specific density of cow's milk ranges from 1.029 to 1.039 (15 °C). Defatted (skim) milk has a higher specific density than whole milk. From the relationships given by *Fleischmann:*

$$m = 1.2f + \frac{266.5(s-1)}{s} \tag{10.1}$$

and by *Richmond:*

$$m = 0.25s + 1.21f + 0.66 \tag{10.2}$$

the dry matter content of milk, m, as a percent, can be calculated from the percent fat content (f), knowing the specific density (s).

The freezing point of milk is -0.53 to -0.55 °C. This rather constant value is a suitable test for detection of watering of milk.

The pH of fresh milk is 6.5–6.75, while the acid degree according to *Soxhlet-Henkel* (°SH) is 6.5–7.5.

The refractive index (n_D^{20}) is 1.3410–1.3480, and the specific conductivity at 25 °C is $4\text{–}5.5 \times 10^{-3}$ $ohm^{-1}\ cm^{-1}$.

The measurement of redox potentials of milk and its products can also be of value. The redox potential is +0.30 V for raw and +0.10 V for pasteurized milk, +0.05 V for processed cheese, − 0.15 V for yoghurt and − 0.30 V for Emmental cheese.

10.1.2 Composition

The composition of dairy cattle milk varies to a fairly significant extent. Table 10.4 provides some data. In all cases water is the main ingredient of milk at 63–87%. In the following sections, only cow's milk will be dealt with in detail since it is the main source of our dairy foods.

Table 10.4. Composition of human milk and milk of various mammals (%)

Milk	Protein	Casein	Whey protein	Sugar	Fat	Ash
Human	1.2[a]	0.5	0.7	7.0	3.7	0.2
Cow (bovine)	3.6	3.0	0.6	5.0	3.7	0.7
Buffalo	5.3	3.5	0.5	4.9	8.0	0.7
Mare	2.7	1.8	0.9	6.1	1.6	0.5
Goat	3.7	2.9	0.8	4.3	4.3	0.9
Sheep	5.3	4.5	0.8	4.9	6.3	0.9
Reindeer	10.3	8.7	1.6	2.5	22.5	1.4
Cat	7.0	3.8	3.2	4.8	4.8	0.6
Dog	7.4	4.8	2.6			
Rabbit	10.4					

[a] After 15-th day of breast feeding period the protein content is increased to 1.6%.

10.1.2.1 Proteins

In 1877 *O. Hammarsten* distinguished three proteins in milk: casein, lactalbumin and lactoglobulin. He also outlined a procedure for their separation: skim milk is diluted then acidified with acetic acid. Casein flocculates, while the whey proteins stay in solution. This established a specific property of casein: it is insoluble in weakly-acidic media. It was later revealed that the milk protein system is much more complex. In 1936 *Pedersen* used ultracentrifugation to demonstrate the nonhomogeneity of casein, while in 1939 *Mellander* used electrophoresis to prove that casein consists of three fractions, i.e. α-, β- and γ-casein.

The most important proteins of milk are listed in Table 10.5. The casein fraction forms the main

Table 10.5. Bovine milk proteins

Fraction	Genetical variant	Portion[a]	Isoionic point	Molecular weight[b] (kdal)	Phosphorus content (%)
Caseins		75–85	–	–	0.9
α_{s1}-Casein	A, B, C, D	39–46	4.92–5.35	23.6	1.1
α_{s2}-Casein	A, B, C, D	8–11		25.2	1.4
κ-Casein	A, B	8–15	5.37	19	0.2
β-Casein	A^1, A^2, A^3, B, C, D, E, B_Z	25–35	5.20–5.85	24	0.6
γ-Casein		3–7	5.8–6.0	12–21	0.1
γ_1-Casein	A^1, A^2, A^3, B			20.5	
γ_2-Casein	A^1/A^2, A^3, B			11.8	
γ_3-Casein	$A^1/A^2/A^3$, B			11.6	
Whey proteins		15–22	–	–	
β-Lactoglobulin	A, B, C, D, Dr	7–12	5.35–5.41	18.3	
α-Lactalbumin	A, B	2–5	4.2–4.5[e]	14.2	
Serum albumin		0.7–1.3	5.13	66.3	
Immunoglobulin		1.9–3.3			
IgG1		1.2–3.3	5.5–6.8	162	
IgG2		0.2–0.7	7.5–8.3	152	
IgA		0.2–0.7	–	400[c]	
IgM		0.1–0.7	–	950[d]	
FSC(s)		0.2–0.3		80	
Proteose-Peptone		2–6	3.3–3.7	4–41	

[a] As % of skim milk total protein, [b] monomers, [c] dimer, [d] pentamer, [e] isoelectric point.

portion. Major constituents of whey proteins, β-lactoglobulin A and B and α-lactalbumin, can be differentiated genetically. Other protein constituents, e. g. enzymes, are present in much lower quantities; they are not listed in Table 10.5.

10.1.2.1.1 Casein fractions

The main constituents of this milk protein fraction have been fairly well investigated. Their amino acid sequences are summarized in Table 10.6. Data showing the genetic variations are provided in Table 10.7.

α_s-caseins. The B variant of α_{s1}-casein consists of a peptide chain with 199 amino acid residues and has a molecular weight of 23 kdal. The sequence contains 8 phosphoserine residues, 7 of which are localized in positions 43–80, and these positions have an additional 12 carboxyl groups. Thus these positions are extremely polar acidic segments along the peptide chain. Proline is uniformly distributed along the chain and apparently to a great extent hinders the formation of a regular structure. A portion of the chain, up to 30%, is assumed to have regular conformations. Amino acid residues 100–199 are distinctly apolar and are responsible for strong association tendencies, which are limited by the repulsing forces of phosphate groups. In the presence of Ca^{2+} ions, in the levels found in milk, α_{s1}-casein forms an insoluble Ca-salt. In the A variant of the molecule, amino acid residues 14–26 are missing; in the C variant the glutamic acid in position 192 (Glu-192) is replaced by Gly-192; and in the D variant Pth-53 (phosphothreonine) replaces Ala-53.

β-caseins. The A^2 variant is a peptide chain consisting of 209 residues and has a molecular weight of 24.5 kdal. Five phosphoserine residues are localized in the 1–40 positions; these positions contain practically all of the ionizing sites of the molecule. Positions 136-209 contain the predominant portions of apolar branched chains. The molecule has a structure with a "polar head" and an "apolar tail", thus resembling a "soaplike" molecule. A proportion approximately 20% of regular conformations is discussed for the β-casein. The self-association of the β-casein is an endothermic process. The protein precipitates in

Table 10.6. Amino acid sequences of bovine milk caseins. (1): α_{s1}-casein B, (2): β-casein A^2, (3): ϰ-casein B

(1)	Arg	Pro	Lys	His	Pro	Ile	Lys	His	Gln	Gly
(2)	Arg	Glu	Leu	Glu	Glu	Leu	Asn	Val	Pro	Gly
(3)	Pyg	Glu	Gln	Asn	Gln	Glu	Gln	Pro	Ile	Arg
(1)	Leu	Pro	Gln	Glu	Val	Leu	Asn	Glu	Asn	Leu
(2)	Glu	Ile	Val	Glu	Pse	Leu	Pse	Pse	Pse	Glu
(3)	Cys	Glu	Lys	Asp	Glu	Arg	Phe	Phe	Ser	Asp
(1)	Leu	Arg	Phe	Phe	Val	Ala	Pro	Phe	Pro	Gln
(2)	Glu	Ser	Ile	Thr	Arg	Ile	Asn	Lys	Lys	Ile
(3)	Lys	Ile	Ala	Lys	Tyr	Ile	Pro	Ile	Gln	Tyr
(1)	Val	Phe	Gly	Lys	Glu	Lys	Val	Asn	Glu	Leu
(2)	Glu	Lys	Phe	Gln	Pse	Glu	Glu	Gln	Gln	Gln
(3)	Val	Leu	Ser	Arg	Tyr	Pro	Ser	Tyr	Gly	Leu
(1)	Ser	Lys	Asp	Ile	Gly	Pse	Glu	Pse	Thr	Glu
(2)	Thr	Glu	Asp	Glu	Leu	Gln	Asp	Lys	Ile	His
(3)	Asn	Tyr	Tyr	Gln	Gln	Lys	Pro	Val	Ala	Leu
(1)	Asp	Gln	Ala	Met	Glu	Asp	Ile	Lys	Glu	Met
(2)	Pro	Phe	Ala	Gln	Thr	Gln	Ser	Leu	Val	Tyr
(3)	Ile	Asn	Asn	Gln	Phe	Leu	Pro	Tyr	Pro	Tyr
(1)	Glu	Ala	Glu	Pse	Ile	Pse	Pse	Pse	Glu	Glu
(2)	Pro	Phe	Pro	Gly	Pro	Ile	Pro	Asn	Ser	Leu
(3)	Tyr	Ala	Lys	Pro	Ala	Ala	Val	Arg	Ser	Pro
(1)	Ile	Val	Pro	Asn	Pse	Val	Gln	Glu	Lys	His
(2)	Pro	Gln	Asn	Ile	Pro	Pro	Leu	Thr	Gln	Thr
(3)	Ala	Gln	Ile	Leu	Gln	Trp	Gln	Val	Leu	Ser
(1)	Ile	Gln	Lys	Glu	Asp	Val	Pro	Ser	Glu	Arg
(2)	Pro	Val	Val	Val	Pro	Pro	Phe	Leu	Gln	Pro
(3)	Asp	Thr	Val	Pro	Ala	Lys	Ser	Cys	Gln	Ala
(1)	Tyr	Leu	Gly	Tyr	Leu	Glu	Gln	Leu	Leu	Arg
(2)	Glu	Val	Met	Gly	Val	Ser	Lys	Val	Lys	Glu
(3)	Gln	Pro	Thr	Thr	Met	Ala	Arg	His	Pro	His
(1)	Leu	Lys	Lys	Tyr	Lys	Val	Pro	Gln	Leu	Glu
(2)	Ala	Met	Ala	Pro	Lys	His	Lys	Glu	Met	Pro
(3)	Pro	His	Leu	Ser	Phe	Met	Ala	Ile	Pro	Pro
(1)	Ile	Val	Pro	Asn	Pse	Ala	Glu	Glu	Arg	Leu
(2)	Phe	Pro	Lys	Tyr	Pro	Val	Gln	Pro	Phe	Thr
(3)	Lys	Lya	Asn	Gln	Asp	Lys	Thr	Glu	Ile	Pro
(1)	His	Ser	Met	Lys	Glu	Gly	Ile	His	Ala	Gln
(2)	Glu	Ser	Gln	Ser	Leu	Thr	Leu	Thr	Asp	Val
(3)	Thr	Ile	Asn	Thr	Ile	Ala	Ser	Gly	Glu	Pro
(1)	Gln	Lys	Glu	Pro	Met	Ile	Gly	Val	Asn	Gln
(2)	Glu	Asn	Leu	His	Leu	Pro	Pro	Leu	Leu	Leu
(3)	Thr	Ser	Thr	Pro	Thr	Ile	Glu	Ala	Val	Glu
(1)	Glu	Leu	Ala	Tyr	Phe	Tyr	Pro	Glu	Leu	Phe
(2)	Gln	Ser	Trp	Met	His	Gln	Pro	His	Gln	Pro
(3)	Ser	Thr	Val	Ala	Thr	Leu	Glu	Ala	Pse	Pro
(1)	Arg	Gln	Phe	Tyr	Gln	Leu	Asp	Ala	Tyr	Pro
(2)	Leu	Pro	Pro	Thr	Val	Met	Phe	Pro	Pro	Gln
(3)	Glu	Val	Ile	Glu	Ser	Pro	Pro	Glu	Ile	Asn
(1)	Ser	Gly	Ala	Trp	Tyr	Tyr	Val	Pro	Leu	Gly
(2)	Ser	Val	Leu	Ser	Leu	Ser	Gln	Ser	Lys	Val
(3)	Thr	Val	Gln	Val	Thr	Ser	Thr	Ala	Val	
(1)	Thr	Gln	Tyr	Thr	Asp	Ala	Pro	Ser	Phe	Ser
(2)	Leu	Pro	Val	Pro	Glu	Lys	Ala	Val	Pro	Tyr
(1)	Asp	Ile	Pro	Asn	Pro	Ile	Gly	Ser	Glu	Asn
(2)	Pro	Gln	Arg	Asp	Met	Pro	Ile	Gln	Ala	Phe
(1)	Ser	Glu	Lys	Thr	Thr	Met	Pro	Leu	Trp	
(2)	Leu	Leu	Tyr	Gln	Gln	Pro	Val	Leu	Gly	Pro
(2)	Val	Arg	Gly	Pro	Phe	Pro	Ile	Ile	Val	

the presence of Ca^{2+} ions at the levels found in milk. However, at temperatures at or below 1 °C the calcium salt is quite soluble.

ϰ-caseins. The B variant consists of a peptide chain with 169 residues and has a molecular weight of 18 kdal. This monomer is accessible only under reducing conditions. Normally, ϰ-casein occurs as a trimer or as a higher oligomer in which the formation of disulfide bonds is probably involved. The protein contains varying amounts of carbohydrates (average values: 1% galactose, 1.2% galactosamine, 2.4% N-acetyl neuramic acid) that are bound to the peptide chain through Thr-133. ϰ-Casein is separated electrophoretically into various components that have the same composition of amino acids, but differ in their carbohydrate moiety, e.g. per

Table 10.7. Amino acid sequences of genetical variants of bovine casein

Reference variant[a]	Variant	Changes in the amino acid sequence
α_{s1}-B	α_{s1}-A	Residues 14–26 are absent
	α_{s1}-C	Glu-192 → Gly-192
	α_{s1}-D	Ala-53 → Pth-53
β-A^2	β-C	Pse-35 → Ser-35, Glu-37 → Lys-37, Pro-67 → His-67
	β-A^1	Pro-67 → His-67
	β-B	Pro-67 → His-67, Ser-122 → Arg-122
	β-A^3	His-106 → Gln-106
ϰ-B	ϰ-A	Ile-136 → Thr-136, Ala-148 → Asp-148

[a] Amino acid sequences are presented in Table 10.6.

protein molecule they contain 0–3 moles N-acetyl neuramic acid, 0–4 moles galactose and 0–3 moles galactosamine. A tri- and a tetrasaccharide with the following structures (ϰ-casein A, from cow's milk; NeuNAc = N-acetyl neuramic acid; GalNAc = N-acetyl galactosamine) could be isolated:

$$\text{NeuNAc} \xrightarrow{\alpha\text{-}2,3} \text{Gal} \xrightarrow{\beta\text{-}1,3} \text{GalNAc} \xrightarrow{\beta\text{-}1} \text{Thr}^{133}$$

$$\text{NeuNAc} \xrightarrow{\alpha\text{-}2,3} \text{Gal} \xrightarrow{\beta\text{-}1,3} \text{GalNAc} \xrightarrow{\beta\text{-}1} \text{Thr}^{133}$$
$$\uparrow \alpha\text{-}2,6$$
$$\text{NeuNAc} \qquad (10.3)$$

ϰ-Casein is the only main constituent of casein which remains soluble in the presence of Ca^{2+} ions in the concentrations found in milk. Aggregation of α_{s1}- and β-caseins with ϰ-casein prevents their coagulation in the presence of Ca^{2+} ions. This property of ϰ-casein is of utmost importance for formation and maintenance of stable casein complexes and casein micelles, as occur in milk.

Rennin (cf. 1.4.5.2) selectively cleaves the peptide chain of ϰ-casein at $-\text{Phe}^{105}-\text{Met}^{100}-$ into two fragments: para-ϰ-casein and a glycopeptide (Pyg = pyroglutamic acid, i.e. pyrrolidone carboxylic acid):

$$\begin{array}{llll} 1 & 105 & 106 & 169 \\ \text{Pyg} \ldots\ldots & \text{Phe} - & \text{Met} \ldots\ldots\ldots & \text{Val} \\ & \varkappa\text{-Casein} & & (10.4) \\ 1 & 105 & 106 & 169 \\ \xrightarrow{\text{Lab}} \text{Pyg} \ldots\ldots & \text{Phe} + & \text{Met} \ldots\ldots\ldots & \text{Val} \\ \text{para-}\varkappa\text{-Casein} & & \text{Glycopeptide} & \end{array}$$

The released glycopeptide is soluble, while para-ϰ-casein precipitates in the presence of Ca^{2+} ions. In this way ϰ-casein loses its protective effect; the casein complexes and casein micelles coagulate (curdle formation) from the milk. The specificity of rennin is high, as is shown by data given in Table 10.8. The sugar moiety of ϰ-casein is not essential for rennin action, nor for the stabilizing property of its protein portion. However, the sugar moiety delays protein cleavage by rennin. Also, it appears that the stability of α_s- and ϰ-casein mixtures in the presence of Ca^{2+} ions is influenced by the carbohydrate content of ϰ-casein.

Table 10.8. Rennet specificity: relative rate of hydrolysis of ϰ-casein and peptides

Substrate	k_{rel}
ϰ-Casein	
(Pyg——Leu-Ser-Phe-Met-Ala——Val) (105 106 169)	1
Phe-Met	0
Phe-Met-OMe	0
Z-Phe-Met-OMe	0
Leu-Ser-Phe-Met-Ala-OMe	0.001 to 0.01

γ-caseins. These proteins are degradation products of the β-caseins, formed by milk proteases, e.g. γ_1-casein is obtained by cleavage of the residues 1–28. The peptide released is identical with the protease-peptone PP8F which has been found in milk. Correspondingly, γ_2- and γ_3-caseins are formed by hydrolysis of the amino acid residues 1–105 and 1–107, respectively.

λ-caseins. The λ-casein fraction consists mainly of fragments of the α_{s1}-caseins. In vitro the λ-caseins are formed by incubation of the α_{s1}-caseins with bovine plasmin.

All casein forms contain phosphoric acid, which always occurs in a tripeptide sequence pattern (Pse = phosphoserine):

Pse-X-Glu or Pse-X-Pse (10.5)

in which X is an amino acid, including phosphoserine and glutamic acid. Examples are:

α_{s1}-Casein:	Pse-Glu-Pse
	Pse-Ile-Pse-Pse-Glu
	Pse-Val-Glu
	Pse-Ala-Glu (10.6)
β-Casein:	Pse-Leu-Pse-Pse-Pse-Glu
	Pse-Glu-Glu
ϰ-Casein:	Pse-Pro-Glu

Most probably this regularity pattern originates from the action of a specific protein kinase.

The various distribution of polar and apolar groups of the individual proteins outlined above are summarized in Table 10.9. The hydrophobicity values listed are average hydrophobicity values $\bar{H}$ of the amino acid side chains present in the sequence of the given segments, and are calculated as follows:

A measure of the hydrophobicity of a compound is the apparent free energy, F, needed to transfer the compound from water into an organic solvent, and is given as the ratio of the compound's solubility in water (N_w, as mole fraction) and in an organic solvent (N_{org}, as mole fraction), involving the activity coefficients (γ_w, γ_{org}):

$$\Delta F_t = RT \ln \frac{N_w \cdot \gamma_w}{N_{org} \cdot \gamma_{org}}$$

The corresponding apparent free energy of transfer of the side chain of an amino acid $H\Phi_i$ is obtained from the following relationship:

$$H\Phi_i = \Delta F_t \text{ (amino acid i)} - \Delta F_t \text{ (glycine)}$$

The average hydrophobicity of a sequence segment of a polypeptide chain with n amino acid residues is then:

$$\bar{H} = \frac{\Sigma H\Phi_i}{n}$$

Table 10.9. Distribution of amino acid residues with ionizing side chains (net charge) and with nonpolar side chains (hydrophobicity) in α_{s1}-casein and β-casein

Residue	α_{s1}-Casein		Residue	β-Casein	
	1	2		1	2
1–40	+3	1,340	1–43	−16	783
41–80	−22.5	641	44–92	−3.5	1,429
81–120	0	1,310	93–135	+2	1,173
121–160	−1	1,264	136–177	+3	1,467
161–199	−2.5	1,164	178–209	+2	1,738

1 Net charge,
2 Hydrophobicity $\bar{H}$ (Cal/mole; cf. text).

Monomers (soluble caseins)

−H″, −Ca″, +citrate +phosphate temperature decrease	↑ ↓	+H″, +Ca²″, −citrate −phosphate temperature increase

Casein complex

−H″, −Ca²″ +citrate +phosphate temperature decrease	↑ ↓	+H″, +Ca²″ −citrate −phosphate temperature increase

Micella
(calcium caseinate + calcium phosphate)

Fig. 10.1. Casein complex and casein micelle formation

The higher $H\Phi_i$, i.e. $\bar{H}$, the higher is the hydrophobicity of individual side chains, i.e. the sequence segment. Data provided in Table 10.9 are related to the ethanol/water system.

Only up to 10% of the total casein fraction is present as monomers. They are usually designated as serum caseins and the concentration ratio $c_\beta > c_\varkappa > c_{\alpha s1}$ is quite valid. However, the main portion is aggregated to casein complexes and casein micelles. This aggregation is regulated by a set of parameters, as presented in Figure 10.1. Dialysis of casein complexes against a chelating agent might shift the equilibrium completely to monomers, while against high Ca^{2+} ion concentrations the shift would be to large micelles.

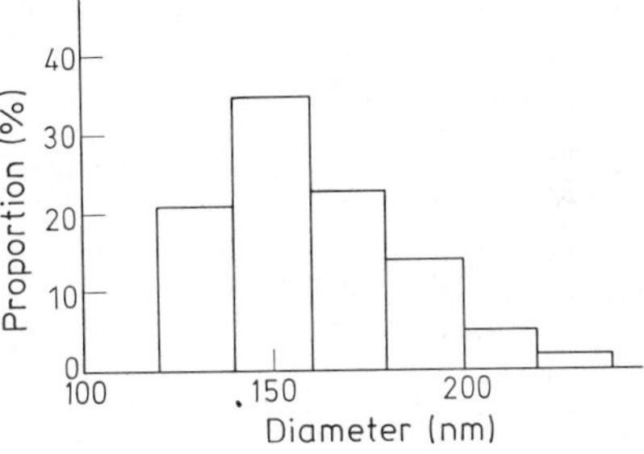

Fig. 10.2. Particle size distribution of casein micelles in skim milk (fixation with glutaraldehyde)

From Figure 10.2 it follows that the diameter of the micelles in skim milk varies from 50–300 nm, with a particle distribution peak at 150 nm. Using an average diameter of 140 nm, the micelle volume is 1.4×10^6 nm³ and the particle weight is 10^7–10^9 dal. This corresponds to 25,000 monomers per micelle. Casein micelles are substantially smaller than fat globules, which have diameters between 0.1–10 µm. Scanning electron micrographs of micelles are shown in Figure 10.3. and compositional data are provided in Table 10.10.

Table 10.10. Composition of casein micelles (%)

Casein	93.2	Phosphate (organic)	2.3
Ca	2.9		
Mg	0.1	Phosphate (inorganic)	2.9
Na	0.1		
K	0.3	Citrate	0.4

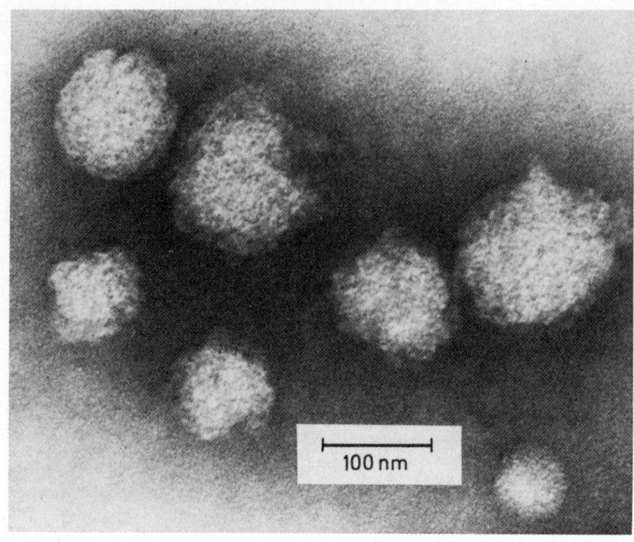

Fig. 10.3. Electron micrograph of the casein micelles in skim milk. (After *Webb,* 1974.) The micelles are fixed with glutaraldehyde and then stained with phosphomolybdic acid

Table 10.11. Typical distribution of components in casein micelles

Component	Ratio numbers			
α_{s1}	3	6	9	12
β	1	1	4	4
γ		1	1	1
$\varkappa$	1	3	3	3

The ratio of monomers in micelles varies to a great extent (Table 10.11), depending on dairy cattle breed, season and fodder, and is influenced also by micellular size (Table 10.12).

The micelles are not tightly packed and so are of variable density. They are strongly solvated (1.9 g water/g protein) and hence are porous. The monomers are kept together with:

- Hydrophobic interactions that are minimal at a temperature less than 5 °C.
- Electrostatic interactions, mostly as calcium or calcium phosphate bridges between phosphoserine and glutamic acid residues (Figure 10.4).
- Hydrogen bonds.

On a molecular level different micelle models have been proposed which to a certain extent explain the experimental findings. The most probable model is shown in Figure 10.5. Small submicelles which consists of α_{s1}-, β-, γ- and $\varkappa$-caseins are aggregated to a large micelle by involvement of calcium phosphate bridges. Thus, in a porous structure rennin enzyme action might occur not only on the surface of the micelle, but also in its interior. Therefore, glycopeptide cleavage results in an increase in hydrophobicity within the micelle, associated with dehydration of the micelle and the consequent collapse of its structure.

Fig. 10.4. Peptide chain bridging with calcium ions

Table 10.12. Composition and size of casein micelles isolated by centrifugation

Centrifugation time (min)[a]	Composition of the sediment (%)			
	α_{s1}	β	ϰ	Others
0[b]	50	32	15	3
0–7.5	47	34	16	3
7.5–15	46	32	18	4
15–30	45	31	20	4
39–60	42	29	26	3
Serum casein	39	23	33	5

[a] Centrifugation speed $10^5 \times g$.
[b] Isoelectric casein.

10.1.2.1.2 Whey proteins

β-*lactoglobulin* occurs in genetic variants A, B and C of the Jersey dairy cattle breed, and variant D of the Montbeliarde dairy cow. Two other A and B variants, of Australian drought master cows, are identical to variants A and B apart from the carbohydrate content. Table 10.13 shows the corresponding changes in amino acid composition.

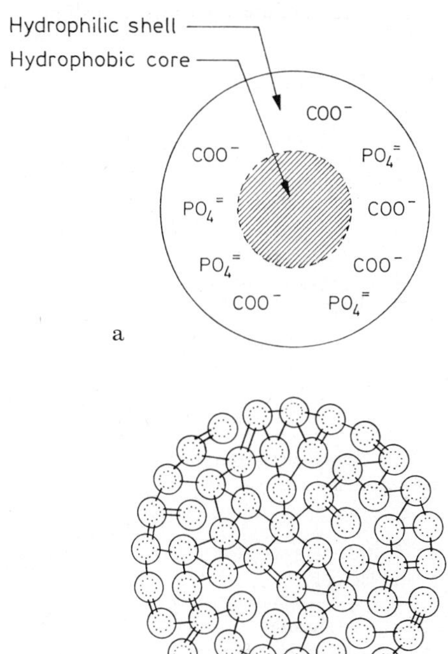

Abb. 10.5. Schematic model of a casein micelle; (**a**) a submicelle consisting of a_{s1}-, b-, c-, ϰ-caseins, (**b**) submicelles bound by calcium phosphate bridges. (After *Webb*, 1974)

Table 10.13. Changes in amino acid composition for genetical variants of β-lactoglobulin

Variant:	A	→ B	→ C
	Asp	→ Gly	
	Val	→ Ala	
		Gln	→ His

The sequence of amino acids in β-lactoglobulin has been elucidated. The monomeric β-lactoglobulin, with a molecular weight of 18 kdal, shows a reversible, pH-dependent oligomerization, as represented by the equation:

$$\underset{pH < 3.5}{A \leftrightarrows A_2} \leftrightarrows \underset{3.7 < pH < 5.1}{(A_2)_4} \leftrightarrows \underset{pH < 7.5}{A_2 \leftrightarrows A} \quad (10.7)$$

Hence, the monomer is stable only at a pH less than 3.5 or above 7.5. The octamer occurs with variant A, but not with variants B and C. Irreversible denaturation occurs at a pH above 8.6 as well as by heating or at higher levels of Ca^{2+} ions. β-lactoglobulin contains one thiol-group that in the native protein is masked within the molecule. This group becomes reactive by partial denaturation and then can participate in protein dimerization through disulfide bond formation.

α-*lactalbumin*. This protein exists in two genetic forms, A and B (Gly → Arg). Its amino acid sequence, which is similar to that of lysozyme, has been elucidated. α-lactalbumin has a biological function since it is the B subunit of the enzyme lactose-synthetase. The enzyme subunit A is a nonspecific galactosyltransferase; the subunit B promotes selective galactose transfer to glucose in lactose biosynthesis.

10.1.2.2 Carbohydrates

The main sugar in milk is lactose, an O-β-D-galactopyranosyl-(1 → 4)-D-glucopyranose, which is 4–6% of milk.
The most stable form is α-lactose monohydrate, $C_{12}H_{22}O_{11} \cdot H_2O$. Lactose crystallizes in this form

from a supersaturated aqueous solution at T < 93.5 °C. The crystals may have a prism- or pyramid-like form, depending on conditions. Vacuum drying at T > 100 °C yields a hygroscopic α-anhydride. Crystallization from aqueous solutions above 93.5 °C provides water-free β-lactose (β-anhydride, cf. Formula 10.8). Rapid drying of a lactose solution, as in milk powder production, gives a hygroscopic and amorphous equilibrium mixture of α- and β-lactose.

(10.8)

β-Lactose

Some physical data of lactose are summarized in Table 10.14. The ratio of anomers is temperature dependent. As temperature increases, the β-form decreases. The mutarotation rate is temperature ($Q_{10} = 2.8$) and pH dependent (Figure 10.6). The rise in mutarotation rate at pH < 2 and pH > 7 originates from the rate-determining step of ring opening, which is catalyzed by both H^+ and OH^- ions (Reaction 10.9).

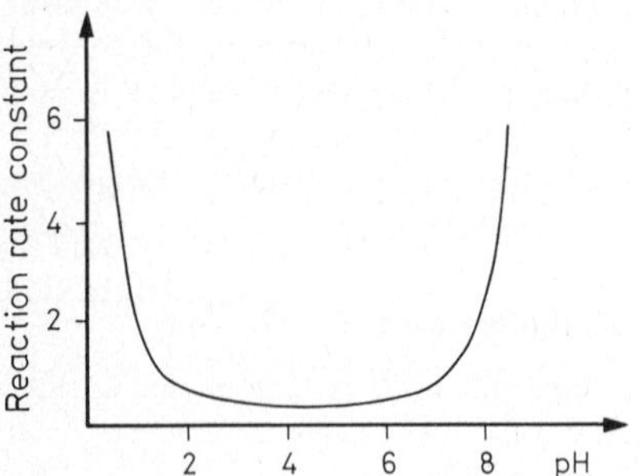

Fig. 10.6. Mutarotation rate of lactose as affected by pH

(10.9)

The great solubility difference between the two anomers is noteworthy. The sweetness of lactose is significantly lower than that of fructose, glucose or sucrose (Table 10.15). Glucose and some other amino sugars and oligosaccharides are present in small amounts in milk.

Table 10.14. Some physical characteristics of lactose

	α-Lactose	β-Lactose	Equilibrium mixture
Melting point (°C)	201.6[a]	252.2	
Spec. rotation $[\alpha]_D^{20}$	89.4	35.0	
Equilibrium in aqueous solution[b]			
0 °C	1.00	1.80	
20 °C	1.00	1.68	
50 °C	1.00	1.63	
Solubility in water[c]			
0 °C	5.0	45.1	11.9
25 °C	8.6		21.6
39 °C	12.6		31.5
100 °C	70	94.7	157.6

[a] Hydrate. [b] Relative concentration.
[c] g Lactose/100 g water.

Table 10.15. Relative sweetness of saccharose, glucose, fructose and lactose[a]

Saccharose	Glucose	Fructose	Lactose
0.5	0.9	0.4	1.9
5.0	8.3	4.2	15.7
10.0	12.7	8.7	20.7
20.0	21.8	16.7	33.3

[a] Results are expressed as concentration % for isosweet aqueous sugar solutions.

Table 10.16 a. Milk lipids

Lipid fraction	Percent of the total lipid
Triacylglycerols	95–96
Diacylglycerols	1.3–1.6
Monoacylglycerols	0.02–0.04
Keto acid glycerides	0.9–1.3
Hydroxy acid glycerides	0.6–0.8
Free fatty acids	0.1–0.4
Phospholipids	0.8–1.0
Sphingolipids	0.06
Sterols	0.2–0.4

10.1.2.3 Lipids

The composition of milk fat is presented in Table 10.16a. Milk fat contains 95–96% triglycerols. Its fatty acid composition is given in Table 10.16b. The relatively high content of low molecular weight fatty acids, primarily of butyric acid, is characteristic of milk. The content of unsaturated acids varies with season and fodder. The level can be increased by incorporating highly unsaturated fats, in protected encapsulated forms, into feed. The disadvantage of such a nutritionally/physiologically-interesting approach is the changed physico-chemical properties of the dairy product, e. g. an increased susceptibility to oxidation and the formation of unsaturated lactones (γ-dodec-cis-6-enolactone from linoleic acid) which influences the flavor of milk and meat.

In addition to the main straight-chain fatty acids, small amounts of odd-C-number, branched-chain and oxo-fatty acids (cf. 3.2.1.3) are present.

Phospholipids are 0.8–1.0% in milk fat and sterols, mostly cholesterol, are 0.2–0.4%.

Butterfat melting properties, as affected by season and fodder, are listed in Table 10.17.

Milk fat is very finely distributed in plasma. The diameter of fat globules is 0.1 – 10 μm. Homogenization, during which milk at 50–75 °C is forced through small passages under pressure of 25–350 bar, the diameter of the globules lowers to < 1 μm, depending on the pressure. The fat droplets are surrounded by a membrane (Figure 10.7) that consists of phospholipids and a double layer of proteins, both about 2% of the total weight of the globule. Membrane compositional data are given in Table 10.18.

Table 10.16 b. Fatty acid composition of milk fat[a]

Fatty acid	Weight-%
Saturated, straight chain	
Butyric acid	2.79
Caproic acid	2.34
Caprylic acid	1.06
Capric acid	3.04
Lauric acid	2.87
Myristic acid	8.94
Pentadecanoic acid	0.79
Palmitic acid	23.8
Heptadecanoic acid	0.70
Stearic acid	13.2
Nonadecanoic acid	0.27
Arachidic acid	0.28
Behenic acid	0.11
Saturated, branched chain	
12-Methyltetradecanoic acid	0.23
13-Methyltetradecanoic acid	0.14
14-Methylpentadecanoic acid	0.20
14-Methylhexadecanoic acid	0.23
15-Methylhexadecanoic acid	0.36
3,7,11,15-Tetramethylhexadecanoic acid	0.12–0.18
Unsaturated	
9-Decenoic acid	0.27
9-cis-Tetradecenoic acid	0.72
9-cis-Hexadecenoic acid	1.46
9-cis-Heptadecenoic acid	0.19
8-cis-Octadecenoic acid	0.45
Oleic acid	25.5
11-cis-Octadecenoic acid	0.67
9-trans-Octadecenoic acid	0.31
10-trans-Octadecenoic acid	0.32
11-trans-Octadecenoic acid	1.08
12-trans-Octadecenoic acid	0.12
13-trans-Octadecenoic acid	0.32
14-trans-Octadecenoic acid	0.27
15-trans-Octadecenoic acid	0.21
16-trans-Octadecenoic acid	0.23
Linoleic acid	2.11
Linolenic acid	0.38

[a] Only acids with a content higher than 0.1% are listed.

Membrane proteins are specific in nature. Casein proteins enter and participate in membrane formation when the fat globule surface area is 4- to 6-fold expanded during homogenization of milk. In addition to lipoproteins, the membrane also contains enzymes, including a very active lipo-

Table 10.17. Melting characteristics of butterfat

Temperature (°C)	Solid content (%)	Temperature (°C)	Solid content (%)
5	43–47	30	6–8
10	40–43	35	1–2
20	21–22	40	0

Table 10.18. Membrane composition of milkfat globules

Constituent	Proportion (%)
Protein	41
Phospho- and glycolipids	30
Cholesterol	2
Neutral glycerides	14
Water	13

protein lipase (a glycoprotein, 8.3% carbohydrates, molecular weight 48.3 kdal). However, if the milking and storage procedures are appropriate, the raw milk can be kept for several days without the development of a rancid off-flavor. It is likely that the membranes of the fat globules prevent lipolysis. Disruption of the organized structure of the membrane, for instance by homogenization, allows the lipase to bind to the fat globules and to hydrolyze the triacylglycerols with a high rate (1 μmole fatty acid per min per ml milk, pH 7, 37 °C). The milk becomes unpalatable within a few minutes. Therefore the lipoprotein lipase has to be inactivated by pasteurization prior to milk homogenization.

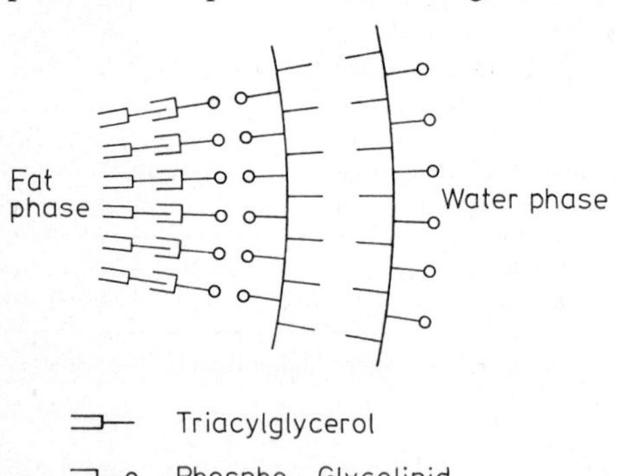

Fig. 10.7. Structure of a milk fat globule (schematic representation)

10.1.2.4 Organic Acids

Citric acid (1.8 g/l) is the predominant organic acid in milk. During storage it disappears rapidly as a result of the action of bacteria. Other acids (lactic, acetic) are degradation products of lactose. The occurrence of orotic acid (73 mg/l), an intermediary product in biosynthesis of pyrimidine nucleotides, is specific for milk:

Dihydro orotic acid → Orotic acid

→ Uridine– / Cytidine– / Thymidine– } 5′-phosphate (10.10)

10.1.2.5 Minerals

Minerals, including trace elements, in milk are compiled in Table 10.19.

10.1.2.6 Vitamins

Milk contains all the vitamins in variable amounts (Table 10.20). During processing, the fat-soluble vitamins are retained by the cream, while the water-soluble vitamins remain in skim milk or whey.

Table 10.19. Mineral composition of milk

Constituent	mg/l	Constituent	μg/l
Potassium	1,500	Zinc	4,000
Calcium	1,200	Aluminum	500
Sodium	500	Iron	400
Magnesium	120	Copper	120
		Molybdenum	60
Phosphate	3,000	Manganese	30
Chloride	1,000	Nickel	25
Sulfate	100		
		Silicium	1,500
		Bromine	1,000
		Boron	200
		Fluorine	150
		Iodine	60

Table 10.20. Vitamin content of milk

Vitamin	mg/l	Vitamin	mg/l
A (Retinol)	0.3	Nicotinamide	1
D (Calciferol)	0.001	Pantothenic acid	3
E (Tocopherol)	1.4	C	20
B_1 (Thiamine)	0.4	Biotin	0.04
B_2 (Riboflavin)	1.7	Folic acid	0.05
B_6 (Pyridoxine)	0.5		
B_{12} (Cyanocobalamine)	0.005		

10.1.2.7 Enzymes

Milk contains a great number of enzymes of analytical significance for detection of heat-treated milk. The rate of heat inactivation of the enzymes indicates the type and extent of heating (cf. 2.6.4 and Figure 2.35). Hydrolases identified include: amylases, lipases (cf. 10.1.2.3), esterases, proteinases and phosphatases. Proteinase inhibitors have also been found.
Important oxidoreductases in milk are aldehyde dehydrogenase (xanthine oxidase), lactoperoxidase and catalase.

10.1.3 Processing of Milk

The processing of raw milk for drinking or into other dairy products requires stringent control to obtain a product that is hygienic and of high organoleptic quality.
Milk is produced on the dairy farm under good hygienic conditions, primarily as a result of using milking machines. The warm milk (30 °C) is strained in the milk house, cooled and stored in cans or storage tanks to avoid sunlight. Milk is usually transported to the processing plant in cans or in the tank of a milk truck. Transportation of milk from mountains to valleys by pipelines made of polyethylene or PVC was first introduced in Austria, France and Switzerland and has been used since 1956 in Allgau, Germany. In-plant milk treatments are depicted schematically in Figure 10.8.
The milk is first purified with a continuously-operated clarifier (centrifuge). Contaminants are removed in the sediment. Cream separation is often achieved simultaneously. The process is conducted at 40 °C at 5,500–6,500 rpm. Such centrifuges have a flow production capacity of up to 20,000 kg/h. Back-mixing allows the milk fat content to be adjusted as desired.

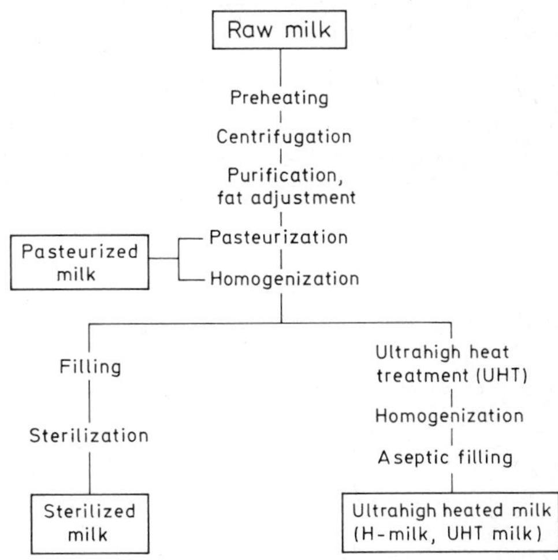

Fig. 10.8. Treatment of milk

The fluid milk is heated after clarification to improve its durability and to kill disease-causing microorganisms. Heat treatments used are (cf. Figure 10.9):

- Pasteurization.
  The milk is treated: at high temperature (85 °C for 2 sec); in a short-time, flash process (71–74 °C for 15–40 sec) in plate heaters; or by the low temperature or holder process, in which it is heated at 62–65 °C for at least 30 min, with stirring, and is then cooled.
- Ultrahigh temperature (UHT) treatment.
  The process involves indirect heating by coils or plates at 135–140 °C for 6–10 sec, or direct heating by live steam injection at 140–150 °C for 2–4 sec, followed by aseptic packaging.
- Sterilization.
  Milk in retail packages is heated in autoclaves at 110–120 °C for 10–20 min.

Heat treatment affects several milk constituents. Casein, strictly speaking, is not a heat-coagulable protein; it coagulates only at very high temperatures (cf. Figure 10.9). Heating at 120 °C for 5 h dephosphorylates sodium or calcium caseinate solutions (100% and 85%, respectively) and releases 15% of the nitrogen in the form of low molecular weight fragments.
However, temperature and pH strongly affect casein association and cause changes in micellular structure. An example of such a change is the pH-dependent heat coagulation of skim milk.

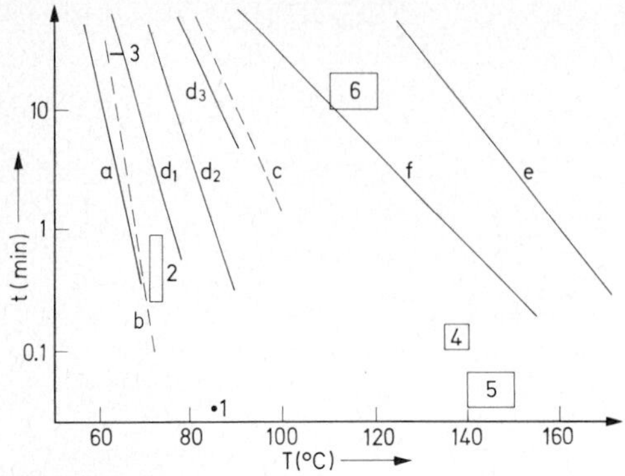

Fig. 10.9. Heating of milk. 1–3 Pasteurization: 1 high temperature treatment, 2 short time and 3 long time heat treatment. 4 and 5 UHT treatment: 4 indirect and 5 direct. 6 sterilization. a: killing pathogenic microorganisms (*Tubercle bacilli* as labelling organism), b/c: inactivation of alkaline/acid phosphatase. d_1, d_2, d_3 denaturation (5, 40, 100%) of whey proteins. e: casein heat coagulation, f: start of milk browning

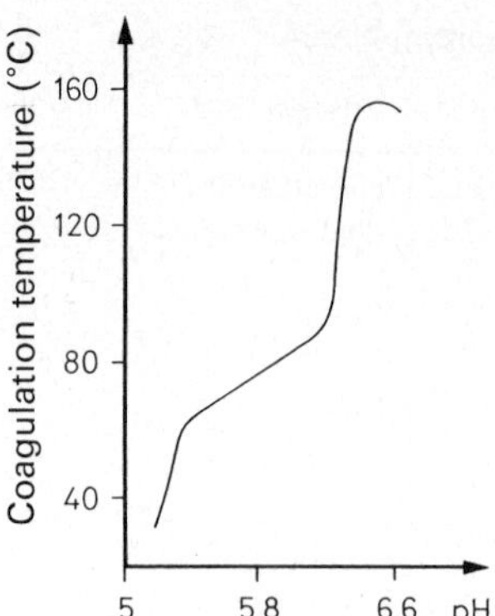

Fig. 10.10. Thermal coagulation of skim milk

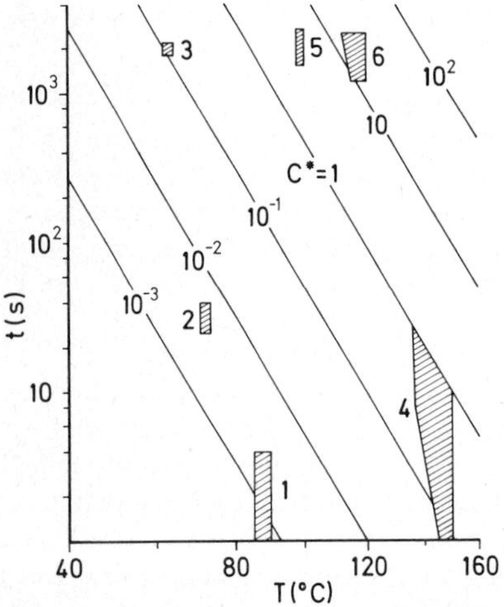

Fig. 10.9 a. Chemical reactions in heat-treated milk. ("Chemical effect" C* = 1: losses of approx. 3% thiamine and approx. 0.7% lysine and formation of approx. 0.8 mg/l HMF); commonly used heat treatments: 1 high heat, 2 short time heating, 3 prolonged heating, 4 UHT treatment, 5 boiling, 6 sterilization. (After *Kessler*, 1983)

The coagulation temperature drops with decreasing pH (Figure 10.10). Salt concentration also has an influence, e.g. the heat stability of milk decreases with a rise in the content of free calcium.

All pasteurization processes supposedly kill the pathogenic microorganisms in milk. The inactivation of the alkaline phosphatase is used in determining the effectiveness of pasteurization. At higher temperatures or with longer heating time, the whey proteins start to denature – this coincides with the complete inactivation of acid phosphatase. Denatured whey proteins, within the pH-range of their isoelectric points, cease to be soluble and coagulate together with casein due to souring or rennin action of the milk. Such coprecipitation of the milk proteins is of importance in some milk processing (as in cottage cheese production). The thermal stability of whey proteins is illustrated in Figure 10.11.

Heat treatment of milk inactivates thiol groups; e.g. a thiol-disulfide exchange reaction occurs between ϰ-casein and β-lactoglobulin.

Further changes induced by heating of milk are:

- Calcium phosphate precipitation on casein micelles.
- *Maillard* reactions between lactose and amino groups (e.g. lysine) which, in a classical sterilization process, causes browning of milk and formation of hydroxymethyl furfural (HMF).
- δ-Lactone and methyl ketone formation from glycerides esterified with hydroxy- or keto-fatty acids.
- Thiamine degradation.
- Changes in membranes of milk fat globules, which affect the cream separation property of the globules.

Some flavor defects of heated milk and the volatile compounds which are involved in their formation are listed in Table 10.20a. For example,

Table 10.20 a. Volatiles contributing to heat-induced off-flavors of milk. Volatile compounds (μg/kg)[a]

Milk	Flavor	H_2S	CH_3SH	CH_3CHO	CH_3SCH_3	Methyl-butanal	2-Pen-tanone	Hexanal	CH_3SSCH_3	2-Hep-tanone
Pasteurized	normal	*0.5*	0	15	*7*	1	1	6	0	–
Pasteurized	malty	–	0	–	7	*70*	1	8	<1	1
Pasteurized	unclean	–	0	–	*55*	1	–	–	–	–
Pasteurized	spoilt		*15*		*40*	*57*	–	–		–
Pasteurized	"sunlight"	–	*2*	*250*	3	4		–	*4*	
UHT	"UHT"	*50*	<1	–	14	3	*40*	–	*4*	*21*
Sterilized	"sterilized"	0	*3*		20	1	*120*	–	*4*	*82*

[a] 0 = not detected; – = not determined; figures in italics are related to the (off) flavors mentioned.

the "malty" off-flavor of pasteurized milk is mainly due to methylbutanal. The UHT-flavor is caused by H_2S which originates mostly from thermally induced reactions of the proteins of the fat globule membrane. Furthermore, 2-pentanone and 2-heptanone, which both are formed upon thermal decarboxylation of β-keto acids (liberated by hydrolysis of milk fat), contribute to the UHT-flavor and to the "sterilized" flavor. Quantitation of these volatiles for product control can be performed by an automatic headspace analysis.

Detailed studies have shown that the rate of several reactions which occur during heating of milk, e.g. thiamine and lysine degradations, formation of HMF and nonenzymic browning, can be calculated over a great temperature-time range (including extended storage) by application of a second-order rate law. Assuming an average activation energy, $E_a = 102$ kJ/mole, a "chemical effect" $C^* = 1$ has been defined which gives a straight line in a log t/T^{-1} diagram, from which the thiamine loss is seen to be approx. 0.8 mg/l (Figure 10.9 a). Other lines in Figure 10.9 a represent a power of ten of heat treatment and chemical reactions ($C^* = 10^{-1}, 10^{-2}, \ldots$, or $10^1, 10^2, \ldots$). The pigments formed in a browning reaction become visible only in the range of $C^* = 10$.

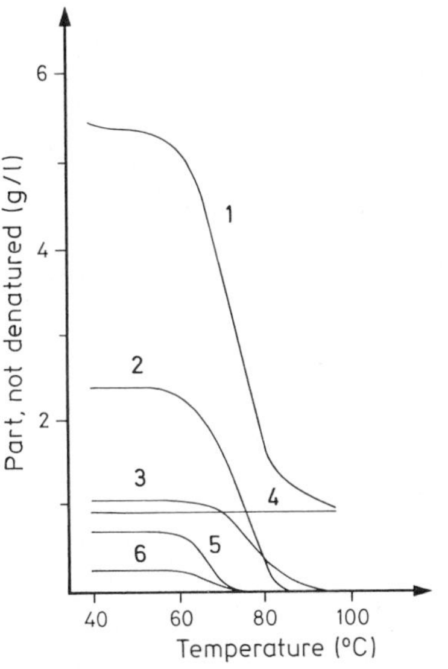

Fig. 10.11. Denaturation of whey proteins by heating at various temperatures for 30 min. 1 Total whey protein, 2 β-lactoglobulin, 3 α-lactalbumin, 4 proteose peptone, 5 immunoglobulin, 6 serum albumin

Quality deterioration in the form of nutritional degradation, changes in color or development of off-flavor have also been predicted for other foods by application of a suitable mathematical model. In most cases the loss of quality fits a zero- or first-order rate law. Knowledge of the rate constant allows one to predict the extent of reaction for any time.

The influence of temperature on the reaction rate follows the *Arrhenius* equation (cf. 2.6.2). Thus by studying a reaction and measuring the rate constants at two or three high temperatures, one could then extrapolate with a straight line to a lower temperature and predict the rate of the reaction at the desired lower temperature. However, these data allow only a prediction of the shelf-life, when the physical and chemical properties of the components of a food do not alter with temperature. For example, as temperature rises a solid fat goes into a liquid state. The reactants may be mobile in the liquid fat and not in the solid phase. Thus, shelf-life will be underestimated for the lower temperature.

10.1.4 Types of Milk

Milk is consumed in the following forms:

- *Raw fluid milk* (high quality milk), which has to comply with strict hygienic demands.
- *Whole milk* is heat-treated and contains at least 3% fat. It can be a standardized whole milk adjusted to a predetermined fat content, in which case the fat content has to be at least 3.5%.
- *Defatted milk* is heat-treated and the cream is separated. The fat content is 1.5–2%.
- *Skim milk* is heat-treated and the fat content is less than 0.3%.
- *Reconstituted milk* is most common in regions where milk production is not feasible (e.g. many Japanese cities). It is made from whole or skim milk powders, with the addition of cream or butter fat.
- *Filled* or *imitation milk* is a fat-substituted milk produced from nonfat milk solids by addition of vegetable fats (coconut, safflower or corn oil).
- *Toned milk* is a blend of a fat-rich fresh milk and reconstituted skim milk in which the nonfat solids are "toned up". Addition of water "tones down" the fat and nonfat solids.

10.2 Dairy Products

Milk processing is illustrated schematically in Figure 10.12.

10.2.1 Fermented Milk Products

10.2.1.1 Sour Milk (Buttermilk)

Sour milk is the product of fluid milk fermentation, either by spontaneous souring effected by various lactic acid-forming bacteria (e.g. *Streptococcus lactis* or *cremoris*) or by the much less numerous aroma-forming bacteria (e.g. *Streptococcus diacetylactis, Betacoccus citrovorus*) or by inoculation of heated milk with pure bacterial cultures (cultured buttermilk). As fermentation proceeds, lactose is transformed into lactic acid, which coagulates casein at pH 4–5. The thick, sour-tasting curdled milk is called, depending on the region, thick-, set-, slime- or silt-curdled or coagulated milk. Sour milk is manufactured from whole milk (at least 3.5% milk fat), often by blending with skim milk powder to increase the total solids content and to improve the resultant protein gel structure. In some countries sheep, water buffalo, reindeer or mare's milk are also processed. Sour cream is produced by a process very similar to that used in sour milk manufacture except that coffee grade cream is used as the raw material.

10.2.1.2 Yoghurt

Yoghurt (also spelled yogurt) is made from milks of various fat contents by fermentation using special cultures. It has a fine, curdled, gel-like consistency, and has a clear sour and a pleasant aromatic flavor that is readily distinguishable from a normal sour milk. Yoghurt is produced by inoculating homogenized pasteurized milk with about 1.5–3% thermophylic lactic acid bacteria (mixed cultures of one or more strains of organisms such as *Streptococcus thermophilus* or *Lactobacillus bulgaricus*) and then incubating at 42–45 °C for approx. 3 h. Curdled gel consistency develops during this time. The final product is acid in flavor (sour degree 40–60° SH). A significant portion of the specific aroma of yoghurt

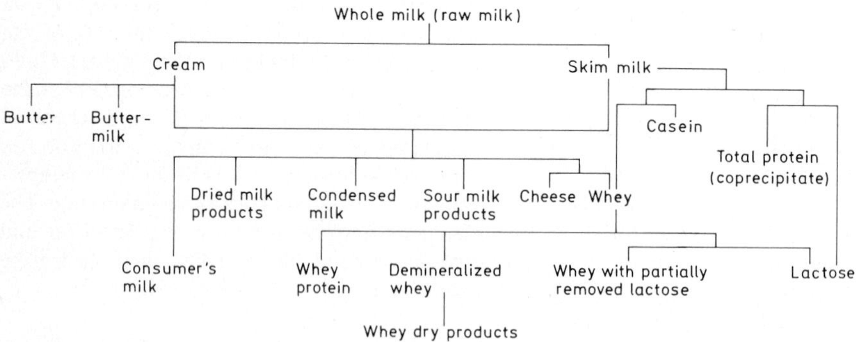

Fig. 10.12. Schematic presentation of milk processing

consists of carbonyl compounds, with diacetyl and acetaldehyde predominant. A special yoghurt, "acidophilus milk", is produced by *Lactobacillus acidophilus*. Addition of sugar and fruit or fruit pastes provides special products (fruit yoghurts) which have lead to a significant rise in yoghurt consumption.

10.2.1.3 Kefir and Kumiss

Kefir and kumiss are sparkling, carbonated alcoholic beverages. The microflora of kefir include *Torula* yeast (responsible for alcoholic fermentation) and of *Streptococcus lactis* and *Lactobacillus caucasicus* (responsible for lactic acid fermentation). The kefir bacillus causes a build-up of "kefir grains", which resemble cauliflower heads when wet and brownish seeds when dry, and are particles of clotted milk plus the kefir organisms. Their addition to fluid milk produces kefir. Kumiss is made of mare's or goat's milk fermented by the obligatory pure kumiss culture [*Thermobacterium bulgaricus* and yeast *Candida (Torula)* kefir].
Both dairy beverages are indigenous to the Caucasus and steppes of Turkestan. Kefir contains lactic acid (0.5–1.0%), noticeable amounts of alcohol (0.5–2.0%) and carbon dioxide, and some products of casein degradation resulting from proteolytic action of yeast.

10.2.1.4 Taette Milk

Taette (Lapp's milk) is a specially fermented, sour cow's milk product consumed in Sweden, Norway and Finland. Its thread-like, viscous structure is due to a symbiotic growth of lactic acid bacteria and yeast. The principal microorganism involved is *Streptococcus lactis* var. *hollandicus*.

10.2.2 Cream

Homogenized or heat-treated cream is obtained form whole milk (by removal of skim milk) and contains at least 10% milk fat. Sour cream is a product of lactic acid fermentation (see above). Whipping cream contains at least 30% milk fat, coffee cream at least 10% and butter cream 25–82%. Cream is utilized in many ways, either by direct consumption or for production of butter and ice creams.
Whippability and stability of the whipped foam products are necessary whipping cream properties. For the best quality cream, a volume increase of at least 80% is excepted and a standard cone with 100 g load must penetrate 3 cm deep in > 10 sec. No serum separation should occur at 18 °C after 1 h. Fat droplets accumulate during whipping on the surface of large air bubbles which form the froth. An increased build-up of smaller bubbles tears apart the membrane of the droplet and enlarges the fat interphase area, thus resulting in gel setting of the lamella separating the individual air bubbles
Whipping of the cream may give a slight increase in the autoxidation of the unsaturated acyl lipids. In very low concentrations the volatile compounds which arise from the antoxidation process improve the flavor. 4-cis-Heptenal in particular, if present in the ppb-range, contributes to the full flavor of cream.

10.2.3 Butter

Butter is a water-containing fat produced from milk. According to its manufacturing process, three types exist:

- Butter from sour cream (cultured-cream butter).
- Butter from nonsoured, sweet cream (sweet cream butter).
- Butter from sweet cream, which is soured in a subsequent step (soured butter).

Butter contains 81–85% fat, 14–16% water and 0.5–2.0% fat-free solids. The composition generally must meet legal standards. Butter is an emulsion with a continuous phase of liquid milk fat in which are trapped crystallized fat grains, water droplets and air bubbles. Butter consistency is determined by the ratio of free fluid fat to that of solidified fat. Due to seasonal variations in the unsaturated fatty acid content of milk fat, the solid/fluid fat ratio fluctuates at 24 °C between 1.0 in summer and 1.5 in winter. Equalization of these ratios is achieved by a preliminary cream-tempering step in a cream-ripening process, then churning and kneading the cream, which influences the extent of fluid fat inclusion into the solidified "fat grains".
The flavor of sweet cream butter is primarily determined by compounds that are present in the lipid fraction of milk or are formed from precursors during processing. Alkanoic acids (C_{10}, C_{12}), δ-lactones ($C_{8,10,12}$), dimethyl sulfide, 4-cis-heptenal, indole and skatole are present in

quantities high enough to contribute to the flavor.

The flavor of cultured-cream butter is composed of aroma compounds also present in sweet-cream butter and others formed by the starter culture. Diacetyl seems to be the most important component which, together with acids (mainly lactic and acetic acids), sweet-cream butter constituents, and dimethyl sulfide (which balances the diacetyl flavor), creates the flavor of sour-cream butter.

The important processing steps in butter production are: cream separation, cream ripening and souring, and cream churning.

10.2.3.1 Cream Separation

Cream is separated from whole milk by high-efficiency separators (flow-through capacity 20,000 kg/h at 5,000–6,500 rpm) at an elevated temperature of 40–45 °C in order to increase the sharpness of the separation. The cream, depending on the subsequent churning process, might contain 25–82% milk fat. The residual fat level in the skim milk by-product is 0.03–0.06%. The cream is then pasteurized at 90–110 °C (smaller creameries use batch methods; larger plants use flash pasteurization at 73–85 °C for 16–25 sec or vacuum pasteurization, heating the cream to 94 °C with steam).

10.2.3.2 Cream Ripening and Souring

Cream ripening and souring are the most important steps in the production of sour cream butter. The process is performed in a cream ripener or vat, with suitable mixing and temperature control. Soon after the cream has filled the ripener, a "starter culture" is added to the cream to achieve souring. The "starter culture" consists of various strains of lactic acid bacteria (primarily *Streptococcus lactis* or *Str. cremoris* mixed with *Streptococcus diacetylactis* and/or *Betacoccus citrovorus*). The subsequent ripening at 8–19 °C proceeds for up to 24 h.

In addition to build-up of acids and some aroma substances, suitable temperature control gives a partial crystallization of fat. This has a significant influence on butter consistency. The souring step is omitted in sweet cream butter production. Accelerated fat crystallization is achieved by additional storage of cream at 6–10 °C for 3–29 h.

10.2.3.3 Churning

Churning is essentially strong mechanical cream shearing which tears the membranes of the fat globules and facilitates coalescence of the globules. The cream "breaks" and tiny granules of butter appear. Prolonged churning results in a continuous fat phase. Foam build-up is also desirable since the tiny air bubbles, with their large surface area, attract some membrane materials. Some membrane phospholipids are transferred into the aqueous phase. Buttermilk, a milky, turbid liquid, separates out initially (it is later drained off), followed by the butter granules of approx. 2 mm diameter. These granules still contain 30% of the aqueous phase. This is reduced to 15–19% by churning. The finely distributed water droplets (diameter 10 μm or less) are retained by the fat phase.

Churning may be a discontinuous, batch-type process in which the cream utilized has a fat content of 28–30% (sour cream) or 40–50% (sweet cream), respectively. The stainless steel, mechanical churns can have many forms and revolve in a non-symmetrical way. In a steady-flow, continuous-type process, the cream is 30–38% fat (sour cream butter) or 40–50% fat (sweet cream butter) and is produced, by the *Fritz-Eisenreich* process, in a cylinder equipped with rotational kneading rollers. In the continuous *Alfa*-process the phase conversion is achieved in a screw-type cooler, using an 82% cream and repeated chilling, without the aqueous phase being separated.

The *Booser* process and the *NIZO* process allow a subsequent souring of butter from sweet cream. Both processes are of economic interest, because they yield a more aromatic sour butter and sweet buttermilk, which is a more useful by-product than sour buttermilk.

During the *Booser* process 3–4% of starter cultures are incorporated into the butter granules (water content: 13.5–14.5%) obtained from sweet cream.

Lactic acid and a flavor concentrate are obtained by separate fermentations during the first step of the *NIZO* process. In a second step they are mixed and incorporated into the butter granules from sweet cream.

Lactobacillus helveticus cultivated on whey produces the lactic acid, which is then separated by ultrafiltration and concentrated in vacuum up to about 18%. The flavor concentrate is obtained by growing starter cultures and *Streptococcus diacetylactis* on skim milk of about 16% dry matter.

10.2.3.4 Packaging

After the butter is formed, it is cut by machine into rectangular blocks and is wrapped in waxed or grease-proof-paper or metallic (aluminium) foil laminates (coated within with polyethylene).

10.2.3.5 Products Derived from Butter

- Melted butter consists of at least 99.3% milk fat. The aqueous phase is removed by decantation of the melted butter or by evaporation.
- Fractionated butterfat. The butter is separated by fractional crystallization into high- and low-melting fractions, and is utilized for various purposes (e.g. consistency improvement of whipping creams and butters).
- Spreadable blends with vegetable oils ("butterine").

10.2.4 Evaporated Milk, Sweetened Condensed Milk

Evaporated milk is manufactured from milk by removal of about 60% of the water and it is used as fluid milk, perhaps in diluted form. It is made from milk of the desired fat content and from which the albumins have been removed. To reduce the bacterial count and to stabilize the milk so that it will not coagulate in the sterilizer or after sterilization, the milk is first heated at 85–100 °C for 10–25 min and, only then, evaporated at 40–80 °C in a continuous vacuum evaporator. During the concentration process, changes occur to a greater extent than those mentioned for heat-treated fluid milk. The volatile fraction of evaporated milk contains 2-alkanones, benzaldehyde, acetophenone, 2-furfural, furfuryl alcohol, naphthalene, δ-decalactone, benzothiazole and o-amino acetophenone. During the storage period "stale" flavors may develop from an increased concentration of a number of these compounds. o-Amino acetophenone seems to be the most relevant compound for this flavor defect at concentrations of 1 $\mu g \times kg^{-1}$ and above. A rubbery aroma results from increased concentrations of benzothiazole. Lactulose was found in the fraction of carbohydrates. The evaporated milk, with a solid content of 25–33% or more, is homogenized, poured into lacquer (enamel)-coated cans made of white metal sheets, and is sterilized in an autoclave at 115–120 °C for 20 min. Continuous flow sterilization followed by aseptic packaging is also used. To prevent coagulation during processing and storage, Na-hydrogen carbonate, disodium phosphate and trisodium citrate are incorporated into the condensed milk. These additives have a dual effect: pH correction and adjustment of free Ca^{2+} ion concentration, both aimed at preventing casein aggregation (cf. Figure 10.1).

In production of *sweetened condensed milk,* after a preheating step (short-time heating at 110–130 °C), sucrose is added to a concentration of 45–50% of the weight of the end-product. Homogenization and sterilization steps are omitted.

To avoid grainyness caused by lactose crystallization – the solubility limit of lactose is exceeded after sucrose addition – the condensed milk is cooled rapidly, then seeded with finely-pulverized α-lactose hydrate. Seeding ensures that the lactose crystal size is 10 μm or less.

10.2.5 Dehydrated Milk Products

Milk or milk concentrate with 30–55% solids is dried to a residual moisture content of 3–4% by dripping or spraying onto heated rollers (100–130 °C, 2–3 sec), or by spraying in a hot stream of air (air inlet temperature 180–220 °C, product exposure temperature 80–100 °C for 0.5–1 sec). Due to its lower thermal exposure, spray-dried milk powder is more soluble than the roller (drum)-dried product. The solubility of the powder can be further improved by tailoring it into an instant product with improved dispersing and reliquefication characteristics. This is done by moistening, agglomerization and post-drying of the powder.

Table 10.21. Composition of dried milk products (%)

	1	2	3	4	5
Water	3.5	4.3	4.0	3.1	7.1
Protein	25.2	35.0	21.5	33.4	12.0
Fat	26.2	1.0	40.0	2.3	1.2
Lactose	38.1	51.9	29.5	54.7	71.5
Minerals	7.0	7.8	5.0	6.5	8.2

1: Whole milk powder, 2: skim milk powder, 3: cream powder, 4: buttermilk powder, and 5: whey powder.

Other dehydrated dairy products, in addition to whole milk or skim milk powders, are manufactured by similar processes. Products include dehydrated malted milk powder, spray- or roller-dried creams with at least 42% fat content of their solids and a maximum 4% moisture, and butter or cream powders with 70–80% milk fat. Dehydrated buttermilk and lactic acid-soured milk are utilized as children's food.

Infant milk food product formulation to approximate mothers' milk can be achieved, for example, by addition of whey proteins, sucrose, whey or lactose, vegetable oil, vitamins and trace elements and by reduction of minerals, i.e. by a shift of the Na/K ratio.

The compositions of some dehydrated dairy products are illustrated in Table 10.21.

10.2.6 Ice Creams

Ice cream is a frozen mass which can contain whole milk, skim milk products, cream or butter, sugar, vegetable oil, egg products, fruit and fruit ingredients, coffee, cocoa, aroma substances and approved food colors. Polysaccharides are added to bind the free water and to retard ice crystal growth during freezing and storage, while emulsifiers are incorporated to stabilize oil droplets. In ice cream production the mixed ingredients are chilled, frozen to about −10 °C by air trapping and, lastly, hardened at −15 to −25 °C (soft or hard ice creams).

10.2.7 Cheese

Cheese is obtained from curdled milk by removal of whey and by curd ripening in the presence of special microflora. The great abundance of cheese varieties, in many different sizes and shapes, can be classified from many viewpoints, e.g. according to:

- Milk utilized (cow, goat or sheep milk).
- Curd formation (using acids, rennet extract or a combination of both).
- Texture or consistency, or water content (%) in fat-free cheese. Following the latter criterion, the more important cheese groups are (water content, %, in brackets):
  Very hard cheese for grating (< 47);
  Hard cheese (< 56);
  Cutting (slicing) cheese (54–63);
  Semi-fat slicing cheese (61–69);
  Soft cheese (78–87).
- Fat content (% dry matter). By this criterion, the more important groups are:
  Double cream cheese (60–85% fat);
  Cream cheese (≥ 50);
  Whole-fat cheese (≥ 45);
  Fat cheese (≥ 40);
  Semi-fat cheese (≥ 20);
  Skim cheese (max. 10)

Within each group, individual cheeses are characterized by aroma. A small selection of the more important cheese varieties is listed in Table 10.22. Cheese manufacturing essentially consists of curd formation and ripening.

Table. 10.22. Cheese Varieties

Unripened Cheeses (F: < 10–70, T: 39–44, R: unripened)
Quark, Neuchâtel, Petit Suisse, Demi Sel, Cottage Cheese
Schichtkäse (layers of different fat content)
Rahm-, Doppelrahmfrischkäse, Demi Suisse, Gervais, Carré-frais, Cream Cheese
Mozzarella (plastic curd by heating to > 60 °C within the whey), Scamorze
Ripened Cheese
Hard Cheeses (F: 30–50, T: 58–63; R. 2–8 M)
Chester, Cheddar, Cheshire, Cantal
Emmental, Alpkäse, Bergkäse, Gruyère, Herrgårdskäse, Samsoe
Gruyère, l'Emmental française, Beaufort, Gruyère de Comte, Geyerzer
Parmigiano-Reggiano (granular structure, very hard, grating type), Grana, Bagozzo, Sbrinz
Provolone (plastic curd by heating to > 60 °C within the whey: Pasta filata), Cacciocavallo
Semi-soft Cheeses (F: 30–60, T: 44–57, R: 3–5 W)
Edam, Geheimratskäse, Brotkäse, Molbo, Thybo Gouda, Fynbo, Naribo
Pecorino (from ewe's milk), Aunis Brinsenkäse
Port Salut, St. Paulin, Esrom, Jerome, Deutscher Trappistenkäse
Tilsiter, Appenzeller, Danbo, Steppenkäse, Svecia-Ost
Butterkäse, Italico, Bel Paese, Klosterkäse
Roquefort (from ewe's milk), Bleu d'Auvergne, Bresse Bleu, Bleu du Haut-Jura, Gorgonzola, Stracchino, Stilton, Blue Dorset, Blue Cheese, Danablue
Steinbuscher
Weißlacker, Bierkäse

Soft Cheeses (F: 20–60, T 35–52, R: 2 W)
Chevre (from goat's milk), Chevret, Chevretin, Nicolin, Cacciotta, Rebbiola, Pinzgauer Käse
Brie, Le Coulommiers
Camembert, veritable Camembert, Petit Camembert
Limburger, Backsteinkäse, Allgäuer Stangenkäse
Münsterkäse, Mainauer, Mondseer, Le Munster, Gérômè
Pont l'Eveque, Angelot, Maroilles
Romadour, Kümmelkäse, Weinkäse
Low-fat Cheeses (F: < 10, T: 35, R: 1–2 W)
Harzer Käse, Mainzer Käse (ripened with *Bact. linens,* different cocci and yeasts)
Handkäse, Korbkäse, Stangenkäse, Spitzkäse (ripened with *Bact. linens,* different cocci and yeasts, or which Penicillium camemberti), Gamelost
Kochkäse (from Cottage Cheese by heating with emulsifying agents, F: < 10–60)

[a] Related types are grouped together. For the classes average values are given for
- fat content in the dry matter: F (%)
- dry matter: T(%)
- ripening time: R in months (M) or weeks (W)

10.2.7.1 Curd Formation

The milk fat content is adjusted to a desired level and, when necessary, the protein content is also adjusted. Additives include calcium salts to improve protein coagulation and cheese texture, nitrates to inhibit anaerobic spore-forming microflora, and color pigments. The prepared raw or pasteurized milk is mixed at 15–50 °C in a vat with a starter culture (lactic acid or propionic acid bacteria; molds, such as *Penicillium camemberti, P. candidum, P. roqueforti;* red- or yellow-smearing cultures, such as *Bacterium linens* with *cocci* and yeast) and the milk coagulates into a soft, semi-solid mass, the curd, after lactic acid fermentation, followed by addition of rennet, or some other combination, the most common being combined acid and rennet treatment. This protein gel is cut into cubes while being heated and is then gently stirred. The whey is drained off while the retained fat-containing curd is subjected to a firming process (syneresis). The firming is more intense as the mechanical input and the applied temperature are higher. The process and the starter culture (pH) determine the curd properties. When the desired curd consistency has been achieved, curd and whey separation is accomplished either by draining off the whey or by pressing off the curd while simultaneously molding it. After salting, the fresh cheese is left to ripen or "cure".
Cheesemaking is becoming increasingly mechanized and automated.

10.2.7.2 Curing

Curing, or ripening, is dependent on cheese mass composition, particularly the water content, the microflora and the external conditions, such as temperature and humidity in the curing rooms.
The ripening of soft cheeses proceeds inwards, so in the early stages there is a ripened rind and an unripe inner core. This nonuniform ripening is due to the pH gradient. The pH is low inside the cheese mass due to the presence of lactic acid. In the rind special molds that grow more favorably at higher pH values contribute to a pH increase by decarboxylating amino acids.
Ripening in hard cheeses occurs uniformly throughout the whole cheese mass. Rind formation is the result of surface drying, so it can be avoided by packaging the cheese mass in suitable plastic foils before curing commences. The duration of curing varies and lasts several days for soft cheeses and up to several months or even a couple of years for hard cheeses. The yield per 100 kg fluid milk is 8 kg for hard cheeses and up to 12 kg for soft cheeses.
All cheese ingredients are degraded biochemically to varying extents during curing.
Lactose is degraded to lactic acid by homofermentation. In cheddar cheese, for example, the pH drops from 6.55 to 5.15 from the addition of the starter culture to the end of mold pressing. In the presence of propionic acid bacteria (as in the case of Emmental cheese), lactic acid is converted further to propionic and acetic acids and CO_2, according to the reaction:

$$3\,CH_3CHOHCOOH \rightarrow 2\,CH_3CH_2COOH + CH_3COOH + CO_2 + H_2O \qquad (10.11)$$

The ratio of propionic to acetic acid is influenced by the redox potential of the cheese, and, in the presence of nitrates, for example, the ratio is lower. Propionic acid fermentation is shown in

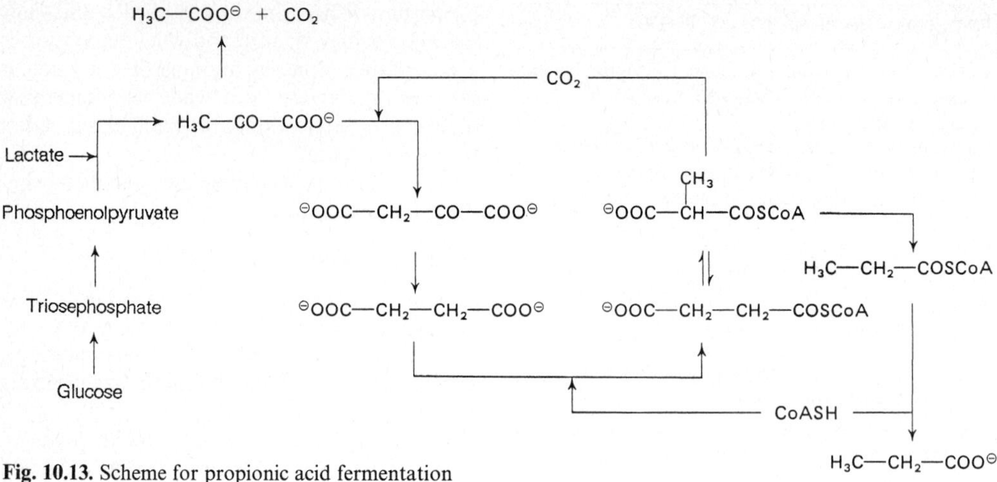

Fig. 10.13. Scheme for propionic acid fermentation

Figure 10.13. The crucial step is the reversible rearrangement of succinyl-CoA into methylmalonyl-CoA:

$$H_3\overset{*}{C}-CH(CO-SCoA)(CO_2^{\ominus}) \longrightarrow CO-SCoA-\overset{*}{C}H_2-CH_2-CO_2^{\ominus} \qquad (10.12)$$

The catalysis is mediated by adenosyl-B_{12}, which is a coenzyme for transformations of the general type:

$$\begin{matrix} | & & | \\ X-C-H & & H-C-H \\ | & \rightarrow & | \\ Y-C-H & & Y-C-X \\ | & & | \end{matrix} \qquad (10.13)$$

Based on a study of a coenzyme B_{12}-analogue, it is obvious that a nonclassical carbanion mechanism is involved (cf. Reaction 10.14).

(10.14) [Reaction scheme: Adenine–ribose–CH_2–Co (adenosyl-cobalamin); CH_2–Co + $H_3C-CH(C(=O)SCoA)(CO_2^{\ominus})$ → $CH_3^{\oplus}$–Co + $H_2\overset{\ominus}{C}-CH(C(=O)SCoA)(CO_2^{\ominus})$ → H_2C–$C(\overline{O}|)(SCoA)$–$CH(CO_2^{\ominus})$ (cyclic) → $H_2C(C(=O)SCoA)-CH_2-CO_2^{\ominus}$]

The mode and extent of milk *fat* degradation depend on the microflora involved in cheese ripening. Lipolysis is strongly enhanced by homogenization of the milk (Figure 10.14). The release of fatty acids, especially those that affect cheese aroma, depends on the specificity of the lipases (Table 10.23). In addition to free fatty acids, 2-alkanones and 2-alkanols are formed as by-products of the β-oxidation of the fatty acids (cf. 3.7.5).

Table 10.23. Substrate specificity of a lipase from *Penicillium roqueforti*

Substrate	Hydrolysis (V_{rel})
Tributyrin	100
Tripropionin	25
Tricaprylin	75
Tricaprin	50
Triolein	15

Table 10.24. 2-Alkanones in blue cheese

2-Alkanone n[a]	mg/100 g cheese dry matter
3	0.5–0.8
5	1.4–4.1
7	3.8–8.0
9	4.4–17.6
11	1.2–5.9

[a] Number of C-atoms.

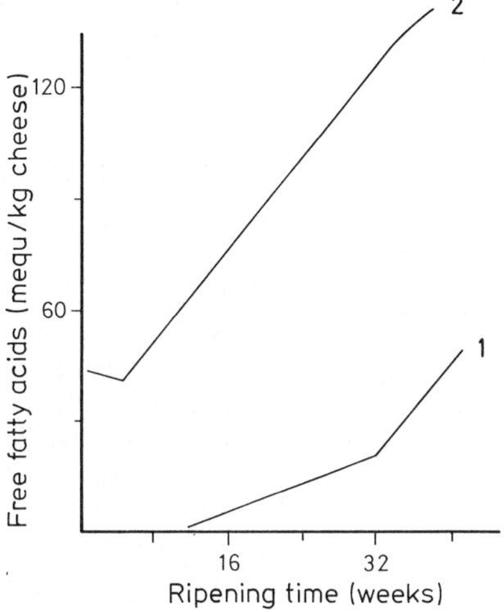

Fig. 10.14. Lipolysis during ripening of blue cheese: 1 untreated milk, 2 homogenized milk

Molds, particularly *Penicillium roqueforti,* utilize β-ketoacyl-CoA deacylase (thiohydrolase) and β-ketoacid decarboxylase to provide the compounds typical for the aroma of semi-soft cheeses e.g. of the blue-veined cheese (Roquefort, Stilton, Gorgonzola, cf. Table 10.24).

Protein degradation to amino acids occurs through peptides as intermediary products. Depending on cheese variety, 20–40% of casein is transformed into soluble protein derivatives, of which 5–15% are amino acids. A pH range of 3–6 is optimum for the activity of proteinases from *Penicillium roqueforti.* Proteolysis is strongly influenced by salt content of the cheese. The amino acid content is 2.6–9% of the cheese solids and is of importance to cheese aroma. Ripening defects can produce bitter-tasting peptides.

The amino acids are transformed further. In early stages of cheese curing, with a lower pH, they are decarboxylated to amines. In later stages, with a higher pH, oxidation reactions prevail:

$$\text{R—CH(NH}_2\text{)—COOH} \rightarrow \begin{cases} \text{R—CH}_2\text{—NH}_2 + \text{CO}_2 \\ \text{R—CO—COOH} \rightarrow \text{R—CHO} + \text{CO}_2 \end{cases} \quad (10.15)$$

Proteolysis contributes not only to aroma formation, but it affects cheese texture. In overripening of soft cheese, proteolysis can proceed almost to liquefaction of the entire cheese mass.

Many compounds are responsible for cheese aroma. Table 10.25 illustrates several blue cheese recipes that approximate or simulate the characteristic blue cheese aroma.

Table 10.25. Formulation of blue cheese aroma (mg/kg cheese) 1: amounts found in blue cheese (+: detected), 2–4: artificial formulations

Compound	1	2	3	4
Acetic acid	825	8	550	
Butyric acid	1,500	6	950	
Propionic acid				126
Isobutyric acid		2		
Valeric acid		3		
Isovaleric acid		3		
Caproic acid	900	265	600	
Heptanoic acid		5.4		
Caprylic acid	770	89	515	
Pelargonic acid		4.5		
Capric acid	2,000	133		
Lauric acid	3,000			
Glyoxylic acid		71		
Cinnamic acid		0.03		
Pyruvic acid		20		
2-Keto isovaleric acid		17		
2-Keto isocaproic acid		43		
Oxalacetic acid		51		
2-Ketoglutaric acid		42		
Butyric acid ethyl ester	+	0.5	1.5	
Caproic acid methyl ester			6	
Caproic acid ethyl ester		0.5		

Table 10.25. (Continued)

Compound	1	2	3	4
Caprylic acid methyl ester			6	
Caprylic acid ethyl ester		0.5		
Cinnamic acid methyl ester				4
2-Pentanol	0.5	6	1	
2-Heptanol	6.0	1	12	28
2-Nonanol	3.5	10	7	
1-Octen-3-ol		22		4
Phenylethanol	+		2	
Ethanal		5		
Propanal		2		
Butanal		2		
Pentanal		2		
Phenylethanal		2		
Methional		0.1		
Acetone	3		6	
2-Pentanone	15	8	30	
2-Heptanone	35	65	70	1.6
2-Nonanone	33	116	65	36
2-Undecanone	8	10	17	
2-Tridecanone		2		
δ-Decalactone		0.7		
δ-Dodecalactone		0.7		
Indole		0.1		

A major difference in flavor between Camembert and blue cheese seems to be that the concentration of 1-octen-3-ol in the former cheese is higher. This allyl alcohol is responsible for a mushroom-like aroma note. The flowery note of Camembert cheese is due to 2-phenylethanol and phenylacetaldehyde; the hazelnut note to 1,3-dimethoxybenzene and cinnamic acid methyl ester. The ripe Camembert garlic note originates from sulfur compounds, such as 2,4-dithiapentane, 2,4,5-trithiahexane and 3-methylthio-2,4-dithiapentane.

The important aroma compounds of cheeses with bacterial surface ripening, e. g. Pont l'Eveque, are phenol, cresol, acetophenone and methylthio esters (methyl thioacetate, -propionate, -butyrate) and methyl butyrate.

Methyl thioacetate also occurs in cheeses fermented with propionic acid, e.g.Emmental and Gruyère. In addition to propionic acid, alcohols and esters have a role in the build-up of aroma. The bitter taste of cheeses is probably derived not only from peptides, but also from other amides. The presence of bitter N-isobutyl acetamide has been confirmed in Camembert cheese.

10.2.7.3 Processed Cheese

Processed (or melted) cheese is made from natural, very hard grating or hard cheeses by shredding and then heating the shreds to 75–95 °C in the presence of 2–3% melting salts (lactate, citrate, phosphate) and, when required, utilizing other ingredients, such as milk powder, cream, aromas, seasonings and vegetable and/or meat products. The cheese can be spreadable or made firm and cut as desired. Processed cheese shelf life is long due to thermal killing of microflora.

10.2.8 Other Products

10.2.8.1 Casein, Caseinates, Coprecipitates

Coagulation and separation of casein from milk is possible by souring the milk by lactic acid fermentation, or by adding acids such as HCl, H_2SO_4, lactic acid or H_3PO_4. Another way to achieve coagulation is to add proteinase enzymes, such as rennin and pepsin. The acid coagulation is achieved at 35–50 °C and pH 4.2–4.6 (isoelectric point for casein is pH 4.6–4.7). Casein separates as a coarse, grainy coagulate which is then washed and dried (whirlwind drier). The curd and whey are heated to 65 °C in the presence of enzyme. Increasing the level of Ca^{2+} ions (addition to milk of 0.2% $CaCl_2$) causes casein and whey proteins to coagulate when the temperature is at 90 °C. Joint coagulation of proteins can also be achieved by first heat-denaturing the whey proteins, then acidifying the milk. Washing followed by drying of the curd gives a coprecipitate which contains up to 96% of the total proteins of the milk. When casein dispersions, 20–25%, are treated with alkali [NaOH or $Ca(OH)_2$, alkali or earth-alkali carbonates or citrates] at 80–90 °C and pH 6.2–6.7, and then the solubilized product is spray-dried, a soluble or readily dispersable casein product is obtained (caseinate, disintegrated milk protein).

Casein and caseinate are utilized as food and also have nonfood uses. In food manufacturing they are used for protein enrichment and/or to achieve stabilization of some physical properties of processed meats, baked products, candies, cereal products, ice creams, whipping creams, coffee

Table 10.26. Protein, lactose and mineral contents of whey products[a]

Product	DM[b] (%)	Protein (%)	Lactose (%)	Minerals (%)
Skim milk	9.0	36	53	7
Whey (from coagulating with rennet)	6.0–6.4	13	75	8
Whey (from coagulating with acid)	5.8–6.2	12	67	14
Demineralized whey powder		12–13	85	1–2
Whey protein powder[c]				
I		47	44	9
II		74	20	6

[a] Average values are expressed as % of dry matter.
[b] Dry matter.
[c] After one (I) and two (II) ultrafiltrations.

whiteners, and some dietetic food products and drugs.

The nonfood uses involve wide application of casein/caseinate as a sizing (coating) for better quality papers (for books and journals, with a surface suitable for fine printing); in glue manufacturing, as a type of waterproof glue (alkali caseinate with calcium components as binder); in the textile industry (dye fixing, water-repellent impregnations); and for casein paints and production of some plastics (knobs, piano keyboards, etc.).

10.2.8.2 Whey Products

Whey is a by-product in cheese and casein manufacturing. In 1975 in FR Germany, 4.3×10^6 tonnes of whey were obtained, i.e. 2.7×10^5 t of dry matter. Of this amount, 29% was utilized as animal feed, 38% for production of whey powder, and close to 14% for lactose recovery. The rest, 19%, was industrial waste. The important products of whey processing, as listed in Table 10.26, are:

- Whey powder. To produce the powder, the whey is first concentrated to 45–50% solids, and then dried on a drum dryer or spray-dried. A nonhygroscopic product is obtained by predrying of whey to about 12% moisture and hydration of amorphous lactose to α-lactose monohydrate, followed by final drying.
- Whey powder, made of the residual whey from the lactose recovery process.
- Demineralized whey powder, obtained from whey treated with ion-exchangers or, preferentially, by electrodialysis. The content of various ions in whey versus demineralization extent is presented in Figure 10.15.
- Whey protein powder. The whey proteins can be precipitated by heating of the whey (95 °C, 30 min) at pH 4.5. A gentler process involves protein enrichment by ultrafiltration; this approach provides soluble and biologically-valuable products.

Additional uses for whey and whey products are in animal nutrition, dietetic food products (infant nutrition), baking products, confections and beverages.

10.2.8.3 Lactose

For lactose production the whey is evaporated to 55–65% solid content, and the concentrate is then seeded and cooled slowly to induce sugar crystallization. The raw lactose (food quality) is recrystallized to yield a raffinade (pharmaceutical-grade lactose). Lactose is used in manufacturing of drugs (tablet filler), dietetic food products, baked products, dehydrated foods, cocoa products, beverages and ice creams.

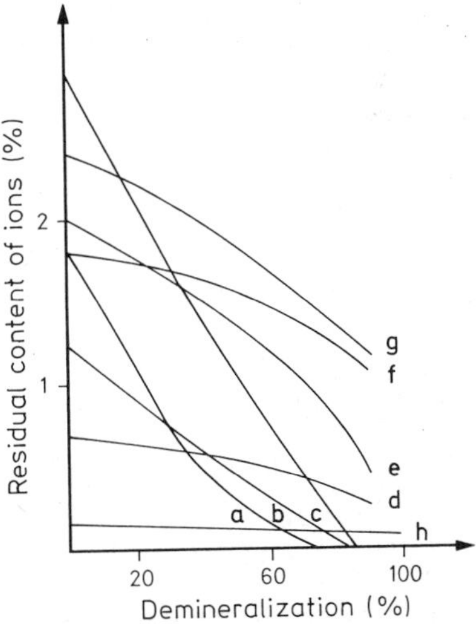

Fig. 10.15. Whey demineralization
Ions of **a** chloride, **b** sodium, **c** potassium, **d** calcium, **e** phosphate, **f** lactate, **g** citrate, and **h** magnesium

10.3 Literature

Adda, J., Roger, S., Dumont, J.-P.: Some recent advances in the knowledge of cheese flavor. In: Flavor of foods and beverages (Eds.: Charalambous, G., Inglett, G. E.), p. 65, Academic Press: New York–London. 1978

Aimutis, W. R., Eigel, W. N.: Identification of λ-casein as plasmin-derived fragments of bovine α_{-s1}-casein. J. Dairy Sci. *65,* 175 (1982)

Andrews, A. T.: The formation and structure of some proteose-peptone components. J. Dairy Res. *46,* 215 (1979)

Badings, H. T., De Jong, C., Dooper, R. P. M., De Nijs, R. C. M.: Rapid analysis of volatile compounds in food products by purge- and cold-trapping/capillary gas chromatography. In: Progress in Flavor Research 1984 (Ed.: Adda, J.), p. 523, Elsevier Science Publ., Amsterdam. 1985

Creamer, L. K., Richardson, T., Parry, D. A. D.: Secondary structure of bovine α_{s1}- and β-casein in solution. Arch. Biochem. Biophys. *211,* 689 (1981)

Davies, F. L., Law, B. A.: Advances in the microbiology and biochemistry of cheese and fermented milk. Elsevier Appl. Science Publ., London. 1984

Jolles, P., Fiat, A.-M.: The carbohydrate portion of milk glycoproteins. J. Dairy Res. *46,* 187 (1979)

Kessler, H. G.: Lebensmittelverfahrenstechnik – Schwerpunkt Molkereitechnologie. Kessler-Verlag: München-Weihenstephan. 1976

Kessler, H. G.: Leidet der ernährungsphysiologische Wert industriell bearbeiteter Lebensmittel – am Beispiel Milch? Deutsche Molkerei-Zeitung *104,* 482 (1983)

Kiermeier, F., Lechner, E.: Milch und Milcherzeugnisse. Verlag Paul Parey: Berlin–Hamburg. 1973

Kinsella, J. E., Hwang, D. H.: Enzymes of Penicillium roqueforti involved in the biosynthesis of cheese flavor. Crit. Rev. Food Sci. Nutr. *8,* 191 (1977)

Labuza, T. P., Riboh, D.: Theory and application of Arrhenius kinetics to the prediction of nutrient losses in foods. Food Technol. *36* (10), 66 (1982)

Nemitz, G.: Technologie der Milch-Nebenprodukte. Landwirtschaftliche Forschung, Sonderheft 33/II, S. 300, Kongreßband 1976 Oldenburg, J. D. Sauerländer's Verlag: Frankfurt/Main. 1977

Olivecrona, T., Bengtsson, G.: Lipase in milk. In: Lipases (Eds.: Borgström, B., Brockman, H. L.), Elsevier Science Publ., Amsterdam. 1984

Payens, T. A. J.: Casein micelles: the colloid-chemical approach. J. Dairy Res. *46,* 291 (1979)

Scott, R.: Cheesemaking Practice. Applied Science Publ., London. 1981

Stein, J. L., Imhof, K.: Milch und Milchprodukte. In: Ullmanns Encyklopädie der technischen Chemie, 4. Aufl., Bd. 16, S. 689, Verlag Chemie: Weinheim. 1978

Swaisgood, H. E.: The Caseins. Crit. Rev. Food Technol. *3,* 375 (1972/73)

Walstra, P., Jenness, R.: Dairy chemistry and physics. John Wiley & Sons, New York. 1984

Webb, B. H., Johnson, A. H., Alford, J. A. (Eds.): Fundamentals of dairy chemistry. 2nd edn., AVI Publ. Co.: Westport, Conn. 1974

11 Eggs

11.1 Foreword

Eggs have been a human food since ancient times. They are one of nature's most nearly perfect protein foods and have other high quality nutrients. Eggs are readily digested and can provide a significant portion of the nutrients required daily for growth and maintenance of body tissues. They are utilized in many ways both in the food industry and the home. Chicken eggs are the most important. Those of other birds (geese, ducks, plovers, seagulls, quail) are of lesser significance. Thus, the term "eggs", without a prefix, generally relates to chicken eggs and is so considered in this chapter. Table 11.1 gives some data for production and utilization of eggs.

Table 11.1. World production of eggs, 1981 (1,000 t)[a]

Continent	Chicken egg	Other eggs
World	29,210	403
Africa	959	7
America, North-, Central-	5,423	1
America, South-	1,715	9
Asia	9,716	257
Europe, West-	5,217	6
Europe, East- + USSR	5,899	107
Australia + New Zealand	282	17

Country	Chicken egg	Country	Other eggs
China	4,902	Indonesia	89
USA	4,122	USSR	65
USSR	3,893	Vietnam	63
Japan	1,999	Bangladesh	36
France	875	China	29
Brasil	850	Rumania	19
FR Germany	800	Czechoslovakia	18
India	776	Australia	15
UK	763	Philippines	13
Spain	690	Thailand	10
Italy	665		
Mexico	627	Σ (%)[b]	89
Holland	564		
Poland	494		
Σ (%)[b]	75		

[a] Including eggs for hatching.
[b] World production ≙ 100%.

11.2 Structure, Physical Properties and Composition

11.2.1 General Outline

The egg (Figure 11.1) is surrounded by a 0.2–0.4 mm thick calcareous and porous shell. Shells of chicken eggs are white-yellow to brown, duck's are greenish to white, and those of most wild birds are characteristically spotted. The inside of the shell is lined with two closely-adhering membranes (inner and outer). The two membranes separate at the large end of the egg to form an air space, the so-called air cell. The air

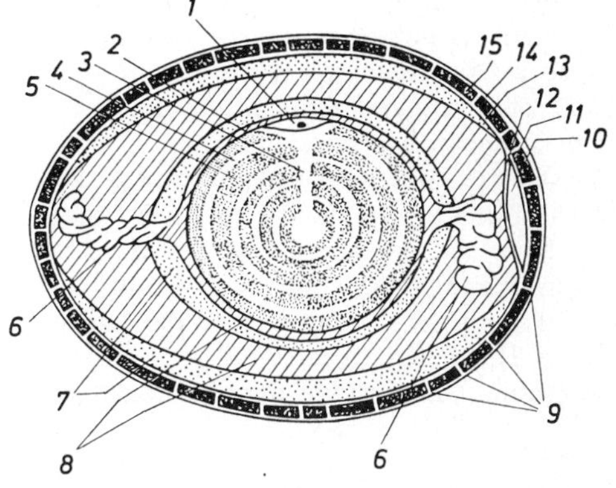

Fig. 11.1. Cross-section of a chicken egg – a schematic representation. Egg yolk: 1 germinal disc (blastoderm), 2 yolk membrane, 3 latebra, 4 a layer of light colored yolk, 5 a layer of dark colored yolk, 6 chalaza, 7 egg white (albumen) thin gel, 8 albumen thick gel, 9 pores, 10 air cell, 11 shell membranes, the outer one the ammilary layer adhering to the shell, 12 inner egg membrane, 13 shell surface cemented to the mammilary layer, 14 cuticle, and 15 the spongy calcareous layer

cell is approx. 5 mm in diameter in fresh eggs and increases in size during storage, hence it can be used to determine the age of eggs. The egg white (albumen) is an aqueous, faintly straw-tinted, gel-like liquid, consisting of four fractions that differ in viscosity. The inner portion of the egg, the yolk, is surrounded by albumen. A thin but very firm layer of albumen (chalaziferous layer) closely surrounds the yolk and it branches on opposite sides of the yolk into two chalazae that extend into the thick albumen. The chalazae resemble two twisted rope-like cords, twisted clockwise at the large end of the egg and counter-clockwise at the small end. They serve as anchors to keep the yolk in the center. In an opened egg the chalazae remain with the yolk. The germinal disc (blastoderm) is located at the top of a club-shaped latebra on one side of the yolk. The yolk consists of alternate layers of dark- and light-colored material arranged concentrically.
The average weight of a chicken egg is 58 g. The proportions of the three main egg parts, yolk, white and shell, and the major ingredients are listed in Table 11.2. Table 11.2a gives the amino acid composition of whole egg, white and yolk.

Table 11.2a. Amino acid composition of whole egg, egg white and yolk (g/100 g edible portion)

Amino acid	Whole egg	Egg white	Egg yolk
Ala	0.71	0.65	0.82
Arg	0.84	0.63	1,13
Asx	1.20	0.85	1.37
Cys	0.30	0.26	0.27
Glx	1.58	1.52	1.95
Gly	0.45	0.40	0.57
His	0.31	0.23	0.37
Ile	0.85	0.70	1.00
Leu	1.13	0.95	1.37
Lys	0.68	0.65	1.07
Met	0.40	0.42	0.42
Phe	0.74	0.69	0.72
Pro	0.54	0.41	0.72
Ser	0.92	0.75	1.31
Thr	0.51	0.48	0.83
Trp	0.21	0.16	0.24
Tyr	0.55	0.45	0.76
Val	0.95	0.84	1.12

11.2.2 Shell

The shell consists of calcite crystals embedded in an organic matrix or framework of interwoven protein fibers and spherical masses (protein-mucopolysaccharide complex) in a proportion of 50 : 1. There are also small amounts of magnesium carbonate and phosphates.
The shell structure is divided into four parts: the cuticle or bloom, the spongy layer, the mammillary layer and the pores. The outermost shell coating is an extremely thin (10 μm), transparent, mucilaginous protein layer called the cuticle, or bloom. The spongy, calcareous layer, i.e. a matrix comprising two-thirds of the shell thickness, is below the thin cuticle. The mammilary layer consists of a small layer of compressed, knob-like particles, with one side firmly cemented to the spongy layer and the other side adhering closely to the outer surface of the shell membrane. The shell membrane is made of two layers (48 and 22 μm, respectively), each an interwoven network of protein polysaccharide fibers. The outer layer adheres closely to the mammilary layer. Tiny pore canals which extend through the shell are seen as minute pores or round openings (7,000–17,000 per egg). The cuticle protein partially seals the pores, but they remain permeable to gases while restricting penetration by microorganisms.

Table 11.2. Average composition of chicken eggs

Fraction	Percent of the total weight	Dry matter (%)	Protein (%)	Fat (%)	Carbohydrates (%)	Minerals (%)
Shell	10.3	98.4	3.3[a]			95.1
Egg white	56.9	12.1	10.6	0.03	0.9	0.6
Egg yolk	32.8	51.3	16.6	32.6	1.0	1.1

[a] A protein mucopolysaccharide complex.

11.2.3 Albumen (Egg White)

Albumen is a 10% aqueous solution of various proteins. Other components are present in very low amounts. The thick, gel-like albumen differs from thin albumen (cf. 11.1) only in its approx. four-fold content of ovomucin. Albumen is a pseudoplastic fluid. Its viscosity depends on shearing force (Figure 11.2). The surface tension (12.5% solution, pH 7.8, 24 °C) is 49.9 dynes cm^{-1}. The pH of albumen of a freshly-laid egg is 7.6–7.9 and rises to 9.7 during storage due to diffusion of solubilized CO_2 through the shell.

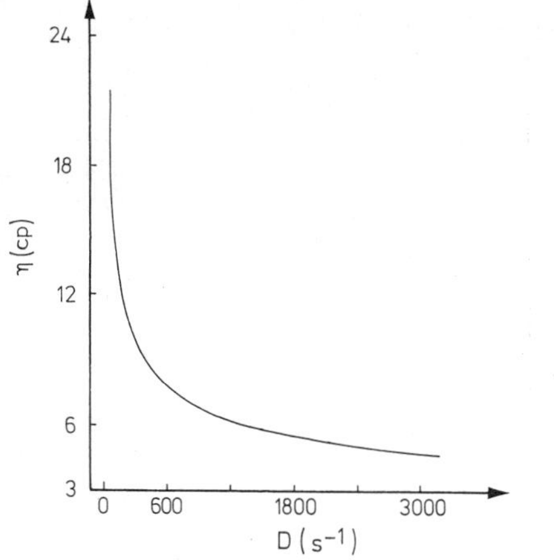

Fig. 11.2. Egg white viscosity, η, as affected by shear rate D, at 10 °C. (After *Stadelman*, 1977)

This rise is time and temperature dependent. For example, a pH of 9.4 was recorded after 21 days of storage at 3–35 °C.

11.2.3.1 Proteins

Table 11.3 lists the most important albumen proteins in order of their abundance in egg white. The carbohydrate moieties of the glycoprotein constituents are presented in Table 11.4. Several albumen proteins have biological activity (Table 11.3), i.e. as enzymes (e.g. lysozyme), enzyme inhibitors (e.g. ovomucoid, ovoinhibitor) and complex-forming agents for some coenzymes (e.g. flavoprotein, avidin). The biological activities may be related to protection of the egg from microbial spoilage. Egg white protein separation is relatively easy: the albumen is treated with an equal volume of saturated ammonium sulfate; the globulin fraction precipitates together with lysozyme, ovomucin and other globulins; while the major portion of the egg white remains in solution. This albumen fraction consists of ovalbumin, conalbumin and ovomucoid. Further separation of these fractions is achieved by ion-exchange chromatography.

Table 11.4. Carbohydrate composition of some chicken egg white glycoproteins

Protein	Carbohydrate (%)	Components (moles/mole protein)				
		Gal	Man	GlcN	GalN	Sialic acid
Ovalbumin	3.2		5	3		
Ovomucoid	23	2	7	23		1
α-Ovomucin[a]	13	21	46	63	6	7
Ovoglycoprotein	31	6	12	19		2
Ovoinhibitor (A)	9.2	10[b]		14		0.2
Avidin[c]	10		4 (5)	3		

[a] In addition to carbohydrate it contains 15 moles of esterified sulfuric acid per mole protein.
[b] Sum of Gal + Man.
[c] Data per subunit (16 kdal).

Table 11.3. Proteins of egg white

Protein	Percent of the total protein[a]	Molecular weight (kdal)	Isoelectric point (pH)	Comments
Ovalbumin	58	45	4.6	
Conalbumin (Ovotransferrin)	13	80	6.6	binds metal ions
Ovomucoid	11	28	3.9	proteinase inhibitor
Lysozyme (Ovoglobulin G_1)	3.5	14.6	10.7	N-acetylmuramidase
Ovoglobulin G_2	4	30–45	5.5	good foam builders
Ovoglobulin G_3	4		5.8	good foam builders
Ovomucin	1.5	210		inhibits viral hemagglutination
Flavoprotein	0.8	35	4.1	binds riboflavin
Ovoglycoprotein	0.5	24		
Ovomacroglobulin	0.5	760–900	4.5–4.7	
Ovoinhibitor	0.1	49	5.2	proteinase inhibitor
Avidin	0.05	67[b]	9.5	binds biotin

[a] Average values are presented.
[b] Four times 15.6 kdal + approx. 10% carbohydrate.

11.2.3.1.1 Ovalbumin

This is the main albumen protein, crystallized by *Hofmeister* in 1890. It is a glycophosphoprotein with 3.2% carbohydrates (Table 11.4) and 0–2 moles of serine-bound phosphoric acid per mole of protein (ovalbumin components A_3, A_2 and A_1, approx. 3, 12 and 85%, respectively). Ovalbumin contains 4 thiol and 1 disulfide groups. During storage of eggs, heat-stable S-ovalbumin is formed from native protein, probably by a thiol-disulfide exchange reaction.

The carbohydrate moiety is bound to an asparagine residue in the sequence:

–Glu–Lys–Tyr–Asn–Leu–Thr–Ser–

with a probable structure as follows:

$$(\beta GlcNAc)_{0\text{–}1} \rightarrow \alpha Man \rightarrow (\alpha Man)_3(1 \rightarrow$$

$$\uparrow$$

$$(\alpha Man)_{0\text{–}1}$$

$$\rightarrow 3)\beta GlcNAc(1-4)\beta GlcNAc \rightarrow Asn$$

$$4$$

$$|$$

$$1$$

$$\beta Man$$

$$\uparrow$$

$$(\beta GlcNAc)_{0\text{–}2} \quad (11.1)$$

Ovalbumin is relatively readily denatured, for example, by shaking or whipping its aqueous solution. This is an interphase denaturation which occurs through unfolding and aggregation of protein molecules.

11.2.3.1.2 Conalbumin (Ovotransferrin)

Conalbumin and serum transferrin are identical in the chicken. This protein, unlike ovalbumin, is not denatured at the interphase but coagulates at lower temperatures.

Binding of metal ions (2 moles Mn^{3+}, Fe^{3+}, Cu^{2+} or Zn^{2+} per mole of protein) at pH 6 or above is a characteristic property of conalbumin. Table 11.5 lists the absorption maxima of several complexes. The occasional red discoloration of egg products during processing originates from a conalbumin-iron complex. The complex is fully dissociated at a pH less than 4. Tyrosine and histidine residues are involved in metal binding. Alkylation of 10 to 14 histidine residues with bromoacetate or nitration of tyrosine residues with tetranitromethane removes its iron-binding ability. Conalbumin has the ability to retard growth of microorganisms.

Table 11.5. Metal complexes of conalbumin

Metal ion	λ max (nm)	ε ($cm^2 \times M^{-1}$)	Complex color
Fe^{3+}	470	3,280	pinkish
Cu^{2+}	440	2,500	yellow
	670	350	
Mn^{3+}	429	4,000	yellow

11.2.3.1.3 Ovomucoid

Ion-exchange chromatography or electrophoresis reveals 2 or 3 forms of this protein, which apparently differ in their sialic acid contents. The carbohydrate moiety (Table 11.4) consists of three oligosaccharide units bound to protein through asparagine residues. The protein has 9 disulfide bonds and, therefore, stability against heat coagulation. Hence, it can be isolated from the supernatants of heat-coagulated albumen solutions, and then precipitated by ethanol or acetone. Ovomucoid inhibits bovine but not human trypsin activities.

11.2.3.1.4 Lysozyme (Ovoglobulin G_1)

Lysozyme is widely distributed and is found not only in egg white but in many animal tissues and secretions, in latex exudates of some plants and in some fungi. This protein, with three known components, is an N-acetylmuramidase enzyme that hydrolyzes the cell walls of Gram-positive bacteria (murein; AG = N-acetyl-glucosamine; AMA = N-acetylmuramic acid; → = lysozyme attack):

$$\downarrow \qquad\qquad \downarrow$$

$$-\beta AG(1-4)\beta AMA(1-4)\beta AG(1\rightarrow$$

$$|$$

$$\text{Peptide}$$

$$|$$

$$\rightarrow 4)\beta AMA(1-4)\beta AG-$$

$$|$$

$$\text{Peptide}$$

$$| \quad (11.2)$$

Lysozyme consists of a peptide chain with 129 amino acid residues and four disulfide bonds. Its primary (Table 11.6) and tertiary structures have been elucidated.

11.2.3.1.5 Ovoglobulins G_2 and G_3

These proteins are good foam builders.

Table 11.6. Amino acid sequences of avidin (1) and lysozyme (2)

1)			Ala	Arg	Lys	*Cys*	Ser	*Leu*	Thr	Gly	Lys	Trp
2)	Lys	Val	Phe	Gly	Arg	*Cys*	Glu	*Leu*	Ala	Ala	Ala	Met
1)			Thr	Asn	Asp	Leu	Gly	Ser	*Asn*[a]	Met	Thr	Ile
2)			Lys	Arg	His	Gly	Leu	Asp	*Asn*[a]	Tyr	Arg	Gly
1)			Gly	Ala	Val	Asn	Ser	Arg	Gly	Glu	Phe	Thr
2)			Tyr	Ser	Leu	Gly	Asn	Trp	Val	Cys	Ala	Ala
1)			Gly	Thr	Tyr	Ile	Thr	Ala	Val	*Thr*	Ala	Thr
2)			Lys	Phe	Glu	Ser	Asn	Phe	Asn	*Thr*	Glu	Ala
1)			Ser	*Asn*	Glu	Ile	Lys	Glu	Ser	Pro	Leu	His
2)			Thr	*Asn*	Arg	Asn	Thr	Asp	Gly	Ser	Thr	Asp
1)			Gly	Thr	Glu	Asn	Thr	*Ile*	*Asn*	Lys	*Arg*	Thr
2)			Tyr	Gly	Ile	Leu	Glu	*Ile*	*Asn*	Ser	*Arg*	Trp
1)			Gln	Pro	Thr	Phe	*Gly*	Phe	*Thr*	Val	Asn	Trp
2)			Trp	Cys	Asn	Asp	*Gly*	Arg	*Thr*	Pro	Gly	Ser
1)			Lys	Phe	Ser	Glu	Ser	Thr	Thr	Val	Phe	Thr
2)			Arg	Asn	Leu	Cys	Asp	Ile	Pro	Cys	Ser	Ala
1)			Gly	Gln	Cys	Phe	Ile	Asp	Arg	Asn	Gly	Lys
2)			Leu	Leu	Ser	Ser	Asp	Ile	Thr	Ala	Ser	Val
1)			Glu	Val	Leu	*Lys*	*Thr*	*Met*	Trp	Leu	Leu	Arg
2)			Asn	Cys	Ala	*Lys*	*Lys*	*Ile*	Val	Ser	Asp	Gly
1)			Ser	Ser	Val	*Asn*	*Asp*	*Ile*	Gly	Asp	Asp	*Trp*
2)			Asp	Glu	Met	*Asn*	*Ala*	*Trp*		Val	Ala	*Trp*
1)			Lys	Ala	Thr	*Arg*	*Val*	*Gly*	Ile	Asn	Ile	Phe
2)			Arg	Asn	Arg	*Cys*	*Lys*	*Gly*	Thr	Asp	Val	Gln
1)			Thr	Arg	Leu	*Arg*	*Thr*	Gln	Lys	Glu		
2)			Ala	Trp	Ile	*Arg*	*Gly*	Cys	Arg	Leu		

[a] Binding site for carbohydrate.

11.2.3.1.6 Ovomucin

This protein, of which three components are known, can apparently form fibrillar structures and so contribute to a rise in viscosity of albumen, particularly of the thick, gel-like egg white (see egg structure, Figure 11.1), where it occurs in a four-fold higher concentration than in fractions of thin albumen. The compositions of its carbohydrate moieties are given in Table 11.4. Ovomucin is heat stable. It forms a water-insoluble complex with lysozyme. The dissociation of the complex is pH dependent. Presumably it is of importance in connection with the thinning of egg white during storage of eggs.

11.2.3.1.7 Flavoprotein

This protein binds firmly with riboflavin and probably functions to facilitate transfer of this coenzyme from blood serum to egg.

11.2.3.1.8 Ovoinhibitor

This protein is, like ovomucoid, a proteinase inhibitor. It inhibits the activities of trypsin, chymotrypsin and some proteinases of microbial origin. Its carbohydrate composition is given in Table 11.4.

11.2.3.1.9 Avidin

Avidin is a basic glycoprotein (Table 11.4). Its amino acid sequence has been determined. Noteworthy is the finding that 15 positions (12% of the total sequence, Table 11.6) are identical with those of lysozyme. Avidin is a tetramer consisting of four identical subunits, each of which binds one mole of biotin. The dissociation constant of the avidin-biotin complex at pH 5.0 is $k_{-1}/k_1 = 1.3 \times 10^{-15}$M, i.e. it is extremely low. The free energy and free enthalpy of complex formation are $\Delta G = -85$ kJ/mole and $\Delta H = -90$ kJ/mole, respectively. Avidin, in its form in egg white, is practically free of biotin, and presumably fulfills an antibacterial role. Of interest is the occurrence in *Streptomyces* spp. of a related biotin-binding protein (streptavidin) with antibiotic properties.

11.2.3.2 Other Constituents

11.2.3.2.1 Lipids

The lipid content of albumen is negligible (0.03%).

11.2.3.2.2 Carbohydrates

Carbohydrates (approx. 1%) are partly bound to protein (approx. 0.5%) and partly free (0.4–0.5%). Free carbohydrates include glucose (98%) and mannose, galactose, arabinose, xylose, ribose and desoxyribose, totaling 0.2–2.0 mg/100 g egg white. There are no free oligosaccharides or polysaccharides. Bound carbohydrates were covered previously with proteins (cf. 11.2.3.1, Table 11.4). Mannose, galactose and glucosamine are predominant, and sialic acid and galactosamine are also present.

Table 11.7. Mineral composition of eggs

	Egg white (%)	Egg yolk (%)
Sulfur	0.195	0.016
Phosphorus	0.018	0.543–0.980
Sodium	0.161–0.169	0.070–0.093
Potassium	0.145–0.167	0.112–0.360
Magnesium	0.009	0.032–0.128
Calcium	0.008–0.02	0.121–0.262
Iron	0.0009	0.0053–0.011

11.2.3.2.3 Minerals

The mineral content of egg white is 0.6%. Its composition is listed in Table 11.7.

11.2.3.2.4 Vitamins

Data on vitamins found in egg white are summarized in Table 11.11.

11.2.4 Egg Yolk

Yolk is a fat-in-water emulsion with about 50% dry matter content, and consisting of proteins (one-third) and lipids (two-thirds). Water transfer from egg white drops the solid content of the yolk by 2–4% during storage for 1–2 weeks. Yolk contains particles of differing size that can be classified into two groups:

- *Yolk droplets* of highly variable size, with a diameter range of 20–150 μm. They resemble fat droplets, consist mostly of lipids, and some have protein membranes. It is a mixture of lipoproteins with a low density (LDL, cf. 3.5.1.2).
- *Granules* that have a diameter of 0.3–1.6 μm, i.e. they are substantially smaller than yolk droplets, and are more uniform in size but less uniform in shape. They have a substructure and consist of proteins but also contain lipids and minerals.

Older methods of yolk separation, which included at least a partial defatting with various solvents (ether, ethanol, butanol), led to lipoprotein destruction and through it to artifacts of varying composition. Yolk studies are now based on ultracentrifugation, when necessary in the presence of electrolytes, which provides native yolk fractions.

Figure 11.3 schematically presents such a fractionation. The granules are separated from the plasma by centrifugation of diluted yolk solution. After NaCl addition, the granules are separated further into a low density lipoprotein fraction (LDL granules' fraction) and a lipovitellin-phosvitin complex. The latter can be separated into its constituents by chromatographic techniques. In the presence of NaCl the plasma can be further separated by centrifugation into a floating, low density lipoprotein fraction (LDL-fraction, lipovitellenin, cf. Table 3.14) and a water-soluble livetin fraction.

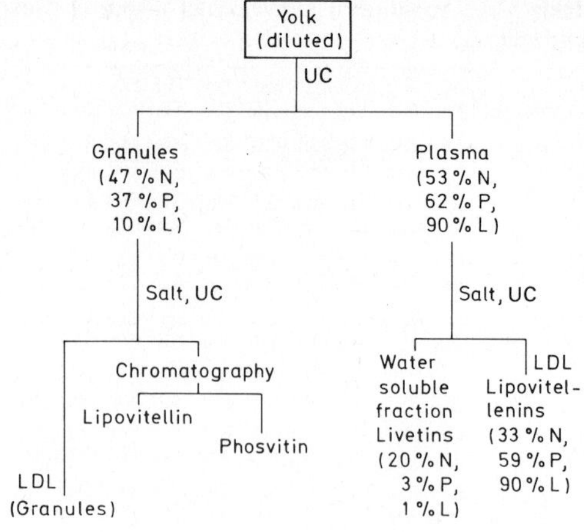

Fig. 11.3. A fractionation scheme for egg yolk. All percentages refer to the total content of the yolk. N: nitrogen, P: phosphorus, L: lipids, UC: ultracentrifugation, LDL: low density lipoprotein

Table 11.8 provides compositional data for granules, plasma and some of their constituents. The values given are calculated from different literature data and so should be considered only as guiding values. The data available deviate significantly due to methodological difficulties with yolk separation.

Egg yolk is a pseudoplastic non-*Newton*ian fluid with a viscosity which depends on the shear forces applied. Its surface tension is 44 dynes cm^{-1} (25 °C), while its pH is 6.0 and, unlike egg

Table 11.8. Composition of egg yolk granules and plasma fractions

Fraction	Lipid	Protein	Minerals
Egg yolk	63.5	32.4	2.1
Granules	*6.9*	*16.1*	*1.4*
Lipovitellins (HDL)[b]	3.5	12.3	
Phosvitin		4.6	
LDL[c]	2.5	0.3	
Plasma	*59.3*	*13.9*	*1.5*
Livetins		10.6	
LDL[c]	59.4	6.6	

[a] Data are expressed in percentage of egg yolk dry matter.
[b] High density lipoprotein.
[c] Low density lipoprotein.

white, increases only slightly (to 6.4–6.9) even after prolonged storage.

11.2.4.1 Proteins of Granules

11.2.4.1.1 Lipovitellins

The lipovitellin fraction represents high density lipoproteins (HDL). Its lipid moiety is 22% of dry matter and consists of 35% triglycerides, approx. 60% phospholipids and close to 5% cholesterol and cholesterol esters (cf. Table 3.14). The lipovitellins can be separated by electrophoretic and chromatographic methods into their α- and β-components, which differ in their protein-bound phosphorus content (0.50 and 0.27% P, respectively). The protein components probably also contain carbohydrates. At a pH < 7 lipovitellins occur as dimers with a molecular weight of 400 kdal. The amino acid composition is provided in Table 11.9. The protein moiety of this lipoprotein fraction has not been well elucidated. It can be assumed that protein subunits of 30 kdal are present. Lipovitellins occur in the yolk as a complex with phosvitin.

11.2.4.1.2 Phosvitin

Phosvitin is a glycophosphoprotein with an exceptionally high amount of phosphoric acid bound to serine residues. It consists of 2 components, α- and β-phosvitin. Its amino acid composition is given in Table 11.9. The molecular weight, based on sedimentation-diffusion measurements, is 36 kdal. The partial specific volume of 0.545 ml/g is very low, probably due to the large repulsive charges of the molecule. The frictional ratio suggests the presence of a long, mostly stretched molecular form.

A partial review of its amino acid sequence shows that sequences of 6–8 phosphoserine residues, interrupted by basic and other amino acid residues, are typical for this protein:

$$
\begin{aligned}
&\cdots\text{Asp}-\text{Pse}-\text{Pse}-\text{Pse}-\text{Pse}-\\
&\qquad-\text{Pse}-\text{Pse}-\text{Arg}-\text{Asp}\cdots\\
&\cdots\text{His}-\text{His}-\text{Arg}-\text{Pse}-\text{Pse}-\\
&\qquad-\text{Pse}-\text{Pse}-\text{Pse}-\\
&\qquad-\text{Pse}-\text{Arg}-\text{His}-\text{Lys}\cdots
\end{aligned}
\tag{11.3}
$$

The carbohydrate moiety is a branched oligosaccharide, consisting of mannose (3 residues), galactose (also 3 residues), N-acetylglucosamine (5) and N-acetylneuramic acid (2). The oligosaccharide is bound by an N-glycosidic linkage to asparagine. The amino acid sequence in the vicinity of the linkage position is:

$$
\begin{aligned}
&\cdots\text{Ser}-\underset{\substack{|\\ \text{Carbohydrate}}}{\text{Asn}}-\text{Ser}-\text{Gly}-(\text{Pse})_8-\\
&\qquad-\text{Arg}-\text{Ser}-\text{Val}-\\
&\qquad\qquad-\text{Ser}-\text{His}-\text{His}\cdots
\end{aligned}
\tag{11.4}
$$

There are indications that phosvitin contains a phosphothreonine residue and that 5–7 serine residues per mole are in free rather than esterified form.

Phosvitin efficiently binds metal ions. Intermolecular complexes are formed through cross linkages in the presence of Ca^{2+} and Mg^{2+}. The Fe^{3+} ion forms a monomeric complex and phosvitin is saturated with iron at a molar ratio of Fe/P = 0.5; this strongly suggests formation of a chelate complex involving two phosphate groups from the same peptide chain per iron. It can be assumed that metal complexing is one of the biological roles of phosvitins.

Table 11.9. Amino acid composition of phosvitin and α- and β-lipovitellins (mole %)

Amino acid	Phosvitin[a]	α-Lipovitellin	β-Lipovitellin
Gly	2.7	5.0	4.6
Ala	3.6	8.0	7.5
Val	1.3	6.2	6.6
Leu	1.3	9.2	9.0
Ile	0.9	5.6	6.2
Pro	1.3	5.5	5.5
Phe	0.9	3.2	3.3
Tyr	0.5	3.3	3.0
Trp	0.5	0.8	0.8
Ser	54.5	9.0	9.0
Thr	2.2	5.2	5.6
Cys	0.0	2.1	1.9
Met	0.5	2.6	2.6
Asx	6.2	9.6	9.3
Glx	5.8	11.4	11.6
His	4.9	2.2	2.0
Lys	7.6	5.7	5.9
Arg	5.3	5.4	5.6

[a] The phosphoric acid content amounts to 50–55 mole-%.

11.2.4.2 Plasma Proteins

11.2.4.2.1 Lipovitellenin

Lipovitellenin is obtained as a floating, low density lipoprotein (LDL) by ultracentrifugation of diluted yolk. Several components with varying densities can be separated by fractional centrifugation. The lipid moiety represents 84–90% of the dry matter and consists of 74% triglycerides and 26% phospholipids. The latter contain predominantly phosphatidyl choline (approx. 75%), phosphatidyl ethanolamine (approx. 18%) and lysophospholipids (approx. 8%). The molecular weight of lipovitellenin is several million dal. The individual components of this plasma protein are not well characterized.

11.2.4.2.2 Livetin

The water-soluble globular protein fraction can be separated electrophoretically into α-, β- and γ-livetins. These have been proven to correspond to chicken blood serum proteins, i.e. serum albumin, α_2-glycoprotein und γ-globulin.

Table 11.10. Egg yolk lipids

Lipid fraction	a	b
Triacylglycerols	66	
Phospholipids	28	
Phosphatidyl choline		73
Phosphatidyl ethanolamine		15.5
Lysophosphatidyl choline		5.8
Sphingomyelin		2.5
Lysophosphatidyl ethanolamine		2.1
Plasmalogen		0.9
Phosphatidyl inositol		0.6
Cholesterol, cholesterol esters and other compounds	6	

[a] As percent of total lipids.
[b] As percent of phospholipid fraction.

Table 11.11. Vitamin content of whole egg, egg white and yolk (mg/100 g edible portion)

Vitamin	Whole egg	Egg white	Egg yolk
A	0.22	0	1,12
Thiamine	0.11	Trace	0.29
Riboflavin	0.30	0.27	0.44
Niacin	0.1	0.1	0.1
B_6	0.12	Trace	0.3
Pantothenic acid	1.59	0.14	3.72
Biotin	0.025	0.007	
Folic acid	0.051	0.016	0.15
Tocopherols	1.0	0	3.0
α-Tocopherol	0.46		

11.2.4.3 Lipids

The lipid composition of egg yolk is given in Table 11.10. These lipids occur as the lipoproteins described above and, as such, are closely associated with the proteins occurring in yolk.

11.2.4.4 Other Constituents

11.2.4.4.1 Carbohydrates

Egg yolk carbohydrates are about 1% of the dry matter, with 0.2% bound to proteins. The free carbohydrates present in addition to glucose are the same monosaccharides identified in egg white (cf. 11.2.3.2.2).

11.2.4.4.2 Minerals

The minerals in egg yolk are listed in Table 11.7.

11.2.4.4.3 Vitamins

The vitamins in egg yolk are presented in Table 11.11.

11.3 Storage of Eggs

A series of changes occurs in eggs during storage. The diffusion of CO_2 through the pores of the shell, which starts soon after the egg is laid, causes a sharp rise in pH, especially in egg white. The gradual evaporation of water through the shell causes a decrease in density (initially approx. 1.086 g/cm^3; the daily reduction coefficient is about 0.0017 g/cm^3) and the air cell enlarges. The viscosity of the egg white drops. The yolk is compact and upright in a fresh egg, but it flattens during storage. After the egg is cracked and the contents are released onto a level surface, this flattening is expressed as yolk index, the ratio of yolk height to diameter. Furthermore, the vitellin membrane of the yolk becomes rigid and tears readily once the egg is opened. Of importance for egg processing is the fact that several properties change, such as egg white whipping behavior and foam stability. In addition, a "stale" flavor develops.

These changes are used for determination of the age of an egg, e.g. floating test (change in egg density), flash candling (egg yolk form and position), egg white viscosity test, measurement of air cell size, refractive index, and sensory assay of the "stale" flavor (performed mostly with soft-boiled eggs). The quality loss during storage of eggs is lower as the storage temperature is lower, as are the losses of CO_2 and water. Therefore, cold storage is an important part of egg preservation. A temperature of 0 to $-1.5\,°C$ (common chilled storage or subcooling at $-1.5\,°C$) and a relative humidity of 85–90% are generally used. A coating (oiling) of the shell surface with light paraffin-base mineral oil quite efficiently retards CO_2 and vapor escape, but a tangible benefit is derived only if oil is applied soon (1 h) after laying, since at this time the CO_2 loss is the highest. Controlled atmosphere storage of eggs (air or nitrogen with up to 45% CO_2) has been shown to be a beneficial form of egg preservation. Cold storage preserves eggs for 6–9 months, with a particularly increased shelf life with subcooled storage at $-1.5\,°C$. Egg weight loss is 3.0–6.5% during storage.

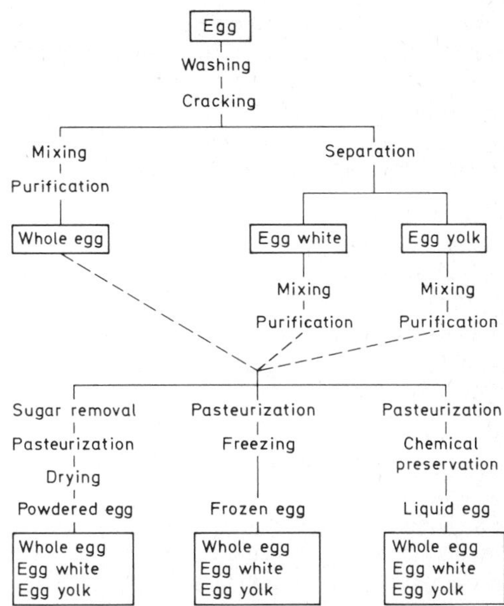

Fig. 11.4. Egg product manufacturing

11.4 Egg Products

11.4.1 General Outline

Egg products, in liquid, frozen or dried forms, are made from whole eggs, white or yolk. They are utilized further as semi-end products in the manufacturing of baked goods, noodles, confectionery, pastry products, mayonnaise and other salad dressings, soup powders, margarine, meat products, ice creams and egg liqueurs. Figure 11.4 gives an overview of the main processing steps involved in manufacturing of egg products.

11.4.2 Technically-Important Properties

The many uses of egg products are basically a result of three properties of eggs: coagulation when heated; foaming ability (whippability); and emulsifying properties. The coloring ability and aroma of egg should also be mentioned.

11.4.2.1 Thermal Coagulation

Egg white begins to coagulate at 62 °C and egg yolk at 65 °C. The coagulation temperature is influenced by pH. At a pH at or above 11.9 egg white gels or sets even at room temperature, though after a while the gel liquifies. All egg proteins coagulate, except ovomucoid and phosvitin. Conalbumin is particularly sensitive, but can be stabilized by complexing it with metal ions. Due to their ability to coagulate, egg products are important food-binding agents.

11.4.2.2 Foaming Ability

Whipping of egg white builds a foam which entraps air and hence is used as a leavening agent in many food products (baked goods, angel cakes, biscuits, soufflés, etc.).

Due to a large surface area increase in the liquid/air interphase, proteins denature and aggregate during whipping. In particular, ovomucin forms a film of insoluble material between the liquid lamella and air bubble, thereby stabilizing the foam. Egg globulin also contributes to this effect by increasing the fluid viscosity and by decreasing the surface tension, both effects of importance in the initial stage of the whipping process. In angel cake, egg white without ovomucin and globulins leads to long whippability times and cakes with reduced volumes. An excessive ovomucin content degrades the elasticity of the ovomucin film and thus decreases the thermal stability (expansion of air bubbles) of the foam.

The whippability of egg white can be assayed by measurement of foam volume and foam stability (amount of liquid released from the foam in a given time).

11.4.2.3 Emulsifying Effect

The emulsifying effect of whole egg or egg yolk alone is utilized, for example, in the production of mayonnaise (creamy salad dressings, made by beating a mixture of egg yolk, olive oil, lemon juice or vinegar, and seasonings; cf. 14.4.6). Lipoproteins and proteins are responsible for the emulsifying action of eggs.

11.4.3 Dried Products

The liquid content ("mélange") of eggs is mixed or churned either immediately or only after egg white and yolk separation. This homogenization is followed by a purification step using centrifuges (separators), and then by a pasteurization step (Figure 11.4).

The sugars are removed prior to egg drying to prevent reaction between amino components (proteins, phosphatidyl colamines) and reducing sugars (glucose), thereby avoiding undesired brown discoloration and faulty aroma.

Sugars are removed from egg white after pasteurization (cf. 11.4.5), usually by microbiological sugar fermentation. The pasteurized egg liquid is adjusted to pH 7.0–7.5 using citric or lactic acid, and then is incubated at 30–33 °C with suitable microorganisms (*Streptococcus* spp.; *Aerobacter* spp.). The sugar in whole egg homogenate or yolk is removed in part by yeasts (e.g. *Saccharomyces cerevisiae*) or mainly by glucoseoxidase/catalase enzymes (cf. 2.8.2.1.1 and 2.8.2.1.2), which oxidize glucose to gluconic acid. Addition of hydrogen peroxide releases oxygen and accelerates the process.

Spray drying with a jet or centrifugal spray drier is the most important egg drying process. The dispersed pulp meets a current of warm air, which enters at 120–230 °C. This rapidly reduces the pulp moisture to 5% or less. Whole egg or egg yolk powders are then rapidly cooled. Other egg drying processes, e.g. freeze drying, are rarely applied commercially.

Dried instant powder can be made in the usual way: rewetting and additionally drying the agglomerated particles. Egg white agglomerization is facilitated by addition of sugar (sucrose or lactose).

Table 11.12. Composition of dried egg products (values in %)

Constituent	Whole egg	Egg white	Egg yolk
Moisture[a]	5.0	8.0	5.0
Fat[b]	40.0	traces	57.0
Protein[b]	45.0	80.0	30.0
Ash	3.7	5.7	3.4
Reducing sugars[a]	0.1	0.1	0.1

[a] Maximum values.
[b] Minimum values.

The shelf life of dried egg white is essentially unlimited. Whole egg powder devoid of sugar has a shelf life of approx. 1 year at room temperature, while sugarless yolk lasts 8 months at 20–24 °C and more than a year in cold storage. The shelf-life of powders containing egg yolk is limited by aroma defects which develop gradually from oxidation of yolk fat. The compositions of dried egg products are given in Table 11.12.

11.4.4 Frozen Egg Products

The eggs are pretreated as described above (cf. 11.4.3 and Figure 11.4). The homogenate is pasteurized at 63 °C for 1 min (cf. 11.4.5) to lower the microflora count and is then frozen between −23 and −25 °C. The shelf-life of the frozen eggs is 8–10 months at a storage temperature of −15 to −18 °C.

Frozen egg white thickens negligibly after thawing, while the viscosity of egg yolk rises irreversibly when freezing and storage temperatures are below −6 °C (Figure 11.5). The egg yolk has a gel-like consistency after thawing, which hampers further utilization by dosage metering or mixing. Thawed whole egg gels can cause similar problems, but to a lesser extent than yolk.

Pretreatment of yolk with proteolytic enzymes, such as papain, and with phospholipase A prevents gel formation. Mechanical treatments after thawing of yolk can result in a drop in viscosity. Gel formation can also be prevented by adding 2–10% common salt or 8–10% sucrose to egg yolk. Although salted and sugar-sweetened yolk is of limited acceptability to some manufacturers, this process is of great importance (Figure 11.6).

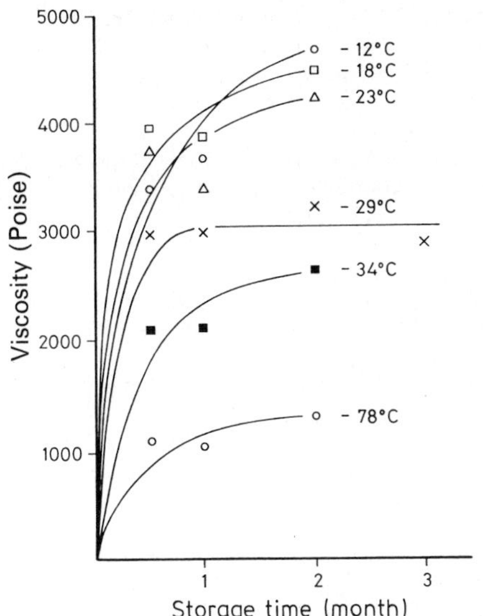

Fig. 11.5. Egg yolk viscosity after frozen storage. (After *Palmer* et al., 1970)

The consistency of the frozen egg products is influenced by the temperature gradients during freezing and thawing, and also by storage duration and temperature. Rapid freezing and thawing are best.

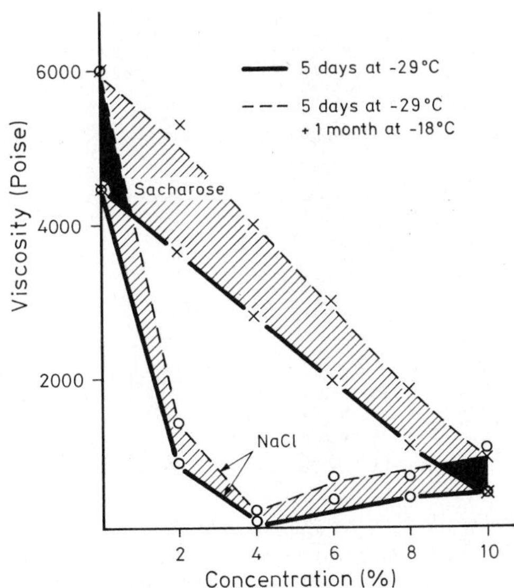

Fig. 11.6. Egg yolk viscosity after addition of NaCl or saccharose and after frozen storage. (After *Palmer* et al., 1970)

Table 11.13. Composition of frozen and liquid egg products (values in %)

Constituent	Whole egg	Egg white	Egg yolk
Moisture	75.3	88.0	57.0
Fat	11	<0.03[a]	27.2
Protein	12	10.5	13.5
Reducing sugars	0.7	0.8	0.7

[a] As percent of egg yolk weight.

The molecular events leading to gel formation by freezing are poorly understood. Apparently, the formation of ice crystals causes a partial dehydration of protein, coupled with a rearrangement of lipoprotein. This probably induces formation of entangled protein strands.

The whippability of egg white can be enhanced by various additives, such as glycerol, starch syrup and triethyl citrate. Typical compositional data for frozen egg products are provided in Table 11.13.

11.4.5 Liquid Egg Products

Eggs are pretreated as described earlier (cf. 11.4.3 and Figure 11.4). Despite sanitary conditions at plants, eggs cannot be entirely protected from microorganisms. Pasteurization is difficult due to the heat sensitivity of egg protein and the need to kill the pathogens under specific conditions. It is especially important to eliminate *Salmonella* spp., which have varying resistances to heat. The most resistant are *S. senftenberg, S. oranienburg* and *S. paratyphi B.* Inactivation of α-amylase occurs as the temperature lethal to *S. senftenberg* is approached; hence, this enzyme can be used as an indicator to monitor the sufficient heating of eggs. The heating conditions differ for different liquid egg products.

A generally adequate heating regimen for whole egg and egg yolk is 64–65 °C for 2.5–3 min, though milder or more drastic conditions are suggested. The conditions for egg white can be milder, since the heat resistance of *Salmonella* spp. is lower at higher pH's. However, at a higher pH protein coagulation occurs at a lower temperature. Addition of salt or sucrose to liquid eggs increases the resistance of microflora to heat.

Most of the egg white proteins are relatively stable at pH 7, so normal pasteurization conditions

do not negatively affect processing properties such as whippabililty. An exception is conalbumin, but addition of metal ions (e.g. Al-lactate) can stabilize even this protein. Addition of Na-hexametaphosphate can also improve the stability of conalbumin.
Pasteurized liquid egg products are generally also preserved by chemical means, e.g. addition of sorbic or benzoic acid.
The compositions of liquid egg products are presented in Table 11.13.

11.5 Literature

Chang, P.K., Powrie, W.D., Fennema, O.: Effect of heat treatment on viscosity of yolk. J. Food Sci. *35,* 864 (1970)

Green, N.M.: Avidin. Adv. Protein Chem. *29,* 85 (1975)

Janssen, H.J.L.: Neuere Entwicklungen bei der Fabrikation von Eiprodukten. Alimenta *10,* 121 (1971)

Maga, J.A.: Egg and egg product flavour. J. Agric. Food Chem. *30,* 9 (1982)

Palmer, H.H., Ijichi, K., Roff, H.: Partial thermal reversal of gelation in thawed egg yolk products. J. Food Sci. *35,* 403 (1970)

Spiro, R.G.: Glycoproteins. Adv. Protein Chem. *27,* 349 (1973)

Stadelmann, W.J., Cotterill, O.J. (Eds.): Egg science and technology. 2nd edn., AVI Publ. Co.: Westport, Conn. 1977

Taborsky, G.: Phosphoproteins. Adv. Protein Chem. *28,* 1 (1974)

Tunmann, P., Silberzahn, H.: Über die Kohlenhydrate im Hühnerei. I. Freie Kohlenhydrate. Z. Lebensm. Unters. Forsch. *115,* 121 (1961)

12 Meat

12.1 Foreword

Much evidence from many civilizations has verified that the meat of wild and domesticated animals has played a significant role in human nutrition since ancient times. In addition to the skeletal muscle of warm-blooded animals, which in a strict sense is "meat", other parts are also used: fat tissue, some internal organs and blood. Definitions of the term "meat" can vary greatly, corresponding to the intended purpose. From the aspect of food legislation for instance, the term meat includes all the parts of warm-blooded animals, in fresh or processed form, which are suitable for human consumption. In the colloquial language the term meat means skeletal muscle tissue containing more-or-less adhering fat. Some data concerning meat production and consumption are compiled in Tables 12.1–12.3.

12.2 Structure of Muscle Tissue

12.2.1 Skeletal Muscle

Skeletal muscle tissue consists of long, thin, parallel cells arranged into fiber bundles. Each of these muscle fibers exists as a separate entity surrounded by connective tissue, the endomysium. Numbers of these primary muscle fibers are held together in a bundle which is surrounded by a larger sheet of thin connective tissue, the perimysium. Many such primary bundles are then held together and wrapped by an outer, large, thick layer of connective tissue called the epimysium. Figure 12.1 shows a cross-section of rabbit *Psoas major* muscle in which the endomysium and perimysium are readily recognized.

The membrane surrounding each individual muscle fiber is called the sarcolemma. It consists of three layers: the endomysium, a middle amorphous layer and an inner plasma membrane. The individual myofibrils, the contractile units of the muscle fiber, are within the muscle fiber and are surrounded and imbedded in a homogeneous matrix, the sarcoplasm, as are other subcellular particles, such as nuclei, mitochondria and the sarcoplasmic reticulum.

White muscle (birds, poultry), which has a high ratio of myofibrils to sarcoplasm, contracts rapidly but tires quickly. It can be distinguished from red muscle, which is poor in myofibrils but rich in sarcoplasm. Red muscles are used for slow, long-lasting contractions and do not tire quickly. Figure 12.2 shows a cross-section of a muscle fiber with numerous myofibrils.

The organization of the muscle contractile apparatus is revealed in a longitudinal section of the muscle fiber. The characteristic cross-bondings ("striations"; Figure 12.3) of skeletal muscle are due to the regularly-overlapping anisotropic A bands, which double refract polarized light, and the isotropic I bands. The dark bands, the Z line, are in the middle of the light I bands and perpendicular to the axis of the fiber. The dark A bands are crossed in the middle with light H bands, while the dark M line is situated in the middle of the H bands (Figure 12.4). A single contractile unit of a myofibril, called the sarcomere, stretches from one Z line to the next and consists of thick and thin filaments. The thick filaments are formed from the protein myosin. They stretch through the entire A band and are kept in the correct position by the bulge at the center (M line). Thin filaments consist mainly of actin. They originate from the Z line and pass across the I band and between the thick filaments to the edge of the H zone, where they penetrate the A bands. During muscle contraction, the mechanism of which is explained in section 12.3.2.1.4, the thick filaments penetrate into the H zones and the Z lines move closer to each other. Thus, the width of the I band gradually decreases and finally disappears. Figure 12.5 schematically presents these changes which take place during muscle contraction.

Table 12.1. World meat production in 1981 (1,000 t)[a]

Continent	Beef/Veal	Buffalo	Mutton/Lamb	Goat	Pork	Horse	Poultry	Meat, grand total
World	45,548	1,304	5,984	2,049	55,195	569	28,696	142,303
Africa	2,870	121	731	561	374	60	1,241	6,740
America, North-, Central-	12,546		181	30	8,754	157	8,400	30,328
America, South-	7,124		262	65	1,762	87	2,526	11,942
Asia	3,952	1,167	1,679	1,258	20,102	115	6,514	35,373
Europe, West	8,174		882	85	12,216	122	5,337	27,601
Europe, East-, + USSR	8,889	16	1,055	41	11,684	28	4,335	26,440
Australia + South Pacific Islands	1,993		1,193	2	303	1	343	3,879

Country	Beef/Veal	Country	Buffalo	Country	Mutton/Lamb
USA	10,330	China	513	USSR	816
USSR	6,690	Pakistan	181	New Zealand	605
Argentina	3,000	India	130	Australia	588
Brazil	2,250	Egypt	121	China	405
France	1,835	Thailand	73	Turkey	302
China	1,691	Vietnam	65	UK	256
FR Germany	1,520	Philippines	44	Iran	240
Australia	1,481	Indonesia	36	France	174
Italy	1,125	Turkey	25	USA	153
UK	1,035	Laos	22	Pakistan	146
Canada	1,020			Sudan	133
Mexico	601	Σ (%)[b]	93	South Africa	133
Colombia	576			Spain	129
South Africa	510			India	125
Poland	500			Argentina	116
				Afghanistan	112
Σ (%)[b]	75			Mongolia	88
				Σ (%)[b]	76

Country	Goat	Country	Pork (swine)	Country	Horse
China	355	China	16,584	USA	79
India	280	USA	7,220	China	62
Pakistan	194	USSR	5,250	Mexico	54
Nigeria	130	FR Germany	2,700	Italy	44
Turkey	125	France	1,890	Ethiopia	42
Ethiopia	55	Japan	1,396	Argentina	40
Somalia	54	Poland	1,350	Mongolia	38
Greece	47	German DR	1,320	Brazil	37
Bangladesh	46	Holland	1,195	France	30
Iran	45	Italy	1,105	Poland	18
Sudan	45	Spain	1,030		
USSR	41	Denmark	995	Σ (%)[b]	78
Indonesia	40				
Yemen	38	Σ (%)[b]	76		
Niger	30				
South Africa	29				
Σ (%)[b]	76				

[a] Data refer to slaughtered animals irrespective of the possibility of being imported as live animals.
[b] World production ≙ 100%.

Table 12.1 (continued)

Country	Poultry	Country	Meat, grand total
USA	7,000	USA	25,031
China	3,206	China	23,112
USSR	2,300	USSR	15,367
Brazil	1,416	France	5,506
France	1,240	Brazil	4,734
Japan	1,135	FR Germany	4,701
Italy	1,020	Argentina	3,822
Spain	820	Italy	3,554
UK	745	Japan	3,022
Mexico	524	UK	2,989
Canada	516	Australia	2,628
Poland	454	Spain	2,538
Rumania	428	Canada	2,418
FR Germany	400	Poland	2,357
Hungary	378	Holland	2,020
		German DR	1,919
Σ (%)[b]	75	Rumania	1,824
		Σ (%)[b]	76

[a] Data refer to slaughtered animals irrespective of the possibility of being imported as live animals.
[b] World production ≙ 100%.

Table 12.2 Annual meat consumption in FR Germany (kg/person)

Year	Beef/Veal	Pork	Poultry	Total
1960	18.7	29.3	4.2	59.0
1964/65	19.2	33.9	6.0	66.5
1970	22.9	36.1	7.4	72.0
1972/73	20.5	42.0	9.0	79.0
1974/75	21.0	44.6	8.8	82.5
1976/77	21.6	45.5	9.1	84.9

Table 12.3. Meat consumption in selected countries (kg/person/year)

Country/ region	Year	Beef/ Veal	Pork	Poultry	Total
USA	1960	41.5	29.5	17.2	95.3
	1970	48.6	30.6	18.9	99.9
EEC (European Economic	1960	19.9	19.2	5.2	52.2
Community)	1970	25.2	23.7	8.9	65.7
France	1960	29.2	19.8	8.6	74.9
	1970	35.9	24.8	11.3	89.3
Italy	1960	12.9	7.2	3.6	30.0
	1970	18.9	9.1	9.2	43.5

12.2.2 Heart Muscle

The structure of heart muscle is similar to striated skeletal muscle but has significantly more mitochondria and sarcoplasm.

12.2.3 Smooth Muscle

The smooth muscle cells are distinguished by their centrally-located cell nuclei and optically uniform myofibrils which do not have cross-striations. Smooth muscles occur in mucous linings, the spleen, lymphatic glands, epidermis and intestinal tract. Smooth muscle fibers are useful in the examination of meat products; preferentially for the detection of pharynx (esophagus), stomach or calf pluck (heart, liver and lungs).

12.3 Muscle Tissue: Composition and Function

12.3.1 Overview

Muscles freed from adhering fat contain on the average 76% moisture, 21.5% N-substances, 1.5% fat and 1% minerals. In addition, variable amounts of carbohydrates (0.05–0.2%) are pre-

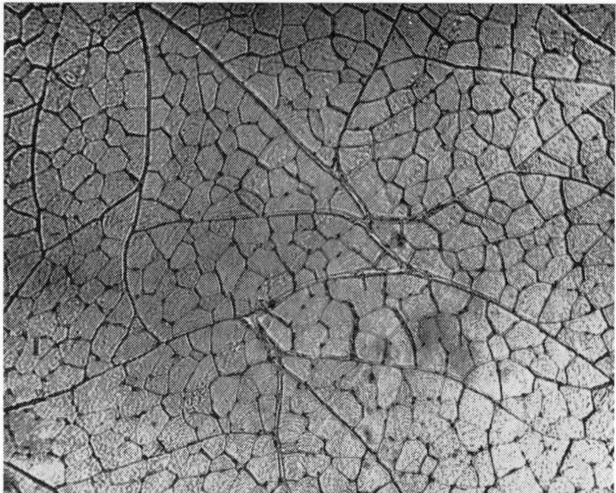

Fig. 12.1. A cross-section of *M. psoas* rabbit muscle. (From *Schultz, Anglemier,* 1964)

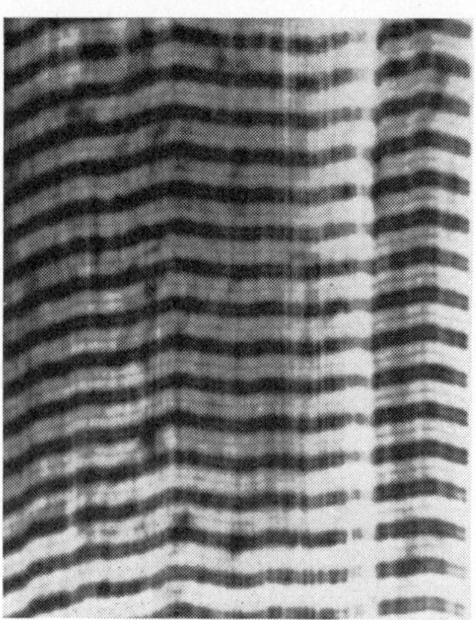

Fig. 12.3. A longitudinal section of two adjacent muscle fibers. (From *Schultz* and *Anglemier,* 1964)

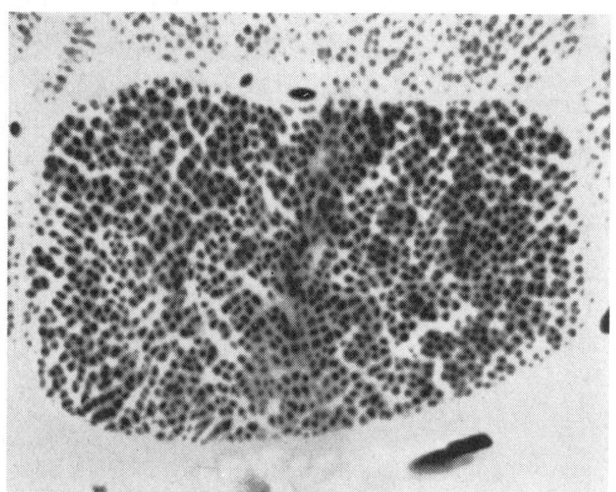

Fig. 12.2. A cross-section of a muscle fiber. (From *Schultz, Anglemier,* 1964)

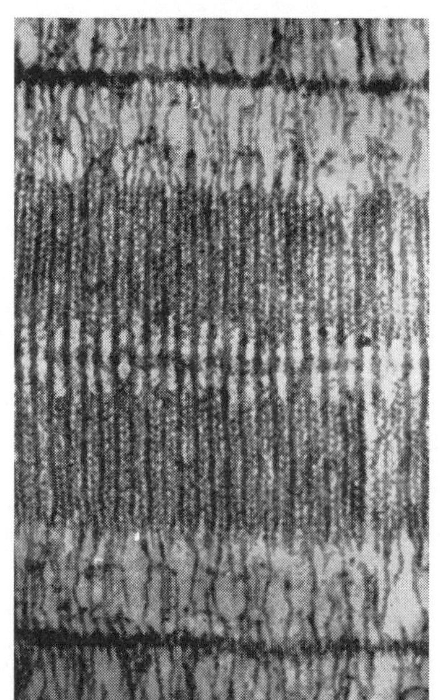

Fig. 12.4. A longitudinal section of a sarcomere. (From *Schultz, Anglemier,* 1964)

sent. Table 12.4 provides data for the average composition of some cuts of beef, pork and chicken.

12.3.2 Proteins

Muscle proteins can be divided into three large groups:

- Proteins of the contractile apparatus, extractable with concentrated salt solutions (actomyosin, together with tropomyosin and troponin).
- Proteins soluble in water or dilute salt solutions (myoglobin and enzymes).
- Insoluble proteins (connective tissue and membrane proteins).

12.3.2.1 Proteins of the Contractile Apparatus and Their Functions

12.3.2.1.1 Myosin

Myosin molecules form the thick filaments and make up 50–60% of the total proteins present in the contractile apparatus. Myosin can be isolated from muscle tissue with a high ionic strength buffer, for example, 0.3 M KCl/0.15 M phosphate buffer, pH 6.5. The molecular weight of myosin is approx. 500 kdal. Myosin consists of

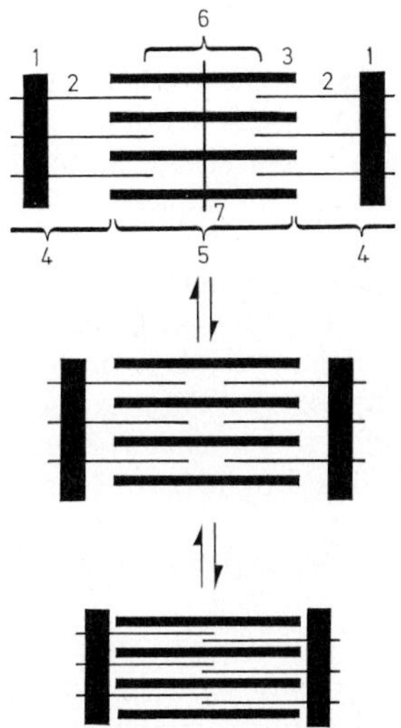

Fig. 12.5. A schematic representation of a sarcomere in a relaxed and contracted state. 1 Z-line, 2 thin filament, 3 thick filament, 4 I-band, 5 A-band, 6 H-zone, and 7 M-line

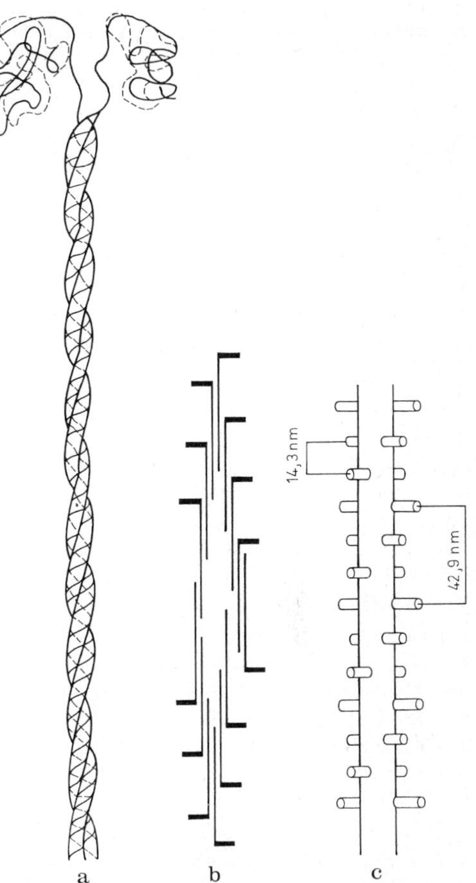

Fig. 12.6. Schematic representations of (**a**) a myosin molecule (after *Lehninger,* 1975), (**b**) arrangement of myosin molecules in a thick filament (after *Huxley,* 1963), and (**c**) a thick filament (after *Lehninger,* 1975)

Table 12.4. Average composition of meat (%)

Meat	Cut	Moisture	Protein	Fat	Ash
Pork	Boston butt				
	(M. subscapularis)	74.9	19.5	4.7	1.1
	Loin				
	(M. psoas maior)	75.3	21.1	2.4	1.2
	Cutlets, chops[a]	54.5	15.2	29.4	0.8
	Ham	75	20.2	3.6	1.1
	Side cuts	40	11.2	48.2	0.6
Beef	Shank	76.4	21.8	0.7	1.2
	Sirloin steak[a]	74.6	22.0	2.2	1.2
Chicken	Hind leg (thigh + drum stick)	73.3	20.0	5.5	1.2
	Breast	74.4	23.3	1.2	1.1

[a] With adhering adipose tissue.

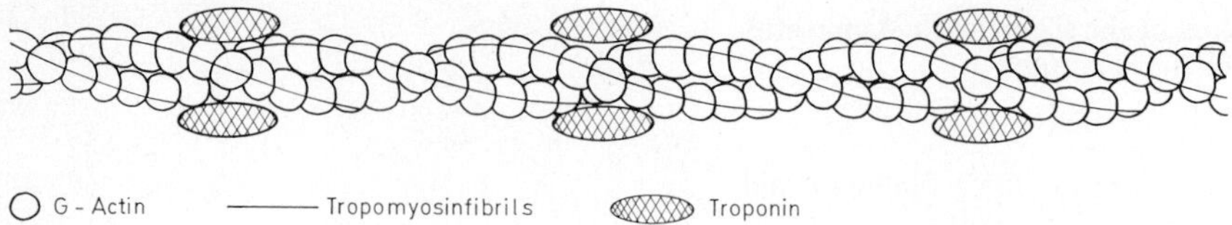

Fig. 12.7. A schematic representation of a thin filament. (After *Karlsson*, 1977)

two very long, identical peptide chains (2 × 140 nm). The two peptide chains form a long, double-stranded α-helical rod with a double head of globular protein, both heads being joined at the same end of the coil (head dimensions, 5 × 20 nm). The myosin ATPase activity is localized in the heads and is required for the interaction of the heads with actin, the protein constituent of the thin filaments (Figure 12.6,a). Myosin is cleaved by trypsin into two fragments: light (LMM) and heavy meromyosin (HMM). The HMM fraction contains the globular-headed region and has the ATPase activity and the ability to react with actin.

The individual myosin molecules in the thick filaments are arranged as presented schematically in Figure 12.6 (b, c). By bringing the tails together, a major cord is formed and on its surface the heads are spirally located. The distance between two adjacent heads on such a spiral is 13.3 nm, and that between the two repeating heads in the same row or line is 42.9 nm. Each filament contains approx. 400 myosin molecules. Their association is reversible under certain conditions.

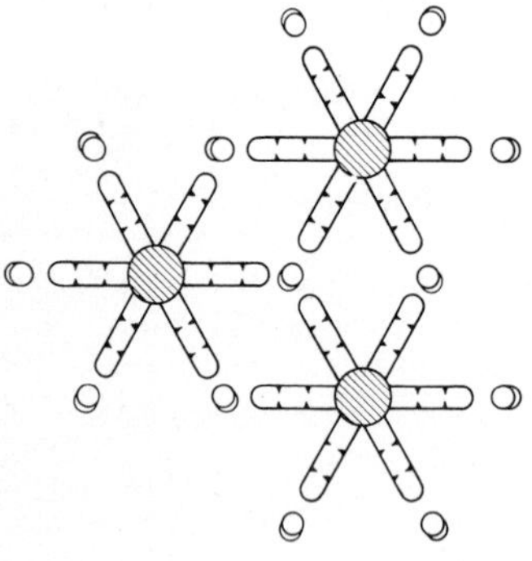

Fig. 12.8. Arrangement of thick and thin filaments in myofibrils (a schematic cross-section). (After *Schultz* and *Anglemier*, 1964)

12.3.2.1.2 Actin

Actin is the main constituent of the thin filament. It makes up 15–30% of the total protein of the contractile apparatus. It is substantially less soluble than myosin, probably because it is fixed to substances in the Z line. Actin can be isolated, for example, by extraction of pulverized, acetone-dried muscle tissue with an aqueous ATP solution.

The actin monomer has a globular shape and hence is designated as G-actin. It has a molecular weight of approx. 46 kdal and binds to myosin. In the presence of salts or of ATP and Mg^{2+} (ATP hydrolyzes to ADP which remains bound to actin), actin polymerizes into its fibrous form, F-actin.

F-actin in the thin filaments is in the form of a double-stranded helix in which the G-actin beads are stabilized by tropomyosin fibrils (cf. 12.3.2.1.3), as shown in Figure 12.7. Six F-actin strands surround a thick filament; consequently, each F-actin strand adheres to the heads of three thick filaments (Figure 12.8).

12.3.2.1.3 Tropomyosin and troponin

Tropomyosin is a highly elongated molecule (2 × 45 nm) with a molecular weight of about 70 kdal, and is assumed to be a double-stranded α-helix. The monomer readily forms polymeric fibrils which are bound to F-actin on the thin filament.

Troponin is a complex of three proteins: Tn-C binds Ca^{2+} ions, Tn-I binds to actin and Tn-T binds to tropomyosin. The troponin complex has a molecular weight of about 79 kdal and is located on the thin filaments at intervals set by the length of tropomyosin.

12.3.2.1.4 Contraction and relaxation

Muscle stimulation by a nerve impulse triggers depolarization of the outer membrane of the muscle cell and thus release of Ca^{2+} ions from the

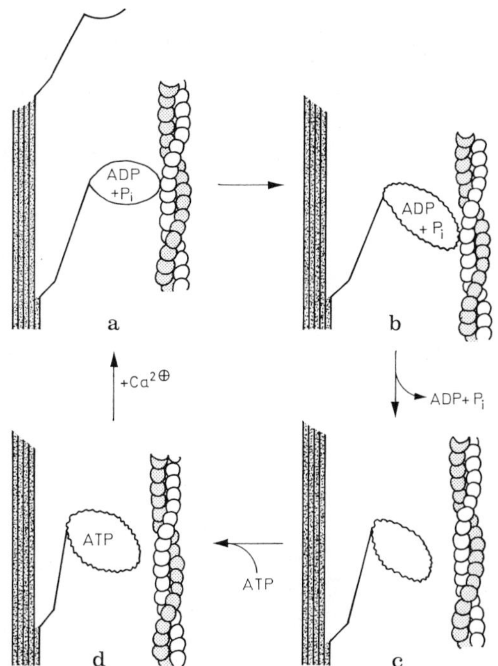

Fig. 12.9. Molecular processes involved in muscle contraction (see text; after *Karlsson,* 1977)

sarcoplasmic reticulum. The Ca^{2+} concentration in the sarcoplasm of the resting muscle increases quickly from 10^{-7} to 10^{-5} mole/l. The binding of this Ca^{2+} to the Tn-C component of the troponin complex causes a conformational change of this protein. As a consequence, displacement of the tropomyosin fibrils occurs along the F-actin filament. Thus, the sterically hindered sites on the actin units are exposed for interaction with the myosin heads. The energy required for the shifting of the unbound myosin heads is obtained from the hydrolysis of ATP. The hydrolysis products of ATP splitting, ADP and inorganic phosphate (P_1), remain on the myosin heads, which then bind to the actin monomers (Figure 12.9,a). Consequently, the myosin heads, now bound to actin, are forced to undergo a conformational change, which forces the thin filament to move relative to the thick filament (Figure 12.9,b).

The polarity of the thin filaments and the heads of the thick filaments reverses half way between the Z lines. Therefore, the two thin filaments which interact with one thick filament are drawn toward each other, resulting in a shortening of the distance between the Z lines.

When the myosin heads release ADP and P_1 and become detached from the thin filaments (Figure 12.9,c), the heads are ready to take up a fresh charge of ATP (Figure 12.9,d). If the Ca^{2+} concentration in the sarcoplasm remains high, the ATP will again hydrolyze and the interaction of the myosin heads with the thin filament is repeated (Figure 12.9,a). However, if the Ca^{2+} concentration drops in the meantime, no ATP hydrolysis occurs, tropomyosin again blocks the access of myosin heads to the actin binding sites and the muscle returns to its resting state.

The decrease in Ca^{2+} concentration when muscle excitation has ceased, as well as the increase in Ca^{2+} during stimulation, i.e. the flow of calcium ions, is controlled by the sarcoplasmic reticulum. The Ca^{2+} concentration is low in the sarcoplasm of the resting muscle, while it is high within the sarcoplasmic reticulum. When the ATP level is low, detachment of the myosin and actin filaments does not occur. The muscle remains in a stiff, contracted state called rigor mortis (cf. 12.4). Hence, relaxation of muscle depends on the presence of regenerated ATP.

12.3.2.1.5 Actomyosin

Solutions of F-actin and myosin at high ionic strength ($\mu = 0.6$) *in vitro* form a complex called actomyosin.The formation of the complex is reflected by an increase in viscosity and occurs in a definite molar ratio: 1 molecule of myosin per 2 molecules of G-actin, the basic unit of the double-helical F-actin strand. It appears that a spike-like structure is formed, which consists of myosin molecules embedded in a "backbone" made of the F-actin double helix. Addition of ATP to actomyosin causes a sudden drop in viscosity due to dissociation of the complex. When this addition of ATP is followed by addition of Ca^{2+}, the myosin ATPase is activated, ATP is hydrolyzed and the actomyosin complex again restored after the ATP concentration decreases.

Upon spinning of the actomyosin solution into water, fibers are obtained which, analogous to muscle fibers, contract in the presence of ATP. Glycerol extraction of muscle fibers removes all the soluble components and abolishes the semipermeability of the membrane. Such a model muscle system shows all the reactions of *in vivo* muscle contraction after the readdition of ATP and Ca^{2+}.

This and similar model studies demonstrate that the muscle contraction mechanism is understood in principle, although some molecular details are still not clarified.

12.3.2.2 Soluble Proteins

Soluble proteins make up 20–30% of the total protein in muscle tissue. They consist mostly of enzymes and myoglobin. The high viscosity of the sarcoplasm is derived from a high concentration of solubilized proteins.

12.3.2.2.1 Enzymes

Sarcoplasm contains most of the enzymes needed to support the glycolytic pathway and the pentosephosphate cycle. Glyceraldehyde-3-phosphate dehydrogenase makes up 10% of the total soluble protein. A series of enzymes involved in ATP metabolism, e.g. creatine phosphokinase and ADP-desaminase (cf. 12.3.6 and 12.3.8) are also present.

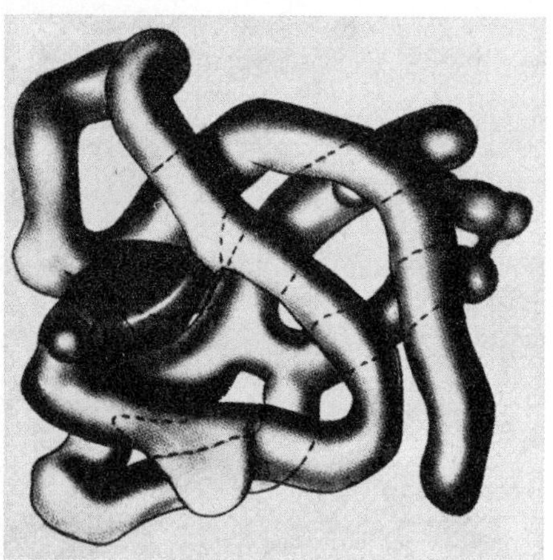

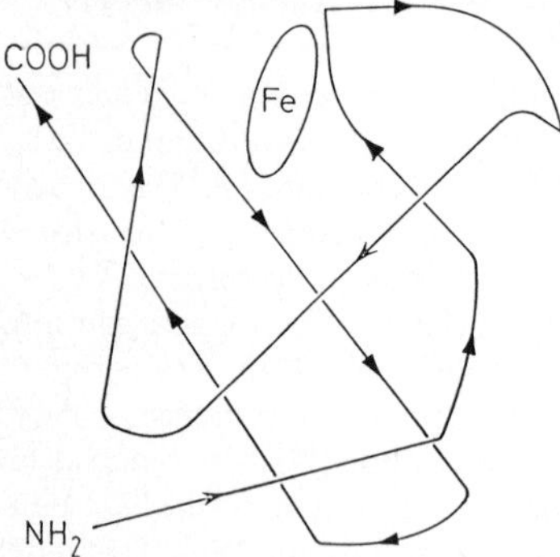

Fig. 12.10. Molecular model of myoglobin (**a**) and a schematic representation of peptide chain course (**b**). (From *Schormueller,* 1965)

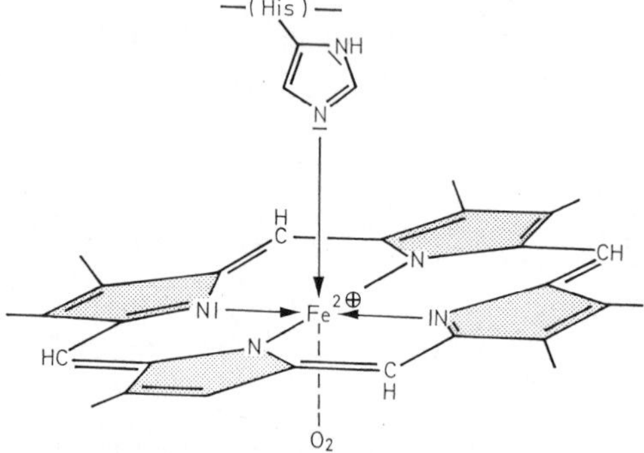

Fig. 12.11. Fe^{2+}-protoporphyrin, formula and its binding to the peptide chain. (After *Karlsson,* 1977)

12.3.2.2.2 Myoglobin

Muscle tissue dry matter contains an average of 2.5% of the purple-red pigment myoglobin. However, the amount varies considerably between white and red meat.

Myoglobin consists of a peptide chain (globin) of molecular weight of 16.8 kdal. It has known primary and tertiary structures (Figure 12.10). The pigment component is present in a hydrophobic pocket of the globin and is bound to a histidyl (His^{93}) residue of the protein. The pigment, heme, is the same as that in hemoglobin (blood pigment), i.e. Fe^{2+}-protoporphyrin (Figure 12.11).

Myoglobin supplies oxygen because of its ability to bind oxygen reversibly. Comparison of the oxygen binding curves for hemoglobin and myoglobin (Figure 12.12) shows that at low p_{O_2}, such as exists in muscle, hemoglobin releases oxygen

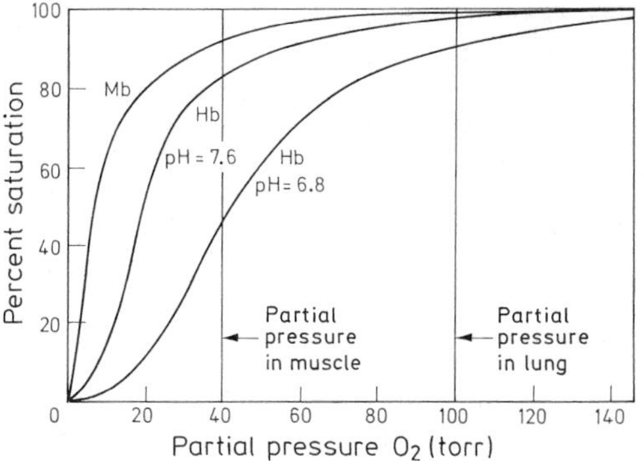

Fig. 12.12. Oxygen binding curves of myoglobin and hemoglobin

to myoglobin. The sigmoidal shape of the O_2-binding curve for hemoglobin is due to its quaternary structure. It consists of four polypeptide chains, with one pigment molecule bound to each. The binding of O_2 on the four pigment molecules occurs cooperatively because of allosteric effects. Therefore, the degree of saturation, S, is expressed by the following equation (p_{O_2} = oxygen partial pressure; k = dissociation constant for the O_2-protein complex):

$$S = \frac{k \cdot p^n_{O_2}}{1 + k \cdot p^n_{O_2}} \quad (12.1)$$

For hemoglobin, n ~ 2.8 (sigmoidal saturation curve), and for myoglobin, n = 1 (hyperbolic saturation curve). The efficiency of O_2 transfer from hemoglobin to myoglobin is further enhanced by a decrease in pH since oxygen binding is pH-dependent (the *Bohr* effect).

While in the living animal approx. 10% of the total iron is bound to myoglobin, 95% of all the iron in well-bled beef muscle is bound to myoglobin. Unlike myoglobin, hemoglobin contributes little to the color of meat, while the contribution of other pigments, such as the cytochromes, is negligible.

However, attention must be paid to the fact that the visual appearance of a cut of meat is influenced not only by the light absorption of pigments, i.e. primarily myoglobin, but also by light scattering by the surface matrix of muscle fiber. A bright red color is obtained when the coefficient of absorption is high and that of light scattering is low.

Myoglobin (Mb) is purple (λ_{max} = 555 nm); oxymyoglobin (MbO_2), a covalent complex of ferrous Mb and O_2, is bright red (λ_{max} = 542 and 580 nm); and the oxidation product of Mb to the ferric state, metmyoglobin (MMb^+), is brown (λ_{max} = 505 and 635 nm). Some other ligands, such as electron pair donors (e.g. CO, NO, N_3^-, CN^-), like O_2, bind covalently, giving rise to low-spin complexes with similar absorption spectra and hence to a color similar to MbO_2. Figure 12.13 shows several absorption spectra of myoglobins.

Heme devoid of globin (free heme, Fe^{2+}-protoporphyrin) does not form the O_2-adduct, but oxidizes rapidly to hemin (Fe^{3+}-protoporphyrin). A prerequisite for reversible O_2 binding is the presence of an effective donor ligand on the iron's axial site, which is bound under formation of a quadratic-pyramidal complex. The imidazole side chain of the His^{93} of the myoglobin has this function. Upon interaction with this fifth ligand, iron is raised above the heme plane by about 0.05 nm:

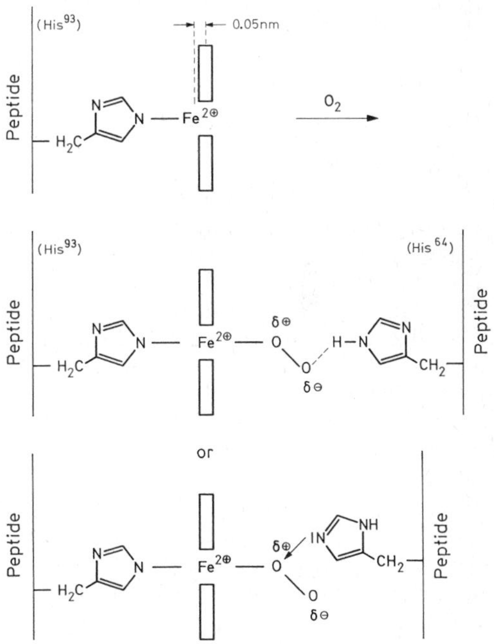

(12.2)

Binding of the sixth ligand moves the iron to its original position in the heme plane. Since the Fe–N bond distance (His^{93}) remains constant, dislocation of the fifth ligand occurs (His^{93},

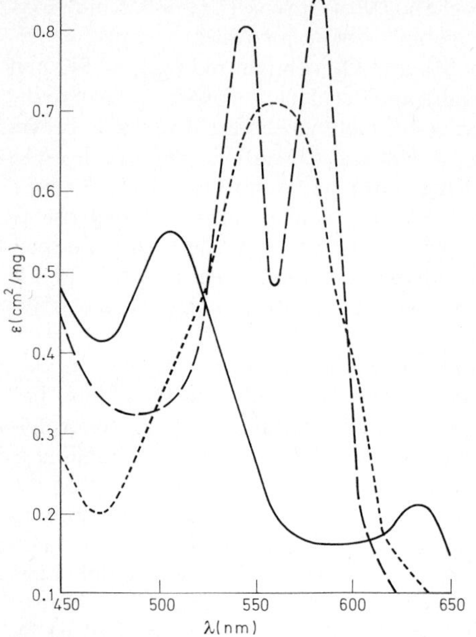

Fig. 12.13. Absorption spectra of myoglobin (----), oxymyoglobin (– – – –) and metmyoglobin (——). (After *Fennema*, 1976)

proximal His), i.e. a conformational change of the globin takes place.

The basicity of the fifth ligand affects the binding of the sixth ligand. The imidazole ring of His^{93} is a good π-donor and, hence, stabilizes the O_2-adduct. A weaker base would enhance oxidation of the iron rather than adduct formation, while a stronger base would increase the stability of the adduct and diminish the probability of iron oxidation. From a biochemical viewpoint, the latter effect is rated as (O_2 supplier) negative; while from a food science point of view, it is desirable and positive (stable, bright red meat color).

As mentioned above, His^{93} is located in a hydrophobic pocket of the myoglobin molecule. The electron density and, therefore, the oxidation state of the iron, is regulated by protonation and deprotonation of the imidazole ring. With an increase in pH, there is an increase in basicity and, hence, an increase in binding of O_2 (the *Bohr* effect; cf. Figure 12.12).

A second histidine residue of myoglobin, His^{64} (distal His), contributes to heme-O_2-complex stabilization by formation of a hydrogen bridge or ionic bond formation between N and O (cf. Diagram 12.2).

The color of fresh meat is determined by the ratios of myoglobin (Mb), oxymyoglobin (MbO_2) and metmyoglobin (MMb^+):

$$Mb \underset{-O_2}{\overset{+O_2}{\rightleftarrows}} MbO_2, \quad Mb \rightleftarrows MMb^{\oplus} \leftarrow MbO_2 \tag{12.3}$$

Stable MbO_2 is formed at a high partial pressure of oxygen. Fresh cuts of meat, to a depth of about 1 cm, acquire a bright cherry-red color which is considered a mark of quality. A slow and continuous oxidation to MMb^+ occurs at a low partial pressure of O_2. The change of $Fe^{2+} \rightarrow Fe^{3+}$ is reflected in the change in color from red to brown. MMb^+ does not form an O_2-adduct, since Fe^{3+} appears to be a less efficient π donor than Fe^{2+}. With better donor ligands than O_2 (CN^-, NO, N_3^-), low-spin complexes are formed, the spectra of which are similar to those of MbO_2.

The change of $Fe^{2+} \rightarrow Fe^{3+}$ is designated as autoxidation:

$$Fe^{2\oplus}{-}O{=}O \rightleftarrows Fe{\equiv}O{-}O \rightleftarrows Fe^{3\oplus}{-}O{-}O^{\ominus} \tag{12.4}$$

"Dioxygen-Fe" "Fe^{3+}-superoxide" (ionic)

The oxygen molecule dissociates from the heme, taking along an electron from the iron, after protonation of its outer, more negative oxygen atom to form a hydroperoxy radical, the conjugate acid of the superoxide anion (cf. 3.7.2.1.4). The proton may originate from the distal histidine residue or other globin residues or from the surrounding medium. Autoxidation is accelerated by a drop in pH. The reason is the increased dissociation of the protein-pigment complex:

$$\text{Globin} + \text{Heme} \underset{k_{-1}}{\overset{k_1}{\rightleftarrows}} \text{Myoglobin} \tag{12.4a}$$

Soon after slaughter, the meat has a pH at or near 7, at which the equilibrium constant of the above reaction is $K = k_1/k_{-1} = 10^{12}\text{-}10^{15}$ mole^{-1}. Since, during post-rigor, glycolysis decreases the pH of the meat to 5–6, myoglobin becomes increasingly susceptible to autoxidation.

The stability of MbO_2 is also highly dependent on temperature. Its half-life, τ, at pH 5 is 2.8 h at 25 °C and 5 days at 0 °C. Fresh meat has a system which can reduce MMb^+ back to Mb. This system appears to be related to that which reduces methemoglobin in erythrocytes. The slow formation of MMb^+ can be reversed at the low partial pressure of O_2 which is found inside the cut of meat or in packaged, sealed meat. Therefore, for color stability, packaging of meat in O_2-permeable materials is not suitable since, after a time, its reduction capacity is fully exhausted. A non-O_2-permeable material is suitable for packaging meat. All of the pigment is present as Mb and is transformed to the bright red MbO_2 only when the package is opened.

Stabilization of the color of meat is also possible under controlled atmosphere packaging. A gaseous mixture of CO and air appears to be advantageous.

Copper ions promote autoxidation of heme to a great extent, while other metal ions, such as Fe^{3+}, Zn^{2+} or Al^{3+}, are less active.

Color stabilization by the addition of nitrate or nitrite (meat curing) plays an important role in meat processing. Nitrite initially oxidizes myoglobin to metmyoglobin:

$$Mb + NO_2^{\ominus} \longrightarrow MMb^{\oplus} + NO \qquad (12.4b)$$

The resultant NO forms bright-red, highly stable complexes with Mb and MMb^+, MbNO and MMb^+NO:

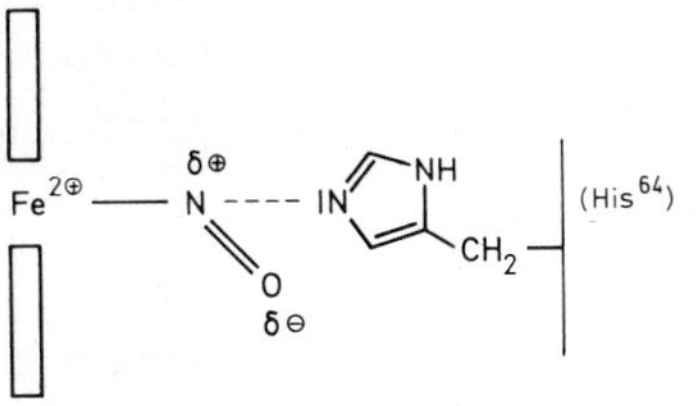

$$Fe^{2\oplus} — \overset{\delta\oplus}{N}{=}\underset{\delta\ominus}{O} \;\text{----}\; \text{N(imidazole)}—CH_2— \;(\text{His}^{64}) \qquad (12.5)$$

Reducing agents, such as ascorbate, thiols or NADH, accelerate the formation of a red color by reducing MMb^+ to Mb. It appears that MMb^+NO, after reduction of iron, is partially converted to MbN^+O. It is possible that this nitrosylmyoglobin is the sole end product of NO interaction with Mb and MMb^+.

MbNO is highly stable when O_2 is absent. However, in the presence of O_2, the NO released by dissociation of MbNO is oxidized to NO_2.

The color of cured meat is heat stable. Denatured nitrosylmyoglobin is present in heated meat, or, due to dissociation of the protein-pigment complex, heme occurs with NO ligands present in both axial binding sites of the heme molecule:

$$O^{\oplus}N—Fe^{2\oplus}—NO^{\ominus} \qquad (12.6)$$

A color change to brown is observed when non-cured meat is heated. Here, an Fe^{3+}-complex is present and presumably contains a nitrogen base (imidazole) as its fifth ligand and water as its

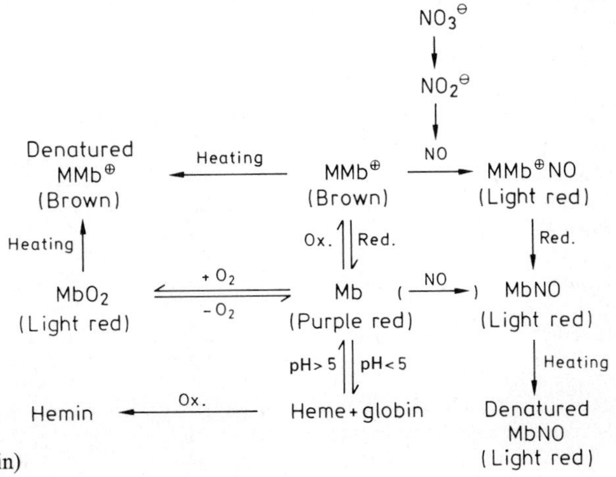

Fig. 12.14. Myoglobin reactions (Mb: myoglobin, MMb^+: metmyoglobin, MbO_2: oxymyoglobin, MbNO: nitrosomyoglobin, MMb^+NO: nitrosometmyoglobin)

sixth ligand; water may readily be replaced by stronger nucleophilic ligands such as NO, CO or CN^-.
The myoglobin reactions relevant to meat color are presented schematically in Figure 12.14.

12.3.2.3 Insoluble Proteins

The main fraction of proteins insoluble in water or salt solutions are the proteins of connective tissue. Membranes and the insoluble portion of the contractile apparatus are included in this group.
Connective tissue contains various types of cells. These cells synthesize a great many intercellular amorphous substances (carbohydrates, lipids, proteins) in which the collagen fibers are embedded.
Lipoproteins are present mostly in membranes. The lipids make up 3–4% of muscle tissue and are located in membranes. They consist of phospholipids, triacylglycerols and cholesterol. The phospholipid portion varies greatly: it makes up 50% of the plasma membrane and 90% of the mitochondrial membrane.

12.3.2.3.1 Collagen

Collagen constitutes 20–25% of the total protein in mammals. Table 12.5 shows data about its amino acid composition. The high contents of glycine and proline and the occurrence of 4-hydroxyproline and 5-hydroxylysine are characteristic. Since the occurrence of hydroxyproline is confined to connective tissue, its determination may provide quantitative data about the extent of connective tissue incorporation into a meat product.
Collagen also contains carbohydrates (glucose and galactose). These are attached to hydroxylysine residues of the peptide chain by O-glycosidic bonds. The presence of 2-O-α-D-glucosyl-O-β-D-galactosyl-hydroxylysine and of O-β-D-galactosyl-hydroxylysine has been confirmed.
The collagen chain consists of three polypeptide chains and, in the most prevalent species (collagen type I), two chains are the same (designated as α^1-chains), while the third chain differs (α^2-chain). A part of the amino acid sequence on an α^1-chain is presented in Table 12.6. Every third residue in this sequence is glycine. Moreover, the

Table 12.5. Amino acid composition of muscle proteins (values are in g/16 g N)

Amino acid	Beef muscle total	Poultry muscle total[a]	Myosin	Actin	Collagen (calves skin)	Elastin
Aspartic acid	9.7–9.9	9.7–11.0	10.9	10.4	5.4	1.0
Threonine	4.8	3.5–4.5	4.7	6.7	2.1	1.1
Serine	4.1–4.5	–	4.1	5.6	2.9	0.9
Glutamic acid	15.8–16.2	16–18	21.9	14.2	9.7	2.4
Proline	3.0–4.1	–	2.4	4.9	13.0	11.6
Hydroxyproline					10.5	1.5
Glycine	4.6–6.1	4.6–6.7	2.8	4.8	22.5	25.5
Alanine	6.1–6.3	–	6.7	6.1	8.2	21.1
Cystine	1.3–1.5	–	1.0	1.3	0	0.30
Valine	4.8–5.5	4.7–4.9	4.7	4.7	2.9	16.5
Methionine	4.1–4.5	–	3.1	4.3	0.7	Trace
Isoleucine	5.2	4.6–5.2	5.3	7.2	4.8[b]	3.7
Leucine	8.1–8.7	7.3–7.8	9.9	7.9		8.6
Tyrosine	3.8–4.0	–	3.1	5.6	1.2	1.3
Phenylalanine	3.8–4.5	3.7–3.9	4.5	4.6	2.2	5.9
Lysine	9.2–9.4	8.3–8.8	11.9	7.3	3.9	0.5
Hydroxylysine					1.1	–
Histidine	3.7–3.9	2.2–2.3	2.2	2.8	0.7	0.1
Arginine	5.3–5.5	5.7–6.1	6.8	6.3	7.6	1.2
Tryptophan	–	–	0.8	2.0	0	–

[a] Chicken, duck, turkey: average values.
[b] Sum of isoleucine and leucine.

Table 12.6. Amino acid sequence of α^1-chain of collagen in rat skin (a segment is presented with residues positioned between 56 to 109)

Sequence
·····-Gly-Pro-Arg-Gly-Pro-Hyp-Gly-Pro-Hyp-Gly-Lys-Asn-Gly-Asp-Asp-
60 65 70
Gly-Gln-Ala-Gly-Lys-Pro-Gly-Arg-Hyp-Gly-Glu-Arg-Gly-Pro-Hyp-
75 80 85
Gly-Pro-Gln-Gly-Ala-Arg-Gly-Leu-Hyp-Gly-Thr-Ala-Gly-Leu-Hyp-
90 95 100
Gly-Met-Hyl-Gly-His-Arg-Gly-Phe-Ser-········
(Hyl) │ R   105

R: αDGlc$p$(1-2)βDGal$p$-, Hyp: Hydroxyproline, Hyl: Hydroxylysine.

sequence glycine (Gly), proline (Pro) and hydroxyproline (Hyp) recurs frequently:

– Gly – Pro – Hyp –

Proline hydroxylation in vertebrates is a highly specific enzymic reaction. Proline can be hydroxylated only if it is located next to a glycine residue. Hence, hydroxyproline is always next to glycine.

The three peptide chains, each of which has a helical structure, occur together as a triple-stranded helix. The three strands wind around each other and are held together by hydrogen bonds. Their skeletal model corresponds to that of polyglycine II.

Two arrangements are possible for the three strands. They are designated as collagen I and II (Figure 12.15).

The basic structural unit of collagen fibers is called tropocollagen. It has a molecular weight of approx. 30 kdal. With a length of approx. 280 nm and a diameter of 1.4–1.5 nm, collagen is one of the longest proteins. Tropocollagen fibers associate in a specific way to form collagen fibers, as presented schematically in Figure 12.16. The association of adjacent rows is not in register, but is displaced by about one-fourth of the tropocollagen length (a "quarter staggered" array). This is responsible for cross-striations in the collagen fibers.

During maturation or aging, collagen fibers strengthen and are stabilized, primarily by covalent cross-linkages. Thus, cross-links confer mechanical strength to collagen fibers.

Cross-link formation involves the following reactions:

- Enzymatic oxidation of lysine and hydroxylysine to the corresponding ω-aldehydes.
- Conversion of these aldehydes to aldols and aldimines.
- Stabilization of these primary products by additional reduction or oxidation reactions.

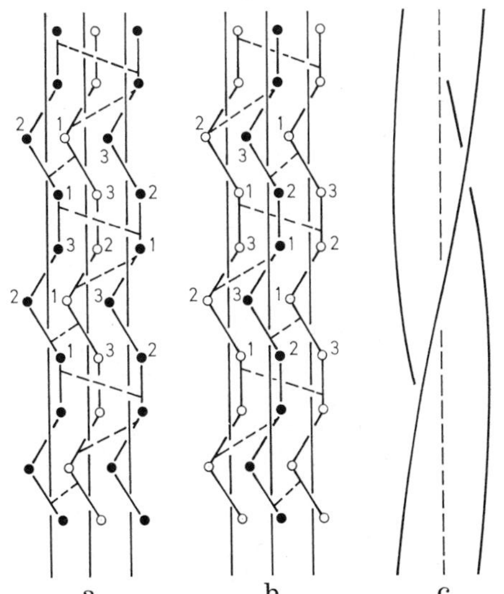

Fig. 12.15. Collagen conformation **a** Collagen I, **b** collagen II, **c** axes of three intertwined helices. In **a** and **b** the carbon α-atoms are denoted as • or ○. The tripeptide sequence is numbered starting with glycine (1). H-bridges: ----. (After *Rich, Davies* and *Crick*, 1961)

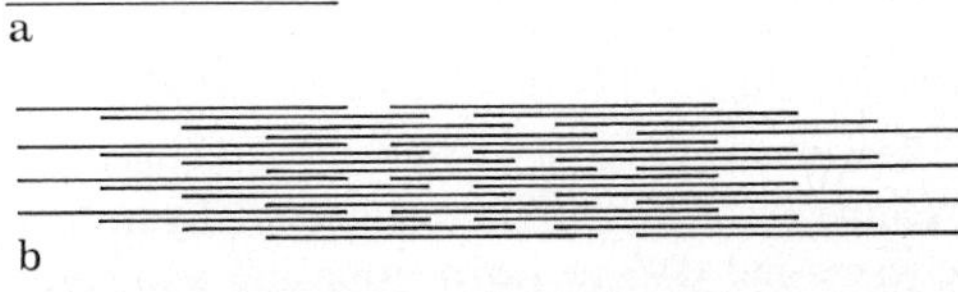

Fig. 12.16. Build up of a collagen fiber (**b**) from tropocollagen (**a**) molecules

Lysine and hydroxylysine residues within the peptide chain are oxidized by an enzyme that requires Cu^{2+} and pyridoxal phosphate for its activity and which is related to amine oxidase. This reaction yields an α-aminoadipic acid semi-aldehyde residue bound to the existing peptide chain (R = H or OH):

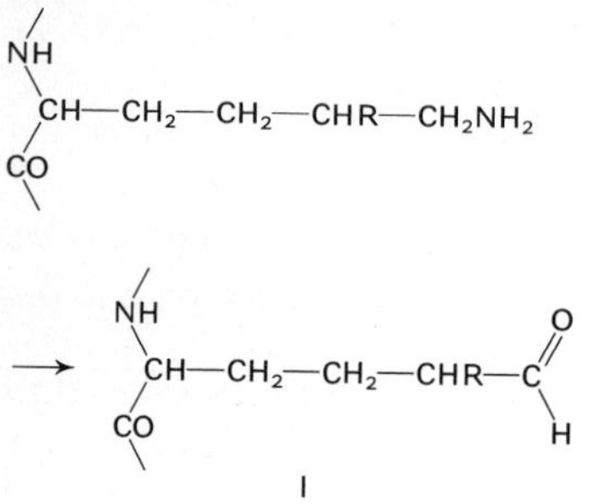

$$\text{NH—CH(CO)—CH_2—CH_2—CHR—CH_2NH_2} \rightarrow \text{NH—CH(CO)—CH_2—CH_2—CHR—CHO} \quad \text{(I)} \qquad (12.7)$$

The two aldehyde-containing chains may interact through an aldol condensation followed by elimination of water, forming a cross-link:

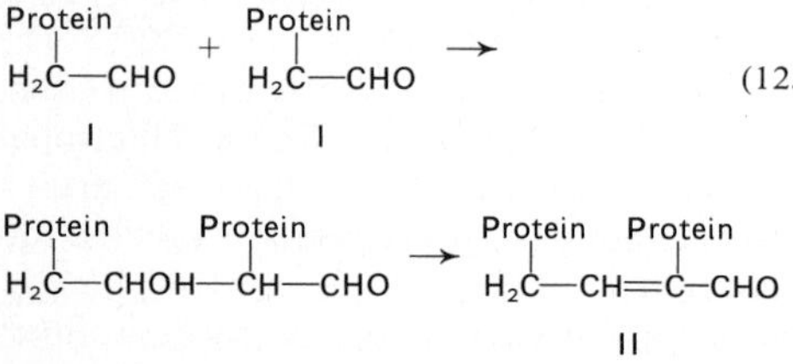

$$\text{Protein—H_2C—CHO (I)} + \text{Protein—H_2C—CHO (I)} \rightarrow \qquad (12.8)$$

$$\text{Protein—H_2C—CHOH—CH(Protein)—CHO} \rightarrow \text{Protein—H_2C—CH=C(Protein)—CHO (II)}$$

A polypeptide chain with an aldehyde residue (I) can interact with a lysine residue of the adjacent chain to form an aldimine, which can be further reduced to peptide-bound lysinonorleucine (III):

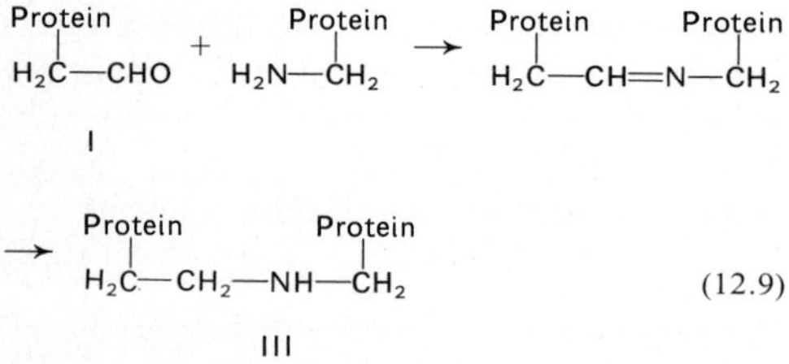

$$\text{Protein—H_2C—CHO (I)} + \text{H_2N—CH_2—Protein} \rightarrow \text{Protein—H_2C—CH=N—CH_2—Protein}$$

$$\rightarrow \text{Protein—H_2C—CH_2—NH—CH_2—Protein (III)} \qquad (12.9)$$

Likewise, aldehyde II can interact with one lysine residue through the intermediary dehydromerodesmosine (IV) to merodesmosine and, thus, provide cross-links between the three adjacent polypeptide chains:

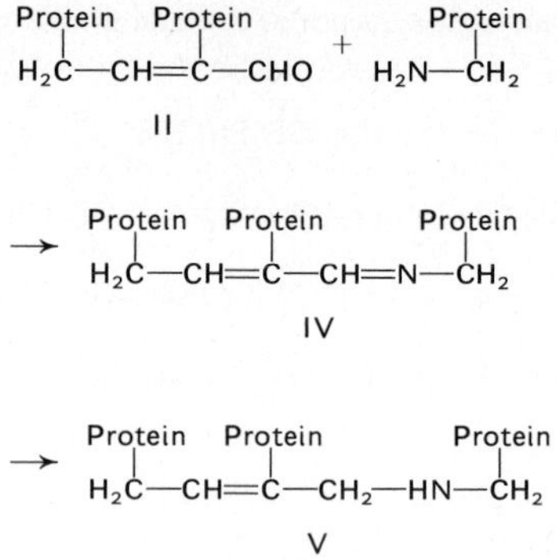

$$\text{Protein—H_2C—CH=C(Protein)—CHO (II)} + \text{H_2N—CH_2—Protein}$$

$$\rightarrow \text{Protein—H_2C—CH=C(Protein)—CH=N—CH_2—Protein (IV)}$$

$$\rightarrow \text{Protein—H_2C—CH=C(Protein)—CH_2—HN—CH_2—Protein (V)} \qquad (12.10)$$

During the reaction of three aldehyde molecules of type I with a lysine residue (actually a total of four lysine side chains are involved), a pyridine derivative is formed which, depending on the extent of reduction, yields desmosine (VI), dihydro- (VII) and tetrahydrodesmosine (VIII) (cf. Reactions 12.11).

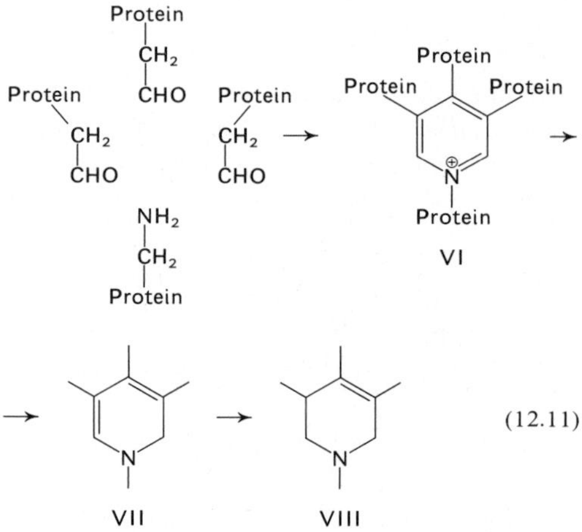

$$3\ \text{Protein—CH_2—CHO} + \text{Protein—CH_2—NH_2} \rightarrow \text{VI} \rightarrow \text{VII} \rightarrow \text{VIII} \qquad (12.11)$$

Depending on the kind of condensation, in addition to desmosine VI, designated as an A-type condensation product, rings with other substitution patterns are observed, i.e. B- and C-type condensation products:

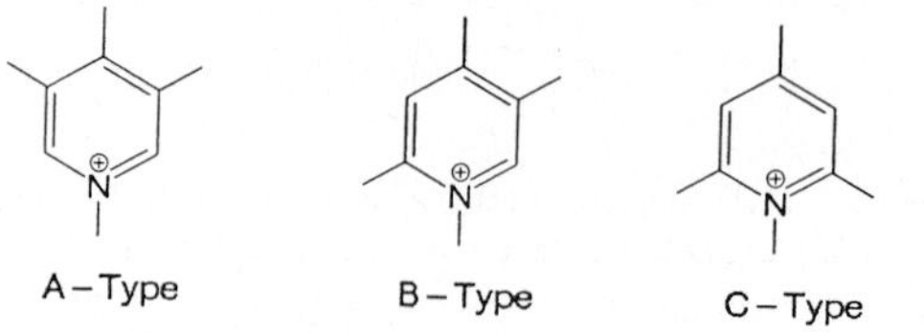

A-Type B-Type C-Type

(12.12)

The outlined reactions can also occur with hydroxylysine residues present on collagen fibers. Collagen biosynthesis (Figure 12.7,a-h) involves first the synthesis of pro-α^1- and pro-α^2-precursor chains. The N-terminus of these precursors contains up to 25% of extended α^1- and α^2-chains (a). Immediately after the chains are released from polysomes, hydroxylation of the proline and lysine residues occurs (cf. reactions under 12.13).

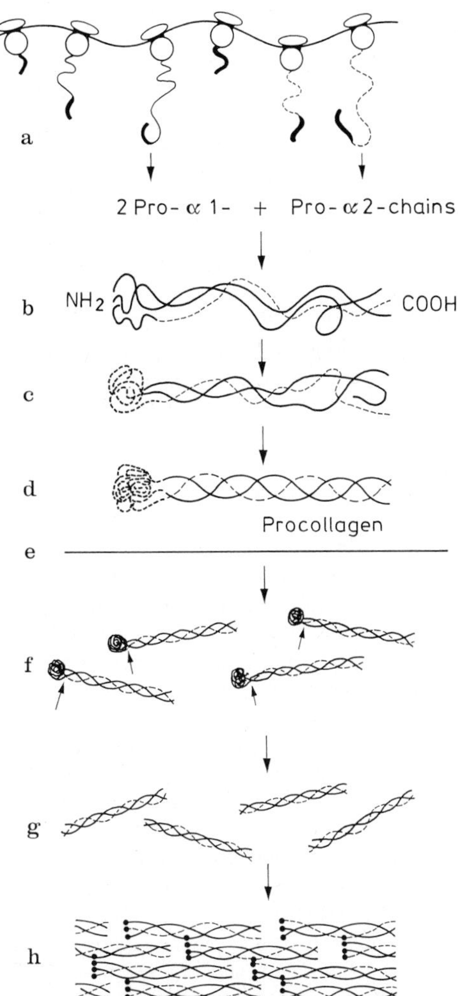

Fig. 12.17. Collagen biosynthesis. (After *Bornstein*, 1974.) **a** Polysome, **b** hydroxylation, **c** chain straightening, **d** disulfide bond formation, **e** cell membrane, **f** segregation, **g** a limited hydrolysis to tropocollagen, **h** collagen fiber formation, cross-linking

$$\text{(N–R)}\cdots COR^1 + O_2 \xrightarrow[\text{(Fe}^{2\oplus}\text{,ascorbate)}]{\text{Proline hydroxylase}} H\text{—}O\text{—}O\text{–(N–R)}\cdots COR^1 \xrightarrow{+2-\text{oxoglutarate}}$$

$$HO\text{—}C(\text{—}C(=O)O^{\ominus})(CH_2CH_2COO^{\ominus})\text{—}O\text{—}O\text{–(N–R)}COR^1$$

$$\longrightarrow {}^{\ominus}OOC\text{—}CH_2\text{—}CH_2\text{—}COO^{\ominus} + CO_2 + HO\text{–(N–R)}\cdots COR^1$$

$$H_2C\text{—}NH_2,\ H_2C,\ R \xrightarrow[\text{(Fe}^{2\oplus}\text{,ascorbate)}]{\text{Lysine hydroxylase}} H_2C\text{—}NH_2,\ CHOH,\ R \qquad (12.13)$$

Realignment of the chains follows: two strands of pro-α^1 and one chain of pro-α^2 are joined to form a triple-stranded helix (b–d). The extended peptides at the N-terminus appear to play a distinct role in these reactions. As well, disulfide bridging occurs between the strands at this stage in order to stabilize the structure. The procollagen thus formed will cross the membrane of the cell in which it was synthesized (e). The N-terminal peptides are removed by limited proteolysis (f) and the procollagen is converted to tropocollagen (g). Finally, the tropocollagen is realigned to form collagen fibers (h). At this stage, collagen maturation, which coincides with strengthening of collagen fibers by covalent cross-linking along the peptide strands, begins. The maturation is initiated by oxidation of lysine and is followed by the reactions described above.

Collagen swells but does not solubilize. Enzymatically, it can be hydrolyzed to various extents with a series of collagenases from different sources and with different specificities. A vertebrate animal collagenase, which is a metal proteinase, splits a special bond in native collagen while the collagenase from *Clostridium histolyticum,* also a metal proteinase, cleaves collagen preferentially at glycine residues, forming tripeptides:

$$—Pro—X\overset{\downarrow}{—}Gly—Pro—X\overset{\downarrow}{—}Gly—Pro— \quad (12.13a)$$

Collagenase enzymes which are serine proteinases are also known.

One characteristic of the intact collagen fiber is that it shrinks when heated (cooking or roasting). The shrinkage temperature (T_s) is different for different species. For fish collagen, the T_s is 45 °C and for mammals, 60–65 °C. When native or intact collagen is heated to $T > T_s$, the triple-stranded helix is destroyed to a great extent, depending on the cross-links. The disrupted structure now exists as random coils which are soluble in water and are called gelatin. Depending on the concentration of the gelatin solution and of the temperature gradient, a transition into organized structures occurs during cooling. Figure 12.18 schematically summarizes these transitions. At low concentrations, intramolecular back-pleating occurs preferentially with single-strands. At higher concentrations and slow rates of cooling, a structure is rebuilt which resembles the original native structure. At even higher concentrations and rapid cooling, structures are obtained in which the helical segments alternate with randomly coiled portions of the strand. All these structures can immobilize a large amount of water and form gelatin gels.

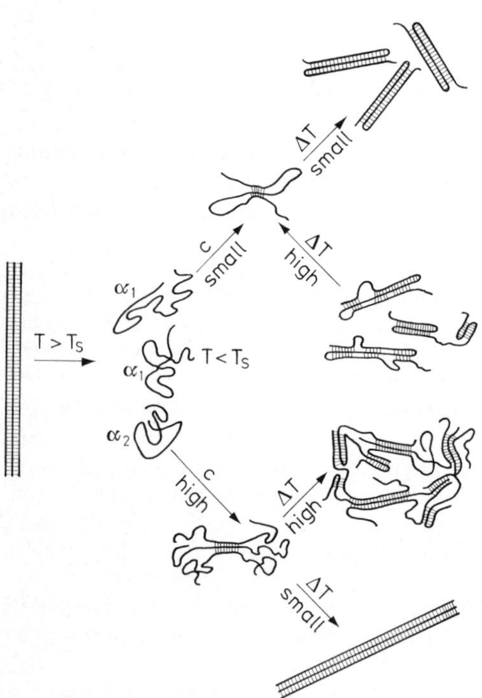

Fig. 12.18. Collagen conversion into gelatin. (After *Traub* and *Piez,* 1971.) T_s: shrinkage temperature, T: temperature, c: concentration; see text)

The transition of collagen to gelatin outlined above occurs during the cooking and roasting of meat. The extent of gelatinization is affected by the collagen cross-linking as determined by the age of the animal and the amount of heat applied (temperature, time, pressure).

Gelatin plays a role as a gelling agent. It is produced on a large scale from animal bones or skin by treatment with alkali or acid, followed by a water extraction step. Depending on the process, products are obtained which differ in molecular weight and, consequently, in their gelling properties. Some brands are used as food gelatins, others play an important role in industry (film emulsions, glue manufacturing).

12.3.2.3.2 Elastin

Elastin is found in lower amounts in connective tissue along with collagen. It is a nonswelling, highly stable protein which forms elastic fibers. The protein has rubber-like properties. It can stretch and then return to its original length or shape. Large amounts of elastin are present in ligaments and the walls of blood vessels. The ligament located in the neck of grazing animals is an exceptionally rich source of this protein. Table 12.5 shows that the amino acid composition is different from that of collagen. The amount of basic amino acids is low, while the content of acids rich in nonpolar aliphatic residues (alanine, valine, leucine, isoleucine) is markedly higher. This may account for the lack of swelling of elastin when it is heated in water. Elastin is hydrolyzed by the serine proteinase elastase, which is excreted by the pancreas. This enzyme preferentially cleaves peptide bonds at sites where the carbonyl residue has a nonaromatic, nonpolar side chain.

12.3.3 Free Amino Acids

Fresh beef muscle contains 0.1–0.3% free amino acids (fresh weight basis). All amino acids are detectable in low amounts (<0.005%), with alanine (0.01–0.05%) and glutamic acid (0.01–0.05%) being most predominant.
The free amino acid fraction also contains 0.02–0.1% taurine (I). As such, taurine should be regarded as a major constituent of this fraction. It is obtained biosynthetically from cysteine through cysteic acid and/or from a side pathway involving cysteamine and hypotaurine (II):

$$\begin{array}{ccccc} CH_2{-}SH & & CH_2{-}SO_2H & & CH_2{-}SO_3H \\ CH{-}NH_2 & \longrightarrow & CH{-}NH_2 & \longrightarrow & CH{-}NH_2 \\ COOH & & COOH & & COOH \\ \downarrow -CO_2 & & \downarrow -CO_2 & & \downarrow -CO_2 \\ CH_2{-}SH & & CH_2{-}SO_2H & & CH_2{-}SO_3H \\ CH_2{-}NH_2 & \longrightarrow & CH_2{-}NH_2 & \longrightarrow & CH_2{-}NH_2 \\ & & II & & I \end{array} \quad (12.14)$$

The biochemical role of taurine is as a constituent of bile acids (taurocholic and taurodesoxycholic acids). A neurotransmitting function has also been ascribed to this compound.

12.3.4 Peptides

The characteristic β-alanyl histidine peptides, carnosine, anserine and balenine, of muscle are described in section 1.3.4.2 (cf. Table 1.15).

12.3.5 Amines

Methylamine in fresh beef muscle is present at 2 mg/kg, while the other volatile aliphatic amines (dimethyl-, trimethyl-, ethyl-, diethyl- and isopropylamine) are detected only in trace amounts. Biogenic amines, such as histamine (from histidine), tyramine (tyrosine), tryptamine (tryptophan) and colamine (serine), are obtained by amino acid decarboxylation. These amines are present at 10 mg/kg fresh muscle. Putrescine (from ornithine), cadaverine (from lysine), spermine [1,4-bis-(3′-aminopropylamino)-butane] and spermidine (3′-aminopropyl-1,4-diaminobutane) are present in very low amounts. There is a significant increase in the content of biogenic amines (1–20 g/kg) in muscle due to autolysis, to be described later, and, initially, to bacterial degradation.

12.3.6 Guanidine Compounds

Creatine and creatinine (I and II, respectively; cf. Formula 12.15) are characteristic constituents of muscle tissue and their assay is used to detect the presence of meat extract in a food product. Creatine is present in fresh beef at 0.3–0.6% and creatinine at 0.02–0.04%.
In living muscle, 50–80% of creatine is in the phosphorylated form, creatine phosphate (III), which is in equilibrium with ATP. The reaction rate (cf. Formula 12.15) is highly influenced by the enzyme creatine phosphokinase. Creatine phosphate serves as an energy reservoir (free energy of hydrolysis, $\Delta G^0 = -42.7$ kJ/mole; of ATP, $\Delta G^0 = -29.7$ kJ/mole). Creatine phosphate has a higher phosphoryl group transfer potential than ATP. Hence, when muscle is stimulated for a prolonged period in the absence of glycolysis or respiration, the supply of creatine phosphate will become depleted within a couple of hours by maintaining the ATP concentration (Formula 12.15). This is especially the case in

$$\underset{\text{I}}{HN{=}C(NH_2){-}N(CH_3){-}CH_2{-}COOH} + ATP \rightleftharpoons \underset{\text{III}}{HN{=}C(NH{-}PO_3H_2){-}N(CH_3){-}CH_2{-}COOH} + ADP$$

$$\text{I} \rightleftharpoons \underset{\text{II}}{\text{creatinine (cyclic: HN=C, NH, C=O, CH}_2\text{, N–CH}_3)} \quad (12.15)$$

post-mortem muscle, when the ATP supply has declined significantly through oxidative respiration.

12.3.7 Quaternary Ammonium Compounds

Choline and carnitine are present in muscle tissue at 0.02–0.06% and 0.05–0.2%, respectively (on a fresh weight basis). Choline is synthesized from serine with colamine as an intermediary product (cf. Reactions 12.16a) and carnitine is obtained from lysine through ε-N-trimethyllysine and butyrobetaine (cf. Reactions 12.16b).

$$\begin{array}{l} CH_2OH \\ | \\ CHNH_2 \\ | \\ COOH \end{array} \xrightarrow{-CO_2} \begin{array}{l} CH_2OH \\ | \\ CH_2NH_2 \end{array} \longrightarrow \begin{array}{l} CH_2OH \\ | \\ CH_2N^{\oplus}(CH_3)_3 \end{array} \qquad (12.16a)$$

$$\begin{array}{l} CH_2{-}NH_2 \\ | \\ CH_2 \\ | \\ CH_2 \\ | \\ CH_2 \\ | \\ CHNH_2 \\ | \\ COOH \end{array} \longrightarrow \begin{array}{l} CH_2N^{\oplus}(CH_3)_3 \\ | \\ CH_2 \\ | \\ CH_2 \\ | \\ CH_2 \\ | \\ CHNH_2 \\ | \\ COOH \end{array} \longrightarrow \begin{array}{l} CH_2N^{\oplus}(CH_3)_3 \\ | \\ CH_2 \\ | \\ CH_2 \\ | \\ COOH \end{array} \longrightarrow$$

$$\xrightarrow[Fe^{2\oplus},\ \text{Ascorbate}]{O_2,\ \text{2-Ketoglutarate}} \begin{array}{l} CH_2N^{\oplus}(CH_3)_3 \\ | \\ CHOH \\ | \\ CH_2 \\ | \\ COOH \end{array} \qquad (12.16b)$$

The carnitine fatty acid ester, which is in equilibrium with long chain acyl-CoA molecules in living muscle tissue, is of biochemical importance. The carnitine fatty acid ester, but not the acyl-CoA ester, can traverse the inner mitochondrial membrane. After the fatty acid is oxidized within the mitochondria, carnitine is instrumental in transporting the generated acetic acid out of the mitochondria.

12.3.8 Purines and Pyrimidines

The total content of purines in fresh beef muscle tissue is 0.1–0.25% (on a fresh weight basis). ATP, present predominantly in living tissue, breaks down to inosine-5′-monophosphate (IMP) in the post-mortem stages. The breakdown rate is influenced by the condition of the animal and by temperature. IMP is then slowly

Table 12.7. Purines and pyrimidines in fresh-beef muscle

Compound	Content (%)
Inosine-5′-phosphate	0.02–0.2[a]
Inosine	Trace
Hypoxanthine	0.01–0.03
Adenosine-5′-phosphate	0.001–0.01
Adenosine-5′-diphosphate } Adenosine-5′-triphosphate }	<0.3[b]
Nicotinamide-adenine-dinucleotide	0.1
Guanosine-5′-phosphate	0.002
Cytidine-5′-phosphate	0.001
Uridine-5′-phosphate	0.002

[a] Until approx. 1 h post-mortem no IMP is found in muscle.

[b] There is a fairly rapid decrease in post-mortem concentration influenced by cooling and other muscle handling conditions.

decomposed through successive steps to hypoxanthin, with inosine as an intermediary product (cf. also Table 12.7):

$$\begin{array}{ll} \text{ATP} & \text{ADP} + P_{an}\ \text{(Myosin-ATPase)} \\ \text{2 ADP} & \text{ATP} + \text{AMP (Myokinase)} \\ \text{AMP} & \text{IMP} + NH_3\ \text{(Adenylate deaminase)} \\ \text{IMP} & \text{Inosine} + P_{an}\ \text{(5′-Nucleotidase)} \\ \text{Inosine} & \text{Hypoxanthine} + \text{Ribose (Nucleosidase)} \end{array} \qquad (12.17)$$

Post-mortem data on the *Psoas major* rabbit muscle are given in Table 12.8. They relate to nucleotide breakdown and to other important muscle tissue constituents.

The changes in water holding capacity of meat resulting from ATP transition to IMP are dealt with in section 12.5. Unlike purines, pyrimidine nucleotide content in muscle is very low (Table 12.7).

12.3.9 Organic Acids

The predominant acid in muscle tissue is the lactic acid formed by glycolysis (0.2–0.8% on a fresh meat weight basis), followed by glycolic (0.1%) and succinic acids (0.05%). Other acids of the *Krebs* cycle are present in negligible amounts.

12.3.10 Carbohydrates

The glycogen content of muscle varies greatly (0.2–1.0% on a fresh tissue weight basis) and is influenced by the age and condition of the animal prior to slaughter. The rate of the post-mortem decrease in glycogen is different.

Sugars are only 0.1–0.15% of the weight of fresh muscle, of which 0.1% is shared by glucose-6-phosphate and other phosphorylated sugars. The free sugars present are glucose (0.009–0.09%), fructose and ribose.

Table 12.8. Post-mortem changes in the concentration of some constituents of rabbit muscle *(M. psoas)*

Compound	µmole/g Fresh tissue	
	living muscle	post-rigor muscle
Total acid-soluble phosphorus	68	68
Inorganic phosphorus	< 12	> 48
Adenosine triphosphate (ATP)	9	< 1
Adenosine diphosphate (ADP)	1	< 1
Inosine monophosphate	< 1	9
Creatine phosphate	20	< 1
Creatine	23	42
NAD/NADP	2	1
Glycogen	50	< 10
Glucose-1-phosphate	< 1	< 1
Glucose-6-phosphate	5	6
Fructose-1,6-diphosphate	< 1	< 1
Lactic acid	10	100

Table 12.9. Vitamins in beef muscle

Compound	mg/kg Fresh tissue
Thiamine	0.6–1.6
Riboflavin	1–3
Nicotinamide	40–120
Pyridoxine, pyridoxal, pyridoxamine	1–4
Pantothenic acid	4–10
Folic acid	0.1–0.3
Biotin	4
Cyanocobalamine (B_{12})	0.01–0.02

Table 12.10. Minerals in beef muscle

Element	% Fresh tissue	Element	% Fresh tissue
K	0.25–0.4	Zn	0.001–0.008
Na	0.07–0.2	P (as P_2O_5)	0.30–0.55
Mg	0.015–0.035	Cl	0.04–0.1
Ca	0.005–0.025		
Fe	0.001–0.005		

12.3.11 Vitamins

Table 12.9 provides data about water-soluble vitamins in beef muscle.

12.3.12 Minerals

Table 12.10 provides data about minerals in meat. The trace elements, which are 1 mg/kg fresh meat tissue, are not listed individually.

12.4 Post-Mortem Changes in Muscle

12.4.1 Rigor Mortis

Cessation of blood circulation ends the O_2 supply to muscle. Anaerobic conditions start to develop. The energy-rich phosphates, such as creatine phosphate, ATP and ADP, are degraded. The glycolysis process, which is pH and temperature

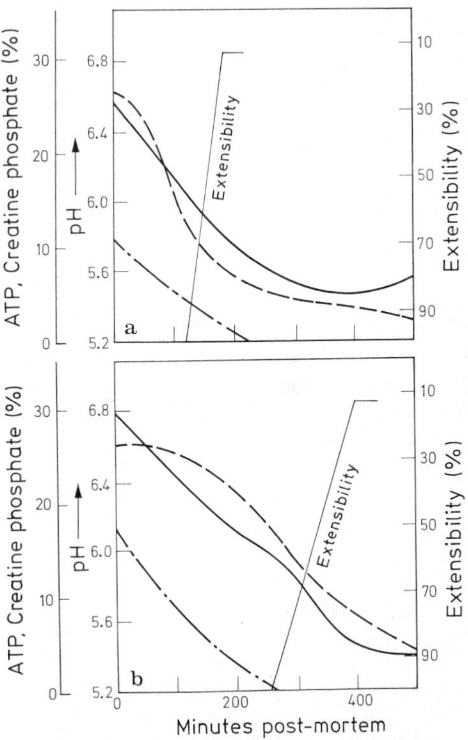

Fig. 12.19. Post-mortem changes in beef muscle. **a** *M. longissimus dorsi;* **b** *M. psoas;* ——: pH value; – – –: ATP as % of the total acid-soluble phosphate; –·–·–: creatine phosphate as % of the total acid soluble phosphate. (After *Hamm,* 1972)

dependent and which is influenced by the presence of glycogen, is the sole remaining energy source. The lactic acid formed remains in the muscle, thereby decreasing the muscle pH from 6.5 to less than 5.8.

Table 12.8 gives an example of post-mortem changes in rabbit *Psoas major* as related to concentrations of some of the more important muscle tissue constituents. The data shown in Figure 12.19 illustrate the post-mortem decreases in pH, creatine phosphate and ATP in beef *Longissimus dorsi* and *Psoas major* muscles and emphasizes that the changes are dependent on the type of muscle.

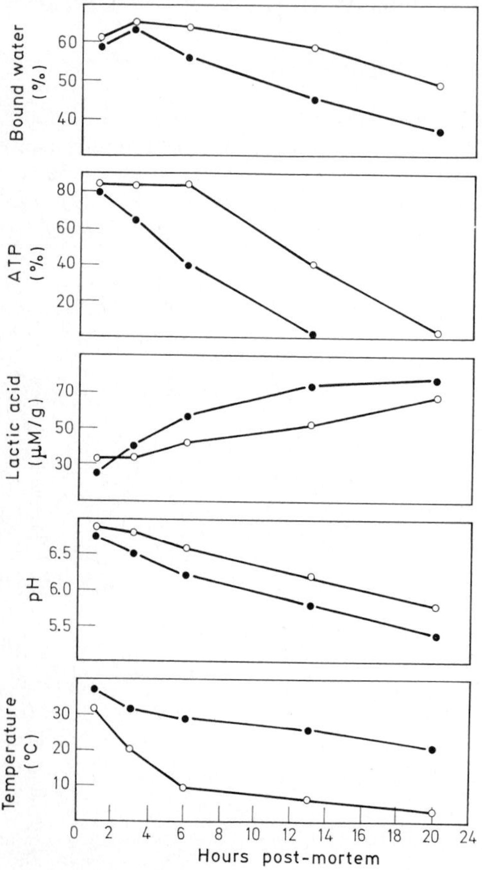

Fig. 12.20. The effect of temperature on post-mortem changes in beef muscle.
M. semimembranosus •–•: normal cooling, animal carcass kept for the first hour post-mortem at 2–4 °C then posterior hind quarters cut and kept at 14 °C for 10 h followed by 2 °C; ○–○: cooling in ice, hind quarters 11 h in crushed ice, followed by 2 °C. Temperature measurement of the meat at 4 cm depth; bound water as percent of total water; lactic acid results are on fresh weight basis and ATP expressed as percent of total nucleotides. (After *Disney* et al., 1967)

Although muscle tissue is soft and flexible and dry on the surface immediately following death, its flexibility or extensibility is lost very rapidly. ATP breaks down (Figure 12.19). The muscle tissue becomes stiff and rigid (death's stiffening, rigor mortis; cf. 12.3.2.1.4 and 12.3.2.1.5) and, as the rigor proceeds, the muscle tissue surface becomes wetter (the drip or muscle exudate increases).

The onset of rigor mortis occurs in beef muscle within 10–24 h; in pork, 4–18 h; and in chicken, 2–4 h.

The rate of decrease in pH and the final pH value of meat are of significance for water holding capacity and, therefore, for meat quality. Figure 12.20 shows that a more rapid and intensive cooling of the post-mortem muscle results in meat with a noticeably higher water holding capacity than that of muscle cooled slowly.

12.4.2 Defects (PSE and DFD Meat)

Rapid drops in ATP and pH (Figure 12.21) cause pork muscle to become pale and soft and to undergo extensive drip loss because of lowered water holding capacity (PSE meat: pale, soft, exudative). PSE meat has a low tensile strength and loses a substantial amount of weight when hung and, when thawed, drip losses occur. Such defects are typical for hogs with an inherited

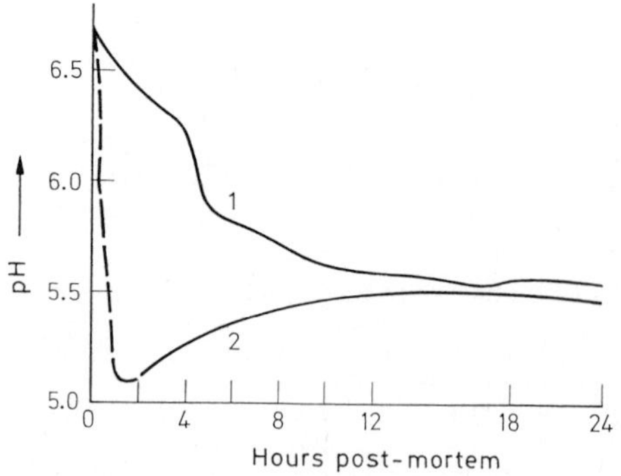

Fig. 12.21. Post-mortem pH-fall in normal pork meat (1) and a PSE-meat (2); *M. longissimus dorsi.* (After *Briskey* and *Wismer-Pedersen,* 1961)

sensitivity towards stress, such as fear prior to slaughter, anxiety during transport, or exposure to temperature changes, etc. Immediately prior to or during slaughtering, an abnormally rapid ATP breakdown occurs and, consequently, the rate of glycolysis is accelerated. A lower pH is thus achieved at a relatively high temperature. As a result, precipitation of sarcoplasmic proteins occurs on the surface of myofibrils and, hence, their swelling properties are changed (Figure 12.22).

The phenomena described above occur preferentially with light-colored muscle tissue, for example, *Longissimus dorsi* or *Glutaeus medius,* while in dark muscles of the same animal, e.g. *Rectus abdomini,* normal conditions may pertain.

The occurrence of dark and firm pork meat (DFD meat: dark, firm and dry) is likewise characteristic of a stress-impaired hog. In contrast to the PSE effect, it seems that lactate and hydrogen ions move from DFD muscle to blood prior to or soon after slaughtering. Thus, DFD meat, in contrast to PSE meat, has a high pH and a low level of lactic acid 45 min post-mortem.

Data relating to normal and faulty cuts of meat are summarized in Table 12.11. Both defects mentioned may occur in different muscles of the same animal. The PSE effect is not significant in beef muscle tissue since energy is available from fat oxidation so glycogen breakdown can occur slowly. These meat defects may be avoided in hog muscles by careful handling of stress-sensitive animals and by rapid cooling of carcasses.

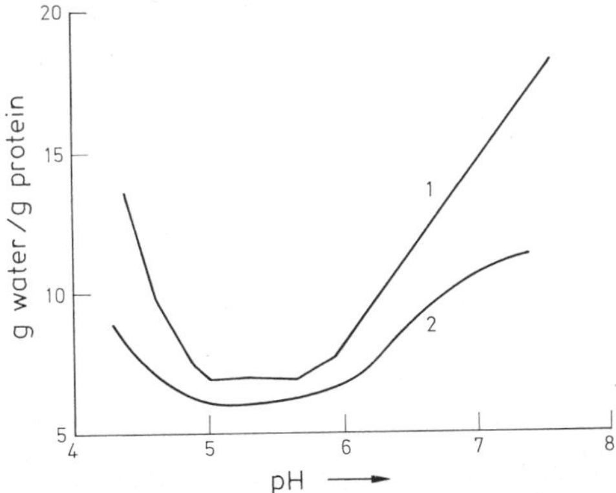

Fig. 12.22. Water holding capacity of washed myofibrils from normal (1) and exudative (2) pork meat (PSE-meat). (After *Bendall* and *Wismer-Pedersen,* 1962)

Table 12.11. Some differences between normal and faulty meat[a]

	Quality	pH (1 h)	pH (24 h)	ATP	Glycogen	Lactate
Normal meat		6.5	5.8	2.2	6.2	4.7
PSE-Meat	Pale, exudative, loose soft texture	5.6	5.6	0.3	1.9	9.0
DFD-Meat	Dark, sticky, firm texture	6.5	6.3	1.1	1.5	4.0

[a] Pork meat: *M. longissimus dorsi.* Values are averages expressed as mg/g muscle 1 h post-mortem; pH 1 h (initial) and pH 24 h (final) values post-mortem.

12.4.3 Aging of Meat

Rigor mortis in beef muscle tissue is usually resolved 2–3 days post-mortem. By this time, the meat again becomes soft and tender. Further aging of the meat to improve tenderness and to form aroma requires various amounts of time, depending on the temperature. Beef requires 14 days at 0 °C, 6 days at 8–10 °C and 4 days at 16–18 °C. A slight rise in pH is observed with aging, the water holding capacity is increased somewhat and, also, fluid loss from heat-treated meat is slightly decreased. Tenderizing involves autolysis of Z line-bound substances by endogenous proteinases.

12.5 Water Holding Capacity of Meat

Some of the previous sections are relevant to the water holding capacity of meat (cf. 1.4.3.1 and 1.4.3.3). Muscle tissue contains 20–25% protein and approx. 74–76% water, i.e. 350–360 g water per 100 g protein. Of this total water not more than 5% is bound directly to hydrophilic groups on the proteins. The rest of the water in the muscle tissue, i.e. 95%, is held by capillary forces between the thick and thin filaments. When a larger amount of water is bound to the network, the muscle is more swollen and the meat is softer and juicier. Hence, water holding capacity, protein swelling and meat consistency are intimately interrelated. The extent of water holding by the protein gel network depends on the abundance of cross-links among the peptide chains. These links may be hydrogen or ionic bonds and may involve divalent metal ions. A decrease in the number of these cross-linkages results in swelling, whereas an increase in the number of cross-linkages results in shrinkage (syneresis) of the protein gel.

The water holding capacity of meat is of great practical importance for meat processing and is affected by pH and the ion environment of the proteins.

The total charge on the proteins and, hence, their electrostatic interactions are the highest at their isoelectric points. Therefore, meat swelling is minimal in the pH range of 5.0–5.5 (Figure 12.23).

Addition of salt to meat shifts the isoelectric point and, hence, the corresponding swelling minimum to lower pH values, due to the preferable binding of the anion. This means that, in the presence of salts, water holding is increased at all pH's higher than the isoelectric point of the unsalted meat (Figure 12.24).

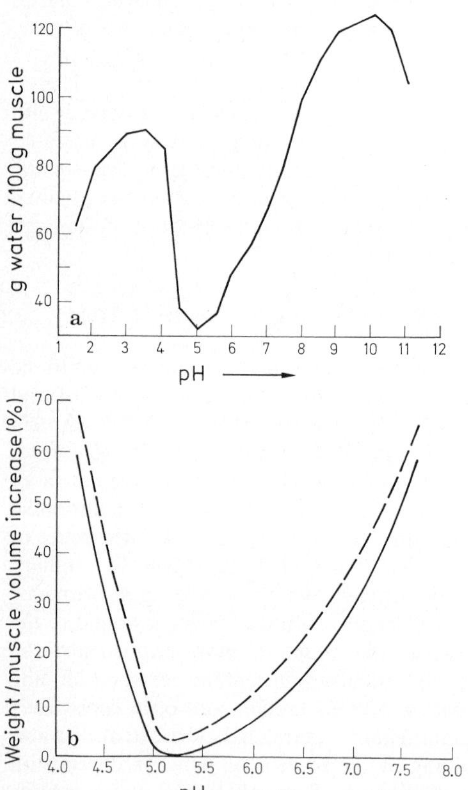

Fig. 12.23. Swelling of meat as affected by pH. **a** Beef muscle homogenate, 5 days post-mortem, **b** beef muscle cut in cubes 3 mm edge length, --- weight increase, — volume increase. (After *Hamm*, 1972)

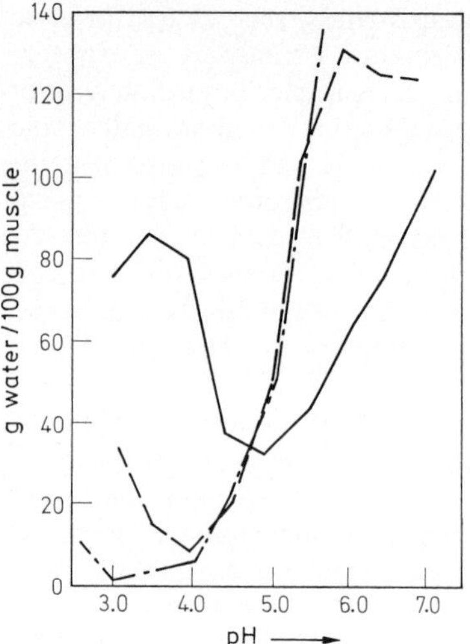

Fig. 12.24. Swelling of meat as affected by salts. Beef muscle homogenate; the ionic strength of the salt added to homogenate is $\mu = 0.20$; — control, --- NaCl, -·-·- NaSCN. (After *Hamm*, 1972)

The water holding capacity of muscle tissue soon after slaughter, while the muscle is still warm and before the onset of rigor mortis, is high and is due to the presence of high concentrations of ATP. As ATP breakdown continues, the rigidity

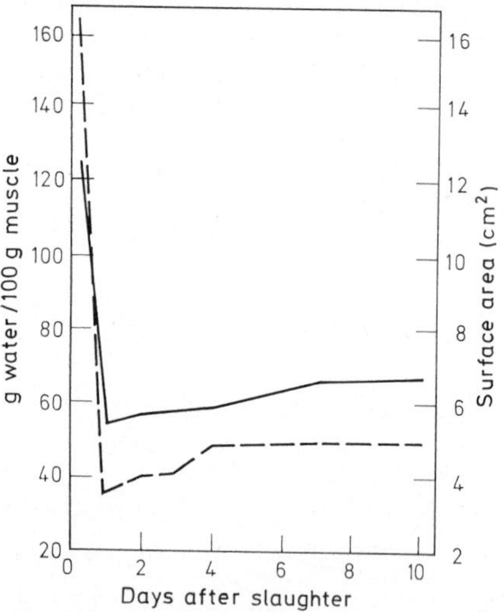

Fig. 12.25. Water holding capacity and rigidity of beef muscle. — Water holding capacity, --- rigidity (stiffness) expressed as the surface area acquired by homogenate after being pressed between filter papers, under standardized conditions. (After *Hamm*, 1972)

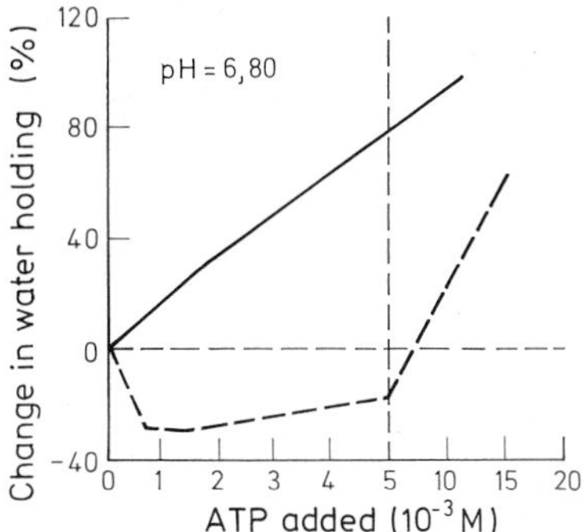

Fig. 12.26. Swelling of meat as affected by ATP addition. Beef muscle homogenate; pH 6.8; —— 2 h post-mortem, --- 4 days post-mortem. (After *Hamm*, 1972)

of the tissue increases and the water holding capacity starts to decrease (Figure 12.25). Addition of ATP to muscle tissue homogenates prior to the onset of rigor mortis brings about a rise in tissue swelling (Figure 12.26). Addition of low levels of ATP (to about 1×10^{-3} molar) during post-rigor brings about tissue contraction or shrinkage, while higher levels of ATP cause tissue swelling (Figure 12.26). This influence on swelling, however, is of short duration since, as ATP breaks down, contraction and shrinkage take place. Nevertheless, these studies amply illustrate the softening effect of ATP and, as already mentioned, the ability of ATP to dissociate actin-myosin complexes (cf. 12.3.2.1.4 and 12.3.2.1.5). Thus, because of high ATP levels and high pH, the slaughtered muscle which is still warm has a high water holding capacity, whereas post-rigor meat, with low ATP and low pH, has a low water holding capacity.

12.6 Kinds of Meat; Storage; Processing

12.6.1 Kinds of Meat; By-Products

12.6.1.1 Beef

Depending on the age of the animal and breed of beef cattle, meat is bright cherry red or dark red and lean or marbled with fat deposits between the connective tissue membranes of the muscle fibers. The best beef originates from 4–6 year old oxen (castrated bulls), nonlactating cows up to 8 years old, or 1½ to 2 year old steers (young bulls). The meat of old cows or discarded dairy cattle more than 12–14 years old is of low value. The average amount of waste from slaughterhouse oxen is 40–55%; that from cows, 42–66%. Beef carcasses are hung for 4–8 days before being cut up for soup meat, and 10–14 days for roasts or steaks.

12.6.1.2 Veal

Veal is white to pale red, with fine, tender fibers. The muscle tissue is generally limp and sticky. The characteristic odor of veal is sour and aromatic and is influenced by the presence of lactic acid. The best age at which to slaughter the calf for veal is 4–14 weeks. Older calves are sold as yearling beef. Younger calves yield meat with a water content exceeding 80%. The meat is hung for 8 days before use.

12.6.1.3 Mutton and Lamb

Mutton (sheep older than 1 year) and lamb (young sheep) meats are interspersed with fat tissue. The best quality meat is obtained from 2–4 year old sheep slaughtered in autumn. The flavor and taste are characteristic and are milder for lamb than mutton.

12.6.1.4 Goat Meat

Goat meat is lighter in color than mutton. The male goat (ram) readily gives off a sex aroma if slaughtered improperly.

12.6.1.5 Pork

Pork is the meat from pigs. It has fine fibers, a soft, tender consistency and is pale pink or pink, or greyish-white in color, thus making it different from all other meats. The meat is interspersed with and surrounded by fat, which may be as much as 20–40%. The carcasses are hung for 3–4 days before butchering.

12.6.1.6 Horse Meat

The meat of a young horse is bright red, whereas that of older horses is dark or reddish-brown or, when exposed to air, darkens to a reddish-black color. The consistency of the meat is firm and compact and the muscle tissue is not marbled with fat. During cooking, the white fat (melting point 30 °C) appears as yellow droplets on the surface of the broth. The characteristic sweet flavor and taste of the meat are derived from the high glycogen content. In addition to the determination of glycogen, sensitized rabbit serum and, particularly, fatty acid analysis can be used to detect horse meat. Horse fat is characterized by a higher content of linolenic acid than beef or pork lard. In the United States, unlike Europe, horse meat is seldom used for human consumption.

12.6.1.7 Poultry

The color of poultry meat differs according to age, breed and body part (breast meat is light, thighs and drumsticks are dark). Species of poultry which have dark meat (geese, ducks, pigeons) can be distinguished from those with light meat (chickens, turkeys, peacocks). The age, breed and feeding of the bird influence meat quality. Poultry fat tends to become rancid because of its high content of unsaturated fatty acids.

12.6.1.8 Game

Wild game can be divided into fur-bearing animals: deer (antelope, caribou, elk, white-tailed deer), wild boars (wild pigs) and other wild game (hare, rabbit, badger, beaver, bear); and birds or fowl (heathcock, partridge, pheasant, snipe, etc.). The meat of wild game consists of fragile fibers with a firm consistency and, because of less drainage of blood, the meat remains red to reddish-brown in color. It has low amounts of connective and adipose tissues. The taste and flavor of each type of wild meat is characteristic. Aging of the meat requires a longer time than meat from domestic animals because of the thick and compact muscle tissue structure. The flavor of the meat then becomes pleasantly piquant or pungent and the meat is colored dark-brown to black-red.

12.6.1.9 Variety Meats

Meats of various animal organs are called variety meats. They include tongue, heart, liver, kidney, spleen, brains, retina, intestines, tripe (the first and second stomachs of ruminants), bladder, pork crackling (skin), cow udders, etc. Many of these variety meats, such as liver, kidney or heart, are highly-valued foods because they contain vitamins and trace elements as well as high quality protein. Liver contributes the specific aroma of liver sausage and pastes (goose liver). Liver is also consumed as such. Heart, kidney, lungs, pork or beef stomach, calf giblets and cow's udders are incorporated into less expensive sausages; spleen is also made into sausage. Tongues are cooked, pickled and smoked, are used for the production of better-quality sausages, and are canned or sold as fresh meat. Calf brain and sweetbreads (thymus glands) are especially valued as food for patients. The compositions of some variety meats are shown in Table 12.12.

Table 12.12. Average composition of some internal organs and blood (g/100 g edible portion)

Organ	Moisture	Protein	Fat	Carbo-hydrate	Caloric value (kJ)
Heart					
beef	75.5	16.8	6.0	0.56	556
pork	76.8	16.9	4.8	0.4	510
Kidney					
beef	76.1	16.6	5.1	–	510
pork	76.3	16.5	5.2	0.80	523
Liver					
beef	69.9	19.7	3.1	5,90	590
pork	71.8	20.1	5.7	1,14	615
Spleen					
beef	76.7	18.5	2.9		456
pork	77.4	17.2	3.6	–	464
Tonque, beef	66.8	16.0	15.9	0.4	933
Lung, pork	79.1	13.5	6.7	–	510
Brain, veal	79.4	9.8	8.6	0.8	536
Thymus, veal	77.7	17.2	3.4	–	452
Blood					
beef	80.5	17.8	0.13	0.065	335
pork	79.2	18.5	0.11	0.06	372

Intestines, with their high content of elastin, make excellent natural sausage casings. These and beef stomach are specialty dishes.

Pork cracklings play a role in the production of jellied meat and blood sausage, as is the case with collared pork (head); these are also consumed directly and are a good source of vitamin D. Cartilage and bones contain tendons and ligaments which are collagen- and elastin-type proteins. Cartilage and bones are similar in composition, with the exception of their mineral content; the former contains 1% minerals and the latter averages 22% minerals, ranging from 20–70%. The fat content of bones can be as high as 30% and commonly varies between 10–25%. Spinal cord and ribs, when boiled in water, release gelatin-type substances and fat and, therefore, both are used in soup preparations (bouillon, clear broth or bouillon cubes or concentrated stock).

12.6.1.10 Blood

The blood which drains from a slaughtered animal is, on the average, about 5% of the live weight (oxen, cows, calves) but is particularly high for horses (9.98%) and low for hogs (3.3%). Blood has been used since ancient times for making blood and red sausages and other food products.

Blood consists of protein-rich plasma in which the cells or corpuscles are suspended. They are the red and white blood cells (erythrocytes and leucocytes, respectively) and the platelets (thrombocytes). The red blood cells do not have nuclei and are flexible round or elliptical discs with indented centers. The diameters of red blood cells vary (in μm: 4 in goat; 6 in pig; 10 in whale; and up to 50 or more in birds, amphibians, reptiles and fish). Red blood cells contain hemoglobin, the red blood pigment. White blood cells contain nuclei but no pigments, are surrounded by membranes, are 4–14 μm in diameter and are fewer in number than red blood cells. In addition to salts (potassium phosphate, sodium chloride and lesser amounts of Ca-, Mg- and Fe-salts), various proteins, such as albumins, globulins and fibrinogen, are present in blood.

The N-containing low molecular weight substances ("residual nitrogen") of blood contain primarily urea and lesser amounts of amino acids, uric acid, creatine and creatinine.

During coagulation or clotting of blood, the soluble fibrinogen in the plasma is converted to insoluble fibrin fibers which separate as a clot. Coagulation is a complex reaction catalyzed by the enzyme thrombin, the precursor of which is prothrombin. Thrombin reacts with fibrinogen to form insoluble fibrin. The mesh of long fibrin fibers traps and holds blood cells (platelets, erythrocytes and leucocytes). Hence, the clot is colored red. The remaining fluid, which contains albumins and globulins, is the serum. Blood plasma contains 0.3–0.4% fibrinogen and 6.5–8.5% albumin plus globulin in the ratio of 2.9:2.0.

The composition of blood is given in Table 12.12. Blood clotting requires the presence of Ca^{2+} ions. Hence, Ca^{2+}-binding agents, such as citrate, phosphate, oxalate and fluoride, prevent blood coagulation. In the processing of blood into food, coagulation is occasionally retarded by stirring the blood with metal rods onto which the fibrin deposits. Currently, blood clotting is inhibited by using Ca^{2+}-complexing salts. In this way the stabilized blood yields about 70% of plasma after centrifugation, containing 7–8% protein. The proteins can be processed further by spray-drying into powdered plasma. Recovery of liquid plasma is permitted only from the blood of cattle (excluding calves) and hogs. Addition of dried and liquid plasma to processed meats is legal. Citrate and/or phosphate are used as calcium-binding agents.

12.6.1.11 Glandular Products

Animal glands, such as the adrenal, pancreas, pineal, mammary, ovary, pituitary and thyroid glands, provide useful by-products for pharmaceutical use. Some of these products are adrenalin, cortisone, epinephrine, insulin, progesterone, trypsin and thyroid gland extract.

12.6.2 Storage and Preservation Processes

12.6.2.1 Cooling

Refrigeration (cooling or freezing the meat) is an important process for prolonged preservation of fresh meat of the carcasses in the form of sides or quarters. Cooling is performed slowly (e.g. with a blast of air at 0.5 m/sec and 4 °C) or quickly (e.g. stepwise for 3 h with a 3.5 m/sec blast of air at −10 °C, for 19 h with a blast of air at 1.2 m/sec at 2 °C, and over 18 days with air at 4 °C). The shelf-life of meat at 0 °C is 3 to 6 weeks. Weight loss due to moisture evaporation is low at high relative humidities, and decreases as the water holding capacity increases. As long as the meat is stored in the cold in large cuts, lipid oxidation is very slow. Only chopping or mincing or warming of the muscle tissue causes a high rate of peroxidation. Muscle disintegration results in a low but significant release of highly unsaturated membrane phospholipids and Fe^{2+} ions from myoglobin. Even after short cold storage, an off-flavor ("warmed over flavor") may develop. Pickling of meat inhibits such quality deterioration.

12.6.2.2 Freezing

The shelf-life of meat is substantially lengthened by freezing. Freezing can be performed in a single step (direct freezing) or in a two step process (initial cooling followed by freezing) using an air blast freezer with an air temperature of −40 °C and an air stream velocity of 3–10 m/sec. The shelf-life for storage at −18 °C to −20 °C and 90% relative humidity is 9 to 15 months. The

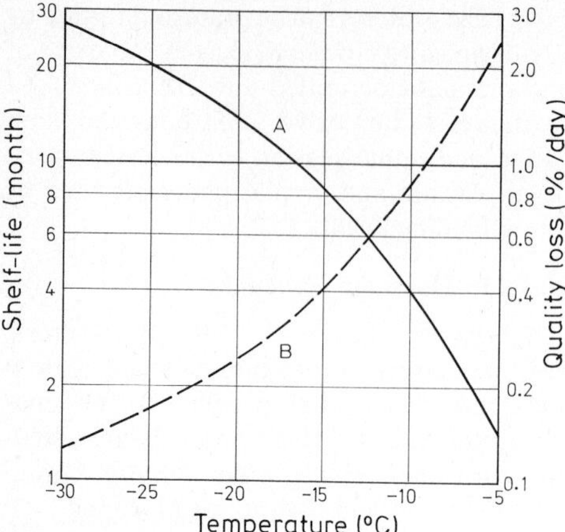

Fig. 12.27. Shelf-life of frozen chicken as affected by storage temperature. —— Shelf-life A, --- quality loss B = 100/A. (After *Gutschmidt*, 1974)

shelf-life of frozen chicken, as affected by storage temperature, is presented in Figure 12.27, while Table 12.13 shows the deterioration of frozen chicken as it is shipped from producer to consumer. The shelf-life is largely determined by oxidative changes affecting the lipids, which take place more readily in poultry (ducks, geese chickens) and pork than in beef or mutton.

The water holding capacity of frozen meat increases as the freezing temperature decreases. Water holding capacity also remains high when freezing is performed rapidly. Under these conditions the formation of large ice crystals is supressed and damage to membranes and the irreversible change in myofibrillar proteins caused by temporary high salt concentrations are avoided. Rupturing of muscle cells has never been observed as a result of rapid freezing.

Freezing meat immediately after slaughter while the carcass is still warm results in meat which, after thawing, loses a large amount of fluid. However, the freezing of meat prior to onset of rigor mortis is possible only with small cuts. The fluid loss is the result of a high rate of ATP breakdown and its associated decrease in water holding capacity. The sudden high rate of ATP breakdown is initiated by release of Ca^{2+} ions from the sarcoplasmic reticulum, which triggers the high activity of myosin-ATPase ("thaw rigor"). This "thaw rigor" can be avoided if the warm meat is frozen and then minced in the frozen state after addition of NaCl. Thaw rigor may also be avoided by disintegrating warm meat in the presence of NaCl and then freezing it.

Freezing aged meat results in negligible fluid losses. Nevertheless, this process is not widely used because of economic reasons.

Long storage of frozen meat results in a decrease in water holding capacity. Solubility changes and shifts in isoelectric points of proteins of the sarcoplasm and contractile apparatus are observed.

Slow thawing of frozen meat is generally considered more favorable than rapid thawing, although some opposing data exist. Obviously, freezing, storage and thawing should be considered as related process steps, which should be coordinated.

Table 12.13. Loss of quality of frozen chicken from producer to consumer[a]

Frozen food chain	Average storage temperature (°C)	Shelf-life (day)	Quality loss (% per day)	Average storage time (day)	Quality loss (%)
Producer	−23	540	0.186	40	7.5
Transport	−20	420	0.239	2	0.5
Wholesaler	−22	520	0.196	190	37.1
Transport	−16	370	0.370	1	0.4
Retailer	−20	420	0.239	30	7.2
	−14[b]	210	0.476	3	1.4
Transport	− 7	60	1.67	0.17	0.3
Consumer	−12	150	0.666	14	9.3
Σ				280	63.7

[a] For definition cf. Fig. 12.27.
[b] A temperature estimate for food storage on the surface of open freezers.

12.6.2.3 Drying

Drying is an ancient method of meat preservation. Some processes are: drying in a stream of hot air (40–60 °C), drying in vacuum under variable conditions, e.g. in hot fat, and freeze-drying, the most gentle process. The moisture content of the end product is usually 3–10%. Important quality criteria of such dried meat products are the rehydration capacity, which can be determined by water uptake under standard conditions, and the fraction of firmly-bound water. The drying process should not affect the water holding and aroma characteristics of the meat. The shelf-life of dried meat products is limited by the development of off-flavors due to fat oxidation and by discoloration due to the *Maillard* reaction. Dried beef and chicken are important ingredients of many soup powders. In addition to pieces of meat, minced meat, with or without binders, and processed meats, e.g. meat balls or dumplings, are also dried for this purpose.

12.6.2.4 Salt and Pickle Curing

Salt in high concentrations inhibits the growth of microorganisms and curtails activity of meat enzymes, hence, salt is considered as a meat preservative. Salting meat at a level up to 5% NaCl causes swelling (cf. Figure 12.24). Higher salt concentrations (10–20%) induce shrinkage in meat and its products, causing a decrease in moisture to a level below that of untreated meat. The meat retains its natural color, usually dark red, since the myoglobin concentration increases due to the moisture loss. The color of such meat changes upon cooking to grayish-brown.

Salting by the addition of sodium nitrite and/or nitrate (curing or pickling) produces products with highly stable color (cf. 12.3.2.2.2). Since nitrite reacts faster and less is required for color stabilization, it is widely used in place of nitrate. Salt curing is done either by rubbing salt on the meat surface (dry curing or pickling), by submerging the meat in 15–20% brine (wet pickle curing), or by injection of brine into blood vessels (pumping into arteries is limited to hams; multiple needle stitch pumping to boneless bellies). The latter type of salt curing or pickling is known as rapid pickling.

Additives, such as sugar or spices, which favorably affect the red color and formation of meat aroma, are often added to pickling salts. The aroma of pickled meat is specific and differs from that of nonpickled meat. Aroma formation is enhanced by the microflora (*Micrococcus* spp. and *Achromobacter* spp.) of pickling brine, which are simultaneously involved in reduction of nitrate (NO_3^-) and nitrite (NO_2^-) ions and thereby contribute to the stabilization of the pinkish or red color of pickled meat.

12.6.2.5 Smoking

Smoking of meat is usually associated with salting. Depending on the smoking procedure, the moisture drops 10–40%. Compounds present in smoke with bactericidal and antioxidative properties are deposited on and penetrate into the meat. Important smoke ingredients include phenols and carbonyl compounds. Polycyclic aromatic hydrocarbons are also present in smoke. Their content depends on the sawdust used and the smokehouse operation. Proper choice of the conditions of smoke generation can keep the presence of undesirable compounds to a minimum.

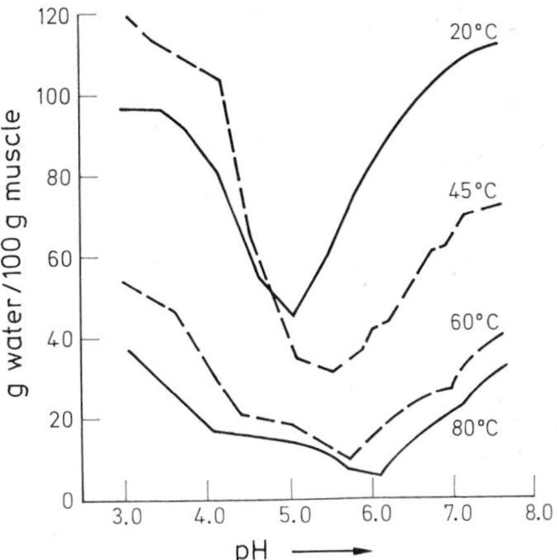

Fig. 12.28. Water holding capacity of beef muscle versus heat treatment and pH. (After *Hamm*, 1972)

[1] At natural pH of the muscle (pH 5.5)

12.6.2.6 Heating

Heat treatment is an important finishing process and also serves for the production of canned meat. Typical changes involved in heat treatment are: development of grayish-brown color, protein coagulation, release of juices due to decrease of water holding capacity (Figure 12.28), an increase in pH, development of a typical cooked or roasted meat aroma and, finally, softening induced by the shrinking and partial conversion of collagen to gelatin (cf. 12.3.2.3).

12.7 Meat Products

12.7.1 Canned Meat

Examples of canned meat are beef and pork in their own juice, corned beef, luncheon meat, cooked sausages, jellied meat, and cured and pickled hams. In order to achieve a sterile canned meat, the required heating time and temperature depend on the size and content of the can since

Table 12.14. Effect of can size and product on required heating time for canned meat (time in min to reach 121 °C at the center of the can)

Canned meat	400 g	850 g	2,500 g
Beef	47	57	80
Pork	58	98	120
Liver sausage	90	130	
Blood sausage	106	113	130

heat penetration is highly variable (Table 12.14). Prolonged heating leads to inferior aroma, texture and appearance (separation of fat and jelly), hence the current processing trend for canning of meat is to use high temperatures and short times.

12.7.2 Ham, Sausages, Pastes

12.7.2.1 Ham, Bacon

Raw smoked hams

The salt-cured and pickled, smoked hind leg of the pig is known as ham. The hind leg section, which extends from the knee to the hip, is cut either long or short. The long cut includes the whole hip and a long part of the shank (part of the leg above the knee). The short cut ham has only part of the hip and a short section of shank. Ham is marketed as halves or as prime cuts called butt (part of the hip), center (bone-in) and shank (bone-out). The halves are obtained by cutting the ham longitudianally into a half called the cushion (practically deboned) and the knuckle or calp half, which retains the bones. Most all bones are removed in processing of ham.

Fresh or "green" ham is unprocessed ham (not cooked, boiled, cured or smoked). It is used for dry, hard country ham and production of other uncooked hams with ethnic flavor: Westphalian, Scottish and the dry-cured Italian ham called Prosciutto.

Hams are divided into bone-in, semi-boneless and boneless hams. Boneless ham is currently an important cured pork product. It is often sold canned.

Some bone-in hams sold in Europe are initially dried and then wet cured for 4–7 weeks, then matured for 2–3 weeks by dry storing, followed by washing, drying and exposure to cold smoke for 4–7 weeks.

In rolled boneless hams, the curing, smoking and maturation steps are similar, except that the pickling immersion time is shorter. Lightly-salted lean hams are made from cutlet or chop meats by a mild curing process, then are filled into casings and additionally smoked.

Cooked hams (or boiled hams, if they were boiled in water) are rarely smoked and are sold sliced and used for sandwiches. They are made of boneless cured and matured hams which are stuffed into metal molds and heated in water tanks at 74–83 °C. When smoked, they are exposed to warm smoke. Praha ham is a type of cooked ham.

Bacon

Bacon is produced from green pork fat bellies either by dry curing or pickling. In the former process, the curing salt mixture is rubbed on all surfaces, while in the latter process the pickling solution is injected into bellies by multiple needle stitch pumping. The bellies are then matured for 1–2 weeks before final smoking and/or cooking. Cooking is usually done in a smokehouse. Often, dry-cured bacon is dried and exposed to cold smoke after washing. Variety bacons are Wiltshire and Canadian bacon, fat backs and those of heavy bellies.

The better quality bacon in Germany is made only from back fat. It is salt-cured, washed and cold smoked.

12.7.2.2 Sausages

Sausage manufacturing consists of grinding, mincing or chopping the muscle tissue and other organs and blending them with fat, salts, seasonings (herbs and spices) and, when necessary, with binders or extenders. The sausage mix or dough is then stuffed into cylindrical synthetic or cellulose casings or tubings of traditional sausage shape or, often, natural casings, such as hog or sheep intestines or the hog's bun (for liver sausage) are used. They are sold as raw, precooked or cooked, and/or smoked sausages. Sausages are classified solely according to the type of processing: coarsely ground or fermented and emulsified, or raw (made of pork or beef), water-cooked (luncheon and breakfast sausage) or heat-treated or precooked [e.g. frankfurters (hot dogs), wieners or mortadella].

Table 12.15. Types of Sausages

Sausage Variety	Formulation/Production[a]
Raw Fermented Sausages	Coarsely-ground, semi-dry or dry sausages often added to meat mixture starter cultures and cured only with nitrate.
Coarse mettwurst	Beef (24%), pork trimmings, lean (48%), pork backfat (24%), salt (37%)[b], sodium nitrate (0.03%), white pepper (0.4%). Dried and exposed to cold smoke until slicing texture is achieved.
Tea Sausage	Veal (30%), light beef (30%), pork trimmings (30%), unsalted pork backfat (10%), salt, sodium nitrate, white pepper, nutmeg. Stored until curing is finished and then smoked.
Salami (Italian style)	Beef trimmings (84%), pork backfat (16%), salt, sodium nitrate, ginger, cloves. Dried and moderately-warm smoked.
Salami (Hungarian style)	Beef trimmings (84%), pork backfat (25%), salt, sugar, pepper, paprika. Dried and smoked.
Cervelat Sausage	Beef trimmings (20%), lean pork trimmings (50%), pork backfat (30%), salt, sodium nitrate and whole white pepper. Dried and then exposed to cold smoke.
Plockwurst	Beef trimmings (50%), pork trimmings (25%), pork backfat (25%), salt, sodium nitrate, sugar, ground and whole white pepper. Dried and smoked until slicing consistency is achieved.
Landjager Sausage	Lean pork trimmings (40%), pork backfat (30%), beef trimmings (30%), salt, sodium nitrate, sugar, pepper, mustard seed. Molded into flat rod-shaped matching pairs of small sausages.
Saxony Knackwurst	Beef (50%), pork (33%), pork backfat (17%), salt, sodium nitrate, sugar, pepper, caraway.
Thuringian Knackwurst	Marbled pork (100%), salt, caraway, pepper, marjoram. Dried and exposed to cold smoke.
Frying Sausage	
Thuringian Bratwurst	Marbled pork (70%), steer[c] meat (30%), salt, pepper, nutmeg, caraway. To be roasted on a grate over fire (barbecued).
Nuremberg Bratwurst	Lean pork trimmings (70%), pork backfat (30%), salt, pepper, marjoram.
Munich Wollwurst	Steer meat (70%), milk (30%), salt, pepper, lemon peel.
Emulsion-Type Sausages	For fast color stabilization, these sausages usually contain nitrite pickle brine.
Wieners	Steer meat (49%), lean pork (19.5%), pork backfat (20%), nitrite pickle brine (2%), sugar (0.2%), white pepper (0.2%), nutmeg (0.03%), coriander (0.02%). Stuffed into sheep casings and warm smoked until a yellow-pink color is attained.
Halberstaedter Sausages	Steer meat (60%), lightly-marbled pork (10%), pork backfat (30%), nitrite pickle brine, sugar, white pepper, nutmeg, ginger, grainy meat broth. Stuffed into animal casings and smoked.
Frankfurters	Lean pork (70%), pork backfat (30%), salt, nitrite pickle brine, sugar, nutmeg, pepper, cardamon. Stuffed into animal casings and warm smoked.
Hunter's Sausage	Lean pork (60%), beef (30%), pork backfat (10%), nitrite pickle brine, sugar, pepper, coriander, nutmeg, grainy meat broth. Ripened to a red color, and hot smoked.

Table 12.15 (continued)

Sausage Variety	Formulation/Production[a]
Mortadella	Steer meat (50%), marbled pork (30%), pork backfat (20%), nitrite pickle brine, sugar, pepper, cardamon, ginger, pistachio nuts, grainy meat broth. Smoked hot and, after cooling, further exposed to cold smoke.
Goettingers	Lean beef (40%), lean pork (30%), pork backfat (30%), nitrite pickle brine, sugar, white pepper, ginger, rum, meat broth. Left to ripen until red colored, then hot smoked, cooled and exposed further to cold smoke.
Gelbwurst (yellow sausage)	Steer meat (70%), marbled pork (30%), salt, sugar, pepper, mace; stuffed and heated to about 67 °C and colored yellow with saffron.
Munich Weisswurst (white sausage)	Steer meat (50%), marbled pork (30%), veal head trimmings (20%), salt, sugar, pepper, mace, ginger, lemon, peel, onions, parseley.
Bavarian Sausage Loaf	Steer meat (50%), marbled pork (40%), pork backfat (10%), some beef liver, nitrite pickle brine, sugar, pepper, mace, ginger, onions. It is then roasted in molds.
Cooked Sausages	
Veal Liver Sausage, economy grade (top grade product is Braunschweiger)	Pork liver (34%), cooked veal (19%), cooked marbled pork (44%), salt (2.5%)[b], pepper (0.3%), mace (0.1%), cloves (0.04%), ginger (0.04%). Cooked in boiling water, cooled, dried and moderately smoked.
Homemade Liver Sausage	Liver (30%), meat and fat trimmings (50%), fresh onions (5%), pork backfat (15%), salt, pepper, pimentos, marjoram, thyme. Cooked in boiling water, cooled, dried and moderately smoked.
Goose Liver Sausage (Strasbourg Style)	Pork liver (22%), pork pouch (32%), liver of fattened geese (35%), tongue (6%), truffle garnish (5%), salt, onion, mace, ginger, white pepper, pistachio. Cooked, cooled, dried and moderately smoked.
Liver Paste (Pâté)	Pork liver (50%), pork backfat (50%), salt, mushrooms, nutmeg, ginger, white pepper, onion. Cooked in molds or baked in the oven.
Rotwurst (Blood Sausage)	Diced pork backfat (50%), meat trimmings, heart (15%), cooked pork skin (15%), pork blood (20%), salt, black pepper, marjoram, clove, pimentos. Cooked, cooled, dried and smoked.
Meat-Blood Sausage	Marbled pork (50%), meat trimmings, heart, kidney (25%), pork skin (15%), pork blood (10%), salt, white pepper, clove, marjoram. Cooked, then cold smoked.
Blood and Tongue sausage	Pork, veal and beef tongue (40%), pork backfat (28%), cooked pork skin (14%), pork blood (18%), salt, black pepper, clove, pimentos, thyme, onions, marjoram. Cooked, dried and cold smoked.
Hannover Cooked Mettwurst (white sausage)	Fat pork (100%), salt, white pepper, mace, onions. Stuffed and then cooked.
Bulk Headcheese	Pork knuckles, pork back, other meat, pork skin, salt, pepper, clove, caraway. The cooked broth is chilled in molds, then wrapped in cellulose or rectangular plastic casings.

[a] The formulations listed should be considered only as guides. They may deviate greatly within a group or between related sausage groups, by country, region, district or market situation, and by degree of quality.

[b] Data for the amount of salt and seasonings (herbs and spices) used are given for only one sausage variety within a group.

[c] A steer is a young castrated bull.

Raw sausages are made of raw skeletal muscle tissue, fat and seasonings. Typical products are Cervelat sausages, many semihard or hard salamis of European origin (semi-dry or summer salami, hard and dry winter salami) and the German Mettwurst sausage. After stuffing with the sausage mix, the sausage is dried and smoked. The specific aroma is formed by ripening in the presence of microorganisms (*Micrococci* and *Lactobacilli*). The pH-fall caused by lactic acid accumulation is reflected in shrinkage of the product. The sausages become firm and suitable for slicing after vaporization of the released water. Accelerated ripening is made possible by souring with the help of glucono-δ-lactone.

Cooked sausages are made by stuffing the cooked ingredients and heating the sausages for the shortest possible time. Cooking is performed in water or in a stream of hot air. The internal temperature of the sausages reaches 69–71 °C. Typical cooked sausages are liver, blood and jellied meat sausages.

Emulsion-type sausages (German "Brühwurst") are usually cooked together with other food. Typical products are bockwurst, frankfurters, wieners (these are similar to frankfurters but shorter), white and hunter's sausages and mortadella. Disintegration of the meat is done in a grinder or in a chopper with rotating knife blades with the addition of water. When slaughtered meat is processed after rigor mortis, the water holding capacity is increased by addition of common salt and of ingredients which enhance the chopping process (condensed phosphates and salts such as lactates, acetates, tartrates or citrates). The swelling resulting from added salts is caused by an increase in the pH of the meat slurry and by the complexing of divalent cations which, in free form, suppress swelling. The temperature during grinding/chopping has to be kept low (addition of ice or ice-cold water) since higher temperatures decrease the water holding capacity. Water retention goes up also, as the fat component of the meat slurry is increased, as long as the fat:protein ratio does not exceed 2.8 to 1. As a consequence the salt concentration is increased. The slurry is actually an emulsion (the frankfurter or hot dog is a typical emulsion product). The fat droplets are coated with myosin and are suspended in a continuous phase of aqueous protein. The ice is needed to prevent heat coagulation of myosin at the emulsifying stage. To sufficiently coat all the fat droplets, without exhausting available protein, the droplet surfaces are not allowed to increase through further disintegration. Hence, overchopping in the process is also avoided.

After chopping and stuffing, the sausages may be smoked at a temperature less than 80 °C (internal product temperature of 68 °C). At this temperature, coagulation of protein gel, which holds the water, forms the broken texture so typical of emulsion-type sausages.

Table 12.16. Protein and fat contents of ham and sausage products

Product	Moisture %	Protein %	Fat %	Caloric value (kJ) (kJ/100 g)
Salami				
(German style)	27.7	17.8	49.7	2303
Cervelat sausage	34.8	16.9	43.2	2028
Knackwurst	50.1	11.9	33.7	1559
Bratwurst (pork)	52.7	12.7	32.4	1522
Hunter's sausage	52.5	15.6	29.2	1448
Gelbwurst	53.1	11.8	32.7	1520
Munich Weisswurst (white sausage Munich style)	65.2	11.1	21.7	1067
Bockwurst	59.1	12.3	25.3	1232
Liver sausage	42.9	12.4	41.2	1881
Rotwurst	45.5	13.3	38.5	1775
Ham, raw	43.4	18.0	33.3	1665
Ham, cooked	62.0	21.4	12.8	903
Bacon, marbled	20.0	9.1	65.0	2751

12.7.2.3 Meat Paste (Pâté)

Meat pastes are delicately cooked meat products made primarily from meat and fat of calves and hogs and, often, from poultry (e.g. goose liver paste) or wild animal meat (hare, deer or boar). Unlike sausages, pastes contain quality meat and are free of slaughter scrapings or other inferior by-products. A portion of meat or the whole meat used is present as finely comminuted spreadable paste.

12.7.3 Meat Extracts and Related Products

12.7.3.1 Beef Extract

Meat extract is a concentrate of water-soluble beef ingredients devoid of fat and proteins. Its preparation dates back to *Liebig's* work in Mu-

Table 12.17. Chemical composition of beef extract

	%
Organic matter	56–64
Amino acids, peptides	15–20
Other N-compounds	10–15
Total creatinine	5.4–8.2
Ammonia	0.2–0.4
Urea	0.1–0.3
N-free compounds	10–15
Total lipids	> 1.5
Pigments	10–20
Minerals	18–24
Sodium chloride	2.5–5
Moisture	15–23
pH-value of a 10% aqueous solution	approx. 5.5

nich in 1847. Comminuted beef is countercurrently extracted with water at 90 °C. After removal of fat by separators and subsequent filtration, the extract, containing 1.5–5% solids, is concentrated to 45–65% solids in a multiple stage vacuum evaporator which operates in a decreasing temperature gradient (a range of 92 to 46 °C). The final evaporation to 80–83% solids is then carried out under atmospheric pressure at a temperature of 65 °C or higher.

In the same way, the cooking water recovered during the production of corned beef can be processed into meat extract. Only this latter source of meat extract is of economic significance. The yield is 4% of fresh meat weight. The composition of the extract is given in Table 12.17. For addition to soup powders and sauce powders, the thick pasty meat extract is blended with a carrier substance and vacuum- or spray-dried.

12.7.3.2 Whale Meat Extract

This product is obtained from meat of various whales (blue, finback, sei, humpback and sperm) in a process similar to that of beef extract.

12.7.3.3 Poultry Meat Extract

Chicken extract is obtained by evaporation of chicken broth or by extraction of chicken halves with water at 80 °C, followed by a concentration step in a vacuum to an end-product of 70–80% solids.

12.7.3.4 Yeast Extract

Yeast cells (*Saccharomyces* and *Torula* spp.) are forced to undergo shrinking of protoplasm by addition of salt, which causes loss of cell water and solutes (plasmolysis), or the cells are steamed or subjected to autolysis. Cells treated in this way are extracted with water and the extract is concentrated to yield a brown paste. Yeast extract is rich in the B-vitamins. The concentrations of thiamine and thiamine diphosphate are above their taste threshold values and may contribute to the product's unpleasant flavor. On the other hand, the spicy flavor of the paste is essentially due to 5′-nucleotides freed during hydrolysis and to amino acids, particularly glutamic acid.

12.7.3.5 Hydrolyzed Vegetable Proteins

Plant proteins (wheat gluten, corn zein and the coarsely-ground peanut and soybean) are hydrolyzed by hydrochloric acid. The hydrolysate is then adjusted with sodium hydroxide or carbonate to pH 4.8–5.8, the brown insoluble pigments are removed by filtration and, after treatment with activated carbon to bleach the product, the hydrolysate is left to mature. This product, with 40–50% solids, is utilized as a food seasoning or, with 80% solids, as a paste for further food processing. Both products have a meat- or bouillon-like odor and flavor.

12.8 Meat Aroma

Raw meat has only a weak aroma. Numerous intensive aroma variations arise from heating, the character of the aroma being dependent on the type of meat and the method of preparation (stewing, cooking, pressure cooking, roasting or broiling-barbecuing). The kind of meat is effective mainly via lipid degradation products, while preparation effects are based on reaction temperatures and reactant concentrations. Thus, a carefully dried, cold aqueous meat extract provides a roasted meat aroma when heated, while an extract heated directly, without drying, provides a bouillon aroma. When meat is fried in hot oil, the frying oil is an additional reactant.

Meat aroma consists of: (a) nonvolatile flavor substances, (b) flavor enhancers and (c) volatile aroma constituents. The latter compounds or their precursors originate essentially from the water-soluble fraction. Amino acids and peptides and some acids such as lactic acid are significant flavoring substances. Flavor enhancers are glutamic acid and inosine 5′-monophosphate. The volatile aroma substances are formed by *Maillard* and *Strecker* degradation reactions. On the one hand are reactants such as amino acids and peptides and, possibly, whole proteins and glycoproteins, while on the other hand, nucleic acids and nucleotides supply the sugars. In addition, an aroma role is ascribed to thiamine and related compounds. The volatiles from heated meat identified so far belong to the following classes of compounds: alcohols, aldehydes, ketones, carboxylic acids, esters, benzene derivatives, furans, γ- and δ-lactones, pyrans, oxazolines, pyrroles, pyrimidines, pyrazines, thiols, sulfides, thiophenes, thiazoles, trithiolanes, trithianes, dithiazines and dithianes.

Although a great number of aroma compounds have been identified, their sensory contribution has not yet been sufficiently elucidated. It is known that sulfur compounds play a significant role in meat aroma. It is assumed that their occurrence is due to the interaction of carbonyls with H_2S and ammonia.

Several compounds have a meat-like aroma note. Among them are methional, 2-formylthiophene, 5-thiomethyl furfural and 3,5-dimethyl-1,2,4-trithiolane. Nevertheless, the typical meat aroma in its many variations is not the result of a few but of many compounds present in the correct concentrations and proportions. The generation of a meat-like aroma is also possible by heating cystine, cysteine, methionine or thiamine with reducing sugars, by reaction of H_2S with alkenals and hydroxydihydrofurans, or by other reactions. A more or less satisfactory imitation of meat aroma is also made possible by utilizing identified aroma constituents or replacing them by synthetic analogues and using them either singly or in a mixture.

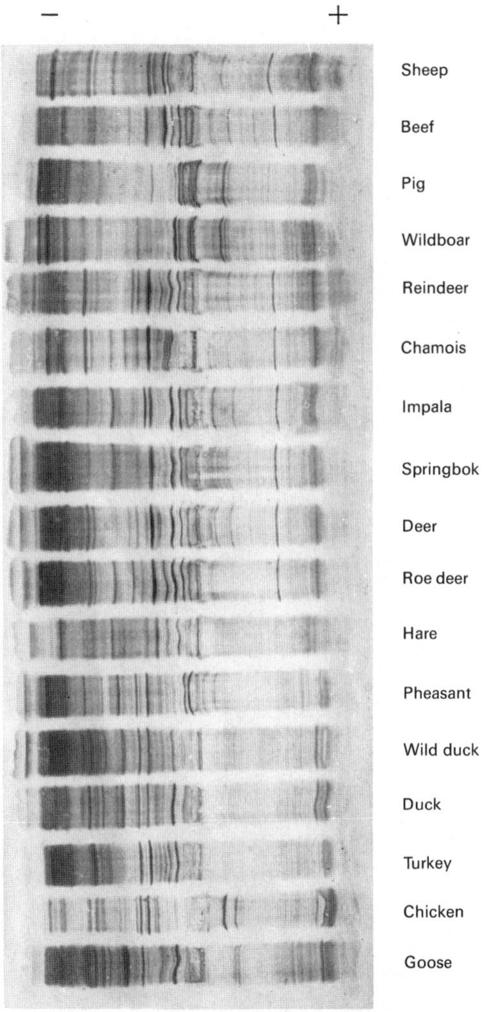

− +

Sheep
Beef
Pig
Wildboar
Reindeer
Chamois
Impala
Springbok
Deer
Roe deer
Hare
Pheasant
Wild duck
Duck
Turkey
Chicken
Goose

Fig. 12.29. Separation of sarcoplasm proteins of various warm blooded animals (mammals and fowl) by isoelectric focusing on polyacrylamide gels (PAGIF, PAGplate, pH range 3.5–9.5). (After *Kaiser,* 1980 a)

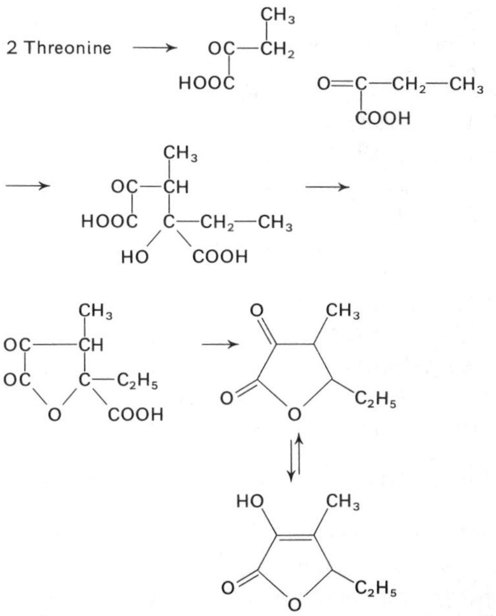

2 Threonine ⟶ $HOOC{-}CO{-}CH_2{-}CH_3$ $\quad O{=}C(COOH){-}CH_2{-}CH_3$

⟶ $HOOC{-}CO{-}CH(CH_3){-}C(OH)(COOH){-}CH_2{-}CH_3$ ⟶

(cyclic anhydride: $OC{-}CH(CH_3){-}C(C_2H_5)(COOH){-}O{-}CO$) ⟶ 3-methyl-4-oxo-5-ethyl-dihydrofuran-2-one ⇌ 3-hydroxy-4-methyl-5-ethyl-2(5H)-furanone (HO, CH_3, C_2H_5)

(12.18)

The meat-like aroma note of plant protein hydrolysates has been outlined. Also, numerous volatile compounds in these preparations have been confirmed; among them, an abundance of heterocyclic compounds, carbonyls and phenols. Their formation also follows the reactions mentioned above. In addition, lignin degradation occurs in plant tissue, providing ferulic acid and related compounds.

A typical meat-like aroma compound is 2-hydroxy-3-methyl-4-ethyl-2-butene-1,4-olide, obtainable from threonine through α-ketobutyric acid (cf. Reactions 12.18).

Homocysteic acid, cysteine-S-sulfonic acid, tricholomic and ibotenic acids are flavor enhancers with an effect similar to that of glutamic acid.

12.9 Meat Analysis

12.9.1 Meat

The determination of the kind of animal, the origin of meat, differentiation of fresh meat from that kept frozen and then thawed, and the control of veterinary medicines is of interest. The latter include antibiotics (penicillin, streptomycin, tetracyclines, etc.) used to treat dairy cattle infected with mastitis, and other chemicals, including diethyl stilbestrol, used for cattle to increase the efficiency of conversion of feed into meat.

12.9.1.1 Animal Origin

The animal origin of the meat can be determined by immunochemical and/or electrophoretic methods of analysis.

Serological differentiation. This analysis is performed in the same way as described for soya and milk proteins (cf. 12.9.2.3.2), i.e. by an antibody-antigen reaction, when antibody is available from the animal being investigated.

Electrophoresis. To determine the animal or plant origin of the food, electrophoretic procedures have often proved to be valuable when the electropherograms of the protein extracts reveal protein zones or bands specific for the protein source. Thus, in meat analysis, such a method allows for the differentiation between beef, pork, horse, sheep and wildlife meat (Figure 12.29). From the zone intensity of an electropherogram, it is possible to estimate the proportion of one kind of meat in a meat mix. This is illustrated in Figure 12.30 using a mixture of beef and pork as an example.

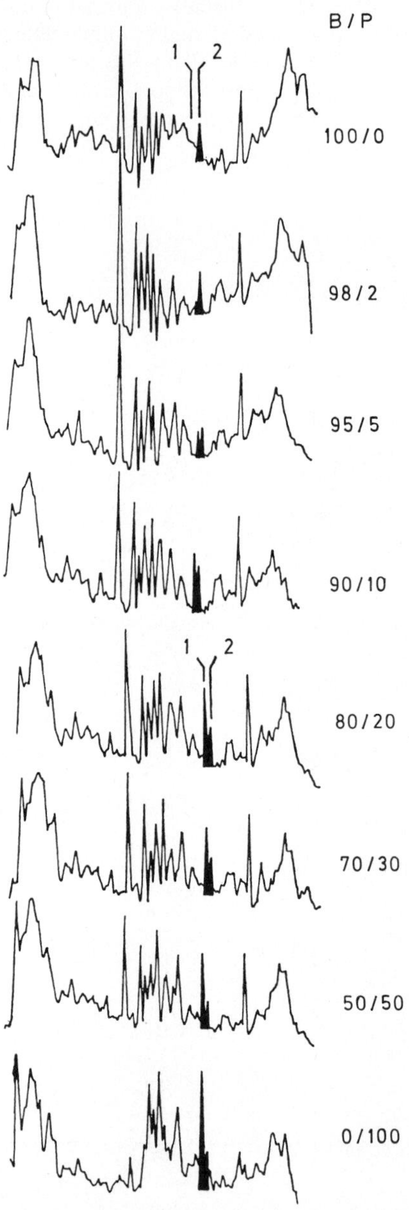

Fig. 12.30. Blended beef and pork meat: Densitograms of sarcoplasm proteins separated by PAGIF and PAG-plate; pH range 3.5–9.5. B/P (beef/pork) blend ratios in weight %. (After *Kaiser,* 1980b)

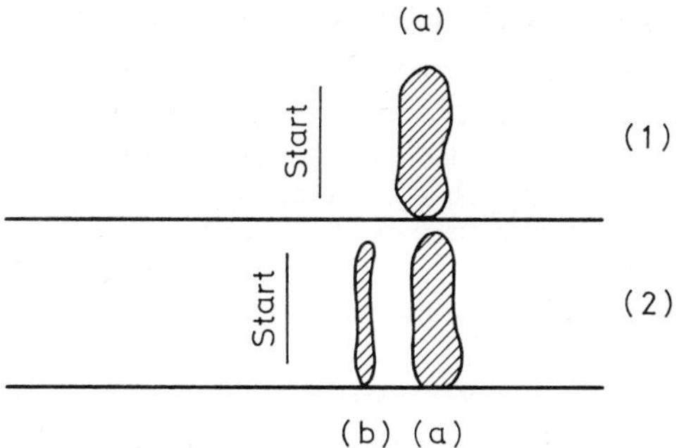

Fig. 12.31. Differentiation of fresh liver (1) from frozen and thawed liver (2) by electrophoretic separation of glutamate-oxalacetate transaminases (a) GOT sarcoplasm, (b) GOT mitochondria. (After *Hamm* and *Mašić,* 1975)

In order to carry out an analysis, the proteins should be extracted in the presence of urea or sodium dodecyl sulfate (SDS), which increase the protein solubility. The electrophoretic separation is achieved in starch, agarose or polyacrylamide gels at a constant pH or by applying a pH gradient within the gel (isoelectric focusing). In many cases protein patterns are obtained which provide useful information. Protein denaturation occurs in heat-treated meat samples. The extent is proportional to the temperature and length of heating; as a consequence, immunochemical and electrophoretic identification procedures become less applicable and less reliable.

12.9.1.2 Differentiation of Fresh and Frozen Meat

The isoenzyme patterns of cell organelles, for instance mitochondria and microsomes, differ often from those of cytoplasm. When the organelle membranes are damaged by a physical or chemical process, isoenzyme blending will occur in the cytoplasm.

Such membrane damage has been observed by freezing and thawing of tissue, for example, of muscle tissue, in which the isoenzymes of glutamate oxalacetate transaminase (GOT) bound to mitochondrial membranes are partially released and found in the sarcoplasm. The pressed sap collected from fresh unfrozen meat has only sarcoplasm enzymes, while the frozen and thawed meat has, in addition, the isoenzymes derived from mitochondria. The GOT isoenzymes can be separated by electrophoresis (Figure 12.31). This procedure is also applicable to fish.

12.9.1.3 Pigments

Pigment analysis is carried out for the evaluation of meat freshness. The individual pigments, such as myoglobin (purple-red), oxymyoglobin (red) and metmyoglobin (brown), are determined.

12.9.1.4 Treatment with Proteinase Preparations

Proteinases injected intramuscularly or through blood vessels degrade the structural proteins and, hence, proteolytic enzymes can be used to soften or tenderize meat. The enzymes are of plant or microbial origin and are used in the meat and poultry industries, while some are also used in the household as meat tenderizers. Analytical determination of proteinases is relatively difficult.

A possible assay may be based on disc gel electrophoresis of meat extracts, prepared in the presence of urea and SDS. The band intensities of the lower molecular weight collagen fragments increase in proteinase-treated meat.

12.9.1.5 Anabolic Steroids

Anabolic compounds present in animal feed as an additive increase muscle tissue deposition. Owing to a potential health hazard, some of these compounds are banned in many countries. Their detection can be achieved by the mouse uterus

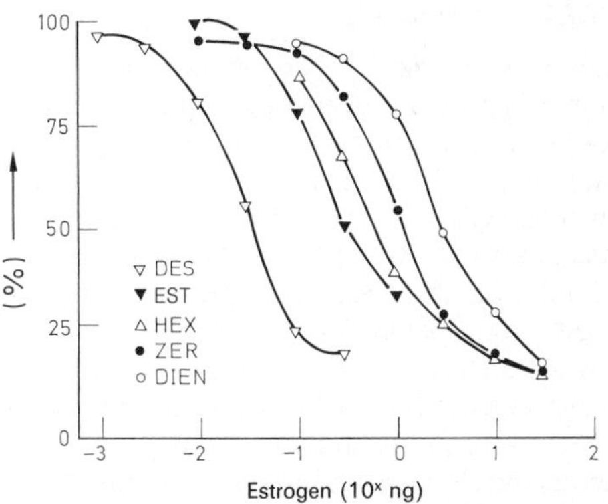

Fig. 12.32. Relative binding affinity of estrogen compounds to estrogen receptor. 50% Binding achieved by: 0.034 ng diethylstilbestrol (DES), 0.33 ng 17–β-estradiol (EST), 0.6 ng hexestrol (HEX), 1.2 ng zeranol (ZER), 2.9 ng dienestrol (DIEN). (After *Ingerowski* and *Stan,* 1978)

test or by a radioimmunoassay. Special receptor proteins which have the property of binding strongly to estrogens are isolated from rabbit or cattle uterus. The hormone-receptor complex is in equilibrium with its components:

Receptor + estrogen ⇄ Receptor-estrogen complex

The nonlabelled estrogens bound to receptor in the test sample will be competitively displaced by the addition of 17-β-estradiol labelled with tritium for radiochemical assay.

To reach equilibrium, a suitable amount of receptor protein and a constant amount of labelled estradiol are incubated together with the test sample. The amount of the radioactive ^{3}H-estradiol receptor complex will decrease in the presence of competitive estrogens from the meat extract. The binding affinity of the estrogen receptor depends on the type of estrogen present (Figure 12.32). Hence, detection limits differ and range from 0.3 to 50 ppb (mg per metric ton).

Anabolic compounds can be further separated by gas-liquid chromatography after derivatization of the polar functional groups, and be identified by mass spectrometry. This method allows the determination of weak or nonestrogenic components too, but in the past it suffered from high losses in sample preparation and could not compete with radioimmunoassay in sensitivity. In the meantime disadvantages of the method have been eliminated.

12.9.1.6 Antibiotics

Antibiotics are used as part of therapy to treat animal diseases and, sometimes, in low concentrations, as constitutents of animal feed to increase feed utilization and thus to accelerate animal growth.

Detection of antibiotics is usually achieved by the inhibition of the growth of bacteria ("inhibitor test"). *Bacillus subtilis,* strain BGA, is one of the recommended test organisms.

Chemical methods must be used in order to identify and quantify the antibiotics and other veterinary medical residues. The principal method is chromatographic separation coupled with mass spectrometry. The tetracyclines, which are common antibiotics, can be determined relatively easily by fluorometric measurement of meat extracts which are adequately prepared and purified.

12.9.2 Processed Meats

Besides the estimation of the animal species and the control of additives, the analysis of processed meats is associated with verifying composition. Here the emphasis is on the content of extraneous added water, carbohydrate-containing thickeners and binders, nonmeat protein additives and fat. In addition, the determination of nitrites, nitrates, nitrosamines and, for enhancing the pinkish-red color of processed meat, L-ascorbic acid is of importance in pickle-cured meat products. Other analytical problemes involve the detection of condensed phosphates, citric acid and glucono-δ-lactone, as well as the detection of polycyclic aromatic compounds in smoked meats and, also, of mycotoxins in products with desirable or undesirable growth of molds.

12.9.2.1 Main Ingredients

The first insight into the composition of processed meat, i.e. whether it contains an excess of fat or carbohydrate, which would lower the protein content and thus lower the value of the processed meat, is obtained by proximate analysis of the product's main ingredients: moisture, raw protein, fat and ash content. If their sum is less than $100 \pm 0.5\%$ of the sample weight, then the presence of carbohydrate binders should be verified. A positive finding should be further investigated since incorporation of liver into processed meat may provide glycogen. Hence, thorough carbohydrate analysis is required.

12.9.2.2 Added Water

Moisture content is related to protein content and is relatively constant. *Feder's* method of analysis of water added to chopped or ground meat or to emulsion-type sausages is based on these findings. The method uses the empirical equation:

$$\text{Water added (\%)} = \text{Moisture (\%)} - \text{Protein (\%)} \times F \qquad (12.19)$$

F for beef, emulsion-type sausage = 4.0;
F for pork = 4.5

This indirect method for assessing the amount of added water has been repeatedly criticized. In spite of this, no better method has yet been developed. Moreover, the calculated water content is never used alone to evaluate a meat product. Other significant data, such as muscle protein content and the proportion of fat to protein, are also included.

12.9.2.3 Lean Meat Free of Connective Tissue

A measure of meat quality is expressed as the amount of lean meat free of connective tissue, which corresponds to meat proteins devoid of connective tissue protein (MPDCP). To obtain this value, the meat sample is analyzed for connective tissue proteins (CP), added extraneous protein (EP) and nonprotein-nitrogen (NPN), e.g. glutamate, purine and pyrimidine derivatives, urea. These values are then deducted from the value for total protein (TP):

$$MPDCP = TP - (CP + EP + NP)$$

Another method still being tested is based on drastic treatment (heating to 130 °C at pH 9) of a meat sample, under which conditions extraneous proteins, collagen and blood plasma proteins solubilize, while the residual protein is calculated using a constant factor as MPDCP.

12.9.2.3.1 Connective tissue protein

The amino acid 4-hydroxyproline is a marker compound for connective tissue. It occurs only in connective tissue protein. The amount of 4-hydroxyproline is determined in the acid hydrolysate of the sample or the isolated protein using an amino acid analyzer, or colorimetrically using a specific color reaction. The latter, accepted widely in practice, is a direct photometric procedure based on the oxidation of hydroxyproline in alkaline solution by H_2O_2 or N-chloro-p-toluenesulfonamide (chloramine-T). The oxidation yields a pyrrole derivative which is then condensed with p-dimethylaminobenzaldehyde to form a red pigment. The connective tissue content of meat is calculated by multiplying the hydroxyproline value by a factor of 8, which corresponds to an average of 12.4% hydroxyproline content of connective tissue.

12.9.2.3.2 Added protein

In order to extend or improve the water holding capacity of processed meat, the product may contain milk, egg or soy proteins. These proteins can be detected immunochemically with high sensitivity using a simple and rapid diffusion test on agar gel (Figure 12.33).

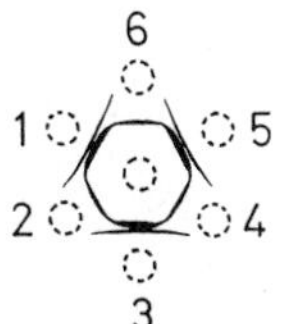

Fig. 12.33. Immunochemical detection of nonmeat proteins in a emulsion-type sausage made at 85 °C. (After *Guenther,* 1969.) Agar-agar gel-coated glass plate has reservoir cavities (wells). A diffusion process occurs against antibody solution in centrally-located well with soya protein (reference) from wells 1, 3, 5 and of emulsion sausage extract from wells 2, 4, 6. After staining with Amido Black two precipitation zones are revealed for soya protein and one for emulsion-type sausage extract

Intensive heat treatment as, for example, in emulsion-type canned meat (Table 12.18), may cause severe changes in the solubility and the antigenic properties. In such cases, no quantitative assessment is possible. An additional difficulty arises when the antibody reacts with several related proteins. This so-called "cross reaction" has been observed, for example when egg white antibody precipitated not only egg white proteins, but also those of milk. In such cases, instead of an agar gel diffusion, the immunoelectrophoresis method is suitable. In addition to precipitation, the positions of the protein bands in an electropherogram provide the qualitative data needed to unequivocally identify the extraneous protein in the processed meat.

However, a simple electrophoretic separation of protein is suitable for some problems. The addition of soya proteins to emulsion-type sausages is readily revealed by disc gel electrophoresis. Two soya protein bands are clearly separated from sausage meat proteins and other added proteins, such as those of milk or egg white.

Table 12.18. Determination of soya protein in emulsion type sausages

Soya protein (%)		Temperature (°C)[a]
added	found	
0.75	0.5 ± 0.1	115
1.50	1.4 ± 0.1	115
3.0	2.0	115
0.75	0.6 ± 0.1	121
1.50	1.0 ± 0.1	121
3.0	0.9 ± 0.1	121

[a] Temperature recorded in center of can.

12.9.2.4 Nitrosamines

Not only does the question of the content of nitrite or nitrate in pickle-cured meat arise, but also whether nitrosamines are formed and to what extent they occur in meat (cf. 9.8).

In general, nitrosamines arise only in very low concentrations. Since some of these compounds are a great health hazard, they should be detectable below 0.1 ppm in food for human consumption. The same procedures are available for identifying volatile nitrosamines which had been described earlier for the analysis of aroma constituents (cf. 5.2). However, precautions should be taken during the isolation step. Isolation of nitrosamines should not proceed at low pH since an acid medium in the presence of residual meat nitrites promotes further *de novo* synthesis of nitrosamines. Since the isolated fraction of neutral volatile compounds, which also includes nitrosamines, is highly complex in composition, reliable nitrosamine identification by gas chromatographic retention data is not possible. Additional mass spectrometric data are needed to verify the chemical structure.

The limit of detection of nitrosamines can be further improved when quantitative analysis involves prior derivatization to a product which is readily detected by the highly sensitive electron capture detector (ECD). Two methods are outlined in Figure 12.34: nitrosamine is oxidized to nitramine, or the nitroso group is eliminated, and the corresponding amine is converted into heptafluorobutyric acid amide.

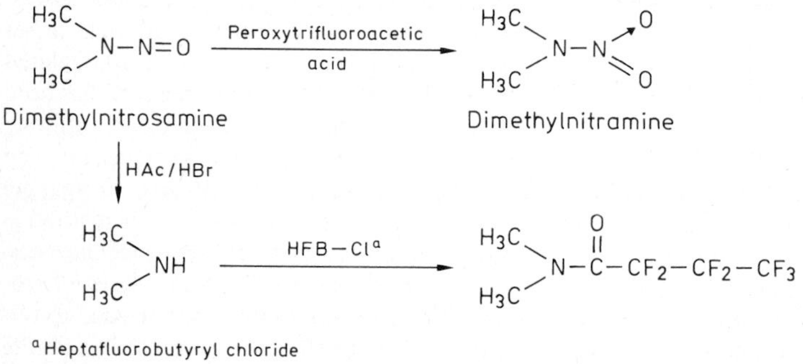

$(H_3C)_2N-N=O$ (Dimethylnitrosamine) —Peroxytrifluoroacetic acid→ $(H_3C)_2N-NO_2$ (Dimethylnitramine)

Dimethylnitrosamine —HAc/HBr→ $(H_3C)_2NH$ —HFB−Cl[a]→ $(H_3C)_2N-C(=O)-CF_2-CF_2-CF_3$

[a] Heptafluorobutyryl chloride

Fig. 12.34. Gas chromatographic detection of nitrosamines after derivatization (electron capture detector)

12.10 Literature

Beck, G., Mantel, Th.: Der biologische Hemmstofftest, ein Weg zur Klärung des Rückstandsproblems in Fleisch. Fleischwirtschaft *52,* 1605 (1972)

Bendall, J. R., Wismer-Pedersen, J.: Some properties of the fibrillar proteins of normal and watery pork muscle. J. Food Sci. *27,* 144 (1962)

Bornstein, P.: The biosynthesis of collagen. Ann. Rev. Biochem. *43,* 567 (1974)

Briskey, E.J., Wismer-Pedersen, J.: Biochemistry of pork muscle structure; I. Rate of anaerobic glycolysis and temperature change versus the apparent structure of muscle tissue. J. Food Sci. *26,* 297 (1961).

Disney, J.G., Follet, M.J., Ratcliff, P.W.: Biochemical changes in beef muscle post mortem and the effect of rapid cooling in ice. J. Sci. Food Agric. *18,* 314 (1967)

Fennema, O.R. (Ed.): Principles of food science. Part I: Food chemistry. Marcel Dekker, Inc.: New York 1976

Flament I., Willhalm, B., Ohloff, G.: New developments in meat aroma research. In: Flavor of foods and beverages (Eds.: Charalambous, G., Inglett, G.E.), p. 15, Academic Press: New York–London. 1978

Giddings, G.G.: The basis of color in muscle foods. Crit. Rev. Food Sci. Nutr. *9,* 81 (1977)

Grau, R.: Fleisch und Fleischwaren. Verlag Paul Parey: Berlin–Hamburg. 1969

Grüttner, F., Lienhop, E.: Taschenbuch der Fleischwarenherstellung. 6. Aufl., Verlag Günter Hempel: Braunschweig. 1962

Günther, H.: Bestimmung von Fremdeiweiß in Fleischwaren. Arch. Lebensmittelhyg. *20,* 97 (1969)

Gutschmidt, J.: The storage life of frozen chicken with regard to the temperature in the cold chain. Lebensm. Wiss. Technol. *7,* 139 (1974)

Hamm, R.: Kolloidchemie des Fleisches. Verlag Paul Parey: Berlin–Hamburg. 1972

Hamm, R., Maic, D.: Routinemethode zur Unterscheidung zwischen frischer Leber und aufgetauter Gefrierleber. Fleischwirtschaft *55*, 242 (1975)

Harrington, W.F., Rao, N.V.: Collagen structure in solution. I. Kinetics of helix regeneration in single-chain gelatins. Biochemistry *9*, 3714 (1970)

Herrmann, Ch., Merkle, Ch., Kotter, L.: Zur Problematik serologischer Nachweisreaktionen für Fremdeiweiße in erhitzten Fleischerzeugnissen. Fleischwirtschaft *53*, 97 und 249 (1973)

Herrmann, Ch., Thoma, H., Kotter, L.: Zur direkten Bestimmung von Muskeleiweiß in Fleischerzeugnissen. Fleischwirtschaft *56*, 87 (1976)

Huxley, H.E.: Electron microscope studies on the structure of natural and synthetic protein filaments from striated muscle. J. Mol. Biol. *7*, 281 (1963)

Ingerowski, G.H., Stan, H.-J.: Nachweis von Östrogen-Rückständen in Fleisch mit Hilfe des cytoplasmatischen Östrogenrezeptors aus Rinderuterus. Dtsch. Lebensm. Rundsch. *74*, 1 (1978)

Kaiser, K.-P., Matheis G., Kmita-Dürrmann, Ch., Belitz, H.-D.: Identifizierung der Tierart bei Fleisch, Fisch und abgeleiteten Produkten durch Proteindifferenzierung mit elektrophoretischen Methoden. I. Rohes Fleisch und roher Fisch. Z. Lebensm. Unters. Forsch. *170*, 334 (1980a)

Kaiser, K.-P., Matheis, G., Kmita-Dürrmann, Ch., Belitz, H.-D.: Proteindifferenzierung mit elektrophoretischen Methoden bei Fleisch, Fisch und abgeleiteten Produkten. II. Qualitative und quantitative Analyse roher binärer Fleischmischungen durch isoelektrische Fokussierung in Polyacrylamidgel. Z. Lebensm. Unters. Forsch. *171*, 415 (1980b)

Karlson, P.: Kurzes Lehrbuch der Biochemie. 10. Aufl., Georg Thieme Verlag: Stuttgart. 1977

Lawrie, R.A.: The conversion of muscle to meat. In: Recent advances in food science (Eds.: Hawthorn, J., Leitch, J.M.), Vol. I, p. 68, Butterworths: London. 1962

Lawrie, R.A. (Ed.): Developments in meat science-1, Applied Science Publ.: London. 1980

Lehninger, A.L.: Biochemie. 2. Aufl., Verlag Chemie: Weinheim–New York. 1977

MacLeod, G., Seyyedain-Ardebili, M.: Natural and simulated meat flavors (with particular reference to beef). Crit. Rev. Food Sci. Nutr. *14*, 309 (1981)

Ohloff, G., Flament, I: Heterocyclic constituents of meat aroma. Heterocycles *11*, 663 (1978)

Pearson, A.M., Gray, J.I.: Mechanism responsible for warmed-over flavor in cooked meat. In: The Maillard reaction in foods and nutrition (Eds.: Waller, G.R., Feather, M.S.), ACS Symposium Series 215, p. 287, American Chemical Society: Washington, D.C. 1983

Potthast, K., Hamm, R.: Biochemie des DFD-Fleisches. Fleischwirtschaft *56*, 978 (1976)

Price, J.F., Schweigert, B.S. (Eds.): The science of meat and meat products. 2nd edn., W.H. Freeman: San Francisco. 1971

Rich, A., Davies, D.R., Crick, F.H.C., Watson, J.D.: The molecular structure of polyadenylic acid. J. Mol. Biol. *3*, 71 (1961)

Schormüller, J. (Hrsg.): Handbuch der Lebensmittelchemie. Bd. I, Springer-Verlag: Berlin–Heidelberg 1965

Schultz, H.W., Angelmier, A.F. (Eds.): Proteins and their reactions. AVI Publ. Co.: Westport, Conn. 1964

Scanlan, R.A.: N-Nitrosamines in foods. Crit. Rev. Food Technol. *5*, 357 (1974)

Sulser, H.: Die Extraktstoffe des Fleisches. Wissenschaftliche Verlagsgesellschaft mbH: Stuttgart. 1978

Tóth, L.: Chemie der Räucherung. Verlag Chemie: Weinheim. 1982

Traub, W., Piez, K.A.: The chemistry and structure of collagen. Adv. Protein Chem. *25*, 243 (1971)

13 Fish, Whales, Crustaceans, Mollusks

13.1 Fish

13.1.1 Foreword

Fish and fish products fulfill an important role in human nutrition as a source of biologically-valuable proteins, fats and fat-soluble vitamins. Fish can be categorized in many different ways, such as:

- By environment in which the fish lives: sea fish (herring, cod, saithe) and freshwater fish (pike, carp, trout), or those which can live in both environments, e.g. eels and salmon. Sea fish can be subdivided into groundfish and pelagial fish.
- By body form: round (cod, saithe) or flat (common sole, turbot or plaice).

Commercial fishing takes place in the open sea, coastal and freshwater areas. Conservation programs and hatcheries to rebuild stocks play important roles in management of fresh and saltwater fish resources.
The fishing industry catch has risen sharply in tonnage during this century. In 1900 the catch was approx. 4 million t, while it had increased to 74 million t by 1977. Table 13.1 shows the catch in tonnage of the leading countries engaged in fishing. This includes shellfish products, i.e. lobsters and crustaceans such as crabs, crayfish and shrimps, and mollusks such as clams, oysters, scallops, squid, etc., which are not true fish but are harvested from the sea by the fishing industry. Table 13.2 lists the catch in the same year of the chief kinds of fish and shellfish. A review of the forms of fish products entering the market is given in Table 13.3.

Table 13.1. World catch of fish, crustaceans and mollusks (1981)

Continent	1,000 t	Country	1,000 t
World	74,760	USA	3,767
Africa	4,036	Chile	3,393
America, North-,		Peru	2,751
Central-	7,220	Norway	2,552
America, South-	8,551	India	2,415
Asia	31,489	South Korea	2,366
Europe, West-	11,201	Indonesia	1,863
Europe, East-		Denmark	1,814
+ USSR	10,788	Philippines	1,651
Australia		Thailand	1,650
+ South Pacific		Mexico	1,565
Islands	355	North Korea	1,500
		Greenland	1,421
Country	**1,000 t**	Canada	1,362
		Spain	1,264
Japan	10,657		
USSR	9,546	Σ (%)[a]	75
China	4,605		

[a] World production ≙ 100%.

13.1.2 Food Fish

Table 13.4 shows the important food fish. In general a predatory fish is better tasting than a nonpredaceous fish, a fatty fish better than a nonfatty fish: The fishbone-rich species, such as carp, perch, pike and fench, are often less in demand than fish with fewer bones.
Some important food fish will be described in more detail.

13.1.2.1 Sea Fish

13.1.2.1.1 Sharks

Dogfish (*Squalus acanthias*) about 1 m long are marinated or smoked before marketing. In North America fish of the family *Squalidae* are generally referred to as dogfish sharks. Other names are spiny, spring or piked dogfish and rock salmon. Trade names used in th U.K. are flake, huss or rig. In Germany the name used is Dornhai (Dornfisch), and the smoked dorsal muscle is sold as "Seeaal", while the hot-smoked, skimmed belly walls are called "Schillerlocken". Mackerel sharks of the family *Lamnidae* are also in this group. The main species of this family are: (a) porbeagle, blue dog or Beaumaris shark (*Lamna nasus*); and (b) salmon, (c) mako and (d)

Table 13.2. Catch of fish, crustaceans and mollusks by species (1981)

	1,000 t
Freshwater fish[a]	
Carps, barbels etc.	706
Tilapias	416
Sturgeons	29
River eels	87
Salmons, trouts, smelts, etc.	856
Others	5,442
Sea fish	
Flounders, halibuts, sole, etc.	1,089
Cods, hakes, haddocks etc.	10,601
Redfishes, basses, congers, etc.	5,213
Jacks, mullets, sauries, etc.	7,991
Herrings, sardines, anchovis	17,456
Tunas, bonitos, billfishes, etc.	2,453
Mackerels, snoeks, cutlassfishes, etc.	3,665
Sharks, rays, chimaeras, etc.	601
Others	8,546
Crustaceans	
Fresh water crustaceans	100
Sea-spiders, crabs, etc.	831
Lobsters	106
Shrimps	1,698
Other sea crustaceans	68
Mollusks	
Freshwater mollusks	253
Abaloues, winkles, couchs, etc.	90
Oysters	946
Octopuses, squids, cuttlefishes	1,304
Other sea mollusks	131
Turtles	6

[a] Includes fish species journeying between the sea and freshwater lakes or rivers.

Table 13.3. World market for fish and fish products (1981)

Marketed	Amount (%)[a]
Fresh fish	19.4
Frozen fish	22.5
Salted, smoked, marinated	14.5
Canned	14.0
	70.4
Fish meal, fish oil	28.2
Others	1.4

[a] As % of total catch.

white sharks. The blue shark is found in the Atlantic Ocean and the North Sea as an escort of herring schools. It possesses a meat similiar to veal and is known in the trade as sea or wild sturgeon, or calf-fish. Due to a high content of urea (cf. 13.1.3.3.6), the meat of these fish is often tainted with a mild odor of ammonia. Endeavors to popularize shark and related fish meat are well justified since the meat is highly nutritious; however consumer acceptance would be hampered by the word "shark", therefore other terms commonly are preferred in the trade. Shark fins are a favorite dish in China and are imported into Europe as a luxury food.

13.1.2.1.2 Herring

The herring (*Clupea harengus*) is one of the most processed and most important food fish and is a source of raw materials for meal and marine oil. Herring are categorized by the season of the catch (spring or winter herring); spawning time or stage of development (e.g. matje, the young fatty herring with roe only slightly developed, is cured and packed in half barrels); or by the ways in which the fish were caught: drag or drift nets, trawling nets, gill and trammel nets or by seining (purse seining), the most important form of snaring. Electrofishing methods have proven to be particularly economical on the high seas. The main fishing time for the German fleet is July, and for the English and Norwegian fishing fleets it is October to December.

The herring averages 12–35 cm in length and migrates in large swarms or schools throughout the nothern temperate and cold seas.

Herring is marketed cold or hot smoked (kippers, buckling), frozen, salt-cured, dried and spiced, jellied, marinated and canned in a large variety of sauces, creams, vegetable oils, etc.

Sprat (*Sprattus sprattus phalericus*), called brisling in Scandinavia and Sprotte in Germany, is processed into an "Appetitsild" (skinned fillets or spice-cured sprats packed in vinegar, salt, sugar and seasonings). Canned brisling is packed in edible oil, tomato and mustard sauces, etc. and is sold as brisling sardines. Brisling is often lightly smoked and marketed as such. Sprats are also processed into a delicatessen product called "Anchosen", which consists mostly of small sprats, sometimes mixed with cured matje, and preserved in salt and sugar, with or without spices and sodium nitrate.

Table 13.4. Major commercial fish species – quality and utilization

Name	Family	Genus, sp.	Comments for quality and processing
Sea fish			
Pleurotremata (sharks)			
Dogfish	Squalidae	*Squalus acanthias (Acanthias vulgaris)*	
Rajiformes (skates)			
Skates, e.g. thornback, common skate	Rajidae	*Raja clavata, R. batis*	used are the wing shaped body widening, and the pectoral part and breast fins as a delicacy; it is fried, smoked or jellied
Acipenseriformes (sturgeons)			
Sturgeon	Acipenseridae	*Acipenser sturio*	exceptionally delicate when smoked, from its roe caviar is made
Clupeiformes (herrings)			
Herring	Clupeidae	*Clupea harengus*	valuable fish with fine white meat, fried and grilled; industrially processed, for example into Bismarck herrings, rollmaps and brat-herring
Sprat	Clupeidae	*Sprattus sprattus*	mostly cold or warm smoked; anchovies
Sardine	Clupeidae	*Sardina pilchardus*	mostly steam cooked and canned in oil; along sea coast grilled and fried
Anchovy (Anchovis)	Engraulidae	*Engraulis encrasicolus*	pleasant, aromatic fragrant, cured in brine, made into rings and paste
Lophiiformes (anglers)			
angler, allmouth	Lophiidae	*Lophius piscatorius*	white, good and firm meat, poached
Gadiformes (cods)			
Ling	Gadidae	*Molva molva*	tasty firm white meat
Cod	Gadidae	*Gadus morrhua*	meat is prone to fracturing, used fresh, filleted, salted and frozen, dried (stock- and klipfish), cooked, poached; from liver oil is produced
Haddock	Gadidae	*Melanogrammus aeglefinus*	very fine in taste, processed as fresh; pickled, or marinated, smoked, fried, roasted, cooked or poached, or used for fish salad
Coalfish pollack, black cod or Boston bluefish	Gadidae	*Pollachius virens, P. pollachius*	meat is lightly tinted grayish-brown, it is filleted, smoked, sliced as cutlets or chops and processed in oil (used for salmon substitute)
Whithing	Merlangius	*Merlangius merlangius*	good meat, easily digested, but very sensitive, fried or deep fried roasted or smoked, used for fish stuffings
Hake	Merluccidae	*Merluccius merluccius*	fresh of frozen, all processing methods are used

Table 13.4. (Continued)

Name	Family	Genus, sp.	Comments for quality and processing
Scorpaeniformes			
Red fish, ocean perch	Scorpaenidae	*Sebastes marinus*	tasty meat it is more fatty than cod, it is filleted or smoked
Gurnard, sea robin (gray gurnard, red gurnard)	Triglidae	*Trigla gurnardus, T. lucerna*	white firm meat (red sp. is of higher quality), used freshly or as smoked
Lumpfish, sea hen	Cyclopteridae	*Cyclopterus lumpus*	smoked, its roe is processed into caviar substitute
Perciformes (percoid fishes)			
Red mullet	Mullidae	*Mullus barbatus*	white fine and a piquant delicius meat, mostly grilled
Catfish, wolffish	Anarhichadidae	*Anarhichas lupus, A. minor*	fine white fragrant meat, poached grilled, dough type crust coated
Mackerel	Scombridae	*Scomber scombrus*	highly valued fish, tasty reddish meat, fried, grilled, smoked or canned; its meat is not readily digestible
Tuna	Scombridae	*Thunnus thynnus*	reddish meat of exceptional taste, it is fried, roasted, smoked, or canned in oil or processed into paste sausages or rolls
Pleuronectiformes (flat fishes)			
Turbot (butt or britt)	Scopthalmidae	*Psetta maxima (Rhombus maximum)*	beside the common sole the highest valued flat fish, meat is snow-white firm and piquant, it is cooked, grilled or poached
Halibut	Pleuronectidae	*Hippoglossus h. hippoglossus*	tasty meat, it is poached, fried or smoked
Plaice	Pleuronectidae	*Pleuronectes platessa*	tasty meat fried or filleted and poached
Flounder	Pleuronectidae	*Platichthys flesus (Pleuronectes flesus)*	good white meat, it is poached, fried or smoked
Common sole	Soleidae	*Solea solea*	it is the finest flat fish, it is poached, fried, grilled or roasted
Freshwater fish			
Petromyzones (lampreys)			
Lamprey	Petromyzonidae	*Lampetra fluviatilis*	it is industrially processed
Anguilliformes (eel sp.)			
Eel	Angullidae	*Anguilla anguilla*	tasty meat, good quality when not exceeding 1 kg in weight, as fresh it is fried, roasted, or it is smoked, marinated of jellied
Salmoniformes (salmons)			
Salmon	Salmonidae	*Salmo salar*	high quality fish (5–10 kg), it is poached, grilled, cured or smoked, also pickled

Table 13.4. (Continued)

Name	Family	Genus, sp.	Comments for quality and processing
River trout	Salmonidae	*Salmo trutta*	high quality fish, fishbone poor, bluish tinted when cooked, roasted a la menuiere
Rainbow trout (lake- or steelhead trout)	Salmonidae	*Salmo gairdnerii*	
Brook trout	Salmonidae	*Salvelinus fontinalis*	a worthy fish, meat is pale pinkish, processed as trout but mostly fried
Whitefish	Salmonidae	*Coregonus sp.*	processed as trout
Coregonus, whitefish	Salmonidae	*Coregonus sp.*	white tender and tasty meat, though somewhat dry, it is fried or deep fried
Smelt	Osmeridae	*Osmerus eperlanus*	a fishbone rich meat which is mostly deep fried
Pike (jackfish)	Esocidae	*Esox lucius*	young pikes (best quality 2–3 kg) are soft tender and tasty the meat is well rated though it is bone rich, steam cooked, cooked or fried
Cypriniformes (carps)			
Roach	Cyprinidae	*Rutilus rutilus*	it has a tasty meat though fishbone rich
Bream	Cyprinidae	*Abramis brama*	tasty meat but fishbone rich
Tench	Cyprinidae	*Tinca tinca*	tender fatty meat, tasty, bluish when cooked, mostly steam cooked
Carp	Cyprinidae	*Cyprinus carpio*	soft meat, readily disgestible, a valuable fish food, bluish when cooked
Crucian	Cyprinidae	*Carassim carassim*	a good food fish, but not as carp, the meat is bone rich
Perciformes (perchoid fishes)			
Perch	Percidae	*Perca fluviatilis*	firm, white and very tasty meat, best quality is below 1 kg (25–40 cm), it is fried, filleted and/or steam cooked
Zander	Percidae	*Stizostedion lucioperca*	white soft juicy and tasty meat, 40–50 cm, fried or steam cooked, it is the best quality freshwater fish
Ruffle	Percidae	*Gymnocephalus cernua*	exceptionally tasty meat

Anchovies (*Engraulis encrasicholus*; German term "Sardellen"; found in Atlantic and Pacific Oceans) should also be included in the herring group. Anchovies are usually salted (cured in brine in barrels until the flesh has reddened). They are also canned in glass jars, marketed as a paste or cream, smoked or dried. Sardines (*Sardina pilchardus*), from the Mediterranean or Atlantic (France, Spain, Portugal) or from Africa's west coast, are often marketed steamed, fried or grilled, or canned in oil or tomato sauce. The fully-grown sardine is known in the trade as pilchard (Californian, Chilean, Japanese). It is also salt-cured and pressed in barrels or canned in edible oil or in sauce. "Russian sardines" or "Kronsardine" are actually marinated small herrings or sprats caught in the Baltic Sea. Also in the herring group is the allis shad (*Alosa alosa* or *Clupea alosa*), which is sold fresh, smoked or canned.

13.1.2.1.3 Cod fish

These nonfatty fish (from the North Sea, Iceland or Greenland) are usually marketed fresh, whole

and gutted, and may have the head and/or skin removed or be filleted. The Atlantic cod (*Gadus morhua*) is considered the most important food fish of Northern Europe. Classified by increasing size, designations are: small codling, codling, sprag and cod in the U.K. and Iceland; and scrod, market, large and whole, the latter being over 11 kg and more than 1 m long, in the United States.
Saithe is also known as coalfish or pollack (*Pollachius virens*) or by names such as black cod or Boston bluefish. After saltcuring and slicing, it is lightly smoked and packed in edible oil. Saithe is marketed in Germany as a salmon substitute called "Seelachs". Rolled in balls and canned, it is called "side boller" in Norway.
Whiting (*Merlangius merlangius*), known as merlan in France and Germany, is a North Atlantic, North Sea fish, marketed in many forms.
Haddock (*Melanogrammus aeglefinus*) is a North Atlantic and Arctic Sea fish. Small haddock are called gibbers or pingers, and large ones are jumbos. The annual haddock catch is lower than those of the above-mentioned fish, i.e. anchovy, herring, cod, sardine and pollack.
Hake (*Merluccius merluccius*) is an Atlantic and North Sea fish. Its various subspecies are the Cape, Chilean, Northeast Pacific, Mediterranean and North American east coast white hake. The annual catch is somewhat higher than that of haddock. Even higher than both is the catch of menhaden (*Brevoortia tyrannus*), which accounts for almost 38% of the fish tonnage in the United States.

13.1.2.1.4 Scorpaemids

Red fish of the North Atlantic and arctic regions (*Sebastes marinus* and other species), which are known as red fish or ocean perch (U.K.) or rosefish or Norway haddock (U.S.A.), have gained in importance in recent decades. In Europe the meat is sold minced or as ready-to-serve pieces garnished with rice and potato. Red fish meat is rich in vitamins and fat. It is marketed fresh or frozen, whole or as fillets; as cold or hot smoked steaks; and roasted or cooked.

13.1.2.1.5 Perch-like fish

The bluefin tuna (*Thunnus thynnus*) is one of the several *Thunnus* spp. It has a red, beef-like muscle tissue and is caught primarily in the North and Mediterranean Seas and the Atlantic Ocean. It is marketed salted and dried, smoked, or canned in edible oil, brine or tomato sauce. Tuna meat is also a common delicatessen item (tuna paste, sausages, rolls, etc.).
The Atlantic mackerel (*Scomber scombrus*) is of great importance, as are the chub or Pacific mackerel (*S. japonicus*) and the blue Australian mackerel (*S. australasicus*). Mackerel are sold whole, gutted or ungutted; or filleted, frozen, smoked, salted, pickle-salted (Boston mackerel), etc.

13.1.2.1.6 Flat fish

This group includes: plaice or hen fish (*Pleuronectes platessa*); flounder (*Platichthys flesus*, also known as fluke); Atlantic halibut or butt (*Hippoglossus hippoglossus*); common dab (*Pleuronectes limanda*); brill (*Rhombus laevis*); Atlantic and North Sea common sole (*Solea solea*, "Dover" sole); and turbot (*Psetta maxima*, also called butt or britt). These fish and haddock (cf. 13.1.2.1.3) are the sea fish most popular with consumers.

13.1.2.2 Freshwater Fish

Some important freshwater fish are: eels; carp; tench; roach; silver bream; pike, jackfish or pickerel; perch, pike-perch or blue pike; salmon; rainbow, river or brown trout; and pollan (freshwater herring or white fish). Unlike sea fish, the catch of freshwater fish is of little economic importance (cf. Table 13.2), although they do offer an important source of biologically-valuable proteins.

13.1.2.2.1 Eels

Freshwater and sea eels (*Anguilla anguilla, A. rostrata, Conger conger*, etc.) are sold as fresh, marinated, jellied, frozen or smoked unripe summer (yellow or brown eel) or ripe winter eels (bright or silver eel). Due to their high fat content (approx. 25% fresh weight), eels are not readily digestible.

13.1.2.2.2 Salmon

Salmon (*Salmo salar*) and sea trout (*Salmo trutta*) are migratory. Salted or frozen fish are supplied to the European market by Norway and by imports from Alaska and the Pacific Coast of Canada. Also included in this groups are: river trout (*Salmo trutta f. fario*) and lake trout (*Salmo gairdnerii*), which is commonly called steelhead trout in North America when it is journeying between the sea and inland lakes.

13.1.3 Skin and Muscle Tissue Structure

As in other backboned animals, fish skin consists of two layers: the outer epidermis, and below it the derma (cutis or corium). The outer epidermis is not horny, as in other animals, but is rich in water, has numerous gland cells and is responsible for the slimy surface. Mucopolysaccharides are major components of this mucous, with galactosamine and glucosamine as the main sugars. The derma is permeated with connective tissue fibers and has various pigment cells, among them guanophores, which contain silvery-white glistening guanine crystals. Scales protrude from the derma. Their number, size and kind differ from species to species. This is of importance in fish processing since it determines whether a fish can be processed with or without skin. The nature and state of fish skin affects shelf life and flavor. The spread of skin microflora after death is the main cause of rapid decay of fish. The skin contains numerous spores resistant to low temperatures; they can grow even at $< -10\,°C$ (psychrophiles or psychrotolerant microorganisms). The decay is also enhanced by bacteria present in fish intestines.

The fish body is fully covered by muscle tissue. It is divided in the dorsoventral by spinous processes and by fin rays; and in the horizontal direction by septa. Corresponding to the number of vertebra, the rump muscle tissue is arranged into muscle sections (myomeres) which are separated from each other by connective tissue envelopes. The transversal envelopes are called myocommata, the horizontal ones myosepta. While myosepta are arranged in a straight line, myocommata are pleated in a zig-zag fashion. Since cooking gelatinizes the connective tissue, the muscle tissue is readily disintegrated into flake-like segments.

Table 13.5. Average chemical composition of fish

Fish	Moisture[a]	Protein[a]	Fat[a]	Minerals[a]	Edible portion[b]
Freshwater fish					
Eel	61	13	26	1.0	70
Perch	80	18	0.8	1.3	38
Zander	78	19	0.7	1.2	50
Carp	72	19	7	1.3	55
Tench	77	18	0.8	1.8	40
Pike	80	18	0.9	1.1	55
Salmon	66	20	14	1.0	64
River trout	78	19	2	1.2	50
Smelt	80	17	1.7	0.9	48
Sea fish					
Cod	82	17	0.3	1.0	56
Haddock	81	18	0.1	1.1	57
Ling	79	19	0.6	1.0	68
Hake	81	17	0.9	1.1	58
Red fish (ocean perch)	78	19	3	1.4	52
Catfish	80	16	3	1.1	52
Plaice	81	17	0.8	1.4	56
Flounder	81	17	0.7	1.3	45
Common sole	80	18	1.4	1.1	71
Halibut (butt)	75	19	5	1.0	75
Turbot (britt)	80	17	1.7	0.7	46
Herring (Northern Sea)	63	17	18	1.3	67
Herring (Baltic Sea)	71	18	9	1.3	65
Sardine	74	19	5		59
Mackerel	68	19	12	1.3	62
Tuna	62	22	16	1.1	61

[a] As % of edible portion.
[b] As % of the whole fish weight.

As in mammals, fish muscle fibers are striated. Depending on myoglobin content, fish flesh is dark or light colored. In some fish, such as herring and mackerel, the portion of darkcolored tissue is very high (10%), while in others, such as cod, it is limited to a thin layer immediately below the derma.

13.1.4 Composition

13.1.4.1 Review

The edible portion of a fish body is less than in warm-blooded animals. The total waste might approach 50%, with only 10–15% after head removal. Fish meat and that of land animals are readily digestible, but fish is digested substantially faster and has therefore a much lower nutritive saturation value. The cooking loss is approx. 15% with fish, and meat shrinkage is significantly less than that of beef. The biological value of fish proteins is similar to that of land animals. While the crude protein content of fish is about 17–20%, the fat and water contents vary widely. Some are distinctly nonfatty, with fat contents of only 0.1–0.3% (haddock or cod), while some are very fatty (eels, herring or tuna), with fat contents of 16–26%. Many fish species have fat contents between these extreme values. Table 13.5 provides a review.

13.1.4.2 Proteins

The protein-N content of fish muscle tissue is between 2–3%. The amino acid composition, when compared to that of beef or milk casein (Table 13.6), reveals the high nutritional value of fish proteins. The sarcoplasma protein accounts for 16–22% of the muscle tissue total protein. The contractile apparatus is 75% protein; the connective tissue of teleosts is 3%; and of elasmobranchs, such as sharks and rays (skate or rocker), is up to 10%. The individual protein groups and their functions in muscle tissue of a mammal (cf. 12.3.2) also apply to fish.

13.1.4.2.1 Sarcoplasma proteins

Fish sarcoplasma proteins consist largely of enzymes. The enzymes correspond to those of mammalian muscle tissue. When these proteins are separated electrophoretically, specific patterns are obtained for each fish species. This is a helpful chemical means of fish taxonomy. The content of pigments (myoglobin, cytochromes) varies greatly, but is never as high as in mammalian muscle. In strongly pigmented fish (e.g. tuna), pigment degradation reactions can induce meat discoloration (e.g. observable "greening" in canned tuna meat).

Table 13.6. Amino acid composition of fish muscle, beef muscle and casein (amino acid-N as % of total-N)

	Casein	Beef muscle	Cod muscle
Aspartic acid	4.7	4.0	6.8
Threonine	3.6	3.7	3.4
Serine	5.3	4.6	3.6
Glutamic acid	13.3	9.3	8.8
Proline	7.5	4.3	3.4
Glycine	3.2	6.0	5.8
Alanine	3.0	4.9	5.9
Cystine	0.2	0.8	2.5
Valine	5.4	3.7	2.5
Methionine	1.8	2.2	2.0
Isoleucine	4.1	4.2	2.7
Leucine	6.1	5.1	5.1
Thyrosine	3.0	2.1	1.7
Phenylalanine	2.7	2.7	2.1
Tryptophan	1.0	1.2	1.1
Lysine	9.8	9.8	11.7
Histidine	5.3	4.9	3.5
Arginine	8.2	14.5	13.2

13.1.4.2.2 Contractile proteins

The proportion of contractile proteins in a fish total protein is higher than in mammalian muscle tissue, however the proportions between individual components are similar. The heat stability of fish proteins is lower than that of mammals, the protein denaturation induced by urea occurs more readily, and protein hydrolysis by trypsin and chymotrypsin is faster (Figure 13.1). These properties provide additional evidence of the good digestibility of fish proteins.

13.1.4.2.3 Connective tissue protein

The content of connective tissue protein in fish muscle is lower than in mammalian flesh. The shrinkage temperature, T_5, is about 45 °C in fish collagen, i.e. much lower than in mammalian collagen (60–65 °C). These two factors make fish meat softer and more tender than mammalian meat.

13.1.4.2.4 Serum proteins

The congealing temperature of the blood serum of polar fish of Arctic or Antarctic regions (e.g.

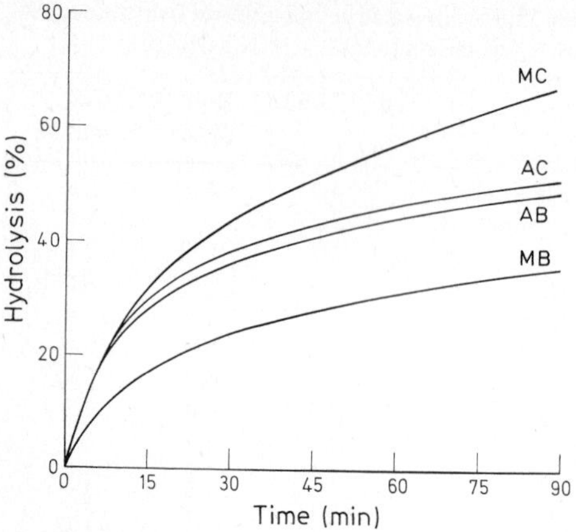

Fig. 13.1. Tryptic hydrolysis of myofibrils (M) and actin (A) from cod fish (C) and beef (B) under the same conditions. (After *Connell*, 1964)

Trematomus borchgrevinki, Dissostichus mawsoni, Boreogadus saida) is about $-2\,°C$ and thus is significantly lower than that of other fish (-0.6 to $-0.8\,°C$). Antifreeze glycoproteins account for such low values. The amino acid sequence of this class of proteins is characterized by high periodicity:

$$[\text{Ala—Ala—Thr}]_n\text{—Ala—Ala}^a \qquad (13.1)$$

[a] C-terminal has one or two alanine residues

The molecular weight range is 10.5–27 kdal, while the conformation is generally stretched, with several α-helical regions. These glycoproteins are hydrated to a great extent in solution. The antifreezing effects are attributed to the disaccharide residues as well as to the methyl side chains of the peptide moiety.

13.1.4.3 Other N-Compounds

The nonprotein-N content is 9–18% of the total nitrogen content in teleosts and 33–38% in elasmobranchs.

13.1.4.3.1 Free amino acids, peptides

Histidine is the predominant free amino acid in fish with darkcolored flesh (tuna, mackerel). Its content in the flesh is 0.6–1.3% fresh weight and can even exceed 2%. During bacterial decay of the flesh, a large amount of histamine is formed from histidine. Fish with light flesh color contain only 0.005–0.05% free histidine. Free 1-methylhistidine is also present in fish muscle tissue. Anserine and carnosine contents are 25 mg/kg fresh tissue. Taurine content is high (500 mg/kg).

13.1.4.3.2 Amines, amine oxides

Sea fish contain 40–120 mg/kg of trimethylamine oxide (TMAO), which is reduced to trimethylamine after death. The TMAO content of freshwater fish is low (0–5 mg/kg), though higher amounts have been detected. TMAO was found to be part of the main metabolic pathway for nitrogen elimination in elasmobranchs. In addition to trimethylamine, the amine fraction contains dimethyl- and monomethylamines and ammonia, and some other biogenic amines derived from amino acid decarboxylation. The concentration of volatile nitrogen bases increases after death, the increase being influenced by storage duration and conditions. This concentration can be used as an objective measure of fish freshness (Figure 13.2).

13.1.4.3.3 Guanidine compounds

Imino urea compounds, such as creatine, are 600–700 mg/kg fresh fish muscle tissue. In crustaceans, the role of creatine is taken over by arginine.

13.1.4.3.4 Quaternary ammonium compounds

Glycine betaine and γ-butyrobetaine are present in low amounts in fish flesh.

13.1.4.3.5 Purines

The purine content in fish muscle tissue is about 300 mg/kg.

13.1.4.3.6 Urea

The fairly high content of urea in muscle tissue (1.3–2.1 g/kg) is characteristic of elasmobranchs (rays, sharks). The compound is decomposed to ammonia by bacterial urease during fish storage.

13.1.4.4 Carbohydrates

Glycogen is the principal carbohydrate. Its content (up to 0.3%) is generally lower than in mammalian muscle tissue.

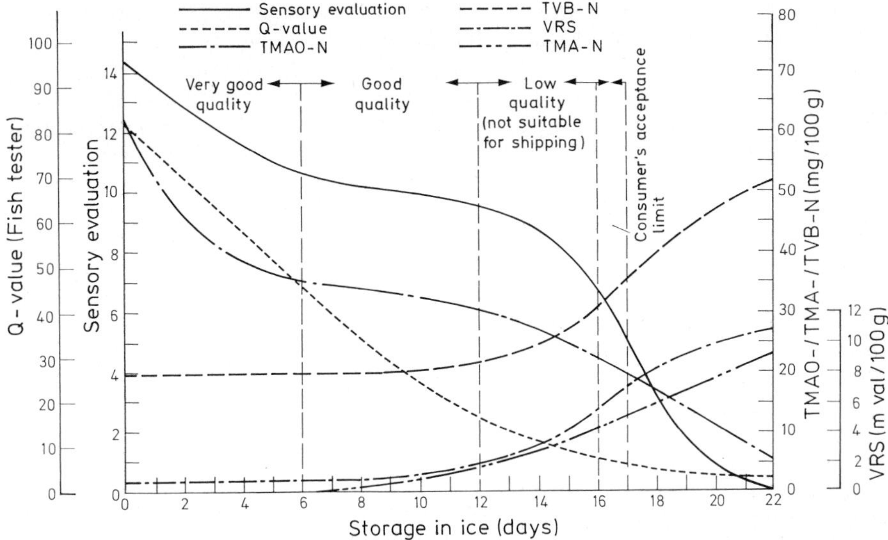

Fig. 13.2. Cod fish quality change during storage. (After *Ludorff*, 1973.) Sensory evaluation: in total 15 points are given, 5 for visual appearance and 10 for odor, taste and texture; Q-value: electric resistance of the fish tissue as recorded by a "fish tester"; Q40: quality class S, Q = 30–40: A, Q = 20–30: B, Q20: C and worse; TMAO-N: trimethylamine oxide-N; TVB-N: total volatile base-N; VRS: volatile reducing substances, TMA-N: trimethylamine-N

13.1.4.5 Lipids

The fat (oil) content of fish is highly variable. It is influenced not only by the kind of fish but by the maturity, season, food availability and feeding habit. Fat deposition occurs in muscle tissue (e.g. carp, herring), in liver (cod, haddock, saithe) or in intestines (blue pike, pike, perch).
The high content of polyenic acids with 4–6 double bonds (cf. 14.3.1.2) is highly characteristic of a fish fat (oil), while there is a relatively low level of tocopherols. The fat portion of fish is the major problem in preservation, particularly with refrigeration

13.1.4.6 Vitamins

Fish fat and liver (liver oil) are significant sources of fat-soluble vitamins, A and D. Also present are vitamins E (tocopherol) and K. The water-soluble vitamins, thiamine, riboflavin and niacin, occur in relatively high amounts, while others are present only in low amounts.

13.1.4.7 Minerals

The average contents of major minerals in fish muscle tissue are compiled in Table 13.7.

13.1.4.8 Other Constituents

More than 500 tropical fish species (barracuda, sting ray, fugu, globefish), including some valuable food fish, are known to be passively poisonous. Poisoning can occur as a result of their consumption. The toxicity can vary with the season, and can extend to the whole body or be localized in individual organs (gonads, i.e. ovaries and testicles, liver, intestines, blood). Cooking can destroy some of these toxic substances. They consist of peptides, proteins and other compounds. Some of their structures have been elucidated. There are also actively-poisonous fish, with prickles or tiny needle-like spines used as the poisoning apparatus. These are triggered as a weapon in defence or attack. This group of fish includes the species *Dasytidae*, *Scorpaenidae* and

Table 13.7. Minerals in fish muscle

Element	Content (mg/kg)	Element	Content (mg/kg)
Ca	48–420	Fe	5–248
Mg	2 40–310	Cu	0.4–1.7
P	1,730–2,170	J	0.1–1.0

Trachinidae. The latter, known as lesser weever (*Trachinus vipera*), is a fish of the Atlantic Ocean and the Mediterranean Sea.

13.1.5 Post-mortem Changes

After death, fish muscle tissue is subjected to practically the same spontaneous reactions as mammalian muscle tissue. Due to the low glycogen content of fish muscle, its pH drop is low. Generally, pH values of 6.2 are obtained. Rigor mortis in cold-blooded animals is much shorter than in warm-blooded animals, a fact of great importance for preservation of fish quality.

Fish exhausted by lengthy struggle in a trawling net give meat of low keeping quality. The duration of rigor mortis is shortened as under these conditions the pH remains high. Therefore, to have an extended rigor the current trend in fishing is to avoid fish exhaustion. Fish muscle tissue differs basically from that of a land animal in that its maturation time is short and a land animal's is prolonged.

Because of the particular structure of fish muscle, the tendency to generate an alkaline pH reaction in muscle and a high probability of microbial infection during fishing and fish dressing, conditions are highly favorable for rapid spoilage of fish. Therefore, bacteriological supervision and control, from the market to processing plants and during distribution, are of utmost importance. Fish muscle autolysis is low, since fish proteinases, the cathepsins, have a pH optimum at 4.3, so they are not active at the fish muscle pH of 6.7.

There are various physical and chemical criteria for assessing fish meat freshness:

The pH of fresh fish is 6.0–6.5 The suitability limit for consumption is pH 6.8, while spoiled fish meat has a pH of 7.0 or above.

The specific resistance of fish muscle changes with storage duration. Soon after catching it is 440–460 ohms, after 4-days approx. 280 ohms, and after 12 days it drops to 260 ohms. The suitability limit for consumption is reached after 16 days, when the resistance is 220 ohms.

The refractive index (n) of fish eye fluid is affected by storage duration. In cod of very good quality, n ranges from 1.3347 to 1.3366. Fish with n 1.3394 or higher is not suitable for marketing.

The decrease in TMAO concentration and a concomitant increase in volatile N-containing substances, such as trimethylamine and several volatile reducing compounds, are chemical criteria for fish quality assessment.

Figure 13.2 provides data on the usefulness of some quality criteria, with cod stored in ice taken as an example. In addition to chemical and physical data, sensory evaluation data are included. A method is considered to be more suitable if it is highly sensitive during the initial 12 days of storage, since it is that time period in which fish quality changes from very good to tolerable generally occur.

13.1.6 Storage and Processing of Fish and Fish Products

13.1.6.1 General Remarks

The ready decomposition or spoilage of fish flesh is the result of the special structure of the muscle tissue and the diverse ways in which microbial contamination is possible while handling fish, from catching, through processing and during distribution. From the earliest times, fish handling methods, like those for land animals, have been designed to increase the shelf life or storage ability.

Fish are usually initially cooled or frozen, or are dried, salted and smoked, followed by pickling in vinegar or in gelatin with vinegar added. They may also be deep fried in oil, or pickled with or without vinegar and soaked in a sauce in an airtight, sealed container. The expected shelf life of such products determines if they are considered fully-preserved, canned or semi-preserved products. Semi-preserves may contain additives against microbial spoilage. The composition of some fish products are given in Table 13.8.

13.1.6.2 Cooling and Freezing

Preservation of freshness by refrigeration is the most modern and effective way to retain the wholesomeness and nutritional value of food. Refrigeration also enables fishing fleets to range the oceans for months in search of fish. Refrigeration permits stockpiling of fish, thus making fish processing plants more economical and better able to respond to market demand and supply.

Fish deteriorates rapidly even at temperatures only slightly above 0 °C. Therefore, immediately after catching fish are packed in ice on board the ship. The ice used may be sprinkled with a

Table 13.8. Average chemical composition of processed fish

Product	Moisture[a]	Protein[a]	Fat[a]	NaCl[a]	Edible portion[b]
Salted fish					
Matje herring	54	18	18	10	68
Salt cured herring	48	21	16	15	68
Dried fish					
Stockfish	15	79	25	2	64
Klipfish	34	45	0.7	13	99
Smoked fish					
Buckling	58	23	16	3	62
Smoked sprats	62	17	20	2	60
Eel	53	19	26	1	73
Mackerel	61	21	16	1	70
Schillerlocken (smoked haddock filet)	53	21	24		100
Semi preserved fish					
Bismarck herring	60	20	17	3	95
Bratherring (fried and pickled herring)	62	17	15	4	92
Herring, jellied	56	29	13		55
Anchovies	69	13	5	1	100
Herring tidbit	62	15	10	3	100

[a] As % of edible portion. [b] As % of the whole fish weight.

bactericidal substance. Freezing, which may also be used on ships, is suitable for whole fish (gutted or ungutted, with or without head or skin removal), as is the case with flat fish, tuna, mackerel or herring, or for fish fillets (cod, haddock, saithe, red fish).

Only quick freezing is used (−30 to −40 °C; cf. Figure 13.3), so the critical temperature range of −0.5 to −5 °C is rapidly passed over. As a protection against oxidation, the whole fish is often sprinkled with water to form a glaze of ice, or it is frozen in an alginate jelly (the "protan" method) or coated with a latex film. Sealed glass jars, various plastic films (polyethylene,.polyester or saran) or wax-impregnated cartons are used for packaging of fillets.

Fish are frozen at sea in flat or round pieces. Fillet sides (strips of flesh cut parallel to the central bone, from which main bones, fins, belly flap and, sometimes, skin are removed) of these fish are cut into thin slates or slabs then are sprinkled with egg, bread crumbs, etc. or are left untreated to reach the consumer as raw fillet. If fillet cutting was performed hygienically, then the cutting residues (8–12%) can be collected and processed into fish meat balls. Fish products

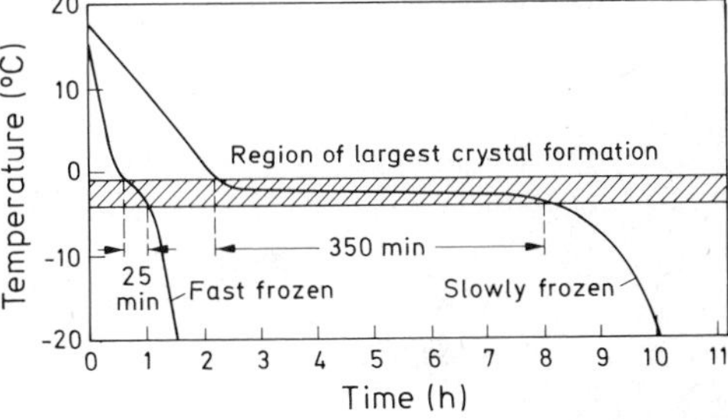

Fig. 13.3. Temperature course during fish fillet freezing process

frozen on land include deep-frozen fish and its end-product dishes, frozen young fatty herring (matje), pickled or sauced herring, etc. Immersion in liquid nitrogen or in dichlorodifluoromethane are recommended freezing processes.

During freezing of fish problems associated with drip or sap losses, discoloration and rancidity or fat oxidation and, consequently, fish weight loss, poor visual appearance and flat taste may arise and must be avoided by suitable processing. Cold storage is augmented by high air humidity (90%) and by stationary, noncirculating air.

Table 13.9. Shelf life of frozen fish, crustaceans and mollusks as influenced by storage temperature

Product	Shelf life (months) at		
	−18 °C	−25 °C	−30 °C
Fatty fish	4	8	12
Nonfatty fish	8	18	24
Lobsters, cray- and crawfish	6	12	15
Crabs	6	12	12
Oysters	4	10	12

Data on the storage properties of some frozen fish are provided in Table 13.9.

Thawing of fish in the home often is done either at room temperature (20 °C) or under a stream of running tap water (15 °C). It is important that the refrigeration chain from processor to distributor to consumer be maintained. Fish must be consumed shortly after thawing, otherwise juices are rapidly lost by dripping and the meat begins to decay. Fish muscle tissue enzymes have noticeable activity even at −10 °C. Excessively long storage or insufficient cooling, especially of fatty fish, results in a rancid off-flavor and an unattractive, yellow-colored fish surface. Antioxidants and associated synergistic compounds, such as ascorbic and citric acids, are used to inhibit fat deterioration. Any change in muscle texture is primarily a result of a change in protein solubility (Figure 13.4).

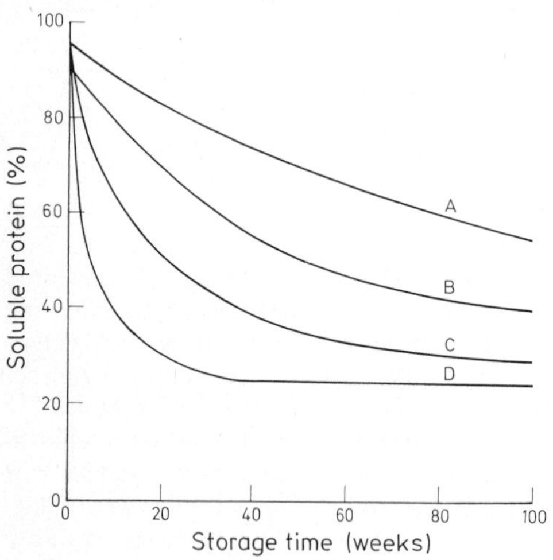

Fig. 13.4. Change in fish muscle solubility as a result of cold storage (−14 °C). A: plaice, B: halibut, C: dogfish, D: cod. (After *Connell*, 1964)

13.1.6.3 Drying

Fish can be preserved by drying naturally in the sun or in drying installations.

Stock fish, primarily nonfatty fish (cod, saithe, haddock, ling or tuck, which is often called cusk in North America) with head removed, split and gutted, is spread outdoors to dry in air. It is an unsalted fish product that is consumed in Southern Europe and tropical countries.

Alternatively, machine-cut, headless and tailless fish which has been belly-clipped ("clipped" fish) is salted, either directly or in brine, and then put through a drying process. This is most often done with cod or other nonfatty fish species. The main consumers of dried salted fish are Italy, Portugal, Spain and South American countries.

13.1.6.4 Salting

Salted fish (whole or parts) are obtained by salting fresh, deep frozen or frozen fish, which aids preservation. Salt is the most important and oldest preservative for fish. Rubbing or sprinkling of fish with salt or immersion in brine, often followed by smoking, is called fish curing. Pickling in vinegar might be an additional preservation step. Salted products include: herring, anchovies, saithe, cod, salmon, tuna, and roe or caviar.

In heavily-salted fish there are at least 20 g of salt in 100 g tissue fluid; in a medium-salted fish the salt content is 12–20 g.

Salting of herring is of special importance. There are the mildly-salted matje (8–10% NaCl), the medium-salted "scotch cure", and the heavily-salted herring, i.e. dry salting to 25% NaCl. Herring are also dressed, salted and packaged, both at sea and on land. The shelf life of salted herring is several months. Matje herring (immature sea herring, often wrongly called "sardines") must be consumed soon after they are removed from refrigeration. Salting might provide a finished end-product, but it is often used as a form of fast, temporary preservation, yielding semi finished products which are later to be processed further. After salting, herring pass through a maturation process which generates a typical flavor. The proteolytic enzymes of the fish are involved in such "gibbed" herring maturation. During gibbing (i.e. a process of removing gills, long gut and stomach), the milt (male fish) or roe (female fish) and some of the pyloric cacea are left in the fish.

These ingredients contribute to the enzymatic maturation of the fish. If all organs are removed, no maturation occurs. Salting causes protein denaturation and cell shrinkage, wherein the glossy, sticky and nearly transparent fish muscle tissue starts to ripen. Similar in importance to herring are salted cod (Atlantic, Pacific and Greenland cod), which are salted dry or in brine as split or boneless fillets, and salted saith, pollack (Dover hake) and some other saltwater fish of *Gadus* species. Salted anchovies (Mediterranean or Scandinavian) are also of importance. They are salted headless, gutted or whole ungutted, packed in barrels and ripened for several months, or are just salted and sold as a semi-preserved product.

13.1.6.5 Smoking

Smoked fish are obtained from fresh, deep frozen or frozen fish which have been dressed in various ways and are dried directly, with or without being salted. The whole fish body or fish portions are exposed to freshly-generated sawdust smoke.

Cold smoking is performed for 2–4 days below 30 °C, generally at 18–26 °C, and is most often used with salt-cured fish (large size herring, salmon, cod, tuna). A product called "Buckling" is a large, fatty herring, sometimes nobbed (with head), that has been smoked. Delicatessen Buckling are made from gutted herring. Kippered herring (Newcastle Kipper) are obtained from fresh, fatty herring with the back split from head to tail. The dressed fish are then lightly brined and cold smoked. In the United States the term "kipper" corresponds to products hot smoked on trays (e.g. kippered black cod). Salted or frozen salmon are smoked in North America.

Hot smoking is performed for 2–4 h at temperatures above 60 °C (usually 100–120 °C). It is used with whole, gutted or descaled fish such as herring (Buckling), sprat ("Kieler Sprotten" in Germany), plaice, flounder, halibut (with or without skin), eels, mackerel, tuna, haddock, whiting (merlan), saithe, cod, red fish (ocean perch), dogfish, sturgeon and shad. Unlike cold-smoked fish, hot-smoked fish have only a limited shelf life, 3–4 days, which can be extended only by cold storage. Hot-smoked caviar (cod or saithe) is available. Smoking of fish in traditional kilns or in batch-type smoke houses is being increasingly replaced by large continuous smoking installations.

13.1.6.6 Marinated, Fried and Cooked Fish Products

Marinade is vinegar or wine, or a mixture of both, usually spiced and salted, in which fish are soaked or steeped before use or before being pickled and stored for a longer time. The fish used might be fresh, deep frozen or frozen, or salted whole fish or fish portions. Marinating tenderizes muscle tissue without heat treatment. Fish preservation by pickling in this manner is based on the combined action of salt and vinegar. Vinegar-packed herring, called simply marinades, are a popular German fish food packaged in retail glass jars. Pickled fish can be packaged together with some plant extracts, sauces, gravy, creams, mayonnaise or related products, or they can be immersed in an edible oil (although oil-packed fish are not called pickled fish). Some of these products might contain chemical preservatives.

Marinated fish are packaged in cans, jars, etc., and may be handled without packaging. Fish marinades have only a limited shelf life (they are semi-preserves), and even chemical preservatives can not prevent their eventual decay.

Marinated fish considered as delicatessen items are "Kronsardines", Bismarck herring, rollmops and pickled herring.

Fried fish products are prepared from variously dressed fresh, deep frozen or frozen whole fish or fish portions, with or without further dressing in eggs and bread crumbs (batter formulations, such as "shake and bake"). They are then made tender by frying, baking, roasting or barbecuing. These products may be packaged or canned in the presence of vinegar, sauces, gravy or an edible oil, often with a chemical preservative added. Examples of these products are fried marinated fish sticks, "Brathering", "Bratrollmops", balls, etc.

Cooked fish products are processed in a similar manner. Tenderization is achieved by cooking or steaming. Processing also involves the use of vinegar or wine, addition of salt and the use of a preservative. The cooked products can be solidified, with or without plant ingredients, into a jelly (herring in jelly) or packaged with other extracts, sauces or gravy. Cooked fish products include herring marinades, rollmops, bacon rollmops in jelly, sea eel (dogfish) in jelly, or broths made from disintegrated saltwater fish meat. The occasional liquefaction of cooked fish jelly ("jelly disease") indicates microbial proteolysis.

13.1.6.7 Saithe

Saithe, often called coalfish, coley, pollack or Boston bluefish (trade name "Dover hake"), are processed into fillets, saltcured, dyed or tinted, and smoked, then are cut into slices or cutlets and covered with edible oil. The product has a good shelf life.

13.1.6.8 Anchosen

Anchosen are made from fresh, frozen or deep-frozen small sprats and herring, preserved with salt in the presence of added sugar or of sugars derived from starch saccharification, spiced and biologically ripened in the presence of sodium nitrate. Flavors are also added.
Anchosen can be packed in sauce (gravy), creams or in edible oils, garnished with plant ingredients and a chemical preservative may be added. Examples for Anchosen are appetitsild, cut spiced herring and spiced herring. Appetitsild is a product consisting of skinned fillets of spice-cured sprats, cured and packed in vinegar, salt, sugar and spices.

13.1.6.9 Pasteurized Fish Products

Pasteurized fish products made of fresh, deep-frozen or frozen fish or fish portions have shelf lives, even without cold storage, of at least 6 months. These products are prepared by prolonged heat treatment of fish at temperatures below 100 °C. They are then tightly sealed in a container. Such products are salted or soaked in vinegar prior to pasteurization.

13.1.6.10 Fish Products with an Extended Shelf Life

Canned fish products of extended shelf life are made by steam retorting of fresh or frozen whole fish or fish portions, followed by packaging in vacuum-sealed, air-tight containers. Their shelf life, without special cold storage, is at least 1 year. Cans are usually wrapped in paper for labelling. Special can materials have to be chosen when the fish is canned with corrosive ingredients such as tomato or mustard sauce, vinegar or lemon juice. The can is usually made of a lacquer-coated tin-plate or inert aluminum.
Products with extended shelf lives are in their own juice or in added oil, or are filled with some sauce or cream (e.g. "sardine" pilchards, *Sardina pilchardus*, packed in olive or soya oil, tomato mustard, or lemon juice). Also available are fish paste, meat balls or "Frikadellen" (Germany), i.e. flesh of white fish made into rissoles using flour, eggs and spices, which are then roasted, deep fried and used ready-to-serve, hors d'œuvres, and fish salad. The latter products are canned or packed in glass jars, and may be packed under controlled atmosphere.

13.1.6.11 Other Fish Products

These include ready-to-serve or instant fish foods, usually garnished with vegetables and fruit, e.g. fish cakes, dumplings, sausages (including cod liver sausage), pastes of anchovy, herring or cod liver, or salmon paste which may include prawn or shrimp and, occasionally, butter. There are also many fish salads and butters, such as anchovy and salmon.

13.1.6.12 Fish Eggs and Sperm

Specially prepared sturgeon eggs (roe) are called caviar. The roe ("hard roe") are detached from the fish ovary gland. The roe are washed in cold water, salted and left to ripen until they become transparent. They are then drained from the brine slime and are marketed for the wholesale market in small metal or glass containers or in barrels. Occasionally, the caviar is pasteurized. Two basic types are marketed: grainy caviar, where eggs are readily detached from roe, and pressed caviar, where the ovarian membrane and the excess fluid are removed by gentle pressing.
Caviar is made from various sturgeon species (beluga, stoer or sevruga). The roe of these sturgeon species caught in winter, when mildly salted (below 6% NaCl), gives a high quality caviar called "Malossol". The beluga (the largest of the three sturgeons mentioned) provides the most valuabel caviar.
Pressed caviar is obtained from all species. Salmon caviar (such as Amur and Keta caviars from Siberian salmon roe) is processed using less than 8.5% salt. American whitefish caviar is a mixture of roe from salmon, whitefish, carp and some other fish. Scandinavian caviar is from cod and lumpfish.
Sturgeon caviar is gray or brown to black in color. Salmon caviar is yellow-red or red. Most caviar is imported from the Soviet Union and Iran (Caspian Sea caviars). It readily decays and

so must be kept refrigerated. A medium-size beluga sturgeon can provide 15–20 kg caviar.
Caviar substitutes are made of roe of various sea and freshwater fish. Germany produces the dyed caviar of lumpfish (lumpsuckers), and also cod and herring caviars. The roe are soured, salted, spiced, dyed black, treated with traganth gum and, occasionally, a preservative is added.
Fish sperm are a product of the gonads of male fish and are often called milt or soft roe. Salted sperm from sea and freswater fish, particularly, herring, are most commonly marketed.

13.1.6.13 Some Other Fish Products

These include the nutritional products and seasonings derived from fish protein hydrolysates; insulin from shark pancreas; proteins recovered from saltwater fish fillet cutting; fish meal used as feed for young animals, poultry and pod fish; and, lastly, fish fat (oil), as mentioned in 14.3.1.2. Of increasing importance is the production of fish protein concentrates and, when necessary, their modified products (cf. 1.4.6.3.2 and Table 1.34).

13.2 Whales

Although a whale is in a true sense a mammal and not a fish, it will be covered here. The blue (*Balaenoptera muculus*) and the finback whale (*B. physalus*) are the two most important whales, each growing up to 30 m in length and up to 150 tonnes in weight. Also caught are the humpback (*Megaptera nodosa*), the sperm (*Physeter macrocephalus*) and the sei whale (*Balaenoptera borealis*). Whale meat is similar to big game meat or beef. It has long and coarse muscle fibers arranged in bundles and colored gray-reddish. The color of the meat is affected by age of the whale, and may be bright red or dark red, while frozen whale meat becomes dark black-brown in color. Freezing imparts a rough, firm texture to the meat.
The fresh meat has a pleasant flavor but, due to the fast rate of fat oxidation, the shelf life is very short. For this reason bulk whale meat is not readily accepted by food wholesalers and the retail market. Whale meat extracts are also produced (cf. 12.7.2.4.2).

Table 13.10. Average chemical composition of crustaceans and mollusks

Crustaceans/ mollusks	Moisture	Protein[a]	Fat[a]	Minerals[a]	Edible portion[b]
Shrimps	78	19	2		41
Lobsters	80	16	2	2.1	36
Crayfish	83	15	0.5	1.3	23
Oysters	83	9	1.2	2.0	10
Scallop	80	16	0.1	1.4	44
Mussel	83	10	1.3	1.7	18

[a] As % of edible portion.
[b] As % of the whole fish weight.

13.3 Crustaceans

Crustaceans have no backbone; their body is divided into sections, each bearing a pair of jointlegs. An armor-like shell covers and protects the body. Included are shrimp, crayfish (also called crabfish), crabs (e.g. freshwater, edible green shore crab) and lobster. Compositional data are provided in Table 13.10.

13.3.1 Shrimps

The most important shrimps are the common or brown shrimp from the North Sea (*Crangon crangon*), the Baltic Sea shrimp (*Palaemon adspersus fabricii*), the deep sea shrimp (*Pandalus borealis*) and the larger species in tropical waters, such as blue Brasilian (*Penaeus* spp.) or the royal red shrimps (*Hymenopenaeus robustus*). Larger species are called prawn.
Shrimps are marketed soon after catch as: live, fresh with shell, with or without head, cooked in brine, cooked without shell. The foregoing have very short shelf lives. Shrimps are also sold canned, deep frozen or as an extract or a salad ingredient. Canned shrimp are heated (pasteurized) at just 80–90 °C so as not to affect their flavor; hence, they are semi-preserves, with a limited shelf life.

13.3.2 Crabs

Crabs live in shallow or deep water along the sea coast or in freshwater. Blue crab (*Callinectes sapidus*) is the most common crab of the Atlantic coast of North America. Other important species are the common shore crab (also called green shore crab); the edible crab of Europe (*Cancer pagurus*), which lives in sandy, shallow water; the king crab of Alaska (*Pralithodes camchaticus*), also called Japanese crab; and the dungeness crab (*Cancer magister*) from the shallow waters from California to Alaska. These crabs differ in shape and size of their big claws, but all have no tail (hence the German term: "Kurzschwanz Krebse"). The color and shape of the body varies, as does the ability to swim or to run sideways.

When crabs shed their shell and the new shell has not yet hardened, they are at their tastiest and are marketed as "extra choice soft" crabs. The forms sold are: live, fresh, frozen and canned. Crab paste, canned soup and crab cakes similar to deepfried fish cakes are delicatessen sea foods. In the trade the term crab meat means white muscle meat, colored red only in leg muscles and chelae, and is distinguished from brown crab meat obtained from crab liver and gonads. The latter are usually processed into crab paste. All crab products are of limited shelf life.

13.3.3 Lobsters

The European lobster (*Homarus gammarus*) caught in the Atlantic is the largest in Europe. It reaches a length of 35–90 cm and a maximum weight of 10 kg. The major area of catch is Helgoland, the north and west sea coast of Europe, the Mediterranean and the Black Sea. The tastiest lobster meat is that from the breast shell. The American or northern lobster (*Homarus americanus*) is closely related to the European lobster. Lobsters are marketed live (remain alive up to 36 h after catch), whole boiled, or canned as cooked meat in its own juice or as soup (cream of lobster, lobster chowder). Lobster paste is also available. Cooking of lobster changes its color to red. The color change involves the release of astaxanthin from ovoverdin, a brown-green chromoprotein (cf. 3.8.4.7.2).

The Norway lobster (*Nephrops norvegicus*; also called Langoustine) also belongs to the lobster family. It is marketed fresh, frozen, semi-preserved, as in salad, or canned, as soup, paste or mildly-brined meat in its own juice.

13.3.4 Crayfish, Crawfish

Crayfish are freshwater crustaceans considered as a delicacy in Europe. The major crayfish of Europe belongs to *Astacus* spp. (*Astacus astacus* or *fluviatilis*). Its meat is the most tasteful in May–August when it sheds its shell and the new shell is still soft. The eastern part of North America has the freshwater crayfish of *Cambarus* spp. The Australian crayfish belongs to *Enastacus serratus*.

The cray(craw)fish die when they are dropped into boiling water. Their tail curls up – this is a sign that they were cooked fresh or alive. For the color change during cooking see above (cf. 13.3.3).

The seawater species of crayfish are called crawfish. They include *Palinurus, Panulirus* and *Jasus* spp. The most important crawfish are the European spiny lobster (*Palinurus vulgaris*), the Pacific North American counterpart (*Panulinus interruptus*) and others from Africa, Australia and Japan. The European spiny lobster is 30–40 cm long, up to 6 kg in weight, has rudimentary front legs shaped into sharp claws and has a knobby shell covering the body. It is often caught in the Mediterranean Sea, the west and south coasts of England, and along the coast of Ireland. The rock lobster (*Jasus lalandii*) and the Mediterranean crawfish (*Palinurus elephas*) are also available on European markets. The meat of these crawfish is rather coarse and fiberlike and is colored yellow to yellow-red.

The cray- and crawfishes are marketed fresh live, raw or cooked, and canned in different forms: meat, butter (precooked meat mixed with butter), soup and soup powders, soup extracts (these are crawfish butter, spiced, salted and blended with flour) and crayfish bisque (French purée or thick soup of crayfish and lobsters).

13.4 Mollusks

13.4.1 Mollusks (Bivalves)

The bivalve mollusks include clams, oysters, mussels and scallops. The common oyster (also called flat native or European oyster) and the blue or common mussel are the most often processed molluscan shellfish.

Oysters (*Ostreidae*, e.g. the European oyster, *Ostrea edulis*) live in colonies along the sea coast or river banks, or are cultivated in ponds ("oyster farms") which are often connected with the sea. Oysters are usually sold unshelled. Only the adductor muscle is consumed; the pleated gills and the digestive system are discarded.

In addition to the common oyster, the Portuguese oyster (*Gryphea angulata*) and the American blue point oyster, (*Crassostrea virginica*), used most commonly for canning, are of importance. The best meat is obtained from oysters harvested when they are 3–5 years old, with the top quality harvested between September and April (an old saying is: oysters should be eaten in months which have"r" in their names).
The blue or common mussel (*Mytilus edulis*) lives in shallow, sandy freshwater, while the sea mussel lives in ocean water or is cultivated in ponds or lakes. The shell, 7–15 cm long, is bluish black and the body meat is yellowish. The meat is rich in protein (16.8%) and also in vitamin A and B-vitamins; hence it is a nourishing food. The meat is eaten cooked, fried or marinated. The major mussel growing areas in Germany are the Kiel Bay and the East Friesian Islands.
In addition to common mussel, numerous other mussels are eaten, mostly canned in vegetable oil, e.g. Pacific Bay or Cape Cod scallops (*Pectinidae*) and cockles (*Cardidae*).
Due to rapid spoilage, mussels are marketed live or canned, i.e. processed. They are eaten soon after being caught or after the can is opened, and are avoided in warm seasons. Moreover, they should originate from uncontaminated clear waters.

13.4.2 Snails

Snails are univale mollusks, i.e. they have only a single, coiled shell. They are eaten preferentially in Italy, France and Germany, and are nearly exclusively the large Helix garden snail (*Helix pomatia*). Snails are sometimes collected wild in South or Central Germany and in France, but most are supplied by snail gardens and feeding lots where lettuce and cabbage leaves are the food source, or in damp shady cellars, where wheat bran and leafy vegetable leaves (e.g. cabbage) are used as a feed. The meat is considered a great delicacy. Since the shelf life of the meat is very limited, snails are marketed live (with the shell plugged) or canned. Marine snails of various kinds are fried, steamed, baked or cooked in soups, and are also considered a great delicacy.

13.4.3 Octopus, Sepia, Calmar

Octopus, sepia and calmar (*Cephalopoda*) are softbodied mollusks with eight or ten arms, and without an outside shell.
The sepia or cuttlefish (*Sepia officinalis*), the squid or calmar (*Loligo loligo*) and the octopus or devilfish (*Octopus vulgaris*) are caught in the Mediterranean region, mostly in Italy, and other parts of the world (Atlantic and Pacific Oceans, e.g. the North American poulp, Japanese *Polypus* spp., etc.). They are consumed deep fried in oil, baked, cooked in wine, pickled in vinegar after being boiled, cooked in soups, in salads, stewed or canned.

13.4.4 Turtles

Turtles, tortoises or terrapin (for American freshwater turtles) are reptiles with a shell used as a "house". The logger head and green sea turtles are caught commercially for their meat. In Germany turtle is mostly eaten in soup or stew. The meat of the so-called soup turtle (*Chelonia mydas*) is faintly red to bright red, and is marketed canned. An imitation or mock turtle soup is prepared from edible parts of heads of calves and has no connection to turtles except for the name.

13.4.5 Frogdrums

The thigh portion (frogdrum) of a frog's hinged leg is sold as a delicacy. Frogs providing frogdrums are the common bullfrog (*Rana catesboniana*), the leopard frog (*Rana pipiens*) and others (*Rana arvalis, Rana tigrena, Rana esculenta*). The meat is soft in texture, white in color and tasty; however, it has a very limited shelf life as it readily deteriorates. Frogdrums are eaten cooked, roasted or stewed.

13.5 Literature

Borgstrom, G. (Ed.): Fish as food. Vol. I–III. Academic Press: New York–London. 1961–1965

Connell, J. J.: Fish muscle proteins and some effect on them of processing. In: Proteins and their reactions (Eds.: Schultz, H. W., Anglemier, A. F.), p. 255, AVI Publ. Co.: Westport, Conn. 1964

Connell, J. J. (Ed.): Advances in fish science and technology. Fishing News Books Ltd: Farnham, Surrey, UK. 1980

Feeney, R. E., Yin Yeh: Antifreeze proteins from fish bloods. Adv. Protein Chem. *32,* 191 (1978)

Habermehl, G.: Gift-Tiere und ihre Waffen. 2. Aufl., Springer-Verlag: Berlin–Heidelberg–New York. 1977

Ludorff, W., Meyer, V.: Fische und Fischerzeugnisse. 2. Aufl., Verlag Paul Parey: Berlin–Hamburg. 1973

Multilingual dictionary of fish and fish products (prepared by the OECD), 2nd ed., Fishing News Books Ltd., Farnham, Surrey, UK. 1978

14 Edible Fats* and Oils

14.1 Foreword

Most fats and oils consist of triacylglycerides (recently also denoted as triacylglycerols; cf. 3.3.1) which differ in fatty acid composition to a certain extent. Other constituents which make up less than 3% of fats and oils, are the unsaponifiable fraction (cf. 3.8) and a number of acyl lipids; e.g. traces of free fatty acids, mono- and diacylglycerols.

The term "fat" generally designates a solid at room temperature and "oil" a liquid. The designations are rather imprecise, since the degree of firmness is dependent on climate and, moreover, many fats are neither solid nor liquid, but are semi-solid. Nevertheless, in this chapter, unless specifically emphasized, these terms based on consistency will be retained.

It should be noted that the fatty acid composition of individual fat samples may deviate greatly. The fat composition of land animals is affected by the kind and breed of animal and by the feed. The composition of plant fats depends on the cultivar and growth environment, i.e. climate and location of the oilseed or fruit plant (cf. Figure 3.9). Therefore only average values are given in the following tables dealing with fatty acid composition.

14.2 Data on Production and Consumption

Data on the production of oilseeds and other crops are summarized in Table 14.0. The world production of these crops has doubled (Table 14.1) since the Second World War. There has been a significant rise in production since 1964 for soybean, palm and sunflower oils, while marine oil production has declined steadily. Soybean oil, butter and edible beef fat and lard are most commonly produced in FR Germany (Table 14.1)

The per capita consumption of edible fats (Table 14.2) in FR Germany more than doubled when compared to the world average. The slight drop recorded from 1970–75 was due to a temporary reduction in butter consumption.

14.3 Origin of Individual Fats and Oils

14.3.1 Animal Fats

14.3.1.1. Land Animal Fats

The deposited fats and organ fats of domestic animals, such as cattle and hogs, and milk fat, which was covered in Chapter 10, are important animal sources of fat production. The role of sheep fat, however, is not significant. The major fatty acids of these three sources are oleic, stearic and palmitic (Table 14.3).

In contrast to oil from plant tissue, the recovery of animal fat is not restricted by rigid cell walls or sclerenchyma supporting tissue. Only heating is needed to release fat from adipose tissue (dry or wet rendering with hot water or steam). The fat expands with heating, tearing the adipose tissue cell membrane and flowing freely. Further fat separation is simple and does not pose a technical problem.

14.3.1.1.1 Edible beef fat

Edible beef fat is obtained from bovine adipose tissue covering the abdominal cavity and surrounding the kidney and heart and from other compact, undamaged fat tissues. The fat is light-yellow due to carotenoids derived from animal feed. It is of a friable, brittle consistency and melts between 45 and 50 °C.

The fatty acid composition of beef fat (Table 14.3) is not influenced greatly by feed intake, but that of hog fat (lard) is. The composition of edible beef fat triacylglycerols is given in Table 3.8.

The following commercial products are prepared from beef fat: *Prime Beef Fat ("premier jus")* is obtained by melting fresh and selected fat trim-

---

* Butter is dealt with in Chapter 10.2.3

Table 14.0. World production of major oilseeds 1981 (1,000 t)[a]

Continent	Castor-bean	Sunflower seed	Rape-seed	Sesame seed	Linseed	Safflower seed	Cotton-seed
World	810	13,765	12,147	1,959	2,274	889	29,337
Africa	44	652	22	459	54	31	2,033
America, North-, Central-		2,301	1,797	127	681	464	6,664
America, South-	312	1,352	62	43	628	1	1,950
Asia	403	1,642	6,466	1,327	548	341	11,871
Europe, West-		844	2,546	2	37	27	347
Europe, East- + USSR	43	6836	1,230		311	4	6,311
Australia + South Pacific Islands		138	25		14	21	161
Oil content	45–55	40–65	40–50	45–63	38–44	50	16–24

Continent	Copra	Palm kernel	Palm oil	Olives	Olive oil
World	5,054	1,891	5,384	8,403	1,579
Africa	175	739	1,365	1,318	204
America, North-, Central-	188	21	46	73	3
America, South-	34	343	162	143	25
Asia	4,335	770	3,756	1,066	175
Europe, West-				5,788	1,169
Europe, East- + USSR				13	2
Australia + South Pacific Islands	322	19	56	3	
Oil content (%)	63–70	40–52	100	20	100

Country	Castor-bean	Country	Sunflower seed	Country	Rape-seed	Country	Sesame seed
Brazil	278	USSR	4,600	China	3,803	India	500
India	210	USA	2,098	India	2,247	China	401
China	120	Argentina	1,260	Canada	1,794	Sudan	200
USSR	40	China	1,000	France	1,023	Burma	162
Thailand	26	Rumania	824	Poland	486	Mexico	86
Paraguay	23	Hungary	622	FR Germany	363	Nigeria	73
Philippines	20	Turkey	575	Sweden	353	Syria	63
Pakistan	19	South Africa	495	Germany DR	330	Uganda	46
Ethiopia	12	Bulgaria	444	UK	325	Afghanistan	40
Equador	11	France	424	Denmark	310	Ethiopia	35
Σ (%)[b]	94	Σ (%)[b]	90	Σ (%)[b]	91	Σ (%)[b]	82

[a] Soybean and peanuts are presented in Table 16.1. [b] World production ≙ 100%.

mings in water heated to 50–55 °C. The acid value resulting from lipolytic action (cf. 14.5.3.1) is not allowed to exceed 1.3 (corresponding to approx. 0.65% free fatty acid). This beef fat, when heated to 30–34 °C, yields two fractions: oleomargarine (liquid) and oleostearine (solid). Oleomargarine is a soft fat with a consistency similar to melted butter. It is used by the margarine and baking industries. Oleostearine (pressed tallow) has a high melting point of 50–56 °C and is used in the production of shortenings (cf. Table 14.15). *Edible Beef Fat (secunda beef fat)* is obtained by melting fat in water at 60–65 °C, followed by a purification step. It has a typical beef fat odor and taste and a free fatty acid content not exceeding 1.5%. Lower quality tallow has only industrial or technical importance, for example as raw material for the soap and detergent industries.

Table 14.0. (Continued)

Country	Linseed	Country	Safflower seed	Country	Cotton-seed	Country	Copra
Argentina	598	Mexico	372	USSR	6,300	Philippines	2,275
Canada	477	India	340	China	6,000	Indonesia	1,254
India	428	USA	92	USA	5,673	India	370
USSR	200	Ethiopia	31	India	2,720	Malaysia	208
USA	198	Spain	25	Pakistan	1,500	Papua, New Guinea	144
China	85	Australia	21	Brazil	1,206	Mexico	143
Poland	49	Portugal	2	Egypt	860	Sri Lanka	123
Rumania	40	Argentina	1	Turkey	785	Mozambique	70
Egypt	37	Israel	1	Mexico	530	Thailand	52
France	25	Turkey	1	Guatemala	249	Vietnam	40
Σ (%)[b]	94	Σ (%)[b]	100	Σ (%)[b]	88	Σ (%)[b]	93

Country	Palm seed	Country	Palm oil	Country	Olives	Country	Olive oil
Malaysia	588	Malaysia	2,822	Italy	2,800	Italy	566
Nigeria	350	Indonesia	722	Greece	1,350	Spain	281
Brazil	275	Nigeria	675	Spain	1,348	Greece	280
Indonesia	131	China	190	Tunisia	700	Tunisia	140
Benin	75	Ivory Coast	190	Turkey	650	Turkey	107
Zaire	65	Zaire	155	Marocco	350	Syria	51
China	46	Colombia	88	Syria	297	Marocco	38
Cameroon	46	Cameroon	80	Portugal	220	Portugal	33
Guinea	35	Sierra Leone	50	Libya	162	Argentina	24
Ivory Coast	30	Equador	42	Argentina	115	Libya	16
Ghana	30	Guinea	42	Σ (%)[b]	95	Σ (%)[b]	97
Sierra Leone	30	Σ (%)[b]	94				
Σ (%)[b]	90						

[b] World production • 100%

14.3.1.1.2 *Sheep tallow*

The unpleasant odor adhering to sheep tallow is difficult to remove, hence it is not used as an edible fat. Sheep tallow is harder and more brittle or friable than beef tallow. The fatty acid composition of sheep tallow is presented in Table 14.3.

14.3.1.1.3 *Hog fat (lard)*

Hog (swine) fat, called lard, is obtained from fat tissue covering the belly (belly trimmings) and other parts of the body. The back fat is mainly utilized for manufacturing bacon. After tallow and butter lard is currently the animal fat which is consumed the most (Table 14.1). Its grainy and oily consistency is influenced by the breed and feeding of hogs.

Some commercial products are:

Lard obtained exclusively from belly trimmings (abdominal wall fat). This is the highest quality *neutral lard.* It has a mild flavor, is white in color and its acid value is not more than 0.8.

Lard from other organs and from back is rendered using steam. The maximum acid value is 1.0.

Lard obtained from all the dispersed fat tissues, including the residues left after the recovery of neutral lard, is rendered in an autoclave with steam (120–130 °C). This type of lard has a maximum acid value of 1.5.

In contrast to the composition of triacylglycerols found in beef fat (Table 3.8), lard contains fewer triacylglycerols of the type SSS and more of the types SUU, USU and UUU (S = saturated; U = unsaturated fatty acid). As a consequence, lard

Table 14.1. World production of fats and oils (in 1,000 t)

Fat/oil	1935/39	1965	1981
Soya oil	1,229	4,860	12,495
Sunflower oil	562	2,375	4,515
Cottonseed oil	1,560	2,570	3,245
Peanut oil	1,506	3,165	2,945
Rapeseed (canola) oil	1,207	1,465	3,750
Palmkernel- and palm oil	1,334	1,595	6,155
Coconut oil	1,932	2,225	2,925
Olive oil	871	1,951[a]	1,325[b]
Other oils of plant origin	95	1,720	1,940
Butter (butter fat)	3,611	4,735	5,515
Lard	2,495	4,375	4,615
Edible tallow	1,442	4,305	5,980
Marine oils	975	1,075	1,195
	18,819	36,416	56,000

[a] Production data for 1964.
[b] An estimate for 1982.

Table 14.2. Consumption of edible fats and oils in FR Germany (kg per capita per year)

Year	Butter[a]	Animal fat	Oils[b]	Total
1969/70	7.3	6.1	12.8	26.2
1970/71	7.0	6.4	13.2	26.6
1971/72	6.2	6.3	13.4	25.9
1972/73	6.2	6.1	13.4	25.7
1973/74	6.0	6.1	13.3	25.4
1974/75	5.9	6.3	12.9	25.1
1975/76	5.5	6.4	13.2	25.1
1976/77	5.3	6.3	14.0	25.6

[a] Butter fat; butter weight is higher by 19.5%.
[b] It includes vegetable and marine oils.

Table 14.3. Average fatty acid composition of some animal fats (weight-%)

Fatty acid	Beef tallow	Sheep tallow	Lard	Goose fat
12:0	0	0.5	0	0
14:0	3	2	2	0.5
14:1 (9)	0.5	0.5	0.5	0
16:0	26	21	24	21
16:1 (9)	3.5	3	4	2.5
18:0	19.5	28	14	6.5
18:1 (9)	40	37	43	58
18:2 (9, 12)	4.5	4	9	9.5
18:3 (9, 12, 15)	0	0	1	2[a]
20:0	0	0.5	0.5	0
20:1, 20:2	0	0.5	2	
Others	3	3	0	0

[a] It includes fatty acid 20:1.

melts at lower temperatures and over a range of temperatures rather than sharply at a single temperature, and its shelf life is not particularly long.

14.3.1.1.4 Goose fat

As the only kind of poultry fat produced, goose fat is a delicacy. Its production is insignificant in quantity. The fatty acid composition of goose fat is given in Table 14.3.

14.3.1.2 Marine Oils

Sea mammals, whales and seals, and fish of the herring family serve as sources for marine oils. These oils typically contain highly unsaturated fatty acids with 4–6 allyl groups (Table 14.4). The following acids are predominant (double bond positions are given in brackets): 18:4 (6, 9, 12, 15); 20:5 (5, 8, 11, 14, 17); 22:5 (7, 10, 13, 16, 19); and 22:6 (4, 7, 10, 13, 16, 19). Since these acids are readily susceptible to autoxidation, marine oils are not utilized directly as edible oils, but first their double bonds are hydrogenated, followed by refining.

Of analytical interest is the occurrence in marine oils of about 1% branched methylated fatty acids, for example, 12-methyl and 13-methyl-tetradecanoic acids or 14-methyl-hexadecanoic acid. These acids are also readily detectable in hardened marine oils.

Table 14.4. Average fatty acid composition of some marine oils (weight-%)

Fatty acid	Blue whale	Seal	Herring (*Clupea harengus*)	Pilchard[a]	Menhaden (*Brevoortia tyrannus*)
14:0	5	4	7.5	7.5	8
16:0	8	7	18	16	29
16:1	9	16	8	9	8
18:0	2	1	2	3.5	4
18:1	29	28	17	11	13
18:2	2	1	1.5	1	1
18:3	0.5		0.5	1	1
18:4	0.4		3	2	2
20:1	22	12	9.5	3	1
20:4	0.5		0.5	1.5	1
20:5	2.5	5	9	17	10
22:1	14	7	11	4	2
22:5	1.5	3	1.5	2.5	1.5
22:6	3	6	7.5	13	13

[a] Trade name of grown sardines *(Sardinóps caerulea)*

14.3.1.2.1 Whale oil

There are two suborders of whales: Baleen whales which have horny plates rather than teeth, and whales which have teeth. The blue and the finback whales, which both live on plankton, belong to the Baleen suborder. The oils from these whales do not differ substantially in their fatty acid composition.

A blue whale, weighing approx. 130 t, yields 25–28 t of oil, which is usually recovered by a wet rendering process. In the past two decades the ruthless and unrestricted exploitation of the sea has nearly wiped out the whale population, hence their raw oil has become a rare product.

14.3.1.2.2 Seal oil

The composition of seal oil is similar to that of whale oil (Table 14.4).

14.3.1.2.3 Herring oil

The following members of the herring fish family are considered to be satisfactory sources of oil: herring, sardines (Californian or Japanese pilchards, etc.), sprat or brisling, anchovies (German Sardellen or Swedish sardell) and the Atlantic menhaden. The fatty acid compositions of the various fish oils differ from each other (Table 14.4).

14.3.2 Oils of Plant Origin

With regard to the processing used to recover plant oils, it is practical to divide them info fruit and oilseed types. While only two fruits are of economic importance in oil production, the number of oilseed sources is enormous.

All the edible oils (with the sole exception of oleomargarine-type products) are of plant origin. They are sold and consumed as pure oil from a single oilseed plant or fruit plant, for example, olive, sunflower or corn oils, or are marketed and used as blended oils, which are generally designated as edible, cooking, frying, table or salad oil.

14.3.2.1 Fruit Pulp Oils

The oils obtained from the fruits of the olive tree and several oil palm species are of great economic importance. The fatty acid compositions of the oils of these fruits are summarized in Table 14.5.

Table 14.5. Characteristics of olives (fruits/oil) and oil palm

	Olive *(Olea europaea, ssp. europaea)*	Oil palm *(Elaeis guineensis)*
Fruits		
Length (cm)	2–3	3–5
Width (cm)	2–3	2–4
Fruit pulp (weight-%)	78–84	35–85
Fruit seed (weight-%)	14–16	65–15
Fruit pulp (mesocarp)		
Oil (weight-%)	38–58	30–55
Moisture (weight-%)	to 60	35–45
Fruit pulp oil		
Solidification point (°C)	−5 to −9	27–38
Average fatty acid composition (weight-%)		
	Olive oil	Palm oil
14:0	0	1
16:0	11.5	40
16:1	1.5	0.5
18:0	2.5	5
18:1 (9)	75.5	40.5
18:2 (9, 12)	7.5	12
18:3 (9, 12, 15)	1.0	0.5
20:0	0.5	0.5

Due to the high enzymatic activity in fruit pulp, particularly of lipases, the shelf life of fruit oil is severely limited.

14.3.2.1.1 Olive oil

Olive oil is obtained from the pulp of the stone fruit of the olive tree (*Olea europaea* ssp *europaea*). More than 90% of the world's olive harvest takes place in the Mediterranean region, primarily in Italy (cf. Table 14.0). Olive tree plantations are found to a smaller extent in Japan, Australia, California and South America. Altogether, olive oil production is declining (Table 14.1). The laborious, painstaking harvesting, which has not yet been mechanized, is especially to blame for this situation.

Oil production: The disintegrated fruit is kneaded to release the oil droplets from the pulp, occasionally by adding common salt. The oil is then pressed out or separated by gravity decantation. The initial cold pressing provides virgin oil (provence oil). This is then followed by oil recovery using warm pressing at about 40 °C.

In addition to the conditions used for oil recovery, the quality of olive oil is affected by ripeness of the fruit (not overripe fruit is preferred) and length of storage. In virgin oils there is a relationship between sensory properties and the content of free fatty acids:

- Prime virgin oil: pleasant aromatic flavor with the free fatty acid content not exceeding 1% (calculated as oleic acid).
- Fine virgin oil: slightly less aromatic in flavor, with a free fatty acid content not exceeding 1.5%.
- Semi-fine virgin oil (courante grade): cold pressed oil, less aromatic in flavor and with a free fatty acid content up to 3%.

Crude olive oil, with a higher content of free fatty acids and which includes the hot pressed oils, is not suitable for direct consumption. Crude olive oil is either partially refined or is utilized for technical purposes.
The fact that the fatty acid composition of tea seed oil is very similar to that of olive oil is of concern. However, these two oils can be differentiated by using the *Fitelson* Test (cf. Table 14.18).

14.3.2.1.2 Palm oil

This oil is obtained from oil palm, the utilization of which is constantly increasing (cf. Table 14.1). Palm plantations are found primarily in western Malaysia, Nigeria and Indonesia (cf. Table 14.10). The fruits provide two different oils, the first from the pulp and the second from the seeds.
Oil production. The fruit cluster, which contains about 3,000–6,000 fruits, is first steam-treated to inactivate the high lipase activity and to separate the pulp from the seed. The oil is recovered by pressing the disintegrated pulp. The crude oil is then clarified by centrifugation. Washing with hot water, followed by drying, provides a crude oil product that has a high carotene content (cf. 14.4.1) and, hence, the color of the oil is yellow to red. During refining (cf. 14.4.1), the palm oil color is destroyed by bleaching and the free fatty acids are removed. Palm fruit characteristics and oil composition are given in Table 14.5.

14.3.2.2 Seed Oils

Some oilseeds have acquired great significance in the large-scale industrial production of edible oils. After a general review of their production, some individual oils, grouped together according to characteristics of the fatty acid composition will be discussed.

14.3.2.2.1 Production

Conditioning. The ground or flaked seeds are heated with live steam of about 90 °C to facilitate oil recovery. This treatment ruptures all the cells and partly denatures the proteins and inactivates most of the enzymes. The temperature is regulated to avoid formation of undesirable colors and aromas.
After conditioning and moisture adjustment to about 3%, the oil is obtained by pressing and/or solvent extraction. The choice between these two processes depends on the oil content of the seed. Solvent extraction is the only economic choice for seeds with an oil content below 25%.
Pressing. The oil is removed by pressure from an expeller or screw press. The residual oil in the resultant meal flakes is 4–7%. It is, however, more economical to apply lower pressures and to leave 15–20% of the oil in the flakes, and then to remove this oil by a solvent extraction process ("prepress solvent extraction" process).
Extraction. The ground seeds are rolled into thin flakes by passing them between smooth steel rollers. This flaking step provides the enlarged surface area needed for efficient solvent extraction. The extraction is performed using petroleum ether, i.e. technical hexane, as a solvent (boiling point 60–70 °C). In addition to n-hexane, it contains 2- and 3-methylpentane and 2,3-dimethylbutane and is free of aromatic compounds.
Solvent removal from the raw oil-solvent mixture, called miscella, is achieved by distillation. The maximum amount of solvent remaining in the oil is 0.1%. The oil-free flakes are then steamed to remove the solvent ("desolventizing") and then, after dry heating ("toasting"), they are cooled and sold as protein-enriched feed meal for cattle.
The crude oil obtained either by pressing or solvent extraction contains suspended plant debris, protein and mucous substances. These impurities are removed by filtration.

14.3.2.2.2 Oils high in lauric and myristic acids

The most important representatives of this group of oils are coconut, palm seed and babassu oils. The acceptable shelf life stability of these oils is reflected in their fatty acid compositions (Table 14.6). Since linoleic acid is present in negligible amounts, autoxidative changes in these oils do not occur. However, when these oils are used in preparations containing water, microbiological deterioration may occur; this involves release of free C_8-C_{12} fatty acids and their partial degradation to methyl ketones ("perfume scent rancidity", cf. 3.7.5).

Coconut and palm seed oils are important ingredients of vegetable margarines which are solid at room temperature. However, they melt in the mouth with a significant heat uptake, as reflected by a cooling effect.

Coconut oil is obtained from the stone fruit of the coconut palm, which grows throughout the tropics. The moisture content of the oil-containing endosperm, when dried, decreases from 50% to about 5–7%. Such crushed and dried coconut endosperm is called "copra" and is sold under this name as a source for oil production around the world.

Palm kernel oil is obtained from the kernels of the fruit of the oil palm. The kernels are separated from the fruit pulp, then removed from the stone shells and dried prior to recovery of the oil. Babassu oil is obtained from seeds of the babassu palm, which is native to Brazil. This oil is rarely found on the world market and is mainly consumed in Brazil.

14.3.2.2.3 Oils high in palmitic and stearic acids

Cocoa butter and fats (solid at room temperature) belong to this group, with the latter referred to as cocoa butter substitutes ("cocoa butter interchangeable fats"). They are relatively hard and can crystallize in several polymorphic forms (cf. 3.3.1.2). Their melting points are between 30 and 40 °C. The relatively narrow range of fat melting points for cocoa butter, as well as for some other types of butter, is to be expected (Table 14.7). When cocoa butter melts in the mouth, a pleasant, cooling sensation is experienced (cf. 14.3.2.2.2), which is characteristic of only a few types of triacylglycerols present in fats which contain predominantly palmitic, oleic and stearic acids. This fatty acid composition is also reflected by the resistance of these fats to autoxidation and microbiological deterioration (Table 14.7). These fats are utilized preferentially in the manufacturing of chocolates, candy and confections.

Table 14.6. Characteristics of palm kernel oils

	Oil palm (*Elaeis guineensis*)	Coconut palm (*Cocos nucifera*)	Babassu palm (*Orbignya speciosa*)
Kernel oil content (weight-%)	40–52	63–70	67–69
Fat/oil melting point range (°C)	23–30	20–28	22–26
Average fatty acid composition (weight-%)			
8:0	6	8	4.5
10:0	4	6	7
12:0	47	47	45
14:0	16	18	16
16:0	8	9	7
18:0	2,5	2.5	4
18:1 (9)	14	7	14
18:2 (9, 12)	2.5	2.5	2.5

Table 14.7. Fatty acid composition of cocoa butter and cocoa butter substitutes

Trade name	Cocoa butter	Illipè butter (Mowrah butter)	Borneo tallow (Tencawang fat, Illipè butter)	Shea butter (Kerité fat)
Source	Cacao tree (*Theobroma cacao*)	*Madhuca longifolia*	*Shorea stenoptera*	*Butyrospermum parkii*
Fat, melting point range (°C)[a]	28–36	24.5–28.5	28–37	23–42
Average fatty acid composition (weight-%)				
16:0	25	28	20	7
18:0	37	14	42	38
18:1 (9)	34	49	36	50
18:2 (9, 12)	3	9	< 1	5

[a] The melting points range reflects a pronounced fat polymorphysm (cf. 3.3.1.2); the highest temperature given represents the melting point of the stable fat modification.

Cocoa butter is the fat from cocoa beans. The seed germ contains up to 50–58% of the fat, which is recovered as a by-product during cocoa manufacturing (cf. 21.3.2.7). It is light yellow and has the pleasant, mild odor of cocoa.

The denotation of the "cocoa butter interchangeable fats" may be confusing since fats from diverse sources are sometimes marketed under a collective name such as Illipè butter. Confusion can be avoided by providing the Latin name of the plant, i.e. the source of the fat.

Shea butter (kerité fat) is obtained from seeds of a tree which grows in western Africa and the cultivation of which appears to be uneconomic.

The high content of unsaponifiable matter (up to 11%) in this kind of butter is of interest.
Borneo tallow (Illipè butter) is obtained from the seeds of a plant native to Java, Borneo, the Philippines and India. It serves as a valuable edible fat in the Tropics. *Mowrah butter* (often marketed as Illipè butter) is derived from a different plant *(Madhuca longifolia)* and is also indigenous to the Asian tropics.

14.3.2.2.4 Oils high in palmitic acid

Oils in this group contain more than 10% palmitic acid along with oleic and linoleic acids (cf. Table 14.8).

Cottonseed oil is obtained from seeds of many cotton plant cultivars. The plant is widely cultivated (cf. Table 14.0). The raw oil is dark, usually dark red, and has a unique odor. It contains a poisonous phenolic, gossypol,

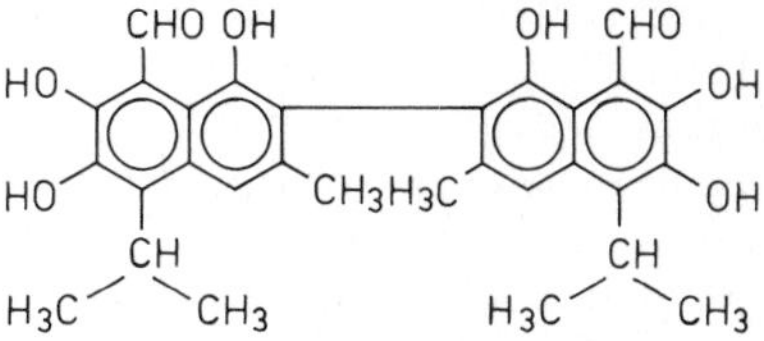

CHO OH OH CHO; HO, HO, OH, OH; CH_3H_3C; CH, CH; H_3C CH_3, H_3C CH_3 (14.1)

which is removed during refining. Another substance present in this oil is malvalia acid,

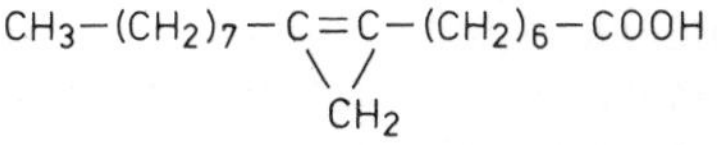

$$CH_3-(CH_2)_7-\underset{\diagdown\ \diagup \atop CH_2}{C=C}-(CH_2)_6-COOH \qquad (14.2)$$

which survives refining, but not hydrogenation of the oil. This substance is responsible for detection or identification of the oil by the *Halphen* reaction (cf. Table 14.18).
At temperatures below +8 °C, cotton seed oil becomes turbid due to crystallization of high melting point triacylglycerols. Such undesirable low temperature characteristics are avoided using a "winterization" process (cf. 14.4.4).
Cereal germ oils. All cereals contain significant amounts of oil in the germ. It is available after the germ is separated during grain processing. Corn (maize) oil is the most important. Germ separation is achieved during dry or wet processing of the kernels into corn meal and starch (cf. 4.4.4.12). The oil is recovered from the germ collected as a by-product by pressing and solvent extraction. After crude oil refining, the corn waxes remaining in the oil and which originate from the skin-like layer coating of the epidermis (the cuticle), are removed by a winterization process (cf. 14.4.4).
Corn oil is suitable for manufacture of margarine and mayonnaise (creamy salad dressing), but is used preferentially as salad and cooking oil.
The oil present in wheat and rice is also concentrated in the germ. This oil can be recovered by pressing and/or solvent extraction of the germ. Wheat germ oil has a high content of tocopherol and, therefore, additional nutritive value. Rice germ oil is consumed to a minor extent in Asia.
Pumpkin oil is obtained by pressing dehulled pumpkin seeds. In southern Europe it is utilized as an edible oil. It is brown in color and has a nut-like taste.

Table 14.8. Oils high in palmitic acid

	Cottonseed *(Gossypium)*	Corn germ *(Zea mays)*	Wheat germ[a] *(Triticum aestivum)*	Pumpkin seed *(Cucurbita pepo)*
Seed oil content (weight-%)	22–24	3,5–5%[b]		35%
Solidification point (°C)	0 to +4	−10 to −18		−15 to −16
	Average fatty acid composition (weight-%)			
14:0	1.5	0	0	0
16:0	22	10.5	7	16
18:0	5	2.5	1	5
20:0	1	0.5	0	0
16:1 (9)	1.5	0.5	0	0.5
18:1 (9)	19	32.5	20	24
18:2 (9, 12)	50	52	52	54
18:3 (9, 12, 15)	0	1	10	0.5

[a] Oil content in germ amounts to 8–11 weight-%.
[b] Of the seed oil content 80% is located in germ and the rest in seed endosperm.

14.3.2.2.5 Oils low in palmitic acid and high in oleic and linoleic acids

A large number of oils from diverse plant families belong to this group (cf. Table 14.9). These oils are important raw materials for manufacturing margarine.
The sunflower is the most cultivated oilseed plant in Europe. Data about the production of the sunflower by regions and countries are given in Table 14.0. Prepressing of dehulled sunflower seeds yields a light yellow oil with a mild flavor. The oil is suitable for consumption once it is clarified mechanically. Refined oils are used in large amounts as salad oil or as frying oil and as a source of margarine production. The refining of the oil includes a wax-removal step.

Table 14.9. Oils low in palmitic acid and high in oleic and linoleic acids

	Sunflower (*Helianthus annuus*)	Soya (*Glycine max.*)	Peanut (*Arachis hypogaea*)	Rapeseed[b] (*Brassica napus*)	Sesame (*Sesamum indicum*)	Safflower (*Carthamus tinctorius*)	Linseed (*Linum usitatissimum*)
Seed oil content (weight-%)	25–30	18–23	42–52	ca. 40	45–55	25–37	32–43
Solidification point (°C)	−18 to −20	−8 to −18	−2 to −3	0 to −2	−3 to −6	−13 to −20	−18 to −27
			Average fatty acid composition (weight-%)				
16:0	6.5	10	10	4	8.5	6	6.5
18:0	5	3.5	3	1.5	4.5	2.5	3.5
20:0	0.5	0.5	1.5	0.5	0.5	0.5	0
22:0	0	0	3	0	0	0	0
18:1 (9)	23	21	55	63	42	12	18
18:2 (9, 12)	63	56	25	20	44.5	78	14
18:3 (9, 12, 15)	0.5	8	0	9	0	0.5	58
20:1 u. 20:2	1	0.5	1	1	0	0.5	0
22:1 (13)	0	0	–	0.5	0	0	0

[a] African peanut oil. [b] Canola-type oil (practically free of erucic acid).

Two legume oils, soybean and peanut (or ground nut), are of great economic significance (cf. Table 14.1). Soybean oil is currently at the top of the world production of edible oils of plant origin. It is cultivated mostly in the United States, Brazil and China. The refined oil is light yellow and has a mild flavor. A "green seed" off-flavor, designated as a reversion flavor (cf. 3.7.2.1.5), develops from autoxidation, particularly of the linolenic acid component (cf. fatty acid composition, Table 14.9), which occurs as a consequence of improper packaging or storage. The shelf life of the oil is improved significantly be selective hydrogenation to give a melting point range of 22–28 °C or 36–43 °C. Such oils are utilized as raw materials for the manufacture of margarine and shortening (semi-solid vegetable fats used in baked products, such as pastry, to make them crisp or flaky).

The fatty acid composition of peanut oil is greatly influenced by the region in which the peanuts are grown. In contrast to the peanut oils produced in Africa (Senegal or Nigeria), the peanut oils from South America are enriched in linoleic acid (41% vs 25%, w/w; see fatty acid composition, Table 14.9) at the expense of oleic acid (37% vs 55%, w/w). The contents of arachidic (20:0, eicosenoic (20:1), behenic (22:0), erucic (22:1) and lignoceric (24:0) acids are characteristic of peanut oil. Their glycerols readily crystallize below 8 °C.

Peanut butter is a spreadable paste made from roasted and ground peanuts by the addition of peanut oil and, occasionally, hydrogenated peanut oil.

Rapeseed oil. This oil is produced from seeds of two *Brassica* species: *B. napus,* known as Argentinian rape or, in Europe, as summer or Swedish rape, and *B. campestris* (Polish or summer turnip rape). The latter plants yield slightly less oil, are shorter (approx. 80 cm), but mature more quickly. They are more tolerant to frost and have improved resistance to pests and diseases. Old rape and turnip rape cultivars contained high levels of erucic acid (45–50% by weight), which is hazardous in human nutrition. "Zero"erucic acid cultivars (22:1 < 5% by weight), called *Canola,* have been developed and, recently, "double zero" cultivars, with low levels of erucic in the oil and goitrogenic compounds in the seed meal, have been developed. The major rapeseed-cultivating regions and countries are listed in Table 14.0.

The above-mentioned plants, such as *Brassicacea,* contain mustard oil glucosides (glucosinolates, cf. 17.1.2.6.5) which, immediately after seed crushing, are hydrolyzed to esters of isothiocyanic acid ("mustard oils", cf. Figure 14.1). The hydrolysis is dependent on seed moisture and is catalyzed by a thioglucosidase enzyme called myrosinase (EC 3.2.3.1). In the presence of the enzyme, some of the isocyanates are isomerized into thiocyanates (esters of normal thiocyanic acid, or rhodanides) and, in part, are decomposed into nitrile compounds which do not contain sulfur. All these compounds are volatile and, when dissolved in oil, are hazardous to health and detrimental to oil flavor. Moreover, they interfere with hydrogenation of the oil by acting as Ni-catalyst poisons (cf. 14.4.2.2). Therefore,

in the production of rapeseed oil, a dry conditioning step is used (without live steam) to inactivate the myrosinase enzyme and, only then, the seed is ground and subjected to prepress and solvent extraction processes.
Despite these precautions, small amounts of volatile sulfur compounds are formed. However, they are removed during the refining process. Irrespective of technical achievements in rapeseed production and processing, the selection and breeding of rapeseed "double zero" cultivars is being continued.
Rapeseed (Canola) oil is used as an edible oil. It is saturated by hydrogenation to a melting point of 32–34 °C and, with these melting properties, resembles coconut oil.
Turnip rape oil has practically the same composition as the *B. napus* oil.
Sesame oil is obtained from an ancient oilseed crop (*Sesamum indicum,* L.), which is widely cultivated in India, China, Burma and east Africa (cf. Table 14.0). In its refined form the oil is nearly crystal clear and has a good shelf life. In addition to a considerable amount of tocopherols, it contains another phenolic antioxidant, sesamol, which is derived from hydrolysis of sesamolin (Figure 14.2).
Sesame oil can be readily identified with great reliability (cf. Table 14.8). Therefore, in some countries, blending this oil into margarine is required by law in order to identify the product as margarine.
Safflower oil is obtained from a thistle-like plant *(Carthamus finctorius)* grown in the arid regions of North America and India (cf. Table 14.0). New cultivars have been bred with oil compositions which deviate greatly from those listed in Table 14.9. These new cultivars contain 80% by weight oleic acid (18:1) and 15% by weight linoleic acid (18:2; 9, 12).
Linseed oil. Flax, for fiber and seed production and the subsequent processing of the seed into linseed oil, is grown mainly in North America, India, the U.S.S.R. and Argentina (cf. Table 14.0). Due to its high content of linolenic acid (cf. Table 14.9), the oil readily autoxidizes, one of the processes by which some bitter substances are created. Since autoxidation involving polymerization reactions proceeds rapidly, the oil solidifies ("fast drying oil"). Therefore, it is used as a base for oil paints, varnishes and linoleum manufacturing, etc. A comparatively negligible amount, particularly of the cold-pressed oil, is utilized as an edible oil.

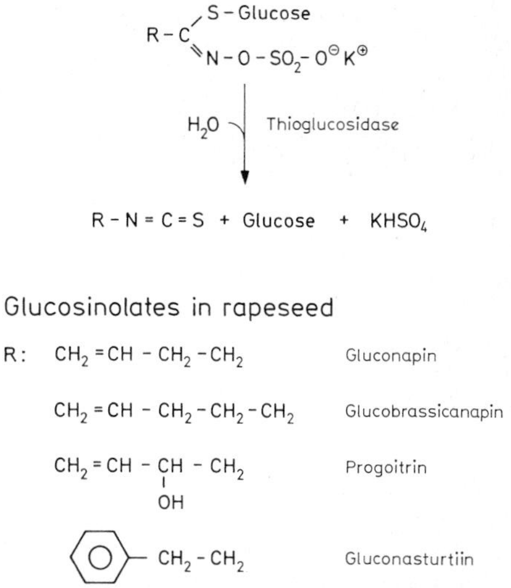

Fig. 14.1. Enzymatic degradation of isothiocyanate glucosides (glucosinolates) during rapeseed crushing

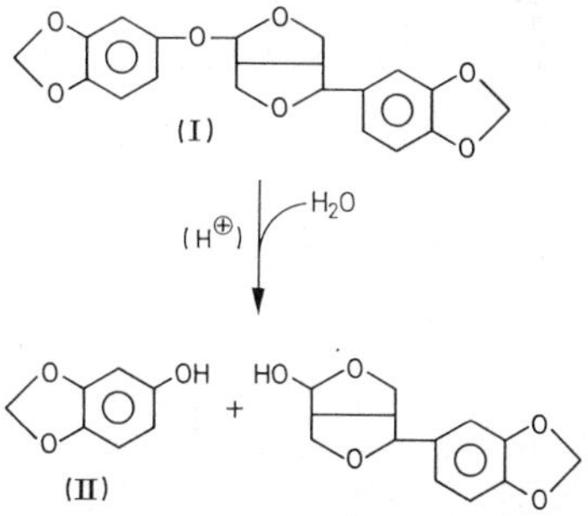

Fig. 14.2. Sesame oil: sesamol (II) formation by sesamolin (I) hydrolysis

14.4 Processing of Fats and Oils

14.4.1 Refining

Apart from some oils obtained by cold pressing (examples in 14.3.2.1), most of the oils obtained using expeller, screw or hydraulic presses, solvent extraction or by melting at elevated temperatures are not suitable for immediate consumption. Depending on the raw material and the oil recovery process, the oil contains polar lipids, especially phospholipids, free fatty acids, some odor- and taste-imparting substances, waxes, pigments

(chlorophyll, carotenoids and their degradation products), sulfur-containing compounds (e.g. thioglucosides in rapeseed oils), phenolic compounds, trace metal ions, contaminants (pesticides or polycyclic hydrocarbons) and autoxidation products.
In a refining process comprising the following steps:

- Vegetable lecithin removal
- Degumming
- Free fatty acid removal
- Bleaching
- Deodorization,

all the undesired compounds and contaminants are removed. In practice the refining steps used depend on the quality of the crude oil and its special constituents (e.g. carotene in palm oil or gossypol in cottonseed oil). The following precautionary measures are taken during refining in order to avoid undesirable autoxidation and polymerization reactions:

- Absence of oxygen (also required during transport or storage)
- Avoidance of heavy metal contamination
- Maintaining the processing temperatures as low and duration as short as possible.

14.4.1.1 Removal of Vegetable Lecithin

This processing step is of special importance for rapeseed and soybean oils. Water (2–5%) is added to crude oil, thereby enriching the phospholipids in the oil/water interface. The emulsion thus formed is separated or clarified by centrifugation. The "crude lecithin" (cf. 3.4.1.1) is isolated from the aqueous phase and is recovered as crude vegetable lecithin after evaporating the water in a vacuum.

14.4.1.2 Degumming

Finely-dispersed protein and carbohydrates are coagulated in oil by addition of phosphoric acid (0.1% of oil weight). A filtering aid is then added and the oil is clarified by filtration. This also removes the residual phospholipids from the previous processing step.

14.4.1.3 Removal of Free Fatty Acids (Deacidification)

Several methods exist for deacidification of fat or oil. The choice depends on the amount of free fatty acids present in crude fat or oil.

The removal of fatty acids with 15% sodium hydroxide (alkali refining) is the most frequently used method. Technically, this is not very simple since fat hydrolysis has to be avoided and, moreover, the sodium soap (the "soapstock"), which tends to form stable emulsions, has to be washed out by hot water. After vacuum drying, the fat or oil may contain only about 0.05% free fatty acids and 60 to 70 ppm of sodium soaps. When the fat or oil is treated with diluted phosphoric acid, the content of sodium soaps decreases to 20 ppm and part of the trace heavy-metal ions is removed.
Fats (oils) with 5% or more free fatty acids require relatively high amounts of alkali for extraction, resulting in an unavoidably high loss of neutral fats (oils) due to alkaline hydrolysis. Removal of free fatty acids from fats (oils) which contain a low content of unsaturated lipids, e.g. palm or coconut oils, is possible by steam distillation (a form of physical refining). This process is efficient and economical at a vacuum of 1–10 mbar and a temperature of 210–270 °C. Carotenes are destroyed during distillation, e.g. thermal bleaching occurs in palm oil. The fatty acids collected as by-products are of higher quality than the so-called "refining fatty acids" obtained by alkali treatment. Physical refining is often combined with an oil deodorization step.
In special cases, a selective fluid/fluid extraction is of interest. Ethanol extracts free fatty acids (above a level of 3%) from triacylglycerols in crude oils – this is a suitable way to treat oils with exceptionally high amounts of free acids. At a given temperature, furfural can extract the polyunsaturated triacylglycerols. On the other hand, propane under pressure preferentially solubilizes the saturated triacylglycerols and leaves behind the unsaturated ones, together with unsaponifiable matter. Pressurized propane is utilized in marine oil fractionation, e.g. in the production of vitamin A concentrates.

14.4.1.4 Bleaching

In order to remove the plant pigments (chlorophyll, carotenoids) and autoxidation products, the fat or oil is stirred for 30 min in the presence of an Al-silicate (bleaching or *Fuller's* earth) in a vacuum at 90 °C. The silicate has to be activated prior to use – a suspension in water is treated with hydrochloric acid, followed by thorough washing with water, then drying. The amount of silicate used is 0.5–2% of the fat (oil) weight. It is often used together with 0.1–0.4% activated charcoal.

The bleached oil is removed from the adsorbent by filtration. The oil retained by the adsorbent can be recovered by hexane extraction and be recycled into the refining process.The residual alkali soaps, gums, part of the unsaponifiable matter and the heavy metal ions are also removed during the bleaching process.

After bleaching, some oils or fats which contain polyunsaturated fatty acids show an increase in absorbance at 270 nm. This is due to decomposition of hydroperoxides, formed by autoxidation, into oxo-dienes (I) and fatty acids with three double bonds (II):

$$
\begin{array}{l}
R_1-CH=CH-CH=CH-\underset{\|}{C}(=O)-R_2 \quad \text{(I)} \\
\uparrow \; -H_2O \\
R_1-CH=CH-CH=CH-CH(OOH)-R_2 \quad \text{(Hydroperoxide)} \\
\downarrow \; -{}^{\bullet}OH \\
R_1-CH=CH-CH=CH-CH(O^{\bullet})-R_2 \\
\downarrow \; +RH, \; -R^{\bullet} \\
R_1-CH=CH-CH=CH-CH(OH)-R_2 \\
\downarrow \; -H_2O \\
R_1-CH=CH-CH=CH-CH=CH-R_3 \quad \text{(II)}
\end{array}
\tag{14.3}
$$

14.4.1.5 Deodorization

Deodorization is essentially vacuum steam distillation (190–210 °C, 0.5–10 mbar). The volatile compounds, together with undesirable off-flavors present in the fat or oil, are separated in this refining step. Deodorization takes from 20 min to 6 h, depending on the type of fat or oil and the content of volatile compounds.

The processing loss of this refining step is 0.2%. This is negligible since the fat or oil droplets carried by the steam are caught by baffles or are intercepted by an external trap system.

14.4.1.6 Product Quality Control

In addition to sensory evaluation, free fatty acid analysis (the content is usually below 0.05%) and analysis of possible contaminants are carried out. The data given in Table 14.10 illustrate the amounts of pesticides and polycyclic aromatic compounds removed by deodorization. To be sure, this refining step also removes the highly desirable aroma substances which are so standard and valuable for the individuality of cold-pressed oils such as olive oil.

14.4.2 Hydrogenation

14.4.2.1 General Remarks

Liquid oils are supplied mostly from natural sources. However, a great demand exists for fats which are solid or semi-solid at room temperature. To satisfy this demand, *W. Normann* developed a process in 1902 to convert liquid oil into solid fat, based on the hydrogenation of unsaturated triacylglycerols using Ni as a catalyst; a process designated as "fat hardening". The process rapidly gained great economic importance; even marine oils became suitable for human consumption after the hardening process. Today, in excess of 4×10^6 tonnes/year of fat are produced worldwide by hydrogenation of oils; most is consumed as food.

The unsaturated triacylglycerols can be fully hydrogenated, providing high melting point cooking, frying and baking fats, or partially hydrogenated, providing products such as:

- Oils rich in fatty acids with one double bond. They are stable and resistant to autoxidation and have a shelf life similar to olive oil. They are used as salad oil or as shortening.
- Products in which the linolenic acid is hydrogenated, but the essential fatty acid, linoleic

Table 14.10. Endrin and polycyclic hydrocarbons removal during edible oil refining (μg/kg)

Compound	Content in raw oil	Content in oil after		
		deacidification	bleaching	steaming
Endrin	620[a]	590	510	< 30
Anthracene	10.1[b]	5.8	4.0	0.4
Phenanthrene	100[b]	68	42	15
1,2-Benz-anthracene	14[b]	7.8	5.0	3.1
3,4-Benzpyrene	2.5[b]	1.6	1.0	0.9

[a] Soybean oil. [b] Rapeseed oil.

acid, is left intact. An example is soybean oil, hydrogenated selectively to increase its stability against oxidation.

- Fats that melt close to 30 °C and have a plastic or spreadable consistency at room temperature.

Fully or partially hydrogenated oils are important raw materials for margarine manufacturing.

14.4.2.2 Catalysts

The principle of the heterogenous catalytic hydrogenation of unsaturated acylglycerols was outlined under 3.2.3.2.4. As catalysts, nickel is mostly used and, next to it, nickel subsulfide (Ni_3S_2) and copper are important. The choice of catalyst is made according to:

- Reaction specificity.
- Extent of trans-isomer formation
- Duration of activity and cost.

To determine the specificity of a catalyst, the reaction rates for each individual hydrogenation step must be determined. Simplified, there are three reaction rate constant (k) involved (AG = acylglycerol):

$$\text{Triene-AG} \xrightarrow{k_3} \text{Diene-AG} \xrightarrow{k_2} \text{Monoene-AG} \xrightarrow{k_1} \text{Saturated-AG} \tag{14.4}$$

The catalytic reactions considered here require that $k_3 > k_2 > k_1$. The following equations determine the specificity"S":

$$S_{32} = \frac{k_3}{k_2}; \quad S_{21} = \frac{k_2}{k_1}; \quad S_{31} = \frac{k_3}{k_1}; \tag{14.5}$$

That is, the greater the value of "S", the faster the hydrogenation at this step. Therefore, specificity (or selectivity) is proportional to the value of "S". For the three catalysts mentioned, Table 14.11 shows that the hydrogenation of diene → monoene by Ni_3S_2 and the hydrogenation of triene → monoene by copper become accelerated, with marked specificity. Copper is particularly suitable for decreasing the linolenic acid content in soybean and rapeseed oils. However, copper catalysts are not sufficiently economical, since they can not be used more than twice. Therefore, an Ag/Ni catalyst with similar hydrogenation properties is more advantageous.

Table 14.11. Properties of hydrogenation catalysts

Catalyst	Selectivity		trans-Fatty acids
	S_{32}	S_{21}	(weight-%)[a]
Nickel-contact	2–3	40	40
Ni_3S_2-contact	1–2	75	90
Copper-contact	10–12	50	10

[a] trans-Fatty acids as monoenoic acids total content is calculated as elaidic acid.

From an economic point of view, the nickel catalyst is the best since it can be used repeatedly for up to 50 times under the following conditions: the plant oil must be deacidified, freed of gum ingredients and contain no sulfur compounds (cf. rapeseed oil, 14.3.2.2.5). The favorable ratio of duration of activity to cost places the nickel catalysts ahead with advantages not readily surpassed by any other catalyst. Two methods exist for the production of this catalyst:

a. A suitable carrier (kieselguhr, silicic acid, aluminum oxide) impregnated with nickel hydroxide or nickel carbonate is "roasted" to form nickel oxide, followed by reduction with hydrogen to metallic Ni.
b. A carrier is impregnated with nickel formate which is then decomposed by heating to active nickel:

$$Ni(HCOO)_2 \xrightarrow[(200-250\,^\circ C)]{} Ni + 2\,CO_2 + H_2 \tag{14.6}$$

The Ni produced in this way is a finely-dispersed pyrophoric metal. For this reason it is embedded in fat and handled and marketed in this form.
The earlier catalyst, Ni_3S_2, gained importance because of its resistance to poisons and its long life span. It was manufactured by precipitation of nickel sulfide onto a carrier, followed by reduction using hydrogen.
Calculation programs were developed for the determination of the actual selectivity of a catalyst based on the fatty acid composition of the starting material and of the hydrogenated product.
During hydrogenation, linolenic acid yields, among others, isolinoleic and isooleic acids (cf. Reaction 14.7).
The diversity of the reaction products present in partially hydrogenated fat is increased further by the positional- and stereoisomers of the double bonds. Hydrogenation of soybean oil in the pres-

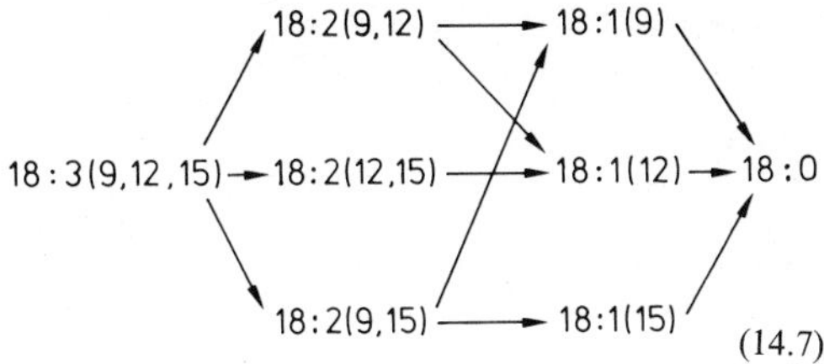

(14.7)

ence of a copper catalyst gives, for example, a number of trans-monoene fatty acids (Table 14.12). The extent of isomerization is affected, among other factors, by the type of catalyst used in hydrogenation.

Although extensive nutritional/physiological studies have provided no evidence about the possible deleterious effects of elaidic or isoelaidic acids in fat metabolism, the current processing trend is still to lower the accumulation of trans-fatty acids in hydrogenated fats. Hydrogenated fat is readily distinguished from nonhydrogenated fat by the presence of trans-fatty acids. The trans-acids are routinely revealed and quantitatively determined by infra-red (IR) spectroscopy or by gas chromatography (cf. 14.5.2.3).

A further drawback of the partial hydrogenation of an oil is the pattern of linoleic acid isomers formed. The two isomers formed during hydrogenation, linol-elaidic acid 18:2 (9 trans, 12 cis) and 18:2 (9 cis, 12 trans), unlike linoleic acid, are not essential fatty acids (cf. 3.2.1.2).

14.4.2.3 The Process

The hydrogen required can be obtained by electrolysis of dilute aqueous KOH, through water-to-gas conversion:

$$\begin{aligned} H_2O + C &\longrightarrow H_2 + CO \\ CO + H_2O &\longrightarrow H_2 + CO_2 \end{aligned} \qquad (14.8)$$

or by the decomposition of natural gas with steam:

$$\begin{aligned} CH_3(CH_2)_xCH_3 + H_2O &\longrightarrow H_2 + CO \\ CO + H_2O &\longrightarrow H_2 + CO_2 \end{aligned} \qquad (14.9)$$

In the latter two processes, the poisonous by-products, H_2S and CO, have to be completely removed.

Oil hydrogenation is performed in an autoclave equipped with a stirrer under hydrogen gas pressure of 1–3 bar and a temperature of 150–250 °C. A newer hydrogenation process uses a recycling

Table 14.12. Fatty acid composition of a soya oil before and after hydrogenation with copper catalyst

Fatty acid	Hydrogenation	
	before (weight-%)	after (weight-%)
16:0	10,0	10.0
18:0	4.2	4.2
18:1 (9)	26.0	30.4
18:1[a]	0	5.5
18:2 (9, 12)	52,5	42.5
18:2 (conjugated)[b]	0	0.7
18:2[c]	0	5.2
18:3 (9, 12, 15)	7.3	0.7

[a] This fraction contains eight trans fatty acids: 18:1 (7 tr) – 18:1 (14 tr); major components are 18:1 (10 tr) and 18:1 (11 tr).
[b] It consists of various conjugated fatty acids.
[c] Isolinoleic and isolinolelaidic acids.

hydrogenation unit equipped with a spraying nozzle, external heat exchanger and recycling pump.

The process conditions have a significant effect on the composition and therefore on the consistency of the end-product. Selective hydrogenation of double bonds is favored by a high concentration of catalyst (which, depending on the Ni activity grade, is 200–800 g Ni/t fat), a higher temperature and lower pressure of hydrogen gas. After hydrogenation, the fat is filtered, then deacidified, bleached and deodorized during further refining (cf. 14.4.1.3–14.4.2.5).

Some constituents of the unsaponifiable matter are also affected by the hydrogenation process. Carotenoids, including vitamin A, are hydrogenated extensively. Some of the chlorine-containing pesticide contaminants are hydrogenated. Sterols, under the usual operating conditions, are not affected. The ratios and levels of tocopherols are essentially unchanged.

14.4.3 Interesterification

Natural fats and oils are subjected to extensive interesterification during processing. This involves a rearrangement or randomization of acyl residues in triacylglycerols and thus provides fats or oils with new properties. By choosing the raw material and processing parameters, the interesterification can be controlled to obtain a fat with melting characteristics and consistency which match the aim of its use ("tailored fats").

The basics of the interesterification process are outlined under 3.3.1.3. Sodium methylate is used almost exclusively as the catalyst. The dried and deacidified fat (or oil) is stirred at 80–100 °C in the presence of alcoholate (0.1–0.3% of fat weight). When the reaction is completed, the catalyst is destroyed by adding water, then the degraded catalyst is removed, together with the resultant soaps, from the fat (oil) by repeated washing with water. The interesterified product is then bleached (cf. 14.4.1.4) and deodorized (cf. 14.4.1.5).

Table 14.13 illustrates the extent of change in triacylglycerols brought about by the process and its influence on fat melting points.

The properties of lard needed for baking (volume and softness of the baked goods) are improved by interesterification. The uniform distribution of palmitic acid in the triacylglycerols accounts for such an improvement.

Furthermore, interesterification is of significant importance in the manufacturing of different varieties of margarine with a given composition, for example:

- Vegetable margarine with 30% w/w of 18:2 (9, 12) fatty acid may be produced by interesterification of partially hardened sunflower oil blended with its natural liquid oil.
- Interesterification of palm oil with palm seed or coconut oil (2:1), then interesterification of this product with 4 parts of sunflower oil, provides a margarine which contains 20–25% w/w of linoleic acid and does not contain hydrogenated fat.

Table 14.13. Changes in the pattern of triacylglycerols in a partially hydrogenated palm oil by interesterification

Melting point	Prior to interester-ification	Single phase interester-ification	Directed interester-ification
Melting point (°C)	41	47	52
	Triacylglycerols[a] in mole-%		
S_3	7	13	32
S_2U	49	38	13
SU_2	38	37	31
U_3	6	12	24

[a] S: Saturated, U: unsaturated fatty acids.

14.4.4 Fractionation

The undesirable fat ingredients are removed or the desirable triacylglycerols (TG) are enriched by fractional crystallization. The rising demand of food processors for special fats with standardized properties has led to large-scale isolation of special fractions, particularly from palm oil and the fats and oils listed under 14.3.2.2.2. The following procedures are used for the fractional crystallization of fats: The melted fat is slowly cooled until the high melting TG selectively crystallize, i.e. without forming mixed crystals of low and high melting TG. A sharp separation into two or more fractions is assumed to be satisfactory when the melting points of the fractions differ by at least 10 C^0. The separated crystals are either removed by filtration or are washed out with a tenside solution. In the latter case, the fat crystals adsorb a water soluble surface-active agent , such as sodium dodecyl sulfate, and thus acquire hydrophilic properties. The crystals are then transferred into the aqueous phase. The isolated aqueous suspension is then heated and the TG recovered as liquid fat.

An even sharper fractional crystallization procedure may be achieved by solubilization of fat in hexane or some other suitable solvent. However, solvent distillation and recovery are rather time consuming, so the use of this procedure is justified only in very special cases.

In the processing step of "winterization" of rapeseed (Canola), cottonseed or sunflower oil, small amounts of higher melting TG or waxes are removed which would otherwise cause turbidity during refrigeration. The basis of winterization is the fractional crystallization by slowly cooling the oil, as outlined above. Other procedures for the production of cold-stable oils are based on the use of crystallization inhibitors. These are mono- and diacylglycerols, esters of succinic acid, etc.

The application of the fractional extraction of fat or oil, instead of crystallization, has been outlined under 14.4.1.3.

14.4.5 Margarine – Manufacturing and Properties

The inventor of margarine, *H. Mège Mouries,* described in his patent issued in 1869 a process for the production of spreadable fat from beef fat which would substitute for and imitate the scarce and costly dairy butter. Based on the assumption that margaric acid (17:0) is the predominant fatty acid of beef fat, the name "margarine" was suggested for the new product. The assumption was, however, proven to be incorrect (cf. Table 14.3). Nevertheless, the name remained. Table 14.14 provides an overview of margarine production in some industrialized nations.

14.4.5.1 Composition

The properties of margarine, such as nutritional value, spreadability, plasticity, shelf life and melting properties, resemble those of butter and are influenced essentially by the varieties and properties of the main fat ingredients. Since choice of ingredients is large, numerous varieties of margarine are produced (cf. Table 14.15).

The fat in margarine, which by regulation is 80% by weight (diet margarine is 39–41% fat), contains about 18% emulsified water. The W/O emulsion is stabilized by a mixture of mono- and diacylglycerols (approx. 0.5%) and crude lecithin (approx. 0.25%). Diet margarines have higher levels of emulsifiers. Skim milk or skim milk powder suspended in water (milk proteins, 1%; 2% in diet brands) is added in the production of high quality retail brands of margarine. The casein assists the action of the emulsifiers and, during heating, together with lactose, provides the desired browning.

The aqueous phase of the margarine acquires a pH of 4.2–4.5 by addition of citric and lactic acids. This not only affects the flavor, but protects against microbial spoilage. In addition, trace heavy metal ions are complexed. Margarine also contains aroma substances, the same ones present in butter and which can be produced by microbiological souring (cf. 10.2.3.2). Readily-available synthetic compounds, such as diacetyl, butyric acid, lactones of C_8–C_{14} hydroxy-fatty acids (cf. 5.3.1.4) and 4-cis-heptenal, may also be used for aromatization. Common salt (0.1–0.2%) is used to round-off the flavor. Margarine is colored with β-carotene or with slightly-refined, unbleached palm oil. Attention is also given to maintaining the presence of 1 mg of α-tocopherol per g of linoleic acid. High quality products are vitaminized by the addition of about 25 IE/g vitamin A and 1 IE/g vitamin D_2. The authenticity of margarine is verified in some countries by an indicator substance added to it. This is required by legislation. Slightly-refined sesame oil (for its detection, see Table 14.18) is one of these substances.

Table 14.14. Production of margarine in 1975 (1,000 t)

Country	
FR Germany	509
German DR	180
USA	1,088
USSR	1,000
Great Britain and Northern Ireland	292
Poland	183
France	158
Japan	157
Belgium	148

14.4.5.2 Manufacturing

Margarine manufacturing is done continuously by a process consisting essentially of three steps:

- Emulsification of water within the continuous oil phase.
- Chilling and mechanical handling of the emulsion.
- Crystallization, which preserves the type of w/o emulsion by efficient removal of the released heat of crystallization.

The triacylglycerols should preferentially crystallyze in their β′-forms (cf. 3.3.1.2). The higher melting β-forms are not desired since they cause a "sandy" texture. The transition β′ → β-form is inhibited by additions of 1% saturated diacylglycerols.

14.4.5.3 Varieties of Margarine

The characteristic features of some varieties of margarine are summarized in Table 14.15.

14.4.6 Mayonnaise

Mayonnaise is an "oil in water", or o/w, emulsion (cf. 8.15.1) consisting of 50–85% edible oil, 5–10% egg yolk, vinegar, salt and seasonings (cf. 11.4.2.3). The emulsion is stabilized by egg yolk phospholipids. Products with a lower oil content (< 50%) may contain thickening agents such as starch, pectin, traganth, agar-agar, alginate, carboxymethylcellulose, milk proteins or gelatin. Sorbic acid, benzoic acid or the ethyl ester of p-hydroxybenzoic acid are added as preservatives.

Table 14.15. Examples of margarine types

Type	Comments
A. Household margarine	
Standard product	At least 50% of fat portion is vegetable oil, the rest being animal fat.
Vegetable margarine	At least 98% of fat portion is vegetable oil; contains at least 15% linoleic acid.
Linoleic acid enriched margarine	At least 30% linoleic acid, otherwise as vegetable margarine.
B. Hemi-fat margarine	The fat content is halved. This type is not suitable for baking and frying.
C. Molten or fused margarine	Practically free of water and protein. It is aromatized with diacetyl and butyric acid; mellow consistency; with large TG crystals it has a grainy structure; applied in cooking, frying and baking.
D. Special types for industrial processing	
Baking margarine	Strongly aromatized with heat stable compounds that contribute to baked products aroma; high melting point TG's are embedded in oil phase.
Margarine for pastry production	This margarine is strongly aromatized; its high melting point TG's are embedded in oil phase; suitable for dough extension into thin sheets ("strudel dough") used in flaky pastry production.
Creamy margarine	It is not or is just slightly aromatized; has a mellow consistency; contains high content of coconut oil and approx. 10 vol-% of air.

TG: Triacylglycerol

14.4.7 Fat Powder

In contrast to fats and oils, fat powders have better stability against autoxidation and, in some food products such as dehydrated soup powders or broths, are easier to handle. They are manufactured from natural or hardened plant fats, sometimes with the addition of emulsifiers and protein carriers. Butter and cream powders are also produced.

In a cold-spray process, the melted fat is sprayed under high pressure into a cooled (−35 °C) air-blast crystallization chamber, where the fat particles solidify. After being recrystallized, the particles are coated to avoid clumping.

In a spray-drying process, the fat is homogenized with emulsifiers, water and skim milk and dried and additionally crystallized in the form of a concentrated spray.

14.5 Analysis

14.5.0 Scope

The problems and scope of fat or oil analysis include identification of the variety, determination of the composition of the blend, detection of additives, antioxidants, color pigments, and extraneous contaminants (solvent residues, pesticides, trace metals, mineral oils, plasticizers). In addition, the scope of analysis encompasses determination of other quality parameters, such as the extent of lipolysis, autoxidation or thermal treatments. Also of interest is the extent of refining which the fats and oils have been subjected to as well as detection of hardened fat and products which were interesterified.

14.5.1 Determination of Fat in Food

The methods used for determination of fat or oil in food are often based on extraction with either ethyl ether or petroleum ether and gravimetric determination of the extraction residue. These methods may provide unreliable or incorrect results, particularly with food of animal origin. As shown in Table 14.16, where a corned beef sample was analyzed, the amount and composition of fatty acids in the fat residue were influenced greatly by the analytical methods used. In addition to the accessible, free lipids, the emulsifiers present and the changes induced by autoxidation affect the content of extracted lipids and the lipid-to-nonlipid ratio of the residue. The use of a standard method still does not eliminate the disadvantages shown by methods of fat analysis.

Table 14.16. Determination of the fat content of canned corned beef

Analytical method	Fat content (%)[a]	Fatty acid composition (g/100 g)			
		Saturated acids	18:1 (9)	18:2 (9, 12)	18:3 (9, 12, 15)
1. Dried sample is extracted with ethyl ether	7.9	3.98	2.06 2.60	0.05 0.77	0.08 0.32
2. Sample is homogenized in 95% ethanol and then extracted with ether	15.8	4.0	2.60	0.17	0.32
3. Sample is hydrolyzed with 4N HCl (at 60 °C for 30 min), then extracted with ether	12.3	5.66	3.94	0.95	0.71
4. Sample is hydrolyzed with conc. HCl (at 100 °C for 1 h), methanol added and then extracted with carbon tetrachloride	13.9	2.45	1.68	0.34	0.21
5. Sample is homogenized in chloroform methanol mixture (2:1 v/v), washed with water and then the chloroform phase recovered	11.2	4.89	3.31	0.85	0.39

[a] The fat is determined gravimetrically after the solvent is evaporated.

Therefore, in questionable cases, quantitative determination of fatty acids and/or glycerol is recommended.

A rapid and accurate determination of fats or oils in food is achieved by ^{1}H–NMR spectrometry. The method is based on the fact that hydrogen nuclei in fluids respond to substantially higher magnetic resonance effects than do immobilized hydrogen atoms of solid substances. Thus, the ^{1}H–NMR signal of a fluid, such as an oil, differs from that of a nonoil matrix, such as carbohydrate, protein or firmly-bound water. The intensity of the signal is directly proportional to the oil content. This method is also of great value in oilseed selection or breeding research, since it permits determination of the oil content in a single kernel without damaging it by grinding or drying, i.e. retaining its ability to germinate.

The proportion of solid to fluid triacylglycerols in fat can also be determined using ^{1}H–NMR spectrometry.

14.5.2 Identification of Fat

14.5.2.1 Chemical Constants

For both, the identification and the determination of the quality of a fat or oil, the older lipid chemistry defines a series of so-called chemical constants in which the reagent uptake is used to quantitatively estimate the selected functional groups or calculate the constituents of a fat or oil. The introduction of new analytical methods, such as gas chromatography of fatty acids and the HPLC of triacylglycerols (cf. 3.3.1.4), has made many of these constants obsolete. The constants which are still used to differentiate the kinds of fats or oils are:

Saponification number (SN). This is the number of milligrams of KOH needed to hydrolyze 1 g of fat or oil under standardized conditions. The higher the SN, the lower the average molecular weight of the fatty acids in the triacylglycerols (for examples, see Table 14.17).

Table 14.17. Iodine (IN) and saponification numbers (SN) of various edible fats and oils

Oil/fat	IN	SN	Oil/fat	IN	SN
Coconut	256	9	Rapeseed (turnip)	225	30
Palm kernel	250	17	Sunflower	190	132
Cocoa	194	37	Soya	192	134
Palm	199	55	Butter	225	30
Olive	190	84			
Peanut	192	156			

Iodine number (IN). This number is the number of grams of halogen, calculated as iodine, which binds to 100 g fat (cf. 3.2.3.2.1). The halogen uptake by fat or oil is affected by the contents of oleic (IN: 89.9), linoleic (IN: 181) and linolenic (IN: 273) acids. Examples of iodine numbers are provided in Table 14.17.

Hydroxyl number (OHN). This number reflects the content of hydroxy fatty acids, fatty alcohols, mono- and diacylglycerols and free glycerol.

14.5.2.2 Color Reactions

Some oils give specific color reactions caused by particular ingredients. Examples are summarized in Table 14.18. Since many specific nonfat components are removed from oils by refining, these tests are negative when applied to refined oils.

14.5.2.3 Gas Chromatographic Analysis of Fatty Acid Composition

The acyl residues of an acylglycerol are freed as methyl esters (cf. 3.2.3.1) and are analyzed as such by gas chromatography. However, free fatty acid analysis is also possible by using specially selected stationary solid phases. Capillary-column gas chromatography should be used to differentiate between cis and trans fatty acids, which is required for the detection of partially-hydrogenated fats. The fatty acids indicative of the identity or type of fat or oil are summarized in Table 14.19. An enrichment step must precede gas chromatographic separation when fatty acids of analytical significance are present as minor constituents.

Prior to the enrichment step, specific techniques such as "argentation" chromatography (cf. 3.2.3.2.3) or fractionation by urea-adduct formation (cf. 3.2.2.3) are involved in addition to the usual preparative chromatographic procedure. The methyl branched fatty acids in marine oils are an example. These acids are enriched by the urea-adduct inclusion method since, unlike straight-chain acids, they are unable to form inclusion compounds. These branched-chain fatty acids do not change during hydrogenation, hence they can be used as marine oil indicators, i.e. to reveal the presence of marine oil in hydrogenated vegetable oils such as margarine.

Table 14.18. Color reactions for fat and oil identification

Reaction after[a]	Identification of
Baudouin (furfural and hydrochloric acid)	Sesame oil
Halphen (sulfur and carbondisulfide)	Cottonseed oil
Fitelson[b] (sulfuric acid and acetic acid anhydride)	Teaseed oil

[a] Reagents are listed in brackets.
[b] It is a modification of *Liebermann-Burchard* reaction for sterols (cf. 3.8.2.4).

Table 14.19. Fatty acid indicators suitable for determination of fat and oil origin

Fatty acid	Content (%)[a]	Indicator for
4:0	3.7	Milk fat
12:0	45	Coconut-, palm kernel-, and babassu fat
18:1 (9)	65–85[b]	Teaseed-, olive- and hazelnut oil
22:1 (13)	45	Rapeseed oil
18:3 (9, 12, 15)	9	Soya-, rapeseed (also erucic acid free) oil
18:2 (9, 12)	50–70	Sunflower-, corn germ-, cottonseed-, wheat germ-, and soya oil
22:0	3	Peanut oil
20:4 (5, 8, 11, 14)	0.1–0.6	Animal fat
18:1 (9, 12-OH)	80	Castor bean oil
Trans-fatty acids		Partially or fully hydrogenated oil/fat[c]
Methyl-branched fatty acids	0.2–1.6	Animal fat[d]

[a] When value range is omitted fatty acid percentage composition is given as an average value.
[b] A high percentage of this acid is a characteristic indicator.
[c] Here precautions are needed: animal fat might contain up to 10% trans fatty acids.
[d] It is relatively high in marine oils (approx. 1%).

14.5.2.4 Unsaponifiable Constituents

Some fats or oils which can not be unequivocally identified by their fatty acid composition may be identified by analysis of their unsaponifiable constituents. Examples are given in Table 14.20. If this possibility fails, then the only alternative is the analysis of the distribution of triacylglycerols (cf. 3.3.1.4).

14.5.2.5 Melting Points

In addition to specific density, index of refraction, color and viscosity, the melting properties can be used to identify fats and oils.

The composition and the crystalline forms (cf. 3.3.1.2) of triacylglycerols present in fat determine the melting points and the temperature range over which melting occurs. The onset, flow

Table 14.20. Fat or oil identification by analysis of unsaponifiable constituents

Analysis	Identification
Squalene	Olive or rice oil and fish liver oil
Campesterol/stigmasterol[a] (cf. 3.8.2.3.1)	Cocoa butter substitutes
Cholesterol[c] (cf. 3.8.2.2.1)	Animal fat
Carotene (cf. 3.8.4.5)	Raw palm oil
γ-/β-Tocopherol[b] (cf. Table 3.41)	Corn oil
γ-Tocopherol (cf. Table 3.41)	Wheat germ oil
α-/γ-Tocopherol[b] (cf. Table 3.41)	Sunflower oil
γ-/δ-Tocopherol[b] (cf. Table 3.41)	Soybean oil
Alkoxylipids (cf. 3.6.2)	Lard and edible tallow

[a] Concentration ratios are characteristics.
[b] Concentration of individual compounds and their concentration ratios are characteristic.
[c] Cholesterol concentration must exceed by 5% the total sterol fraction.

point and end point of melting are of interest. They are determined by standardized procedures.

The melting properties of fat are more accurately determined by differential thermal analysis. The temperature difference is measured between the fat sample and a blank, i.e. a thermally-inert substance, versus the heating temperature (Figure 14.3). In this way the temperatures at which polymorphic transitions of fat occur are detectable. In addition, the content of solid triacylglycerols can be assessed from the heat absorbed during melting at various temperatures. Thus, the solid triacylglycerol (TG) portion of coconut oil at $-3\,^{\circ}\mathrm{C}$ can be calculated using data from the recorded curve (Figure 14.3) and the following formula:

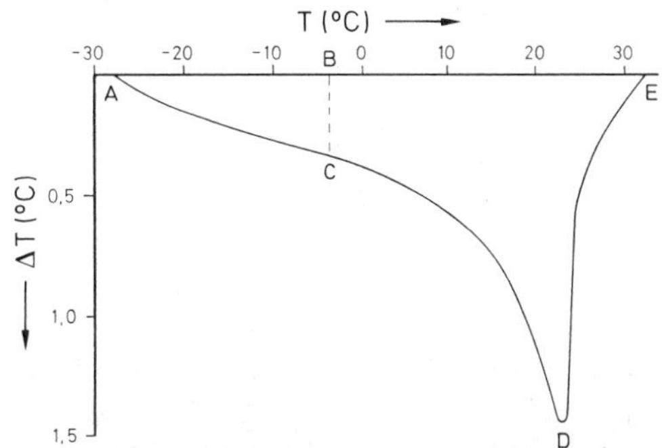

Fig. 14.3. Differential thermal analysis of a coconut fat

$$\%\ \text{Coconut (solid TG)} = \frac{\text{Area (BCDE)}}{\text{Area (AEDA)}} \times 100 \tag{14.10}$$

The solid: liquid ratio of acylglycerols is of importance in fat hydrogenation and interesterification processes (cf. 14.4.3). This ratio can also be assessed using the Solid Fat Index by measuring the expansion of the fat, i.e. the volume increase of a fat during its transition from solid to liquid.

14.5.3 Determination of Quality

Processes used in recovery and refining and subsequent storage conditions are the main factors affecting the quality of edible fat or oil. A number of analytical methods are available for assessing the quality and deterioration of fat or oil.

14.5.3.1 Lipolysis

The extent of lipolysis (cf. 3.7.1) is determined by the free fatty acid content (FFA or Acid Value). Oils with FFA content exceeding 1% are commonly designated as crude oils, while lard with this level of free acids is considered spoiled. An exception is olive oil, which is still considered suitable for direct consumption even with a 3% FFA content. The FFA content is lowered to less than 0.1% by refining of oil or fat.

A relationship has been established between the sensory perception of quality deterioration and the levels of FFA in fats or oils that contain low molecular weight acyl residues. It is valid for fats or oils such as those of milk and coconut and palm seed kernels.

14.5.3.2 Oxidative Deterioration

Fats and oils deteriorate rather rapidly by autoxidation of their unsaturated acyl residues (cf. 3.7.2.1). A number of analytical methods have been developed to determine the extent of such deterioration and to predict the expected shelf life of a fat or oil.

14.5.3.2.1 Oxidation state

Peroxide value. The method for determination of peroxide concentration is based on the reduction of the hydroperoxide group with HI or Fe^{2+}. The result of the iodometric titration is expressed as the peroxide value. The Fe^{2+} method is more suitable for determining a low hydroperoxide concentration since the amount of the resultant Fe^{3+}, in the form of the ferrithiocyanate (rhodanide) complex, is determined photometrically with high sensitivity (Fe-test in Table 14.21). The peroxide concentration reveals the extent of oxidative deterioration of the fat, nevertheless, no relationship exists between the peroxide value and aroma defects due to rancidity (already existing or anticipated). This is because hydroperoxide degradation into volatile off-flavors is influenced by so many factors (cf. 3.7.2.1.5) which make its retention by fat or oil or its further conversion into volatiles unpredictable.

Carbonyl compounds. The analysis of the compounds responsible for the rancid aroma defect of a fat or oil is of great value. Volatile carbonyls (cf. 3.7.2.1.5) are among such compounds.

In a simple test, such as benzidine, anisidine or heptanal values, the volatile aldehydes are not separated from fat or oil, rather the reaction with the group-specific reagents is carried out in the fat or oil. In addition to the odorous aldehydes, the flavorless oxo-acylglycerols and oxo-acids can be determined. Since the proportion of aroma-active and sensory-neutral carbonyls is not known, any correlation found between the carbonyl value and aroma defects is clearly coincidental.

The *thiobarbituric acid test* (TBA) is a preferred method for detecting lipid peroxidation in biological systems. However, the reaction is nonspecific since a number of primary and secondary products of lipid peroxidation form malonaldehyde which in turn reacts in the TBA test. In food containing oleic and linoleic acids, the TBA-test is not as sensitive as the Fe^{2+}-test outlined above.

The gas chromatographic determination of individual carbonyl compounds appears to be a method suitable for comparison with findings of sensory panel tests. Data in Table 14.22 illustrate that, in deteriorating soybean or peanut oils, pentanal or hexanal formation is correlated with sensory findings. However, such a correlation could not be established for cottonseed oil.

The example of peanut oil/citric acid (Table 14.22) is given to emphasize the effect of an additive – because of its presence, the spectrum of volatile compounds changed to such an extent that the above correlation ceased to exist. This may be clarified by a model system. Degradation of 9-hydroperoxy-octadeca-trans-10,cis-12-dienoic acid in the presence of a copper-palmitate salt is modified by the presence of an antioxidant (Table 14.23). Thus, in the presence of an antioxidant, the portion of 2-trans,4-cis-decadienal is increased. Since this aldehyde has a substantially lower threshold value than hexanal (cf. Table 3.24), such concentration shifts are desirable in the control of the aroma of an oxidized fat or oil.

Table 14.21. Analytical aspects related to the determination of the extent of oxidation of unsaturated fatty acids: relative sensitivities of spectrophotometric procedures[a]

Method		Autoxidized fatty acid methyl esters	
		18:2 (9, 12)[b]	18:3 (9, 12, 15)[c]
UV-Absorption	234 nm	1.0	1.0
	270 nm	0.1	0.3
Fe^{2+}/Thiocyanate (rhodanide)		9.4	6.3
Thiobarbituric	452 nm	0.1	0.5
acid test	530 nm	0.1	1.0
Kreis-test		0.1	0.1
Anisidine value		0.3	0.75
Heptanal value		0.1	0.1

[a] Related to UV absorption at 234 nm.
[b] Peroxide value: 475.
[c] Peroxide value: 450.

Table 14.22. Oxidative fat or oil deterioration: correlation between the rancid odor assay and pentanal or hexanal levels

Oil	Correlation coefficient	
	Pentanal	Hexanal
Soya	0.99[a]	0.99[a]
Soya + 0.01% citric acid + 0.02% TBHQ[d]	0.94[b]	0.90[b]
Cottonseed	0.71[c]	0.93[c]
Peanut	0.96[b]	0.97[b]
Peanut + 0.01% citric acid	0.06[c]	0.02[c]

[a] Statistically significant at a level of 99%.
[b] Statistically significant at 95% level.
[c] Not significant.
[d] tert-Butyl-hydroxyhydroquinone.

14.5.3.2.2 Shelf life prediction test

To estimate susceptibility to oxidation, the fat or oil is subjected to an accelerated oxidation test under standardized conditions so that the signs of deterioration are revealed within several hours or days. Examples of such tests are the *Schaal test* (fat maintained at 60 °C) and the *Swift stabil-*

Table 14.23. Degradation of hydroperoxy-octadeca-trans-10, cis-12-dienoic acid (75µmoles) into volatile aldehydes in presence of Cu-palmitate and antioxidants

Added antioxidant	Aldehydes (in mole %) after 3 h at 38 °C				
	Amount (µmole)	Hexanal	2-trans-Octenal	2,4-Nonadienal	2,4-Decadienal
None	2.5	78	20	1.0	1.0
α-Tocopherol	2.0	73.5	22	1.0	3.5
γ-Tocopherol	1.85	77	19.5	1.0	2.5
δ-Tocopherol	1.6	73.5	22.5	1.2	2.5
BHA	1.8	69.5	24	1.0	5.5
BHT	1.25	73.5	23	1.0	2.5

ity test (fat kept at 97.8 °C and aerated continuously). The extent of oxidation is then measured by sensory and chemical tests such as peroxide value (cf. 3.2.1.1), ultraviolet absorption (suitable for fats and oils containing linoleic or linolenic acids) or oxygen uptake. There are also methods based on the fact that in the process of triglycerol oxidation, when the initiation period is terminated, large amounts of low molecular weight acids are released. They are then determined electrochemically. During oxidation of a given fat or oil sample, a good correlation exists between the length of the induction period and the shelf-life.

14.5.3.3 Heat Stability

The behavior of a frying oil, when heated, is assessed from the content of oxidized fatty acids which are insoluble in petroleum ether and from the smoke point (cf. 3.7.4.2) of the fat or oil. The smoke point of a fat or oil is the temperature at which its triacylglycerols start to decompose in the presence of air. Smoke is the sign of decomposition. The smoke point of a fat or oil is normally in the range of 200–230 °C during prolonged frying, and it decreases in the presence of decomposition products. When it falls below 170 °C, the fat is considered to be spoiled. At this point, the amount of fatty acids which are insoluble in petroleum ether exceeds 0.7%. However, this petroleum ether method is not reproducible. Fat separation by column chromatography is more reliable. The heated fat or oil is separated into a polar and a nonpolar fraction using silicic acid as an adsorbent. The value of 0.7% oxidized petroleum ether-insoluble fat corresponds, in this chromatographic separation, to 73% nonpolar and 27% polar fractions.

14.6 Literature

Baltes, J.: Gewinnung und Verarbeitung von Nahrungsfetten. Verlag Paul Parey: Berlin–Hamburg. 1975

Bernardini, E.: Oilseeds, Oils and Fats Vol I–II., 2nd ed., B. E. Oil Publ. House, Rome, 1985.

Christie, W. W.: Lipid analysis. 2nd edn. Pergamon Press: Oxford. 1982.

DGF-Einheitsmethoden. Hrsg. Deutsche Gesellschaft für Fettwissenschaft e. V., Münster/Westf., Wissenschaftliche Verlagsgesellschaft mbH: Stuttgart. 1950–1979

Grosch, W., Tsoukalas B.: Analysis of fat deterioration – Comparison of some photometric tests. J. Am. Oil Chem. Soc. *54,* 490 (1977)

Guhr, G., Waibel, J.: Untersuchungen an Fritierfetten; Zusammenhänge zwischen dem Gehalt an petroletherunlöslichen Fettsäuren und dem Gehalt an polaren Substanzen bzw. dem Gehalt an polymeren Triglyceriden. Fette Seifen Anstrichm. *80,* 106 (1978)

Gunstone, F. D., Norris, F. A.: Lipids in foods. Pergamon Press: Oxford. 1983

Kroll, S.: Margarine und Backfette. In: Ullmanns Encyklopädie der technischen Chemie, 4. Aufl., Bd. 16, S. 481, Verlag Chemie: Weinheim–New York. 1978

Official Methods and Recommended Practices of the American Oil Chemists' Society. 3rd edn. (R. C. Walker, Ed.) 1984

Pardun, H.: Analyse der Nahrungsfette. Verlag Paul Parey: Berlin–Hamburg. 1976

Sheppard, A. J., Hubbard, W. D., Prosser, A. R.: Evaluation of eight extraction methods and their effects upon total fat and gas liquid chromatographic fatty acid composition analyses of food products. J. Am. Oil Chem. Soc. *51,* 417 (1974)

Stansby, M. E.: Fish oils. AVI Publ. Co.: Westport, Conn. 1967

Swern, D. (Ed.): Bailey's industrial oil and fat products. 4th edn., Vol. 1, John Wiley and Sons: New York 1979

Thomas, A.: Fette und Öle. In: Ullmanns Encyklopädie der technischen Chemie, 4. Aufl., Bd. 11, S. 455, Verlag Chemie: Weinheim 1976

Weiss, T. J.: Food Oils and Their Uses 2nd ed., Avi Publ. Co., Westport, CT, 1983.

15 Cereals and Cereal Products

15.1 Foreword

15.1.1 Introduction

Cereal products are amongst the most important staple foods of mankind. Nutrients provided by bread consumption in industrial countries include close to 50% of the carbohydrates, one-third of the proteins and 50–60% of the daily requirement of vitamin B. Moreover, cereal products are also a source of minerals and trace elements.
The major cereals are wheat, rye, rice, barley, millet and oats. Wheat and rye have a special role since only they are suitable for breadmaking.

15.1.2 Origin

The genealogy of the cereals begins with wild grasses *(Poaceae)*, as shown in Figure 15.1. Barley *(Hordeum vulgare)*, probably one of the first cereals grown systematically, was known as early as 5000 B.C. in Egypt and Babylon. Also, the bearded wheat cultivars from the groups Einkorn *(Triticum monococcum)* and Emmer *(T. dicoccum)*, with diploid ($2n = 14$) and tetraploid ($2n = 28$) sets of chromosomes, were found among cultivated plants that were widely spread in temperate zones of Euroasia during the neolithic period. These cultivars are becoming extinct. Only the durum form of Emmer (*T. durum*, hard wheat), at 10% of the total wheat grown, has a significant role. The hexaploid ($2n = 42$) bare wheats (soft, bread wheats), derived from spelt wheat (*T. aestivum* L.), are grown worldwide.
Rice *(Oryza sativa)* and corn *(Zea mays)* have been cultivated for 5,000 years, first in tropical Southeast Asia and then in Central and South America. Cereals designated as millet have had a role from antiquity in subtropical and tropical regions of Asia and Africa. True millet from the subfamilies *Eragrostoideae* and *Panicoideae*, to which many regionally important cultivars belong (for instance, *Eragrostis tef, Eleusine coracan, Echinochloa frumentacea, Pennisetum glaucum, Setaria italica*), is distinguished from sorghum *(Sorghum bicolor)*, which belongs to the subfamily *Andropogonoideae*, and is cultivated worldwide.
Rye *(Secale cereale)* and oats *(Avena sativa)* are so-called secondary culture plants. Initially hardy and unwanted escorts of cultivated plants,

Table 15.1. Cultivated land area of a cereal crop as % of the world total area under cereals (1979: 7.6×10^8 ha)

Cereal	1966	1976	1979	1981
Wheat	30.6	31.5	31.4	32.3
Rice	18.8	19.2	19.1	19.6
Corn	15.5	15.9	15.8	18.1
Millet	15.4	15.6	13.7	12.3
Barley	12.2	11.9	12.8	10.8
Oats	4.5	3.8	3.5	3.6
Rye	2.4	2.1	1.9	2.1

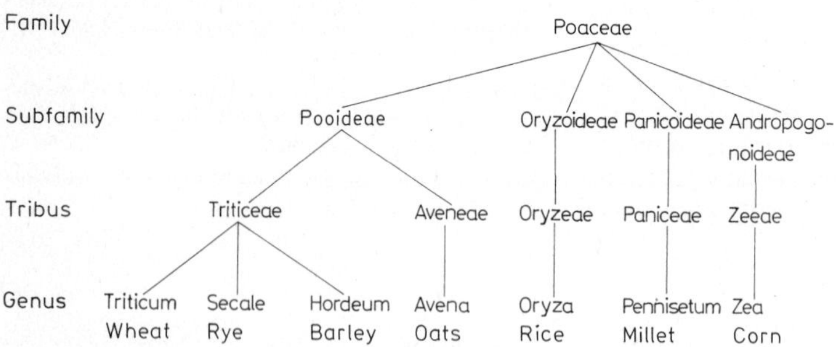

Fig. 15.1. Evolutionary development (phylogeny) of cereals

they prospered and established themselves in northern regions with unfavorable climates. Their high tolerance for unfavorable climates surpassed both wheat and barley. Rye and barley have been cultivated for millenia.
Breeders have for many years attempted to combine the baking quality of wheat with the hardiness of rye. *Triticale*, the man-made hybrid of wheat and rye, does not yet fulfill this aim, hence its economic significance is low.

15.1.3 Production

Cereals are of great importance as raw materials for production of food and feed. Accordingly, they are grown on close to 60% of cultivated

Table 15.2. Cereal production in 1981 (1,000 t)

Continent	Wheat	Rice	Barley	Corn	Rye	Oats	Millet	Sorghum	Cereals, grand total
World	458,195	413,785	158,488	451,704	24,473	44,024	29,653	71,984	1,663,828
Africa	8,584	8,562	3,197	32,860	10	210	10,282	11,174	76,329
America, North-, Central-	103,781	10,736	24,358	231,937	1,437	11,050		29,174	414,473
America, South-	11,743	13,258	806	38,406	231	771	238	9,146	74,725
Asia	137,371	376,232	17,216	86,570	1,561	1,082	17,588	20,566	661,439
Europe, West-	61,014	1,597	49,829	22,769	3,363	10,163	18	644	150,941
Europe, East- + USSR	118,928	2,611	59,446	38,858	17,854	19,180	1,514	183	261,721
Australia and South Pacific Islands	16,773	789	3,636	304	17	1,568	13	1,098	24,201

Country	Wheat	Country	Rice	Country	Barley
USSR	88,000	China	146,087	USSR	43,000
USA	76,026	India	82,000	Canada	13,384
China	57,003	Indonesia	33,000	USA	10,414
India	36,460	Bangladesh	20,000	France	10,180
Canada	24,519	Thailand	19,000	UK	10,149
France	22,782	Burma	14,636	FR Germany	8,687
Turkey	17,040	Japan	12,824	Denmark	6,010
Australia	16,400	Vietnam	12,570	Turkey	5,900
Pakistan	11,340	USA	8,408	Spain	4,709
Italy	8,921	Brasil	8,261	German DR	3,500
				China	3,400
Σ (%)[a]	78	Σ (%)[a]	86		
				Σ (%)[a]	75

Country	Corn	Country	Rye	Country	Oats
USA	208,314	USSR	8,500	USSR	15,000
China	61,601	Poland	6,789	USA	7,375
Brazil	21,098	German DR	1,800	Canada	3,570
Mexico	14,766	FR Germany	1,729	Poland	2,768
South Africa	14,650	China	1,000	FR Germany	2,678
Argentina	13,500	Canada	964	France	1,754
Rumania	11,200	Czechoslovakia	500	Sweden	1,732
Yugoslavia	9,800	Turkey	500	Australia	1,520
France	9,100	USA	473	Finland	1,008
USSR	8,000	France	342	UK	622
Σ (%)[a]	82	Σ (%)[a]	92	Σ (%)[a]	86

[a] World production ≙ 100%.

Table 15.2. (Continued)

Country	Millet	Country	Sorghum	Country	Cereals, grand total
India	10,500	USA	22,360	USA	333,748
China	5,903	India	11,500	China	286,104
Niger	3,230	Argentina	7,550	USSR	167,306
USSR	1,500	China	7,510	India	149,702
Nigeria	1,117	Mexico	6,296	Canada	50,257
Mali	930	Nigeria	3,835	France	45,004
Senegal	750	Sudan	2,800	Indonesia	36,991
Egypt	643	Australia	1,090	Brazil	32,023
Chad	580	Burkine Faso (Upper Volta)	750	Argentina	30,672
Sudan	500	Egypt	687	Mexico	25,574
Σ (%) [a]	87	Σ (%)[a]	89	Turkey	25,322
				Australia	23,242
				Thailand	23,084
				FR Germany	22,826
				Σ (%)[a]	75

[a] World production ≙ 100%.

Table 15.3. World production of cereals 1948–1981 (10^6 t)

Year	Amount	Year	Amount
1948	683	1976	1,456
1956	789	1978	1,596
1964	1,019	1979	1,553
1968	1,180	1981	1,664

land in the world. Wheat production takes up the greatest part of land cultivated in cereals (Table 15.1) and wheat is produced in the largest quantity (Table 15.2). Wheat surplus producers are the USA, Canada, Argentina, Australia, and France. In the Federal Republic of Germany (FRG), winter wheat (92%) and spring wheat (8%) are both cultivated. The rise of cereal production in the world is shown in Table 15.3. The yields per hectare vary greatly from one country to another (Table 15.4). Due to an intensive effort in breeding and crop production programs, the yields per hectare in the FRG are very high and are surpassed by only a few countries, e. g. Holland. The FRG utilized 25.7×10^6 tonnes of cereals in 1976/77, of which 38% was bread and 62% feed cereals.

Table 15.4. Yields per hectare in year 1978/1979/1981 [in dt (100 kg/ha)]

	Wheat	Rye	Corn	Rice
FR Germany	50.1/49.5/51.0	37.8/37.6/35.7	53.0/60.0/64.6	
German DR	45.9/43.6/41.1	29.1/26.9/26.9	34.3/33.3/20.0	
Argentina	17.3/17.1/14.1	8.1/ 8.7/ 8.2	36.5/31.1/38.6	
Australia	17.8/13.9/13.6			
China	14.4/15.0/19.5		28.6/31.1/30.0	36.8/37.2/42.4
France	50.3/47.7/47.9	31.2/29.3/29.2	52.9/51.4/58.0	
India	14.8/15.7/16.5		10.8/ 9.1/12.1	20.1/17.9/20.5
Turkey	17.9/19.0/18.1	13.1/12.4/11.9	22.4/22.6/19.0	
USSR	19.2/15.6/14.9	17.6/12.5/11.3	35.3/31.5/22.6	36.1/39.3/37.9
USA	21.2/23.0/23.2		63.3/68.7/69.0	50.3/51.4/54.6
World	19.1/17.8/19.1	19.8/16.0/16.0	30.8/32.7/33.7	26.6/26.2/28.6

15.1.4 Anatomy – Chemical Composition, a Review

Cereals, in contrast to forage grasses, form a relatively large fruit, termed a caryopsis, in which the fruit shell is strongly bound to the seed shell. The kernel size, which is expressed as grams per 1000 kernels (Table 15.5), is not only dependent on the kind of cereal but on cultivar and crop production techniques, hence it varies widely.
In oats, barley, and rice the front and back husks are fused together with the fruit. In contrast, threshing separates wheat and rye kernels from the husks as bare seed.
The major constituents of seven kinds of cereal are fairly uniform (Table 15.6). Noteworthy variations are the higher lipid content of oats and a lower starch content in oats, barley and rye. Instead of starch, in these cereals the "other carbohydrates" are increased and consist mainly of nonstarchy polysaccharides (cf. 15.2.3.2). These cereals also differ in vitamin B content (Table 15.6).
Fruit and seed coats enclose the nutrient tissue, endosperm and germ, in the kernel (Figure 15.2). Botanically the endosperm consists of the starchy endosperm (70–83% of the kernel; Table 15.7) and the aleurone layer, which, with exception of barley, is a single cell layer. The aleurone layer is rich in protein and also contains high amounts of fat, enzymes and vitamins (Tables 15.8 and 15.9). The proteins, of which half are water-soluble, appear as granules in the aleurone cells. They have no influence on the baking properties of wheat. Millers regard the aleurone layer as part of the bran.
The starchy endosperm is the source of flour. Its thin-walled cells are packed with starch granules which lie imbedded in a matrix which is largely protein. A portion of these proteins, the gluten proteins, is responsible for the special baking properties of wheat. The concentrations of the proteins and some other constituents (vitamins and minerals) decrease from outer to inner cells

Table 15.5. Average thousand kernel weight of cereals (g)

Wheat	37	Oats	32
Rye	21	Barley	37
Corn	285	Millet	23
Rice	27		

Table 15.6. Chemical composition of cereals (average values)

	Wheat	Rye	Corn	Barley	Oats	Rice	Millet
	weight %						
Moisture	13.2	13.7	12.5	11,7	13.0	13.1	12.1
Protein (N × 6.25)	11.7	11.6	9.2	10.6	12.6	7.4	10.6
Lipids	2.2	1.7	3.8	2.1	5.7	2.4[a]	4.05
Starch	59.2	52.4	62.6	52.2	40.1	70.4	64.4
Other carbohydrates	10.1	16.6	8.4	19.6	22.8	5.0	6.3
Crude fiber	2.0	2.1	2.15	1.55	1.56	0.67	1.1
Minerals	1.5	1.9	1.30	2.25	2.85	1.2	1.6
	mg/kg						
Thiamine	5.5	4.4	4.6	5.7	7.0	3.4	4.6
Niacin	63.6	15.0	26.6	64.5	17.8	54.1	48.4
Riboflavin	1.3	1.8	1.3	2.2	1.8	0.55	1.5
Pantothenic acid	13.6	7.7	5.9	7.3	14.5	7.0	12.5

[a] Polished rice: 0.8%

Table 15.7. Fractions from various cereals separated by milling (average weight-%)

Cereal variety	Husk	Bran	Germ	Endosperm
Wheat	0	15.0	2.0	83.0
Corn	0	7.2	11.0	81.8
Oats	20	8.0	2.0	70.0
Rice	20	8.0	2.0	70.0
Millet	0	7.9	9.8	82.3

of the endosperm. The germ is separated from the endosperm by the scutellum. The germ is rich in enzymes and lipids (Table 15.8). Table 15.9 shows that wheat milling, when starchy endosperm cells are separated from germ and bran, results in a substantial loss of B-vitamins and minerals.

15.1.5 Special Role of Wheat – Gluten Formation

After addition of water a viscoelastic cohesive dough can be kneaded only from wheat flour. The resulting gluten, which can be isolated as a residue after washing out the dough with water, removing starch and other ingredients, is responsible for plasticity and dough stability.
Gluten consists of 90% protein (cf. 15.2.1.3), 8% lipids and 2% carbohydrates. The latter are primarily the water-insoluble pentosans (cf. 15.2.3.2.1), which are able to bind and hold a significant amount of water, while the lipids (cf. 15.2.4) form a lipoprotein complex with certain

Table 15.8. Chemical composition of anatomical parts of a wheat kernel (average weight-% on dry weight basis)

	Ash	Crude protein (N × 6.25)	Lipids	Crude fiber[a]	Cellulose	Pentosans	Starch
Longitudinal cells	1.3	3.9	1.0	27.7	32.1	50.1	–
Cross- and tube cells	10.6	10.7	0.5	20.7	22.9	38.9	–
Fruit and seed coatings	3.4	6.9	0.8	23.9	27.0	46.6	–
Aleurone cells[b]	10.9	31.7	9.1	6.6	5.3	28.3	–
Germ[b]	5.8	34.0	27.6	2.4	–	–	–
Starchy endosperm	0.6	12.6	1.6	0.3	0.3	3.3	80.4

a Crude fiber includes parts of cellulose and pentosans [b] Data for carbohydrates are incomplete.

Table 15.9. Mineral and vitamin distribution as % in kernel fractions of wheat

Fraction	Minerals	Thiamine	Riboflavin	Niacin	Pyridoxal phosphate	Pantothenic acid
Fruit coat	7	1	5	4	12	9
Germ	12	64	26	2	21	7
Aleurone layer	61	32	37	82	61	41
Starchy endosperm	20	3	32	12	6	43

gluten proteins. In addition, enzymes such as proteinases and lipoxygenase are detectable in freshly-isolated gluten.

The gluten proteins, in association with lipids, are responsible for the cohesive and viscoelastic flowing properties of dough. Such rheological properties provide the dough's gas-holding capacity during leavening and provide a porous, spongy product with an elastic crumb after baking.

Unlike wheat, rye and other cereals can not form gluten. The baking quality of rye is due to pentosans and to some proteins which swell after acidification (cf. 15.4.2.2) and contribute to gas-holding properties.

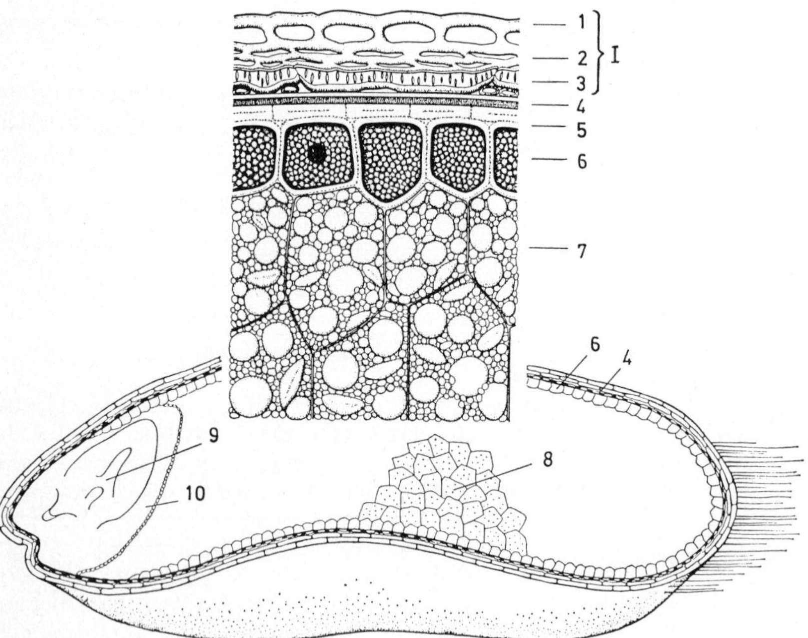

Fig. 15.2. Longitudinal section of a wheat grain. I Pericarp, 1 epidermis (epicarp), 2 hypodermis, 3 tube cells, 4 seed coat (testa), 5 nucellar tissue, 6 aleurone layer, 7 outer starchy endosperm cells, 8 inner starchy endosperm cells, 9 germ and 10 scutellum

15.1.6 Celiac Disease

Wheat, rye and barley can cause celiac disease (celiac sprue, or gluten-induced enteropathy); the role of oats in this disease is uncertain. Celiac disease affects both infants and adolescents, but rarely adults. It is associated with a loss of villous structure of the intestinal mucosa; epithelial cells exhibit degenerative changes; and nutrient absorption functions are severly impaired. Incidence of the disease varies, e.g. 0.05% of the children are affected in central Europe and 0.33% in Ireland. The prolamin fractions of wheat, barley or rye are the cause of the disease, which is therefore eliminated by a change of diet to rice, millet or corn.

15.2 Individual Constituents

The role of constituents is of particular interest in the processing of wheat and rye into bakery products.

15.2.1 Proteins

15.2.1.1 Differences in Amino Acid Composition

The proteins of different cereal flours vary in their amino acid composition (Table 15.10). Lysine content is low in all cereals. Methionine is also low, particularly in wheat, rye, barley, oats and corn. Both amino acids are significantly lower in flour than in muscle, egg or milk proteins. By breeding, attempts are being made to improve the content of all essential amino acids. This approach has been successful in the case of high-lysine barley and several corn cultivars.

15.2.1.2 A Review of the *Osborne* Fractions of Cereals

In 1907 *T. B. Osborne* separated wheat proteins, on the basis of their solubility, into four fractions. Sequential extraction of a flour sample yielded: water-soluble albumins; salt-soluble (e.g. 0.4 M NaCl) globulins; and 70% aqueous ethanol-soluble prolamins. The glutelins remained in the flour residue. The latter can be partly solubilized in dilute acids, but completely solubilized only after reduction of disulfide bonds in the presence of detergents such as sodium dodecyl sulfate. The *Osborne* fractionation is still used in addition to

Table 15.10. Amino acid composition of total proteins (mole-%) from flours of various cereals

Amino acid	Wheat	Rye	Barley	Oats	Rice	Millet	Corn
Asx	4.2	6.9	4.9	8.1	8.8	7.7	5.9
Thr	3.2	4.0	3.8	3.9	4.1	4.5	3.7
Ser	6.6	6.4	6.0	6.6	6.8	6.6	6.4
Glx	31.1	23.6	24.8	19.5	15.4	17.1	17.7
Pro	12.6	12.2	14.3	6.2	5.2	7.5	10.8
Gly	6.1	7.0	6.0	8.2	7.8	5.7	4.9
Ala	4.3	6.0	5.1	6.7	8.1	11.2	11.2
Cys	1.8	1.6	1.5	2.6	1.6	1.2	1.6
Val	4.9	5.5	6.1	6.2	6.7	6.7	5.0
Met	1.4	1.3	1.6	1.7	2.6	2.9	1.8
Ile	3.8	3.6	3.7	4.0	4.2	3.9	3.6
Leu	6.8	6.6	6.8	7.6	8.1	9.6	14.1
Tyr	2.3	2.2	2.7	2.8	3.8	2.7	3.1
Phe	3.8	3.9	4.3	4.4	4.1	4.0	4.0
His	1.8	1.9	1.8	2.0	2.2	2.1	2.2
Lys	1.8	3.1	2.6	3.3	3.3	2.5	1.4
Arg	2.8	3.7	3.3	5.4	6.4	3.1	2.4
Trp	0.7	0.5	0.7	0.8	0.8	1.0	0.2
Amide group	31.0	24.4	26.1	19.2	15.7	22.8	19.8

separation procedures by gel chromatography (cf. Figure 15.14). It is used as an initial step in the separation of cereal proteins.

In the literature, *Osborne* fractions derived from different cereals are often designated by special names (cf. review Table 15.11). The various designations may result in confusion and incorrect conclusions with regard to protein homogeneity. Therefore, it may be better to use the general designations of the *Osborne* fractions and specify the protein source, e.g. wheat glutelin instead of glutenin.

Albumins and globulins are derived mostly from cytoplasmic residues and other subcellular fractions which are part of the kernel. Thus, enzymes are present in the first two *Osborne* fractions. Prolamins and glutelins, on the other hand, are storage proteins.

Cereals contain variable levels of *Osborne* fractions (Figure 15.3). Wheat has the highest content of prolamin; corn has the second highest. The albumin fraction is the highest in rye and the lowest in corn (Table 15.12). The content of albumin in oats is comparable to that in rye (Figure 15.3). Oats and rice have a higher content of glutelin than wheat, while rye, millet and corn have a much lower glutelin content (Figure 15.3). The amino acid composition of only the prolamins (Figure 15.4) can be correlated to the botanical genealogy of cereals as shown in Figure 15.1. In general, the amino acid composition is

Table 15.11. Designations of *Osborne*-fractions

Fraction	Wheat	Rye	Oats	Barley	Corn	Rice	Millet
Albumins	Leukosin						
Globulins	Edestin		Avenalin				
Prolamins	Gliadin	Secalin	Gliadin	Hordein	Zein	Oryzin	Cafirin
Glutelins	Glutenin	Secalinin	Avenin	Hordenin	Zeanin	Oryzenin	

Table 15.12. Protein distribution in *Osborne*-fractions (%)[a]

Fraction	Wheat	Rye	Barley	Oats	Rice	Millet	Corn
Albumins	14.7	44.4	12.1	20.2	10.8	18.2	4.0
Globulins	7.0	10.2	8.4	11.9	9.7	6.1	2.8
Prolamins	32.6	20.9	25.0	14.0	2.2	33.9	47.9
Globulins	45.7	24.5	54.5	53.9	77.3	41.8	45.3

[a] Ash content of the flours (% dry weight basis), amounted in wheat to 0.55; rye, 0.97; barley, 0.96; oat, 1.87; rice, 1.0; millet, 1.10 and corn 0.33.

similar for wheat, rye and barley. The prolamin composition of oats is intermediate between *Triticeae* and the other cereals. The amount of glutamic acid in oat prolamins is similar to that of the *Triticeae,* whereas amounts of proline and leucine in oat prolamins are lower and higher, respectively, than those found in the *Triticeae;* this is also the case in comparison with rice, millet and corn. The amino acid compositions of rice, millet and corn are not related to the *Pooideae.*

The *Triticeae,* in which the prolamin amino acid compositions are closely related, can also cause *Celiac* disease (cf. 15.1.6). In comparison to other cereals, *Triticeae* prolamins contain significantly higher levels of glutamic acid and proline. This suggest that the difference in prolamin composition, induced by these amino acids, may be responsible for *Celiac* disease.

15.2.1.3 Gluten Proteins

Wheat protein fractionation by the *Osborne* method provides prolamins and glutelins in a ratio of 2:3. Both fractions, in hydrated form, have different effects on the rheological characteristics of dough: prolamins are responsible, preferentially, for viscosity, and glutelins for dough elasticity.

The genes for the gluten proteins occur at nine different complex loci in the wheat genome. The high molecular weight glutenin subunits are coded by the loci Glu-A1, Glu-B1 and Glu-D1, which are carried on the long arms of the chromosomes 1A, 1B and 1D. The low molecular weight glutenin subunits, the ω-gliadins and the γ-gliadins are coded by the loci Gli-A1, Gli-B1 and Gli-D1, which occur on the short arms of the chromosomes 1A, 1B and 1D. The α- and β-gliadins are coded by the loci Gli-A2, Gli-B2 and Gli-D2 on the short arms of the group G chromosomes. It is presumed that the variation seen in different varieties is due to the presence of allelic genes at each of the nine storage protein loci. The relative importance of different alleles for gluten quality seems to be Glu-1 > Gli-1 > Gli-2. A fractionation of total endosperm protein is possible by two-dimensional electrophoresis (isoelectric focussing followed by SDS-polyacrylamide gel electrophoresis).

Studies on wheat prolamin and glutelin structures, being performed in several laboratories, aim to clarify on a molecular basis the distinct rheological properties of wheat dough. As a result of these studies, answers are expected as to

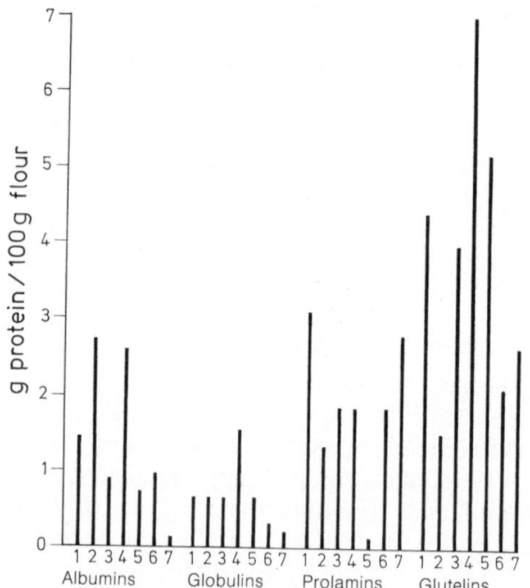

Fig. 15.3. The occurrence of *Osborne*-fractions in different cereal flours. (After *H. Wieser* et al., 1980.) 1 Wheat, 2 rye, 3 barley, 4 oats, 5 rice, 6 millet, 7 corn

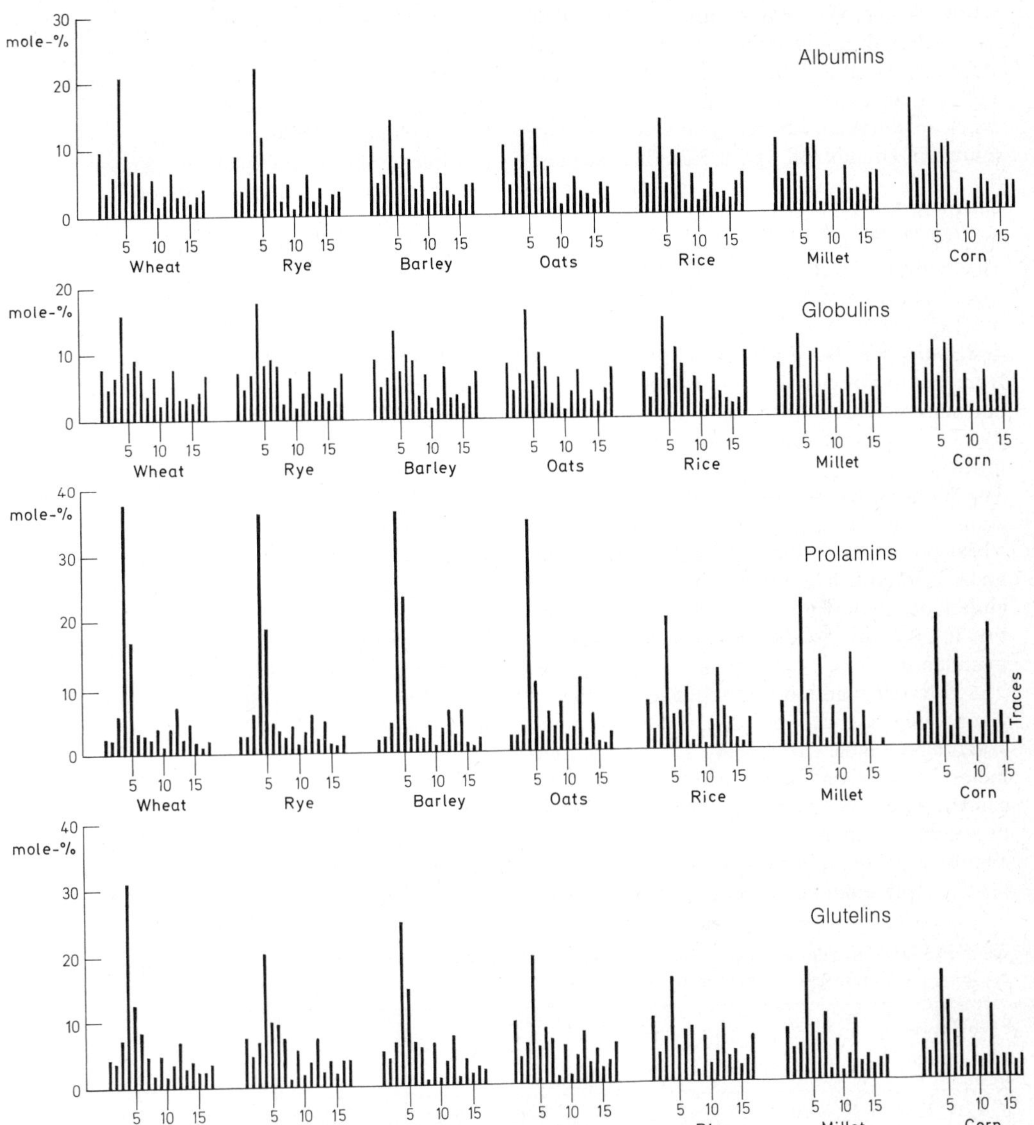

Fig. 15.4. Amino acid compositions (mole-% without Trp) of *Osborne* fractions. (After *H. Wieser* et al., 1980.) The following amino acids are given on the abscissa: 1 Asx, 2 Thr, 3 Ser, 4 Glx, 5 Pro, 6 Gly, 7 Ala, 8 Cys, 9 Val, 10 Met, 11 Ile, 12 Leu, 13 Tyr, 14 Phe, 15 His, 16 Lys and 17 Arg.

what extent individual proteins of both *Osborne* fractions contribute to gluten build-up.

15.2.1.3.1 Wheat prolamins

The *Osborne* fraction which is soluble in 70% aqueous ethanol (gliadin fraction) can be separated by gel chromatography into three major fractions: high molecular weight (MW) gliadins, MW >100 kdal; ω-gliadins, MW 60–80 kdal; and low molecular weight gliadins, MW 30–40 kdal.

The high molecular weight gliadin fraction contains predominantly components of MW 140 kdal; components of exceptionally high MW of

800 kdal; and low MW components of 20–60 kdal. After disulfide bond reduction, the high molecular weight gliadin decreases to 40 kdal and thus corresponds to those glutenin subunits which are obtainable by reduction and which are soluble in ethanol (cf. 15.2.1.3.2). This suggests that high molecular weight gliadin consists mostly of low molecular weight subunits joined by disulfide bonds.

The ω-gliadin fraction is eluted in gel filtration together with dimer forms of low molecular weight gliadins (α-, β- and γ-gliadins). The designations α-, β-, γ- and ω-gliadin are derived from electrophoretic separations in acidic media, and mark the fractions by their decreasing mobility. The gliadin fraction contains at least 50 proteins.

The amino acid compositions of the various gliadins are similar and deviate little from that of wheat prolamins as shown in Figure 15.4. Glutamic acid, which is present predominantly as glutamine, makes up more than one-third of the total amino acids, with proline next in abundance.

The N-terminal amino acid sequences of the α-, β- and γ-prolamins studied so far show many homologies (Table 15.13); only γ_2- and γ_3-prolamines deviate significantly. From Table 15.13 it can also be concluded that N-terminal sequences of wheat prolamins do not reflect the overall amino acid composition of these proteins. Hence, representative sequences (cf. Fig. 15.5a, Fig. 15.5b) are located further away from the N-terminus.

Recently the complete amino acid sequence of A-gliadin, an α-type gliadin, has been derived from a cloned cDNA-sequence and in part also by direct amino acid sequencing (Figure 15.5b). The sequence determined is a composite, because the A-gliadin separates into about five components upon two-dimensional electrophoresis. On the basis of this sequence information, the mature α-type gliadins may be divided into five different sections or domains. Section I (residues 1–95) is essentially composed of only Pro, Gln, Phe/Tyr in the ratio 3:2:1. Tandem repeats are characteristic features of this section. Sections II and IV (96–113, 183–190) are, with a few exceptions, only polyglutamine. Sections III and V (114–182, 191–266) have somewhat less Pro and Gln than does section I and contain most of the other amino acids of α-type gliadins. No obvious repeating sequences occur within these sections.

The only thing known about prolamin conformation is that it is 38% helical (in trifluoroethanol as a solvent), which is somewhat greater than in glutelins.

15.2.1.3.2 Wheat glutelin

Wheat glutelin (glutenin) constitutes about 40% of the total proteins of hexaploid wheat and contributes significantly to the elastic properties of flour dough. It consists of very high molecular

Table 15.13. N-Terminal sequence of various prolamins of wheat cultivar “Ponca” (for one letter code designation cf. Table 1.1; differences revealed in several positions provide a proof for protein heterogeneity)

Prolamin	Position																													
	1				5					10					15					20					25					30
α_8	V	R	V	P	V	P	Q	L	Q	P	Q	N	P		Q	Q	Q	P	Q	E	Q	V	P	L	V	Q	Z	P	Z	
α_9	V	R	V	P	V	P	Q	L	Q	P	Q	N	P		Q	Q	Q	P	Q	E	Q	V	P	L	V					Q
α_{10}	V	R	V	P	V	P	Q	L	Q	P	Q	N	P	S	Q	Q	Q	P	Q	E	Q	V	P	L	V	Z	Z			
α_{11}	V	R	V	P	V	P	Q	L	Q	P	Q	N	P	S	Q	Q	Q	P	Q	E	Q	V	P	L	V	Q	Z			
α_{12}	V	R	V	P	V	P	Q	L	Q	P	Q	N	P	S	Q	Q	Q	P	Q	E	Q	V	P	L	V	Q	Q			
β_5	V	R	V	P	V	P	Q	L	Q	P	Q	N	P		Q	Q	Q	P	Q	E	Q	V								
γ_1	V	R	V	P	V	P	Q	L	Q	P	Q	N	P	S	Q	Q	Q	P	Q	Z	Q	V	P	L	V	Z	Z	I	Z	
γ_2	N	I	G	V	D	P	W	G	Q	V	Q	W	L	P	Q	Q	Q	V	P	Q	L	Z	Q	P		Q				
	P		Q		V	Q		L				I	V	Q			L		Q		Q			Q		R				
					Q																									
γ_3	N	M	G	V	D	P	W	G	Q	V	Q	W	V	P	Q	Q	L	Q		Q	Q	Q		Z			V	Q		
	P		Q	Q	V	Q	Q					L	P	Q			Q	H												
α_2	V	R	V	P	V	P	Q	L	Q	P	Q	N	P	S	Q	Q	Q	P	Q	E	Q	V	P	L	V					
γ	V	I	V	Q	V	R	Q	L	Q	V	Q	Q																		

PQPQPFPSQQPY
LQPQQPFPQQPQQPFP
TQQPQQPFPQQPQQPFPQ
TQQPQQPFPQQPQQPFPQQPQQP.
SQQQQPPF
AQIPQQL
IQPSLQ
GIIQPQQPAQQL
VQQQQFPGQQQPFPPQQPYPQPQ.
VPVPQ
GQQPQQQQL
TQQPQQPF
LQLQPFPQPQLP
SQQPQQPFPQPQQP
VQGQGIIQPQQPA
SQPQQPI
XQQPQQPFPQP[a]
VNVPL
PQPQPQY
LQPQQ
AQGSVQPQ
GSVQPQQ

[a] X: Unspecified amino acid

Fig. 15.5a. Amino acid sequence of typical peptides of a chymotrypsin hydrolyzate of prolamin of wheat cultivar "Kolibri"

```
      L G Q G Q P G Y Y P T S   Q Q P G Q
K Q Q A G Q
G Q Q S G Q G Q Q G Y Y P T S P Q Q T G Q
G Q Q P G Q G Q P G Y Y P T S P Q Q S G Q
W Q Q P G Q
G Q Q P G Q
G Q Q S G Q G Q Q
G Q Q P
G Q R P G Q G Q Q G Y Y P I S P Q Q P G Q
G Q Q S G Q G Q P G Y Y P T S L R Q P G Q
W Q Q P G Q
G Q Q P G Q G Q Q
G Q Q P G Q
G Q Q S G Q G Q Q G Y Y P T S L Q Q P G Q
G Q Q L G Q G Q P G Y Y P T S P Q Q S G Q
G Q Q L G Q G Q P G Y Y P T S P Q Q S G Q
G Q Q S G Q G Q Q G Y Y P T S P Q Q S G Q
G Q Q P
```

Fig. 15.5b. Partial amino acid sequence of a high molecular weight subunit of glutenin of wheat cultivar "Chinese spring". The sequence was postulated in accord with DNA sequence of the initial 730 nucleotides of the corresponding gene. The sequence is arranged in an order to readily reveal the existing homology

weight aggregates which are composed of subunits joined together by noncovalent and covalent bonds.

Native glutenin is only partially soluble (25 to 85%) in dilute acetic acid, concentrated urea, guanidine hydrochloride or sodium dodecyl sulfate. When analyzed by agarose gel chromatography, glutenin gives a complex elution profile of proteins with molecular weights ranging from 30–20,000 kdal.

Studies of various wheat cultivars have shown that good breadmaking quality is apparently related to large amounts of the high molecular weight glutenin fraction which is insoluble in urea or acetic acid.

Glutenin, after disulfide bond reduction followed by subsequent alkylation of the resulting SH-groups (in order to prevent their reoxidation), provides monomers (subunits) with molecular weights of about 100 (high molecular weight subunits), 80 and 40 kdal. These subunits vary with wheat cultivars, hence the different properties of glutenin may be related to different combinations of these subunits. The high molecular weight subunits have been investigated in more detail, since they are found only in glutenin and have therefore had ascribed to them a special role in breadmaking.

The location of the genes on the corresponding chromosomes responsible for biosynthesis of the high molecular weight subunits is well known. By determining the base sequence of the corresponding nucleic acid, the partial sequence of amino acids was deduced for a high molecular weight subunit of the wheat cultivar "Chinese spring" (Figure 15.5b). Typical sequence segments appear to be –YPTS– and –GQGQQ–. Obviously, glycine has a special role in this high molecular weight glutenin subunit. The amount of a hydrophilic high-molecular weight peptide fraction, with a high glycine content, obtained by chymotryptic hydrolysis of glutenin of various wheat cultivars, correlates with the baking quality of these cultivars.

The association of glutenin subunits into native glutenin and its interaction with other wheat protein components is not well understood. It is obvious that quality differences among wheat cultivars can not be explained only by the qualitative and quantitative differences in the glutenin subunit composition, but must depend among others on the molecular weight pattern of the native glutenin.

15.2.1.3.3 Gluten formation

An hypothesis (which indeed needs experimental support) is that gluten formation takes place in a stepwise process. The low molecular weight subunits of glutenin (ethanol soluble), together with particular gliadin subunits, form high molecular weight gliadin. The high molecular weight glutenin subunits polymerize and then combine with high molecular weight gliadin. Lastly, glutenin, together with low molecular weight gliadin and possibly with albumins and globulins, form gluten. Such a large complex is held together by intermolecular disulfide, hydrogen, hydrophobic and ionic bonds.

15.2.2 Enzymes

Of the enzymes present in cereal kernels, those which play a role in processing or are involved in the reactions which are decisive for the quality of a cereal product have been thoroughly investigated.

15.2.2.1 Amylases

α- and β-amylases (for their reactions, see 2.8.2.2.2 and 2.8.2.2.3) are present in all cereals. Wheat and rye amylases are of particular interest; their optimum activities are desirable in dough making in the presence of yeast (cf. 15.4.1.4.8). In mature kernels, α-amylase activity is minimal, while it increases abruptly during sprouting or germination. Unlike the situation with wheat, dormancy in rye is not very pronounced. Unfavorable harvest conditions (high moisture and temperature) favor premature germination ("sprouting"), not visible externally. During this time, α-amylase activity rises, resulting in extensive starch degradation during the baking process. Bread faults appear, as mentioned under 15.4.1.2.

15.2.2.2 Proteinases

Acid proteinases with pH optima of 4–5 occur in wheat, rye and barley. Their substrate specificity has been determined. The possibility that the wheat proteinases are involved in cleavage of gluten bonds, thereby affecting softening or mellowing of gluten during baking, is still disputed.

15.2.2.3 Lipases

These enzymes occur in various concentrations in all cereals. Carboxylester hydrolase, readily isolated from wheat germ, is not considered a lipase but an esterase (cf. 3.7.1.1). In addition to the esterase a lipase occurs in wheat, which is enriched in the bran. A rise in free fatty acids observable during flour storage also involves lipases from metabolism of microorganisms present in flour.

In comparison to other cereals, oats contain a significant level of lipase. Its high activity is released once the oat kernel is disintegrated, crushed or squeezed. Linoleic acid is released from the acyl lipids that are present. It then is changed into hydroxy fatty acids by lipoxygenase and hydroperoxidase enzymes, giving rise to off-flavors (Figure 15.6.). All these enzymes are inactivated by heat treatment and thus quality deterioration can be avoided.

15.2.2.4 Phytase

Close to 70% of the phosphorus in wheat is bound to phytin, which is 1% of the kernels. A major portion of this is hydrolyzed during dough making by microbial phytases:

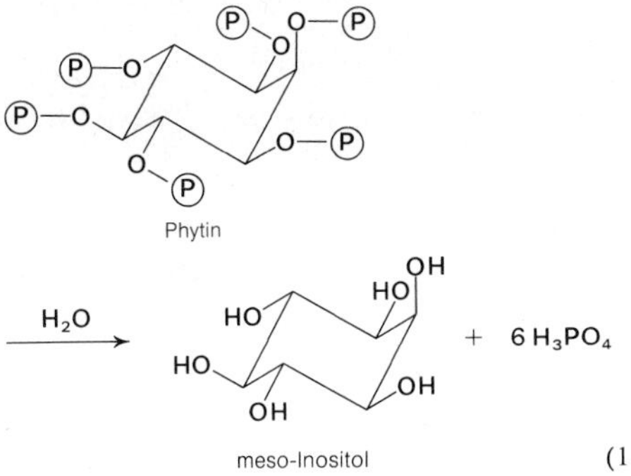

Phytin

$\xrightarrow{H_2O}$ meso-Inositol $+ 6\,H_3PO_4$ (15.1)

The reaction is nutritionally and physiologically desirable since phytin inhibits, by formation of water insoluble salts, the intestinal absorption of calcium and iron ions.

15.2.2.5 Lipoxygenases

Cereals contain lipoxygenases (cf. 3.7.2.2) which, with the exception of rye, oxidize linoleic acid to, preferentially, 9-hydroperoxy acids. The rye lipoxygenase forms mainly the 13-hydroperoxide isomer. Though the enzyme from wheat belongs to type I (cf. Table 3.26) and thus cooxidizes carotenoids at a slow rate, it can still bring about a loss of yellow color in pasta products. This is the reason for heat inactivation of wheat lipoxygenase during preparation of pasta products (cf. 15.5.3).

The involvement of endogenous lipoxygenase in the baking process when wheat flour is an ingredient is not clear. However, by addition of lipoxygenase-active soy flour, a significant improvement of the flour quality is achieved (cf. 15.4.1.4.3). As shown in Figure 15.6, oats contain a lipoxygenase with lipoperoxidase activity. This activity reduces the hydroperoxides initially formed, in the presence of phenolic compounds as H-donors, to corresponding hydroxy acids:

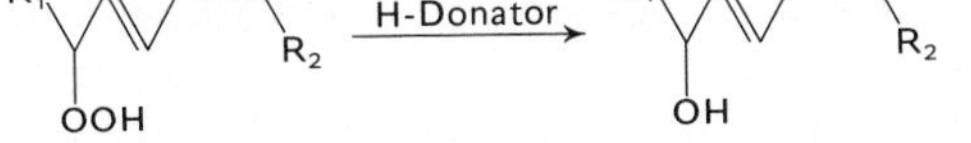

R_1–CH(OOH)–CH=CH–CH=CH–R_2 $\xrightarrow{\text{H-Donator}}$ R_1–CH(OH)–CH=CH–CH=CH–R_2

R_1: $HOOC—(CH_2)_7$; R_2: $H_3C—(CH_2)_4$ (15.2)

15.2.2.6 Peroxidase, Catalase

Both enzymes are widely distributed among cereals. At the pH of a wheat dough the catalase is more active than the peroxidase.

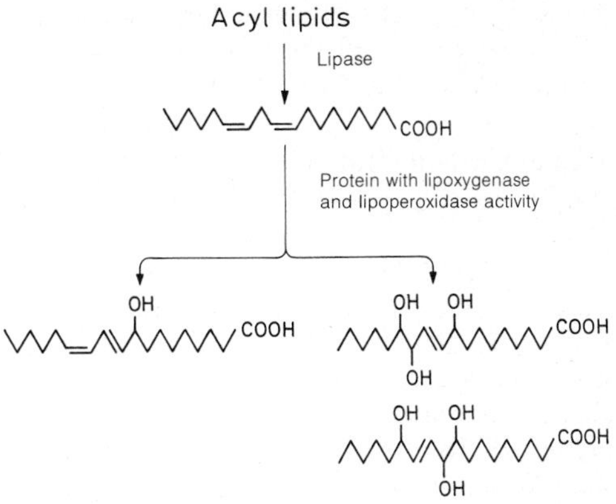

Acyl lipids → (Lipase) → COOH → (Protein with lipoxygenase and lipoperoxidase activity) → OH … COOH; OH OH … OH … COOH; OH OH … OH … COOH

Fig. 15.6. Formation of bitter tasting compounds in oats. (Taste threshold values in Table 3.28)

As heme catalysts they accelerate the non-enzymatic oxidation of ascorbic acid to the dehydro form. Therefore the involvement of both enzymes in the action of ascorbic acid as an improver (cf. 15.4.1.4.1) is discussed.

15.2.2.7 Glutathione Dehydrogenase

This enzyme catalyzes the oxidation of glutathione (GSH) in the presence of dehydroascorbic acid as an H-acceptor:

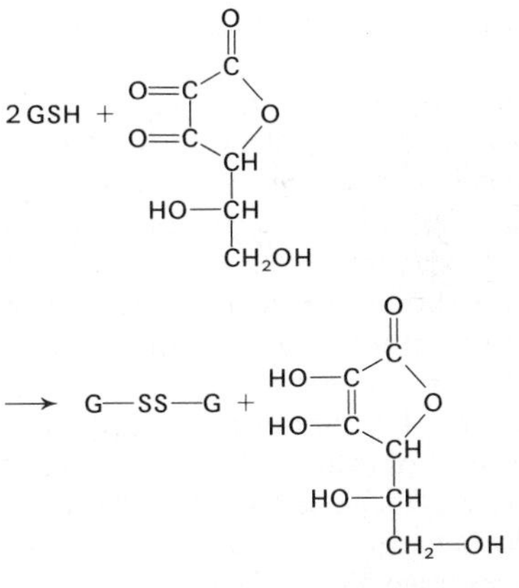

$2\,GSH +$ dehydroascorbic acid $\longrightarrow G—SS—G +$ ascorbic acid (15.3)

It has been purified from wheat flour in which its activity is relatively high. The enzyme is more specific for the H-donor (cysteine is not oxidized) than for the H-acceptor. From the four dehydroascorbic acid (DHAsc) diastereomers it reduces the L-threo-DHAsc (oxidation product of vitamin C) faster than both the D- and L-erythro-DHAsc, which in turn are reduced faster than the D-threo-diastereomer. These and other experimental findings strongly suggest the involvement of the enzyme in flour improvement by addition of ascorbic acid (cf. 15.4.1.4.1).

15.2.3 Carbohydrates

15.2.3.1 Starch

The major carbohydrate storage form of cereals, starch (cf. Table 15.6), occurs only in the endosperm cells. The size and form of the starch granules is specific for different cereals. The polysaccharide molecules in starch granules are radially organized. Due to the presence of alternating water-deficient and water-enriched layers, differences in indices of refraction can be observed under a microscope.

Starch granules swell when heated in water suspension. At the end of swelling, they lose their form; i.e. they gelatinize. The temperature range in which these changes occur and also the extent of swelling at a given temperature are specific (cf. Table 4.17) and may be used for starch source identification.

Cereal starches consist of about 25% amylose and 75% amylopectin (cf. Table 4.17). The chemical structures of these polysaccharides are presented in 4.4.4.12.2 and 4.4.4.12.3. Starch granules in some cultivars, for instance waxy corns, contain only amylopectin, while some cultivars are rich in amylose (cf. Table 4.17). Waxy corn starch swells considerably on heating, while granules with amylose swell only slightly (cf. Table 4.17 and Figure 4.20 b).

Lipids (Table 15.14) and proteins (about 0.5%) are among the heterogenous constituents of starch granules. Lipids are enclosed within the amylose helices. In wheat starch, they consist predominantly of lysolecithins (Table 15.14). They are extractable from partially gelatinized starch by using hot water-saturated butanol. During extraction, the lipid in the amylose helix is replaced by butanol.

The lipids complexed within the starch granules retard swelling and increase their gelatinization temperatures; thus they influence the baking behavior of cereals.

Table 15.14. Lipids in various cereal starches

	Wheat	Corn[a] (maize)	Amylo-maize[a]	Waxy maize[a]
	(% or mg/100 g)[b]			
Nonpolar lipids	6%	60%	73%	88%
Sterol esters	2	3	9	7
Triacylglycerols	15	5	16	12
Diacylglycerols	7	3	16	6
Monoacylglycerols	8	12	13	5
Free fatty acids	27	380	650	105
Glycolipids	5%	1%	5%	6%
Sterol glycosides	3	7	13	3
Monogalactosyldiacylglycerols	4			1
Monogalactosylmonoacylglycerols	10		18	
Digalactosyldiacylglycerols	11			2
Digalactosylmonoglycerols	24		17	3
Phospholipids	89%	39%	22%	6%
Lyso-phosphatidyl ethanolamines	104	17	16	1
Lyso-phosphatidyl glycerols	23	6	7	trace
Lyso-phosphatidyl cholines	783	226	183	8
Lyso-phosphatidyl serines Lyso-phosphatidyl inositols	26	8	6	trace
Total lipids	1,047	667	964	153

[a] Amylose content in starch amounts to 23% (corn), 70% (amylomaize) and 5% (waxy maize cultivars).
[b] Results for lipid classes are expressed as % of total lipids present in starch, and for individual lipid compounds as mg/100 g starch dry matter.

Table 15.15. Distribution of carbohydrates in wheat (%)

	Endosperm	Germ	Bran
Pentosans and hemicelluloses	2.4	15.3	43.1
Cellulose	0.3	16.8	35.2
Starch	95.8	31.5	14.1
Sugars	1.5	36.4	7.6

15.2.3.2 Polysaccharides other than Starch

Cereals contain polysaccharides other than starch. In endosperm cells their content is much less than that of starch (cf. Table 15.15). They include hemicelluloses, pentosans, cellulose, β-glucans and glucofructans. These designations are not uniformly applied in the literature. This is partly due to the fact that analytical criteria are not yet reliable for unequivocal classification. These polysaccharides are primarily constituents of cell walls, and are more abundant in the outer portions than the inner portions of the kernel. In flours with increased flour extraction grade, their content is higher (cf., e.g. rye in Table 15.20).

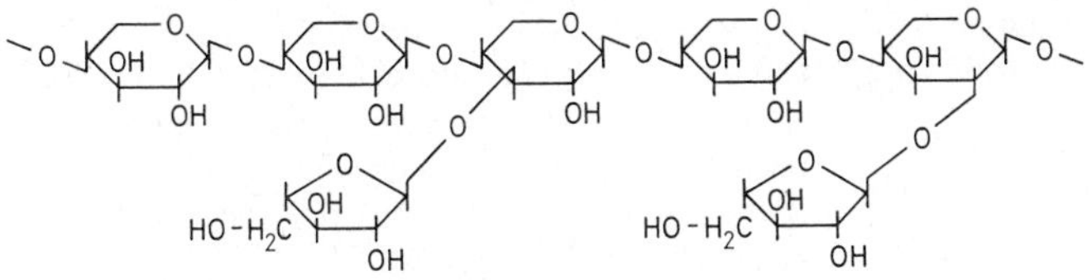

Fig. 15.7. Structural segment of a water-soluble pentosan of wheat flour

15.2.3.2.1 Pentosans

The pentosan content of cereals varies. Rye flour is exceptionally rich (6–8%) in comparison to wheat flour (2–3%). A portion of pentosans, 1–1.5% for wheat and 15–25% for rye, is water-soluble. Some authors have designated the water-insoluble pentosans as hemicelluloses.

Unlike the water-soluble proteins of cereals, the soluble pentosans are able to absorb 15–20 times more water and thus can form highly viscous solutions.

The insoluble portion of pentosans swells extensively in water. This portion is responsible for rheological properties of dough and the baking behavior of rye, and increases the crumb juiciness and chewability of baked products. An optimum starch-pentosan ratio is 16:1 for rye flour. Pentosans also play an important role in wheat baking quality, since they also participate in gluten formation (cf. 15.1.5).

Insoluble pentosans are solubilized by alkali treatment. Pentosan preparations from rye, when prepared and additionally purified by electrodialysis and added to wheat flour to 2%, improve flour baking quality, in particular bread volume and crumb texture, but decrease the freshness. Additionally, pentosans allow wheat flour to be blended with 5–10% of nonbread cereals (corn, millet, soy) without affecting its baking quality. Such flour mixtures are designated as "composite flours".

Pentosans are a mixture of polysaccharides and glycoproteins. The building constituents of the polysaccharides are D-xylose (50–60%), L-arabinose (30–35%) and glucose (6–7%). The structure is typified by a chain of D-xylopyranose units with HO-groups in positions 2 or 3 which are bound by glycosidic linkage to L-arabinofuranose (Figure 15.7).

The glycoproteins also contain branched arabinoxylanes. The composition of the peptide component and the nature of its binding to the carbohydrate moiety are unknown. The importance of the peptide moiety is still unclear.

Pentosan solutions gel when treated with hydrogen peroxide/peroxidase. This is due to the presence of low levels of ferulic acid (ca. 0.2%). An enzymic phenol oxidation occurs (cf. Figure 15.8), which causes polymerization. This results in build-up of a network which is, along with the low content of branched arabinofuranose, responsible for the lack of solubility of most pentosans. However, whether this phenol oxidation takes place in dough where normally the catalase activity is preferred in comparison to the peroxidase activity (cf. 15.2.2.6) is an open question since both enzymes use H_2O_2 as substrate.

15.2.3.2.2 β-Glucan

Oat and barley kernels contain 6–8%, wheat and rye kernels only 0.5–2% of slimy mucous substances, the β-glucans. These are linear polysaccharides with D-glucopyranose units joined by β-1,3 and β-1,4 linkages. Polysaccharides of the β-glucan type are also called lichenins. The slimy mucous substances provide a high viscosity to water solutions in barley processing and can interfere in wort filtration in beer production.

15.2.3.2.3 Glucofructans

Wheat flour contains 1% water-soluble, nonreducing oligosaccharides of molecular weight up to 2 kdal. They consist of D-glucose and D-

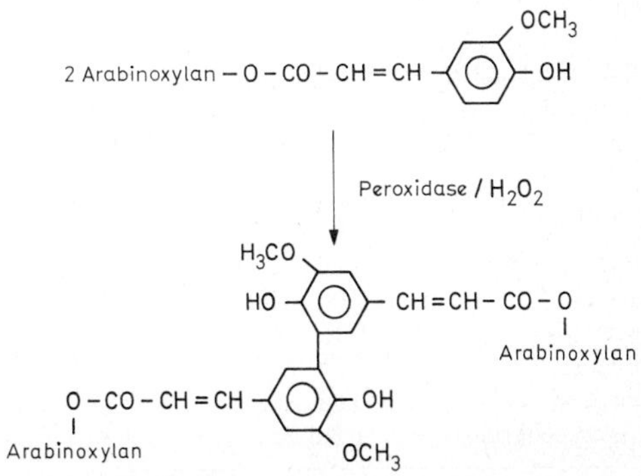

Fig. 15.8. Oxidative crosslinking of cereal pentosans

fructose. Glucofructan, which predominates in durum wheats, probably has the following structure:

$$\alpha\text{-D-Glc}p(1\rightarrow2)\beta\text{-D-Fru}f(6\rightarrow2)\beta\text{-D-Fru}f(6\rightarrow2)\beta\text{-D-Fru}f \quad \begin{matrix} 1 \\ \downarrow \\ 2\beta \\ \text{Fruc}f \end{matrix} \qquad (15.4)$$

15.2.3.2.4 Cellulose

Cellulose is a minor constituent of the carbohydrate fraction obtained from starchy endosperm cells (cf. Table 15.15).

15.2.3.3 Sugars

Mono-, di- and trisaccharides, as well as other low molecular weight degradation products of starch, occur in wheat and other cereals in relatively low concentrations (Table 15.16). When starch degradation occurs during dough making, their levels increase (cf. 1.5.4.2.5). Mono-, di- and trisaccharides are of importance for dough leavening in the presence of yeast (cf. 15.4.1.6.1).

15.2.4 Lipids

Cereal kernels contain relatively low levels of lipids; nevertheless, differences occur among cereals (cf. Table 15.6.). The endosperm cells of oats contain a higher level of lipids (6–8%) than wheat (1.6%; cf. Table 15.8). For this reason, the overall lipid content of oat is higher than in wheat and in other cereals.
The lipids are preferentially stored in the germ which, in the case of corn and wheat, serves as a source of oil production (cf. 14.3.2.2.4). Lipids are stored to a smaller extent in the aleurone layer.

Table 15.16. Mono- and oligosaccharides in wheat flour

Compound	(%)
Raffinose	0.05–0.17
Glucodifructose	0.20–0.30
Maltose	0.05–0.10
Saccharose	0.10–0.40
Glucose	0.01–0.09
Fructose	0.02–0.08
Oligosaccharides[a]	1.2–1.3

[a] Fraction soluble in 80% ethanol.

Cereal lipids do not differ significantly in their fatty acid composition (Table 15.17). Linoleic acid always predominates. Close attention has been given to wheat lipids since they greatly influence baking quality and they have therefore been studied thoroughly.
Wheat flour contains 1.5–2.5% lipids, depending on milling extraction rate. Part of this lipid is nonstarch lipid. This portion is extracted with a polar solvent, water-saturated butanol, at room temperature. Nonstarch lipid comprises about 75% of the total lipid of flour (Figure 15.9). The residual lipids (25%) are bound to starch (cf. 15.2.3.1).

Table 15.17. Average fatty acid composition of acyl lipids of cereals (weight-%)

	14:0	16:0	16.1	18:0	18:1	18:2	18:3
Wheat		20	1.5	1.5	14	55	4
Rye		18	< 3	1	25	46	4
Corn		17.7		1.2	29.9	50.0	1.2
Oats	0.6	18.9		1.6	36.4	40.5	1.9
Barley	2	22	< 1	< 2	11	57	5
Millet		14.3	1.0	2.1	31.0	49.0	2.7
Rice	1	< 28	6	2	35	39	3

Nonstarch- and starch-bound lipids in wheat differ in their composition (cf. Table 15.18 and Table 15.14). In nonstarch-bound lipids the major constituents are the triacylglycerides and digalactosyl diacylglycerides, while in starch-bound lipids, the major constituents are lysophosphatides in which the acyl residue is located primarily in position 1. A decrease of amylose content is accompanied by a decrease in the lipid content (Table 15.14). The ratios of nonstarch-bound lipid classes are dependent on flour extraction

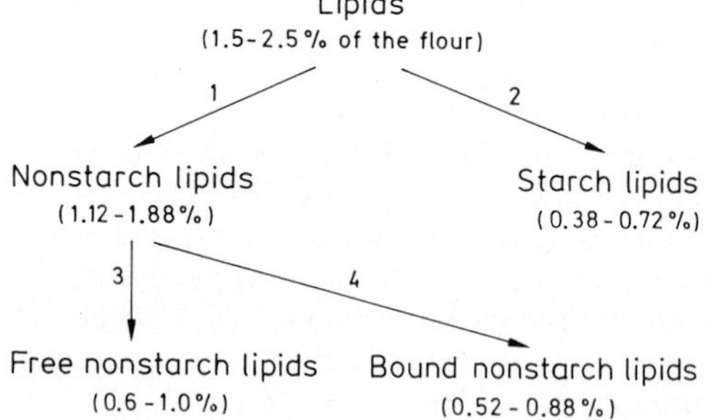

Fig. 15.9. Differentiation of wheat flour lipids by their solubility. 1 Flour extraction with water-saturated butanol (WSB) at room temperature, 2 with WSB at 90–100 °C, 3 with petroleum-ether, and subsequently, 4 WSB

grade. An increase in extraction grade increases the triacylglyceride content, since more of the germ is transferred into the flour.
The rheological dough properties are affected by nonstarch-bound lipids. The starch-bound lipids appear to affect the properties of the baked products.
Nonstarch-bound lipids are separated into free and bound lipids when extracted with solvents of different polarity. The free lipid fraction contains 90% of the total nonpolar lipids and 20% of the total polar lipids listed in Table 15.18. By kneading the flour into dough, the glycolipids become completely bound to gluten, while other lipids are only 70–80% bound. The extent of binding of triacylglycerides depends on dough handling. Intensive oxygen aeration and, particularly, addition of lipoxygenase (cf. 15.4.1.4.3) increase the fraction of free lipids.
The nature of lipid bonds in flour and gluten is still unclear. Separations by gel chromatography and sedimentation show the existence of lipoproteins. This is the basis for the claim stated in 3.5.1.1 that noncovalent interactions among lipids and proteins are involved in gluten formation.

Table 15.18. Nonstarch lipids in wheat flour

	mg/100 g[a]
Nonpolar lipids (59%)	
Sterol lipids	43
Triacylglycerols (TG)	909
Diacylglycerols (DG)	67
Monoacylglycerols (MG)	53
Free fatty acids (FFA)	64
Glycolipids (26%)	
Sterol glycosides	18
Monogalactosyldiacylglycerols (MGDG)	115
Monogalactosylmonoacylglycerols (MGMG)	17
Digalactosyldiacylglycerols (DGDG)	322
Digalactosylmonoacylglycerols (DGMG)	52
Phospholipids (15%)	
N-Acyl-phosphatidyl ethanolamines	95
N-Acyl-lyso-phosphatidyl ethanolamines	33
Phosphatidyl ethanolamines Phosphatidyl glycerols	19
Phosphatidyl cholines	96
Phosphatidyl serines Phosphatidyl inositols	9
Lyso-phosphatidyl glycerols	5
Lyso-phosphatidyl cholines	29

[a] Based on dry matter

In baking tests with flours from different wheat cultivars, the addition of nonpolar, nonstarch lipids results in unsatisfactory bread volume (Figure 15.10). In contrast, the polar lipids provide a positive volume increase. The major flour glycolipid, digalactosyl diacylglyceride, is particularly effective.
Models supported by some physical-chemical data have been proposed to explain the effect of digalactosyl diacylglycerides. As presented in Figure 15.11, one hypothesis claims that the glycolipids are confined between the starch and gluten, thus improving starch adherence to glutelin and prolamine constituents of the gluten complex. The lipid hydrophobic acyl moiety is attached to protein in such a complex, while the disaccharide moieties H-bond to starch.
Carotenoids belong to the minor components of the cereal lipid fraction. Wheat flour has a carotenoid content averaging 5.7 mg/kg. In durum wheats, which have a more intense yellow color, the carotenoids are 7.3 mg/kg of flour.
The major carotenoid, lutein (cf. 3.8.4.1.2), is present in free or esterified form (either mono- or diester) with the fatty acids listed in Table 15.17. The following carotenoid pigments are also present: β-carotene, β-apo-carotenal, cryptoxanthin, zeaxanthin and antheraxanthin (for structures see 3.8.4.1 and Table 3.44). Carotenoid content of corn, depending on the cultivar, is 0.6–57.9 mg/kg, with lutein and zeaxanthin being the major constituents.

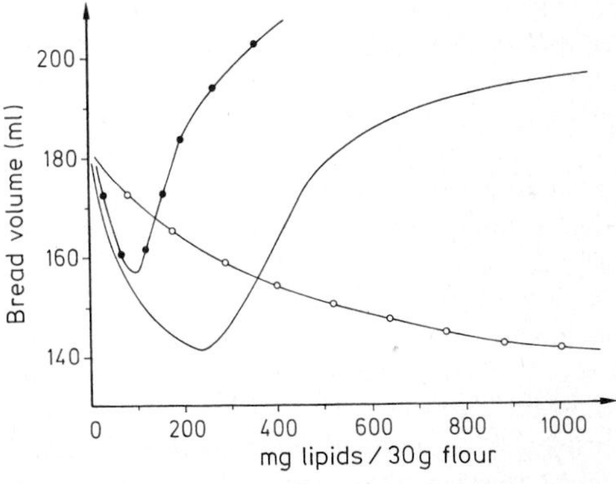

Fig. 15.10. The effect of free nonstarch lipids on the baking quality of defatted wheat flour. (After *W.R. Morrison*, 1976.) —— Lipids (total), –○–○– nonpolar lipids, –●–●– polar lipids

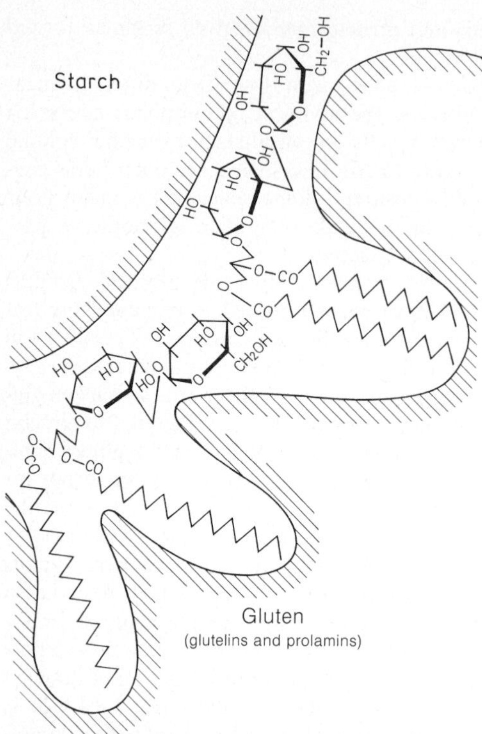

Fig. 15.11. A model to explain the role of digalactosyldiglycerides in wheat. (After *O. K. Chung* et al., 1978)

15.3 Cereals – Milling

15.3.1 Breadmaking Cereals

15.3.1.1 Storage

Cereals can be stored without loss of quality for 2 to 3 years, provided that the kernel moisture content, which is 20–24% after threshing, is reduced to at least 14%. The low moisture content prevents microbial spoilage, especially by mycotoxin-forming organisms, and it also lowers kernel respiration, i.e. metabolism.

The water is slowly removed from grains by ripple-type dryers in a stream of hot air or burned gas at 60–80 °C (to the extent of 4% per passage) to avoid damage to kernels by uncontrolled shrinkage. Grains with high moisture content can be stored for short periods of time in the cold without quality deterioration. Stored grains are fumigated for pest control. Aluminum and magnesium phosphides are introduced. At 20 °C and 75% relative humidity, they decompose into gaseous PH_3. HCN or ethylene oxide fumigants are also used.

15.3.1.2 Milling

The aim of milling is to obtain preferentially a flour in which the constituents of the endosperm cells predominate. The outer part of the kernel, including the germ and aleurone layer (cf. Figure 15.2) is removed. Such a requirement is not easy to accomplish since the kernel's groove and the unequal sizes of aleurone cells in cereals do not facilitate simple dehulling. Therefore, the grain has to be carefully broken, the particles sorted and separated by size and, only then, further disintegrated.

In a preliminary step to milling, the grain is cleaned of impurities such as weed seeds, straw, soil particles, spoiled decayed grains, dust, etc. This cleaning step is based on the cereal's kernel size and specific gravity. Washing with water is rarely done, since it promotes the growth of microorganisms.

The next step is grain wetting or steeping in water for 3–24 h, since an increased moisture content to 15–17% facilitates the separation of starchy endosperm cells from germ and hull. An alternative procedure is wheat conditioning at elevated temperatures, up to 65 °C; it is faster than steeping and also favorably affects the baking quality.

The kernels are disintegrated stepwise. Each passage through rollers involves particle size reduction by pressure and shear forces followed by flour separation by particle size using sieves in the form of flat sifters. Rollers are matched to the product needed. Their size, surface flutes, rotation velocity, gap between pairs of rollers rotating in opposite directions at dissimilar speeds – all can be selected or adjusted. Wheat and rye are milled differently because of structural differences in the kernels. The wheat kernel is rather brittle; the rye kernel is gluey or sticky. Therefore, rye is less suitable for coarse grist milling than wheat. The wheat milling process can be adjusted so that the first passages provide the grist and the following ones provide the flour.

The germ of the rye kernel, because of its loose attachment, falls off readily during the cleaning step, while the wheat germ is removed only on sifters. The hull and a substantial part of the aleurone layer is removed in the form of bran.

15.3.1.3 Milling Products

A miller distinguishes the end-products of milling on the basis of particle size or diameter, e.g.: > 500 μm for grist; 200–500 μm for semolina from durum or farina from bread wheats; 120–200 μm for "dunst"; and 14–120 μm for flour. The larger flour particles can be felt between the fingers (graspable flour), as opposed to smooth or polished flours in which the average particle size is 40–50 μm.

Differently milled flours vary considerably in baking quality. Flours obtained also differ greatly from cultivar to cultivar. This is especially the case with wheat cultivars (cf. 15.4.1.1). In addition, quality depends on whether the milled flour is enriched by the inner or outer parts of endosperm. Therefore, milled flour is controlled in the plant for its baking properties and blended or mixed to yield a commercial product based on preset market standards (see also below). The characteristics of a few milling products and their applications are listed in Table 15.19.

The chemical composition of the flour depends on the milling extraction grade, e.g. flour weight obtained from 100 weight of grains. Examples are given in Table 15.20. Increasing the grade of flour extraction decreases the proportion of starch and increases amounts of kernel-coating constituents such as minerals, vitamins and crude fiber (cf. Tables 15.8 and 15.9). On the basis of the same extraction grade, rye flour contains higher proportions of both minerals and vitamins which occur in the kernels than wheat flour (Figure 15.12). It should be pointed out that in the case of some B-vitamins, such as niacin, this difference is well-balanced by the higher concentrations in wheat in comparison to rye kernels (cf. Table 15.6). In consequence the concentrations of such vitamins are similar in rye and wheat flour.

Table 15.19. Wheat and rye milling products

All purpose flour	Commercially available (retail market) flour for household preparations of baked products.
Special flour	It is used for special baked products, e.g. strong gluten wheat flour for toast bread, wheat flour with weak gluten for baked goods of loose tender or crispy structure as pastry etc.
Compounded (ready to use) flour	Special flour that contains other ingredients as milk or egg powder, sugar etc., required by formulation of a selected baked product.
Groats (grist)	Coarsely ground dehulled cereal (devoid of germ and seed hull).
Whole grain groats	Ground from whole kernel (including germ).

Bread flours are standardized on the basis of ash content in Europe and, particularly, Germany. The type of flour = ash content (weight %) × 1000 corresponds to the extraction grade. Examples are provided in Table 15.20 for wheat and rye flours and their chemical composition is detailed. Protein and starch contents are also related to flour particle size (cf. Table 15.21).

Because of the variable particle sizes and densities of protein and starch granules, a flour sample can be separated by air classification into a fraction enriched in protein and starch. These are the so-called special purpose flours.

The commercial product semolina ("griess") is made from endosperm cells of hard durum wheats. Semolina keeps its integrity during cooking and is used mostly for pasta production. Since semolina is a milled flour of low extraction grade, it contains few minerals and vitamins.

15.3.1.4 Starch Flours

Starch flours should be considered as practically pure starch. They are made from cereal grains, potato tubers, sago palms (*Metroxylon sagu*), arrowroot (*Maranta arundinaceae*) and cassava root (*Manihot esculenta*). The starch isolation process used depends on the raw material.

Wheat starch is obtained from wheat dough by washing out and collecting the raw aqueous starch suspension. After fiber removal, the starch is fractionated and sedimented by centrifugation. This provides starch fractions such as "secunda" (finer granules, in part contaminated with pentosans) and "prima", a more purified starch. Starch is then dried and further graded. In such starch isolation processes, the remaining gluten (cf. 15.1.5) is collected and used as an ingredient in seasoning or production of glutamic acid. Gentle drying preserves the native baking quality of gluten. As a "vital gluten" it is used as a flour ingredient in order to improve flour quality. Potato starch is also readily isolated by washing and centrifugation, while isolation of corn and rice starches requires additional steps. As an exam-

Table 15.20. Average composition of wheat- and rye flours[a]

A. Wheat flour					
	Type				
	405	550	812	1,050	1,700[b]
	Flour extraction rate[c]				
	40–56%	64–71%	76–79%	82–85%	100%
Starch	84.2	81.8	78.4	78.2	66
Protein (N × 6.25)	11.7	12.3	13.0	13.3	14.8
Lipids	1.0	1.2	1.5	1.9	2.3
Dietary fiber[d]	3.7	3.7		4.9	10.9
Minerals (ash)	0.41	0.55	0.81	1.05	1.7
B. Rye flour					
	Type				
	815	997	1,150	1,370	1,740
	Flour extraction rate[c]				
	69-72%	75–78%	79–83%	84–87%	90–95%
Starch	77.5	74.6	72.2	69.3	62.8
Protein (N × 6.25)	9.6	10.1	10.6	11.2	12.4
Lipids	1.1	1.1	1.3	1.5	1.5
Insoluble pentosans	3.8	4.3	4.8	5.2	6.5
Soluble pentosans	1.4	1.5	1.6	1.7	1.9
Minerals (ash)	0.82	1.0	1.15	1.37	1.74

[a] Weight-% per dry matter of wheat and rye flours. Flour average moisture content is 13 weight-%.
[b] Whole wheat flour.
[c] Approximate data.
[d] Indigestible carbohydrates (water soluble and insoluble), lignin.

Table 15.21. Protein content of wheat flours as affected by flour particle size

Particle size (μm)	As portion of flour (weight %)	Protein content (weight %)
0–13	4	19
13–17	8	14
17–22	18	7
22–28	18	5
28–35	9	7
> 35	43	11.5

ple, ground, germ-free corn has to be first steeped in 0.1–0.2% sulfurous acid at 50 °C for 40–60 h (cf. 4.4.4.12).

Corn starch is the major food starch and is used for pastries, puddings, sauces, etc., and also as a source of starch for conversion into syrups and glucose (cf. 2.8.2.2.4 and 19.1.4.3). Starch isolation from rye is not easy to perform, due to a high content of constituents that swell in water. In tropical countries starch isolated from cassava tubers is marketed in a partially gelatinized form under the name "tapioca". True *sago* and *tacca* starches are the corresponding products of pulps of sago palms and arrowroot tubers.

15.3.2 Non-Breadmaking Cereals

15.3.2.1 Corn

Corn endosperm, with the germ removed, is ground to grist for corn porridge (Polenta) and into corn flour for flat cakes (tortillas). Corn flakes are made from cooked and sweetened corn slurry, by drying, flaking and toasting. Similar products are made from millet, rice and oats.

15.3.2.2 Hull Cereals

Dehulling of rice, oats and barley requires special processes (cf. 15.1.4).

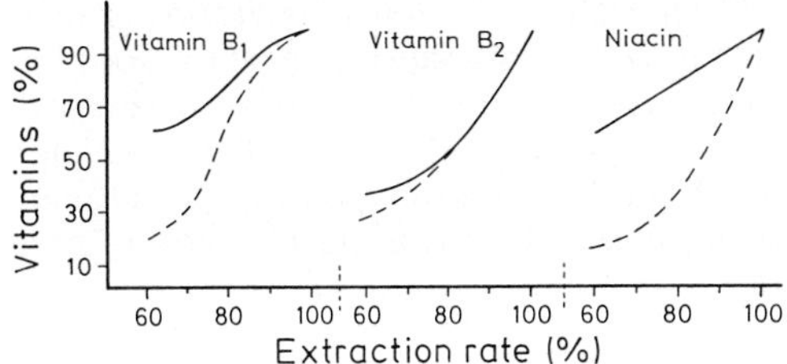

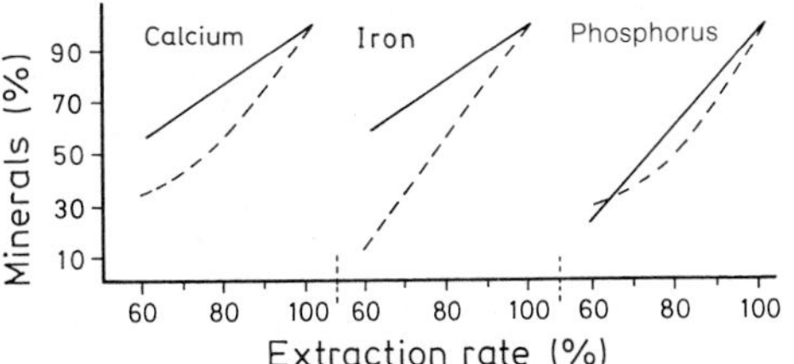

Fig. 15.12. Content of B vitamins and minerals in flour as affected by milling extraction rate. (After Lebensmittellexikon, 1979.) —— rye, --- wheat

[a] Calculated as percent of the total content present in grain

15.3.2.2.1 Rice

Rice milling involves the following processing steps: rough rice (paddy rice) → hull removal → brown rice → polishing to remove the bran coats (fruit and seed coats), the silvery cuticle, the germ and the aleurone layer → rubbing-off or rice polishing to obtain the end-product, white rice. Undamaged rice (45–55%), broken kernels or flour (20–35%) and a husk/hull fraction (20–24%) are obtained.

Polished white rice is made from this cleaned rice by additional treatment of the kernels with talc (a magnesium silicate) and 50% glucose solution. This imparts a glossy, transparent coating to the kernels.

White rice, in comparison to rough or brown rice, is low in vitamin content (cf. Table 15.22) and in minerals. A nutritionally-improved product may be obtained by a parboiling process, originally developed to facilitate seed coat removal. About 25% of the world's rice harvest is treated by the following process: raw rice → steeping in hot water, steaming in autoclaves, followed by drying and polishing → parboiled rice. Minerals and vitamins migrate from the outer layers of the kernels to the inner endosperm, while the aleurone layer is lost (Table

Table 15.22. Vitamin content of raw-, white- and parboiled rice

	B-vitamins (mg/kg)		
	Thiamine	Riboflavin	Niacin
Raw rice	3.4	0.55	54.1
White rice	0.5	0.19	16.4
Parboiled rice	2.5	0.38	32.2

15.22). In addition starch gelatinization occurs, thus lessening subsequent cooking time (converted or instant rice).

15.3.2.2.2 Oats

Oat flakes are produced by the following processing steps: the kernels are steamed and then the moisture content is decreased to 5% by heating at 75 °C for 60–90 min. The hull (fruit and seed coats) is removed, i.e. the kernel is polished. This is followed by repeated steaming, squeezing between drum rollers, and drying of the moist flakes. The yield is 55–65%. This hydrothermic process also inactivates the oat enzymes involved in off-flavor development.

15.3.2.2.3 Barley

Removal of hull (fruit and seed coatings) yields groats which, after grinding, provide marketable products of large or fine particle size.

15.4 Baked Products

Baked products (for a review, see Table 15.23) are made from milled wheat, rye and, to a lesser extent, other cereals by the addition of water, salt, a leavening agent and other ingredients (shortening, milk, sugar, eggs, etc.). The process involves selection and preparation of ingredients, dough making and handling, baking and measures for quality preservation.

15.4.1 Raw Materials

Among the ingredients involved in a formulation, only flour and those additives which affect dough rheological and/or baking properties will be covered. Flour improvers and dough leavening agents will be emphasized.

Characterization of the raw materials and additives is, in practice, made by assessing the dough

Table 15.23. Classification of baked products

Bread including small baked products (rolls, buns)	Made entirely or mostly from cereal flours; moisture content on average 15%. Addition of sugar, milk and/or shortenings amounts to a total less than 10%. Small baked products differ from bread only by their size, form and weight.
Fine baked goods, including long term or extended shelf-life products such as biscuits, crackers, cookies etc.	Made of cereal flours with at least 10% shortening and/or sugar, as well as other added ingredients. In baked goods for long shelf-life the moisture content is greatly reduced.

rheological properties and by baking tests. Basic research endeavors to understand the nature of flour constituents and the reactions which affect their behavior in dough handling and baking.

15.4.1.1 Wheat Flour

A flour of optimal baking properties is required and chosen to match the quality of the desired product (cf. Table 15.19). The baking quality of wheat is strongly influenced by the cultivar (cf. Table 15.25) and also by conditions of growth and cultivation (climate, location), and subsequently by flour storage conditions and duration. Prior quality control is of importance to assess the overall baking quality of wheat flour. Flour particle size and color are assessed by sensory analysis. Graspable flours (cf. 15.3.1.3) are made from hard gluten-rich cultivars. Water uptake is slow when compared to smooth flour, and they make dry doughs. The color difference is important, and is assessed with a wetted flour sample on a black background (*Pekar*-test).

15.4.1.1.1 Chemical assays

Flour acidity (ml of 0.1 N NaOH/10 g, titrated in the presence of phenolphthalein) depends upon the extraction grade of the flour and ranges between 2.0 ml/g (flour type 450) and 5.5 ml/g (flour type 1800). Too low acidity often reflects poorly aged flour. Acidity above 7.0 suggests microbial spoilage.

The content of gluten, which is the residue left after the dough is washed (10 g flour kneaded into a dough with 6 ml of 2% NaCl, then washed with tap water), provides an indication of flour quality. A very low gluten content (<20%) frequently results in dough deterioration when machine-handled and also in baking faults. A higher content of gluten will not guarantee good baking quality (see "Maris Huntsman" cultivar, Table 15.25). Gluten swelling power is assessed by a *sedimentation value* as recommended by *Zeleny*. The flour is mixed in a solution of lactic acid in aqueous isopropanol and is left to stand. The higher the volume of the sedimented gluten, the better should be the baking quality of the flour.

For a given wheat cultivar, grown under similar climatic and soil conditions, the baking volume correlates with the protein content of the flour (Figure 15.13). A similar linear relationship is not readily attainable for flours from different cultivars, as evidenced by the very different slopes of the regression lines.

Various methods have been developed to assess flour baking quality by the determination of the glutelin content. Figure 15.14 shows, as an example, a gel chromatographic procedure; clearly, the glutelin levels are higher as the baking quality improves.

Wheat cultivars differ in the content of their thiol and disulfide groups (Table 15.24). This implies that the stability of a dough may be strongly influenced by a disulfide-exchange between a low molecular weight SH-peptide and gluten proteins (cf. 15.2.1.3.2). This also implies that a positive correlation between the contents of SH- and -SS-groups in flour, or their ratios, would be reflected in baking quality. However, low correlation coefficients of about 0.6 have been found, suggesting that such relationships are much more complex

Table 15.24. Concentration of SH- and SS-groups in flour of different wheat cultivars

Cultivar	SH	SS	SS/SH
	µmole per g flour		
Kolibri	1.15	12.5	10.9
Caribo topfit	0.88	12.2	13.9
Strong Canadian wheats	0.95	13.4	14.1
Inland wheat I[a]	0.75	10.2	13.6
Inland II[a]	1.05	12.6	12.0
Canadian Red Spring Wheat (CWRS)	1.26	12.9	10.2

[a] Marketed flour blendings.

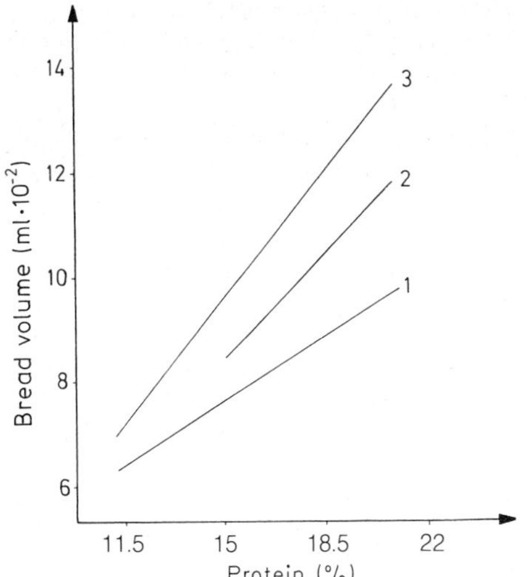

Fig. 15.13. Examples of relationship between protein content and bread volume. (After *Y. Pomeranz*, 1977.) United States winter wheat cultivars: 1 Chikan, 2 Blackhull, 3 Nebred. The regression lines are based on numerous sample analyses

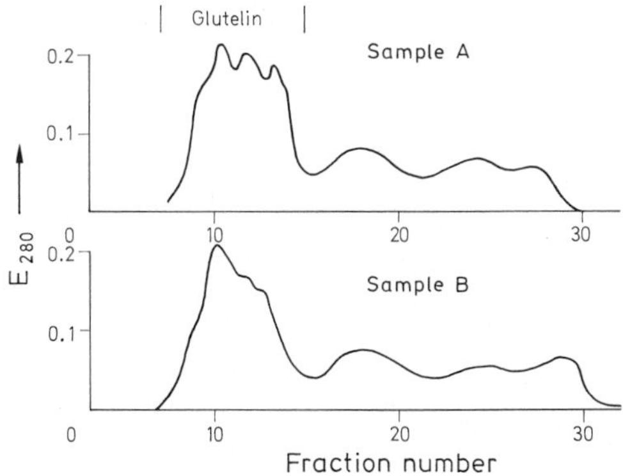

Fig. 15.14. Wheat flour proteins separation by Sephadex G-150 gel chromatography. Eluents are aqueous solutions of 0.1 M acetic acid, 3 M urea, and 0.01 M cetyltrimethylammonium bromide. (After *K. Khan* and *W. Bushuk*, 1978.) A, flour of good and B poor baking quality

than assumed in a model system. Hence, SH- and -SS- data, obtainable by simple analysis, are not suitable for flour quality assessment.

Of all enzymes in flour, quality control is aimed at determination of amylase activity. The Falling Number test (*Hagberg* and *Perten*) serves this aim. A piston-type mixer falls through an aqueous flour paste. The falling time of the piston is measured for a given distance under standard conditions. The results are related, among other things, to starch granule stability in the presence of amylase enzymes. Dextrin values should be determined to assess amylase activity specifically. In a method developed by *Lemmerzahl,* the extent of standard dextrin hydrolysis in the presence of flour extract is measured. The fermentation power of a flour (cf. 15.4.1.6.1) involves determination of the maltose value (diastatic activity). This is a quantitative determination of reducing sugars prior to and after incubating a flour suspension at 27 °C for 1 h. Flours with a maltose content of $<1.0\%$ are regarded as weak fermentation promoters; values above 2.5% are flours from sprouted kernels. They provide poor baking quality.

15.4.1.1.2 Physical assays

Assays with farinographs and extensigraphs are of the utmost importance for estimation of rheological properties of flour doughs. Dough development is followed with a farinograph (Figure 15.15), with which measurement is made of the volume of water absorbed by the flour in order to make a dough of predetermined consistency (normal consistency). A plot of dough consis-

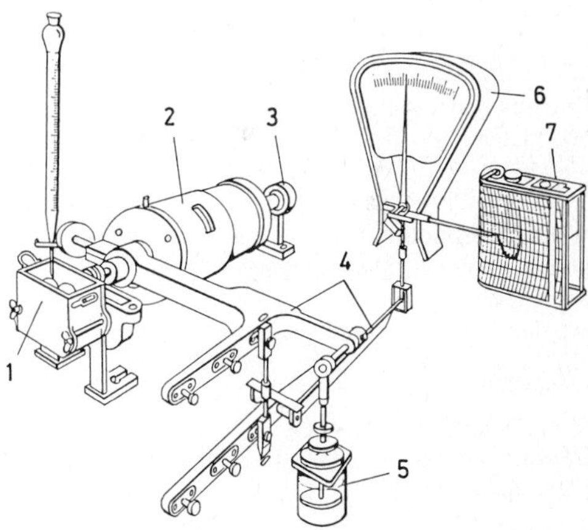

Fig. 15.15. Farinograph. (After *M. Rohrlich* and *B. Thomas,* 1967.) The apparatus consists of a thermostated mixer or kneader (1), its blades are driven by an electromotor (2). The reaction torque acts through a lever system (4) of analytical balance precision on the indicator scale (6) simultaneously recorded on a strip chart recorder (7). The movement of the lever system is damped by an oil dash pot (5). The farinogram is a diagram of force versus time

tency versus time is recorded, as shown in Figure 15.16.

In addition to the water-absorption, the shape of the farinogram is used to characterize a flour. Various indices have been defined (cf. Figure 15.16); usually they refer to doughs with a maximum consistency of 500 FU.

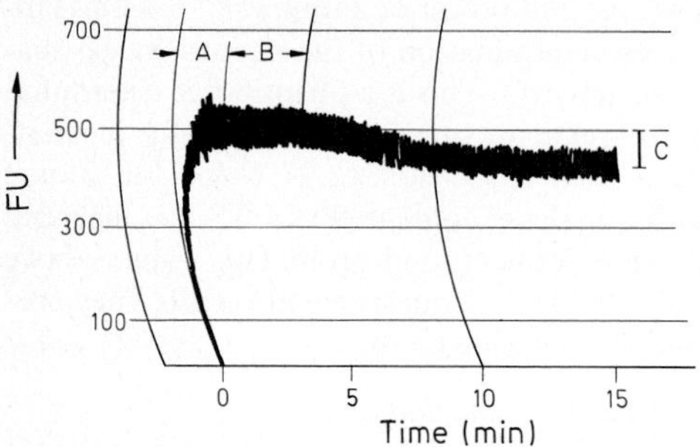

Fig. 15.16. Farinogram. The following data are pertinent for quality assessment of flour: A dough development time, B dough stability (dough consistency does not change), C decrease in dough consistency after a given time, here 12 min. FU: farinogram units

Flours with strong gluten absorb more water and show longer dough development and stability times than do flours with weak gluten (Table 15.25).

A standardized piece of dough is stretched with the hook of an extensigraph until the piece breaks (Figure 15.17). As shown in Figure 15.18, a graph of force (resistance to extension) versus stretching distance (extensibility) provides information about the stability of a dough, its gas-holding capacity and fermentation tolerance. Of the examples given in Table 15.25, the "Monopol" cultivar obviously has strong gluten. The "Nimbus" cultivar has short gluten, as reflected by its low extensibility. The "Maris Huntsman" cultivar has a very weak gluten, as shown by the low resistance of its dough to extension and also by its low extensibility, and thus its extension area is very small.

15.4.1.1.3 Baking tests

Straightforward and reliable information about the baking quality of a flour is obtained from baking tests under standardized conditions.

Table 15.25. Baking quality data of some wheat flours

	Wheat cultivar[a]		
	Monopol	Nimbus	Maris Huntsman
Protein (% dry matter)[b]	13.2	11.6	11.8
Wet gluten (%)	35.1	24.7	34.3
Farinogram[c]			
Water absorption	59.2	54.8	59.8
Dough development time (min)	5.0	1.0	2.0
Dough stability (min)	5.0	1.5	0.5
Mixing Tolerance Index[d] (FU)	30	80	130
Extensigram[e]			
Area (Dough strength, cm^2)	143	75	17
Resistance of the dough to extension	700	680	110
Extensibility (mm)	170	92	100
Baking test			
Dough surface	somewhat wet to normal	normal	wet, gluey
Dough elasticity	normal	somewhat short	weak
Baking volume (ml)	738	630	510

[a] Wheat cultivars are with breadmaking quality corresponding to very good ("Monopol"), average ("Nimbus") and poor ("Maris Huntsman").
[b] Factor N × 5.7.
[c] Explanation in Fig. 15.16; dough consistency: 500 BU.
[d] Measured after 10 min in Farinogram units (FU).
[e] Explanation in Fig. 15.18.

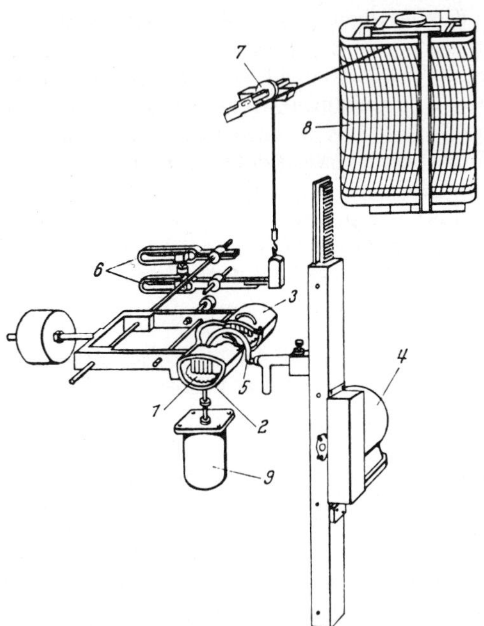

Fig. 15.17. Extensigraph. (After *M. Rohrlich* and *B. Thomas*, 1967.) The cylindrical piece of dough (1) is fixed by dough clamps (3) and placed on the balance fork (2). The motor (4) of the stretching unit (5) is then started. The arm moves downward into the dough and extends it at constant speed. Simultaneously, the forces opposing the stretching action are transmitted through the lever system (6) to the balance system (7). This is coupled to a recording arm of the strip chart recorder (8). The fork of the balance system is coupled to an oil damper (9) to reduce the recoil.

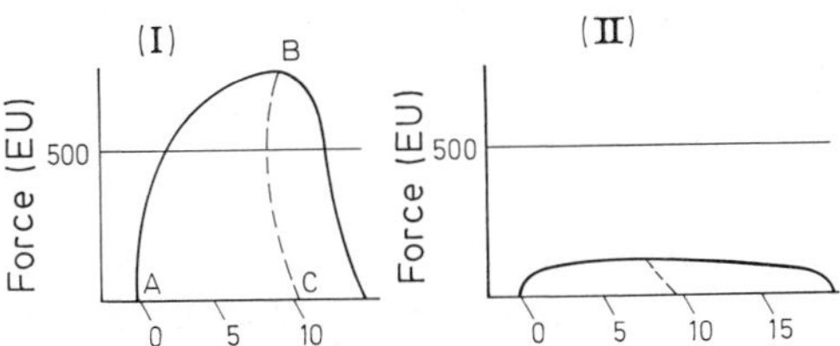

Fig. 15.18. Extensigrams of a normal (I) and weak dough (II). For quality assessment the following parameters are determined: resistance to extension, height of the curve at its peak (B–C) given in extensigram units (EU); extensibility, abscissa length between A–C in mm; extensibility area (A–B–C–A, cm²) is related to energy input required to reach the maximum resistance; extensigram number (overall dough quality) is the ratio of extension resistance to extensibility

Baking volume (cf. Table 15.25), form, crumb structure and elasticity, and other tests of the baked product are evaluated. A baking test is performed with at least 100 g flour for each product.

When the effects of expensive and not readily available flour constituents and/or additives are tested or a new cultivar is assessed, of which only several hundred kernels are available, a "micro baking test" is used, with 10 g flour for each baked product (cf. Figure 15.22).

15.4.1.2 Rye Flour

The Falling Number test (cf. 15.4.1.1.1) and an amylographic assay are the most important tests to assess the baking properties of rye flour. These tests depend to a great extent on gelatinization properties of starches and the presence of α-amylase. The higher the α-amylase activity, the lower the Falling Number.

An amylograph is a rotational torsion viscometer. It measures the viscosity change of an aqueous suspension of flour as a function of temperature. The recorded curve, called an amylogram (Figure 15.19), shows that with increasing temperature there is an initial small fall followed by a steep rise in viscosity to a maximum value. The steep rise is due to intensive starch gelatinization. The viscosity value and temperature at

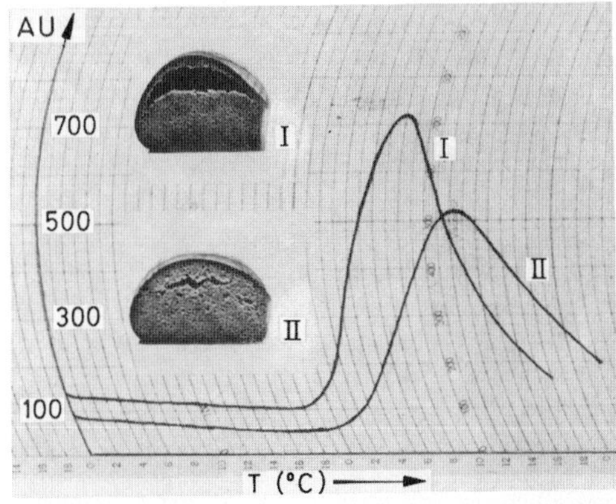

Fig. 15.19. Amylograms of two rye flours. (After *H. Stephan*, 1976)

	Gelatinization maximum (peak)	Gelatinization temperature	α-Amylase
Flour I	720 AU	67 °C	high
Flour II	520 AU	73.5 °C	low

AU: amylogram units

maximum viscosity (i.e. the temperature reflecting the end of gelatinization) are then read.

In rye flour with balanced baking properties, an optimal relation should exist between α-amylase activity and starch properties, since the extent of enzymatic starch degradation influences the stability of the gas-cell membranes which are formed by gas released in the dough and which consolidate during baking into an elastic crumb structure. These membranes contain pentosans, proteins and intact starch granules in addition to gelatinized and partially hydrolyzed starch. High α-amylase activity in rye or a large difference between the temperatures needed for enzyme inactivation (close to 75 °C) and those required for termination of starch gelatinization will produce poor bread, since too much starch will be degraded during breadmaking. The gas-cell membranes are liquified to a great extent; so the gas can escape. This gas will then be trapped in a hollow space below the bread crust (Figure 15.19,I). Low α-amylase activity, especially in conjunction with low starch gelatinization, leads to a firm and brittle crumb structure.

15.4.1.3 Storage

Rye flour acquires optimal baking properties after 1–2 weeks of storage after milling. Wheat flour requires 3–4 weeks. This storage period is the flour "maturation time". In wheat the time is needed for oxidative processes to occur and thus provide a stronger (shorter) gluten. Also, the concentration of endogenous glutathione will be lowered by oxidation to its disulfide (Table 15.26), hence gluten stability during dough handling will be enhanced.

Flour with a moisture content of $< 12\%$ may be stored at 20 °C and a relative humidity of $< 70\%$ for more than 6 months without significant change in baking quality.

Table 15.26. Occurrence of reduced and oxidized forms of glutathione (GSH and GSSG) in wheats flours

Wheat cultivar	Flour type	GSH	GSSG
		µmole/g flour	
Caribo topfit	550	0.29	0.19
Kolibri	405	0.27	0.26
Kolibri	550	0.32	0.27
Bussard	550	0.37	0.29
Kranich	550	0.40	0.19
Benno	405	0.46	0.37
Maris Huntsman	550	0.72	0.28

Flour fumigation with Cl_2, ClO_2, NOCl, N_2O_4 or NO, or treatment with dibenzoyl- or acetone peroxide results in carotenoid destruction. The flour becomes bleached. Other reactions, not yet elucidated, are involved with Cl_2, NOCl, ClO_2 and acetone peroxide treatment since they provide simultaneous improvement in baking quality of flours which have weak gluten.

15.4.1.4 Influence of Additives/Minor Ingredients on Baking Properties of Wheat Flour

The baking properties of wheat flours differ widely (cf. Table 15.25). In small traditional plants, a baker can use his experience to compensate for changes in the quality of raw materials: flexibility in formulations, dough handling and baking – all these parameters can be adjusted in order to obtain the desired end-product.

In a large-scale automated bakery, economic production demands uniform raw materials, with uniform properties. Additives are used when necessary to adjust the flour characteristics to match the baking process (for instance, shortened dough handling time and with low energy input). Additives are also used to ensure that the end-product meets existing standards. Incorporation of ascorbic acid, alkali bromates or enzyme-active soy flour improves the quality of weak gluten flour – this is done in bread or bun baking. In these cases the dough becomes drier and there are increases in dough resistance to extensibility, mixing tolerance and fermentation stability. In addition, baking volume will increase and the crumb structure will improve. Ascorbic acid and lipoxygenase require oxygen for their actions; hence their beneficial role is very dependent on the intensity of dough mixing, which traps oxygen from the air.

In contrast, opposite effects may be observed by adding cysteine or proteinases, the result being gluten softening. Biscuits are made from such mellowed, softened doughs, which are made with little energy input. Additives which affect the rheological quality of the dough and/or the quality of baked products include emulsifiers, shortenings, salt, milk, soy flour, α-amylase preparations and starch syrups.

15.4.1.4.1 Ascorbic acid

The improver effect of ascorbic acid (Asc) was recognized by *Jørgensen* as early as 1935. He found that small amounts (2–6 g Asc per 100 kg flour) caused a pronounced increase of both dough strength and bread volume. The oxidation product of Asc, dehydroascorbic acid (DHAsc) is also effective, but its use would be uneconomical. The improver action in conventional doughmaking is different from that in the continuous process, in which Asc lowers requirement for mixing and the DHAsc is inactive. The four diastereomers of Asc are differently active as improvers in conventional doughmaking: L-threo-Asc (vitamin C) enhances most strongly the handling and baking characteristics, both D- and L-erythro Asc are less active while D-threo-Asc is inactive. This ranking was also shown in rheological measurements (cf. Figure 15.21). The dehydro forms of the four ascorbic acids show the same ranking as improvers. This specificity, suggests that at least one enzyme is involved in the action of ascorbic acid in the conventional process.

The effect of Asc or DHAsc is related to a rapid reaction with endogenous glutathione (GSH), converting it into its disulfide form during dough kneading (Table 15.27). Gluten becomes softer in the presence of glutathione since its protein molecules (Pr – SS – Pr) are depolymerized via a thiol/disulfide interchange reaction:

$$\text{Pr} - \text{SS} - \text{Pr} + 2\,\text{GSH} \rightarrow 2\,\text{Pr} - \text{SH} + \text{G} - \text{SS} - \text{G}.$$

By adding Asc, part of this glutathione is removed from the reaction, resulting in a stronger gluten and consequently a stronger dough.

The effect of Asc takes place in two steps (Figure 15.20). First, Asc is oxidized into DHAsc by oxygen present in the dough. The reaction is catalyzed by trace heavy metal ions or heme-containing compounds such as the enzymes peroxidase or catalase. Involvement of an Asc-oxidase is possible. The second step is the oxidation of glutathione to its disulfide by DHAsc, a reaction catalyzed by glutathione dehydrogenase. The enzyme, with relatively high activity in wheat flour,

Table 15.27. Changes in GSH and GSSG-content during dough making from wheat flours

Cultivar		Kneading time (min)	Found (μmole/g) GSH	GSSG	Calculated (μmole/g) 2 × GSSG + GSH
Benno	Flour	0	0.46	0.37	1.29
	Dough	2	0.29	0.48	1.25
	Dough	5	n.d.	0.66	1.32
Kolibri	Flour	0	0.27	0.26	0.79
	Dough	5	n.d.	0.44	0.88

n.d. not detected.

I. Ascorbic acid

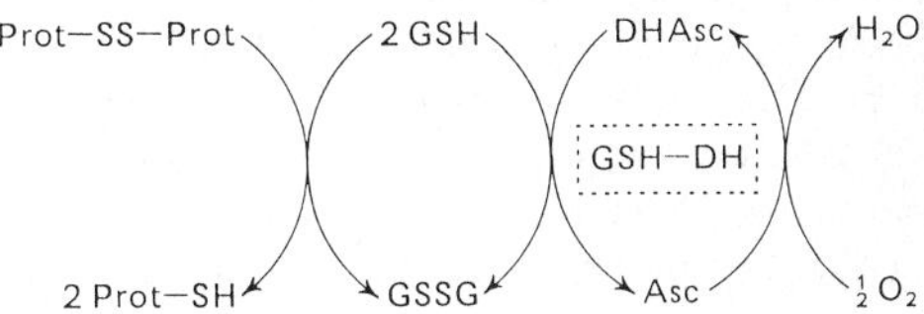

II. Potassium bromate

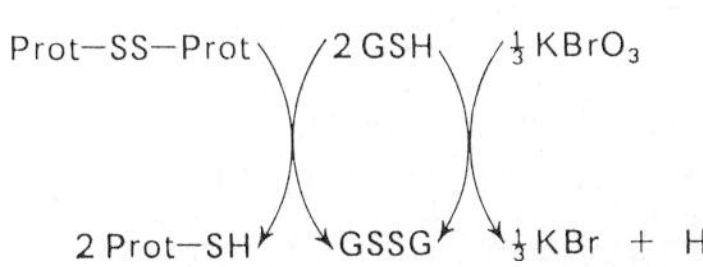

Fig. 15.20. Reactions involved in flour improvement by ascorbic acid or potassium bromate. (After *G. Mair* and *W. Grosch*, 1979.) Asc L-threo-ascorbic acid, DHAsc L-threo-dehydroascorbic acid, GSH reduced glutathione, GSSG oxidized glutathione, Prot-SS-Prot gluten proteins (compare 15.2.1.3.2), GSH-DH glutathione dehydrogenase (see 15.2.2.7)

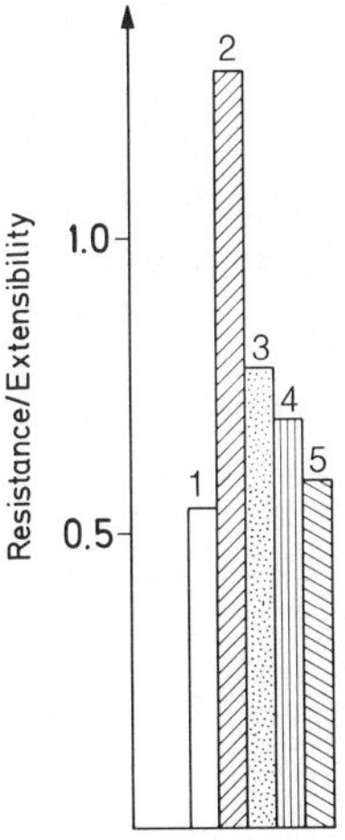

Fig. 15.21. Action of ascorbic acid diastereomers on rheological properties of wheat flour dough. The dough contains 0.23 μmol ascorbic acid (Asc)/g flour: *1* control (no addition), *2* L-threo-Asc, *3* D-erythro-Asc, *4* L-erythro-Asc, *5* D-threo-Asc. The extensigrams are recorded by a micro-scale assay. The y-axis shows the ratio of the resistance to mixing to the extensibility (extensigram number)

reduces the four diastereomers of DHAsc with different velocities (cf. 15.2.2.7). This substrate specificity agrees with the improver effect of the four Asc diastereomers (cf. Figure 15.21).

15.4.1.4.2 Bromates, azodicarbonamide

Addition of alkali bromates to flour also prevents excessive softening of gluten during dough making. The reaction involves oxidation of endogenous glutathione to its disulfide (Figure 15.20). During baking, bromates are completely reduced to bromides, with no bromination of flour constituents.

Azodicarbonamide is of interest as a flour improver,

$$H_2N—CO—N{=}N—CO—NH_2 \quad (15.5)$$

since it improves not only the dough properties of weak gluten flour, but also lowers the energy input in dough mixing (cf. Figure 15.27). Details of the reactions involved are unknown.

15.4.1.4.3 Lipoxygenase

The addition of a small amount of enzyme-active soy flour to a wheat dough increases the mixing tolerance, improves the rheological properties and may increase the bread volume. The effect on the dough rheology is shown only with high power mixing in the presence of air. The carotenoid pigments of the wheat flour are bleached by the addition of enzyme-active soy flour. This is desirable in the production of white bread. The amount of enzyme-active soy flour is restricted to approximately 1% since higher levels catalytically generate off-flavors (cf. 3.7.2.2).

It was demonstrated that the type II lipoxygenase (cf. 3.7.2.2) is responsible for the improver action (Figure 15.22) and the bleaching effect caused by the enzyme-active soy flour. This enzyme, in contrast to endogenous wheat flour lipoxygenase, releases peroxy radicals which cooxidize carotenoids and other flour constituents.

15.4.1.4.4 Cysteine

Cysteine, in its hydrochloride form, softens gluten due to a thiol/disulfide interchange with gluten proteins, as outlined for GSH under 15.4.1.4.1. Decreases of dough development time and dough stability, as shown in farinograms (Figure 15.23), clearly reveal the addition of cysteine. Flours with strong gluten and with optimum levels of cysteine also show a favorable increase in baking volume since, prior to baking, the gas trapped within the dough can develop a more spongy dough. The action of sodium sulfite is similar to that of cysteine.

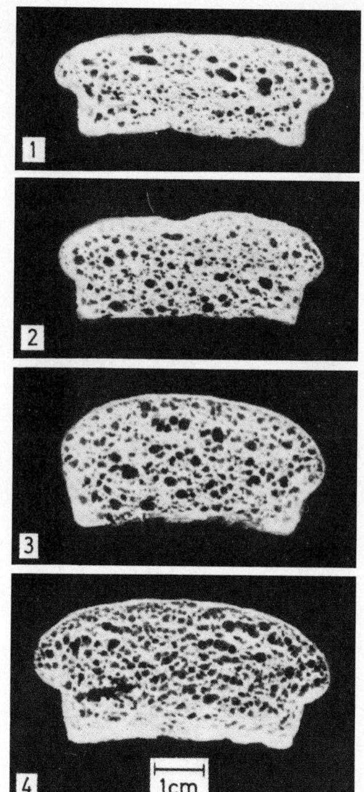

Fig. 15.22. Wheat flour quality improvement by lipoxygenase type-II enzyme of soybean. (After *R. Kieffer* and *W. Grosch*, 1979.) Additions: **1** control (no addition, bread volume 31 ml), **2** extract of defatted soya meal in which lipoxygenase was thermally inactivated (31 ml), **3** extract of a defatted soya meal with 290 units of lipoxygenase (35 ml), **4** purified type-II enzyme with 285 activity units (37 ml)

[a] Results in small-scale baking, 10 g flour cv. Clement.
[b] One enzyme unit = 1 μmole · min^{-1} oxygen uptake.

15.4.1.4.5 Proteinases

Proteinase preparations of microbial or plant origin are used for dough softening (cf. 2.8.2.2.1). Their action involves protein hydrolysis, i.e. gluten-protein endo-hydrolysis. Their effect on dough rheology, therefore, depends on the nature of the enzymes and the activity of the preparations towards gluten proteins. This is shown in Figure 15.24. Despite equal hydrolase activities

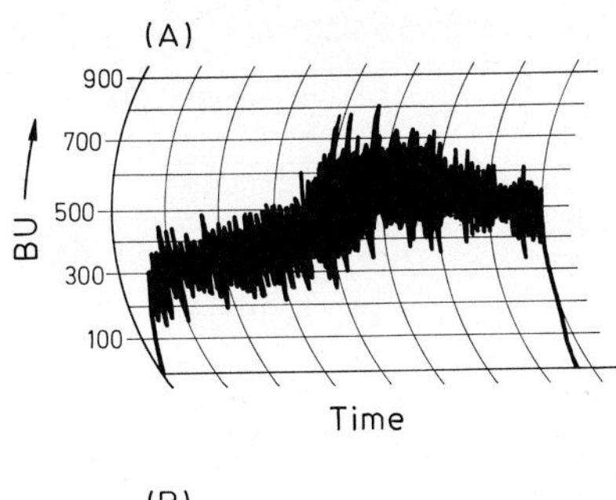

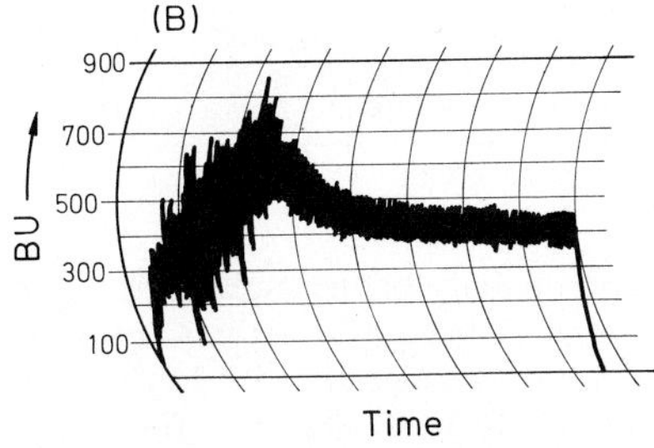

Fig. 15.23. Farinograms. Effect of L-cysteine hydrochloride on a flour with strong gluten. (After *K.F. Finney* et al., 1971.) **A** control (no addition), **B** cysteine added (120 ppm)

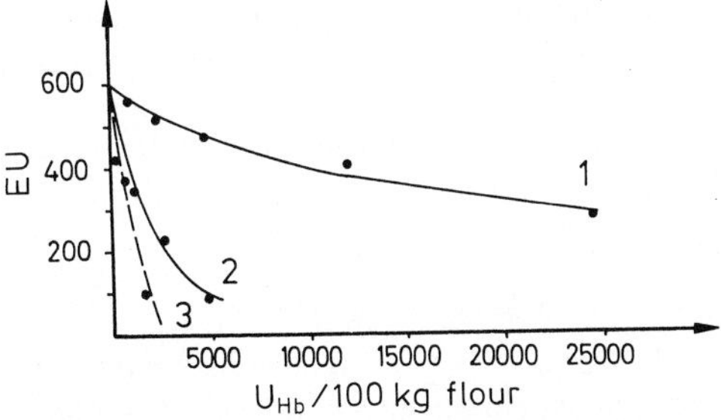

Fig. 15.24. The effect of a proteinase preparation on resistance to extension (in extensigram units) of a wheat flour dough. (After *B. Sproessler,* 1980.) Proteinase preparation: 1 fungal, 2 papain, and 3 bacterial. U_{HB} proteinase activity units determined with hemoglobin as a substrate

with hemoglobin as a test substrate, a fungal proteinase degrades gluten to a lesser extent and consequently causes a smaller decrease in dough resistance to extension in comparison to a bacterial enzyme preparation. Also, the latter is more effective than papain.

Fungal proteinases, because of their low enzyme activity and, therefore, high dosage tolerance, are suitable for optimization of bread- and bun-type flours containing strong gluten. However, bacterial enzymes are preferred in production of biscuits and wafers since they degrade gluten to a greater extent, providing accurate flat dough pieces with high form stability. Bacterial enzymes are also preferred for the desirable end product qualities of porosity and breaking strength.

Data are shown in Table 15.28 for white bread prepared with and without papain. There is a rise in the content of both free amino acids in crumb and volatile carbonyl compounds in the crust when proteinase is used. As long as proteinases are active in a baking process, they release amino acids from flour proteins, which are then changed via *Strecker* degradation into volatile carbonyl compounds in the crust. Bread aroma is enhanced, as is the crust color, by a build-up of melanoidin compounds from nonenzymatic browning reactions.

15.4.1.4.6 Salt

The taste of bread is rounded-off by the addition to dough of about 1.5% NaCl. As with other salts with small cations (e.g. sodium fumarate or phytate), the addition of NaCl increases dough stability. It is assumed that this is due to the ions masking the repulsion of one charged gluten protein molecule for another of like charge. This allows a sufficiently close approach of one molecule to another, thus hydrophobic and hydrophilic interactions can occur.

15.4.1.4.7 Emulsifiers, shortenings

Flour baking quality is positively correlated to the content of polar lipids, particularly glycolipids (cf. 15.2.4). Further improvements in dough properties, baking results and end-product freshness or shelf-life (cf. 15.4.4) are gained by adding emulsifiers to the dough, e.g. crude lecithin (cf. Table 3.13), mono- and diacylglycerides or their derivatives in which the OH-

Table 15.28. Effects of papain addition in white bread making (values in µmole/g dry matter)

Constituent		Without papain	With papain
Free amino acids	Dough	183	186
	Crumb	182	272
	Crust	10	15
Volatile carbonyl compounds	Crust	158	217

group(s) is esterified with acetic, tartaric, lactic, monoacetyl or diacetyl tartaric acid (cf. 3.3.2 and 8.15). A plausible explanation of positive effects of emulsifiers in the baking process is similar to that ascribed to glycolipids (cf. Figure 15.11); that is, emulsifiers enhance the adherence of starch to protein and vice-versa.

Addition of triacylglycerides (shortenings) generally reduces the end-product volume, but there are exceptions. As illustrated by flour type I in Table 15.29, addition of 3% shortening provides a substantial increase in baking volume.

15.4.1.4.8 α-Amylases

Flours contain very small amounts of sugars which are metabolizable by yeast (cf. Table 15.16). Addition of sucrose or starch syrup at 1–2% to dough is advisable to maintain favorable growth of yeast and therefore to provide CO_2 needed for dough leavening. Uniform leavening over an extended time improves the quality of many baked end-products; the crumb structure acquires finer and more uniform porosity, while the crust has greater elasticity.

Flours derived from wheat without sprouted grains have some β- but very little α-amylase activity (cf. 15.2.2.1). Thus, only a small amount of starch is degraded into fermentable maltose by handling dough. An insight into the extent of starch degradation is provided by the maltose value (cf. 15.4.1.1.1). Addition of α-amylase in the form of malt flour or as a microbial preparation increases the flour capacity to hydrolyze the starch.

The activity of α-amylase as well as the levels of maltose and glucose increase in the germination of cereals; hence, addition of flour from malted grains enhances the growth of yeast in dough. However, the addition of malt to flours with weak gluten may be not expedient because of the proteolytic activity of the malt. α-Amylase preparations free of proteolytic activity are available from microorganisms (cf. 2.8.2.2.2).

Table 15.29. The effect of shortening on baking volume

Wheat flour	Baking volume (ml)[a]	
	Without shortening	With 3% shortening
I	64.5	81.0
II	73.3	71.8
III[b]	51.6	46.3

[a] Baking test performed on a small scale (10 g flour).
[b] Flour of poor baking quality.

Table 15.30. The effect of α-amylase preparations on baking results

α-Amylase preparation		White bread		
Origin	Activity[a] (units)	Volume (ml)	Crumb	
			pores	structure
Without addition		2,400	average	average
Wheat malt	140	2,790	good	good
	560	3,000	good	good
	1,120	2,860	average	good
Aspergillus oryzae	140	2,750	very good	very good
	560	2,900	good	good
	1,120	2,950	average	average
Bacillus subtilis	7	2,600	good	good
	35	2,600	good	average
	140	2,640	poor	very poor

[a] α-Amylase units in 700 g flour.

Examples in Table 15.30 illustrate the effects of α-amylase from various sources on baking quality. While malt and fungal amylases show similar effects, the heat-stable α-amylase from *Bacillus subtilis,* with its prolonged activity even in the oven, may be easily used to excess.

Maltose obtained from the combined action of α- and β-amylases is also available as a reactant for nonenzymatic browning reactions. This favorably affects the aroma and color of the crust.

15.4.1.4.9 Milk and soy products

Dairy products such as skim milk, buttermilk, whey and casein are added to flour in combination with the ingredients or additives mentioned so far. These dairy products are used in either powdered or liquid form; and, as well, either whole or in the form of defatted flour. In such cases, dough supplemented by added proteins increases its water binding capacity and provides a mellowed juicy crumb.

15.4.1.5 Rye Flour Baking Quality as Influenced by Some Additives

Rye flour often requires an improved water binding capacity. For this purpose, 2–4% of pregelatinized flour is added. In addition, artificial acidification of the rye dough is practiced; hence both aspects will be covered.

15.4.1.5.1 Pregelatinized flour

Pregelatinized flour is made from ground cereals such as wheat, rye, rice, millet, etc. by cooking and steaming in autoclaves followed by drying and repeated milling. Such pregelatinized flours are sometimes blended with guar flour or alginates.

15.4.1.5.2 Acids

Rye flour is used in bread baking after sour dough fermentation (cf. 15.4.2.2).
Artificial acidification can be achieved by the addition of lactic, acetic, tartaric or citric acid to rye dough or by adding acidic forms of sodium and calcium salts of ortho- and/or pyrophosphoric acids.
Other preparations for acidification, the so-called dry or instant acids, consist of pregelatinized flour blended with a sour dough concentrate or of cereal mash prefermented by lactic bacteria. The acid values (for definitions see 15.4.1.1.1) vary from 100–1000.

15.4.1.6 Dough Leavening Ingredients

Dough consisting only of flour and water gives a dense flat cake. Baked products with a porous crumb, such as bread, are obtained only after the dough is leavened. This is achieved for wheat dough by addition of yeast while, for fine baked products, baking powders are used. Rye dough leavening is achieved by a sour dough formulation which includes lactic and acetic acid bacteria.

15.4.1.6.1 Yeast

A given amount (Table 15.31) of surface-fermenting yeast, *Saccharomyces cerevisiae,* is used. While normal yeasts preferentially degrade sucrose rather than maltose, special rapidly-fermenting yeasts are used which metabolize both disaccharides at the same rate, shortening the fermentation time.
Yeasts differ in their growth temperature optima (24–26 °C) and their fermentation temperature optima (28–32 °C). The optimum pH for growth is 4.0–5.0. In addition to CO_2 needed for dough leavening, the yeast forms a set of aroma compounds (cf. 5.3.2.1). Whether other compounds released by the growth of yeast would affect the dough rheology is unclear; there appears to be no effect of yeast proteinase.

Table 15.31. Amount of yeast used in bread and other baked products

Baked product	Yeast added[a] (%)
Rye bread	0.5–1.5
Rye mix bread	1.0–2.0
Wheat mix bread	1.5–2.5
Wheat bread	2.0–4.0
Breakfast rolls	4.0–6.0
Rusk ("Zwieback")	6.0–10.0

[a] Based on flour content.

15.4.1.6.2 Chemical leavening agents

The interaction of water, acid, heat and chemical leavening agents (baking powders) releases CO_2. The release of gas may occur in the dough prior to or during oven baking. The agents consist of a CO_2-generating source, usually sodium bicarbonate, and of an acid constituent; tartaric, citric or adipic acid or their sodium or calcium salts, disodium hydrogen phosphate or aluminum sulfate. Glucono delta lactone is recommended since it hydrolyzes quickly in the dough to produce gluconic acid. In the baking powder the two reactive constituents are blended with a divider which consists mostly of corn, rice, wheat or tapioca starch or just wheat flour. The divider content in a powder is up to 30%. The role of the divider is to prevent premature release of CO_2. The market also offers baking powders flavored with vanillin or ethyl vanillin.
For every 500 g of flour, baking powder should develop 2.35–2.85 g CO_2, equivalent to about 1.25 liters.
$NaHCO_3$ alone is used for some flat shelf-stable cookies. Ginger beer and honey cookies are sometimes leavened by ammonium carbonate together with potassium carbonate. The former salt is often a mixture of ammonium bicarbonate and ammonium carbamate ($H_2NCOONH_4$). Both decompose above 60 °C to NH_3, CO_2 and water.

15.4.2 Dough Preparation

15.4.2.1 Addition of Yeast

15.4.2.1.1 Direct addition

Flour, water, yeast, salt and other ingredients are directly mixed into the dough.

15.4.2.1.2 Indirect addition

Yeast is propagated at 25–27 °C in a well-aerated liquid pre-ferment which contains flour, water and some sugar. After a given time, the liquid is blended with the bulk of flour and water and other ingredients and then made into a dough in a mixer.

For continuous indirect addition of yeast, special liquid starters (sponges) with a pH of 5.0–5.3 are also used, with incubation at 38 °C to develop aroma. Such matured fermented sponge is then metered continuously into a kneader which handles the dough.

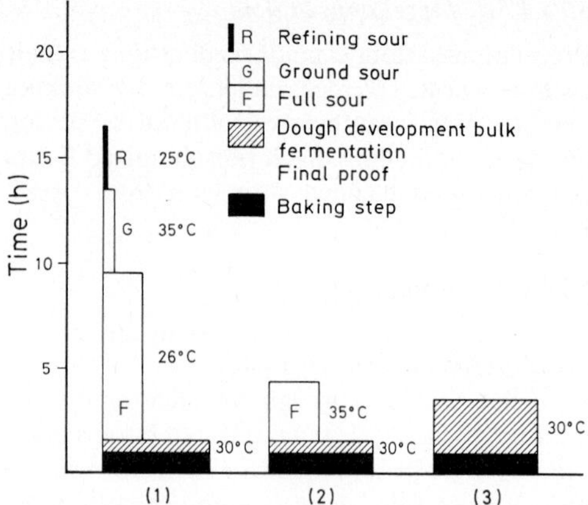

Fig. 15.25. Time requirement for various sour dough development methods. (After *M. Rothe,* 1974.) 1 A three step process, 2 short sour, 3 dough souring agents used

15.4.2.2 Sour Dough Making

In sour dough making (lowering the pH to 4.0–4.3) rye flour acquires the aroma and taste properties so typical of rye bread (cf. 15.1.5).

Yeast (*Saccharomyces cerevisae, Saccharomyces minor* and others), which are mainly responsible for dough leavening, and a complex bacterial flora in which lactic acid-forming organisms dominate (*Lactobacillus plantarum* and *Lactobacillus brevis*) are present in sour dough.

Sour dough is prepared by various procedures which differ considerably in length of time required (Figure 15.25). A multiple-stage procedure takes into account the optimum temperature and humidity needs of yeast and bacteria. Yeast prefer to grow at 26 °C, while the bacteria of interest grow best at 35 °C.

In setting-up a three-stage process, initially an aqueous flour suspension is inoculated. This is the first "full sour" build-up stage (Figure 15.25). After maturation, further amounts of flour and water are added and the process is continued with a second "ground sour" stage at 35 °C and then, in a similar way, continued with an additional "full sour" third stage at 26 °C.

The incubation conditions given in Figure 15.25 are only the essential outline. Temperature deviations influence the spectrum of fermentation products. At warmer temperatures (30–35 °C) lactic acid is preferentially formed (Figure 15.26), while at cooler temperatures (20–25 °C) more acetic acid is produced. The desirable lactic acid: acetic acid ratio, called the "fermentation ratio", is close to 80:20. A ratio with a higher acetic acid concentration gives too sharp an acid taste. The portion of rye flour in the end-product affects the amount of rye sour (full sour) to be

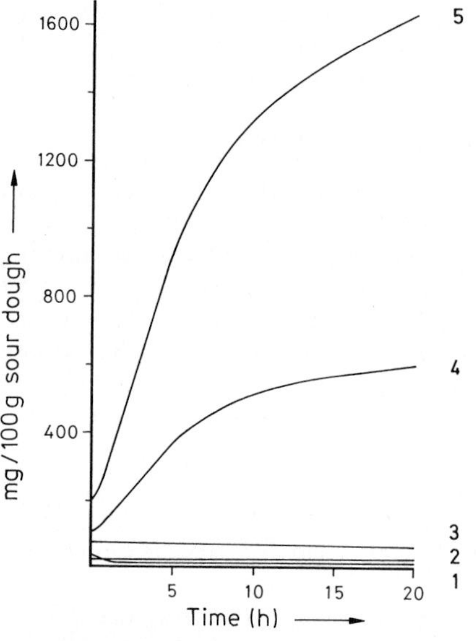

Fig. 15.26. Acid formation in sour dough versus time. (After *E. Rabe,* 1980.) 1 Malate, 2 pyruvate, 3 citrate, 4 acetate, and 5 lactate

added to the dough in the preparation stage. Thus, for rye bread the sour dough to be added is 35–45%, while for a rye mix bread it is 40–60% (on the basis of rye flour). In the short sour method the growth of yeast is negligible. Only a single sour stage, which lasts about 3 h, is involved and yeast is added (Figure 15.25). However, this short method requires a relatively high content of starter saved from a previous ripe sour. Additional time can be saved by using dough acidifiers (cf. 15.4.1.5.2 and Figure 15.25). In short sour processes all the organic acids needed for the sour taste of the rye end-product are present. However, there is a lack of aroma compounds and precursors from which baking can generate aroma. In a multiple-stage rye sour procedure, part of the flour proteins is hydrolyzed by proteinases of the microflora into free amino acids which then participate in *Maillard* reactions during baking, providing the more intense aroma.

15.4.2.3 Kneading

The kneading process is characterized by the following stages: mixing of the ingredients and seasonings; dough development and dough plastification.

The energy input into dough kneading, the dough properties and baking volumes are interrelated. For each dough the baking volume passes through an optimum which is dependent on kneading energy input (Figure 15.27). This optimum shifts towards lower energy input with a flour of weak gluten content and towards higher energy input with flours of strong gluten content; and, as expected, the position of the optimum can be influenced by flour improvers. Increased additions, especially of azodicarbonamide, to the dough result in a successive drop in kneading energy input (Figure 15.27).

As the kneading energy moves away from the optimum, the dough becomes wetter, it starts to stick to trough walls and its gasholding ability drops. Dough development of wheat flours requires close to double the kneading time of rye flours.

The machines used for kneading are grouped by their performance based on kneading time: fast, intensive, and high power kneaders and mixers (Table 15.32). However, the groups are not sharply divided. As the kneading speed increases, the temperature of the dough rises (Table 15.32). Hence, cooling must be used during kneading to keep the temperature at 22–30 °C or, with high speed mixers, at 26–33 °C. The mixer, in a true sense, does not knead the dough, but rips or ruptures it. This could reduce the stability of the dough to such an extent that it could be baked only as a panbread (in which case the pan walls

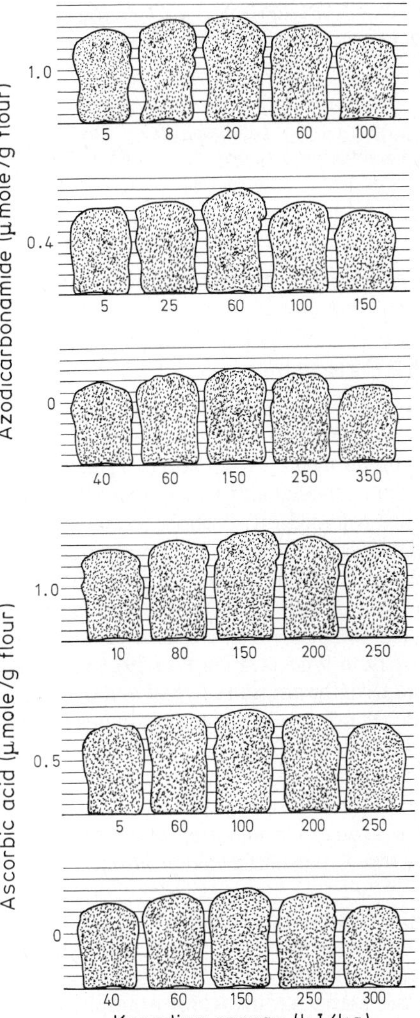

Fig. 15.27. Bread volume as affected by kneading energy input. (After *P. J. Frazier* et al., 1979)

Table 15.32. Examples for kneading conditions in white bread dough making

Dough mixer/ kneader	Speed (rpm)	Kneading time (min)	Dough heat[a] ΔT (°C)
Rapid kneader	60–75	20	2
Intensive kneader	120–180	10	5
High power kneader	450	3–5	
Mixer	1,440	1	9
Mixer	2,900	0.75	14

[a] Temperature rise during kneading time.

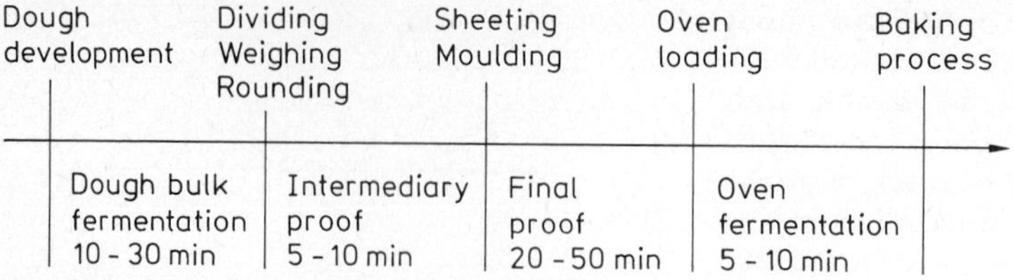

Dough development | Dividing Weighing Rounding | Sheeting Moulding | Oven loading | Baking process

Dough bulk fermentation 10 - 30 min | Intermediary proof 5 - 10 min | Final proof 20 - 50 min | Oven fermentation 5 - 10 min

Fig. 15.28. Fermentation process for biologically-leavened dough; temperature 26–32 °C (After *H. Bueskens,* 1978.)

support the dough), but not as bread made from self-supporting dough.

15.4.2.4 Fermentation

Dough passes through several stages of fermentation in the presence of growing yeast, a biological leavening agent. After initial fermentation, the dough is divided and scaled, then the dough pieces are rounded-off. A short fermentation is followed by sheeting and moulded dough fermentation. The dough acquires its enlarged end-volume by CO_2 and steam action in the oven. The length of time of the fermentation varies. It depends on flour type (cf. Figure 15.29), seasonings incorporated, the amount of yeast and oven temperature. The flour character determines the fermentation tolerance, i.e. the time period within which the fermentation break starts and stops. During this time the dough has to be loaded into the oven. Dough fermentation of a weak gluten flour is rapid, but its fermentation tolerance is low.

The main dough fermentation step (cf. Figure 15.28) can be substantially shortened by kneading the dough energetically and/or by incorporating fast-acting additives (for example, a mixture of bromates, ascorbic acid and cysteine) into the dough. This provides a favorable dough structure, able to accommodate large amounts of yeast. This is the basis for "no-time" dough making procedures, which provide a continuous flow of dough.

15.4.2.5 Events Involved in Dough Making

Bread dough is prepared by mixing water and flour (70 : 30 w/w). Water uptake, which depends on flour type, predetermines most of the subsequent reactions. A high water uptake favors the

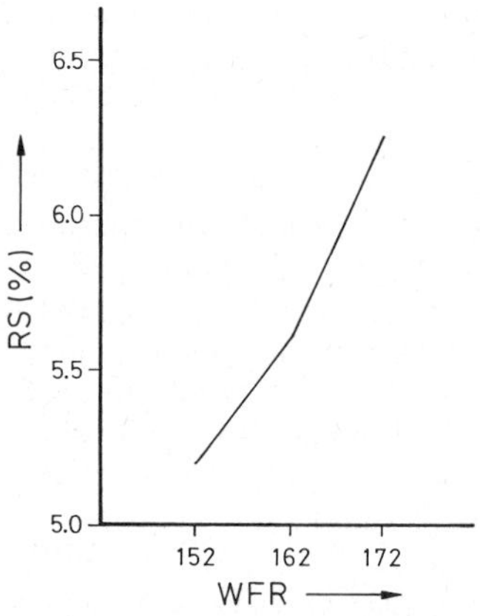

Fig. 15.30. The reducing sugar content in wheat bread crumb as affected by water content of dough. (After *L. Wassermann* and *H. H. Doerfner,* 1971.) Water flour ratio (WFR) $= \frac{(\text{flour} + \text{water}) \times 100}{\text{flour}}$. RS: Reducing sugar expressed as maltose

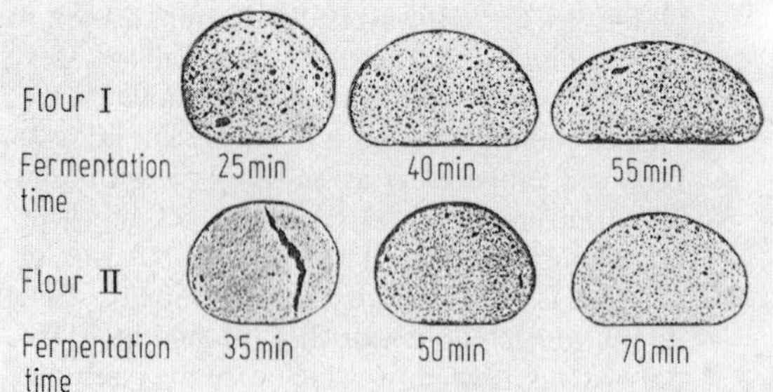

Fig. 15.29. The effect of fermentation time on baking results. (Rye-mix bread with two wheat flours which differ in baking quality. After *H. Bueskens,* 1978)

mobility of all the constituents involved in reactions, e.g. enzymatic degradation of starch into reducing sugars (Figure 15.30).

Observation of wheat dough development by light or scanning electron microscopy reveals that a sequence of forceful changes occurs in the arrangement of the flour proteins:

The proteins are present in a platelet-sheet form, with a thickness of about 100 Angstroms. Contact with water causes these sheets to rapidly develop cavities (Figure 15.31 a), and the cavity rims swell (Figure 15.31 b). Ultimately, the sheet structure breaks down into a fibrillar network (Figure 15.31 c). The microfibrils, with a di-

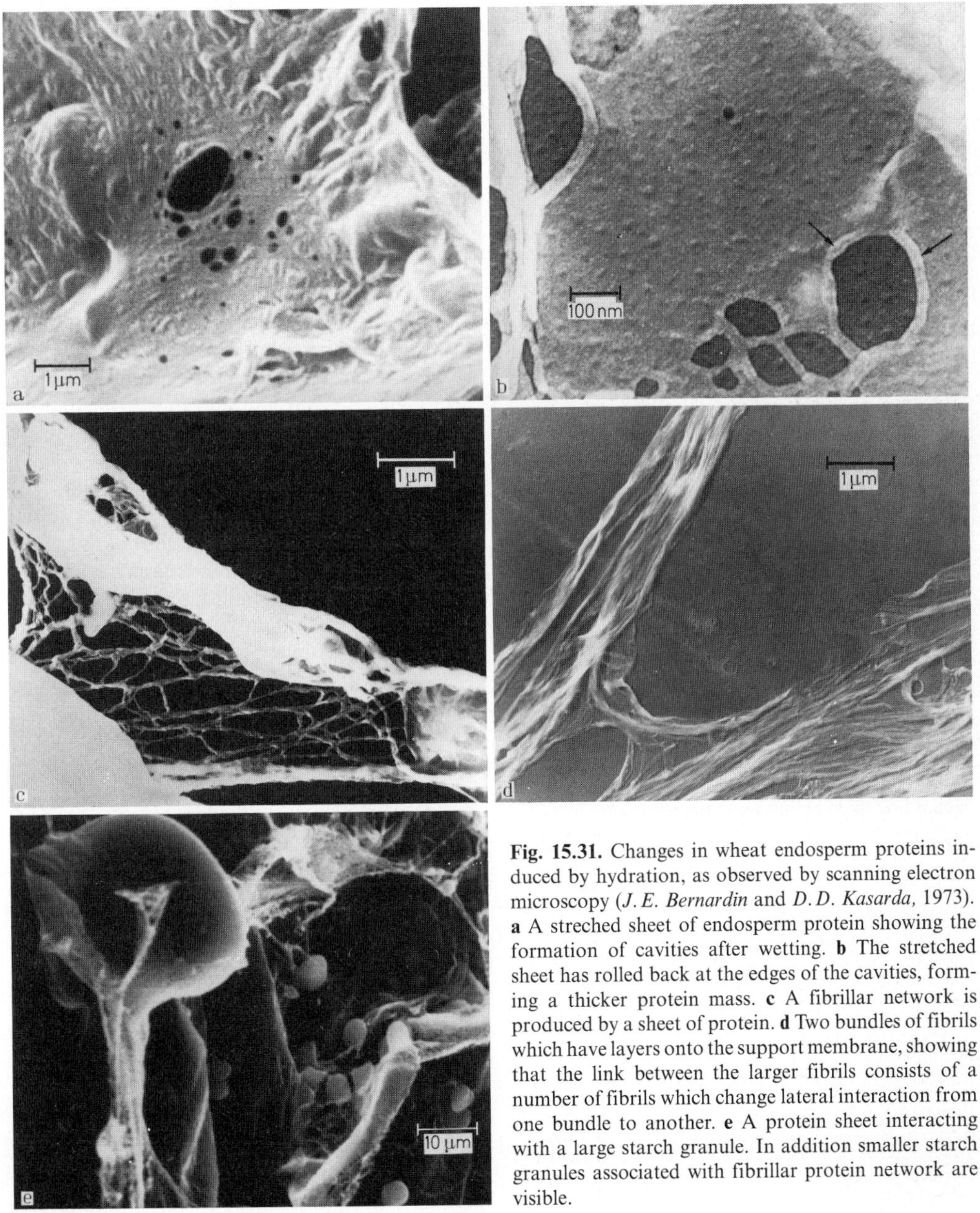

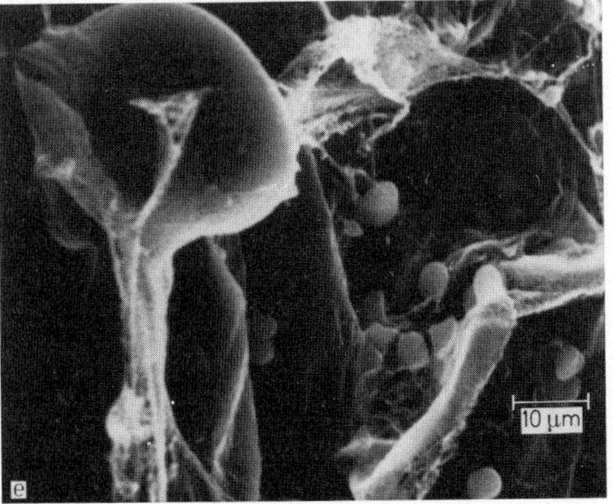

Fig. 15.31. Changes in wheat endosperm proteins induced by hydration, as observed by scanning electron microscopy (*J. E. Bernardin* and *D. D. Kasarda,* 1973). **a** A streched sheet of endosperm protein showing the formation of cavities after wetting. **b** The stretched sheet has rolled back at the edges of the cavities, forming a thicker protein mass. **c** A fibrillar network is produced by a sheet of protein. **d** Two bundles of fibrils which have layers onto the support membrane, showing that the link between the larger fibrils consists of a number of fibrils which change lateral interaction from one bundle to another. **e** A protein sheet interacting with a large starch granule. In addition smaller starch granules associated with fibrillar protein network are visible.

ameter of 50–100 Angstroms, extend or spread out further with increased hydration. During kneading, the microfibrils associate, forming bundles (Figure 15.31 d), which then merge into a network in which the starch granules are embedded (Figure 15.31 e).

Such transformation of platelet-sheet structures into microfibrils is observable with rye to a restricted extent, but not with corn, rice or barley proteins. In wheat flour the gluten-forming proteins, prolamines and glutelins (cf. 15.2.3.2.1), are involved, with some participation of lipids (cf. 15.2.4) and pentosans (cf. 15.2.3.2.1). The gluten gives the dough viscoelastic properties and gas-holding capacity.

Kneading can initiate additional chemical reactions. Cleavage of disulfide bonds into thiyl radicals is just one that has been reported.

15.4.3 Baking Process

15.4.3.1 Conditions

The oven temperature and time of baking for some baked products are summarized in Table 15.33. Conditions for baking of rye and rye mix bread sometimes deviate from these values. They are prebaked at higher temperatures, for instance at 400 °C for 1–3 min, and then post-baked at 150 °C (for the effect on quality see Table 15.36). In an oven with the temperatures given in Table 15.33, since heat transfer occurs slowly in dough, there is a steep temperature gradient, 200 → 120 °C, inward from the crust of the dough piece. By the end of baking, a temperature of 96 °C is attained within the product. Higher temperatures up to 106 °C are found when the crust is able to resist the rise of inner steam pressure. The water evaporates only in the crust region during dough baking. Water diffusion towards the center of the bread can give the fresh crumb a higher moisture content than the dough. The steam concentration in the oven also affects the baking results. A steam header is provided in most oven designs to regulate oven moisture.

A baking weight loss is experienced as a result of water evaporation during crust formation. The extent of the loss is related to the form and size of the baked bread and is 8–14% of the fresh dough weight.

Table 15.33. Baking times and temperatures

Baked product	Weight (g)	Baking time (min)	Oven temperature (°C)
Buns, rolls and other small baked products	45	18–20	250–240
Wheat bread (self-supported dough)[a]	500	25–30	240–230
Wheat bread (pan-baked)[b]	500	35–40	240–230
Wheat bread (self-supported dough)	1,000	40–50	240–220
Rye mix bread (self-supported dough)	1,500	55–65	250–200
Rye bread (self-supported dough)	1,500	60–70	260–200
Pumpernickel (pan-baked)	3,000	16–14 hrs.	180–100

[a] Hearth bread
[b] Pan bread

15.4.3.2 Chemical and Physical Changes

15.4.3.2.1 Texture

A rise in temperature causes protein denaturation and starch swelling. In white bread dough this solidifies the crumb framework, which contains pores up to 3 μm in diameter, and is built-up of gluten foam and entrapped or adhering swollen starch granules. The gluten foam arises during dough fermentation. The extent of starch swelling depends on the available water. The water in dough is preferentially bound by prolamines, glutelins and pentosans. Part of this water becomes available to swell the starch during baking. Limited starch swelling results in a brittle crumb, whereas extensive swelling makes the crumb greasy or gluey.

In contrast to the crumb, the starch granules of the crust surface gelatinize almost completely. This is especially the case when the oven humidity is high, e.g. when baking occurs below a steam header. Investigations involving gluten and starch mixtures to which the emulsifier stearyl-2-lactylate was added revealed that lipid transfer occurs from gluten to starch during heating of the mixture above 50 °C (Table 15.34). Apparently, the high swelling and gelatinization of the starch granules, which occurs above 50 °C (cf. Table 15.17), promotes lipid binding.

The specific volume of white bread is higher than that of rye bread (Table 15.35). The rye crumb is stronger and less elastic, suggesting that the pentosans can not fully compensate for the lack of texturizing quality of rye proteins (cf. 15.1.5).

Heating of a dough accelerates enzymatic reactions, e.g. starch degradation (cf. 15.4.1.2). Above the "temperature optimum" (cf. 2.6.3) the

reactions are inhibited by denaturation of the enzymes.
Starch degrades to dextrins, mono- and disaccharides at the relatively high temperatures to which the outer part of the dough is exposed. Caramelization and nonenzymatic browning reactions also occur, providing the sweetness and color of the crust. The thickness of the crust is dependent on temperature and baking time (Table 15.36) and type of baked product (Table 15.37).

15.4.3.2.2 Aroma

Aroma is generated in two stages in the bread-making process. The yeast or the sour dough procedures provide the initial fermentation aroma, while baking produces the crust aroma. The major volatile compounds, formed primarily by yeast metabolism, are acetaldehyde, dimethyl sulfide and ethanol (Table 15.38). Due to their high aroma values (definition in 5.2.4.1), they are considered to be responsible for the fermented aroma of white bread.
However, the fermented aroma is covered by the typical aroma released in *Maillard* (cf. 4.2.4.4) and caramelization reactions (cf. 4.2.4.3.2) which occur mainly in the crust during baking. The following of the numerous volatile compounds arising from such thermal reactions have been identified as major carriers of crust aroma:

Table 15.34. The effect of temperature on stearyl-2-lactylate (SSL) binding in a blend of gluten and starch[a]

T (°C)	SSL free[b]	SSL bound[b] on	
		gluten	starch
30	22.0	64	14
40	20.0	66	14
50	22.0	62	16
60	20.0	6	74
70	16.0	6	78
80	12.0	8	80
90	12.0	2	86

[a] Blends from 17.9 g starch, 2.7 g gluten and 0.103 g SSL.
[b] Values in % of total SSL.

Table 15.35. Specific volumes[a] of bread

Bread variety	ml/g
Toast bread	3.5–4.0
White bread	3.3–3.7
White mix bread[b]	2.5–3.0
Rye mix bread[b]	2.1–2.6
Rye bread	1.9–2.4

[a] Specific volume = volume/weight.
[b] cf. Table 15.41.

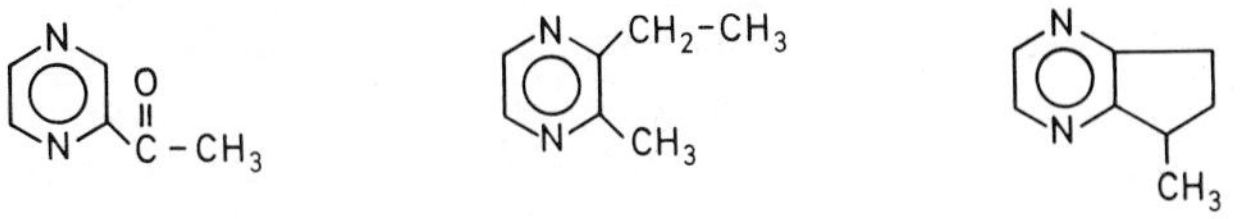

(I; 62 ppb) (II; 30 ppb) (III; 15-25 ppb)

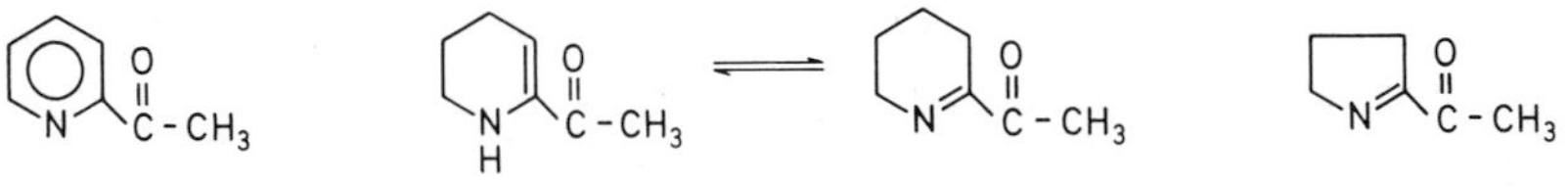

(IV; 19 ppb) (Va) ⇌ (Vb) (1.6 ppb) (VI; 0.1 ppb)

(15.6)

The flavor thresholds (water as solvent) given in brackets indicate that 2-acetyl-1-pyrroline (VI) is the most potent flavor compound with a crusty note. From 2-acetyl-1,4,5,6-tetra hydropyridine the two tautomers (Va and Vb) were found in gas chromatograms of bread crust volatiles. They do not differ in flavor.
Several volatile carbonyls and alcohols present in bread can be assayed quantitatively by rather simple methods. Though they are not typical

Table 15.36. The effect of baking time and temperature on the quality of rye whole meal bread

Baking time (min)	90	180	270
Baking temperature (°C)	240–210	210–185	185–160
Bread yield (ml)	142	142	140
Crust strength (mm)	4	5	6
Taste	raw, slightly aromatic	aromatic	strongly aromatic

aroma constituents, their determinations provide an insight into aroma build-up as affected by baking ingredients and processes.

A study of different bread types (Table 15.39) reveals that volatile compounds, with the exception of ethanol, are in higher concentration in the crust than in the crumb. The volatile constituents in rye bread crust are higher than in the crust of white bread. This agrees with sensory perception; rye bread having a more intensive aroma than white bread.

Table 15.37. Crumb and crust ratios in different bread varieties

Bread variety	Crumb (%)	Crust (%)
Buns, rolls (50 g)	72.5	27.5
Stick (French) white bread	68.5	31.5
White bread, pan-baked, 500 g	75.0	25.0
White bread (self-supported dough, 500 g)	73.8	26.2
Rye bread (self-supported dough) 1000 g	73.3	26.7
Rye mix bread (pan-baked, 1000 g)	84.5	15.5

Table 15.38. Volatile compounds contributing to "fermentation aroma" of a white bread

Compound	Aroma value[a]
Acetaldehyde	57
Dimethyl disulfide	15.6
Ethanol	11.1
2-Methylpropanal	10
3-Methyl-1-butanol	9.7
Dimethyl sulfide	5
1,1-Diethoxyethane	3.6
3-Methylbutanal	2.9
2-Methyl-1-propanol	2.2

[a] Definition cf. 5.2.4.1.

15.4.4 Changes During Storage

Bread quality changes rapidly during storage. Due to moisture adsorption, the crust loses its crispiness and glossyness. The aroma compounds of freshly-baked bread evaporate or are entraped preferentially by amylose helices which occur in the crumb. Repeated heating of aged bread releases these compounds. Labile aroma compounds are also present (e.g. VI in 15.4.3.2.2). These compounds also change rapidly by oxidation and other reactions.

The crumb structure also changes, although at a lower rate. The crumb becomes firm, its elasticity and juiciness are lost, and it crumbles more easily. The so-called staling defect of the crumb is basically a starch retrogradation phenomenon (cf. 4.4.4.12.2). It involves the increased realignment of starch molecules from an amorphous to a semi-crystalline form.

The retrogradation is reflected in a decrease in content of polysaccharides which can be extracted from aged bread. As shown in Table 15.40, the solubility drop of amylose with time is even more pronounced than that of amylopectin, implying that amylose retrogradation occurs at a faster rate.

Increasing storage temperature from 21 to 35 °C decreases the rate of amylopectin retrogradation by a factor of 4 and improves freshness of the crumb, while aroma is dissipated. Increased protein or pentosan content slows retrogradation. A choice – actually a rule – to extend the shelf-life or freshness of the baked product is the use of emulsifiers, such as monoacylglycerides or stearyl-2-lactylate. During baking the emulsifier will be complexed with both starch constituents,

Table 15.39. Differences in content (ppm) of volatile compounds in three bread varieties

Compound	*White bread*		*Rye bread*		*Rye whole meal bread*	
	Crumb	Crust	Crumb	Crust	Crumb	Crust
Ethanol	3,900	1,800	3,400	1,100	2,300	1,000
5-Hydroxymethylfurfural	9	40	12	300	20	400
Acetaldehyde	4.3	12.8	4.7	22.6	4.6	26.2
Isopentanol	1.2	4.7	2.7	15.2	1.9	19.0
Furfural	0.3	5.5	1.5	12.4	2.3	28.7
Methylglyoxal	0.7	0.8	1.5	8.9	1.9	13.5
Isobutanal	0.3	2.6	0.9	6.0	0.8	12.9
Acetone	0.7	4.5	1.4	5.6	2.0	6.5
Acetoin	0.9	1.0	0.2	1.1	0.3	0.7
Diacetyl	0.2	0.9	0.2	1.3	0.2	1.3

Table 15.40. Soluble starch in white bread stored at 21 °C

Time (days)	Soluble starch[a]	Amylose[a]	Amylo-pectin[a]
0	3.34	0.52	2.82
1	2.16	0.19	1.97
2	1.72	0.14	1.58
5	1.22	0.10	1.12

[a] In g per 100 g dry matter.

though to a different extent (Table 15.34 and Figure 15.32). Such complexes retard starch retrogradation.

Less carbohydrates can be extracted from starch-monoglyceride complexes than from starch alone. This property probably contributes to the increased cooking strength of pasta products which contain monoglycerides (cf. 15.5).

The simplified events involved in bread staling can be summarized as follows: During dough baking the swelling of the starch granules is limited since the available water is limited. Swelling solubilizes some of the amylose, which diffuses from the granules and concentrates in the aqueous phase in the free space between the swollen granules. In the early cooling period after baking, amylose molecules start to form a gel structure in the crumb because of their ability to form micelles and to retrograde. In other words, the crumb of freshly baked bread is a network of amylose chains and embedded swollen starch granules, in which the hydrated amylopectin moiety is retained. During bread staling, changes occur in the swollen starch granules. Amylopectin molecules which are separated by hydration start to aggregate via their branches, thus reinforcing the swollen starch granules. When baking involves the use of emulsifiers, amylose diffusion from granules is prevented. It remains within the granules as a fixed lipid complex. No gel setting occurs and the crumb softness is retained. It is also assumed that polar lipids (phospho- and glycolipids) affect amylopectin aggregation.

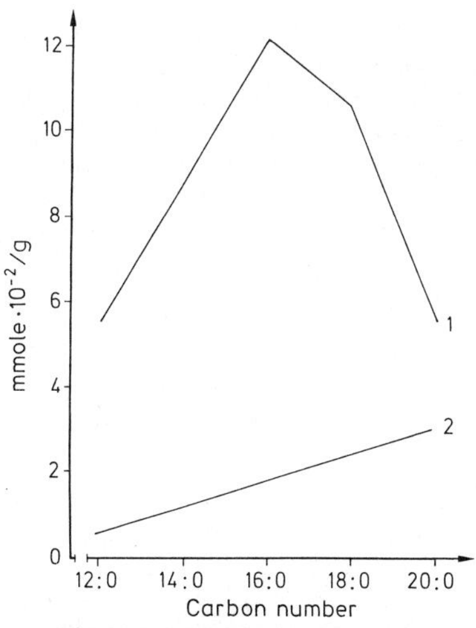

Fig. 15.32. Complex formation between monoglycerides and amylose (1) or amylopectin (2). (After *W. H. Knightly*, 1977.) x-axis: carbon number of the saturated acyl residue. y-axis: tendency of a monoglyceride (MG) to form a complex with amylose or amylopectin (mmole $\cdot 10^{-2}$ MG/g polysaccharide)

15.4.5 Bread Types

Only those bread types of significant economic importance are listed in Table 15.41. Corresponding data on chemical composition are given in Table 15.42.

Crisp bread (Knaeckebrot) and Pumpernickel are special rye breads.

Table 15.41. Bread varieties

No.	Bread variety	Formulation
1.	Wheat bread (white bread)	At least 90% wheat; middlings less than 10%; occasionally with addition of dairy products, sugar, shortenings (ref. Table 15.25).
2.	Wheat mix bread	50-89% wheat, the rest rye milling products and other ingredients as under 1.
3.	Rye mix bread	50–89% rye, the rest wheat milling products and others as under 1.
4.	Rye bread	At least 90% rye flour, up to 10% wheat flour; other ingredients as under 1.
5.	Rye whole grain bread	From whole rye meal including also whole kernels, other rye and wheat products less than 10%.

Table 15.42. Bread composition

Constituent	%	Constituent	%
Water	35–43	NaCl	1.0–1.5
Carbohydrates	45–58	Other minerals	0.4–1.1
Protein	6–16	Crude fiber	1.1–1.2
Fat	0.5–1.4		

The flat crisp bread is produced mostly from whole rye meal with low α-amylase activity. Under ice cooling the dough is mixed using compressors until foaming occurs, then sheeted and baked for 8–10 min in a tunnel-type oven. Additional drying reduces moisture by 10–20% to a level of 5%. In addition to this mechanically-leavened bread, made by mixing air or nitrogen into the dough, there are crisp breads in which biological leavening (yeast or rye sour) is used. Flat bread production can be a fully automated process if cooker-extruders are used.
Pumpernickel bread originates from Westphalia. The sour rye dough, heated in unsealed ovens, is more steam-cooked than baked (cf. Table 15.33). Prolonged heating considerably degrades the starch into dextrins and maltose, which are responsible for the sweet taste. The increased build-up of melanoidin pigments accounts for the dark color.

15.5 Pasta Products

15.5.1 Raw Materials

Pasta products are made of wheat semolina and grist (cf. 15.3.1.3), in which the flour extraction grade is less than 70%, and may incorporate egg. The preferred ingredient is durum wheat semolina rather than the soft wheat counterpart (farina) since the former has better cooking and biting strengths and also has a higher content of carotenoids (cf. 15.2.4) which provide the yellow color of pasta products. In wheat mixtures, the soft wheat characters emerge when the soft wheat content is higher than 30%. In egg-pasta products (chemical composition in Table 15.43), 2–4 eggs/kg semolina provide a pasta with improved cooking strength and color.

15.5.2 Additives

Cysteine hydrochloride (about 0.01%) lowers the mixing/kneading time by 15–20% (cf. 15.4.1.4.4). The cysteine also inhibits melanoidin build-up due to nonenzymatic browning, and suppresses the greyish-brown pigmentation. Addition of monoglycerides (about 0.4%) brings about amylose and amylopectin complexing, thereby increasing cooking strength (cf. 15.4.4). Through competitive inhibition, ascorbic acid prevents the action of lipoxygenase (Figure 15.33). Though the enzyme is of the lipoxygenase I type (cf. Table 3.26) and so only slowly cooxidizes carotenoids, the low enzyme activity can still destroy the pigments when pasta production is a prolonged process. Addition of ascorbic acid inhibits this cooxidation (Figure 15.34).

Table 15.43. Composition of pasta products containing eggs (4 eggs per 1 kg flour)

Constituent	%	Constituent	%
Water	11.1	Carbohydrates	70.0
Protein	14.5	Crude fiber	0.5
Fat	2.9	Minerals	1.0

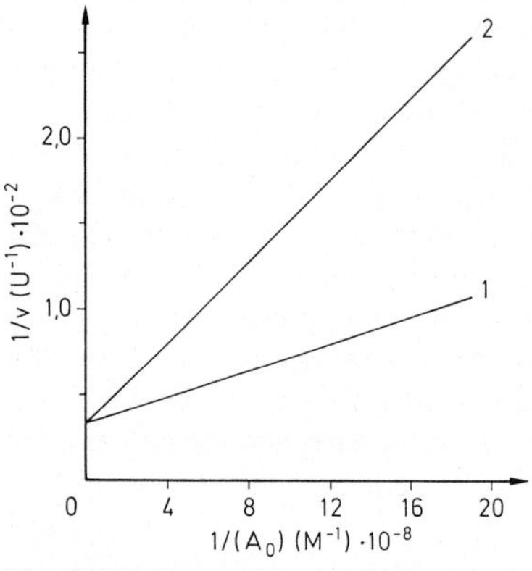

Fig. 15.33. Competitive inhibition of wheat lipoxygenase by ascorbic acid. (After *D.E. Walsh* et al., 1970.) Activity assay with linoleic acid as a substrate (1) without, and (2) in the presence of ascorbic acid ($2 \cdot 10^{-6}$ M)

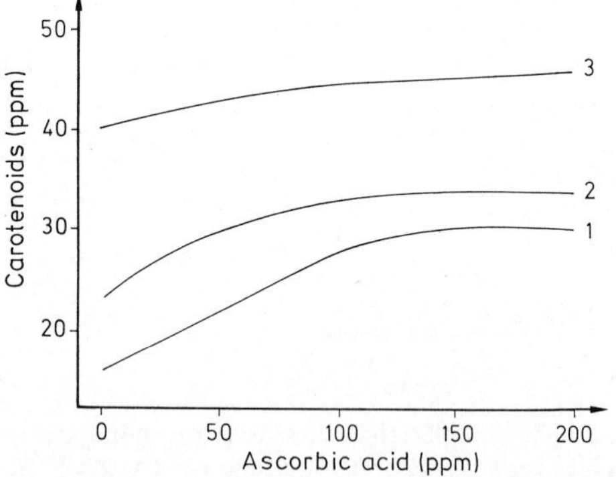

Fig. 15.34. Carotenoid stability in pasta products made of three Durum wheat cultivars, as affected by the added ascorbic acid. (After *D.E. Walsh* et al., 1970.) 1–3 wheat cv. Durum

15.5.3 Production

Pasta products are manufactured continuously by a vacuum extruder, which consists of a mixing trough and press segments. The vacuum is used to retard oxidative degradation of carotenoids. The semolina and added water (30%) and, when necessary, egg or egg powder, are mixed in a mixing trough to form a crumb dough (diameter 1–3 mm), pressed at 150–200 bar into a uniform paste and then pressed through an extruder pressure head die, to provide the familiar pasta strings.

Drying is the most demanding stage of pasta manufacturing. The surface of a pasta product must not be allowed to harden prior to the interior core, otherwise cracks, fractures or bursts develop. The freshly-extruded strings are initially dried from the outside until they are no longer sticky, then drying is continued at 45–60 °C, either very slowly or stepwise. The moisture content drops to 20–24% after such a predrying process. The moisture is then allowed to equilibrate between inner and outer parts, which brings the content of the final dried product to 11–13%.

15.6 Literature

Acker, L.: Die Lipide der Stärken – ein Forschungsgebiet zwischen Kohlenhydraten und Lipiden. Fette Seifen Anstrichm. *79,* 1 (1977)

Acker, L., Schmitz, H.J., Hamza, Y.: Über die Lipide des Weizens. Getreide Mehl *18,* 45 (1968)

Ali, M.R., D'Appolonia, B.L.: Einfluß von Roggenpentosanen auf die Teig- und Broteigenschaften. Getreide Mehl Brot *33,* 334 (1979)

Autorenkollektiv: Lebensmittel-Lexikon. VEB Fachbuchverlag: Leipzig. 1979

Barnes, P.J. (Ed.): Lipids in Cereal Technology. Academic Press: New York–London. 1983

Bernardin, J.E.: Gluten protein interaction with small molecules and ions – the control of flour properties. Bakers Dig. *52,* Aug. 1978, p. 20

Bernardin, J.E., Kasarda, D.D.: Hydrated protein fibrils from wheat endosperm. Cereal Chem. *50,* 529 (1973)

Bernardin, J.E., Kasarda, D.D.: The microstructure of wheat protein fibrils. Cereal Chem. *50,* 735 (1973)

Bietz, J.A., Huebner, F.R., Sanderson, J.E., Wall, J.S.: Wheat gliadin homology revealed through N-terminal amino acid sequence analysis. Cereal Chem. *54,* 1070 (1977)

Bietz, J.A., Huebner, F.R., Wall, J.S.: Glutenin the strength protein of wheat flour. Bakers Dig. *47,* Febr. 1973, p. 26

Bietz, J.A., Shepherd, K.W., Wall, J.S.: Single-kernel analysis of glutenin: use in wheat genetics and breeding. Cereal Chem. *52,* 513 (1975)

Bietz, J.A., Wall, J.S.: Identity of high-molecular weight gliadin and ethanol-soluble glutenin subunits of wheat: relation to gluten structure. Cereal Chem. *57,* 415 (1980)

Bietz, J.A., Wall, J.S.: Wheat gluten subunits: molecular weights determined by sodium dodecyl sulfate polyacrylamide gel electrophoresis. Cereal Chem. *49,* 416 (1972)

Bolling, H., Meyer, D.: Zur Verarbeitungsqualität neu zugelassener Weizensorten 1976. Getreide Mehl Brot *30,* 291 (1976)

Bushuk, W. (Ed.): Rye: Production, chemistry, and technology. American Association of Cereal Chemists: St. Paul, Minn. 1976

Büskens, H.: Die Backschule – Fachkunde für Bäcker. 7. Aufl., Verlag W. Girardet: Essen. 1978

Chung, O.K., Pomeranz, Y., Finney, K.F.: Wheat flour lipids in breadmaking. Cereal Chem. *55,* 598 (1978)

D'Appolonia, B.L., Kim, S.K.: Wheat flour pentosans. Bakers Dig. *50,* June 1976, p. 45

Finney, K.F., Tsen, C.C., Shogren, M.D.: Cysteine's effect on mixing time, water absorption, oxidation requirement, and loaf volume of Red River 68. Cereal Chem. *48,* 540 (1971)

Fox, P.F., Mulvihill, D.M.: Enzymes in wheat, flour, and bread. In: Advances in cereal science and technology (Ed.: Pomeranz, Y.), Vol. V, p. 107, American Association of Cereal Chemists: St. Paul, Minn. 1982

Frazier, P.J.: Lipoxygenase action and lipid binding in breadmaking. Bakers Dig. *53,* Dec. 1979, p. 8

Frazier, P.J., Brimblecombe, F.A., Daniels, N.W.R., Russell Eggitt, P.W.: Besseres Brot aus schwächerem Weizen – rheologische Überlegungen. Getreide Mehl Brot *33,* 268 (1979)

Hoseney, R. Carl: Principles of Cereal Science and Technology. American Association of Cereal Chemists St. Paul, Minn. 1986

Houston, D.F. (Ed.): Rice, chemistry and technology. American Association of Cereal Chemists: St. Paul, Minn. 1972

Huebner, F.R., Wall, J.S.: Fractionation and quantitative differences of glutenin from wheat varieties varying in baking quality. Cereal Chem. *53,* 258 (1976)

ICC–Standards: Standard-Methoden der Internationalen Gesellschaft für Getreidechemie (ICC). Verlag Moritz Schäfer: Detmold.

Inglett, G.E. (Ed.): Maize, recent progress in chemistry and technology. Academic Press: New York–London. 1982

Inglett, G.E., Munck, L. (Eds.): Cereals for food and beverages, recent progress in cereal chemistry and technology. Academic Press: New York–London. 1980

Jackson, E. A., Holt, L. M., Payne, P. I.: Characterization of high-molecular weight gliadin and low-molecular weight glutenin subunits of wheat endosperm by two-dimensional electrophoresis and the chromosomal location of their controlling genes. Theor. Appl. Genet. *66,* 29 (1983)

Kasarda, D. D., Bernardin, J. E., Nimmo, C. C.: Wheat proteins. In: Advances in cereal science and technology (Ed.: Pomeranz, Y.), Vol. I, p. 158, American Association of Cereal Chemists: St. Paul, Minn. 1976

Kasarda, D. D., Okita, T. W., Bernardin, J. E., Baecker, P. A., Nimmo, C. C., Lew, E. J.-L., Dietler, M. D., Greene, F. C.: Nucleic acid (cDNA) and amino acid sequences of α-type gliadins from wheat *(Triticum aestivum).* Proc. Natl. Acad. Sci. USA *81,* 4712 (1984)

Khan, K., Bushuk, W.: Glutenin: Structure and functionality in breadmaking. Bakers Dig. *52,* April 1978, p. 14

Kieffer, R., Belitz, H.-D.: Molekulargewichte von Kleberproteinen des Weizens. Z. Lebensm. Unters. Forsch. *172,* 96 (1981)

Kieffer, R., Grosch, W.: Verbesserung der Backeigenschaften von Weizenmehlen durch die Typ II-Lipoxygenase aus Sojabohnen. Z. Lebensm. Unters. Forsch. *170,* 258 (1980)

Knightly, W. H.: The staling of bread. Bakers Dig. *51,* Oct. 1977, p. 52

Kosmina, N. P.: Biochemie der Brotherstellung. VEB Fachbuchverlag: Leipzig. 1977

Lawrence, G. J., Shepherd, K. W.: Variation in glutenin protein subunits of wheat. Aust. J. Biol. Sci. *33,* 221 (1980)

Lorenz, K.: The history, development, and utilization of Triticale. Crit. Rev. Food Technol. *5,* 175 (1974)

MacRitchie, F.: Physicochemical aspects of some problems in wheat research. In: Advances in cereal science and technology (Ed.: Pomeranz, Y.), Vol. III, p. 271, American Association of Cereal Chemists: St. Paul, Minn. 1980

Maga, J. A.: Bread flavour. Crit. Rev. Food Technol. *5,* 55 (1974)

Maga, J. A.: Bread staling. Crit. Rev. Food Technol. *5,* 443 (1974)

Mair, G., Grosch, W.: Changes in glutathione content (reduced and oxidised form) and the effect of ascorbic acid and potassium bromate on glutathione oxidation during dough mixing. J. Sci. Food Agric. *30,* 914 (1979)

Menger, A.: Einfluß von Rohstoffen und Prozeßfaktoren auf die Teigwarenqualität. Getreide Mehl Brot *30,* 149 (1976)

Morrison, W. R.: Lipide in Mehl, Teig und Brot. Getreide Mehl Brot *30,* 244 (1976)

Morrison, W. R.: Cereal lipids. In: Advances in cereal science and technology (Ed.: Pomeranz, Y.), Vol. II, p. 221, American Association of Cereal Chemists: St. Paul, Minn. 1978

Orth, R. A., Bushuk, W.: A comparative study of the proteins of wheat of diverse baking qualities. Cereal Chem. *49,* 268 (1972)

Payne, P. I., Corfield, K. G., Blackman, J. A.: Identification of a high-molecular-weight subunit of glutenin, whose presence correlates with bread-making-quality in wheats of related pedigree. Theor. Appl. Genet. *55,* 153 (1979)

Payne, P. I., Harris, P. A., Law, C. N., Holt, L. M., Blackman, J. A.: The high-molecular-weight subunits of glutenin: structure, genetics and relationship to bread-making-quality. Ann. Technol. Agric. *29,* 309 (1980)

Payne, P. I., Corfield, K. G., Holt, L. M., Blackman, J. A.: Correlation between the inheritance of certain high-molecular-weight subunits of glutenin and bread-making quality in progenies of six crosses of bread wheat. J. Sci. Food Agric. *32,* 51 (1981)

Payne, P. I., Holt, L. M., Jackson, E. A., Law, C. N.: Wheat storage proteins: their genetics and their potential for manipulation by plant breeding. Phil. Trans. R. Soc. Lond. B *304,* 359 (1984)

Pomeranz, Y.: Dispersibility of wheat proteins in aqueous urea solution – a new parameter to evaluate bread-making potentialities of wheat flours. J. Sci. Food Agric. *16,* 586 (1965)

Pomeranz, Y. (Ed.): Wheat: Chemistry and technology. American Association of Cereal Chemists: St. Paul, Minn. 1971

Pomeranz, Y.: Industrial uses of cereals. Symp. Proc., American Association of Cereal Chemists: St. Paul, Minn. 1973

Pomeranz, Y.: Theorie und Praxis der Bestimmung unterschiedlichen Einflusses einzelner Weizenmehlkomponenten auf das Backverhalten. Getreide Mehl Brot *31,* 147 (1977)

Pomeranz, Y., Shellenberger, J. A.: Bread Science and Technology. AVI Publ. Co.: Westport, CT. 1971

Pyler, E. J.: Baking Science and Technology. IIIed. Vol. I–II. Siebel Publ. Co.: Chicago, Ill. 1982

Rabe, E.: Organische Säuren in unterschiedlich geführten Broten. Getreide Mehl Brot *34,* 90 (1980)

Rohrlich, M.: Kleberforschung – Ein historisch-wissenschaftlicher Versuch. Verlag Moritz Schäfer: Detmold. 1969

Rohrlich, M.: Getreideenzyme. Verlag Paul Parey: Berlin–Hamburg. 1969

Rohrlich, M., Brückner, G.: Das Getreide. I. und II. Teil, 2. Aufl., Verlag Paul Parey: Berlin–Hamburg. 1966/1967

Rohrlich, M., Eßner, W.: Menge und Verteilung der SH-Gruppen und SS-Bindungen in Weizenmahlprodukten. Brot Gebäck *20,* 4 (1966)

Rohrlich, M., Thomas, B.: Getreide und Getreidemahlprodukte. In: Handbuch der Lebensmittelchemie

(Hrsg.: Schormüller, J.), Bd. V/1. Teil, Springer-Verlag: Berlin–Heidelberg. 1967

Rothe, M.: Aroma von Brot. Akademie-Verlag: Berlin. 1974

Schieberle, P., Grosch, W.: Identification of the volatile flavour compounds of wheat bread crust – Comparison with rye bread crust. Z. Lebensm. Unters. Forsch. *180,* 474 (1985)

Shukla, T. P.: Cereal proteins: Chemistry and food applications. Crit. Rev. Food. Sci. Nutr. *6,* 1 (1975)

Spicher, G.: Brot und andere Backwaren. In: Ullmanns Encyklopädie der technischen Chemie, 4. Aufl., Bd. 8, S. 702, Verlag Chemie: Weinheim. 1974

Sprößler, B.: Wirkung von Proteinasen beim Zusatz zum Mehl. Getreide Mehl Brot *35,* 60 (1981)

Stephan, H.: Roggenmehl. Getreide Mehl Brot *30,* 77 (1976)

Tan, S. L., Morrison, W. R.: The distribution of lipids in the germ, endosperm, pericarp and tip cap of amylomaize, LG-11 hybrid maize and waxy maize. J. Am. Oil Chem. Soc. *56,* 531 (1979)

Thompson, R. D., Bartels, D., Harberd, N. P., Flavell, R. B.: Characterization of the multiogen family coding for HMW glutenin subunits in wheat using cDNA clones. Theor. Appl. Genet. *67,* 87 (1983)

Walsh, D. E., Youngs, V. L., Gilles, K. A.: Inhibition of durum wheat lipoxygenase with L-ascorbic acid. Cereal Chem. *47,* 119 (1970)

Wassermann, L.: Getreide und Getreideprodukte. In: Ullmanns Encyklopädie der technischen Chemie, 4. Aufl., Bd. 12, S. 253, Verlag Chemie: Weinheim. 1976

Wassermann, L., Dörffner, H. H.: Der Einfluß des Wasser-Mehl-Verhältnisses in Brotteigen auf die Zusammensetzung und Eigenschaften der daraus hergestellten Brote. Brot Gebäck *25,* 148 (1971)

Webster, F. H. (Ed.): Oats: Chemistry and Technology. American Association of Cereal Chemists: St. Paul, Minn. 1986

Wieser, H., Seilmeier, W., Belitz, H.-D.: Vergleichende Untersuchungen über partielle Aminosäuresequenzen von Prolaminen und Glutelinen verschiedener Getreidearten. Z. Lebensm. Unters. Forsch. *170,* 17 (1980)

Wieser, H., Stempfl, Ch., Belitz, H.-D.: Peptidmuster von Glutelinfraktionen verschiedener Weizensorten. Z. Lebensm. Unters. Forsch. 181, 1 (1985)

Wood, P. J., Siddiqui, I. R., Paton, D.: Extraction of high-viscosity gums from oats. Cereal Chem. *55,* 1038 (1978)

Youngs, V. L., Peterson, D. M., Brown, C. M.: Oats. In: Advances in cereal science and technology (Ed.: Pomeranz, Y.), Vol. V, p. 49, American Association of Cereal Chemists: St. Paul, Minn. 1982

16 Legumes

16.1 Foreword

Ripe seeds of the plant family *Fabaceae*, known commonly as "legumes" or "pulses", are an important source of proteins for much of the world's population*. The extent of the production of major legumes is illustrated in Table 16.1. Legumes contain relatively high amounts of protein (Table 16.2). Hence, they are an indispensable supply of protein for the "third world". Soybeans and peanuts are oil seeds (cf. 4.3.2.2.5) and, even in industrialized countries, are used as an important source of raw proteins.
With regard to biological value, legume proteins are somewhat deficient in the S-containing amino acids (Tables 16.3 and 1.8).
Some legumes contain toxic substances (e.g. cyanogenic glycosides and nonprotein amino acids) and antinutritive factors (e.g. proteinase inhibitors, lectins) which, when necessary, are destroyed by suitable procedures, for example, heating.

16.2 Individual Constituents

16.2.1 Storage Proteins

About 80% of the proteins from soybean can be extracted at pH 6.8. A large number of these proteins can be precipitated by acidification at pH 4.5. This pH-dependent solubility is used in large-scale preparations of soy proteins.
Fractionation of legume proteins using solubility procedures, as applied to cereals by *T.B. Osborne*, yields three fractions: albumins, globulins, and glutelin, with globulins being predominant (Table 16.4).
The high content of globulins in seeds indicates that their function is mostly as storage proteins which are mobilized during the course of germination.

* Semi-ripe peas and beans are considered as vegetables (cf. Chapter 17).

The globulin fraction can be separated by ultracentrifugation or chromatography into two major components present in all the legumes: *vicilin* (~7S) and *legumin* (~11S). Legumin from soybeans is called glycinin and from peanuts is called arachin. Molecular weights and sedimentation coefficients for the 7S and 11S globulins isolated from various legumes are presented in Table 16.5.
The 11S globulin obtained from soybeans has been studied in detail. It contains six subunits per molecule (Figure 16.1). Each subunit consists of one basic and one acidic protein moiety, which are linked together by one or more disulfide bonds. The acidic protein is larger (pI ~ 5; MW ~ 37 kdal) than the basic one (pI ~ 8.2; MW ~ 22 kdal). Hydrophobic bonds are involved in the stabilization of the quaternary structure of the 11S globulin.
When the ionic strength at pH 7.6 is decreased from $\mu = 0.5$ to $\mu < 0.1$, the 11S globulin dissociates stepwise into subunits:

$$11\ S(6\ AB) \rightarrow 7.5\ S(3\ AB) \rightarrow 3\ S(AB) \qquad (16.1\,a)$$

Complete dissociation occurs when the disulfide

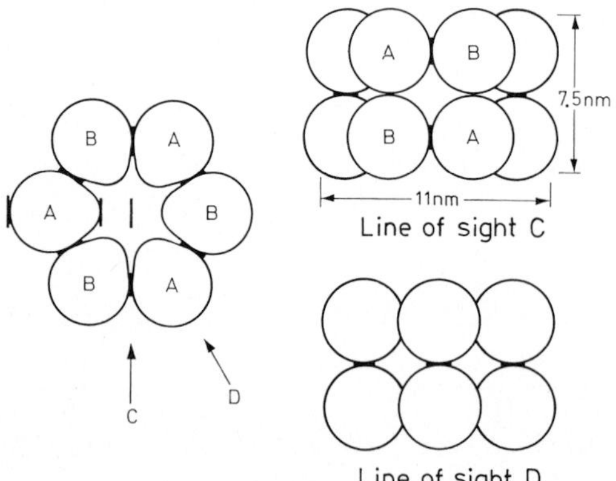

Fig. 16.1. 11 S-Globulin from soybean: model of the quaternary structure. (After *R.A. Badley* et al., 1975.) Acidic (A) and basic (B) proteins

Table 16.1. World production of seed legumes, 1979 (1,000 t)

Continent	Legumes total[a]	Beans[b]	Broad beans	Peas	Chick peas	Lentils	Soy-beans	Peanuts[c]
World	42,403	14,053	4,178	8,215	6,292	1,128	87,941	19,368
Africa	5,076	1,353	716	327	185	446	319	5,201
America, North-, Central-	4,109	3,328	97	251	260	94	56,605	1,980
America, South-	3,294	2,984	115	111	24	39	19,575	700
Asia	22,241	5,672	2,791	2,764	5,748	889	10,320	11,408
Europe, West-	1,395	294	361	268	71	51	41	22
Europe, East- + USSR	6,033	419	82	4,360	3	9	1,011	8
Australia and South Pacific Islands	255	2	15	136			70	48

Country	Legumes, grand total	Country	Beans	Country	Broad beans	Country	Peas
India	11,207	India	2,700	China	2,700	USSR	4,000
China	6,856	Brazil	2,339	Ethiopia	277	China	2,300
USSR	5,027	China	1,856	Egypt	262	India	400
Brazil	2.401	Mexico	1,469	Italy	206	USA	144
Mexico	1.820	USA	1,443	Czechoslovakia	71	Ethiopia	130
USA	1,673	Thailand	275	France	65	Hungary	125
Nigeria	903	Uganda	240	Morocco	65	France	120
Turkey	858	Argentina	221	Brazil	62	Canada	97
Pakistan	731	Burma	195	Tunisia	54	UK	88
Ethiopia	632	Ruanda	178	Turkey	53	Czechoslovakia	77
Σ (%)[d]	76	Σ (%)[d]	78	Σ (%)[d]	91	Σ (%)[d]	91

Country	Chick peas	Country	Lentils	Country	Soy-beans	Country	Peanuts
India	4,652	India	440	USA	55,260	India	6,000
Pakistan	532	Turkey	220	Brazil	14,978	China	3,513
Turkey	280	Syria	94	China	8,016	USA	1,791
Mexico	260	USA	86	Argentina	3,770	Senegal	900
Burma	98	Bangladesh	49	Mexico	712	Indonesia	855
Ethiopia	82	Pakistan	37	Indonesia	653	Sudan	800
Syria	53	Iran	28	Canada	631	Nigeria	580
Iran	43	Ethiopia	27	Paraguay	600	Burma	476
Bangladesh	38	Spain	23	India	500	South Africa	374
Nepal	34	France	22	USSR	500	Brazil	355
Spain	34	Σ (%)[d]	91	Σ (%)[d]	97	Σ (%)[d]	81
Σ (%)[d]	97						

[a] Without soybeans and peanuts.
[b] Without broad beans.
[c] With hull included.
[d] World production ≙ 100%.

Table 16.2. Chemical composition of legumes[a]

Name	Systematic name	Crude protein (%)	Lipid (%)	Carbo-hydrate (%)	Crude fiber (%)	Minerals (%)
Soybeans	*Glycine max*	39.0	19.6	35.5	4.7	5.5
Peanuts	*Arachis hypogaea*	24.8	47.9	24.6	3.1	2.7
Peas	*Pisum sativum*	25.7	1.6	68.6	1.6	3.0
Garden beans	*Phaseolus vulgaris*	24.1	1.8	65.2	4.5	4.4
Runner beans	*Phaseolus coccineus*	23.1	2.1	70.7	5.5	3.9
Black gram	*Phaseolus mungo*	26.9	1.6	66.9	1.0	3.6
Green gram (mungo beans)	*Phaseolus aureus*	27.2	1.3	66.6	0.9	3.8
Lima beans	*Phaseolus lunatus*	25.0	1.6	70.3	4.9	3.9
Chick peas	*Cicer arietinum*	22.7	5.0	66.3	3.0	3.0
Broad beans	*Vicia faba*	26.7	2.3	64.0	7.2	3.6
Lentils	*Lens culinaris*	28.6	0.8	67.3	0.8	2.4

[a] The results are average values given as weight-%/dry matter.

Table 16.3. Essential amino acids in legumes (g/16 g N)

Amino acid	Soybean	Broad bean
Cystine	1.3	0.8
Methionine	1.3	0.7
Lysine	6.4	6.5
Isoleucine	4.5	4.0
Leucine	7.8	7.1
Phenylalanine	4.9	4.3
Tyrosine	3.1	3.2
Threonine	3.9	3.4
Tryptophan	1.3	n. a.
Valine	4.8	4.4

n. a.: not analyzed.

Table 16.4. Legumes: protein distribution (%) by *Osborne* fractions

Fraction	Soy-beans	Peanuts	Peas	Mungo beans	Broad beans
Albumin	10	15	21	4	20
Globulin	90	70	66	67	60
Glutelin	0	10	12	29	15

Table 16.5. Molecular weight and sedimentation coefficient of the 7S and 11S globulins from legumes

Legume	7S globulin		11S globulin	
	Sedimentation coefficient	Mol. weight (kdal)	Sedimentation coefficient	Mol. weight (kdal)
Soybeans	7.9 ($S_{20,w}$)	193	12.3 ($S_{20,w}$)	320
Peanuts	8.7 (S_{20})	190	13.2 ($S_{20,w}$)	340
Peas	8.1 (S_{20})		13.1 (S_{20})	398
Garden beans	7.6 ($S_{20,w}$)	140	11.6 ($S_{20,w}$)	340
Broad beans	7.1 ($S_{20,w}$)	150	11.4 ($S_{20,w}$)	328

bonds are reduced in the presence of protein-unfolding agents, such as urea or SDS:

$$(AB) \rightarrow A + B \qquad (16.1\,b)$$

A and B represent acidic and basic proteins, respectively.

The 7S globulin has similar properties, as illustrated in Figure 16.2 for soybean vicilin. Hence, its molecular weight is also strongly dependent on pH and ionic strength.

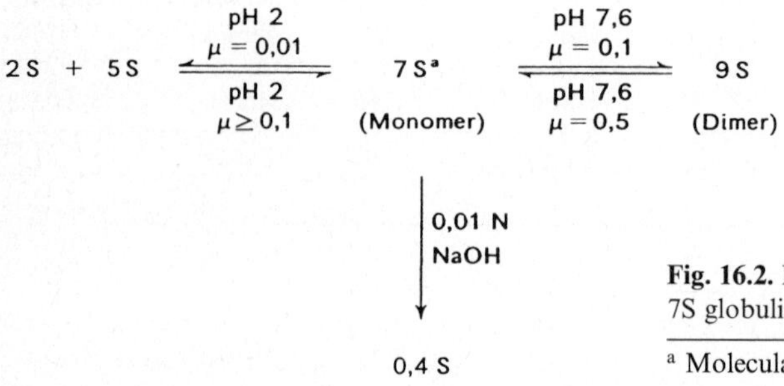

$$2S + 5S \underset{\text{pH 2},\ \mu \geq 0,1}{\overset{\text{pH 2},\ \mu = 0,01}{\rightleftharpoons}} \underset{\text{(Monomer)}}{7S^{a}} \underset{\text{pH 7,6},\ \mu = 0,5}{\overset{\text{pH 7,6},\ \mu = 0,1}{\rightleftharpoons}} \underset{\text{(Dimer)}}{9S}$$

$$7S^{a} \xrightarrow{\text{0,01 N NaOH}} 0,4\ S$$

Fig. 16.2. Dissociation and aggregation of the soybean 7S globulin

[a] Molecular weight: 193 kdal

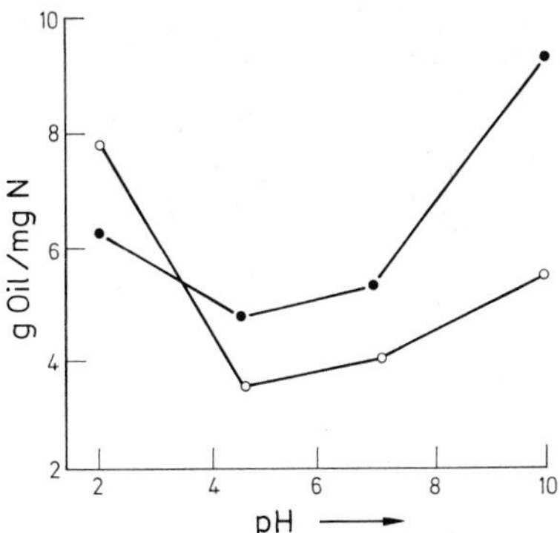

Fig. 16.3. Soybean globulin as an emulsifier. (After *H. Aoki* et al., 1980.) The capacity of an o/w-emulsion after addition of 11S globulin (-○-) and 7S globulin (-●-) is plotted versus pH

The amino acid compositions of both major soybean proteins, with the exception of methionine, are very similar (Table 16.6). However, large differences exist in their carbohydrate contents. The 7S globulin contains 5% carbohydrate and the 11S globulin less than 1% carbohydrate. In the pH range of 4–10, the 7S globulin is a better emulsifier than the 11S globulin, when the capacity (Figure 16.3) and the stability of an o/w emulsion are compared.

16.2.2 Enzymes

Various forms of lipoxygenase (cf. 3.7.2.2) are of interest in food chemistry since they strongly affect the legume aroma.
Urease, which hydrolyses, urea,

$$NH_2-CO-NH_2 + H_2O \rightarrow CO_2 + 2NH_3 \quad (16.2)$$

occurs in soybeans in relatively high concentration. Heat treatments of soy preparations can be detected by measuring the activity of this enzyme.

16.2.3 Proteinase Inhibitors

Proteinase inhibitors are found in animal and plant tissues. They are particularly widespread in legumes and also occur in cereals and potatoes. They are proteins with molecular weights ranging from 6–46 kdal. These inhibitors combine with proteinases to yield an inactive complex which has a low dissociation constant. Also, covalent bonding occurs to a certain extent between the enzyme and its inhibitor.
Often, several different inhibitors are found in plant materials. They differ in their isoelectric points and also in their specific activities and thermal stabilities. For example, in the more than 30 legumes analyzed so far, nine inhibitors have been identified and five partially purified.

Table 16.6. Amino acid composition of 7S and 11S globulins from soybeans

Amino acid	g amino acid/100 g protein	
	7S globulin	11S globulin
Asx	11.18	13.10
Thr	3.14	3.37
Ser	4.79	4.16
Glx	17.54	18.03
Pro	5.21	5.40
Gly	3.37	3.97
Ala	3.66	3.55
Cys	1.52	1.44
Val	4.68	5.05
Met	0.43	1.84
Ile	4.99	4.69
Leu	8.15	7.17
Tyr	3.51	4.05
Phe	5.55	5.73
His	2.32	2.22
Lys	6.26	4.88
Arg	7.37	7.75

Table 16.7. Thermal stability of proteinase inhibitors of pea and bean seeds[a]

Seed	Heat treatment						
		Duration (min)					
	Temperature (°C)	10		30		60	
		Residual activity[b] (%)					
		T	CT	T	CT	T	CT
Pisum	85					78	24
sativum	95	47	4			0	0
Phaseolus	85			90	94	89	94
vulgaris	95	86		58	66	34	52
Phaseolus	85			44	79	32	19
coccineus	95	31	5	12	0		
Vicia faba	95			70	26	30	18

[a] Ground seed aqueous extract.
[b] Activity of the extract after heat treatment against trypsin (T) and chymotrypsin (CT), respectively; control = 100%.

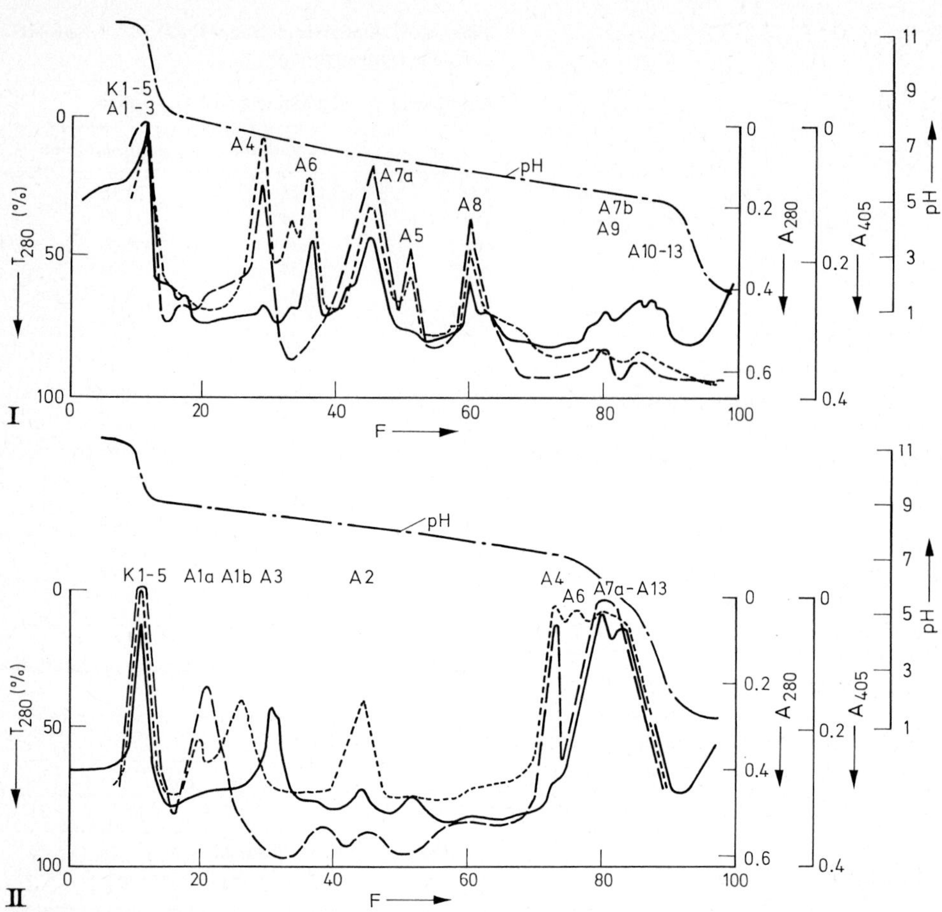

Fig. 16.4. Isoelectric focusing of proteins from potato tubers. (After *H.-D. Belitz* et al., 1971.) **I**: pH range 5–8, **II**: pH range 7–10, F: 1 ml-fractions: —— protein, transmittance scanning at 280 nm; ---- trypsin inhibition activity: inhibition of 20 µg bovine trypsin by 0.025 ml aliquots, substrate N$^{\alpha}$-benzoyl-L-arginine-p-nitroanilide (BAPA). Absorbance at 405 nm; — — — inhibition of chymotrypsin activity: inhibition of 12 µg bovine chymotrypsin by 0.025 ml aliquots, substrate casein. Absorbance 280 nm; –·–·– pH gradient

Figure 16.4 provides an example of raw protein separation from potato sap by isoelectric focusing. More than ten inhibitors for bovine trypsin and chymotrypsin have been found. The thermal stability of proteinase inhibitors depends on molecular weight and on the extent of stabilization of the active conformation by disulfide bonds. An example of a highly thermostable inhibitor is the *Bowman-Birk* inhibitor from soybean. It inhibits trypsin and chymotrypsin, has a molecular weight of 7.9 kdal, and has seven disulfide bonds (cf. 1.4.2.3.2 and Figure 1.19).

A second inhibitor from soybean, isolated by *Kunitz,* is substantially less thermostable. Table 16.7 shows data for the thermal stability of several inhibitors from several cultivars of peas and beans.

The specificity of proteinase inhibitors varies. In addition to inhibitors for trypsin, chymotrypsin and other proteinases, there are proteins which possess reaction centers for two enzymes, e.g. for trypsin and chymotrypsin, and as such are designated as "double-headed" inhibitors. The *Bowman-Birk* inhibitor from soybean (cf. Figure 1.19) is an example from this group. The aforementioned inhibitors from potatoes also inhibit serine proteinases, but differ individually in their specificity (Table 16.8). It is also known that inhibitors for the same enzyme but from various sources, respond differently. For example, hen ovomucoid inhibits bovine trypsin but not human trypsin. Figure 16.5 provides some additional examples of differing inhibitions of animal and human enzymes by proteinase inhibitors.

Food which contains inhibitors might cause nutritional problems. For example, feeding rats and chickens with raw soymeal leads to reversible pancreatic blistering.

A consequence of excessive secretion of pancreatic juice is increased secretion of nitrogen in the feces. Furthermore, growth inhibition occurs which can be eliminated by incorporating methionine, threonine and valine into the diet. These findings indicate that the poor growth rate might be due to some amino acid deficiencies, which are a result of increased N-excretion. All the possible effects of proteinase inhibitors are not fully understood.

Table 16.8. Inhibition of serine proteinases by proteinase inhibitors from potatoes[a]

Enzyme	Inhibitor[c]									
	K-group	A1a	A1b	A2	A4	A5	A6	A7a	A7b	A8
Trypsin[b]	+	+	+	+	+	+	+	+	+	+
α-Chymotrypsin[b]	+	+	0	0	+	+	0	+	+	+
B. subtilis[d]	+	−	0	0	+	+	0	0	+	0
Asp. oryzae[d]	+	−	0	0	+	+	0	0	+	0
Pronase E	+	−	0	0	+	+	0	0	+	0
Proteinase K	+	−	0	0	+	+	0	0	+	0

[a] cv. Maritta. [b] Bovine. [c] +: inhibition, 0: no inhibition, −: not analyzed.
[d] Proteinase of the given microorganism.

In food and feed processing or preparation, proteinase inhibitors are mostly or completely inactivated by heat. The extent of inactivation is dependent on the stability of the inhibitors (cf.

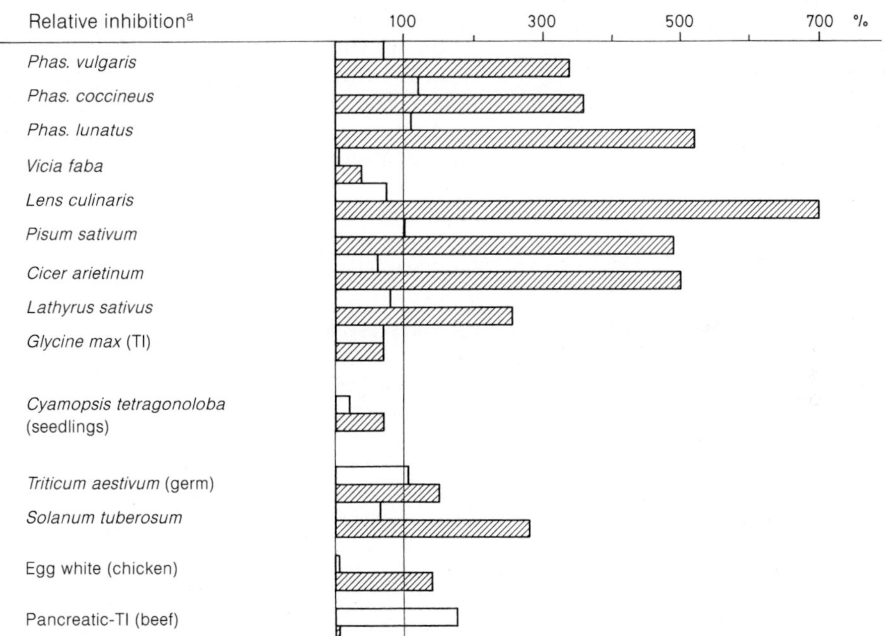

Fig. 16.5. The extent of inhibition of enzymes present in human intestine exudates[b] as compared to those of bovine trypsin and α-chymotrypsin

[a] Related to

Inhibition of bovine trypsin = 100%

Inhibition of bovine α-chymotrypsin = 100%

[b] Substrate: N^{α}-benzoyl-L-arginine-p-nitroanilide (BAPA) for trypsin and N-succinyl-L-phenylalanine-p-nitroanilide (Suphepa) for chymotrypsin

Table 16.9) and/or on processing parameters, e.g. temperature and moisture content of the treated food. Table 16.9 shows that, in texturized soy protein products, an appreciable residual activity of the inhibitors is still present.

16.2.4 Hemagglutinins (Lectins)

Some proteins or glycoproteins which are able to attach themselves to erythrocytes and cause their precipitation or agglutination are widely distributed in foods of plant origin. These substances are therefore designated as hemagglutinins or by the selectivity in their reactions with red cells of human blood groups as lectins (Latin *legere* = to choose, select, pick out). In erythrocytes, certain glycan sites function as receptors for lectins. Accordingly, any biopolymer will be precipitated which possesses a glycosidic residue recognizable by lectins (cf. data for specificity in Table 16.10).

Table 16.9. Inhibition of bovine trypsin activity by some soya products[a]

Product	Extracted with	
	0.25 N H_2SO_4	0.01 N NaOH
Untreated soybean (cv. Caloria)	51.5	33.7
Supro G 10	6.8	15.6
Soyflour	1.1	8.7
TVP U 110 chunks	0	4.1
Flocosoya	0	1.9

[a] A 50% inhibition of mg trypsin/g product; substrate: N^{α}-benzoyl-L-arginine-p-nitroanilide.

Most lectins are glycoproteins. When their molecular weight exceeds 30 kdal, they consist of several subunits (Table 16.10), which readily dissociate by a change in pH or ionic strength. A characteristic feature of their amino acid composition is the high content of acidic and hydroxy amino acids and the absence or low content of methionine.

Among the well-characterized lectins is concanavalin A, a protein isolated from jack beans (*Canavalia ensiformis*). This lectin interacts preferentially with polysaccharides containing D-mannose, but is able to bind to biopolymers consisting exclusively of α-D-glucopyranose, such as dextrins.

Lectins occur mostly in legume seeds. Animal tests have demonstrated that their toxicity often does not parallel hemagglutination activity. Thus, both lectins from soybeans and garden beans are toxic, but not that from peas. These and other observations suggest that it is not the hemagglutination activity but other activities of lectins which are responsible for their toxicity. Most observations point to the binding of lectins to epithelial cells on the intestinal wall, causing a deleterious nutritional effect by interfering with nutrient absorption, while some other lectins act as inhibitors of protein biosynthesis.

After prolonged cooking or dry heating, the activities of legume lectins and the associated toxic effects are destroyed.

16.2.5 Carbohydrates

The carbohydrates which are present in legumes are listed in Table 16.11. The major carbohydrate

Table 16.10. Occurrence of phytohemagglutinins (lectins) in food

Source	Molecular weight (kdal)	Subunits	Glycan-component		Specificity[a]
			% carbohydrate	Building blocks	
Soybean	110	4	5.0	D-Man, D-GlcNAc	D-GalNac
Garden beans	98–138	4	4.1	GlcN, Man	D-GalNAc
Jack beans[b]	112	4	0		α-D-Man
Lentils	42–69	2	2.0	GlcN, Glc	α-D-Man
Potato	20		5.2	Ara	D-GlcNAc
Wheat	26		4.5	Glc, Xyl, Hexosamine	D-GlcNAc

[a] Biopolymers precipitation occurs that contain the given building blocks (polysaccharides, glycoproteins, lipopolysaccharides).
[b] *Canavalia ensiformis*.

is starch, amounting to 75–80%. In soybeans, the presence of starch has not been proven. Instead, the presence of arabinoxylane and galactane (3.6 and 2.3%, respectively) has been confirmed. In peanuts, about one-third of the total carbohydrate is starch.

Oligosaccharides in legumes are present in higher concentration than in cereals. Predominant in this fraction are sucrose, stachyose and verbascose (Table 16.11).

After legume consumption, oligosaccharides might cause flatulency, a symptom of gas accumulation in the stomach or intestines. It is a result of the growth of anaerobic microorganisms in the intestines, which hydrolyze the oligo- into monosaccharides and cause their further degradation to CO_2, CH_4 and H_2. Model feeding tests have demonstrated that phenolic ingredients, such as ferulic and syringic acids, inhibit microorganism metabolism and the related flatulency.

Table 16.11. Carbohydrates in legume flours[a]

Flour	Glucose	Saccharose	Raffinose	Stachyose	Verbascose	Starch
Garden beans	0.04	2.23	0.41	2.59	0.13	51.6
Broad beans	0.34	1.55	0.24	0.80	1.94	52.7
Lentils	0.07	1.81	0.39	1.85	1.20	52.3
Green gram (mungo beans)	0.05	1.28	0.32	1.65	2.77	52.0
Soybean[b]	0.01	4.5	1.1	3.7		

[a] Weight-% of the dry matter.
[b] Defatted flour

16.2.6 Cyanogenic Glycosides

Cyanogenic glycosides (Table 16.12) are present in lima beans and in some other plant foods. Precursors of cyanogenic glycosides are the amino acids listed in Table 16.12. As in the biosynthesis of glucosinolates (cf. 17.1.2.6.5), an aldoxime is initially formed, which is then transformed into a cyanogenic glycoside by means of the postulated reaction pathway shown in Figure 16.6.

Seeds are ground and moistened in order to detoxify them. This initiates glycoside degradation with formation of HCN (cf. Table 16.13) which, after incubation, is expelled by heating.

The cyanogenic glycoside degradation is initiated by β-glucosidase (Figure 16.7) which in the cells is separated from its substrate. Once the cell structure is ruptured by seed grinding, the enzyme and the substrate are brought together and the reaction starts.

The substrate specificity of β-glucosidase is governed by an aglycon moiety. Thus, the enzymes present in "emulsin", a glycosidase mixture from bitter almonds, hydrolyze not only amygdalin but also other cyanogenic glycosides which are derived from phenylalanine or tyrosine, but not linamarin.

Table 16.12. Cyanogenic glycosides in fruit and some field crops

```
R2  O-β-sugar
  \ /
   C
  / \
R1  CN
```

Glycoside Name	Structure R_1	Structure R_2	Sugar	Amino acid precursor	Occurrence (seeds)
Linamarin	CH_3	CH_3	Glucose	Val	Lima bean Linseed (flax) Cassava
(R)-Lotaustralin	C_2H_5	CH_3	Glucose	Ile	(as Linamarin)
(R)-Prunasin	Phenyl	H	Glucose	Phe	Prunes
(R)-Amygdalin	Phenyl	H	Gentiobiose	Phe	Bitter almond Apricots Peaches Apple
(S)-Dhurrin	HO–Phenyl	H	Glucose	Tyr	Sorghum sp.

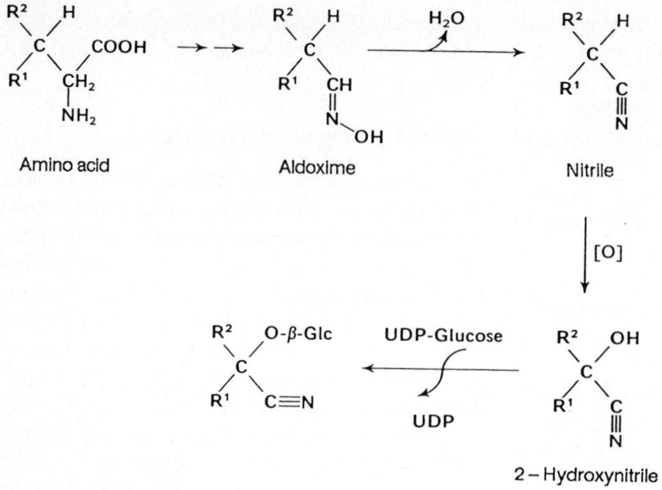

Amino acid → → Aldoxime → (H_2O) Nitrile → [O] 2-Hydroxynitrile → (UDP-Glucose, UDP) cyanogenic glucoside

Fig. 16.6. Biosynthesis of cyanogenic glucosides

As shown in Figure 16.7, β-glucosidase hydrolysis produces an unstable hydroxynitrile which slowly degrades into the corresponding carbonyl compound and HCN. However, most legume seeds contain an hydroxynitrile lyase which accelerates this reaction.

16.2.7 Lipids*

With exception of soybeans and peanuts the lipid content of most legumes is so low (cf. Table 16.1) that they can not be considered as a source of fats or oils. Examples of their fatty acid composition are listed in Table 16.14.

16.2.8 Vitamins and Minerals

Vitamin and mineral content of some legumes is presented in Table 16.15. In addition to B-vitamins, the two oilseeds are rich in vitamin E (cf. Table 16.15).

16.2.9 Coumestrol

Among the phenolic constituents of legumes, coumestrol is of interest because of its estrogenic effect:

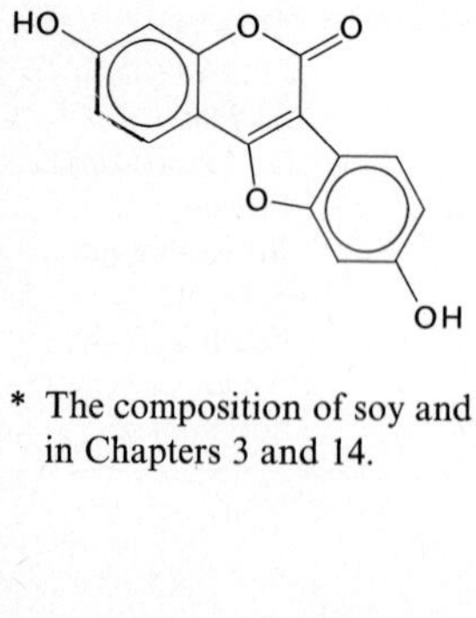

(16.3)

* The composition of soy and peanut lipids is covered in Chapters 3 and 14.

Table 16.13. Amount of glycoside-bound hydrocyanic acid in food

Food	HCN (mg/100 g)
Lima bean[a]	210–310
Bitter almond	250
Sorghum sp.	250
Cassava	110
Pea	2.3
Bean	2.0
Chick pea	0.8

[a] In the United States new cultivars have been developed that contain only 10 mg HCN/100 g seed.

Table 16.14. Fatty acid composition of legume lipids (weight-%)[a]

Fatty acid	Garden beans	Chick peas	Broad beans	Lentils
14:0	0.22	1.3	0.6	0.85
16:0	21.8	8.9	9.3	23.2
18:0	4.7	1.6	4.9	4.6
20:0	0.53	0.03	0.7	2.3
22:0	2.9	0	0.42	2.7
24:0	1.1	0	0	0.85
16:1 (9)	0.21	0.05	0	0.15
18:1 (9)	11.6	35.4	33.8	36.0
18:2 (9, 12)	29.8	51.1	42.1	20.6
18:3 (9, 12, 15)	27.4	1.7	6.4	1.6
20:1	0.02	0	0.7	1.9

[a] In Table 14.9 the fatty acid compositions are provided for soya oil and peanut butter.

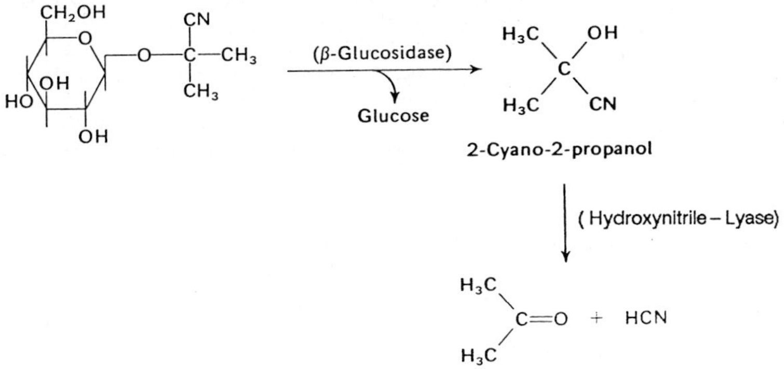

Fig. 16.7. Lima beans: linamarin degradation, resulting in a release of hydrocyanic acid

It occurs mostly in soybean hulls at levels of 0.05–30 ppm.

16.2.10 Saponins

Legumes contain a number of saponins. Saponin has been isolated and identified from soy and other beans (cf. Formula 16.4). The suggestion that these compounds impart a bitter taste and might develop during storage of soy products has been proven wrong. On the other hand, the insecticidal and or fungicidal effect of saponins is still of practical interest.

16.3 Processing

16.3.1 Soybeans and Peanuts

Preparation and storage of products from both oilseeds is often inhibited by rancidity and bitter aroma defects caused mostly by volatile carbonyl compounds, particularly hexanal derived from oxidized fatty acids. The rancidity-causing compounds are formed through lipid peroxidation, accelerated by the enzyme lipoxygenase and/or by hem(in) proteins (cf. 3.7.2.2). One way to increase quality is to thermally inactivate enzymes or hem(in) catalysts. Table 16.16 illustrates steam heating of peanuts for a prolonged time in order to inactivate peroxidase activity. Lipoxygenase denaturation, under the conditions given in Table 16.16, occurs after 2 min, but this alone does not yield a satisfactory storage stability. Peroxidase and probably other catalysts should be excluded as well (Figure 16.8).

Table 16.15. Vitamin and mineral composition of legumes[a]

	Soybeans	Peas	Garden beans	Peanuts
Vitamins				
E	127	12		202
B_1	8.2	1.2	4.5	9.0
B_2	4.3	0.64	1.6	1.5
Nicotinamide	20.8	9.5	20.8	153
Pantothenic acid	15.9	2.9	9.7	26
B_6	9.9	0.64	2.8	3.0
Minerals				
Na	33	8.0	20	52
K	1.4×10^4	1.2×10^3	1.3×10^4	7.1×10^3
Mg	2.1×10^3	132	1.3×10^3	1.6×10^3
Ca	$2{,}1 \times 10^3$	96	1.1×10^3	590
Fe	71	7.4	60.4	21.1
Zn	8.3	10.6		30.7
P	4.9×10^3	432	4.3×10^3	3.7×10^3
Cl	58	160	248	70

[a] Results are given in mg/kg.

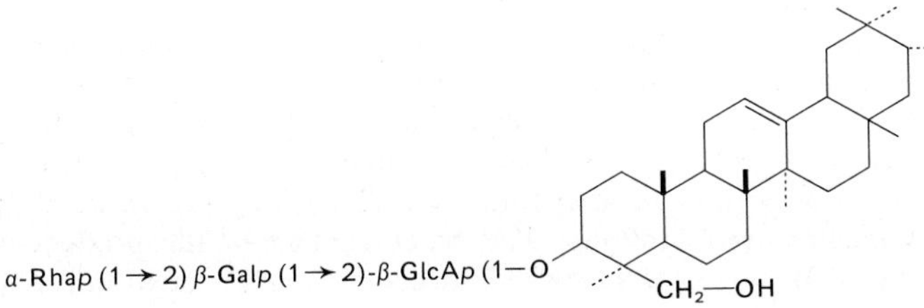

(16.4)

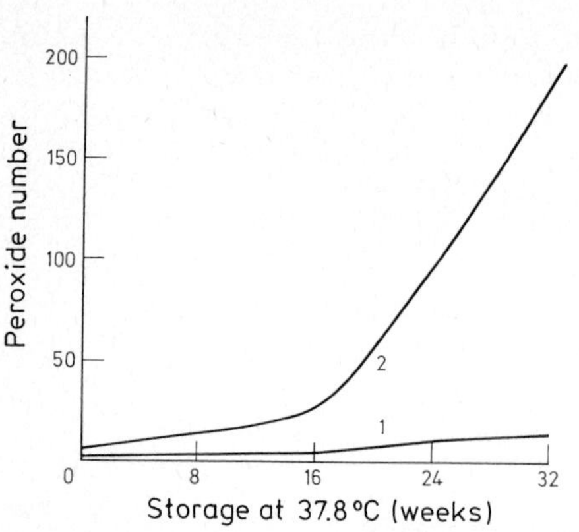

Fig. 16.8. Storage stability of peanut flakes. (After *J. H. Mitchel* and *R. K. Malphrus*, 1977.) Peanut flakes treated with steam at 100 °C for 30 min (1) and 5 min (2)

The complete removal of lipids is used as an additional precautionary measure in order to obtain an off-flavor-free product, particularly in the case of production of protein isolates. For example, the lipid residue which remains in soy flakes after hexane solvent extraction (cf. 14.3.2.2.1) is removed by extraction with hexane-ethanol 82:18 v/v.

16.3.1.1 Individual Products

Protein preparations and milk-like products are processed from soybeans and peanuts. The following products are made from soybeans.

16.3.1.1.1 Soy proteins

Soy protein concentrate is usually obtained from the flaked and defatted soy meal that is left after oil extraction (cf. 14.3.2.2.1). The process involves soaking of flakes in water, acidification of the aqueous extract to pH 4–5 (cf. 16.2.1) and separation of the precipitate from solubilized ingredients by centrifugation followed by washing and drying of the sediment collected.
Soy meal isolates enriched in protein are obtained by a preliminary extraction of soluble soy constituents with water or diluted alkali, pH 8–9, followed by protein precipitation from the aqueous extract by adjusting the pH to 4–5. Such protein isolates, texturized and flavored (cf. 1.4.7), are used as meat substitutes.
The compositions of protein concentrates and isolates are compared in Table 16.17. For both products, the essential amino acid content corresponds to that of soybeans (cf. Table 16.3).

Table 16.16. Thermal inactivation of lipoxygenase and peroxidase in peanuts

Heat treatment			*Enzyme activity* (%)	
type	°C	time (min)	per-oxidase	lipoxy-genase
Control			100	100
Dry heat	110	60	48	7
Steam	100	2	35	0
Steam	100	6	8	0
Steam	100	30	1	0

Table 16.17. Composition of soya protein concentrate and isolate (%)

Product	Protein	Crude fiber	Ash
Concentrate	72	3.5	5.5
Isolate	95.6	0.2	4.0

Soy protein is added as an ingredient to baked and meat products and to baby food preparations to raise their protein level and to improve their processing qualities, such as increased water binding capacity or stabilization of o/w emulsions. These properties are required for processing at higher temperatures. The addition of soy protein to beverages at a pH of 3 results in better solubility of beverage constituents. Soy protein market value may be increased by its partial hydrolysis with papain (cf. 2.8.2.2.1).

16.3.1.1.2 Soy milk

Soybeans are swollen and ground in the presence of a 10-fold excess of water. Heating the suspension close to its boiling point for 15–20 min pasteurizes the suspension and inactivates lipoxygenase enzyme and proteinase inhibitors. A soy milk preparation enriched with calcium and vitamins is of importance in infant nutrition as a replacement for cow's milk, which close to 7% of infants in the USA are unable to tolerate.

16.3.1.1.3 Tofu

When calcium sulfate (3 g salt/kg milk) is added to soy milk at 65 °C, a gel (called soy "curd") slowly precipitates. The curd is separated from excess fluid by gentle squeezing in a special wooden filter box. A washing procedure then follows. The water content of the product is about 88%. Tofu contains 55% protein and 28%

fat dry weight. In China and some other Asian countries, tofu is the largest source of food protein. It is consumed fresh or dried, or fried in fat and seasoned with soy sauce.
Tofu inoculated with *Actinomucor elegans* provides, after a ripening period, a product called "sufu".

16.3.1.1.4 Soy sauce (Shoyu)

Defatted soy meal is used as a starting material in the production of this seasoning sauce. The meal is moistened, then mixed with roasted and crushed wheat and heated in an autoclave for 45 min. The mix ratio in Japan is fixed at 1:1, while in China it varies up to 4:1. Increasing the amount of soy decreases the quality of the end-product. The mix, with a water content of 26%, is then inoculated with *Aspergillus oryzae* and *Aspergillus soyae*. Initial incubation is at 30 °C for 24 h and then at 40 °C for an additional 48 h. This fermentation starter, called "koji", is then salted to 18% by addition of 22.6% NaCl solution. Inoculation with *Lactobacillus delbrueckii* and with *Hansenula* yeast species provides lactic acid fermentation, which proceeds under gentle aeration in order to prevent the growth of undesired anaerobic microorganisms. It is a long and tedious fermentation carried out in stepwise fashion: for example, starting at 15 °C for one month, followed by 28 °C for four months, and finishing at 15 °C for an additional month. Highly-valued products ripen for several years. After the fermentation is completed, the soy sauce of pH 4.6 is filtered, pasteurized at 65–80 °C and preserved with benzoic acid for the export market.
During fermentation the microorganisms produce extracellular hydrolases which decompose the main components of the raw material: proteins, carbohydrates and nucleic acids. Soy sauce contains 1.5% N (of which 60% corresponds to amino-N) and 4.4% reducing sugar. The N-containing fraction consists of 40–50% amino acids (glutamic acid predominates as 1.2% of the product), 40–50% peptides with 10–15% ammonia and less than 1% protein. In addition, soy sauce contains by-products of microorganism metabolism, such as ethanol (1.2%) and lactic, succinic and acetic acids.
Soy sauce products of lower quality are blended with spices and are prepared by acid hydrolysis of the above mix of raw materials (cf. 12.7.2.4.5).

16.3.2 Peas and Beans

Peas and beans are consumed only when cooked. In order to shorten the cooking time which, even after a preliminary soaking in water overnight (preliminary swelling), is several hours, the legumes are precooked or parboiled by the process described in 15.3.2.2.1.
Additionally, seed hull removal provides about a 40% reduction in cooking time which, for peas, involves seed steaming at 90 °C, followed by drying and subsequent dehulling.

16.4 Literature

Angelo, A. J. St., Ory, R. L.: Effects of lipoperoxides on proteins in raw and processed peanuts. J. Agric. Food Chem. *23*, 141 (1975)

Aoki, H., Taneyana, O., Inami, M.: Emulsifying properties of soy protein: characteristics of 7S and 11S proteins. J. Food Sci. *45*, 534 (1980)

Badley, R. A., Atkinson, D., Hauser, H., Oldani, D., Green, J. P., Stubbs, J. M.: The structure, physical and chemical properties of the soybean protein glycinin. Biochim. Biophys. Acta *412*, 214 (1975)

Belitz, H.-D.: Vegetable proteins as human food. FEBS 11th Meeting Copenhagen 1977, Vol. 44, Symposium A3, Pergamon Press: Oxford–New York. 1978

Belitz, H.-D., Kaiser, K.-P., Santarius, K.: Trypsin and chymotrypsin inhibitors from potatoes: isolation and some properties. Biochem. Biophys. Res. Commun. *42*, 420 (1971)

Boulter, D., Derbyshire, E.: The general properties, classification and distribution of plant proteins. In: Plant proteins (Ed.: Norton G.), p. 3, Butterworths: London. 1978

Derbyshire, E., Wright, D. J., Boulter, D.: Legumin and vicilin, storage proteins of legume seeds. Phytochemistry *15*, 3 (1976)

Liener, I. E.: Phytohemagglutinins: their nutritional significance. J. Agric. Food Chem. *22*, 17 (1974)

Liener, I. E.: Protease inhibitors and other toxic factors in seeds. In: Plant proteins (Ed.: Norton, G.), p. 117, Butterworths: London. 1978

Lookhart, G. L., Jones, B. L., Finney, K. F.: Determination of coumestrol in soybeans by high-performance liquid and thin-layer chromatography. Cereal Chem. *55*, 967 (1978)

Mitchell, J. H., Malphrus, R. K.: Lipid oxidation in spanish peanuts: the effect of moist heat treatments. J. Food Sci. *42*, 1457 (1977)

Naivikul, O., D'Appolonia, B. L.: Comparison of legume and wheat flour carbohydrates. I. Sugar analysis. Cereal Chem. *55*, 913 (1978)

Pernollet, J.-C., Mossé, J.: Structure and location of legume and cereal seed storage proteins. In: Seed proteins (Eds.: Daussant, J., Mossé, J., Vaughan, J.), p. 155, Academic Press: New York–London. 1983

Rackis, J. J., Sessa, D. J., Steggerda, F. R., Shimizu, T., Anderson, J., Pearl, S. L.: Soybean factors relating to gas production by intestinal bacteria. J. Food Sci. *35,* 634 (1970)

Smith, A. K., Circle, S. J. (Eds.): Soybeans: Chemistry and technology. Vol. 1, AVI Publ. Co.: Westport, Conn. 1972

Whitaker, J. R., Feeney, R. E.: Enzyme inhibitors in foods. In: Toxicants occurring naturally in foods, 2nd edn., p. 276, National Academy of Sciences: Washington, D. C. 1973

Wolf, W. J.: Chemistry and technology of soybeans. In: Advances in cereal science and technology (Ed.: Pomeranz , Y.), Vol. I, p. 325, American Association of Cereal Chemists: St. Paul, Minn. 1976

17 Vegetables and Their Products

17.1 Vegetables

17.1.1 Foreword

Vegetables are defined as the fresh parts of plants which, either raw, cooked, canned or processed in some other way, provide suitable human nutrition. Fruits of perennial trees are not considered to be vegetables. Dried seeds are also excluded (peas, beans, cereal grains, etc.). From a botanical point of view, vegetables can be divided into algae (seaweed), mushrooms, root vegetables (carrots), tubers (potatoes, yams), onions, and stem or stalk (kohlrabi, parsley), leafy (spinach), inflorescence (broccoli), seed (green peas) and fruit (tomato) vegetables. The most important vegetables, with data relating to their botanical classification and use, are presented in Table 17.1. Information about vegetable production follows in Table 17.2.

17.1.2 Composition

The composition of vegetables can vary significantly depending upon the cultivar and origin. Table 17.3 shows that the amount of dry matter in most vegetables is between 10 and 20%. The nitrogen content is in the range of 1–5%, carbohydrates 3–20%, fats 0.1–0.3%, crude fiber about 1%, and minerals close to 1%. Some tuber and seed vegetables have a high starch content and therefore a high dry matter content. Vitamins, minerals, flavor substances and dietary fibers are important secondary constituents.

17.1.2.1 Nitrogen Compounds

Vegetables contain an average of 1–3% nitrogen compounds. Of this, 35–80% is protein, the rest is amino acids, peptides and other compounds.

Table 17.1. List of some important vegetables

Number	Common name	Latin name	Class, order, family	Consumed as
Mushrooms (cultivated or wildly grown edible species)				
1	Ringed boletus	*Suillus luteus*	Basidiomycetes/Boletales	Steamed, fried, dried, pickled or salted
2	Saffron milk cap	*Lactarius deliciosus*	Basidiomycetes/Agaricales	
3	Field champignon	*Agaricus campester*	Basidiomycetes/Agaricales	
4	Garden champignon	*Agaricus hortensis*	Basidiomycetes/Agaricales	
5	Cep	*Xerocomus badius*	Basidiomycetes/Boletales	
6	Truffle	*Tuber melanosporum*	Ascomycetes/Tuberales	
7	Chanterelle	*Cantharellus cibarius*	Basidiomycetes/ Aphyllophorales	
8		*Xerocomus chrysenteron*	Basidiomycetes/Boletales	
9	Morel	*Morchella esculenta*	Ascomycetes/Pezizales	
10	Edible boletus	*Boletus edulis*	Basidiomycetes/Boletales	
11	Goat's lip	*Xerocomus subtomentosus*	Basidiomycetes/Boletales	
Algae (seaweed)				
12	Sea lettuce	*Ulva lactuca*		Eaten raw as a salad, cooked in soups (Chile, Scotland, West Indies)
13	Sweet tangle	*Laminaria saccharina*		Eaten raw or cooked (Scotland)
14		*Laminaria sp.*		Eaten dried ("combu") or as a vegetable (Japan)
15		*Porphyra laciniata*		Eaten raw in salads, cooked as a vegetable (England, America)
16		*Porphyra sp.*		Dried or cooked ("nari" product, Japan and Korea)
17		*Undaria pinnatifida*		Eaten dried ("wakami") and as a vegetable (Japan)

Table 17.1. (Continued)

Number	Common name	Latin name	Class, order, family	Consumed as
Rooty vegetables				
18	Carrot	*Daucus carota*	Apiaceae	Eaten raw or cooked
19	Radish (white elongated fleshy root)	*Raphanus sativus var. niger*	Brassicaceae	The pungent fleshy root eaten raw, salted
20	Viper's grass, scorzonera	*Scorzonera hispanica*	Asteraceae	Cooked as a vegetable
21	Parsley	*Petroselinum crispum ssp. tuberosum*	Apiaceae	Long tapered roots cooked as a vegetable, or used for seasoning
Tuberous vegetables (sprouting tubers)				
22	Arrowroot	*Tacca leontopetaloides*	Taccaceae	Cooked or milled into flour for breadmaking
23	White (Irish) potato	*Solanum tuberosum*	Solanaceae	Cooked, fried or deep fried in many forms, or unpeeled baked, also for starch and alcohol production
24	Celery tuber	*Apium graveolens, var. rapaceum*	Apiaceae	Cooked, as salad, and cooked and fried as a vegetable
25	Kohlrabi, turnip cabbage	*Brassica oleracea convar. acephala var. gongylodes*	Brassicaceae	Eaten raw or cooked as a vegetable
26	Rutabaga	*Brassica napus var. naprobrassica*	Brassicaceae	Cooked as a vegetable
27	Radish (reddish round root)	*Raphanus sativus var. sativus/var. niger*	Brassicaceae	The pungent fleshy root is eaten raw, usually salted
28	Red beet, beetroot	*Beta vulgaris ssp. vulgaris var. conditiva*	Chenopodiaceae	Cooked as a salad
Tuberous (rhizomatic) vegetables				
29	Sweet potatoes	*Ipomoea batatas*	Convolvulaceae	Cooked, fried or baked
30	Cassava (manioc)	*Manihot esculenta*	Euphorbiaceae	Cooked or roasted
31	Yam	*Dioscorea*	Dioscoreaceae	Cooked or roasted
Bulbous rooty vegetables				
32	Vegetable fennel	*Foeniculum vulgare var. azoricum*	Apiaceae	Eaten raw as salad, cooked as a vegetable
33	Onion	*Allium cepa*	Liliaceae	Eaten raw, fried as seasoning, cooked as a vegetable
34	Leek	*Allium porrum*	Liliaceae	The pungent succulent leaves and thick cylindrical stalk are cooked as a vegetable
Stem (shoot) vegetables				
35	Bamboo roots	*Bambusa vulgaris*	Poaceae	Cooked for salads
36	Asparagus	*Asparagus officinalis*	Liliaceae	Young shoots cooked as a vegetable or eaten as salad
Leafy (stalk) vegetables				
37	Celery	*Apium graveolens var. dulce*	Apiaceae	Leafy crispy stalks eaten raw as salad, or are cooked as vegetable
38	Rhubarb	*Rheum rhabarbarum, Rheum rhaponticum*	Polygonaceae	Large thick and succulent petioles are cooked as preserves of baked; used as a pie filling
Leafy vegetables				
39	Watercress	*Nasturtium officinale*	Brassicaceae	Moderately pungent leaves are eaten raw in salads or used as garnish
40	Endive (escarole, chicory)	*Cichorium intybus L. var. foliosum*	Cichoriaceae	Eaten raw as a salad, or is cooked as a vegetable
41	Chinese cabbage	*Brassica chinensis*	Brassicaceae	Eaten raw in salads, or is cooked as a vegetable
42	Lamb's salad (lettuce or corn salad)	*Valerianella locusta*	Valerianaceae	Eaten raw in salads
43	Garden cress	*Lepidium sativum*	Brassicaceae	Eaten raw in salads
44	Kale (borecole)	*Brassica oleracea convar. acephala var. sabellica*	Brassicaceae	Cooked as a vegetable
45	Head lettuce	*Lactuca capitata var. capitata*	Cichoriaceae	Juicy succulent leaves are eaten raw in salads

Table 17.1. (Continued)

Number	Common name	Latin name	Class, order, family	Consumed as
46	Mangold (mangel-wurzel, beet root)	*Beta vulgaris ssp. vulgaris var. vulgaris*	Chenopodiaceae	Cooked as a vegetable
47	Chinese (Peking) cabbage	*Brassica pekinensis*	Brassicaceae	Cooked as a vegetable
48	Brussels sprouts	*Brassica oleracea convar. oleracea var. gemmifera*	Brassicaceae	Cooked as a vegetable
49	Red cabbage	*Brassica oleracea convar. capitata var. capitata f. rubra*	Brassicaceae	Eaten raw in salads or is cooked as a vegetable
50	Romaine lettuce	*Lactuca capitata var. crispa*	Cichoriaceae	Eaten raw as a salad
51	Spinach	*Spinacia oleracea*	Chenopodiaceae	Cooked as a vegetable or is eaten raw as a salad
52	White (common) cabbage	*Brassica oleracea convar. capitata var. capitata f. alba*	Brassicaceae	Juicy succulent leaves are eaten raw in salads, or are fermented (sauerkraut), steamed or cooked as a vegetable
53	Winter endive	*Cichoricum endivia*	Cichoriaceae	Eaten raw as a salad
54	Savoy cabbage	*Brassica oleracea convar. capitata, var. sabauda*	Brassicaceae	Cooked as a vegetable
Flowerhead (calix) vegetables				
55	Artichoke	*Cynara scolymus*	Asteraceae	Flowerhead is cooked as a vegetable
56	Cauliflower	*Brassica oleracea convar. botrytis var. botrytis*	Brassicaceae	Cooked as a vegetable or used in salads (raw or pickled)
57	Broccoli	*Brassica oleracea convar. botrytis var. italica*	Brassicaceae	The tight green florets are cooked as a vegetable
Seed vegetables				
58	Chestnut	*Castanea sativa*	Fagaceae	Cooked as a vegetable, roasted, or milled into a flour and used in soups and bread doughs
59	Green beans	*Phaseolus vulgaris*	Fabaceae	The immature pod is cooked as a vegetable or is steamed or pickled for salads
60	Green peas	*Pisum sativum ssp. sativum*	Fabaceae	The rounded smooth or (wrinkled) green seeds are cooked as a vegetable or are steamed/cooked for salads
Fruity vegetables				
61	Eggplant	*Solanum melongena*	Solanaceae	Steamed as a vegetable
62	Garden squash	*Cucurbita pepo*	Cucurbitaceae	Cooked as a compote or as a vegetable
63	Green bell pepper	*Capsicum annuum*	Solanaceae	Eaten raw in salads, or is cooked, steamed or baked
64	Cucumber	*Cucumis sativus*	Cucurbitaceae	Eaten raw in salads, cooked as a vegetable or pickled
65	Okra	*Abelmoschus esculentus*	Malvaceae	Its mucilaginous green pods are cooked as a vegetable in soups or stewed, or eaten as a salad
66	Tomato	*Lycopersicon lycopersicum*	Solanaceae	The reddish pulpy berry is eaten raw, in salads, cooked as a vegetable, used as a paste or seasoned puree; immature green tomatoes are pickled and then eaten as salad
67	Zucchini	*Cucurbita pepo convar. giromontiina*	Cucurbitaceae	The cylindrical dark green fruits are peeled and cooked as a vegetable

17.1.2.1.1 Proteins

The protein fraction consists to a great extent of enzymes which may have either a beneficial or a detrimental effect on processing. They may contribute to the typical flavor or to formation of undesirable flavors, tissue softening and discoloration. Enzymes of all the main groups are present in vegetables:

- *Oxidoreductases* such as lipoxygenases, phenoloxidases, peroxidases;

- *Hydrolases* such as glycosidases, esterases, proteinases;
- *Transferases* such as transaminases;
- *Lyases* such as glutamic acid decarboxylase, alliinase;
- *Ligases* such as glutamine synthetase.

Enzyme inhibitors are also present, e.g. potatoes contain proteins which have an inhibitory effect on serine proteinases, while proteins from beans and cucumber inhibit pectolytic enzymes. Protein and enzyme patterns, as obtained by electrophoretic separation, are often characteristic for species or cultivars and can be used for analytical differentiation. Figure 17.1 shows typical protein and proteinase inhibitor patterns for several potato cultivars (see also Figures 18.1 and 18.2).

17.1.2.1.2 Free amino acids

In addition to protein-building amino acids, non-protein amino acids occur in vegetables as well as in other plants. Tables 17.4 and 17.5 present data on the occurrence and structure of these amino acids. Information about their biosynthetic pathways is given as follows:
The higher homologues of amino acids, such as homoserine, homomethionine and aminoadipic acid, are generally derived from a reaction sequence which corresponds to that of oxalacetate to ketoglutarate in the *Krebs* cycle:

$$\mathrm{R{-}CH(NH_2){-}COOH \rightarrow R{-}CO{-}COOH \rightarrow R{-}C(OH)(COOH){-}CH_2{-}COOH \rightarrow R{-}C(COOH){=}CH{-}COOH \rightarrow R{-}CH(COOH){-}CH(OH){-}COOH \rightarrow R{-}CH(COOH){-}CO{-}COOH \rightarrow R{-}CH_2{-}CO{-}COOH \rightarrow R{-}CH_2{-}CH(NH_2){-}COOH} \quad (17.1)$$

Table 17.2. Production of vegetables in 1981 (1,000 t)

Continent	Vegetables + melons, grand total	Cabbages	Artichokes	Tomato	Cauliflower	Squash
World	351,961	35,093	1,160	50,396	4,555	5,257
Africa	24,163	647	91	5,287	148	1,009
America, North-, Central-	32,292	1,807	51	8,817	291	275
America, South-	11,311	563	68	2,795	70	633
Asia	190,382	15,430	13	12,557	1,801	2,218
Europe, West-	46,205	3,868	938	10,706	1,790	582
Europe, East- + USSR	45,869	12,675		9,946	365	461
Australia + South Pacific Islands	1,738	104		287	112	80

Continent	Cucumbers	Eggplants	Chilies[a]	Onions, air dried	Garlic
World	10,782	4,612	7,205	19,780	2,202
Africa	316	427	1,096	1,522	176
America, North-, Central-	1,068	70	757	1,796	120
America, South-	45	7	194	1,670	153
Asia	5,348	3,535	2,884	8,844	1,344
Europe, West-	1,234	547	1,180	2,798	270
Europe, East- + USSR	2,755	25	1,095	2,958	131
Australia + South Pacific Islands	17	–	–	192	2

[a] Data include other Capsicum species. [b] World production • 100%

Table 17.2. (Continued)

Continent	Green beans	Green peas	Carrots	Water-melons	Cantaloups and other melons (muskmelons)
World	2,448	4,478	10,555	25,014	6,625
Africa	313	129	401	2,122	449
America, North-, Central-	206	1,187	1,336	1,861	1,144
America, South-	115	119	418	993	323
Asia	678	608	2,730	14,096	3,148
Europe West-	912	1,663	2,306	2,001	1,421
Europe, East- + USSR	180	619	3,231	3,893	141
Australia + South Pacific Islands	44	153	135	48	–

Country	Vegetables + melons, grand total	Country	Cabbages	Country	Artichokes	Country	Tomato
China	81,474	USSR	8,400	Italy	565	USA	6,339
India	40,017	China	5,986	Spain	246	USSR	6,150
USSR	28,389	Japan	3,667	France	95	Italy	4,457
USA	24,868	Korea, South-	3,457	Argentina	59	China	4,302
Japan	15,052	Poland	1,600	USA	51	Turkey	3,900
Italy	13,574	USA	1,494	Morocco	34	Egypt	2,632
Turkey	13,529	Rumania	1,000	Greece	32	Spain	2,074
Korea, South-	9,099	UK	883	Egypt	30	Greece	1,666
Spain	8,283	Yugoslavia	767	Algeria	14	Rumania	1,600
Egypt	7,356	Turkey	588	Tunisia	13	Brazil	1,495
France	6,763					Mexico	1,370
Brazil	4,423	Σ (%)[b]	80	Σ (%)[b]	98	Japan	1,000
Poland	4,357					Bulgaria	911
Rumania	4,357						
UK	4,006					Σ (%)[b]	75
Σ (%)[b]	75						

Country	Cauli-flower	Country	Squash	Country	Cucumbers	Country	Eggplants
China	787	China	956	China	2,739	China	1,568
India	650	Egypt	461	USSR	1,600	Japan	620
Italy	508	Rumania	406	Japan	1,030	Turkey	659
France	481	Turkey	361	USA	795	Italy	337
UK	307	Italy	324	Turkey	511	Egypt	313
USA	222	Argentina	274	Poland	400	Syria	190
Spain	201	Japan	252	Netherlands	375	Iraq	126
Poland	143	South Africa	220	Syria	285	Philippines	110
German DR	129	Syria	192	Egypt	283	Spain	105
Egypt	99	Mexico	161	Spain	260	Sudan	79
		Chile	130				
Σ (%)[b]	77	Spain	115	Σ (%)[b]	77	Σ (%)[b]	89
		Bangladesh	96				
		Σ (%)[b]	75				

[a] Data include other Capsium species. [b] World production ≙ 100%

Table 17.2. (Continued)

Country	Chilies[a]
China	1,455
Nigeria	635
Turkey	596
Spain	584
Italy	466
Mexico	415
Yugoslavia	381
Bulgaria	280
USA	266
Rumania	210
Indonesia	200
Σ (%)[b]	76

Country	Onions, air dried
China	2,687
India	1,650
USA	1,599
USSR	1,450
Japan	1,067
Spain	1,045
Turkey	1,000
Brazil	777
Italy	530
Egypt	527
Netherlands	527
Pakistan	435
Korea, South-	407
Poland	400
Rumania	340
Yugoslavia	310
Iran	265
Colombia	265
Σ (%)[b]	77

Country	Garlic
China	541
India	200
Thailand	188
Egypt	162
Korea, South-	154
Spain	153
Turkey	80
USA	75
Argentina	69
Italy	53
Σ (%)[b]	76

Country	Green beans
China	363
Italy	260
Egypt	255
Spain	222
USA	118
Japan	96
UK	94
Chile	83
France	82
Greece	74
Netherlands	62
Syria	52
Canada	50
Rumania	46
Hungary	46
Σ (%)[b]	78

Country	Green peas
USA	1,050
UK	700
France	419
USSR	260
India	255
Italy	242
China	231
Hungary	197
Australia	110
Belgium, Luxembourg	93
Σ (%)[b]	79

Country	Carrots
USSR	2,000
China	1,852
USA	942
Japan	630
UK	620
Poland	549
France	501
Italy	355
German DR	304
Canada	260
Σ (%)[b]	76

Country	Water-melons
Turkey	4,457
China	4,325
USSR	2,800
Egypt	1,267
USA	1,183
Japan	999
Iran	946
Syria	923
Italy	733
Greece	710
Iraq	683
Σ (%)[b]	76

Country	Cantaloups and other melons
China	1,536
USA	771
Spain	770
Iran	494
Italy	320
Japan	305
Mexico	302
Syria	210
France	205
Morocco	155
Σ (%)[b]	77

[a] Data include other Capsium species. [b] World production $\hat{=}$ 100%

$$H_3C{-}CO{-}COOH + H_3C{-}CO{-}COOH \longrightarrow H_3C{-}C(OH)(COOH){-}H_2C{-}CO{-}COOH$$
$$\longrightarrow H_3C{-}C(OH)(COOH){-}CH_2{-}CH(NH_2){-}COOH$$
$$\longrightarrow H_2C{=}C(COOH){-}H_2C{-}CH(NH_2){-}COOH \qquad (17.2)$$

4-methyleneglutamic acid (Table 17.4: XXXI) is formed from pyruvic acid (cf. Reaction 17.2). Reactions leading to the important precursors for onion flavor, the S-alkylcysteine sulfoxides, are presented in Reaction 17.3.

$$H_2C(SH){-}CH(NH_2){-}COOH \longrightarrow H_2C(SR){-}CH(NH_2){-}COOH$$
$$\longrightarrow H_2C(O{=}SR){-}CH(NH_2){-}COOH \qquad (17.3)$$

Table 17.2a. Production of starch containing roots, rhizomes and tubers in 1981 (1,000 t)

Continent	Tubers + rhizomes grand total	Potato	Sweet potatoes	Cassava (manioc)
World	561,567	256,978	145,045	127,261
Africa	84,990	5,127	5,151	47,818
America, North-, Central-	21,947	18,938	1,347	954
America, South-	43,125	10,531	1,404	30,677
Asia	225,271	37,929	137,108	47,584
Europe, West-	45,808	45,663	130	
Europe, East- + USSR	137,610	137,607		
Australia + South Pacific Islands	2,816	1,182	620	229

Country	Potato	Country	Sweet potatoes	Country	Cassava (maniok)
USSR	72,000	China	125,680	Brazil	25,050
Poland	42,600	Vietnam	2,400	Thailand	17,900
USA	15,135	Indonesia	2,079	Indonesia	13,726
China	15,039	India	1,500	Zaire	13,000
German DR	10,500	Japan	1,317	Nigeria	11,000
India	9,599	Korea, South-	1,108	India	5,817
FR Germany	8,045	Philippins	1,100	Tansania	4,650
France	6,480	Ruanda	938	Vietnam	3,400
Netherlands	6,445	Burundi	928	China	3,276
UK	6,108	Brazil	800	Mozambique	2,850
Σ (%)[a]	75	Σ (%)[a]	95	Σ (%)[a]	79

[a] World production ≙ 100%

2,4-diaminobutyric acid and some other compounds are derived from cysteine (cf. Reaction 17.4).

$$\text{Cysteine} \xrightarrow[H_2S]{HCN} NC{-}CH_2{-}CH(NH_2){-}COOH$$

$$\xrightarrow{-CO_2} NC{-}CH_2{-}CH_2{-}NH_2 \rightarrow \gamma\text{-Glutamyl-}\beta\text{-aminopropionitrile}$$

$$\xrightarrow{+H_2O} H_2NOC{-}CH_2{-}CH(NH_2){-}COOH$$

$$\xrightarrow{\text{Red.}} H_2N{-}CH_2{-}CH_2{-}CH(NH_2){-}COOH \rightarrow OHC{-}CH_2{-}CH(NH_2){-}COOH \rightarrow \text{azetidine-2-carboxylic acid (XXI)} \quad (17.4)$$

The aspartic acid semi-nitrile formed initially can be decarboxylated to β-amino propionitrile which, just as its γ-glutamyl derivative, is responsible for osteolathyrism in animals. Hydrolysis of the semi-nitrile yields aspartic acid, hydrolysis and reduction yield 2,4-diaminobutyric acid, the oxalyl derivative of which, like oxalyldiaminopropionic acid, is a human neurotoxin. The main

Table 17.3. Average composition of vegetables (as % of fresh edible portion)

Vegetable	Dry matter	N-Compounds	Carbohydrates	Lipids	Crude fiber	Ash
Mushrooms						
Champignon (cultivated *Agaricus arvensis, campestris*)	10.0	4.8	3.5	0.2	0.8	0.8
Chanterelle	8.5	2.6	3.5	0.8	1.0	0.7
Edible boletus (*Boletus edulis*)	13.0	5.4	5.2	0.4	1.0	1.0
Rooty vegetables						
Carrots	11.8	1.1	8.7	0.2	1.0	0.8
Radish (*Raphanus sativus,* elongated white fleshy root)	5.5	1.0	2.9	0.2	0.7	0.8
Viper's grass, *scorzonera*	21.4	1.4	16.3	0.4	2.3	1.0
Parsley	12.0	2.9	2.3	0.6		1.6
Tuberous vegetables (sprouting tubers)						
White (Irish) potato	22.2	2.0	18.9[a]	0.2	0.8	1.1
Celery (root)	11.6	1.8	7.2	0.3	1.3	1.0
Kohlrabi	9.7	2.0	5.6	0.1	1.0	1.0
Rutabaga	13.0	1.1	9.9	0.2	1.1	0.8
Radish (*Raphanus sativus,* reddish fleshy root)	5.6	1.1	3.5	0.1		0.9
Red beet, beetroot	12.7	1.6	9.1	0.1	0.8	1.1
Tuberous root vegetables						
Sweet potato	30.8	1.6	26.6[b]	0.6	0.9	1.1
Cassava (manioc)	35.0	0.9	32.0	0.4	0.8	0.4
Yam	28.0	1.8	23.8	0.2	0.7	1.0
Bulbous root vegetables						
Onion	10.9	1.5	8.1	0.3	0.6	0.6
Leek	14.6	2.2	9.9	0.3	1.3	0.9
Vegetable fennel	14.0	2.4	9.1	0.3	0.5	1.7
Stem (shoot) vegetables						
Asparagus	8.3	2.5	4.3	0.1	0.7	0.8
Leafy (stalk) vegetables						
Rhubarb	5.2	0.6	3.0	0.2	0.7	0.8
Leafy vegetables						
Endive (escarole)	5.6	1.3	2.3	0.2	0.9	1.0
Kale (curly cabbage)	17.3	6.0	7.5	0.9		1.5
Head lettuce	5.1	1.6	1.7	0.3	0.7	0.9
Brussels sprouts	14.8	4.9	6.7	0.6	1.6	1.2
Red cabbage	9.8	2.0	5.9	0.2	1.0	0.7
Spinach	9.3	3.2	3.7	0.4	0.6	1.5
Common (white) cabbage	7.6	1.3	4.6	0.2	0.8	0.7
Flowerhead (calix) vegetables						
Artichoke	14.5	2.9	8.2	0.1	2.4	0.8
Cauliflower	9.0	2.7	4.2	0.3	1.0	0.9
Broccoli	10.9	3.6	4.4		1.5	1.1

[a] Starch content 14.1%. [b] Starch and saccharose contents 19.6 and 2.8%, respectively.

Table 17.3. (Continued)

Vegetable	Dry matter	N-Compounds	Carbohydrates	Lipids	Crude fiber	Ash
Seed vegetables						
Chestnut	49.9	2.9	42.8	1.9	1.4	1.2
Green beans	9.9	1.9	6.1		1.0	0.7
Green peas	22.0	6.3	12.4		2.0	0.9
Fruity vegetables						
Eggplant	7.6	1.2	4.7	0.2	0.9	0.6
Squash	8.7	1.1	5.5	0.1	1.2	0.8
Green bell pepper	6.6	1.2	3.4	0.3	1.4	0.4
Cucumber	5.9	0.9	2.8	0.2	0.6	0.5
Tomato	6.5	1.1	4.2	0.2	0.5	0.5

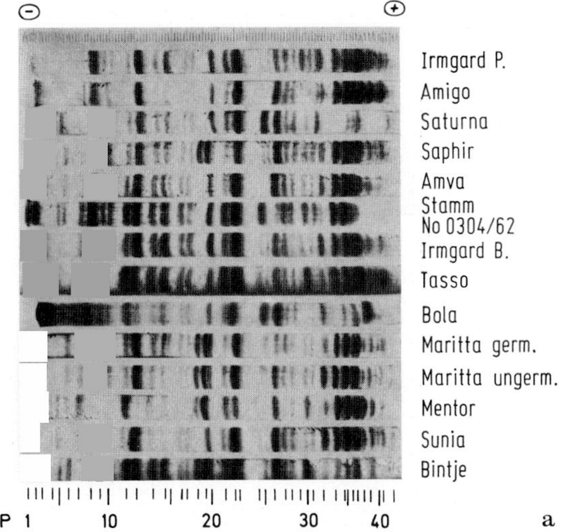

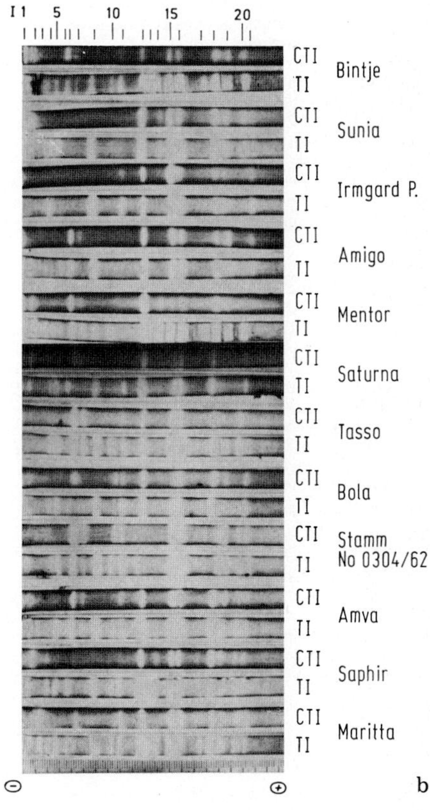

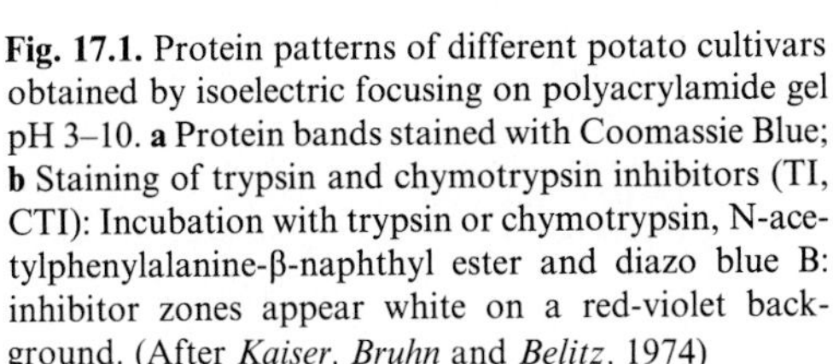

Fig. 17.1. Protein patterns of different potato cultivars obtained by isoelectric focusing on polyacrylamide gel pH 3–10. **a** Protein bands stained with Coomassie Blue; **b** Staining of trypsin and chymotrypsin inhibitors (TI, CTI): Incubation with trypsin or chymotrypsin, N-acetylphenylalanine-β-naphthyl ester and diazo blue B: inhibitor zones appear white on a red-violet background. (After *Kaiser, Bruhn* and *Belitz,* 1974)

symptoms of neurolathyrism are paralysis of the limbs and muscular rigidity. 2,4-diaminobutryic acid can be converted via the aspartic acid semialdehyde into 2-azetidine carboxylic acid (XXI), which occurs, for example, in sugar beets (Table 17.4).

Freshly harvested mushrooms contain approx. 0.1% agaritin, β-N-(γ-L(+)-glutamyl)-4-hydroxymethylphenylhydrazine. The enzymes present can hydrolyze agaritin and the released 4-hydroxymethyl-phenylhydrazine oxidize to the diazonium salt.

Table 17.4. Occurrence of nonprotein amino acids in plants (The Roman numerals refer to Table 17.5)

	Amino acid	Plant		Family
Neutral aliphatic amino acids				
I	2-(Methylenecyclopropyl)-glycine	litchi	*Litchi chinensis*	Sapidaceae
II	3-(Methylenecyclopropyl)-L-alanine (Hypoglycine A)	akee	*Bligia sapida*	Sapidaceae
III	3-Cyano-L-alanine	common vetch	*Vicia sativa*	Fabaceae
IV	L-2-Aminobutyric acid	garden sage	*Salvia officinalis*	Lamiaceae
V	L-Homoserine	garden pea	*Pisum sativum*	Fabaceae
VI	O-Acetyl-L-homoserine	garden pea		
VII	O-Oxalyl-L-homoserine	vetchling	*Lathyrus sativum*	Fabaceae
VIII	5-Hydroxy-L-norvaline	jackbean	*Canavalia ensiformis*	Fabaceae
IX	4-Hydroxy-L-isoleucine	fenugreek	*Trigonella foenum-graecum*	Fabaceae
X	1-Amino-cyclopropane-1-carboxylic acid	apple	*Malus sylvestris*	Rosaceae
		pear	*Pyrus communis*	Rosaceae
Sulfurcontaining amino acids				
XI	S-Methyl-L-cysteine	garden bean	*Phaseolus vulgaris*	Fabaceae
XII	S-Methyl-L-cysteinesulfoxide	radish, cabbage cauliflower, broccoli	*Brassica oleracea*	Brassicaceae
XIII	S-(Prop-1-enyl)cysteine	garlic	*Allium sativum*	Liliaceae
XIV	S-(Prop-l-enyl)cysteinesulfoxide	onion	*Allium cepa*	Liliaceae
XV	γ-Glutamyl-S-(prop-1-enyl)cysteine	chive	*Allium schoenoprasum*	Liliaceae
XVI	S-(Carboxymethyl)cysteine	radish	*Raphanus sativus*	Brassicaceae
XVII	3,3′-(Methylenedithio)dialanine (Djenkolic acid)	djenkol bean	*Pithecolobium lobatum*	Fabaceae
XVIII	3,3′(-2-Methylethenyl-1,2-dithio)-dialanine (as γ-Glutamyl derivative)	chive	*Allium schoenoprasum*	Liliaceae
XIX	S-Methylmethionine	jackbean	*Canavalia ensiformis*	Fabaceae
		white cabbage	*Brassica oleracea*	Brassicaceae
		asparagus	*Asparagus officinalis*	Liliaceae
XX	Homomethionine	white cabbage	*Brassica oleracea*	Brassicaceae
Imino acids				
XXI	Azetidine-2-carboxylic acid	sugar beet	*Beta vulgaris ssp.*	Chenopodiaceae
XXII	tr-4-Methyl-L-proline	apple	*Malus sylvestris*	Rosaceae
XXIII	cis-4-Hydroxymethyl-L-proline	apple peel	*Malus sylvestris*	Rosaceae
XXIV	trans-4-Hydroxymethyl-L-proline	loquat	*Eriobotrya japonica*	Rosaceae
XXV	trans-4-Hydroxymethyl-D-proline	loquat	*Eriobotrya japonica*	Rosaceae
XXVI	4-Methylene-D,L-proline	loquat	*Eriobotrya japonica*	Rosaceae
XXVII	cis-3-Amino-L-proline	morel	*Morchella esculenta*	Ascomycetes
XXVIII	Pipecolic acid	many plants		
XXIX	3-Carboxy-6,7-dihydroxy-1,2,3,4-tetrahydroisoquinoline	cowage	*Mucuna sp.*	Fabaceae
XXX	1-Methyl-3-carboxy-6,7-dihydroxy-1,2,3,4-tetrahydroisoquinoline	cowage	*Mucuna sp.*	Fabaceae
Acidic amino acids and related compounds				
XXXI	4-Methyleneglutamic acid	peanut	*Arachis hypogaea*	Fabaceae
XXXII	4-Methyleneglutamine	peanut	*Arachis hypogaea*	Fabaceae
XXXIII	N^5-Ethyl-L-glutamine (L-Theanine)	tea	*Thea sinensis*	Theaceae
XXXIV	L-threo-4-Hydroxyglutamic acid			
XXXV	3,4-Dihydroxyglutamic acid	garden cress	*Lepidium sativum*	Brassicaceae
		rhubarb	*Rheum rhabarbarum*	Polygonaceae
		carrot	*Daucus carota*	Apiaceae
		currant	*Ribis rubrum*	Saxifragaceae
		spinach	*Spinacia oleracea*	Chenopodiaceae
		longwort	*Angelica archangelica*	Apiaceae
XXXVI	L-2-Aminoadipic acid	many plants		
Basic amino acids and related compounds				
XXXVII	N^2-Oxalyl-diaminopropionic acid	vetchling	*Lathyrus sativus*	Fabaceae
XXXVIII	N^3-Oxalyl-diaminopropionic acid	vetchling	*Lathyrus sativus*	Fabaceae
XXXIX	2,4-Diaminobutyric acid (as N^4-Lactyl compound)	sugar beet	*Beta vulgaris ssp.*	Chenopodiaceae
XL	2-Amino-4-(guanidinooxy)butyric acid (Canavanine)	jackbean	*Canavalia ensiformis*	Fabaceae
		soybean	*Glycine max*	Fabaceae
XLI	4-Hydroxyornithine	common vetch	*Vicia sativa*	Fabaceae
XLII	L-Citrulline	watermelon	*Citrullus lanatus*	Cucurbitaceae

Table 17.4. (Continued)

Amino acid		Plant		Family
XLIII	Homocitrulline	horse bean	*Vicia faba*	Fabaceae
XLIV	4-Hydroxyhomocitrulline	horse bean	*Vicia faba*	Fabaceae
XLV	4-Hydroxyarginine	common vetch	*Vicia sativa*	Fabaceae
XLVI	4-Hydroxylysine	garden sage	*Salvia officinalis*	Lamiaceae
XLVII	5-Hydroxylysine	lucern	*Medicago sativa*	Fabaceae
XLVIII	N^6-Acetyl-L-lysine	sugar beet	*Beta vulgaris*	Chenopodiaceae
XLIX	N^6-Acetyl-allo-5-hydroxy-L-lysine	sugar beet	*Beta vulgaris*	Chenopodiaceae
Heterocyclic amino acids				
L	3-(2-Furoyl)-L-alanine	buck wheat	*Fagopyrum esculentum*	Polygonaceae
LI	3-Pyrazol-1-ylalanine	watermelon	*Citrullus lanatus*	Cucurbitaceae
LII	1-Alanyluracil (Willardin)	cucumber	*Cucumis sativus*	Cucurbitaceae
		garden pea	*Pisum sativum*	Fabaceae
LIII	3-Alanyluracil (Isowillardin)	garden pea	*Pisum sativum*	Fabaceae
LIV	3-Amino-3-carboxypyrrolidine	musk melon	*Cucurbita monlata*	Cucurbitaceae
LV	3-(2,6-Dihydroxypyrimidine-5-yl)-alanine	garden pea	*Pisum satium*	Fabaceae
LVI	3-(Isoxazoline-5-one-2-yl)alanine	garden pea	*Pisum sativum*	Fabaceae
LVII	3-(2-β-D-Glucopyranosyl-isoxazoline-5-one-4-yl)alanine	garden pea	*Pisum sativum*	Fabaceae
Aromatic amino acids				
LVIII	N-Carbamoyl-4-hydroxy-phenylglycine	horse bean	*Vicia faba*	Fabaceae
LIX	L-3,4-Dihydroxyphenylalanine	horse bean	*Vicia fabea*	Fabaceae
		cowage	*Mucuna sp.*	Fabaceae
Other amino acids				
LX	γ-Glutamyl-L-β-phenyl-β-alanine	adzuki bean	*Phaseolus angularis*	Fabaceae
LXI	Saccharopine	yeast	*Saccharomyces cerevisiae*	Saccharomycetaceae

17.1.2.1.3 Amines

The presence of amines has been confirmed in various vegetables; e.g. histamine, N-acetylhistamine and N,N-dimethylhistamine in spinach; and tryptamine, serotonin and tyramine in tomatoes and eggplant (3–4 mg/100 g dry matter).

17.1.2.2 Carbohydrates

17.1.2.2.1 Mono- and oligosaccharides, sugar alcohols

The predominant sugars in vegetables are glucose and fructose (0.3–4%) as well as sucrose (0.1–12%). Other sugars occur in small amounts; e.g. glycosidically-bound apiose in *Umbelliferae* (celery and parsley); 1^F-β- and 6^G-β-fructosylsaccharose in the allium group (onions, leeks); raffinose, stachyose and verbascose in *Fabaceae;* and mannitol in *Brassicaceae* and *Cucurbitaceae.*

17.1.2.2.2 Polysaccharides

Starch occurs widely as a storage carbohydrate, and is present in large amounts in some root and tuber vegetables. In *Compositae* (e.g. artichoke, viper's grass, bot. *Scorzonera*), inulin, rather than starch, is the storage carbohydrate.

Other polysaccharides are cellulose, hemicelluloses and pectins. The pectin fraction has a distinct role in tissue firmness of vegetables. Tomatoes become firmer as the total pectin content and the content of some minerals (Ca, Mg) increases, and also as the degree of esterification of the pectin decreases. In processing cauliflower (cf. 17.2), 70 °C is favorable for preserving tissue firmness. The reason for this effect is the presence of pectinmethylesterase which, in vegetables, is fully inactivated only at temperatures above 88 °C, while at 70 °C it is active and provides a build-up of insoluble pectates. For the conversion of protopectin to pectin during plant tissue maturation or ripening see 18.1.3.3.1.

17.1.2.3 Lipids

The lipid content of vegetables is generally low (0.1–0.9%). In addition to triacylglycerids, glyco- and phospholipids are present.

Carotenoids are occasionally found in large

Table 17.5. Structures of nonprotein amino acids in plants (structures and Roman numerals refer to Table 17.4)

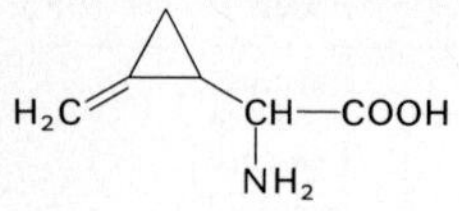

$H_2C{=}$(cyclopropyl)$-CH(NH_2)-COOH$

I

$H_2C{=}$(cyclopropyl)$-CH_2-CH(NH_2)-COOH$

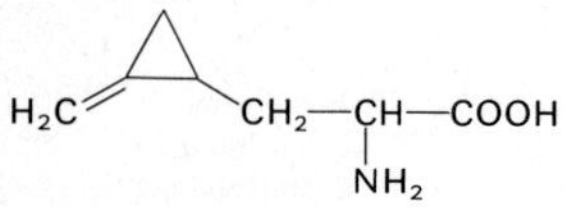

II

$NC-CH_2-CH(NH_2)-COOH$

III

$H_3C-CH_2-CH(NH_2)-COOH$

IV

$ROCH_2-CH_2-CH(NH_2)-COOH$

V: R = H
VI: R = H_3C-CO
VII: R = $HOOC-CO$

$HO-CH_2-CH_2-CH_2-CH(NH_2)-COOH$

VIII

$H_3C-CH(OH)-CH(CH_3)-CH(NH_2)-COOH$

IX

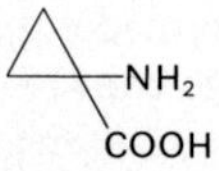

(cyclopropane)$(NH_2)(COOH)$

X

$R-S-CH_2-CH(NH_2)-COOH$

XI: R = CH_3
XIII: R = $H_3C-CH{=}CH$
XVI: R = $HOOC-CH_2$

$R-S(=O)-CH_2-CH(NH_2)-COOH$

XII: R = CH_3
XIV: R = $H_3C-CH{=}CH$

$H_3C-CH{=}CH-S-CH_2-CH(COOH)-NH-CO-(CH_2)_2-CH(NH_2)-COOH$

XV

$HOOC-CH(NH_2)-CH_2-S-CH_2-S-CH_2-CH(NH_2)-COOH$

XVII

$HOOC-CH(NH_2)-CH_2-S-CH_2-CH(CH_3)-S-CH_2-CH(NH_2)-COOH$

XVIII

$H_3C-S^{\oplus}(CH_3)-CH_2-CH_2-CH(NH_2)-COOH$

XIX

$H_3C-S-CH_2-CH_2-CH_2-CH(NH_2)-COOH$

XX

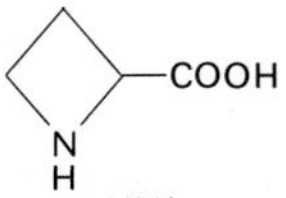

(azetidine, N–H)$-COOH$

XXI

(4-R-pyrrolidine, N–H)$-COOH$

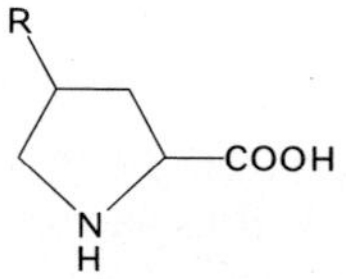

XXII: R = CH_3
XXIII, XXIV, XXV: R = $HOCH_2$

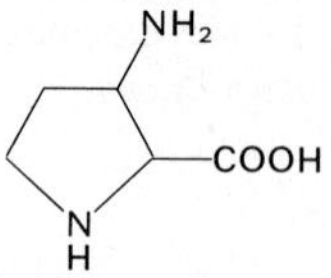

(3-NH_2-pyrrolidine, N–H)$-COOH$

XXVII

(4-methylene ($H_2C{=}$) pyrrolidine, N–H)$-COOH$

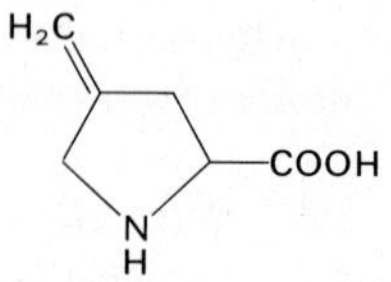

XXVI

Table 17.5. (Continued)

XXVIII (piperidine-2-carboxylic acid)

XXIX: R = H
XXX: R = CH_3

$H_2N-C(=O)-NH-(CH_2)_n-CH(R)-CH_2-CH(NH_2)-COOH$

XLII: n = 1, R = H
XLIII: n = 2, R = H
XLIV: n = 2, R = OH

$H_2N-C(=NH)-NH-CH_2-CH(OH)-CH_2-CH(NH_2)-COOH$

XLV

$ROC-C(=CH_2)-CH_2-CH(NH_2)-COOH$

XXXI: R = OH XXXII: R = NH_2

$R^2-HN-CH_2-CH(R^1)-CH(R)-CH_2-CH(NH_2)-COOH$

XLVI: R = OH, R^1, R^2 = H
XLVII: R^1 = OH, R, R^2 = H
XLVIII: R, R^1 = H, R^2 = CH_3CO
XLIX: R = H, R^1 = OH, R^2 = CH_3CO

$H_5C_2-NHOC-CH_2-CH_2-CH(NH_2)-COOH$

XXXIII

$HOOC-CH(R)-CH(R^1)-CH(NH_2)-COOH$

XXXIV: R = HO, R^1 = H
XXXV: R, R^1 = HO

(2-furyl)$-CO-CH_2-CH(NH_2)-COOH$

L

$HOOC-(CH_2)_3-CH(NH_2)-COOH$

XXXVI

(pyrazol-1-yl)$-CH_2-CH(NH_2)-COOH$

LI

$H_2C(NHR^1)-CH(NHR)-COOH$

XXXVII: R = HOOC—CO, R^1 = H
XXXVIII: R^1 = HOOC—CO, R = H

(uracil-1-yl)$-CH_2-CH(NH_2)-COOH$

LII

$H_2C(NH_2)-CH_2-CH(NH_2)-COOH$

XXXIX

(uracil-3-yl)$-CH_2-CH(NH_2)-COOH$

LIII

$H_2N-C(=NH)-NH-O-CH_2-CH_2-CH(NH_2)-COOH$

XL

3-amino-pyrrolidine-3-carboxylic acid (NH_2, COOH)

LIV

$H_2C(NH_2)-CH(OH)-CH_2-CH(NH_2)-COOH$

XLI

Table 17.5. (Continued)

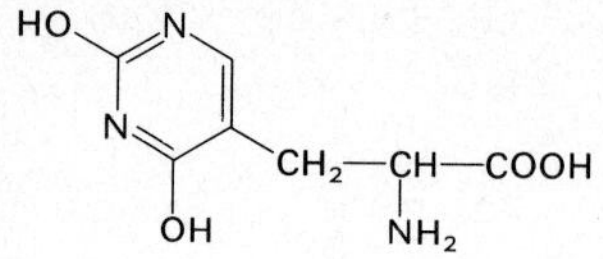

LV

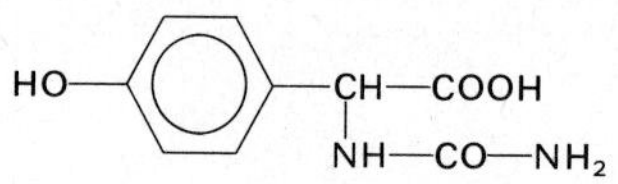

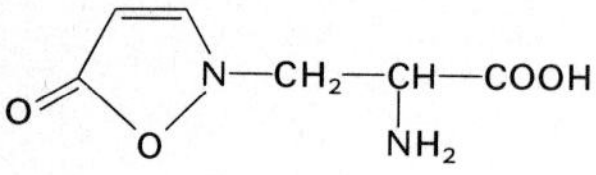

LVI

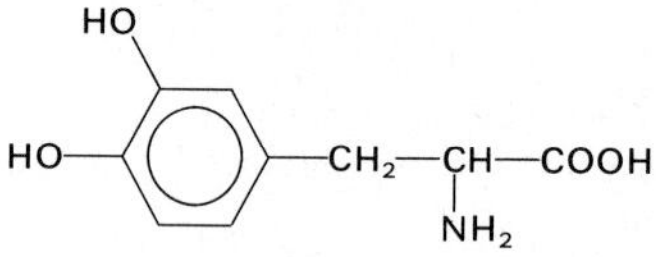

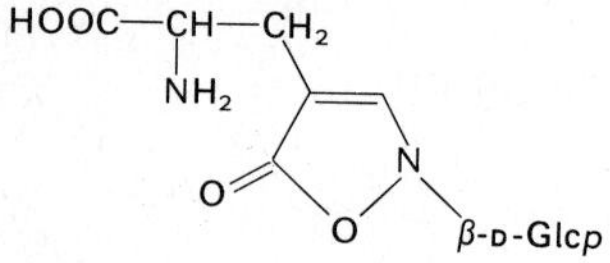

LVII

LVIII

LIX

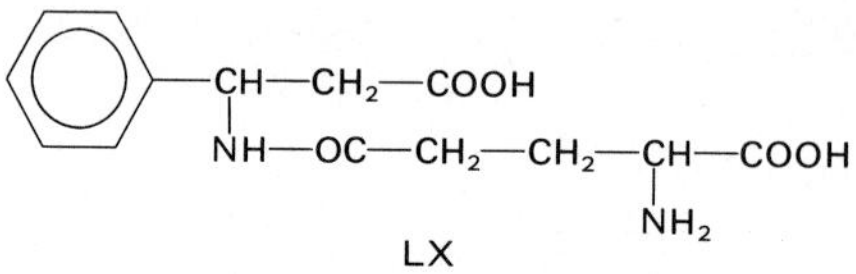

LX

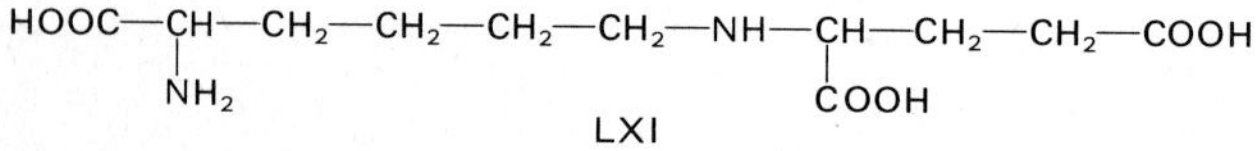

LXI

amounts (cf. 18.1.2.3.2). Table 17.6 provides data about carotenoid compounds in green bell and paprika peppers, tomato and watermelon. For the occurrence of bitter cucurbitacins in *Cucurbitaceae,* see 18.1.2.3.3.

Table 17.6. Carotenoids in vegetables[a]

	Green bell pepper	Red pepper (paprika)	Tomato	Water-melon
Total carotenoids[b]	0.9–1.1	12.7–28.4	5.1–6.3	2.5
Phytoene (I)	1.4	1.7	5.3	2.1
Phytofluene (II)	0.2	1.1	2.8	1.4
α-Carotene (VI)	0.4	0.2	0.03	0.06
β-Carotene (VII)	13.4	11.6	3.7	4.1
γ-Carotene (V)			1.2	0.4
ζ-Carotene (III)	0.4	1.5	0.9	1.6
Lycopene (IV)			78.7	81.3
α-Cryptoxanthin	1.2	1.0		
β-Cryptoxanthin	0.5	6.7		
Lutein (IX)	40.8		2.0	
Zeaxanthin (VIII)	0.6	2.3	0.08	
Violaxanthin (XIII)	13.8	9.9	0.8	
Luteoxanthin (XIV)	6.9	0.9	0.1	
Capsanthin (X)		34.7		
Neoxanthin (XX)	15.1	0.7	0.7	

[a] Roman numeral refers to structural formula presented in Chapter 3.8.4.
[b] mg carotene/100 g fresh weight.

17.1.2.4 Organic Acids

The organic acids present in highest concentration in vegetables are malic and citric acids (Table 17.7). The content of free titratable acids is 0.2–0.4 g/100 g fresh tissue, an amount which is low in comparison to fruits. Accordingly, the pH, with several exceptions such as tomato or rhubarb, is relatively high (5.5–6.5). Other acids of the citric acid cycle are present in negligible amounts.

Oxalic acid occurs in larger amounts in some vegetables (Table 17.7).

17.1.2.5 Phenolic Compounds

The phenolic compounds in plant material are dealt with in detail in 18.1.2.5. Hydroxybenzoic and hydroxycinnamic acids, flavones and flavo-

Table 17.7. Organic acids in vegetables (mg/100 g fresh weight)

Vegetable	Malic acid	Citric acid	Oxalic acid
Artichoke	170	100	
Eggplant	170	–	
Cauliflower	390	210	
Green beans	112	34	20–45
Broccoli	120	210	
Green peas	75	142	
Kale	50	350	13–125
Carrot	240	90	0–60
Leek		59	0–89
Rhubarb	910	137	230–500
Brussels sprouts	200	240	37
Red beet	–	110	30–138
Sorrel			360
White common cabbage	100	140	
Onion	170	20	

nols also occur in vegetables. Table 17.8 provides data on the occurrence of anthocyanins in some vegetables.

17.1.2.6 Aroma Substances

Vegetable aromas, just as fruit aromas (cf. 18.1.2.6), are derived either from one single characteristic (impact) compound or from a more or less large number of compounds. Table 17.9 gives examples of both. Several vegetables will be dealt with in more detail. The number following each vegetable corresponds to that given in Table 17.1. For aroma biosynthesis see 5.3.2.

Table 17.8. Anthocyanins in vegetable

Vegetable	Anthocyanin
Eggplant	Delphinidin-3-(p-coumaroyl-L-rhamnosyl-D-glucosyl)-5-D-glucoside
Radish	Pelargonidin-3-[glucosyl(1→2)-6-(p-coumaroyl)-β-D-glucosido]-5-glucoside
	Pelargonidin-3-[glucosyl(1→2)-6-(feruloyl)-β-D-glucosido]-5-glucoside
Red cabbage	Cyanidin-3-sophorosido-5-glucoside (sugar moiety esterified with sinapic acid, 1–3 moles)
Onion (red shell)	Cyanidin glycoside
	Peonidin-3-arabinoside

Table 17.9. Aroma compounds in vegetables

Vegetable	Compound[a]
Group 1[b]	
Watercress	*2-Phenylethyl isothiocyanate*
Mushroom (champignon, *Agaricus campestris*)	*1-Octen-3-ol*
Cucumber	*trans-2-cis-6-Nonadienal*
	cis-3-cis-6-Nonadienal
Potato (raw)	*2-Isopropyl-3-methoxypyrazine*
Garlic	*Di-2-propenyldisulfide*
Red beet	*Geosmin*
Group 2[c]	
Potato (cooked)	*2-Ethyl-3,6-dimethylpyrazine*
	Methional
Cabbage (cooked)	*Dimethyldisulfide*
	2-Propenyl isothiocyanate
	Dimethyltrisulfide
Celery	*3-Isobutyliden-3a,4-dihydrophthalide*
	3-Isovaliden-3a,4-dihydrophthalide
	cis-3-Hexen-1-yl-pyruvate
	2,3-Butandione
Tomato	Hexanal
	trans-2-Hexenal
	cis-3-Hexenal
	trans-2,trans-4-Decadienal
	β-Ionone
	2-Isobutylthiazole
	Several less volatile compounds
Onion (raw)	*Thiopropanal-S-oxide* (lachrymatory factor)
	Thiosulfinate
	Thiosulfonate
Onion (cooked)	*Dipropyldisulfide*
	Di-1-propenyldisulfide

[a] Character impact compounds (cf. 5.1.3) are given in italics.
[b] The aroma character is determined by the presence of one compound (character impact compound).
[c] The aroma is characterized by the presence of more compounds, of which one might play the role of a character impact compound.

17.1.2.6.1 Fungi (4)

The aroma in mushrooms originates from 1-octen-3-ol, derived from enzymatic oxidative degradation of linoleic acid (cf. 3.7.2.3). Heating the fungus results in oxidation to 1-octen-3-one, which has a fungal-metallic aroma. The fungus

Lentium ediodes, which is widely consumed in China and Japan, has a very intensive aroma. The presence of 1,2,3,5,6-pentathiepane (lenthionine) has been confirmed, and it is a typical impact compound:

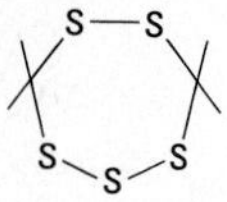

(17.5)

Its threshold values are 0.27–0.53 ppm (in water) or 12.5–25 ppm (in edible oil). It is derived biosynthetically from an S-alkyl cysteine sulfoxide, lentinic acid. Truffles, edible potato-shaped fungi, contain approx. 50 ng/g 5α-androst-16-ene-3α-ol, which has a musky odor that contributes to the typical aroma.

17.1.2.6.2 Parsley roots (21)

Parsley aroma is derived from various monoterpene hydrocarbons, including myrcene, α-pinene, β-pinene, α-thujone, camphene, sabinene, 3-carene, α- and β-phellandrene (-)-limonene, γ-terpinene, *p*-cymene and terpinolene.

17.1.2.6.3 Potatoes (23)

Of great importance for the aroma of raw potatoes is 2-isopropyl-3-methoxypyrazine. Addition of this compound or of the related compound 2-methoxy-3-ethylpyrazine to potato products (e.g. dehydrated mashed potatoes) in amounts of 0.1–0.2 ppm significantly enhances the typical potato aroma. 2,5-dimethylpyrazine, which possesses an earthy aroma, likewise plays a significant role. Numerous carbonyl compounds and alcohols have also been identified.

17.1.2.6.4 Celery tubers (24)

Celery aroma is due to the occurrence of phthalides and dihydrophthalides, namely 3-*n*-butylphthalide (I) and 3-*n*-butyl-4,5-dihydrophthalide (II), in leaves, root, tuber and seeds. Their content in essential oil from celery seeds is 3–10%.

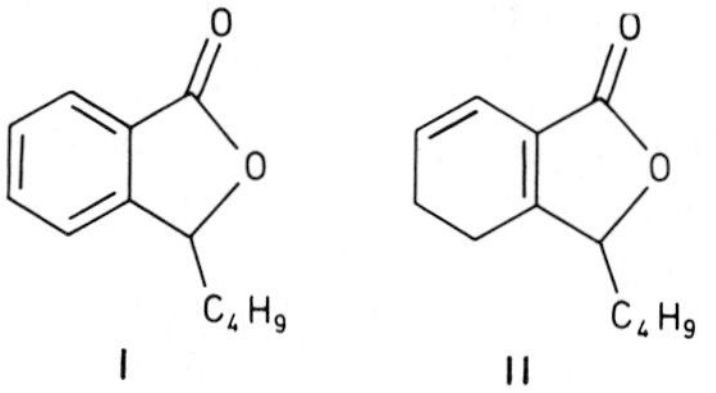

(17.6)

Table 17.9 a. Examples of glucosinolates and their corresponding mustard oils or oxazolidin-2-thiones

Glucosinolate	Occurrence	Biosynthetized from	Mustard oil
Sinigrin	Seeds of black mustard *(Brassica nigra)*	Homomethionine	Allyl-
Sinalbin	Seeds of white mustard *(Sinapis alba)*	Tyrosine	p-Hydroxybenzyl-
Gluconapin	Rapeseed *(Brassica napus)*		3-Butenyl-
Glucobrassicanapin	Rapeseed		4-Pentenyl-
Glucotropeolin	Garden cress *(Lepidium sativum),* Nasturtium *(Tropaeolum majus)*	Phenylalanine	Benzyl-
Gluconasturtiin	Watercress *(Nasturtium officinale),* Winter cress *(Barbaraea praecox)*	Homophenylalanine	Phenylethyl-
Glucoibervirin		Homomethionine	ω-Methylthiopropyl-
Glucobrassicin		Tryptophan	Indolylmethyl-
			Oxazolidin-2-thione
Glucoconringiin	*Conringia orientalis*	Leucine/4-hydroxyleucine	5,5-Dimethyl-
Glucobarbarin		2-Amino-4-phenyl-4-hydroxybutyric acid	5-Phenyl-

[a] Many glucosinolates are widely spread; the listed sources are just a few examples.

17.1.2.6.5 Radishes (27)

The sharp taste of the radish is due to 4-methylthio-trans-3-butenyl-isothiocyanate, which is released from the corresponding glucosinolate after the radish is sliced.

Glucosinolates are widely distributed among *Brassicaceae, Capparaceae* and some other plant families. Table 17.9a provides a review. Glucosinolates are hydrolyzed by myrosinase, a thioglucosidase enzyme, into the corresponding isothiocyanates (mustard oils) by disintegration of tissue. The 'R' radical represents a branched or unbranched alkyl-, alkenyl- or an ω-methylthioalkyl-, monoketoalkyl-, aryl-alkyl- or heterocyclic moiety of the glucosinolate molecule:

$$R{-}C({-}S{-}Glc){=}N{-}OSO_3^{\ominus} \xrightarrow[\text{Thioglucosidase}]{H_2O} R{-}C(S^{\ominus}){=}N{-}OSO_3^{\ominus} + \text{Glucose} \rightarrow \left[R{-}C(=S){-}\ddot{N}\right] + SO_4^{2\ominus} \rightarrow R{-}N{=}C{=}S \qquad (17.7)$$

The decomposition corresponds to a *Lossen's* rearrangement of an hydroxamic acid. In addition to isothiocyanates, rhodanides and nitriles have been observed among the reaction products.

The isothiocyanates can react further, e.g. with hydroxy compounds or thiols, to form thiourethanes or dithiourethanes. In the presence of amines, thioureas result; while hydrolysis yields the corresponding amines and releases CO_2 and H_2S:

$$R{-}N{=}C{=}S \begin{cases} \xrightarrow{R^1OH} R{-}NH{-}CS{-}OR^1 \\ \xrightarrow{R^1NH_2} R{-}NH{-}CS{-}NHR^1 \\ \xrightarrow{H_2O} R{-}NH_2 + CO_2 + H_2S \end{cases} \qquad (17.8)$$

Biosynthesis of glucosinolates (reaction 17.9) originates from the corresponding amino acids, and occurs via an oxime (I) and thiohydroximic acid (III).

The intermediate reactions between steps I and III are not yet clarified. Tests with ^{14}C- and ^{35}S-labelled compounds suggest that the aci-form of

$$R{-}CH(NH_2){-}COOH \rightarrow R{-}CH(NHOH){-}COOH \rightarrow \underset{\text{I}}{R{-}CH{=}NOH} \rightarrow \left[R{-}CH_2{-}NO_2 \rightleftharpoons \underset{\text{II}}{R{-}CH{=}N^{\oplus}(OH){-}O^{\ominus}}\right]$$

$$\xrightarrow[OH^{\ominus}]{R^1S^{\ominus}} R{-}C(=NOH){-}SR^1 \xrightarrow{R^1} \underset{\text{III}}{R{-}C(=NOH){-}S^{\ominus}}$$

$$\xrightarrow[UDP]{UDPGlc} R{-}C(=NOH){-}SGlc \xrightarrow[PAP]{PAPS} R{-}C(=NOSO_3^{\ominus}){-}SGlc \qquad (17.9)$$

the corresponding nitro-compound (II) functions as a thiol acceptor. Cysteine may be involved as a thiol donor. The sulfation is achieved by 3′-phosphoadenosine-5′-phosphosulfate (PAPS). The biosynthetic pathway for cyanogenic glycosides branches at the aldoxime (I) intermediate (cf. 16.2.6).

17.1.2.6.6 Red beets (28)

Geosmin is the character impact compound of the red beet:

HO (17.10)

17.1.2.6.7 Onions (33) and garlic

The compound which causes tears (the lachrymatory factor) is (Z)-propanethial-S-oxide (II) which, once the onion bulb is sliced, is derived from trans-(+)-S-(1-propenyl)-L-cysteine sulfoxide (I) by the action of the enzyme alliinase. Alliinase has pyridoxalphosphate as its coenzyme (cf. reaction sequence 17.11).

Alkylthiosulfonates (III) are responsible for the aroma of raw onions, while propyl- and propenyl disulfides (IV) and trisulfides are predominant in the aroma of cooked onions. The aroma of fried onions is derived from dimethylthiophenes.

Precursors of importance for the aroma of onions, other than compound I, are S-methyl- and S-propyl-L-cysteine sulfoxide. Precursor I is biosynthesized from valine and cysteine (cf. reaction sequence 17.12).

(17.11)

$$\begin{array}{l} (CH_3)_2CH{-}CH(NH_2){-}COOH \longrightarrow (CH_3)_2CH{-}CO{-}COOH \\ \longrightarrow (CH_3)_2CH{-}CO{-}CoA \longrightarrow H_2C{=}C(CH_3){-}CO{-}CoA \\ \longrightarrow H_2C{=}C(CH_3){-}COOH \xrightarrow{CySH} \\ HOOC(H_2N)CH{-}CH_2{-}S{-}CH_2{-}CH(CH_3){-}COOH \\ \longrightarrow HOOC(H_2N)CH{-}CH_2{-}S{-}CH{=}CH{-}CH_3 \\ \longrightarrow HOOC(H_2N)CH{-}CH_2{-}S(=O){-}CH{=}CH{-}CH_3 \quad (I) \end{array} \qquad (17.12)$$

The key precursor for garlic aroma is S-allyl-L-cysteine sulfoxide (alliin) which, as in onions, occurs in garlic bulbs together with S-methyl- and S-propyl-compounds. The allyl- and propyl-compounds are assumed to be synthesized from serine and corresponding thiols (rection 17.13).

$$\begin{array}{l} R{-}SH + HO{-}CH_2{-}CH(NH_2){-}COOH \\ \longrightarrow R{-}S{-}CH_2{-}CH(NH_2){-}COOH \\ \longrightarrow R{-}S(=O){-}CH_2{-}CH(NH_2){-}COOH \end{array} \qquad (17.13)$$

Diallylthiosulfinate (allicin) and diallyldisulfide are formed from the main component by means of the enzyme alliinase. Both are character impact compounds.

17.1.2.6.8 Watercress (39)

Phenylethylisothiocyanate is responsible for the aroma of this plant of the mustard family (*Brassicaceae*). Decomposition of the corresponding glucosinolate gives phenylpropionitrile, the main component, and some other nitriles, e.g. 8-methylthiooctanonitrile and 9-methylthiononanonitrile.

17.1.2.6.9 White cabbage, red cabbage and Brussels sprouts (52, 49, 48)

Mustard oil is more than 6% of the total volatile fraction of cooked white and red cabbages. Major constituents are 2-propenyl-, 3-butenyl- and 2-phenylethyl lisothiocyanates. A great number of other sulfur-containing compounds have been identified, including dimethylsulfide and dimethyltrisulfide. It also appears that 3-alkyl-2-methoxypyrazine plays a role in cabbage aroma.

The total impact of the aroma in cooked frozen Brussels sprouts is less satisfactory than in cooked fresh cabbage. In the former case, analysis has revealed comparatively less allyl mustard oil and more allylnitrile. Isothiocyanates in low levels are pleasant and appetite-stimulating, while nitriles are reminiscent of garlic odor. The shift in the concentration ratio of the two compounds is attributed to myrosinase enzyme inactivation during blanching prior to freezing. As a consequence of this, allylglucosinolate in frozen Brussels sprouts is thermally-degraded only in subsequent cooking, preferentially forming nitriles.

17.1.2.6.10 Cauliflower (56)

Aroma compounds of importance in cooked cauliflower include sulfur-containing compounds such as hydrogen sulfide, methane-, ethane- and propanethiols and dimethylsulfide, carbonyls such as ethanal, and the alcohol 2-methylpropanol.

17.1.2.6.11 Green peas (60)

The aroma of green peas is derived from aldehydes and pyrazines (3-isopropyl-, 3-*sec*-butyl- and 3-isobutyl-2-methoxypyrazine).

17.1.2.6.12 Cucumbers (64)

The following aldehydes play an important role in cucumber aroma: cis-3,cis-6-nonadienal, trans-2,cis-6-nonadienal and cis-3-nonenal. Linoleic and linolenic acids, as shown in Figure 3.28, are the precursors for these and other aldehydes (cis-3-hexenal, trans-2-hexenal, trans-2-nonenal). In addition, nonanal, cis-6-nonenal and all the alcohols which could be derived from the listed aldehydes have been identified. Also, 3-alkyl-2-methoxypyrazines are present.

17.1.2.6.13 Tomatoes (66)

A great number of volatile compounds are known to occur in tomatoes: carbonyl compounds, alcohols, esters, lactones, acetals, ketals, sulfur compounds. The compounds listed in Table 17.9 are of exceptional importance for tomato aroma. As with all fruits, the stage of maturity and the method applied for ripening greatly influence aroma. Figure 17.2 shows gas chromatograms of (a) artificially ripened, (b) field ripened and (c) overripe tomatoes; they differ significantly. In artificially ripened fruits, the concentrations (peak numbers from Figure 17.2 are in brackets) of butanol (10), 3-pentanol (13), 2-methyl-3-hexanol (14), isopentanal (4), 2,3-butanedione (30), propyl acetate (6) and isopentyl butyrate (18) are higher than in field ripened tomatoes. However, the latter have higher concentrations of nonanal (21), decanal (26), dodecanal (35), neral (34), benzaldehyde (25), citronellyl propionate (57), citronellyl butyrate (58), geranyl acetate (59) and geranyl butyrate (60) than artificially ripened tomatoes.
Unlike field ripened tomatoes, overripe tomatoes show increased levels of isopentyl acetate (18), citronellyl butyrate (58) and geranyl butyrate (60).
The aroma of cooked tomatoes is influenced by increased concentrations of dimethylsulfide and linalool.

17.1.2.7 Vitamins

Table 17.10 provides data on the vitamin content of some vegetables. The values given may vary significantly with vegetable cultivar and climate. In spinach, for example, the ascorbic acid content varies from 40–155 mg/100 g fresh weight.

17.1.2.8 Minerals

Table 17.11 reviews the mineral contents of some vegetables. Potassium is by far the most abundant constituent, followed by calcium, sodium and magnesium. The major anions are phosphate, chloride and carbonate. All other elements are present in much lower amounts. For nitrate content see Table 9.9.

Table 17.10. Vitamin content in vegetables (mg/100 g fresh weight)

Vegetable	Ascorbic acid	Thiamine	Riboflavin	Nicotinic acid	Folic acid
Artichoke	12	0.08	0.05	1.0	
Eggplant	5	0.05	0.05	0.6	
Cauliflower	78	0.11	0.10	0.7	0.02
Broccoli	113	0.10	0.23	0.9	
Kale	186	0.16	0.26	2.1	
Cucumber	11	0.03	0.04	0.2	0.02
Head lettuce	10	0.06	0.09	0.4	
Carrot	8	0.06	0.05	0.6	
Green bell pepper	128	0.08	0.08	0.5	
Leek	17	0.11	0.06	0.5	
Radish	26	0.03	0.03	0.3	
Brussels sprouts	102	0.10	0.16	0.9	
Red beet	10	0.03	0.05	0.4	
Red cabbage	61	0.09	0.06	0.4	0.04
Celery	8	0.05	0.06	0.7	
Asparagus	33	0.18	0.20	1.5	
Spinach	51	0.10	0.20	0.6	0.08
Tomato	23	0.06	0.04	0.7	0.01

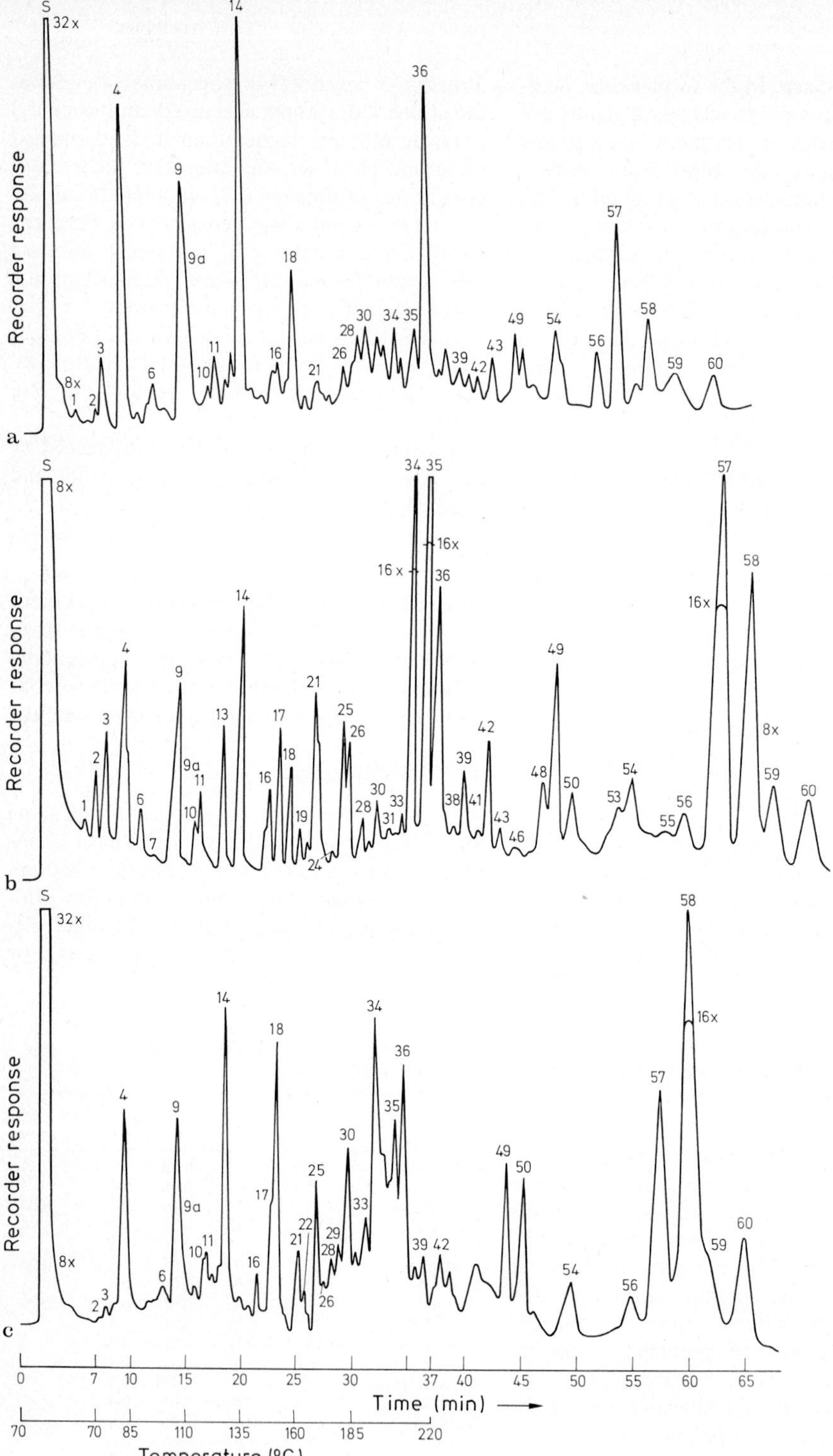

Fig. 17.2. Gaschromatograms of artificially ripened (**a**), field ripened (**b**), and over ripened (**c**) tomatoes. (After *Shah* et al., 1969)

Table 17.11. Minerals in vegetables (mg/100 g fresh weight)

Vegetable	K	Na	Ca	Mg	Fe	Mn	Co	Cu	Zn	P	Cl	F	I
Cauliflower	328	16	20	17	0.6	0.2		0.1	0.2	54	29	0.01	
Green beans	256	1.7	51	26	0.8	0.5		0.1		37	36	0.01	
Green peas	296	2	26	33	1.9	0.7	0.003	0.2	3	119	40	0.02	0.004
Cucumber	141	8.5	15	8	0.5	0.2		0.1	0.2	23	37	0.02	0.003
Red beet	336	86	29	1.4	0.9	1.0	0.01	0.2	0.6	45	0.4	0.02	
Tomato	297	6.3	14	20	0.5	0.1	0.01	0.1	0.2	26	60	0.02	0.002
White common cabbage	227	13	46	23	0.5	0.1	0.01	0.1	0.8	28	37	0.01	0.005

17.1.2.9 Other Constituents

Plant pigments other than carotenoids and anthocyanins, e.g. chlorophyll and betalains, are also of great importance in vegetables and are covered in this section together with goitrogenic compounds occurring in *Brassicaceae*.

17.1.2.9.1 Chlorophyll

The green color of leaves and unripe fruits is due to the pigments chlorophyll a (blue-green) and chlorophyll b (yellow-green) occurring together in a ratio of about 3:1 (see formula 17.14). Figure 17.3 shows the absorption spectra of chlorophylls a and b. Removal of magnesium from the chlorophylls gives pheophytins a and b, both of which are olive-brown. Replacing magnesium by metal ions such as Sn^{2+} or Fe^{3+} likewise yields greyish-brown compounds, while copper or zinc ions retain the green color. Upon removal of the phytol group, for example by the action of the chlorophyllase enzyme, the chlorophylls are converted into chlorophyllides a and b, while the hydrolysis of pheophytins yields pheophorbides a and b.

Chlorophylls and pheophytins are lipophylic due to the presence of the phytol group, while chlorophyllides and pheophorbides, without phytol, are hydrophylic. Conversion of chlorophylls to pheophytins, which is accompanied by a color change, occurs readily upon heating plant material in weakly acidic solutions and, less readily, at pH 7. Color changes are encountered most visibly in processing of green peas, green beans, kale, Brussels sprouts and spinach. Table 17.12 shows that higher temperatures and shorter heating times provide better color retention than prolonged heating at lower temperatures.

Chlorophyllase is mostly inactivated when vegetables are blanched, hence chlorophyllides and pheophorbides are rarely detected. However, in the fermentation of cucumbers chlorophyllase is

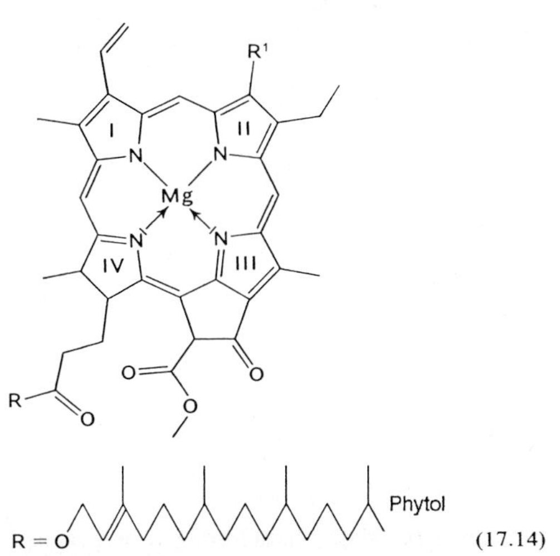

(17.14)

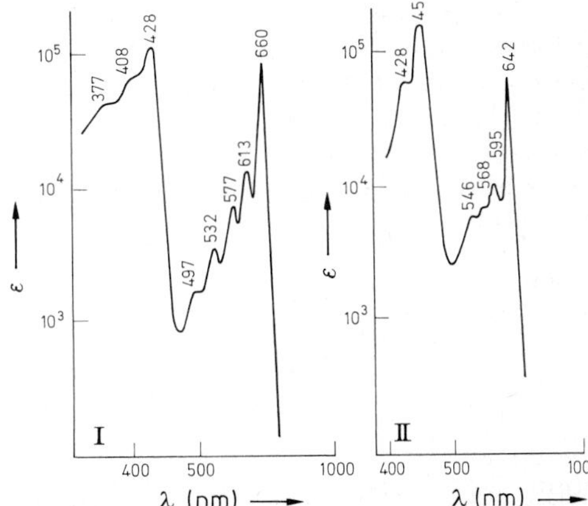

Fig. 17.3. Absorption spectra of chlorophylls a (I) and b (II). Solvent: diethyl ether (I) or diethyl ether + 1% CCl_4 (II)

Table 17.12. Changes in chlorophyll fractions as affected by processing (values in % of the total pigment content of unprocessed vegetables)

Vegetable	Process	Chlorophylls		Chlorophyllides		Pheophytins		Pheophorbides	
		a	b	a	b	a	b	a	b
Green beans	Untreated	49	25	0	0	18	8	0	0
	Blanched, 4 min/100 °C	37	24	0	0	19	10	0	0
Cucumbers	Untreated	51	30	0	0	15	5	0	0
	Blanched, 4 min/100 °C	34	24	6	3	22	1	5	7
Cucumbers	Untreated	67	33	0	0	0	0	0	0
	Fermented (pickled), 6 days	4	7	3	5	10	3	47	15
	Fermented (pickled), 24 days	0	0	0	0	16	7	57	28
Spinach puree	Untreated	48	20			26	7		
	Heated in oilbath for								
	19.5 min/116 °C	4	4			82	27		
	3.2 min/127 °C	21	13			55	15		
	1.6 min/138 °C	28	16			48	11		

active. The result is a color change from dark-green to olive-green, caused by large amounts of pheophorbides.

A change in color occurs during drying of vegetables, its extent being dependent on water activity. The conversion of chlorophylls to pheophytins continues in blanched vegetables even during frozen storage. In beans and Brussels sprouts immediately after blanching (2 min at 100 °C), the pheophytin content amounts to 8–9%, while after storage for 12 months at −18 °C it increases to 68–83%. Pheophytin content rises from 0% to only 4–6% in paprika peppers and peas under the same conditions.

17.1.2.9.2 Betalains

Pigments known as betalains occur in centrospermae, e.g. in red beet and also in some mushrooms (the red cap of fly amanita). They consist of red-violet betacyanins ($\lambda_{max} \sim 540$ nm) and yellow betaxanthins ($\lambda_{max} \sim 480$ nm). They have the general structure:

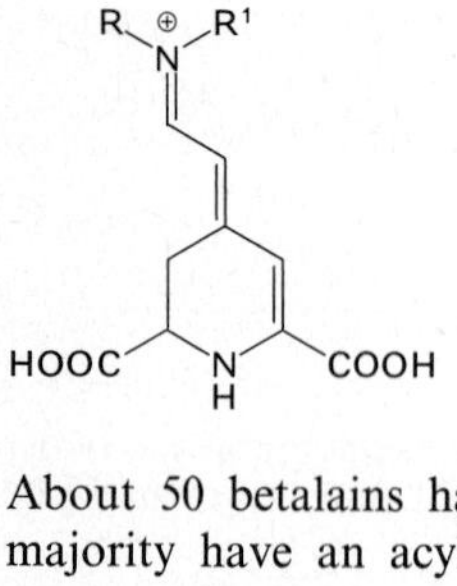

(17.15)

About 50 betalains have been identified. The majority have an acylated sugar moiety. The acids involved are sulfuric, malonic, caffeic, sinapic, citric and *p*-coumaric acids. All betacyanins are derived from two aglycones: betanidin (I) and isobetanidin (II), the latter being the C-15 epimer of betanidin:

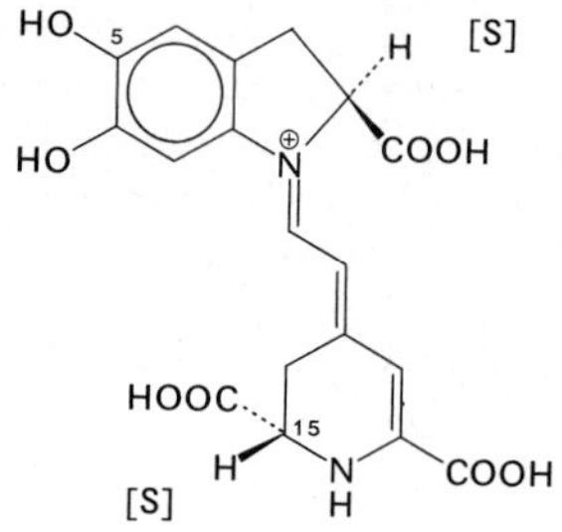

I

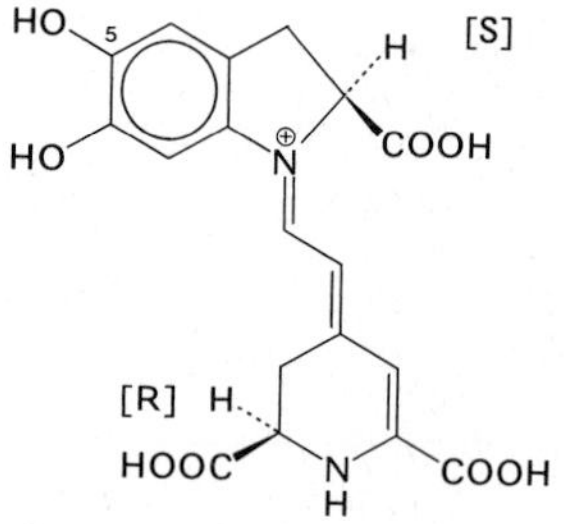

II

(17.16)

Betanin is the main pigment of red beet. It is a betanidin 5-0-β-glucoside. The betaxanthins have only the dihydropyridine ring in common.

The other structural features are more variable than in betacyanins. Examples of betaxanthins are natural vulgaxanthins I and II, also from red beet (*Beta vulgaris*):

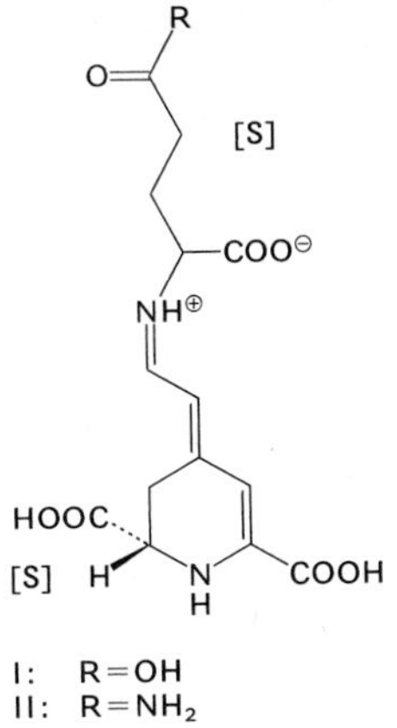

I: R=OH
II: $R=NH_2$

(17.17)

Betalain biosynthesis starts with dopa by opening of its benzene ring followed by cyclization to a dihydropyridine. The (S)-betalamic acid which is formed undergoes condensation with (S)-cyclodopa to betacyanins or with some other amino acids to betaxanthins (cf. reaction sequence 17.18).

17.1.2.9.3 Goitrogenic substances

Brassicaceae contain glucosinolates which decompose enzymatically, e.g. into thiocyanates. For example, in savoy cabbage thiocyanate content is 30 mg/100 g fresh weight, while in cauliflower it is 10 mg and in kohlrabi 2 mg. Since thiocyanates interfere with iodine uptake by the thyroid gland, large amounts of cabbage together with low amounts of iodine in the diet may cause goiter.

Oxazolidine-2-thiones are also goitrogenic. They occur as secondary products in the enzymatic

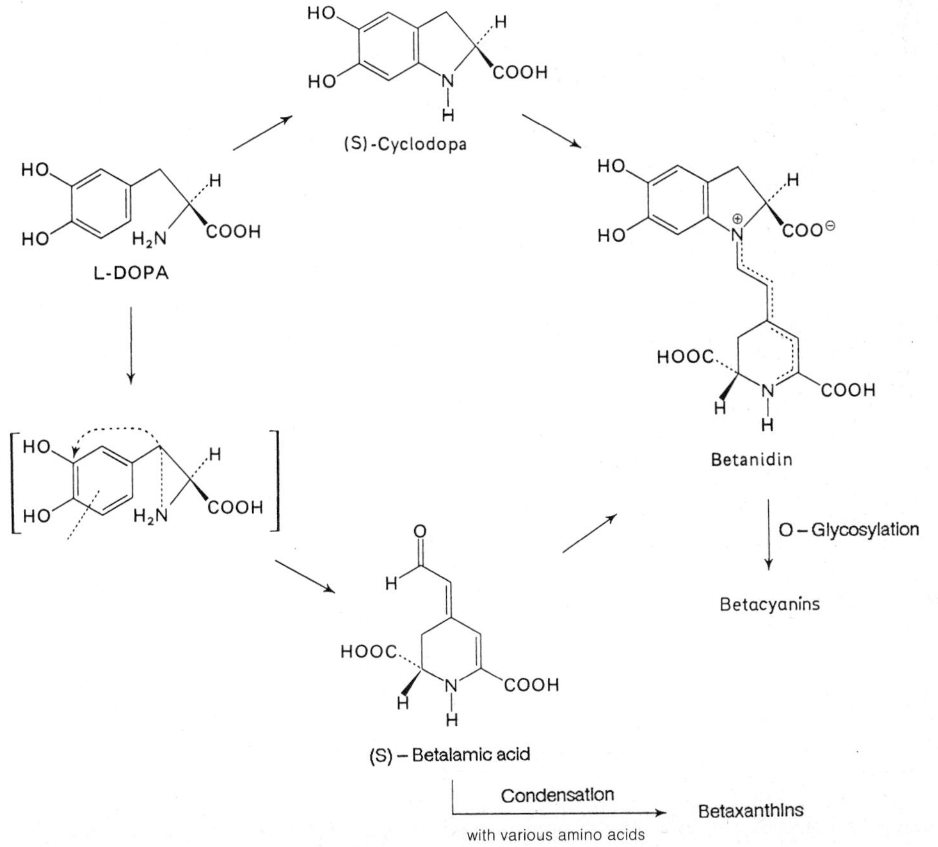

L-DOPA

(S)-Cyclodopa

Betanidin

O-Glycosylation

Betacyanins

(S)-Betalamic acid

Condensation with various amino acids

Betaxanthins

(17.18)

hydrolyzate of glucosinolates when the initially formed isothiocyanates contain an hydroxy group in position 2:

$$\text{Glucosinolate} \xrightarrow[\text{sinase}]{\text{Myro-}} R{-}\underset{\displaystyle OH}{\underset{|}{C}H}{-}CH_2{-}N{=}C{=}S$$

$$\longrightarrow \text{5-R-oxazolidine-2-thione} \qquad (17.19)$$

The levels of the corresponding glucosinolates are up to 0.02% in yellow and white beets and up to 0.8% in seeds of *Brassicaceae* (all members of the cabbage family; kohlrabi, turnip; rapeseed). The leaves contain only negligible amounts of these compounds.

There are 3–15 mg/kg of 5-vinyloxazolidine-2-thione in sliced turnips. Direct intake of thioxazolidones by humans is unlikely since the vegetable is generally consumed in cooked form. Consequently, the myrosinase enzyme is inactivated and there is no release of goitrogenic compounds. An indirect intake is possible through milk when such plants are used as animal feed, resulting in a goitrogenic compound content of 50–100 μg/l of milk. The oxazolidine-2-thiones inhibit the iodination of tyrosine, an effect unlike that of thiocyanates, which may be offset not by intake of iodine but only by intake of thyroxine.

17.1.3 Storage

The storability of vegetables varies greatly and depends mostly on type but also on vegetable quality. While some leafy vegetables, such as lettuce and spinach, and beans, peas, cauliflower, cucumbers, asparagus and tomatoes have limited storage time, root and tuber vegetables, such as carrots, potatoes, kohlrabi, turnips, red table beets, celery, onions and late cabbage cultivars, can be stored for months. Cold storage at high air humidity is the most appropriate. Table 17.13 lists some common storage conditions. The relative air humidity has to be 80–95%. The weight loss experienced for these storage times is 2–10%. Ascorbic acid and carotene contents generally decrease with storage. Starch and protein degradation also occurs and there can be a rise in free acid content of vegetables such as cauliflower, lettuce and spinach.

Table 17.13. Effect of cold storage temperature on vegetables shelf-life

Vegetable	Temperature range (°C)	Shelf-life (weeks)
Cauliflower	−1/0	4–6
Green beans	+3/+4	1–2
Green peas[a]	−1/0	4–6
Kale	−2/−1	12
Cucumber	+1/+2	2–3
Head lettuce	+0.5/+1	2–4
Carrot	−0.5/+0.5	8–10
Green bell pepper	−1/0	4
Leek	−1/0	8–12
Brussels sprouts	−3/−2	6–10
Red beet	−0.5/+0.5	16–26
Celery	−0.5/+1	26
Asparagus	+0.5/+1	2–4
Spinach	−1/0	2–4
Tomato	+1/+2	2–4
Onion	−2.5/−2	40

[a] Kept in pods.

17.2 Vegetable Products

A number of processing techniques provide vegetable products which have a substantially longer storage capability compared to fresh vegetables, and are readily converted into a consumable form. As is the case with dairy products, unique vegetable products can be produced by fermentation.

17.2.1 Dehydrated Vegetables

Vegetable dehydration reduces the natural water content of the plant below the level critical for the growth of microorganisms (12–15%) without being detrimental to important nutrients. Also, it is aimed at preserving flavor, aroma and appearance, and the ability to regain the original shape or appearance by swelling when water is added. The dehydration process is accompanied by significant changes. First, there is a concentration of major ingredients such as proteins, carbohydrates and minerals. This occurs along with some chemical changes. Fats are oxidatively degraded and, although present in low amounts in vegetables, this oxidation often diminishes odor and flavor. Amino compounds and carbohydrates interact in a *Maillard* reaction, resulting in a darker color and development of new aroma substances. Vitamin levels may also drop sharply. The original volatile aroma and flavor compounds are lost to a great extent.

In the production of the dehydrated product, the vegetable is first washed, peeled or cleaned, and may be sliced or diced. Blanching for 2–7 min to inactivate the enzymes is then done in hot water or steam. Vegetables may also be treated with SO_2.

Dehydration is performed in a conveyor or tube dryer at 55–60 °C to a residual moisture content of 4–8%. Liquid or paste forms, such as tomato or potato mash, are dried in a spray or drum dryer or, in the case of some special products, in a fluidized bed dryer. Dehydration by freeze-drying provides high quality products (good shape retention) with a spongy and porous structure that is readily rehydrated. Some vegetables used in soup powders, e.g. peas and cauliflower, are prepared in this way. For production of dehydrated potato products, tubers are peeled, cleaned, sliced into strings or chips or diced and, after steam-cooking, dried. For production of dehydrated mashed potatoes or flakes, the steamed slices are squeezed between rollers into a mash with the least possible damage to cell walls. Cell wall damage allows the gelatinized starch to escape from the ruptured cells and to later impart a gluey-sticky texture to the reconstituted mash.

Dehydrated vegetables are light, air and moisture sensitive and therefore require careful packaging. Wax-impregnated paper or cardboard, multilayer foils, metal cans or glass containers are commonly used and, occasionally, the packaging is done under nitrogen or vacuum. Also, the dehydrated product may be pressed prior to packaging.

17.2.2 Canned Vegetables

Canning, which involves heat sterilization, is one of the most important processes in vegetable preservation. The selected and sorted freshly-harvested products are trimmed and blanched as outlined with dehydrated vegetables. Blanching here serves not only to inactivate the enzymes, but to remove both undesirable flavoring compounds, such as occur in cabbages, and the air present in plant tissue, and to induce shrinkage or softening of the product, thereby increasing packaging density.

Brine (1–2% NaCl solution) often serves as a filling liquid. Sugar (peas, red table beets, tomato, sweet corn), citric acid (up to 0.05%, used for example for celery, cauliflower and horse beans), calcium salts for firming the plant tissue (tomato, cauliflower) or monosodium glutamate (100–150 mg per kg filling) are also added to round-off the flavor.

Sterilization is performed in autoclaves. Temperature and duration of heating are adjusted to match the type and texture of the vegetable and the size and type of container. For sterilization, a kilogram of vegetable in a sealed container requires 18 min at 115 °C for asparagus, 18 min at 118 °C for peas, 15 min at 116 °C for cauliflower and 20 min at 118 °C for chanterelle mushrooms. Heating and cooling require an additional 6–10 min. As with other foods, vegetable sterilization processes tend toward higher temperatures and shorter times since, in this way, the products retain a better quality (texture, aroma, color).

The nutritional/physiological value of the main constituents of vegetables (proteins and carbohydrates) is not diminished by this common heat sterilization process. Damage due to interaction of amino acids with reducing sugars, which occurs to a small extent, is also negligible. However, there is often a negative effect on vitamins (cf. Tables 6.1 and 6.2). Carotene, a fat-soluble provitamin A, is not affected by the washing and blanching steps, but it is moderately destroyed (5–30%) during actual canning. Vitamin B_1 in carrots and tomatoes does not decrease significantly, while losses are 10–50% for other vegetables (green beans, peas and asparagus). Vitamin B_2 losses are high in spinach (66%) due to large surface area. Vitamin B_2 is lost (5-25%) by leaching during blanching, but not significantly during further processing. Nicotinic acid losses are similar. Vitamin C losses are due to its water solubility and its enzymatic and chemical degradation, particularly in the presence of traces of heavy metal ions. Vitamin C retention is 55–90% during the canning of asparagus, peas and green beans. Storage of canned vegetables for several years generally results in an additional 20% vitamin loss.

17.2.3 Deep-Freezing of Vegetables

Beans, peas, paprika peppers, Brussels sprouts, edible mushrooms (*Boletus edulis*), tomato pulp and carrots are particularly suitable for freezing. Radishes, lettuce or whole tomatoes are unsuitable. High quality fresh vegetables are treated in boiling water for 1.5–4 min, or by steam for 2–5 min for enzyme inactivation. The blanching time is generally shorter than that used in canning, and varies according to type, ripeness and size of vegetable. It is kept as short as possible to prevent leaching. Steam blanching is generally more advantageous. A peroxidase enzyme test is suitable to verify the correct blanching time required for enzyme inactivation (cf. 2.6.4).

Immediately after blanching, the vegetable is cooled, frozen at −40 °C or lower in a plate or air freezer, then stored at −18 to −20 °C.

Freezing preserves vegetable nutrients to a great extent. Vitamin A and its provitamin, carotene, are well preserved in spinach, peas and beans, or are moderately lost as in asparagus, after proper blanching, during freezing and deep-freeze storage and even after thawing to room temperature. Losses in the Vitamin B group depend mostly on the conditions of the primary processing steps (washing, blanching). The other steps have no effect on B vitamins. Vitamin C leaching by water or steam is detrimental. It is generally preserved during freezing and thawing. Careful blanching and low temperature storage are critical for vitamin C preservation (Figures 17.4 and 17.5).

Irreversible textural changes can occur in deep-frozen vegetables. Typical symptoms are softening, ductile stickiness, or looseness or flaccidity (beans, cucumbers, carrots); build-up of a sticky, ductile, gum-like structure (asparagus), or pasty, soggy structure (celery, kohlrabi); or hull hardening (peas).

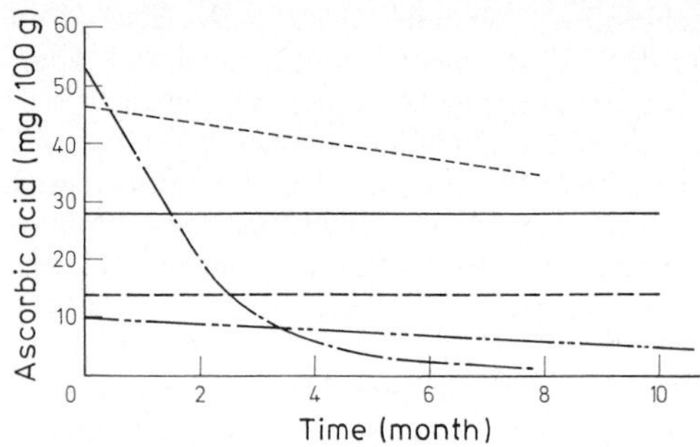

Fig. 17.4. Changes in vitamin C content in frozen vegetables kept at −21 °C. —— Peas precooked, – – – beans precooked, –··–··– beans raw, –·–·– spinach raw, ----- spinach precooked. (After *W. Heimann*, 1958)

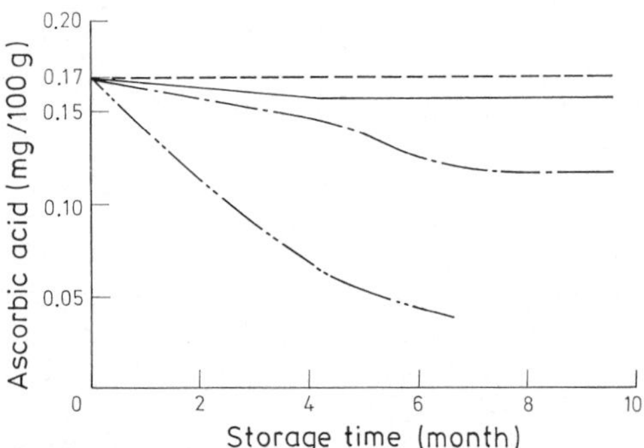

Fig. 17.5. Ascorbic acid losses in frozen peas as influenced by storage temperature. – – – −40 °C, —— −18 °C, –·–·– −12 °C, –··–··– −9 °C. (After *Schormueller*, 1966)

17.2.4 Pickled Vegetables

Pickled vegetables are produced by spontaneous lactic acid fermentation (white cabbage, green beans, cucumbers, etc.). The fermentation lowers the pH, inhibits the growth of undesirable acid-sensitive microorganisms and, simultaneously, affects the enzymatic softening of cells and their tissues, thus improving digestibility and wholesomeness. The use of salt also has a preservative effect. The acidic pH of the medium stabilizes vitamin C.

While the preservation techniques outlined in earlier sections were aimed at retention of the original odor and flavoring substances of the raw material, including regeneration of lost aroma constituents, this is not important in pickled vegetables since a new typical aroma is developed.

17.2.4.1 Pickled Cucumbers (Salt and Dill Pickles)

Unripe cucumbers, after addition of dill herb and, if necessary, other flavoring spices (vine leaf, garlic or bay leaf), are placed into 4–6% NaCl solution or are sometimes salted dry. Usually, the salt solution is poured on the cucumbers in a barrel and then allowed to ferment and, if necessary, glucose is added. Fermentation takes place at 18–20 °C and initially is heterofermentative (*Leuconostoc* ssp.). In the main phase, fermentation becomes homofermentative (*Lactobacillus plantarum*, etc., including yeasts), and lactic acid, CO_2, some volatile acids, ethanol and small amounts of various aroma substances are formed. The lactic acid (0.5–1%) initially formed is later metabolized partly by film yeast or oxidative yeasts that grow on the surface of the brine. Thus, the original pH value of the fermenting medium (3.4–3.8) is slightly increased.

17.2.4.2 Other Vegetables

Green beans, carrots, kohlrabi, celery, asparagus, turnips and others are processed similarly to cucumbers. Sliced green beans, for example, are treated with salt (2.5–3%), subjected to lactic acid fermentation at about 20 °C, and marketed in barrels, cans or glass jars ("Rheinische Brühbohnen"). Some pickled vegetables, mostly those that were not blanched or precooked, will not soften during later cooking.

17.2.4.3 Sauerkraut

Lactic acid fermentation has been used for millenia for the production of sauerkraut. It was also customary earlier to place the cabbage into acidified wine or vinegar. White cabbage heads are cut into 0.75–1.5 mm thick shreds, then mixed with salt at 1.5–2.5% by weight. The shreds are then packed into tanks of wood or reinforced concrete, coated with synthetics. After the shreds have been packed in layers, they are tamped and weighted down so that a layer of expressed brine juice covers the surface. The lactic acid fermentation occurs spontaneously at 18–24 °C for 3–6 weeks. The acid formed inhibits the growth of competing interfering microorganisms. The amount of acid formed depends on the initial sugar content of the cabbage. Hence, sugar is sometimes added (to 1%) to cabbage which does not ferment readily. In addition to *Lactobacillus* ssp., yeasts are also involved in fermentation. The products are lactic and acetic acids (in ratios of 4 : 1 to 6 : 1), ethanol (0.2–0.8%), CO_2, mannitol (from fructose) and, most importantly, aroma substances which appear in the prefermentation phase. After fermentation is complete, the sauerkraut pH is about 3.6. Lactic acid values less than 6 g/l indicate unsatisfactorily fermented cabbage. The end-product is kept in barrels under brine. The sauerkraut is also packaged or canned in retail containers. The cans are filled at 70 °C, then exhausted, sealed and sterilized at 95–100 °C. In addition, sauerkraut is packed and distributed in plastic foils and containers. Mildly acidic sauerkraut, preferred in South Germany, is produced by stopping the fermentation before all the sugar is degraded. After pasteurization, the product can be stored for a longer time and still retains a clearly sour taste. Sauerkraut is flavored and spiced to some extent by addition of sugar, juniper berries, caraway or dill seeds. For wine sauerkraut at least 1 liter of wine per 50 kg sauerkraut is added after fermentation.

Drained sauerkraut contains on the average 90.7% water, 1.5% nitrogen compounds, 0.3% crude fat, 3.9% carbohydrates, 1.1% crude fiber, 0.6% minerals (excluding NaCl), 0.8–3.3% NaCl, 1.4–1.9% titratable acid (calculated as lactic acid; 0.28–0.42% is acetic acid) and 0.29–0.61% ethanol. There are small amounts of formic, *n*-heptanoic and *n*-octanoic acids, methanol, and compounds important for palatability, i.e. dextran and mannitol. Vitamin C content (10–38 mg/100 g) is not changed when sauerkraut is heated in a pressure cooker. However, after several reheatings about 30% is destroyed.

17.2.4.4 Faulty Processing of Pickles

Pickled cucumbers are often softened due to the effects of their own or microbial pectolytic enzymes. Brown-to-black discoloration is caused by iron sulfide build-up or by black pigments formed by microorganisms (*Bacillus nigrificans*). Hollowness is caused by gas-forming microorganisms, i.e. gaseous fermentation, and can be prevented readily by pickling in the presence of sorbic acid.

Sauerkraut is darkened by chemical or enzymatic oxidations when the brine does not cover the surface of the pickles. Reddish color is caused by yeasts. Sauerkraut softening occurs when fermentation takes place at too high a temperature; the kraut is exposed to air; too little salt is added; or by faulty fermentation when the lactic acid content remains too low. In addition to faulty fermentation, the kraut can be ruined by infections caused by molds and other flora of the surface film and by rottening (insufficient brine for full protection).

17.2.5 Vinegar-Pickled Vegetables

These products are prepared by pouring preboiled and still hot vinegar onto the vegetables. Vegetables used are cucumbers, red table beets, pearl and silver onions, paprika peppers, mixed

vegetables, which also include cauliflower, carrots, onions, peas, mushrooms (in particular the table mushroom, *Boletus edulis*), asparagus, tender corncobs, celery, parsley root, parsnip, kohlrabi, pumpkin and pepperoni peppers.
Only unblemished raw material is used. The vegetable is covered with a solution of 2.5% vinegar. Salt, spices and herbs, herb extracts, sugar and chemical preservatives are usually added. Depending on the vegetable and its preparation method, there are "single pickles" in vinegar (vinegar cucumbers, chili pepper-flavored cucumbers or gherkins, mustard cucumbers, sterilized deli and spiced garlic, dill-flavored cucumbers) and "mixed pickles" in vinegar, which are made partly from fresh and partly from precanned vegetables (unsliced cucumbers, cauliflower, onions, delicate and tender corncobs, paprika peppers). When an infusion of mustard paste is added, they are marketted as "Piccalilly" mixed pickles.
Mixed salads pickled in vinegar are marketed under various trade names. They are made of onions, green or red bell peppers (pimientos), and/or tender corncobs. Such products are partially pasteurized.

17.2.6 Stock Brining of Vegetables

Salting is a practical method for preserving some vegetables in bulk until further processing. Usually the vegetable is salted with table salt after being blanched. Brined vegetables are kept for the production of other products. Salted asparagus, for example, is obtained by addition of ~20% by weight of salt and used for the preparation of "Leipzig medley" and mixed fresh vegetables. Stock brining of beans is also important. Blanched or nonblanched beans are soaked in salt brine or are treated with dry salt to 10–20% by weight (added by hand or by machine spreading or dusting) and kept in brine prior to the manufacture of other products. As with other vegetables, the beans are thoroughly drained of brine and rinsed in a stream of hot water before further processing. In the same way, vegetables such as cauliflower, cabbage, carrots, pearl onions and gherkins are stock brined. Mushrooms and morels are also salted; a practice primarily found in Poland and the USSR.

17.2.7 Vegetable Juices

The vegetable is cleaned, washed, then blanched and disintegrated in a mill. In some instances, e.g. the tomato, it is first disintegrated and the slurry heated to > 70 °C for some time. The juice is then separated in presses or by centrifuging and salt is usually added at 0.25–1%. Nonsour juices are mixed with lactic or citric acid. For storage stability, such products are subjected to pasteurization by plate heat exchangers. Mostly tomatoes and occasionally other vegetables such as cucumbers, carrots, red beets, radishes, sauerkraut, celery or spinach are used for processing into juice.

17.2.8 Vegetable Purée (Paste)

A vegetable purée or paste is a finely dispersed slurry from which skins and seeds have been removed by passing the slurry through a pulper or finisher. The most widely used of these products is tomato purée which, depending on the brand, has a dry matter content of 14–36% and contains 0.8–2% NaCl. Some other vegetable purées are important primarily as baby foods.

17.2.9 Vegetable Powders

Vegetable powders are obtained by drying the corresponding juice with or without addition of a drying enhancer, such as starch or a starch degradation product, to a residual moisture content of about 3%. Drying processes used are spray-drying; vacuum drum drying; and freeze-drying. The most important product is tomato powder. Other powders, such as those of spinach or red beets, are in part used in food colorings.

17.3 Literature

Adler, G.: Kartoffeln und Kartoffelerzeugnisse. Verlag Paul Parey: Berlin–Hamburg. 1971

Bötticher, W.: Technologie der Pilzverwertung. Verlag Eugen Ulmer: Stuttgart. 1974

Hadziyev, D., Steele, L.: Dehydrated mashed potatoes chemical and biochemical aspecty. Adv. Food Res. 25, 55 (1979)

Herrmann, K.: Gemüse und Gemüsedauerwaren. Verlag Paul Parey: Berlin–Hamburg. 1969

Herrmann, K.: Tiefgefrorene Lebensmittel. Verlag Paul Parey: Berlin–Hamburg. 1970

Kaiser, K.-P., Bruhn, L. C., Belitz, H.-D.: Protease inhibitors in potatoes. Protein-, trypsin- and chymotrypsin inhibitor patterns by isoelectric focusing in polyacrylamidegel. A rapid method for identification of potato varieties. Z. Lebensm. Unters. Forsch. *154,* 339 (1974)

Ross, A. E., Nagel, D. L., Toth, B.: Evidence for the occurrence and formation of diazonium ions in the Agaricus bisporus mushroom and its extracts. J. Agric. Food Chem. *30,* 521 (1982)

Salunkhe, D. K., Do, J. Y.: Biogenesis of aroma constituents of fruits and vegetables. Crit. Rev. Food Sci. Nutr. *8,* 161 (1977)

Schobinger, U.: Frucht- und Gemüsesäfte, Verlag Eugen Ulmer: Stuttgart. 1978

Schormüller, J.: Die Erhaltung der Lebensmittel. Ferdinand Enke Verlag: Stuttgart. 1966

Shah, B. M., Salunkhe, D. K., Olson, L. E.: Effects of ripening processes on chemistry of tomato volatiles. J. Am. Soc. Hort. Sci. *94,* 171 (1969)

Stamer, J. R.: Lactic acid fermentation of cabbage and cucumbers. In: Biotechnology (Eds.: Rehm, H.-J., Reed, G.), Vol. 5, p. 365, Verlag Chemie: Weinheim 1983.

Whitaker, J. R.: Development of flavor odor and pungency in onion and garlic. Adv. Food Res. *22,* 73 (1976)

18 Fruits and Fruit Products

18.1 Fruits

18.1.1 Foreword

Fruits include both true fruits and spurious fruits, as well as seeds of cultivated and wild perennial plants. Fruits are commonly classified as pomaceous fruits, stone fruits, berries, tropical and subtropical fruits, hard-shelled dry fruits and wild fruits. The most important fruits are presented in Table 18.1 with pertinent data on botanical classification and use. Table 18.2 provides data about fruit production.

18.1.2 Composition

Fruit composition can be strongly influenced by the variety and ripeness, thus data given should be used only as a guide. Table 18.3 shows that the dry matter content of fruits (berries and pomme, stone, citrus and tropical fruits) varies between 10–20%. The major constituents are sugars, polysaccharides and organic acids, while N-compounds and lipids are present in lesser amounts. Minor constituents include pigments and aroma substances of importance to organoleptic quality, and vitamins and minerals of nutritional importance. Nuts are highly variable in composition (Table 18.4). Their moisture content is below 10%, N-compounds are about 20% and lipids are as high as 50%.

18.1.2.1 N-Containing Compounds

Fruits contain 0.1–1.5% N-compounds, of which 35–75% is protein. Free amino acids are also widely distributed. Other nitrogen compounds are only minor constituents. The special value of nuts, with their high protein content, has already been outlined.

18.1.2.1.1 Proteins

The protein fraction varies greatly with fruit variety and ripeness. This fraction is primarily enzymes. Besides those involved in carbohydrate metabolism (e. g. pectinolytic enzymes, cellulases, amylases, phosphorylases, saccharases, enzymes of the pentose phosphate cycle, aldolases), there are enzymes involved in lipid metabolism (e. g. lipases, lipoxygenases, enzymes involved in lipid biosynthesis), and in the citric acid and glyoxylate cycles, and many other enzymes such as acid phosphatases, ribonucleases, esterases, catalases, peroxidases, phenoloxidases and O-methyl transferases.

Protein and enzyme patterns, which can be obtained, for example, by electrophoretic separation, are generally highly specific for fruits, and can be utilized for analytical differentiation of the species and variety. Figure 18.1 shows protein patterns of various grape species and Figure 18.2 presents enzyme patterns of various species and cultivars of strawberries.

18.1.2.1.2 Free amino acids

Free amino acids are an average of 50% of the soluble N-compounds. The amino acid pattern is typical for a fruit and hence can be utilized for the analytical characterization of a fruit product. Table 18.5 provides some relevant data.

In addition to common protein-building amino acids, there are nonprotein amino acids present in fruits, as in other plant tissues. Examples are the toxic 2-(methylene cyclopropyl)-glycine (I) in litchi fruits *(Litchi sinensis)*, the toxic hypoglycine A (II) in akee *(Blighia sapida)*, 1-aminocyclopropane-1-carboxylic acid (X) in apples and pears, trans-4-methylproline (XXII), 4-hydroxymethylprolines (XXIII–XXV) and 4-methyleneproline (XXVI) in apples and in loquat fruits *(Eviobotrya japonica)*, 3,4-dihydroxyglutamic acid (XXXV) in red currants, 4-methyleneglutamic acid (XXXI) and 4-methyleneglutamine (XXXII) in peanuts and 3-amino-3-carboxypyrrolidine) (LIV) in cashew. The nonprotein amino acids are discussed in more detail in Section 17.1.2.1.2. The Roman numerals given in brackets above correspond to Tables 17.4 and 17.5.

Table 18.1. Edible fruits: a classification

Number	Common name	Latin name	Family/subfamily	Form of consumption
Pomme fruits				
1	Apple	*Malus sylvestris*	Rosaceae	Fresh, dried, purée, jelly, juice, apple cider, brandy
2	Pear	*Pyrus communis*	Rosaceae	Fresh, dried, compote, brandy
3	Quince apple shaped pear shaped	*Cydonia oblonga* *var. maliformis* *var. pyriformis*	Rosaceae	Jelly, ingredient of apple purée
Stone fruits				
4	Apricot	*Prunus armeniaca*	Rosaceae	Fresh, dried, compote, marmalade, juice, seed for percipan, brandy
5	Peach	*Prunus persica*	Rosaceae	Fresh, compote, juice, brandy
6	Prune/plum	*Prunus domestica*	Rosaceae	Fresh, dried, compote, jam, marmalade, brandy
7	Sour cherry	*Prunus cerasus*	Rosaceae	Fresh, compote, marmalade, juice, brandy
8	Sweet cherry	*Prunus avium*	Rosaceae	Fresh, candied, compote
Berry fruits				
9	Blackberry	*Rubus fruticosus*	Rosaceae	Fresh, marmalade, jelly, juice, wine, liqueur
10	Strawberry	*Fragaria vesca*	Rosaceae	Fresh, compote, marmalade, brandy
11	Bilberry	*Vaccinium myrtillus*	Ericaceae	Fresh, compote, marmalade, brandy
12	Raspberry	*Rubus idaeus*	Rosaceae	Fresh, marmalade, jelly syrup, brandy
13	Red currant	*Ribes rubrun*	Saxifragaceae	Fresh, jelly, juice, brandy
14	Black currant	*Ribes nigrum*	Saxifragaceae	Fresh, juice, liqueur
15	Cranberry	*Vaccinium vitis-idaea*	Ericaeae	Compote
16	Gooseberry	*Ribes uva-crispa*	Saxifragaceae	Unripe: compote; ripe: fresh, marmalade, juice
17	Grapes	*Vites vinifera* *ssp. vinifera*	Vitaceae	Fresh, dried (raisins) juice, wine, brandy
Citrus fruits				
18	Orange	*Citrus sinensis*	Rutaceae	Fresh, juice
19	Grapefruit	*Citrus paradisi*	Rutaceae	Fresh juice
20	Kumquat	*Fortunella margarita*	Rutaceae	Fresh, compote, marmalade
21	Mandarine	*Citrus reticulata*	Rutaceae	Fresh compote
22	Pomelo	*Citrus maxima*	Rutaceae	Fresh, juice
23	Seville orange	*Citrus aurantium* *ssp. aurantium*	Rutaceae	Candied, marmalade
24	Lemon	*Citrus limon*	Rutaceae	Juice
25	Citron	*Citrus medica*	Rutaceae	Peel candied (citronat)
Other tropical/subtropical fruits				
26	Acerola	*Malpighia emarginata*	Malpighiaceae	Fresh, compote, juice
27	Pineapple	*Ananas comosus*	Bromeliaceae	Fresh, compote, marmalade, juice
28	Avocado	*Persea americana*	Lauraceae	Fresh
29	Banana	*Musa*	Musaceae	Fresh, dried, cooked, baked
30	Cherimoya	*Annona cherimola*	Annonaceae	Fresh
31	Date	*Phoenix dactylifera*	Arecaceae	Fresh, dried

Table 18.1. (Continued)

Number	Common name	Latin name	Family/subfamily	Form of consumption
32	Fig	*Ficus carica*	Moraceae	Fresh, dried, marmalade, dessert wine
33	Indian fig	*Opuntia ficus-indica*	Cactaceae	Fresh
34	Guava	*Psidium guajava*	Myrtaceae	Compote, juice
35	Persimmon	*Diospyros kaki*	Ebenaceae	Fresh, candied, compote
36	Kiwi	*Actinidia chinensis*	Actinidiaceae	Fresh, compote
37	Litchi	*Litchi chinensis*	Sapindaceae	Fresh, dried, compote
38	Mango	*Mangifera indica*	Anacardiaceae	Fresh, compote, juice
39	Melons			
	cantaloups	*Cucumis melo*	Cucurbitaceae	Fresh
	watermelon	*Citrullus lanatus*	Cucurbitaceae	Fresh
40	Papaya	*Carica papaya*	Caricaceae	Fresh, compote, juice
41	Passion fruit	*Passiflora edulis*	Passifloraceae	Fresh, juice
42	Golden shower	*Cassia fistula*	Caesalpiniaceae	Fresh
Shell(nut) fruits				
43	Cashew nut	*Anacardium occidentale*	Anacardiaceae	Roasted
44	Peanut	*Arachis hypogaea*	Fabaceae	Roasted, salted
45	Hazel-nut (Filbert)	*Corylus avellana*	Betulaceae	Fresh, baked and confectionary products (nougat, crocant)
46	Almond	*Prunus dulcis*	Rosaceae	Baked and confectionary products (marzipan); flavoring of baked and confectionary products
	sweet	*var. dulcis*		
	bitter	*var. amara*		
47	Brazil nut	*Bertholletia excelsa*	Lecythidaceae	Fresh
48	Pistachio	*Pistacia vera*	Anacardiaceae	Fresh, salted, sausage flavoring, decoration of baked products
49	Walnut	*Juglans regia*	Juglandaceae	Fresh, baked and confectionary products, unripe fruits; vinegar and sugar-containing preserves
Wild fruits				
50	Rose hips	*Rosa sp.*	Rosaceae	Marmalade, wine
51	Elderberry	*Sambucus nigra*	Caprifoliaceae	Juice, jam
52	Seabuckthorn	*Hippophae rhamnoides*	Elaeagnaceae	Marmalade, juice

18.1.2.1.3 Amines

A number of aliphatic and aromatic amines are found in various fruits (Tables 18.6 and 18.7). They are formed in part by amino acid decarboxylation such as in apples, or by amination (cf. Reaction 18.1 a) or transamination of aldehydes (cf. Reaction 18.1 b).

$$R{-}CHO \xrightarrow[2[H]]{NH_3} R{-}CH_2{-}NH_2 \qquad (18.1a)$$

$$R{-}CHO + R^1{-}\underset{NH_2}{\underset{|}{CH}}{-}COOH \longrightarrow R{-}CH_2{-}NH_2 + R^1{-}\underset{O}{\underset{\|}{C}}{-}COOH \qquad (18.1b)$$

Some amines are derived from tyramine (e.g. hordenine, synephrine, octopamine, dopamine and noradrenaline; cf. Formula 18.2).

18.1.2.2 Carbohydrates

18.1.2.2.1 Monosaccharides

In addition to glucose and fructose, the ratios of which vary greatly in various fruits (Table 18.8), other monosaccharides occur only in trace amounts. For example, arabinose and xylose have been found in several fruits. An exceptional case is avocado in which a number of higher sugars are present at 0.2 to 5.0% of the fresh

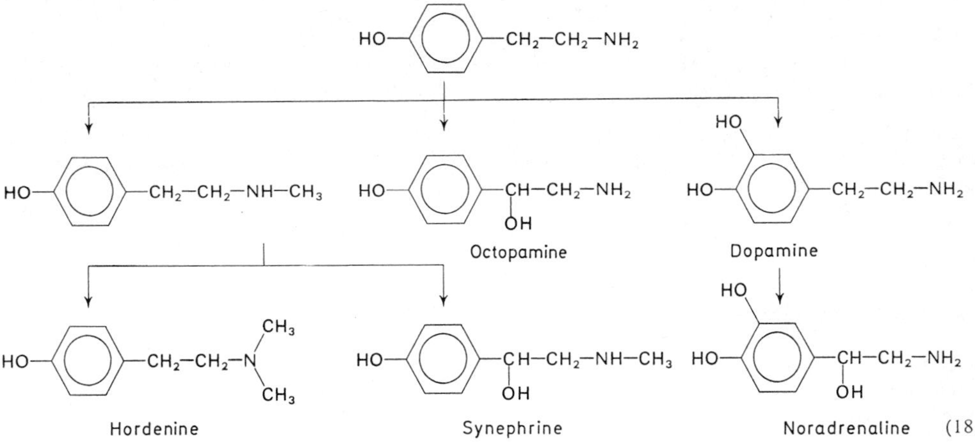

$HO-C_6H_4-CH_2-CH_2-NH_2$

$HO-C_6H_4-CH_2-CH_2-NH-CH_3$

$HO-C_6H_4-CH(OH)-CH_2-NH_2$ Octopamine

$(HO)_2C_6H_3-CH_2-CH_2-NH_2$ Dopamine

$HO-C_6H_4-CH_2-CH_2-N(CH_3)_2$ Hordenine

$HO-C_6H_4-CH(OH)-CH_2-NH-CH_3$ Synephrine

$(HO)_2C_6H_3-CH(OH)-CH_2-NH_2$ Noradrenaline (18.2)

Table 18.2. World production of fruits in 1981 (1,000 t)

Continent	Fruits[a], grand total	Nuts, grand total	Grapes	Raisins	Dates
World	282,526	3,999	61,739	960	2,718
Africa	33,128	321	2,119	24	1,054
America, North-, Central-	44,380	747	4,666	184	24
America, South-	40,686	141	4,590	8	
Asia	78,028	1,561	7,279	506	1,628
Europe, West-	54,018	1,009	30,368	168	12
Europe, East- + USSR	28,621	212	11,976	10	
Australia + South Pacific Islands	3,665	9	741	60	

Continent	Apples	Pears	Peaches + nectarines	Plums/ prunes	Oranges	Manda-rins[b]	Lemons
World	31,915	8,535	7,309	5,210	37,544	7,046	5,403
Africa	474	218	222	61	3,422	574	239
America, North-, Central-	4,185	900	1,636	777	12,035	730	1,791
America, South-	1.433	243	560	101	11,773	837	680
Asia	8,239	2,841	1,213	848	5,858	3,714	1,259
Europe, West-	7,260	3,007	2,974	773	3,809	1,154	1,387
Europe, East- + USSR	9.760	1,167	597	2,625	254		
Australia + South Pacific Islands	564	167	108	25	394	36	47

Continent	Grape-fruit	Apricots	Avocado	Mango	Pine-apple	Bananas	Floury-bananas	Papaya
World	4,450	1,550	1,534	13,444	8,866	39,925	22,410	1,891
Africa	290	172	121	852	1,247	4,442	13,639	220
America, North-, Central-	2,865	93	990	1,316	1,405	7,266	1,631	425
America, South-	320	27	338	866	1,070	11,856	4,659	541
Asia	922	427	81	10,401	5,003	14,788	2,477	689
Europe, West	19	495	2		1	475		
Europe, East- + USSR		303						
Australia + South Pacific Islands	35	34	3	9	140	1,100	4	17

[a] Without melons and nuts. [b] Inclusive of tangerines, clementines and satsumas.

Table 18.2. (Continued)

Continent	Strawberries	Raspberries	Currants	Almonds	Pistachio-nuts	Hazelnuts	Cashew nuts	Sweet chestnuts	Walnuts
World	1,674	218	420	1,064	86	552	454	542	824
Africa				51			171		6
America, North-, Central-	437	22		310	6	13	3		209
America, South-	13						85	10	7
Asia	316			134	76	368	196	370	305
Europe, West-	515	3	199	535	4	166		150	123
Europe, East- + USSR	386	180	220	32		5		12	174
Australia + South Pacific Islands	8	2	2	3					

Country	Fruits, grand total	Country	Nuts, grand total	Country	Grapes
USA	25,912	USA	709	Italy	12,400
Italy	20,479	Turkey	591	France	8,800
Brazil	19,659	Italy	421	USSR	6,700
India	18,510	China	385	Spain	5,239
USSR	15,030	Spain	370	USA	4,018
France	12,019	India	206	Turkey	3,600
Spain	11,009	Iran	126	Argentina	2,700
China	9,029	Brazil	116	Rumania	1,755
Turkey	7,852	USSR	92	Greece	1,603
Mexico	7,443	Greece	87	South Africa	1,000
Philippines	6,607				
Japan	6,155	Σ (%)[c]	78	Σ (%)[c]	77
Argentina	5,578				
Thailand	4,221				
Colombia	4,058				
Equador	3,987				
Uganda	3,953				
Greece	3,636				
Nigeria	3,200				
Indonesia	3,112				
Rumania	3,067				
Yugoslavia	2,970				
South Africa	2,858				
Iran	2,618				
Egypt	2,563				
FR Germany	2,507				
Zaire	2,493				
Σ (%)[c]	75				

[c] World production ≙ 100%.

Table 18.2. (Continued)

Country	Raisins	Country	Dates	Country	Apples
Turkey	366	Saudi Arabia	429	USSR	6,000
USA	180	Egypt	428	USA	3,468
Greece	164	Iraq	405	China	3,068
Afghanistan	70	Iran	301	France	1,840
Australia	60	Algeria	206	Italy	1,750
Iran	50	Pakistan	205	Turkey	1,479
South Africa	18	Sudan	119	Hungary	1,100
Rumania	10	Morocco	105	Spain	1,064
Syria	8	Libya	88	Argentina	905
China	7	Yemen	84	Japan	847
				India	822
Σ (%)[c]	97	Σ (%)[c]	87	Poland	800
				FR Germany	772
				Σ (%)[c]	75

Country	Pears	Country	Peaches	Country	Plums/ prunes
China	1,660	Italy	1,550	Yugoslavia	809
Italy	1,160	USA	1,430	USA	695
USA	810	France	480	Rumania	590
USSR	620	Greece	448	USSR	520
Japan	522	Spain	441	China	433
Spain	520	China	387	Poland	210
France	435	Japan	287	Italy	180
Turkey	339	Turkey	250	FR Germany	171
FR Germany	276	USSR	250	Hungary	170
Australia	150	Argentina	222	Turkey	161
Poland	100	Σ (%)[c]	79	Σ (%)[c]	76
Σ (%)[c]	76				

Country	Oranges	Country	Mandarins	Country	Lemons
USA	9,547	Japan	2,841	USA	1,139
Brazil	9,315	Spain	725	Italy	750
Italy	1,600	USA	564	Mexico	530
Mexico	1,600	Brazil	470	India	490
Spain	1,500	Italy	340	Spain	431
India	1,180	Morocco	280	Argentina	300
Egypt	1,137	China	267	Turkey	300
Israel	920	Argentina	208	Greece	185
Turkey	723	Pakistan	200	Brasil	115
Morocco	720	Turkey	174	Peru	91
Σ (%)[c]	75	Σ (%)[c]	86	Σ (%)[c]	82

[c] World production ≙ 100%.

Table 18.2. (Continued)

Country	Grapefruit	Country	Apricots	Country	Avocado
USA	2,503	Spain	174	Mexico	474
Israel	536	Turkey	166	USA	210
Mexico	163	USSR	125	Brazil	140
Argentina	156	Italy	113	Dominican Rep.	134
China	147	Greece	104	Peru	64
Cuba	110	France	85	Haiti	59
South Africa	110	USA	81	Venezuela	48
Cyprus	89	Morocco	65	Indonesia	46
Sudan	59	Iran	55	Equador	30
Paraguay	58	Syria	51	El Salvador	30
Σ (%)[c]	88	Hungary	50	Σ (%)[c]	81
		South Africa	38		
		Afghanistan	37		
		Pakistan	35		
		Σ (%)[c]	76		

Country	Mango	Country	Pineapple	Country	Bananas
India	8,516	Thailand	1,800	Brazil	6,696
Mexico	620	Philippines	1,200	India	4,500
Brazil	600	Brazil	625	Philippines	4,000
Pakistan	550	India	593	Equador	2,275
Philippines	380	USA	590	Thailand	2,021
Indonesia	357	Mexico	568	Indonesia	1,622
Haiti	330	Ivory Coast	350	Mexico	1,562
China	289	Vietnam	350	Honduras	1,330
Bangladesh	210	China	298	Colombia	1,155
Dominican Rep.	180	Indonesia	266	Costa Rica	1,144
Σ (%)[c]	90	Σ (%)[c]	75	Panama	1,082
				Burundi	983
				Venezuela	980
				Papua/New Guinea	932
				Σ (%)[c]	76

Country	Floury-bananas	Country	Papaya	Country	Strawberries
Uganda	3,550	Brazil	380	USA	336
Colombia	2,400	Mexico	322	Japan	204
Nigeria	2,250	Indonesia	277	Italy	184
Ruanda	2,100	India	270	Poland	157
Sri Lanka	1,800	Zaire	159	France	80
Zaire	1,473	Philippines	65	USSR	80
Cameroon	1,026	Peru	47	Korea, South-	78
Ghana	900	Venezuela	45	Mexico	77
Ivory Coast	830	Cuba	38	Spain	51
Peru	805	Mozambique	38	UK	51
Σ (%)[c]	76	Σ (%)[c]	87	Σ (%)[c]	78

[c] World production ≙ 100%.

Table 18.2. (Continued)

Country	Raspberries	Country	Currants	Country	Almonds
USSR	80	Poland	117	Spain	310
Poland	21	FR Germany	100	USA	310
Hungary	20	USSR	52	Italy	180
FR Germany	19	German DR	29	Iran	50
UK	19	Austria	27	Greece	40
USA	13	Czechoslovakia	22	Tunisia	39
Yugoslavia	13	UK	20	Turkey	25
Canada	9	Norway	17	Syria	18
France	8	Hungary	17	USSR	9
Bulgaria	7	France	6	Lebanon	7
Σ (%)[c]	96	Σ (%)[c]	97	Σ (%)[c]	93

Country	Sweet chestnuts	Country	Cashew-nuts	Country	Walnuts
China	212	India	190	USA	205
Italy	63	Brazil	85	China	152
Turkey	60	Mozambique	75	Turkey	126
Japan	50	Tanzania	72	USSR	70
Korea, South-	42	Kenya	15	Italy	45
France	26	Madagascar	4	Rumania	35
Portugal	25	Philippines	4	France	27
Spain	21	Guinea-Bissau	3	Greece	21
Greece	14	Angola	1	Yugoslavia	19
Bolivia	10	Dominican Rep.	1	Bulgaria	18
Σ (%)[c]	96	Σ (%)[c]	99	Σ (%)[c]	87

[c] World production ≙ 100%.

Table 18.3. Average chemical composition of fruits (as % of fresh edible portion)

Fruit	Dry matter	Total sugar	Titratable acidity[a]	Insoluble matter	Pectin[b]	Ash	pH
Apple	16.0	11.1	0.6 (M)	2.1	0.6	0.3	3.3
Pear	17.5	9.8	0.2 (M)	3.1	0.5	0.4	3.9
Apricot	12.6	6.1	1.6 (M)	1.6	1.0	0.6	3.7
Sour cherry	14.7	9.4	0.7 (M)	1.6	0.3		3.4
Sweet cherry	18.7	12.4	0.7 (M)	2.0	0.3	0.6	4.0
Peach	12.9	8.5	0.6 (M)			0.5	3.7
Plum/prune	14.0	7.8	1.5 (M)	1.3	0.9	0.5	3.3
Blackberry	19.1	5.0	0.6 (C)	9.2	0.7	0.5	3.4
Strawberry	10.2	5.7	0.9 (C)	2.4	0.5	0.5	
Currant, red	16.4	5.1	2.3 (C)	5.9	0.7	0.6	3.0
Currant, black	19.7	6.3	3.2 (C)	5.9	1.1	0.6	3.3
Raspberry	13.9	4.5	1.8 (C)	5.1	0.4	0.5	3.4
Grapes	17.3	14.8	0.4 (T)			0.5	3.3

[a] Calculated as malic (M), citric (C), or tartaric acid (T).
[b] Results are expressed as calcium pectate.

Table 18.3. (Continued)

Fruit	Dry matter	Total sugar	Titratable acidity[a]	Insoluble matter	Pectin[b]	Ash	pH
Orange	13.0	7.0	0.8 (C)			0.5	3.3
Grapefruit	11.4	6.7	1.3 (C)			0.4	3.3
Lemon	11.7	2.2	6.0 (C)			0.5	2.5
Pineapple	15.4	12.3	1.1 (C)	1.5		0.4	3.4
Banana	26.4	18.0	0.4 (M)	4.6	0.9	0.8	4.7
Cherimoya	19	13	0.2			0.9	
Date	80	61				1.8	
Fig	22	16	0.4 (C)		0.6		
Guava	22	4.9			0.7		
Mango	19	14	0.5		0.5		
Papaya	11	9	0.1		0.6		

[a] Calculated as malic (M), citric (C), or tartaric acid (T).
[b] Results are expressed as calcium pectate.

Table 18.4. Proximate composition of shell-nut fruit (as % of fresh edible portion)

Fruit	Moisture	N-Compounds	Lipids	Carbohydrates	Ash	Crude fiber
Cashewnut		16	45.5	21.5		
Peanut	5.0	28.5	47.5	20	2.9	2.8
Hazelnut	7.1	17.4	62.6	9.5	2.5	3.2
Pistachio		21	53.5	16.5		
Almond	4.7	20.5	53.5	16.5	2.3	3.7
Walnut	3.3	15.0	64.4	15.6	1.7	2.1

weight (D-manno-heptulose, D-talo-heptulose, D-glycero-D-galacto-heptose, D-glycero-D-manno-octulose, D-glycero-L-galacto-octulose, D-erythro-L-gluco-nonulose and D-erythro-L-galacto-nonulose). Small amounts of heptuloses have been found in the fruit flesh of apples, peaches and strawberries, and in peels of grapefruit, peaches and grapes.

18.1.2.2.2 Oligosaccharides

Saccharose (sucrose) is the dominant oligosaccharide. Other disaccharides do not have quantitative importance. Maltose occurs in small amounts in grapes, bananas and guava *(Myrtaceae)*. Melibiose, raffinose and stachyose have also been detected in grapes. 6-Kestose has been identified in ripe bananas.

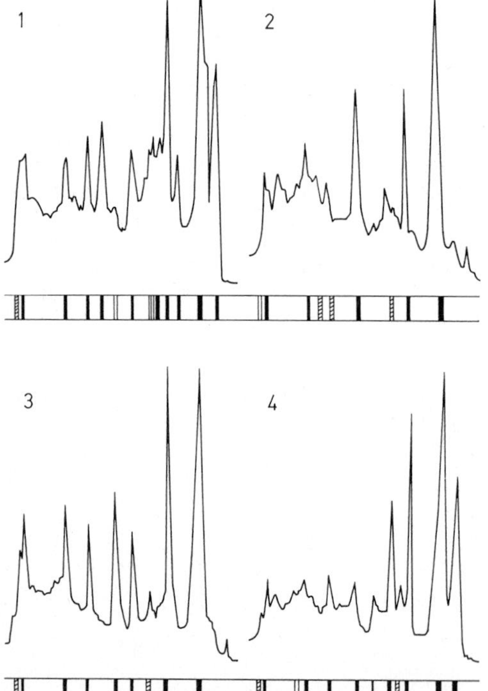

Fig. 18.1. Protein patterns of various wine cultivars obtained by isoelectric focusing (pH 3–10) using Sephadex G-75 as a gel support medium. Staining was done by Coomassie Blue. The figures show the electrophoretograms and the corresponding densitograms. Cultivation region South Palatinate; 1 Morio Muscat, 2 Mueller-Thurgau, 3 Rulaender, 4 Sylvaner. (After *Drawert* and *Mueller,* 1973)

Table 18.5. Free amino acids in fruit (as % of total free amino acids)

Fruit	Asp	Asn	Glu	Gln	Ser	Thr	Pro	Ala	Abu[b]	His	Arg	Pip[c]
Apple (juice)	21	17	15		10	3	2	7	5			
Pear (juice)	10	9	10		11	2	14	9	3			
Grapes	3		13			6	31	9	6		27	
Currant, black		7	17	24	5		8	17	12			
Orange[a]	7–115	20–188	6–71	3–63	4–37		6–295	3–26	4–73		23–150	
Grapefruit[a]	470		280		310	10				76		
Lemon[a]	19–60		6–35		12–28			1–31	4–20		25–106	
Banana	5–10	15		10–15					5–10	10–15		5–10

[a] Values in mg/100 ml juice.
[b] γ-Aminobutyric acid.
[c] Pipecolinic acid.

Other oligosaccharides occur only in trace amounts. The proportion of reducing sugars to saccharose can vary greatly (Table 18.8). Some fruits have no saccharose (e.g. cherries, grapes and figs), while in some the saccharose content is significantly higher than the reducing sugar content (e.g. apricots, peaches and pineapples).

Table 18.6. Amines in fruit

Fruit	Amines
Apple	Methylamine, ethylamine, propylamine, butylamine, hexylamine, octylamine, dimethylamine, spermine, spermidine
Plum/prune	Dopamine
Orange	Feruoylputrescin, methyltyramine, synephrine
Grapefruit	Feruoylputrescine
Lemon	Tyramine, synephrine, octopamine
Pineapple	Tyramine, serotonin
Avocado	Tyramine, dopamine
Banana	Methylamine, ethylamine, isobutylamine, isoamylamine, dimethylamine, putrescine, spermidine, ethanolamine, propanolamine, histamine, 2-phenylethylamine, tyramine, dopamine, noradrenaline, serotonin

Table 18.7. Amines in peel and flesh of banana (μg/g fresh weight)

Amine	Peel	Flesh
Serotonin	50–150	28
Tyramine	65	7
Dopamine	700	8
Noradrenaline	122	2

Table 18.8. Sugar content in various fruits (as % of the edible portion)

Fruit	Glucose	Fructose	Saccharose
Apple	1.8	5.0	2.4
Pear	2.2	6.0	1.1
Apricot	1.9	0.4	4.4
Cherry	5.5	6.1	0.0
Peach	1.5	0.9	6.7
Plum/prune	3.5	1.3	1.5
Blackberry	3.2	2.9	0.2
Strawberry	2.6	2.3	1.3
Currant, red	2.3	1.0	0.2
Currant, black	2.4	3.7	0.6
Raspberry	2.3	2.4	1.0
Grapes	8.2	8.0	0.0
Orange	2.4	2.4	4.7
Grapefruit	2.0	1.2	2.1
Lemon	0.5	0.9	0.2
Pineapple	2.3	1.4	7.9
Banana	5.8	3.8	6.6
Date	32.0	23.7	8.2
Fig	5.5	4.0	0.0

18.1.2.2.3 Sugar alcohols

D-sorbitol is abundant in *Rosaceae* fruits (pomme fruits, stone fruits). For example, its concentration is 300–800 mg/100 ml in apple juice. Since fruits such as berries, citrus fruits, pineapples or bananas do not contain sorbitol, its detection is of analytical importance in the evaluation of wine and other fruit products. Meso-inositol also occurs in fruits; in orange juice it ranges from 130–170 mg/100 ml.

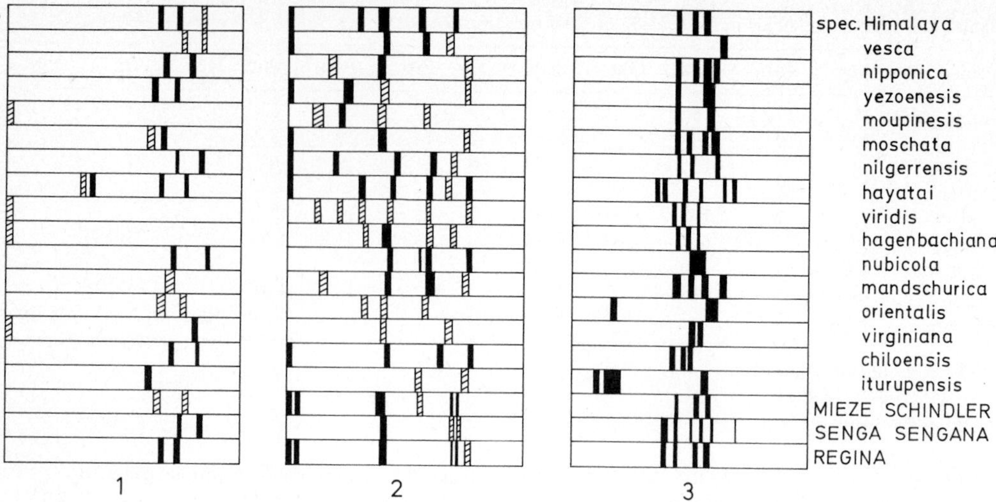

Fig. 18.2. Enzyme patterns of some strawberry species *(Fragaria sp.)* and *(Fragaria ananas)* obtained by PAGE disc gel electrophoresis. Large pore concentrating gel pH 6.7, small pore separating gel, pH 8.9. **1** Peroxidase: incubation with o-toluidine/H_2O_2 at pH 7. **2** Esterase: incubation with α-naphthylacetate at pH 7, the released α-naphthol is diazotized and then coupled with p-chloroaniline. **3** Malate-dehydrogenase: incubation with malate, nitro-blue-tetrazolium chloride and NAD at pH 7.5. (After *Drawert* et al., 1974)

18.1.2.2.4 Polysaccharides

All fruits contain cellulose, hemicellulose, pentosans and pectins. The building blocks of these polysaccharides are glucose, galactose, mannose, arabinose, xylose, rhamnose, fucose and galacturonic and glucuronic acids. The pectin fractions of fruits are particularly affected by ripening. A decrease of insoluble pectin is accompanied by an increase in the soluble pectin fraction. The total pectin content can also decrease. Starch is present primarily in unripe fruits and its content decreases to a negligible level as ripening proceeds. Exceptions are bananas, in which the starch content can be 3% or more even in ripe bananas, and various nuts such as cashew and Brazil nuts.

18.1.2.3 Lipids

The lipid content of fruits is generally low, 0.1–0.5% of the fresh weight. Only fruit seeds and nuts contain significantly higher levels of lipids (cf. Table 18.4). The fruit flesh of avocado is also enriched in fat. The lipid fraction from fruits consists of triacylglycerols, glyco- and phospholipids, carotenoids, triterpenoids and waxes.

18.1.2.3.1 Fruit flesh lipids (other than carotenoids and triterpenoids)

Table 18.9 presents the lipid fractions of apple flesh. Phospholipids, about 50% of the lipid frac-

Table 18.9. Lipids of apple flesh (as % of the total lipids)

Triacylglycerols	5	Sterols	15
Glycolipids	17	Sterol esters	2
Phospholipids	47	Sulfolipids	1
		Others	15

Table 18.10. Fatty acid composition of some fruit flesh lipids (as % of the total fatty acid)

Fatty acid	Avocado	Apple	Banana
12:0	+[a]	0.6	+
14:0	+	0.6	0.6
16:0	15	30	58
16:1	4	0.5	8.3
18:0	+	6.4	2.5
18:1	69	18.5	15
18:2	11	42.5	10.6
18:3	+	1	3.6

[a] Traces.

tion, are predominant. The most abundant fatty acids are palmitic, oleic and linoleic acids (Table 18.10).

18.1.2.3.2 Carotenoids

Carotenoids are widespread in fruits and, in a number of fruits, such as citrus fruits, peaches and sweet melons, their presence is the main factor determining color. The most important carotenoids found in fruits are compiled in Table 18.11, while Table 18.12 gives the carotenoid composition of some fruits.

Fruits can be divided into various classes according to the content and distribution pattern of carotenoids:

a) Fruits with low content of carotenoids (occurring mostly in chloroplasts) such as β-carotene, lutein, violaxanthin, neoxanthin; found in e.g. pineapples, bananas, figs and grapes.
b) Fruits, such as peaches, with relatively high contents of lycopene, phytoene, phytofluene, ζ-carotene and neurosporene.
c) Fruits with relatively high contents of β-carotene, cryptoxanthin and zeaxanthin. This class includes oranges, pears, peaches and sweet melons.
d) Fruits, e.g. oranges and pears, with high amounts of epoxides.
e) Fruits, e.g. oranges, which contain unusual carotenoids.

The compositional pattern of carotenoids which can be readily analyzed by HPLC is important for analytical characterization of fruit products.

18.1.2.3.3 Triterpenoids

This fraction contains bitter compounds of special interest, limonoids and cucurbitacins. Limonoids are found in the flesh and seeds of *Rutaceae* fruits. For example, limonin (II) is present in seed juice, and fruit flesh of oranges and grapefruit. The limonin content decreases with fruit ripening in oranges but remains constant in grapefruit. Development of a bitter taste in heated orange juice is a processing problem. Limonin monolactone (I), a nonbitter compound which is stable in the neutral pH range, is present in orange albedo and endocarp. During production of orange juice it is transferred into the juice in which, due to the lower pH, it is transformed

Table 18.11. Carotenoids occurring in fruit (Roman numerals refer to their structures under 3.8.4)

Number	Carotenoid
1	Phytoene (I)
2	Phytofluene (II)
3	ζ-Carotene (III)
4	Lycopene (IV)
5	α-Carotene (VI)
6	β-Carotene (VII)
7	β-Zeacarotene (V a)
8	Lycoxanthin (16-hydroxylycopene)
9	α-Cryptoxanthin (3-hydroxy-α-carotene)
10	β-Cryptoxanthin (3-hydroxy-β-carotene)
11	β-Carotene-5,6-epoxide
12	Mutatochrome (β-carotene-5,8-epoxide)
13	Lutein (IX)
14	Zeaxanthin (VIII)
15	Cryptoflavin (α-cryptoxanthin-5,8-epoxide)
16	β-Carotene-5,6,5′,6′-diepoxide
17	Antheraxanthin (zeaxanthin-5,6-epoxide)
18	Lutein-5,6-epoxide
19	Mutatoxanthin (XVI)
20	Lutein-5,8-epoxide
21	Cryptoxanthin-5,8,5′,8′-diepoxide
22	Violaxanthin (XIII)
23	Luteoxanthin (XIV)
24	Auroxanthin (zeaxanthin-5,8,5′,8′-diepoxide)
25	Neoxanthin (XX)
26	Capsanthin (X)

Table 18.12. Carotenoid patterns of various fruits

Fruit	Carotenoid	
	Content[a]	Compounds[b]
Pineapple		6, 13
Orange	24	1, 2, 3, 4, 6, 10, 11, 12, 15, 17, 20, 21, 22, 23, 24
Banana		6, 13
Pear	0.3–1.3	2, 3, 6, 7, 8, 11, 12, 13, 14, 16, 18, 20, 24, 25
Fig	8.5	1, 2, 5, 6, 13, 14, 22, 23, 25
Guava		5, 6
Peach	27	1, 2, 3, 5, 6, 9, 10, 13, 14, 17, 18, 19, 22, 23, 24
Plum/prune		1, 2, 3, 5, 6, 9, 10, 12, 14, 15, 17, 18, 19, 20, 22, 23, 25
Grapes	1.8	1, 2, 4, 5, 13, 14, 22, 23
Cantaloupe	20–30	1, 2, 3, 5, 6, 13, 14, 22, 23

[a] mg/kg fresh weight.
[b] The numerals refer to Arabic numerals in Table 18.11.

into the bitter tasting dilactone, limonin (II; cf. Formula 18.3).

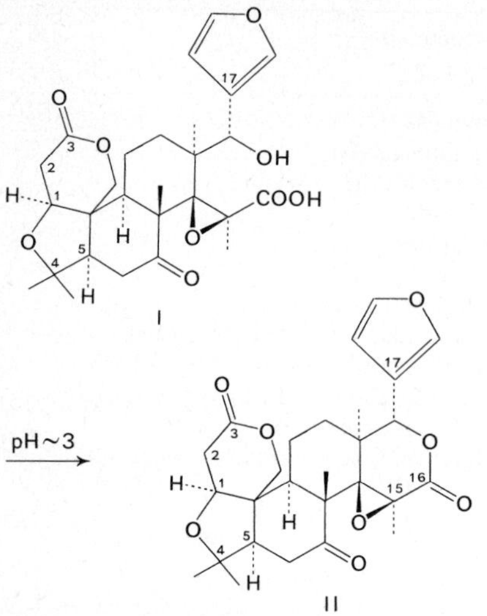

(18.3)

Bitter and nonbitter forms of the many *Cucurbitaceae* are known. The bitter forms contain cucurbitacins (III) in fruits and seeds. For example, *Citrullus lanatus* (watermelon) contains IIIE in glycosidic form; while *Cucumis sativus* (cucumber) contains IIIC and *Cucurbita* spp. (pumpkin) contain IIIB, D, E and I (cf. Formula 18.4).

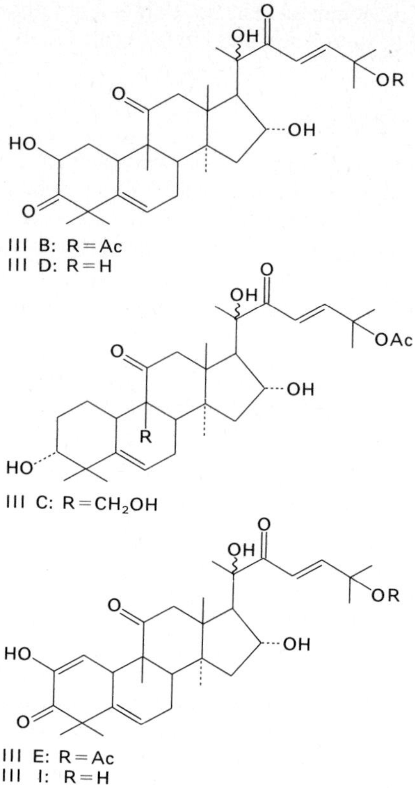

III B: R = Ac
III D: R = H
III C: R = CH_2OH
III E: R = Ac
III I: R = H

(18.4)

Table 18.13. Organic acids in various fruits (milliequivalents 100g fresh weight)

Fruit	Major acid	Other acids
Apple	Malic 3–19	Quinic (in unripe fruits)
Pear	Malic 1–2	Citric
Apricot	Malic 12	Citric 12, quinic 2–3
Cherry	Malic 5–9	Citric, quinic, and shikimic
Peach	Malic 4	Citric 4
Plum/prune	Malic 4–6	Quinic (especially in unripe fruits)
Strawberry	Citric 10–18	Malic 1–3, quinic 0.1, succinic 0.1
Raspberry	Citric 24	Malic 1
Currant, red	Citric 21–28	Malic 2–4, succinic, oxalic
Currant, black	Citric 43	Malic 6
Gooseberry	Citric 11–14	Malic 10–13, shikimic 1–2
Grapes	Tartaric 1.5–2	Malic 1.5–2
Orange	Citric 15	Malic 3, quinic
Lemon	Citric 73	Malic 4, quinic
Pineapple	Citric 6–20	Malic 1.5–7
Banana	Malic 4	
Fig	Citric 6	Malic, acetic
Guava	Citric 10–20	Malic

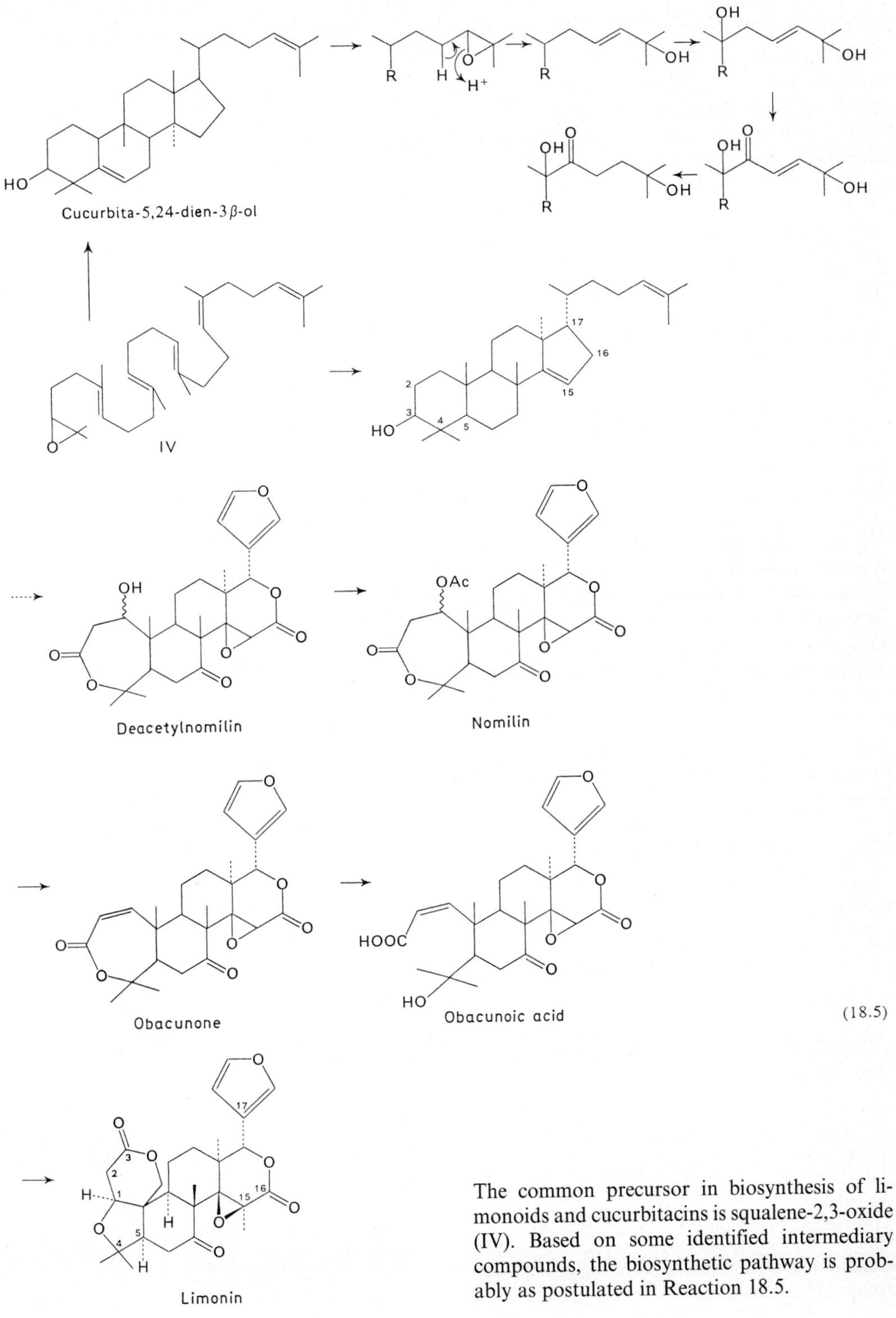

Cucurbita-5,24-dien-3β-ol

IV

Deacetylnomilin

Nomilin

Obacunone

Obacunoic acid

(18.5)

Limonin

The common precursor in biosynthesis of limonoids and cucurbitacins is squalene-2,3-oxide (IV). Based on some identified intermediary compounds, the biosynthetic pathway is probably as postulated in Reaction 18.5.

18.1.2.3.4 Fruit waxes

The fruit peel is often coated with a waxy layer. In addition to the esters of higher fatty acids with higher alcohols, these waxes contain hydrocarbons, free fatty acids, free alcohols, ketones and aldehydes. The ester fraction in apples and grapes predominantly consists of alcohols of 24, 26 and 28 carbons, but their fatty acid patterns differ. Apples contain mostly 18:1, 18:2, 16:0 and 18:0 fatty acids, while grapes contain 20:0, 18:0, 22:0 and 24:0 fatty acids.

18.1.2.4 Organic Acids

L-malic and citric acids are the major organic acids of fruits (Table 18.13). Malic acid is predominant in pomme and stone fruits, while citric acids is most abundant in berries, citrus and tropical fruits. (2R:3R)-Tartaric acid occurs only in grapes. Many other acids, including the acids in the citric acid cycle, occur only in low amounts. Examples are cis-aconitic, succinic, pyruvic, citramalic, fumaric, glyceric, glycolic, glyoxylic, isocitric, lactic, oxalacetic, oxalic and 2-oxoglutaric acids. Important phenolic acids, dealt with in Section 18.1.2.5.1, are quinic, caffeic, chlorogenic and shikimic acids. Galacturonic and glucuronic acids are also found.

Tartaric acid biosynthesis in *Vitis* spp. starts from glucose or fructose and probably proceeds through 5-oxogluconic acid or ascorbic acid, respectively:

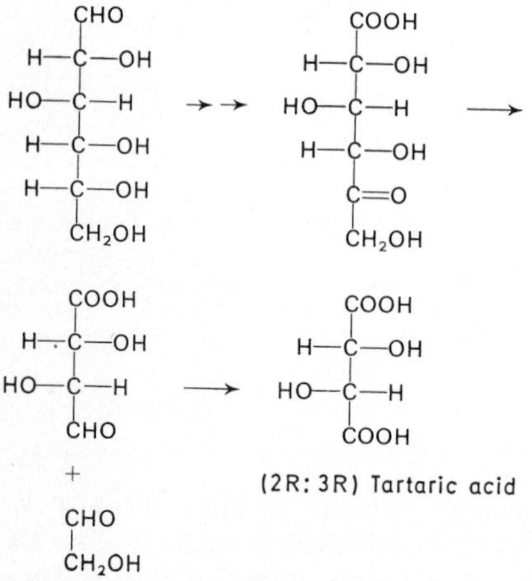

CHO / H—C—OH / HO—C—H / H—C—OH / H—C—OH / CH_2OH →→ COOH / H—C—OH / HO—C—H / H—C—OH / C=O / CH_2OH →

COOH / H—C—OH / HO—C—H / CHO + CHO / CH_2OH → COOH / H—C—OH / HO—C—H / COOH

(2R: 3R) Tartaric acid (18.6)

Human and animal metabolism oxidatively degrade (2R:3R)-tartaric and meso-tartaric acids into glyoxylic and hydroxypyruvic acids, respectively:

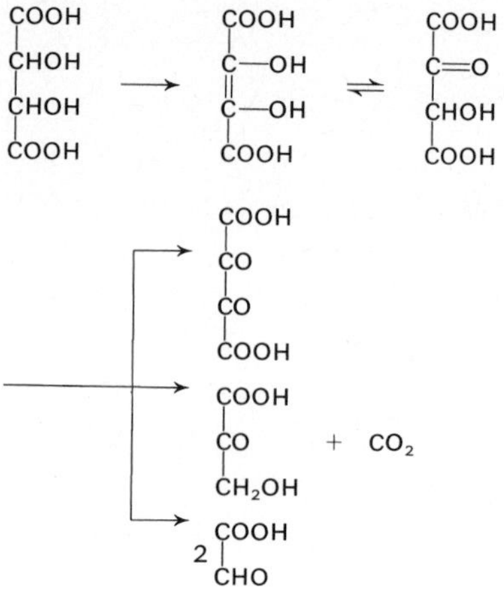

COOH / CHOH / CHOH / COOH → COOH / C—OH / ‖ / C—OH / COOH ⇌ COOH / C=O / CHOH / COOH

→ COOH / CO / CO / COOH

→ COOH / CO / CH_2OH + CO_2

→ 2 COOH / CHO (18.7)

18.1.2.5 Phenolic Compounds

These compounds occur in most fruits and most of them contribute to color and taste. They can form metal complexes during fruit processing, resulting in discoloration of fruit pulp. Table 18.14 provides data for the total phenol content of some selected fruits.

18.1.2.5.1 Hydroxycinammic acids, hydroxycoumarins and hydroxybenzoic acids

p-Coumaric (I), ferulic (II), caffeic (III) and sinapic (IV) acids are widespread in fruits and vegetables:

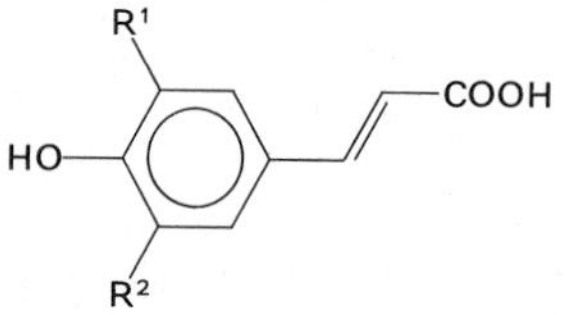

R^1, HO, COOH, R^2

I: R^1, R^2=H,
III: R^1=H, R^2=OH,
II: R^1=H, R^2=OCH_3
IV: R^1, R^2=OCH_3

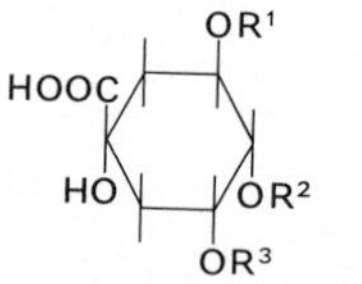

OR^1, HOOC, HO, OR^2, OR^3

V: R^1=Caffeoyl-, R^2, R^3=H
VI: a) R^1, R^2=Caffeoyl-, R^3=H,
b) R^1, R^3=Caffeoyl-, R^2=H,
c) R^2, R^3=Caffeoyl-, R^1=H,
VII: R^1, R^2=H, R^3=Caffeoyl-,
VII a: R^1, R^3=H, R^2=Caffeoyl- (18.8)

Table 18.14. Phenolic compounds in fruit

Fruit	Total phenol (g/100 g fresh tissue)
Apple	0.1–1
Pear	0.4
Cherry	0.2
Peach	0.03–0.14
Plum/prune	0.2–1.4
Grapes	0.1–1

Table 18.15. Hydroxycoumarins in fruit (VIII)[a]

Compound	Substitution pattern		
	R^1	R^2	R^3
Coumarin	H	H	H
Umbelliferone	H	OH	H
Herniarin	H	OCH_3	H
Aesculetin	OH	OH	H
Scopoletin	OCH_3	OH	H
Fraxetin	OCH_3	OH	OH

[a] See Formula 18.9.

These hydroxycinnamic acids are rarely free but are present mainly as acid derivatives. For example, caffeic acid is esterified with quinic acid, giving rise to chlorogenic acid (V), isochlorogenic acids a, b and c (VI), neochlorogenic acid (VII) and cryptochlorogenic acid (VII a).

Esters of other hydroxycinnamic acids with quinic acid are also widespread. Other alcoholic components are shikimic, malic and tartaric acids and meso-inositol. Sinapine, which is found in mustard seed, is the counter ion of glucosinolate sinalbin and is the choline ester of sinapic acid. Hydroxycinnamic acid glycosides and amides are also found in plants.

Scopoletin in esterified form is the only hydroxycoumarin (VIII, Table 18.15) which has been found, in small amounts, in plums and apricots:

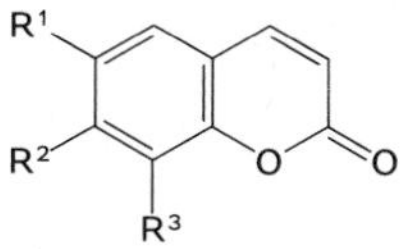

R¹ R² R³ O O

VIII

(18.9)

Hydroxybenzoic acids found in various fruits include: salicylic acid (2-hydroxybenzoic acid), 4-hydroxybenzoic acid, gentisic acid (2,4-dihydroxybenzoic acid), protocatechuic acid (3,4-dihydroxybenzoic acid), gallic acid (3,4,5-trihydroxybenzoic acid), vanillic acid (3-methoxy-4-hydroxybenzoic acid) and ellagic acid (IX):

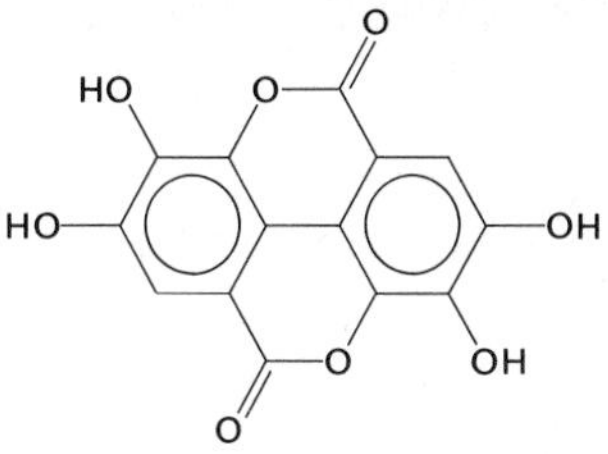

HO O O HO OH O O OH O

IX

(18.10)

Table 18.16 lists the occurrence of phenolic acids in various fruits, and Table 18.17 shows their changes in apples and strawberries during ripening and storage.

Hydroxycinnamic acid biosynthesis starts with phenylalanine [cf. Reaction 18.11: a) phenylalanine-ammonia lyase; b) cinnamic acid 4-hydroxylase; c) phenolases; d) methyl transferases, R: OH and OCH_3 in various positions]:

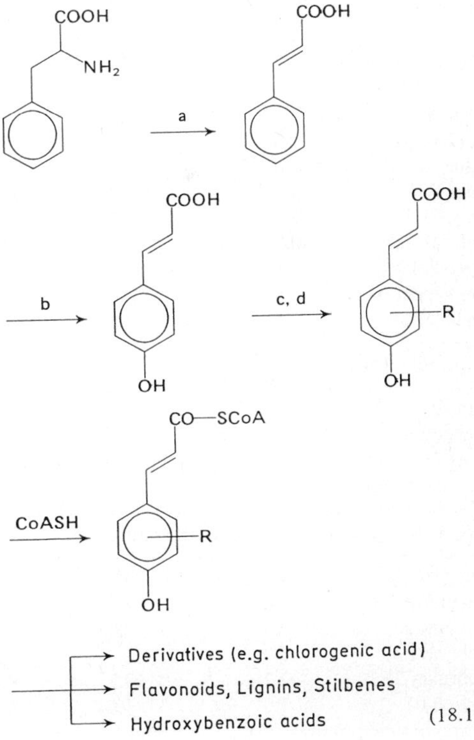

COOH NH_2 → a COOH

b → COOH OH → c, d COOH R OH

CoASH → CO—SCoA R OH

→ Derivatives (e.g. chlorogenic acid)
→ Flavonoids, Lignins, Stilbenes
→ Hydroxybenzoic acids

(18.11)

Table 18.16. Occurrence of phenolic acids in fruit

Fruit	1[a]	2	3	4	5	6	7	8	9	10	11	12	13	14	15	16	17
Pear	+	+	+	+													
Quince	+	+	+		+	+											
Cherry	+	+	+	+		+											
Plums/prunes	+	+	+	+	+	+											
Peach	+	+	+				+										
Blackberry	+	+			+												
Strawberry	+	+				+	+				+	+	+	+	+	+	+
Raspberry	+	+			+												
Currant	+	+			+	+											
Gooseberry	+	+			+	+				+							
Grapes	+	+	+	+			+		+	+	+	+					
Orange					+	+	+	+			+		+				
Grapefruit					+	+	+	+		+	+		+				
Lemon					+	+	+	+			+		+				

[a] 1 = Chlorogenic acid, 2 = neochlorogenic acid, 3 = isochlorogenic acid, 4 = cryptochlorogenic acid, 5 = ferulic acid, 6 = p-coumaric acid, 7 = caffeic acid, 8 = sinapic acid, 9 = cinnamic acid, 10 = quinic acid, 11 = salicylic acid, 12 = p-hydroxybenzoic acid, 13 = gentisic acid, 14 = gallic acid, 15 = ellagic acid, 16 = protocatechuic acid, 17 = vanillic acid.

Table 18.17. Phenolic acids in apple and strawberry during ripening and storage

	p-Coumaric acid	Caffeic acid	Ferulic acid	Gallic acid	p-Hydroxy-benzoic acid	Vanillic acid
Ripening of apples[a]						
June 01		640				
June 07	250	1,000	60			
June 12	460	1,270	95			
June 23	425	1,010	50			
July 03	250	665	29			
July 14	147	470	22			
August 04	51	214	12			
September 07	28	137	6.4			
October 05	15	85	4.0			
Ripening of strawberries[a]						
June 05	69	15		80		3
June 20	110	34		110	19	23
June 27	119	30		111	87	25
July 01 (ripe)	175	39		121	108	34
Apple storage at 4 °C, picking date: October 19[b]						
October 28	4.4	42	1.0			
December 16	3.0	30.1	0.6			
January 20	2.3	28.5	0.4			
February 01	2.2	19.7	0.4			
February 15	2.2	19.7	0.4			
March 01	1.9	10.4	0.4			

[a] Picking date. [b] Analysis date.

The hydroxybenzoic acids are derived from hydroxycinnamic acids by a pathway analogous to β-oxidation of fatty acids:

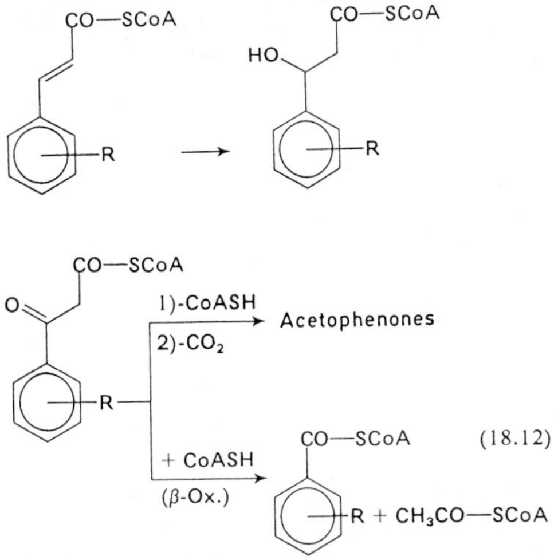

CO—SCoA → HO, CO—SCoA; R

CO—SCoA; 1)-CoASH, 2)-CO_2 → Acetophenones; + CoASH (β-Ox.) → CO—SCoA, R + CH_3CO—SCoA (18.12)

Reduction of the benzoic acid carboxyl groups yields the corresponding aldehydes and alcohols, as for instance vanillin and vanillyl alcohol (X and XI, respectively):

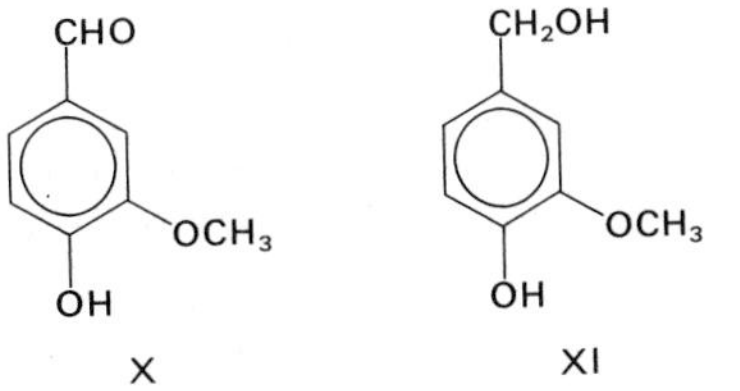

CHO, OCH_3, OH (X); CH_2OH, OCH_3, OH (XI) (18.13)

The glucosides of cis-o-coumaric acids are the precursors of coumarins. Disintegration of plant tissue releases the free acids from the glucosides. The acids then close spontaneously to the ring forms (R: OH and OCH_3 in various positions):

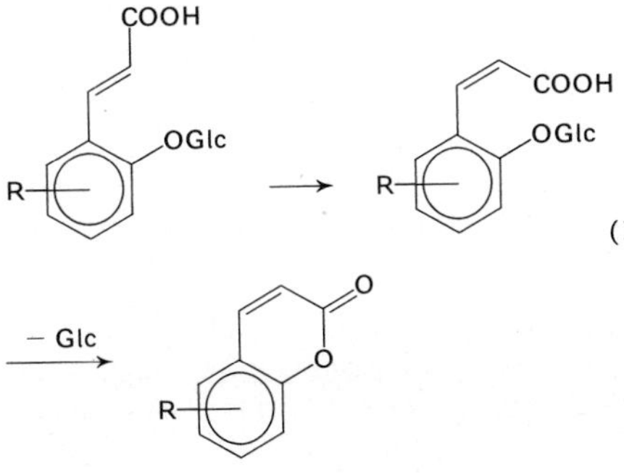

COOH, OGlc, R → COOH, OGlc, R (18.14)

– Glc → O, O, R

18.1.2.5.2 Catechins (3-hydroxyflavanes) and leucoanthocyanidins (3,4-dihydroxyflavanes)

These colorless compounds (Table 18.23) occur in all commonly grown fruits as intermediary products in flavanoid biosynthesis (cf. 18.1.2.5.6) [R, R^1 = H: a) catechin, b) epicatechin; R = H, R^1 = OH: gallocatechin; R = OH: leucoanthocyanidin]:

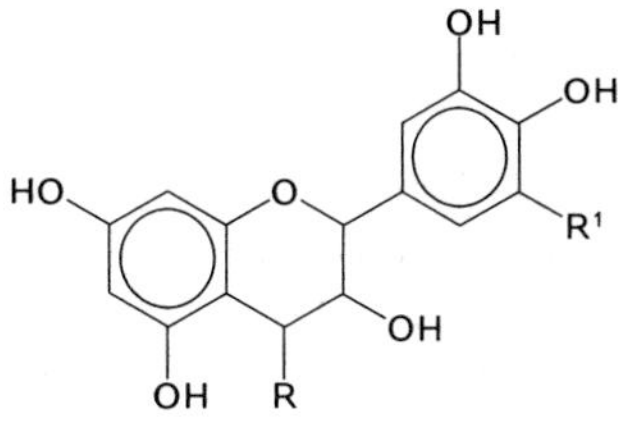

OH, OH, HO, O, R^1, OH, OH, R

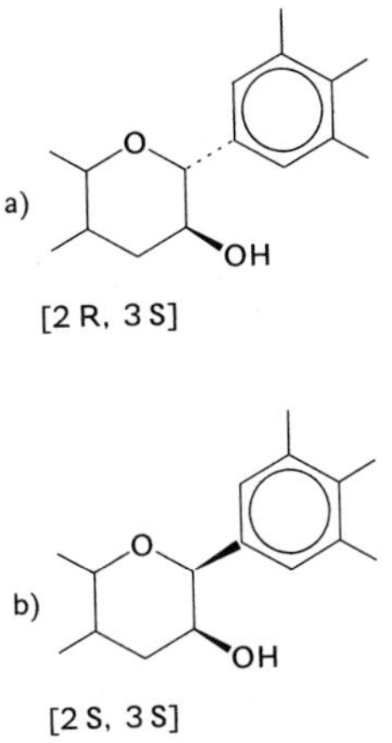

a) O, OH [2 R, 3 S]

b) O, OH [2 S, 3 S] (18.15)

These compounds are converted oxidatively into tannins of various degrees of polymerization. Up to a molecular weight of 3 kdal they have an astringent taste (R^1: various OH-substituents; R = H: dimers; R = other flavane monomers: oligomer):

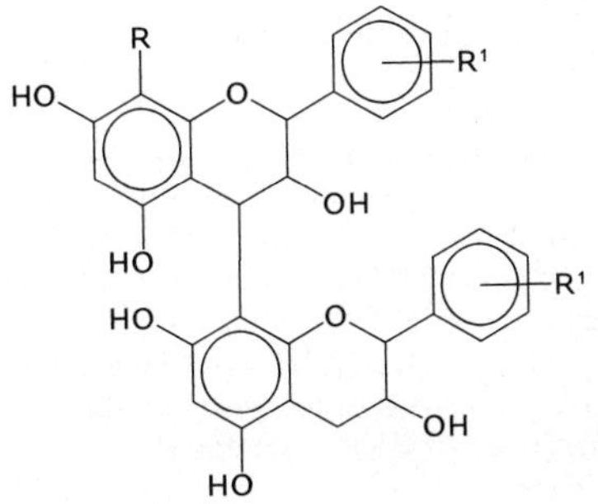

R, R^1, HO, O, OH, HO, HO, O, R^1, OH, HO (18.16)

18.1.2.5.3 Anthocyanidins

These red, blue or violet compounds occur in the form of glycosides in most commonly grown fruit varieties and also in some citrus and tropical fruits (Table 18.18):

(18.17)

Table 18.18. Anthocyanins in various fruits (the major constituents in italics)

Fruit	Anthocyanin
Apple	*Cy-3-gal,* Cy-3-ara, Cy-7-ara
Pears	Cy-3-gal
Peach	Cy-3-glc
Plums/prunes	*Cy-3-glc,* Cy-3-rut, Peo-3-glc, Peo-3-rut
Sour cherry	*Cy-3-sop,* Cy-3-rut, Cy-3-glc-rut, Cy-3-glc
Sweet cherry	Cy-3-glc, Cy-3-rut
Blackberry	*Cy-3-glc,* Cy-3-rut
Strawberry	*Pg-3-glc,* Pg-3-gal, Cy-3-glc
Bilberry	Pel-3-gly, Cy-3-gly, Pet-gly, Del-3-glc, Del-3-gal, Mv-3-glc
Raspberry	*Cy-3-glc,* Cy-3-glc-rut, Cy-3-rut, Cy-3-sop, Cy-3-glc-sop
Currant, red	Del-3-glc, Cy-3-rut, Cy-3-xyl-rut, Cy-3-glc-rut, Cy-3-sop, Cy-3-sam
Currant, black	Cy-3-glc, Cy-3-rut, Del-3-glc
Grapes (*Vitis vinifera, V. labrusca, V. riparia,* including hybrids)	Cy-, Del-, Peo-, Pet-, Mv-3-glc, Mv-3,5-diglc, Mv-3-p-cumaroylglc-5-glc, Mv-3-p-caffeoylglc-5-glc, Peo-3-p-cumaroylglc-5-glc
Orange	Cy-3-glc, Del-3-glc
Banana	Pet-3-gly
Fig	Cy-3-gly
Passion fruit	Del-3-glc, Del-3-glc-glc

Cy: Cyanidin, Del: delphinidin, Mv: malvidin, Peo: peonidin, Pet: petunidin, Pg: pelargonidin, ara: arabinoside, gal: galactoside, glc: glucoside, gly: glycoside, rut: rutinoside, sam: sambubioside, sop: sophoroside, xyl: xyloside, glc-rut: glucosyl-rutinoside etc. (Sophorose: β-D-Glc*p*(1 → 2)-D-Glc*p*, sambubiose: β-D-Xyl*p*(1 → 2)-D-Glc*p*).

Table 18.19. Anthocyanidins: absorption maxima of the visible spectrum

Compound	R^1	R^2	R^3	λ_{max} (nm)[a] R = H	λ_{max} (nm)[a] R = Glc[b]
Pelargonidin	H	OH	H	520	506
Cyanidin	OH	OH	H	535[c]	525[c]
Peonidin	OCH_3	OH	H	532	523
Delphinidin	OH	OH	OH	544[c]	535[c]
Petunidin	OCH_3	OH	OH	543[c]	535[c]
Malvidin	OCH_3	OH	OCH_3	542	535

[a] In methanol with 0.01% HCl.
[b] 3-Glucoside.
[c] $AlCl_3$ shifts the absorption towards blue region of spectrum by 14 to 23 nm.

Table 18.19 provides data about the structure and absorption maxima of the most important anthocyanidins. An increased extent of hydroxylation results in a shift towards blue color (pelargonidin → cyanidin → delphinidin), whereas glycoside formation and methylation results in a shift towards a red color (pelargonidin → pelargonidin-3-glucoside; cyanidin → peonidin).

The color of an anthocyanin changes with the pH of the medium (R = sugar moiety; cf. Formula 18.18).

The flavylium cation (I) is stable only at very low pH. As the pH increases it is transformed into colorless chromenol (II). Figure 18.3 shows the absorption decrease in the visible spectrum at

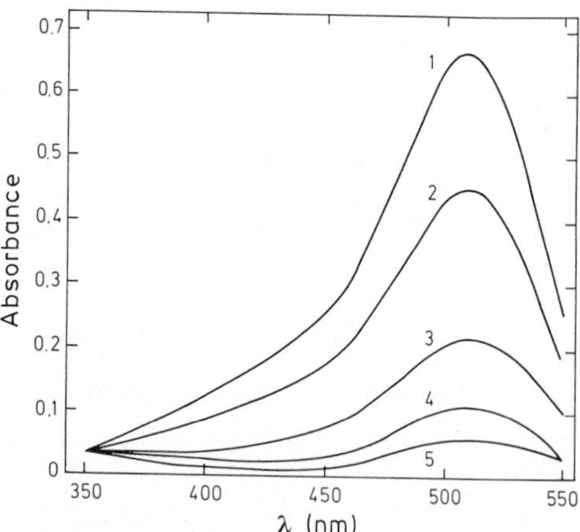

Fig. 18.3. Absorption spectra of cyanidin-3-rhamnoglucoside (16 mg/l) in aqueous buffered solution at pH 0.71 (1), pH 2.53 (2), pH 3.31 (3), pH 3.70 (4), and pH 4.02 (5). (After *Jurd,* 1964)

I: pH ≤ 1, red

II: pH = 4 - 5, colorless

III: pH = 6 - 7, purple

IV: pH = 7 - 8, deep blue

V: pH = 7 - 8, yellow

(18.18)

various pH's, reflecting these transformations. Formation of a quinoidal (III) and ionic anhydro base (IV) at pH 6–8 intensifies the color. At pH 7–8 structure IV is transformed through ring opening to yellow chalcone (V). At higher pH's the color can be stabilized by the presence of multivalent metal ions (Al^{3+}, Fe^{3+}). The complexes formed are deep blue (cf. Formula 18.19).

VI

(18.19)

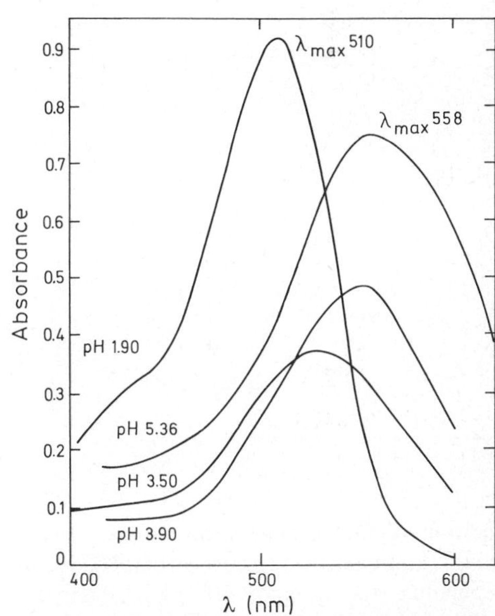

Fig. 18.4. Absorption spectra of cyanidin-3-glucoside (35 µmole/l + 830 µmole/l $AlCl_3$) in aqueous buffered solutions at pH 1.90, pH 3.50, pH 3.90, and pH 5.36. (After *Jurd* and *Asen,* 1966)

Figure 18.4 illustrates the shift in absorption maximum from 510 to 558 nm for cyanidin-3-glucoside over the pH range of 1.9–5.4. Readings were taken in the presence of aluminium chloride.

At higher pH's free anthocyanidins (VII) are degraded through chromenols (VIII) and α-diketones (IX) to aldehydes (X) and carboxylic acids (XI):

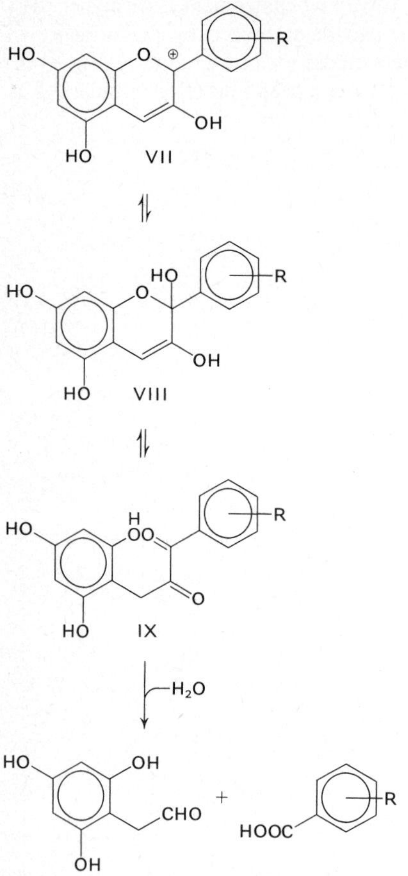

(18.20)

Addition of SO_2 bleaches anthocyanins. The flavylium cation reacts to form a carbinol base corresponding to compounds XII or XIII. The color is restored by acidification to pH 1 or by addition of a carbonyl compound (e.g. ethanal). Since compounds of type XIV ($R^1 = CH_3, C_2H_5$) are not affected by SO_2, it appears that compound XIII is involved in such bleaching reactions (cf. Formula 18.21).

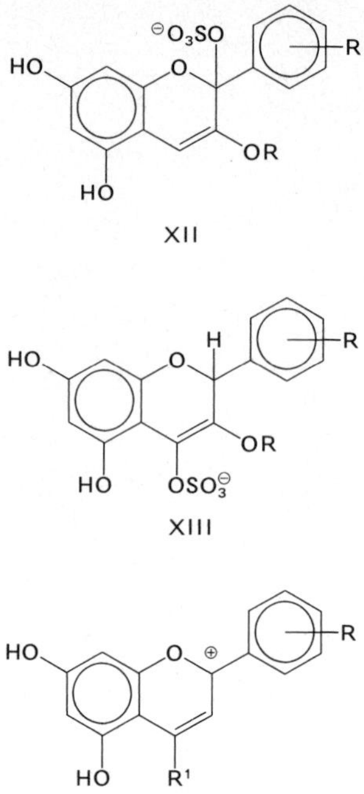

(18.21)

18.1.2.5.4 Flavanones

Flavanones ($R^1 = H, R^2 = OCH_3$: isosakuranetin; $R^1 = H, R^2 = OH$: naringenin; $R^1 = OH, R^2 = OCH_3$: hesperetin; $R^1, R^2 = OH$: eriodictyol) occur mostly as glycosides in citrus fruits (Table 18.20):

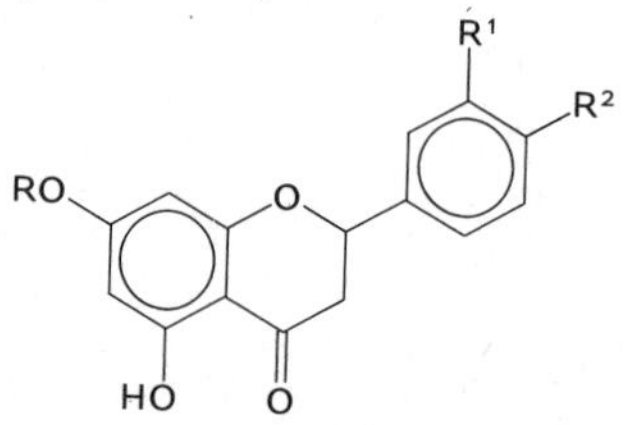

(18.22)

Table 18.21 shows that flavanone-7-rutinosides are usually nonbitter, whereas flavanone-7-neohesperidosides are generally bitter. The intensity of the bitter taste is influenced by the substitution pattern. Compounds with $R^1 = H, R^2 = OH, OCH_3$ (e.g. naringin, poncirin) are an order of magnitude more bitter than those with $R^1 = OH, R^2 = OH, OCH_3$ (e.g. neohesperidin, neoeriocitrin). Naringenin-7-neohesperidoside (naringin) is the bitter constituent of grapefruit. Hesperetin-7-neohesperidoside (neohesperidin) is the bitter compound of bitter oranges *(Citrus*

Table 18.20. Flavanones, flavones and flavonols in citrus fruit

Fruit	Compound
Orange	
flesh	Hes-7-rut
peel	Hes-7-rut, Nob, Nav-7-neo, Isa-7-rha-glc
Bitter orange	Hes-7-neo
Grapefruit	Nar-7-neo, Isa-7-neo, Api-7-rut
Lemon	
peel	Hes-7-rut, Dio-7-rut, Lin, Eri-7-rut, Lol, Api, Lut, Chr, Que, Irh

Api: Apigenin, Chr: chrysoeriol, Dio: diosmetin, Eri: eriodictyol, Hes: hesperitin, Irh: isorhamnetin, Isa: isosacuranetin, Lin: limocitrin, Lol: limocitrol, Nar: naringenin, Nob: nobiletin, Que: quercetin, glc: glucoside, neo: neohesperidoside, rha: rhamnoside, rut: rutinoside.

Table 18.21. Taste of flavanone glycosides[a]

Compound	R	R^1	R^2	Taste quality	Taste intensity[b]
Naringenin-rutinoside	rut[c]	H	OH	neutral	–
Naringin	neo[d]	H	OH	bitter	20
Isosacuranetin-rutinoside	rut	H	OCH_3	neutral	–
Poncirin	neo	H	OCH_3	bitter	20
Hesperidin	rut	OH	OCH_3	neutral	–
Neohesperidin	neo	OH	OCH_3	bitter	2
Eriocitrin	rut	OH	OH	neutral	–
Neoeriocitrin	neo	OH	OH	bitter	2

[a] Data for R, R^1 and R^2 refer to Formula 18.22.
[b] Relative bitterness refers to quinine hydrochloride = 100
[c] Rutinosyl.
[d] Neohesperidosyl.

auranticus). The nonbitter isomer, hesperetin-7-rutinoside (hesperidin) occurs in oranges *(Citrus sinensis)*.

Removal of the bitter taste of citrus juices and citrus fruit pulps is possible by enzymatic cleavage of the sugar moiety using a mixture of α-rhamnosidase and β-glucosidase. These enzymes are isolated from microorganisms such as *Phomopsis citri, Cochliobolus niyabeanus* or *Rhizoctonia solanii:*

$$\text{Naringin} \longrightarrow \text{Naringenin} + \text{Rhamnose} + \text{Glucose} \quad (18.22\text{ a})$$

A number of neutral or bitter flavanone glycosides can be converted through ring opening to sweet chalcones (II) which, by additional hydrogenation, can be stabilized as sweet dihydrochalcones (III):

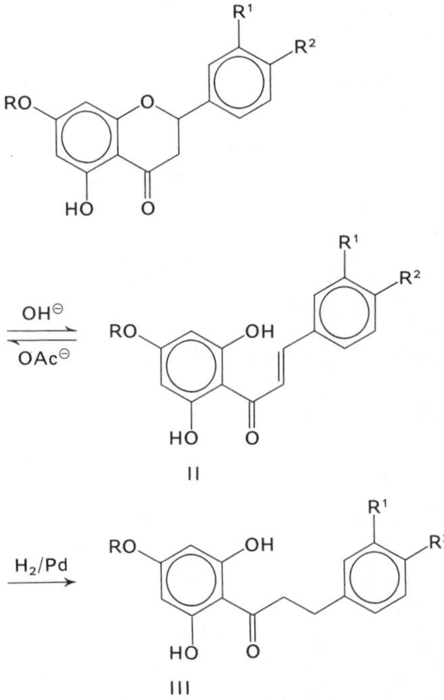

RO, HO, O, R^1, R^2 (flavanone)

$\xrightarrow[\text{OAc}^{\ominus}]{\text{OH}^{\ominus}}$ II (chalcone: RO, OH, HO, O, R^1, R^2)

$\xrightarrow{H_2/Pd}$ III (dihydrochalcone: RO, OH, HO, O, R^1, R^2) (18.23)

Table 18.22. Taste of dihydrochalcones

Dihydrochalcone from	Taste quality	Taste intensity[a] (μmole/l)	Relative intensity[b]
Naringin	sweet	200	1
Neohesperidin	sweet	10	20
Neoeriocitrin	slightly sweet	–	–
Poncirin	slightly bitter	–	–
Saccharin (Sodium salt)	sweet	200	1

[a] Concentration of iso-sweet solutions.
[b] Related to saccharin.

The presence of a free OH-group in position R^1 or R^2 is necessary for a sweet taste. Table 18.22 shows that the dihydrochalcone of naringin corresponds to saccharin in sweetness intensity, whereas the dihydrochalcone of neohesperidin is sweeter than saccharin by a factor of 20.

Conversion of naringin to highly sweet neohesperidin dihydrochalcone (VII) is possible by alkali fragmentation to a methylketone (IV), condensation with isovanillin (V) to the corresponding chalcone (VI), then hydrogenation:

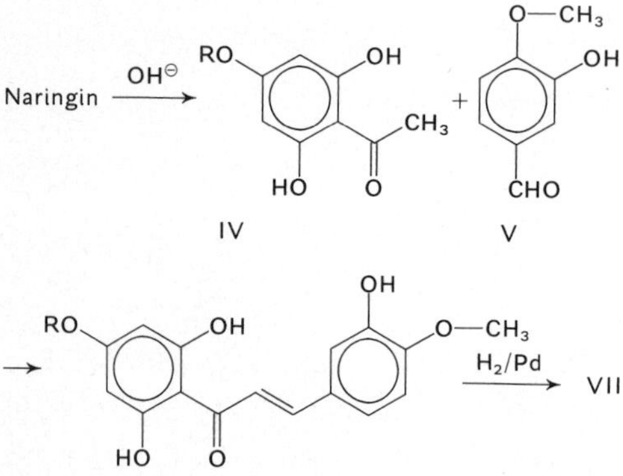

(18.24)

A sweet compound can be obtained from the neutral-tasting hesperidin of oranges by first converting hesperidin to another neutral-tasting compound, hesperidin dihydrochalcone. The latter can then be hydrolyzed, by acidic or enzymatic catalysis, to remove the rhamnose residue, yielding hesperetin dihydrochalcone glucoside, which is sweet.

Dihydrochalcones show potential as new sweeteners (cf. 8.8.13).

18.1.2.5.5 Flavones, Flavonols

Flavones (Formula 18.25: I, R = H; R^1, R^2 = H, R^3 = OH: apigenin; R^1, R^2 = OH, R^3 = H: luteolin; R^1 = OH, R^2 = H, R^3 = OCH_3: diosmetin; R^1 = OCH_3, R^2 = H, R^3 = OH: chrysoeriol; II: nobiletin) and flavonols (Formula 18.25: I, R, R^3 = OH; R^1, R^2 = H: kaempferol; R^1 = OH, R^2 = H: quercetin; R^1, R^2 = OH: myricetin; R^1 = OCH_3, R^2 = H: isorhamnetin) occur in all common fruits and citrus and tropical fruits as the 3-glycosides and, less frequently, as the 7-glycosides (Tables 18.20 and 18.23):

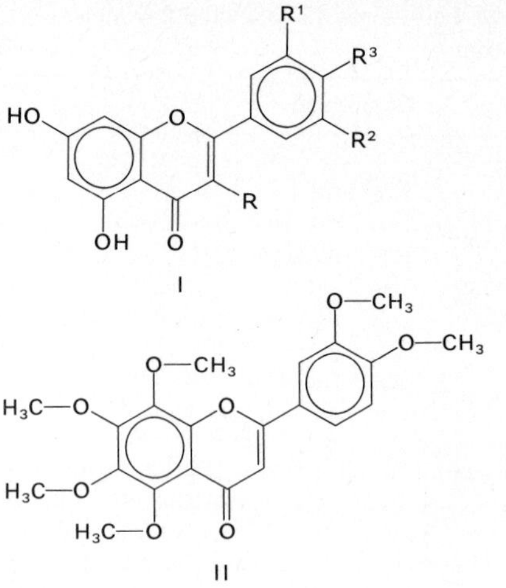

(18.25)

They are faintly-yellow compounds.

Table 18.23. Occurrence of catechins and flavonols in fruit (major constituents in italics)

Fruit	Catechins	Flavonols
Apple	Cat, *Epi*, Gal	Que-3-gal, Que-3-glc, Que-3-rha, Que-3-rha-glc, Que-3-ara, Que-3-xyl
Pear	Cat, *Epi*	Que-3-glc
Peach	Cat, *Epi*	Que-3-glc
Apricot	Cat	Que-3-glc, Kaem-gly
Plum/prune	*Cat*, Epi, Gal	Que-3-glc, Que-3-rha, Que-3-ara
Sour cherry	Cat, *Epi*	
Sweet cherry	Cat, *Epi*	Que-3-glc
Blackberry	Cat, *Epi*	Que-3-glc
Strawberry	*Cat*, Gal	
Currant, black	Cat, Gal	Que-3-glc, Kaem-3-glc, Myr-3-glc, further Que-glc, Kaem-glc
Raspberry	Cat, *Epi*	
Grapes	*Cat*, Epi, Gal	Que-3-rha, Que-3-glc, Que-3-rha-glc

Cat: Catechin, Epi: epicatechin, Gal: gallocatechin. Kaem: kaempferol, Myr: myricetin, Que: quercetin. ara: arabinoside, gal: galactoside, glc: glucoside, gly: glycoside, rha: rhamnoside, and xyl: xyloside.

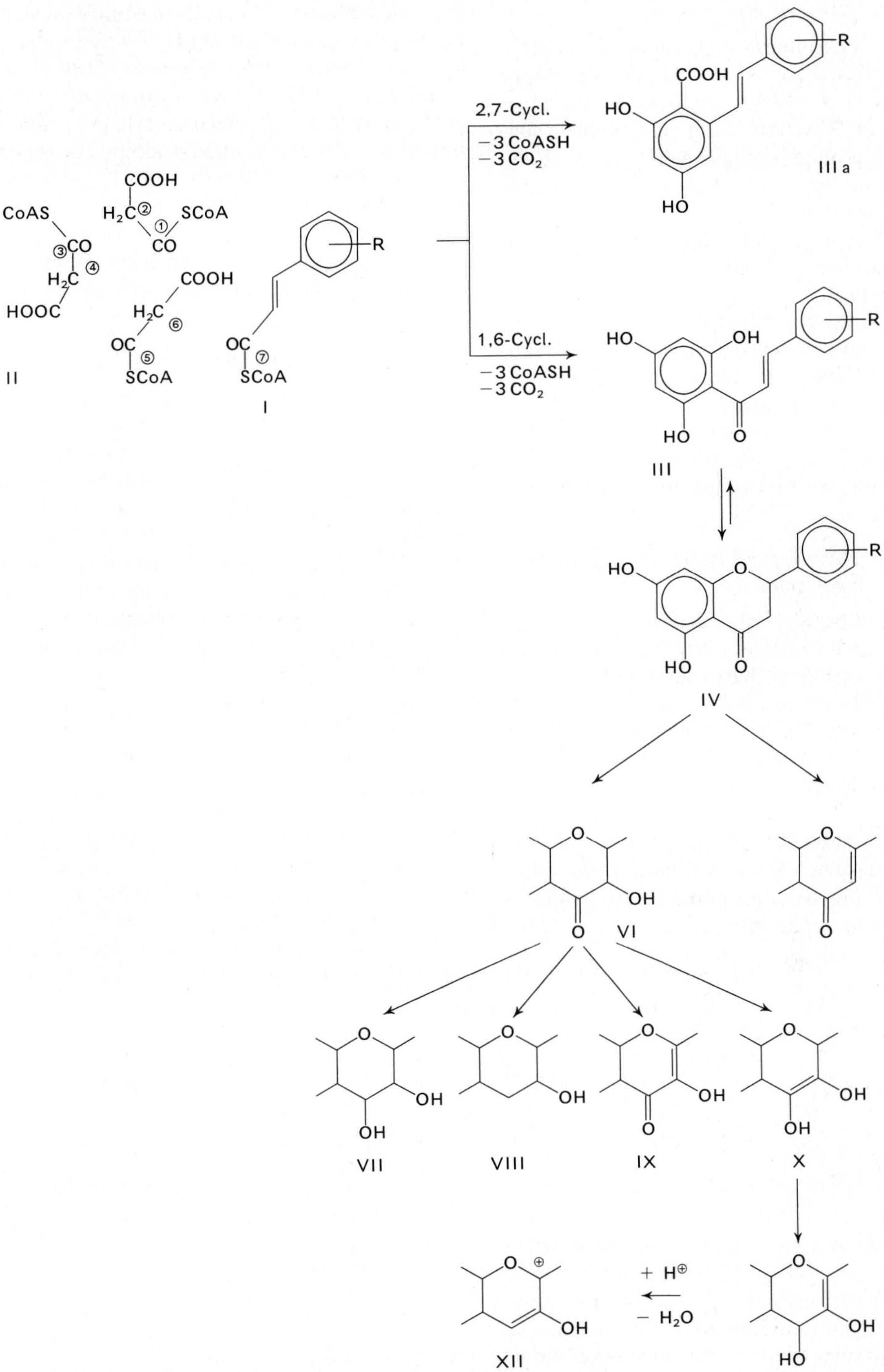

2,7-Cycl.
−3 CoASH
−3 CO_2
III a

1,6-Cycl.
−3 CoASH
−3 CO_2

II I III IV

(18.26)

VI V

VII VIII IX X

+ $H^{\oplus}$
− H_2O

XII XI

18.1.2.5.6 Flavonoid biosynthesis

Flavonoid biosynthesis occurs through the stepwise condensation of activated hydroxycinnamic acid (I) with three activated malonic acid molecules (II; cf. Reaction 18.26). The primary condensation product, a chalcone (III), is in equilibrium with a flavanone (IV) with the equilibrium shifted toward product IV. Recent research suggests that condensation directly yields a flavanone, hence chalcone is not an obligatory intermediary product.
A 2,7-cyclization yields stilbenes (III a).
One pathway converts flavanones (IV) to flavones (V) and, through another pathway, flavanones are converted to flavanonols (VI). The latter compounds are converted to flavandiols (VII), flavanols (VIII) and flavonols (IX), as well as anthocyanidins (XII) via endiols (X) and enols (XI).

18.1.2.5.7 Technological importance of phenolic compounds

The taste of fruits is influenced by phenolic compounds. Condensation of catechins into tannins yields an astringent, harsh taste, similar to an unripe apple (or an apple variety suitable only for processing). Table quality apples are low in phenolic compounds. Flavanones (naringin, neohesperidin) are the bitter compounds of citrus fruits.
Phenolic compounds are substrates for phenolase enzymes. These enzymes hydroxylate monophenols to o-diphenols and also oxidize o-diphenols to o-quinones:

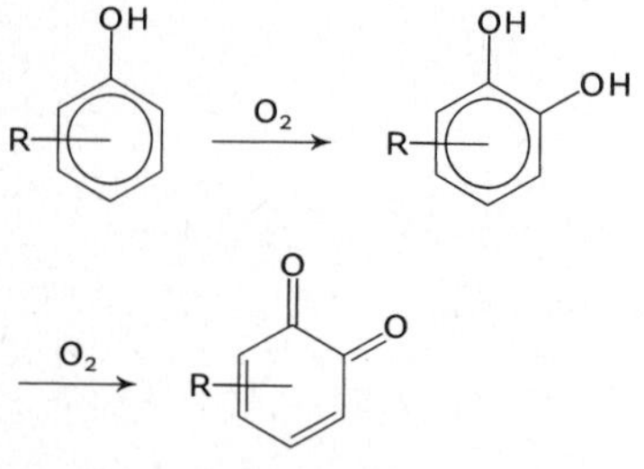

(18.27)

o-Quinones can enter into a number of other reactions, thus giving the undesired brown discoloration of fruits and fruit products. Protective measures against discoloration include inactivation of enzymes by heat treatment, use of reductive agents such as SO_2 or ascorbic acid, or removal of available oxygen.

Polyvalent phenols form colored complexes with metal ions. For example, at pH > 4, Fe^{3+} forms complexes which are bluish-gray or bluish-black in color. Al^{3+} and Sn^{2+} also form intensely colored complexes. Leucoanthocyanins, when heated in the presence of an acid, are converted into anthocyanins. The red color of apples and pears, which is formed during cooking, is derived from leucoanthocyanins.
Phenolic compounds can also form complexes with proteins. These complexes increase the turbidity of fruit juices.

18.1.2.6 Aroma Compounds

Aroma compounds contribute significantly to the importance of fruits in human nutrition. Foods can be divided into four groups depending on the presence of character impact compounds in their aromas. This is explained in more detail in Section 5.1.3.
Table 18.24 gives examples of these four groups of fruits, based on the above-mentioned criterion. Aroma biosynthesis is covered in Section 5.3.2. Selected fruits will be outlined below in more detail.

18.1.2.6.1 Bananas

The characteristic aroma compound of bananas is isopentyl acetate. Some esters of pentanol, such as those of acetic, propionic and butyric acids, also contribute to the typical aroma of bananas, while esters of butanol and hexanol with acetic and butyric acids are generally fruity in character. An important contribution to the complete, mild banana aroma is provided by eugenol (I), O-methyleugenol (II) and elemicin (III):

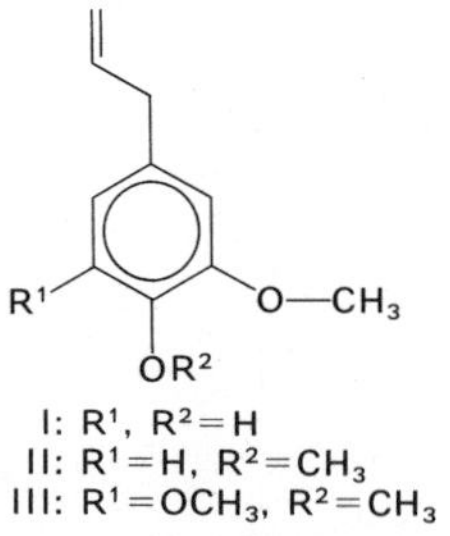

I: R^1, R^2=H
II: R^1=H, R^2=CH_3
III: R^1=OCH_3, R^2=CH_3

(18.28)

18.1.2.6.2 Grapes

Methyl anthranilate is the character impact compound of grapes. Other aroma compounds are

Table 18.24. Aroma compounds of various fruits

Fruit	Compound[a]
Group 1: The fruit aroma resides largely in one compound, a so-called character-impact compound.	
Banana	*Isopentyl acetate*
Pear	*trans-2-cis-4-Decadienoate*[b]
Currant, black	*(+)-p-Menthane-3-one-8-thiol*
Almond	*Benzaldehyde*
Grapes (cv. Concord)	*Methyl anthranilate*
Lemon	*Citral*
Group 2: The fruit aroma is essentially due not to a single compound, but to a mixture of a small number of compounds. A character-impact compound may be present.	
Apple (Delicious)	*Ethyl-2-methylbutyrate,* hexanal, trans-2-hexenal
Bilberry	Ethyl-2-methylbutyrate, ethyl-3-methylbutyrate, trans-2-hexenal
Raspberry	*1-p-Hydroxyphenyl-3-butanone,* cis-3-hexene-1-ol, damascenone, α-ionone, β-ionone
Mandarin	Methyl-N-methylanthranilate, thymol, γ-terpinene, α-pinene
Group 3[c]: The fruit aroma can only be reproduced reasonably faithfully by the use of quite a large number of compounds. A character-impact-compound is unlikely to be present.	
Pineapple[c]	
Apricot	Myrcene, limonene, p-cymene, terpinolene, trans-2-hexenal, α-terpineol, geranial, geraniol, linalool, linalool oxide, 2-methylbutyric acid, caproic acid, γ-lactones (C_8, C_{10}, C_{12}), δ-lactones (C_8,C_{10}), benzyl alcohol
Passion fruit	Ethyl butyrate, ethyl hexanoate, hexyl butyrate, hexyl hexanoate
Peach	γ-Lactones (C_6, C_8, C_{10}), δ-lactones (C_{10}), various esters, alcohols, acids, benzaldehyde
Walnut[c]	
Group 4: The fruit aroma can not be reasonably reproduced even by a complex mixture of specific compounds.	
Strawberry[d]	

[a] Character-impact compounds are in italics.
[b] Methyl-, ethyl-, and higher esters.
[c] For illustration, the major aroma compounds are presented only for some fruits
[d] A comprehensive survey of volatile compounds of strawberry is presented in Table 18.25.

responsible for the specific aromas of some grape varieties.

18.1.2.6.3 Citrus fruits

Terpenes are of great importance in citrus fruits. Limonene (IX in Table 5.22) at 80–90%, is the major component of citrus oils. (+)-Limonene, with its low odor threshold of 10 ppb, contributes to the aroma of the citrus fruits.
The typical aroma note of grapefruit is due to 1-p-menthene-8-thiol (IV) and to nootkatone (V), which is about 5% of the grapefruit oil. Compound IV is an exceptionally powerful aroma constituent (cf. Table 5.1), probably obtained by addition of hydrogen sulfide to limonene. Traces of hydrogen sulfide and dimethyl sulfide are present in all citrus fruit juices and also contribute to their aromas.

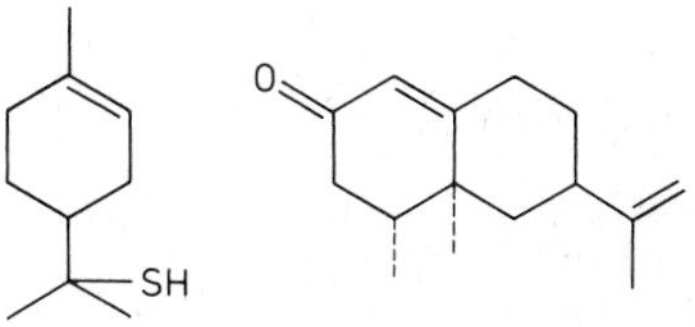

SH O
IV V (18.29a)

Citral, which is actually a mixture of two stereoisomers, geranial (VI) and neral (VII), is the character-impact compound of lemon oil:

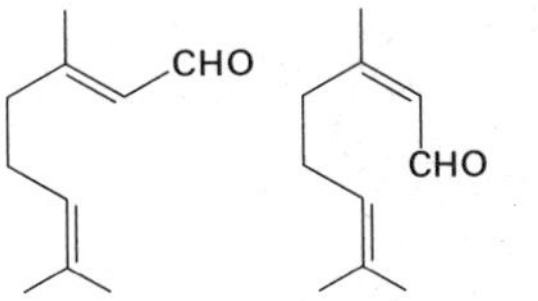

CHO CHO
VI VII 18.29b

The aroma of mandarin oranges is due to: thymol (0.18 weight % of the essential oil), N-methylanthranilic acid methyl ester (0.65), γ-terpinene (14.0) and α-pinene (1.8). Orange aroma is more complicated, but typical are aldehydes such as octanal, nonanal, decanal and dodecanal.

18.1.2.6.4 Apples

Ethyl-2-methylbutyrate, with an odor threshold of 10^{-4} ppm, occurs in apples (variety "Delicious"). This compound provides the "ripe" note to apple aroma. An excess of hexanal and cis-3- and trans-2-hexenal imparts the green, unripe aroma note.

18.1.2.6.5 Raspberries

The character-impact compound is the "raspberry ketone", i.e. 1-(p-hydroxyphenyl)-3-butanone. Additional aroma notes are provided by cis-3-hexenol, α- (VIII) and β-ionone (IX) and α-irone (X):

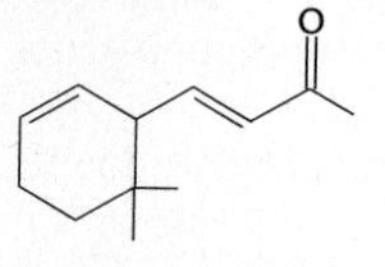

VIII 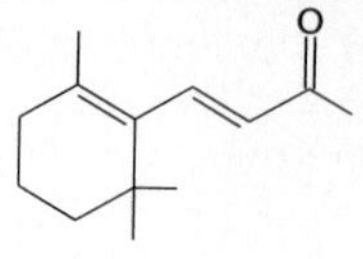 IX (18.30)

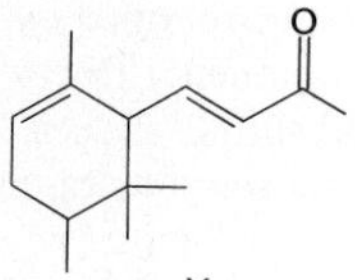

X

18.1.2.6.6 Apricots

Apricots belong to the third group, i. e. a number of compounds in the right proportion is required to satisfactorily reproduce the aroma. Compounds involved here (cf. Formula 18.31) include: terpenes, e. g. myrcene (XI), limonene (IV), p-cymene (XII), terpinolene (XIII), α-terpineol (XIV), geranial (VI), geraniol and linalool (XV); acids, e. g. acetic and 2-methylbutyric acids; alcohols, e. g. trans-2-hexenol; and a number of γ- and δ-lactones, e. g. γ-caprolactone (XVI), γ-octalactone (XVII), γ-decalactone (XVIII), γ-dodecalactone (XIX), δ-octalactone (XX) and δ-decalactone (XXI):

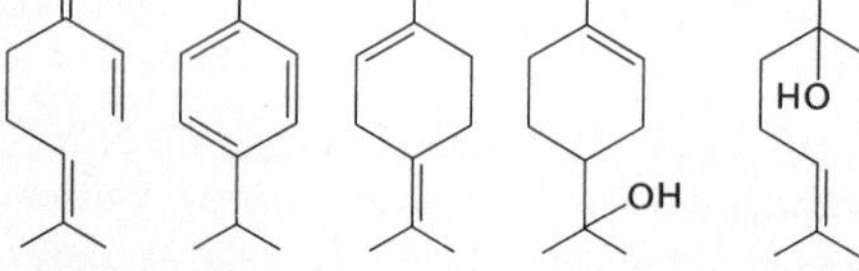

HO OH

XI XII XIII XIV XV (18.31)

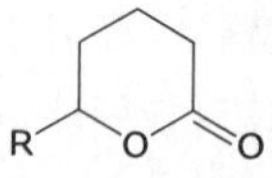

R O O   R O O

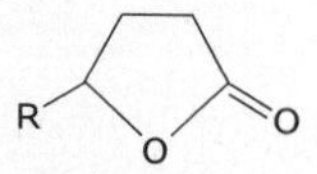

XVI: $R=C_2H_5$
XVII: $R=C_4H_9$
XVIII: $R=C_6H_{13}$
XIX: $R=C_8H_{17}$

XX: $R=C_3H_7$
XXI: $R=C_5H_{11}$

18.1.2.6.7 Peaches

The aroma of peaches is also characterized by various γ- (C_6, C_8, C_{10}) and δ-lactones (C_{10}). Various esters and acids and benzaldehyde are also important.

18.1.2.6.8 Passion fruit

2-Methyl-4-propyl-1,3-oxathiane contributes to the aroma of the yellow passion fruit. In addition the following esters are involved in the formation of the aroma of passion fruit: ethyl butyrate (1.4% of the volatile fraction), ethyl hexanoate (9.7), hexyl butyrate (13.9) and hexyl hexanoate (69.6).

18.1.2.6.9 Strawberries

Strawberries belong to the fourth group of fruits, i.e. satisfactory reproduction of their aroma has not yet been achieved, though more than 300 volatile compounds have been identified in strawberry aroma; among them, more than 100 esters and lactones (Table 18.25).

Synthetic strawberry aroma may contain 3-methyl-3-phenylglycidic acid ethyl ester or related compounds, none of which occur in natural strawberry aroma. Of the four possible isomers, only one (2R:3R; XXII) has a characteristic strawberry odor, while the others have weak, nonspecific aromas:

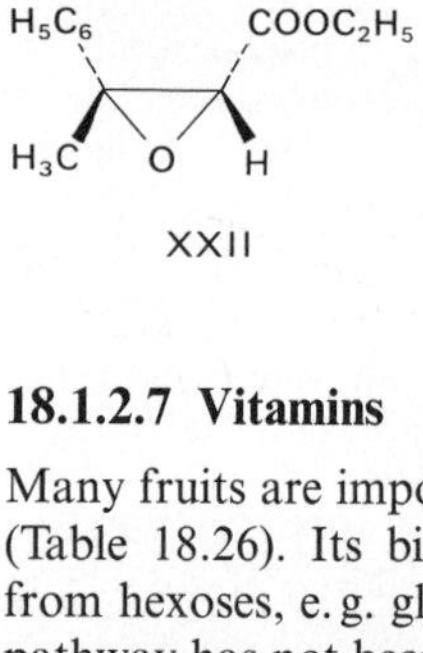

H_5C_6 $COOC_2H_5$

H_3C O H (18.32)

XXII

18.1.2.7 Vitamins

Many fruits are important sources of vitamin C (Table 18.26). Its biosynthesis in plants starts from hexoses, e. g. glucose, but the biosynthetic pathway has not been fully elucidated. It is postulated that, following C-1 oxidation and cyclization to 1,4-lactone (II), the 5-keto compound (III) appears as an intermediary product which is oxidized to the 2,3-endiol (IV) then reduced stereospecifically to L-ascorbic acid (V; cf. Reaction 18.33).

Industrial-scale production of ascorbic acid also starts with glucose. The sugar is first reduced to sorbitol (VI) and then oxidized with *Acetobacter suboxidans* to L-sorbose (VII). The diisopropyli-

Table 18.25. Volatile compounds of strawberries

Hydrocarbons

Ethane, Ethylene, Hexane, Heptane, Octane, Nonane, Decane, Undecane, Dodecane, Methylcyclohexane, Limonene, α-Pinene, β-Pinene, Benzene, Toluene, 1,2-Dimethylbenzene, 1,3-Dimethylbenzene, 1,4-Dimethylbenzene, 1,1,6-Trimethyl-1,2-dihydronaphthalene

Alcohols

Methanol, Ethanol, 1-Propanol, 2-Propanol, 2-Methylpropan-1-ol, 1-Butanol, 2-Butanol, 2-Methyl-1-butanol, 3-Methyl-1-butanol, 2-Methyl-2-butanol, 3-Methyl-2-butanol, 1-Pentanol, 2-Pentanol, 3-Pentanol, 1-Penten-3-ol, 1-Hexanol, 2-Hexanol, 3-Hexanol, trans-2-Hexen-1-ol, cis-3-Hexen-1-ol, 1-Hexen-3-ol, 2-Ethyl-1-hexanol, 1-Heptanol, 2-Heptanol, 3-Heptanol, 1-Octanol, 2-Octanol, 3-Octanol, 3-Octen-1-ol, 1-Octen-3-ol, 1-Nonanol, 2-Nonanol, 1-Nonen-3-ol, 1-Decanol, 2-Decanol, 2-Undecanol, 1-Dodecanol, 2-Dodecanol, 2-Tridecanol, 2-Pentadecanol, Linalool, Benzyl alcohol, 2-Phenylethanol, 2-(4-Hydroxyphenyl)ethanol, α-Terpineol, Terpin, Borneol, Isofenchylalcohol

Carbonyl Compounds

Acetaldehyde, Propanal, 2-Propenal, Butanal, 2-Butenal, 2-Pentenal, Hexanal, trans-2-Hexenal, cis-3-Hexenal, Heptanal, Nonanal, Benzaldehyde, 2-Propanone, 2-Butanone, 3-Methyl-2-butanone, 2,3-Butanedione, 2-Pentanone, 3-Pentanone, 2-Hexanone, 2-Heptanone, 2-Octanone, 2-Nonanone, 2-Decanone, 2-Undecanone, 2-Tridecanone, 2-Pentadecanone, 2-Heptadecanone, Acetophenone

Acids

Formic, Acetic, Propionic, 2-Methylpropionic, Butanoic, 2-Methylbutanoic, 3-Methylbutanoic, 2-Methyl-2-butenoic, Pentanoic, 4-Methylpentanoic, 2-Methyl-2-pentenoic, 2-Methyl-3-pentenoic (cis and trans), Hexanoic, 2-Hexenoic, 5-Methylhexanoic, 3-Hydroxyhexanoic, Heptanoic, Octanoic, 2-Octenoic, 3-Hydroxyoctanoic, Nonanoic, 3-Nonenoic, Decanoic, Decenoic, Undecanoic, Dodecanoic, Tridecanoic, Benzoic, 2-Hydroxybenzoic, 4-Methylbenzoic, 2-Phenylacetic, 3-Phenylpropionic, Cinnamic

Esters

Methyl formate, Ethyl formate, Butyl formate, Isoamyl formate, Methyl acetate, Ethyl acetate, Propyl acetate, Isopropyl acetate, Butyl acetate, Isobutyl acetate, Amyl acetate, Isoamyl acetate, 2-Methylbutyl acetate, 2-Pentyl acetate, Hexyl acetate, 2-Hexyl acetate, trans-2-Hexenyl acetate, cis-3-Hexenyl acetate, 1-Hexen-3-yl-acetate, Hexenyl acetate, 2-Heptyl acetate, 1-Hepten-3-ylacetate, Octyl acetate, Decyl acetate, Benzyl acetate, Phenethyl acetate, Methyl propanoate, Ethyl propanoate, Hexenyl propanoate, Methyl-2-methyl propanoate, Ethyl-2-methyl propanoate, Methyl butanoate, Ethyl butanoate, Propyl butanoate, Isopropyl butanoate, Butyl butanoate, Isobutyl butanoate, Amyl butanoate, Isoamyl butanoate, 2-Pentyl butanoate, Pentenyl butanoate, Hexyl butanoate, trans-2-Hexenyl butanoate, 2-Heptyl butanoate, Octyl butanoate, 2-Nonyl butanoate, Phenethyl butanoate, Ethyl-2-butenoate, Methyl-2-methylbutanoate, Ethyl-2-methylbutanoate, Isopropyl-2-methylbutanoate, Butyl-2-methylbutanoate, Isobutyl-2-methylbutanoate, 2-Methylbutyl-2-methylbutanoate, Isoamyl-2-methylbutanoate, Hexyl-2-methylbutanoate, Octyl-2-methylbutanoate, Ethyl-3-methylbutanoate, Ethyl-3-oxobutanoate, Ethyl pentanoate, Methyl-4-methylpentanoate, Methyl hexanoate, Ethyl hexanoate, Butyl hexanoate, Amyl hexanoate, Isoamyl hexanoate, 2-Pentyl hexanoate, Hexyl hexanoate, 2-Hexenyl hexanoate, trans-3-Hexenylhexanoate, 2-Heptyl hexanoate, Octyl hexanoate, Decyl hexanoate, Methyl heptanoate, Ethyl heptanoate, Methyl octanoate, Ethyl octanoate, Isopropyl octanoate, Butyl octanoate, Isoamyl octanoate, Hexyl octanoate, Hexenyl octanoate, Methyl nonanoate, Isobutyl nonanoate, Isoamyl nonanoate, Methyl decanoate, Ethyl decanoate, Isopropyl decanoate, Hexyl decanoate, Methyl dodecanoate, Ethyl dodecanoate, Methyl benzoate, Ethyl salicylate, Methyl cinnamate, Ethyl cinnamate, 4-Hydroxyhexanoic acid lactone, 4-Hydroxyheptanoic acid lactone, 5-Hydroxyheptanoic acid lactone, 4-Hydroxyoctanoic acid lactone, 5-Hydroxyoctanoic acid lactone, 4-Hydroxydecanoic acid lactone, 4-Hydroxydodecanoic acid lactone

Sulfur Compounds

Methanethiol, Hydrogen sulfide, Ethylthioethane, Ethyldithioethane

Other Compounds

1,1-Dimethoxymethane, 1,1-Diethoxymethane, 1,1-Dimethoxyethane, 1-Ethoxy-1-methoxyethane, 1-Butoxy-1-methoxyethane, 1-Methoxy-1-pentoxyethane, 1,1-Diethoxyethane, 1-Ethoxy-1-propoxyethane, 1-Butoxy-1-ethoxyethane, 1-Ethoxy-1-pentoxyethane, 1-Ethoxy-1-hexoxyethane, 1-Ethoxy-1-(3-hexenoxy)ethane, 1,1-Dihexoxyethane, 1,1-Diethoxypentane, Furfural, 2,5-Dimethyl-4-hydroxy-(2H)-furan-3-one, 2-Furane carboxylic acid, 3-Hydroxy-2-methyl-4-pyrone

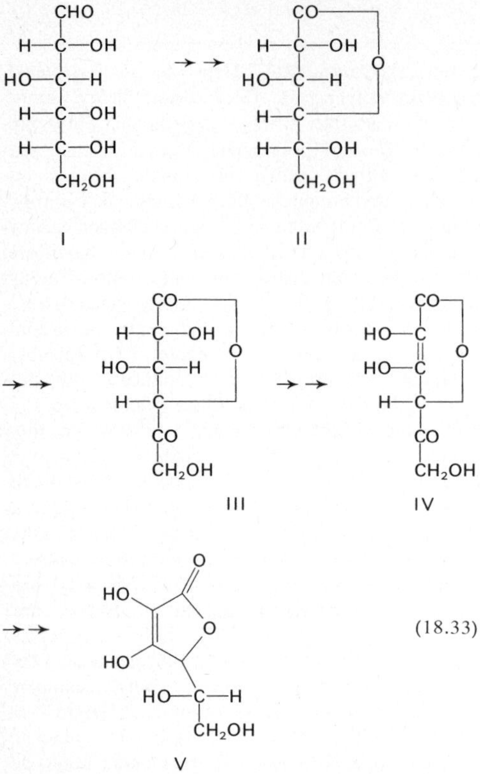

(18.33)

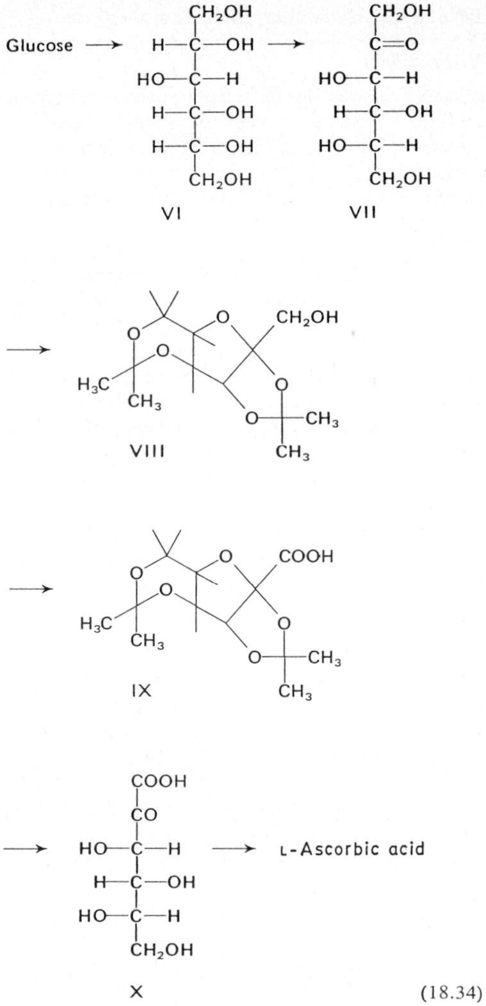

dene derivative (VIII) of the latter is oxidized to the corresponding derivative of L-2-ketogulonic acid (IX). After removal of the protecting isopropylidene groups, L-ascorbic acid (vitamin C) is obtained via L-2-ketogulonic acid (X cf. Reaction 18.34).

(18.34)

Table 18.26. Ascorbic acid in various fruits (mg/100 g edible portion)

Fruit	Ascorbic acid	Fruit	Ascorbic acid
Apple	2–10	Currant, black	210
Pear	4		
Apricot	7–10	Orange	50
Cherry	5–8	Grapefruit	40
Peach	7	Lemon	50
Plum/prune	3		
Blackberry	15	Acerola	1,300
Strawberry	60	Pineapple	25
Raspberry	25	Banana	10–30
Currant, red	40	Guava	300
		Melons	25–35

β-Carotene (provitamin A) occurs in large amounts in apricots, cherries, cantaloups and peaches. B-vitamins present in some fruits (apricots, citrus fruits, figs, black currants and gooseberries) are pantothenic acid and biotin. Other B-group vitamins occur at levels of no nutritional significance. Vitamins B_{12} and D and tocopherols are found in no more than trace amounts.

18.1.2.8 Minerals

Table 18.27 gives the composition of the ash of orange juice and apples. The most important cation is potassium and the most important inorganic anion is phosphate.

Table 18.27. Minerals in fruit

Element	Orange juice (% in ash)	Apple (mg/100 g dry matter)
Potassium	40	840
Sodium	1.8	
Calcium	2.8	90
Magnesium	1.7	40
Iron	0.06	2.5
Aluminium	0.12	
Phosphorus	6.8	90
Sulfur	0.8	
Chlorine	0.7	
Zinc, titanium, barium, copper, manganese, tin	≤ 0.03	Zinc 4.8
		Manganese 0.6
Boron	≤ 0.01	Copper 1.2

18.1.3 Chemical Changes during Ripening of Fruit

Ripening of fruit involves highly complex changes in physical and chemical properties. Softening, increasing sweetness, aroma and color changes are among the most striking phenomena related to ripening. Some changes will be outlined below in more detail.

18.1.3.1 Changes in Respiration Rate

The respiration rate is affected by the development stage of the fruit. A rise in respiration rate occurs with growth. This is followed by a slow decrease in respiration rate until the fruit is fully ripe. In a number of fruits ripening is associated with a renewed rise in respiration rate, which is often denoted as a climacteric rise. Maximal CO_2 production occurs in the climacteric stage. Depending on the fruit, this can occur before or after harvesting. Figures 18.5 and 18.6 show that such a rise occurs a short time after harvest for apples and tomatoes and is accompanied by increased ethylene production.

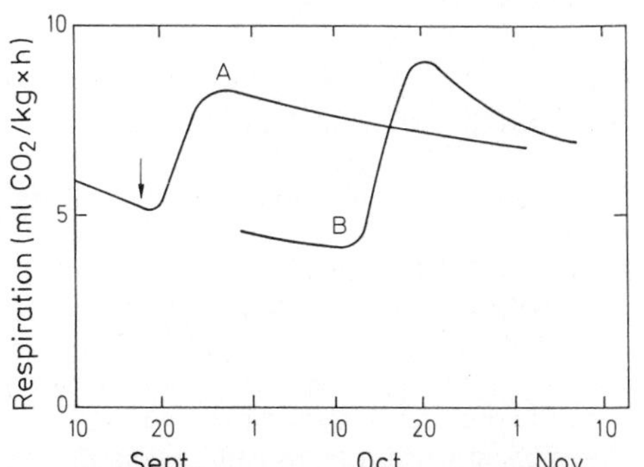

Fig. 18.5. Respiration rise in apples, Bramsley's seedlings. (After *A. C. Hulme,* 1963.) A, apple picked →, B, left on tree to ripen

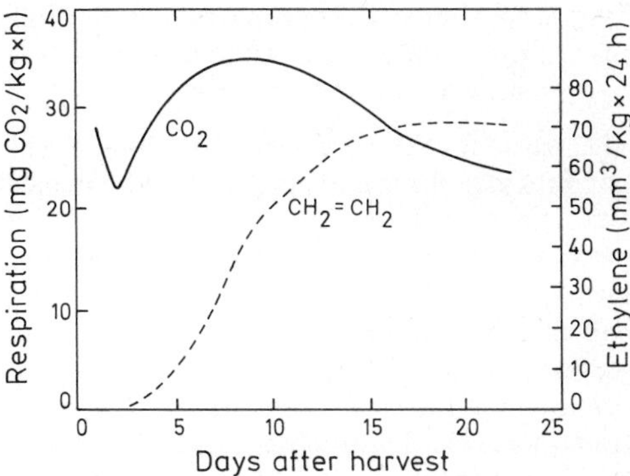

Fig. 18.6. Respiration rise in tomatoes —: CO_2, ----: Ethylene

The climacteric respiration rise is so specific that fruits can be classified into:

- *Climacteric types* of fruits, such as apples, apricots, avocados, bananas, pears, mangoes, papaya, passion fruit, peaches, plums/prunes and tomatoes; and
- *Nonclimacteric types* of fruits, which include pineapples, oranges, strawberries, figs, grapefruit, cucumbers, cherries, cantaloupes, melons, grapes and lemons.

It should be emphasized that nonclimacteric fruits generally ripen on the plants and contain no starch. The differing effects of ethylene on the two types of fruits are covered in Section 18.1.4.2.

Fruits can also be classified according to respiration behavior after harvesting. Three fruit types are distinguished:

Type 1: A slow drop in CO_2 production during ripening (as illustrated by citrus fruits).

Type 2: A temporary rise in CO_2 production. The fruits are fully ripe after this increase reaches a maximum (e.g. avocados, bananas, mangoes or tomatoes).

Type 3: Maximum CO_2 production in the fully ripe stage until the fruit is overripe (e.g. strawberries and peaches).

The reason for the increase in CO_2 production is not yet fully elucidated. Physical and chemical factors are involved. For example, a change in permeability for gases occurs in fruit peels. With increasing age the peel cuticle becomes thicker and is more strongly impregnated with fluid waxes and oils. Thus, the total permeability drops, while the CO_2 concentration within the fruit increases. Three possibilities are usually considered for the rise in CO_2 production. The first is related to increased protein biosynthesis coupled with increased ATP consumption thus stimulating enhanced respiration. Secondly, since the respiratory quotient (RQ) increases from 1 to 1.4–1.6, it is assumed that the additional CO_2 source is not due to respiration but to decarboxylation of malate and pyruvate, i.e. there is a switch from the citric acid cycle to malate degradation. Another possibility is the partial uncoupling of respiration from phosphorylation by an unknown decoupler.

New concepts involving structural factors suggest that fruit flesh possesses marked photosynthetic activity which is then associated with CO_2 uptake. With the onset of ripening, an increased disorganization occurs in chloroplasts and other cell organelles. Photosynthetic activity decreases and finally stops completely. The same is the case for other synthetic activities. Catabolic processes, catalyzed by cytoplasmic enzymes, become dominant. Based on such a perception (*Phan et al.*, 1975) the "climacteric is seen as an indication of the natural end of a period of active synthesis and maintenance, and the beginning of the actual senescence of the fruit".

18.1.3.2 Changes in Metabolic Pathways

Metabolic shifts may occur in several fruits during ripening. For example, during ripening of bananas, there is a marked rise in aldolase and carboxylase activities and thus it appears that at this stage the *Embden-Meyerhoff* pathway becomes dominant and the pentose-phosphate pathway is suppressed.

An increase in malate and pyruvate decarboxylase activities is observed in apples during the climacteric stage. The activities drop as CO_2 production decreases. This provides an explanation for the change in RQ during the climacteric stage. CO_2 production increases more rapidly than O_2 uptake, thus the RQ is greater than 1. The shift from citric acid cycle to malate degradation in apples is also reflected by the effect of citrate and malate on succinate production. As ripening proceeds, production of succinate from citrate drops to zero. An increase in succinate content after addition of malate in the initial stage of ripening is probably a feedback reaction. In this case, a decrease is also observed later on, suggesting a greater change in metabolic patterns.

18.1.3.3 Changes in Individual Constituents

18.1.3.3.1 Carbohydrates

During ripening of fruits, significant changes occur in the carbohydrate fraction. For example, between picking and onset of decay in apples about 20% of the available carbohydrates have been utilized.

During the growth of apples on trees, the starch content rises and then drops to a negligible level by the time of harvest. This drop appears to be related to the increase of climacteric respiration. Unlike starch, the sugar content rises. Other sources in addition to starch should be available for conversion to sugars. A decrease in hemicelluloses suggests that they are a possible source. Organic acids may also be an additional source of sugars.

A marked decrease in starch in bananas parallels an increase in the contents of glucose, fructose and saccharose. Biosynthesis of the latter occurs by two pathways:

$$1)\ \text{UDPG} + \text{Fru-6-P} \longrightarrow \text{UDP} + \text{Sac-6}^{F}\text{-P} \longrightarrow \text{Sac} + \text{Pin}$$

$$2)\ \text{UDPG} + \text{Fru} \longrightarrow \text{UDP} + \text{Sac} \quad (18.34\,a)$$

The content of hemicelluloses drops from 9% to 1–2% (relative to fresh weight), hence they act as a storage pool in carbohydrate metabolism. There is also a drop in sugar content in bananas during the post-climacteric stage.

Differences in various fruits can be remarkable. In oranges and grapefruits the acid content drops during ripening while the sugar level rises. In lemons, however, there is an increase in acids.

Decreases in arabinans, cellulose and other polysaccharides are found in pears during ripening. Cellulase enzyme activity has been confirmed in tomatoes.

Remarkable changes occur in the pectin fractions during ripening of many fruits (e.g. bananas, citrus fruits, strawberries, mangoes, cantaloupes and melons). The molecular weight of pectins decreases and there is a decrease in the degree of methylation. Insoluble protopectin is increasingly transformed into soluble forms. Protopectin is tightly associated with cellulose in the cell wall matrix. Its galacturonic acid residues are acetylated at OH-groups in positions 2 and 3 or are bound to polysaccharides as lignin (R^1 = H, CH_3, polysaccharides: arabinan, galactan and possibly cellulose; R^2 = H, CH_3CO, polysaccharides, lignin):

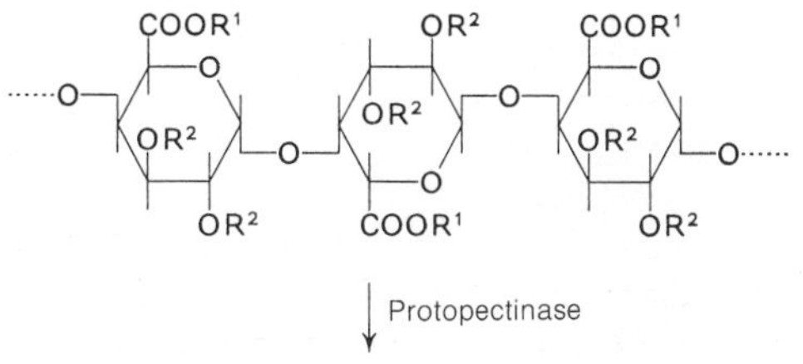

COOR¹ OR² COOR¹ / OR² OR² OR² / OR² COOR¹ OR²

Protopectinase

Polysaccharides + soluble Pectins (18.35)

After prolonged ripening there is a decrease in soluble pectins in apples. This drop is associated with a mealy, soft texture. Similar events occur in pears, but much more rapidly and with more extensive demethylation of pectin. Generally, the degree of pectin esterification drops from 85% to about 40% during ripening of pears, peaches and avocados. This drop is due to a remarkable increase in activities of polygalacturonases and pectin esterases. The rise in free galacturonic acid is negligible; therefore it appears that the release of uronic acid is associated with its simultaneous conversion through other reactions.

18.1.3.3.2 Proteins, enzymes

During ripening of some fruits and with a constant total content of nitrogen, there is an increase in protein content, an increase assigned primarily to increased biosynthesis of enzymes. For example, during ripening of fruit there is increased activity of hydrolases (amylases, cellulases, pectinolytic enzymes, glycolytic enzymes, enzymes involved in the citric acid cycle, transaminases, peroxidases and catalases). Proteinaceous enzyme inhibitors which inhibit the activities of amylases, peroxidases and catalases are found in unripe bananas and mangoes. The activities of these inhibitors appear to decrease with increasing ripeness.

The ratios of $NADH/NAD^+$ or $NADPH/NADP^+$ pass through a maximum during ripening of fruit. For example, the values for mangoes are 0.32–0.67 in the unripe stage, 1.44–6.50 in the semi-ripe stage and 0.57–0.93 in the ripe stage. During ripening of fruit, shifts also occur in the amino acid and amine fractions. The shifts are not uniform and are affected by type and ripening stage of fruits.

18.1.3.3.3 Lipids

Little is known about changes in the lipid fraction. Shifts in composition and quantity have been found, especially in the phospholipid fraction.

18.1.3.3.4 Acids

There is a drop in acid content during ripening of fruits. Lemons, as already mentioned, are an exception. The proportion of various acids can change. In ripe apples malic acid is the major acid, while in young, unripe apples, quinic acid is dominant. In the various tissues of any single fruit, various acids can be dominant. For example, apple peels contain citramalic acid (I) which is formed from pyruvic acid, and can produce acetone through acetoacetic acid. Acetone is formed abundantly during ripening:

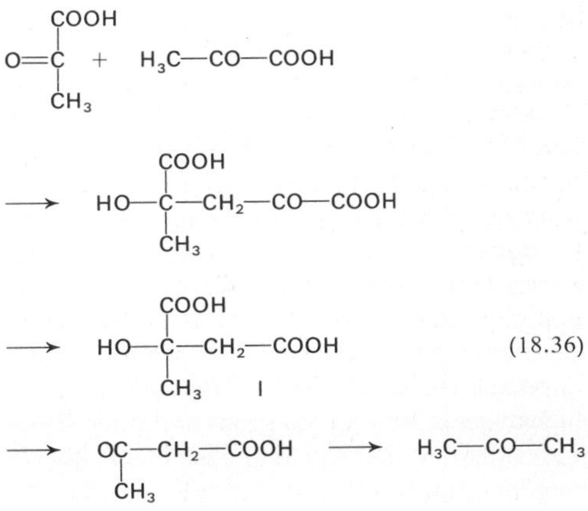

$$O{=}C(COOH)(CH_3) + H_3C{-}CO{-}COOH$$

$$\longrightarrow HO{-}C(COOH)(CH_3){-}CH_2{-}CO{-}COOH$$

$$\longrightarrow HO{-}C(COOH)(CH_3){-}CH_2{-}COOH \quad (\text{I}) \qquad (18.36)$$

$$\longrightarrow OC(CH_3){-}CH_2{-}COOH \longrightarrow H_3C{-}CO{-}CH_3$$

The synthesis of ascorbic acid is also of importance. It takes place in many fruits during ripening (cf. 18.1.2.7).

18.1.3.3.5 Aroma compounds

The formation of typical aromas takes place during the ripening of fruit. In bananas, for example, noticeable amounts of volatile compounds are formed only 24 h after the climacteric stage has passed. The aroma build-up is affected by external factors such as temperature and day/night variations. Bananas, with a day/night rhythm of 30 °C/20 °C, produce about 60% more volatiles than those kept at a constant temperature of 30 °C. Biosynthesis of volatile compounds is outlined in Section 5.3.2.

18.1.4 Ripening as influenced by Chemical Agents

The regulation of plant growth is achieved with phytohormones. They can be divided into five classes: auxins, gibberellins, cytokinins, abscisic acid and ethylene. There are many synthetic compounds which imitate the activities of phytohormones or act as their antagonists. These compounds are used in practice for various reasons: to induce blossom, to induce parthenocarpy, to enhance yields, to facilitate harvesting, to promote fruit ripening, to prevent sprouting or germination, and to retard aging and thus to increase the shelf life or storage stability of fruits. Some important examples are outlined in the following sections.

18.1.4.1 Compounds with Retarding Effects

A number of compounds retard the ripening and aging of fruits and vegetables and, thus, extend their storage life. Kinetin (I, cf. Formulas 18.37) and N^6-benzyladenine (II), which are cytokinins, retard chlorophyll degradation and aging of leafy vegetables (e.g. spinach, beans, cucumbers). Compound II affects the storability of fruits (e.g. sweet cherries and strawberries) and vegetables (e.g. cauliflower, chicory, various cabbage cultivars, radishes, celery and asparagus). It also improves the setting of the fruit and provides higher yields for grapes, melons and other fruits. Gibberellin A_3 (III) can be applied before harvest (e.g. to oranges or lemons) or after harvest (e.g. to bananas, guava and tomatoes) to retard fruit ripening. Generally, quality improvements have been reported for many fruits and vegetables. For example, larger grapes are obtained, with a looser arrangement of berries. Parthenocarpy is stimulated in tomatoes and grapes.

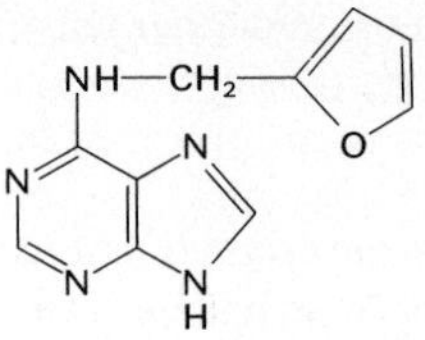

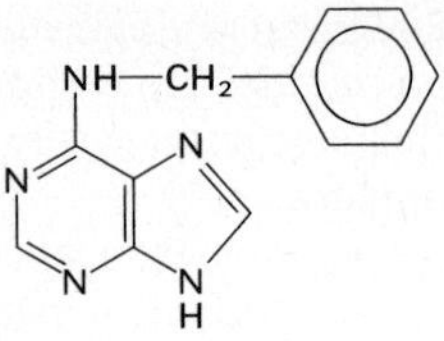

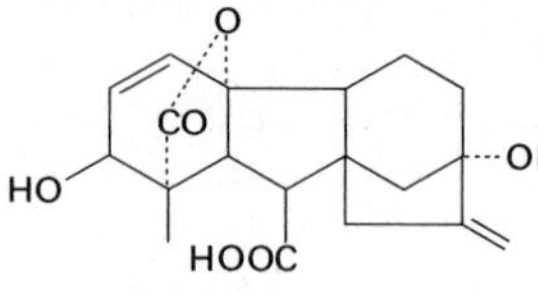

NH—CH₂ (furfuryl) adenine — I; NH—CH₂ (benzyl) adenine — II; gibberellin (HO, CO, O, OH, HOOC) — III

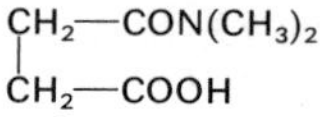

CH_2—$CON(CH_3)_2$ / CH_2—COOH — IV

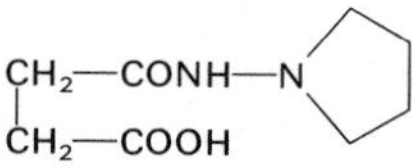

CH_2—CONH—N(pyrrolidine) / CH_2—COOH — V (18.37)

Application of gibberellin antagonists, e.g. N,N-dimethylsuccinic acid amide (IV), favorably affects the firmness and color of fruits, accelerates ripening of cherries, peaches and nectarines, and improves formation of grapes. An even more positive influence has been found with N-pyrrolidinyl succinic acid amide (V).

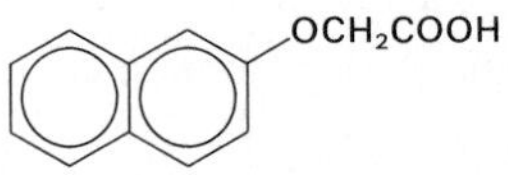

OCH_2COOH — VI; CH_2COOH — VII; O—CH(CH_3)—COOH, Cl — VIII; COOH, J, J, J — IX (18.38)

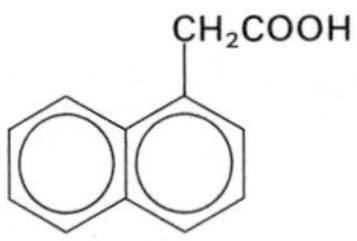

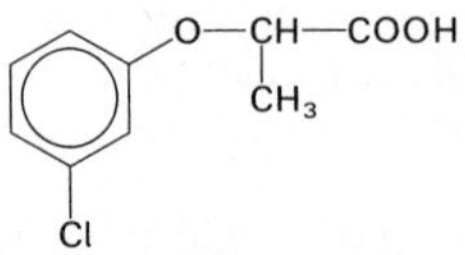

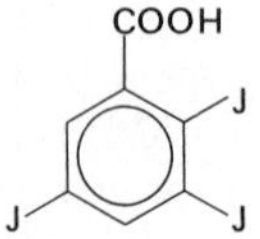

β-Naphthoxyacetic acid (VI, cf. Formulas 18.38) is a synthetic compound of the auxin group. Applied before harvest, it has a favorable affect on the storage properties of mandarin oranges. Fruit weight loss is diminished and the content of ascorbic acid remains high. α-Naphthylacetic acid (VII) is used for the induction of blossoms in pineapples and for thinning of apples and oranges on the trees. 2-(3-Chlorophenoxy)-propionic acid (VIII) is utilized for peach thinning, while 2,3,5-triiodobenzoic acid (IX) induces blossoming of apple trees and retards tomato growth.

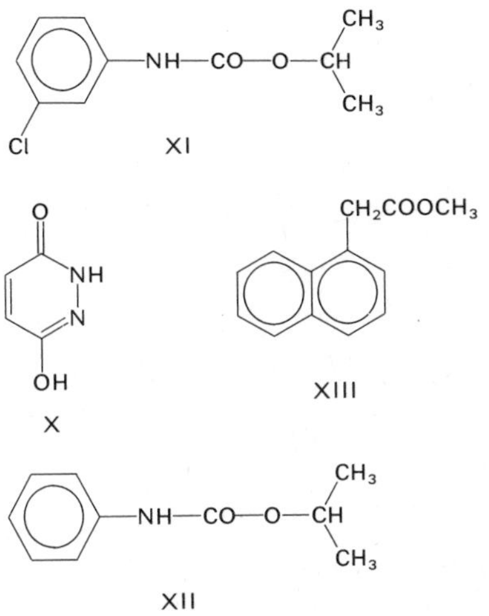

XI, X, XIII, XII

(18.39)

Maleic acid hydrazide (X, cf. Formulas 18.39) interferes with cell division and also affects mobilization of storage polysaccharides in favor of mono- and oligosaccharides. It is used as a sprouting inhibitor (e.g. with potatoes, carrots or onions). Compound X is also used to retard ripening of various fruits (mangoes and tomatoes) and to maintain the firm texture of apples during storage.

Various carbamates are also active sprouting inhibitors. Examples are isopropyl-(3-chlorphenyl)-carbamate (XI), isopropyl-N-phenylcarbamate (XII) and α-naphthylacetic acid methyl ester (XIII).

Retardation of ripening of various fruits has also been observed by using cycloheximide (XIV). However, this compound facilitates the detachment of citrus and olive fruits from trees.

Actinomycin D, ethylene oxide (e.g. for mango fruits) and dehydroacetic acid (XV; e.g. for strawberries) are also able to retard ripening.

Compounds which can bind ethylene (a ripening promoter) also have retarding activities. For example, bananas sealed in polyethylene bags have a prolonged shelf life in the presence of a silica carrier impregnated with alkaline $KMnO_4$.

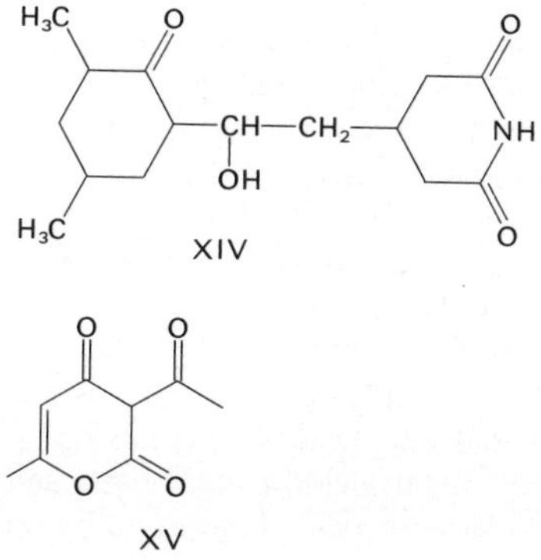

XIV, XV

(18.39a)

$$CH_3-S-CH_2-CH_2-CH(NH_3^{\oplus})-COO^{\ominus} \xrightarrow{ATP \rightarrow PP_i + P_i} CH_3-\overset{\oplus}{S}(Ad)-CH_2-CH_2-CH(NH_2)-COO^{\ominus} \xrightarrow{R-CHO} CH_3-\overset{\oplus}{S}(Ad)-CH_2-CH_2-CH(COO^{\ominus})-N=CH-R \rightarrow \xrightarrow{-CH_3-S-Ad} \text{cyclopropane}(COO^{\ominus})(N=CH-R) \xrightarrow{H_2O,\ -R-CHO} \text{cyclopropane}(COO^{\ominus})(NH_3^{\oplus}) \xrightarrow{Ox.} \text{cyclopropane}(COO^{\ominus})(NH-OH) \xrightarrow{H_2O} \{CH_2=CH_2,\ CO,\ NH_3,\ HCOOH\}$$

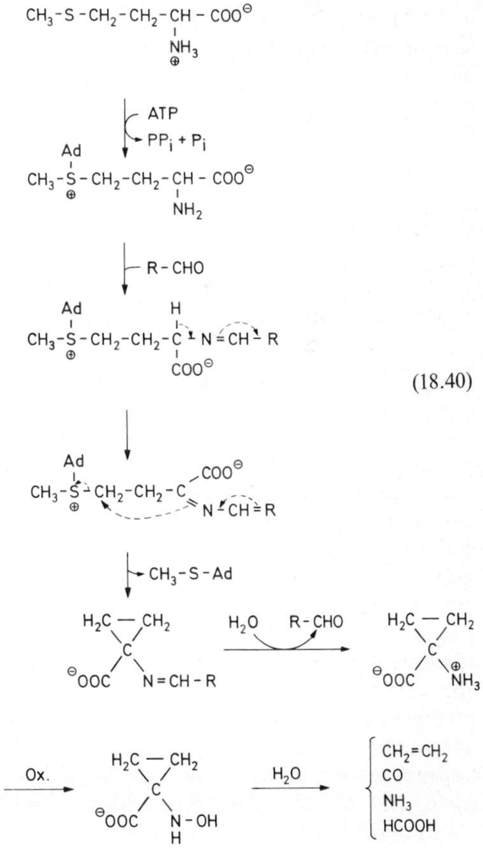

(18.40)

18.1.4.2 Compounds promoting Fruit Ripening

Fruit ripening is coupled with ethylene biosynthesis. This gaseous compound increases membrane permeability and, probably, thereby accelerates metabolism and fruit ripening. With mango fruits, for example, it has been demonstrated that, before the climacteric stage, ethylene stimulates oxidative and hydrolytic enzymes (catalase, peroxidase and amylase) and inactivates inhibitors of these enzymes.

Climacteric and nonclimacteric fruits respond differently to external ethylene (Figure 18.7). Depending on the ethylene level, the respiratory increase sets in earlier in unripe climacteric fruits, but its height is not influenced. In contrast, in nonclimacteric fruits there is an increase in respiration rate at each ripening stage which is clearly dependent on ethylene concentration.

The reaction pathway under 18.40 is sugqested for the biosynthesis of ethylene (R – CHO; pyridoxal phosphate; Ad: adenosine).

Ethylene and compounds able to release ethylene under suitable conditions are utilized commercially for enhancing the ripening process. A number of such compounds are known, e.g. 2-chloroethylphosphonic acid (ethephon; R = H or CH_2-CH_2Cl):

$$Cl-CH_2-CH_2-\overset{\overset{\displaystyle O}{\|}}{\underset{\underset{\displaystyle OR}{|}}{P}}-OH \xrightarrow{pH>4} H_2C{=}CH_2 + H_2PO_4^{\ominus} + Cl^{\ominus} \quad (18.41)$$

The use of ethylene before picking fruit (as with pineapples, figs, mangoes, melons, cantaloups and tomatoes) results in faster and more uniform ripening. Its utilization after harvesting accelerates ripening (e.g. with bananas, citrus fruits and mangoes). Ethylene can induce blossoming in the pineapple plant and can accelerate detachment of stone fruits and olives. Vine defoliation can also be achieved.

The activity of propylene is only 1% of that of ethylene. Acetylene also accelerates ripening but only at substantially higher concentrations.

Methionine affects the ripening of apples, bananas and mangoes by stimulating ethylene biosynthesis. Other stimulants are auxins, abscisic acid (XVI), ethanedial-dioxime (XVII; used for citrus fruits) and 5-chloro-3-methyl-4-nitropyrazole (XVIII). Application of the latter facilitates harvesting of ripe oranges.

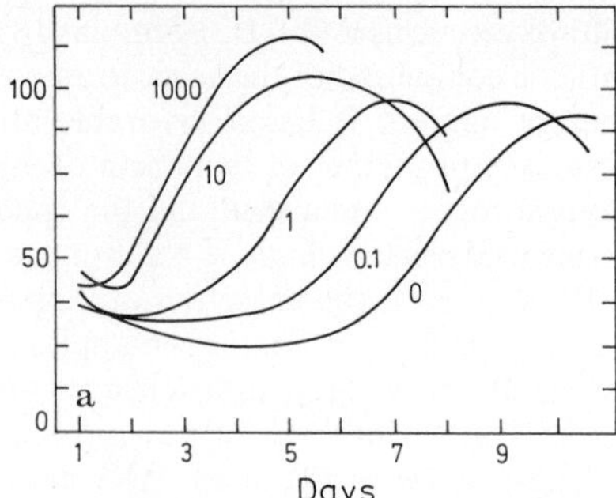

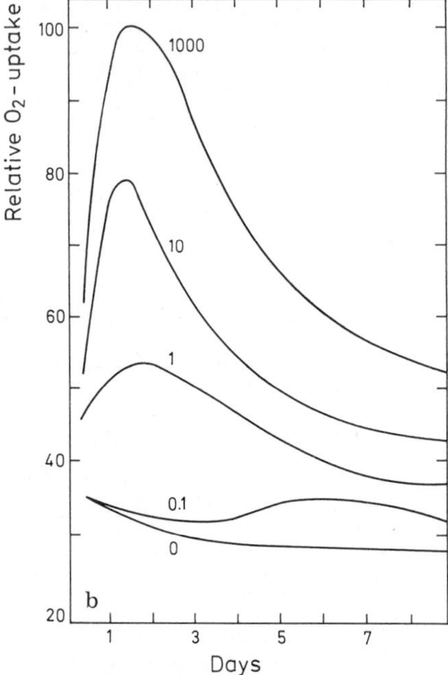

Fig. 18.7. The effect of ethylene on fruit respiration. (a) climacteric, (b) nonclimacteric. Numerals on the curves: ethylene in air, ppm. (After *Biale*, 1964)

Ethylene biosynthesis from methionine is inhibited by rhizobitoxine (XIX), which is found in the nodules of a symbiotic association of *Rhizobium japonicum* with the root system of soybeans. Carbon dioxide is an ethylene antagonist and, thus, a retardant of fruit ripening.

N,N-bis-phosphonomethyl glycine (XX), which interferes in protein synthesis, accelerates ripening and increases sugar yield in sugar beets and sugar cane. The sucrose yield is increased by up

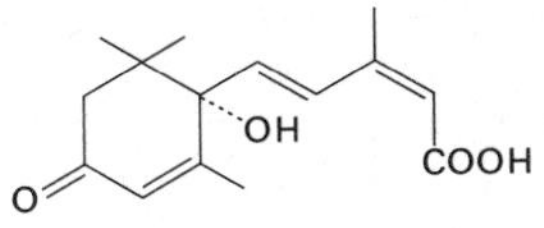

XVI

HO—N=CH—CH=N—OH

XVII

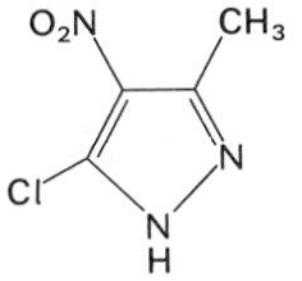

XVIII

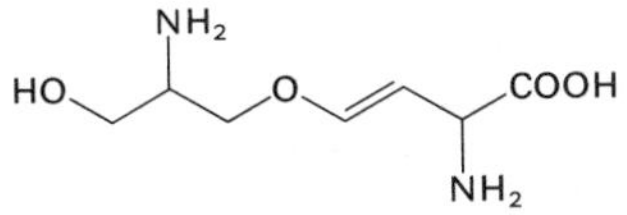

XIX

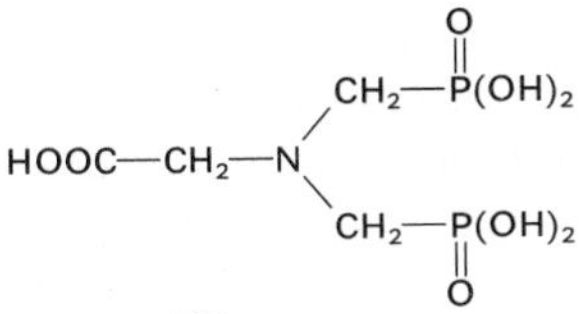

XX

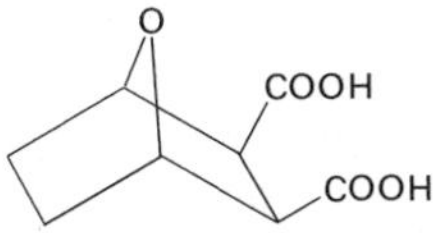

XXI

XXII

(18.42)

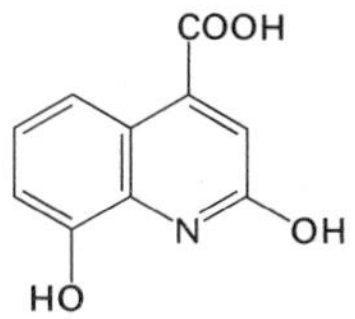

to 10%. A sucrose yield increase of up to 25% is recorded in sugar cane after application of 7-oxabicyclo[2,2,1]heptane-2,3-dicarboxylic acid (XXI). This compound appears to interfere with lipid biosynthesis and probably causes changes in the permeability of cell wall membranes, and it apparently retards the cleavage of sucrose.

Zeaninic acid (XXII) is a growth promoter with an unknown mechanism of activity. It is recovered from wastes of corn starch production. It facilitates formation of the grape berry.

18.1.5 Storage of Fruits

18.1.5.1 Cold Storage

The suitability, duration and required conditions of fruit storage are dependent on variety and quality. Commonly used conditions are −1° to +2° C at 80–90% relative humidity. The storage time varies from 4–8 months for apples, 2–6 months for pears, 2–3 months for grapes, 1–2 weeks for strawberries and raspberries, and 4–5 days for cherries. Efficient aeration is required during storage. Air circulation is often combined with purging to remove ethylene, the volatile promoter of fruit ripening. Weight losses occur during fruit storage due to moisture losses of 3–10%.

18.1.5.2 Storage under a Controlled (Modified) Atmosphere

This term is applied to an atmosphere which, in comparison to air, has a lowered oxygen concentration and an increased CO_2 concentration. Common conditions for storage of many fruits are temperature of 0–5 °C, O_2 concentration of about 3% and CO_2 level of about 0–5%. For each fruit variety it is important that optimal conditions for controlled atmosphere storage be maintained. For example, a high O_2 concentration accelerates ripening, while an overly low O_2 concentration results in high production of CO_2. The result can be aroma defects (apples, oranges, bananas and strawberries) and discoloration of fruit.

18.2 Fruit Products

The short shelf life for most fruits and the frequent need to store and spread out the surplus of a harvest for a prolonged period of time has brought about a number of processes which provide more durable and stable fruit products.

18.2.1 Dried Fruits

Like many other food products, moisture removal from fruits by a suitable drying process results in a product in which microbial growth is retarded and, with a suitable pretreatment, the enzymes present are largely inactivated. Fruit drying is probably the oldest procedure for preservation. It was originally performed in a rather primitive way (spreading the fruit in the hot air of a fireplace or hearth, kitchen stove or oven), thus providing dark "baked products". Solar drying is still a common process in southern and tropical countries for obtaining dried apple slices, apricots, peaches or pears or tropical fruits such as dates, figs or raisins. Predrying is often achieved in sunshine and additional drying by artifical heat in drying installations. The temperature in drying chambers, flat or tunnel dryers is between 75 °C (incoming air) and 65 °C (temperature of the exit air) at a relative humidity of 15–20%. Vacuum drying at about 60 °C is particularly gentle.

Carefully washed and trimmed fruits of suitable varieties are pretreated in various ways:

Pomme fruits (apples, pears) are initially peeled mechanically and freed from the core and calix (seed compartment). Apples are then cut preferentially into 5–7 mm thick slices, and dried in rings (a yield of 10–20% of the unpeeled fresh weight). Sulfite treatment is used to prevent browning during processing and storage. The sulfurous acid prevents both enzymatic and nonezymatic browning reactions, stabilizes vitamin C and prevents microbial contamination during storage of the end product. The utilization of dilute solutions of citric acid is also suitable for preventing browning. Whole or sliced pears are heated with steam to achieve a translucent appearance and then are dried at 60–65 °C. The yield is 13–14% of the fresh weight.

The *stone fruits* which are usually dried are plums/prunes, apricots and peaches. Plums are first dipped for 5–15 s into a hot, diluted solution of sodium hydroxide, or into 0.7% aqueous K-carbonate and then rinsed and dried at 70–75 °C or dried in the sun. Plum peels are often fissured to facilitate drying. In order to clean and to provide a black, glossy surface, dried plums are steamed additionally at 80–85 °C for a short time. The plum yield is 25–30% at a moisture content of not more than 19%. Apricots and peaches are treated alternately with cold and hot water, then are halved, the stone seed is removed and the fruit is dried in the sun or in drying installations at 65–70 °C. The yield, depending on fruit size, is 10–15%. SO_2 (sulfurous acid) treatment is common for apricots and peaches. Cherries play a less important role as dried fruit. To avoid substantial aroma losses, cherries are dried slowly and with a number of precautions.

Grapes are the most commonly dried *berry fruits*. Raisins are dark-colored, dried grapes which contain seeds, whereas sultana raisins are seedless, light-colored, dried grapes. Currants, with or without seeds, are dark and are much smaller in size than the other two raisin products. The surface treatment of raisins, with the exception of the currants, involves the use of acetylated monoglycerides to prevent caking or sticking.

The compositions of some dried fruits are presented in Table 18.28. Dried fruits are exceptionally rich in calories and they supply significant amounts of minerals. Of the vitamins found in fruit, β-carotene and the B-vitamins remain intact. Vitamin C is lost to a great extent. Sulfite treatment destroys vitamin B_1. However, fruit color and vitamin C content are retained and stabilized.

Table 18.28. Composition of some dried fruits (g/100 g edible portion)

Fruit	Moisture	N-containing compounds	Fat	Carbohydrates	Crude fiber	Minerals	Vitamin C
Apricots	15–24	5.0	0.4	68	3.2	3.5	0.011
Dates	20.2	1.9	0.5	73.2	2.4	1.8	0.003
Figs	24.6	3.5	1.3	61.5	6.7	2.4	0–0.005
Peaches	24.0	3.0	0.6	65.9	3.5	3.0	0.017
Plums/ prunes	24.0	2.3	0.6	69.4	1.6	2.1	0.004
Raisins	24.2	2.3	0.5	64.2	7.0	1.86	0.001

18.2.2 Canned Fruits

Since the middle of the last century, heat sterilization in cans and glass jars has been the most important process for fruit preservation.

Undamaged, aroma-rich and not overripe fruits are suitable for heat sterilization. Aseptic canning is applicable only for fruit purées. Canned fruits used are primarily stone fruits, pears, pineapples and apples (usually apple purée). Strawberries and gooseberries are canned to a lesser extent.

Canned fruits are produced in a large volume by the food industry and also in individual households. Cherries are freed from stone seeds and

stems, plums/prunes, apricots and peaches are halved and the stone seeds are removed, strawberry calix is removed, gooseberry and red currant stems are removed, apples and pears are peeled and sliced. Specialized equipment has been developed for these procedures.
With a few exceptions (raspberries and blackberries) all fruits are washed or rinsed. Apricots are readily peeled after alkali treatment at 65 °C. Fruits sterilized unpeeled, e.g. prunes or yellow plums, are first fissured to prevent later bursting. To avoid aroma loss and to prevent floating in the can, fruits which shrink considerably (such as cherries, yellow plums, strawberries and gooseberries) are dipped prior to canning into a hot 30% sugar solution and then covered with sugar solution in cans, with a sugar concentration approximately twice the desired final concentration in the can. Finally, the can is vacuum sealed at 77–95 °C for 4–6 min and, according to the fruit species, heat sterilized under the required conditions. For example, a 1 liter can of strawberries is sterilized in a boiling water bath at 100 °C for 18 min, while pears, peaches and apricots are heated at 100 °C for 22 min. Additions of ascorbic and citric acids for stabilization of color and calcium salts for the preservation of firm texture have been accepted as standard procedures for canned fruits consumed as desserts.
Canned fruits used for bakery products, confections or candies are produced like canned dessert fruits, however, the fruits are covered by water instead of sugar solutions.

18.2.3 Deep-Frozen Fruits

Fruits are frozen and stored either as an end product or for further processing. The choice of suitable varieties of fruit at an optimal ripening stage is very important. Pineapples, apples, apricots, grapefruit, strawberries and dark-colored cherries are highly suitable. Light-colored cherries, plums, grapes and numerous subtropical or tropical fruits are of low suitability.
Rapid chilling is important (air temperature ≤ -30 °C, freezing time about 3 h) to avoid microbial growth, large concentration shifts in fruit tissues, and formation of large ice crystals which damage tissue structure. A blanching step prior to freezing is commonly used only for few fruits, such as pears, and occasionally for apples, apricots and peaches. Some fruits are covered, prior to freezing, with a 30–50% sugar solution or with solid granulated sugar (1 part per 4–10 parts by weight) and are left to stand until the sap separates. In both instances oxygen is eliminated, enzymatic browning is prevented, and the texture and aroma of the fruit are better preserved. Addition of ascorbic acid or citric acid is also common.
Frozen fruit which is stored at -18 to -24 °C is stable for two to four years.

18.2.4 Rum Fruits, Fruits in Sugar Syrup, etc.

Rum fruits are produced by steeping the fruit in dilute spirits in the presence of sufficient sugar.
Fruits preserved in vinegar, mostly pears and plums, are prepared by poaching in wine vinegar sweetened with sugar and spiced with cinnamon and cloves.
Fruits in sugar syrup are prepared by treating raw or precooked fruits or fruit portions (may be precooked under a vacuum) with highly concentrated sucrose solutions which also contain starch syrup. The latter is added to enhance translucency, smoothness and tractability of the product. Candied lemon or orange peels are products of this kind.
Other varieties provide intermediary products processed further into fruit confections: glazed fruits (these are washed fruits treated with a sugar solution containing gum arabic and then subsequently dried at 30–35 °C) or candied fruits in which the dried, glazed fruit is also immersed in a concentrated sugar solution and then dried to form a candied hull. Another product is crystallized fruit in which the dried, glazed fruits are rolled over icing or granulated sugar (sucrose), then dried additionally and, to achieve a shiny, glossy appearance, are exposed to steam for a short time.

18.2.5 Fruit Pulps and Slurries

Fruit pulp is not suitable for direct consumption. The pulp is in the form of slurried fresh fruit or pieces of fruit either split or whole, and, when necessary, stabilized by chemical preservatives. The minimum dry matter content of various pulps is 7–11%. For pulp production the fruit, which has been washed in special machines, is lightly steamed in steam conduits or precooking retorts.

The fruit slurry is an intermediary product and is not suitable for direct consumption. The production steps are similar to those for pulp. However, there is an additional step: slurrying and straining, i.e. passing the slurry through sieves. Both the pulp and the slurry can be stored frozen.

18.2.6 Marmalades, Jams and Jellies

Marmalades are thick, spreadable fruit slurries obtained by boiling (thickened by boiling) fresh or fresh stored fruits, or are obtained from fruit pulps or slurries by the addition of sugar. Addition of unpeeled fruits, fruit pectin, starch syrup, and tartaric, citric or lactic acids is common. Marmalades can be made from a single fruit or a mixture of fruits, or can be blends of several marmalades.

For the production of marmalade, the fresh fruits or intermediary products, such as fruit pulps or slurries, are boiled in an open kettle with the addition of sugar (usually added in two batches). Other ingredients (gelling agents, starch syrup and acids) are added before the thickening is completed by boiling. The end of boiling is determined by refractometer readings (the total boiling time is usually 15–30 min). Boiling in a vacuum decreases the time and the required temperature. In addition, it enhances color and aroma retention. However, marmalade boiled in a vacuum does not have the characteristic flavor of marmalade boiled in an open kettle since the flavor of the latter is imparted by a limited extent of sucrose inversion and sugar caramelization. The hot marmalade is then poured into appropriate containers.

Jams and marmalades which have not been boiled are pasteurized at temperatures under 85 °C. This assures retention of fruit aroma and color in the product.

Jams are produced similarly to marmalades but usually from one kind of fruit. They are thickened by boiling and constant stirring of the whole or sliced fresh or fresh stored raw material, or of fruit pulp. Unlike marmalades, whole fruits or pieces can be found in ready-made jams. Table 18.29 provides compositional data for some commercial jams.

For the production of jelly the sap is mixed with half its weight of sugar and, when necessary, with fruit pectin and other constituents, and is thickened by boiling in an open kettle or in a vacuum kettle. The scum is carefully skimmed off and the mixture is boiled further until a moisture content of about 42% is reached.

18.2.7 Plum Sauce (Damson Cheese)

Plum/prune sauce is produced by thickening through boiling of fresh fruit pulps or fruit slurries. The use of dried plums is also common. Normally, the product has no added sugar, although sweetened products or products with other ingredients added are also produced.

18.2.8 Thickened Fruit Syrups

Thickened fruit syrups are products from fresh apples or pears which are obtained by steaming or boiling, followed by pressing and evaporation of the extract or by evaporation of the juices with or without the addition of sugar. The dry matter content of the end product is at least 65%.

Table 18.29. Composition of various jams (average values in %)

Jam from	Moisture	Water soluble extract	Total sugar	Sugar-free extract	Titratable acidity	Ash	Pectin as Ca-pectate
Strawberries	32.2	66.2	57.7	8.5	0.49	0.30	0.34
Apricots	33.1	66.2	51.3	5.0	0.71	0.36	0.50
Cherries	28.6	70.8	61.6	9.3	0.55	0.38	0.42
Blackberries	34.2	64.8	58.0	6.8	0.71	0.32	0.34
Raspberries	30.0	67.2	60.3	6.8	0.90	0.30	0.38
Bilberries	30.1	68.0	60.0	8.0	0.78	0.22	0.37
Plums/prunes	31.1	68.0	59.5	8.3	0.42	0.24	0.43

Details of the production are as follows: fresh sweet apples or pears are boiled in water or steamed until a soft, mealy consistency is achieved. The fruit is then subjected to hydraulic pressing. The sap collected is boiled in an open kettle under constant stirring to a thick consistency with a moisture content not exceeding 35%. The same method is used to produce a sweetened product from apples (with a sucrose content of 25% of the weight of the end product), a blend of apples and pears, and thickened sugar beet syrup.

18.2.9 Fruit Juices

Fruit juices are usually obtained by mechanical means, and also from juice concentrates (cf. 18.2.11) by dilution with water. The solid matter content is generally 5–20%. The juices are consumed as is or are used as intermediary products, e.g. for the production of syrups, jellies, lemonades, fruit juice liqueurs or fruit candies. Fruit juice production is regulated in most countries.
Juices from acidic fruits are usually sweetened by adding sucrose, glucose or fructose. Juices used for further processing usually contain chemical preservatives to inhibit fermentation.
Some juices from berries and stone fruits, because of their high acid content, are not suitable for direct consumption. Addition of sugar and subsequent dilution with water provides fruit nectars or sweet musts (cf. 18.2.10). Table 18.30 lists data on the composition of some juices and nectars.
Production of fruit juice involves several processing steps: fruit preparation and the removal, treatment and preservation of the juice.
Preparation of the fruit involves washing, rinsing and trimming, i.e. the faulty and unripe fruits are removed. The stone seeds and stalks, stems or calyx are then removed. Disintegration is accomplished mechanically in mills, thermally by heating (thermobreak at about 80 °C) or by freezing (less than − 50 °C). The yield can be increased by enzymatic pectin degradation ("mash fermentation", particularly of stone fruits and of berries) or by applying procedures such as ultrasound or electroplasmolysis.
Separation of the juice is achieved using continuous or discontinuous presses or processes such as vacuum filtration or extraction.
Newer approaches involve liquefaction of fruit tissue by pectinolytic and cellulolytic enzymes. Such an approach is very convenient for soft tropical fruits. A schematic production line, fruit preparation → fruit mashing-milling → enzyme treatment → filtration → pasteurization → container filling (omitting dilution with water), provides fruit juices suitable for direct consumption.
The juice treatment step involves clarification, i.e. removal of turbidity, and stabilization to prevent additional turbidity. This step commonly involves treatment with enzymes, mostly pectinolytic, and, if necessary, removal of starch and polyphenols using gelatin, alone or together with colloidal silicic acid or tannin, or polyvinylpyrrolidone. Finally, proteins are removed by adsorption on bentonite.
Clarification of juice is achieved by filtration through porous pads or layers of cellulose, asbestos or kieselguhr, or by centrifugation.
Since juice production provides juices which are well-saturated with air, oxygen-sensitive products are also deaerated. This is achieved by an evacuation step or by purging the juice with an inert gas such as N_2 or CO_2.
Fruit juices (with the exception of citrus juices) are produced as transparent, clarified products,

Table 18.30. Composition of some fruit juices (g/l)

	Extract	Total sugar	Volatile acids	Sugar-free extract	Ash	Titratable acidity[a]	Vitamin C
Apple	97–130	72–102	0.15–0.25	14–34	2.2–3.1	4.1–10.4 (M)	0–0.03
Grape	145–195	120–180	0.08–0.25	21.6–35	2.1–3.2	3.6–11.7 (T)	0.017–0.02
Currant, black[b]	120–165	95–145	0.12–0.25	13.3–44.5	2.25–3.2	9.15–12.75 (T)	0.1–0.56
Cherry[b]	126.4–166.4	104.3–138.4	0.08–0.12	17.8–32.6	1.99–3.02	8.0–10.1 (T)	–
Raspberry	45–100	2.7–69.6	–	22.8–64.8	3.5–5.4	13.5–27.8 (T)	0.12–0.49
Orange	87–148	60–110	–	15.2–41.0	2.2–4.0	5–18 (C)	0.28–0.86
Lemon	71–119	7.7–40.8	–	–	1.5–3.5	42–83.3 (C)	0.37–0.63
Grapefruit	76–126	50–83	–	10.3–53	2.5–5.6	5–27 C)	0.25–0.5

[a] Calculated as malic (M), tartaric (T) or citric acid (C).
[b] Diluted and sweetened.

although some turbid juices are produced. In the latter case, measures are required to obtain a stable, turbid suspension. This is achieved with stone fruit juices by a short treatment with polygalacturonase preparations which have a low pectin esterase activity and which then partially degrade and, thus, stabilize the ingredients required for turbidity. Citrus juices (lemons, oranges, grapefruits) are heat-treated to inactivate the endogenous pectinesterase, which would otherwise provide pectic acid which can aggregate and flocculate in the presence of Ca^{2+} ions. However, since heat treatment damages fruit aroma, the use of polygalacturonase is preferred. This enzyme degrades the pectic acid to such an extent that flocculation does not occur in the presence of divalent cations. Finally, the fruit juice preservation step involves pasteurization, preservation by freezing, storage under an inert atmosphere, or concentration (cf. 18.2.11) and drying (cf. 18.2.13).

Pasteurization kills the microflora and inactivates the enzymes, particularly the phenol oxidases. Since a longer heating time is detrimental to the quality, a short, high-temperature heat treatment is the preferred process, using plate heat exchangers (82–90 °C for 15–150 s) with subsequent rapid cooling. The juice is stored in germ-free tanks. Filling operations for the retail market can lead to reinfection, hence a second pasteurization is required. It is achieved by filling preheated containers with the heated juice, or by heating the filled and sealed containers in chambers or tunnel pasteurizers.

Preservation by freezing generally involves transforming the juice or juice concentrate into an ice slurry (at −2.5 °C to −6.5 °C), then packing and cooling to the retail market storage temperature. The product is stable for 5–10 months in a temperature range of −18 °C to −23 °C.

Storage under an inert atmosphere makes use of the fact that filtered, sterilized juices are microbiologically stable at temperatures below 10 °C and under an atmosphere of more than 14.6 g CO_2/l. To attain such a concentration of CO_2, the filled storage tank has to be at a pressure of 5.9 bar at 10 °C, or 4.7 bar at 5 °C.

Fruit juices are poured into retail containers, i.e. glass bottles, synthetic polyethylene pouches, aluminum cans, or aluminum-lined cardboard containers.

Pomace is the residue from the production of fruit juices. Citrus fruits and apple pomace are used for the recovery of pectins. Other fruit residues are used as animal feed, as organic fertilizer, or are incinerated.

18.2.10 Fruit Nectars

Fruit nectars are produced from fruit slurries or whole fruits by homogenization in the presence of sugar, water and, when necessary, citric and ascorbic acids. The fruit portion (as fresh weight) is 25–50% and is regulated in most countries, as is the minimum total acid content. Apricots, pears, strawberries, peaches and sour cherries are suitable for nectar production. The fruits are washed, rinsed, disintegrated and heated to inactivate the enzymes present. The fruit mash is then treated with a suitable mixture of pectinolytic and cellulolytic enzymes. The treatment degrades protopectin and, thus, separates the tissue into its individual intact cells ("maceration").

High molecular weight and highly esterified pectin formed from protopectin provides the high viscosity and the required turbidity for the nectar. Finally, the disintegrated product is filtered hot, then saturated with the usual additives, homogenized and pasteurized.

Fruit products from citrus fruits (comminuted bases) are obtained by autoclaving (2–3 min at 3 bars) and then straining the fruits through sieves, followed by homogenization.

Fruit nectars also include juices or juice concentrates from berries or stone fruits, adjusted by addition of water and sugar. Such products are commonly denoted as sweet musts.

18.2.11 Fruit Juice Concentrates

Fruit juice concentrates are chemically and microbiologically more stable than fruit juices and their storage and transport costs are lower. The solid content (dry matter) of the concentrates is 60–75%. Intermediary products, less stable concentrates with a dry matter content of 36–48%, are also produced. Fruit juice concentration is achieved by evaporation, freezing, or by a process involving high pressure filtration. Initially, the pectin is degraded to eliminate high viscosity and gel setting (undesired properties).

Concentration by evaporation is the preferred industrial process. Since the process leads to losses in volatile aroma constituents, it is combined with an aroma recovery step. The aroma of the juice is enriched by 100- to 200-times by a counter-current distillation. This aroma is stored and recombined with the juice only at the dilution stage. In order to maintain quality, the residence time in evaporators is as short as possible. In a high-temperature, short-time heating installation, e.g. in a 3- to 4-fold stepwise gradient-type evaporator, the residence time is 3-8 min at an evaporating temperature of 100 °C in the first step and at about 40 °C in the fourth step. The concentrate is then cooled to 10 °C. Recovery of the aroma is achieved by rectifying the condensate of the first evaporation stage. A short-time treatment of juices is also possible in thin-layer falling film evaporators. These are particularly suitable for concentrating highly viscous products such as fruit slurries.

Concentration of juice by freezing is less economical than evaporation. Hence, it is utilized mostly for products containing sensitive aroma constituents, e.g. orange juice. The juice is cooled continuously below its freezing point in a scraper-type cooler. The ice crystals are separated from the resultant ice slurry by pressing or by centrifugation. The obtainable solid content of the end product is 40–50%. This content is a function of freezing temperature, as illustrated with apple juice in Figure 18.8.

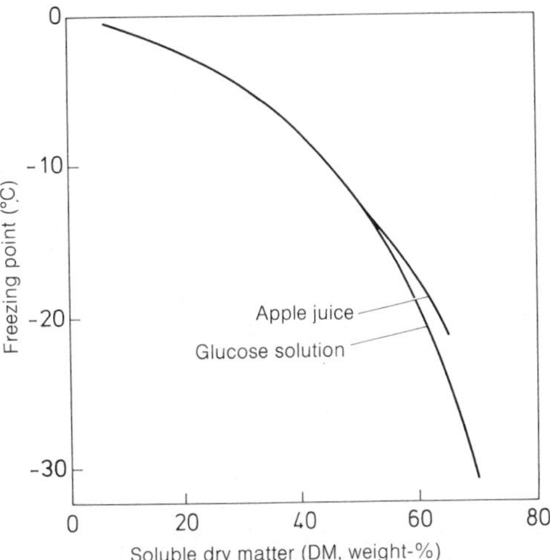

Fig. 18.8. Freezing points of apple juice and glucose solution as affected by soluble dry matter (DM). (After *Schobinger*, 1978)

Concentration of juice by filtration using semi-permeable membranes and high pressure (1–10 bars) is known as ultrafiltration. When the membrane is permeable for water and only to a limited extent for other small molecules, the process is called reverse osmosis (small molecules are of less than 500 dal and are salts, sugars or aroma compounds). Concentration of juice is possible only to about 25% dry matter content. The reverse osmosis process is still under development, but may gain in importance as a pretreatment step in combination with other concentrating procedures.

18.2.12. Fruit Syrups

Fruit syrups are thick, fluid preparations made from boiling one kind of fruit with an excess of sugar. They are sometimes prepared without heating by directly treating fresh fruit or fruit juice with sugar, occasionally also using small amounts of tartaric or lactic acids. Fruit syrups from citrus fruits often contain small amounts of peel aromas.

Fruit syrups are rapidly cooled to avoid aroma losses and caramelization of sugar. The boiling process partially inverts sucrose, preventing subsequent sucrose crystallization. Low-acid fruits are treated with tartaric or lactic acid. Boiling in closed kettles permits recovery of vaporized aroma compounds which can be added back to the end product. As in marmalade production, the boiling is occasionally done under vacuum (50 °C starting temperature, 65–70 °C final temperature) in order to retain the aroma. Syrup production by a cold process is particularly gentle. The raw juice flows over the granulated sucrose in the cold until the required sugar concentration has been achieved. Aroma-sensitive syrups which contain turbidity-causing substances, e.g. citrus fruit syrups, are made by adding sugar to the mother liquor with vigorous stirring.

18.2.13 Fruit Powders

Fruit powders are produced by drying of juices, juice concentrates or slurries. The hygroscopic powders contain no more than 3–4% moisture. Addition of sugars that enhance drying (such as glucose, maltose or starch syrup) in amounts greater than 50% of the dry matter can efficiently control clumping or caking due to the presence of fructose in the drying process. Freeze-drying, vacuum foam-drying (1–10 torr, 40–60 °C) and spray-drying are suitable drying processes. The last two are of industrial significance.

18.3 Alcohol-Free Beverages

18.3.1 Fruit Juice Beverages

These drinks are prepared from fruit juices or their mixtures or from fruit concentrates, with or without addition of sucrose or glucose, and are diluted with water or soda or mineral water. Fruit juice refreshments have mainly citrus fruits as their base and contain at least 6% fruit juice ingredients. These drinks are occasionally mixed with juices of Seville oranges, mandarin oranges, tangerines, lemons or limes to round-off the base flavor.

18.3.2 Lemonades, Cold and Hot Beverages

These drinks are prepared from natural fruit essences and sugar (sucrose or glucose), fruit acids and soda or mineral water. They are also consumed without added carbon dioxide, either cold or warmed. The drinks are usually colored.

Tonic water is also considered a lemonade. It contains about 80 mg quinine/l to provide the characteristic bitter taste.

18.3.3 Caffeine-Containing Beverages

These are also considered as "lemonades" (particularly in Europe). The most popular are the cola drinks, which contain extracts from the cola nut *(Cola nitida)* or aromatic extracts from ginger, orange blossoms, carob and tonka beans or lime peels. Caffeine is often added (6.5–25 mg/100 ml). Phosphoric acid is sometimes used as an acidulant (70 mg/100 ml). The sugar content of cola drinks averages 10–11%. The deep-brown color of the drink is adjusted with caramel.

18.3.4 Other Pop Beverages

Some effervescent pop drinks are imitations of fruit juices and lemonade-type drinks, however, their sugar content is fully or partially replaced by artificial sweeteners and the natural essence or flavoring ingredients are replaced by artificial or artificially-enhanced essences. Coloring substances are usually added.

18.4 Literature

Bell, E. A., Charlwood, B. Y. (Eds.): Secondary plant products. Springer-Verlag: Berlin–Heidelberg. 1980

Biale, J. B.: Growth, maturation, and senescence in fruits. Science *146,* 880 (1964)

Demole, E., Enggist, P., Ohloff, G.: 1-p-menthene-8-thiol: a powerful flavor impact constituent of grapefruit juice *(Citrus paradisi McFayden).* Helv. Chim. Acta *65,* 1785 (1982)

Drawert, F., Müller, W.: Über die elektrophoretische Differenzierung und Klassifizierung von Proteinen. II. Dünnschichtisoelektrische Fokussierung von Proteinen aus Trauben verschiedener Rebsorten. Z. Lebensm. Unters. Forsch. *153,* 204 (1973)

Drawert, F., Görg, A., Staudt, G.: Über die elektrophoretische Differenzierung und Klassifizierung von Proteinen. IV. Disk-Elektrophorese und isoelektrische Fokussierung in Poly-Acrylamid-Gelen von Proteinen und Enzymen aus verschiedenen Erdbeerarten, -sorten und Artkreuzungsversuchen. Z. Lebensm. Unters. Forsch. *156,* 129 (1974)

Friend, J., Rhodes, M. J. C. (Eds.): Recent Advances in the Biochemistry of Fruits and Vegetables. Academic Press: London–New York. 1981

Fowden, L., Lea, P. J., Bell, E. A.: The nonprotein amino acids of plants. Adv. Enzymol. *50,* 117 (1979)

Herrmann, K.: Obst, Obstdauerwaren und Obsterzeugnisse. Verlag Paul Parey: Berlin–Hamburg. 1966

Hulme, A. C.: Problems in the biochemistry of fruits. In: Biochemical principals of the food industry (Eds.: Kretovich, V. L., Pijanowski, E.), p. 143, Pergamon Press: Oxford. 1963

Hulme, A. C. (Ed.): The biochemistry of fruits and their products. Vol. 1, Academic Press: London–New York. 1970

Jurd, L.: Reactions involved in sulfite bleaching of anthocyanins. J. Food Sci. *29,* 16 (1964)

Jurd, L., Asen, S.: Formation of metal and copigment complexes of cyanidin 3-D-glucoside. Phytochemistry *5,* 1263 (1966)

Maarse, H., Visscher, C. A. (Eds.): Volatile compounds in food. Qualitative data. 5th edn. with supplements 1 and 2, TNO-CIVO Food Analysis Institute: Zeist, the Netherlands. 1985

Maarse, H., Visscher, C. A. (Eds.): Volatile compounds in food. Quantitative data. Vol. 4, TNO-CIVO Food Analysis Institute: Zeist, the Netherlands. 1985

Maga, J. A.: Amines in foods. Crit. Rev. Food Sci. Nutr. *10,* 373 (1978)

Mosandl, A.: Struktur und Geruch substituierter Glycidsäureester. Habilitationsschrift, Würzburg 1981

Mosel, H. D., Herrmann, K.: Changes in catechins and hydroxycinnamic acid derivatives during development of apples and pears. J. Sci. Food Agric. *25,* 251 (1974)

Nagy, S., Attaway, J. A. (Eds.): Citrus nutrition and

quality. ACS Symposium Series 143, American Chemical Society: Washington, D.C. 1980

Pantastico, E.B. (Ed.): Postharvest physiology, handling and utilization of tropical and subtropical fruits and vegetables. AVI Publ. Co.: Westport, Conn. 1975

Phan, C.T., Pantastico, E.B., Ogata, K., Chachin, K.: Respiration and respiratory climacteric. In: Postharvest physiology, handling and utilization of tropical and subtropical fruits and vegetables (Ed.: Pantastico, E.B.), p. 86, AVI Publ. Co.: Westport, Conn. 1975

Salunkhe, D.K.: Biogenesis of aroma constituents of fruits and vegetables. Crit. Rev. Food Sci. Nutr. *8*, 161 (1977)

Salunkhe, D.K., Desai, B.B.: Postharvest Biotechnology of fruits Vol. I–II. CRC Press, Inc. Boca Raton, FL. 1984

Schobinger, U. (Hrsg.): Frucht- und Gemüsesäfte. Verlag Eugen Ulmer: Stuttgart. 1978

Schormüller, J.: Die Erhaltung der Lebensmittel. Ferdinand Enke Verlag Stuttgart. 1966

Shaw, P.E., Wilson, C.W., III: Importance of nootkatone to the aroma of grapefruit oil and the flavor of grapefruit juice. J. Agric. Food Chem. *29*, 677 (1981)

Spencer, Mary.: Ethylene in nature. In: Progress in the chemistry of organic natural products (Ed.: Zechmeister, L.), p. 31, Springer Verlag, Vol. XXVII, 1969.

Sulc, D.: Fruit juice concentration and aroma separation. Confructa 28, 258 (1984)

Stöhr, H., Herrmann, K.: Die phenolischen Inhaltsstoffe des Obstes, VI. Die phenolischen Inhaltsstoffe der Johannisbeeren, Stachelbeeren und Kulturheidelbeeren. Veränderungen der Phenolsäuren und Catechine während Wachstum und Reife von schwarzen Johannisbeeren. Z. Lebensm. Unters. Forsch. *159*, 31 (1975)

Wegler, R. (Ed.): Chemie der Pflanzenschutz- und Schädlingsbekämpfungsmittel. Bde. 1 bis 8, Springer-Verlag: Berlin–Heidelberg–New York. 1970 to 1982

19 Sugars, Sugar Alcohols and Honey

19.1 Sugars, Sugar Alcohols and Sugar Products

19.1.1 Foreword

Only a few of the sugars occurring in nature are used extensively as sweeteners. Other important sugars besides sucrose (saccharose) are: glucose (the starch sugar or starch syrup); invert sugar (equimolar mixture of glucose and fructose); maltose; lactose; and fructose. In addition, some other sugars and sugar alcohols (polyhydric alcohols) are used in diets or for some technical purposes. These include sorbitol, xylitol, mannitol, maltulose, isomaltulose, maltitol, isomaltitol, lactulose and lactitol. Some are used commonly in food and pharmaceutical industries, while applications for others are being developed. Table 19.1 reviews production data, while Table 19.2 lists data on relative sweetness, source and means of production, and Table 19.3 gives nutritional and physiological properties. Whether one of the compounds discussed will be successful as a sweetener depends on nutritional, physiological and processing properties, cariogenicity as compared to sucrose, economic impact, and the quality and intensity of the sweet taste.

Table 19.1 Approximate world production of sweeteners of carbohydrate origin (1978)

Sweetener	Amount (1,000 t)
Saccharose	90,000
Glucose syrup (80%)	4,000
Glucose	2,000
Isoglucose (72%)	1,500
Sorbitol (70%) and sorbitol, crystalline	300
Lactose	150
Fructose	20
Mannitol	8
Xylitol	5
Hydrogenated glucose syrup (75%)	4

19.1.2 Processing Properties

The potential of a compound for use as a sweetener depends upon its physical, processing and sensory properties. Important physical properties are solubility, viscosity of the solutions, and hygroscopicity. Figure 19.1 shows that the solubility of sugars and their alcohols in water is variable and affected to a great extent by temperature.

There are similar temperature and concentration influences on the viscosity of aqueous solutions of many sugars and sugar alcohols. As an example, Figure 19.2 shows viscosity curves for sucrose as a function of both temperature and concentration.

The viscosity of glucose syrup depends on its composition. It increases as the proportion of the high molecular weight carbohydrates increases (Figure 19.3).

Figure 19.4 shows the water absorption characteristics of several sweeteners. Sorbitol and fructose are very hygroscopic, while other sugars absorb water only at higher relative humidities. Chemical reactions of sugars were covered in detail in Chapter 4. Only those reactions important from a technological viewpoint will be emphasized here.

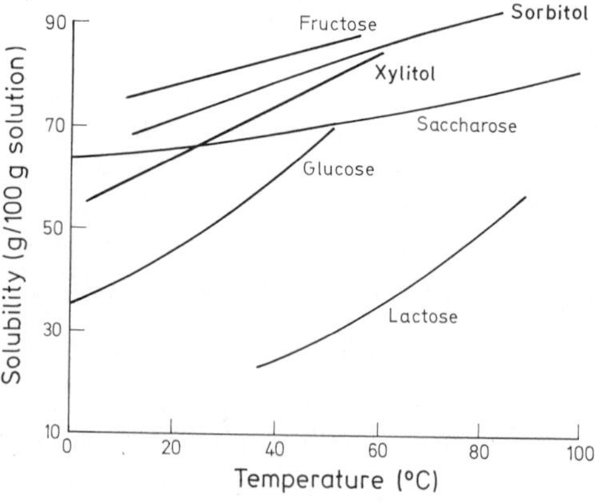

Fig. 19.1. Solubility of sugars and sugar alcohols in water. (After *Koivistoinen*, 1980)

Table 19.2. Sweeteners of carbohydrate origin

Name	Relative sweetnes[a]	Starting material, applied process
Products of economic importance		
Saccharose	1.00	Isolation from sugar beet and sugar cane
Glucose	0.5–0.8	Hydrolysis of starch with acids and/or enzymes (α-amylase + glucoamylase)
Fructose	1.1.–1.7	a) Hydrolysis of saccharose followed by separation of hydrolysate by chromatography. b) Hydrolysis of starch to glucose, followed by isomerization and separation by chromatography
Lactose	0.2–0.6	Isolation from whey
Mannitol	0.4–0.5	Hydrogenation of fructose
Sorbitol	0.4–0.5	Hydrogenation of glucose
Xylitol	1.0	Hydrogenation of xylose
Glucose syrup (starch syrup)	0.3–0.5[b]	Hydrolysis of starch with acids and/or enzymes; hydrolysate composition is strongly affected by process parameters (percentage of glucose, maltose, maltotriose and higher oligosaccharides)
Maltose syrup		As glucose syrup; process parameters adjusted for higher proportion of maltose in hydrolysate (amylase from *Aspergillus oryzae*)
Glucose/fructose syrup (isoglucose, high fructose syrup)	0.8–0.9	Isomerization of glucose to glucose/fructose mixture with xylose isomerase; conversion degree 45–50%
Invert sugar		Hydrolysis of saccharose
Hydrogenated glucose syrup	0.3–0.8	Hydrogenation of starch hydrolysate (glucose syrup); composition is highly dependent on starting material (content of sorbitol, maltitol and hydrogenated oligosaccharides)
Maltitol syrup		Hydrogenation of maltose syrup
Products of potential economic importance		
Arabinitol	approx. 1.0	Hydrogenation of arabinose
Galactitol		Hydrogenation of galactose
Galactose	0.3–0.5	Hydrolysis of lactose, followed by separation of hydrolysate
Isomaltitol	0.5	Hydrogenation of isomaltulose (palatinose)
Lactitol	0.3	Hydrogenation of lactose
Lactulose	approx. 0.6	Isomerization of lactose
Maltose	0.3–0.6	Hydrolysis of starch
Maltitol	approx. 0.9	Hydrogenation of maltose
L-Sorbose	0.6–0.8	From glucose microbiologically
D-Xylose	approx. 0.5	Hydrolysis of hemicellulose
Palatinit		Isomerization of saccharose to isomaltulose (palatinose) followed by hydrogenation to a mixture of glucopyranosido-sorbitol and glucopyranosido-mannitol

[a] Sweetness is related to saccharose sweetness (= 1); the values are affected by sweetener concentration.
[b] Sweetness value is strongly influenced by syrup composition.

Table 19.3. Nutritional/physiological properties of carbohydrate-derived sweeteners

Sweetener	Resorption	Utilization in metabolism	Effect on blood sugar level and insulin secretion	Other properties
Sucrose	Effective after being hydrolyzed	Hydrolysis to fructose and glucose	Moderately high	Cariogenic
Glucose	Effective	Insulin-dependent in all tissues	High	Less cariogenic than sucrose
Fructose	Faster than by diffusion process	In liver to an extent of 80%	Low	Accelerates alcohol conversion in liver
Lactose	Effective after being hydrolyzed	Hydrolysis to glucose and galactose	High	Intolerance by humans lacking lactase enzyme; laxative effect
Sorbitol	Diffusion	Oxidation to fructose	Low	Slightly cariogenic and laxative
Mannitol	Diffusion	Partially utilized by liver	Low	Slightly cariogenic and laxative
Xylitol	Diffusion	Utilized preferentially by liver and red blood cells	Low	Not cariogenic, available data indicate an anticariogenic effect; mildly laxative
Hydrogenated glucose syrup	After hydrolysis glucose effective; sorbitol by diffusion	Variable depending on composition	Variable, composition dependent	Slightly cariogenic; mildly laxative
Arabinitol	Diffusion	Not metabolized by humans	None	Side effects unknown; probably laxative
Galactose	Effective	Isomerization to glucose	High	Forms cataract of the eyes in feeding trials with rats; probably laxative
Isomaltitol	None	Probably not metabolized	None	Side effects unknown; strongly laxative
Lactitol	None	No hydrolysis	None	Side effects unknown; strongly laxative
Lactulose	None	No hydrolysis	None	Effects the N-balance; strongly laxative
Maltitol	Effective as glucose after hydrolysis; sorbitol by diffusion	Hydrolysis to glucose and sorbitol	Probably slight	Side effects unknown; laxative
Maltose	Effective after hydrolysis	Hydrolysis to glucose	High	Cariogenic; intravenously given it appears to be utilized directly and as glucose it is insulin-dependent
L-Sorbose	Diffusion	Utilized preferentially by liver	Probably slight	Feeding trials with dogs revealed hemolytic anaemia at a higher dosage intake; probably laxative

Table 19.3. (Continued)

Sweetener	Resorption	Utilization in metabolism	Effect on blood sugar level and insulin secretion	Other properties
D-Xylose	Diffusion	Not metabolized by humans	None	Forms cataract of the eyes in feeding trials with rats; probably laxative
Palatinit		Partial hydrolysis to glucose, sorbitol, and mannitol	Probably slight	Side effects unknown

All sugars with free reducing groups are very reactive. In mildly acidic solutions monosaccharides are stable, while disaccharides hydrolyze to yield monosaccharides. Fructose is maximally stable at pH 3.3; glucose at pH 4.0. At lower pH's dehydration reactions prevail, while the *Lobry de Bruyn-van Ekenstein* rearrangement occurs at higher pH's. Reducing sugars are unstable in mildly alkaline solutions, while nonreducing disaccharides, e.g. sucrose, have their stability maxima in this pH region.

The thermal stability of sugars is also quite variable. Sucrose and glucose can be heated in neutral solutions up to 100 °C, but fructose decomposes at temperatures as low as 60 °C. Sugar alcohols are very stable in acidic or alkaline solutions. Relative taste intensity values for various sweeteners are found in Table 19.2. Taste intensity within a food can be affected by a series of parameters, e.g. aroma, pH or food texture. Creams and gels with equal amounts of sweetener are often less sweet than the corresponding sweetener in pure solution. The sweet taste intensity may also depend on temperature (Figure 19.5), an effect which is particularly pronounced with fructose – hot fructose solutions are less sweet than cold ones. The cause of such effects is the mass equilibrium of sugar isomers in solution. At higher temperatures the concentration of the very sweet β-D-fructopyranose drops in favor of both the less sweet α-D-fructopyranose and the β-D-fructofuranose (Figure 19.6). Such strong shifts in isomer concentrations do not occur with glucose, hence its sweet taste intensity is relatively unchanged in the range of 5–50 °C.

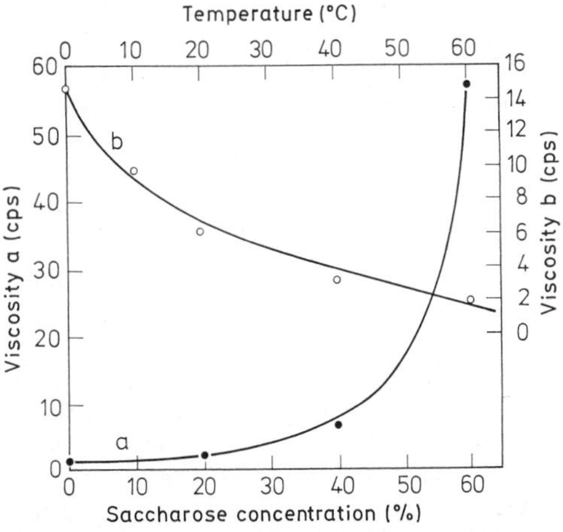

Fig. 19.2. Viscosity of aqueous saccharose solution as affected by (a) saccharose concentration (20 °C) and temperature (40% saccharose). (After *Shallenberger* and *Birch*, 1975)

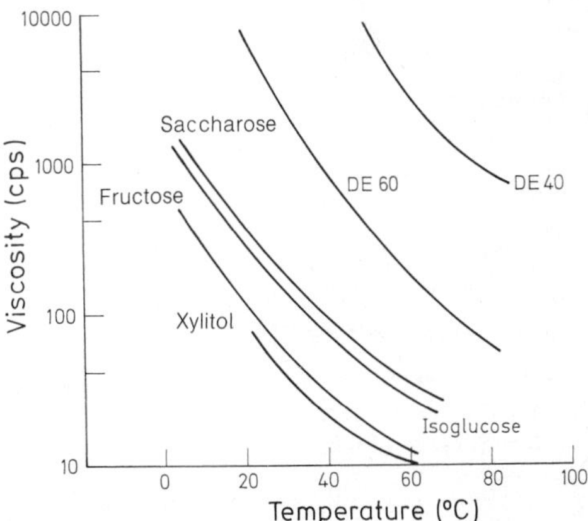

Fig. 19.3. Viscosity of some sugar solutions. Glucose syrup DE 40: 78 weight-%; glucose syrup DE 60: 77 weight-%; all other sugar solutions: 70 weight-%. (After *Koivistoinen*, 1980)

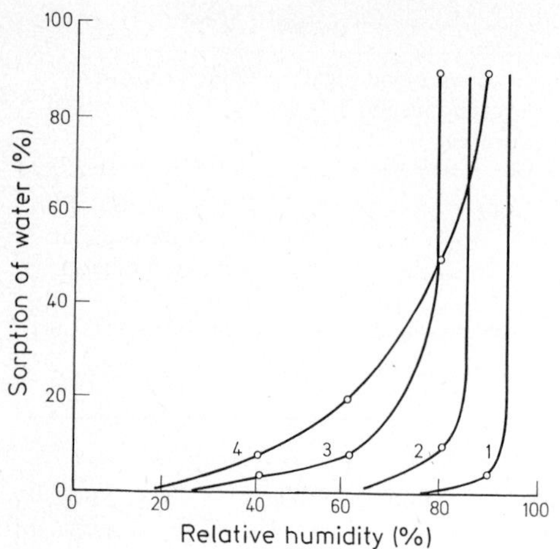

Fig. 19.4. Sorption of water by sugars at room temperature. 1 Saccharose, 2 xylitol, 3 fructose, 4 sorbitol (After *Koivistoinen,* 1980)

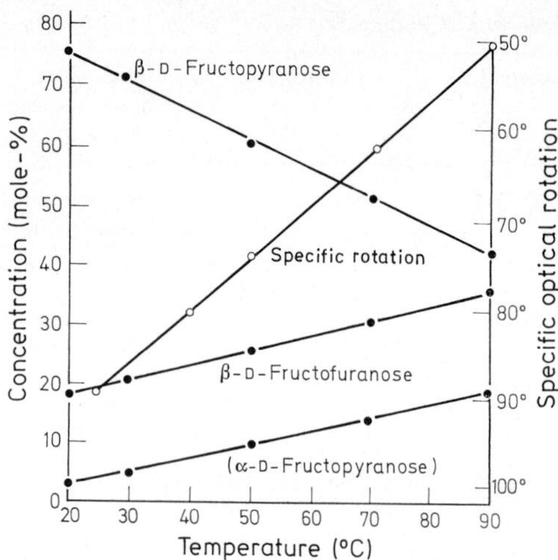

Fig. 19.6. Fructose mutarotation equilibrium as affected by temperature. (After *Shallenberger,* 1975)

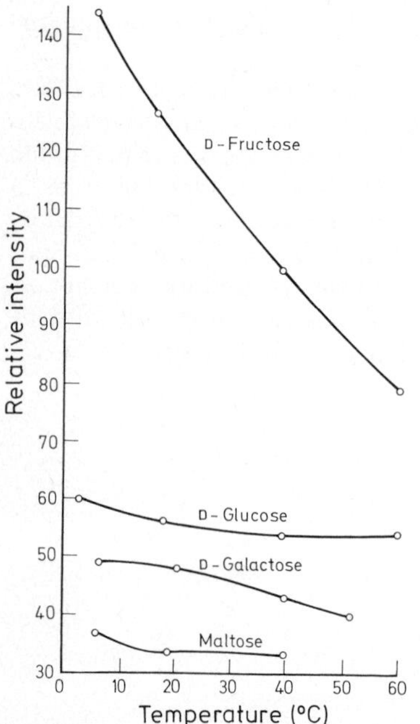

Fig. 19.5. Sugar sweetness intensity versus temperature. At all temperatures the saccharose taste intensity is 100. (After *Shallenberger,* 1975)

19.1.3 Nutritional/Physiological Properties

The role of carbohydrates in metabolism is primarily determined by the ability of disaccharides to be hydrolyzed in the gastrointestinal tract and by the mechanisms of monosaccharide absorption.

The human organism hydrolyzes sucrose, lactose and oligosaccharides of the maltose and isomaltose type. The enzyme lactase, which is responsible for lactose hydrolysis, is lacking in some adults. Glucose and galactose are actively transported, while all other monosaccharides are transported only by diffusion. Sugar phosphorylation occurs preferentially in the liver. All monosaccharides which are metabolized can be interconverted. Sugar alcohols are oxidized: sorbitol → fructose, xylitol → xylulose. However, only glucose can enter the insulin-regulated and -dependent energy metabolism and be utilized by all tissues. Galactose is rapidly transformed into glucose and is therefore nutritionally equal to glucose. Oral intake of glucose and galactose causes a rapid increase in blood sugar levels and, as a consequence, insulin secretion. All other monosaccharides are primarily metabolized by the liver and do not directly affect glucose status or insulin release. After fructose intake, insulin secretion is only 50% of that after glucose intake. Sugars to be avoided by diabetics are, therefore, glucose, galactose, lactose and maltose. Fructose, xylitol, sorbitol and mannitol can be well tolerated by diabetics and sucrose, invert sugar and hydrogenated glucose syrup only moderately so.

19.1.4 Individual Sugars and Sugar Alcohols

19.1.4.1 Sucrose (Beet Sugar, Cane Sugar)

19.1.4.1.1 General outline

Sucrose is widely distributed in nature, particularly in green plants, leaves and stalks (sugar cane, 12–26%; sweet corn, 12–17%; sugar millet, 7–15%; palm sap, 3–6%); in fruits and seeds (stone fruits, such as peaches; core fruits, such as sweet apples; pumpkins; carobs or St. John's bread; pineapples, coconuts; walnuts; chestnuts); and in roots and rhisomes (sweet potatoes, 2–3%; peanuts, 4–12%; onions, 10–11%; beet roots and selected breeding forms, 3–20%). The two most important sources for sucrose production are sugar cane (*Saccharum officinarum*) and sugar beet (*Beta vulgaris ssp. vulgaris* var. *altissima*).

Sucrose is the most economically significant sugar and is produced industrially in the largest quantity. Table 19.4 provides an overview of yearly world production of beet and cane sugar. Table 19.5 lists the main producers.

Honey is the oldest known sweetener and has relatively recently been displaced by cane sugar. Cane sugar was brought to Europe from Persia by the Arabs. After the Crusades, it was imported by Cyprus and Venice and, later, primarily by Holland, from Cuba, Mexico, Peru and Brazil. In 1747 *A. S. Marggraf* discovered sucrose in beets and in 1802 *F. Achard* was the first to produc sucrose commercially from sugar beets. The new sugar source had great economic impact; the more so when sucrose accumulation in the beets was increased by selection and breeding.

Table 19.4. World production of sugar (beet/cane) 1901–1981

Year	Total production 10^6 t	Beet sugar 10^6 t	Cane sugar 10^6 t	Beet sugar %	Cane sugar %
1900/01	11.3	6.0	5.3	53.0	47.0
1920/21	16.4	4.8	11.6	29.5	70.5
1940/41	30.9	11.6	19.2	37.7	62.3
1960/61	61.1	24.3	36.8	39.7	60.3
1965/66	71.1	27.2	43.9	38.2	61.8
1970/71	82.3	29.5	52.8	35.8	64.2
1975/76	92.2	32.7	59.5	35.4	64.6
1980/81	98.4	32.8	65.6	33.4	66.6
1981/82	108.5	36.6	71.9	33.8	66.2

Table 19.5. Production of sugar beet, sugar cane and saccharose 1981/82 (1,000 t)

Continent	Sugar cane	Sugar beet	Saccharose[a]
World	775,285	281,485	96,546
Africa	64,687	2,464	6,955
America, North-, Central-	171,625	25,951	19,379
America, South-	219,249	1,808	13,681
Asia	290,357	24,675	21,265
Europe, West-	384	117,318	19,671
Europe, East- + USSR		109,269	11,616
Australia + South Pacific Islands	28,983		3,979

Country	Sugar cane	Country	Sugar beet
Brazil	153,858	USSR	60,600
India	150,522	France	31,800
Cuba	67,000	USA	24,740
Mexico	35,461	FR Germany	24,090
China	33,000	Italy	16,800
Pakistan	32,359	Poland	15,800
USA	27,076	Turkey	11,000
Colombia	25,900	Belgium – Luxembourg	8,000
Australia	25,160	German DR	7,980
Philippines	20,450	Spain	7,890
Thailand	18,600	Czechoslovakia	7,770
Σ (%)[b]	76	Σ (%)[b]	77

Country	Saccharose[a]	Country	Saccharose[a]
Brazil	8,500	Italy	2,170
Cuba	7,359	South Africa	2,050
USSR	6,100	Poland	1,872
USA	5,771	Thailand	1,641
France	5,600	Argentina	1,624
India	5,587	Indonesia	1,449
China	4,031	Turkey	1,300
FR Germany	3,600	UK	1,200
Australia	3,450	Colombia	1,185
Mexico	2,518		
Philippines	2,394	Σ (%)[b]	75

[a] As raw (centrifuged) sugar.
[b] World production ≙ 100%.

19.1.4.1.2 Production of beet sugar

Prolonged selection efforts have led to sugar beets which reach their maximum sucrose content of 15–20% in the middle of October. The average for the past 5 years in FR Germany is

16.3%. The early yield achieved by *F. Achard* of 4.5 kg/100 kg beets has been increased to about 14 kg. Currently, beet varieties have a high sugar content and small amounts of nonsugar substances. Anatomically they have a favorable shape, i.e. are small and slim with a smooth surface, and have a firm texture. Since the sugar accumulation in beets peaks in October and since sugar decomposition due to respiration occurs during subsequent storage of beets, they are rapidly processed from the end of September to the middle of December.

The beet sugar extract contains about 17% sucrose, 0.5% inorganic and 1.4% organic nonsucrose matter. Invert sugar and raffinose content is 0.1% (in molasses this may be as high as 2%). The trisaccharide kestose (cf. Table 4.13), which is present in the extract, is an artifact generated in the course of beet processing. In addition to pectic substances, beet extract contains saponins which are responsible for foaming of the extract and which bind with sugars. N-containing, nonsugar constituents of particular importance are proteins, free amino acids, and their amides, (e.g. glutamine) and glycine betaine ("betaine"). These constituents are 0.3% of beets and about 5% of molasses. Beet ash averages 28% potassium, 4% sodium, 5% calcium and 13% phosphoric acid, and contains numerous trace elements. The nonsugar constituents of the sugar extract also include steam-distillable odorous compounds, phenolic acids, e.g. ferulic acid, and numerous beet enzymes which are extensively inactivated during extract processing. These enzymes, e.g. polyphenol oxidase, can induce darkening through melanin build-up, with the color being transferred during beet extraction into the raw sugar extract.

Beet processing involves the following steps:

- *Flushing and cleaning* in flushing chutes and whirlwashers.
- *Slicing* with machines into thin shreds (cossettes) with the shape of "shoestrings" 2–3 mm thick and 4–7 mm wide.
- *Extraction* by leaching of beet slices. This was once performed in a so-called diffusion battery of 12–14 bottom sieve-equipped cylindrical containers (diffusers) connected in series and operating discontinuously on a countercurrent principle. Today this battery operation has been replaced to a great extent by a continuous and automatically-operated extraction tower into which the shreds are introduced at the bottom while the extraction fluid flows from the top. The extracted shreds (pulp) are discharged at the top. The pulp, extracted at 70–75 °C contains residual sugar (approx. 0.2% of the beet dry weight). The dried, pressed and/or pelleted or acidified pulp is used as cattle feed. In order to avoid sucrose inversion, the extraction is done under mildly alkaline conditions.
- *Raw sugar extract purification* (lime treatment followed by purification). This step removes proteins, pectins, organic and phosphoric acids as well as cellulose fibers or cell debris. The raw extract flowing from the extraction tower is turbid and colored dark-greyish to bluish-black by melanins (resulting from polyphenol oxidase action). This extract is first mechanically clarified by filtration, then is lime-treated (calcium hydroxide, lime milk) in two steps: a pre- and then a main liming step at 85 °C. A number of organic acids and phosphate ions precipitate as calcium salts at this stage and colloids flocculate, while part of the added lime goes into solution as calcium saccharate. In order to remove excess calcium and to decompose calcium saccharate, thus recovering additional sucrose, and to transform the previously-obtained turbid coagulate into a more filtrable form, the solution is gassed with carbon dioxide. This converts the excess calcium into calcium carbonate. Carbonation is also performed in two steps. The turbid solution is then clarified by decantation and filtration, yielding a clean, lightly-colored, thin syrup with 12–15% dry matter content. Sugar content in the sludge is 0.2–0.6%. This purification step, however, does not remove the alkali salts, amino acids and other N-containing substances. The resultant thin syrup may also be treated with SO_2.

Ion-exchange resins have become important in raw sugar extract purification. They soften the thin syrup and avoid subsequent salt precipitation on evaporating heating coils, remove substances which retard sugar crystallization, and partially bleach the thin syrup. Complete desalting of the thin syrup appears to be detrimental since a temporary drop in the pH might induce sucrose loss by inversion. A substitution of alkali with earth alkali ions (Mg^{2+}) is beneficial since it decreases the sugar bound and lost in molasses by approx. 30%. Bleaching of thin syrup with large-pore ion-exchangers is also possible, wherein the pigments are largely removed by adsorption.

- *Evaporation of the thin syrup* is achieved in multiple-stage evaporators (falling film evaporators, natural or forced circulation evaporators). Mildly alkaline conditions (pH 9) are maintained to prevent sucrose inversion. The boiling temperature does not exceed 135 °C. The resultant thick syrup (yield of 25–30 kg/100 kg beet) is once more filtered.. The syrup contains 60–70% solids and sucrose content is 56–65%. The raw, thin and thick syrups have purity quotients of approx. 89, 92 and 93, respectively, i.e. the percentage sucrose on a dry matter basis.
- *Crystallization.* To obtain white sugar, the washed raw sugar crystals from the affination step (see below) are solubilized, then mixed with thick syrup and evaporated (cf. Figure 19.7). This mixture is further concentrated in a vacuum at 70–80 °C until the bulk of the sugar is crystallized from the oversaturated solution (evaporation to seeding). Seeding is done with a small amount of ground white sucrose dispersed in isopropanol. Centrifugation at 40–45 °C provides raw sugar crystals 2–4 mm in size, and a mother liquor ("green syrup") which is subjected to two additional crystallization steps (Figure 19.7). The last mother liquor is molasses, a highly viscous, brown syrup. Modern technology usually makes crystallization a fully-continuous operation (crystallization by evaporation, followed by centrifugation).
- *Affination.* The bulk raw sugar produced in raw sugar plants is shipped to a refinery. The raw sugar crystals are sticky and moist and yellow-brown in color due to the adhering dark molasses. The crystals have a raw, beet-like flavor and contain numerous microorganisms. Therefore, before the raw sugar is solubilized, it is subjected to a preliminary purification or affination process. The raw sugar is mixed in a mingler with enough of the thick syrup of suitable purity to form a thick paste or magma. The magma is heated in a mixer to reduce its viscosity and is then discharged into centrifugal baskets where the liquid portion is spun off. The crystals which remain in the basket are, if required for white sugar production, spray washed with a small volume of water or steamed to solubilize only the outer surface of the crystals. Such cleaned sugar is called affinade or washed, raw or affinated sugar. The resultant affination syrup is recycled for further crystallization (or, as in the USA, may be processed in part to an edible liquid syrup). A simple crystallization scheme for white sugar production, consisting of three stages, is given in Figure 19.7.
- *Raffination.* In this processing step the affinade is solubilized to a clear liquid and, again under reduced pressure, "evaporated to seeding", i.e. until the raffinade filling mass is separated as crystallized granulated sugar. Raffination provedes the purest sugar form which is screened for crystal size. Table sugar is made from fine-sized grains. Sugar is also molded into cubes or tablets by forming a mixture of sugar crystals and white sugar syrup under pressure, followed by drying. The first mother liquor (green syrup) of raffination is naturally much cleaner than that spun off from raw

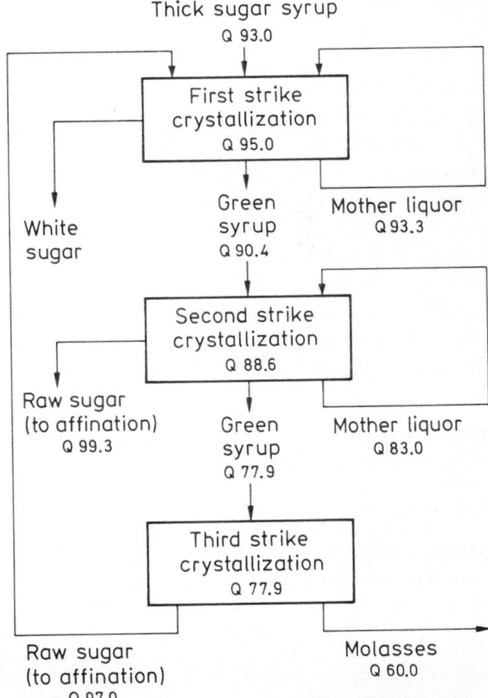

Fig. 19.7. White sugar evaporation and crystallization

sugar filling mass. It is added to thick syrup and processed into white sugar (see above). Occasionally, the first mother liquor is evaporated to seeding into a pale-yellow product called "sugar berry" which is used by the baking industry.

Processing losses in sucrose recovery from beets in 1974 were 0.4–0.9% (sugar determined polarimetrically; and based on processed beet weight) and, when compared to 1950, represent a significant improvement of the sucrose yield (Table 19.6). This technological progress is also reflected in a rise of work productivity (work min/t beets), which was 130–150 in 1950 but only 12–30 in 1974.

19.1.4.1.3 Production of cane sugar

Sugar cane processing starts with squeezing out of the sweet sap from thoroughly-washed cane. For this purpose, the cane moves to a shredding machine where knives shred the stalks and then moves to crushing machines where a series of revolving heavy steel rollers squeeze the cane under high pressure. After the first roller, more than 60% of the cane weight is removed in the form of sap which contains 70% or more of the cane sucrose content. Repeated squeezing provides a sucrose yield of 93–97.5%. The squeezing may be combined with extraction by mixing the "bagasse" (the pressed cane) with hot water or dilute hot cane juice, followed by a final pressing. The experience gained in continuous thin beet syrup production is applied to sugar cane production, with a resultant energy saving and a rise in sugar yields.

Clarification and neutralization of the mildly acidic, raw extract (pH 4.8–5.0) is done by treatment with lime or lime and carbon dioxide. Further processing of the clarified pure syrup parallels that of sugar beet processing. The yield of raw cane sugar is 6–11% of the cane weight. The "bagasse" is used as fuel, made into wallboard or used as insulation.

Table 19.6. Production losses[a] during saccharose recovery from sugar beet

Processing step	1950	1974
Beet slice extraction	0.4–0.5	0.15–0.25
Sugar extract purification	0.1–0.2	0.02–0.05
Other steps	0.6–0.8	0.25–0.90
Total process	1.1–1.5	0.42–0.60

[a] Sugar amount in % based on the processed beet weight.

19.1.4.1.4 Other sources for sucrose production

Some plants other than sugar beet or sugar cane can serve as sources of sucrose:

Date sugar is obtained from sweet, fleshy fruit of the date palm (Algeria, Iraq), which contains up to 81% sucrose in its solids.

Palm sugar originates from various palm species, e.g. palmyra, saga or Toddy palm, coconut and Nipa palm grown in India, Sri Lanka, Malaysia and the Philippines, respectively.

Maple sugar is obtained from the maple tree (*Acer saccharum*), found solely in North America (USA and Canada) and Japan. The sap, which drips from holes drilled in the maple tree trunk, flows down metal spiles to metal pails. This sap contains about 5% sucrose, minute amounts of raffinose and several other oligosaccharides of unknown structures. It is marketed either as maple syrup or as maple sugar. Aroma substances are important constituents of these products. The syrup also contains various acids, e.g. citric, malic, fumaric, glycolic and succinic acids. The main component of maple sugar is sucrose (88–99% of the total solids). Aroma constituents include vanillin, syringic aldehyde, dihydroconiferyl alcohol, vanilloyl methyl ketone and furfural.

Sorghum sugar. Sugar sorghum (*Sorghum dochna*) stalks contain 12% sucrose. This source was important earlier in the USA. Sugar sorghum is processed into sorghum syrup on a small-scale on individual farms in the Midwestern United States.

19.1.4.1.5 Packaging and storage

Sucrose is packaged in paper, jute or linen sacks, in cardboard boxes, paper bags or cones, in glass containers and in polyethylene foils; the latter serving as lining in paper, jute or wooden containers.

Sugar is stored at a relative humidity of 65–70% in loose form in bins or by stacking the paper or jute sacks. The unbagged, loose or bulk sugar is distributed to industry and wholesalers in bins on trucks or rail freight cars.

19.1.4.1.6 Types of sugar

Sucrose is known under many trade and popular names. These may be related to its purity grade (raffinade; white, consumer's berry, raw or yellow sugar), to its extent of granulation or crystal size (icing, crystal, berry and candy sugar, and cube and cone sugar) and to its use (canning, confectionery or soft drink sugar). Liquid sugar is a sucrose solution in water with at least 62% solids (of which a maximum of 3% is invert sugar). The invert sugar content is high in liquid invert sugars and invert sugar syrups. Such solutions are easily stored, handled and transported. They are best metered by pumps and are widely used by the beverage industry (soft drinks and spirits), the canning industry and ice cream makers, confectionery and baking industries, and in production of jams, jellies and marmalades. Use of liquid sugar avoids the additional crystallization steps of sugar processing and problems associated, with packaging of sugar.

Criteria for the analytical determination of sugars are: (a) color; (b) color extinction coefficient (absorbance) of a 50% sugar solution, expressed in ICUMSA-units; (c) ash content determined from conductivity measurements of 28% aqueous sugar solutions; (d) moisture content; (e) optical rotation; and (f) criteria based on the content of invert sugar.

19.1.4.1.7 Composition of some sugar types

The chemical composition of a given type of sugar depends on the extent of sugar raffination. A raffinade, as mentioned above, consists of practically 100% sucrose. Washed raw beet sugar has about 96% sucrose, <1.4% moisture, 0.9% ash and 1.5% nonsugar organic substances. Berry sugar consists of 98.8% sucrose, 0.70% moisture, 0.20% ash and 0.29% nonsugar organic substances. The presence of raffinose, a trisaccharide, is detected by high optical rotation readings or by the presence of needle- or spear-like crystals.

19.1.4.1.8 Molasses

The molasses obtained after sugar beet processing contains about 60% sucrose and 40% other components (both on dry basis). The nonsucrose substances, expressed as percent weight of molasses, include: 10% inorganic salts, especially those of potassium; raffinose (about 1.2%); the trisaccharide kestose, an artifact of processing; organic acids (formic, acetic, propionic, butyric and valeric); and N-containing compounds (amino acids, betaine, etc.). The main amino acids are glutamic acid and its derivative, pyrrolidone carboxylic acid. Molasses is used in the production of baker's yeast; in fermentation technology for production of ethanol and citric, lactic and gluconic acids, as well as glycerol, butanol and acetone; as an ingredient of mixed feeds; or in the production of amino acids.

The residual molasses after cane sugar processing contains about 4% invert sugar, 30–40% sucrose, 10–25% reducing substances, a very low amount of raffinose and no betaine, but, unlike beet molasses, contains about 5% aconitic acid. Cane sugar molasses is fermented to provide arrack and rum.

19.1.4.2 Sugars Produced from Sucrose

Hydrolysis of sucrose with acids, cation-exchange resins or enzymes (invertase or saccharase) results in invert sugar which, after chromatographic separation, can provide *glucose* and *fructose*. Invert sugar syrup is a commercially-available liquid sugar. Invert sugar also serves as a raw material for production of sorbitol and mannitol. Enzymatic isomerization of sucrose results in *isomaltulose* (palatinose, I) which, after hydrogenation at pH>7 via isomaltose (II), yields mainly *isomaltitol* (III); while hydrogenation at pH=7 yields a mixture *(palatinit)* of isomaltitol and glucopyranosido-1,6-mannitol (IV) [cf. Reaction 19.1].

This mixture of sugar alcohols can be separated by fractional crystallization. Palatinit is a potential sugar substitute.

19.1.4.3 Starch Degradation Products

19.1.4.3.1 General outline

In principle, either starch or cellulose could be used as a source for saccharification, but only starch hydrolysis is currently of economic importance. Improvements in cellulose saccharification are being sought.

19.1.4.3.2 Starch syrup (glucose or maltose syrups)

Starch saccharification is achieved by either acidic or enzymatic hydrolysis. Controlled processing conditions yield products of widely dif-

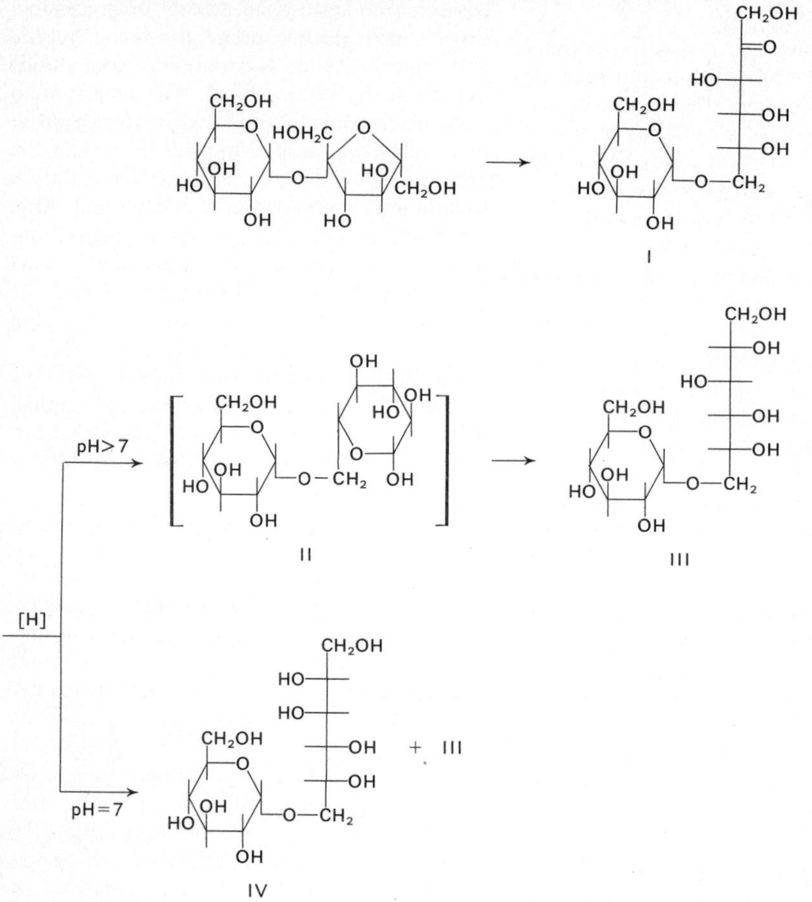

(19.1)

ferent compositions to suit the diversified fields of application. The enzyme most commonly used is α-amylase isolated from, for example, *Bacillus subtilis* or *B. licheniformis*. Optimal pH and temperature are 6.5 and 70–90 °C, respectively. The enzyme from *B. licheniformis* is active even at 100 °C. Hydrolysis can be carried out to obtain a product consisting mostly of maltose and in addition maltotriose and small amounts of glucose. The amylase enzyme from *Aspergillus oryzae*, with an optimum pH of 5 and temperature of 50–55 °C, is suitable for production of such a maltose syrup.

The extent of starch conversion into sugars is generally expressed as dextrose equivalents (DE value), i.e. the amount of reducing sugars produced, calculated as glucose.

The sweet taste intensity of the starch hydrolysates depends on the degree of saccharification and ranges from 25–50% of that of sucrose. Table 19.7 provides data for some hydrolysis products. The wide range of starch syrups starts with those with DE value of 10–20 (maltodextrins) and ends with those with DE value of 96.

Starch syrups are used in sweet commodity products. They retard sucrose crystallization (hard caramel candies) and act as softening agents, as in soft caramel candies, fondants and chewing gum. They are also used in ice cream manufacturing, production of alcoholic beverages and soft drinks, canning and processing of fruits and by the baking industry.

19.1.4.3.3 *Dried starch syrup (dried glucose syrup)*

Dried starch syrups with a moisture content of 3–4% are produced by spray drying of starch hydrolysates. The products are readily soluble in water and dilute alcohol and are used, for example, in sausage production as a red color enhancer. The average composition of dried starch syrups is 50% dextran, 30% maltose and 20% glucose.

19.1.4.3.4 Glucose (Dextrose)

The raw source for glucose production is primarily starch isolated from corn, potatoes or wheat. Saccharification is achieved enzymatically by α-amylase and/or by microbial amyloglucosidase, or by amyloglucosidase after the starch has been subjected to partial acid hydrolysis. The enzyme from *Aspergillus niger,* at pH 4.5 and 60 °C, provides a hydrolysate with 94–96% glucose. After a purification step, the hydrolysate is evaporated and crystallized. Glucose crystallizes as α-D-glucose monohydrate. Water-free α-D-glucose is obtainable from the monohydrate by drying in a stream of warm air or by crystallization from ethanol, methanol or glacial acetic acid. Dextrose, due to its great and rapid resorption, is used as an invigorating and strengthening agent in many nourishing formulations and medicines. Like dried glucose syrup, crystalline dextrose is used as a red color enhancer of meat and frying sausages.

19.1.4.3.5 Glucose-fructose syrup (high fructose syrup)

A glucose-fructose isomerase enzyme (cf. 2.8.2.3) occurs in some species of *Bacillus megaterium, B. coagulans* and *Lactobacillus brevii.* In a neutral or weakly alkaline reaction medium (pH 8.2) and at 35–60 °C, the isomerase converts glucose into fructose. Using immobilized enzymes, large amounts of glucose syrup can be converted into glucose-fructose syrup with a fructose content of 40–50%. When coupled with chromatographic separation techniques, products up to 90% fructose are obtainable. Glucose-fructose syrups are used as invert sugars derived from sucrose.

19.1.4.3.6 Starch syrup derivatives

Hydrogenation of glucose syrups results in products which, since they are nonfermentable and are less cariogenic, are used in manufacturing of sweet commodity products. Alkaline isomerization of maltose gives *maltulose,* which is sweeter than maltose, while hydrogenation yields *maltitol* in a mixture with maltotriit. This mixture of sugar alcohols is not crystallizable but, after addition of suitable polysaccharides (alginate, methylcellulose), can be spray-dried into a powder.

Table 19.7. Average composition of starch hydrolysates[a]

DE-Value	Glucose	Maltose	Maltotriose	Higher oligosaccharides
Acid hydrolysis				
30	10	9	9	72
40	17	13	11	59
60	36	20	13	31
Enzymatic hydrolysis[b]				
20	1	5	6	88
45	5	50	20	25
65	39	35	11	15
97	96	2	–	2

[a] All values expressed as % of starch hydrolysate dry weight basis.
[b] Occasionally it involves a combined acid/enzymatic hydrolysis.

19.1.4.4 Milk Sugar (Lactose) and Derived Products

19.1.4.4.1 Milk sugar

Lactose is produced from whey concentrates and whey, a by-product of cow milk clotted with rennet or acids. For milk sugar (lactose) isolation, the whey is adjusted to pH 4.7 and then heated directly by steam at 95–98 °C to remove milk albumins. The deproteinated filtered fluid is further concentrated by a multi-stage evaporator and then the separated salts are removed. The desalted concentrate yields a yellow raw sugar with a moisture content of 12–14%. The remaining mother liquor still contains an appreciable amount of lactose, so it is recirculated through the process or is used for the production of ethanol or lactic or propionic acids. The raw lactose is raffinated by solubilization, filtration and several crystallizations. The snow-white α-lactose monohydrate is pulverized in a pin mill and separated by particle size in a centrifugal classifier. Spray drying of lactose is gaining in importance.

To increase lactose digestibility, sweetness and solubility, a 60% lactose solution can be heated to 93.5 °C and the crystallizate discharged to a vacuum drum dryer. β-Lactose is formed. Its moisture content is not more than 1% and it is more soluble than α-lactose. Uses of β-lactose include: a nutrient for children; a filler or diluter in medicinal preparations (tablets); and an ingredient of nutrient solutions used in microbial production of antibiotics.

19.1.4.4.2 Products from lactose

Enzymatic or acidic hydrolysis of lactose provides a glucose-galactose mixture which is twice as sweet as lactose. A further increase in taste intensity is achieved by enzymatic isomerization of glucose. Such enzyme-treated products contain about 50% galactose, 29% glucose and 21% fructose.

Lactulose is obtained by isomerization of lactose. It is sweeter than lactose. Hydrogenation of lactose yields *lactitol,* while hydrogenation of lactulose yields a mixture of lactilol and β-D-galactopyranosido-1,4,-mannitol.

19.1.4.5 Fruit Sugar (Fructose, Levulose)

Fructose is obtainable from its natural polymer, inulin, which occurs in: topinambur tubers (India) or in its North American counterpart, Jerusalem artichoke (*Helianthus tuberosus*); chicory; tuberous roots of dahlia plants; and in flowerheads of globe or true artichoke *(Cyanara scolymus)*, grown extensively in France. Fructose is obtained by acidic hydrolysis of inulin or from chromatographic separation of a glucose-fructose mixture (invert sugar, isomerized glucose syrup). Only the latter process has commercial significance. Sweeter than sucrose, fructose is used as a sugar substitute for diabetics. Fruclose yields a mixture of lactitol and β-D-galactopyranosido-1,4-mannitol.

19.1.4.6 Sorbitol

Sorbitol, a hygroscopic alcohol, is approximately half as sweet as sucrose. It is used as a sweetener for diabetics and in food canning. It is also used as a softener and to retain moisture in candies (humectant), to which it is added in a 70% syrup form in amounts of 5–15%. Sorbitol can be produced by electrochemical reduction, but the bulk is still made by catalytic hydrogenation of glucose.

19.1.4.7 Sorbose

Sorbitol is oxidized by *Acetobacter xylium* into L-sorbose, an intermediary product for commercial synthesis of ascorbic acid (cf. 18.1.2.7). Sorbose is used as a sucrose substitute for diabetics and as an ingredient with neglible cariogenicity in low calorie foods.

19.1.4.8 Xylitol

Xylose is obtained by hydrolysis of hemicelluloses. Catalytic hydrogenation of xylose yields xylitol. Like fructose and sorbitol, xylitol is a sugar substitute for diabetics, has a role in parenteral (intravenous or intramuscular) nutrition and is used as an ingredient of low cariogenicity in production of "sugar-free" chewing gum.

19.1.4.9 Mannitol

Mannitol is obtained from fructose hydrogenation and, due to its laxative effect, it is used mainly in chewing gum manufacturing.

19.1.5 Candies

19.1.5.1 General Outline

Candies represent a subgroup of sweet commodities generally called confectionery. Such products as long-storage cookies, cocoa and chocolate products, ice cream and artificial "honey" (to be discussed elsewhere) are also confections.

Candies are manufactured from all forms of sugar and may also incorporate other foods of diverse origin (dairy products, honey, fat, cocoa, chocolate, marmalade, jellies, fruit juices, herbs, spices, malt extract, seed kernels, rigid or elastic gels, liqueurs or spirits, essences, etc.). The essential and characteristic component of all types of candy is sugar, not only sucrose, but also other forms of sugar such as starch sugar, starch syrup, invert sugar, maltose, lactose, etc.

The diversity of products classified as candy is great indeed. Nevertheless some main groups can be listed: (1) Hard candy is made of sugar and water (simple syrup) and is flavored or colored. There are clear candies with less water, and pulled hard candies with more water, and filled hard candies with a hard candy coating surrounding the fillings. (2) Candies with other ingredients, but not exceeding 5% of the candy's sugar content, e.g. marshmallows. (3) Candies containing large amounts of ingredients other than syrup, e.g. caramels, starch jellies and cream-filled chocolates. Other examples are caramels, made of sugar syrup, fat, color and flavorings (hard or soft caramel, the latter also called toffee); malt, honey, cream or milk bonbons; chocolates filled with cream (candy bars) or fondant, a soft filling made of small sugar crystals and milk); cocoflakes, made with coconut, i.e. candy with a water-containing and partially crystallized sugar mass. (4) Candies with flexible swelling ingredients (starch, pectins or other natural gums or gelatin), such as jellies, cotton candy and Turkish delight, a jelly-like candy covered with icing sugar. (5) Candied fruits (fruit or fruit paste impregnated and covered with sugar). (6) Licorice and licorice-like products. (7) Effervescent lemonade-type powders. (8) Dragees (sugar-coated candies) and pastilles (rolled, shaped and filled candy; lozenges). (9) Fancy cookies or cakes. (10) Icing products. (11) Products from almonds, nuts, apricot seeds, peanuts and other protein-rich oilseeds. (12) Marzipan and marzipan-like (persipan) products. (13) Nougat, croquant and similar fillings. Many candies with different fillings are dipped in molten chocolate to provide a candy with better flavor and smoother and glossier surface.

An additional classification of various candies within the listed tentative main groups is nearly impossible. Only a few important representative products listed above will be outlined in the following sections.

19.1.5.2 Marzipan

Marzipan is produced initially as a raw filler. Dehulled, moistened sweet almonds, together with not more than 35% sucrose, are disintegrated and refined into a fine paste using rubber rolls on a porcelain roller frame. The paste is briefly roasted in open pans until the water content is decreased to 17% and made into a homogeneous "plastic" filler. This raw filler is then processed into commercial marzipan by blending with an equal weight of sucrose. A portion of the sucrose may be replaced by starch syrup and/or sorbitol.

19.1.5.3 Persipan

As with marzipan, a raw paste is initially prepared, but instead of almonds, the seed kernels of apricots, peaches or bitter almonds (with bitterness removed) are used. Commercial persipan is a mix or raw persipan filler and sucrose, the latter not more than half of the mix weight. Sucrose can be partially replaced by starch syrup and/or sorbitol.

19.1.5.4 Other Raw Candy Fillers

These are produced from dehulled nuts such as cashews (kidney-shaped nuts from an evergreen tree grown in Brazil) or from peanuts. They correspond in composition to raw persipan paste. They are designated according to the oilseed component.

19.1.5.5 Nougat Fillers

Nougat paste serves as a soft or firm candy filling. It contains up to 2% water and roasted dehulled filberts (hazelnuts) or roasted dehulled almonds, finely ground in the presence of sugar and cocoa products. Cocoa products used are cocoa beans; cocoa liquor and butter; pulverized defatted cocoa; chocolate; baking, cream and milk chocolate; chocolate icing; cream and milk chocolate icings; and chocolate powders. The filler may contain a small amount of flavoring and/or lecithin. Also, part of the sugar may be replaced by cream or milk powder. Sweet nougat fillings can also be produced without cocoa ingredients and cream or milk powders. The kneaded nougat paste is often designated just as nougat or noisette.

Recently the trans- and cis-isomers of 5-methyl-4-hepten-2-one have been detected as character impact compounds for the flavor of filberts. The aroma threshold of the trans-isomer is extremely low: 5 ng/kg (water as solvent).

19.1.5.6 Croquant

Croquant serves generally as a filling for candy. It is made of molten sucrose which has been at least partly caramelized, and ground and roasted almonds or nuts. It is occasionally mixed with marzipan, nougat, stable dairy products, fruit constituents and/or starch syrup. Croquant can be formulated to a brittle or soft consistency.

19.1.5.7 Licorice and its Products

To manufacture licorice products, flour dough is mixed with sugar, starch syrup, concentrated flavoring of licorice herb root and gelatin, and the mix is evaporated to a thick consistency. It is then molded into sticks, bands, figurines, etc.

and dried further. The characteristic and flavor-determining ingredient derived from the perennial licorice herb is the diglucuronide of β-glycyrrhetinic acid (cf. 8.8.11).
Simple licorice products contain starch (30–45%), sucrose (30–40%) and at least 5% licorice extract. Better quality products have an extract content of at least 30%. The aroma is enhanced, usually, with anise seed oil in conjunction with low amounts of ammonium chloride.

19.1.5.8 Roasted Almonds

Raw or roasted almonds are coated with hot, saturated caramelized sugar syrup, i.e. made into dragee, a sugar-coated candy. The coating is puffed up by forced hot air into a curly, crinkled, crispy surface. The coating contains herbs or other flavoring additives such as vanillin. The weight ratio of sugar to almond generally does not exceed 4:1. The sugar coating may be colored.

19.1.5.9 Chewing Gum

Chewing gum is made of a natural or a synthetic gum base impregnated with nutrients and flavoring constituents, mostly sugars and aroma substances, which are gradually released by chewing. The gum base is a blend of latex products from rubber trees that grow in tropical forests or plantations. The most important sources are chicle latex from the *Sapodilla* tree of Mexico, Indonesia and Malaysia; jelutongs; and rubber latex. Natural (mastic tree) and synthetic resins and waxes are also used. Synthetic thermoplastic resins are polyvinyl esters and ethers, polyethylene, etc. The gum base may also contain cellulose as a filler and a break-up agent. The gum is flattened into oblong sticks or formed into pellets which are coated with sugar or candy. The aroma and flavor carriers used are sucrose, invert sugar, starch syrup or other sweetener substitutes and essential oils (spearmint, peppermint, etc.). In recent years extruders have been used increasingly to produce chewing gum in a continuous process.

19.1.5.10 Effervescent Lemonade Powders

The powder or compressed tablets (effervescent bonbons) are used for preparation of artificial sparkling lemonades. They contain sodium bicarbonate and an acid component (lactic, tartaric or citric acid). When dissolved in water, they generate carbon dioxide. Other constituents of the product are sucrose or another sweetener, and natural or artificial flavoring substances. Sodium bicarbonate and acids are often packaged and marketed separately in individual capsules or in two separate containers.

19.2 Honey

19.2.1 Honey

19.2.1.1 Foreword

Honey is produced by honeybees. They suck up nectar from flowers or other sweet saps found on living plants, store the nectar in their honey sac, and enrich it with some of their own substances to induce changes. When the bees return to the hive, they deposit the nectar into honeycombs for storage and ripening.
Honey production starts immediately after the flower pollen, nectar and honeydew are collected and deposited in the bee's pouch (honey sac). The mixture of raw materials is then given to worker bees in the hive to deposit it into the six-sided individual cells of the honeycomb. The changing of nectar into honey proceeds in the cell in the following stages: water evaporates from the nectar, which then thickens; the content of invert sugar increases through sucrose hydrolysis by acids and enzymes derived from bees, while an additional isomerization of glucose to fructose occurs in the honey sac; absorption of proteins from plant and bees, and acids from

Table 19.8. Production of honey in 1979 (1,000 t)

Continent		Country	
World	1,031	China	257
Africa	90	USSR	220
America, North-, Central-	210	USA	108
		Mexico	56
America, South-	44	Canada	30
Asia	301	Argentina	28
Europe, West-	82	Turkey	21
Europe, East- + USSR	280	Ethiopia	20
		France	20
Australia + South Pacific Islands	24	Australia	18
		Σ (%)[a]	75

[a] World production ≙ 100%.

the bee's body; assimilation of forage minerals, vitamins and aroma substances; and absorption of enzymes from the bees' salivary glands and honey sacs. When the water content of the honey drops to 16–19%, the cells are closed with a wax lid and ripening continues, as reflected by a continued hydrolysis of sucrose by the enzyme invertase and in addition by the synthesis of new sugars.

Table 19.8 provides information about honey production in some countries. The honey consumption in FR Germany in 1970 was 0.98 kg per capita.

19.2.1.2 Production and Types

In the production and processing of honey, it is important to preserve the original composition, particularly the content of aroma substances, and to avoid contamination. The following kinds of honey are differentiated according to recovery techniques:

Comb honey (honey with waxy cells), i.e. honey present in freshly-built, closed combs devoid of brood combs (young virgin combs). Such honey is produced in high amounts, but is not readily found in Germany. In other countries, primarily the USA, Canada and Mexico, it is widely available. Darker colored honey is obtained from covered virgin combs not more than one year old and from combs which include those used as brood combs.

Extracted honey is obtained with a honey extractor, i.e. by centrifugation at somewhat elevated temperatures of brood-free comb cells. This recovery technique provides the bulk of the honey found on the market. Gentle warming up to 40 °C facilitates the release of honey from the combs.

Pressed honey is collected by compressing the brood-free honey combs in a hydraulic press at room temperature.

Strained honey is collected from brood-free, pulped or unpulped honey combs by gentle heating followed by pressing.

Beetle honey is recovered by pulping honey combs which include brood combs. This type of honey is used only for feeding bees.

Based on its use, honey is distinguished as:

Honey for domestic use. This is the highest quality product, and is consumed and enjoyed in pure form.

Baking honey. This type of honey is not of high quality and is used in place of sugar in the baking industry. Such honey has to a certain degree spontaneously fermented, has absorbed or acquired other foreign odors and flavors, or was overheated. This category includes caramelized honey.

According to the recovery (harvest) time, honey is characterized as: early (collected until the end of May); main (June and July); and late (August and September).

Honey can be classified by geographical origin, e.g. German (Black Forest or Allgäu honey), Hungarian, Californian, Canadian, Chilean, Havanan, etc.

The flavor and color of honey are influenced by the kinds of flowers from which the nectar originates. The following kinds of honey are classified on the basis of the type of plant from which they are obtained.

Flower honey, e.g. from: heather; linden; acacia; alsike, sweet and white clovers; alfalfa; rape; buckwheat and fruit tree blossoms. When freshly manufactured, these are thick, transparent liquids which gradually granulate by developing sugar crystals. Flower honey is white, light-to-dark, greenish-yellow or brownish. Maple tree honey is light amber; alfalfa honey, dark-red; clover honey, light amber-to-reddish; and meadow flower honey, amber-to-brown. Flower honey has a typical sweet and highly aromatic flavor that is dependent on the flavor substances which together with the nectar are collected by the bees. Honey sometimes has a flavor reminiscent of molasses. This is especially true of honey derived from heather (alfalfa and buckwheat honeys).

Honedew honey (pine, spruce or leaf honeydew). This type of honey solidifies with difficulty. It is less sweet, darkly colored, and may often have a resinous terpene-like odor and flavor.

19.2.1.3 Processing

Honey is marketed as a liquid or semisolid product.

It is usually oversaturated with glucose, which granulates, i.e. crystallizes, within the thick syrup in the form of glucose hydrate. To stabilize liquid honey, it has to be filtered under pressure to remove the sugar crystals and other crystallization seeds. Heating of honey decreases its viscosity during processing and filling, and provides complete glucose solubilization and pasteurization. Heating has to be gentle since the low pH

of honey and its high fructose content make it sensitive to heat treatment. As with other foods, continuous, high temperature-short time processing (e.g. 65 °C for 30 s followed by rapid cooling) is advantageous.
Processing of honey into a semisolid product involves seeding of liquid honey with fine crystalline honey to 10% and storing for one week at 14 °C to fully crystallize. This product is marketed as creamed honey.

19.2.1.4 Physical Properties

Honey density (at 20 °C) depends on water content and may range from 1.4404 (14% water) to 1.3550 (21% water). Honey is hygroscopic and hence is kept in airtight containers. Viscosity data at various temperatures are given in Table 19.9. Most honeys behave as *Newtonian* fluids. Some, however, such as alfalfa honey, show thixotropic properties which are traceable to the presence of proteins, or dilating properties (as with opuntia cactus honey) due to the presence of trace amounts of dextran.
The specific heat (20 °C; 17.4% water) is 2.26 J/g/°C. Because of poor heat conductivity, the possibility of heating honey with microwaves is a viable approach. Heating 1 L honey for 1 h from 30–55 °C requires 25 kW of energy.

19.2.1.5 Composition

Honey is essentially a concentrated aqueous solution of invert sugar, but it also contains a very complex mixture of other carbohydrates, several enzymes, amino and organic acids, minerals, aroma substances, pigments, waxes, pollen grains, etc. Table 19.10 provides compositional data. The analytical data correspond to honey from the USA, nevertheless, they basically represent the composition of honey from other countries.
Water. The water content of honey should be less than 20%. Honey with higher water content is readily susceptible to fermentation by osmophilic yeasts. Yeast fermentation is negligible when the water contents is less than 17.1%, while between 17.1 and 20% fermentation depends on the count of osmophilic yeast buds.
Carbohydrates. Fructose (averaging 38%) and glucose (averaging 31%) are the predominant sugars in honey. Other monosaccharides have not been found. However, more than 20 di- and oligosaccharides have been identified (Table 19.11), with maltose predominating, followed by kojibiose (Table 19.12). The content of sucrose varies appreciably with honey ripening stage. The composition of disaccharides depends largely on the plants from which the honey was derived, while the effects of regions or season of the year are negligible.
Enzymes. The most prominent enzymes in honey are α-glucosidase, invertase or saccharase, α- and β-amylases (diastase), glucose oxidase, catalase and acid phosphatase. Average enzyme activities are presented in Table 19.13. Invertase and diastase activities, together with hydroxymethyl fur-

Table 19.9. Viscosity of honey at various temperatures

	Temperature (°C)	Viscosity (Poise)
Honey 1[a]	13.7	600.0
	20.6	189.6
	29.0	68.4
	39.4	21.4
	48.1	10.7
	71.1	2.6
Honey 2[b]	11.7	729.6
	20.2	184.8
	30.7	55.2
	40.9	19.2
	50.7	9.5

[a] Melilot honey (*Melilotus officinalis;* 16.1% moisture).
[b] Sage honey (*Salvia officinalis;* moisture content 18.6%).

Table 19.10. Composition of honey (%)

Constituent	Average value	Variation range
Moisture	17.2	13.4–22.9
Fructose	38.2	27.3–44.3
Glucose	31.3	22.0–40.8
Saccharose	1.3	0.3–7.6
Maltose	7.3	2.7–16.0
Higher sugars	1.5	0.1–8.5
Others	3.1	0–13.2
Nitrogen	0.04	0–0.13
Minerals (ash)	0.17	0.02–1.03
Free acids[a]	22	6.8–47.2
Lactones[a]	7.1	0–18.8
Total acids[a]	29.1	8.7–59.5
pH-value	3.9	3.4–6.1
Diastase-value	20.8	2.1–61.2

[a] mequivalents/kg

Table 19.11. Sugars identified in honey

Common name	Systematic name
Glucose	
Fructose	
Saccharose	α-D-glucopyranosyl-β-D-fructofuranoside
Maltose	O-α-D-glucopyranosyl-(1 → 4)-D-glucopyranose
Isomaltose	O-α-D-glucopyranosyl-(1 → 6)-D-glucopyranose
Maltulose	O-α-D-glucopyranosyl-(1 → 4)-D-fructose
Nigerose	O-α-D-glucopyranosyl-(1 → 3)-D-glucopyranose
Turanose	O-α-D-glucopyranosyl-(1 → 3)-D-fructose
Kojibiose	O-α-D-glucopyranosyl-(1 → 2)-D-glucopyranose
Laminaribiose	O-β-D-glucopyranosyl-(1 → 3)-D-glucopyranose
α,β-Trehalose	α-D-glucopyranosyl-β-D-glucopyranoside
Gentiobiose	O-β-D-glucopyranosyl-(1 → 6)-D-glucopyranose
Melezitose	O-α-D-glucopyranosyl-(1 → 3)-O-β-D-fructofuranosyl-(2 → 1)-α-D-glucopyranoside
3-α-Isomaltosylglucose	O-α-D-glucopyranosyl-(1 → 6)-O-α-D-glucopyranosyl-(1 → 3)-D-glucopyranose
Maltotriose	O-α-D-glucopyranosyl-(1 → 4)-O-α-D-glucopyranosyl-(1 → 4)-D-glucopyranose
1-Kestose	O-α-D-glucopyranosyl-(1 → 2)-β-D-α-fructofuranosyl-(1 → 2)-β-D-fructofuranoside
Panose	O-α-D-glucopyranosyl-(1 → 6)-O-α-D-glucopyranosyl-(1 → 4)-D-glucopyranose
Isomaltotriose	O-α-D-glucopyranosyl-(1 → 6)-O-α-D-glucopyranosyl-(1 → 6)-D-glucopyranose
Erlose	O-α-D-glucopyranosyl-(1 → 4)-α-D-glucopyranosyl-β-D-fructofuranoside
Theanderose	O-α-D-glucopyranosyl-(1 → 6)-α-D-glucopyranosyl-β-D-fructofuranoside
Centose	O-α-D-glucopyranosyl-(1 → 4)-O-α-D-glucopyranosyl-(1 → 2)-D-glucopyranose
Isopanose	O-α-D-glucopyranosyl-(1 → 4)-O-α-D-glucopyranosyl-(1 → 6)-D-glucopyranose
Isomaltotetraose	O-α-D-glucopyranosyl-(1 → 6)-[O-α-D-glucopyranosyl-(1 → 6)]$_2$-D-glucopyranose
Isomaltopentaose	O-α-D-glucopyranosyl-(1 → 6)-[O-α-D-glucopyranosyl-(1 → 6)]$_3$-D-glucopyranose

fural content, are of significance for assessing whether or not the honey was heated.

For α-glucosidase, 7–18 isoenzymes are known. In a wide pH optimum between 5.8–6.5 the enzyme hydrolyzes maltose and other α-glucosides. The K_M with sucrose as substrate is 0.030 M. It also possesses transglucosylase activity. During

Table 19.12. Oligosaccharide composition of honey

Sugar	Content[a] (%)
Disaccharides	
Maltose	29.4
Kojibiose	8.2
Turanose	4.7
Isomaltose	4.4
Saccharose	3.9
Maltulose (and two unidentified ketoses)	3.1
Nigerose	1.7
α-, β-Trehalose	1.1
Gentiobiose	0.4
Laminaribiose	0.09
Trisaccharides	
Erlose	4.5
Theanderose	2.7
Panose	2.5
Maltotriose	1.9
1-Kestose	0.9
Isomaltotriose	0.6
Melezitose	0.3
Isopanose	0.24
Gentose	0.05
3-α-Isomaltosylglucose	+[b]
Higher Oligosaccharides	
Isomaltotetraose	0.33
Isomaltopentaose	0.16
Acidic fraction	6.51

[a] Values are based on oligosaccharide total content (= 100%, which in honey averages 3.65%. Only the most important sugars are presented.
[b] Traces.

the first stage of sucrose hydrolysis the trisaccharide erlose (α-maltosyl-β-D-fructofuranoside) plus other oligosaccharides are formed (E = enzyme, S = sucrose, G = glucose, F = fructose):

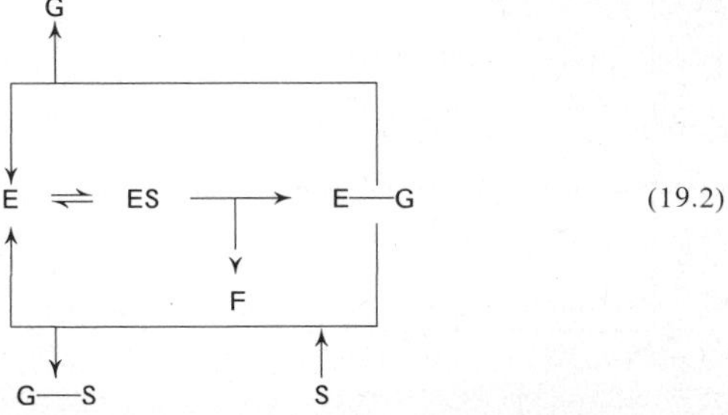

(19.2)

As the hydrolysis proceeds, most of these oligosaccharides are cleaved into monosaccharides.

Table 19.13. Average enzyme activity in honey

Number	Enzyme	Activity[a]
1	α-Glucosidase (saccharase)	7.5–10
2	Diastase (α- and β-amylase)	16–24
3	Glucose oxidase	80.8–210
4	Catalase	0–86.8
5	Acid phosphatase	5.07–13.4

[a] **1:** g saccharose hydrolyzed by 100 g honey per hour at 40 °C; **2:** g starch degraded by 100 g honey per hour at 40 °C; **3:** μg H_2O_2 formed per g honey/h; **4:** catalytic activity/g honey, and **5:** mg P/100 g honey released for 24 h.

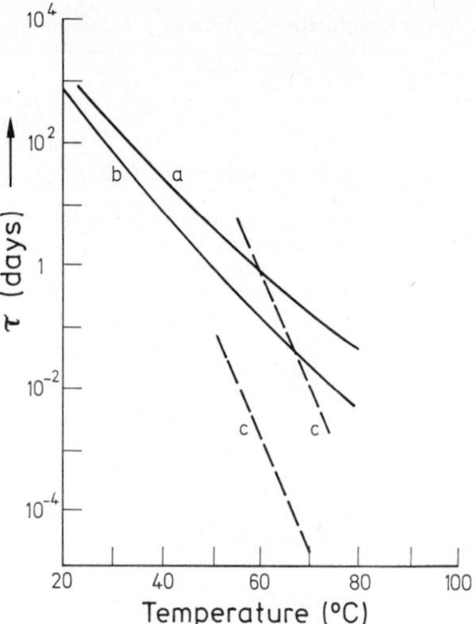

Fig. 19.9. Half-life activity ("τ") for enzymes diastase (a), invertase (b), and glucose oxidase (c) in honey at various temperatures. (After *White*, 1978)

Thermal inactivation of invertase in honey and its half-life values at various temperatures have been thoroughly investigated. These data are presented in Figures 19.8 and 19.9. Practically all invertase activity is derived from bees.

Honey *α-* and *β-amylases (diastase)* likewise originate from bees. Their pH optimum range is 5.0–5.3. Diastase activity is somewhat more thermally stable than invertase activity (Figures 19.8 and 19.9).

Glucose oxidase presence in honey is also derived from bees. Its optimum pH is 6.1. The enzyme oxidizes glucose (100%) and mannose (9%). The enzymatic oxidation by-product, hydrogen peroxide, is responsible for a bacteriostatic effect of nonheated honey, an effect earlier ascribed to a so-called "inhibine". The enzymatic oxidation yields gluconic acid, the main acid in honey. Glucose oxidase activity and thermal stability in honey vary widely (limit values were given in Table 19.9), hence the enzyme is not a suitable indicator for thermally-treated honey.

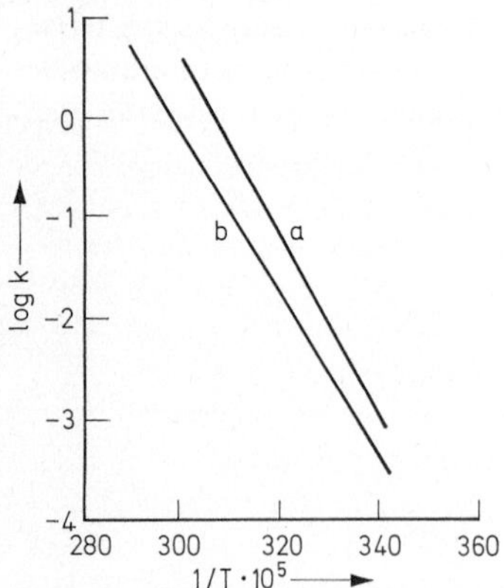

Fig. 19.8. Inactivation rate of (a) invertase and (b) diastase enzymes in honey. (After *White*, 1978)

Catalase in honey most probably originates from pollen which, unlike flower nectar, has a high activity of this enzyme. Similarly, honey *acid phosphatase* originates mainly from pollen, although some activity comes from flower nectars.

Proteins. Honey proteins are derived partly from plants and partly from honeybees. Figure 19.10 shows that bees fed on sucrose provide proteins with less complex patterns than, for example, cottonflower honey.

Amino acids. Honey contains free amino acids at a level of 100 mg/100 g solids. Proline, which might originate from bees, is the prevalent amino acid and is 50–85% of the amino acid fraction (Table 19.14). Based on several amino acid ratios, it is possible to identify the geographical or regional origin of honeys (Figure 19.11).

Organic acids. The principal organic acid in honey is gluconic acid, which results from glucose oxidase activity. In honey gluconic acid is in equilibrium with its gluconolactone. The acid level is mostly dependent on time elapsed between nectar collection by bees and time when the final thickness of honey is achieved in honeycomb cells. Glucose oxidase activity drops to a negligible level in thickened honey. Other acids

Table 19.14. Free amino acids in honey

Amino acid	mg/100 g honey (dry weight basis)	Amino acid	mg/100 g honey (dry weight basis)
Asp	3.44	Tyr	2.58
Asn + Gln	11.64	Phe	14.75
Glu	2.94	β-Ala	1.06
Pro	59.65	γ-Abu	2.15
Gly	0.68	Lys	0.99
Ala	2.07	Orn	0.26
Cys	0.47	His	3.84
Val	2.00	Tyr	3.84
Met	0.33	Arg	1.72
Met-O	1.74	Unidentified AA's (6)	24.53
Ile	1.12		
Leu	1.03	Total	118.77

present in honey only in small amounts are: acetic, butyric, lactic, citric, succinic, formic, maleic, malic and oxalic acids.

Aroma substances. About 120 volatile compounds are present in honey, and more than 80 have been identified. There are esters of aliphatic and aromatic acids, aldehydes, ketones and alcohols. A significant constituent is the ester of

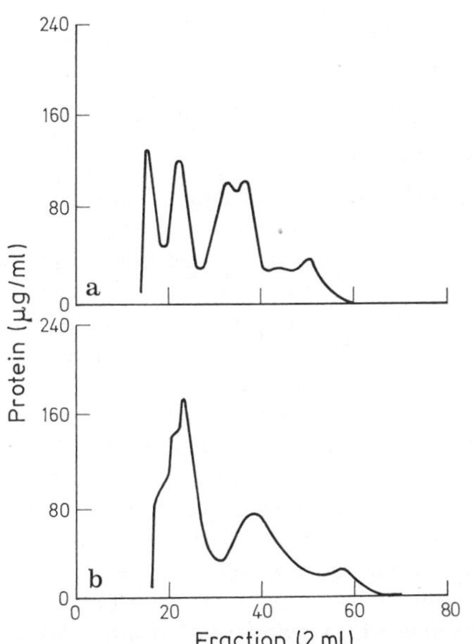

Fig. 19.10. Protein profiles of two honey varieties as revealed by gel filtration on Sephadex G-200. (a) Cottonflower honey (b), honey from sugar-fed bees. (After *White,* 1978)

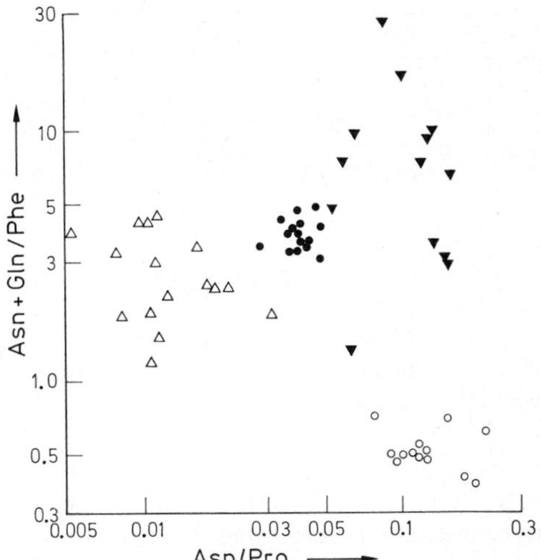

Fig. 19.11. Regional origin of honey as related to its amino acid composition. (After *White,* 1978)
Honey origin: △ Australia, ● Canada, ▼ United States (clover), ○ Yucatan

phenylacetic acid, which possesses a typical honey-like odor and flavor. The occurrence of anthranilic acid is typical for honey from lemon and lavender blossoms.

Pigments. Relatively little is known about honey color pigments. The amber color appears to originate from phenolic compounds and from products of the nonenzymic browning reactions between amino acids and fructose.

Toxic constituents. Poisonous honey (pontius or insane honey) has been known since the time of the Greek historian and general, *Xenophon,* and the Roman writer, *Plinius. Xenophon* described a mass poisoning of an expedition by *Kyros* to Asia Minor in 401 B.C. Recent intoxications have been reported in the USSR, the USA and Japan. Toxic honey comes mostly from bees collecting their nectar from: rhododendron species (Asia Minor, Caucasus Mountains); some plants of the family *Ericacea;* insane ("mad") berries; *Kalmia* evergreen shrubs; *Eurphorbiaceae;* and honey collected from other sweet substances, e.g. honeydew exudates of grasshoppers. Rhododendrons contain two poisonous compounds, andromedotoxin (an acetylandromedol) and graianotoxin (a tetracyclic diterpene) used in medicine as a muscle relaxant (see Formula 19.3). The poisonous nature of New Zealand honey is a result of tutin and hyenanchin (mellitoxin)

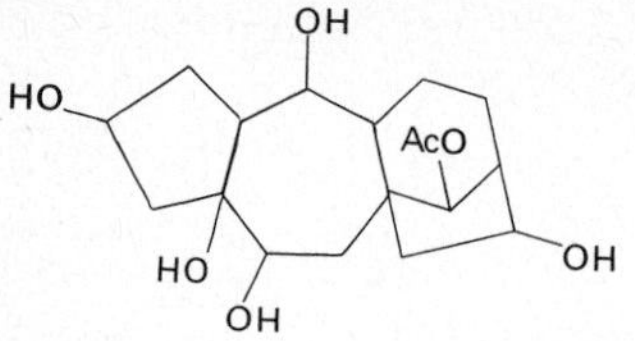

(19.3)

toxins from the tutu shrub (tanner shrub plant, *Coriaria arbora*). Poisonous flowers of tobacco, oleander, jasmine, henbane *(Datura metel)* and of hemlock *(Conium maculatum)* provide non-poisonous honeys. The yield of these honeys is negligible in Europe.

19.2.1.6 Storage

Honey color generally darkens on storage, the aroma intensity decreases and the content of hydroxymethyl furfural increases, depending on pH, storage time and temperature (Figure 19.12). The enzymatic inversion of sucrose also continues at a low level even when honey has reached its final density.

Honey should be protected from air moisture and kept at temperatures lower than 10 °C when stored. The desired temperature range for use is 18–24 °C.

19.2.1.7 Utilization

Honey use goes back to prehistoric times. Beeswax and honey played an important role in ancient civilizations. They were placed into tombs as food for deceased spirits, while the Old Testament describes the promised land as "a land flowing with milk and honey". In the Middle Ages honey was used as an excellent energy food and, up to the introduction of cane sugar, served as the only food sweetener. Besides being enjoyed as honey, it is used in baking (honey cookies, etc.) or in the manufacturing of alcoholic beverages by mixing with alcohol (honey liqueur, "beartrag") or by fermentation into honey flavored wine (Met). Preparations containing honey, in combination with milk an cereals, are processed for children. Tobacco products are occasionally flavored with honey. In medicine, honey is used in pure form or prescribed in preparations such as honey milk, fennel honey and ointments for wounds. It is incorporated into cosmetics in glycerol-honey gels and tanning cream products. The importance of honey as a food and as a nutrient is based primarily on its aroma constituents and the high content and fast absorption of its carbohydrates.

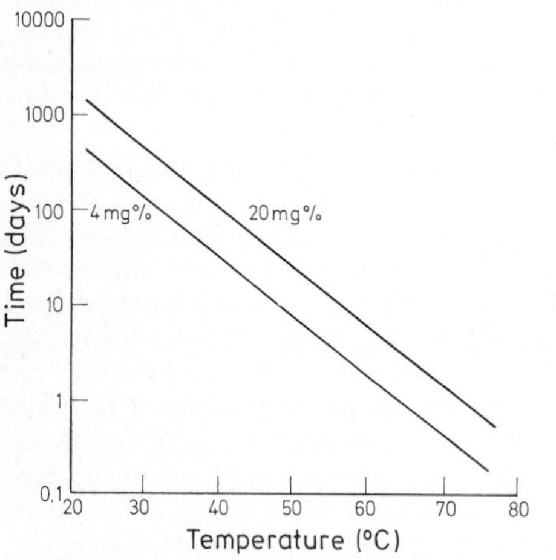

Fig. 19.12. Hydroxymethyl furfural formation in honey versus temperature and time. (After *White*, 1978)

19.2.2 Artificial Honey

19.2.2.1 Foreword

Artificial honey is mostly inverted sucrose from beet or cane sugar and is produced with or without starch sugar or starch syrup. It is adjusted in appearance, odor and flavor to imitate true honey. Depending on production method, such creams contain nonsugar constituents, minerals, sucrose and hydroxymethyl furfural.

19.2.2.2 Production

Sucrose (75% solution) is cleaved into glucose and fructose by acidic hydrolysis using hydrochloric, sulfuric, phosphoric, carbonic, formic, lactic, tartaric or citric acid or, less frequently, enzymatically using invertase. The acid used for inversion is then neutralized with sodium carbonate or bicarbonate, calcium carbonate, etc. The inverted sugar is then aromatized, occasionally with strongly-flavored natural honey. To facilitate crystallization, it is seeded with an invert sugar mixture already solidified, then packaged with automated machines. During inversion, an oligosaccharide (a "reversion dextrin") is also formed, mostly from fructose. Overinversion by prolonged heating results in dark coloring of the product and in some bitter flavor. Moreover, glucose and fructose degradation forms a noticeable level of hydroxymethyl furfural – this could be used for identification of artificial honey.

Liquid artificial honey is made from inverted and neutralized sucrose syrup. To prevent crystallization, up to 20% of a mildly-degraded, dextrin-enriched starch syrup is added (the amount added is proportional to the end-product weight).

19.2.2.3 Composition

Artificial honey contains invert sugar ($\geq 50\%$), sucrose ($\leq 38.5\%$) water $\leq 22\%$), ash ($\leq 0.5\%$) and, when necessary, saccharified starch products ($\leq 38.5\%$). The pH of the mixture should be ≥ 2.5. The aroma carrier is primarily phenylacetic acid ethyl ester and, occasionally, diacetyl, etc. Hydroxymethyl furfural content is 0.08–0.14%. The product is often colored with certified food colors.

19.2.2.4 Utilization

Artificial honey is used as a sweet spread for bread and for making Printen (honey cookies covered with almonds), gingerbread and other baked products.

19.3 Literature[a]

Birch, G.G., Green, L.F. (Eds.): Molecular structure and function of food carbohydrates. Applied Science Publ.: London. 1973

Birch, G.G., Parker, K.J. (Eds.): Sugar: Science and Technology. Applied Science Publ.: London. 1979

Crane, E. (Ed.): Honey. Heinemann. London. 1979

Hough, C.A.M., Parker, K.J., Vlitos, A.J. (Eds.): Developments in sweeteners. Applied Science Publ.: London. 1979

Jeanes, A., Hodge, J. (Eds.): Physiological effects of food carbohydrates. ACS Symposium Series 15, American Chemical Society: Washington, D.C. 1975

Koivistoinen, P., Hyvönen, L. (Eds.): Carbohydrate sweeteners in foods and nutrition. Academic Press: New York. 1980

Pancoast, H.M., Junk, W.R.: Handbook of sugars, 2nd edn., AVI Publ. Co.: Westport, Conn. 1980

Schiweck, H.: Disaccharidalkohole. Süßwaren *22* (14), 13 (1978)

Shallenberger, R.S., Birch, G.G.: Sugar chemistry. AVI Publ. Co.: Westport, Conn. 1975

White jr., J.W.: Honey. Adv. Food Res. *24*, 287 (1978)

---

[a] cf. 4.5.

20 Alcoholic Beverages

Alcoholic beverages are produced from sugar-containing liquids by alcoholic fermentation. Sugars, fermentable by yeasts, are either present as such or are generated from the raw material by processing, i.e. by hydrolytic cleavage of starches and dextrins, yielding simple sugars. The most important alcoholic beverages are beer, wine and brandy. Beer and wine were known to early civilizations and were produced by a well-developed industry. The distillation process for liquor production was introduced much later.

Figure 20.1 illustrates the *Embden-Meyerhoff-Parnas* scheme of alcoholic fermentation and glycolysis. For related details about the reactions and enzymes involved, the reader is referred to a textbook of biochemistry.

20.1 Beer

20.1.1 Foreword

Beer making or brewing involves the use of germinated barley (malt), hops, yeast and water. In addition to malt from barley, other starch- and/or sugar-containing raw materials have a role, e.g. other kinds of malt such as wheat, unmalted cereals called adjuncts (barley, wheat, corn, rice), starch flour, starch degradation products and fermentable sugars. The use of additional raw materials may necessitate in part the use of microbial enzyme preparations.

Beer owes its invigorating and intoxicating properties to ethanol; its aroma, flavor and bitter taste to hops, kiln-dried products and numerous aroma constituents formed during fermentation; its nutritional value to the content of unfermented solubilized extracts (carbohydrates, protein); and, lastly, its refreshing effect to carbon dioxide, a major constituent. Data about beer production and consumption are given in Table 20.1.

Table 20.1. World production and consumption of beer (1978)

Continent	Production (10^6 hectoliters)
World	873.1
Africa	27.4
America, North-, Central-	237.2
America, South-	58.3
Asia	67.6
Europe, West-	275.3
Europe, East- + USSR	152.1
Australia + South Pacific Islands	23.9

Country	Production (10^6 hectoliters)	Consumption (l/capita)
USA	190.3	88.6
FR Germany	91.7	145.6
UK	66.4	122.1
USSR	65.0	24.0[a]
Japan	44.3	32.4[b]
Brazil	26.5	
German DR	23.0	
France	22.8	45.9
Czechoslovakia	22.1	130.7
Mexico	22.0	34.0
Canada	20.4	84.9[b]
Australia	19.5	137.7
Spain	18.7	52.1
Holland	14.7	84.9
Belgium	13.4	124.0
	Σ 75.7%[c]	

[a] 1977.
[b] 1976.
[c] World production ≙ 100%.

20.1.2 Raw Materials

20.1.2.1 Barley

Barley is the most important of the raw materials used for beer production. Different cultivars of the spring barley (*Hordeum vulgare convar. disti-*

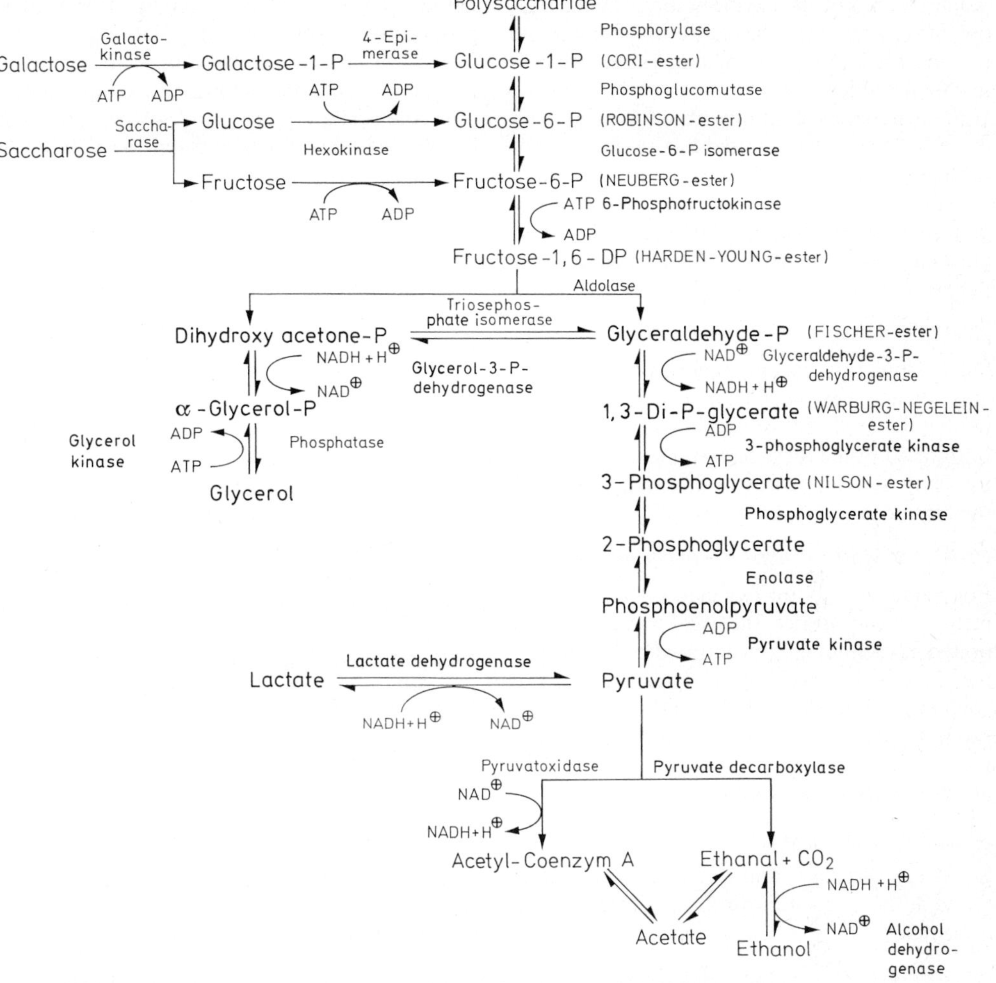

Fig. 20.1. *Embden-Meyerhoff-Parnas*-scheme of glycolysis and alcoholic fermentation

chon) with exceptionally suitable properties are used as brewing and malting barley in Germany. In addition, six-row winter barley has an increasing role. Barley of high brewing value provides an ample content of extract from the resultant malt, and has high starch and moderate protein (9–10%) content, a high degree of germination (over 95% of kernels), high germination vigor and good swelling ability. Sensory assay (hand appraisal) should also be included in evaluation of barley.

20.1.2.2 Other Starch- and Sugar-Containing Raw Materials

20.1.2.2.1 Wheat malt

Wheat malt is used in a mix with barley malt in a ratio of 40 : 60 in the production of top fermented beer.

20.1.2.2.2 Adjuncts

In addition to barley malt, supplementary sources of starch are used in the form of unmalted cereals (adjuncts) in order to dilute the mash by 15–50%. The adjuncts are barley, wheat, corn and rice (cracked rice) grains in the form of whole meal, grits, flakes or flour.

Adjuncts are low in enzyme activity, hence their use may necessitate the addition of microbial enzyme preparations with α-amylase and proteinase activities.
Unmalted barley contains about threee times more β-glucans than malted barley. In order to decrease the viscosity of unmalted barley extract to values similar to those of malted barley, β-glucans must be degraded with the enzyme β-glucanase, which is present in microbial enzyme preparations.

20.1.2.2.3 Syrups, extract powders

Since adjunct processing may result in undesirable changes, extracts from enzyme- or acid-treated barley, wheat or corn have recently been introduced in the form of syrup or powder. The use of syrup from barley to as much as 45% of the total mash is possible.

20.1.2.2.4 Malt extracts, wort concentrates

For production of hop-free malt extracts or hopped wort concentrates, the usual worts are evaporated in vacuum or concentrated by freeze drying. Such concentrate is diluted prior to use. The content of bitter substances and the tendency to produce cloudiness or turbidity are decreased in such concentrates, since tannins and proteins are removed during the evaporation step.

20.1.2.2.5 Brewing sugars

Sucrose, invert sugar and starch-sugar are introduced at the stage of hopping or before the beer is bottled.

20.1.2.3 Hops

20.1.2.3.1 General outline

Hops are a very important and indispensable ingredient in beer production. They act as a clarifier, since they precipitate the proteins in wort, change the wort character to give a specific aroma and bitter taste and, together with ethanol and carbon dioxide, their active antibiotic properties contribute to stability of beer. Lastly, the pectin content of hops enhances the foam-building ability of beer.
The hop (*Humulus lupulus*) is a tall, hardy, perennial climbing vine. The flowers of the female plants, though lacking pollination, grow well and cluster into a conical blossom which has large thin scales or bracts. This cone, when ripe, is harvested and used commercially. The plant is propagated vegetatively by planting cuttings from fleshy roots. The hop cones are picked in August or September and are dried and pressed into bales. The lupulin gland in the upper and lower portion of bracts contains, in addition to essential oils, bitter constituents. Data about hop production are given in Table 20.2

20.1.2.3.2 Composition

Table 20.3 presents data on the composition of hops. The constituents of utmost importance are the bitter substances. In fresh hops they occur mostly in the form of α-acids (cf. Formula 20.1): humulon (I), cohumulon (II), adhumulon (III); and in the form of β-acids: lupulon (IV), colupulon (V) and adlupulon (VI). These compounds are susceptible to changes during drying, storage and processing of hops. The changes usually involve isomerization, oxidation and/or polymerization. As a consequence, a great number of secondary products are found.

Table 20.2. Production of hops in 1981 (1,000 t)

Continent	Hops	Country	Hops
World	116	USA	36
America, North-, Central-	37	FR Germany	31
Asia	2	Czechoslovakia	10
Europe, West-	46	UK	10
Europe, East- + USSR	29	USSR	8
Australia + South Pacific Islands	2	Yugoslavia	4
		Australia	2
		Belgium–Luxemburg	2
		German DR	2
		Japan	2
		Poland	2
		Spain	2
		Σ (%)[a]	96

[a] World production ≙ 100%.

Table 20.3. Composition of hops

Constituent	Content (%)[a]	Constituent	Content (%)[a]
Bitter compounds	18.3	Crude fiber	15.0
Essential oil	0.5	Ash	8.5
Polyphenols	3.5	N-free extractable matter	34.0
Crude protein	20.0		

[a] As % dry matter; moisture content approx. 11%.

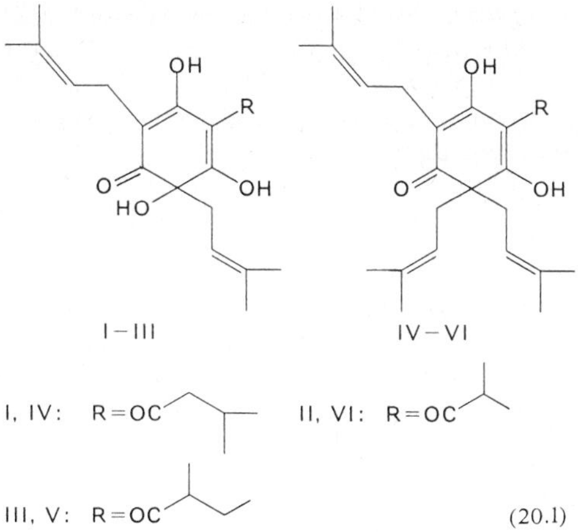
OH
R
O
HO
OH
OH
R
O
OH
I–III
IV–VI
I, IV: R=OC
II, VI: R=OC
III, V: R=OC

(20.1)

The quality and intensity of the bitter taste derived from these secondary products are different. Evaluation of hops is therefore based on a determination of composition of individual α- and β-acids, rather than of the total content of bitter substances. As seen from Table 20.4, the composition varies greatly with hop origin. During the boiling of hops, humulons isomerize into isohumulons (cis-compounds, VII; trans-compounds, VIII; cf. Formula 20.2), which are more soluble and bitter than the initial compounds. The isohumulons can be further transformed into humulinic acids (IX, X), which have only about 30% of the bitterness of isohumulons.

Hulupons (XI) and luputrions (XII) are the secondary products of the lupulons. They possess an exceptionally pleasant and mild bitter taste which is much less bitter than the compounds from which they are derived. Hence the bitter taste of beer is primarily due to compounds of the humulon fraction.

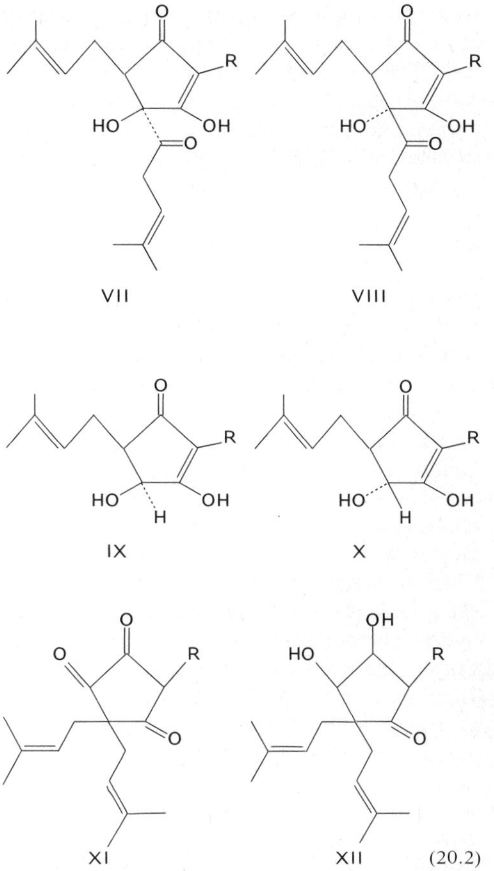
O
R
HO
OH
O
VII
VIII
IX
X
H
XI
XII
HO
OH

(20.2)

More than 150 compounds have been identified in essential oil from hops. Among the most important are the terpenes, such as myrcene, humulene (XIV), which is obtained biosynthetically from farnesyl pyrophosphate (XIII), and caryophyllene (XV), an isomer of humulene:

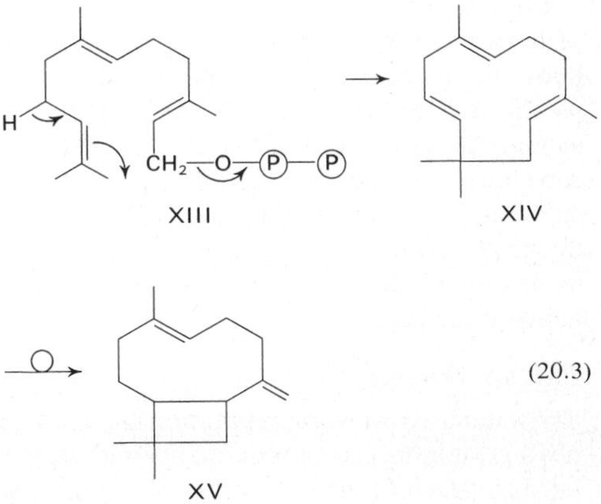
H
CH_2–O–Ⓟ–Ⓟ
XIII
XIV
XV

(20.3)

Table 20.4. Content of humulon and lupulon of hops from various sources (values in %)

Hops	α-Acids			β-Acids		
	humulon	cohumulon	adhumulon	lupulon	colupulon	adlupulon
Japan	46	41	13	21	68	11
America	54	34	12	32	57	11
Hallertau	59	27	14	45	43	12
Northern Brewer	64	24	12	46	43	11
Saaz	67	21	12	51	37	12

Based on terpene constituents, hops can be classed as myrcene- and humulene-enriched cultivars. The latter (such as cv. Spalt, Hallertauer, Mittelfrueher, Saaz) have a particularly pleasant aroma. The essential oil content decreases during storage and a shift in composition occurs: the hydrocarbons decrease, while the oxygen-containing terpenes increase.

20.1.2.3.3 Processing

Freshly-harvested hops are dried in a hop kiln in a stream of warm air (30–65 °C) to 8–10% moisture, followed by a readjustment of moisture content to 11–12%. The dried hops are also fumigated with sulfur to increase their stability.
In addition to hop cones, which are prone to quality loss even under proper storage conditions, processed products from hops are acceptable and utilized.
Hop powder is obtained by grinding the cones, which makes the active aroma ingredients more extractable. Prior to grinding, part of the inert material is separated and thus lupulin-enriched concentrates are obtained.
Hops are extracted with a mixture of water and an organic solvent (e.g. dichloromethane), giving extracts of varying compositions. Recently, a hops extraction process using supercritical carbon dioxide is of increasing importance. Such extracts can replace 20–60% of the hops.
Isomerized extracts, in which humulon has been converted into isohumulon by heat treatment, are suitable for a cold hopping procedure. In traditional beer hopping this conversion is achieved by boiling the wort for a long time. Isomerized extracts are used in the main fermentation or at a later step of brewing. Boiling results in the loss of a large portion of oil constituents of hops by steam distillation. The addition of hops shortly before the end of boiling or the use of hop resins or concentrates may greatly enhance the hop aroma of the product. Phenolic constituents in hops contribute to protein coagulation during wort boiling. A part of protein-tannin complexes formed may precipitate at low temperatures after long storage, resulting in turbidity in the beer.

20.1.2.4 Brewing Water

The water used for wort preparation in a brewery has a great influence on beer quality and character. The salt constituents of water can change the pH of the mash and wort. Bicarbonate ions cause a pH increase, while Ca^{2+} and Mg^{2+} ions cause a pH decrease. Heating of water which contains bicarbonates increases the alkalinity according to the equation:

$$HCO_3^{\ominus} + H^{\oplus} \rightleftharpoons CO_2 + H_2O \quad (20.4a)$$

in which the equilibrium is shifted to the left since, during heating, the CO_2 component escapes as a gas. Ca and Mg ions react with secondary phosphates in wort to form insoluble tertiary phosphates, releasing protons which add to the acidity of the water:

$$3\,Ca^{2\oplus} + 2\,HPO_4^{2\ominus} \rightleftharpoons Ca_3(PO_4)_2 + 2\,H^{\oplus} \quad (20.4b)$$

Magnesium sulfate in high concentrations imparts an unpleasant bitter taste to beer. Manganese and iron salts induce turbidity, discoloration and taste deterioration. Nitrates and silicates interfere with fermentation.
The unique character of different kinds of beer (Pilsen, Dortmund, Munich, Burton-on-Trent), without doubt, can historically be ascribed to the brewing water used in those places, with residual alkalinity playing the major role. Water, low in soluble bicarbonates of calcium, magnesium, sodium or potassium, and soluble carbonates and hydroxides, is suitable for strongly-hopped light beers, such as Pilsener, while high alkalinity water is suitable for dark beers, such as those from Munich.
Preparation of brewing water is mainly directed to the removal of carbonates. Precipitation by heating with lime is customary. Furthermore, when lime water is used without heating, water softening occurs. Removal of excess salt by ion-exchange resins is also advantageous. Today any water can be treated to match the requirement of a desired type of beer.

20.1.2.5 Brewing Yeasts

Brewing yeasts are exclusively from strains of *Saccharomyces*. Two types are recognized: top fermenting yeasts for temperatures > 10 °C, and bottom fermenting yeasts, used down to 0 °C. The top fermenting yeasts, e.g. *Saccharomyces cerevisiae Hansen,* rise to the surface during fermentation in the form of large budding ("sprouting") associations. They ferment only ¾ of raffinose, since they lack the enzyme melibiase. The bottom fermenting yeasts, e.g. *Saccharomyces carlsbergensis Hansen,* settle to the bottom during fermentation and completely ferment all sugars, including raffinose. There are yeasts with high fermentation ability which remain suspended for a long time, giving a high fermentation rate. Yeasts with low fermentation ability flocculate early and settle to the bottom (superflocculent yeasts) and hence are unable to continue active fermentation. Pure cultures of many yeast strains currently in use are derived from a single yeast cell and are used as "starter yeast" in plant operations. After the main fermentation, a part of the yeast is harvested for use in freshly-prepared worts, until the yeast becomes useless due to contamination or degeneration. In this way, it is possible to continuously select suitable yeast strains for a defined goal.

20.1.3 Malt Preparation

The cereals are soaked (steeped) in water, then are allowed to germinate. The product, green malt, is dried and mildly roasted into a more or less dark and aroma-rich kiln-dried malt. During processing, the rootlets are removed from the malt. The loss due to malting is 11–13% of the dry weight. Prior to use, the malt is stored for 4–6 weeks.

20.1.3.1 Steeping

Cereal kernels are steeped in water to raise their moisture content to induce germination. The water content is 42–44% for light and 44–46% for dark malt. Usually the steeping water is alternately added and removed. In this way, in contrast to a single-wetting process, the steeping time is shortened and better germination and enzyme build-up are achieved. Good aeration is needed in all phases to remove the CO_2 produced by respiration. The normal steeping temperature is 10–15 °C. Warm (20–30 °C) or hot (35–38 °C) water steeping accelerates germination and thus shortens the process. Alkali treatment (CaO, NaOH) of steeping water serves to reduce microbial contamination and to remove undesirable polyphenols from the hulls.

20.1.3.2 Germination

When the cereals reach the desired moisture content, they are allowed to germinate in germinator compartments, on floors, or in chests or drums. The removal of CO_2 and heat is achieved by turning and mixing in traditional floor malting, whereas a stream of moistened air is used in the newer chest or drum procedures. The optimal germination temperature range is 15–20 °C. The process for light malt takes about 7 days; for dark malt, 9 days. Germination can be stimulated by the use of gibberellic acid.

20.1.3.3 Kilning

The germinated cereals, termed green malt, contain 42–45% moisture. They are dried in a kiln into a malt with 2–3% moisture. Simultaneously, the color and roasted aroma of the malt are formed from *Maillard* reactions.

Initially, the green malt is dried at 35–40 °C for about 12 h. For light malt production the moisture is decreased rapidly to about 10%. For dark malt, where more extensive hydrolytic degradation of starch is needed, the moisture content is maintained for a longer time at about 20%. The malt is then heated to the roasting temperature within 2 h and is roasted for 4–5 h at 80 °C for light malt or at 105 °C for dark malt. Finally, the kiln-dried malt is freed of rootlets, cleaned and polished.

20.1.3.4 Continuous Processes

Several kinds of installations have been developed which provide continuous steeping, germination and, occasionally, also kilning; offering substantial savings in time. Steeping in this case is performed as a single washing followed by water spraying and continuous transferring to the germination stage. The process conditions are regulated by means of forced air. Green malt is obtained within 5–7 days. In some installations the malt is moved for different stages, while in others it remains in the same container from steeping to kilning.

20.1.3.5 Special Malts

Special malts are prepared for many purposes. Dark caramelized malt is held briefly at 60–80 °C to saccharify its starch and is then roasted at 150–180 °C for the desired color grade. Such color-rich malt is free of diastase enzyme activity, is a good foam builder, and is mostly used for aromatizing of malt beers and strong bock beers. Light caramelized malt is made in a similar way, but is treated at lower temperatures after the saccharification step. This preserves the enzyme activities. It is lightly colored and when used gives beer an increased foaming capacity and full-bodied properties. Colored malt is obtained by roasting the kiln-dried malt at 190–220 °C, omitting the prior saccharification step. It can be used to intensify the color of dark beers.

20.1.4 Wort Preparation

The coarsely-ground malt is dispersed in water. During this time, the malt enzymes hydrolyze starch and other ingredients. A clear fermentable solution, the so-called wort, is obtained by filtration. When boiled with added hops, the wort takes on the typical beer flavors.

20.1.4.1 Ground Malt

Malt is disintegrated by passing it through several grinding rolls and sifters. The ground products, hull, middlings and flour, are then combined in the desired proportions. By using finely-ground meal, the extraction yield increases, but problems arise in wort filtration. Wet milling is commonly preferred for better filtration as it yields a higher proportion of intact hulls. In addition, it provides the desired high extraction yield. For wet milling the water content of the malt is adjusted to 25–30%. It is then ground by a set of rollers and processed immediately into wort.

20.1.4.2 Mashing

For 100 kg of malt 8 hectoliters of water are needed. This amount of water is divided into a major portion for production of the mash, and into one or several post-mashing rinses used to wash out extract from the hulls. The course of pH and temperature during mashing are of utmost importance for determining wort composition and, hence, the type and quality of beer. The optimum activity of malt α-amylases is from 72–76 °C at pH 5.3–5.8, and of malt β-amylases from 60–65 °C at pH 4.6, while that for malt proteinases is from 55–65 °C at pH 4.6. Hence, wort with a pH near 6 will not, without prior pH adjustment, provide optimal conditions for the action of enzymes. The methods used for temperature control in mashing are of two types: decoction and infusion. In the decoction method the temperature of the total mash from start-up to end temperature is raised by removing an aliquot of mash, heating this to boiling and then returning it to the main mash in the mash tun. According to the number of such boiled mash returns, one-, two- or three-mash return procedures are being used commercially. The latter is used exclusively for dark beer brewing; the two-mash return for light beer; and the one-mash return procedure for brewing all types of beer. The three-mash return procedure will be briefly described as an example. The crushed malt is mixed in the mash tun with water at 37 °C; the first aliquot is drawn off, heated to boiling and returned to the mash tun. In this way the total mash temperature is raised to 52 °C. Two repetitions raise the total mash temperature stepwise to 64 and then to 75 °C. The mashing process is completed at a terminal mash temperature of 74–78 °C.

For improving the enzymatic degradation and the extraction yield a brief stop at 47–50 °C is advisable before further temperature increase. On the other hand, when a low alcohol beer is desired, the malt mashed at 37 °C is drained into boiling water, increasing the temperature to 70 °C and resulting in extensive enzyme inactivation.

In infusion mashing, used mostly in England for brewing top fermented beer, the terminal mashing temperature is achieved not by stepwise increases, but by live steam injection or addition of hot water As in the decoction method, the temperature programme used can vary greatly.

20.1.4.3 Lautering

The separation of wort from hulls and insoluble residues of the grain is done by a classical procedure in a lauter tun, a vessel with a slotted false bottom. The hull and other residues form a 35 cm deep layer in the bottom which acts as a filter through which the extract, or wort, is strained. The initial turbid liquid (turbid wort) with 16–20% extract is pumped back to the tun. Finally, to obtain more wort, the spent grains are rinsed or sparged 3 to 4 times with water.

Modern installations for lautering use strain masters or discontinuous or continuous mash filters. The draff, the lautering residue, is used for animal feed.

20.1.4.4 Wort Boiling and Hopping

Wort boiling with hops or hop products is done in a brew kettle (hop kettle) in which the initial and subsequent worts from the lautering step are collected. Addition of hops is adjusted according to the type and quality of beer desired. The quantity (in hop cones/hectoliter) for light lager beer is 130–150 g; for Dortmund-type beer, 180–220 g; for Pilsener beer, 250–400 g; for dark Munich beer, 130–170 g; and for malt beer and dark bock beer, 50–90 g. The critical factor is the content of bitter substances in the hops selected. Boiling for 2 h concentrates the wort, coagulates protein ("break forming"), solubilizes hop ingredients and converts the bitter components to their isoforms and, lastly, inactivates enzymes. The hot wort is then chilled, filtered, aerated and, finally, "pitched" with yeast.

20.1.4.5 Continuous Processes

Processes are now available which allow continuous wort production followed by fermentation in a fermentation tower. Installation and operating costs are considerably below those of discontinuous operations.

20.1.5 Fermentation

20.1.5.1 Bottom Fermentation

Bottom fermentation involves a primary and a secondary step. In the primary fermentation step, the cooled wort with about 6.5–18% dry mass extracted from malt ("stemwort") is pumped into fermenting tanks, located in fermentation cellars cooled to 5–6 °C. In the past the tanks were made of lacquered, microparaffin, wax or asphalt base-treated oakwood, but today the tanks are made of plastic-lined concrete, enamel-coated steel, aluminum or V_2A steel. The wort is inoculated ("pitched") with yeast in the form of a thick yeast slurry (0.5–1 l/hl) and fermented at 4–11 °C until 85–90% of the fermentable extract has been converted. The primary fermentation is completed in 8–10 days, at which point the yeast "breaks", i.e. flocculates and settles to the bottom. The beer is transferred to large clean tanks. The middle layer of yeast, the "core yeast", is removed from the bottom of the fermentation tank and is filtered, washed and reused in the next fermentation.

The young "green" beer is stored for 1–4 months in tanks at 1.5–2 °C for secondary fermentation. During this time, extensive fermentation of residual sugars occurs, accompanied by carbon dioxide enrichment, beer clarification and maturation. The storage time is 6–12 weeks for Munich beer containing 11–14% stemwort, or 14–18 weeks for Dortmund beer with 13–14% stemwort.

20.1.5.2 Top Fermentation

Primary fermentation proceeds in fermentation tanks, but at higher temperatures (15–20 °C) than with bottom fermentation, and requires a total time of 2–7 days. The yeast builds a solid cap at the top of the tank. It is skimmed off into individual fractions (hops flock, yeast flock, post-flock). The secondary fermentation is a very slow process and may continue in tanks or bottles. Top fermentation is used mostly in England and Belgium, while in Germany it is used in the production of Weiss beer, a light tart ale made from wheat.

20.1.5.3 Continuous Processes, Rapid Methods

Several continuous processing methods provide accelerated fermentation. They make use of thermophyle yeasts, higher fermentation temperatures and more intensive wort aeration.

20.1.6 Bottling

After ageing, beer is filtered through filter pads of cotton and some silicates, often having been preclarified through a kieselguhr pad or by centrifugation. Then, with the aid of a special cask/keg filling apparatus, it is foamlessly filled into transportable casks or metal cisterns. In addition to impregnated oakwood casks, specially-lined iron, aluminum or V_2A steel containers are also acceptable. Bottle filling proceeds from a "bottle tank" in a fully automated process. Tin-plated or aluminum cans are also used.

Pasteurization gives the beer biological stability for overseas export. To avoid cloudiness due to protein precipitation and changes in flavor, the beer is heated to 60–70 °C in a water bath or by steam. The beer is often pasteurized at 62 °C for 20 min. For sterile filling the beer is heated to 70 °C for 30 sec or is passed through microfilters (with pore size less than the size of bacteria) and then poured into sterilized bottles or cans.

Temperature fluctuations during storage and transport must be avoided if beer quality is to be preserved.

20.1.7 Composition

20.1.7.1 Ethanol

The ethanol content is 1.0–1.5% by weight for a low fermented extract-rich beer, 1.5–2.0% for a weak or thin beer, 3.5–4.5% for a full beer, and

4.8–5.5% for a strong beer. Higher alcohols, such as 2-methylbutanol, 3-methylbutanol, 2-methylpropanol and 2-phenylethanol, are also present in very small quantities.

20.1.7.2 Extract

The nonalcoholic constituents of beer vary within a wide range of from 2–3% for plain beers up to 8–10% for strong beers. These constituents are the beer solids and consist of up to 80% carbohydrate, mostly dextrins. It is possible to calculate the solids content of the original wort before fermentation from the solids content (E, weight %) and alcohol content (A, weight %) of the beer product. The calculation is based on the fermentation equation: 2 parts by weight of sugar equal 1 part by weight of alcohol. The initial solids content of wort, which actually represents a measure of malt utilization, is designated as "stemwort" (St) and can be calculated by the formula:

$$St = \frac{100\,(E + 2{,}0665\,A)}{100 + 1{,}0665\,A} \quad (20.5a)$$

Thus, for example, if the solids content (E) of a beer is 3% (w/v) and the alcohol content (A) is 5.0% (v/v), then the solids content of the wort before fermentation was 12.6% (w/v).
The stemwort content in Germany is 2–5.5% for plain beers, 7–8% for draft beers, 11–14% for full beers and above 16% for strong beers.

20.1.7.3 Acids

Carbon dioxide is responsible to a substantial extent for the refreshing value and stability of beer. CO_2 is 0.36–0.44% in bottom fermented beers, while in Weiss beer the CO_2 content is up to 0.6–0.7%. A CO_2 content below 0.2% gives flat and dull beers. Small amounts of lactic, acetic, formic and succinic acids are found in all beers. The pH is between 4.7 (dark, strong beers) and 4.1 (Weiss beers).

20.1.7.4 Nitrogen Compounds

The N-compounds in beer (0.15–0.75%) originate primarily from proteins in the raw materials and from yeast. They consist mainly of proteins plus high molecular weight protein degradation products; both being responsible for cloudiness in beer during cold storage. The free amino acids found in malt are also present in beer. It appears that glutamic acid contributes to beer taste. The presence of volatile amines has also been confirmed.

20.1.7.5 Carbohydrates

The carbohydrate content is approximately 3–5%, while in some strong beers or malt beers it may be considerably higher. Pentosans are also present in addition to dextrins, mono- and oligosaccharides (maltotriose, maltose, etc.). Glycerol normally is 0.2–0.3% in beer.

20.1.7.6 Minerals

Minerals make up 0.3–0.4% of beer and consist mostly of potassium and phosphate. Calcium, magnesium, iron, chloride, sulfate and silicates are also present.

20.1.7.7 Vitamins

Vitamins of the B-group (vitamins B_1 and B_2, nicotinic acid, pyridoxine and pantothenic acid) are present in various beers, often in significant amounts.

20.1.7.8 Aroma Substances

Beer aroma is determined by the presence of higher alcohols (fusel oil), volatile organic acids, esters and carbonyl compounds. The aroma of a young beer typically reflects the presence of sulfur compounds such as H_2S, thiols and thiocarbonyl compounds which, during secondary fermentation, are released along with CO_2. The total ester content is 40–50 mg/l, of which 10–20 mg/l is ethyl acetate. Other esters, for example isobutyl-, isoamyl- and phenylethyl acetate and ethyl capronate, are also present. Carbonyl compounds include ethanal, 2,3-butanedione and its reduction products acetoin and 2,3-butanediol, and 2,4-pentanedione.
The ethanal concentration is 3–14 mg/l of beer. Levels of ethanal exceeding 25 mg/l give an undesirable aroma. The content of ethanal increases when fermentation is conducted at higher temperatures and/or when an excessive amount of yeast is used. Concentrations of diacetyl above 0.35 mg/l also give an undesirable aroma to beer. A well-selected active yeast reduces diacetyl to acetoin which, in concentrations >1 mg/l, provides an undesirable cellar-type flavor. Beer flavor is negatively affected by the presence of as little as 0.13 mg diacetyl, 1.5–1.7 mg acetoin and 140–165 mg butanediol per liter. Accelerated fermentation brought about by more vigorous stir-

ring of wort results in an increase in the content of higher alcohols and diacetyl and a decrease in the content of esters and free acids. Consequently, the beer aroma becomes unpleasant.
Light-induced flavor defects are primarily due to 3-methyl-2-buten-1-thiol (cf. Table 5.3). Reactions related to a decrease in beer aroma are primarily oxidative in nature. For example, a "cardboard" aroma defect is due to the formation of trans-2-nonenal, an oxidative degradation product of linoleic acid. Hence, bottling beer under O_2-depleted conditions is very important. Bottled beer should contain no more than 1 mg O_2/l.

20.1.7.9 Foam Builders

The foam building properties of beer are due to proteins, polysaccharides and bitter constituents. The β-glucans stabilize the foam through their ability to increase viscosity. Addition of semi-synthetic polysaccharides, e.g. propyleneglycol alginate (4 g/hectoliter), to beer provides a very stable foam although the addition is judged as unfavorable.

20.1.8 Kinds of Beer

There is a distinction between top and bottom fermented beers.

20.1.8.1 Top Fermented Beers

Selected examples of top fermented beers from Germany are: Berlin weiss beer, brewed from a wort having 7–8% solids from barley and wheat malts and inoculated at fermentation with yeast and lactic acid bacteria; Bavarian weiss beer brewed from weakly-smoked barley malt with a little wheat malt and fermented only with yeast; Graetzer beer made from wheat malt with a smoky flavor and with a stemwort content of 7–8%; malt beer (caramel beer), a dark, sweet and slightly hop-flavored full beer; the bitter beers such as those from Cologne or Duesseldorf (Altbier) which are strongly hop-flavored full beers; top fermented plain beers (Jungbier or Frischbier) with a low stemwort content and often artificially sweetened; Braunschweig's mumme, an unfermented, non-hop-flavored malt extract, hence not a true beer or a beer-like beverage. English beers have a stemwort content up to 11–13%. Stout is a very darkly colored and alcohol-rich beer made from concentrated boiled wort (up to 25% stemwort; alcohol content >6.5%). Milder varieties of stout are known as Porter beer.
Pale ale is strongly-hopped light beer, whereas mild ale is mildly-hopped dark beer. Incorporation of ginger root essence into these beers yields ginger-flavored ale.
Top fermented beers from Belgium, which are stored for a longer time, are called Lambic and Faro beers.

20.1.8.2 Bottom Fermented Beers

These beers show a significantly increased storage stability and are brewed as light, mildly colored or dark beers.
Pilsener beer, an example of a light colored beer, is typically hop flavored, containing 11.8–12.7% stemwort. In contrast is the Dortmunder-type beer made from a more concentrated wort which is fermented longer and thereby has a higher alcohol content. Lager beer (North German Lager) is similar to Dortmunder in hop flavoring, while the stemwort content is close to a Pilsener beer. Munich beers are dark, lightly hop flavored and contain 0.5–2% colored malt and often a little caramel malt. They taste sweet, have a typical malt aromatic flavor, and are fermented with a stemwort content of 11–14%. Beers with a high content of extract are designated as export beers. Traditional beers, dark, and currently produced also as a special light beer, are the bock beers (Salvator, Animator, etc.). They are also strong beers, with more than 16% stemwort. The dark Nuernberg and Kulmbacher beers are even higher in colored malt extracts and thereby are darker than Munich beers. An example of mildly colored beer is the Maerzen beer (averaging 13.8% stemwort). It is produced from malt of Munich in which the use of colored malt is omitted.

20.1.8.3 Diet Beers

Diet beers are high fermentation-grade beers which contain practically no residual carbohydrates, which are a burden for diabetics. They are brewed by special fermentation processes and may contain relatively high levels of alcohol.

20.1.8.4 Export Beers

These originate from widely different kinds of beer. They are mostly pasteurized and additionally-treated with flocculating or adsorption agents (tannin, bentonite) or with proteolytic enzyme preparations to remove most of the proteins. The proteolytic enzymes split the large protein molecules into soluble products. Such beers are free of cloudiness or turbidity (chill-proofed beers) even after prolonged transport and cold storage.

20.1.9 Beer Flavor and Beer Defects

The pleasant flavor of a good beer is derived from carbon dioxide, the tannin content, bitter compounds of hops, esters, amino acids and some other normal beer constituents. Body depends on stemwort content. Foaming is also considered an important component of beer flavor assessment. Foam is perceived by its volume, which is related to the amount of carbon dioxide, by its density and, especially, by its stability. The stability is related to protein degradation products, the presence of bitter hop compounds and pentosans. Short-chain fatty acids, which are present in beer bouquet, act as natural antifoaming agents.

Beer defects (a term for faulty beer) involve odor and taste and arise primarily as a result of improper processing and/or storage. An example of a taste defect is the very harsh bitter taste caused by oxidized polyphenols and some hop ingredients. A flat taste, as already mentioned, originates from a low content of carbon dioxide. Beer is markedly light and oxidation sensitive (cf. 20.1.7.8). To overcome color and flavor deterioration related to beer oxidation, addition of ascorbic acid or glucose oxidase/catalase enzymes is recommended (cf. 2.8.2.1.1). Cloudiness, which can settle as a sediment, may develop during storage of beer. Proteins and polypeptides are 40–75% of this sediment. They become insoluble during storage due to reactions which form intermolecular disulfide bonds, or to polyphenol complex formation or the involvement of heavy metal ions (Cu, Fe, Sn). Other components of the sediment or of the turbidity are carbohydrates (2–15%), primarily α- and β-glucans, and polyphenols. For measures used to prevent cloudiness, see 20.1.8.4.

Infection of beer can result in total loss of the product. Contaminating organisms include: *Pediococcus cerevisiae* ("sarcina sickness"); lactic acid and acetic acid bacteria, which cause souring of beer; and contamination by cocci such as *Streptococcus mucilaginosus* or, even more probably, a pediococcus (*Pediococcus viscosus*), causing sliminess or ropiness. Mild contamination with these microorganisms may give beer an acidic, butter-like diacetyl flavor.

Table 20.5. Wine production (1,000 t) and wine consumption (l/capita)

Continent	Production		
	1969–71	1979	1981
World	28,884	37,618	31,335
Africa	1,568	1,081	997
America, North-, Central-	1,239	1,691	1,556
America, South-	2,658	3,614	3,004
Asia	158	223	238
Europe, West-	18,148	25,033	19,031
Europe, East- + USSR	4,835	5,600	6,109
Australia + South Pacific Islands	278	377	400

Country	Production (1981)	Consumption (1971)
Italy	7,650	111
France	5,791	107
Spain	3,331	60
USSR	3,200	
Argentina	2,075	85
USA	1,450	5
Rumania	1,020	23
FR Germany	750	18
Yugoslavia	670	27
Portugal	640	91
Σ (%)	85	

[a] World production ≙ 100%.

20.2 Wine

20.2.1 Foreword

Wine is a beverage obtained by full or partial alcoholic fermentation of fresh, crushed grapes or grape juice (must). The woody vine grape has thrived in the Mediterranean region since ancient times and Italy, France and Spain are still among the leading wine-producing countries of the world. Other major producers are Argentina, the USSR and the USA. Table 20.5 provides data for wine production and consumption in some countries.

20.2.2 Grape Cultivars

Among the cultivated species of *Vitis*, the most important is the grapevine *Vitis vinifera*, L. ssp. *vinifera* in its many forms; more than 8,000 cultivars are known. The size, shape and color of the grapes vary: there are round, elongated, large or small grape clusters. Grapes are either wine-type grapes, for white or red wine making, or table grapes, which are even grown in greenhouses in some northern countries. The cultivars are different in sugar content and aroma. Table 20.6 provides information about the major grape cultivars of Germany, with some of their characteristics. Table 20.7 shows the share of the major cultivars in German wine growing areas. Table 20.8 gives data on the grape cultivars of some other countries. The European *V. vinifera* and the American vines (*V. labrusca*) have been crossed in order to produce pest-resistant forms (hybrids, "direct producers"), giving plants with pest resistance and good quality must production, although the hybrids still leave much to be desired. The wines are considered rather ordinary, with less character and a more obtrusive flavor than the parent plants.

Grape cultivars providing top quality white wines are:

- *Riesling* – native to Germany; a hardy cultivar grown in the Pfalz (Rhine Palatinate) and along the Mosel (Moselle), Rhine and Nahe rivers.
- *Traminer* – cultivated extensively in Alsace, Baden and Pfalz, and in Austria.
- *Rulaender* (gray Burgundy) – from Alsace and Burgundy regions in the Kaiserstuhl district, and from Hungary.
- *Semillon Blanc* – together with Sauvignon and sometimes with Muscatel, provides Sauternes from the Bordeaux region.
- *Sauvignon* – used for Sauternes, and processed into its own types of wine, such as in the Loire region.
- *White Burgundy* (Pinot blanc) – yields the white wines from Burgundy (Chablis, Meursault, Puligny-Montrachet).
- *Chardonnay* – related to white burgundy, cultivated for example in Champagne.
- *Auxerrois* – also related to white burgundy.

Grape cultivars providing good white wines are:

- *Muscatel* and *Muscat-Ottonel* – cultivars with an exceptionally rich bouquet.
- *Furmint* – the grape cultivar of Hungarian Tokay wines.
- *Sylvaner* – grown in Pfalz, Rheinhessen and Franken regions of Germany.
- *Mueller-Thurgau* – grown widely in east Switzerland and in Germany; it is a cross between Riesling and Sylvaner.
- *Gutedel* (Chasselas, Fendant, Dorin) – often found in Baden, Alsace, West Switzerland, and France.
- *Scheurebe* – a favored cultivar in Germany, obtained by crossing Sylvaner and Riesling.
- *Morio-Muscat,* a cultivar of exceptional bouquet.
- *Veltliner* – of significance in Austria, as is
- *Zierfandler.*

Grape cultivars providing top quality red wines are:

- *Pinot noir* – the famous red vine cultivated in the Cote d'Or region of Burgundy, and also in Germany along the river Ahr and in Baden.
- *Cabernet-Sauvignon,*
- *Cabernet-Franc,* and
- *Merlot* – are cultivated together and provide the famous red wines of the Bordeaux region.

Other red grape cultivars are:

- *Gamay* – from the southern part of Burgundy and from Beaujolais and Maconnais.
- *Pinot meunier* – black Riesling; of importance in Champagne, Wuerttemberg and Baden.
- *Portuguese* – found in Pfalz, Rheinhessen, and Wuerttemberg.
- *Trollinger* (Vernatsch) – cultivated in south Tyrol and in Wuerttemberg.
- *Limberger* – found in Wuerttemberg and Austria.
- *Blue Aramon* – the cultivar which provides the wines from Midi, France.
- *Rossara* – widely cultivated in south Tyrol.

Grape vine cultivation requires an average annual temperature of 10–12°C. The average monthly temperature from April to October should not fall below 15°C. The northern limit for cultivating the grape vine is close to 50° latitude. The permissible altitude for cultivation is dependent on climate (plains in Italy, Spain and Portugal; sunny slopes of Germany; up to 1,300 m on Mt. Aetna in Sicily; up to 2,700 m in the Himalayas). Soil cultivability and quality and weather are of decisive importance.

Table 20.6. Important German grape cultivars

Cultivar	Area under cultivation		Wine type[a]	Acid[b]	Must weight[c]	Maturation characteristics[d]	Yield[e]	Comments about wine[f]
	(ha)	(%)						
White wine cultivars								
Auxerrois	109	0.1				4		A vivacious wine with distinquished bouquet
Bacchus	1,157	1.3	M	2	2	3	3	Flowery with a muscat note fragrance
Burgundy, white	802	0.9	S	3	2	4		A full-bodied wine, pleasantly aromatic and more neutral as Rulaender
Ehrenfelser	171	0.2	R	2	2	5	2	Fruity, mildly acidic, a riesling-like wine
Elbling, white	1,196	1.4	S	3	1	6	3	A light wine, devoid of rounded body and bouquet
Faber	1,157	1.3	M		2	2	3	A refined, refreshing and fruity flavored wine
Gutedel, white	1,221	1.4	M	1	1	3	3	Light wine, pleasing and captivating, mildly aromatic
Huxelrebe	746	0.8	T	2	2	2	3	A mellow wine with muscat-like bouquet
Kerner	2,116	2.4	R	2	2	4	2	A refreshing wine with a fine riesling-like bouquet
Morio-Muscat	2,777	3.2	B	2	1	3	3	A wine with strong captivating muscat aroma
Mueller-Thurgau	24,116	27.5	M		1	2	3	Mild and refreshing wine with fine muscat flavor
Muscatel-yellow	19	0.02	B	2				A superior wine with a strong muscat-like aroma
Muscat-Ottonel	25	0.03	B	2				A pleasing wine with a strong refined muscat bouquet
Nobling	105	0.1	S	2	2	2	2	A full-bodied wine with a fruity flavor and fine bouquet
Optima	216	0.2	A	2	3	2	2	A refined, captivating wine with a fragrant aroma
Ortega	350	0.4	B	2	3	1	1	A wine with refined peach-like aroma
Perle	276	0.3	T	1	2	3	3	A mellow wine with flowery bouquet
Riesling, white	18,351	20.9	R	3	1	5	2	A superior refreshing and pleasing wine, with a fruity and flowery flavor
Rulaender (gray Burgundy)	3,221	3.7	T	2	2	3		A body-rich wine with burning and passionate perception, and a pleasing bouquet
Scheurebe	2,529	2.9	T	3	2	4	3	A strong fruity flavored body-rich wine with a bouquet reminiscent of black currants
Siegerrebe	195	0.2	B	1	3	1	1	A wine with highly intensive refined bouquet

Table 20.6. (Continued)

Cultivar	Area under cultivation (ha)	(%)	Wine type[a]	Acid[b]	Must weight[c]	Maturation characteristics[d]	Yield[e]	Comments about wine[f]
Sylvaner, green	14,111	16.1	S	2	1	4		A mellow pleasing wine with a delicately fruity flavor
Traminer, reddish (Clevner)	881	1.0	T	2	2	4	1	A wine with an exceptionally strong persisting bouquet
Others		1.3						
White wine grand total	76,950	87.6						
Red wine cultivars								
Burgundy, blue, late-	3,086	3.5						Full-bodied, strongly flavored with a rounded bouquet, dark red mellow wine
Heroldrebe	165	0.2						A superior neutral with a tannin-like astrigency accentuated wine
Limberger, blue	376	0.4						Characteristically fruity, a somewhat herbaceous, tarty and finely astrigent wine
Muellerrebe (black riesling)	903	1.0						Reminiscent of late Burgundy, but of lesser quality
Portuguese, blue	4,062	4.6						A neutral mellow wine with a bouquet deficiency
Trollinger, blue	1,866	2.1						A mellow refreshing light wine with a pungent flavor and light-red in color
Others		0.6						
Red wine grand total	10,891	12.4						

[a] By quality German wines are classified into table wines (Tafelwein, Oechsle degrees less than 60), quality wines (with all the required characteristics of the growing region and an Oechsle degree of at least 60) and the high quality wines with a predicate (Oechsle degrees at least 73). The latter are denoted by increasing quality as Kabinett, Spaetlese, Auslese, Beerenauslese and for the top quality as Trockenbeerenauslese. In addition to a predicate the label might carry a designation as Eiswein (ice-wine, see text).
R: Riesling group of wine (superior, fruity wine with distinct acidity)
S: Sylvaner group (neutral wine devoid of a distinct bouquet)
M: Mueller-Thurgau group (light, flowery with discreet bouquet)
T: Traminer group (wine with a fine bouquet)
B: Bouquet group of wine (wine with exceptionally strong and rich bouquet)
A: Auslese group of wines (thick, fullbody great wines).

[b] 1: Low (approx. 5 g/l), 2: medium (approx. 5–10 g/l), and 3: high acidity (10–15 g/l).

[c] 1: 60–70 Oechsle degrees, 2: 70–85°, and 3: > 85 Oechsle degrees.

[d] 1: Very early maturing (beginning-middle of September), 2: early (middle-end of September, 3: early-medium (end of September, beginning of October), 4: medium late (beginning-middle of October), 5: late (middle-end of October), and 6: very late maturing cultivar (end of October beginning of November).

[e] 1: Low (60 hl/ha), 2: average (60–80 hl/ha), and 3: high yielding cultivar ($\geq$ 90 hl/ha).

[f] The wine organoleptic quality description has its own wine dictionary. Terms are classified and refer to wine (1) aroma or bouquet, (2) body, (3) sweetness and acids, (4) variety or cultivar, (5) age and (6) wine taste harmony (i.e. to which extent are the constituents of wine agreeably blended or related).

Table 20.7. Distribution of major grape cultivars as percent of total vineyard area in Germany (1968)

Vineyard area[a]	1[b]	2	3	4	5	6	7	8	9	10	11	12
Rheinpfalz	14	35	23				82		16			18
Rheinhessen	6	41	36				92		7			8
Mosel–Saar–Ruwer	77		12	11			100					0
Baden	7	8	24		13	13	77	22				23
Wuerttemberg	23	12	6				47		11	27	6	53
Rheingau	77	7	12				98	2				2
Nahe	27	33	28				99					1
Franken	4	48	38				99					1
Mittelrhein	85	4	8				99					1
Ahr	23		17				43	24	31			57

[a] By existing wine law in FR Germany eleven grape growing regions are established for production of quality wines. They are given on the label of the wine bottle. They are listed in this Table with the exception of the region "Hessische Bergstrasse".

[b] 1: Riesling, 2: Sylvaner, 3: Mueller-Thurgau, 4: Elbling, 5: Gutedel, 6: Rulaender, 7: white grapes grand total, 8: Blue late Burgunder, 9: Portuguese, 10: Trollinger, 11: Limberger, 12: red grapes grand total.

20.2.3 Grape Must

20.2.3.1 Growth and Harvest

After blooming and fruit formation, the grape berry continues to grow until the middle or the end of August, but remains green and hard. The acid content is high, while the sugar content is low. As ripening proceeds, the berry color changes to yellow-green or blue-red. The sugar content rises abruptly, while both the acid and water contents drop (Figure 20.2).

The harvest (picking the berry clusters from the vines) is performed as nearly as possible when the grape is fully ripe, about the middle of September until the end of November, or it may be delayed until the grapes are overripe. Terms which relate to the time of harvest include "vorlese", early harvest, "normallese", normal harvest, and "spaetlese", late harvest. The latter term, when applied to German wines, identifies excellent, top quality wines. Particularly well-developed grapes of the best cultivars from selected locations are picked separately and processed into a wine called "Auslese". When the grapes are left on the vine stock, they become overripe and dry – this provides the raisins or dried berries for "Beerenauslese", "Trockenbeerenauslese", or "Ausbruch" wine (fortified wine). In some districts, such as Tyrol and Trentino, the grapes are spread on straw or on reed mats to obtain shrivelled berries – this provides the so-called straw wines. Grapes that are botrytised (a state of "dry rot" caused by the mold *Botrytis cynerea,* the noble rot) have a high sugar content and a must of superior quality, consequently producing a superior, fortified wine. Frozen grapes left on the vine stock provide ice-must which, because of freezing, is enriched in sugar and, as such, is a source of high quality wines (ice-wines).

20.2.3.2 Must Production and Treatment

The grape clusters cut from vine stocks using grape shears are cleaned of rotten and dried berries and then, as fast as possible, separated from the stems. This is done in a roller crusher which consists of two fluted horizontal rolls by which the berries are crushed without breaking the seeds or grinding the stems. The latter are separated out by a stemmer. The crushed grapes are then subjected to pressing to release their juice, the must. The mechanical and partly continuously-operated presses are basket-type screw-presses (extruder-like tapered screw), hydraulic or pneumatic presses. The free-running must is collected prior to pressing (first run) and, after mild pressing, the major portion (pressed-must) is produced. The remaining grape skins and seeds (pomace) are loosened or shaken-up and pressed again. This provides the second or post-extract. In red wine making the crushed berries (the mash) are fermented without prior removal of the pomace, i.e. the must is fermented together with the skin. This is done in order to extract the red pigments localized in the skin and which are released only during fermentation. When blue grapes are processed in the same manner as white

Table 20.8. Major grape cultivars of selected countries

Country	Grape cultivar	Comments about cultivation area and quality
France	*White wine cultivars*	
	Aligote	Bourgogne, a "vin ordinaire", modest quality wine
	Chardonnay	Cultivated in Champagne and Bourgogne area (Chablis, Montrachet, Pouilly). A very good quality wine
	Chemin blanc	Cultivated in regions of Tourraine, Anjou and Loire
	Folle blanche	Wine used for brandy production in Cognac and Armagnac
	Grenach blanc	Midi
	Melon blanc (Muscadet)	Mellow refreshing wine with a slight muscat bouquet
	Muscadelle	Cultivated in Bordeaux and Charente regions, 5–10% blended into Sauternes and Graves wines
	Pinot blanc	Cultivated in Alsace, Champagne, Loire and Cote d'Or
	Pinot gris	Alsace wine
	Roussane (Rouselle)	Cultivated in Rhone region, a full-bodied, pleasing fragrancy wine
	Sauvignon	Wine of Bordeaux, Loire and Cher regions, a full-bodied fragrant wine, with Semillon used for production of Sauternes wine
	Semillon blanc	As Sauvignon, used for production of Sauternes wine
	Red wine cultivars	
	Cabernet Franc	Spread in Bordeaux and Loire regions, a superior, strong pleasing wine, with Cabernet Sauvignon and Merlot is an ingredient of Bordeaux wines
	Cabernet Sauvignon	As Cabernet Franc, aroma rich, a superior quality wine
	Carignan	Grown in Rhone, Midi and Provence regions
	Cot (Malbec)	Bordeaux, one of the best grape cultivars
	Cinsault	Grown in Southern France
	Grenach noir	Grown in Southern France
	Gamay noir	Beaujolais, Maconnais; fruity pleasant, refreshing wine
	Merlot	A Bordeaux wine, full-bodied, rich and mellow; as Cabernet Franc and Cabernet Sauvignon is an ingredient of Bordeaux wines
	Petite Verdot	Grown in Bordeaux region, component of Bordeaux wines
	Pinot noir	Bourgogne, and wine of Cote d'Or
	Syrah	Grown in Southern France
Italy	*White wine cultivars*	
	Malvasia, bianca	An important cultivar across Italy
	Mascato bianco	Grown mostly in Northern Italy, a wine of Asti region
	Trebbiano	Widely grown across Italy
	Vermentino	Grown along Italian riviera, a very good white wine
	Weissterlaner	Wine of South Tyrol
	Red wine cultivars	
	Aleatico	Widely grown in Italy
	Barbera	One of the most important cultivars
	Freisa	Grown in Piemont and Vercelli regions, one of the best Italian cultivars
	Gross-Vernatsch (Trollinger)	The wine of Bolzano, Trento and Como
	Lagrein	Grown in South Tyrol

Table 20.8. (Continued)

Country	Grape cultivar	Comments about cultivation area and quality
	Merlot	
	Nebbiolo	A prefered cultivar of Piemont and Lombardy regions
	Pinot Nero	Grown in Northern Italy with Rome as Southern limit
	San Giovese	Spread from Toscana till Latium; major constituent of Chianti wine
Austria	*White wine cultivars*	
	Mueller-Thurgau	
	Muscat-Ottonel	
	Neuburger	A pronounced cultivar bouquet, pleasantly acidic
	Rheinriesling	
	Rotgipfler	Fruity, aroma rich, full-bodied; together with Zierfandler an ingredient of Gumpoldskirchner wines
	Sylvaner	
	Traminer	
	Veltliner, green	A pleasant pleasing bouquet refreshing wine
	Veltliner, red	Fruity wine with a fine bouquet
	Veltliner (early red, Malvasier)	
	Welschriesling	A mellow wine with fine bouquet
	Zierfandler, red	A wine with burning and passionate perception, fragrant aromatic, with a cultivar specific bouquet
	Red wine cultivars	
	Burgundy, blue, late	
	Blaufraenkisch (Limberger)	
	Portuguese, blue	
	Sankt Laurent	A strong wine, dark red colored, with a fine Bordeaux-like bouquet
Switzerland	*White wine cultivars*	
	Gutedel (Chasselas, Fendant, Dorin)	A Swiss major grape cultivar
	Marsanne Blanche (Hermitage)	A mellow wine with a refined bouquet
	Riesling	
	Mueller-Thurgau	Major cultivar of Eastern Switzerland
	Red wine cultivars	
	Burgundy, blue	
	Gamay	Grown in Western Switzerland
	Merlot	The wine of Tessin (Ticino)
Hungary	*White wine cultivars*	
	Furmint, yellow	Used for Tokay wine production
	Red wine cultivars	
	Kadarka	The most important Hungarian red wine cultivar

ones, or blends of blue and white grapes are combined and then processed, pink wines are obtained. They are designated as rose wines. In red wine making the extraction of red pigments is sometimes facilitated by raising the temperature to 50 °C prior to fermentation of the mash, or to 30 °C after the main fermentation, followed by a short additional fermentation.

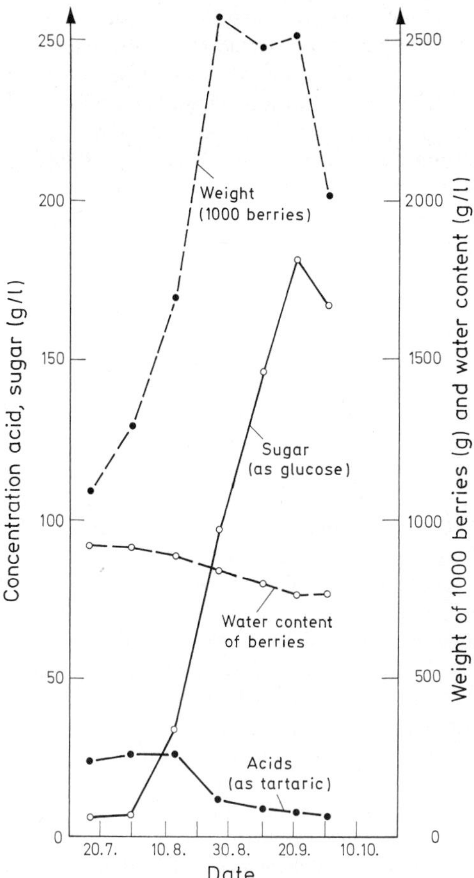

Fig. 20.2. Sylvaner wine grape ripening with measurement of the content of acid (as tartaric), sugar (as glucose), weight of 1000 berries and water content of the berries

The leftover stems, skins and seeds provide the pomace. It is used as feed or fertilizer, or is fermented to provide pomace wine. This is consumed as a homemade drink and is not marketed. Pomace brandy is obtained by distillation of fermented pomace. The average must yield is 75 l/100 kg grapes. Of this, 60% is free juice (must), 30% press-must and 10% must from the second pressing.

The fresh, sweet must is treated with sulfur dioxide to suppress oxidative discoloration and the growth of undesirable microorganisms. In order to remove undesirable odors or off-tastes, the must is treated with activated charcoal and, when necessary, is clarified by separators or filters. If required, the must is pasteurized by a short heat treatment (87 °C/2 min).

20.2.3.3 Must Composition

Table 20.9 provides data for the average composition of grape-musts. For the quality assessment of grape-must, its specific density has to be determined at 20 °C (must weight M). It is then usually expressed in *Oechsle* degrees (OD):

$$OD = (M-1) \times 10^3$$

The sugar contents (g/l) can be calculated from the following relationship (X is the acid content in g/l; 25 for a good vintage, and 30 for an average year):

$$S = 2.5\,OD - X$$

Carbohydrates. Ripe grapes contain equal amounts of glucose and fructose, while fructose predominates in overripe or botrytised berries. Saccharose content is much lower (10–12 g/l) and pectins (0.12–0.15%) and small amounts of pentosans are present.

Acids. The major acids of must are L-tartaric and L-malic acids. Succinic, citric and some other acids (cf. 18.1.2.4 and 20.2.6.5) are minor constituents. In a good vintage, tartaric acid is 65–70% of the titratable acidity, but in years when unripe grapes are fermented, its content is 35–40% and malic acid predominates. The good vintage year of 1911, for example, yielded grapes with 3.1 g/l malic acid and 6.4 g/l tartaric acid; in the inferior vintage year of 1912, on the other hand, malic acid was 10.7 g/l and tartaric acid 6.0 g/l.

Nitrogen compounds. Proteins, which include various of enzymes, peptides and amino acids, are present in low amounts (cf. 18.1.2.1).

Lipids. The lipid content of must is about 0.01 g/l (cf. 18.1.2.3).

Table 20.9. Average composition of grape must

Constituent	Content (g/l)
Water	780–850
Sugar (as glucose)	120–250
Acids (as tartaric acid)	6–14
N-Compounds	0.5–1
Minerals	2.5–3.5

Phenolic compounds (cf. 18.1.2.5). Tannins occur primarily in stems, skin and seeds. In a carefully prepared white must, the tannin content is no more than 0.2 g/l. In contrast, red wines contain high levels of tannin, 1–2.5 g/l or even higher. The pigments in must are anthocyanins.

Minerals. Must contains predominantly potassium, followed by calcium, magnesium, sodium and iron. Important anions are phosphate, sulfate, silicate and chloride.

Aroma substances. The must aroma substances will be discussed together with wine aroma substances (cf. 20.2.6.9).

20.2.4 Fermentation

Wine fermentation may occur spontaneously due to the presence of various desirable wine yeasts and wild yeasts found on the surface of grapes. Fermentation can also be conducted after must pasteurization by inoculation of the must with a pure culture of a selected strain of wine yeast. Wild yeast include *Saccharomyces apiculatus* and *exiguus,* while the pure selected yeast are derived from *Saccharomyces cerevisiae var. ellipsoides* or *pastorianus.* The pure wine yeast possesses various desirable fermentation properties. High fermenting strains are used to give high alcohol wines (up to 145 g/l) and those which are resistant to tannin and high alcohol levels are used in red wine fermentation. Other types of yeast are "sulfite yeast" with little sensitivity to sulfurous acid (sulfur dioxide solutions), "cold fermentation yeast", which are active at low temperatures and, finally, special yeast for sparkling wines, which are able to form a dense, coarse-grained cloudiness that is readily removed from the wine. Such yeast are added to must held in fermenters (vats made of oak, or chromium-nickel steel tanks lined with glass, enamel or plastic). The must is then fermented slowly for up to 21 days at 12–14 °C for white wines or 20–24 °C for red wines. When must has not been treated extensively with sulfur dioxide, the primary fermentation starts within a day and reaches its maximum after 3–4 days.

As a safeguard against air (discoloration), bacterial spoilage (acetic acid bacteria) and also to retain carbon dioxide, the liquid loss in the fermenter is compensated for by topping up with the same wine. After the end of main (primary) fermentation, which lasts 5–7 days, the sugar has been largely converted to alcohol while the protein, pectin and tannins, along with tartrate and cell debris, settle along with yeast cells to the bottom of the fermenter. This sediment is called bottom mud, dregs or lees. Partial precipitation of tartaric acid in the form of potassium acid tartrate (cream of tartar) is affected by temperature, alcohol content and pH (Figure 20.3). Tartaric acid also precipitates to a small extent as calcium tartrate. The unfermented residual sugar (residual sweetness) may be retained when necessary, if the secondary fermentation is suppressed by addition of sorbic acid or diethylpyrocarbonate or dimethylpyrocarbonate. Fermentation stops at an ethanol concentration of 12–15% (v/v).

The young wine, which is drunk in some regions of Germany and Austria ("Federweisser" or "Sauser"), is kept in casks for a few weeks up to several months, during which time secondary fermentation occurs. In such wines the residual sugar is degraded, valuable aroma substances (the bouquet) develop and some additional dregs, such as yeast cells and cream of tartar, sediment out.

Red wine mash is fermented at somewhat higher temperatures by using various procedures, often in closed double-walled enamel-lined tanks and often by treating extensively with sulfur dioxide. The wine initially drawn off is the better quality free-run wine, followed by the pressed wine, an astringent and dry fraction ("press-wine"). The young wine should not stay on the pomace longer than necessary to extract the pigments, otherwise it will become tannin-enriched and hence harsh and astringent. In industrial production the ex-

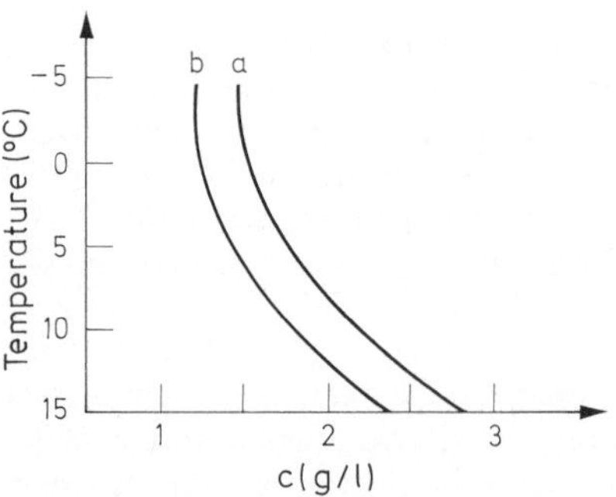

Fig. 20.3. The effect of temperature and ethanol concentration on cream of tartar solubility in wine. (a) 8 vol-%, (b) 12 vol-%. (After *Vogt,* 1974)

traction of the red pigments is not done by fermentation of the mash but by heat treatment of the mash, as outlined above.

The fermentation residue, or pomace, is processed into yeast-pressed wine or yeast-brandy, into wine oil (for brandy essence) and into tartaric acid. The left-over pomace is used as an excellent feed or fertilizer. Pomace wine, obtained by fermenting a sugar solution containing the dispersed pressed-out pomace, is made only into a household drink and is not marketed.

20.2.5 Cellar Operations after Fermentation; Storage

The following cellar operations develop a particular character in the wine and give it stability and durability.

20.2.5.1 Racking, Storing and Aging

Racking of young wine is required to get rid of the sediment. The wine is drawn-off or decanted into large sulfur-treated vats, with or without aeration. The time for racking is determined by the cellar master's experience. The wine racking is repeated as is required. It is good to do it as early as possible. When necessary, 5–10% of unfermented sterile grape must is blended with the young wine to round-off and sweeten its body.

The objective of wine aging/storage is to further build up the aroma and flavor constituents. Aging requires various lengths of time. In general the wine is removed from vats after 3–9 months and poured into bottles in which aging continues. Duration of aging and storability differ and depend on wine quality. The great Burgundy and Bordeaux wines require at least 4–8 years in order to develop, while for an average German wine maximum development is achieved within 5–7 years or the maximum may even be passed within that time. Only great quality wines endure aging lasting 10–12 years or more without quality loss.

During wine aging or maturation, one of the more important reactions which contribute to flavor is the partial degradation of malic acid to lactic acid, i.e. malo-lactic fermentation under the influence of diverse organisms such as *Bacterium gracile, Micrococcus malolactius, M. acidovorax* and *M. variococcus:*

$$\mathrm{HOOC{-}CHOH{-}CH_2{-}COOH} \longrightarrow \mathrm{HOOC{-}CHOH{-}CH_3} + \mathrm{CO_2} \quad (20.6)$$

This degradation reduces wine acidity and mellows the wine. Table 20.10 provides information on the time course of this reaction.

Changes induced during wine maturation are not yet well understood. Reactions between wine ingredients, such as ethanol, acids and carbonyl compounds, which form the typical aroma components of wine, are covered in 20.2.6.9.

20.2.5.2 Sulfur Treatment

Crushed grapes (mash) or must are treated with sulfur immediately after grape crushing to preserve the constituents that are sensitive to oxidation, prevent enzymatic browning via phenol oxidation and suppress the growth of undesirable microorganisms (acetic acid bacteria, wild yeast, molds). Sulfur treatment of wines prior to the first racking serves the same purpose: wine stabilization. Sulfur treatment is achieved by burning sulfur sticks, by the addition of sulfites in the form of a 6% aqueous solution of sulfurous acid or its salts, or by adding liquid SO_2. Only a part of the added sulfurous acid remains as free acid. A portion is oxidized to sulfate, while another binds to sugars and carbonyl compounds. Use of the right amount of SO_2 is important for fermentation and aging and hence for the quality of the wine.

20.2.5.3 Clarification

Wine clarification is usually achieved by precipitation reactions, filtration or centrifugation (so-called "fining"). In blue-fining the excess metal ions which are responsible for metal-induced cloudiness (iron, copper and zinc) are precipitated by precisely calculated amounts of potas-

Table 20.10. Lactic acid formation during aging of white wine

Date	Total acid[a] (g/l)	Lactic acid[a] (g/l)
June	11.3	0.0
July	11.0	0.4
September	10.5	0.8
February	9.3	2.0
July	8.1	2.8

[a] Calculated as tartaric acid

sium ferrocyanide. The blue turbidity formed helps to eliminate the persistent protein turbidity (grayish and black casse). Wine fining with an excess of potassium ferrocyanide causes the formation of hydrocyanic acid. In other fining procedures, gelatin, isinglass (sturgeon bladder gelatin), egg albumen, tannin, iron-free bentonite, kaoline, agar-agar and purified or activated charcoals are added to the wine. This results in adsorption or precipitation of the substances causing cloudiness, the interaction products all being quick-settling coagulums. Iron salts are removed from wine by phytates, phenolic compounds by polyvinylpyrrolidone (detanninizing) and undesirable sulfur compounds by colloidal silver chloride.

The clarification by filtering involves pads of asbestos, cellulose, infusorial earth, and filter aids such as Hyflo Super Cel and Filter Cel. The filters are built either as leaf filters or as washable filter presses. Sterilizing filters (which remove microorganisms) are suitable for stopping fermentation and thus retaining a desired level of unfermented sugar at a selected stage of fermentation. High efficiency centrifuges (separators) are being increasingly used for rapid clarification in industrial wineries.

20.2.5.4 Amelioration

Must and wine amelioration is required when unfavorable weather in some years results in grapes with an excess of acids and a low sugar content. Such grapes would provide a must which could not be processed directly into a drinkable, palatable wine. Amelioration may be considered as a type of wine adulteration. The ameliorated wine should not contain more alcohol nor less acid than the wine of the same type and origin from a good vintage year. The usual procedures involved are the addition of sugar, de-acidification and wine blending.

Addition of sugar, for which regulations exist in most countries, is done before, during or after fermentation, though it is usually done at the must or grape crushing stage. Sucrose (dry sweetening), sucrose solutions (wet sweetening) or grape must concentrate are added. The bouquet (aroma) is not improved. Poor or inferior wine is not improved by amelioration.

De-acidification is achieved primarily by adding calcium carbonate, which may give either a precipitate of calcium tartrate or a mixture of calcium tartrate and calcium malate. Ion-exchange resins can also be used for de-acidification.

Wine blending is a suitable way of rectifying defects, refreshing old wines, deepening the color of red wines (table wines) and enhancing the bouquet or readjusting the low acid content, thus producing a uniform quality wine for market.

Other wine treatments involve, for example, the addition of tartaric or citric acid to low acid wines of southern European countries and addition of glycerol to enhance the body. Addition of gypsum or phosphate treatment, used to increase the color of red wines, is based on color improvement caused by a decrease in pH by $CaSO_4$ or $CaHPO_4$. Saturation with carbon dioxide does not provide substantial amelioration to flat or bland wines.

20.2.6 Composition

The chemical composition of wine varies over a wide range. It is affected by environmental factors, such as climate, weather and soil, as well as by cultivar and by storage and handling of the grapes, must and wine.

Within the scope of wine analysis, wine extract, alcohol, sugar, acids, ash, tannins, color pigments, nitrogen compounds and bouquet-forming substances are important. Hence, the value and quality of a wine is assessed through the content of ethanol, extract, sugar, glycerol, acids and bouquet substances.

20.2.6.1 Extract

The extract includes all the components of wine mentioned earlier, except the volatile, distillable ones. Many of the extract components are present in must and are described in that section; others are typical fermentation and degradation products. The extract content of 85% of all German white wines is about 20–30 g/l (average about 22 g/l), while the extract content of red wines is somewhat higher – German "Auslese" wines contain about 60 g/l; other sweet wines, 30–40 g/l.

20.2.6.2 Carbohydrates

Carbohydrates (0.03–0.5%) present in fully-fermented wine are small amounts of the hexoses glucose and fructose and of nonfermentable pentoses. Incompletely fermented wines contain higher concentrations of both hexoses, but substantially more of the slower fermenting fructose. The average ratio of glucose to fructose in the residual sugar of a wine is 0.58 : 1, but it varies to a great extent. Wines with less than 4 g/l of residual sugar are designated as dry. The pentose sugars which are present in fermented wines consist of 0.05–0.13% arabinose, 0.02–0.04% rhamnose, and xylose in trace amounts.

20.2.6.3 Ethanol

The ethanol content of wine varies over a wide range. Normal grape wines, according to vintage and cultivar, have an ethanol content of 55–110 g/l. Light wines contain 55–75 g/l; average wines, 75–90 g/l; and stronger wines 90–110 g/l or more. The ethanol content of strong wines from southern European countries is 110–130 g/l. An alcohol level above 144 g/l indicates addition of ethanol.

20.2.6.4 Other Alcohols

Methanol occurs in wines at a very low level (38–200 mg/l), but much more is present in the fermentation of pomace as a product of pectin hydrolysis. Brandy distilled from pomace often contains 1–2% methanol. Higher alcohols in wine are propyl, butyl and amyl alcohols which, together, constitute 99% of the wine fusel oil. Hexyl, heptyl and nonyl alcohol and other alcohols are present in small amounts. The average butylene glycol (2,3-butanediol) content is 0.4–0.7 g/l and is derived from diacetyl by yeast fermentation. Glycerol, 6–10 g/l, originates from sugars and gives wine its body and round taste. Sorbitol is found in very low amounts. D-Mannitol is not present in healthy wines, but is present in spoiled, bacteria-infected wines at levels up to 35 g/l.

20.2.6.5 Acids

The pH of grape wine is between 2.8 and 3.8. Titratable acidity in German wines is between 5.5 and 8.5 g/l (expressed as tartaric acid). Acid degradation and cream of tartar precipitation decrease the acid content of ripe wines. Red wines generally contain more acids than do white wines. The wines from Mediterranean countries are low in acid content. Wine acids from grapes are tartaric, malic and citric acids; from fermentation and acid degradation they are succinic, carbonic (carbon dioxide) and lactic acids and low amounts of some volatile acids. The presence of acetic and propionic acids and longer chain fatty acids, as well as an anomalous amount of lactic acid, is an indication of diseased wine.

20.2.6.6 Phenolic Compounds

The phenolic compounds were described under grape must and in the chapter "fruits" (cf. 18.1.2.5).

20.2.6.7 Nitrogen Compounds

The nitrogen compounds in must precipitate to a lesser extent by binding to tannins during grape crushing and mashing, while most (70–80%) of them are metabolized by the growing yeast during fermentation. Free amino acids (about 200–800 mg/l) are the major compounds which remain in wine.

20.2.6.8 Minerals

The mineral content of wine is lower than that of the must since part of the minerals is removed by the yeast and by precipitation as salts of tartaric acid. The ash content of wines is about 1.8–2.5 g/l, while that of must is 3–5 g/l. The average composition of ash, in %, is: K_2O, 40; MgO, 8; CaO, 4; Na_2O, 2; Al_2O_3, 1; CO_2, 18; P_2O, 16; SO_3, 10; Cl, 2; SiO_2, 1.
The iron (as Fe_2O_3) content of wine is 5.7–13.4 mg/l, but it can rise to much higher levels (20–30 mg/l) through improper processing of grapes.

20.2.6.9 Aroma Substances

In general, the volatile constituents of young wine are designated as aroma substances. The aroma changes into bouquet in a good wine during storage and maturation. The bouquet can be imprinted with a special character, e.g. through oxidation reactions, depending on maturation conditions. This is the case with southern wines such as Madeira, the bouquet of which is very typical due to the accumulated high levels of aldehydes and acetals. In contrast, the table wines of higher quality, when stored in bottles for longer periods of time, appear to develop their bouquet by reduction reactions. All in all, little is known about the development of bouquet.
The volatiles of in wine, about 400–600 compounds, with a total concentration of 0.8–1.2 g/l, originate in part from the grapes (primary aroma) and are formed in part during fermentation (secondary aroma). Most of the grape cultivars

Table 20.11. Free and glycosidically bound monoterpenes in Muscat of Alexandria grapes[a]

Monoterpene	Free (µg/kg fruit)	Bound (µg/kg fruit)	Odor threshold (µg/l; water)
Linalool	100	390	6
Geraniol	<5	330	130
Nerol	<5	170	400
α-Terpineol	<5	40	350
trans-Furan linalool oxide[b]	5	130	7,000
cis-Furan linalool oxide[b]	<5	<5	6,000
trans-Pyran linalool oxide	100	20	3,600
cis-Pyran linalool oxide	30	5	5,400
3,7-Dimethylocta-1,*trans*-5-dien-3,7-diol[b]	2,200	300	odorless

[a] The quantitative data relate 120-th day of the survey.
[b] Chemical structure: cf. 5.3.2.3.

used for wine production are neutral in aroma. However, there are aroma-rich grapes, e.g. the Muscatel, Pinot and Sauvignon cultivars.
It has been demonstrated that free and glycosidically bound monoterpenes occur in grapes. The latter have been elucidated as a mixture of disaccharide glycosides of several mono-, di- and polyhydroxylated monoterpenes (examples cf. 5.3.2.3).
The concentrations of the free volatile terpenes are estimated in Muscatel grapes (Table 20.11). In this fraction linalool, *trans*- and *cis*-pyran linalool oxides (I and II in formula 20.6a)

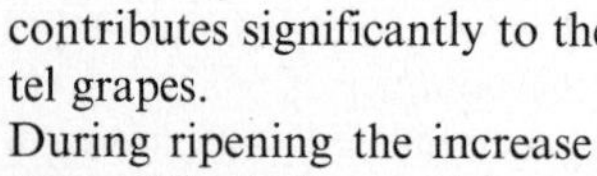

HO O (I) HO O (II) (20.6a)

are predominant (Table 20.11). A comparison of the aroma values (definition cf. 5.2.4.1) of these three compounds indicated that only linalool contributes significantly to the flavor of Muscatel grapes.
During ripening the increase in the concentrations of both free and bound monoterpenes is delayed in comparison to that of sugar and acids (Fig. 20.4). This demonstrates that a higher concentration of flavor compounds could be formed in the berries by leaving the fruit on the vine for extended periods.
A greater amount of the total monoterpenes in the grape juice is present as odorless glycosides, di- and polyols (Table 20.11). Juice processing techniques such as pH adjustment and heating enhances the hydrolysis of the glycosides to yield higher concentrations of linalool, geraniol, nerol and α-terpineol. In addition a broad pattern of aroma-active monoterpenes (e.g. nerol oxide, hotrienol) is formed by cyclization and dehydration reactions of di- and polyhydroxylated monoterpenes (examples, cf. 5.3.2.3). From the limited number of monoterpenes present in the grape juice numerous and/or larger quantities of aroma compounds are formed during the production of wines especially that of Muscatel cultivar.

Table 20.11 a. Free volatile and potential volatile terpene (T) contents of ripe grapes of some selected varieties.

Variety	Terpene content[a] (mg/l)	
	Free volatile T.	Potential volatile T.
Muscat of Alexandria	1.10	5.6
Traminer	0.54	1.9
Riesling	0.28	0.88
Chardonnay	0.16	0.22
Doradillo	0.18	0.58
Canada Muscat	0.96	4.1
Red Labrusca	1.2	1.4
Concord	0.84	1.3
Delaware	0.46	1.6

[a] Steam distillation of juice, first at neutrality, which yields free aroma compounds of the grape, and then at low pH, monoterpenes derived from the polyols and glycosidically bound forms are collected. Reaction of these distillates with a vanillin-sulphuric acid reagent gives a color, the intensity of which is proportional to the concentration of monoterpenes in the two fractions.

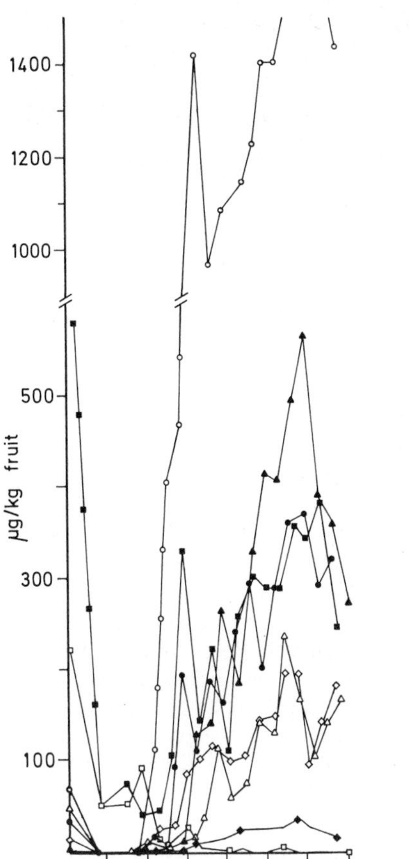

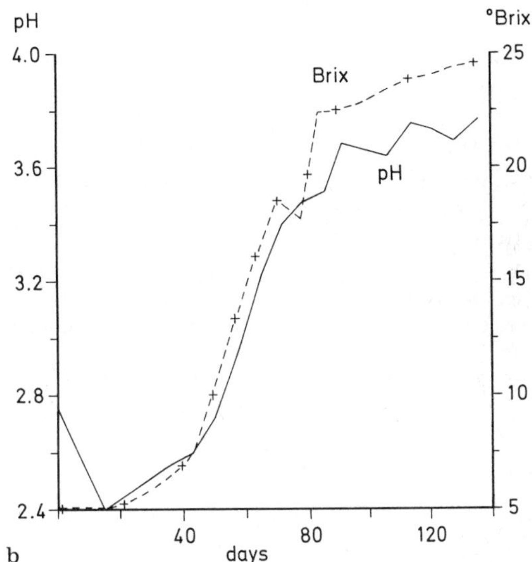

Fig. 20.4. Changes in concentration of monoterpenes, pH and sugar levels in the juice of developing Muscat of Alexandria grapes (After *Wilson et al.*, 1984)
a Concentration of free (open symbols) and glycosidically bound (closed symbols) monoterpenes.
○–○, ●–● 3,7-dimethylocta-1, *trans*-5-dien-3,7-diol; □–□, ■–■ geraniol; △–△, ▲–▲ linalool; ◇–◇, ◆–◆ trans-pyran-linalool oxide
b Juice pH and sugar levels
—— pH; –+–+ °Brix.

A simple procedure has been developed for the determination of the free volatile terpenes and the potential volatile terpenes in grape cultivars. The data listed in Table 20.11 a indicate the high flavor potential of two Muscatel cultivars.

The aroma during fermentation is due to the formation of higher alcohols and esters. Assessment of individual compounds for their sensory relevance is very difficult since the aroma is a complex mixture in which the sensory effect of one compound is influenced by the presence of the others.

Moreover, other factors, such as pH and sugar content (Table 20.12) are very important. Table 20.13 provides information about the concentration ranges of the higher alcohols and esters found in a number of white and red wines.

Higher alcohols are present in amounts which barely exceed their odor thresholds. At such low levels they are desired constituents of wine aroma. Higher levels (>400 mg/l) are perceived as being unpleasant.

Ethyl acetate is the most prominent ester con-

Table 20.12. Effect of medium on odor threshold values of some wine aroma compounds

Compound	Threshold value in solution[a]			
	1	2	3	4
Ethyl caproate	36	37	56	850
Ethyl acetate	25,000	40,000		160,000
Amyl alcohol[b]	1,900	12,500	60,000	180,000

[a] 1: Water, 2: 12% ethanol, with tartaric acid adjusted to pH 3, 3: solution under 2 + 100 g/l saccharose, 4: dry white wine.
[b] Blend (1:3) of 2-methyl-1-butanol with 3-methyl-1-butanol.

Table 20.13. Volatile compounds in wine

Compound	White wine (mg/l)	Red wine (mg/l)
Methanol	38–118	43–222
1-Propanol	9–48	11–52
2-Methyl-1-propanol	28–170	45–140
1-Butanol	1.4–8.5	2.1–2.3
2-Methyl-1-butanol	17–82	48–150
3-Methyl-1-butanol	70–320	117–490
1-Hexanol	3–10	3–10
Ethyl formate	0.02–0.84	0.03–0.20
Methyl acetate	0–0.11	0.08–0.15
Ethyl acetate	4.5–180	22–190
Ethyl propionate	0–7.50	0.07–0.25
Propyl acetate	0–0.04	0–0.08
Ethyl-2-methyl-propionate	0–0.60	0.03–0.08
2-Methylpropyl acetate	0.03–0.60	0.01–0.08
Ethyl butyrate	0.04–1.0	0.01–0.20
Ethyl-2-methyl-butyrate	0–0.02	0–0.08
Ethyl-3-methyl-butyrate	0–0.04	0–0.09
3-Methylbutyl acetate	0.04–6.10	0.04–0.15
Ethyl capronate	0.06–0.60	0.06–0.13
Hexyl acetate	0–0.63	0–0.60
Ethyl lactate	3.80–15	9–17
Ethyl caprylate	1.10–5.10	1.0–6.0
Ethyl caprinate	0.90–3.50	0.60–4.0
Diethyl succinate	0.01–0.80	
2-Phenylethyl acetate	0.20–5.10	
Ethyl laurate	0.10–1.20	
Ethyl myristate	0.10–1.20	
Ethyl palmitate	0.10–0.85	

stituent of aroma. Its threshold value in water is 25 mg/l. At a low level (50–80 mg/l) it contributes greatly to the pleasant bouquet of red wines. Ethyl succinate and ethyl lactate have threshold values of 100 mg/l and probably play no role in the aroma. Ethyl esters of caproic, caprylic, capric and lauric acids, each present in aroma at a concentration of one power of ten above their threshold values, are particularly important for the aroma of white wine. While the ethyl ester of caproic acid is fruity, the aroma quality decreases with increasing number of carbons of the acid, changing finally to a soap-like aroma. The content and composition of the ester fraction is greatly affected by fermentation conditions. The higher the temperature and the lower the pH during fermentation, the lower the ester concentration (Table 20.14).

Table 20.14. Effect of fermentation conditions on formation of higher alcohols and esters

Temperature (°C)	pH	Higher alcohols total (mg/l)	Fatty acid esters total (mg/l)
20	3.4	201	10.8
20	2.9	180	9.9
30	3.4	188	7.8
30	2.9	148	5.4

20.2.7 Spoilage

As with beer, defects in wine are reflected in appearance, odor and taste and, if not controlled, result in complete spoilage. A full explanation of all defects is beyond the scope of this book; hence only a general outline will be provided.

Of importance is browning due to oxidative reactions of phenolic compounds which, in red wine, may result in complete flocculation of the color pigments. This oxidative darkening process is as much chemical as enzymatic (polyphenol oxidases). Sulfurous acid is the preferred agent to prevent browning. Once the wine is affected by browning, it may be lightened by treatment with activated charcoal. The charcoal treatment can also remove other defects, such as the taste of mash or rotten grapes. Iron-induced turbidity (white or greyish casse) appears as a white, greyish-white or greyish haze or cloudiness and consists mostly of ferric phosphate ($FePO_4$). It is formed from the oxidation of ferrous compounds in wine. Proteins, tannins or pectins can participate in the build-up of such cloudiness (black casse). The so-called copper casse or turbidity is based on the formation of Cu_2S and other compounds with monovalent copper. It originates from the Cu^{2+} ions present in wine and their reduction in the presence of excess SO_2. The so-called "boeckser" of young wine is caused by the smell of hydrogen sulfide. The very unpleasant, rotten and yeasty "boeckser" (= mercaptan) odor is most objectionable and lingers for a long time. It is due to ethyl mercaptan, which can be removed by activated charcoal. The volatile sulfur compounds originate from sulfite which is reduced to H_2S by yeast, later reacting with ethanol to form ethyl mercaptan. Additional wine taste defects are the odd and disagreeable cork tastes which is due to the formation of 2,4,6-trichloroanisole (cf. 5.1.4). Above concentrations of 15–20 ng/l, it contributes to the cork flavor of wine.

Additional yeast spoilage is induced by species of the genera *Candida (Mycoderma)*, *Pischia* and *Hansenula (Willia)*. Other microorganisms are involved in the formation of viscous, moldy and ropy wine flavor defects. Bacterial spoilage may involve acetic acid and lactic acid bacteria. In this case vinegar or lactic acid souring is detectable. It has usually been associated with mannitol fermentation which may result in considerable amounts of mannitol are detected.

A "mousy" taint is occasionally detected in fruit and berry wines and, less often, in grape wines. It is discussed that 2-acetyl-1,4,5,6-tetrahydropyridine which has been also identified as important flavor compound of the bread crust (cf. 15.4.3.2.2) contributes to the "mousy" taint. The compound might be formed by microorganisms in wine.

Likewise, red wines, particularly color-deficient wines, show a microbiologically-induced change reflected in a substantial increase of volatile acids and the degradation of tartaric acid and glycerol. The bitter taste of red wines is caused by bacteria, mold and yeast. The bitter taste is usually a result of glycerol conversion to divinyl glycol. Cloudiness of red wines appears due either to bacterial or yeast spoilage or to physical reasons alone, such as the precipitation of cream of tartar. The latter occurs frequently and mostly in bottled wines. Cream of tartar precipitates as a result of oversaturation of the salt solution, as appears to be the case with protein-tannin interaction products. With oversaturation, they sediment as a fine greyish-yellow haze. Cloudiness caused by mucic acid salts also occurs.

20.2.8 Dessert Wines

Dessert wines are those with a high alcohol content which is not derived directly from the fermentation of fresh grape must, and/or are wines with high sugar content. They are made principally by two processes, which can to a certain extent be combined:

- The fermentation of grape juice of exceptionally high sugar concentration, or blending of concentrated grape juices with normal wines (concentrated dessert wine).
- When the must is sufficiently fermented, alcohol or concentrated must, mixed with alcohol is added (spirit-enriched dessert wine), stopping further fermentation.

Among the great number of dessert wines coming from southern European countries, several are worthy of mention: Hungarian dessert wines, Szamorodny, fortified Tokay and Tokay essence; and the French Haute Sauternes – they are concentrated dessert wines. Malaga and sherry wines originate from Spain; Port and Madeira from Portugal; and from Italy, Marsalla from Sicily, Calabrese from Syracuse and sweet Muscatel from Syracuse and Catania. Greek dessert wines include Samos and vino santo (from the island of Santorin). The extract, alcohol and sugar contents of dessert wines are given in Table 20.15.

Table 20.15. Composition of some dessert wines and top quality wines

	Extract (g/l)	Alcohol (g/l)	Sugar (g/l)	Glycerol (g/l)	Tritratable acid (g/l)[a]
Malaga	159.2	143.4	135.8	5.0	5.3
Portwine	67.6	166.5	47.0	2.8	4.5
Madeira	129.0	149.5	107.5		
Marsala	81.0	150.4	52.2	6.2	5.9
Samos	119.0	152.0	82.0	7.5	6.8
Tokay essence	257.5	84.4	225.3	4.1	6.5
Rheingauer top quality	140.6	107.7	99.4	14.3	10.2
Pfaelzer (Palatinate) top quality	171.6	86.7	121.3	10.5	11.6
Sauternes top quality	127.8	101.2	82.7		0.3

[a] Expressed as tartaric acid.

At least 2–5 years are needed to make dessert wines. In the production of sherry the wine is stored in partially filled butts, i.e. in the presence of excess air. Flor yeast develop on the wine surface in the form of a continuous film or wine cover (sherry yeast). The typical sherry flavor is derived from the aerobic conditions of maturation. During this time the concentrations of the following compounds increase at the expense of alcohol and volatile acids: ethanal, acetals, esters, acetoin and 2,3-butylene glycol. In port wine production the wine is drawn off to casks before the end of fermentation and is fortified with wine distillates. The fortifying procedure is repeated several times ("multiple addition") until the desired alcohol content is reached.

20.2.9 Sparkling Wine

Experience has shown that carbon dioxide imparts a refreshing, prickling and lively character to wine (as already mentioned for young wines). Hence, the production of a refined form of wine, enriched with carbon dioxide (sparkling wine) was developed and used in the early 18th Century, originally in the Champagne region of France ("Champagne" wine).

20.2.9.1 Bottle Fermentation ("Méthode Champenoise")

In the production of sparkling wine the young wines from suitable regions are used since fermentation of their grape juice in casks provides the special, fresh, fruity bouquet ("cuvé") desired. Blending of wines ("coupage") from different localities, often with older wines, is aimed at obtaining a uniform end-product. In this way clarified wine is then converted into an effervescent beverage by subjecting it to a second fermentation. Sugar is added (about 20–25 g/l) to wine, together with a pure yeast culture, for the purpose of attaining the desired final alcohol content (85–108 g/l) and carbon dioxide pressure (4.5 bar at 20 °C). Special yeast are selected which, in addition to being good fermenters and insensitive to carbonic acid, sediment as a firm, grainy precipitate ("depot") after fermentation is complete. The wine is bottled ("tirage") in such a way as to leave a small headspace of air and is then corked with a natural or plastic cork or, very often, with "crown" caps. The cork is finally and firmly secured with an iron clamp ("agrafe"). The bottles are stacked in cellars at normal temperatures (9–12 °C). The fermentation lasts several months, while the build-up of carbonation may go on longer, perhaps for up to 3–5 years. During this time, the carbon dioxide pressure within the bottle rises considerably. The sparkling wines are classified in France depending on the pressure: "grand mousseux", high pressure (4.5–5 bar); "mousseux", intermediate pressure (4–4.5 bar) and "cremant", low pressure (below 4 bar).

At this stage the sparkling wine is ready for yeast removal (disgorging). The bottles are restacked upside-down. Then the contents are repeatedly shaken until the yeast are lossened and settle on the cork. After 6–8 weeks the bottles are placed upright and the cork is removed using disgorging pliers. Simultaneously, the yeast are pushed out by the pressure from within the bottle. In order to simplify this production step, which is considered the most difficult step, the neck of the bottle is frozen to about −20 °C and the yeast are forced out as an "ice" plug. Because of this time consuming, costly procedure, the loss of wine, and other problems inherent in clearing the bottle of its yeast deposit, a "transfer system" has been introduced. The raw wine which has fermented in the bottle is emptied into a tank. The measured wine is filtered from the tank under pressure into a shipping/export bottle. The sparkling, yeast-free wine ("vin brut", dry wine) is then, depending on the market demand, supplemented with "liqueur" (dosage), quite commonly now a plain solution of candy or cane sugar. The bottle, with a headspace volume of 15 ml, is then corked and the cork is wired down. For further build-up of CO_2, the finished sparkling wine needs to be stored for additional 3–6 months. Wine with a small amount of liqueur added is designated as dry ("sec"); with more liqueur added as semi-sweet ("demi-sec"); and with more than 12% as sweet ("doux"). Sparkling wine for diabetics is sweetened with sorbitol.

20.2.9.2 Tank Fermentation Process ("Produit en Cuve Close")

With the aim of simplifying the costly and time consuming classical process, much of the sparkling wine production is now based on fermentation of wine in pressurized steel tanks instead of in bottles. The carbon dioxide-saturated wine is clarified and filtered and then chilled thoroughly and bottled. Fermentation is carried out at a pressure of about 7 bar over a 3 to 4 week period.

20.2.9.3 Carbonation Process

The carbonation of wine ("vin mousseux gacéifié") involves artificial saturation of wine with carbon dioxide, instead of the natural CO_2 developed during fermentation. Thus, the process is identical to the production of carbonated mineral water. The second fermentation, sugar addition and disgorging are omitted. However, sweetening with liqueur, corking and cork wiring are all retained.

20.2.9.4 Various Types of Sparkling Wines

Champagne is obtained by the bottle fermentation of wine from the French grape which grows in the region of Champagne. Sparkling wines produced in this region are the only ones that may be sold under the name of "Champagne". German sparkling wines are called "Schaumwein" and are commonly sold as "Sekt"; such Italian wines are "Spumante"; while in Spain and Portugal they are "Espumante".

Sparkling wines are also made from fruit and berry wines (apple, pear, white and red currant, bilberry). The process is that described above for carbonation.

Perlwine is a prematurely bottled wine which is left to ferment in the bottle.

20.2.10 Wine-Like Beverages

Compositions of some typical wine-like products are given in Table 20.16

20.2.10.1 Fruit Wines

For the production of fruit wine, pressed juice (fruit must) is made from apples, pears, cherries, plums, peaches, red currants, gooseberries, bilberries, cranberries, raspberries, hip berries and rhubarb. A general outline of this process is the same as making wine from grapes. Apple and pear mash are first pressed and the pressed juice (must) is fermented, while berry mash is fermented directly in order to extract the color pigments. Natural fermentation is suppressed by inoculation with pure, cultured yeast (cold-fermenting yeast). The vigor of the fermentation of berry musts, which are nitrogen deficient, is increased by addition of small amounts of ammonium salts. Lactic acid (3 g/l) is added to acid-deficient musts, such as that from pears, in order to achieve a clear ferment and, often, sucrose solutions are added to berry and fruit musts to alleviate acidity. The yield and quality of pome fruit must is improved by mixing 9 parts of fruit residue with 1 part of water and adding sucrose to raise the density of the must to 55 Oechsle degrees.

Fruit wines are produced industrially in many countries, e.g. apple wine, which is called cider in France, the UK and the USA, and pear wine, known as "poiré" in France. In Germany fruit wine is made along the Mosel river, around Frankfurt and in the state of Baden-Wuerttemberg. It is a popular beverage and is commonly called "plain must".

Table 20.16. Composition of some wine-like beverages[a]

Beverage	Alcohol	Extract	Acids[b]	Sugar	Minerals
Apple cider	58.4	23.4	3.8+	1.7	2.8
Cidre	51.0	29.7	2.8+	10.4	2.6
Pear wine	49.3	53.7	6.5+	9.0	4.1
Red currant cider	62.1	39.8	18.6*	1.8	4.0
Gooseberry cider	96.3	78.6	7.5*	55.8	1.8
Sour cherry cider	101.4	62.7	11.7*	3.8	3.61
Malt wine	70.6	24.5	4.6+	4.9	1.36
Malton sherry	123.0	115.2	8.1+	55.9	2.3
Mead	51.4	242.4	3.9+	208.0	1.34
Sake	121.2	28.6	5.7+	5.5	1.0

[a] Results are given in g/l.
[b] Acids are calculated as malic(+) or citric acid(*).

20.2.10.2 Malt Wine; Mead

Malt wine is made from fermented malt extract (the hot water extract of whole meal malt). Malton wine is made in the same way, except that sucrose is added at 1.8-times the amount of malt in order to increase the sugar and alcohol content of the wine. The wort is then soured by the action of lactic acid bacteria (0.6–0.8% lactic acid, final concentration). The acid fermentation is stopped by heating the wort to 78 °C and, after inoculation with a pure yeast culture, the wort is fermented to an alcohol content of 10–13%. The beverage thus formed has the character of a dessert wine, but is different because of its high content of lactic acid and its malt extract flavor.

Mead is an alcoholic liquor made of fermented honey, malt and spices, or just of honey and water (not more than 2 l water per kg of honey). Since early times, mead has been widely consumed in Europe and, even today, it is enjoyed the most of all the wine beverages in eastern and northern Europe.

20.2.10.3 Other Products

Other wine-like products include palm and agave wines ("Pulque"), maple and tamarind (Indian date) wines, and sake, the Japanese alcoholic drink made from fermented rice, which resembles sherry and is enjoyed as a warm drink.

20.2.11 Wine-Containing Beverages

Wine-containing beverages are made with wine, liquor wines and sparkling wines and, hence, they are alcoholic beverages.

20.2.11.1 Vermouth

Vermouth was first produced in the late 18th century in Italy (Vermouth di Torino, Vino Vermouth) and later in Hungary, France, Yugoslavia (Slovenia) and Germany. For the production of vermouth, wormwood (*Artemisia absynthium*) is extracted with the fermenting must or with wine, or it is made from a concentrate of plant extracts added to wine. Other herbs or spices are additionally used, such as seeds, bark, leaves or roots, as is the case with thyme, gentian or calamus, the sweet flag plant.

20.2.11.2 Aromatic Wines

These wines are similar to vermouth aperitif wines. They are flavored by different herbs and spices. Ginger-flavored wine is an example of this type of wine.

20.2.11.3 Prescription or Medicinal Wines

Such wines are pharmaceutical preparations. Some are quite bitter, especially those made from Cinchona bark, a source of the alkaloid quinine. Quinine-based wine is used in Europe and South America as a bitter aperitif. Kondurango, camphor and pepsin wines also belong to this class.

20.2.11.4 May Wines and Punches

May wine (a German summer drink) has wine as its base, sweetened with sugar and aromatized with the German Waldmeister plant (*Asperula odorata,* genus Woodruff) or its essence. May punch may also contain sparkling wine and mineral water or fruit wines.
Wine is the basis for other similar drinks, e.g. punches (mixtures of wine, sparkling wine, soda water and fruit such as pineapples, strawberries or peaches), drink mixes with soda water ("Spritzer", "Schorle-Morle"), or cold punch or Cardinal (a mixture of sparkling and red or white wine, mixed with sugar, orange, lemon or pineapple).

Table 20.17. Production of various alcoholic beverages in FR Germany (1983)

Spirits	Amount (10^6 l)	Spirits	Amount (10^6 l)
Grain distillates	84.4	Geneva gin	12.4
Wine brandy	83.4	Whiskey	4.3
Fruit brandy	8.2	Vodka	5.4
Liqueurs	54.2	Other distilled beverages	29.5
Rum and arrack	29.6		

20.2.11.5 Wine Punch

Wine punch is a mixed drink consisting of wine and rum or arrack, lemon juice, sugar, water and aromatic substances.

20.3 Spirits

20.3.1 Foreword

Spirits or liquors are alcoholic beverages in which the high alcohol concentration is achieved by distillation of a fermented sugar-containing liquid. Examples are distilled wines (brandies), liqueurs, punch extracts and alcohol-containing mixed drinks. Statistics on the production of liquor are presented in Table 20.17.

20.3.2 Liquor

The term liquor includes all liquids, even pure alcohol, which are obtained by fermentation followed by distillation. Some types of liquors contain flavorings.

20.3.2.1 Production

Liquors are produced by removing alcohol from an alcohol-containing liquid by distillation. Such liquids may already contain the alcohol, or alcohol is produced by the fermentation of a sugar-containing mash. The mash may include fermentable forms of sugars (D-glucose, D-fructose, D-mannose and D-galactose), or those forms are prepared by prior hydrolysis of di- and oligosaccharides (sucrose, lactose, raffinose, gentianose, melecitose, etc.) or polysaccharides.
The main raw materials are:

- alcohol-containing liquids (wine, beer, fruit wines, fermented milk);
- sugar-containing sources, such as sugar cane and beet, molasses, fruit and fruit products, fruit pomace, whey, palm extract and sugar-rich parts of tropical plants;
- starch- and inulin-containing raw materials (fruit, cereal, potato, topinambur, sweet potato, cassava, tapioca or chicory);

Saccharification of the starch-containing material is achieved with malt (green malt or kiln-dried malt), or by microbial amylases e.g. from the molds *Aspergillus niger* and *A. oryzae*. Distillation is performed in various ways, depending on the source and desired end-product. For the distillation of rum, arrack, fruit brandies and cereals, and brandy from wine, the apparatus is often a relatively simple still, used in such a way as to obtain a distillate which contains several other products of fermentation besides ethanol, or which contains the aroma substances of the starting raw material. These aroma substances are alcohols, esters, aldehydes, acids, essential oils and hydrogen cyanide. Repeated distillation is needed to obtain an alcohol-enriched distillate. In the production of pure or absolute alcohol the aim is the opposite: the final product being free from materials other than ethanol.

20.3.2.2 Alcohol Production

Alcohol used for drinks is made primarily from potatoes, cereals and molasses. Distiller's yeast, especially the top fermenting culture, is used for fermentation. Since the fermentation proceeds in an unsterilized mash and at elevated temperatures and since the growth of yeast occurs in mash acidified with lactic or sulfuric acid, the yeast must be highly fermentative, tolerant of elevated temperatures and resistant to acids and alcohol. In addition to saccharification by malt which contains mainly β-amylase, high-activity microbial α-amylases are also used. Molasses does not require saccharification. The saccharified mash is cooled to 30 °C and then inoculated with a yeast starter which has been cultured on a sulfuric or lactic acid medium of the mash. After 48 h of fermentation, the ethanol present at 6–10% by volume in the mash is distilled off along with the other volatile constituents. This step and the following rectification of the crude alcohol are achieved by continuous processes.
To facilitate the removal of the fusel oils, the crude alcohol is diluted to 15% by volume prior to rectification. The head product obtained from the rectification column consists nearly of pure ethanol (96.6% by volume) which is used for production of alcohol-fortified beverages. Large amounts of acetaldehyde, methanol and low boiling esters are present in the first runnings of the distillate, while the last runnings contain primarily fusel oil, other high alcohols, furfural and esters. These runnings combined with other intermediate fractions provide technical alcohol. The fusel oil, obtained in amounts of 0.1–0.5 l per 100 l alcohol, is used for technical purposes, while the distillation residue (the wash or stillage) is frequently used as animal feed. The yield of alcohol from 100 kg of mash starch is 62–64 l, i.e. about 89% of the theoretical value.
Technical alcohol is denatured or embittered to prevent its use for other than technical purposes, e.g. for drinking. Burning alcohol is denatured by addition of a mixture of methylethylketone and pyridine, alcohol for industrial use with other solvents, such as petroleum ether, camphor, diethyl ether or dyes.

20.3.2.3 Liquor from Wine, Fruit, Cereals and Sugar Cane

These beverages have a distinct taste and odor and contain at least 38% ethanol by volume. They are called natural, genuine or true liquors. The distillate resulting from a single distillation has a low alcohol content and often contains the specific odor and taste components of the starting material (harsh raw grain or harsh raw juniper liquor-gin).
In the production of liquor, the ultimate aim is to collect most of the desirable, specific fragrance and aroma substances (esters, essential oils) or to develop them (hydrogen cyanide, fermentation products, yeast oil) by using suitable mashing, fermentation and distillation processes. The freshly-distilled liquor has a hard, burning taste and unpleasant odor. It is improved by aging, which gives it a new, desirable aroma and flavor. Therefore, aging of liquor is of the utmost importance.

20.3.2.3.1 Wine liquor (brandy)

Brandy is distilled wine which contains at least 38% by volume of alcohol. Brandy to which alcohol is added is designated as a brandy blend or adulterated brandy.
The term "cognac" is restricted to brandy made in France in the region of Charente. The brandy produced in southern France, called Armagnac, is close in quality to cognac. Brandy production originated in France. Fermented grape juices (must) are distilled in very simple copper-pot stills on an open fire, often without prior removal of the yeast. The primary distillate (sectionnement) with a harsh, unpleasant odor is refined by repeated distillations ("repasse"). Brandy production soon spread to other countries (Germany, Russia, Spain, Hungary, the USA, Australia) and today brandy is frequently distill-

ed by a continuous process and its production has become a large-scale industry. In Germany imported wines which are fortified with wine distillate serve as starting material. The primary wine distillate contains 52–86% by volume ethanol and is considered as an intermediate product. It is used as the raw ingredient in the production of adulterated brandy by aging from 6 months to several years in wooden casks. Hard oak wood is used predominantly (barrels are made from "limousin" wood, holding about 300 l). Wild chestnut and other woods are also used. During aging, the wine distillate extracts phenolic compounds and colors of the wood, thus acquiring the typical golden-yellow and, occasionally, greenish-yellow color of brandy. Simultaneously, oxidation and esterification reactions mellow and polish the flavor and aroma. In order to improve quality, it is common to add an essence prepared by extraction of oakwood, plums, green walnuts or deshelled almond with a distillate obtained from wine. And also to add sugar, burnt sugar ("couleur") and 1% dessert wine to sweeten the brandy. In addition, treatment of brandy with clarifying agents and by filtering agents is also common. The desired alcohol content is obtained by dilution of brandy with water. The possibility of shortening the long and costly aging in casks has been repeatedly investigated. Attempts to age brandy artificially have included treatment with ozone or ion-exchange resin or application of ultrasound. None of them were successful.

20.3.2.3.2 Fruit liquor (fruit brandy)

Fruit liquors are also called cherry or plum waters or bilberry or raspberry spirits. Production of fruit liquor will be illustrated by cherry and plum liquors. Kirschwasser is made mostly in southern Germany (Black Forest's cherry water), France and Switzerland (Chriesiwasser). Whole fruits of the various sweet cherry cultivars are partly crushed together with the seeds and are pounded into a pulp. The fruit is left to ferment for several weeks, using a pure yeast culture. The fermented mash is then distilled in a copper still on an open fire or is heated with steam. During distillation the first and last fractions are separated. The main distillate contains 60% by volume or more alcohol. It is usually diluted with water to about 40–50% by volume alcohol and is marketed as clear, colorless brandy. The low levels of benzaldehyde and hydrogen cyanide which both contribute to the flavor are derived from the enzymatic cleavage of seed amygdalin. Kirschwasser, as is the case with Marasca from Dalmatia or Italy, is often used as an admixture in liqueur or cordial production, as with curacao, cherry brandy, maraschino, etc.

Plum brandy is produced from fully-ripe plums in a similar way to Kirschwasser, though no seed crushing is involved. It is often marketed under the name "Slivovitz" (from the old Slavic word sliva = plum). Besides Germany and Switzerland (Pfluemli water), major producers are Yugoslavia and the Balkan states, Czechoslovakia, Hungary and France. In addition to the common plum, the highly aromatic yellow plum, mirabelle, is also fermented. Mirabelle liquor is a desirable admixture to liqueurs containing fruit extract.

Fruit spirits are obtained from fresh or frozen fruit pulp or juice to which alcohol has been added prior to distillation. Fruits and berries used for this purpose are apricot, peach, bilberry, raspberry, strawberry, red currant, etc. Pome fruit liquor is obtained from freshly fermented apple or other pome fruits, either whole or crushed, or their juices, without prior addition of sugar-containing materials, sucrose or alcohol of some other origin. The alcohol content of liquor from pome fruits is at least 38% by volume. Hydrogen cyanide plays an important role in the chemical composition of fruit liquors of either stone or pome fruit. The cherry liquor sold on the market contains about 0.3–60 mg of hydrogen cyanide per liter of alcohol. In the same range are the concentrations of benzaldehyde (at least 20 mg/l) and the bouquet substances (about 7–15 mg/100 ml). Plum brandy contains less hydrogen cyanide (0.6–21.3 mg/l).

20.3.2.3.3 Gentian liquor ("Enzian")

Gentian brandy is a product obtained by distilling the fermented mash of gentian roots, or in which gentian distillate is used. The raw materials are the roots of many plants of the gentian family which, in the fresh state, contain substantial amounts of sugars (6–13%) in addition to the bitter glycoside-type compounds, such as gentiopicrin, amarogentin and others. The major production regions are the Alps (Tyrol, Bavaria, Switzerland) as well as the French and Swiss Jura mountains.

20.3.2.3.4 *Juniper liquor (brandy)*

Juniper brandy is obtained from pure alcohol and/or grain distillate by the addition of juniper distillate or its harsh, raw brandy. The use of juniper oil is uncommon. Juniper spirit is made exclusively from the distillate of whole juniper berries or from a fermented aqueous extract of juniper. The berries of *Juniper communis* are processed into brandy in Germany, Hungary, Austria, France and Switzerland. Pure juniper brandy is also used as an intermediate product; for the production of alcoholic beverages with a juniper flavor as, for example, in Geneva gin. The alcohol of this gin is obtained by distillation of a cereal mash prepared from kiln-dried smoked malt. Juniper brandy also flavors the Bommerlunder from the state of Schleswig-Holstein and the Doornkaat of East Friesland in Germany. Common gin is made from juniper distillates and spices, and contains at least 38% by volume alcohol. Dry gin has an alcohol content of at least 40% by volume.

20.3.2.3.5 *Rum*

Major rum-producing countries are in the West Indies (Jamaica, Cuba, Barbados, Puerto Rico, Guyana and Martinique) and also Brazil and Mauritius.

Rum production in sugar cane-cultivating regions uses sugar syrup and freshly pressed extract, often by the addition of such by-products as foam skimmings, molasses, press-skimmings and their extracts, and of distiller's wash ("dunder"), the residue left from a previous distillation. The sugar-containing solutions are diluted and allowed to ferment spontaneously at a maximum temperature of 36 °C and then are usually distilled in simple pot stills. Parts of aromatic plants are occasionally added to increase the aroma of the fermenting mash. This results in rum brands with different aromas. The quality of individual products fluctuates greatly. Especially highly regarded is Jamaican rum, which is marketed in various quality grades. A general classification divides them into drinking and blending types. Export rums have an alcohol content of about 76–80% by volume ("original rum"). Rum has the most intense aroma of all the distilled spirits enjoyed as drinks. This is acquired only after long aerobic aging in casks, by absorption of extracted substances from oakwood, and by formation of esters and other aroma constituents during aging.

Original rum contains about 80–150 mg acids per 100 ml, calculated as acetic acid. A great part occurs in free form as acetic and formic acids, the rest, along with other low molecular weight fatty acids, is esterified. The ester content and composition are of utmost importance for the assessment of aroma quality.

20.3.2.3.6 *Arrack*

Arrack is made from rice, sugar cane molasses, or sugar-containing plant juices (primarily from sweet coconut palm extract or its bloom spadix) by fermentation and subsequent distillation. Dates are used for the same purpose in the Middle East.

Countries which produce arrack are Indonesia (Java), Sri Lanka, India (Malabar coast) and Thailand. Well-known brands are Batavia and Goa arracks. In comparison to rum, arrack is not available in very many varieties; it is imported as the "original arrack" with an alcohol content of 56–60% by volume, from which "true arrack" is obtained by dilution with water to 38–50% by volume alcohol. At least a tenth of the alcohol in arrack blends must be from genuine arrack. Arrack is used for hot drink preparations, for Swedish punch, as an admixture for liqueurs, and in baking and as a seasoning ingredient in candy manufacturing. Batavia brand arrack, with an alcohol content of about 57% by volume, contains on the average 92 mg acids, 189 mg esters, 21 mg aldehydes and 174 mg higher alcohols per 100 ml of ethanol.

20.3.2.3.7 *Liquors from cereals*

Typical products are grain alcohol and whiskey (American and Irish brands are usually spelled with an "e", while Scottish and Canadian brands tend to use "whisky"). Different cereals (rye, wheat, buckwheat, oats, barley, corn, millet) are used. The cereals are first ground, mixed with acidified water, and made into an uniform mash by starch gelatinization. Saccharification is then accomplished by incorporating 15% kiln-dried malt in a premashing vat and stirring constantly at 56 °C. Saccharification proceeds rapidly through the action of malt diastase enzymes. The enzymes are inactivated by heating the mash to 62 °C. This step is followed by rapid cooling of the mash to 19–23 °C. The sweet mash is fermented by a special yeast and is then distilled. Grain liquors ar obtained by distilling the mash, while malt liquors commonly are produced by distillation of the wort. Simple stills are used for distillation in small plants, while both distillation and rectification are achieved on highly efficient, continuously-run column stills in industrial-scale production. According to the process used, the yield is 30–35 l of alcohol per 100 kg of cereal (e.g. rye), while the quality and character of the spirits vary greatly. Simple stills, with an unsophisticated separation of head and tail fractions, provide characteristic products rich in grain fusel oils. A modern distillery is able to remove the fusel oils to a great extent, yielding a high percentage grain alcohol, from which it is then possible to make a mellow, tasty, pure grain brandy with a subtle aroma. The final flavor of all these products is dependent on well-conducted aging in wooden casks.

Whiskey, depending on the kind, is made by different processes. The raw material for Scotch malt whiskey is barley malt which has been exposed to peat moss smoke during kiln drying. Such smoked malt is mashed at 60 °C and filtered. The resulting wort is then fermented at 20–32 °C. The distillation is conducted in two steps, sometimes in simple pot stills. The harsh, raw liquor is collected in the first distillation step. The undesirable harsh components are removed in the head and tail fractions in the second distillation.

In the production of Scotch grain whiskey the saccharified starch is distilled in continuous column stills. The character of the distillate is neutral, with less aroma than malt whiskey. In both Scotch whiskey processes, the distillates, with about 63% by volume ethanol, have to be stored/aged in order to develop their full aroma. This is best achieved by aging in old sherry casks or in charred casks. At the end of processing, the alcohol content is reduced to a drinkable level, about 43% by volume. Depending on the desired flavor or current preferences, the malt whiskey might be blended with grain whiskey ("blended whiskey").

American whiskey is made from corn, rye or wheat by saccharification with malt enzymes, fermentation of the wort, followed by double-distillation in column stills and aging, usually in charred oakwood casks. The corn distillate content of bourbon whiskey is at least 51% by volume and that of corn whiskey is at least 80% by volume. Rye whiskey contains at least 51% by volume distillate from rye, while wheat whiskey must be mostly distillate from wheat.

20.3.2.3.8 Volatile ingredients in liquor

In addition to ethanol, distilled spirits contain a great number of volatile constituents which originate from the raw material or arise as by-products of fermentation. Additional compounds may be formed during aging or maturation of liquors by interactions between the ingredients. Many of these volatile compounds are of great importance to an individual product and may fluctuate greatly in their nature and content, depending on the raw material and the process used. Table 20.18 gives a review of selected volatile compounds and their contents in some liquors, as determined by gas chromatography by direct sample injection, i.e. without a prior enrichment step.

Important groups of volatile compounds are:

- Methanol and higher alcohols.

  Methanol is found primarily in pectin-rich fruit and pomace wine liquors. Grain spirits or liquors are low in methanol. Higher alcohols are commonly present, though in highly variable amounts. Average values (in g/l) are, for example: cognac, 1.5; whiskey, 1.0; and rum, 0.6. Higher alcohols originate either from amino acids which, through oxidation or transamination, give keto acids; through decarboxylation, give the corresponding aldehydes; and through reduction give higher alcohols:

$$R{-}\underset{NH_2}{\underset{|}{C}H}{-}COOH \longrightarrow R{-}CO{-}COOH \longrightarrow R{-}CHO \longrightarrow R{-}CH_2OH \qquad (20.7)$$

  Higher alcohols are also generated during the biosynthesis of amino acids (cf. 5.3.2.1).

- Carbonyl compounds, acetals.

  Acetaldehyde is the most important carbonyl compound formed during alcoholic fermentation. Together with diethylacetal, acetaldehyde affects the drinkable quality of liquor even in very low amounts. Other carbonyls present in liquor are: propanal, isobutanal, pentanal, isopentanal, hexanal, diacetyl, 2,3-pentadione, acrolein, furfural, various ketones, vanillin, coniferyl- and *p*-hydroxybenzaldehyde. Some of these compounds leach from wooden casks during the aging of liquor.

Table 20.18. Volatile compounds in distilled spirits (average value in mg/100 ml pure ethanol)

Compound	Plum brandy	Cherry liquor	Pear (Williams) brandy	Cognac	German wine brandy	Grain spirit	Blended whiskey (Scotland)	Bourbon whiskey (USA)
Methanol	1,137	681	1,408	69	97	30	23	26
1-Propanol	146	806	134	52	41	0.4	44	29
1-Butanol	16	2.5	33	0.6	5	–	0.4	0.7
2-Butanol	23	44	16	2.5	13			
Isobutanol	86	47	56	112	63	+	69	81
2-Methyl-1-butanol	46	28	37	60	53	+	25	129
3-Methyl-1-butanol	143	98	119	218	150	0.8	50	207
1-Pentanol	0.4	+	0.6	–	+	–	+	+
1-Hexanol	2	0.8	11	2	1.5	–	+	0.4
1-Octanol	+	+	+	+	+	–	+	+
Benzyl alcohol	3	4.5	–					
2-Phenylethanol	3	1.5	0.7	3.5	2	+	2.2	7.5
Ethyl formate	0.5	0.5	1.5	3.5	2	+	1.1	2.0
Methyl acetate	9	7	33					
Ethyl acetate	204	295	151	50	55	0.9	26	71
Propyl acetate	–	4.5	–					
Isoamyl acetate	0.7	0.7	0.7	+	0.4	–	1.0	0.5
Hexyl acetate	+	–	–			–	+	+
Benzyl acetate	+	+	–					
2-Phenylethyl acetate	+	–	–	+	+	–	0.4	–
Ethyl propionate	4	4.5	0.6	1	3	–	0.4	0.3
Ethyl lactate	57	100	34	15	9	–	1.1	2.4
Isoamyl lactate	0.4	0.4	+	0.4	+	–	–	–
Diethyl succinate	2	2	+	0.8	0.6	–	+	+
Ethyl butyrate	0.5	+	+	+	1			
Ethyl capronate	0.7	0.4	0.3	0.9	0.8	–	+	0.4
Ethyl caprylate	1.5	0.8	0.5	2	2	–	0.9	1.5
Ethyl pelargonate	+	–	–			–	+	+
Ethyl caprinate	3	2	1	4	4	+	2.9	2.8
Ethyl-trans-2-decenoate	–	–	+					
Ethyl laurate	2	0.6	0.4	2.5	2	+	2.5	1.5
Ethyl-trans-2-cis-4-decadienoate	–	–	11					
Ethyl-trans-2-trans-4-decadienoate	–	–	5					
Ethyl benzoate	0.8	0.6	–					
Ethanal	18	16	17	21	53	1.5	7.0	8.6
Benzaldehyde	2.5	1	+	+ +	+ +	–	+	+
Furfural				+ +	+ +	–	0.8	2.0
Acetone	1	1	0.4	1	0.9	+	+	+
Diethylacetal	7	6	11	7	15	0.9	4.4	6.5
cis-Linalool oxide	+	+	+					
trans-Linalool oxide	0.3	+	+					
Terpineol	+	+	–					

- Organic acids.
  Acetic acid (40–95% of the total acids) is the predominant organic acid found in liquor. In addition, the following acids have been detected: propionic, isobutyric, isovaleric, valeric, caproic, caprylic, capric and lauric acids.

A characteristic constituent of rums appears to be 2-ethyl-3-methyl butyric acid.
The total acid content in mg/l is 200 for cognac, 100 for Scotch whiskey, 400 for bourbon whiskey and 600 for a rum with a good aroma.

- Esters.
  Esters, especially those derived from short chain acids and from aliphatic alcohols ("fruit" esters) play an important role in the odor and taste of distilled spirits. Ethyl acetate predominates, followed by the ethyl, isobutyl and 3-methylbutyl esters of lower fatty acids. Also, there are ethyl esters of higher fatty acids such as caprylic, capric and lauric and, in Scotch whiskey, palmitic acid. The effect of the type of process on the composition of the volatile fraction is illustrated by the fact that in brandy distillate the amount of higher esters of fatty acids is considerably greater when the distillation is run in the presence of yeast.
  The ester content of cherry and apricot distillates is 1.1–4.3 g/l. The various qualities of rum are based on the content of esters. The ester value gives the mg of esters, calculated as acetic acid ethyl ester, present in 100 ml of ethanol. According to aroma intensity, the ester values of Jamaican rums vary from 80 to 1,600.
- Other compounds.
  This group includes various phenols (*p*-methyl and *p*-ethylguaiacol, guaiacol, etc.), terpenes derived from essential oils, the bitter glycosidic compounds of gentian brandy (gentiamarin, etc.) and, finally, the nitrogen compounds (e.g. pyridines, picolines and pyrazines) found in rum and whiskey.

20.3.2.4 Miscellaneous Alcoholic Beverages

Many liquors are made "cold" by simply mixing the purified alcohols of various brands with water and are named according to the place of origin: Klarer, Weisser, East-German, etc. Such mixes often contain flavorings (seasonings, spices), e.g. freshly-distilled or aged grain liquor, extracts of caraway, anise, fennel, etc., as well as sugar, essence, essential oils or other flavoring substances. These products are designated as aromatized liquors. Some examples are:

Vodka (in Russian = diminutive of water) is made of alcohol and/or grain distillate by a special process. In all cases the characteristic smoothness and flavor must be achieved. The flavor should be neutral. The extract content is 0.3 g/100 ml and the alcohol content is at least 40% by volume.

Aquavit is a liquor flavored primarily with caraway. It is made from a distillate of herbs, spices or drugs and contains at least 35% by volume alcohol (potato alcohol or grain distillate). It is a favorite type of liquor in the Scandinavian countries.

Bitters are made from alcohol and bitter and aromatic plant or fruit extracts and/or their distillates, fruit saps and natural essential oils, with or without sugar, i.e. starch syrup. This group of products includes Boonekamp, bitter drops, English and Spanish bitters, and Angostura. The so-called "Aufgesetzter" is made of black currants and spirit or grain alcohol.

Absinthe is a liqueur flavored with aromatic constituents of wormwood and other aromatic plants. It becomes turbid after dilution with water.

Other products. Some special liquors of regional importance should be mentioned: tequila and mescal from Mexico and South America, made from fermented sap of the agave cactus; and liquors from the Middle East, made of sultana raisins, figs or dates.

20.3.3 Liqueurs (Cordials)

Liqueurs are alcoholic beverages with 20–35% by volume alcohol and 220–500 g/l sucrose or starch syrup, and flavored with fruit, spices, extracts or essences.

20.3.3.1 Fruit Sap Liqueurs

Fruit liqueurs contain the sap of fruits and this gives the liqueur its name. The lowest concentration of sap is 20 l per 100 l of end-product (25% by volume alcohol). Addition of natural aroma substances, caramel and some other colors is quite common. Examples of fruit liqueurs are pineapple, strawberry, cherry, blackberry liqueurs, etc. Cherry brandy, a special type of cherry liqueur, consists of cherry sap, cherry-water, sucrose or starch syrup, wine essence and water.

20.3.3.2 Fruit Aroma Liqueurs

These liqueurs are alcoholic beverages made of natural fruit essences, distillates or extracts. Use of synthetic aroma substances (with the exception of vanillin) is uncommon. Liqueurs of this type include apricot, barberry, rose hip, plum, lemon, etc. The designations "triple" or "triple sec" are used only for citrus liqueurs with at least 38% by volume alcohol.

20.3.3.3 Fruit Brandies

Fruit brandies are fruit sap and fruit aroma liqueurs with at least 5 l of fruit brandy (40% by volume ethanol) per 100 l of end-product. The fruit brandy selected designates the liqueur's name.

20.3.3.4 Other Liqueurs

Other liqueurs include:

Crystal liqueur, which contains sugar crystals (e.g., "crystal caraway").

Allasch, a special aromatic alcohol- and sugar-rich caraway liqueur with at least 40% by volume alcohol.

Ice liqueur, which is mixed and drunk with ice (e.g. lemon ice liqueur), and has an extract content of at least 30 g/100 ml and a minimum alcohol content of 35%. by volume.

Medoc cordial, which contains at least 35% by volume alcohol, at least 20% of which comes from wine distillate or wine liquor.

Gold water, a spice liqueur containing gold leaf as a characteristic ingredient.

Fragrant vanilla liqueur, the aroma of which is derived exclusively from pod-like vanilla capsules (vanilla beans).

Honey liqueur ("Baerenfang", "Petzfang", the "bear traps"), which has at least 25 kg of honey in 100 l of end-product.

Swedish punch, which is made of arrack and spices and has an alcohol content of at least 25%.

Cocoa, coffee and tea liqueurs, made from the corresponding extracts of raw materials.

Emulsion liqueurs, i.e. chocolate, cream and milk liqueurs. Mocca with cream liqueur, egg liqueur (the egg cream, "Advokat"), egg wine brandy, and other liqueurs with eggs added. The widespread and common egg liqueur is made from alcohol, sucrose and egg yolk.

Herb, spice and bitter liqueurs, made from fruit saps and/or plant parts, natural essential oils or essences, and sugar. Examples are anise, caraway, curacao, peppermint, ginger, quince and many other liqueurs.

20.3.4 Punch Extracts

Punch extracts or punch syrups, known simply as punch, are concentrates which are diluted before they are drunk. Rum or arrack punches contain 5% rum or 10% arrack, calculated relative to the total alcohol content. Aromatization with artificial rum or arrack essences, or with fruit ethers or esters, is not commonly done.

20.3.5 Mixed Drinks

Mixed drinks or cocktails are mixtures of liquors, liqueurs, wines, essences, fruit and plant extracts, etc. They are prepared immediately before their drinking in restaurants or bars, or are marketed as ready-made cocktail mixes or as their separate constituents.

20.4 Literature

Amerine, M.A., Berg, H.W., Cruess, W.V.: The technology of wine making. The AVI Publ. Co.: Westport, Conn. 1972

Buser, H.-R., Zanier, C., Tanner, H.: Identification of 2,4,6-trichloroanisole as a potent compound causing cork taint in wine. J. Agric. Food Chem. *30,* 359 (1982)

Dittrich, H.H.: Mikrobiologie des Weines. Verlag Eugen Ulmer: Stuttgart. 1977

Hillebrand, W.: Taschenbuch der Rebsorten. 5. Aufl., Zeitschriftenverlag Dr. Bilz und Dr. Fraund KG.: Wiesbaden. 1978

Hoffmann, K.M.: Weinkunde in Stichworten. Verlag Ferdinand Hirt: 1970

Horak, W., Drawert, F., Schreier, P., Heitmann, W., Lang, H.: Äthanol und Spirituosen. In: Ullmanns Encyklopädie der technischen Chemie. 4. Aufl., Bd. 8, S. 80, Verlag Chemie: Weinheim. 1974

Jounela-Eriksson, P.: The aroma composition of distilled beverages and the perceived aroma of whisky. In: Flavor of foods and beverages (Eds.: Charalambous, G., Inglett, G.E.), p. 339, Academic Press: New York–San Francisco–London. 1978

Kreipe, H.: Getreide- und Kartoffelbrennerei. 3. Aufl., Verlag Eugen Ulmer: Stuttgart. 1981

Macher, L.: Bier. In: Ullmanns Encyklopädie der technischen Chemie, 4. Aufl., Bd. 8, S. 462, Verlag Chemie: Weinheim. 1974

Molyneux, R.J., Wong, Yen-i.: High-pressure liquid chromatography in the separation and detection of bitter compounds. J. Agric. Food Chem. *21,* 531 (1973)

Narziß, L.: Abriß der Bierbrauerei. 4th Edn., Ferdinand Enke Verlag: Stuttgart. 1980

Nykänen, L., Suomalainen, H.: Aroma of beer, wine and distilled alcoholic beverages. D. Reidel Publ. Co.: Dordrecht–Boston–London. 1983

Palamand, S. R., Aldenhoff, J. M.: Bitter tasting compounds of beer. Chemistry and taste properties of some hop resin compounds. J. Agric. Food Chem. *21,* 535 (1973)

Pieper, H. J., Bruchmann, E.-E., Kolb. E.: Technologie der Obstbrennerei. Verlag Eugen Ulmer: Stuttgart. 1977

Pollock, J. R. A. (Ed.): Brewing Science, Vol. 1, 2. Academic Press: London–New York. 1979/81

Postel, W., Drawert, F., Adam, L.: Aromastoffe in Branntweinen. In: Geruch- und Geschmackstoffe (Hrsg.: Drawert, F.), S. 99, Verlag Hans Carl: Nürnberg. 1975

Rapp, A., Hastrich, H., Engel, L., Knipser, W.: Possibilities of characterizing wine quality and vine varieties by means of capillary chromatography. In: Flavor foods and beverages (Eds.: Charalambous, G., Inglett, G. E.), p. 391, Academic Press: New York–San Francisco–London. 1978

Ribereau-Gayon, P.: Wine flavor. In: Flavor of foods and beverages (Eds.: Charalambous, G., Inglett, G. E.), p. 355, Academic Press: London–New York. 1978

Rinke, W.: Das Bier. Verlag Paul Parey: Berlin–Hamburg. 1967

Strauss, C. R., Heresztyn, T.: 2-Acetyltetrahydropyridines – a cause of the "mousy" taint in wine. Chem. & Ind. *1984,* 109.

Tressl, R., Friese, L., Fendesack, F., Köppler, H.: Gas chromatography – mass spectrometric investigation of hop aroma constituents in beer. J. Agric. Food Chem. *26,* 1422 (1978)

Tressl, R., Friese, L., Fendesack, F., Köppler, H.: Studies of the volatile composition of hops during storage. J. Agric. Food Chem. *26,* 1426 (1978)

Troost, G.: Technologie des Weines. 5. Aufl. Verlag Eugen Ulmer: Stuttgart. 1980

Troost, G., Haushofer, H.: Sekt, Schaum- und Perlwein. Verlag Eugen Ulmer: Stuttgart. 1980

Vogt, E., Jakob, L., Lemperle, E., Weiss, E.: Der Wein. 7. Aufl., Verlag Eugen Ulmer: Stuttgart. 1977

Williams, P. J., Strauss, C. R., Wilson, B., Dimitriadis, E.: Recent studies into grape terpene glycosides. In "Progress in Flavour Research 1984" (Ed.: J. Adda) p. 349. Elsevier Sci. Publ.: Amsterdam. 1985.

Wilson, B., Strauss, C. R., Williams, P. J.: Changes in free and glycosidically bound monoterpenes in developing Muscat grapes. J. Agric. Food. Chem. *32,* 919 (1984).

21 Coffee, Tea, Cocoa

21.1 Coffee and Coffee Substitutes

21.1.1 Foreword

Coffee (coffee beans) includes the seeds of crimson fruits from which the outer pericarp is completely removed and the silverskin (spermoderm) is occasionally removed. The seeds may be raw or roasted, whole or ground, and should be from the botanical genus *Coffea*. The drink prepared from such seeds is also called coffee.

Coffee is native to Africa (Ethiopia). From there it reached Arabia, then Constantinople and Venice. Regardless of the prohibition of use and medical warnings, coffee had spread all over Europe by the middle of the 17th century. The coffee tree or shrub belongs to the family *Rubiaceae*. Depending on the species, it can grow from 3–12 m in height. The shrubs are pruned to keep them at 2–2.5 m height and thus facilitate harvesting. The evergreen shrubs have leathery short-stemmed leaves and white, jasmin-like fragrant flowers from which the stone fruit, cherry-like berries, develop with a diameter of about 1.5 cm. The fruit or berry (Figure 21.1) has a green outer skin which, when ripe, turns red-violet or deep red and encloses the sweet mesocarp or the pulp and the stone-fruit bean. The latter consists of two elliptical hemispheres with flattened adjacent sides. A yellowish transparent spermoderm, or silverskin, covers each hemisphere. Covering both hemispheres and separating them from each other is the strong fibrous endocarp, called the "parchment". Occasionally, 10–15% of the fruit berries consist of only one spherical bean ("peaberry" or "caracol"), which often brings a premium price.

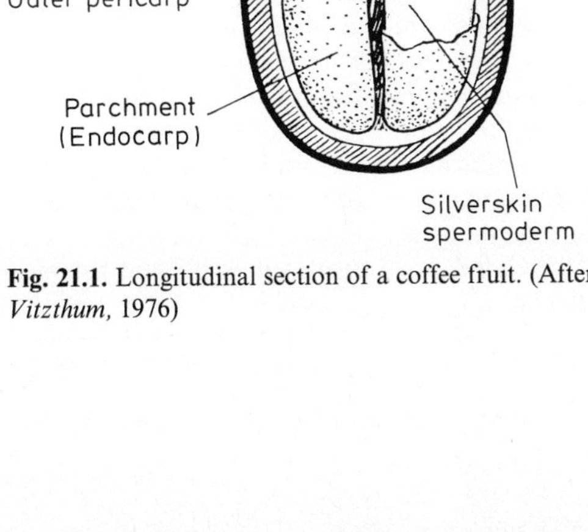

Fig. 21.1. Longitudinal section of a coffee fruit. (After *Vitzthum*, 1976)

The coffee shrub thrives in high tropical altitudes (600–1,200 m) with an annual average temperature of 15–25 °C and of moderate moisture and cloudiness. The shrubs start to bloom 3–4 years after planting and after six years of growth they provide a full harvest. The shrubs can bear fruit for 40 years, but the maximum yield is attained after 10–15 years. Fruit ripening occurs within 8–12 months after flowering. Only 3 of the 70 species of coffee are cultivated: *Coffea arabica,* which provides 75% of the world's production; *C. canephora,* about 25%; and *C. liberica* and others, less than 1%. The quantity (in kg) of fresh coffee cherries which yields 1 kg of marketable coffee beans is for *C. arabica* 6.38, *C. canephora* 4.35, and *C. liberica* 11.5. The most important countries providing the world's coffee harvest in 1981 are listed in Table 21.1.

21.1.2 Green Coffee

21.1.2.1 Harvesting and Processing

The coffee harvest occurs from about December until February from the Equator north to the Tropic of Cancer, while south of the Equator to the Tropic of Capricorn harvest occurs from May until August. Harvesting is done by hand-picking of each ripe berry or by strip-picking all of the berries from tree branches after most of the berries (often present as clusters) have matured. Harvesting may also be done by sweeping under the tree, i.e. collecting the ripe berries from the

Table 21.1. Production of coffee beans in 1981 (1,000 t)

Continent	Raw coffee	Country	Raw coffee
World	5,846	Brazil	1,878
Africa	1,288	Colombia	808
America, North-, Central-	945	Ivory Coast	350
America, South-	2,965	Indonesia	265
Asia	595	Mexico	217
Australia, + South Pacific Islands	53	Ethiopia	198
		Guatemala	173
		El Salvador	150
		Philippines	140
		India	131
		Uganda	130
		Σ (%)[a]	76

[a] World production ≙ 100%.

ground. Processing commences with removal of the fleshy pulp and using one of the two following processes:

The dry or natural process used in Brazil involves rapid transport of the harvested berries to a central processing plant or sun-drying terraces where the whole fruit is spread out and dried until the beans separate by shrinking from the surrounding parchment layer.

Dehulling machines – conical screws with a helical pitch increasing toward the discharge end – remove the dried husks and parchment from the dried berries and, as much as possible, the silverskin. The dehulled and cleaned coffee beans are then classified by size and packed in 60 kg bags. Often, the fresh cherries, instead of being spread on the drying terrace, are piled-up, left for 3–4 days under their own heat to ferment the fruity pulp, and are then processed as outlined below. In both cases unwashed beans are obtained.

The wet (washing) process is more sophisticated than the dry process, and by general consent leads to better quality coffee. The method is generally used for Arabica coffee (except in Brazil) in Central America, Colombia and Africa. The freshly-harvested berries are brought to a pulper in which the soft fruit is squeezed between a rotating cylinder or disc and a slotted plate, the gap of which is adjustable. The passage of the fruit produces a squeezing action which detaches the skin and the pulp from the beans without damaging the seed. The removed pulp is used as fertilizer. The pulped beans still have the silverskin, the parchment and a very adhesive mucilaginous layer (mucilage). Hence, such coffee is carried by the water stream to fermentation tanks made of concrete, the water is drained off and the beans are left to ferment for 12–48 h. During this time, the mucilaginous layer, which consists of 84.2% water, 8.9% protein, 4.1% sugar, 0.91% pectic substances and 0.7% ash, is hydrolyzed by enzymes of the coffee and by similar enzymes produced by microorganisms found on the fruit skins. The mucilage is degraded to an extent which can be readily dispersed by washing with water. The beans are then collected, sun-dried on concrete floors or dried in mechanical dryers in a stream of hot air (65–85 °C). Beans dried in this way are still covered with the parchment shell ("pergament" coffee or "cafe pergamino") and are further processed by dehulling machines as in the dry process. This yields the green coffee beans. Premium-priced coffee beans are often polished to a smooth, glossy surface and the silverskin except that retained in the centre-cut of beans is removed.

21.1.2.2 Green Coffee Varieties

About 80 varieties of the three coffee bean species mentioned above are known. The most important of the species *Coffea arabica* are *typica, bourbon, maragogips* and *mocca;* and of *Coffea canephora* are *robusta* (the most common), *typica, uganda* and *guillon.* All varieties of *Coffea canephora* are marketed under the common name *"robusta".*

The names of green coffees may be characteristic of the place of origin; i.e. the country and the port of export. Important washed Arabica coffees are, for example, Kenyan, Tanzanian, Colombian, Salvadorian, Guatemalan or Mexican.

Unwashed Arabica beans are the mild Santos and the hard Rio and Bahia beans. All three are from Brazil. Robusta coffees, mostly unwashed, are, for example, those from Angola, Uganda, the Ivory Coast and Madagascar.

Arabica coffees, particularly those from Kenya, Colombia and Central America, have a soft, rich, clean flavor or "fine acid" and "good body". The Arabica Santos from Brazil is an important ingredient of roasted coffee blends because of its strong but mellow flavor. Robusta coffee, on the other hand, is stronger but harsh and rough in aroma.

The quality assessment of green coffee is based on odor and taste assays, as well as on the size, shape, color, hardness and cross-sectioning of the bean. Major defects or imperfections are primarily due to objectionable off-flavored blemished beans, which are removed by careful hand sorting. Blemished beans consist of: unripe seeds (grassy beans) which stay light colored during roasting; overfermented beans with an off-flavor due to the presence of acetic acid, acetoin, diacetyl, butanol and isobutanol; frost-bitten and cracked beans; insect- and rainfall-damaged beans; and excessively withered beans. Even a single blemished bean can spoil the whole coffee infusion. Additional imperfections are the moldy, musty flavor of insufficiently-dried and prematurely-sacked coffee and earthy or hay-like off-flavors. Coffee varieties grown at high altitudes are generally more valuable than those from the plains or lowlands.

21.1.2.3 Composition of Green Coffee

The composition of green coffee is dependent on variety, origin, processing and climate. A review is provided in Table 21.2. The constituents will be covered in more detail in the section dealing with roasted coffee.

Table 21.2. Chemical composition of green coffee beans (% dry weight basis)[a]

Constituent	Content	
	Average value	Variation range
Water soluble extract	33	29.0–36.2
Protein		8.7–12.2
Lipids	12.6	8.3–17.0
Reducing sugars[b]		0–0.5
Reducing sugars after inversion[c]		2–9
Saccharose	6–7	
Crude fiber		10–11.7
Citric acid		0.5–1.15
Malic acid		0–0.5
Oxalic acid		< 0.2
Chlorogenic acids		4.5–11.1
Caffeine	1.45	0.9–2.6
Trigonelline	0.63	0.24–1.2
Minerals	4.0	3.0–5.4

[a] Moisture content 9.5% (5.0–12.1).
[b] Calculated as glucose.
[c] Calculated as saccharose.

21.1.3 Roasted Coffee

21.1.3.1 Roasting

Green beans are devoid of coffee aroma, so they must be heat treated in a process called roasting to bring about their truly delightful aroma. Roasting in the temperature range of 200–250 °C causes profound changes to occur. The beans expand their volume (50–80%) and change their structure and color. The green is replaced by a brown color, a 13-20% loss in weight occurs, and there is a build-up of the typical roasted flavor of the beans. Simultaneously, the specific gravity falls from 1.126–1.272 to 0.570–0.694, hence the roasted coffee floats on water and the green beans sink. The horny, tough and difficult-to-crack beans, become brittle and mellow after roasting.

Four major phases are distinguished during the roasting process: drying, development, decomposition and full roasting. The initial changes occur at or above 50 °C when the protein in the tissue cells denatures and water evaporates. Browning occurs above 100 °C due to thermal decomposition and pyrolysis of organic compounds, accompanied by swelling and an initial dry distillation; at about 150 °C there is a release of volatile products (water, CO_2, CO) which results in an increase in bean volume. The decomposition phase, which begins at 180–200 °C, is recognizable by: the beans being forced to pop and burst (bursting by cracking along the groove or furrow); formation of bluish smoke; and the release of coffee aroma. Lastly, under optimum caramelization, the full roasting phase is achieved, during which the moisture content of the beans drops to its final level of 1.5–3.5%.

The roasting process is characterized by a decrease of old and formation of new compounds. This is covered in section 21.1.3.3, which deals with the composition of roasted coffee. The running of a roasting process requires skill and experience to achieve uniform color and optimum aroma development and to minimize the damage through overroasting, scorching or burning.

The most generally used roasting apparatus is that of a horizontal rotating drum, for tumbling the green coffee beans in a current of hot air, either batch wise or continuously operated. Single passage air flow was general, but hot air re-circulation is now more usually adopted. The former is, however, generally associated with further combustion (after-burners) to minimize atmospheric pollution by the discharged gases, so that re-circulation of the gases is a logical extension for fuel economy.

The heat transfer in roasting occurs through contact of beans with the walls of the roasting machine and by convection stream of hot air or combusted gas. The roasting process (roasting time 5–15 min) strives to increase the convection component as much as possible by means of suitable process control. In a novel short-time roasting process (roasting time 2–5 min), the warm-up phase is significantly shortened by improved heat transfer. Water evaporation proceeds by puffing, thereby also providing a greater volume increase for the beans than any customary roasting process. Thus, the volume density of the ground coffee roasted by this process is 15–25% less.

The roasting process is controlled electronically or by sampling roasted beans. The end-product is discharged rapidly to cooling sifters or is sprinkled with water in order to avoid over-roasting or burning and aroma loss. During roasting the vapors formed and cell fragments (silverskin particles) are removed by suction of an exhauster and, in larger plants, incinerated.

There are different roasting grades desired. In the USA. and Central Europe, beans are roasted to a light color (200–220 °C, 3–10 min, weight loss 14–17%), and in France, Italy and the Balkan states, to a dark color (espresso, 230 °C, weight loss 20%). Sugar coating or glazing of coffee beans is becoming obsolete. Not more than 0.5 parts of glazing substance are used per 100 parts of raw coffee.

21.1.3.2 Storing and Packaging

Roasted coffee is freed of faulty beans either by hand picking on a sorting board or, at large plants, automatically, using photo cells. Commercially-available roasted coffee is a blend of 4–8 varieties which, because of their different characteristics, are normally roasted separately. Especially strong blends are usually designated as mocca blends.

While green coffee can be stored for 1–3 years, roasted coffee, commercially packaged (can, plastic bags, pouches, bottles), remains fresh for only 8–10 weeks. The roasting aroma decreases, while a stale, rancid taste or aroma appears.

Table 21.3. Composition of roasted coffee[a]

Constituent	Content (%)
Moisture	2.5
Protein[b]	9
Polysaccharide, water insoluble	24
Polysaccharide, water soluble	6
Saccharose	0.20
Glucose, fructose, arabinose	0.10
Lipids	13
Formic acid	0.10
Acetic acid	0.25
Nonvolatile acids[c]	0.40
Chlorogenic acids	3.7
Caffeine	1.2
Trigonelline	0.4
Nicotinic acid	0.02
Volatile aroma compounds	0.1
Minerals (ash)	4[d]
Unidentified constituents	35[e]

[a] Arabica-coffee of a normal roasting.
[b] Calculated as a sum of amino acids after acid hydrolysis; water soluble fraction amounts to 1.5%.
[c] Lactic, pyruvic, oxalic, tartaric, and citric acids.
[d] Water soluble fraction amounts to 3.5%.
[e] Water soluble fraction amounts to 7.5%.

Ground coffee packaged in the absence of oxygen (vacuum packaging) keeps for 6–8 months but, as soon as the package is opened, this drops to 1–2 weeks. Little is known of the nature of the changes involved in aroma and flavor damage. The changes are retarded by storing coffee at low temperatures, excluding oxygen and water vapor.

21.1.3.3 Composition of Roasted Coffee

Table 21.3 provides information about the composition of roasted coffee. This varies greatly, depending on variety and extent of roasting.

21.1.3.3.1 Proteins

Protein is subjected to extensive changes when heated in the presence of carbohydrates. There is a shift of the amino acid composition of coffee protein acid hydrolysates before and after bean roasting (Table 21.4). The total amino acid content of the hydrolysate drops by about 30% because of considerable degradation. Arginine, aspartic acid, cystine, histidine, lysine, serine, threonine and methionine, because they are especially reactive amino acids, are somewhat decreased in roasted coffee, while the stable amino acids, particularly alanine, glutamic acid and leucine, are relatively increased. Free amino acids occur only in traces in roasted coffee.

21.1.3.3.2 Carbohydrates

Most carbohydrates present, such as cellulose and other polysaccharides, the latter consisting of mannose, galactose and arabinose, are insoluble. During roasting a proportion of the polysaccharides are degraded into fragments which are soluble. Sucrose present in raw coffee is mostly decomposed in roasting coffee, as are monosaccharides.

21.1.3.3.3 Lipids

The lipid fraction appears to be very stable and survives the roasting process with only minor changes. Its composition is given in Table 21.5. Linoleic acid is the predominant fatty acid, followed by palmitic acid. The raw coffee waxes, together with hydroxytryptamide esters of various fatty acids (arachidic, behenic and lignoceric) originate from the fruit epicarp. These compounds are 0.06–0.1% of normally-roasted coffee. The diterpenes present are cafestol (I) and kahweol (II; see Formula 21.1). The diterpenes decline during storage of the green beans and they are extensively degraded during roasting although sufficient kahweol survives for Arabicas to be distinguished from Robustas by a spot test.

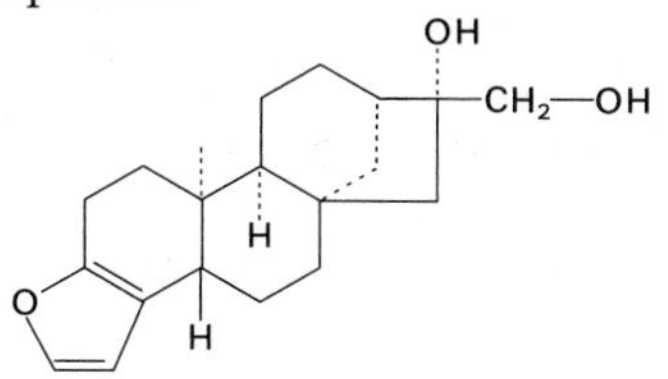

OH
CH_2—OH
H
O
H
I

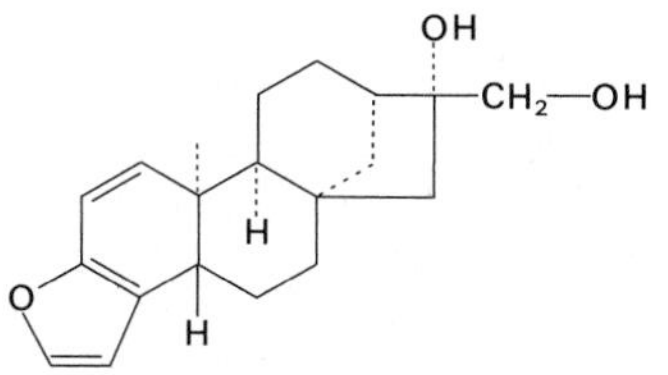

OH
CH_2—OH
H
O
H
II

(21.1)

Diterpene glycosides, including atractyloside (aglycon atractyligenin; cf. formula 21.2), occur in much higher levels in Arabicas (0.07–0.11%) than in Robustas (up to 0.007%). Although progressively degraded during roasting their residue passes into soluble coffee powders and beverage and might contribute to brew bitterness.

H
CH_3
=CH_2
HO
H
OH
H
COOH

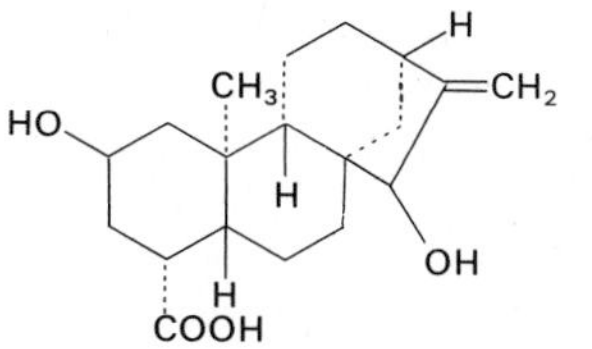

(21.2)

Sitosterol (43–57%), stigmasterol (20–28%) and campesterol (15–19%) are major compounds of the sterol fraction.

21.1.3.3.4 Acids

Formic and acetic acids predominate among the volatile acids, while nonvolatile acids are lactic,

Table 21.4. Amino acid composition of the acid hydrolyzate of Colombia coffee beans prior to and after roasting

Amino acid	Green coffee (%)	Roasted coffee[a] (%)
Alanine	4.75	5.52
Arginine	3.61	0
Aspartic acid	10.63	7.13
Cystine	2.89	0.69
Glutamic acid	19.80	23.22
Glycine	6.40	6.78
Histidine	2.79	1.61
Isoleucine	4.64	4.60
Leucine	8.77	10.34
Lysine	6.81	2.76
Methionine	1.44	1.26
Phenylalanine	5.78	6.32
Proline	6.60	7.01
Serine	5.88	0.80
Threonine	3.82	1.38
Tyrosine	3.61	4.35
Valine	8.05	8.05

[a] A loss due to roasting amounts to 17.6%.

Table 21.5. Lipid composition of roasted coffee beans (coffee oil)

Constituent	Content (%)	Constituent	Content (%)
Triacylglycerols	78.8	Triterpenes (sterols)	0.34
Diterpene esters	15.0	Unidentified compounds	4.0
Diterpenes	0.12		
Triterpene esters	1.8		

tartaric, pyruvic and citric. Higher fatty acids and malonic, succinic, glutaric and malic acids are only minor constituents. Itaconic (I), citraconic (II) and mesaconic acids (III) are degradation products of citric acid, while fumaric and maleic acids are degradation products of malic acid:

HOOC–CH₂–C(=CH₂)–COOH (I)

H_3C–C(COOH)=CH–COOH (II) HOOC–CH=C(CH₃)–COOH (III) (21.3)

Chlorogenic acid is the most abundant acid of coffee. During the normal roasting process, it is decomposed by about 30% and, during stronger roasting, by about 70% (Tables 21.1. and 21.3). The percentages of different isomers of this acid are provided in Table 21.6. Also found are low levels of free quinic, caffeic and ferulic acids and quinic acid esterified with ferulic or coumaric acids.

21.1.3.3.5 Caffeine

The best known N-compound is caffeine (1,3,7-trimethylxanthine) because of its physiological effects (stimulation of the central nervous system, increased blood circulation and respiration). It is mildly bitter in taste (threshold value in water is 0.8–1.2 mmole/l), crystallizes with one molecule of water into silky, white needles, which melt at 236.5 °C and sublime without decomposition at 178 °C. The caffeine content of raw coffee is 0.8–2.5%, while in the Robusta variety, it can be as high as 4%. In contrast there are caffeine-free Coffea varieties. Santos, an Arabica coffee, is on the low side, while Robusta from Angola is at the top of the range given for caffeine content. Caffeine forms, in part, a hydrophobic π-complex with chlorogenic acid in a molar ratio of 1:1. In a coffee drink, 10% of the caffeine and about 6% of the chlorogenic acid present occur in this form. The caffeine level in beans is only slightly decreased during roasting. Caffeine obtained by the decaffeination process and synthetic caffeine are used by the pharmaceutical and soft drink industries. Synthetic caffeine is obtained by methylation of xanthine which is synthesized from uric acid and formamide.

Table 21.6. Chlorogenic acids in roasted coffee beans

Compound	Content (%)
3-Caffeoylquinic acid (chlorogenic acid)	2.0
4-Caffeoylquinic acid (cryptochlorogenic acid)	0.2
5-Caffeoylquinic acid (neochlorogenic acid)	1.0
3,4-Dicaffeoylquinic acid (isochlorogenic acid a)	0.01
3,5-Dicaffeoylquinic acid (isochlorogenic acid b)	0.09
4,5-Dicaffeoylquinic acid (isochlorogenic acid c)	0.01

21.1.3.3.6 Trigonelline, nicotinic acid

Trigonelline (N-methylnicotinic acid) is present in green coffee up tp 0.6% and is 50% decomposed during roasting. The degradation products include nicotinic acid, pyridine, 3-methyl pyridine, nicotinic acid methyl ester, and a number of other compounds.

21.1.3.3.7 Volatile constituents

Molecular distillation of the oils which are obtained by pressing roasted coffee, produces a concentrate of aroma substances (0.1% of roasted coffee). Gas chromatographic separation of this concentrate or of a concentrate obtained by other methods yields numerous compounds and mixtures with very different odors (Figure 21.2). None of the close to 600 compounds identified has, by itself, the typical odor of coffee. The odor is, however, reconstituted when the compounds are blended together. Obviously, numerous individual compounds are responsible for the total sensory perception of coffee aroma. These compounds should be present in the mixture in the correct quantities and proportions.

The compounds identified so far belong to the various classes presented in Table 21.7.

The group of aliphatic compounds includes hydrocarbons, alcohols and, above all, carbonyl compounds which are derived during roasting from carbohydrate fragmentation. Also, numerous alicyclic compounds are found, for

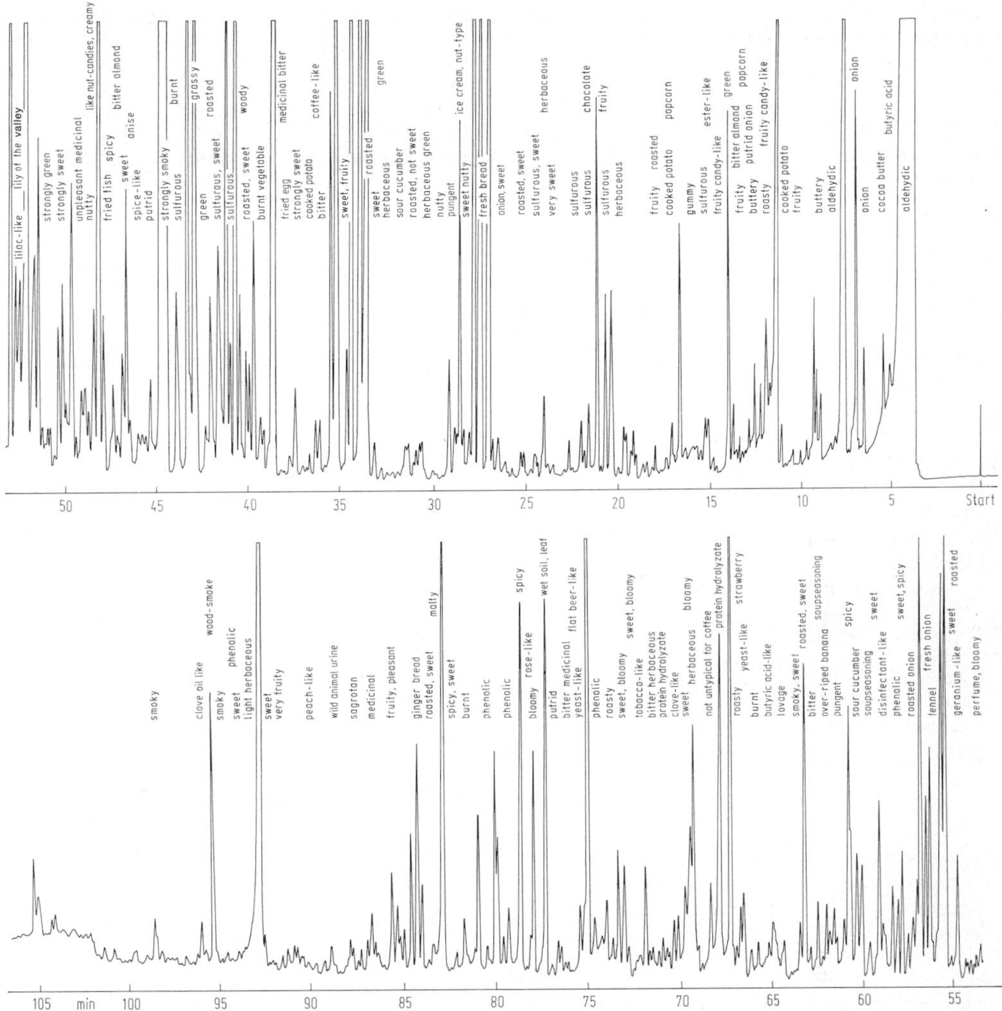

Fig. 21.2. Capillary gas chromatography of a coffee aroma concentrate with sensory evaluation of the effluent in a "sniffing-port". Glass capillary UCON HB 5100, temperature gradient 20–180 °C. (After *Vitzthum,* 1976)

example, cyclopentanone, cyclopenten-2-one, cyclohexen-2-one, cyclopentanedione-(1,2) and cyclohexanedione-(1,2).

Phenols are predominant among the aromatic compounds, and are derived most probably from thermal decomposition of chlorogenic acids (cf. 5.31.11). Phenol ethers, carbonyls, esters and polycyclic compounds are also found.

There is a substantial number of heterocyclic compounds, among which are many 2- and 2,5-substituted furanes, probably derived from the pyrolysis of sucrose and other sugars. In addition, many pyrroles, pyrazines, thiophenes, thiazoles and oxazoles are found. The cyclization of α-amino ketones appears to be responsible for pyrazine formation. The former are derived from the *Strecker* degradation reaction (cf. 5.3.1.10). β-Hydroxyamino acids should also be considered as possible precursors.

Oxazoles can be formed in a similar way (cf. 5.3.1.8). Thiazoles might result from the interaction of α-diketones and cystine, or be derived directly from acetylated cysteamine (cf. 5.3.1.7). The volatiles which are considered as flavor impact compounds of coffee are summarized in Table 21.7a.

Table 21.7. Volatile constituents of roasted coffee (arranged according to the compound classes)

Class of compound	Number	
Aliphatic compounds		148
Hydrocarbons	27	
Alcohols, ketoalcohols	19	
Aldehydes	17	
Acetals	1	
Ketones, diketones	38	
Thioketones	1	
Carboxylic acids	19	
Esters, ethers	13	
N-Compounds (amines, nitriles)	4	
S-compounds	9	
Alicyclic compounds		21
Aromatic compounds		55
Hydrocarbons	20	
Phenols	11	
Alcohols, aldehydes, ketones	11	
Ethers, esters	11	
S-compounds (thioether)	2	
Heterocyclic compounds		317
Furanes	92	
Lactones	8	
Acid anhydrides	4	
Pyrroles	37	
Indoles	3	
Pyridines	9	
Quinolines	2	
Pyrazines	70	
Quinoxalines	8	
Thiophenes	28	
Thiazoles	28	
Oxazoles	28	
Grand total		541

21.1.3.3.8 Minerals

As with all plant materials, potassium is predominant in coffee ash (1.1%), followed by calcium (0.2%) and magnesium (0.2%). The predominant anions are phosphate (0.2%) and sulfate (0.1%). Many other elements are present in trace amounts.

21.1.3.3.9 Other constituents

Brown compounds (melanoidins) are present in the soluble fraction of roasted coffee. They have a molecular weight range of 5–10 kdal and are derived from *Maillard* reactions or from carbohydrate caramelization. The structures of these compounds have not yet been elucidated.

Table 21.7 a. Possible coffee aroma impact compounds

Compound	Flavor description
Diacetyl	buttery
2,3-Pentanedione	buttery
2-trans-Nonenal	woody
1-Octen-3-ol	mushroom-like
3-Methylbutanal	burnt, mocca, slight earthy
Methylmercaptan	sulfur note
Kahweofuran[a]	roasted, smoky
Dimethylsulfide	buttery, creamy
Furfural	earthy, burnt, caramel
Alkyl pyrazines	roasted, nutty
2-Furylmethanethiol[a]	typical, burnt, roasted
Guaiacol	burnt, smoky
2-Methyl-4,5-dihydro-3(2H)-thiophenone	green, burnt
2-Acetyl-4-methylthiazole	toasted, anthranilic, burnt
Maltol	sweet, caramel, burnt
2-Hydroxy-3-methyl-cyclopent-2-en-1-one	sweet, caramel, burnt
4-Hydroxy-2,5-dimethyl-3(2H)-furanone[a]	sweet, caramel, burnt

[a] Chemical structure, cf. 5.3.1.3.

Apparently, chlorogenic acid is also involved in such browning reactions since caffeic acid has been identified in alkali hydrolysates of melanoidins.

Secondary products of the thermolysis of mixtures of carbohydrates and proteins, are probably involved in the formation of the bitter flavors of roasted coffee. This has been demonstrated by model systems consisting of mixtures of sugars and amino acids. Particularly intensive bitter tastes are obtained by heating sucrose and proline together (Table 21.8). Little is yet known about the structure of these bitter substances. Extracts of roasted coffee have been separated by gel chromatography into fractions with coffee taste. These fractions contain carbohydrates, organic and amino acids and trigonelline.

21.1.3.4 Coffee Beverages

In order to obtain an aromatic brewed coffee with a high content of flavoring and stimulant constituents, a number of prerequisites must be fulfilled; the brewing procedure, leaching and filtration, among them, give rise to a variety of combinations.

While in our society brewed coffee is enjoyed as a transparent, clear drink, in the Orient brewed

Table 21.8. Formation of bitter taste from roasted mixture of amino acids and sugars (molar ratio 1:1, roasted at 190 °C for 30 min)

Amino acid	Sugar	Yield[a] (g/100 g)	c_{Sbi}[b] (g/100 ml)	Tb[c]	Aroma description
L-Pro	Sac	75	0.01–0.03	3750	Cracker-like, roasted peanuts
L-Pro	Fru	59	0.02–0.03	2360	Cracker-like, roasted peanuts
L-Pro	Glc	39	0.03–0.04	1114	Cracker-like, roasted peanuts
D-Pro	Sac	69	0.01–0.03	3450	Cracker-like, roasted peanuts
L-Lys	Sac	71	0.04–0.05	1578	Alkaline, burnt
L-Lys	Fru	25	0.04–0.06	500	Alkaline, burnt
L-Lys	Glc	22	0.05–0.06	400	Alkaline, burnt
cyclo-Leu	Sac	60	0.07–0.09	750	Mildly burnt aroma
L-Met	Sac	76	0.11–0.13	633	Potato-like, roasty
L-Met	Fru	6	0.05–0.06	109	Potato-like, roasty
L-Met	Glc	19	0.18–0.22	95	Potato-like, roasty
L-Phe	Sac	63	0.12–0.14	485	Floral with cocoa aroma note
L-Phe	Fru	16	0.12–0.14	123	Floral with slight cocoa aroma
L-Phe	Glc	15	0.12–0.13	120	Floral
L-Val	Sac	78	0.18–0.22	390	Cocoa aroma
L-Val	Fru	20	0.13–0.15	143	Mild cocoa aroma, burnt
L-Val	Glc	28	0.20–0.22	133	Mild cocoa aroma, burnt
L-Thr	Sac	61	0.18–0.19	330	Caramel-like, smoky
L-Thr	Fru	18	0.16–0.18	106	Caramel-like, smoky
L-Thr	Glc	25	0.22–0.26	104	Caramel-like, smoky
L-Ala	Sac	80	0.30–0.34	250	Smoky
L-Ala	Fru	7	0.09–0.10	74	Smoky
L-Ala	Glc	7	0.11–0.12	61	Smoky
L-Leu	Sac	81	0.40–0.50	180	Cocoa aroma, mild sweet note
L-Leu	Fru	17	0.13–0.15	121	Mild cocoa aroma, smoky
L-Leu	Glc	16	0.15–0.16	103	Mild cocoa aroma, smoky
D-Leu	Sac	79	0.40–0.50	176	Cocoa aroma
D-Leu	Fru	16	0.15–0.17	100	Mild cocoa aroma, smoky
D-Leu	Glc	22	0.25–0.35	73	Smoky
Gly	Sac	84	0.45–0.55	168	Caramel-like
Gly	Fru	35	0.40–0.50	78	Mildly burnt, scorched
Gly	Glc	41	0.60–0.70	63	Mildly burnt
L-His	Sac	65	0.35–0.45	163	Caramel-like
L-His	Fru	34	0.25–0.35	113	Smoky
L-His	Glc	38	0.35–0.40	101	Smoky
L-Ile	Sac	69	0.40–0.45	162	Cocoa aroma
L-Ile	Fru	16	0.13–0.15	114	Mild cocoa aroma, smoky
L-Ile	Glc	22	0.20–0.24	100	Mild cocoa aroma, smoky

[a] Yield of water soluble substances is based on reaction system.
[b] Detection threshold value for a bitter taste.
[c] Total bitterness Tb, corresponds to the solution volume with c_{Sbi}-concentration per gram reaction mixture.

coffee is prepared from pulverized beans (roasted beans ground to a fine powder) and water brought to boiling and is drunk as a turbid beverage with the sediment (Turkish mocca). Coffee extract is made by boiling the coffee for 10 min in water and then filtering. In the boiling-up procedure the coffee is added to hot water, brought to a boil within a short time and then filtered. The steeping method involves pouring hot water on a bag filled with ground coffee and occasionally swirling the bag in a pot for 10 min. In the filtration-percolation method, ground coffee is placed on a support grid (filter paper, muslin, perforated plastic filter, sintered glass, etc.) and extracted by dripping or spraying with hot water, i.e. by slow gravity percolation. This procedure, in principle, is the method used in most coffee machines. In an espresso machine, which was developed in Italy, coffee is extracted briefly by superheated water (100–110 °C), while filtration is accelerated by steam at a pressure of 4–5 bar. The exceptionally strong drink is usually turbid and is made of freshly-ground, darkly-roasted coffee. The water temperature should not exceed 85–95 °C in order to obtain an aromatic drink with most of the volatile substances retained. Water quality obviously plays a role, especially water with an unusual composition (some mineral spring waters, excessively hard water, and chlorinated water) might reduce the quality of the coffee brew. Brewed coffee allowed to stand for a longer time undergoes a change of flavor. For regular brewed coffee, 50 g of roasted coffee/l (7.5 g/150 ml cup) is used; for mocca, 100 g/l; and for Italian espresso, 150 g/l. Depending on the particle size and brewing procedure, 18–35% of the roasted coffee is solubilized. The dry matter content of coffee beverages is 1–3%. The composition is presented in Table 21.9.

The taste of coffee depends greatly on the pH of the brew. The pH using 42.5 g/l of mild roasted coffee should be 4.9–5.2. At pH < 4.9 the coffee tastes sour; at pH > 5.2 it is flat and bitter. Coffees of different origins provide extracts with different pH's. Generally, the pH's of Robusta varieties are higher than those of Arabica varieties.

Table 21.9. Composition of coffee beverages[a]

Constituent	Content (% dry weight basis)
Protein[b]	6
Polysaccharides	24
Saccharose	0.8
Monosaccharides	0.4
Lipids	0.8
Volatile acids	1.4
Nonvolatile acids	1.6
Chlorogenic acids	14.8
Caffeine	4.8
Trigonelline	1.6
Nicotinic acid	0.08
Volatile aroma compounds	0.4
Minerals	14
Unidentified constituents (pigments, bitter compounds etc)	29.4

[a] Arabica-coffee, normal roast, 50 g/l.
[b] See footnote b) in Table 21.3.

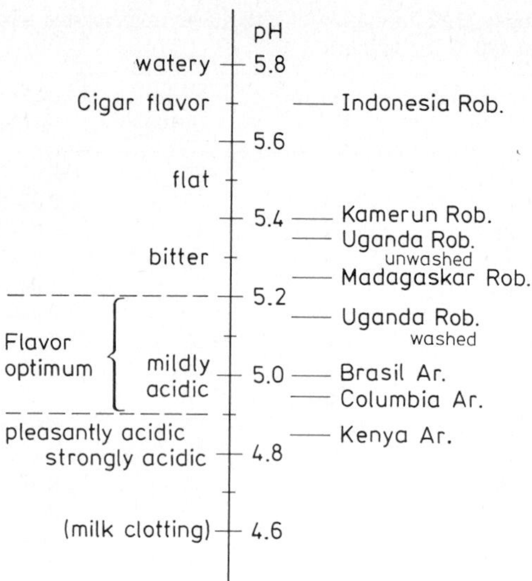

Fig. 21.3. The flavor of roasted coffee brew as related to pH value. (After *Vitzthum*, 1976)

Figure 21.3 shows the relationship between pH and extract taste for some coffees of known origin.

21.1.4 Coffee Products

The coffee products which will be discussed are instant coffee, decaffeinated coffee and those containing additives.

21.1.4.1 Instant Coffee

Instant (soluble) coffee is obtained by the extraction of roasted coffee. The first technically-sound process was developed by *Morgenthaler* in Switzerland in 1938. Ground coffee is batchwise extracted under pressure in percolator batteries or continuously in extractors. The water temperature may be as high as 200 °C, while the temperature of the extract leaving the last extraction cell is 40–80 °C. The dry matter content of the initial extract is 15% and this is concentrated to 35–40%. The yield of solubles amounts 36–46% of the original ground coffee. Further processing involves spray or freeze drying. In the latter method, the liquid extract is foamed and frozen in a stream of cold air or an inert gas, then granulated, sifted and dried in a vacuum in the frozen state. Soluble (instant) coffee is produced from spray-dried products by agglomeration.

Aroma retention is a major problem in an instant coffee process. Since extraction under drastic conditions results in great losses of volatile aroma constituents, methods have been developed for their prior removal and subsequent recovery. Such an aroma concentrate is then added back to aromatize the coffee extract, either before or after the extract has dried.

The resultant extract powder is hygroscopic and unstable. It is packaged in glass jars, vacuum packed in cans, aluminum foil-lined bags, flexible polyethylene, laminated pouches or bags, or packaged in air-tight plastic beakers or mugs, often under vacuum or under an inert gas.

Like roasted coffee, instant coffee is marketed in different varieties, e.g. regular roasted or as a dark, strongly-roasted espresso, or caffeine free. Instant coffee contains 1.0–6.0% moisture. The dry matter consists of 7.6–14.6% minerals, 3.2–13.1% reducing sugars (calculated as glucose), 2.4–10.5% galactomannan, 12% low molecular weight organic acids, 15–28% brown pigments, 2.5–5.4% caffeine and 1.56–2.65% trigonelline.

The products are used not only for the preparation of coffee beverages but also as flavorings for desserts, cakes, sweet cookies and ice cream.

The consumption of instant coffee is rising slowly but steadily; for example, as a percentage of total brewed coffee in the Federal Republic of Germany, it was 10% in 1963–65 and 18% in 1974.

21.1.4.2 Decaffeinated Coffee

The physiological effects of caffeine are not beneficial nor are they tolerated by everyone. Hence, many processes have been developed to remove caffeine from coffee. The first technically usable process was developed by *Roselius* in Bremen in 1908. In this process, which is still used in Europe, the green coffee is treated with superheated steam at high pressure. The swollen beans are then extracted with various organic solvents (dichloromethane, acetic acid ester) with constant stirring for the selective removal of caffeine. The decaffeinated beans are recovered after solvent removal. The moist beans, which, after steaming, acquire a moisture content of about 40%, are then dried under vacuum or in a stream of hot air. In another process, used in the USA, initially all the water-soluble compounds, including caffeine, are removed from the green beans. The aqueous extract is decaffeinated with an organic solvent (e.g. dichloroethylene), and then added back to the green beans and, together with the beans, evaporated to dryness.

A new process uses liquid carbon dioxide as a solvent at 70–90 °C and 100–200 bar. Decaffeinated coffee extract can also be prepared from roasted coffee by this process. The market share of decaffeinated coffee products in Germany in 1974 was 12%.

21.1.4.3 Treated Coffee

The "roast" compounds, the phenolic acids and the coffee waxes, are irritating substances of roasted coffee. Various processes have been developed to separate these constituents to make roasted coffee tolerable for sensitive people. *Lendrich* (1927) investigated the effect of steaming green beans, without caffeine extraction, on the removal of some substances, such as waxes, and hydrolysis of chlorogenic acid. In a process developed by *Bach* (1957), roasted coffee beans are washed with liquid carbon dioxide. In another process, the surface waxes of the raw beans are first removed by a low-boiling organic solvent, followed by steaming, as used by *Lendrich*. The extent of wax removal can be monitored by the analysis of fatty acid tryptamides, which have already been mentioned (cf. 21.1.3.3.3).

21.1.5 Coffee Substitutes and Adjuncts

21.1.5.1 Foreword

Coffee substitutes, or surrogates, are the parts of roasted plants and also other sources which are made into a product which, with hot water, provides a coffee-like brew and serves as a coffee substitute or as a coffee blend.

Coffee adjuncts (coffee spices) are roasted parts of plants or material derived from plants, mixed with sugar, or a blend of all three sources and also, when other ingredients are added, are used as a supplement to coffee or as coffee substitutes. The starting materials for manufacturing such products vary: barley, rye, milo (a sorghum-type grain) and similar starch-rich seeds, barley and rye malts and other malted cereals, chicory, sugar beets, carrots and other roots, figs, dates, locust fruit (St. John's bread) and similar sugar-rich fruits, peanuts, soybeans and other oilseeds, fully- or partially-defatted acorns and other tannin-free plant parts, and, lastly, various sugars.

Coffee substitutes have been known for a long time, as exemplified by the coffee brew made of chicory roots (*Cichoricum intybus* var. *sativum*) or by clear drinks prepared from roasted cereals. The coffee substitute industry during the 2nd World War processed about 315,000 t of such products. In post-war Germany (1966) the amount was 24,000 t, which is about 27 l per capita per year. For comparison, in the same year coffee consumption was 127 l per capita. About ten years later (1977) the production of coffee surrogates dropped to 10,000 t or 10.3 l per capita in comparison to about 159 l of brewed coffee per capita.

21.1.5.2 Processing of Raw Materials

The raw materials are stored as such (all cereals, figs), or are stored until processing as dried slices (e.g. root crops such as chicory or sugar beet). After careful cleaning, steeping, malting and steaming in steaming vats, pots or pressure vats takes place. Roasting follows, with a final temperature of 180–200 °C, and then, like the processing of coffee beans, the grains may be polished or coated with sugar.

During the manufacture of substitutes and adjunct essences, the liquid sugar juice (cane or beet molasses, syrup or starch-sugar plant extracts) is caramelized in a cooker by heating above 160 °C under atmospheric pressure. The dark, brown-black product solidifies to a glassy, strongly hygroscopic mass which is then ground.

Pulverized coffee substitutes are obtained from the corresponding starting materials, as with true coffee, by a spray, drum, conveyor or other drying process.

The starch present in the raw materials is diastatically degraded to readily-caramelized, water-soluble sugars in the manufacture of coffee substitutes during the steeping, steaming and, particularly, the malting steps. This is especially the case with malt coffee. Caramel substances ("bitter roast") formed in the roasting step, which provide the color and aroma of the brew, are derived from carbohydrate-rich raw materials (starch, inulin or sucrose). Since oilseeds readily develop rancidity, processing of carbohydrate-rich materials is preferred to oil- or protein-rich raw materials.

The oils from roasted products, as aroma carriers, especially for malt and chicory coffees, have been analyzed in detail. From the volatiles identified in the coffee aroma numerous constituents are also found in these oils. However, a basic difference appears to be that important sulfur- and nitrogen-containing aroma compounds present in roasted coffee beans are practically absent in coffee substitutes, or are present in negligible amounts.

21.1.5.3 Individual Products

21.1.5.3.1 Barley coffee

Barley (or rye, corn or wheat) coffee is obtained by roasting the cleaned cereal grains after steeping or steaming. The products contain up to 12% moisture and have about 4% ash.

21.1.5.3.2 Malt coffee

Malt coffee is made from barley malt by roasting, with or without an additional steaming step. It contains 4.5% moisture, 2.6% minerals, 74.7% carbohydrates (calculated), 1.8% fat, 10.8% crude protein, 5.6% crude fiber and provides an extract which is 42.4% soluble in water. Polycyclic aromatic hydrocarbons are also detected. Rye and wheat malt coffees are manufactured from their respective malts in the same way.

21.1.5.3.3 Chicory coffee

Chicory coffee is manufactured by roasting the cleaned roots of the chicory plant possibly with addition of sugar beet, low amounts of edible fats or oils, salt and alkali carbonates, followed by grinding of the roasted product, with or without an additional steaming step or treatment with hot water. Chicory contains on the average 13.3% moisture, 4.4% minerals, 68.4% carbohydrates, 1.6% fat, 6.8% crude protein, 5.5% crude fiber and provides an extract which is 64.6% soluble in water.

21.1.5.3.4 Fig coffee

Fig coffee is made from figs by roasting and grinding, with or without an additional steaming step or treatment with hot water. It contains 11.4% moisture, 70.2% carbohydrates and 3.0% fat and provides an extract which is 67.9% soluble in water.

21.1.5.3.5 Acorn coffee

This product is made from acorns, freed from fruit hull and the bulk of the seed coat, by the same process as used for coffee. It contains an average of 10.5% moisture, 73.0% carbohydrates and provides an extract which is 28.9% soluble in water.

21.1.5.3.6 Other products

Coffee substitute blends and similarly-designated products are blends of the above-outlined coffee substitutes, coffee adjuncts and coffee beans. Caffeine-containing coffee substitutes or adjuncts are made by incorporating plant caffeine extracts into substitutes before, during or after the roasting step. The content of caffeine never exceeds 0.2% in such products.

21.2 Tea and Tea-Like Products

21.2.1 Foreword

Tea or tea blends are considered to be the young, tender shoots of tea shrubs, consisting of young leaves and the bud, processed in a way traditional to the country of origin. The tea shrub was cultivated in China and Japan well before the time of Christ. Plantations are now also found in India, Pakistan, Sri Lanka, Indonesia, Taiwan, East Africa, South America, etc. Table 21.10 shows some data on the production of tea.

The evergreen tea shrub (*Camellia sinensis,* synonym *Thea sinensis*) has three principal varieties, of which the Chinese (var. *sinensis,* small leaves) and the Assam varieties (var. *assamica,* large leaves) are the more important and widely cultivated. Grown in the wild, the shrub reaches a height of 9 m but, in order to facilitate harvest on plantations and in tea gardens, it is kept pruned as a low spreading shrub of 1–1.5 m in height. The plant is propagated from seeds or by vegetative propagation using leaf cuttings. It thrives in tropical and subtropical climates with high humidity. The first harvest is obtained after 4–5 years. The shrub can be used for 60 to 70 years. The harvesting season depends upon the region and climate and lasts for 8–9 months per year, or leaves can be plucked at intervals of 6–9 days all year round. In China there are 3–4 harvests per year.

The younger the plucked leaves, the better the tea quality. The white-haired bud and the two adjacent youngest leaves (the famous "two leaves and the bud" formula) are plucked, but plucking of longer shoots containing three or even four to six leaves is not uncommon. Further processing of the leaves provides black or green tea.

Table 21.10. Production of tea, 1981 (1,000 t)

Continent	Tea	Country	Tea
World	1,845	India	550
Africa	196	China	354
America, South-	38	Sri Lanka	210
Asia	1,468	USSR	135
Europe, East- + USSR	135	Japan	103
Australia + South Pacific islands	8	Indonesia	95
		Kenya	91
		Turkey	52
		Bangladesh	40
		Malawi	32
		Σ (%)[a]	91

[a] World production ≙ 100%.

21.2.2 Black Tea

The bulk of harvested tea leaves is processed into black tea. First, the leaves are withered in trays or drying racks in drying rooms, or are drum dried; this involves dehydration, reducing the moisture content of the fresh leaves from about 75% to about 55–65% so that the leaves become flaccid, a prerequisite for the next stage of processing: rolling without cracking of the leaves. Withering at 20–35 °C lasts about 4–18 h. During this time the thinly-spread leaves lose about 50% of their weight in air or in a stream of warm air as in drum drying. In the next stage of processing, the leaves are fed into rollers and are lightly, and without pressure, conditioned in order to attain a uniform distribution of phenoloxidase enzymes. These enzymes are present in epidermis tissue cells, spatially separated from their substrates. This is followed by a true rolling step in which the tea leaf tissue is completely macerated by conventional crank rollers under pressure. The cell sap is released and subjected to oxidation by oxygen from the air. The rolling process is regarded as the first stage of fermentation which proceeds at 35–40 °C for 45 min to 4 h for tea leaves spread thinly in layers 5–7.5 cm thick. The fermentation is stopped when the leaves attain the bright, coppery-red color of a minted copper coin and an odor resembling that of sour apples. Then the fermented tea leaves are heated in large ovens or firing machines or desiccators at

87–93 °C for 20–22 min, or, more recently, by a fully automated process. The firing reduces the moisture content to about 3%, the tea aroma is fixed, and the coppery-red color is changed to black (hence "black tea").

India and Sri Lanka tea factories use both rollers and machines of continuous operation – the so-called CTC machines (cutting, tearing and curling). They provide a simultaneous crushing, grinding, and rolling of the tea leaf, thus reducing the rolling and fermentation time to 1 to 2 hours.

21.2.3 Green Tea

In the green tea manufacture, the development of oxidative processes is regarded as an adverse factor, the fresher the tea leaf used in manufacture, the better the tea produced. Since oxidative processes catalyzed by the leaf enzymes are undesirable, the enzymes are inactivated at an early stage and their reactions are replaced by thermochemical processes. In contrast to black tea manufacture, withering and fermentation stages are omitted in green tea processing.

There are two methods of green tea manufacturing: Japanese and Chinese. The Japanese method involves steaming of the freshly plucked leaf at 95 °C, followed by cooling and drying. Then the leaf undergoes high-temperature rolling at 75 to 80 °C. In the Chinese method the fresh leaves are placed into a roaster which is heated by smokeless charcoal, and roasted. After rolling and sifting, firing is the final step in the production of green tea.

During the processing of green tea the content of tannin, chlorophyll, vitamin C and organic acids decreases only slightly as a consequence of enzyme inactivation.

Green tea provides a very light, clear, bitter tasting beverage. In China and Japan it is often aromatized by flowers of orange, rose or jasmin.

Yellow tea and red tea *(Oolong)* occupy an intermediate position between the black and green teas, yellow tea being closer to green teas, and red tea to black teas.

Yellow tea production does not include fermentation. Nevertheless, in withering, roasting, and firing, a portion of tanning undergoes oxidation, and, therefore, dry yellow tea is darker than green tea.

Red tea is a partially fermented tea. Its special flavor which is free from the grassy note of green tea is formed during roasting and higher-temperature rolling.

21.2.4 Grades of Tea

The numerous grades of tea found in the trade are defined by origin, climate, age, processing method, and leaf grade. They can be classified somewhat arbitrarily:

- According to leaf grade (tea with full, intact leaves), such as Flowery Orange Pekoe and Orange Pekoe (made from leaf buds and the two youngest, hairy, silver leaves with yellowish tips); Pekoe (the third leaf); Pekoe Souchong (with the coarsest leaves, fourth to sixth, on the young twig).
- Broken-tea, with broken or cut leaves similar to the above grades, in which the fine broken or cut teas with the outermost golden leaf tips are distinguished from coarse, broken leaves. Broken/cut tea (loose tea) is the preferred product in world trade since it provides a finer aroma which, because of increased surface area, produces larger amounts of the beverage.
- Fannings and the fluff from broken/cut leaves, freed from stalks or stems, are used preferentially for manufacturing of tea bags.
- Tea dust, which is not used in Europe.
- Brick tea is, likewise, not available on the European market. It is made of tea dust by sifting, steaming and pressing the dust in the presence of a binder into a stiff, compact tea-brick.

All over the world there is blending of teas (e.g. Chinese, Russian, East-Friesen blends, household blends) to adjust the quality and flavor of the brewed tea to suit consumer taste, acceptance or trends and to accommodate regional cultural practices for tea-water ratios. Like coffee, tea extracts are dried and marketed in the form of a soluble powder, often called instant tea.

21.2.5 Composition

The chemical composition of tea leaves varies greatly depending on their origin, age and the type of processing. Table 21.11 provides data about the constituents of fresh and fermented tea leaves. In fermented teas 38–41% of the dry matter is soluble in hot water; this is significantly more than for roasted coffee.

Table 21.11. Composition (%, dry weight basis) of the fresh and fermented tea leaves and of tea brew

Constituent	Fresh flush	Black tea	Black tea brew[a]
Phenolic compounds[b]	30	5	4.5
Oxidized phenolic compounds[c]	0	25	15
Protein	15	15	+[d]
Amino acids	4	4	3.5
Caffeine	4	4	3.2
Crude fiber	26	26	0
Other carbohydrates	7	7	4
Lipids	7	7	+
Pigments[e]	2	2	+
Volatile compounds	0.1	0.1	0.1
Minerals	5	5	4.5

[a] Brewing time 3 min. [b] Mostly flavanols. [c] Mostly thearubigins. [d] Traces. [e] Chlorophyll and carotenoids.

21.2.5.1 Phenolic compounds

Phenolic compounds make up 25–35% of the dry matter content of young, fresh tea leaves. Flavanol compounds (Table 21.12) are 80% of the phenols, while the remainder is leucoanthocyanins, phenolic acids, flavonols and flavones. During fermentation the flavanols are oxidized enzymatically to compounds which are responsible for the color and flavor of black tea. The reddish-yellow color of black tea extract is largely due to theaflavins and thearubigins (cf. 21.2.6), while flavor intensity is correlated with the total content of phenolic compounds and polyphenol oxidase activity. The polyphenol oxidases are inactivated in green tea, hence flavanol oxidation is prevented. The greenish or yellowish color of green tea is due to the presence of flavonols and flavones. Thus, tea which is processed into green or black tea is chemically readily distinguishable mainly by the composition of phenolic compounds.

Table 21.12. Phenolic compounds in fresh tea leaves (% dry matter)

Compound	Content
(−)-Epicatechin	1–3
(−)-Epicatechin gallate	3–6
(−)-Epicatechin digallate	+[a]
(−)-Epigallocatechin	3–6
(−)-Epigallocatechin gallate	9–13
(−)-Epigallocatechin digallate	+
(+)-Catechin	1–2
(+)-Gallocatechin	3–4
Flavonols and flavonolglycosides (quercetin, kaempherol, etc)	+
Flavones (vitexin, etc)	+
Leucoanthocyanins	2–3
Phenolic acids and esters (gallic acid, chlorogenic acids) p-Coumaroylquinic acid, theogallin	~5
Phenols, grand total	25–35

[a] Quantitative data are not available.

Changes in the content of the phenols occur during tea leaf growth on the shrub: the concentration decreases and the composition of this fraction is altered. Therefore, good quality tea is obtained only from young leaves.

Among the remaining phenolic compounds theogallin (3-galloyl quinic acid) plays a special role, since it is found only in tea and is correlated with tea quality:

HO, HO, OH, O, O, COOH, HO, OH, OH (21.4)

The biosynthesis of phenolic compounds occurs via the shikimic acid and the phenylalanine pathway.

21.2.5.2 Enzymes

A substantial part of the protein fraction in tea consists of enzymes.

The *polyphenol oxidases,* which are located mainly within the cells of leaf epidermis, are of great importance for tea fermentation. Their activity rises during the leaf withering and rolling process and then drops during the fermentation stage, probably as a consequence of reactions of some products (e.g. o-quinones) with the enzyme proteins.

Shikimate dehydrogenase which reversibly interconverts dehydroshikimate and shikimate is a key enzyme in the biosynthesis of phenolic compounds via the phenylalanine pathway.

Phenylalanine ammonia-lyase which catalyzes the cleavage of phenylalanine into *trans*-cinnamate and NH_3, is equally important for the biosynthesis of phenols. Its activity in tea leaves parallels the content of catechins and epicatechins.

Proteinases cause protein hydrolysis during withering, resulting in a rise in peptides and free amino acids.
The observed oxidation of linolenic acid to *cis*-3-hexenal, which then isomerizes to *trans*-2-hexenal, is catalyzed by a *lipoxygenase* and a *hydroperoxide lyase* (cf. 3.7.2.3). Both aldehydes and their corresponding alcohols which are formed by enzymatic reduction *(alcohol dehydrogenase)* contribute to the tea flavor
Chlorophyllases participate in the degradation of chlorophyll and *transaminases* in the production of precursors for aroma constituents.
Demethylation of pectins by *pectinesterase* results in the formation of a pectic acid gel, which affects cell membrane permeability, thus resulting in a drop in the rate of oxygen diffusion into leaves during fermentation.

21.2.5.3 Amino Acids

Free amino acids constitute about 1% of the dry matter of the tea leaf. Of this, 50% is theanine (5-N-ethylglutamine) and the rest consists of protein-forming amino acids; β-alanine is also present.
Green tea contains more theanine than black tea. Generally, there is a characteristic difference in amino acid content as well as a difference in phenolic compounds between the two types of tea (Table 21.13).
The contribution of theanine to the taste of green tea is discussed. Theanine biosynthesis occurs in the plant roots from glutamic acid and ethylamine, the latter being derived from alanine. The compound is then transported into the leaves. The analogous compounds, 4-N-ethylasparagine and 5-N-methylglutamine, are present at very low levels in tea leaves.

Table 21.13. Amino acids and phenolic compounds in green and black tea (% dry matter)

Tea	Phenolic compounds	Amino acids
Green tea		
Prime quality (Japan)	13.2	4.8
Consumer quality (Japan)	22.9	2.1
Consumer quality (China)	25.8	1.8
Black tea		
Highlands (Sri Lanka)	28.0	1.6
Plains (Sri Lanka)	30.2	1.7

21.2.5.4 Caffeine

Caffeine constitutes 2.5–5.5% of the dry matter of tea leaves. It is of importance for the taste of tea. Theobromine (0.07–0.17%) and theophylline (0.002–0.013%) are also present but in very low amounts. The biosynthesis of these two compounds involves methylation of hypoxanthine or xanthine:

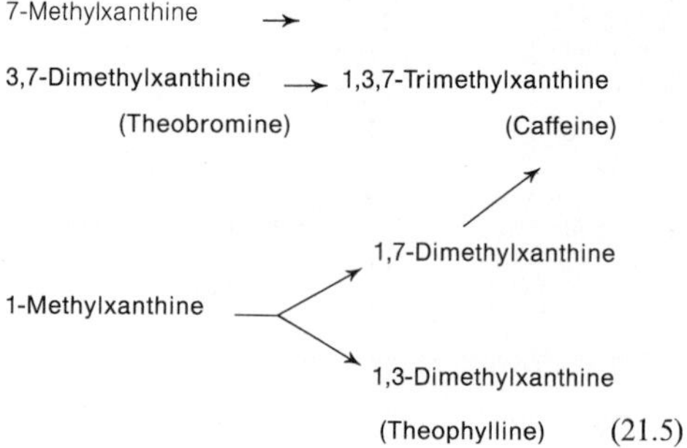

(21.5)

21.2.5.5 Carbohydrates

Glucose (0.72%), fructose, sucrose, arabinose and ribose are among sugars present in tea leaves. Rhamnose and galactose are bound to glycosides. Polysaccharides found include cellulose, hemicelluloses and pectic substances. Inositol occurs also in tea leaves.

21.2.5.6 Lipids

Lipids are present only at low levels. The polar fraction (glycerophospholipids) in young tea leaves is predominant, while glycolipids predominate in older leaves.
Triterpene alcohols, such as β-amyrin, butyrospermol and lupeol are predominant in the unsaponifiable fraction. The sterol fraction contains only Δ^7-sterols, primarily α-spinasterol and Δ^7-stigmasterol.

21.2.5.7 Pigments (Chlorophyll and Carotenoids)

Chlorophyll is degraded during tea processing. Chlorophyllides and pheophorbides (brownish in color) are present in fermented leaves, both being converted to pheophytines (black) during the firing step.
Fourteen carotenoids have been identified in tea leaves. The main carotenoids are xanthophylls, neoxanthin, violaxanthin and β-carotene (cf. 3.8.4.1). The content decreases during the pro-

cessing of black tea. Degradation of β-carotene, as an example, yields β-ionone, a significant contributer to tea aroma (Table 21.14).

Table 21.14. Volatile products formed by oxidation of carotenoids during processing of tea

Carotenoid[a]	Degradation product
β-Carotene (102/61)[a]	1) β-Ionone, 2) dihydroactinidiolide, 3) theaspirone, 4) theaspirane, 5) 3-oxo-β-ionone, 6) 2,2,6-trimethyl-6-hydroxycyclohexanone, 7) 3,7-dimethyl-1,5,7-octatrien-3-ol, 8) 6-methyl-trans-3,5-heptadien-2-one, 9) 2,2,6-trimethylcyclohexanone, 10) 2,4,6-decatrienal, 11) limonene (?), 12) terpinolene (?), 13) linalool (?)
Xanthophyll (260/154)	2), 3), 5), 7), 8), 9), 11)–13)
Neoxanthin (51/23)	2), 3), 5), 7), 8), 9), 11)–13)
Violaxanthin (120/36)	2), 3), 5), 7), 8), 9), 11)–13)

[a] In brackets carotenoid content is given in fresh tea leaves and black tea respectively (μg/g dry matter).

Table 21.15. Volatile compounds in steam distillate of Flavory Ceylon black tea

Compound	Proportion (%)
Linalool	53.6
trans-Furan linalool oxide	16.7
cis-Furan linalool oxide	4.8
trans-2-Hexene-1-al	4.7
Hexanol	1.7
trans-3-Octen-2-one	1.6
Benzaldehyde	1.2
cis/trans-Pyran linalool oxides	0.9
1-Penten-3-ol	0.8
cis-3-Hexen-1-ol	0.6
Furfuryl	0.6
cis-3-Hexenyl acetate	0.5
trans,trans-2,4-Heptadienal	0.5
cis-2-Penten-1-ol	0.4
1-Octen-3-ol	0.4
α-Terpinyl acetate	0.3
Geranyl acetate	0.3
Hexanal	0.3
Octanol	0.3
cis-3-Hexenyl hexanoate	0.3
Heptanal	0.2
3,7-Dimethyl-1,5,7-octatrien-3-ol	0.2
Acetophenone	0.2
Methyl salicylate	0.2
Geraniol	0.2
Pentanal	0.1
trans-2-Octen-1-al	0.1
α-Terpineol	0.1
Neryl acetate	0.1
Nerol	0.1
Benzyl alcohol	0.1
3-Methylbutanal	+[a]
trans-2-Pentenal	+
Butanol	+
3-Methylbutanol	+
Myrcene	+
d-Limonene	+
Pentanol	+
Octanal	+
Terpinolene	+
Nonanal	+
trans,trans-2,4-Hexadienal	+
trans-2-Hexen-1-ol	+
cis-3-Hexenyl butyrate	+
trans-2,*cis*-4-Heptadienal	+
trans,trans-3,5-Octadienone	+
6-Methyl-3,5-heptadienone	+
Nonanal	+
Phenylacetaldehyde	+
1-Ethylformylpyrrole	+
γ-Caprolactone	+
α-Ionone	+
β-Ionone	+
2-Phenylethanol	+

[a] < 0.1%.

21.2.5.8 Volatile Compounds

Volatile compounds constitute approximately 0.01–0.02% of tea dry matter. Black tea has three- to five-times more volatiles than green tea. More than 300 compounds have been identified but not all contribute to the flavor. Obviously, the aroma of tea is based on the proper proportion of a number of key compounds. Also, it should be emphasized that the aroma is greatly affected by the origin and the processing of the tea.

Analysis of an aroma concentrate of a particularly aromatic "Ceylon Flavory Tea" revealed the presence of the compounds summarized in Table 21.15. This tea is harvested in January/February in the Dimbula district of Sri Lanka and in July–September in the Uva district. It is characterized by a sweet, fruity, jasmin-like floral aroma.

Table 21.16. Aroma notes and major constituents of some fractions of aroma concentrate of Flavory Ceylon black tea

Fraction[a]	Aroma note	Major constituents
1 (12,8)[b]	Green, grassy	Pentanol, *cis*-3-hexen-1-ol
2 (43,3)	Floral, pleasant	Furan linalool oxides, 2-phenylethanol
3 (28,3)	Woody	Pyran linalool oxides
4 (15,6)	Peach, apricot-like, sweet, floral	2-methylhept-2-en-6-one (1,8)[b], retro-ionone (2,5), linalool pyran oxides (11,9), geraniol (6,4), 4-octanolide (2,1), *cis*-3-hexenyl-*cis*-3-hexenoate (8,2), 4-nonanolide (4,8), theaspirone (3,0), 2,3-dimethyl-2-nonen-4-olide (1,1), 5-decanolide (5.3), jasminlactone (11,9), methyl jasmonate (4,6), dihydroactinidiolide (15,5)

[a] Fractions 1–4 constitute 2.7% of aroma concentrate and are characterized by a very strong and sweet tea aroma.
[b] Semi-quantitative data for relative percentage composition.

The same tea aroma concentrate, when separated on a silica gel column with ether/hexane as eluent, yielded fractions with various aromas, among them one with an exceptionally sweet tea aroma. When that fraction was further separated by preparative gas chromatography, four subfractions were obtained, the aromas and major constituents of which are presented in Table 21.16.

Based on this and other studies, it has been concluded that the aroma of black tea depends a great deal on the constituents presented in Table 21.16a.

Linalool, its oxides and a number of other compounds are relatively insignificant in green tea aroma; but nerolidol, β-ionone and various other compounds are present in greater abundance than in black tea (Table 21.17). Other as yet unidentified compounds possibly play a role in green tea aroma build-up.

Table 21.16a. Probable black tea aroma impact compounds

Compound	Flavor description
3-Methylbutanol	fatty, burnt
3-*cis*-Hexenol	green, leafy
2-*trans*-Hexenal Hexanal	green, fruity, fatty
2-Phenylethanol	rose, fermented
Methyl salicylate	floral, metallic
α-Terpineol	floral, fruity, earthy
Linalool	leafy, floral
Nerol	fruity, floral
Phenol Guaiacol	phenolic, animal, smoky leather
Furan linalool oxides Pyran linalool oxides	leafy, earthy
β-Damascone[a]	fruity, fermented
Damascenone[a]	floral, fruity, woody
β-Ionone	floral, woody
Dihydroactinidiolide[a]	hay-like
Theaspirone[a]	hay-like

[a] Chemical structures, cf. Table 3.47

Table 21.17. Volatile compounds in green tea

Compound	Relative portion (%)
Linalool	19.9
δ-Cadinene	9.4
Geraniol	5.5
Nerolidol	4.0
Indole	3.9
Benzyl alcohol	3.7
α-Muurolen	3.2
α-Terpineol	3.0
cis-Jasmone	3.0
trans-Furan linalool oxide	2.6
β-Ionone	2.6
Octanol	1.6
cis-Furan linalool oxide	1.4
epi-Cubenol	1.2
Dihydroactinidiolide	1.2
Caryophyllene	1.1
Cubenol	1.1
Calamenene	1.0
cis-3-Hexenyl hexanoate	1.0
2-Acetylpyrrole	0.9
1-Ethylformylpyrrole	0.9
β-Muurolene	0.8
2-Phenylethanol	0.7
Coumarin	0.7

Table 21.17. Volatile compounds in green tea

Compound	Relative portion (%)
α-Cadinol	0.6
3,7-Dimethyl-1,5,7-octatrien-3-ol	0.6
4-Vinylphenol	0.6
Diphenylamine	0.6
Nerol	0.6
cis-3-Hexenyl benzoate	0.6
Furfuryl alcohol	0.5
α-Ionone	0.5
Phenol	0.5
Nonanal	0.5
Limonene	0.5
cis-3-Hexen-1-ol	0.5
α-Copaene	0.5
α-Cubebene	0.4
2′,2″-Dihydro-β-ionone	0.4
Acetophenone	0.4
α-Humulene	0.4
cis-Pyran linalool oxide	0.4
6,10,14-Trimethyl-2-pentadecanone	0.4
m- and p-Cresol	0.3
β-Sesquiphellandrene	0.3
4-Ethylguaiacol	0.3
trans,trans-3,5-Octadien-2-one	0.2
5-Methylfurfural	0.2
α-Terpinyl acetate	0.1
6-Methyl-trans,trans-3,5-heptadien-2-one	0.1
Heptanol	0.1
Unidentified compounds	approx. 14%

21.2.5.9 Minerals

Tea contains about 5% minerals. The major element is potassium, which is half the total mineral content. Some tea varieties contain fluorine in higher amounts (0.015–0.03%).

21.2.6 Reactions Involved in the Processing of Tea

Changes in tea constituents begin during the *withering* step of processing. Enzymatic protein hydrolysis yields amino acids of which a part is transaminated to the corresponding keto acids. Both types of acids provide a precursor pool for aroma substances. The induced chlorophyll degradation has significance for the appearance of the end-product. A more extensive conversion of chlorophyll into chlorophyllide, a reaction catalyzed by the enzyme chlorophyllase (cf. 17.1.2.9.1), is undesirable since it gives rise to pheophorbides (brown) and not the desired olive-black pheophytins. Increased cell permeability during withering favors the fermentation procedure. As already mentioned, a uniform distribution of polyphenol oxidases in tea leaves is achieved during the *conditioning* step of processing.

During *rolling,* the tea leaf is macerated and the substrate and enzymes are brought together; a prerequisite for fermentation. The subsequent enzymatic oxidative reactions are designated as *"fermentation"*. This term is a misnomer and originates from the time when the participation of microorganisms was assumed. In this processing step, the pigments and aroma substances are formed primarily as a result of phenolic oxidation by the polyphenol oxidases. In addition, oxidation of amino acids, carotenoids and unsaturated fatty acids, preferentially by oxidized phenols, is also of importance.

Harler (1963) described tea aroma development during processing: "The aroma of the leaf changes as fermentation proceeds. Withered leaf has the smell of apples. When rolling (or leaf maceration) begins, this changes to one of pears, which then fades and the acrid smell of the green leaf returns. Late, a nutty aroma develops and, finally, a sweet smell, together with a flowery smell if flavor is present."

The enzymatic oxidation of flavanols via the corresponding o-quinones provides condensation reactions leading to theaflavins and epitheaflavic acids and to benzotropolone derivatives which impart the characteristic color to black tea extracts (Formulas 21.6).

A second, obviously heterogenous group of compounds, found in tea after the enzymatic oxidation of flavanols, are the thearubigins, a group of compounds responsible for the characteristic reddish-yellow color and astringent taste of black tea extracts. Their structures have not yet been elucidated; their molecular weights range from 700–40,000 dal (Formula 21.7).

Oxidized Flavanoles ↘
Theaflavines → Thearubigines (21.7)
Epitheaflavic acids ↗

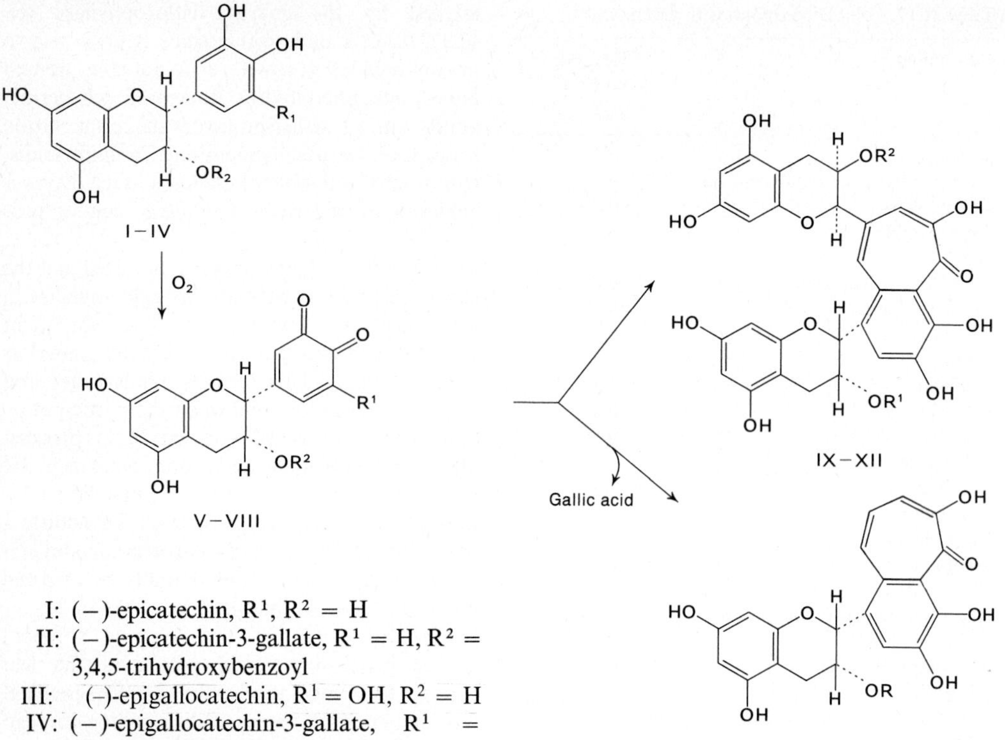

I–IV

$\xrightarrow{O_2}$

V–VIII

Gallic acid

IX–XII

XIII–XIV (21.6)

I: (−)-epicatechin, $R^1, R^2 = H$
II: (−)-epicatechin-3-gallate, $R^1 = H, R^2 =$ 3,4,5-trihydroxybenzoyl
III: (−)-epigallocatechin, $R^1 = OH, R^2 = H$
IV: (−)-epigallocatechin-3-gallate, $R^1 =$ OH, $R^2 =$ 3,4,5-trihydroxybenzoyl
V–VIII: o-quinones of compounds I–IV
IX: theaflavin, $R^1, R^2 = H$
X: theaflavin gallate A, $R^1 = H, R^2 =$ 3,4,5-trihydroxybenzoyl
XI: theaflavin gallate B, $R^1 =$ 3,4,5-trihydroxybenzoyl, $R^2 = H$
XII: theaflavin digallate, $R^1, R^2 =$ 3,4,5-trihydroxybenzoyl
XIII: epitheaflavic acid, R = H
XIV: 3-galloyl epitheaflavic acid, R = 3,4,5-trihydroxybenzoyl.

Aroma development during fermentation is accompanied by an increase in the occurrence of volatile compounds typical for black tea. They are produced by *Strecker* degradation reactions of amino acids with oxidized flavonols (Formula 21.8),
by oxidation of unsaturated fatty acids, and particularly by carotenoid oxidation (Formula 21.9). From the latter reaction results e.g. β-ionone, hydroxy-β-ionone, and possibly limonene as primary oxidation products of β-carotene (R = H) or xanthophyll (R = OH). Secondary reactions would lead to important tea aroma constituents such as dihydroactinidiolide, theaspirone and linalool (Table 21.14).

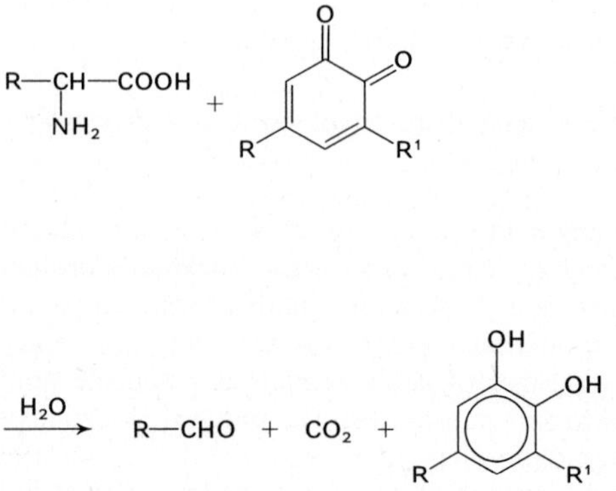

$$R{-}CH(NH_2){-}COOH + \text{o-quinone} \xrightarrow{H_2O} R{-}CHO + CO_2 + \text{catechol} \quad (21.8)$$

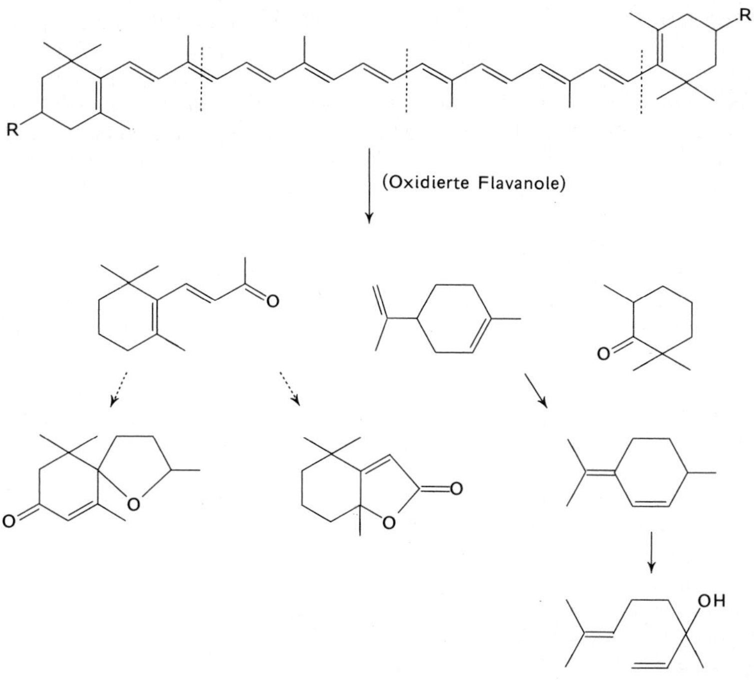

(Oxidierte Flavanole)

OH

(21.9)

In addition to degradation reactions, biosynthetic reactions also appear to contribute to tea aroma formation. For example, the participation of leucine in nerolidol biosynthesis via mevalonic acid has been confirmed using labelled leucine.

During the *firing* step of tea processing, there is an initial rise in enzyme activity (10–15% of the theaflavins are formed during the first 10 min), then all the enzymes are inactivated. The tea acquires its black color from the conversion of chlorophyll into pheophytin. A prerequisite for this reaction is high temperature and an acidic environment. The undesired brown color is obtained at higher pH's. The astringent character of teas is decreased by the formation of complexes between phenolic compounds and proteins.

The firing step also affects the balance of volatiles. On the one hand there is a loss of volatile compounds, on the other hand, at high temperatures, an enhancement of the build-up of typical aroma constituents occurs: β-ionone, theaspirone and dihydroactinidiolide. Pyrazines, pyridines and quinolines are most probably also formed in this step as a result of sugar-amino acid interactions.

21.2.7 Packaging, Storage, Brewing

In the country in which it is grown, the tea is cleaned of coarse impurities, graded by leaf size, and then packed in standard plywood chests of 20–50 kg lined with aluminum, zinc or plastic foils. To preserve tea quality, the foils are sealed, soldered or welded. China, glass or metal containers are suitable for storing tea. Bags made of pergament or filter papers and filled with metered quantities of tea are also very common.

During storage, the tea is protected from light, heat ($T < 30\,°C$) and moisture, otherwise its aroma becomes flat and light. Other sources of odor should be avoided during storage.

To prepare brewed tea, hot water is usually poured on the leaves and, with occasional swirling, left for 3–5 min. An initial tea concentrate or extract is often made, which is subsequently diluted with water. Usually 4–6 g of tea leaves per liter are required, but stronger extracts need about 8 g. The stimulating effect of tea is due primarily to the presence of caffeine.

21.2.8 Maté (Paraguayan Tea)

Maté, or Paraguayan tea, is made from leaves of a South American palm, *Ilex paraguariensis*. The palm grows in Argentina, Brazil, Paraguay and Uruguay, either wild or cultivated, and reaches a height of 8–12 m. To obtain mate, the palm leaves, petioles, flower stems and young shoot tips are collected and charred slightly on an open fire or in a woven wire drum. During such firing, oxidase enzymes are inactivated, the green color is fixed and a specific aroma is formed. The dried product is then pounded into burlap sacks or is ground to a fine powder (maté pulver, maté en pod). Maté may also be prepared by an alternative process: brief blanching of the leaf in boiling water, followed by drying on warm floors and disintegration of the leaves to rather coarse particles. In the countries in which it is grown, maté is drunk as a hot brew (yerva) from a gourd (maté = bulbshaped pumpkin fruit) using a special metal straw called a bombilla, or it is enjoyed simply in a powdered form. Maté stimulates the appetite and, because of its caffeine content (0.5–1.5%), it has long been the most important alkaloid-containing brewed plant product of South America. It contains on the average 12% crude protein, 4,5% ether-soluble material, 7.4% polyphenols and 6% minerals. About one third of the total dry matter of the leaves is solubilized in a maté brew, except for caffeine, which solubilizes to the extent of only 0.019–0.028%, and is 50% bound in leaves.

21.2.9 Products from Cola Nut

Cola (kola) nuts, called guru, goora and bissey nuts by Africans, are not nuts but actually seeds of an evergreen tree of the *Sterculiacea* family, genus *Cola*, species *verticillata*, *nitida* or *acuminata*, which grows wild in West Africa up to a height of 20 m. The tree is indigenous to Africa, but plantations of Cola are found on Madagascar, in Sri Lanka, Central and South America. Each fruit borne by the tree contains several red or yellow-white cola nuts, shaped like horse chestnuts. The nuts change color to brownish-red when dried, with the typical cola-red color resulting from the action of polyphenol oxidase enzymes. The nuts are on the average 5 cm long and 3 cm wide and have a bitter, astringent taste. The fresh nuts, wrapped in cola leaves and moistened with water, are the most enjoyed plant product of Western and Central Africa. They are consumed mostly in fresh form but are also chewed as dried nuts or ground to a powder and eaten with milk or honey. Cola nuts are used in the making of tinctures, extracts or medical stimulants in tablet or pastille form. They are also used in the liqueur, cocoa and chocolate industries and, especially, in the making of alcohol-free soft drinks, cola-wines, etc. The stimulating effect of cola nuts is due to the presence of caffeine (average content 2.16%), the main portion of which is in bound form. In addition, cola nuts contain on the average 12.2% moisture, 9.2% nitrogen compounds, 0,05% theobromine, 1.35% crude fat (ether extract), 3.4% polyphenols, 1.25% red pigments, 2.8% sugar, 43.8% starch, 15% other N-free extractable substances, 7.9% crude fiber and 3% ash.

21.3 Cocoa and Chocolate

21.3.1 Foreword

Cocoa, as a drink, is different from coffee or tea since it is consumed not in the form of an aqueous extract, i.e. a clear brew, but as a suspension. In addition to stimulating alkaloids, particularly theobromine, cacao products contain substantial amounts of nutrients: fats, carbohydrates and proteins. Unlike coffee and tea, cocoa has to be consumed in large amounts in order to experience a stimulating effect.

Cacao beans were known in Mexico and Central America for more than a thousand years before America was discovered by *Columbus*. They were enjoyed originally in the form of a slurry of roasted cocoa beans and corn which was seasoned with paprika, vanilla or cinnamon. In the first half of the 17th century, cacao beans were introduced into Germany. Cacao became popular in the Old World only after sugar was added to the chocolate preparation. Initially, cacao was treated as a luxury item, until the 19th century, when production of pulverized chocolate and defatted cacao was established and they were distributed extensively as a food commodity.

The world production of cacao was 31,000 t in 1870/80, 103,000 t in 1900 and 1,585 million t in 1979. The main cacao-producing countries are listed in Table 21.18.

Table 21.18. Production of cacao beans in 1981 (1,000 t)

Continent	Cacao beans	Country	Cacao beans
World	1,670	Ivory Coast	430
Africa	994	Brazil	345
America, North-, Central-	88	Ghana	230
America, South-	502	Nigeria	160
Asia	52	Cameroon	110
Australia + South Pacific islands	35	Equador	96
		Colombia	41
		Dominican Republic	35
		Mexico	32
		Papua, New Guinea	31
		Σ (%)[a]	91

[a] World production ≙ 100%.

The processing of cacao beans into cocoa powder and chocolate is presented schematically in Figure 21.4.

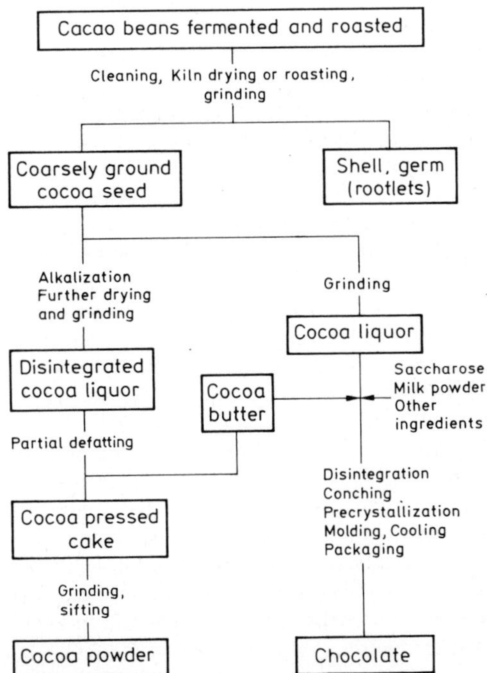

Fig. 21.4. Cocoa powder and chocolate production

21.3.2 Cacao

21.3.2.1 General Information

Cacao beans are the seeds of the tropical cacao tree, *Theobroma cacao,* family *Sterculiaceae.* Originating in the northern part of South America and currently grown within 20° latitude of the Equator, the tree flourishes in warm, moist climates with an average annual temperature of 24–28 °C and at elevations up to 600 m. The tree, because of its sensitivity to sunshine and wind, is often planted and cultivated under shade trees ("cacao mothers"), as are forest trees, coconut palms and banana trees. The perennial tree grows in the wild to a height of 10–15 m, but on plantations it is kept at 2–4 m by pruning. The tree blooms all year round and the small red or white flowers bear 20–50 ripe fruits per tree. The ripe fruit or pod resembles a cantaloupe 15–25 cm long and 7–10 cm wide. The pod is surrounded by a strong 10–15 mm thick shell. Embedded within the pod are pulpa, i.e. a sweet, mucilaginous pulp containing 10% glucose and fructose. The pulp surrounds 20–40 almond-shaped seeds (cacao beans). The seed is oval and flattened, about 2 cm long and 1 cm wide, and weighs close to 1 g after drying. The embryo, with two thick cotyledons (nibs) and a germ rootlet 5 mm long and 1 mm thick, is under a thin, brittle seed coat. The colors in the cross-section of a nib range from white to light brown, to greyish-brown or brownviolet, to deep violet.

The fruit is harvested year round but, preferentially, twice in a year. The main harvest time in Mexico is from March through April; in Brazil, February and, in particular, July. The summer harvest is larger and of higher quality. After the tree is planted (propagation by seed or by vegetative methods), it begins to bear pods after the fifth or sixth years, giving a maximum yield after 20–30 years, while it is nearly exhausted after 40 years of growth. After reaching full beaning capacity, a cacao tree provides only 0.5–2 kg of fermented and dried beans per year. Harvesting at the right time is of great importance for the aroma of cacao and its products. The fruit is harvested fully ripe but not overripe, avoiding damage to the seed during its removal from the fruit.

The tree species *Theobroma cacao* (the only one of commercial importance) is divided into two major groups. The "Criollo" tree (criollo = native) is sensitive to climatic changes and to attack by diseases and pests. It bears highly aromatic beans, hence their commercial name "flavor beans", but they are relatively low yielding. The second group of trees, "Forastero" (forastero = strange, inferior), is characterized by great vigor and the trees are more resistant to climatic changes and to diseases and are higher yielding. The purple-red Forastero beans are less flavorful than Criollo varieties. Nevertheless, the Forastero bean is by far the most important commercial type of cacao and accounts for the bulk of world cacao production (Bahia and Accra cacaos).

Other varieties worthy of mention are the resistant and productive Calabacillo and the Ecuadorian Amelonado varieties.

Cacao beans are differentiated by their geographical origin, grade of cleanliness and the number of preparation steps to which they are subjected prior to shipment. "Flavor beans" come from Ecuador, Venezuela, Trinidad, Sri Lanka and Indonesia, while "commercial beans" are exported by the leading cacao-growing countries of West Africa (Ghana, Nigeria, Ivory Coast and Cameroon), and by Brazil (the port of Bahia) and the Dominican Republic.

21.3.2.2 Harvesting and Processing

At harvest the fully-ripe pods are carefully cut from trees, gathered into heaps, cut open and the beans scooped out with the surrounding pulp. Only rarely are the beans dried in the sun without a prior fermentation step (*Arriba* and *Machala* varieties from South America). The bulk of the harvest is fermented before being dried. In this fermentation step the beans with the adhering pulp are transferred to heaps, ditches or fermentation floors, baskets, boxes or perforated barrels and, depending on the variety, are left to ferment for 2–8 days. From time to time the beans are mixed to make the oxygen in the air accessible to the fermentation process. During this time the temperature of the beans rises rapidly to 45–50 °C, where the germination ability of the seeds is lost and the pH becomes acidic. Flavor and color formation and partial conversion of astringent phenolic compounds also occur. The adhering pulp is decomposed enzymatically and becomes liquid. It drains away as a fermentation juice. In addition, there are reactions between constituents of the beans and pulp. After fermentation is completed, the beans may be washed (Java, Sri Lanka), and are dried to a moisture content of 6–8%.

Well-fermented cocoa provides uniformly-colored, dark-brown beans which are readily separated into their cotyledons. Inadequate or unripe fermented beans are smooth in appearance (violetas) and are of low quality.

The cocoa imported by consuming countries is processed further. The cocoa beans are cleaned by a series of operations and separated by size in order to facilitate uniform roasting in the next processing step. Roasting is being performed more and more as a two-step process. Roasting reduces the moisture content of the beans to 2%, contributes to further oxidation of phenolic compounds and the removal of acetic acid, volatile esters and other undesirable aroma components. In addition the eggs and larvae of pests are destroyed. The aroma of the beans is enhanced, the color deepens, the seed hardens and becomes more brittle and the shell is loosened and made more readily removable because of enzymatic and thermal reactions. The ripeness, moisture content, variety and size of the beans and preliminary processing steps done in the country of origin determine the extent and other parameters of bean roasting processes. Generally, the roasting temperature should not exceed 150 °C; this is substantially lower than in roasting coffee. In the production of cocoa for making chocolate, roasting is more extensive than for beans used for the manufacture of cocoa powder. Losses induced by roasting are 5–8%.

As with coffee, roasted beans are immediately cooled to avoid overroasting. The roasters are batch or continuous. Heat transfer occurs either directly through heated surfaces or by a stream of hot air, without burning the shell of the beans. Roasting lasts 10–35 min, depending on the extent desired.

Roasted beans are transferred, after cooling, to winnowing machines to remove the shells and germ rootlets (these have a particularly unpleasant flavor and impart other undesirable properties to cocoa drinks). During winnowing the beans are lightly crushed in order to preserve the nibs and the shells in larger pieces and to avoid dust formation.

The winnowing process provides on the average 78–80% nibs and yields about 10–12% shells, with a small amount of germ and about 4% of fine cocoa particles as waste. All yields are calculated on the basis of the weight of the raw beans.

The whole nibs, dried or roasted, dehulled and degermed or cracked, are still contaminated with 1.5–2% shell, seed coats and germ. The debris fraction, collected by purifying the cocoa waste, consists of fine nib particles and contains up to 10% shell, seed coating and germ. Although the cocoa shell is considered as waste material of little value, it can be used for recovery of theobromine, production of activated charcoal, or as a feed, cork substitute or tea substitute (cocoa shell tea) and, after extraction of fat, as a fertilizer or a fuel. Detection of shells in a cocoa blend is becoming increasingly more difficult due to ever-improving disintegration techniques. Chemical procedures, such as crude fibre analysis, furfural determination or assay of "mucilage", are far from being satisfactory. Microscopic methods, which involve isolation, enrichment and counting of scleroid cells, are more suitable. However, even here many precautions are necessary. Estimation of galacturonic acid, which is present in shells, and also of pectins which are 4.6–7% (dry weight basis) of the shell (but only a negligible 0.2% in nibs), provides another possibility for distinguishing pure, good quality cocoa powder with a maximum shell content of 1.75% from cocoa adulterated with shells.

21.3.2.3 Composition

The compositions of fermented and air-dried cocoa nib, cocoa shell and germ are presented in Table 21.19.

21.3.2.3.1 Proteins and amino acids

About 60% of the total nitrogen content of fermented beans is protein. The nonprotein nitrogen is found as amino acids, about 0.3%, in amide form, and as ammonia, 0.20%, formed during fermentation of the beans.

Among the various enzymes, α-amylase, β-fructosidase, β-glucosidase, β-galactosidase, pectinesterase, polygalacturonase, proteinase, alkaline and acid phosphatases, lipase, catalase, peroxidase and polyphenol oxidase activities have been detected in fresh cacao beans. These enzymes are inactivated to a great extent during processing.

Table 21.19. Composition (%) of fermented and air dried cacao beans (1), cacao shells (2) and cacao germs (3)

Constituent	1	2	3
Moisture	5.0	4.5	8.5
Fat	54.0	1.5	3.5
Caffeine	0.2		
Theobromine	1.2	1.4	
Polyhydroxyphenols	6.0		
Crude protein	11.5	10.9	25.1
Mono- and oligosaccharides	1.0	0.1	2.3
Starch	6.0		
Pentosans	1.5	7.0	
Cellulose	9.0	26.5	4.3
Carboxylic acids	1.5		
Other compounds	0.5		
Ash	2.6	8.0	6.3

21.3.2.3.2 Theobromine and caffeine

Theobromine (3,7-dimethylxanthine), which is 1.2% in cocoa, provides a stimulating effect, though to a lesser extent than does caffeine in coffee, and therefore is of physiological importance. Caffeine is also present, but in much lower amounts (average 0.2%). A cup of cocoa contains 0.1 g of theobromine and 0.01 g of caffeine. Theobromine crystallizes in the form of small rhombic prisms which sublime at 290 °C without decomposition. In cocoa beans theobromine is often weakly bound to tannins and is released by the acetic acid formed during fermentation of the beans. Part of this theobromine then diffuses into the shell.

21.3.2.3.3 Lipids

Cocoa fat (cocoa butter), because of its abundance and value, is the most significant ingredient of cacao beans, and is dealt with in detail elsewhere (cf. 14.3.2.2.3).

21.3.2.3.4 Carbohydrates

Starch is the predominant carbohydrate. It is present in nibs but not in shells, a fact useful in the microscopic examination of cocoa powders in methods based on the occurrence of starch as a characteristic constituent. Components of the dietary fibre are amongst others pentosans, galactan mucins containing galacturonic acid, and cellulose. Soluble carbohydrates present include stachyose, raffinose and sucrose (0.08–1.5%), glucose and fructose. Sucrose hydrolysis, which occurs during fermentation of the beans, provides the reducing sugar pool important for aroma formation during the roasting process. Mesoinositol, phytin, verbascotetrose, and some other sugars are found in cocoa nib.

21.3.2.3.5 Phenolic compounds

The nib cotyledons consist of two types of parenchyma cells (Figure 21.5). More than 90% of the cells are small and contain protoplasm, starch granules, aleurone grains and fat globules. The larger cells are scattered among them and contain all the phenolic compounds and purines. These polyphenol storage cells (pigment cells) make up 11–13% of the tissue and contain anthocyanins and, depending on their composition, are white to dark purple. Data on the composition of these cells and that of the total tissue are given in Table 21.20.

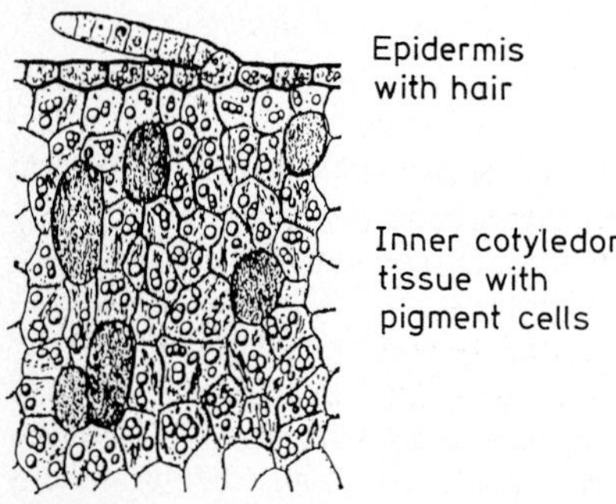

Fig. 21.5. A cross-section of cocoa cotyledon tissue

Table 21.20. Composition of polyphenol storage cells of cacao tissue

Constituent	Polyphenol storage cell (%)	Cotyledons (%)[a]
Catechins	25.0	3.0
Leucocyanidins	21.0	2.5
Polymeric leucocyanidins	17.5	2.1
Anthocyanins	3.0	0.4
Total phenols	66.5	8.0
Theobromine	14.0	1.7
Caffeine	0.5	0.1
Free sugars	1.6	
Polysaccharides	3.0	
Other compounds	14.4	

[a] As % of dry matter.

Three groups of phenols are present: catechins (about 37%), anthocyanins (about 4%) and leucoanthocyanins (about 58%).

The main catechin is (−)-epicatechin, besides (+)-catechin, (+)-gallocatechin and (−)-epigallocatechin. The anthocyanin fraction consists mostly of cyanidin-3-arabinoside and cyanidin-3-galactoside.

Leucoanthocyanins are compounds which, when heated in acidic media, yield anthocyanins and catechins or epicatechins, respectively. The form present in the greatest amount is flavan-3-4-diol (I) which, through $4 \rightarrow 8$ (II) or $4 \rightarrow 6$ (III) linkages, condenses to form dimers, trimers or higher oligomers:

I

II

III (21.10a)

Leucoanthocyanins occur in fruits of various plants in addition to cacao; e.g. apples, pears and cola (kola) nuts.

21.3.2.3.6 Organic acids

Organic acids in cocoa (1.2–1.6%) are formed mainly during cacao fermentation and consist mostly of acetic acid (a flavor constituent), citric acid (0.45–0.75%) and oxalic acid (0.32–0.50%). The amount of acetic acid released by pulp and partly retained by the bean cotyledons depends on the duration of fermentation and on the drying method used. Eight brands of cocoa were found to contain 1.22–1.64% total acids, 0.79–1.25% volatile acids and 0.19–0.71% acetic acid.

21.3.2.3.7 Volatile compounds and flavor substances

Cocoa aroma is crucially dependent on harvesting, fermentation, drying and roasting. The fresh beans have the odor and taste of vinegar. The characteristic bitter and astringent taste and the residual sweet taste of fermented beans might be impaired by various faults, such as processing of unripe or overripe fruit, insufficient aeration, lack of mixing of the fruit, infection with pest organisms and/or smoke damage as a result of improper drying.

There are more than 400 volatile compounds of roasted cocoa which are of more or less importance in reconstituting the roasted cocoa aroma. Precursors of aroma are found among the reaction products of anaerobic fermentation, particularly amino acids and sugars, which will interact during roasting via the mechanisms of *Maillard* and *Strecker* and subsequent reactions. Aldehydes, heterocyclic compounds and terpenes are important cocoa aroma substances (cf. Table 21.20 a). The sweet, flowery green note of cocoa aroma is derived from linalool and a series of aldehydes, such as isopentanal, phenylethanal and their aldol condensation product, 5-methyl-2-phenylhex-2-enal:

CHO + CHO → OH CHO

$-H_2O$ → CHO

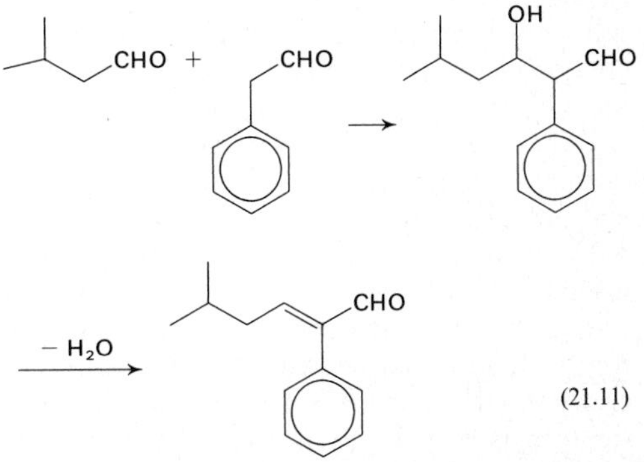

(21.11)

The typical roasted note is derived from pyrazines and pyridine derivatives, e.g. acetylpyridine.

The bitter taste is derived from purines, theobromine and caffeine, and from dioxopiperazines, which are formed during the thermal degradation of proteins during roasting:

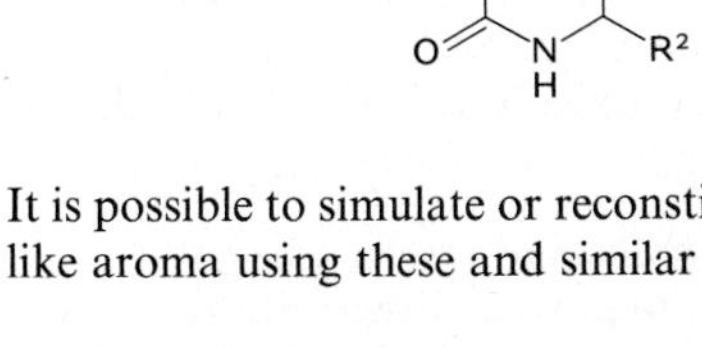

O R¹ H N COOH R N H O R²

→ R OH + R¹ H N O O N H R²

(21.12)

It is possible to simulate or reconstitute a cocoa-like aroma using these and similar compounds.

Table 21.20 a. Possible *cocoa aroma impact* compounds

Compound	Flavor description
2-Methylpropanal	malt, chocolate
2-Methylbutanal	green, cocoa
Dimethylsulfide	buttery, creamy
2-Phenylethanol	rose, fermented
2-Phenylethanal	green, flowery
5-Methyl-2-phenylhex-2-enal	chocolate
Linalool	flowery, sweet
Benzylmethylsufide	burnt, meaty, alliaceous
Methyl anthranilate	fruity
Trimethylpyrazine	sweet, nutty
Maltol	caramel
Trimethyloxazole	sweet, nutty
4-Methyl-5-vinylthiazole	nutty
2-Acetylpyridine	roasted

21.3.2.4 Reactions Occurring During Fermentation and Drying

The reactions occurring within the pulp during fermentation of whole cacao fruit can be distinguished from those occurring in the nibs or co-

tyledons. The pulp sugar is fermented by yeast to alcohol and CO_2 on the first day. Lactic acid fermentation may also occur to a small extent. Pectolytic enzymes and other glycosidases affect the degradation of polysaccharides. This is reflected in the fruit pulp becoming liquid and draining away. This improves aeration, resulting in oxidation of alcohol to acetic acid by acetic acid bacteria during the second to fourth days. The pH drops from about 6.5 to about 4.5 and the temperature increases to 45–50 °C. The seed cell walls become permeable, the living cacao seed is killed and an oxidative process takes over the entire mass. From the fifth to the seventh day, the oxidation and condensation reactions of phenolic compounds predominate. Amino acids and peptides react with the oxidation products of the phenolic compounds, giving rise to water-insoluble brown or brown-violet phlobaphenes (cacao-brown and red), which confer the characteristic color to fermented cacao beans. A decrease in the content of soluble phenols mellows the original harsh and astringent cacao flavor. Finally, the oxidation reactions are terminated by drying the cocoa seeds to a moisture content of less than 8%.
It is extremely important to properly handle the fermentation process for the formation of cocoa aroma. The growth of detrimental microorganisms, such as molds, butyric acid bacteria and putrefaction-inducing bacteria, is thereby prevented.

21.3.2.5 Production of Cocoa Liquor

After roasting and drying, the cocoa nib is disintegrated and milled in order to rupture the cell walls of aggregates and expose the cocoa butter. Knife-hammer mills or crushing rolls usually serve for disintegration, while roller-ball, horizontal "stone" or steel disc or disc attrition mills are used for fine disintegration of cocoa particles. The resultant product is a homogeneous mobile paste, a flowing cocoa mass or cocoa liquor.

21.3.2.6 Production of Cocoa Liquor with Improved Dispersability

The cocoa nib or the cocoa mass is subjected to an alkalization process in order to mellow the flavor by the partial neutralization of free acids and to improve the color and enhance the wettability of cocoa powder, improve dispersability and lengthen suspension-holding ability, thus preventing formation of a sediment in the cocoa drink. The process involves the use of solutions or suspensions of magnesium oxide or hydroxide, potassium or sodium carbonate or their hydroxides. It is occasionally performed at elevated temperature and pressure, usually using steam. In this process, introduced by *C.I. van Houten* in 1828 (hence the term "Dutch cocoa process"), the roasted nibs are treated with a dilute 2–2.5% alkali solution at 75–100 °C, then neutralized, if necessary, by tartaric acid, and dried to a moisture content of about 2% in a vacuum dryer or by further kneading of the mass at a temperature above 100 °C. This treatment, in addition to acid neutralization, causes swelling of starch and an overall spongy and porous cell structure of the cocoa mass. Cocoa so treated is often incorrectly designated as "soluble cocoa" – the process does not increase solubility. Finally, the cocoa is disintegrated with fine roller mills. The "alkalized" cocoa generally contains 52–58% cocoa butter, up to 5% ash and up to 7% alkalized mass or liquor.

21.3.2.7 Production of Cocoa Powder by Cocoa Mass Pressing

To convert the cocoa mass/liquor into cocoa powder, the cocoa fat (54% of nib weight on the average) has to be reduced by pressing, usually by means of a hydraulic, mechanical or, preferentially, horizontally-run expeller press at a pressure of 400–500 bar and a temperature of 90–100 °C. To remove the contaminating cell debris, the hot cocoa butter is passed through a filter press, then molded and cooled. The bulk of the cocoa butter produced is used in chocolate manufacturing. The "stone hard" cocoa press cake, with a residual fat content of 10–24%, is disintegrated by a cook breaker, i.e. rollers with intermashing teeth, then is ground in a peg mill and separated into a fine and a coarse fraction by an air sifter, the coarse fraction being recycled and milled repeatedly. Cocoa powders are divided according to the extent of defatting into lightly defatted powder, with 20–22% residual cocoa butter, and extensively-defatted powder, which contains less than 20% but more than 10% butter. Lightly-defatted powder is darker in color and milder in flavor. Cocoa powder is widely used in the manufacture of other products, e.g. cake fillings, icings, pudding powders, ice creams and cocoa (chocolate) beverages.

21.3.3 Chocolate

21.3.3.1 Foreword

Chocolates were originally made directly from cocoa nibs by grinding them in the presence of sugar. Chocolate is now made from nonalkalized cocoa liquor by incorporating sucrose, cocoa butter, aroma or flavoring substances and, occasionally, other constituents (milk ingredients, nuts, coffee paste, etc.). The ingredients are mixed, refined, thoroughly conched and, finally, the chocolate mass is molded. To obtain a highly aromatic, structurally homogeneous and stable form and a product which "melts in the mouth", a set of chocolate processing steps is required, as described below.

21.3.3.2 Chocolate Production

21.3.3.2.1 Mixing

Mixing is a processing step by which ingredients such as cocoa liquor, high grade crystalline sucrose, cocoa butter and, for milk chocolate, milk powder are brought together in a mixer ("melangeur") or paster. A homogeneous, coarse chocolate paste is formed after intense mixing.

21.3.3.2.2 Refining

The refining step is performed by single or multiple refining rollers which disintegrate the chocolate paste into a smooth-textured mass made up of much finer particles. The rollers are hollow and can be adjusted to the desired temperature lay water cooling. The refined end-product has a particle size of 35–75 ξm. Its fat content should be 23–28%.

21.3.3.2.3 Conching

The refined chocolate mass is dry and powdery at room temperature and has a harsh, sour flavor. It is ripened before further processing by keeping it in warm chambers at 45–50 °C for about 24 h. Ripening imparts a doughy consistency to the chocolate and it may be used for the production of baking or other commercial chocolates. An additional conching step is required to obtain fine chocolates of extra smoothness. It is performed in oblong or round conche pots with roller or rotary conches. The chocolate mass is mixed, ground and kneaded. The step is usually run in two stages. In the first, the mass is heated at 80 °C for 6–12 h. Loss of moisture occurs during heating, a portion of the volatiles is removed (ethanal, acetone, diacetyl, methanol, ethanol, isopropanol, isobutanol, isopentanol and acetic acid ethyl ester) and the fat becomes uniformly distributed such that each cocoa particle is covered with a film of fat. The temperature at this stage is not allowed to rise, since important aroma substances, e.g. pyrazines (cf. 21.3.2.3.7), may be lost. In the second stage, the mass is liquified by the addition of residual cocoa butter and homogenized further. Lecithin is then added to reduce the viscosity of chocolate, or rather to give chocolates a required fluidity by the use of less cocoa butter, and homogenization is continued. Conching is a mixing process which produces a fine flavor and the desired texture, which was not attainable in the previous refining step. Chemical processes involved in conching are only partially understood.

21.3.3.2.4 Tempering and molding

Cocoa butter exists in a number of polymorphic forms (cf. 3.3.1.2) and the nature of the crystalline form depends on the method of cooling the liquid fat.

If chocolate is solidified from the liquid state without any attention to controlled cooling, it will be granular in texture and be of poor color or blotchy in appearance.

To obtain chocolate tablets or covered confectionery of good texture, color and in a stable condition, good tempering and correct cooling are essential. For this purpose, molten chocolate is initially cooled from 50 °C to 18 °C within 10 min with constant stirring. It is kept at this lower temperature for 10 min to form the stable β-modification of cocoa butter (cf. 3.3.1.2). The temperature of the chocolate is then raised within 5 min to 29–31 °C. The process conditions vary according to composition. Regardless of processing variables, tempering serves to provide a great abundance of small fat crystals with high melting points. During the cooling step, the bulk of the molten chocolate develops a solid, homogeneous, finely crystalline, heat-stable fat structure characterized by good melting properties and a nice glossy surface.

Before molding, the chocolate is kept at 30–32 °C and delivered to warmed plastic or metal molds with a metering pump. The filled molds pass over a vibrating shaker to let the trapped air escape. They then pass through a cooling channel where, by slow cooling, the mass hardens and, finally, at 10 °C, the final chocolate product falls out of the mold.

Tempering, metering, filling, cooling and wrapping and packaging machines now provide nearly fully-mechanized and automated production of chocolate.

21.3.3.3 Kinds of Chocolate

In a strict sense, chocolate represents a food commodity which may be molded and which consists of cocoa nibs or nib particles, or of cocoa liquor and sucrose, with or without added cocoa butter, natural herbs or spices, vanillin or ethyl vanillin. Chocolate contains at least 40% cocoa liquor or a blend of liquor and cocoa butter, and up to 60% sugar. The content of coca butter is at least 21% and, when cocoa liquor is blended with cocoa butter, at least 33%.

The composition of the more important kinds of chocolates and confectionery coatings are shown in Table 21.21.

Baking chocolate is made by a special process. Other kinds of chocolates include: cream; full or skim milk or lean; filled; fruit, nut, almond; and those containing coffee or candied orange peels. Cola-chocolate is a caffeine-containing product (maximum of 0.25% caffeine) prepared by mixing with extracts obtained from coffee, cola or other caffeine-containing plants. Diabetic- or diet-chocolates are made by replacing sucrose with fructose, mannitol, sorbitol or xylitol. Information about chocolate coatings is also presented in Table 21.21. Chocolates can also contain nuts and almonds whose oil contents are occasionally reduced by pressing to reach ⅔ of the original amount. This is because the oil has a melting point lower than that of cocoa butter. In filled chocolates, the filler is first placed into a chocolate cup and then closed with a chocolate lid or cover. Fine crumbs of chocolate are made by pressing low-fat chocolate through a plate with orifices. Hollow figures are made in two-part molds, by a hollow press or by gluing together the individually-molded parts.

The term "praline" originates from the name of the French Marshal *Duplessis-Praslin*, whose cook covered sweets with chocolate. Only a few of the many processing options will be mentioned. For pralines with a hard core, the hot, supersaturated sugar syrup (fondant) is poured into molds dusted with wheat powder and left to cool. The congealed core (korpus) is dipped into molten kuverture and, in this way, covered with a chocolate coat (creme-praline). The fondant can be fully or partly replaced by fruit pastes like marzipan, jams, nuts, almonds, etc. (dessert-pralines). Such pralines are prepared with or without a sugar crust. Products with a sugar crust are made from a mixture of thick sugar solution and liqueur by pouring the mixture into mold cavities. The solid crust crystallizes on the outer walls, while the inner portion of the mixture remains liquid. The core so obtained is then dipped into melted chocolate, as described above. For pralines without a sugar crust (brandy or liqueur), the processing involves hollow-body machines in which the chocolate shell is formed, then filled with, e.g. brandy, and covered with a lid in a second machine. The fondant may also contain invertase and, thereby, the praline filling liquifies after several days. Plastic pastes are made by preliminary pulverization of ingredients in a mill and then refiner rollers. The oil content of the ingredients (nuts, almonds, peanuts) provides the consistency of a workable paste after grinding.

Chocolate for beverages or drinks (chocolate powder or flour) is made from cocoa liquor or cocoa powder and sucrose. It is customary to

Table 21.21. Composition of some chocolate products

Product	Cocoa, %	Skim milk powder %	Cocoa butter, %	Total fat, %	Butter fat, %	Sugar %
Baking chocolate	33–50	–	5–7	22–30	–	50–60
Chocolate for coating	35–60	–	to 15	28–35	–	38–50
Milk cream chocolate	10–20	8–16	10–22	33–36	5.5–10	35–60
Whole milk chocolate	10–30	9.3–23	12–20	28–32	3.2–7.5	32–60
Skim milk chocolate	10–35	12.5–25	15–25	22–30	0–2	30–60
Icings	33–65		5–25	35–46		25–50

incorporate seasonings, especially vanillin. The sugar content in chocolate drink powders is at most 65%.

Chocolate syrups are made in the USA by adding bacterial amylase. The enzyme prevents the syrup from thickening or setting by solubilizing and dextrinizing cocoa starch. The fat coating is a glazing on top of chocolate coatings other than cocoa butter (fat from peanuts, coconuts, etc.). It is often used on baked or confectionery products. Tropical chocolates contain high melting fats or are specially prepared to make the chocolate resistant to heat. The melting point of cocoa butter can be raised by a controlled precrystallization procedure. Another option is based on the formation of a coherent sugar skeleton in which the fat is deposited in hollow or void spaces. In this case, in contrast to regular chocolate, there is no continuous fat phase to collapse during heating.

21.3.4 Storage of Cocoa Products

All products, from the raw cacao to chocolate, demand careful storage – dry, cool, well-aerated space, protected from light and sources of other odors. A temperature of 10–12 °C and a relative humidity of 55–65% are suitable. Chocolate products are readily attacked by pests, particularly cacao moths (*Ephestia elutella* and *Cadra cauteila*), the flour moth (*Ephestia kuhniella*) and also beetles (*Coleoptera*), cockroaches (*Dictyoptera*) and ants (order *Hymenoptera*).

Chocolates not properly stored are recognized by a greyish matte surface. Sugar bloom is caused by storage of chocolate in moist conditions (relative humidity above 75–80%) or by deposition of dew, causing the tiny sugar particles on the surface of the chocolate to solubilize and then, after evaporation, to form larger crystals. A fat bloom arises from chocolate fat at temperatures above 30 °C. At these temperatures the liquid fat is separated and, after repeated congealing, forms a white and larger spot. This may also occur as a result of improper precrystallization or tempering during chocolate production. The defect may be prevented or rectified by post-tempering at 30 °C for 6 h.

21.4 Literature

Bokuchava, M. A., Skobeleva, N. I.: The biochemistry and technology of tea manufacture. Crit. Rev. Food Sci. Nutr. *12,* 303 (1980)

Clifford, M. N., Willson, K. C. (Eds.): Coffee, botany, biochemistry and production of beans and beverage. The AVI Publishing Comp. Inc., Westport, Conn. 1985

Flament, I.: Coffee, cacao and tea flavours: a review of present knowledge. In "International Symposium of Food Flavors (J. Adda and H. Richard, Eds.) Technique et Documentation (Lavoisier): Paris, 1983

Forsyth, W. G. C., Quesnel, V. C.: "The mechanism of cacao curing". Adv. Enzymol. *25,* 457 (1963)

Harler, C. R.: "Tea Manufacture". Oxford University Press: London–New York. 1963

Kleinert-Zollinger, J.: Schokolade. In: Ullmanns Encyklopädie der technischen Chemie. 4. Aufl., Bd. 20, S. 673, Verlag Chemie: Weinheim–Basel. 1981

Lange, H., Fincke, A.: "Kakao und Schokolade". In: Handbuch der Lebensmittelchemie, Bd. VI (Hrsg.: Schormüller, J.), S. 210, Springer-Verlag: Berlin–Heidelberg. 1970

Maier, H. G.: Kaffee. Verlag Paul Parey: Berlin–Hamburg. 1981

Sanderson, G. W.: "Black tea aroma and its formation". In: Geruch- und Geschmackstoffe (Hrsg.: Drawert, F.), S. 65, Verlag Hans Carl: Nürnberg. 1975

Vitzthum, O. G.: "Chemie und Bearbeitung des Kaffees". In: Kaffee und Coffein (Hrsg.: Eichler, O.) S. 3, Springer-Verlag: Berlin–Heidelberg. 1976

Wickremasinghe, R. L.: "Tea". Adv. Food Res. *24,* 229 (1978)

Yamanishi, T.: "The aroma of various teas". In: Flavor of foods and beverages (Eds.: Charalambous, G., Inglett, G. E.), p. 305, Academic Press: London–New York. 1978

22. Spices, Salt and Vinegar

22.1 Spices

Some plants with intensive and distinctive flavors and aromas are used dried or in fresh form as seasonings or spices. Table 22.1 lists the most important spice plants together with the part of the plant used for seasoning.

22.1.1 Composition

22.1.1.1 Essential Oils

Table 22.2 lists spices which contain 1–6% essential or volatile oils which, as outlined in 5.5.1.1, are obtainable from plants by steam distillation. The main oil constituents are either mono- and sesquiterpenes or phenols and phenolethers. Examples of the latter two classes of compounds are eugenol (I), carvacrol (II), thymol (III), estragole (IV), anethole (V), safrole (VI) and myristicin (VII):

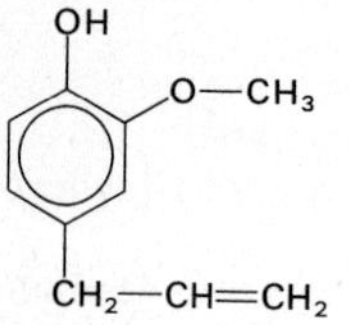

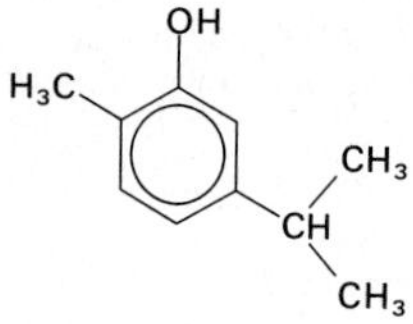

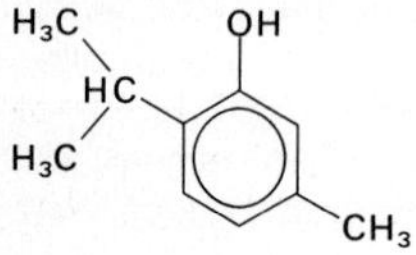

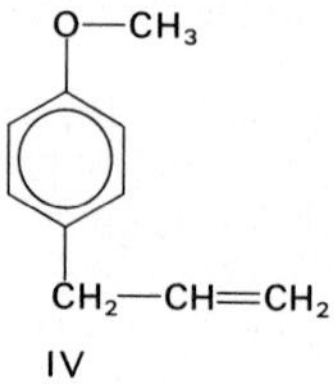

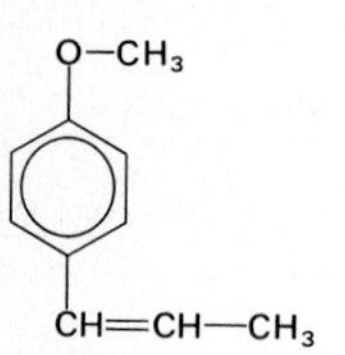

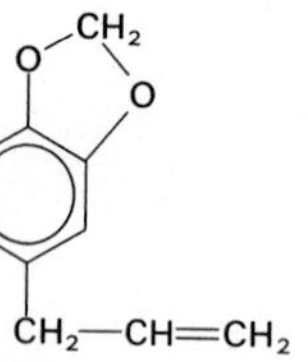

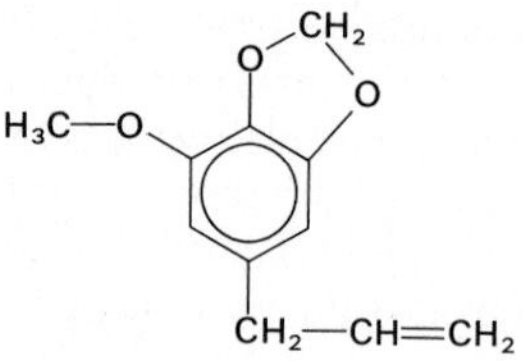

(22.1)

Biosynthesis of cinnamaldehyde (IX), the main constituent of cinnamon bark, and also of eugenol (I) and safrole (VI) originates from phenylalanine (compare biosynthesis of other plant phenols in 18.1.2.5.1). The following reaction sequence is assumed:

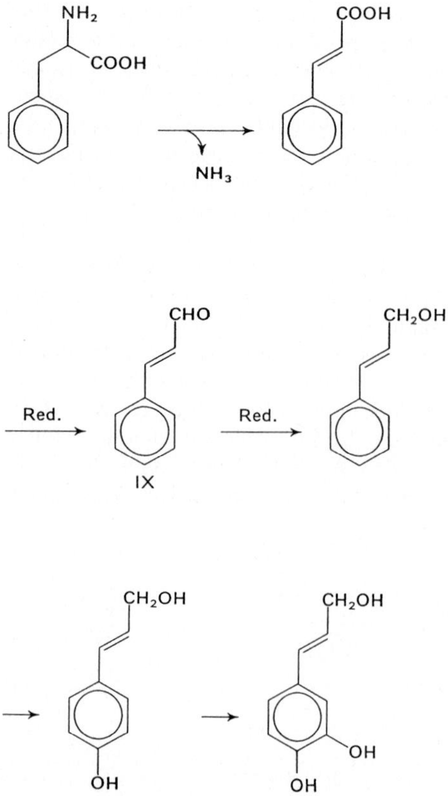

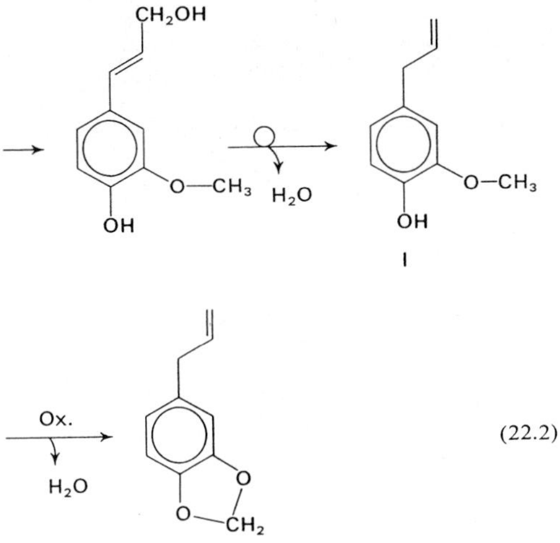

ponents" (cf. 5.1.3) of the distinct aroma. Examples are cinnamaldehyde in cinnamon, anethole in anise, *d*-carvone in caraway, eugenol in cloves, 1,3,8-*p*-menthatriene and 1-methyl-4-isopropylbenzene in parsley leaves and turmerone (XVI) and ar-turmerone (XVII) in curcuma:

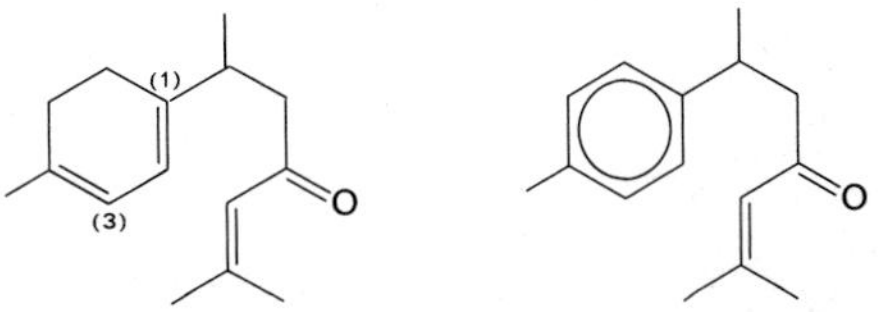

The double bond in position 3 can also be found in position 4 or 5. (22.5)

Some aromatic hydrocarbons are probably generated in spices by terpene oxidation. Examples are: 1-methyl-4-isopropenylbenzene (XI), derived from 1,3,8-Menthatriene (X) and (+)-ar-curcumene (XIV) from zingiberene (XII) or β-sesquiphellandrene (XIII) [cf. Formula 22.3]. The formation of (+)-ar-curcumene from the above-mentioned precursor was detected during storage of ginger oil.

Another aromatic hydrocarbon present in significant amounts in essential oils of some spices (Table 22.2) is *p*-cymene (XV):

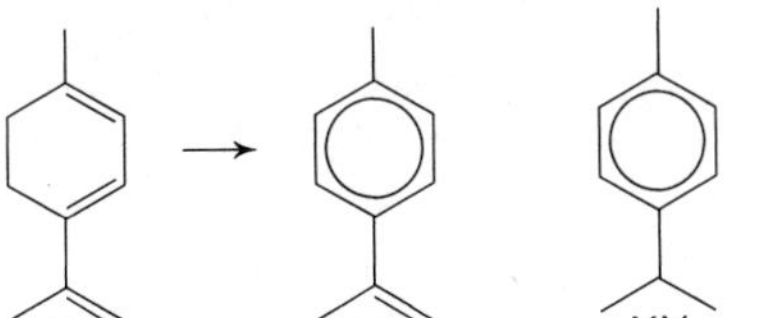

In some spices the main compounds of the essential oils are the "characteristic impact com-

22.1.1.2 Glucosinolates, Pyrazines

Mustard and horseradish contain glucosinolates (Table 22.3) which, after cell rupture, are exposed to the action of a *thioglucosidase* enzymes (cf. 14.3.2.2.5), yielding isothiocyanates (mustard oil; cf. Figure 14.1). Allyl isothiocyanate is obtained from the glucoside sinigrin, a compound responsible for the pungent burning odor and taste of both spices. *p*-Hydroxybenzyl isothiocyanate obtained from sinalbin is only slightly volatile and contributes significantly to the sharp pungent taste of mustard.

The aroma of horseradish is also influenced by methyl, ethyl, isopropyl and 4-pentenyl isothiocyanates which, however, are present only in very small amounts in comparison to allyl isothiocyanate.

The aroma of capsicum pepper plants consists of pyrazines, particularly of 2-isobutyl-3-methoxypyrazine (cf. 5.3.2.5).

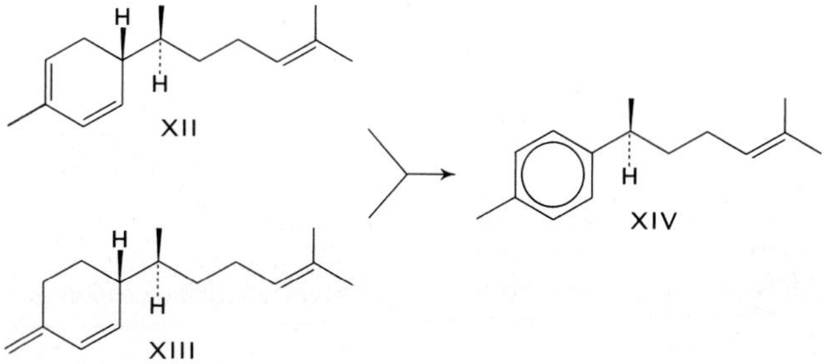

Table 22.1. Spices used in food preparation/processing

Number	Common name	Latin name	Class/order family (bot)	Cultivation region
Fruits				
1	Pepper, black	*Piper nigrum*	Piperaceae	Tropical and subtropical regions
2	Vanilla	*Vanilla planifolia*	Orchidaceae	Madagascar, Comore Island, Mexico, Uganda
3	Allspice	*Pimenta dioica*	Myrtaceae	Caribbean Islands, Central America
4	Paprika (red pepper)	*Capsicum annuum*	Solanaceae	Mediterranean and Balkan region
5	Bay tree[a]	*Laurus nobilis*	Lauraceae	Mediterranean region
6	Juniper berries	*Juniperus communis*	Cupressaceae	Temperate climate region
7	Chili	*Capsicum frutescens*	Solanaceae	Tropical region
8	Aniseed	*Pimpinella anisum*	Apiaceae	Temperate climate region
9	Caraway	*Carum carvi*	Apiaceae	Temperate climate region
10	Coriander	*Coriandrum sativum*	Apiaceae	Temperate climate region
11	Dill[a]	*Anethum graveolens*	Apiaceae	Temperate climate region
Seeds				
12	Mustard	*Sinapsis alba*[b] *Brassica nigra*[c]	Brassicaceae Brassicaceae	Temperate climate region
13	Nutmeg	*Myristica fragrans*	Myristicaceae	Indonesia, Sri Lanka, India
14	Cardamom	*Elettaria cardamomum*	Zingiberaceae	India, Sri Lanka
Flowers				
15	Clove	*Syzygium aromaticum*	Myrtaceae	Indonesia, Sri Lanka, Madagascar
Rhizomes				
16	Ginger	*Zingiber officinale*	Zingiberaceae	South China, India, Japan, Caribbean Islands, Africa
17	Turmeric	*Curcuma longa*	Zingiberaceae	India, China, Indonesia
Barks				
18	Cinnamon	*Cinnamomum zeylanicum, C. aromaticum, C. burmanii*	Lauraceae	China, Sri Lanka, Indonesia, Caribbean Islands
Roots				
19	Horseradish	*Armoracia rusticana*	Brassicaceae	Temperate climate region
Leaves				
20	Parsley	*Petroselinum crispum*	Apiaceae	Temperate climate region
21	Marjoram	*Origanum majorana*	Lamiaceae	Temperate climate region
22	Origano	*Origanum heracleoticum, O. onïtes*	Lamiaceae	Temperate climate region
23	Rosemary	*Rosmarinus officinalis*	Lamiaceae	Mediterranean region
24	Sage	*Salvia officinalis*	Lamiaceae	Mediterranean region
25	Thyme	*Thymus vulgaris*	Lamiaceae	Temperate climate region

[a] Fruits and leaves, [b] white mustard, [c] black mustard.

Table 22.2. Major components of the essential oils of spices

Spice[a]	Components[b]
Pepper (1)	22% α-Pinene (XXIX*), 21% sabinene (XXV*), 17% β-caryophyllene (XLIX*), Δ^3-carene (XXXII*), limonene (IX*), β-pinene (XXX*)
Allspice (3)	70% Eugenol (I), β-caryophyllene (XLIX*), methyleugenol, 1,8-cineol (XXIII*), α-phellandrene (XI*)
Bay leaf (5)	50–70% 1,8-Cineol (XXIII*), α-pinene (XXIX*), β-pinene (XXX*), α-phellandrene (XI*), linalool (IV*)
Juniper berries (6)	36% α-Pinene (XXIX*), 13% myrcene (I*), β-pinene (XXX*), Δ^3-carene (XXXII*)
Aniseed (8)	80–90% Anethole (V)
Caraway (9)	55% Carvone (XXI*), 44% limonene (IX*)
Coriander (10)	Linalool (IV*), linalyl acetate, citral[c]
Dill fruit (11)	35% Carvone (XXI*), 12% dihydrocarvone, 10% limonene (IX*), carveol (XVI*), α-terpinene (X*)
Nutmeg (13)	27% α-Pinene (XXIX*), 21% β-pinene (XXX*), 15% sabinene (XXV*), 9% limonene (IX*), safrole (VI), myristicin (VII)
Cardamom (14)	20–40% 1,8-Cineole (XXIII*), 28–34% α-terpinyl acetate, 2–14% limonene (IX*), 3–5% sabinene (XXV*)
Clove (15)	80–90% Eugenol (I), 9% caryophyllene, eugenol acetate
Ginger (16)	30% (−)-Zingiberene (XLII*), 10–15% β-bisabolene (XLI*), 15–20% (−)-sesqui-phellandrene (XLIII*), (+)-ar-curcumene (XIV), citral[c], citronellyl acetate
Turmeric (17)	30% Turmerone (XVI), 25% ar-turmerone (XVII), 25% zingiberene (XLII*)
Cinnamon (18)	50–80% Cinnamaldehyde (IX), 10% eugenol (I), 0–11% safrole (VI), 10–15% linalool (IV*), camphor (XXXIII*)
Parsley (20)	1,3,8-p-Menthantriene (X), 1-methyl-4-isopropenylbenzene (XI), β-phellandrene (XII*), myrcene (I*)
Majoram (21)	49–65% 1,8-Cineole (XXIII*), 25% esdragole (IV), 15% α-terpineol (XVII*), 11% eugenol (I), linalool (IV*), geranyl acetate, ocimene (II*)
Origano (22)	Carvacrol (II), thymol (III), p-cymene (XV), carvacrol methyl ether
Rosemary (23)	1,8-Cineole (XXIII*), camphor (XXXIII*), β-pinene (XXX*), camphene (XXXI*)
Sage (24)	1,8-Cineole (XXIII*), camphor (XXXIII*), thujone (XXVI*)
Thyme (25)	Thymol (III), p-cymene (XV), carvacrol (II), linalool (IV*)

[a] The number in brackets refers to Table 22.1.
[b] Roman numerals with an asterisk refer to the chemical structures of the terpenes presented in Table 5.22; roman numerals without an asterisk refer to chemical structures shown in chapter 22.
[c] A mixture of neral and geranial.

Table 22.3. The most important glucosinolates of mustard and horseradish

General structure: R—C(=N—O—SO$_2$—O$^{\ominus}$K$^{\oplus}$)—S—Glucose

R	Name	Occurrence
HO—C$_6$H$_4$—CH$_2$—	Sinalbin	Mustard
H$_2$C=CH—CH$_2$—	Sinigrin	Mustard, horseradish
C$_6$H$_5$—CH$_2$—CH$_2$—	Gluconasturtiin	Horseradish

Table 22.4. Compounds present in spices causing a hot burning organoleptic perception

Name	Structure	Occurrence[a]	Relative pungency[b]
Piperine[c]	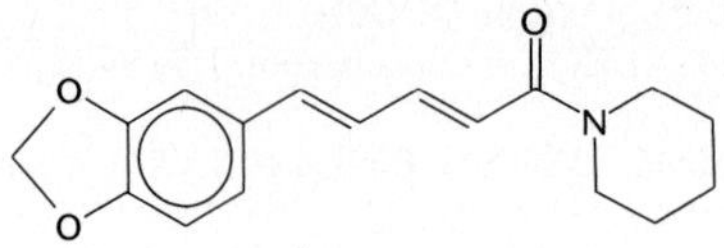	*Pepper* (1)	1.0
Piperanine	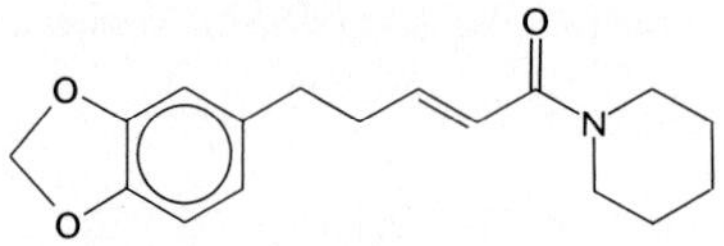	*Pepper* (1)	0.5
Piperylin	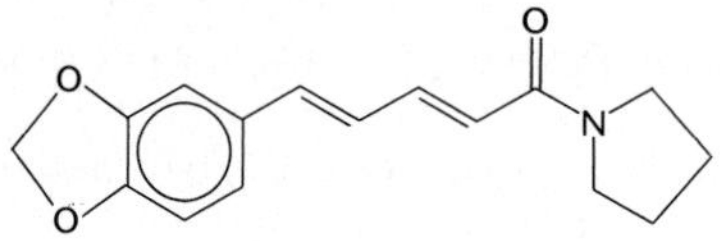	*Pepper* (1)	0···1[d]
Gingerol	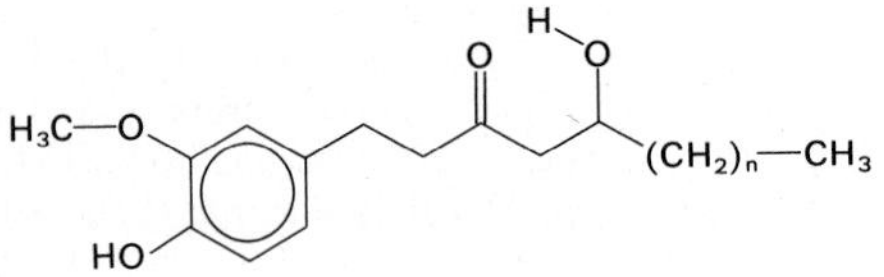 $H_3C—O$, HO, $(CH_2)_n—CH_3$	Ginger (16)	0.8
Shogaol	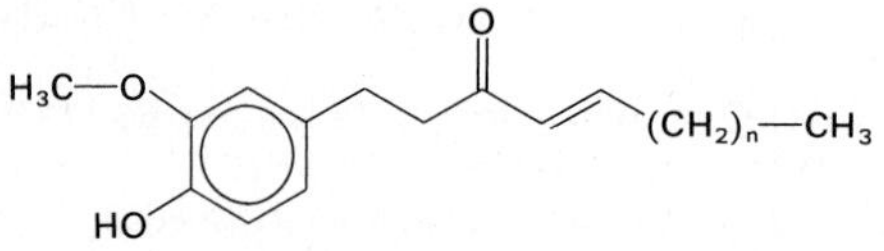 $H_3C—O$, HO, $(CH_2)_n—CH_3$	Ginger (16)	1.6
Zingerone	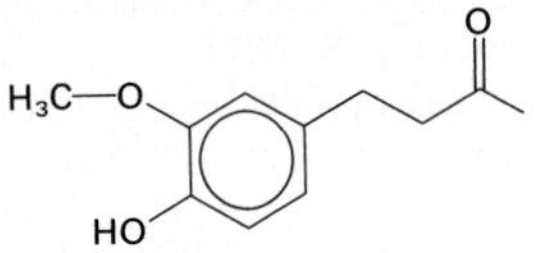 $H_3C—O$, HO	Ginger (16)	0···0.5[d]
Capsaicin	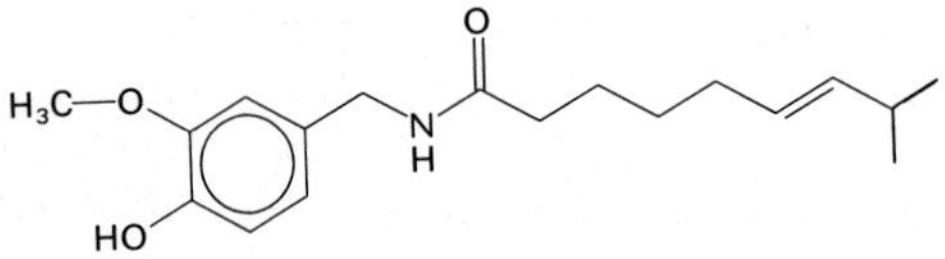 $H_3C—O$, HO	Capsicum (4; 7)	150···300[d]
Dihydro-capsaicin	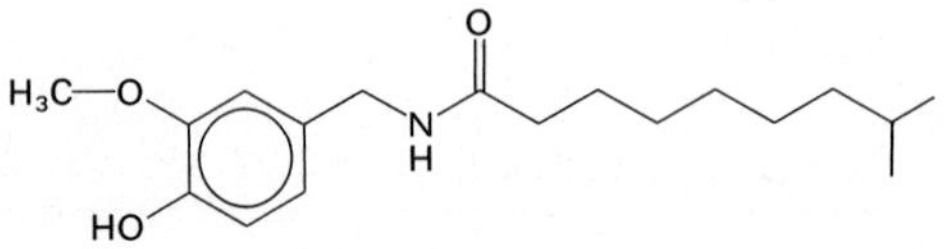 $H_3C—O$, HO	Capsicum (4; 7)	as capsaicin
Nordihydro-capsaicin	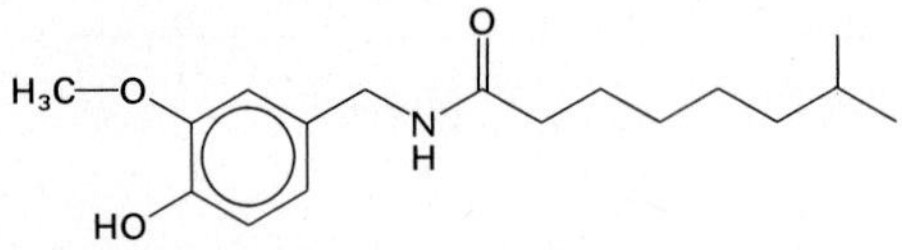 $H_3C—O$, HO	Capsicum (4; 7)	75% capsaicin

[a] The numerals in brackets refer to Table 22.1.
[b] Reference: pungency of piperine = 1.
[c] The corresponding cis,cis-compound is devoid of pungent taste.
[d] Literature data are within the range of values presented.

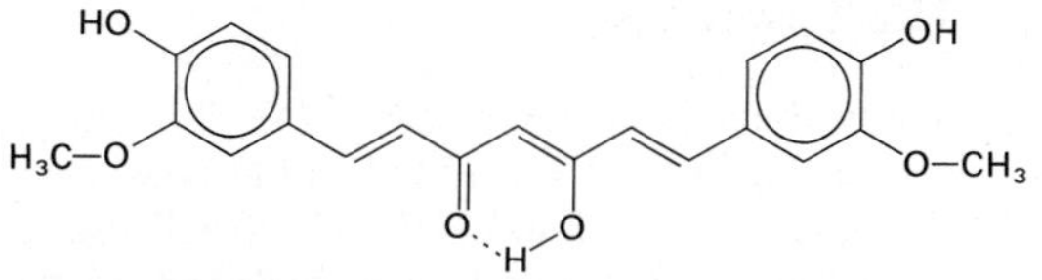

(22.9)

22.1.1.3 Substances with Pungent Flavors

The hot, burning pungent flavor of paprika (red pepper), pepper (black pepper) and ginger are derived from the nonvolatile compounds listed in Table 22.4.
Capsaicin content of the fruits of capsicum or of various other pepper plants depends on the variety, cultivation, drying and storage conditions, and varies between 0.075 and 0.8%. This compound provides a very hot burning sensation (Table 22.4).

22.1.1.4 Pigments

Paprika (red pepper) and curcuma pigments are used as food colorants. Paprika pigments are carotenoids, with capsanthin as the main compound (cf. 3.8.4.1.2 and Figure 3.42). Curcumin is the main pigment of curcuma, a tropical plant of the ginger family (cf. Formula 22.9).

22.1.1.5 Antioxidants

Extracts of several spices, particularly of sage and rosemary, have the ability to prevent the autoxidation of unsaturated triacylglycerols. Among the most effective antioxidant constituents of both spices, the cyclic diterpene diphenols, carnosolic acid (XVIII) and carnosol (XIX) have been identified:

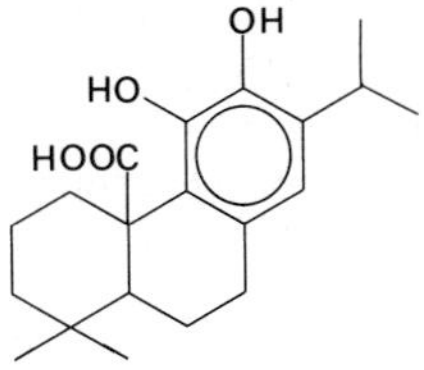

XVIII (22.10a)

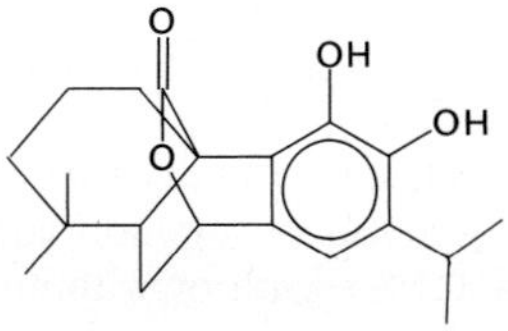

XIX (22.10b)

22.1.2 Products

22.1.2.1 Spice Powders

Spices are marketed unground, or as coarsely or finely ground powder. The flavor is improved when the spices are ground using a cryogenic mill. After grinding the shelf life of the spices is limited. Favorable storage conditions are the absence of air, a relative humidity less than 60% and a temperature less than 20 °C. Crushed spices rapidly lose their aroma and absorb aromas from other sources. The standard plate count of spice powders is often very high, hence their addition to food preparations may accelerate microbial food spoilage.

22.1.2.2 Spice Extracts or Concentrates (Oleoresins)

Spice extracts are being used in increasing amounts in industrial-scale food preparation since they are easier to handle than spice powders and are free of microorganisms. The production of these extracts is outlined in 5.5.1.2. The flavor quality depends on the solvent used and also on the raw material.

22.1.2.3 Blended Spices

Specially blended spices are offered commercially for some food preparations, such as liver sausage which uses a spice blend consisting of sweet marjoram, mace, nutmeg, cardamom, ginger, pepper and a little cinnamon.
Smoked, saveloy sausage spice blend consists of coriander, ginger, mustard kernels, paprika and pepper. Common spices for bread are aniseed, fennel and caraway. Gingerbread spice blend consists of aniseed, clove, coriander, cardamom, allspice and cinnamon.

22.1.2.4 Spice Preparations

Spice preparations are obtained by the addition of spices and blended spices to other substances, such as salt, sugar, glutamate, yeast extract and starch flour.

Curry powder. A spice preparation containing a spice blend of tumeric as the main ingredient and paprika, chili, ginger, coriander, cardamom, clove, allspice and cinnamon, mixed together with up to 10% legume meal, starch and glucose, and with less than 5% salt.

Mustard. A dark yellow paste used as a pungent seasoning for food. It is made from finely ground, often defatted mustard seeds, mixed into a slurry with water, vinegar, salt, oil and some other spices (pepper, clove, coriander, curcuma, ginger, paprika, etc.) and ground repeatedly or refined. During processing, lasting 1–4 h at a temperature not exceeding 60 °C, the mustard oil is released from its glucoside, as outlined in 22.1.1.2. "Extra strong" mustard is primarily made from dehulled black mustard seed, while the "medium hot" or "hot" types are made from seeds with hull, using varying proportions of black and white mustards.

Sambal. A spice preparation derived from Asia and used for seasoning rice dishes. Its base is Sambal oelek, which consists mainly of crushed or pulverized salt-preserved chili.

22.2 Salt (Cooking Salt)

Common salt occupies a special position among the spices. Salt is used in greater amounts than all other spices to enhance the flavor and taste of food. Also, some foods are preserved when salted with large amounts of NaCl (cf. 0.3.2).
Humans require a certain constant level of intake of sodium and chloride ions to maintain their vital concentrations in plasma and extracellular fluids. The daily requirement is about 5 g of NaCl; an excessive intake is detrimental to health.

22.2.1 Composition

Common (cooking or kitchen) salt is nearly entirely NaCl. Impurities are moisture (up to 3%) and other salts, not exceeding 2.5% (magnesium and calcium chloride; magnesium, calcium and sodium sulfates). Salt also contains trace elements.

22.2.2 Occurrence

Salt is abundant in sea water (2.7–3.7%) and in various landlocked seas (7.9% in the Dead Sea; 15.1% in the Great Salt Lake in Utah) and also in salt springs (Lueneburg, Reichenhall) and, above all, in salt beds formed in various geological periods, e.g. the European Zechstein salt deposits.

22.2.3 Production

In FR Germany most of the salt is mined as rock salt. It is selected, crushed and finely ground. Salt springs are also an important source. Brine (at least 4% NaCl) from wells is treated by either direct evaporation, eventually under vacuum, or the brine is preconcentrated by cascade solar evaporation units. These are cradles filled with twigs on the large surface of which the brine is partly evaporated by repeated repumping until a 20% NaCl concentration is achieved. Salt obtained in such a manner is called "boiling" salt.
In warm countries sea water is concentrated in shallow flat basins, by the sun, heat and wind until it crystallizes ("solar salt").
The addition of 0.25–2.0% calcium or magnesium carbonate or 20 ppm of potassium ferrocyanide prevents the formation of lumps in the salt. The latter compound modifies the crystallization pattern of NaCl during the evaporation of salt spring water. In the presence of potassium ferrocyanide, the salt builds dendrites, which have strongly reduced volume, density and inclination to agglomerate.
In 1975 the worldwide production of NaCl was 162.2×10^6 t. In 1974 only 5% of the NaCl produced in FR Germany was used for consumption; the remainder, 95%, was used in industry or trade (raw materials, salt for regeneration of ion-exchange resins, etc.).

22.2.4 Special Salt

Iodized salt is produced as a preventive measure against goitre, a disease of the thyroid gland (cf. 17.1.2.9.3). It contains 5 mg/kg of sodium-, potassium- or calcium iodide.
Nitrite salts are used for pickling and dry curing of meat (cf. 12.6.4). They consist of common salt and sodium nitrite (0.4–0.5%), with or without additional potassium nitrate.

Table 22.5. Substitutes for common salt

Potassium, calcium and magnesium salts of adipic, succinic, glutamic, carbonic, lactic, hydrochloric, tartaric and citric acids;
Monopotassium phosphate, adipic and glutamic acids and potassium sulfate;
Choline salt of acetic, carbonic, lactic, hydrochloric, tartaric and citric acids;
Potassium salt of guanylic and inosinic acids

22.2.5 Salt Substitutes

Some human diseases make it necessary to avoid excessive intake of sodium ions, so attempts have been made to eliminate the use of added salt as a spice or flavoring, without attempting to achieve completely salt-free nutrition. This "low salt" nutrition is actually only related to reduced sodium levels, hence a "low sodium" diet is a more relevant designation.
The compounds listed in Table 22.5 are used as salt substitutes. Their blends are marketed as "diet salts".

22.3 Vinegar

Vinegar was known in old Oriental civilizations and was used as a poor man's drink and later as a remedy in ancient Greece and Rome. Vinegar is the most important single flavoring used to provide or enhance the sour, acidic taste of food (cf. 8.12.5).

22.3.1 Production

Vinegar is produced microbiologically from ethanol or by dilution of acetic acid.

$$CH_3CH_2OH + O_2 \longrightarrow CH_3COOH + H_2O + 494 \text{ kJ} \qquad (22.11)$$

22.3.1.1 Microbiological Production

Acetobacter species are cultivated in aqueous ethanol solution or, to a lesser extent, in wine, fermented apple juice, malt mash or fermented whey. Ethanol, as shown in Figure 22.1, is dehydrogenated stepwise to acetic acid; the reduced cosubstrates are oxidized via the respiratory chain. Part of the energy formed by oxidation is released as heat which has to be removed by cooling during the processing of vinegar. If there is an insufficient supply of oxygen, the microorganisms disproportionate a proportion of the acetaldehyde, the intermediate compound (cf. Figure 22.1) of this aerobic reaction pathway:

$$2\,CH_3CHO \longrightarrow CH_3COOH + CH_3CH_2OH \qquad (22.12)$$

Fermentation of ethanol is conducted as a top fermentation and for 20 years increasingly as a submerged oxidative process. In top fermentation the bacteria are cultivated on spongy, porous laminated carriers (usually beechwood shavings) with the alcoholic solution trickling down over carrier surfaces while a plentiful supply of air is provided from below. The fermentation is stopped at a 0.3% by volume residual ethanol level to avoid overoxidation, i.e. oxidation of acetic acid to CO_2 and water.

22.3.1.2 Chemical Synthesis

Acetic acid is usually synthesized by catalytic oxidation of acetaldehyde:

$$CH_3CHO + \tfrac{1}{2}O_2 \xrightarrow{\text{Kat.}} CH_3COOH \qquad (22.13)$$

Acetaldehyde is obtained by the catalytic hydration of acetylene or by the catalytic dehydrogenation of ethanol. Formic acid and formaldehyde are by-products of acetic acid synthesis. They are removed by distillation. Chemically-pure acetic acid is diluted with water to 60–80% by volume to obtain the vinegar essence. The essence is a

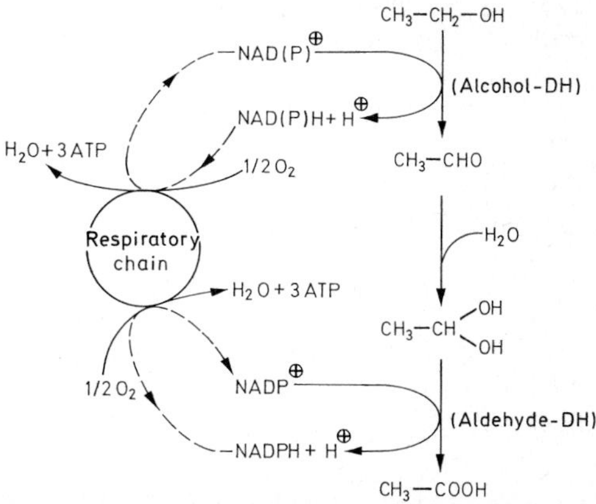

Fig. 22.1. Oxidation of ethanol to acetic acid by *Acetobacter* species. (After *H.J. Rehm*, 1980)

strongly corrosive liquid and is sold with special precautions. It is diluted further with water for production of food grade vinegar.

22.3.2 Composition

There are 5–15.5 g acetic acid in 100 g of vinegar. The blending (or adulteration) of fermented vinegar with synthetic acid can be detected by mass spectrometric determination of the $^{13}C/^{12}C$-isotope ratio; fermented vinegar has 5‰ more ^{13}C isotope than acetic acid synthesized petrochemically. In addition fermented vinegar can be distinguished from synthetic vinegar by analyzing the accompanying compounds. With this method fermented vinegars of different origin can also be distinguished from each other; e.g. spirit vinegar (fermented from aqueous ethanol) from wine, apple, malt and/or whey vinegar. The fermented vinegars contain metabolic by-products of *Acetobacter* strains, such as amino acids, 2,3-butylene glycol and acetyl methyl carbinol, in addition to substances derived from the raw materials used in vinegar production.

22.4 Literature

Brockhaus, R., Förster, G.: Essigsäure. In: Ullmanns Encyklopädie der technischen Chemie, 4. Aufl., Bd. 11, S. 57, Verlag Chemie: Weinheim. 1976

Ebner, H..: Essig. In: Ullmanns Encyklopädie der technischen Chemie, 4. Aufl., Bd. 11, S. 41, Verlag Chemie: Weinheim. 1976

Govindarajan, V. S.: Pepper – chemistry, technology, and quality evaluation. Crit. Rev. Food Sci. Nutr. *9,* 115 (1977)

Govindarajan, V. S.: Tumeric – chemistry, technology, and quality. Crit. Rev. Food Sci. Nutr. *12,* 199 (1980)

Govindarajan, V. S.: Ginger – chemistry, technology, and quality evaluation. Crit. Rev. Food Sci. Nutr. *17,* 1 u. 189 (1982)

Govindarajan, V. S., Narasimhan, S., Raghuveer, K. G., Lewis, Y. S.: Cardamom – production, technology, chemistry and quality. Crit. Rev. Food Sci. Nutr. *16,* 229 (1982)

Herrmann, K.: Übersicht über nichtessentielle Inhaltsstoffe der Gemüsearten. II. Cruciferen (Kohlarten, Radieschen, Rettiche, Speiserüben, Kohlrüben, Meerrettich) sowie Gramineen (Zwiebeln, Porree, Schnittlauch, Knoblauch, Spargel). Z. Lebensm. Unters. Forsch. *165,* 151 (1977)

Maga, J. A.: Capsicum. Crit. Rev. Food Sci. Nutr. *6,* 177 (1975)

Masada, Y.: Analysis of essential oils by gas chromatography and mass spectrometry. John Wiley and Sons: New York. 1976

Melchior, H., Kastner, H.: Gewürze. Verlag Paul Parey: Berlin–Hamburg. 1974

Pruthi, J. S.: Spices and condiments – chemistry, microbiology, technology. Academic Press: London–New York. 1980

Rehm, H.-J.: Industrielle Mikrobiologie. 2. Aufl., Springer-Verlag: Berlin–Heidelberg. 1980

Salzer, U.-J.: The analysis of essential oils and extracts (oleoresins) from seasonings – a critical review. Crit. Rev. Food Sci. Nutr. *9,* 345 (1977)

Schmid, E. R., Fogy, I., Schwarz, P.: Beitrag zur Unterscheidung von Gärungsessig und synthetischem Säureessig durch die massenspektrometrische Bestimmung des $^{13}C/^{12}C$-Isotopenverhältnisses. Z. Lebensm. Unters. Forsch. *166,* 89 (1978)

Wijesekera, R. O. B.: The chemistry and technology of cinnamon. Crit. Rev. Food Sci. Nutr. *10,* 1 (1978)

Appendix: Selected References to the Literature of Food Chemistry and Related Fields

The majority of food chemistry data applied today have been reported as scientific papers in research journals that cover the broad spectrum of the science and technology of food. The titles listed below constitute the primary literature source. Monographs, bibliographies, handbooks, directories, encyclopedias, symposia or convention proceedings and abstracts of published papers and other reference works constitute the secondary sources that assist in tracking down or augmenting specific information. Listed is only a partial selection of important primary and secondary sources related to the field of food science in general and food chemistry in particular. It should be emphasized that it is increasingly difficult to clearly segregate technology, the application of discoveries from the basic science, the acquisition of knowledge. By the same token it is increasingly difficult to address topics in food science without considering the end product of such endeavours – human nutrition. Lastly, for the benefit of readers who use the North American Library of Congress Classification, the classification by letters and numbers of the selected subjects is provided.

Primary Literature

Journals

General Food Science and Technology

Acta Alimentaria, International Journal of Food Science (Akademiai Kiado, Budapest/D. Reidel Publ. Co., Dordrecht/Boston, 1972) TX 341 A18

Biotechnology and Bioengineering (John Wiley and Sons, New York, 1959) TA 166 B616

Canadian Institute of Food Science and Technology, Journal (CIEST, Ottawa, 1968–) TP 368 C213

Chemie, Mikrobiologie, Technologie der Lebensmittel (Verlag Hans Carl, GmbH, Nürnberg, 1971–) TX 541 C52

Critical Reviews in Food Science and Nutrition (CRC Press, Inc., Boca Raton, FL, 1980–) TP 368 C393

CSIRO Food Research Quaterly (CSIRO, Div. Food Research, Sydney, 1941–) TX 599 F68

Food Hydrocolloids (Elsevier Applied Science Publ., London, New York, 1986–)

Food Technology (American Institute of Food Technologists, IFT, Chicago, IL, 1947–) TP 370 A1 F6

Food Technology in Australia (Australian Food Technology Association, Sydney, 1948–) TP 368 F687

Food Technology in New Zealand (Trade Publ. Ltd., Auckland, 1965–) TP 368 F688

Journal of Food Quality (Food and Nutrition Press, Inc., Westport, CT, 1977–) TX 341 J857

Journal of Food Science (Formerly Food Research, IFT, Chicago, IL, 1936/1961–) TX 341 F68

Journal of Food Technology (Blackwell Scientific Publ., 1966–) TX 341 J85

Journal of Food Protection, An International Journal (International Association of Milk Food & Environmental Sanitarians, Inc., Ames, IA, 1977–) SF 221 J862

Lebensmittel-Wissenschaft + Technologie/Food Science + Technology (Forster Verlag AG., Zürich, 1968–) TP 370 L44

Mitteilungen aus dem Gebiete der Lebensmitteluntersuchung und Hygiene/Travaux de Chimie Alimentaire et d'Hygiene (Schweizerischer Verein Analytischer Chemiker, Bern, 1961–) RA 421 M68

Pishchevaia Tekhnologiia, Izvestiia Vysshikh Uchebnyks zavedenii (Krasnodar, USSR, 1958–) TX 341 R96

Cereals, Flour and Breadmaking, Science and Technology

Bakers' Digest (Siebel Publ. Co., Pontiac, IL, 1965–) TX 341 B16

Cereal Chemistry (American Association of Cereal Chemists, AACC, St. Paul, MN, 1924–) TS 2120 C41

Cereal Foods World (AACC, St. Paul, MN, 1975–, formerly Cereal Science Today, 1965–) TS 2120 C422

International Rice Research Newsletter (Intl. Rice Res. Institute, Manila, Philippines, 1981)

Journal of Cereal Science (Academic Press, London, New York, 1983–) SB 188 J86

Khlebopekarnaia i konditerskaia promyshlennost (Moscow, USSR, 1957) TX 761 K45

Dairy Science and Technology

Australian Journal of Dairy Technology (Australian Society of Dairy Technology, Melbourne, 1946–) SF 221 A93

Indian Journal of Dairy Science (Indian Dairy Science Association, Bangalore Press, Bangalore City, 1948–) SF 221 I39

Journal of Dairy Research (Cambridge University Press, London, 1929–) SF 221 J863

Journal of Dairy Science (American Dairy Science Association, Champaign, IL, 1917–) SF 221 J865

Le Lait (INRA, Lucien Declume, Lyon, 1922–) SF 221 L18

Milchwissenschaft, Milk Science International (Organ der Deutschen Gesellschaft für Milchwissenschaft, Volkswirtschaftlicher Verlag, Kempten/Allgäu, 1946–) SF 221 M63

Molochnaia promyshlennost (Moscow, USSR, 1934–) SF 221 M72

Netherlands Milk and Dairy Journal (Association for Adv. of Dairy Science, Wageningen, 1947–) SF 221 N46

New Zealand Journal of Dairy Science and Technology (New Zealand Dairy Research Institute, Palmerston North, 1966–) SF 221 N532

Environmental Science and Technology

Bulletin of Environmental Contamination and Toxicology (Springer-Verlag, New York, Berlin, 1966–) RA 565 A1 B93

Environmental Conservation, International Journal (Elsevier Sequoia, SA, Lausanne, 1974–) QH 540 E62

Environment International, A Journal of Science, Technology, Health, Monitoring and Policy (Pergamon Press, New York, 1984–) TD 169 E615

Food and Chemical Toxicology, International Journal (Pergamon Press, Oxford, 1982–)RA 1190 F68

Journal of Environmental Science and Health (Part B; Pesticides, Food Contaminants, and Agricultural Wastes, Marcel Dekker Inc., New York, 1976–) TD 172 J865

Residue Reviews (Academic Press, New York, 1962–) TX 501 R43

Toxicological and Environmental Chemistry (Gordon and Breach Science Publs., New York, 1981–) QD 415 A1 T752

Water, Air, and Soil Pollution, International Journal of Environmental Pollution (D. Reidel Publ. Co., Dordrecht/Boston, 1971–) TD 172 W32

Fat and Oil Science and Technology

Journal of the American Oil Chemists' Society (Champaign, IL, 1924–) TP 670 A1 A52

Biochimica et Biophysica Acta, Lipids and Lipid Metabolism (Elsevier Science Publ., Amsterdam, 1955–) QP 1 B6

Fette, Seifen, Anstrichmittel (Organ der Deutschen Gesellschaft für Fettwissenschaft e. V. Industrieverlag von Hernhaussen KG, Leinfelden-Echterdingen, formerly with various titles, 1894–) TP 670 F42

Journal of Lipid Research (Bethesda, Md, 1959–) QP 751 J86

Lipids (American Oil Chemist's Society, Champaign, IL, 1966–) QP 501 L76

Maslozhirnovaia promyshlennost (Moscow, USSR, 1952–) TP 1 M39

Progress in Lipid Research (Pergamon Press, Oxford, 1978–) QD 301 P962

Revue Française des Corps Gras (Organe Officiel de l'Institut des Corps Gras, Paris, 1954–) TP 670 A1 R4

Fermentation Technology

Biotechnology and Bioengineering (John Wiley and Sons, New York, 1959–) TA 166 B616

American Journal of Enology and Viticulture (Am. Soc. Enologists, Davis, CA, 1950–) TP 500 A512

Monatsschrift für Brauwissenschaft (Formerly Brauwissenschaft, Verlag Hans Carl, GmbH Co., Nürnberg, 1948–) TP 500 B822

Fermentnaia i spirtovaia promyshlennost (Moscow, USSR, 1923–) TP 500 F35

Journal of the Institute of Brewing (The Institute of Brewing, London, 1895–) TP 500 I59

Process Biochemistry (Wheatland Ltd., Watford, UK, 1966–) QR 53 P962

The Brewers Digest (Siebel Publ. Co., Chicago, IL, 1898–) TP 500 B84

Fish- Water Resources

Aquaculture (Elsevier Science Publ., Amsterdam, 1972–) SH 182 A65

Bulletin of Marine Science (Rosenstiel School of Marine and Atmospheric Sciences, University of Miami, Miami, FL, 1951–) GC 1 B931

Cahiers de Biologie Marine (Edition de la Station Biologique de Roscoff, Paris, 1965–) QH 90 C13

Canadian Journal of Fisheries and Aquatic Sciences (Government of Canada Fisheries and Ocean Sci. Info Publ. Branch, Ottawa, 1980–) SH 1 C2112

Hydrobiological Journal (Translation of Gidrobiologicheskiy Zhurnal USSR, Scripta Technica Inc., John Wiley and Sons, Inc., Publ., New York, 1969–) QH 91 A1 H99

Journal of the Marine Biological Association of the United Kingdom (Cambridge University Press, 1887–) QH 301 H33

Marine Biology, International Journal (Springer-Verlag, Heidelberg, New York, 1967–) QH 91 A1 H33

Marine Ecology, Progress Series (Inter Research, Halstenbek, FRG, 1979–) QH 541.5 S3 M325

Marine Chemistry, International Journal (Elsevier Science Publ., Amsterdam, 1972–) GC 109 M33
Transactions of the American Fisheries Society (American Fisheries Soc., Bethesda, MA, 1914–) SH 1 A5118

Flavor Science and Technology

Chemical Senses (IRL Press, London, 1974–) QP 431 C522
Flavour and Fragrance Journal (John Wiley and Sons, New York, 1986–)
Food Flavourings, Ingredients Packaging & Processing (Formerly The Flavor Industry 1970–74; International Flavours and Food Additives 1974–79; United Trade Press, London, UK, 1980–) TP 450 F5932

Food Chemistry, Biochemistry, Analysis and Related Fields

Agricultural and Biological Chemistry (Agricultural Chemical Society of Japan, Tokyo, 1960–)
Analytical Biochemistry (Academic Press, Inc., New York, 1960–) QP 501 A53
Association of Official Analytical Chemists, Journal (AOAC, Arlington, VA, 1915–) S 583 A842
Chemistry and Industry (Society of Chemical Industry, London, 1944–) TP 1 C518
Food Chemistry, International Journal (Elsevier Applied Science Publ., London, 1976–) TX 501 F68
Food Microstructure, International Journal on the Microstructure and Microanalysis of Food, Feeds and Their Ingredients (SEM, Inc., Chicago, IL, 1982–) TX 543 F685
Journal of Agricultural and Food Chemistry (American Chemical Society, Easton, PA, 1953–) S 583 J86
Journal of Food Biochemistry (Food and Nutrition Press, Inc., Westport, CT, 1977–) TX 501 J86
Journal of Food Science and Technology (Association of Food Scientists & Technologists, Mysore, India, 1964–) TX 341 J858
Journal of the Association of Public Analysts (The Association of Public Analysts; Academic Press, London, 1963–) QP 71 A84
Journal of the Science of Food and Agriculture (Chemical Society, London, Blackwell Sci. Publ., 1950–) TX 341 J861
Nahrung (Akademie Verlag, Berlin, 1957–) TX 341 N15
Zeitschrift für Lebensmittel-Untersuchung und Forschung. International Journal of Food Research and Technology (Springer-Verlag, Berlin, Heidelberg, 1955–) TX 341 2481
The Analyst (London, UK, 1877–) QD 71 A53

Food Engineering

Journal of Food Engineering (Elsevier Applied Science Publ., London, 1982–) TP 368 J86
Journal of Food Process Engineering (Food and Nutrition Press, Inc., Westport, CT, 1977–) TX 341 J855
International Journal of Refrigeration (International Institute of Refrigeration, IPC Science and Technology Press, Guildford, 1978–) TP 490 I625

Food Irradiation

International Journal of Applied Radiation and Isotopes (Pergamon Press, Oxford, 1968–) QC 770 I64
Irradiation des Aliments/Food Irradiation (European Information Centre for Food Irradiation, Saclay, 1960–71) TX 611.5 I72

Food Rheology

Journal of Texture Studies (D. Reidel Publ. Co., Dordrecht/Boston, after vol. 8: Food and Nutrition Press, Inc., Westport, CT, 1969–) TX 341 J864

Fruit and Vegetable Science and Technology

American Potato Journal (The Potato Association USA, Orono, ME, 1941–) SB 211 P8 A51
Confructa (Professional information for the fruit, vegetable-juice and fruit wine-industry; Verlag Flüssiges Obst GmbH, Frankfurt am Main, 1975–) TP 440 C74
Hort Science (American Society for Horticultural Science, Alexandria, VA, 1966–) SB 1 H825
Journal of the American Society for Horticultural Science (Alexandria, VA, 1902–) SB 317.56 U6 A52
Journal of Plant Foods (Newman Publ., London, 1978–) TX 391 P72
Phytochemistry, International Journal of Plant Biochemistry (Pergamon Press, Oxford, 1961–) QK 861 A1 P72
Qualitas Plantarum, Plant Foods for Human Nutrition, International Journal (Junk Publ., The Hague, 1952–) SB 13 Q122

Meat Science and Technology

Fleischwirtschaft (Sponholz, Frankfurt am Main, 1972–) TS 1950 F59
Meat Science (Elsevier Applied Science Publ. Ltd., London, 1977–) TX 373 M48
Miasnaia Industriya (Moscow, USSR, 1930–) TS 1950 M618

Human Nutrition

L'Alimentation et la Vie (Societe Scientifique d'Hygiene Alimentaire, Paris, 1911–) TH 7201 A435

Journal of the American Dietetic Association (The American Dietetic Association, Chicago, IL, 1925–) TX 341 A51

Journal of the Canadian Dietetic Association (CDA, Toronto, 1939–) TX 341 C21

Ernährungsforschung (Akademie Verlag, Berlin, 1955–) TX 341 E72

Human Nutrition: Applied Nutrition (Formerly Journal of Humane Nutrition; British Dietetic Association, London, John Libbey and Co. Ltd., 1976–) QP 141 A1 N9732

Journal of Applied Nutrition (Off. Publ. Intl. College of Appl. Nutrition, La Habra, CA, 1947–) TX 341 J853

Nutrition Reports International (Publ. Geron-X Inc., Los Altos, CA, 1970–) QP 141 A1 N97

Progress in Food and Nutrition Science (Pergamon Press, Oxford, 1975–) QP 141 A1 P96

The British Journal of Nutrition (The Nutrition Society, London, Cambridge University Press, 1947–) TX 341 B86

The Journal of Nutrition (Wistar Institute, Philadelphia, PA, 1928–) TX 341 J86

The Proceedings of the Nutrition Society (Cambridge University Press, Cambridge, 1944–) TX 341 N971

Zeitschrift für Ernährungswissenschaft (Steinkopff, Darmstadt) TX 341 Z48

Starch and Sugar Science and Technology

American Society of Sugar Beet Technologists, Journal (Formerly Proceedings, SSBT, Fort Collins, CO, 1947–) TP 375 A51J

Stärke/Starch, International Journal (VCH Verlagsgesellschaft, Weinheim, 1949–) TP 415 579

UNO-FAO Publications

CERES, FAO Review on Agriculture and Development (Rome, 1968–) S 401 C414

FAO Documentation, Current Bibliography (Rome, 1967–) S 401 F682

Food and Agricultural Legislation (FAO, Rome, 1952–) HD 185 F68

Vitamins

International Journal for Vitamin and Nutrition Research (Formerly Internationale Zeitschrift für Vitaminforschung, Hans Huber Publs., Bern, 1932–) TX 553 V5 I62

Journal of Nutritional Science and Vitaminology (Formerly The Journal of Vitaminology, The Vitamin Society of Japan, Center for Academic Publ., Tokyo, 1954–) TX 553 V5 J862

Water and Waste Water

Wasser und Abwasser (Bundesanstalt für Wassergüte, Vienna, 1958–) TD 741 W32

Water Research (The Journal of the International Association on Water Pollution Research and Control, Pergamon Press, Oxford, 1967–) TD 420 W32

Water Science and Technology (Journal of the International Association. on Water Pollution Research and Control; Formerly Progress in Water Technology, Pergamon Press, Oxford, 1969–) TD 420 D962

Water Pollution Control Federation, Journal (Washington, DC, WPCE, 1951–) TD 511 S51

Secondary Literature

Abstracts

Dairy Science Abstracts (Commonwealth Bureau of Dairy Science and Technology, Reading, 1939–) SF 221 C73

Food Science and Technology Abstracts (Formerly Food Science Abstracts, International Food Information Service, 1969–) TX 341 F694

Microbiology Abstracts, Section A; Industrial and Applied Microbiology (London, 1965–) QR 1 M6222

Nutrition Abstracts & Reviews, Series A (Commonwealth Bureau of Nutrition, Aberdeen, 1931–) RH 214 N972

Approved Official and Tentative Methods of Analysis

Association of Official Analytical Chemists (AOAC): Official Methods of Analysis, 14-th Ed., 1985 S 587 A84

American Association of Cereal Chemists (AACC): Approved Methods, 8-th Ed., 1983 (Supplements, 1985) TX 541 A512

American Oil Chemist's Association (AOCS): Official and Tentative Methods of Analysis, 1985 TP 670 A51

Schweizerisches Lebensmittelbuch. Methoden für die Untersuchung und Beurteilung von Lebensmitteln und Gebrauchsgegenständen (Eidg. Drucksachen und Materialzentrale, Bern, 1964–)

Annual Reviews and Serial Publications

Advances in Biochemical Engineering/Biotechnology (Ed. A. Fiechter, Springer-Verlag, Berlin, 1971–) TP 248.3 A242

Advances in Food Research (Eds. C. O. Chichester, E. M. Mrak, and G. F. Stewart; Academic Press, New York, 1948–) TX 537 A24

Advances in Cereal Science and Technology (Ed. Y. Pomeranz; AACC, St. Paul, MN, 1976–) TS 2120 A24

Advances in Carbohydrate Chemistry and Biochemistry (Eds. R. Stuart Tipson and D. Horton; Academic Press, New York, 1945–) QP 321 A24

Advances in Lipid Research (Eds. R. Paoletti and D.Kritchevsky; Academic Press, New York, 1963–) QP 751 A24

Analytical Methods for Pesticides, Plant Growth Regulators and Food Addivites (Eds. G. Zweig and J. Sherma; Academic Press, New York, 1963–) TX 545 Z97

Annual Review of Biochemistry (Ed. J. M. Luck; Annual Reviews Inc., Palo Alto, CA, 1933–) QP 501 A61

Annual Review of Nutrition (Eds. R. E. Olson, E. Beutler and H. P. Broquist; Annual Reviews Inc., Palo Alto, CA, 1981–) TX 341 A62

Critical Reviews in Food Science and Nutrition (Formerly Critical Reviews in Food Technology, Ed. T. E. Furia; CRC Press Inc., Boca Raton, FL, 1970–) TP 368 C393

Grundlagen und Fortschritte der Lebensmitteluntersuchung und Lebensmitteltechnologie (Ed. F. Kiermeier; Verlag Paul Parey, Berlin, 1953–)

Recent Advances in Food Science (Eds. J. Hawthorn and Jas M. Leitch; Butterworths, 1964–, Pergamon Press, 1948–) TX 345 R29

Food Processing Reviews, Noyes Development Corp (Park Ridge, NJ, 1960–)

Handbuch der Lebensmittelchemie, Vols. I–IX (Eds. L. Acker, K. G. Bergner, W. Diemair, W. Heimann, F. Kiermeier, J. Schormüller, S. W. Souci; Springer-Verlag, Berlin, 1965–) TX 551 H23

Progress in the Chemistry of Organic Natural Products/Fortschritte der Chemie Organischer Naturstoffe (Ed. L. Zechmeister, 1938–) QD 241 F74

Encyclopedias

Encyclopedia of Food Technology (Eds. A. H. Johnson and M. S. Peterson, AVI Publ. Co., Westport, CT, 1971–) TP 368.2 E56 (J67)

Foods and Food Production Encyclopedia (Eds. D. M. Considine and G. D. Considine; Van Nostrand Reinhold Co., New York, 1982 TX 349 F68

Kirk Othmer Encyclopedia of Chemical Technology (Wiley-Interscience Publ. New York, 3-rd Ed., 1978–) TP 9 E56

McGraw-Hill Encyclopedia of Science and Technology, 5-th Ed. (McGraw-Hill, New York, 1982–) Q 121 M14

Ullmann's Encyclopedia of Industrial Chemistry, 5-th Ed. in English (VCH Verlagsgesellschaft, Weinheim, 1985–) TP 9 U42 A1

Patent Literature

Commercial Food Patents, United States (Ed. H. B. North, AVI Publ. Co., Westport, CT, 1969–) TX 341 C73

Subject Index